Geological Survey of Canada

Geology of Canada, no. 8

GEOLOGY OF CANADIAN MINERAL DEPOSIT TYPES

edited by

O.R. Eckstrand, W.D. Sinclair, and R.I. Thorpe

1995

This is volume P-1 of the Geological Society of America's Geology of North America series produced as part of the Decade of North American Geology project.

Available in Canada from

Geological Survey of Canada offices:

601 Booth Street
Ottawa, Canada K1A 0E8

3303-33rd Street N.W.
Calgary, Alberta T2L 2A7

100 West Pender Street
Vancouver, B.C. V6B 1R8

or from

Canada Communication Group – Publishing
Ottawa, Canada K1A 0S9

A deposit copy of this publication is also available for reference
in public libraries across Canada

Cat. No. M40-49/8E
ISBN 0-660-13136-6

Price subject to change without notice
Cette publication est aussi disponible en français

Technical editor
O.E. Inglis

Design and layout
P.A. Melbourne
S. Leslie

Cover:
Upper Left   Native gold and spalerite in vein quartz from the Pamour No. 1 mine, Timmins, Ontario. Quartz-carbonate vein gold (mineral deposit subtype 15.2). Height of specimen is 12 cm. Photograph courtesy of Royal Ontario Museum, ROM specimen M41070.
Upper Right   Tectonically-deformed lead-zinc sulphide ore, Sullivan mine, Kimberley, British Columbia. Sedimentary exhalative sulphides (mineral deposit subtype 6.1). (Width of field of view is 11 cm). GSC 1995-186.
Lower Right   Magnetite-jasper iron ore with cross-cutting quartz veinlets, Sherman mine, Temagami, Ontario. Algoma-type Iron-formation (mineral deposit subtype 3.2). (Width of field of view is 7 cm). GSC 1995-185.
Lower Left   Nickel-copper sulphide ore in noritic host rock, Sudbury, Ontario. Nickel-copper sulphide (mineral deposit subtype 27.1). (Width of field of view is approximately 0.5 m). Natural Resources Canada photograph 8243.

Printed in Canada

PREFACE

The *Geology of North America* series has been prepared to mark the Centennial of The Geological Society of America. It represents the co-operative efforts of more than 1000 individuals from academia, state and federal agencies of many countries, and industry, to prepare syntheses that are as current and authoritative as possible about the geology of the North American continent and adjacent oceanic regions.

This series is part of the Decade of North American Geology (DNAG) Project which also includes eight wall maps at a scale of 1:5 000 00 that summarize the geology, tectonics, magnetic and gravity anomaly patterns, regional stress fields, thermal aspects, seismicity, and neotectonics of North America and its surroundings. Together the synthesis volumes and maps are the first co-ordinated effort to integrate all available knowledge about the geology and geophysics of a crustal plate on a regional scale.

The products of the DNAG Project present the state of knowledge of the geology and geophysics of North America in the 1980s, and they point the way toward work to be done in the decade ahead.

From time to time since its foundation in 1842 the Geological Survey of Canada has prepared and published overviews of the geology of Canada. This volume represents a part of the seventh such synthesis and besides forming part of the DNAG Project series is one of the nine volumes that make up the latest *Geology of Canada.*

J.O. Wheeler
General Editor for the volumes
published by the
Geological Survey of Canada

A.R. Palmer
General Editor for the volumes
published by the
Geological Society of America

ACKNOWLEDGMENTS

Although the *Geology of Canada* is produced and published by the Geological Survey of Canada, additional support from the following contributors through the Canadian Geological Foundation assisted in defraying special costs related to the volume on the Appalachian Orogen in Canada and Greenland.

Alberta Energy Co. Ltd.
Bow Valley Industries Ltd.
B.P. Canada Ltd.
Canterra Energy Ltd.
Norcen Energy Resources Ltd.
Petro-Canada
Shell Canada Ltd.
Westmin Resources Ltd.

J.J. Brummer
D.R. Derry (deceased)
R.E. Folinsbee

CONTENTS

INTRODUCTION

INTRODUCTION

O.R. Eckstrand, W.D. Sinclair, and R.I. Thorpe

OBJECTIVE AND SCOPE

The objective of this volume is to define and summarize in a brief and systematic manner the essential characteristics of all economically significant types of Canadian mineral deposits. These summaries reflect the current general understanding of mineral deposits, and correspond closely to the definitions of mineral deposit types in common use.

Each deposit type summary begins by identifying the main diagnostic geological characteristics, the contained commodities, and examples of deposits of that type, both Canadian and foreign. It subsequently outlines the economic significance, typical size and grade of deposits, and geological features such as geological setting, age, host rocks, associated rocks, form and distribution of mineralization, mineralogy, and alteration. Genetic models, related deposit types, and guides to exploration are also summarized. References and a selected bibliography provide an introduction to the most relevant literature.

Deposits of metallic minerals and some industrial minerals are treated in this volume, but not those of fossil fuels. Each deposit type included in this volume has accounted for at least a moderate or historically significant amount of Canadian production and/or reserves, is represented by mineral occurrences in Canada, or is judged to have potential for significant undiscovered deposits in Canada. Although these guidelines are rather subjective, the resulting collection of deposit types is intended to encompass those that are relevant in the Canadian context.

Eckstrand, O.R., Sinclair, W.D., and Thorpe, R.I.
1996: Introduction; in Geology of Canadian Mineral Deposit Types, (ed.) O.R. Eckstrand, W.D. Sinclair, and R.I. Thorpe; Geological Survey of Canada, Geology of Canada, no. 8, p. 1-7 (also Geological Society of America, The Geology of North America, v. P-1).

Because the emphasis in this volume is on mineral deposit types rather than individual deposits, not all Canadian mineral deposits are described, or even mentioned. The examples described are generally those that have been most studied, commonly because of their economic importance and hence their accessibility, or because they are geologically well preserved.

DEFINITION

Mineral deposits are natural concentrations of one or more mineral commodities. They are the products of various geological processes that have operated in a wide range of geological environments. Within a specified geological setting, or a restricted range of related settings, and under similar conditions (such as temperature, pressure, structural conditions favouring fluid flow, availability of metal sources) a particular genetic process or processes operate to produce mineral concentrations with similar characteristics. Such processes include fractional crystallization of magmas, late stage release of volatiles from crystallizing magmas, magma-country rock interaction, metamorphic dewatering, reduction of oxidizing groundwaters or formation waters, mineral precipitation produced or influenced by organisms, and many others. If more than one ore element is concentrated by a specific process or combination of processes, it is because those elements have similar geochemical properties, and were available in that environment. Most geological processes have recurred repeatedly throughout geological history and around the globe. It is not surprising, therefore, that mineral deposits having similar geological characteristics and suites of commodities occur in comparable settings at numerous locations throughout the world in rocks of different ages.

Mineral deposits that are similar in these ways constitute a mineral deposit type. This leads to the following empirical definition: **a "mineral deposit type" is a** collective term for mineral deposits that (a) share a set of geological attributes, and (b) contain a particular mineral commodity or combination of commodities such that (a) and (b) together distinguish them from other types of mineral deposits.

Two important corollaries follow from this definition. The first is that **mineral deposits of the same type are likely to have a common or similar mode of genesis**. The concept of a mineral deposit type has great importance for geologists concerned with the genesis of mineral deposits. This is because the definition of a mineral deposit type is a convenient summary of the main attributes that any genetic theory must explain.

Perhaps more importantly, the second corollary is that **rock assemblages which contain the geological attributes that are characteristic of a particular mineral deposit type have the best potential for containing mineral deposits of that type**. Thus, a knowledge of the kinds of rocks and structures, and tectonic, sedimentary, and magmatic environments that typify a certain deposit type, aided by an understanding of its genesis, allows the exploration geologist to focus on the geological areas most likely to contain undiscovered deposits of that type.

Recognition of "types" of mineral deposits is a convenience practiced widely by economic geologists in both mineral exploration and research on ore genesis. Hence the concept of mineral deposit types is a fundamental one in the methodology of all practicing economic geologists.

"Model" is another term that is commonly associated with distinct groups of deposits, in somewhat the same manner as is "type". In this volume, however, the term "type" is preferred in order to emphasize the empirical

characteristics that allow groups of deposits to be distinguished from one another, and therefore corresponds closely only to the term "descriptive model". "Genetic models", on the other hand, are considered to be important facets of mineral deposit geology but are not used as criteria for identification of deposit types. This is because an empirically defined type is the main basis on which a genetic model is formulated. It happens commonly that a small addition of new empirical information about a deposit type can lead to a dramatic change in the corresponding genetic model. Thus, genetic models may rise and fall in popularity depending on preferred interpretations, whereas the descriptive models of deposit types represent continually growing databases of information. In this sense, a carefully defined type or descriptive model is more robust, and has a longer useful "life" than its corresponding genetic model.

The usual definition of "ore" is used, namely, material that can be processed to recover useful mineral commodities for purposes of anticipated economic or strategic gain.

MINERAL DEPOSIT TYPES: SOURCES OF MINERAL COMMODITIES

Most mineral deposit types and subtypes that are known or inferred in Canada are listed in the Table of Contents, and additional subtypes are defined in some of the individual mineral deposit type/subtype descriptions. The total number of types and subtypes thus defined is 77. Of these, 21 (shown in bold in the Table of Contents) account for "significant" production in Canada, either currently or in the past.

More than 18 mineral commodities were produced in significant amount in Canada in 1987 (Fig. 1A, Table 1), but two-thirds of the total value was derived from just 4 of these commodities (Au, Zn, Cu, Ni). This mineral production was

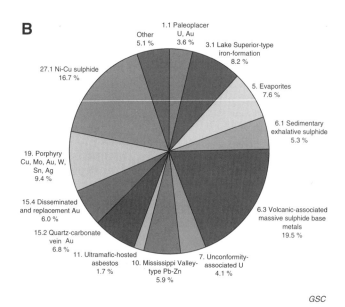

Figure 1. Diagrams illustrating relative proportions of total Canadian mineral production values (1987) by **A)** mineral commodities, and **B)** mineral deposit types.

obtained from more than 25 deposit types (Fig. 1B), but nearly two-thirds of the value of production was obtained from only 5 of these deposit types (6.3 Volcanic-associated massive sulphide base metals; 27.1 Nickel-copper sulphide; 19 Porphyry copper, molybdenum, gold, tungsten, tin, silver; 3.1 Lake Superior-type iron-formation; and 5 Evaporites).

The relationship between mineral commodities and their deposit type sources is illustrated in Table 1. In this table, mineral commodities that were produced in 1987 are listed in the top row, and the deposit types from which they were produced are listed in the left hand column. The numbers listed in the body of the table show, for example, that 300 283 t of zinc valued at CAN$ 382.6 million were produced from sedimentary exhalative sulphide deposits (subtype 6.1), and this represents 17.6% of all zinc produced in Canada in 1987. It may be seen in the bottom row that a total of 1 706 145 t of zinc valued at CAN$ 2147 million were produced in Canada in that year, representing 15.7% of the total value of mineral commodities produced. Similarly, in the right hand column it is seen that sedimentary exhalative sulphide deposits yielded commodities totalling CAN$ 733.6 million in value, representing 5.3% of the total value of all commodities produced in Canada in 1987.

From some mineral deposit types, only a single mineral commodity is produced (e.g., 11 Asbestos); other deposit types produce a number of commodities (e.g., 6.1 Sedimentary exhalative sulphide; 6.3 Volcanic-associated massive sulphide base metals; 19 Porphyry copper, molybdenum, gold, tungsten, tin, silver; and 27.1 Nickel-copper sulphide). Furthermore, some commodities are available from only a single deposit type (e.g., potash), whereas others are recovered from numerous deposit types (e.g., gold, silver, copper, lead, zinc). These associations of commodities in deposit types are the result of (1) the particular processes by which the elements are concentrated, and (2) the similar geochemical behaviors of the associated elements.

DISTINGUISHING MINERAL DEPOSIT TYPES

Many of the mineral deposit types cited here have been recognized and known for a long time; for example, veins of gold (15.2 Quartz-carbonate vein gold) and sedimentary iron deposits (3.3 Ironstone). Others, such as komatiitic nickel deposits (deposit subtype 27.1c), have only been formally recognized as a distinct deposit type for less than 25 years. Still others, such as volcanic redbed copper deposits (deposit Type 9), have received little previous recognition as deposit types. Characterization and understanding of various mineral deposit types evolves continuously as a result of discovery and study of additional mineral deposits, and of advances in research on ore-forming processes and the paleoenvironments in which mineral deposits have formed.

In attempting to classify deposits into types, it is inevitable that some types can usefully be subdivided into two or more subtypes, thus introducing hierarchy into the organizational scheme. This is, for example, clearly the case for the magmatic nickel-copper-platinum group element (PGE) deposit type (27), not only for the purpose of acknowledging the differences between Ni-Cu deposits and PGE deposits, but also for discriminating amongst the various subtypes of Ni-Cu deposits. Subdivision of mineral deposit types into subtypes is a useful way to acknowledge genetic affiliations, while maintaining distinctions that are significant genetically and for exploration purposes. In the opposite sense, the possibility of grouping deposit types can arise; some deposit types that appear rather distinct because of their mining and exploration characteristics can be grouped because they share enough geological similarities that they are better viewed as variants of one type rather than as unrelated types. A good example would be placer gold deposits of the Klondike (1.2 Placer gold, platinum) and paleoplacer uranium deposits of Elliot Lake (1.1 Paleoplacer uranium, gold).

It is inevitable that different choices could have been made in the "lumping and splitting" of types and subtypes. For example, given more emphasis on structure and less on contained metals, the various vein deposit types could have been viewed as subtypes of a single type ("Vein deposits").

Another inherent difficulty in devising a uniform treatment of deposit types is the uneven state of our documentation of mineral deposits. For instance, the many studies on volcanic-associated massive sulphide deposits, porphyry copper deposits, and ultramafic-associated nickel deposits have led to a relatively good appreciation of their characteristics, and have resulted in a relatively clear definition of their types and subtypes. However, other deposits (e.g. some vein deposits) have not received the same attention, and recognition of their specific subtypes seems correspondingly vague.

Names of mineral deposit types

The scheme adopted for the naming of deposit types uses a mixture of traditional and newly proposed terms. Traditional terms such as "placer gold", "Mississippi Valley-type lead-zinc", and "Exhalative base metal sulphide" have been retained. In general, terms are intended to be descriptive, and in most cases consist of two parts. The first part refers to the most noteworthy geological characteristic of that type, which in most cases is the rock association or the related structure. Thus "sandstone-hosted" refers to a type that occurs in sandstone, and "unconformity-associated" refers to one that occurs at or near unconformities. The terms "-hosted" and "-associated" are intended to indicate only a spatial relationship between rock and ore, although in many instances a genetic relationship is also accepted. The second part of the name identifies the main commodities. The commodities linked by hyphens (e.g. Nickel-copper sulphide) consistently occur together, whereas those separated by commas (e.g. Placer uranium, gold) do not occur together in all deposits. The commodities listed first constitute the principal recovered products at one or more deposits of the type in question. However, all the commodities so listed are not necessarily recovered from all deposits of that type. Although this scheme results in some cumbersome names, they are considered useful for the sake of clarity.

Listed order of mineral deposit types

Mineral deposit types in the Table of Contents are listed roughly according to increasing temperature of deposition and/or depth of emplacement. Thus placers as surficial deposits are first and some chromite deposits as mantle-derived

Table 1. Selected mineral commodities produced in Canada (1987) listed by deposit type.

Mineral deposit type	Cu tonnes M$	Ni tonnes M$	Pb tonnes M$	Zn tonnes M$	Mo tonnes M$	Au kg M$	Ag 1000 kg M$	PGE kg M$	Fe tonnes M$
1.1 Paleoplacer U, Au									
1.2 Placer Au, Pt						4 009.5 $76.3 2.9%			
3.1 Lake Superior-type iron-formation									33 923 000 $1133.0 88.3%
3.2 Algoma-type iron-formation									3 244 000 $108.4 8.4%
4.1 Enriched iron-formation									1 173 000 $39.2 3.1%
5 Evaporites									
6.1 Sedimentary exhalative sulphide	6 809 $16.5 0.7%		229 905 $243.2 42.6%	300 283 $382.6 17.6%		513.6 $9.8 0.4%	264.2 $81.5 15.1%		
6.3 Volcanic-associated massive sulphide base metals	360 823 $873.7 38.0%		146 803 $155.3 27.2%	893 446 $1138.4 52.4%		10 922.2 $207.9 7.9%	1 069.5 $330.0 61.0%		
7 Unconformity-associated U									
10 Mississippi Valley-type Pb-Zn			157 966 $167.1 29.3%	510 368 $650.3 29.9%					
11 Ultramafic-hosted asbestos									
14.1 Arsenide vein Ag-Co							19.8 $6.1 1.1%		
15.1 Epithermal Au						3 157.8 $60.1 2.3%	6.8 $2.1 0.4%		
15.2 Quartz-carbonate vein Au	560 $1.4 0.1%	99 $0.7 <0.1%	77 $0.1 <0.1%	360 $0.5 <0.1%		49 015.8 $933.0 35.5%	11.9 $3.7 0.7%		
15.3 Iron-formation-hosted stratabound Au						8 872.8 $168.9 6.4%	2.1 $0.6 0.1%		
15.4 Disseminated and replacement Au	9 494 $20.9 0.9%					42 539.8 $809.7 30.8%	28.5 $8.8 1.6%		
16 Clastic metasediment-hosted vein Ag-Pb-Zn			4 396 $4.7 0.8%	1 688 $2.2 0.1%			89.3 $27.6 5.1%		
17 Vein copper	14 978 $36.3 1.6%					5 151.4 $98.0 3.7%	10.9 $3.4 0.6%		
18 Vein-Stockwork Sn, W									
19 Porphyry Cu, Mo, Au, W, Sn, Ag	397 280 $961.7 41.8%				17 583 $149.0 99.1%	7 421.7 $141.3 5.4%	160.1 $49.4 9.1%		
20.2 Skarn Cu	6 827 $16.5 0.7%				161 $1.4 0.9%	23.4 $0.4 <0.1%	2.0 $0.6 0.1%		
20.3 Skarn Au						1 944.5 $37.0 1.4%	1.1 $0.3 0.1%		
20.4 Skarn Fe									63 000 $2.1 0.2%
24 Carbonatite-associated deposits									
27.1 Ni-Cu sulphide	153 790 $372.4 16.2%	227 887 $1591.6 100%				4 458.2 $84.9 3.2%	87.1 $26.9 5.0%	10 930 $181.8 100%	
Total amount of commodity produced: (T,kg)	950 561	227 986	539 147	1 706 145	17 744	138 030.7	1 753.3	10 930	38 403 000
Total value of commodity produced: M$	$2 299.4	$1 592.3	$570.4	$2 174.0	$150.4	$2 627.3	$541.0	$181.8	$1 282.7
% of total Canadian mineral production	16.6%	11.5%	4.1%	15.7%	1.1%	18.9%	3.9%	1.3%	9.2%

NOTES:
1) Dollar amounts are in 1987 Canadian dollars
2) Ti production was entirely from Allard Lake (26. Mafic intrusion-hosted Ti-Fe); data are confidential
3) W: no production in 1987
4) The production quantities for base and precious metals are based on metal-in-concentrate data; these are the available figures that permit the most reliable identification of the deposit type source of those commodities.
The values of base and precious metal production were calculated by applying average metal prices for that year to the amounts of metals produced.
5) The total value of deposit types does not include the value of minor byproducts such as Te, Se, S, etc.
6) Small, unknown amounts of Au, Cu, and Ag produced from (6.4) Volcanic-associated massive sulphide gold deposits are included under (6.3) Volcanic-associated massive sulphide base metals, (15.3) Iron-formation-hosted stratabound gold, and (15.4) Disseminated and replacement gold.

| Sn | Co | U | Nb_2O_5 | Asbestos | Potash | Salt | Na_2SO_4 | Gypsum | Value of production from deposit type |
tonnes M$	tonnes M$	tonnes M$	tonnes M$	tonnes M$	tonnes M$	tonnes M$	tonnes M$	tonnes M$	M$ % of total Canadian mineral production
		4 214 $499.6 46.8%							$499.6 3.6%
									$76.3 0.5%
									$1 133.0 8.2%
									$108.4 0.8%
									$39.2 0.3%
					7 266 700 $705.8 100%	10 129 053 $238.6 100%	342 076 $26.6 100%	9 093 900 $87.0 100%	$1 058.0 7.6%
									$733.6 5.3%
									$2705.3 19.5%
		8 221 $567.2 53.2%							$567.2 4.1%
									$817.4 5.9%
				664 546 $238.0 100%					$238.0 1.7%
									$6.1 <0.1%
									$62.2 0.4%
									$939.4 6.8%
									$169.5 1.2%
									$839.4 6.0%
									$34.5 0.2%
									$137.7 1.0%
3 439 $31.7 100%									$31.7 0.2%
									$1301.4 9.4%
									$18.9 0.1%
									$37.3 0.3%
									$2.1 <0.1%
			2 630 $17.0 100%						$17.0 0.1%
	2 877 $54.5 100%								$2312.1 16.7%
3 439 $31.7 0.2%	2 877 $54.5 0.4%	12 435 $1 066.8 7.7%	2 630 $17.0 0.1%	664 546 $238.0 1.7%	7 266 700 $705.8 5.1%	10 129 053 $238.6 1.7%	342 076 $26.6 0.2%	9 093 900 $87.0 0.6%	$13885.3 100.0%

Key

Zn

6.1	Sedimentary exhalative sulphide	300 283	←	tonnes of Zn produced from Sedimentary exhalative sulphide deposits
		$382.6	←	value of Zn produced (millions of $ Canadian) from Sedimentary exhalative sulphide deposits
		17.6%	←	% of all Zn produced in Canada that was obtained from Sedimentary exhalative sulphide deposits

Sources:
- Economic and Financial Analysis Branch, Natural Resources Canada
- Canadian Minerals Yearbook, 1987
- Canadian Minerals Yearbook, 1988

deposits are last. Obviously, this ordering is not precise; many types overlap widely with others in regard to temperature-pressure conditions of formation, and for others, these conditions are not well known. Nevertheless, this order provides a crude "profile" of the sequence of deposit types through the crust.

Organization of content

The content of each summary is organized in a format intended to provide a ready guide to the type of information sought by the reader. The introduction states the main geological characteristics and the principal commodities of the deposit type, and lists typical examples, usually both Canadian and foreign. This is followed by statements of the importance to Canadian and/or world mineral supply, and size and grade of deposits (single ore bodies, clusters of ore bodies, or whole mining camps, depending on information available). Most summaries include a map of Canada that shows the distribution of selected mineral deposits of that type, on a background consisting of a simplified version of the geological provinces.

Under geological features are the main geological characteristics, including the geological setting, age, relation of ore to host rocks, form of deposits, nature and zoning of the ore, mineralogy, and alteration. In diagrams showing geological features, ore and mineralization are highlighted in tones of red. A brief review of the definitive characteristics of the deposit type precedes a discussion of the current genetic model (or models). Finally a list of exploration guides summarizes both practical and speculative ways in which the search for such deposits may be focused. In a few cases, areas with potential for undiscovered deposits are suggested.

After acknowledgments, the reader's entry into additional literature is aided by references or selected bibliography. As the present volume is a highly condensed review, only the references considered most useful to the reader are cited, and these are supplemented by other, uncited publications that are deemed particularly informative. The most highly recommended of these are indicated by asterisks.

Mineral deposits map

The map of Canadian mineral deposits (Fig. 2, in pocket) shows the distribution of approximately 1200 significant deposits, according to their mineral deposit type, size, and contained metals or minerals. In order to provide a relatively simple and readily distinguishable graphic display, the mineral deposit types that are described subsequently in this volume have been grouped into seven major classes which are represented on the map by different symbols. Five of the classes are based on the principal host rocks affiliated with the deposits: these are 1) sediment-associated deposits, 2) volcanic-associated deposits, 3) felsic and intermediate intrusion-associated deposits, 4) alkaline intrusion-associated deposits, and 5) mafic and ultramafic volcanic- and intrusion-associated deposits. The remaining two classes are vein and/or replacement (and other miscellaneous deposits), and placer deposits.

The size of the map symbol reflects the size of the deposit as determined by the total contained metals or minerals, including both past production and existing reserves. Four size classifications are shown; the criteria for determining the size of different deposits are given on the map. No distinction is made between producers, past producers, and nonproducers.

The colour of the map symbol reflects the prinicipal metals or minerals contained in each deposit. For example, red represents copper or copper-dominant deposits and yellow represents precious metals, including both gold and silver.

Not all deposits in the list that accompanies the map (in Appendix) are represented by individual symbols. In some places, one symbol may represent a number of deposits in mining districts such as Cobalt, Kirkland Lake, Highland Valley, and Noranda. Such deposits are listed under a single number, but have been assigned different letters to distinguish them.

The list of deposits has been compiled from a variety of sources, in particular CANMINDEX (an unpublished, index-level, computer-based file of Canadian mineral deposits and occurrences compiled by the Mineral Deposits Subdivision of the Geological Survey of Canada) and the volume "Canadian Mineral Deposits Not Being Mined in 1989" (Mineral Bulletin MR 223 of the former Mineral Policy Sector of Energy, Mines and Resources Canada, now the Mining Sector of Natural Resources Canada). Other sources included Provincial government publications and various trade journals, such as *The Northern Miner*. Complete resource data were not available for all deposits; in such cases, estimates of the deposit size were based on limited data, such as the approximate dimensions of the deposit and representative or typical grades.

Acknowledgments

The structure and organization of mineral deposit types that are presented here grew out of wide-ranging discussions over the years with mineral deposits geologists at the Geological Survey of Canada. Exceptional contributions of this nature, as well as overall guidance in determining the scope and structure of this volume, merit the special recognition of Charlie Jefferson and Rod Kirkham. An earlier publication, "Canadian Mineral Deposit Types: A Geological Synopsis" (Geological Survey of Canada, Economic Geology Report 36, 1984) served as a valuable nucleus around which the present volume grew.

Deep appreciation is expressed to all the contributors to this volume. Their efforts have made this undertaking possible. Editorial contributions by Charlie Jefferson, Ian Jonasson, Vlad Ruzicka, and Don Sangster helped bring the volume to completion and are gratefully acknowledged.

Technical help in the preparation of most of the illustrative material was provided by Richard Lancaster. Dave Garson prepared the standardized mineral deposit distribution maps and, together with Brian Williamson, Janet Carrière, and Robert Laramée, helped to maintain the orderly flow of manuscripts. Lo-Sun Jen of the Economic and Financial Analysis Branch, Natural Resources Canada supplied much of the data for Table 1, which was compiled by Andy Douma. Carole Plant and Lara O'Neill contributed word processing services.

A great debt of gratitude is owed to the many reviewers whose contributions added much to the form and content of the summaries. J.O. Wheeler reviewed an early draft of most of the volume. The critical reviewers of deposit type summaries include S.B. Ballantyne, C.T. Barrie, R.T. Bell, T.C. Birkett, A. Bjørlykke, D.R. Boyle, P. Černý, J.A. Donaldson, J.M. Duke, O.R. Eckstrand, J.M. Franklin, M.D. Hannington, L.J. Hulbert, J.L. Jambor, A.J.A. Janse, C.W. Jefferson, J.P. Johnson, I.R. Jonasson, B.A. Kjarsgaard, G. Lynch, A.N. Mariano, R.W. McQueen, L.D. Meinert, A.R. Miller, R.R. Miller, R.H. Mitchell, D. O'Hanley, A. Panteleyev, K.H. Poulsen, V.A. Preto, G.E. Ray, F. Robert, V. Ruzicka, D.F. Sangster, B.H. Scott-Smith, W.D. Sinclair, R.I. Thorpe, R.P. Wares, and D.H. Watkinson.

Authors' addresses

O.R. Eckstrand
Geological Survey of Canada
601 Booth Street
Ottawa, Ontario
K1A 0E8

W.D. Sinclair
Geological Survey of Canada
601 Booth Street
Ottawa, Ontario
K1A 0E8

R.I. Thorpe
Geological Survey of Canada
601 Booth Street
Ottawa, Ontario
K1A 0E8

Printed in Canada

1. PLACER URANIUM, GOLD

INTRODUCTION

Placer deposits represent concentrations of heavy minerals of certain elements, particularly of gold, uranium, and platinum, by sedimentary processes. Depending on the age and the state of their consolidation, the deposits are empirically classified as paleoplacers if they formed in ancient coarse siliciclastic rocks, or as modern placers if they are a part of Pliocene to Recent unconsolidated clastic sediments. According to the main commodities and the host environments the placers are subdivided into 1.1 Paleoplacer uranium, gold, and 1.2 Placer gold, platinum. Paleoplacer deposits are further subdivided into uraniferous and auriferous pyritic quartz pebble conglomerates and sandstones (subtype 1.1.1) which contain detrital pyrite and are older than 2.4 Ga; and auriferous hematitic conglomerates and sandstones (subtype 1.1.2) which are younger, and contain hematite in place of pyrite. The transition from pyrite to hematite with time marks the increase in oxygen in the Earth's atmosphere.

1.1 PALEOPLACER URANIUM, GOLD

S.M. Roscoe

INTRODUCTION

Paleoplacer concentrations of detrital heavy minerals containing elements such as Au, PGEs, Sn, W, REEs, Ti, Zr, Cr, Th, U, and Fe are common in indurated and metamorphosed fluvial to littoral clastic sedimentary rocks of all ages. There are, however, some remarkable differences between heavy mineral suites in Archean and very early Aphebian rocks and those in rocks younger than about 2.4 to 2.2 Ga. In the younger rocks, ferric oxide is present as an important, if not dominant, constituent of minerals in the allogenic suite and also as an ubiquitous authigenic constituent of the host rocks. These younger paleoplacers (subtype 1.1.2) are similar to unconsolidated recent placers except for easily decipherable and expectable modifications incurred during diagenesis and metamorphism. The ancient pyritic paleoplacers (subtype 1.1.1), on the other hand, contain allogenic and authigenic pyrite in lieu of magnetite and hematite. They occur in conglomerates containing clasts of quartz or resistant rocks, or in associated quartz-rich arenite beds, unlike the ferric oxide-bearing paleoplacers which occur in chemically immature as well as in mature clastic sedimentary rocks.

Paleoplacers, like placers, were formed wherever vigorous water or air currents have sorted heavy and light, large and small, mineral and rock clasts, and thus produced lenses or layers containing heavy minerals (and, commonly, relatively large clasts of lighter minerals and rocks) in greater abundance than elsewhere in a clastic sedimentary formation. The term is generally applied only to those heavy mineral accumulations that are notably enriched in particular commodities (e.g., gold paleoplacers) and implies a provenance containing special primary sources of these commodities in addition to sources of the more common heavy minerals. The question of whether gold, uranium minerals, and pyrite in subtype 1.1.1 deposits, pyritic quartz pebble conglomerates, were originally transported and deposited as detrital grains has long been a subject of debate. The empirical classification of the ancient pyritic deposits as paleoplacers is useful, however, because the ore minerals are concentrated – in association with undoubted heavy detrital minerals – in coarse, quartz-rich clastic metasedimentary rocks deposited under high energy conditions especially favourable for development of placer deposits.

Uraniferous pyritic quartz pebble conglomerates (subtype 1.1.1 deposits) in the Huronian Supergroup at Elliot Lake, north of Lake Huron in Canada (Fig. 1.1-1), have

been a major source of uranium since 1956. This is the only district in which this class of paleoplacer contains very little gold and has nevertheless been mined. Auriferous pyritic quartz pebble conglomerates (subtype 1.1.1 deposits), in the Witwatersrand Supergroup and Dominion Group in South Africa (Fig. 1.1-2), have produced far more gold than any other type of gold-bearing mineral deposit. They have also produced very important amounts of byproduct and coproduct uranium. Witwatersrand-type deposits are also mined in Brazil. In contrast to the very limited number of pyritic paleoplacer mining districts, pyritic paleoplacers containing only low concentrations of uranium and gold are common in formations of quartzose arenites more than 2.4 billion years old.

Deposits of subtype 1.1.2 have been mined on a small scale in many places and significant amounts of gold have been produced from hematitic quartz pebble beds in Ghana. In general, however, this most abundant type of paleoplacer is of little importance compared to either the much richer ancient pyritic Witwatersrand gold paleoplacers or unconsolidated placers containing relatively low concentrations of cheaply recoverable particulate gold.

IMPORTANCE

Subtype 1.1.1

Uraniferous pyritic quartz pebble conglomerate (Huronian Supergroup)

From initial production in 1957 until the end of 1992, the Elliot Lake district (also known as the Blind River area) has produced about 140 500 t[1] of uranium from ores grading about 0.09% U. This is 13.8% of total uranium production in the 'Western World' during the same period and 59.9% of total Canadian production. At peak annual production in 1959, 11 mines, with combined total mill capacity of 30 800 t per day, produced 9400 t – at that time 28% of 'Western World' production (Fig. 1.1-3). Elliot Lake ores and tailings contain major potential thorium, yttrium, and rare-earth element resources, and small quantities of these have been produced in the past. Remaining uranium resources in the Elliot Lake district are lower grade than past production. In 1993 only one mine (Stanleigh) was producing uranium.

Roscoe, S.M.
1996:   Paleoplacer uranium, gold; *in* Geology of Canadian Mineral Deposit Types, (ed.) O.R. Eckstrand, W.D. Sinclair, and R.I. Thorpe; Geological Survey of Canada, Geology of Canada, no. 8, p. 10-23 (*also* Geological Society of America, The Geology of North America, v. P-1).

[1]   Production data have been derived from various sources, most importantly: Mining Annual Review; Minerals Yearbook, United States Department of Interior; Mineral Information Bulletin, Mineral Resources Division, Department of Energy, Mines and Resources, Canada (and predecessor Department of Mines and Technical Surveys, Canada); Consolidated Goldfields, P.L.C., annual report on gold; annual reports of Rio Algom Ltd. and Denison Mines Ltd.

Auriferous pyritic quartz pebble conglomerates and quartzites (Witwatersrand Supergroup)

About 40 000 t of gold have been produced (to the end of 1987) from about 4 billion tonnes of pyritic quartz pebble conglomerate ores mined since 1886 when deposits of this type were discovered in the Transvaal, South Africa. Peak annual production of 1000 t reached in 1970, represented 79% of 'Western World' production. For a 38 year period, 1949 to 1987, these deposits yielded 75% of 'Western World'

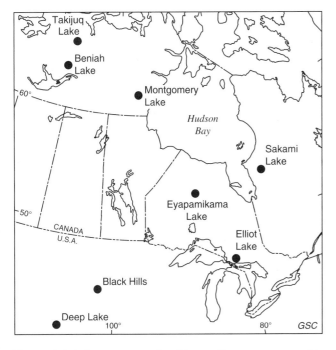

Figure 1.1-1 Occurrences of pyritic uraniferous quartz pebble conglomerates in North America (after Roscoe, 1981, 1990).

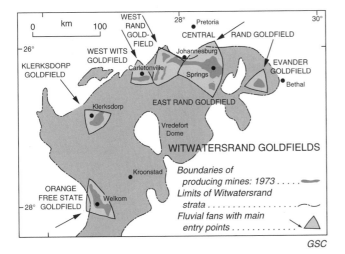

Figure 1.1-2. Areas of gold- and uranium-bearing conglomerates in the Witwatersrand goldfields, South Africa (after Tankard et al., 1982).

gold production. Annual production has decreased in recent years to about 600 t in 1987. This decrease and, more importantly, increasing production elsewhere in the world, notably Nevada, U.S.A., reduced South Africa's share of 'Western World' production to 44% in 1987.

Extraction of uranium from South African gold ores began in 1952 and between 1952 and 1987 has totalled approximately 125 600 t U, about 14% of the uranium produced in this period in the 'Western World'. Most Witwatersrand gold ores and resources contain more than 15 times more uranium than gold (Bourret, 1981) and some, as at Vaal Reefs, as much as 33 times more (Minter, 1981). Clearly, these resources and some 4 billion tonnes of tailings represent a major potential uranium resource. In addition to byproduct uranium, pyrite is recovered from some ores and used to produce sulphuric acid required by uranium extraction plants, and small amounts of osmium, iridium, and diamonds are recovered from South African conglomerate ores.

Subtype 1.1.2

Ghana gold production from hematitic paleoplacers

Production to date of at least 227 t (250 tons) of gold from hematitic conglomerates in Proterozoic quartzites of the Tarkwaian System in Ghana (Vogel, 1987) is substantial, but much smaller than that from vein-type deposits in metamorphic rocks of the underlying Birimian System, an outstanding gold-producing Proterozoic greenstone belt. Uranium is not significantly concentrated in the Tarkwaian conglomerates or in any other hematitic conglomerates.

SIZE AND GRADE OF DEPOSITS

Subtype 1.1.1

Huronian deposits

The main mining areas in the Elliot Lake district are about 10 km apart and extend down dip to depths of about 1200 m below surface on the north and south flanks of a syncline (Fig. 1.1-3). The largest, at Quirke Lake, extends 10 km in an east-southeast direction and is as much as 3.5 km wide. Conglomeratic zones are mined at several stratigraphic levels through a formational thickness of 100 m. The lowermost, and by far the most productive of these, was estimated to contain some 200 million tonnes of drill-indicated ore grading 0.10% U when mining began in 1956 (Roscoe, 1957). The Nordic zone, trending northwest, is 6 km long and as much as 2 km wide. The main zone in this area is overlain by a more extensive, lower grade conglomeratic section. Pronto and Agnew Lake mines, respectively 20 and 75 km south and east of Elliot Lake, mined relatively small remnants of once more extensive conglomerate zones.

Witwatersrand deposits

The Witwatersrand goldfields resemble huge fans containing thin sheets of pebble beds. Mine workings are centered on 'paystreaks' that radiate out from fan apices. Some paystreaks can be followed for more than 10 km. The East Rand fan, the largest of the six goldfields, according to

Pretorius (1974), "extends for 40 km down the central section from the apex of the fan to the base of the fan where it merges with the main lacustrian environment. The mid-fan portion is 50 km wide and the fanbase 90 km wide. The western lobe is 45 km long and the eastern lobe 20 km." Drill exploration has been extended in places to depths of nearly 5 km and mining to nearly 4 km. Some mines exploit two or more reefs (stratiform ore zones) that are widely separated stratigraphically. The uppermost sections of

some of these, truncated at unconformities, are at depths as great as 3 km. Pretorius (1974) cited average ore grades for all Witwatersrand production to 1972 as 9.2 g/t Au, with the especially rich Carletonville Goldfield averaging 19.4 g/t and the most uraniferous Krugersdorp (West Rand) Goldfield, 6.3 g/t Au. Ore processed in uranium extraction plants in the period 1953-1972 contained an average of 213 g/t (0.021%) U for the Witwatersrand as a whole, 188 g/t for Carletonville ores, and 477 g/t for Krugersdorp ores.

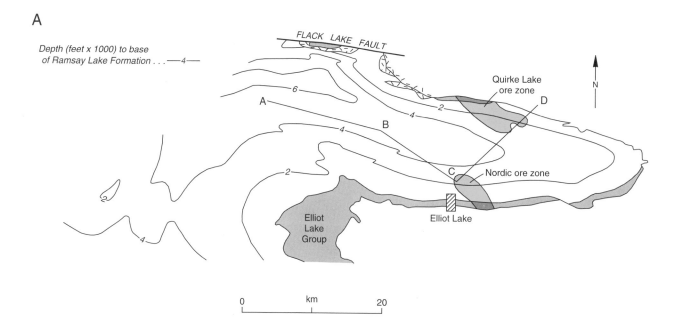

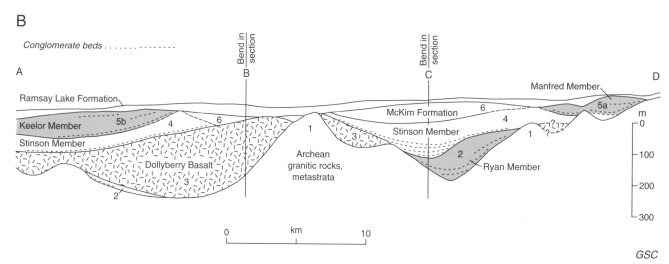

GSC

Figure 1.1-3. Stratigraphic relations in the Huronian Supergroup near Elliot Lake. **A)** Map showing outcrop of Elliot Lake Group (grey shading), depth to the base of the overlying Matinenda Formation, and the Quirke Lake and Nordic ore zones (red). **B)** Schematic cross-section showing relations among Ramsay Lake and McKim Formations and arenaceous members of the Matinenda Formation. Uraniferous members are shown in red, conglomerate beds are represented by dashed lines. Designation of units by numbers and patterns correspond to those shown in Figure 1.1-6. * For lithostratigraphy only <u>see</u> Table 1.1-1. (from Roscoe, 1981).

Subtype 1.1.2

Tarkwaian hematitic paleoplacers, Ghana

Mining operations have been carried out through areas as much as 3 km long and 1 km wide along a 12 km distance. Mining thicknesses have been between 1.2 and 7 m and ore grades between 3.5 and 14 g/t Au (Vogel, 1987).

GEOLOGICAL FEATURES

Pyritic paleoplacers (subtype 1.1.1) and many hematitic paleoplacers (subtype 1.1.2) occur in thick successions of clastic sedimentary rocks dominated by thick quartz arenite units that are chemically, mineralogically, and texturally submature to supermature. Major portions of these variably micaceous and feldspathic quartzite units, including very coarse grained members containing pyritic (or hematitic) quartz pebble beds, consist of channel-type crossbeds that show unimodal dip directions. They are interpreted as having been deposited under fluvial conditions, primarily in braided stream systems within broad valleys and, most importantly, on huge alluvial fans (Minter, 1978; Roscoe, 1981). In addition to such terrestrial deposits, the Huronian Supergroup and the South African host assemblage, which includes the Dominion Group, Witwatersrand Supergroup, and Ventersdorp Supergroup, contain volcanic rocks, quartz arenites with bimodal (herringbone) crossbedding that likely reflects deposition in the intertidal zone, siltstones and argillites deposited in standing water, carbonate rocks, glaciogenic sediments, and iron-formation (in Witwatersrand).

Conglomeratic zones in the Matinenda Formation in the lower part of the Huronian Supergroup have been mined for uranium in two extensive areas near the town of Elliot Lake, north of Lake Huron midway between Sudbury and Sault Ste. Marie, and at two outlying areas south and east of Elliot Lake. Uneconomic occurrences of uraniferous pyritic paleoplacers have been found elsewhere in Canada (Roscoe and Donaldson, 1988; Roscoe et al., 1989; Roscoe, 1990; Fig. 1.1-1): Sakami Lake in Quebec; Eyapamikama Lake in northwestern Ontario; Beniah Lake 130 km northeast, and Takijuq Lake 400 km north, of Yellowknife, Northwest Territories, and in the Montgomery Lake Group near Henik Lakes, Northwest Territories (Roscoe, 1981). Sub-ore grade occurrences have also been discovered at Deep Lake in the United States in the Deep Lake Group and Phantom Lake rocks in the Medicine Bow and Sierra Madre mountains of Wyoming and at Nemo in the Black Hills of South Dakota (Houston and Karlstrom, 1987).

In South Africa, auriferous pyritic quartz pebble conglomerates, with U/Au ratios less than 50, occur in the Witwatersrand Supergroup throughout its 350 by 200 km extent (Fig. 1.1-2). Six gold mining areas, termed 'goldfields' are distributed along an arc 420 km long, concave to the southeast, the direction of sediment transport (Fig. 1.1-2). Mining has also been carried out in much less extensive older strata of the Dominion Reef Group, which contain conglomerate beds with U/Au ratios very much higher (Von Backström, 1981) than Witwatersrand ores, but distinctly lower than the approximate 10 000 U/Au ratios of Elliot Lake ores. Radioactive auriferous pyritic conglomerates are also present in the older Pongola Supergroup

(Saager et al., 1987) 400 km southwest of, and the Pietersburg greenstone belt 280 km northeast of, the Witwatersrand goldfields (Meyer et al., 1987). A pyritic gold paleoplacer is mined near Jacobina, Bahia, Brazil (Gama, 1982), and other pyritic quartz pebble beds near Belo Horizonte, 975 km south of Jacobina, have been explored for uranium (Villaca and Moura, 1981). Similar beds are present at several localities in the Pilbara Block in Western Australia (Carter and Gee, 1987) within the lower Fortesque Group (about 2.7 Ga) and within the Gorge Creek Group (about 3.0 Ga). Small amounts of gold were mined from Fortesque beds early in the century. Auriferous radioactive pyritic quartz pebble beds have also been found in Karnataka State in southern India (Srinivasan and Ojakangas, 1986) and in the Singhbhum craton near Calcutta (Rao et al., 1988).

Occurrences of pyritiferous conglomerates have been reported in Karelia and Ukraine (Salop, 1977), but of these, some in Finland are epigenetic deposits, more akin to sandstone-type or unconformity-type deposits, in conglomerate beds that elsewhere in the vicinity and regionally are hematitic rather than pyritic (Aikas and Sarikkola, 1987). Others are not well enough documented to establish that they are stratiform and pyritic throughout entire formations, as are the Huronian and Witwatersrand deposits.

The most noteworthy auriferous hematitic paleoplacers (subtype 1.1.2) are in the Tarkwaian "System" in Ghana (Vogel, 1987), which was deposited after about 2135 Ma (Davis et al., 1994). Monazite-, zircon-, and rutile-bearing hematitic quartz pebble beds in the Lorrain Formation in the upper part of the Huronian Supergroup, Ontario, although not economically significant, are of special interest because they are the oldest known concentrations of subtype 1.1.2 hematitic, nonpyritic heavy minerals. Their general association with minor redbeds has been considered evidence that the atmosphere changed during deposition of the Huronian sequence to one that, for the first time, was capable of oxidizing iron minerals (Roscoe, 1969).

The Huronian succession (Canada)

The *Huronian Supergroup* comprises 4 groups of formations that form southerly thickening wedges with aggregate maximum thicknesses totalling as much as 15 km according to Roscoe and Card (1992). In ascending order (Table 1.1-1) these are: the *Elliot Lake Group*, up to 4000 m thick; *Hough Lake Group,* up to 3700 m thick; *Quirke Lake Group*, up to 2400 m thick; and *Cobalt Group*, up to 5000 m thick. The Elliot Lake Group contains the only volcanic formations in the Huronian succession and relatively fine grained pelitic sediments, as well as arenaceous units and pyritic quartz pebble conglomerate beds. The basal unit of the Hough Lake Group, the *Ramsay Lake Formation*, consists of polymictic paraconglomerate that disconformably overlies the eroded surface of Elliot Lake Group rocks or, in many northern areas, Archean basement rocks. Similar conglomeratic units, the *Bruce* and *Gowganda formations*, respectively, form the bases of the Quirke Lake and Cobalt groups, and in each of the three upper groups, paraconglomerate is succeeded by fine grained sediments, the *Pecors, Espanola,* and *Firstbrook formations*, which in turn are overlain by great thicknesses of quartz arenites, the *Mississagi, Serpent,* and *Lorrain formations*. At some places west of Elliot Lake, an unconformity at the base of the Gowganda Formation cuts out almost all of the Quirke

13

Table 1.1-1. Stratigraphy of the Huronian Supergroup at Elliot Lake, Canada (from Roscoe, 1969).

Group	Formation			Composite lithological sequence
Cobalt	Bar River			quartzite, red siltstone quartzite
	Gordon Lake			varicoloured siltstone
	Lorrain			quartzite arkose
	Gowganda			reddish argillite, argillite conglomeratic greywacke, grey and pink arkose
Quirke Lake	Serpent			arkose-subgreywacke
	Espanola			dolomite, siltstone siltstone, greywacke limestone
	Bruce			conglomeratic greywacke
Hough Lake	Mississagi			coarse subarkose
	Pecors			argillite, siltstone
	Ramsay Lake			conglomeratic greywacke
Elliot Lake	McKim			subgreywacke, argillite
	Matinenda			gritty subarkose
		Copper Cliff		acid volcanics
	Thessalon	Pater	Stobie	basic volcanics
	Livingstone Creek			subarkose

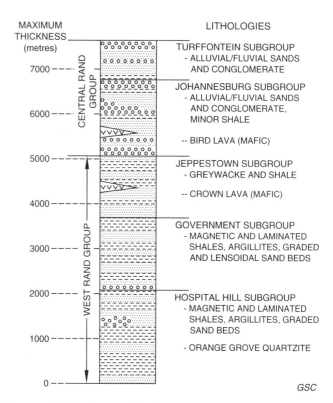

Figure 1.1-4. Generalized stratigraphic column of the Witwatersrand Supergroup, South Africa. Circles = conglomerate; dots = sands, greywacke, quartzite; dashes = shale, argillite; vees = volcanics (after Tankard et al., 1982).

Lake Group. The cyclicity of formations has been interpreted as related to glaciations accompanied by depression, followed by deglaciations resulting in rapid isostatic uplift and resumption of fluvial sand deposition (Frarey and Roscoe, 1970).

South African paleoplacer hosts

The following outline of the major South African paleoplacer host succession (Fig. 1.1-4 and 1.1-5) is taken largely from Pretorius (1976) and Tankard et al. (1982). The major host units include, in ascending order, the 3.07 Ga volcanic *Dominion Group* (2710 m thick), *Witwatersrand Supergroup* (Fig. 1.1-4), including the West Rand Group (4670 m) which contains marine sediments, and the Central Rand Group (2420 m) of dominantly fluvial clastic sediments; and the dominantly volcanic *Ventersdorp Supergroup* (7860 m; Fig. 1.1-5) which has yielded a zircon U-Pb age of 2.71 Ga (Armstrong et al., 1990).

Paleoplacers in the Dominion Group are restricted to a basal unit of submature arenite, as much as 60 m thick, in the *Renosterhoek Formation*. This unit, deposited in southwesterly flowing streams within paleovalleys on granitic basement rocks, is overlain by 1100 m of subaerial mafic to intermediate volcanic rocks containing paleosol zones. The uppermost 1550 m of the group consist of quartz and feldspar phyric lavas that have given a zircon U-Pb age of 3.07 Ga (Armstrong et al., 1990).

The West Rand Group consists mainly of shale and sandstone deposited in both fluvial and marginal marine environments. The lowermost part, the *Hospital Hill Subgroup*, contains a contorted iron-formation unit near its base, and some minor subeconomic paleoplacer beds and tilloidal conglomerate near its top. The overlying *Government Subgroup* contains paleoplacers in the base of upward-fining sequences interpreted as fans that prograded southwest and southeast over tidal deposits. Ten of these deposits produced more than 35 t of gold near the turn of the century. The Coronation Formation, near the middle of the subgroup, contains tillite overlain by magnetic shale. The *Jeppestown Subgroup* at the top of the West Rand Group contains a 250 m sequence of thin amygdaloidal basalt lava flows, *Crown lava*, that extends throughout the preserved part of the West Rand depository. Some minor gold production was obtained from paleoplacers in this subgroup in the early part of the century. The dominantly shaly Roodepoort Formation, as thick as 600 m, caps the subgroup.

The Central Rand Group contains most of the productive gold paleoplacers. The Booysens Formation and equivalent shaly beds mark the top of a lower division, the *Johannesburg Subgroup*, as much as 1500 m thick, which is overlain by an upper division, the *Turffontein Subgroup*, as much as 1800 m thick. Placers at or near the base of the Johannesburg Subgroup, including the *Carbon Leader, Middelvlei, Main,* and *Main Leader placers*, have been

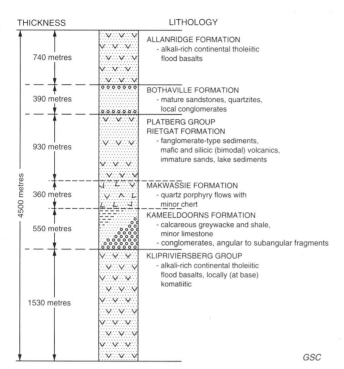

THICKNESS   LITHOLOGY

740 metres
ALLANRIDGE FORMATION
- alkali-rich continental tholeiitic
  flood basalts

390 metres
BOTHAVILLE FORMATION
- mature sandstones, quartzites,
  local conglomerates

930 metres
PLATBERG GROUP
RIETGAT FORMATION
- fanglomerate-type sediments,
  mafic and silicic (bimodal) volcanics,
  immature sands, lake sediments

360 metres
MAKWASSIE FORMATION
- quartz porphyry flows with
  minor chert

550 metres
KAMEELDOORNS FORMATION
- calcareous greywacke and shale,
  minor limestone
- conglomerates, angular to subangular fragments

1530 metres
KLIPRIVIERSBERG GROUP
- alkali-rich continental tholeiitic
  flood basalts, locally (at base)
  komatiitic

4500 metres

GSC

Figure 1.1-5. Generalized stratigraphic column of the Ventersdorp Supergroup, South Africa (after Winter, 1976, and SACS, 1980).

particularly important producers in the northern goldfields. Broad channels filled with diamictite are also present at the base of the subgroup. In the upper part of the subgroup the *Steyn, Basal,* and *Vaal placers* are most important in the Welkom and Klerksdorp goldfields, the *Bird* in the East Rand Goldfield. Four especially uraniferous placers are mined in the upper part of the subgroup in the Krugersdorp Goldfield. Lavas (*Bird amygdaloid)* are present in the central part of the subgroup south and east of Johannesburg. These are about 200 m thick near Johannesburg and thicken easterly to about 1200 m in the Evander area, where an intrusive sill is also present. Sedimentary rocks, on the other hand, thin easterly. The Turffontein Subgroup contains conglomeratic zones and individual massive conglomerate beds which are thicker and contain much larger pebbles than are present lower in the succession, yet numerous placers in this upper subgroup have been less productive than thin placers in the underlying Johannesburg Subgroup. Extensive conglomeratic sections are barren, low grade, or contain patchy gold concentrations. Important Turffontein placers have been mined in the Welkom, Klerksdorp (*Cristalkop, Gold Estates,* and *Orkney),* and Krugersdorp (*Kimberley placers)* goldfields. The only placer mined in the Evander Goldfield is in the Turffontein Subgroup.

Volcanic and sedimentary rocks of the Ventersdorp Supergroup concordantly to unconformably overlie the Witwatersrand Supergroup and overlap its eroded margins onto Dominion Group and older rocks to extend, mainly in the subsurface, over an area 800 km long and as much as 300 km wide. Two unconformities separate three groups of formations collectively comprising as much as 5775 m of strata. The lower one, the *Klipriviersberg Group,* consists

of up to 1830 m of continental tholeiitic basalts with a basal veneer of fluvial sediments, including the very important *Venterpost placer* (formerly called Ventersdorp contact reef or VCR). Overlying formations include fault controlled alluvial fan sediments, andesitic volcanic rocks, and quartz phyric flows as much as 2100 m thick (*Makwassie Formation*) that have given a zircon U-Pb age of 2714 ± 8 Ma (Armstrong et al., 1990).

The youngest auriferous pyritic conglomeratic beds in southern Africa are in the *Black Reef Formation* at the base of the Chuniespoort Group of the Transvaal Supergroup. This group, which consists mainly of dolomite and iron-formation, unconformably overlies Ventersdorp and older rocks and is unconformably overlain by the Pretoria Group. A Rb-Sr whole rock age of 2.2 Ga (Burger and Coertze, 1975) has been obtained for volcanic rocks (*Hekpoort Formation*) in the latter group. Paleosol above the volcanic rocks (Retallack, 1986) shows evidence of development of an oxygen-bearing atmosphere believed by many to be unfavourable for the formation of pyritic placers.

The oldest gold paleoplacers are in the Pongola Supergroup. These lie stratigraphically above volcanic rocks that gave a zircon U-Pb age of 3.09 Ga, and rhyolite within the Pongola Supergroup has yielded a zircon U-Pb age of 2940 ± 22 Ma (Hegner et al., 1984).

Geological setting of Elliot Lake paleoplacers

Pyritic quartz pebble conglomerate uranium ores of the Elliot Lake district occur in the Matinenda Formation of the Elliot Lake Group, which is the lowest Group of the Huronian Supergroup succession (Table 1.1-1). East and south of the main ore zones in the Elliot Lake area, feldspathic, micaceous quartzite, grit, and quartz pebble beds of the Matinenda Formation nonconformably overlie paleosol-mantled Archean basement rocks. The latter are highly deformed metavolcanic rocks, likely about 2.7 Ga in age, that have been intruded by granite believed to be more than 2.65 Ga on the basis of the youngest zircon U-Pb ages obtained for comparable rocks in the region (Krogh et al., 1984). West of the ore zones, Matinenda beds were deposited atop the eroded, weathered surface of the Dollyberry Formation, a succession of subaerial tholeiitic basalt flows that is locally as much as 300 m thick. It too overlies paleosol-mantled Archean basement rocks and, locally in outcrop and drill core 20 km west of Quirke Lake, is separated from basement by a cushion of coarse Huronian clastic sediments including radioactive pyritic conglomerate. This unit, termed Crazy Lake Formation, is poorly sorted but otherwise is not unlike gritty facies of the Matinenda Formation.

Elliot Lake ore zones

The *Nordic* and *Quirke* ore zones and the *Keelor* conglomerate zone (Fig. 1.1-3) are in lenses of Matinenda subarkose termed, respectively, *Ryan, Manfred,* and *Keelor* members. These are characterized by quartz grains that are generally very coarse, but highly varied in grain size and variably sorted, as reflected by greenish colours imparted to the rocks by matrix sericite. Subarkose in the *Stinson Member,* which overlies the Ryan Member and underlies the Manfred and Keelor members, is better sorted, more homogeneous, finer grained, and paler coloured. A distinctive nonpyritic, nonradioactive conglomerate characterized by rounded, well sorted, tightly packed clasts of basalt and subordinate to dominant quartz clasts marks the base of the member. Throughout much of the area, the Stinson conglomerate is merely a layer of single small pebbles, mainly of quartz, but where it occurs above the Nordic ore zone and above Dollyberry Basalt, it contains massive layers of large basalt cobbles.

The Ryan Member is as much as 170 m thick along a northwesterly trending axis that is close to the centreline of the Nordic ore zone (see Pienaar, 1963, Fig. 8). It overlies Archean metavolcanic and metasedimentary rocks and also Huronian basalt and patches of basal conglomerate. The main ore body, *Nordic reef*, the underlying productive *Lacnor reef*, and the overlying less productive *Pardee reef*, occupy a 30 m stratigraphic interval 0 to 30 m above the base of the member. Other pyritic quartz pebble beds and lenses occur throughout overlying coarse Ryan subarkose, but upwards through the pile they become sparser and thinner, contain smaller pebbles, and are less uraniferous. The southeasterly elongated Nordic zone parallels paleocurrent flow as indicated by foreset dips and is bordered on the northeast by a basement ridge of iron-formation against which the Lacnor and Nordic reefs abut. The Pardee reef and overlying Ryan strata, however, cross this ridge and the overall form of the Ryan 'bulge' resembles a fan with a convex upper surface and an apex near the north end of the Nordic ore zone. Deposition may have involved valley filling with upstream (northwesterly) migration of the fall line to a notch in a steeper slope from which alluvial braided distributaries built a broad fan. The original morphology and even the original depositional limits of the member, however, are masked by erosion preceding and accompanying deposition of the Stinson Member and the Ramsay Lake Formation. Thick zones in the Stinson Member mimic those in the underlying Ryan Member and may represent superposed fans (Pienaar, 1963, Fig. 16). Particularly thick sections, as much as 130 m, overlie flanks of the Nordic ore zone and are complementary to thinned sections in the Ryan Member, corresponding to conglomerate-filled channels that cut deeply into the lower member, locally even removing sections of ore zones. The basalt clasts were derived from the Dollyberry Formation, which became a prolific source of such clasts only after the Ryan Member was deposited. This led Roscoe (1981) to suggest that the Dollyberry Formation was erupted and built into a high-standing edifice after deposition of the Ryan Member as well as after deposition of the Crazy Horse Formation. If others, notably Bennett (1979), are correct in believing that only one episode of volcanism occurred in the western part of the Huronian belt and that this preceded deposition of the entire Matinenda Formation, then a structural block containing the Dollyberry Formation must have been uplifted abruptly following deposition of the Ryan Member.

The Manfred Member, hosting the Quirke Lake ore zones, nonconformably overlies Archean basement rocks to the north, and Dollyberry Basalt flows to the west, and to the south it conformably overlies a grey subarkose unit that Robertson and Steenland (1960) and Roscoe (1969) considered correlative with the Stinson Member. The Denison reef zone (C reef in Rio Algom Mines), by far the most important in the district, is at or near the base of the Manfred Member. It is a section of pyritic conglomerate, conglomeratic subarkose, and subarkose lenses and is as much as 10 m thick to the north, but thins and contains reduced proportions of conglomerate towards the south and southeast down the paleoslope. A medial part up to 2.5 m thick, nearly devoid of pebble layers, is traceable throughout the entire mining area and divides the reef into upper and lower sections comparable to the closely spaced Nordic and Lacnor reefs of the Nordic ore zone. In many areas the upper and lower parts have been mined separately or only one has been mined. The Manfred Member, as much as 100 m thick (Pienaar, 1963) in the northwestern part of the Quirke Lake mining zone at the boundary between the Quirke mine and Denison mine properties, contains abundant conglomeratic layers in addition to the Denison reef. Of these, the Quirke reef (also known as A reef and at Denison as E reef), located 30 m above the Denison reef, has been extensively mined. Other sections with abundant pyritic quartz pebble layers, both below and above the Quirke reef, can be followed considerable distances. The upper surface of the Manfred Member was extensively eroded prior to deposition of the Ramsay Lake Formation, a glacial mixtite unit (Fralick and Miall, 1987) at the base of the Hough Lake Group. This unconformity cuts down through the entire member in some areas in both the northern and southern parts of the Quirke Lake mining zone and the basal part of the Ramsay Lake Formation has become enriched in quartz grains and pebbles, pyrite, and radioactive minerals derived from unconsolidated Matinenda sands and gravels (see Fralick and Miall, 1987).

The Keelor Member, and a contained subeconomic conglomeratic zone, 15 to 30 km west of Quirke Lake, overlies Dollyberry Basalt in its proximal northern part and, like the Manfred Member (Fig. 1.1-6), overlies Stinson subarkose in more distal portions to the south (Roscoe, 1981). The most northerly, deepest, most proximal part of the Keelor Member has not yet been delimited by drilling, but it too is truncated by the Ramsay Lake Formation, and lower conglomerate beds abut against rises in the surface of underlying volcanic rocks. Southerly parts of the most conglomeratic section, however, have not been affected by sub-Ramsay Lake erosion and its original fan-like form is apparent in sections and isopachal plans drawn from drill core data (Roscoe, 1957). Deposition of the coarse detritus of the Keelor and Manfred members can be considered to have been initiated at points where streams debouched off Huronian lava fields and Archean uplands onto more gently sloping surfaces of Huronian sands (*Stinson bajadas*). Gravel and coarse sand deposits were built up, evidently both headward and downslope, as alluvial fans and coalescent fans, by distributary braided streams (Roscoe, 1981).

Early investigators of the Elliot Lake deposits noted that the conglomerate orebodies are largely within or overlie depressions in the pre-Huronian surface (e.g., Hart et al., 1955; Roscoe, 1957; Derry, 1960; Robertson and Steenland, 1960; Pienaar, 1963). Most considered this consistent with subaerial deposition of the conglomeratic zones as gravel beds in river valleys. Robertson and Steenland (1960), however, considered that the conglomerates and other Matinenda strata were deposited on marine deltas. This would explain thickness variations of Matinenda units that far exceed sub-Matinenda paleotopographic relief as indicated by variations in thickness of basal units. Construction of alluvial fans from fan heads that migrated headward up valleys (Roscoe, 1969) is a more likely explanation, in the absence of changes in lithologies and sedimentary structures that would suggest transitions from high energy fluvial environments to marginal marine environments.

Uranium concentrations in Huronian strata other than the Matinenda Formation

Beds of uraniferous pyritic quartz pebble conglomerate as much as 1 or 2 m thick are interlayered with thin amygdaloidal basalt flows near the base of the Thessalon Formation, 60 to 115 km west of Elliot Lake, and also in the uppermost quartzite beds of the Livingstone Creek Formation that underlies the volcanic rocks. Correlations of these formations with Matinenda units and volcanic rocks at Elliot Lake are uncertain, but Bennett (1979) considered the Livingstone Creek Formation to predate the latter strata, with the exception of the Crazy Lake Formation. The Mississagi Formation of the Hough Lake Group contains geographically and stratigraphically widespread lenses of small quartz and black chert pebbles with associated pyrite, zircon, and monazite. Some proximal facies of the formation in northern parts of the belt consist of coarse subarkose, grit, and substantial radioactive conglomerate beds lithologically similar to major components of the Matinenda Formation at Elliot Lake.

In the Wanapitei Lake-Temagami Lake area, 25 to 70 km north and northeast of Sudbury, the Mississagi Formation, along with some basal lenses of fine grained clastics, polymictic conglomerate, and quartz pebble conglomerate, nonconformably overlies Archean basement rocks (Long, 1987). Several radioactive conglomerates have been found to contain erratic concentrations of gold much higher than any found in Elliot Lake ores. This is interesting as the Temagami occurrences are only 15 km to 35 km from several gold deposits in basement greenstone rocks, whereas the Elliot Lake area is 140 km south and southeast of greenstone belts that contain known deposits that could have been sources of detrital gold. Several uranium occurrences north of Sudbury, in and near Roberts Township, are noteworthy as they occur in fine grained quartzite and siltstone, rather than conglomerate, and contain abundant ilmenite, as well as pyrite, rutile, and zircon. The heavy minerals, which appear to be detrital grains in hydraulic equilibrium (H.R. Steacy, pers. comm., 1976; Goodwin, 1980 in Long, 1987), occur in graded laminae.

Minerals and mineral zonations

Pyrite occurs as several types of grains in matrices of Elliot Lake and Witwatersrand pyritic conglomerates along with quartz, feldspar, and muscovite. It is the most abundant heavy mineral, averaging about 3% in Witwatersrand ores, and more than this in Elliot Lake ores. Other heavy minerals concentrated in South African and/or Huronian pyritic quartz pebble conglomerates include garnet, chromite, sphene, rutile, monazite, xenotime, apatite, tourmaline, ilmenite, columbite, corundum, cassiterite, osmiridium, and diamond. Modes of occurrence and shapes and sizes of uraninite grains (Liebenberg, 1958; Ramdohr, 1958; Roscoe, 1969) and of some of the gold (Hallbauer, 1981) and pyrite are consistent also with a placer origin, although some pyrite, gold, uraninite (or pitchblende), and other uranium minerals are authigenic or have been remobilized. Feather and Koen (1975) have identified 70 minerals in

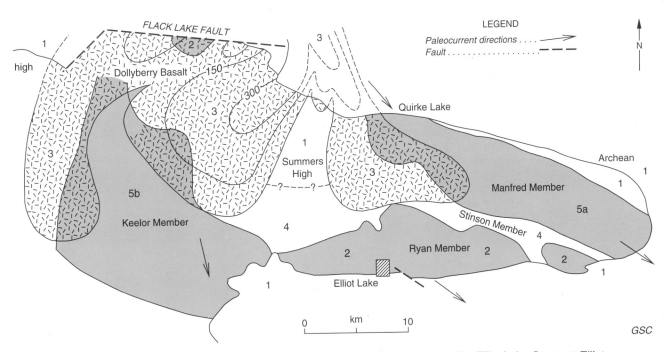

Figure 1.1-6. Subsurface distribution of conglomeratic and volcanic strata in the Elliot Lake Group at Elliot Lake. Designations of units by numbers and patterns are as shown in Figure 1.1-3; red indicates uranium-bearing units; contour lines indicate approximate thickness (m) of unit 3 (from Roscoe, 1981).

Witwatersrand ores. These include common sulphides and sulpharsenides. Kerogen, also referred to as carbon, hydrocarbon, and thucholite, is abundant in some, but not all, South African ores. It occurs as granules and as concordant seams associated with and incorporating heavy minerals. The seams commonly have columnar structure and some are in layers that apparently represent distal pebble-free extensions of thin conglomerate ore beds. They are commonly uranium-rich and may be phenomenally rich in gold.

Origin of kerogen

The character and origin of kerogen in the ancient paleoplacers has been intensely studied and debated (Liebenberg, 1955, 1958; Hallbauer, 1981; Schidlowski, 1981; Zumberge et al., 1981; Ruzicka, 1981; Mossman and Dyer, 1985). Liebenberg considered that the "hydrocarbon" was introduced as a fluid, fixed as a solid by the polymerizing effect of radiation, and that this process was accompanied by dissolution of detrital uraninite and dispersal of uranium and thorium through the resultant 'thucholite'. Others have thought that the kerogen represents algal mats that grew in situ, and that these organic mats trapped uranium and gold both mechanically and chemically. Hallbauer (1975) found extremely delicate, branching cellular uraniferous filaments in organic seams. He considered these to represent microfossils of fungal hyphae, the columnar forms to have been lichen, and spheroidal particles to have possibly been spores. Development of such relatively advanced organisms in Archean time and the preservation of extremely delicate structures in rocks of greenschist facies is difficult to accept. According to Schidlowski (1981), the 'thucolite' might have formed relatively late where fluidal hydrocarbon came in contact with radioactive minerals. Relationships between the character of kerogen and proximity to uraninite grains (Landais et al., 1990) lend support to this hypothesis. It is noteworthy that methane is abundant enough in some South African mines to be a problem. Underlying strata of the West Rand Group offer a credible source.

In the Elliot Lake mines, concentrations of kerogen nodules with fibrous sepiolite, and crystals of calcite, pyrrhotite, pyrite, pale sphalerite, and galena are common in open fissures that are not confined to ore zones. The hydrocarbon is only very weakly radioactive. The galena contains 'future' lead so rich in ^{206}Pb that it must have had a radon parent. Kerogen in these occurrences is only very weakly radioactive. It is noteworthy that gas may be observed bubbling out of water-filled fractures in Denison mine and that methane, radon, and helium have been identified in the gas. Concordant seams of kerogen, such as those described by Ruzicka and Steacy (1976) and Ruzicka (1981), are rare. They appear to be limited to small thin lenses in which they are associated with laminae composed almost entirely of uraninite grains. Kerogen (identified by its volatility under the X-ray beam of a microprobe) also occurs as irregular random replacements within some, but not all, uraninite grains in ores other than uraninite-rich layers, according to Ruzicka (1981). Ruzicka (1981) considered that the concordant kerogen seams were formed in situ during sedimentation by organisms, but it seems reasonable to suppose that both types of kerogen occurrences could be postore, formed from transient methane.

Mineral and grain size zonations in conglomerates

The Denison reef and other conglomeratic sheets in the Quirke Lake mining area show pronounced decreases in pebble size southerly and easterly down the paleoslope, and reduction in U/Th ratios concomitant with reductions in uraninite contents and increases in monazite contents (Robertson and Steenland, 1960). Titanium mineral abundance also increases downstream from the richest uraninite concentrations, and zircon is enriched in the most distal small pebble conglomerate and grit (Roscoe, 1969). Changes in abundance and size of quartz clasts, pyrite, uraninite and other mineral grains have been documented quantitatively by Theis (1979). The zonations can only be interpreted as the results of hydraulic sorting of detrital grains including uraninite. Hydraulic equivalence of ore minerals has also been noted in South Africa (Koen, 1961; Coetzee, 1965; Hiemstra, 1968). Minter (1976) has documented uraninite-gold zonations in Witwatersrand ores. Gold is concentrated with detrital uraninite, but the highest uranium/gold ratios are downstream from the highest gold contents, suggesting that some gold was originally deposited as particles with average equivalent hydraulic sizes larger than those of most of the uraninite grains carried by the streams.

Tectonic settings of pyritic conglomerate-bearing successions

A variety of tectonic settings have been proposed for the major tectonostratigraphic blocks that contain pyritic paleoplacers. Bickle and Eriksson (1982) considered the Witwatersrand Supergroup to have been deposited in a tensional regime. Burke et al. (1985, 1986) proposed that the Dominion Group may be part of an Andean-type volcanic arc, that the Witwatersrand Supergroup filled a foreland basin developed when the Kaapvaal and Zimbabwe cratons collided at 2.7 Ga, and that deposition of the Ventersdorp Supergroup was a result of rifting caused by this collision. Clendenin et al. (1987) have proposed that the Dominion-Witwatersrand, Ventersdorp, and Chuniespoort depositional systems were developed in a series of successor basins formed through a 3.1 to 2.2 Ga interval in response to changes in stresses generated first by subduction of oceanic crust and finally by collision of the Zimbabwe Craton with the Kaapvaal Craton. One wonders whether this time interval is not excessive, and thus whether it might not be more reasonable to consider (a) the Ventersdorp eruptions (Van Niekerk and Burger, 1978) as a prelude to opening of an ocean at ca. 2.6 Ga and deposition of Chuniespoort dolomites and iron-formation as passive margin deposits, and (b) the disconformably overlying fine grained clastic sediments of the Pretoria Group as the fill of post-collision (2.6-2.2 Ga) foreland basins. Models invoking ocean closing and continental collision a few hundred kilometres north of the present erosional edge of the Witwatersand Supergroup perhaps unnecessarily constrain its possible provenance area, as evidence for such a collision during the period postulated is not compelling.

Various models have also been proposed for the tectonic setting for deposition of the Huronian Supergroup. Young (1983) proposed deposition in an aulacogen that opened southeastward into the region that is now the Grenville

Province. Card (1978) considered the Huronian depository to have developed as an elongate intracratonic graben. Zolnai et al. (1984) regarded the Huronian Supergroup as a passive margin sequence deposited along the southern margin of the Superior Province Archean craton and deformed ca. 1.9 Ga when its southern part was overridden by an allochthonous terrane. Fahrig (1987) postulated that deposition began with rifting that was accompanied by intrusion of the Matachewan mafic dyke swarm and extrusion of the early Huronian lavas, then continued on the passive south margin of the Archean craton. A U-Pb zircon age of 2454 ± 2 Ma has been reported (Heaman, 1988) for a dyke of the Hearst-Matachewan swarm. This compares with a U-Pb age of 2450 +25/-10 Ma for felsic volcanic rock (Coppercliff Formation) near the base of the Huronian Supergroup.

A south-facing passive margin model has also been proposed for the Snowy Pass Supergroup in Wyoming (Karlstrom et al., 1983), which contains a succession that is comparable unit for unit with the succession in the western part of the Huronian Supergroup. These similarities are so remarkable that it is inconceivable that accumulation of the two occurred in separate depositories. A reconstruction permitting their deposition as one and the same succession, and their subsequent rifting apart is possible. This reconstruction, with the Snowy Pass belt and the Wyoming craton rotated about 140° (Roscoe and Card, 1993), matches and extends the centripetal pattern of fluvial paleocurrents exhibited in the Huronian belt. It also adds subjacent strata (Phantom Lake suite) to Elliot Lake Group rocks deposited south of Sault Ste. Marie. The reconstructed depository, with effluent direction constrained towards the south or southeast, is not unlike the Witwatersrand depository with its semicircular centripetal inflow. Parts of the Baltic Shield, which contain glaciogenic, hematitic paleoplacers and other strata similar to the Cobalt Group, as well as rocks similar in character and age to those in the Superior Province, may have formed the eastern part of the proposed combined Huronian-Snowy Pass depository. According to this scenario, the early Aphebian deposition of dominantly fluvial 2.45 Ga sediments seems likely to have occurred in a tensionally strained zone that had a northerly trend, like the 2.45 Ga Matachewan dyke swarm, within a large, newly assembled (2.7 Ga) continent that was being uplifted due to trapped heat, including radiogenic heat generated relatively rapidly in the ancient rocks. It can be further postulated that the large continent began to break up at 2.2 Ga when large volumes of gabbro were intruded in the Huronian area, the Lake Superior area, the Wyoming cratonic segment, and the Baltic segment. Strata containing iron-formation and carbonate rocks were deposited subsequently on passive margins, and perhaps in later collision related basins, on the Superior Province, the Wyoming Province, and the Baltic Shield. The Witwatersrand Supergroup, along with pyritic paleoplacer-bearing rocks in Brazil, Australia, and India, might have been involved in a similar sequence of events from 3.0 to 2.6 Ga.

Age of paleoplacers

Pyritic paleoplacers (subtype 1.1.1) occur in strata 3.0 Ga old and older in southern Africa (Saager and Muff, 1978), Australia (Carter and Gee, 1987), and India (Srinivasan and Ojakangas, 1986). Gold-uranium pyritic paleoplacers

in the Dominion Group are 3.07 Ga; those in the Witwatersrand Supergroup, between 2.95 and 2.71 Ga, according to the U-Pb isotopic ages of zircons from associated volcanic formations and intrusive or unconformably underlying felsic igneous rocks. Huronian pyritic paleoplacer uranium ores are nearly coeval with rhyolite of the Coppercliff Formation dated at 2.45 Ga according to U-Pb zircon data of Krogh et al. (1984) and with Matachewan diabase dykes of this age (Heaman, 1988). No succession containing pyritic paleoplacers has been found to be younger than 2.4 Ga, although some have younger possible minimum ages, in the range 2.2 Ga to 1.9 Ga.

Hematitic paleoplacers (subtype 1.1.2) comparable to recently formed placers, on the other hand, have not been found in strata older than 2.4 Ga. Ancient strata containing pyritic conglomerates differ in additional respects from younger strata containing hematitic (black sand) detrital heavy minerals. They lack redbeds, authigenic pyrite is present in lieu of authigenic ferric oxide, and associated paleosols also lack ferric iron oxides. This time-dependent transition in weathering and oxidation of fluvial sediments has been linked to atmospheric evolution and termed *oxyatmoversion* (Roscoe, 1969, 1981). The Huronian Supergroup and the Snowy Pass Supergroup in Wyoming (Karlstrom et al., 1983) contain pyritic conglomerates in lower units, and hematitic conglomerates and redbeds in upper units, so their deposition spans the oxyatmoversion. This event is constrained by U-Pb ages for rhyolite of the Coppercliff Formation (2450 +25/-10 Ma, Krogh et al., 1984) and on Nipissing gabbro (2219.4 +3.6/-3.5 Ma, Corfu and Andrews, 1986), which has intruded folded strata of the uppermost, post-oxyatmoversion Huronian Supergroup. The Huronian sediments were likely deposited in a period considerably less than the 230 Ma between these two events, so the transition may have occurred at about 2.4 Ga. In Africa, it is constrained between 2.6 and 2.2 Ga.

DEFINITIVE CHARACTERISTICS

The prime diagnostic feature of paleoplacers is the presence of concentrations of individual grains of heavy minerals in clastic sedimentary rocks that have been deposited and sorted by strong currents. Hematitic paleoplacers (subtype 1.1.2) due to diagenetic and metamorphic modifications, may differ mineralogically, but not chemically, from unconsolidated placers. Pyritic paleoplacers (subtype 1.1.1) differ from common hematitic paleoplacers in their high sulphur content, absence of magnetite and ferric oxide, and relatively high U/Th ratios. Isolated occurrences may be difficult to distinguish from stratabound epigenetic sulphide-rich gold or uranium concentrations. The epigenetic origin of the latter can be diagnosed if hematitic paleoplacers or redbeds are present in correlative or older strata.

Differences between Huronian (uraniferous) and gold-rich Witwatersrand pyritic paleoplacers

The gold content of the South African ores is two orders of magnitude greater than that of most large exploitable placers, hematitic paleoplacers, and Elliot Lake uranium ores. What differences are there between Witwatersrand and Elliot Lake pyritic paleoplacers that might provide some insight into the reason for the different gold contents?

Some differences of Witwatersrand ores and rocks compared to those of the Huronian are: (1) quartz arenites and conglomerates are texturally and mineralogically more mature; any feldspar not destroyed during weathering and transport was diagenetically altered to phyllosilicates; (2) considerable detritus from underlying sands and gravels were eroded and incorporated in many succeeding beds – in other words, there was likely more reworking; (3) ratios of uraninite to other radioactive minerals and of U/Th ratios in uraninite and in ores are higher; (4) clasts, including pyrite and uraninite grains, are more rounded; ventifacts have been noted; (5) pyrite contains more Ni and less Co; (6) heavy minerals include more chromite and less monazite; (7) kerogen is much more abundant; (8) pyrite 'mudballs' are abundant in some 'Wits' conglomerates, but are rare or absent in the Huronian; (9) arsenopyrite, other sulpharsenides, and probably some sulphides, are more abundant; (10) pyrite (average 3%) is less abundant; (11) it has been inferred that some Witwatersrand ore-bearing zones were likely deposited near shorelines and are intertongued with marine or lacustrine sediments (e.g., Minter, 1981); there is no evidence that Huronian conglomeratic ores were deposited in a littoral environment; (12) it has been suggested that altered zones ('hags') in granitic basement rocks could represent roots of gold deposits that were sources of detrital gold in nearby Witwatersrand goldfields (Robb and Meyer, 1985); and (13) Witwatersrand gold paleoplacers are older (between 3.07 and 2.71 Ga) than Huronian uranium paleoplacers (close to 2.45 Ga).

The hypothesis most commonly advanced for the different U/Au ratios in Witwatersrand and Huronian ores relates them to inferred differences in provenance. This cannot be assessed satisfactorily. Exposures of Witwatersrand source rocks are limited and do not include important gold deposits, whereas the Huronian provenance almost certainly included part of the Abitibi belt, which has produced more gold than any other Archean greenstone belt. It is difficult in any case to imagine any source area capable of supplying quantities of particulate gold comparable to the gold present in Witwatersrand deposits. More gold has been produced from 'Wits' than has been found in all of the 'lode' gold deposits of the world put together (possibly excluding the former U.S.S.R.). The possibility that the most favourable conditions for formation of auriferous paleoplacers might have been attained prior to formation of the Huronian ores must be considered (Hutchinson, 1987). We do not have adequate data, however, to determine whether the age difference between Witwatersrand and Huronian paleoplacers applies generally to gold-rich and gold-poor pyritic paleoplacers. Age certainly could not be the only factor, as older (ca. 3.07 Ga) Dominion paleoplacers have higher U/Au ratios than the considerably younger Venterpost paleoplacers (ca. 2.64 Ga).

GENETIC MODEL

It is doubtful if any genetic model has been more usefully applied to exploration and mine development than the modified placer model, which is almost universally accepted by people involved in these activities with respect to pyritic quartz pebble conglomerate deposits. This does not mean that it is correct in all respects as it is applied; only that no other model has been devised that has comparable predictive value. Many features consistent with the placer

model have been cited above. These include detrital shapes and modes of occurrence of uraninite and evidence of hydraulic sorting.

It has been suggested that peculiarities of the pyritic conglomeratic ores, as compared to hematitic paleoplacers and unconsolidated placers, are the result of an early atmosphere devoid, or nearly devoid, of free oxygen (Roscoe, 1973). A time-dependent change is certainly involved and it seems unlikely that this could have been a change in epigenetic processes. No variation of the placer model, however, provides satisfactory explanations for all features of pyritic conglomeratic ores; nor does any other model. It is particularly difficult to conceive of an adequate source for the immense amount of gold in Witwatersrand strata, regardless of how it may have been concentrated therein. Necessary widespread sources for detrital uraninite and a source for sulphur in ubiquitous authigenic, as well as allogenic, pyrite are also problematical.

EXPLORATION GUIDES

Paleoplacer gold and uranium deposits occur in Archean to Early Proterozoic fluvial to littoral clastic sedimentary sequences, dominated by quartz arenite units, that are older than about 2.4 Ga. The host units are pyritic, mature arenites, and oligomictic (quartz pebble) conglomerates produced by multiple cycles of erosion and redeposition. Deposition of these sequences occurred on and adjacent to stable Archean cratons and at the margins of, and within, intracratonic grabens or aulacogens, or within basins formed by downwarping due to tectonic processes other than rifting. Ore zones are in some cases controlled by basement topography, and paystreaks within the Witwatersrand goldfields radiate from fan apices within huge deltaic fan sequences.

Favourable environments for gold and other paleoplacers in consolidated rocks are the same as those for placers in unconsolidated rocks – high energy environments where heavy detrital mineral grains have been concentrated relative to lighter detrital mineral grains by stream currents, wave action, or wind. To be of economic interest, the concentrations must be much greater than those present in most of the unconsolidated placers that have been exploited. Pyritic paleoplacers (subtype 1.1.1) are far more favourable targets than hematitic, or black sand, paleoplacers (subtype 1.1.2), but they occur only in quartzose arenite formations older than 2.4 Ga. All such formations should be prospected for pyrite-dominated heavy mineral concentrations in quartz pebble layers. These are invariably more radioactive than host quartzites, so a scintillometer is an indispensable field tool. Rusty weathered bands in coarse or pebbly quartzite is a common clue. If analyses reveal slight enrichments in gold and uranium, a search for richer, sourceward concentrations may be considered. This may require special mapping and sedimentological studies to determine the extent of the formation, facies variations within it, foreset dips, and other directional features. Exploration in any extension of the formation 'upstream' from sub-ore grade paleoplacer outcrops will require core drilling and application of subsurface stratigraphic study techniques of various parameters, such as changes from hole to hole in pebble sizes, bed thicknesses, and U/Th ratios, as well as gold and uranium analyses.

SELECTED BIBLIOGRAPHY

References with asterisks (*) are considered to be the best source of general information on this deposit type.

Aikas, O. and Sarikkola, R.
1987: Uranium in lower Proterozoic conglomerates of the Koli area, eastern Finland; in Uranium Deposits in Proterozoic Quartz-pebble Conglomerates, International Atomic Energy Agency, Vienna, IAEA-TECDOC-427, p. 189-234.

Armstrong, R.A., Retief, E., Compston, W., and Williams, I.S.
1990: Geochronological constraints on the evolution of the Witwatersrand Basin, as deduced from single zircon U/Pb ion microprobe studies; Geochemistry, University of Cape Town, v. 27, no. 21, p. 24-27.

Bennett, J.
1979: Huronian volcanism, districts of Sault Ste. Marie and Sudbury; No. 18 in Summary of Field Work, 1978, (ed.) V.G. Milne, O.L. White, R.B. Barlow, and J.A. Robertson; Ontario Geological Survey, Miscellaneous Paper 82, p. 105-110.

Bickle, M.J. and Eriksson, K.A.
1982: Evolution and subsidence of early Precambrian sedimentary basins; Philosophical Transactions Royal Society London, v. 305, p. 225-247.

*Bourret, W.
1981: Investigation of Witwatersrand uranium-bearing quartz-pebble conglomerates in 1944-45; in Genesis of Uranium- and Gold-bearing Precambrian Quartz-pebble Conglomerate; United States Geological Survey, Professional Paper 1161-A-BB, p. A1-7.

Burger, A.J. and Coertze, F.J.
1975: Age determinations - April 1972 to March 1974; Annales Geological Survey South Africa 1973-1974, v. 10, p. 135-141.

Burke, K., Kidd, W.S.F., and Kusky, T.
1985: Is the Ventersdorp rift system of southern Africa related to a continental collision between the Kaapvaal and Zimbabwe cratons at 2.64 Ga. ago; Tectonophysics, v. 115, p. 1-24.

Burke, K., Kidd, W.S.F., and Kusky, T.M.
1986: Archean foreland basin tectonics in the Witwatersrand, South Africa; Tectonics, v. 5, p. 439-456.

Card, K.D.
1978: Geology of the Sudbury - Manitoulin area, districts of Sudbury and Manitoulin; Ontario Geological Survey, Report 166, 238 p.

Carter, J.D. and Gee, R.D.
1987: Geology and exploration history of Precambrian quartz-pebble conglomerates in Western Australia; in Uranium Deposits in Proterozoic Quartz-pebble Conglomerates, International Atomic Energy Agency, Vienna, IAEA-TECDOC-427, p. 387-425.

Clendenin, C.W., Charlesworth, E.G., and Maske, S.
1987: Tectonic style and mechanism of early Proterozoic successor basin development, Southern Africa; Economic Geology Research Unit, University of the Witwatersrand, Information Circular 197, 26 p.

Coetzee, F.
1965: Distribution and grain size of gold, uraninite, pyrite and certain other heavy minerals in gold-bearing reefs of the Witwatersrand basin; Transactions of the Geological Society of South Africa, v. 68, p. 61-89.

Corfu, F. and Andrews, A.J.
1986: U-Pb age for mineralized Nipissing Diabase, Gowganda, Ontario; Canadian Journal of Earth Sciences, v. 23, p. 107-109.

Davis, D.W., Hirdes, W., Schaltegger, U., and Nunoo, E.A.
1994: U-Pb age constraints on deposition and provenance of Birimian and gold-bearing Tarkwaian sediments in Ghana, West Africa; Precambrian Research, v. 67, no. 1-2, p. 89-107.

Derry, D.R.
1960: Evidence of the origin of the Blind River uranium deposits; Economic Geology, v. 55, p. 906-927.

Fahrig, W.F.
1987: The tectonic setting of continental mafic dyke swarms; Failed arm and early passive margin; in Mafic Dyke Swarms, (ed.) H.C. Halls, and W.F. Fahrig; Geological Association of Canada, Special Paper 34, p. 331-348.

Feather, C.E. and Koen, G.M.
1975: The mineralogy of the Witwatersrand reefs; Minerals Science and Engineering, v. 7, p. 189-224.

Fralick, P.W. and Miall, A.D.
1987: Glacial outwash uranium placers? Evidence from the lower Huronian Supergroup, Ontario, Canada; in Uranium Deposits in Proterozoic Quartz-pebble Conglomerates, International Atomic Energy Agency, Vienna, IAEA-TECDOC-427, p. 133-154.

Frarey, M.J. and Roscoe, S.M.
1970: The Huronian Supergroup north of Lake Huron; in Symposium on basins and geosynclines of the Canadian Shield, (ed.) A.J Baer; Geological Survey of Canada, Paper 70-40, p. 143-158.

Gama, H.M.
1982: Serra de Jacobina gold-bearing metasedimentary sequence paleoplacer gold deposits; in International Symposium on Archean and Early Proterozoic Geologic Evolution and Metallogenesis, excursions guidebook, p. 119-133.

Hallbauer, D.K.
1975: The plant origin of the Witwatersrand 'carbon'; Minerals Science and Engineering, v. 7, no. 2, p. 111-131.

***1981: Geochemistry and morphology of mineral components from the fossil gold and uranium placers of the Witwatersrand; in Genesis of Uranium- and Gold-bearing Precambrian Quartz-pebble Conglomerates, United States Geological Survey, Professional Paper 1161-A-BB, p. M1-M22.

Hart, R.C., Harper, H.G., and Algom Field Staff
1955: Uranium deposits of the Quirke Lake Trough, Algoma District, Ontario; Transactions of the Canadian Institute of Mining and Metallurgy, v. 48, no. 517, p. 260-265.

Heaman, L.M.
1988: A precise U-Pb zircon age for a Hearst Dyke; Geological Association of Canada-Mineralogical Association of Canada, Program with Abstracts, v. 13, p. 453.

Hegner, E., Kröner, A., and Hofmann, A.W.
1984: Age and isotope geochemistry of the Archaean Pongola and Usushwana suites in Swaziland, southern Africa; a case for crustal contamination of mantle-derived magma; Earth and Planetary Science Letters, v. 70, no. 2, p. 267-279.

Hiemstra, S.A.
1968: The mineralogy and petrology of the uraniferous conglomerate of the Dominion Reefs Mine, Klerksdorp area; Transactions of the Geological Society of South Africa, v. 71, p. 1-65.

Houston, R.S. and Karlstrom, K.E.
1987: Application of the time and strata bound model for the origin of uranium bearing quartz pebble-conglomerate in southeastern Wyoming, U.S.A.; in Uranium Deposits in Proterozoic Quartz-pebble Conglomerates; International Atomic Energy Agency, Vienna, Austria, IAEA-TECDOC-427, p. 99-131.

*Hutchinson, R.W.
1987: Metallogeny of Precambrian gold deposits: space and time relationships; Economic Geology, v. 82, no. 8, p. 1993-2007.

Karlstrom, K.E., Flurkey, A.J., and Houston, R.S.
1983: Stratigraphy and depositional setting of the Proterozoic Snowy Pass Supergroup, southeastern Wyoming; record of an early Proterozoic Atlantic-type cratonic margin; Geological Society of America, Bulletin, v. 94, p. 1157-1274.

Koen, G.M.
1961: The genetic significance of the size distribution of uraninite in Witwatersrand bankets; Transactions of the Geological Society of South Africa, v. 67, p. 23-54.

Krogh, T.E., Davis, D.W., and Corfu, F.
1984: Precise U-Pb zircon and baddeleyite ages for the Sudbury area; in The Geology and Ore Deposits of the Sudbury Structure; Ontario Geological Survey, Special Volume 1, chap. 20, p. 431-446.

Landais, P., Duessy, J., Robb, L.J., and Nouel, C.
1990: Preliminary chemical analyses and Raman spectroscopy on selected samples of Witwatersrand kerogen; Economic Geology Research Unit, University of the Witwatersrand, Information Circular 222, 8 p.

Liebenberg, W.R.
1955: The occurrence and origin of gold and radioactive minerals in the Witwatersrand System, the Dominion Reef, the Ventersdorp Contact Reef and the Black Reef; Transactions of the Geological Society of South Africa, v. 58, p. 101-254.

1958: The mode of occurrence and theory of origin of the uranium minerals and gold in the Witwatersrand ores; in United Nations, Peaceful Uses of Atomic Energy, Proceedings of the Second International Conference, Geneva, p. 379-387.

Long, D.G.F.
1987: Sedimentary framework of uranium deposits in the southern Cobalt Embayment, Ontario, Canada; in Proterozoic Quartz-pebble Conglomerates; International Atomic Energy Agency, Vienna, IAEA-TECDOC-427, p. 155-188.

Meyer, M., Saager, R., and Köppel, V.
1987: Uranium distribution and redistribution in a suite of fresh and weathered pre-Witwatersrand conglomerates from South Africa; in Uranium Deposits in Proterozoic Quartz-pebble Conglomerates, International Atomic Energy Agency, Vienna, IAEA-TECDOC-427, p. 255-273.

Minter, W.E.L.

1976: Detrital gold, uranium, and pyrite concentrations related to sedimentology in the Precambrian Vaal Reef placer, Witwatersrand, South Africa; Economic Geology, v. 72, p. 157-176.

*1978: A sedimentological synthesis of placer gold, uranium and pyrite concentrations in Proterozoic Witwatersrand sediments; in Fluvial Sedimentology, (ed.) A.D. Miall; Canadian Society of Petroleum Geology, Memoir No. 5, p. 801-829.

1981: Preliminary notes concerning the uranium-gold ratio and the gradient of heavy-mineral size distribution as factors of transport distance down the paleoslope of the Proterozoic Styn Reef placer deposit, Orange Free State, Witwatersrand, South Africa; in Genesis of Uranium- and Gold-bearing Precambrian Quartz-pebble Conglomerates, (ed.) F.C. Armstrong; United States Geological Survey, Professional Paper 1161-A-BB, p. K1-K3.

Mossman, D.J. and Dyer, B.D.

1985: The geochemistry of Witwatersrand-type gold deposits and the possible influence of ancient prokaryotic communities on gold dissolution and precipitation; Precambrian Research, v. 30, p. 303-319.

Pienaar, P.J.

1963: Stratigraphy, petrology, and genesis of the Elliot Group, Blind River, Ontario, including the uraniferous conglomerate; Geological Survey of Canada, Bulletin 83, 140 p.

Pretorius, D.A.

1974: The nature of the Witwatersrand gold-uranium deposits; Economic Geology Research Unit, University of the Witwatersrand, Information Circular 86, 50 p.

*1976: The nature of the Witwatersrand gold-uranium deposits; in Handbook of Strata-bound and Stratiform Ore Deposits II; Regional Studies and Specific Deposits, (ed.) K.H. Wolf; Vol. 7, Au, U, Fe, Mn, Hg, Sb, W and P deposits; Elsevier Publishing Co., New York, p. 29-89.

*1981: Gold and uranium in quartz-pebble conglomerates; Economic Geology, Seventy-fifth Anniversary Volume 1905-1980, (ed.) B.J. Skinner; p. 117-138.

Ramdohr, P.

1958: The uranium and gold deposits of Witwatersrand; Blind River district; Dominion Reef; Serra de Jacobina; microscopic analyses and a geological comparison; in Geochemistry and the Origin of Life, (ed.) K.A. Kvenvolden; Dowden, Hutchison & Ross Inc., Stroudsburg, Pennsylvania, U.S.A., v. 14, p. 188-194, 1974, (reprint from Abhandlungen Deutsche Akademie der Wissenschaften zu Berlin, v. 3, 1958).

Rao, M.V., Sinha, K.K., Misra, B., Balachandran, K., Srinivasan, S., and Rajasekharan, P.

1988: Quartz-pebble conglomerate from the Dhanjori – a new uranium horizon of Singhbhum uranium province; in Uranium and Gold-bearing Quartz-pebble Conglomerates of India; Geological Society of India, p. 89-95.

Retallack, G.

1986: Reappraisal of a 2200 Ma-old paleosol near Waterval Onder, South Africa; Precambrian Research, v. 32, p. 195-232.

Robb, L.J. and Meyer, M.

1985: The nature of the Witwatersrand hinterland; Economic Geology Research Unit, University of the Witwatersrand, Information Circular No. 178, 25 p.

Robertson, D.S. and Steenland, N.C.

1960: The Blind River uranium ores and their origin; Economic Geology, v. 55, p. 659-694.

Robertson, J.A.

1964: Geology of Scarfe, Mackenzie, Cobden, and Striker Townships; Ontario Department of Mines, Geological Report No. 20, p. 16-18.

1986: Huronian geology and the Blind River (Elliot Lake) uranium deposits; in Uranium Deposits of Canada, (ed.) E.L. Evans; The Canadian Institute of Mining and Metallurgy, Special Volume 33, p. 7-43.

Roscoe, S.M.

1957: Geology and uranium deposits, Quirke Lake-Elliot Lake, Blind River area, Ontario; Geological Survey of Canada, Paper 56-7, 21 p.

*1969: Huronian rocks and uraniferous conglomerates in the Canadian Shield; Geological Survey of Canada, Paper 68-40, 205 p.

1973: The Huronian Supergroup, a Paleoaphebian succession showing evidence of atmospheric evolution; in Huronian Stratigraphy and Sedimentation; (ed.) G.M. Young; Geological Association of Canada, Special Paper 12, p. 31-37.

1981: Temporal and other factors affecting deposition of Precambrian quartz-pebble conglomerates; in Genesis of Uranium- and Gold-bearing Precambrian Quartz Pebble Conglomerates, (ed.) F.C. Armstrong; United States Geological Survey, Professional Paper 1161 A-BB, p. W1-W17.

Roscoe, S.M. (cont.)

1990: Quartzose arenites and possible paleoplacers in Slave Structural Province, N.W.T.; in Current Research, Part C; Geological Survey of Canada, Paper 90-1C, p. 231-238.

Roscoe, S.M. and Card, K.D.

1992: Early Proterozoic Tectonics and Metallogeny of the Lake Huron region of the Canadian Shield; Precambrian Research, v. 58, no. 1-4, p. 99-119.

1993: The reappearance of the Huronian in Wyoming: rifting and drifting of ancient continents; Canadian Journal of Earth Sciences, v. 30, no. 12, p. 2475-2480.

Roscoe, S.M. and Donaldson, J.A.

1988: Uraniferous pyritic quartz pebble conglomerate and layered ultramafic intrusions in a sequence of quartzite, carbonate, iron formation and basalt of probable Archean age at Lac Sakami, Quebec; in Current Research, Part C; Geological Survey of Canada, Paper 88-1C, p. 117-121.

Roscoe, S.M., Stubley, M., and Roach, D.

1989: Archean quartz arenites and pyritic paleoplacers in the Beaulieu River supracrustal belt, Slave Structural Province, N.W.T.; in Current Research, Part C; Geological Survey of Canada, Paper 89-1C, p. 199-214.

Ruzicka, V.

1981: Some metallogenic features of the Huronian and post-Huronian uraniferous conglomerates; in Genesis of Uranium- and Gold-bearing Precambrian Quartz Pebble Conglomerates, (ed.) F.C. Armstrong; United States Geological Survey, Professional Paper 1161-A-BB, p. V1- v8.

Ruzicka, V. and Steacy, H.R.

1976: Some sedimentary features of conglomeratic uranium ore from Elliot Lake, Ontario; Geological Survey of Canada, Paper 76-1A, Report of Activities, Part A, p. 343-346.

Saager, R. and Muff, R.

1978: Petrographic and mineragraphic investigations of the Archean gold placers at Mount Robert in the Pietersburg greenstone belt, northern Transvaal; Economic Geology Research Unit, University of the Witwatersrand, Information Circular No. 114, 10 p.

Saager, R., Stupp, H.D., Vorwerk, R., Thiel, K., and Hennig, G.J.

1987: Interpretation of alpha- and gamma-spectrometric data from Precambrian conglomerates: a case study from the Denny Dalton uranium prospect, Northern Zululand, South Africa; in Uranium Deposits in Proterozoic Quartz-pebble Conglomerates, International Atomic Energy Agency, Vienna, IAEA-TECDOC-427, p. 293-311.

SACS (South African Committee for Stratigraphy)

1980: Stratigraphy of South Africa. Part 1; (comp.) L.E. Kent; in Lithostratigraphy of the Republic of South Africa, South West Africa/Namibia, and the Republics of Bophuthatswana, Transkei and Venda; Handbook Geological Survey of South Africa, no. 8, 690 p.

Salop, L.J.

1977: Precambrian of the Northern Hemisphere; Elsevier, New York, 378 p.

Schidlowski, M.

1981: Uraniferous constituents of the Witwatersrand conglomerates: ore microscopic observations and implications for the Witwatersrand metallogeny; in Genesis of Uranium- and Gold-bearing Precambrian Quartz-pebble Conglomerates, (ed.) F.C. Armstrong; United States Geological Survey, Professional Paper 1161-A-BB, p. N1-N29.

Srinivasan, R. and Ojakangas, R.W.

1986: Sedimentology of quartz-pebble conglomerates and quartzites of the Archean Bababudan Group, Dharwar Craton, South India: evidence for early crustal stability; Journal of Geology, v. 94, p. 199-214.

Tankard, A.J., Jackson, M.P.A., Eriksson, K.A., Hobday, D.K., Hunter, D.R., and Minter, W.E.L.

1982: Crustal Evolution of Southern Africa; Springer-Verlag, New York, 523 p.

***Theis, N.J.**

1979: Uranium-bearing and associated minerals in their geological context, Elliot Lake, Ontario; Geological Survey of Canada, Bulletin 304, p. 1-50.

Van Niekerk, C.B. and Burger, A.J.

1978: A new age for the Ventersdorp acidic lavas; Transactions of the Geological Society of South Africa, v. 81, p. 155-163.

Villaca, J.N. and Moura, L.A.M.

1981: Uranium in Precambrian Moeda Formation, Minas Gerais, Brazil; in Genesis of Uranium- and Gold-bearing Precambrian Quartz-pebble Conglomerates, (ed.) F.C. Armstrong; United States Geological Survey, Professional Paper 1161-A-BB, p. T1-T14.

Vogel, W.
1987: The mineralized quartz-pebble conglomerates of Ghana; in Uranium Deposits in Proterozoic Quartz-pebble Conglomerates; International Atomic Energy Agency, Vienna, IAEA-TECDOC-427, p. 235-254.

Von Backström, J.W.
1981: The Dominion Reef Group, western Transvaal, South Africa; in Genesis of Uranium- and Gold-bearing Precambrian Quartz-pebble Conglomerates, (ed.) F.C. Armstrong; United States Geological Survey, Professional Paper 1161-A-BB, p. F1-F8.

Winter, H. de la R.
1976: A lithostratigraphic classification of the Ventersdorp Succession; Transactions of the Geological Society of South Africa, v. 79, no. 1, p. 31-48.

Young, G.M.
1983: Tectono-sedimentary history of early Proterozoic rocks of the northern Great Lakes region; Geological Society of America, Memoir 160, p. 15-32.

Zolnai, A.I., Price, R.A., and Helmstaedt, H.
1984: Regional cross section of the Southern Province adjacent to Lake Huron, Ontario: implications for tectonic significance of the Murray Fault Zone; Canadian Journal of Earth Sciences, v. 21, p. 447-456.

Zumberge, J.E., Nagy, B., and Nagy, L.A.
1981: Some aspects of the development of the Vaal Reef uranium-gold carbon seams, Witwatersrand sequence: organic geochemical and microbiological considerations; in Genesis of Uranium- and Gold-bearing Precambrian Quartz-pebble Conglomerates, (ed.) F.C. Armstrong; United States Geological Survey, Professional Paper 1161-A-BB, p. O1-O7.

1.2 PLACER GOLD, PLATINUM

C.R. McLeod and S.R. Morison

INTRODUCTION

Placer deposits are accumulations of heavy minerals that have been eroded from lode sources and concentrated by sedimentation processes involving gravity, water, wind, or ice. In this description we also include those minerals (especially gold) that may have formed in situ as a result of chemical transport and precipitation. Gold placers commonly occur in stream gravels in uplifted, unglaciated areas containing primary, auriferous source rocks. Silver is recovered from placer gold in the refining process; platinum is a byproduct of placer gold mining in some localities, or, as at Tulameen, British Columbia, platinum was the primary metal recovered and gold was a byproduct (O'Neill and Gunning, 1934).

Numerous placer gold deposits have been and continue to be mined in Canada, particularly in the Cordilleran region. Important examples include the Klondike-Clear Creek area, Yukon Territory; Atlin and Cariboo districts, British Columbia; and the Chaudière River Basin, Quebec. The most significant examples of foreign placer gold districts are Sierra Nevada, California; Victoria, Australia; Lena, Aldan, and Amur rivers, Russia; and Choco, Colombia.

McLeod, C.R. and Morison, S.R.
1996: Placer gold, platinum; in Geology of Canadian Mineral Deposit Types, (ed.) O.R. Eckstrand, W.D. Sinclair, and R.I. Thorpe; Geological Survey of Canada, Geology of Canada, no. 8, p. 23-32 (also Geological Society of America, The Geology of North America, v. P-1).

IMPORTANCE

Historically, placer deposits have contributed more than 7% of the total recorded Canadian gold production of 6.9 million kilograms (222 million fine ounces) (Robinson, 1935; Dominion Bureau of Statistics annual reports, 1923-1949, 1951-1955; Canadian Minerals Yearbooks, 1960-1963, 1965-1990; LeBarge and Morison, 1990; Latoski, 1993; INAC, 1994). Total recorded placer production, about 515 000 kg to 1988, has come chiefly from Yukon Territory (72%) and British Columbia (27%). Minor contributions from the Chaudière region in Quebec, the Saskatchewan River in Alberta, and a small amount from beaches in Nova Scotia make up the remaining 1%. Perhaps as much as 20% of placer production has gone unreported, but this amount is difficult to estimate and is not reflected in the above figures.

In the 1980s, placers contributed 3.5% of the total Canadian gold production; this represents an increase from a low of 0.2% in the early 1970s. In 1900 when the Klondike production was at its peak and British Columbia production was at about one third of the maximum for the province (reached in 1863), placer gold accounted for 85% of the Canadian total. The Chaudière River Basin, Quebec, was the most productive Canadian placer district east of British Columbia. Within this basin about 100 000 fine ounces (more than 3000 kg) of gold was recovered from preglacial, interglacial, and postglacial placers (Boyle, 1979), chiefly between 1860 and 1885.

Platinum and platinum group elements (PGE) have been reported from several Canadian placer gold deposits in British Columbia, Yukon Territory, and the North Saskatchewan River, Alberta (O'Neill and Gunning, 1934).

Platinum production from the Tulameen district, British Columbia, in the latter part of the nineteenth and early part of this century totalled about 20 000 troy ounces (622 000 g). Placers have been a major source of platinum production in Colombia, United States (Goodnews Bay, Alaska), and the former U.S.S.R. (Mertie, 1969).

SIZE AND GRADE OF DEPOSITS

Placer deposits range greatly in size and grade, from small, local, near-source concentrations that may be profitable to work on a small scale, to much larger deposits of sparsely disseminated, fine grained gold in alluvial sediments associated with regional drainage basins. In some instances, as distance from the source increases, the grain size of placer gold decreases. With dilution from nonauriferous tributaries, the gold values in alluvial placers diminish, more or less gradationally, to low grade, uneconomic disseminations.

In the Klondike area, Yukon Territory, more than 200 km of valley bottoms have been worked in Bonanza, Hunker, Dominion, Sulphur, Quartz, and Bear creeks and their tributaries. Richer pay streaks, or parts of them, were originally mined by sinking shafts and drifting along the bedrock surface. Subsequent reworking by open cuts or dredging, in many areas more than once, and the lack of detailed records on the amount of gold recovered, render it difficult to compile accurate "grade-tonnage" relationships. McConnell (1907) estimated production from four miles of pay streak on Eldorado Creek, Yukon Territory, at CAN. $25 000 000, and probable future output at an additional $2 600 000. This would represent more than 6000 g of gold per metre for a distance of 6 km. Production from Eldorado Creek since 1906 has probably equalled or exceeded that during the first 10 years it was worked, making it one of the richest creeks found in Canada.

Placer concentrations of other mineral commodities are also known in Canada. Along the north shore of the St. Lawrence River, concentrations of magnetite, ilmenite, and titaniferous magnetite occur as black sand beds several centimetres thick in postglacial fluvial, deltaic sands which have been partially reworked by wave and tidal action. Drilling near the mouth of the Natashquan River indicated 1.5 billion tonnes of sand containing 3.7% iron (Gross, 1967). Near Steep Rock Lake, Ontario, more than 600 000 t of iron concentrate was produced between 1959 and 1964 from hematite-goethite glacial gravels (Prest, 1970). In the Dublin Gulch area, Yukon Territory, minor quantities of scheelite were recovered from placer workings during the Second World War and again between 1977 and 1981 when more elaborate methods (vibrating sluice, jigs, spiral concentrators, separation tables) were used (Debicki, 1983). However, the scheelite recovery process proved to be uneconomic (LeBarge and Morison, 1990).

GEOLOGICAL FEATURES

Canadian Cordilleran placer deposits occur in both glaciated and unglaciated terrains, and it is assumed that gold was released from bedrock or paleoplacer sources through weathering processes during nonglacial periods or interglacial intervals. These deposits are found in five sedimentary settings (Table 1.2-1; Morison, 1989), which can be summarized as follows: 1) Pliocene to early Pleistocene alluvial placer deposits which are preserved as high level

terraces buried beneath nonauriferous overburden; 2) Pleistocene nonglacial alluvial placer deposits that occur as valley-bottom fill and low to high level terraces in unglaciated terrain; 3) interglacial placer deposits that occur as valley-bottom alluvial fill or low terraces in drainage systems that have escaped the effects of glacial erosion; 4) glacial placer deposits that have formed when gold from regional bedrock or paleoplacer sources was incorporated into some types of glacial drift, such as glaciofluvial or ice contact deposits, moraines, and stratified till deposits; and 5) Recent placer deposits which are found as colluvial deposits, valley-bottom alluvial blankets in gulches and other tributary valleys, bar deposits in major river systems and beach and nearshore marine deposits.

Economic placer gold districts are found throughout the Canadian Cordillera (Fig. 1.2-1) and include all of the preceding placer deposit settings, each with a unique depositional history. For example, in the Clear Creek drainage basin in the Yukon Territory (area 4 of Fig. 1.2-1), placer gold is found in valley-bottom creek and gulch gravel, glacial gravel, and buried preglacial fluvial gravel (Fig. 1.2-2) (Morison, 1985a). In the Mayo district (area 5 of Fig. 1.2-1), small colluvial placers have been developed from local bedrock sources (Boyle, 1979). Pliocene to early Pleistocene White Channel placer deposits in the Klondike district (Fig. 1.2-3; area 4 of Fig. 1.2-1) form high level terraces 50 to 100 m above present day stream levels. White Channel alluvium ranges in thickness from a few metres to more than 35 m, and is characterized by 14 lithofacies types which were deposited in a braided river environment with valley wall alluvial fan and debris flow sedimentation (Morison, 1985b). Lebarge (1993) concluded that placer gold in the Mount Nansen area (Fig. 1.2-1), in central Yukon Territory, is found in valley bottom alluvium, Pleistocene alluvial terraces, and early Pleistocene proglacial gravelly terraces and diamicton that is interpreted as glacial till or resedimented till. In the Cariboo district of British Columbia (area 19 of Fig. 1.2-1), placer deposit settings include Tertiary and interglacial gravel, glacial outwash gravel, and postglacial stream gravel (Boyle, 1979). Clague (1987) identified a previously unknown Pleistocene buried valley near the Bullion mine in the Cariboo district and suggested there may be other such valleys in the region. Levson and Giles (1993) identified the following placer settings in the Cariboo region of British Columbia: Tertiary and pre-late Wisconsinan paleochannel and paleofan settings; Late Wisconsinan glacial and glaciofluvial environments and Holocene high and low terraces, colluvium, and alluvial fan settings. They further concluded that five main auriferous lithofacies types characterize placer deposit settings in this area of British Columbia. Beach placer deposits on the Queen Charlotte Islands (area 14 of Fig. 1.2-1) formed through wind and wave erosion of auriferous glacial sediments (Boyle, 1979).

Placer gold deposits in Alberta and Saskatchewan occur as Recent and modern stream deposits in major drainage basins, such as the North Saskatchewan River (Giusti, 1983; Coombe, 1984). The recovery of placer gold in Alberta is primarily due to gravel washing plants that process late Tertiary, preglacial gravels and sands of river channel deposits in the Edmonton area. In Quebec and Ontario placer gold deposits and occurrences are found in both recent and interglacial stream gravel and glacial drift (Ferguson and Freeman, 1978; LaSalle, 1980). In coastal

Table 1.2-1. Stratigraphy and general characteristics of placer gold deposits in Canada (from Morison, 1989).

Time Period	Tertiary	Quaternary, Pleistocene, preglacial or nonglacial	Interglacial	Glacial	Holocene
Placer environment and geomorphic location	Buried alluvial sediments in benches above valley floors	Buried alluvial sediments in benches above valley floors; valley fill alluvial sediments; alluvial terraces	Valley fill alluvial sediments; alluvial terraces	Benches of proglacial and ice contact deposits; moraines and drifts	Valley bottom alluvial plains and terraces; colluvium and slope deposits; beach and nearshore marine deposits
General sediment characteristics	Mature sediments; well sorted alluvium with a diverse assemblage of sediment types	Locally derived gravel lithology; moderately to well sorted alluvium which is crudely to distinctly stratified	Mixed gravel lithology; moderately to well sorted alluvium which is crudely to distinctly stratified	Regionally derived gravel lithology; variable sorting and stratfication depending upon type of glacial drift	Mixed gravel lithology; moderately to well sorted alluvium which is crudely to distinctly stratified; poorly sorted, massive slope deposits; well sorted beach sand
Gold distribution	Greater concentration with depth	Discrete concentrations throughout to pay streaks at base of alluvium	Discrete concentrations throughout to pay streaks at base of alluvium	Dispersed throughout	Discrete concentrations throughout to pay streaks at base of alluvium; pay streaks follow slope morphology and strandline trend
Mining problems	Thick overburden	Thick overburden; variable grade	Variable grade	Low grade	Variable grade and low volume of auriferous sediment
Examples	"White Channel Gravel" of the Klondike area, Yukon Territory	Preglacial fluvial gravels, Clear Creek drainage basin and unglaciated terrain in Yukon Territory	Interglacial stream gravels in Atlin and Cariboo mining districts, British Columbia	Glaciofluvial gravel, Clear Creek drainage basin, Yukon Territory	Valley bottom creek and gulch placers, Clear Creek drainage basin, colluvial placers in Dublin Gulch area, Yukon Territory; beach placers, Queen Charlotte Islands, British Columbia

areas (Graham Island, British Columbia; southern Nova Scotia; Seward Peninsula, Alaska) placer gold deposits and auriferous sands and gravels occur in submerged, raised, and modern beaches and in submarine drainage channels (Boyle, 1979).

Geological setting

Placer gold deposits are commonly found in deeply weathered unglaciated terrain within a stable craton that contains auriferous source rocks. Deposits occur in the weathered zone of all rock types but are especially common in metamorphic rocks. Sources of placer gold include gold-bearing quartz veins, felsic intrusions, sulphide deposits (skarns, veins, massive sulphides), paleoplacers, and porphyry copper deposits. Both epithermal and mesothermal gold deposits may also give rise to gold placers. However, economic placers may well be derived only from specific types of gold-bearing deposits.

Age of host rocks and mineralization

Unconsolidated placers in Canada are chiefly of Quaternary age, but late Tertiary gravels, particularly in the Klondike district, Yukon Territory, have yielded considerable amounts of placer gold. The geology of these Tertiary deposits has been described in detail by Morison (1985b), Dufresne et al. (1986), and Morison and Hein (1987). Primary and secondary (intermediate collector sediment) sources of gold can range from Precambrian to Tertiary. Boyle (1979) linked the preponderance of rich, productive Tertiary and later placers chiefly to first generation sediments that contain coarse gold. Multiple cycles of erosion and deposition result in comminution and dispersal of gold as fine particles.

Form of deposits and distribution of ore minerals

Coarse placer gold tends to be concentrated in pay streaks, commonly on or near the bedrock surface or on compacted, relatively impervious sediments above bedrock. Factors that control the location of pay streaks are not well understood, but in narrow gulches and creeks they are generally on the bedrock surface in the deepest part of the valleys. Wider, mature valleys contain discontinuous pay streaks and discrete concentrations of placer gold, both on the bedrock surface and within the gravelly sediments. Uplift and downcutting may leave bench placers as stream terraces or may result in the redistribution of gold into creek, river, floodplain, deltaic, and beach placers. In delta and flood plain environments, gold is commonly in very fine flakes and particles, and pay streaks are irregularly distributed as a result of complex channel migration and the large volume of sediments.

Mineralogy

Gold (or electrum) is the prime mineral of interest in Canadian placer deposits, but minor quantities of platinum group metals, scheelite, wolframite, and cassiterite have also been recovered. Common heavy minerals found in black sand include amphibole, pyroxene, tourmaline, topaz, beryl, garnet, chromite, sphene, rutile, goethite, magnetite, ilmenite, pyrite, cassiterite, wolframite, and scheelite. "Uranianpyrochlore", euxenite, and "uranothorite" have been reported from postglacial outwash gravels in Bugaboo Creek, Spillimacheen district, British Columbia (Merrett, 1957). Platinum-iron alloy grains are the most common platinum group minerals found in Yukon Territory, British Columbia, and Alberta placers

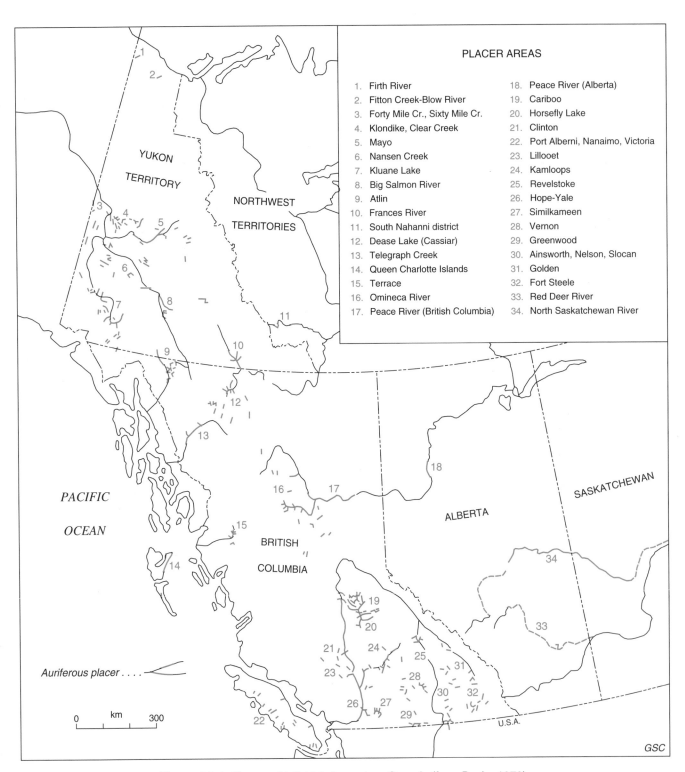

Figure 1.2-1. Placer gold districts in western Canada (from Boyle, 1979).

PLACER AREAS

1. Firth River
2. Fitton Creek-Blow River
3. Forty Mile Cr., Sixty Mile Cr.
4. Klondike, Clear Creek
5. Mayo
6. Nansen Creek
7. Kluane Lake
8. Big Salmon River
9. Atlin
10. Frances River
11. South Nahanni district
12. Dease Lake (Cassiar)
13. Telegraph Creek
14. Queen Charlotte Islands
15. Terrace
16. Omineca River
17. Peace River (British Columbia)
18. Peace River (Alberta)
19. Cariboo
20. Horsefly Lake
21. Clinton
22. Port Alberni, Nanaimo, Victoria
23. Lillooet
24. Kamloops
25. Revelstoke
26. Hope-Yale
27. Similkameen
28. Vernon
29. Greenwood
30. Ainsworth, Nelson, Slocan
31. Golden
32. Fort Steele
33. Red Deer River
34. North Saskatchewan River

Auriferous placer . . . .

(Nixon et al., 1989; Ballantyne and Harris, 1991; S.B. Ballantyne and D.C. Harris, unpub. data, GSC Minerals Colloquium, January 22-24, 1992, Ottawa, Ontario; Harris and Ballantyne, 1994), whereas Os-Ir-Ru grains are found in the Atlin district, British Columbia, and North Saskatchewan River, Alberta (Harris and Cabri, 1973; Harris and Ballantyne, 1994). Minerals that may become economically extractable in the future include monazite, pyrochlore, tantalite, columbite, fergusonite, bastnaesite, xenotime, zircon, baddeleyite, euxenite, samarskite, and cinnabar. Placer minerals are characteristically highly resistant to mechanical breakdown and chemical dissolution in the surface environment.

Placer gold is found in a wide variety of shapes and sizes, but most common are dust and small scales, or particles in the 0.1-2 mm range. The morphology of grains has been characterized by various investigators, e.g., Ballantyne and MacKinnon (1986), Ebert and Kern (1988), and Knight et al. (1994). As a general rule, placer gold near its source is relatively coarse and rough; with transport the particles become smaller, smoother, and more flattened. These effects are the result of both physical and chemical processes (Boyle, 1979), but the natural milling of a high energy stream system appears to play a major role in modifying the size and morphology of gold particles during transport.

Electron microprobe investigations have shown that spongy marginal rims on placer gold commonly have a higher fineness than grain interiors ("fineness" refers to the proportion of gold in the naturally occurring metal; it is calculated and expressed in parts per thousand). Leaching of silver was proposed as the cause of this enrichment by several workers (e.g., Giusti and Smith, 1984; Michailidis, 1989; Knight et al., 1994), but Groen et al. (1990) considered the rims to be formed either by precipitation of gold from the surrounding solution or by "self-electrorefining" of placer electrum grains. Giusti and Smith (1984) recognized a general increase in fineness with decreasing grain size,

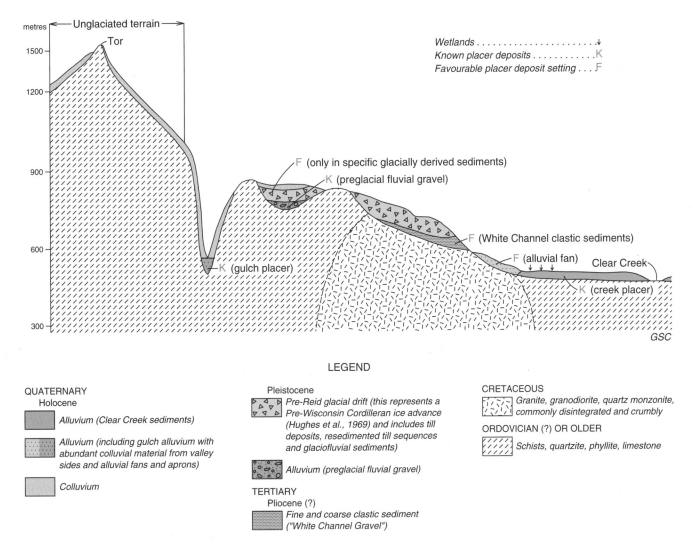

Figure 1.2-2. Schematic profile of the Clear Creek, Yukon Territory drainage basin showing types of placer deposits and Quaternary deposits (from Morison, 1985a, 1989).

27

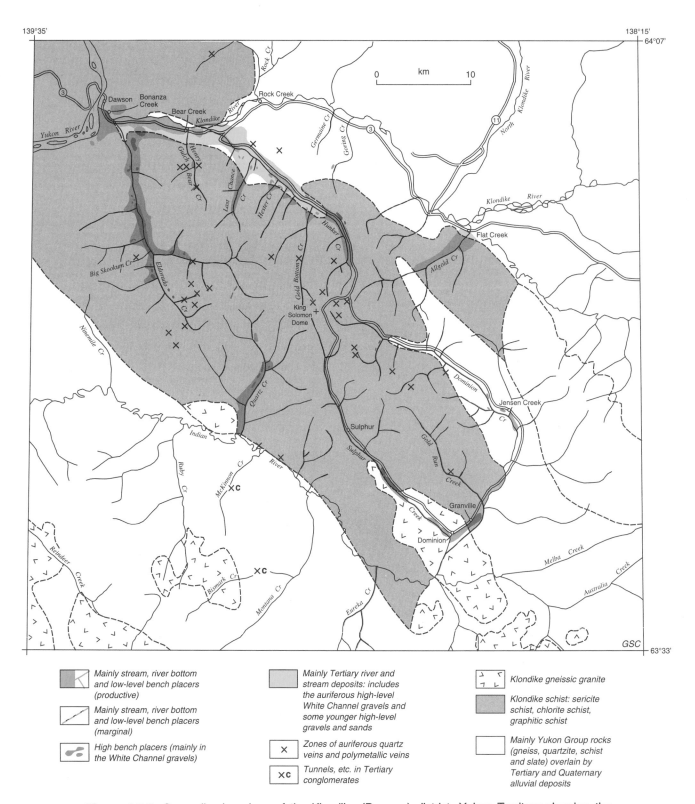

Figure 1.2-3. Generalized geology of the Klondike (Dawson) district, Yukon Territory showing the relationship of the gold placers to the Klondike schist and other rock units (from Boyle, 1979).

attributable to the gold-rich rims; Groen et al. (1990) found the thickest rims were usually on flake-shaped (most transported) grains. S.B. Ballantyne and D.C. Harris (unpub. data, GSC Minerals Colloquium, January 22-24, 1992, Ottawa, Ontario); reported the discovery of "new" gold in the five micrometre size range on the surfaces of placer heavy minerals, such as zircon, monazite, scheelite, and ilmenite, from the alluvial platinum-gold placers of Florence Creek, Yukon Territory. Geochemical variations in placer gold have been related to the depositional environments of the primary gold sources (e.g., Ballantyne and MacKinnon, 1986; Michailidis; 1989: Mosier et al., 1989; Knight et al., 1994). Knight et al. (1994) studied the shape and composition of some 2700 gold particles from lode occurrences and placer deposits in the Klondike area and demonstrated that compositional data may be used to link placer gold with specific lode sources.

Mertie (1969) reported that placer platinoid grains commonly consist of two principal types of PGE alloys; one is dominantly platinum with varying contents of the other five PGEs, whereas the other consists of alloys composed chiefly of iridium and osmium. Both alloy types can contain intimate intergrowths with, and minute inclusions of, the other. More detailed electron microprobe studies (e.g., Cabri and Harris, 1975; Cousins and Kinloch, 1976) confirmed the complex compositions and interrelationships of PGE minerals in placer grains. Compositional zoning in such grains was considered to be primary by Cabri and Harris (1975), but indicative of accretion and secondary growth by Cousins and Kinloch (1976). Rosenblum et al. (1986) found that magnetic concentrates from Goodnews Bay, Alaska, contained discrete grains of PGE minerals, PGE minerals as inclusions in magnetite, and PGEs that had apparently diffused into the magnetite lattice adjacent to PGE minerals. Mardock and Barker (1991) reported that placer platinum group metal minerals from Goodnews Bay exhibit characteristics attributable to both mechanical and accretionary processes of formation. In Canada recent studies have characterized placer platinum group metal minerals in British Columbia, Yukon Territory, and Alberta (see respectively, Nixon et al., 1989; Ballantyne and Harris, 1991; S.B. Ballantyne and D.C. Harris, unpub. data, GSC Minerals Colloquium, January 22-24, 1992, Ottawa, Ontario; S.B. Ballantyne and D.C. Harris, unpub. data, GSC Minerals Colloquium, January 17-19, 1994, Ottawa, Ontario; and Harris and Ballantyne, 1994).

DEFINITIVE CHARACTERISTICS

Placer deposits are commonly found in deeply weathered, unglaciated terrain with subdued topography, represented by broad valleys and accordant rounded hills that have been dissected by a moderately mature drainage system made up of numerous small streams tributary to the main water courses. Auriferous source rocks are a prime prerequisite for the formation of placer deposits. Near the source, gold is coarse and is found in fairly well-defined pay streaks on bedrock; the grain size decreases with transport but fineness of the gold commonly increases. Heavy minerals that accumulate with gold are usually from primary gold deposits or local country rocks. Fine grained gold that is transported to larger stream and river systems is erratically distributed by more complex fluvial processes, but local concentrations may result under favourable depositional

conditions. For example, Hou and Fletcher (1992) reported a (slight) increase in the abundance and size of gold grains downstream in the Harris Creek catchment basin of southern British Columbia.

GENETIC MODEL

The first requirement for the formation of a placer is the presence of auriferous source rocks, which can include gold-bearing quartz veins, large silicified bodies that contain disseminated gold, porphyry copper deposits, disseminated pyritic and massive sulphide deposits, pre-existing placers, and trace contents of gold in otherwise unmineralized rock.

During the weathering cycle, chemical and mechanical processes remove gangue or interstitial host rock material that is far less resistant to structural breakdown and chemical decomposition during weathering than is gold. This can result in low-grade, in situ or residual accumulations that have experienced little, if any, mechanical transport. In Canada such placers are rare and small in size, in part because of extensive glaciation. However, as a first stage of a model for placer formation, the generation of eluvial or residual gold accumulations probably reflects the start of a lengthy and complex process that eventually gives rise to extensive and rich placer deposits and districts.

Deep secular weathering of source rocks, particularly where undisturbed by glaciation, results in the preconcentration of gold in, on, or near the bedrock surface by gravity, creep, frost action, and solifluction processes. Gravity and stream hydrodynamic processes concentrate gold into pay streaks in the alluvium, especially on irregular bedrock surfaces, in much the same manner as that which takes place in a sluice box. These processes can also result in concentrations of gold on impervious layers within the alluvium. Beach placers, such as those on Graham Island (Queen Charlotte Islands), British Columbia, form due to tidal and wave action.

Extremely fine gold particles are more readily transported, and their concentration (or dissemination) in the fluvial or littoral environment is determined by hydraulic sorting (flow separation), which is dependent on such factors as the size, shape, and specific gravity of individual grains in the sediment load, the velocity and degree of turbulence of the water flow, and the nature of the stream bed (Best and Brayshaw, 1985; Reid and Frostick, 1985). Grains of equidimensional character are more effectively trapped and concentrated than scaly or flaky particles. Additional concentration of gold may accompany tectonic uplift, with the resulting incision of stream valleys and the reworking of existing auriferous unconsolidated sediments, and the deposition of new alluvial deposits. In the Klondike area, deep weathering, tectonic uplift, and the absence of glaciation that would disperse accumulations of gold, have combined to contribute to the formation and preservation of Canada's richest placer district. It is evident that the formation of placers is the result of a complex interaction of processes that operate at scales ranging from landform (the abrupt widening of a valley) to the microenvironment (the interstices between sediment clasts in the stream bed that entrap heavy mineral grains) (Slingerland, 1984; Hou and Fletcher, 1992). Time and climatic constraints on these processes are not well understood.

The model summarized above is based on a clastogene process whereby discrete gold grains are released by weathering processes and subsequently concentrated in a stream environment. The role of chemical transport in the formation of placers, and the proportion of gold that is contributed or redistributed in this manner, are not well understood. Whether chemical processes are expressed mainly by dissolution and local reprecipitation, with little transportation, is not known. Boyle (1979) reviewed the criteria for detrital versus chemical origin of nuggets and placer gold and concluded that both processes may be active, but one or the other may predominate, depending on local conditions. Watterson (1985) noted that physical, chemical (particularly organic complexes), and electrical effects, which appear to be connected with freezing action, may influence the weathering of auriferous rocks, the chemical complexing of gold, and its subsequent dissolution, mobilization, and redeposition. These effects, in part, could be responsible for the large number of placer deposits at high latitudes. Knight et al. (1994), however, concluded that the bulk of the gold extracted from the White Channel Gravel deposits in the Klondike area is detrital in origin despite the presence of hydrothermal alteration in both footwall rock and the overlying gravel in some of the White Channel exposures in the Klondike area.

In summary, the distribution of placer deposits is controlled by (a) primary sources and their tectonic setting, and (b) geomorphological processes, both of which are affected by climatic conditions superimposed on them. In discussing geomorphological controls on the global distribution of placer deposits, Sutherland (1985) recognized that the processes responsible for the formation of placer deposits vary markedly throughout the world and that placers in different morphoclimatic regions should exhibit characteristics that are distinctive for, and dependent on, the conditions under which they were deposited.

RELATED DEPOSIT TYPES

Paleoplacer gold deposits (deposit subtype 1.1) presumably formed in much the same manner as modern placers, but may have undergone considerable modification since their deposition. At least part of the gold they contain is detrital, but a considerable portion of it may have been recrystallized or chemically remobilized, transported, and precipitated (Lowey, 1985).

Gold-bearing deposits of almost any type apparently may serve as the source of gold in the formation of placer gold deposits (see "Geological setting"), provided the appropriate conditions exist for release, transport, and concentration of the gold.

Aside from gold and platinum, the only commodities contained in placer deposits that have proven to be of economic interest (only marginally) in Canada are magnetite, hematite-goethite, and scheelite. Elsewhere in the world there are, in addition, important cassiterite (tin), heavy mineral (e.g., rutile, ilmenite, zircon, monazite, magnetite), and diamond placers in stream, beach, and offshore deposits. However, placers or even minor concentrations of heavy minerals may be useful indicators of the presence of, or potential for, primary mineral deposits.

EXPLORATION GUIDES

Guidelines to exploration for placers include:

1. Location of areas of gold-bearing source rocks such as auriferous quartz veins and polymetallic sulphide or porphyry deposits; slightly auriferous quartz stringers and veins in schists and gneisses; slightly auriferous minerals such as pyrite and other sulphides in graphitic schists; and slightly auriferous conglomerates, sandstones, and quartzites, or paleoplacers.

2. Recognition of evidence for deep secular chemical and mechanical weathering of potential source rocks on a comparatively mature topographic surface, or the presence of alteration zones within these rocks.

3. Preservation of a weathered profile in unglaciated areas or in areas protected from glacial advances (e.g., trunk valleys oblique or normal to ice direction). This may indicate potential for preglaciation placers.

4. Identification of areas of uplift in which superposed moderately incised drainage has developed.

5. Location of concentrations of gold in a variety of favourable depositional sites in fluvial systems, e.g., on irregular bedrock surfaces, at the heads of midchannel bars, on point bars, and at abrupt valley widenings. This requires an understanding of the hydraulic conditions which control concentrating processes at all scales.

6. Use of remotely sensed imagery (LANDSAT, RADARSAT, side-looking radar, side-looking airborne radar) taken during different times of the year to enhance contour map and aerial photograph interpretation of features such as denudation, sediment accumulation, aggradation, and neotectonic movements.

7. Use of modern geophysical methods (hammer seismic, ground-penetrating radar, magnetic, electrical) to help delineate buried channels and the profile of the bedrock surface along which pay streaks may be found.

8. Use of stream sediment geochemical surveys for associated elements or gold itself to indicate anomalous areas.

9. Use of testing techniques (e.g., panning, rocker, portable sluice box, systematic pitting or trenching, reverse circulation, or sonic drilling) to establish gold distribution, concentration levels, and reserves.

10. Use of specialized gold separation and concentration techniques, which optimize the retention of fine grained gold, for the treatment of bulk samples (e.g., head grade evaluation).

11. Use of highly sensitive analytical techniques to accurately determine the gold contents of samples.

ACKNOWLEDGMENTS

The authors gratefully acknowledge the critical review and helpful comments of S.B. Ballantyne.

REFERENCES

References with asterisks (*) are considered to be the best source of general information on this deposit type.

Ballantyne, S.B. and Harris, D.C.
1991: An investigation of platinum-bearing alluvium from Florence Creek, Yukon; in Current Research, Part A; Geological Survey of Canada, Paper 91-1A, p. 119-129.

Ballantyne, S.B. and MacKinnon, H.F.
1986: Gold in the Atlin terrane, British Columbia; in Gold '86: An International Symposium on the Geology of Gold Deposits, (ed.) A.M. Chater; Toronto, September 28-October 1, 1986, p. 16-17.

Best, J.L. and Brayshaw, A.C.
1985: Flow separation - a physical process for the concentration of heavy minerals within alluvial channels; Journal of the Geological Society of London, v. 142, p. 745-755.

Boyle, R.W.
1979: The geochemistry of gold and its deposits; Geological Survey of Canada, Bulletin 280, 584 p.
*1987: Gold: History and Genesis of Deposits; Van Nostrand Reinhold Co., New York, 676 p.

Cabri, L.J. and Harris, D.C.
1975: Zoning in Os-Ir alloys and the relation of the geological and tectonic environment of the source rocks to the bulk Pt: Pt + Ir + Os ratio for placers; Canadian Mineralogist, v. 13, p. 266-274.

Clague, J.J.
1987: A placer gold exploration target in the Cariboo district, British Columbia; in Current Research, Part A; Geological Survey of Canada, Paper 87-1A, p. 177-180.

Coombe, W.
1984: Gold in Saskatchewan; in Saskatchewan Geological Survey, Open File Report 84-1, 134 p.

Cousins, C.A. and Kinloch, E.D.
1976: Some observations on textures and inclusions in alluvial platinoids; Economic Geology, v. 71, p. 1377-1398.

Debicki, R.L.
1983: Yukon Placer Mining Industry 1978 to 1982; Indian Affairs and Northern Development, p. 79-80.

Dufresne, M.B., Morison, S.R., and Nesbitt, B.E.
1986: Evidence of hydrothermal alteration in the White Channel sediments and bedrock of the Klondike area, west-central Yukon; Yukon geology, Volume 1; Exploration and Geological Services Division, Yukon, Indian and Northern Affairs Canada, p. 44-49.

Ebert, C. and Kern, H.
1988: Placer gold from the Gevattergraben in the Frankenwald area (Germany) - mineralogical and morphological characteristics and their significance for gold prospecting; Neues Jahrbuch für Mineralogie, Heft 9, p. 405-417.

Ferguson, S.A. and Freeman, E.B.
1978: Ontario occurrences of float, placer gold and other heavy minerals; Ontario Geological Survey, Mineral Deposits Circular 17, 214 p.

Giusti, L.
1983: The distribution, grades and mineralogical composition of gold-bearing placers in Alberta; MSc. thesis, University of Alberta, Edmonton, Alberta, 397 p.

Giusti, L. and Smith, D.G.W.
1984: An electron microprobe study of some Alberta placer gold; Tschermaks Mineralogische und Petrographische Mitteilungen, v. 33, p. 187-202.

Groen, J.C., Craig, J.R., and Rimstidt, J.D.
1990: Gold-rich rim formation on electrum grains in placers; Canadian Mineralogist, v. 28, p. 207-228.

Gross, G.A.
1967: Iron deposits in the Appalachian and Grenville Regions of Canada; Volume II in Geology of Iron Depsits in Canada; Geological Survey of Canada, Economic Geology Report 22, p. 86-87.

Harris, D.C. and Ballantyne, S.B.
1994: Characterization of gold and PGE-bearing placer concentrates from the North Saskatchewan River, Edmonton, Alberta; in Current Research, 1994-E; Geological Survey of Canada, p. 133-139.

Harris, D.C. and Cabri, L.J.
1973: The nomenclature of the natural alloys of osmium, iridium and ruthenium based on new compositional data of alloys from world-wide occurrences; Canadian Mineralogist, v. 12, p. 104-112.

Hou, Z. and Fletcher, W.K.
1992: Distribution and morphological characteristics of visible gold in Harris Creek (82L/2); in Geological Fieldwork 1991, British Columbia Ministry of Energy, Mines and Petroleum Resources, Paper 1992-1, p. 319-323.

Hughes, O.L., Campbell, R.B., Muller, J.E., and Wheeler, J.O.
1969: Glacial limits and flow patterns, Yukon Territory, south of 65°N latitude; Geological Survey of Canada, Paper 68-34, 9 p.

INAC (Indian and Northern Affairs Canada)
1994: 1993 Yukon Mining and Exploration Overview; in Yukon Exploration and Geology 1993; Exploration and Geological Services Division, Indian and Northern Affairs Canada, Yukon Region, 142 p.

Knight, J.B., Mortensen, J.K., and Morison, S.R.
1994: Shape and composition of lode and placer gold from the Klondike district, Yukon, Canada; Bulletin 3, Exploration and Geological Services Division, Indian and Northern Affairs Canada, Yukon Region, 142 p.

LaSalle, P.
1980: L'or dans les sédiments meubles - formation des placers, Extraction et occurrences dans le sud-est du Québec; ministère de l'Energie et des Ressources, DPV-745, 26 p.

Latoski, D.A.
1993: An overview of the Yukon placer mining industry 1991 and 1992; in Yukon Placer Mining Industry 1991-1992; Placer Mining Section 1993, Mineral Resources Directorate, Yukon, Indian and Northern Affairs Canada, p. 1-4.

LeBarge, W.P.
1993: Sedimentology of placer gravels near Mt. Nansen, central Yukon Territory; MSc. thesis, University of Calgary, Department of Geology and Geophysics, 272 p.

LeBarge, W.P. and Morison, S.R.
1990: Yukon placer mining and exploration 1985 to 1988; Exploration and Geological Services Division, Yukon, Indian and Northern Affairs Canada, 151 p.

Levson, V.M. and Giles, T.R.
1993: Geology of Tertiary and Quaternary gold-bearing placers in the Cariboo region, British Columbia (93 A,B,G,H); British Columbia Ministry of Energy, Mines and Petroleum Resources, Mineral Resources Division, Geological Survey Branch, Bulletin 89, 202 p.

Lowey, G.W.
1985: Auriferous conglomerates at McKinnon Creek, west-central Yukon (115O/11): Paleoplacer or epithermal mineralization?; in Yukon Exploration and Geology 1983; Exploration and Geological Services Division, Yukon Region, Indian and Northern Affairs Canada, p. 69-78.

Mardock, C.L. and Barker, J.C.
1991: Theories on the transport and deposition of gold and PGM minerals in offshore placers near Goodnews Bay, Alaska; Ore Geology Reviews, v. 6, p. 211-227.

McConnell, R.G.
1907: Report on gold values in the Klondike high level gravels; Geological Survey of Canada, Publication No. 979, 34 p.

Merrett, J.E.
1957: Spillimacheen, columbium and uranium, Bugaboo (Quebec Metallurgical Industries Ltd.); in Minister of Mines, Province of British Columbia, Annual Report, 1956, p. 142, 143.

Mertie, J.B., Jr.
1969: Economic geology of the platinum metals; United States Geological Survey, Professional Paper 630, 120 p.

Michailidis, K.
1989: Gold chemistry from the Gallikos River placer deposits, Central Macedonia Greece; Chemie der Erde, v. 49, p. 95-103.

Morison, S.R.
1985a: Placer deposits of Clear Creek drainage basin 115P, central Yukon; Yukon Exploration and Geology 1983, Department of Indian Affairs and Northern Development, Exploration and Geological Services Division, Yukon, p. 88-93.
1985b: Sedimentology of White Channel placer deposits, Klondike area, west-central Yukon; MSc. thesis, University of Alberta, Edmonton, Alberta, 149 p.
1989: Placer deposits in Canada; in Chapter 11 of Quaternary Geology of Canada and Greenland, (ed.) R.J. Fulton; Geological Survey of Canada, Geology of Canada, no. 1, p. 687-697 (also Geological Society of America, The Geology of North America, v. K-1).

Morison, S.R. and Hein, F.J.

1987: Sedimentology of the White Channel Gravels, Klondike area, Yukon Territory: fluvial deposits in a confined area; in Recent Developments in Fluvial Sedimentology, (ed.) F.G. Etherige, R.M. Flores, and M.D. Harvey; Society of Economic Paleontologists and Mineralogists, Special Publication No. 39, p. 205-216.

Mosier, E.L., Cathrall, J.B., Antweiler, J.C., and Tripp, R.B.

1989: Geochemistry of placer gold, Koyukuk-Chandalar mining district, Alaska; Journal of Geochemical Exploration, v. 31, p. 97-115.

Nixon, G.T., Cabri, L.J., and Laflamme, J.H.G.

1989: Tulameen placers, 92H/7,10: origin of platinum nuggets in Tulameen placers: a mineral chemistry approach with potential for exploration; in Exploration in British Columbia 1988; British Columbia Ministry of Energy, Mines and Petroleum Resources, p. B83-B89.

O'Neill, J.J. and Gunning, H.C.

1934: Platinum and allied metal deposits of Canada; Geological Survey of Canada, Economic Geology Series No. 13, 165 p.

Prest, V.K.

1970: Quaternary geology of Canada; in Geology and Economic Minerals of Canada, Geological Survey of Canada, Economic Geology Report 1, p. 757.

Reid, I. and Frostick, L.E.

1985: Role of settling, entrainment and dispersive equivalence and of interstice trapping in placer formation; Journal of the Geological Society of London, v. 142, p. 739-746.

Robinson, A.H.A.

1935: Gold in Canada, 1935; Canada Department of Mines, Mines Branch Publication Number 769, 127 p.

Rosenblum, S., Carlson, R.R., Nishi, J.M., and Overstreet, W.C.

1986: Platinum-group elements in magnetic concentrates from the Goodnews Bay district, Alaska; United States Geological Survey, Bulletin 1660, 38 p.

Slingerland, R.

1984: Role of hydraulic sorting in the origin of fluvial placers; Journal of Sedimentary Petrology, v. 54, p. 137-150.

Sutherland, R.

1985: Geomorphological controls on the distribution of placer deposits; Journal of the Geological Society of London, v. 142, p. 727-737.

Watterson, J.R.

1985: Crystalline gold in soil and the problem of supergene nugget formation: freezing and exclusion as genetic mechanisms; Precambrian Research, v. 30, p. 321-335.

Authors' addresses

C.R. McLeod
1128 Cline Crescent
Ottawa, Ontario
K2C 2P2

S.R. Morison
Northern Affairs Program
Indian and Northern Affairs Canada
Whitehorse, Yukon
Y1A 3V1

S.M. Roscoe
(deceased)

Printed in Canada

2. STRATIFORM PHOSPHATE

2. STRATIFORM PHOSPHATE

F.W. Chandler and R.L. Christie

INTRODUCTION

Typical phosphorites are argillaceous to sandy marine sedimentary rocks that contain stratified concentrations of calcium phosphate, mainly as apatite. Depending on the extraction process, the principal commodity obtained from these deposits is elemental phosphorus or phosphoric acid.

Byproducts may include the following commodities: uranium is commonly present in sedimentary phosphate and can be recovered in acid extraction; vanadium is a byproduct of electric furnace production of elemental phosphorus, where it can be recovered from the ferrophosphorus slag; fluorine can be recovered from waste gases and as fluorosilicic acid in the wet, or acid, process. Large quantities of impure calcium sulphate (phosphogypsum) are produced in the wet process; small amounts of this are used directly as a soil conditioner in the U.S.A., but the remainder constitutes a major disposal problem. The coproducts ferrophosphorus and calcium-silicate slag are produced in the thermal reduction of phosphate rock; some of the ferrophosphorus can be used as a ferro-alloy in the steel industry. The calcium-silicate slag, normally a waste product, can be crushed and used as a concrete aggregate, as ballast in highway or railroad construction, or can be foamed to form lightweight aggregate or slag wool (Christie, 1978b; Notholt et al., 1979).

Nomenclature is used in this paper as follows: **phosphate** is an informal term used in the industry to describe a rock, mineral, or salt containing phosphorus compounds.

Chandler, F.W. and Christie, R.L.
1996: Stratiform phosphate; in Geology of Canadian Mineral Deposit Types, (ed.) O.R. Eckstrand, W.D. Sinclair, and R.I. Thorpe; Geological Survey of Canada, Geology of Canada, no. 8, p. 33-40 (also Geological Society of America, The Geology of North America, v. P-1).

The term **phosphate rock** serves both industry and the geological fraternity, but with differing meanings. In the industry, phosphate rock is usually a high grade phosphate product that can be used in a fertilizer plant or put directly on cropland. It may also be beneficiated ore. The term **ore**, is reserved in the industry for the naturally occurring phosphate material that is commonly too low in grade to be used as feed in a fertilizer plant without beneficiation. In a geological sense, phosphate rock contains enough of one or more phosphate minerals, usually apatite, to be used, either directly or after beneficiation, in the manufacture of phosphate products. **Phosphorite** is a sedimentary phosphate rock and is the most widely used ore. The term phosphorite is applied to all sedimentary rocks that contain 10% or more (volumetrically) phosphate grains or matrix (Riggs, 1979a; Notholt et al., 1979). Other phosphate rocks are phosphatized limestones, sandstones, shales, and igneous rocks. **Phosphatic** describes a rock containing 1 to 10% phosphate: e.g., phosphatic quartz sandstone, phosphatic dolomite.

IMPORTANCE

Since recorded production began in 1847 in Suffolk, England, about 2 billion tonnes of phosphate rock has been produced (Notholt et al., 1979, p. 43). Most of this production was derived from marine sedimentary rocks, and three countries – United States, the former Soviet Union, and Morocco – account for about 75%. Sedimentary phosphate deposits, all of which are marine (Cook, 1976), currently provide more than 80% of the world's phosphate rock.

No sedimentary phosphate is mined in Canada, although phosphogenic basins have been recognized (Fig. 2-1). The nearest active mining regions are those of the western United States (the 'Phosphoria Basin') and Florida. About 30% of the phosphate rock for the well developed Canadian fertilizer industry is imported from the western United States and about 70% from Florida.

SIZE AND GRADE OF DEPOSIT

World reserves of phosphate are enormous (some 34 billion tonnes), mostly of the marine sedimentary type (Werner, 1981, p. 32). Published figures for various national reserves of phosphate range from 100 million tonnes to more than 18 billion tonnes (Morocco). One of the largest accumulations of phosphate rock, or phosphorite, is the Western Phosphate Field of the northwestern United States; in this region, the Phosphoria Formation is distributed over an area of about 140 000 km² (Notholt et al., 1979, p. 9).

Phosphate grades as great as 35% P_2O_5 are obtained from some beds; ore of lower grades, e.g., 24 to 29% P_2O_5, is mined and beneficiated to 30% or more. Phosphorite of lower grade (below 24%) can be extracted economically in special circumstances, such as when the deposit is easily mined (e.g., unconsolidated) or when coproducts are available. The once common direct shipping of high grade phosphate rock is now much rarer. Reserves of rock requiring little or no treatment are much reduced, and the need to ship higher grade concentrates has become a greater economic factor, so that beneficiation by means of washing, flotation, or calcining is now widespread.

GEOLOGICAL FEATURES

Lithology

Typical phosphorites are stratified argillaceous to sandy sedimentary deposits containing concentrations of calcium phosphate (apatite). The rocks are white to dark brown or brownish grey, but mainly light coloured. They are not very distinctive and can easily be overlooked or misidentified (e.g., as impure sandstones or cherty argillites). The mining width of individual beds may be as thick as 6 m. Beds contain as much as 35% P_2O_5 (76% phosphate mineral, or "bone phosphate of lime", BPL) (Christie, 1978a, b; Notholt et al., 1979).

The phosphatic beds themselves are variously mudstone, quartz-granulose-pelletal, cherty, or carbonate-rich. They may contain mixed components such as authigenic carbonate (and phosphate), terrigenous debris, and organic matter, and thus are complex deposits that appear to be the product of more than one sedimentary system or process.

Phosphorites commonly have a pelletal structure, and the pellets are rounded to ovoid, usually structureless, and generally range from 0.06 to 4.0 mm in diameter. Some, with a concentric structure, are called "ooliths" or "oolites". Radiating or spherulitic structure is rare. The pellets typically are aggregates of smaller sand or carbonate grains or

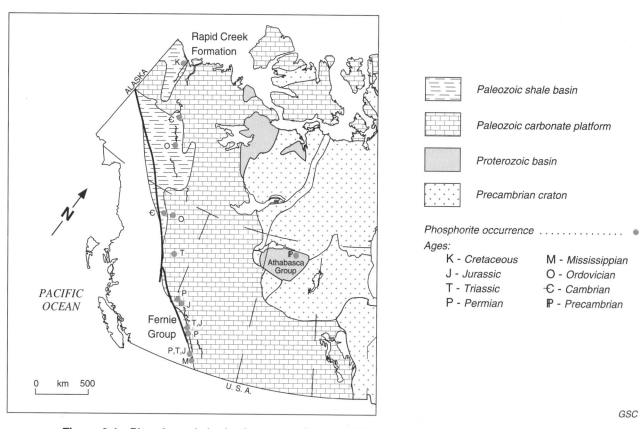

GSC

Figure 2-1. Phosphogenic basins in western Canada (Paleozoic basins after Tipper et al., 1981; Precambrian basins and craton after Douglas, 1969). The Paleozoic regions include overlying Mesozoic syn- and postorogenic basins.

crystallites, and may contain small fossils or fossil debris and recycled clastic phosphorite material (Riggs, 1979a). The fine grained to cryptocrystalline nature of the phosphate minerals, the complex mixtures present, and the widespread alteration (phosphatization, recrystallization) hinder interpretation of the sedimentary conditions or history of development of phosphorites.

Phosphorites are typically associated with either dark carbonaceous shale, chert, and carbonate rock or with cherty carbonate, sandstone, and shale. Some phosphorites, however, are associated with glauconitic sandstone and quartz-siltstone, dolomite, shelly material, or diatomite.

Mineralogy

Most mined phosphate minerals are calcium phosphates of the apatite group. Aluminum phosphates (augelite, crandallite, millisite, and wavellite) are mined in tropical or subtropical regions where they form lateritic, residual, or phosphatized rock deposits. Apatite also occurs in various residual deposits and as replacements of limestone or coralline rock. The apatite series is isomorphous, with three end-members:

fluorapatite $Ca_{10}(PO_4)_6F_2$,

chlorapatite $Ca_{10}(PO_4)_6Cl_2$, and

hydroxyapatite $Ca_{10}(PO_4)_6(OH)_2$.

Fluorapatite is by far the commonest mineral of the series in crustal rocks, but pure fluorapatite is relatively rare in commercial sedimentary deposits due to ionic subsitutions in several positions within the apatite lattice. The most common substitutions are: Mg, Sr, and Na for Ca; OH and Cl for F; As and V for P; and CO_3+F for PO_4 (McClellan, 1980a). The replacement of PO_4 by CO_3 is particularly common and apatite containing appreciable amounts of CO_2 and more than 1% F is named carbonate fluorapatite or francolite, a simplified formula for which (Slansky, 1980) is: $Ca_{10}[(PO_4)_{6-x}(CO_3F)_x]F_2$.

The substituted PO_4^{3-} ions in carbonate fluorapatite are replaced by $(CO_3^{2-}$•$F^-)$; the additional F^- ions help to maintain electroneutrality and cation co-ordination and contribute to the high F content of many sedimentary apatites (McClellan and Lehr, 1969). The P_2O_5 content of pure fluorapatite is 42.2%; the P_2O_5 content of the carbonate apatite decreases with increasing carbonate substitution to a minimum of about 34% (Notholt et al., 1979; McClellan, 1980a, b; Slansky, 1980, p. 10-17).

The name "collophane" has been widely applied to cryptocrystalline phosphate material; it appears amorphous under microscopic or X-ray examination and in many cases occurs as layers resembling opal.

DEFINITIVE CHARACTERISTICS

Phosphorite is a bedded, usually shallow marine, sedimentary rock, typically associated with carbonaceous shale, chert, and carbonate rock or with cherty carbonate, sandstone, and shale. It contains a high proportion of apatite, usually carbonate fluorapatite, which may variously be of authigenic, diagenetic, clastic, and secondary replacement origin. A bluish-white 'phosphate bloom' may be present on weathered surfaces.

GENETIC MODEL

The genesis of phosphorites has been recently reviewed by a significant number of authors (e.g., Cook, 1976; Cook and McElhinny, 1979; Kolodny, 1979; Notholt et al., 1979, p. 10-13; Sheldon, 1980, 1981; Riggs, 1984; Cook and Shergold, 1986; and Cook et al., 1990). The prevailing opinion for the formation of phosphorite (Wang and McKelvey, 1976; Cook, 1976; Kolodny, 1979; Cook and McElhinny, 1979; Riggs, 1979b; Burnett, 1980; Slansky, 1980; Parrish and Barron, 1986) may be stated as follows:

Phosphorus is concentrated in the deep ocean from the dissolution of sinking organisms. Most phosphates form, however, not in the open ocean, but rather in shallow water environments where several important factors may coincide:

a) High biological activity concentrates phosphorus adjacent to zones of oceanic upwelling and local restriction of circulation reduces oxygen levels. Phosphate may be leached from the organic remains below the sediment surface.

b) Terrigenous input is low and phosphate can be concentrated by reworking, as in the dry tropics – hence the suggested association with evaporites.

c) Basement highs are located nearby (Cook and Shergold, 1986). Phosphorite deposits can be enriched by weathering, winnowing, and reworking (Sheldon, 1981) and local topographic depressions form traps (Cook and Shergold, 1986). Secondary deposits formed by these processes are known as residual, phosphatized, or reworked deposits.

The role of tectonics

As outlined below, phosphorites are generally formed in shallow marine environments with access to upwelling nutrient-rich deep oceanic water. Upwelling mostly occurs at certain coastal locations determined by the interplay of major winds and ocean currents. The influence of tectonics is felt in at least three ways. First, the distribution of continents on the globe modifies global winds and ocean currents (Cook and McElhinny, 1979; Parrish and Curtis, 1982; Parrish et al., 1986). Second, major shallow water environments such as on sedimentary prisms of mature rifted continental margins and in epeiric seas (Hays and Pitman, 1973; Hallam, 1977) are directly related to divergent tectonics. Third, world climate which affects both sea level and ocean overturn (Sheldon, 1980), can be related to the position of continents relative to the poles. Phosphogenesis was concentrated in the early Paleozoic, Permian, Cretaceous, and Tertiary (Cook and McElhinny, 1979; Parrish et al., 1986). Apart from the Permian episode, these episodes seem related to first order high stands of sea level.

The role of winds and ocean currents

The planetary wind system is driven by transfer of solar heat from equatorial to polar regions. Because of the rotation of the Earth, the Coriolis effect breaks the atmospheric circulation into six zonal (latitude-parallel) cells, the Hadley cells (Parrish and Curtis, 1982; Parrish et al., 1986). This resultant global wind pattern drives the gyres, large surface currents that circle major oceans (Sheldon, 1980, 1981; Riggs, 1984) (Fig. 2-2). Interaction between the wind pattern and major ocean surface currents, including the gyres, causes surface waters to flow apart or offshore, leading to upwelling of deep, phosphorus-rich ocean water (Parrish, 1982, 1990). This arrangement is modified over geological time by tectonic processes which move the continents.

Upwelling, and the resultant high biological activity, occur primarily at three latitudes: midlatitudes, the equator, and high latitudes (Parrish and Curtis, 1982; Parrish, 1990).

1a. **Wind-driven midlatitude upwelling:** 10-40°N and S; occurs on north-south coasts on the eastern side of wide (3000 km) oceans. The Permian Phosphoria Formation of the western U.S.A. is the classic example of this type of deposit (Cook and Shergold, 1986; Sheldon, 1989). It originally contained at least 1.7×10^{12} tonnes of P_2O_5, equivalent to 18×10^{12} tonnes of rock at 9.6% P_2O_5, more than six times the P in the present world ocean. Coast-parallel northerly trade winds, associated with the arid subtropical high pressure zone (Parrish et al., 1986) interact with the eastern boundary current of the oceanic gyre. The Coriolis effect, which in the northern hemisphere pushes surface water to the right of the wind, drives the surface water offshore, causing deeper water to upwell. Off California, Peru, and southwest Africa (Cook, 1976), this upwelling zone is nowadays related to the deposition of phosphorite (Fig. 2-2, locations A, B, C).

1b. **Dynamic midlatitude upwelling:** occurs on the western coast of wide oceans. It is not related to winds, but to the dynamics of fast ocean currents, being prominent on the western side of oceanic gyres. At present the Atlantic gyre (Gulf Stream), off the southeast North American coast, is in deep water (Parrish, 1982), but in the Miocene, when sea level was higher (Parrish, 1990; Popenoe, 1990), the upwelling impinged on the shelf of the southeastern U.S.A. giving rise to major phosphorite deposition (Fig. 2-2, location D) (Riggs, 1984; Cook et al., 1990).

2. **Equatorial upwelling:** 10°N-10°S; this upwelling is driven by converging easterly winds. The Coriolis effect causes a net transport of water away from the equator. Nowadays it occurs over open ocean but in the past could have occurred over shelves (Parrish et al., 1986; Parrish, 1990). Also, it could have been important in forming phosphates in equatorial east-west seaways (Sheldon, 1980), such as the Tethyan deposits of North Africa and the Middle East, and in forming phosphorites on oceanic islands (Fig. 2-2, location E). This type of upwelling is also caused by friction between the two Pacific gyres and the equatorial counter current (Fig. 2-2) (Sheldon, 1981).

3. **High latitude upwelling:** occurs under the low pressure belts at about 55-60°N and S. At present, in the northern hemisphere continents have broken the high latitude low pressure zone into stable cells. Because winds spiral anticlockwise into these cells the net transport of water is outward. Although this effect occurs at present in the open ocean, coastal upwelling could have occurred in the past (Parrish, 1982, Fig. 4; Parrish and Curtis, 1982).

PALEOGEOGRAPHIC (DEPOSITIONAL) SETTINGS OF CANADIAN PHOSPHORITE DEPOSITS

Marine phosphorite seems to occur in five major settings, discriminated by the factors described above. Examples, including some major deposits of the world and major Canadian occurrences, are:

a) **west coast setting** (Permian Phosphoria Formation, U.S.A.; Fernie Group, Canada),

b) **east coast setting** (Miocene of southeastern U.S.A.),

c) **epicontinental setting** (Cambrian Georgina Basin, Australia; Proterozoic Athabasca Group, Canada),

d) **high latitude setting** (Cretaceous Rapid Creek Formation, Canada), and

e) **the equatorial settings** of east-west coasts and oceanic islands. Although many phosphorites and phosphatic sedimentary rocks have been recorded in Canada (Fig. 2-1) (Christie, 1979, 1980, p. 241) and some have potential for mineable phosphorite (Christie and Sheldon, 1986), none so far has proven economic.

West coast setting: Fernie Group (Jurassic)

Phosphatic shales at the base of the Sinemurian to Bajocian (lower-middle Jurassic) Fernie Group of southeastern British Columbia are preserved in a synformal structure, the Fernie 'basin' or synclinorium (Christie, 1989). The group underlies the Rocky Mountains, adjacent mountains to the west, and the Interior Plains to the east. Reconnaissance exploration has taken place since 1915, when phosphate (in Permian beds) was discovered in the southern Canadian Rockies (de Schmid, 1916). The shales formed at about 43°N (Smith et al., 1981) at a time of low sea level,

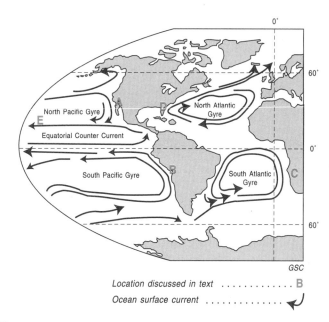

Figure 2-2. Major ocean currents in the present oceans (adapted from New York Times Atlas of the World, 2nd revised edition, 1991).

during which little phosphate was deposited in the world (Sheldon, 1980). They form the lower part of an upward-coarsening cyclic clastic wedge deposited on a marine shelf on the continental side of a foreland basin. The latter formed by accretion of the Intermontane superterrane against the western passive margin of North America (Cant and Stockmal, 1989), at a time when orogenic activity to the west first affected western Alberta (Hall, 1984; Stronach, 1984).

Carbonate, shale, and mature sandstone dominate the sedimentary column. The terrigenous detrital components were derived mainly, if not entirely, from the cratonic interior, a hinterland of low relief to the northeast.

The basal phosphatic bed consists of about 1 to 3 m of dark grey-brown to black, pelletal phosphorite. Assays as high as 25% P_2O_5 have been obtained, but the phosphate content diminishes upward through about 30 m of dark shale (G.L. Webber, pers. comm., 1981). The best potential lies within the Fernie Basin of British Columbia, where 24% P_2O_5 over a thickness of at least 2 m was reported (MacDonald, 1988). Tonnages of sedimentary phosphate rock have been outlined on various commercial holdings (mainly Cominco Ltd.), but no commercial production has taken place because of the high costs of mining thin beds and the medium to low grade of the phosphate rock. Development of an economical method of beneficiation will be a major factor in the establishment of commercial production of phosphate rock in southeastern British Columbia.

East coast setting

This setting, in which phosphate deposition is driven by dynamic midlatitude upwelling, is typified by the Miocene deposits of the southeastern U.S.A., but no Canadian examples are known.

Epicontinental setting: Athabasca Group, (Proterozoic)

Modern data suggest that phosphorites do not form in epicontinental seas, but there is strong evidence from the Cambrian and Proterozoic that they do. Phosphorites may be pelletal, nodular, or structureless, with either a quartzose or carbonate matrix and limestone, dolomite, shale, glauconitic sandstone, and quartz-siltstone are associated. Either oceanic upwelling may not have affected epicontinental seas or may have occurred hundreds of kilometres away and allowed the biomass to be swept into the shallow sea (Cook and Shergold, 1986). Alternatively, it has been suggested that upwelling is possible at depths of greater than 50 m, perhaps driven by surface current divergences (Parrish, 1982). There is strong evidence that the Australian Georgina Basin deposit is related to a form of dynamic upwelling caused by currents moving over basement highs (Cook and Shergold, 1986).

The Athabasca Group of Saskatchewan-Alberta (Fig. 2-1) consists of undeformed and unmetamorphosed east-derived alluvial fanglomerate, siltstone, and diagenetically matured quartzite, with a western marine upper part, deposited, according to Ramaekers (1981), in a tectonically active area. The phosphate within the Athabasca Group has yielded a 1700 Ma U-Pb age, obtained from fluorapatite (Cumming et al., 1987) from the shallow marine Wolverine

Point and Fair Point formations. The Athabaska Group overlies Hudsonian gneisses and probably postdates the 1854 to 1840 Ma Missi Group, a coarse, west-derived molasse that lies east of the Athabasca Group (Andsell et al., 1992).

Crandallite group minerals (hydrous Al phosphates) are present in trace amounts through the Athabasca Group and the underlying regolith and basement, and may be of early diagenetic or detrital origin (Wilson, 1985). Phosphorites are widespread in sandstone to mudstone in the storm- or tide-influenced upper member of the shallow marine to open shelf Wolverine Point Formation in the upper part of the group (Ramaekers, 1981). Apatite occurs as disseminated grains and as a matrix constituent. Phosphatic zones with several per cent P_2O_5 over intervals as much as 6 m contain 30 cm zones with values as great as 20% (Ramaekers, 1979, 1980).

Thick regolith beneath both the Missi and Athabasca groups implies intense weathering (Ramaekers, 1981; Ansdell et al., 1992). Ramaekers (1981) suggested that deposition of the latter occurred under a warm climate, with access to the ocean. Associated with the broadly similar and correlative Thelon and Hornby Bay sequences (Ross and Chiarenzelli, 1986), that lie to the north, are silcretes, eolian quartz arenites, and silicified evaporites, which postdate laterite developed on the granitic basement. These younger features are interpreted (Ross and Chiarenzelli, 1986) to indicate an arid paleoclimate. A paleolatitude of 25°-40°S is suggested by the paleomagnetic data of McGlynn and Irving (1981) for 1750 Ma. With the paleocontinental reconstruction of Moores (1991), the above data indicate to the writers that the phosphorite-bearing Wolverine Point Formation was deposited in an epeiric sea in the continental interior, under a hot dry climate.

High latitude setting: Rapid Creek Formation (Cretaceous)

The Rapid Creek Formation (Young and Robertson, 1984) is a phosphatic ironstone exposed in the Richardson Mountains in the northwestern District of MacKenzie. It is the eastern facies of the upper part of much thicker, fly-schoid units to the west in the Yukon Territory. The phosphatic rock is "compact" microcrystalline siderite, medium grained to pebbly ironstone-intraclast wackestones and packstones, and pyrite-phosphate rock (Young, 1977, p. 73). Phosphate minerals appear within a mud matrix, as sparry replacement cements, and in veinlets and open fractures. Phosphate values range up to 30% P_2O_5 in certain groups of beds, but the phosphate content is highly varied. The composition of the phosphate rock may be described with end members of detrital quartz, phosphate grains, and siderite or clay matrix. High P, Mg, and Mn and low Ca give rise to a suite of unusual phosphorus minerals such as satterleyite, arrojadite, and gormanite. The locality has been of considerable interest because of excellent mineral specimens, and new phosphatic minerals have been described and named from the locality.

From a thickness of 1000 m in the west, the phosphatic sediments thin to 60 m over the Cache Creek High, a former arch within the sedimentary basin, a fairly common relationship in phosphogenic basins. To the east, a carbonaceous grey shale is correlative. Rough estimates of average composition and calculations of possible resources were made by Young (1977). An average estimated composition of Fe_2O_3=33%, P_2O_5=14%, and MnO:5%, with an assumed subsurface (mineable) limit of 1000 m, yielded about 27 billion tonnes of ironstone, of which about 4 billion tonnes would be phosphate equivalent. Remoteness, low phosphate grades, and the metallurgical problems in treating an ironstone make this deposit subeconomic, but further exploration seems inviting from the possibility of coproducts and the large tonnages present.

During Albian time the area became the junction of Pacific Ocean, Boreal Sea, and Interior Seaway. West-derived flysch from the Cordilleran fold belt was channelled to the north along a deepening seaway, on the east flank of which the Rapid Creek Formation was formed on an outer shelf to slope. Pacific Ocean water flowed northeast into the Boreal and Interior seaways, causing cold nutrient-rich ocean water to upwell on the west flanks of the Cache Creek High where dilution of phosphates by clastics was minimal (Young and Robertson, 1984; Dixon, 1992).

The association of marine apatite, iron, and flyschoid sediments is unusual, as is the high paleolatitude of about 75° (Irving, 1979), but modern productive upwelling does occur at high latitude (type 3, above). Further, Parrish and Curtis (1982) suggested that there would have been a mid-Cretaceous upwelling zone west of the interior North American seaway during the winter, at latitude 60-70°N.

Also worth noting for both the Phosphoria and Fernie deposits, offshore terranes did not interfere with the ingress of ocean water.

Equatorial east-west coasts and equatorial oceanic island settings

Oceanic phosphorite of the equatorial Pacific islands formed from the waste of plankton, fish, and birds that were localized by equatorial upwelling (Sheldon, 1980). Some Paleogene-Cretaceous occurrences on seamounts (guyots) are now at depths of 2000-4000 m (Sheldon, 1981) due to thermal subsidence of the ocean floor (Menard, 1984).

The equatorial east-west coast setting (Cook and McElhinny, 1979) and the oceanic setting is unlikely to apply to any Canadian phosphorite and will not be considered further (Cathcart and Gulbrandsen, 1973; Cook, 1976).

RELATED DEPOSIT TYPES

The association: **pyritic black shale**+phosphorite+glauconite+chert, is typical of upwelling zones and related to high biological activity (Sheldon, 1981; Parrish and Barron, 1986; Parrish et al., 1986; Skinner, 1993). Siliceous ooze occurs in the centre of the area of upwelling and is rimmed in turn by phosphorite and glauconite (Parrish et al., 1986). Also, black shales related to coastal upwelling are an important source for petroleum (Parrish, 1982), and some black shales can contain economic concentrations of a number of metals such as PGEs, copper, gold, molybdenum, nickle (see "Sedimentary nickle sulphide", subtype 6.2), uranium and other elements (Coveney and Chan, 1989; Huyck, 1990; Pasava, 1993).

Cook and McElhinny (1979) noted a striking similarity between the distributions of **iron ores** and phosphorites with time. Many ironstones (see "Lake Superior-type iron-formation", subtype 3.1) contain considerable phosphorus, perhaps adsorbed onto the hydrous precursors of the iron oxides (Parfitt, 1989). An upwelling model for iron ores (Holland, 1973) would bring both P and Fe to shallow waters where they could be coprecipitated, especially in the early Proterozoic. In the later Proterozoic and the Phanerozoic an increasingly prolific biota and a decrease in the amount of iron in oceanic upwellings would explain the Phanerozoic dominance of the black shale-phosphate association over the iron-formation-phosphate association (Cook and McElhinny, 1979).

Phosphorites have an average **uranium** content of 190 ppm and are a potential uranium source (Cook, 1976). Weathering of the apatite will cause release and redeposition of the uranium and both elements can be concentrated by weathering on unconformity surfaces (Cook and Shergold, 1986).

EXPLORATION GUIDES

An exploration strategy for phosphorites can be guided by three complementary approaches with decreasing scale:

1. Potential sites of phosphogenesis can be reconstructed on world and smaller scales by modelling winds, ocean currents, paleoclimate, sea level, and continental configuration for the appropriate geological time, as done by Parrish et al. (1986).

2. A regional exploration program should seek:

   - extensive continental prisms that are at the margins of mature oceans, or epicontinental sequences with slow sedimentation rates and reworking of sediments.

   - situations where marine transgression is well defined; the regressive phases are favourable for phosphogenesis.

   - paleolatitudes about 40° or less.

   - black shale-chert-carbonate association (continental shelf) or facies changes such as black, basinal shales oceanwards and carbonates, cherts, saline deposits, and red or light coloured shale or sandstone facies landwards (i.e., at the hinge line along the edge of a basin).

   - a light coloured sandstone-shale-siltstone association (epicontinental), but with chert, calcareous chert, carbonate rock, and siliceous chert(?) also present.

   - presence of synsedimentary arches and basins.

   - anomalous amounts of uranium detected in acid streams that drain potentially phosphatic terranes.

   - secondary (weathered, residual, phosphatized) deposits that may be associated with karstic horizons or weathering profiles.

   - where an older, phosphatic terrane is truncated by unconformities that provide for reworking and concentration of phosphate in younger deposits.

3. At the outcrop scale, the common pelletal or nodular textures suggest sedimentary phosphate, but are not distinctive, and hence phosphorite is easily overlooked. Instead, many phosphatic rocks are discovered through routine chemical analyses. Some guides to exploration include:

- pelletal or nodular texture
- bone fragments, fecal pellets, fish teeth
- glauconite, a common accessory mineral
- bluish-white coating (phosphate bloom) on the weathered surface
- somewhat high specific gravity
- radiation related to associated uranium, especially in marine apatite
- bituminous odour when the rock is struck by a hammer

Suspected phosphorite can be checked in the field using a simple spot-test (ammonium molybdate), but the test is sensitive to low phosphate values.

ACKNOWLEDGMENTS

The authors are grateful to O.R. Eckstrand of the Geological Survey of Canada for his valuable and patient guidance and editing. A review by R.T. Bell helped clarify a significant number of points.

SELECTED BIBLIOGRAPHY

References marked with asterisks (*) are considered to be the best sources of general information on this deposit type.

Ansdell, K.M., Kyser, T.K., and Stauffer, M.R.
1992: Age and source of detrital zircons from the Missi Formation: a Proterozoic molasse deposit, Trans-Hudson Orogen, Canada; Canadian Journal of Earth Sciences, v. 29, p. 2583-2594.

Burnett, W.C.
1980: Fertilizer mineral potential in Asia and the Pacific; in Proceedings of the Fertilizer Raw Materials Resources Workshop, (ed.) R.P. Sheldon and W.C. Burnett; August 20-24, 1979, Honolulu, Hawaii, East-West Resource Systems Institute, East-West Center, Hawaii, p. 119-144.

Cant, D.J. and Stockmal, G.S.
1989: The Alberta foreland basin: relationship between stratigraphy and Cordilleran terrane-accretion events; Canadian Journal of Earth Sciences, v. 26, p. 1964-1975.

Cathcart, J.B. and Gulbrandsen, R.A.
1973: Phosphate deposits; in United States Mineral Resources, United States Geological Survey, Professional Paper 820, p. 515-525.

Christie, R.L.
*1978a: Sedimentary phosphate deposits – an interim review; Geological Survey of Canada, Paper 78-20, 9 p.
1978b: Sedimentary phosphate deposits and the world phosphate industry; Geological Survey of Canada, Open File 533, 47 p.
1979: Phosphorite in sedimentary basins of western Canada; in Current Research, Part B; Geological Survey of Canada, Paper 79-1B, p. 253-258.
1980: Paleolatitudes and potential for phosphorite deposition in Canada; in Current Research, Part B; Geological Survey of Canada, Current Research, Paper 80-1B, p. 241-248.
1989: Jurassic phosphorite of the Fernie Synclinorium, southeastern British Columbia; in Phosphate Deposits of the World, Volume 2, Phosphate Rock Resources (ed.) A.J.G. Notholt, R.P. Sheldon, and D.F. Davidson; International Geological Correlation Programme, Project 156: Phosphorites, Cambridge University Press, p. 79-83.

Christie, R.L. and Sheldon, R.P.
1986: Proterozoic and Cambrian phosphorites – regional review: North America; in Phosphate Deposits of the World, Volume 1, Proterozoic and Cambrian Phosphorites, International Geological Correlation Program, Project 156, Phosphorites, Cambridge University Press, p. 101-107.

Cook, P.J.
1976: Sedimentary phosphate deposits; in Handbook of Strata-Bound Ore Deposits, v. 7, (ed.) Wolfe, K.H.; Amsterdam, Elsevier, p. 505-535.

*Cook, P.J. and McElhinny, M.W.
1979: A reevaluation of the spatial and temporal distribution of sedimentary phosphate deposits in the light of plate tectonics; Economic Geology, v. 74, p. 315-330.

Cook, P.J. and Shergold, J.H.
1986: Proterozoic and Cambrian phosphorites – nature and origin; in Phosphate Deposits of the World, Volume 1, Proterozoic and Cambrian Phosphorites (ed.) P.J. Cook and J.H. Shergold; International Geological Correlation Programme, Project 156: Phosphorites, Cambridge University Press, p. 369-386.

Cook, P.J., Shergold, J.H., Burnett, W.C., and Riggs, S.R.
1990: Phosphorite research: a historical overview; in Phosphate Research and Development, (ed.) A.J.G. Notholt and I. Jarvis; Geological Society Special Publication, no. 52, p. 1-22.

Coveney, R.M., Jr. and Chen, N.
1989: Nickel-molybdenum-platinum-gold deposits in black shales of southern China – a new ore type with possible analogs in Pennsylvanian rocks of the U.S.A.; in Metalliferous Black Shales and Related Ore Deposits - Proceedings, 1989, United States Working Group Meeting, International Geological Congress Correlation Program, Project 254, United States Geological Survey, Circular 1058, p. 9-11.

Cumming, G.L., Krstic, D., and Wilson, J.A.
1987: Age of the Athabasca Group, northern Alberta; Geological Association of Canada–Mineralogical Association of Canada, Program with Abstracts, Saskatoon, May 25-27, 1987, v. 12, p. 35.

de Schmid, H.S.
1916: Investigation of a reported discovery of phosphate in Alberta; Mines Branch, Department of Mines, Canada, Bulletin 12, 38 p.

Dixon, J.
1992: A review of Cretaceous and Tertiary stratigraphy in the northern Yukon and adjacent Northwest Territories; Geological Survey of Canada, Paper 92-9, 79 p.

Douglas, R.J.W.
1969: Geological map of Canada; Geological Survey of Canada, Map 1250, scale 1: 5 000 000.

Hall, R.L.
1984: Lithostratigraphy and biostratigraphy of the Fernie Formation (Jurassic) in the southeastern Canadian Rocky Mountains; in The Mesozoic of Middle North America, (ed.) D.F. Stott and D.J. Glass; Canadian Society of Petroleum Geologists, Memoir 9, p. 361-372.

Hallam, A.
1977: Secular changes in marine inundation of USSR and North America through the Phanerozoic; Nature, v. 269, p. 769-772.

Hays, J.D. and Pitman, W.C.
1973: Lithospheric plate motion, sea level changes and climatic and ecological consequences; Nature, v. 246, p. 18-22.

Holland, H.D.
1973: The oceans: a possible source of iron in iron formations; Economic Geology, v. 68, p. 1169-1172.

Huyck, H.L.O.
1990: Black shales: an economic geologist's perspective; in 1990 Eastern Oil Shale Symposium, Institute for Mining and Minerals Research, University of Kentucky, Lexington, Kentucky, p. 208-217.

Irving, E.
1979: Paleopoles and paleolatitudes of North America and speculations about displaced terrains; Canadian Journal of Earth Sciences, v. 16, p. 669-694.

Kolodny, Y.
1979: Origin of phosphorites in light of recent studies of modern deposits; in Report on the Marine Phosphatic Sediments Workshop, (ed.) W.C. Burnett and R.P. Sheldon; Honolulu, February 1979, East-West Resource Systems Institute, Hawaii, p. 7-8.

MacDonald, D.E.
1988: Sedimentary phosphate rock in Alberta and southeastern British Columbia: resource potential, the industry, technology and research needs; The Canadian Mining and Metallurgical Bulletin, v. 81, p. 45-52.

McClellan, G.H.
1980a: Quality factors of phosphate raw materials; in Fertilizer Mineral Potential in Asia and the Pacific, (ed.) R.P. Sheldon and W.C. Burnett; Proceedings of the Fertilizer Raw Materials Resources Workshop, August, 1979, East-West Resources Systems Institute, Honolulu, Hawaii, p. 359-377.
1980b: Mineralogy of carbonate fluorapatites; Journal of the Geological Society of London, v. 137, p. 675-681.

McClellan, G.H. and Lehr, J.R.
1969: Crystal-chemical investigation of natural apatites; American Mineralogist, v. 54, p. 1374-1391.

McGlynn, J.C. and Irving, E.
1981: Horizontal motions and rotations in the Canadian Shield during the early Proterozoic; in Proterozoic Basins of Canada, (ed.) F.H.A. Campbell; Geological Survey of Canada, Paper 81-10, p. 183-190.

McKelvey, V.E., Williams, J.S., Sheldon, R.P., Cressman, E.R., Cheney, T.M., and Swanson, R.W.
1959: The Phosphoria, Park City and Shedhorn Formations in the western phosphate field; United States Geological Survey, Professional Paper 313-A, 47 p.

Menard, H.W.
1984: Origin of guyots: the *Beagle* to *Seabeam*; Journal of Geophysical Research, v. 89, p. 11, 117-123.

Miller, A.R., Cumming, G.L., and Krstic, D.
1989: U-Pb, Pb-Pb and K-Ar isotopic study and petrography of uraniferous phosphate-bearing rocks of the Thelon Formation, Dubawnt Group, Northwest Territories, Canada; Canadian Journal of Earth Sciences, v. 26, p. 867-880.

Moores, E.M.
1991: Southwest U.S.-East Antarctic (SWEAT) connection: a hypothesis; Geology, v. 19, p. 425-428.

Notholt, A.J.G., Highley, D.E., and Slansky, M.
1979: Dossier on phosphate, Dossier IV of Raw Materials Research and Development; Commission of the European Communities, DG XII-Research, Science, Education.

Parfitt, R.L.
1989: Phosphate reactions with natural allophane, ferrihydrite and goethite; Journal of Soil Science, v. 40, p. 359-369.

Parrish, J.T.
*1982: Upwelling and petroleum source beds, with reference to Paleozoic; American Association of Petroleum Geologists, Bulletin, v. 66, p. 750-774.
1990: Paleoceanographic and paleoclimatic setting of the Miocene phosphogenic episode; in Phosphate Deposits of the World, Volume 3, Neogene to Modern Phosphorites, (ed.) W.C. Burnett and S.R. Riggs; International Geological Correlation Programme, Project 156: Phosphorites, Cambridge University Press, p. 223-240.

Parrish, J.T. and Barron, E.J.
1986: Paleoclimates and economic geology; Lecture notes for Short Course No. 18, Society of Economic Paleontologists and Mineralogists, Tulsa, Oklahoma, 162 p.

***Parrish, J.T. and Curtis, R.L.**
1982: Atmospheric circulation, upwelling and organic-rich rocks in the Mesozoic and Cenozoic eras; Palaeogeography, Palaeoclimatology and Palaeoecology, v. 40, p. 31-66.

***Parrish, J.T., Ziegler, A.M., Scotese, C.R., Humphreville, R.G., and Kirschvink, J.L.**
1986: Proterozoic and Cambrian phosphorites – specialist studies: Early Cambrian paleogeography, palaeoceanography and phosphorites; in Phosphate Deposits of the World, Volume 1, Proterozoic and Cambrian Phosphorites, (ed.) P.J. Cook and J.H. Shergold; International Geological Correlation Programme, Project 156: Phosphorites, Cambridge University Press, p. 280-294.

Pasava, J.
1993: Anoxic sediments – an important environment for PGE: an overview; Ore Geology Reviews, v. 8, p. 425-445.

Popenoe, P.
1990: Paleoceanography and paleogeography of the Miocene of the Southeastern United States; in Phosphate Deposits of the World, Volume 3, Neogene to Modern Phosphorites, (ed.) W.C. Burnett and S.R. Riggs; International Geological Correlation Programme, Project 156, Phosphorites, Cambridge University Press, p. 353-380.

Ramaekers, P.
1979: Reconnaissance geology of the interior Athabasca Basin; Saskatchewan Geological Survey, Summary of Investigations, 1978, Miscellaneous Report 78-10, p. 133-135.
1980: Stratigraphy and tectonic history of the Athabasca Group (Helikian) of northern Saskatchewan; in Summary of Investigations, 1980, Saskatchewan Geological Survey, Miscellaneous Report 80-4, p. 99-106.
1981: Hudsonian and Helikian basins of the Athabasca region, northern Saskatchewan; in Proterozoic Basins of Canada, (ed.) F.H.A. Campbell; Geological Survey of Canada, Paper 81-10, p. 219-233.

Riggs, S.R.
1979a: Petrology of the Tertiary phosphorite system of Florida; Economic Geology, v. 74, p. 195-220.

Riggs, S.R. (cont.)
1979b: Phosphorite sedimentation in Florida – a model phosphogenic system; Economic Geology, v. 74, p. 285-314.
1984: Paleoceanographic model of Neogene phosphorite deposition, U.S. Atlantic continental margin; Science, v. 223, no. 4632, p. 123-131.

Ross, G.M. and Chiarenzelli, J.R.
1986: Paleoclimatic significance of widespread Proterozoic silcretes in the Bear and Churchill provinces of the northwestern Canadian Shield; Journal of Sedimentary Petrology, v. 55, p. 196-204.

Sheldon, R.P.
*1963: Physical stratigraphy and mineral resources of Permian rocks in western Wyoming; United States Geological Survey, Professional Paper 313-B, p. 49-273.
1980: Episodicity of phosphate deposition and deep ocean circulation – a hypothesis; Society of Economic Paleontologists and Mineralogists, Special Publication 29, p. 239-247.
1981: Ancient marine phosphorites; Annual Reviews of Earth and Planetary Science, v. 9, p. 251-284.
1989: Phosphorite deposits of the Phosphoria Formation, western United States; in Phosphate Deposits of the World, Volume 2, Phosphate Rock Resources, (ed.) A.J.G. Notholt, R.P. Sheldon, and D.F. Davidson; International Geological Correlation Programme, Project 156: Phosphorites, Cambridge University Press, p. 53-61.

Skinner, H.C.
1993: A review of apatites, iron and manganese minerals, and their roles as indicators of biological activity in black shales; Precambrian Research, v. 61, p. 209-229.

***Slansky, M.**
1980: Géologie des phosphates sédimentaires; Bureau de recherches géologiques et minières, France, Mémoire no. 114, 92 p.

Smith, A.G., Hurley, A.M., and Briden, J.C.
1981: Phanerozoic Paleocontinental World Maps; Cambridge University Press, Cambridge, New York, 102 p.

Stronach, N.J.
1984: Depositional environments and cycles in the Jurassic Fernie Formation, southern Canadian Rocky Mountains; in The Mesozoic of Middle North America, (ed.) D.F. Stott and D.J. Glass; Canadian Society of Petroleum Geologists, Memoir 9, p. 43-67.

Tipper, H.W., Woodsworth, G.J., and Gabrielse, H.
1981: Tectonic assemblage map of the Canadian Cordillera and adjacent parts of the United States of America; Geological Survey of Canada, Map 1505A, scale 1:2 000 000.

Wang, F.F.H. and McKelvey, V.E.
1976: Marine mineral resources; in World Mineral Supplies, Assessment and Perspective, (ed.) G.J.S. Govett and M.H. Govett; Elsevier, Amsterdam, p. 221-286.

Werner, A.B.T. (comp.)
1981: Phosphate rock: an imported mineral commodity; Energy, Mines and Resources Canada, Mineral Bulletin MR 193 (1982), 61 p.

Wilson, J.A.
1985: Crandallite group minerals in the Helikian Athabasca Group in Alberta, Canada; Canadian Journal of Earth Sciences, v. 22, p. 637-641.

Young, F.G.
1977: The mid-Cretaceous flysch and phosphatic ironstone sequence, northern Richardson Mountains, Yukon Territory; Geological Survey of Canada, Report of Activities, Paper 77-1C, p. 67-74.

Young, F.G. and Robertson, B.T.
1984: The Rapid Creek Formation: an Albian Flysch-related phosphatic iron formation in northern Yukon Territory; in The Mesozoic of Middle North America, (ed.) D.F. Stott and D.J. Glass; Canadian Society of Petroleum Geologists, Memoir 9, p. 361-372.

Authors' addresses

F.W. Chandler
Geological Survey of Canada
601 Booth Street
Ottawa, Ontario
K1A 0E8

R.L. Christie
RR# 5,
Madoc, Ontario
K0K 2K0

Printed in Canada

3. STRATIFORM IRON

3.1 Lake Superior-type iron-formation
3.2 Algoma-type iron-formation
3.3 Ironstone

3. STRATIFORM IRON

G.A. Gross

INTRODUCTION

About 5% of the Earth's crust is composed of iron, the fourth most abundant element after aluminum, silicon, and oxygen, and the most abundant element in the ferride group (atomic numbers 22-28) – iron, titanium, vanadium, chromium, manganese, cobalt, and nickel. The ferride elements are strongly lithophile and form compounds with oxygen or oxyanions. Nickel and cobalt are strongly chalcophile and combine readily with sulphur and, to a lesser extent, with oxygen. Iron and manganese are both chalcophile and lithophile and commonly occur as oxide and sulphide minerals. Iron is also strongly siderophile, but occurrences of native iron are rarely found. Because of its chemical properties and abundance in the Earth, iron is widely distributed in rocks in silicate minerals and concentrated in oxide, sulphide, and carbonate minerals in most of the mineral deposits that occur in, or are associated with, metalliferous sediments or igneous rocks. Iron ore deposits are highly diversified and only the principal types of metallogenetic significance in Canada are considered here.

More than 95% of the iron ore resources of the world occur in iron-formation. The term iron-formation has been used for stratigraphic units of layered, bedded, or laminated rocks of all ages that contain 15% or more iron, in which the iron minerals are commonly interbanded with quartz, chert, and/or carbonate. The bedding of iron-formation generally conforms with the primary bedding in the associated sedimentary, volcanic, or metasedimentary rocks (James, 1954; Gross, 1959a, 1965; Brandt et al., 1972).

Iron-formations consist of minerals formed during diagenesis and metamorphism, and many inferences have been made about the nature of the precursor hydrolithic sediments from which they were derived. Some recent metalliferous sediments on the seafloor are similar in composition to various lithofacies of iron-formation that were deposited by chemical and biochemical processes. These consist of complex mixtures of amorphous particles of iron oxide and silica, hydrated iron oxide minerals (goethite), various forms of quartz and silica gel, carbonate, iron sulphide, and a variety of complex clay minerals, including montmorillionite, illite, and smectite muds. Subsidiary amounts of detrital minerals are commonly present in, or interbedded with, metalliferous sediments.

Gross, G.A.
1996: Stratiform iron; in Geology of Canadian Mineral Deposit Types, (ed.) O.R. Eckstrand, W.D. Sinclair, and R.I. Thorpe; Geological Survey of Canada, Geology of Canada, no. 8, p. 41-54 (also Geological Society of America, The Geology of North America, v. P-1).

Iron-formations are a major part of aqueochemically precipitated (hydrolithic) metalliferous sedimentary rocks classified by Gross (1990) as the stratafer group: "...lithological facies formed by chemical, biogenic, and hydrothermal effusive or exhalative processes, commonly composed of banded chert and quartz interbedded with oxide, sulphide, carbonate, and silicate minerals containing ferrous, nonferrous, and/or precious metals". This group includes the common iron-rich oxide, silicate, carbonate, and sulphide lithofacies of iron-formation, and associated lithofacies units interbedded and associated with them, that evidently formed by similar processes and which may contain economically important amounts of manganese, copper, zinc, lead, gold, silver, rare-earth elements, tungsten, nickel, cobalt, niobium, barite, and/or other elements.

The stratafer group of siliceous hydrolithic sediments has special metallogenetic importance, as they are the source rocks for more than 75% of the metals mined in the world (Gross, 1993a). Iron deposits include the common iron-rich oxide, silicate, carbonate, and sulphide lithofacies of iron-formation which are described in the following papers. Other members of the stratafer group include deposits of manganese and a variety of polymetallic sulphide (stratiform massive sulphide) deposits which are described elsewhere in this volume as volcanic-associated massive sulphide base metals (subtype 6.3) and sedimentary exhalative (subtype 6.1). The polymetallic sulphide deposits are massive and bedded lenses that may contain extensive stratiform concentrations of barium, copper, zinc, lead, gold, silver, rare-earth elements, nickel, tin, tungsten, cobalt, niobium, and/or other associated metals. These are interpreted to have formed by similar processes to the iron-formations and constitute sulphide lithofacies.

Following long established precedent, the term ironstone or clay ironstone is retained for bedded goethite-siderite-chamosite-bearing hydrolithic sedimentary rocks, that commonly have oolitic textures, a relatively high content of alumina and phosphorus, and a low content of silica. Iron-formation and ironstone are not only distinct lithologically, but are also interpreted to have different metallogenetic significance. Modern facies of iron-formation that are forming by hydrothermal-effusive or exhalative processes at many sites on the seafloor provide important analogues for understanding and interpretation of ancient iron-formations and the stratafer sediments associated with them. In contrast, the ironstones appear to have been deposited in sedimentary basins dominated by hydrogenous and chemical-biogenic sedimentary processes. The principal lithofacies and depositional environments for major types of iron-formation and ironstones in North America are indicated in Figure 3-1. The typical compositions of the main types of iron-formation and ironstone are given in Table 3-1.

Deposits of the stratafer group, including the common lithofacies of iron-formation distributed throughout the geological record, mark ancient sites of mineral deposition by hydrothermal-effusive and/or epigenetic processes. Iron-formations are prominent metallogenetic marker beds for the diverse spectrum of lithofacies included in the stratafer group, and have been used effectively for guiding mineral exploration (Frietsch, 1982a, b; Gross, 1986, 1991, 1993a).

SEDIMENTARY LITHOFACIES OF THE STRATAFER GROUP

Understanding the distribution and interrelationships of the various lithofacies in the stratafer group is required for the assessment of their mineral resource potential. Development of the facies concept was a major step in understanding the genesis of chemical sediments and their broader metallogenetic significance. Krumbein and Garrels (1952) demonstrated that pH and Eh in solutions were major factors controlling the composition and mineralogy of hydrolithic sediments deposited in different basin environments. James (1954) applied these concepts in the study of iron-formations in the Lake Superior region and defined four distinct primary facies of iron-formation: oxide, silicate, carbonate, and sulphide. The concept of primary sedimentary facies as demonstrated for iron-formations has provided a basis for understanding the origin and distribution of the variety of other lithofacies in the stratafer group.

Recognition and documentation of facies relationships in iron-formations (Krumbein and Garrels, 1952; James, 1954; Goodwin, 1956; Gross, 1965, 1970a, 1973, 1993a) and in other major groups of stratafer sediments, have been inhibited in many cases because of inadequate stratigraphic and structural data, inadequate documentation of

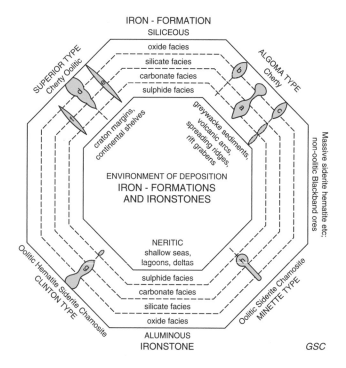

Figure 3-1. Diagram showing major types of chemically precipitated iron-formation and ironstone with their sedimentary facies and depositional environment. Examples are from **a)** Michipicoten, Ontario; **b)** Moose Mountain, Ontario; **c)** Temagami, Ontario; **d)** Knob Lake, Labrador and Quebec; **e)** Iron River, Michigan; **f)** Gunflint and Biwabik iron-formations, Ontario and Minnesota; **g)** Wabana, Newfoundland; **h)** Clear Hills, Alberta (after Gross, 1965).

geological events, and a lack of basin analyses. Nevertheless, multiple facies relationships, including the commonly recognized oxide, carbonate, silicate, and sulphide facies of iron-formation, have now been documented in most of the stratafer group. Stratafer lithofacies are now generally accepted as having formed by similar hydrothermal-sedimentary processes, are interbedded in many stratigraphic successions, and occur throughout the geological record from Early Archean to the modern possible protofacies on the seafloor (Gross, 1987, 1988b).

Mineral exploration for deposits of the stratafer group has been oriented mainly to assessing the resource potential for specific metals or particular mineral groups, ferrous, nonferrous, or precious metals. In formulating guidelines for exploration, some have overlooked, ignored, or not recognized the association of lithofacies in the stratafer group of metalliferous sediments as a whole, and the distribution of metals and interelement relationships between them. Considerable confusion exists in current literature because different facies of iron-formation are not identified or are described under different names, detailed stratigraphic relationships are not reported, primary depositional and diagenetic features are not recognized, and distinctions between isochemical metamorphism and alteration by epigenetic and metasomatic processes are not made.

Discrepancies and inconsistencies in nomenclature and terminology in petrographic descriptions have also contributed to confusion in understanding the relationships of lithofacies and have seriously distracted from our understanding of common genetic processes for the whole assemblage of metalliferous sediments. The development of consistent descriptive nomenclature for local stratigraphic features is of course needed, but it should be placed in the context of established schemes of nomenclature and accepted usage or at least referred to them (cf. Gross, 1965, 1970b, 1990; Brandt et al., 1972).

Studies of genetic processes have tended to be carried out independently on the major lithofacies groups, the iron-formations, the manganese-iron, polymetallic sulphide, and

Table 3-1. Compositions of different types of iron-formation and ironstone (from Gross, 1965).

Type	Iron-formation			Ironstone	
	Algoma	Lake Superior	Rapitan	Clinton	Minette
Location	Temagami	Knob Lake area		Snake River	Wabana	Peace River
Facies	Oxide [1]	Oxide [2]	Silicate Carbonate [3]	Oxide [4]	[5]	[6]
Fe (wt.%)	33.52	33.97	30.23	38.93	51.79	30.9
SiO_2	47.9	48.35	49.41	33.03	11.42	28.0
Al_2O_3	0.9	0.48	0.68	1.08	5.07	5.79
Fe_2O_3	31.7	45.98	16.34	54.71	61.83	29.81
FeO	14.6	2.33	24.19	0.85	11.00	13.0
CaO	1.45	0.1	0.1	3.58	3.32	1.9
MgO	1.8	0.32	2.95	1.66	0.63	1.5
Na_2O	0.2	0.33	0.03	0.05	ND	0.3
K_2O	0.32	0.01	0.07	0.03	ND	0.5
H_2O+	0.47	2.0	5.2	ND		
H_2O-	0.1	0.04	0.38	0.48	1.94	13.1
TiO^2	0.05	0.01	0.01	0.19	0.015	0.1
P_2O_5	0.1	0.04	0.08	0.58	1.96	1.5
MnO	0.3	0.25	0.65	0.17	0.17	0.1
CO_2	ND	0.03	0.22	3.45	2.15	2.8
S	ND	0.013	0.05	0.02	.023	ND
C	ND	0.08	0.15	ND	ND	ND

Notes:
1) Average of 4 analyses for 15 m of section sampled; Algoma-type magnetite-quartz iron-formation, mainly oxide facies, Temagami Lake area, Ontario (analyses, Geological Survey of Canada laboratories).
2) Average of 6 analyses for 100 m of section sampled; Lake Superior-type hematite-magnetite-quartz iron-formation, oxide facies, Knob Lake iron ranges, Quebec and Labrador (analyses, Geological Survey of Canada laboratories).
3) Analyses for 15 m of section sampled; Lake Superior-type silicate-carbonate-chert facies, Knob Lake iron ranges, Quebec and Labrador (analyses, Geological Survey of Canada laboratories).
4) Average of 42 samples collected at random of typical oxide facies, Rapitan-type hematite-quartz iron-formation, Snake River area, Mackenzie Mountains, Yukon Territory (analyses, Geological Survey of Canada laboratories).
5) Analysis of large composite sample; Clinton-type ironstone, Wabana, Newfoundland (courtesy, Wabana Mines).
6) Average of 11 samples, each of 4.5 m of section; Minette-type, oolitic siderite-chamosite-limonite ironstone, Peace River district, Alberta (Mellon, 1962).
7) ND: not determined.

other sedimentary-exhalative facies. Hydrothermal-effusive or volcanogenic processes, proposed initially in Europe and seriously considered about a century ago in the study of iron-formations, were considered independently for the genesis of the other facies in the stratafer group, and have been discussed in voluminous literature. The many distinctive lithofacies of the stratafer group are interbedded in many sedimentary and volcanic successions, have been traced laterally from one group to another, and syngenetic relationships between the facies groups cannot be ignored.

Patterns in the distribution of metalliferous lithofacies in relation to hydrothermal-effusive and volcanic centres that are evident from empirical data, and that are analogous to the patterns for ancient stratafer sediments, have been outlined for sulphide, silicate, carbonate, and oxide facies in recent marine sediments. Descriptive data for these metalliferous sediments from Honnorez et al. (1973); Cronan (1980); Bornhold et al. (1982); Hekinian (1982); Smith and Cronan (1983); Bischoff et al. (1983); Gross (1983b, 1987); Koski et al. (1984); Gross and McLeod (1987); and many others are summarized as follows:

1. Polymetallic "massive" sulphide facies occur within and close to volcanic vents. The thinly banded and layered iron-rich sulphide facies associated with them usually extend to areas distal from the vents.

2. Stacks, mounds, crusts, and gelatinous sediments are developed around vents from which high temperature hydrothermal fluids are discharged. Transitional zoning outward from the vents, from iron and copper-zinc sulphides, to iron and manganese sulphide and oxide facies, evidently follows high-to-low temperature gradients.

3. Sedimentary sulphide facies were deposited close to the higher temperature effusive centres; iron oxide and silicate facies were intermediate, and manganese-iron facies were deposited around cooler hydrothermal vents and in areas distal from active hydrothermal discharge. Overlapping and transitions in facies types are common.

4. Sequential development of different facies of metalliferous sediments during the evolution of a spreading ridge system gave rise to the deposition of iron and manganese oxide facies in the initial and cooler stages of effusive hydrothermal activity along ridge axes, and deposition continued along ridges in the direction of propagation of a ridge system. Facies development extends laterally from polymetallic and iron sulphide facies proximal to vent areas, to silicate facies, and carbonate facies usually in shallow water, to iron and manganese oxide facies distal from hydrothermal centres.

5. Hydrothermal effusive systems are commonly initiated at the intersection of ridge axes and transverse faults and achieve their greatest intensity and maturity in these areas.

6. Metals from hydrothermal effusive sources may be carried in currents and plumes for distances of 100 km or more before they are deposited as iron-dominated metalliferous sediments or manganese nodules.

7. As ridge systems develop, the earlier formed metalliferous deposits are carried on the spreading plate laterally from the axial zone and become covered with clastic sediment, including turbiditic beds, or with lava and

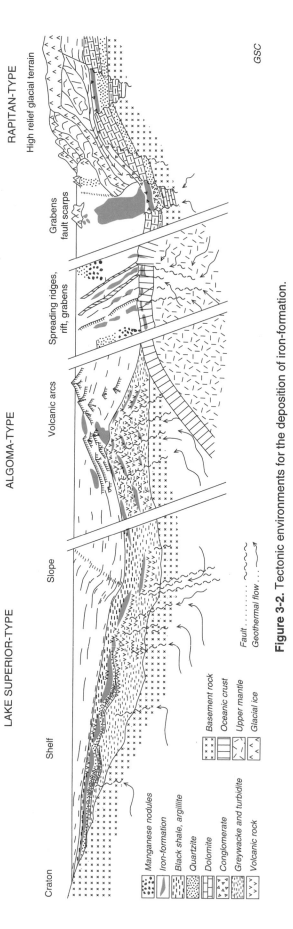

Figure 3-2. Tectonic environments for the deposition of iron-formation.

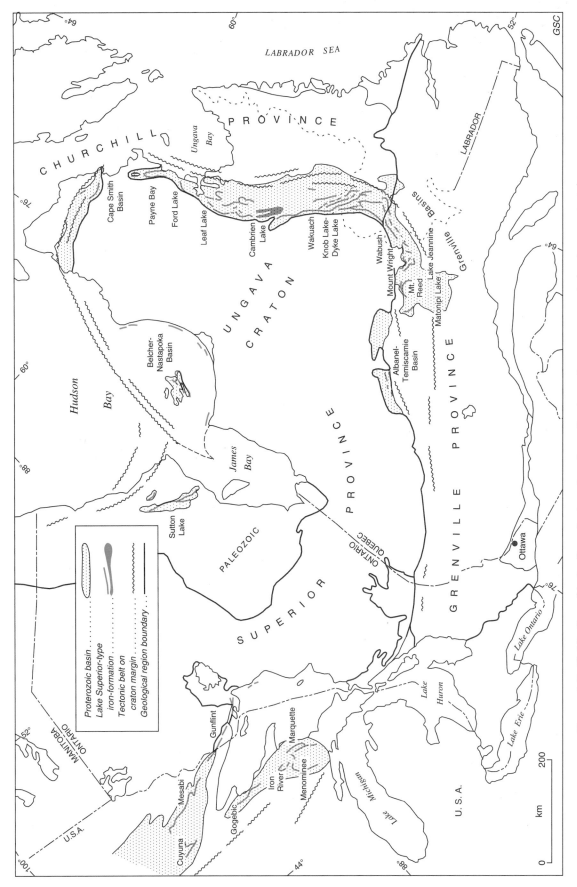

Figure 3-3. Distribution of Lake Superior-type iron-formation in sedimentary-tectonic basins marginal to the Ungava-Superior craton.

45

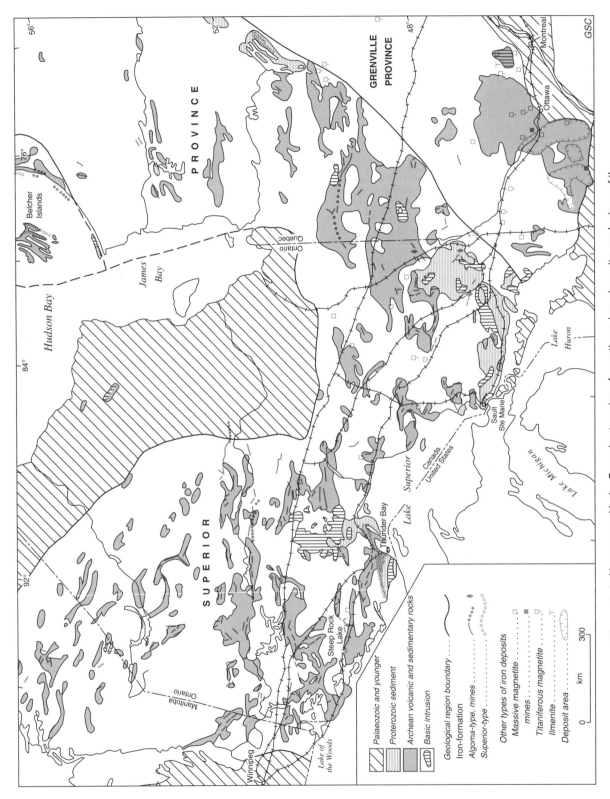

Figure 3-4. Algoma- and Lake Superior-type iron-formations, iron deposits, and mines of the southern Canadian Shield (modified after Lang et al., 1970).

tuff. Deposition of higher temperature polymetallic sulphide minerals takes place in the tectonically active and faster spreading segments in the propagating parts of ridges.

8. Metalliferous sediments, including manganese nodules, accumulate throughout the entire period of ridge development in basins distal from ridge axes where the rate of clastic sedimentation is low.

9. Hydrolithic sediments (e.g., chert) are commonly interspersed with volcanic rocks, clastic sediments, and dolomite, and record hydrothermal-effusive and exhalative processes in the history of an area.

LITHOFACIES OF THE STRATAFER GROUP

Modern concepts for understanding the metallogeny of iron-formations and other members of the stratafer group are based on the recognition of three main factors:

1. their hydrolithic nature and derivation from primary aqueochemically precipitated protofacies;

2. the predominant role of hydrothermal-effusive and volcanogenic processes in their genesis; and

3. the interrelationships and diverse nature of lithofacies formed by hydrothermal-sedimentary processes.

Iron-formation lithofacies

Lithofacies of stratafer sediments that contain more than 15% iron are classified as iron-formation. They are widely distributed throughout the geological record and were deposited in a broad spectrum of depositional environments ranging from continental shelves and platforms, through craton and oceanic plate margins, to tectonic basins and volcanic arcs along subduction zones. The most diverse group of iron-formation lithofacies are proximal to volcanic centres in association with turbidites and volcanic rocks along offshore tectonic-volcanic belts, island arcs, spreading ridges, and rift zones in deep ocean basins (Gross, 1973, 1980, 1983a).

Gross (1959a, 1965) considered the iron-formations and associated hydrolithic sediments to be sensitive indicators of their depositional environments. He recognized two principal types of iron-formation, Lake Superior and Algoma, based on tectonic setting, types of associated rocks, and depositional environment. Both types were considered to have formed by similar hydrothermal-effusive or exhalative processes, but in distinctly different tectonic and depositional environments as depicted in Figure 3-2 (Gross, 1965, 1973, 1983a).

Iron-formations of Lake Superior type have been the principal sources of iron ore throughout the world for more than a century. They formed in shelf and platformal basins along the margins of Early to Middle Proterozoic cratons and are preserved on all of the continents. They are associated with normal shelf-type sedimentary rocks, including dolostone, quartz arenite, arkose, black shale, and conglomerate, and with tuff and other volcanic rocks in linear basins along craton margins. The distribution of Lake Superior-type iron-formations around the margins of the Superior-Ungava craton of the Canadian Shield is outlined in Figure 3-3.

Algoma-type iron-formations occur throughout the geological record in marine depositional environments that are distributed along volcanic arcs, spreading ridges, grabens, fault scarps, and fracture zones, and are interbedded with greywacke, turbidites, metalliferous sediments, and volcanic rocks (Gross, 1959a, 1965, 1973, 1983a). Algoma-type iron-formation and other associated stratafer lithofacies are widely distributed and appear to be consistently present in belts of volcanic and sedimentary rocks such as the Archean greenstone belts of the Superior Province (Fig. 3-4).

Rapitan-type iron-formations have distinctive lithological features, being associated with diamictites (tillite), including dropstones, sandstone, conglomerate, and argillites. Examples, such as the Snake River banded chert-hematite facies in the Mackenzie Mountains in northwestern Canada (Gross, 1964a; Yeo, 1978, 1986) and the Jacadigo iron-formation in Brazil (Dorr, 1973), were deposited in grabens and fault scarp basins along the rifted margins of continents or ancient cratons in sequences of Late Proterozoic and Early Paleozoic rocks.

Evidence for hydrothermal-effusive processes in the genesis of iron-formations, especially the stratigraphic association of Algoma-type iron-formation with volcanic episodes, as pointed out by Goodwin (1962, 1973), Gross (1965, 1983a), Ohmoto and Skinner (1983); and Shegelski (1987), has established a basis for linking their metallogenesis with that of other metalliferous facies of the stratafer groups. Modern iron-formation facies on the seafloor are analogous to common lithofacies of the stratafer group in the rock record (Bischoff, 1969; Honnorez et al., 1973; Bischoff et al., 1983; Smith and Cronan, 1983; Boström and Widenfalk, 1984; Koski et al., 1984; Gross and McLeod, 1987; El Shazly, 1990), supporting the early interpretations of their hydrothermal-sedimentary origins (Gross, 1965, 1991). The history of the development of hydrothermal-effusive, volcanogenic, and exhalative concepts for the origin of iron-formation and associated stratafer lithofacies has been reviewed by Gross (1991) and Stanton (1991).

Manganese iron-formation lithofacies

Manganiferous facies interbedded in iron-formations are the principal host rocks throughout the world for large manganese deposits (Gross, 1983b, 1990). The major deposits consist of primary manganiferous lithofacies and "secondary enrichment deposits" hosted in manganese-rich iron-formation. Roy (1981) comprehensively reviewed manganese deposits throughout the world. An excellent overview of genetic concepts for manganese deposits was presented by Varentsov and Grasselly (1980a, b, c) for Africa, South America, Hungary, India, South Korea, and the former Soviet Union. Manganese carbonate facies in iron-formation on the Nastapoka Islands in Canada were reported by Bell (1879), and manganese oxide facies in iron-formation at Woodstock, New Brunswick were mined as iron ore from 1848 to 1884 (Gross, 1967; Anderson, 1986). Manganese facies in iron-formation in the Lake Superior iron ranges have been described by James (1954), Lepp (1963, 1968), and Schmidt (1963); in high-grade metamorphic terrain in India by Devaraju and Laajoki (1986); in the Karadzhal deposits in Kazakhstan by Sapozhnikov (1963) and Kalinin (1965); at Jalisco, Mexico by Zantop

(1978); in Minas Gerais, Bahia, and Morro do Urucum in Brazil by Park et al. (1951) and Dorr et al. (1956); in South Africa by De Villiers (1956), Beukes (1973), and Sohnge (1977); in Sweden by Frietsch (1982b); and in the former U.S.S.R. by Varentsov and Grasselly (1980b).

Aside from manganese nodules, manganese deposition related to active hydrothermal fields took place in areas distal from hydrothermal vents, in the outer margins of the sedimentary basins, and manganese was generally transported beyond the main depositional areas for siliceous iron-rich facies (Gross, 1987).

Polymetallic sulphide lithofacies

Polymetallic sulphide (massive sulphide) lithofacies were considered separately in North America until the late 1950s when the common syngenetic relationships between oxide, carbonate, silicate, and sulphide facies in iron-formation were recognized by many exploration geologists. The metallogenetic significance of hydrothermal-effusive, exhalative, and volcanogenic processes were emphasized in Canada in the early work of Gross (1959a, 1965, 1973); Stanton (1959); and Goodwin (1962). These concepts were applied successfully in the exploration for massive sulphide deposits (Stanton, 1960; Roscoe, 1965; Hutchinson, 1965; Gilmour, 1965). Hutchinson et al. (1971) pointed out that the distinction between the exhalative iron-formations and the associated sulphide deposits was based on metal content rather than genetic process.

Interbedding of oxide, carbonate, and sulphide facies of iron-formation has been observed throughout the world, for example in the Bathurst area, New Brunswick (McAllister, 1959; Gross, 1967; Van Staal and Williams, 1984); in the Manitouwadge and Michipicoten areas, Ontario (Goodwin, 1962; Franklin et al., 1981); at Broken Hill in New South Wales, Australia (Stanton, 1976); in the Gamsberg base-metal sulphide deposit in Cape Province, South Africa (Rozendaal, 1980); in the Isua iron-formation in Greenland (Appel, 1979); in Sweden (Frietsch, 1980a, b, c, 1982a, b); in Lake Superior-type iron-formations in the Labrador-Quebec fold belt in Canada (Barrett et al., 1988); in the Kuroko and related volcanogenic massive sulphide deposits in Japan (Ohmoto and Skinner, 1983); in the Red Sea metalliferous sediments (Degens and Ross, 1969; El Shazly, 1990); and in many other stratiform sulphide deposits in Canada referred to by Sangster (1972), Franklin et al. (1981), and Franklin and Thorpe (1982).

Gale (1983) discussed the sedimentary and volcanic settings of stratiform Proterozoic exhalative massive sulphide deposits, mainly from the Canadian Shield, and referred to the stratabound nickeliferous sulphide deposits at Outokumpu, Finland and at Thompson, Manitoba, associated with biotite gneiss, quartzite, calc-silicate rock, pelitic schist, greywacke, and banded iron-formation.

The primary nature of many sulphide deposits is obscure because of the extensive recrystallization and remobilization of constituents in sulphide facies in the vent area during their formation, and during metamorphism (see Introduction, deposit Type 6, "Exhalative base metal sulphide"). Element mobility in sulphide facies during diagenesis and metamorphism greatly exceeds that in other lithofacies of the stratafer group and the syngenetic

distribution of elements and primary sedimentary features are not well preserved in many cases, or may be totally destroyed. Depending on the location and configuration of a depositional basin relative to the hydrothermal vents or metal sources, sulphide, oxide, and manganese-iron facies are locally developed as discrete units. Lateral transitions from one facies to another, and the genetic relationships of protofacies, are commonly obscure.

Sedex lithofacies

Sedex deposits (see "Sedimentary exhalative sulphides (Sedex)", subtype 6.1) are another sedimentary-exhalative member of the stratafer group, distinguished by the dominance of zinc and lead, by their predominantly sedimentary depositional environment, and in some cases by the apparent lack of directly associated volcanic rocks. Some Sedex deposits contain significant amounts of copper, antimony, tungsten, tin, gold, silver, and mercury, and relatively minor amounts of iron and manganese in the associated siliceous and carbonate facies (Edwards and Atkinson, 1986). Some Sedex deposits are not clearly distinguished from polymetallic volcanogenic massive sulphide deposits because of similarities in their genesis and in specific lithofacies (see Introduction, "Exhalative base metal sulphide").

The spatial association of iron-formation lithofacies with the Sedex and VMS lithofacies is well established. For example, Schultz (1966) and Edwards and Atkinson (1986) pointed out that oxide facies iron-formation at Tynagh mine, Ireland is the lateral stratigraphic equivalent of the bedded lead-zinc sulphide ore, and they described the association of iron-formation with the sulphide ore beds at Broken Hill, Australia. They considered Howards Pass and Sullivan deposits in Canada; Rammelsberg and Meggen in Germany; Broken Hill, Mount Isa, and McArthur River in Australia; Gamsberg in South Africa; and Tynagh and others in Ireland to be typical sedimentary-exhalative deposits. All are associated with banded chert and carbonate, and some are associated with sulphide and oxide lithofacies of iron-formation.

The tectonic setting and origin of the Jason deposit in the Yukon Territory, described by Winn and Bailes (1987), is considered to be representative of many Sedex deposits that were formed by the expulsion of hydrothermal fluids along a fault zone at the margin of a graben basin. The silicified, carbonatized, and brecciated rocks (stockworks) associated with the main stratiform bedded mineral zone are believed to mark hydrothermal conduits in the area.

The Sedex members of the stratafer group are of special metallogenetic interest because some contain economically significant amounts of the minor or trace elements that are consistently present in the common lithofacies of iron-formation. The following examples illustrate the great variety of metalliferous strata of hydrothermal-sedimentary origin that are included in the stratafer group: phosphorus-rich facies of iron-formation in Finland (Laajoki, 1986) and in Sweden (Frietsch, 1974); zinc-bearing iron-formations in the Grenville Province in southern Quebec (Gauthier et al., 1987); highly metamorphosed zinc deposits related to iron-formation such as the Balmat-Edwards-Pierrepont deposits in New York State, and similar deposits in Sweden (Frietsch, 1982a); stratiform tungsten deposits of San Luis, Argentina

(Brodtkorb and Brodtkorb, 1977); King Island scheelite mine in Tasmania (Burchard, 1977); antimony-tungsten-mercury deposits in the eastern Alps (Höll, 1977); stratiform tin deposits in Bolivia (Schneider and Lehmann, 1977); and stratiform mercury and antimony-tungsten-mercury deposits in Turkey (Sozen, 1977).

Rare-earth elements in iron-formation

Trace amounts of rare-earth elements (REEs) occur in iron-formations in all parts of the world, regardless of their age and facies. Typically the iron-formations are enriched in La and light relative to the heavier REEs and have marked positive Eu anomalies in chondrite-normalized data, except for facies that have a mixed provenance of chemically precipitated and clastic constituents. The REEs, like other minor elements, are not distributed uniformly and their content varies greatly from bed to bed in the cherty iron-formations (Gross, 1993a).

The largest known ore reserves of REEs occur in the Bayan Obo iron-formations in Inner Mongolia, China, where estimates of the iron ore reserves exceed 1000 million tons (907 Mt) containing 30 to 35% Fe, 5.7% REEs, 0.126% Nb, and 2% fluorite (Gross, 1986, 1993b). Oxide and carbonate facies of this highly folded and metamorphosed iron-formation are interbedded with oolitic dolostone in a typical succession of Early to Middle Proterozoic quartz arenite and arkose, dolomite, and black slate-shale that was deposited in a linear basin along the rifted margin of the North China Platform. The distribution of the REEs in the iron-formation and dolomite is interpreted by Gross (1993b) to be syngenetic. For example, although the contents of minor elements range from 10 to 2000 times greater in the Bayan Obo iron-formations than the average contents in most Lake Superior-type iron-formations, the interelement correlation patterns, including those for REEs, are similar to patterns found consistently in similar facies of most other iron-formations (Gross, 1993b). The abnormally high content of REEs and other minor elements in the Bayan Obo iron-formation is attributed to extensive hydrothermal contribution of constituents from the mantle and ideal conditions for rapid precipitation with dolomite and iron-formation and adsorption on the primary iron oxide minerals. Other iron-formations containing abnormally high amounts of REEs have been reported in China (Tu et al., 1985) and these occurrences, along with Bayan Obo, illustrate the metallogenetic importance of REEs in some of the iron-formations.

Gold in iron-formation

The common association of gold with iron-formations and stratafer sediments throughout the world has been recognized for more than a century. Iron-formations are now used widely as marker beds in the exploration for gold in fold-belts of sedimentary-volcanic rocks. Iron-formations and stratafer sediments are good host rocks for gold for the following reasons:

1. they are competent and brittle, providing excellent fracture systems for vein development;

2. their high iron contents in oxide, carbonate, silicate, and sulphide minerals are interpreted to be conducive to the precipitation of gold from hydrothermal solutions; and

3. as hydrolithic sediments they consistently contain minor but significant amounts of syngenetic gold (Gross, 1988a; see subtype 15.3, "Iron-formation-hosted stratabound gold") which may increase to an economically important content in some cases, as in the Homestake mine, South Dakota (Rye and Rye, 1974), in Lupin mine in the Northwest Territories (Kerswill et al., 1983), in the Kolar schist belt of India (Natarajan and Mukherjee, 1986), in many deposits in Canada (Hodgson and MacGeehan, 1982; Hodder and Petruk, 1982; Macdonald, 1990), in Zimbabwe (Saager et al., 1987); and in many other deposits throughout the world.

Element correlation patterns in many iron-formations show that Au distribution is commonly correlated with Sb, Bi, As, S, Zn, and that Au is enriched in mineral assemblages such as the sulphosalts that formed at moderate to lower temperatures in the hydrothermal systems (Gross, 1988a). Hannington and Scott (1988) provided evidence to show that different sulphide mineral equilibria and FeS contents in sphalerite reflect the same physical and chemical conditions which influence gold content. They suggested that petrological indicators of the sulphidation state may be useful guides to gold mineralization in volcanogenic massive sulphide deposits.

TYPES OF IRON ORE DEPOSITS

Iron ore deposits are highly diversified in their characteristics and origin and many different kinds of deposits are recognized. Descriptions of their physical, chemical, and geological features should provide an objective basis for recognizing the different types of ore deposits, primary and secondary genetic processes and genetic models, and an indication of the probable size, grade, and quality of ore that can be extracted from them (Gross, 1965, 1967, 1968).

Gross (1970b) classified iron deposits into five major types; bedded, massive, residual, byproduct, and "other" types. An extension of this classification, which includes six major types of deposits that are defined on the basis of descriptive data and geological features (Gross, 1959a, 1965; Lang et al., 1970), is as follows:

1. iron-formation (see subtype 3.1, "Lake Superior-type iron-formation" and subtype 3.2, "Algoma-type iron-formation"), ironstone (subtype 3.3), and other iron-rich sediments;

2. residual deposits derived from iron-formations by leaching of silica and concentration of iron oxide (see subtype 4.1, "Enriched iron-formation");

3. residual deposits, including chemically and mechanically transported surface deposits, placer sands, bog iron, laterite;

4. deposits in or associated directly with plutonic rocks, including 4a) mafic and ultramafic rocks (see Type 26, "Mafic intrusion-hosted titanium-iron") and 4b) felsic granitoid, and alkaline granitic rocks;

5. skarn, contact metasomatic, vein and structurally-controlled replacement deposits, (see subtype 20.4, "Skarn iron"); and

6. other types of deposits.

About 20 subtypes of these main types are found in Canada. Most of the iron ore resources in Canada occur in siliceous hydrolithic sediments classified as Lake Superior- or Algoma-type iron-formation.

IRON ORE

Iron ore is natural material of suitable grade, composition, and physical quality that can be mined and processed for profit or economic benefit, and similar resources that could be mined under present circumstances if a market and demand for them existed. Iron ore resources include all categories of measured, indicated, and inferred resources, and potential resources that include materials that could provide a source of iron in the future.

Many complex and interrelated factors are considered in the identification and appraisal of iron ore resources. These include the size, location, chemical composition, physical characteristics, amenability of ore to concentration and beneficiation, the quality and composition of other raw materials such as coal and limestone that will be used in smelting the ore, and the kind of smelting and steel-making process being used (blast furnace or direct reduction in rotary kilns; Bessemer, non-Bessemer, open hearth, oxygen smelting, electric-crucible direct reduction, etc.). It is desirable to use natural material that meets the required physical and chemical specifications, or ore that can be easily concentrated and processed to improve its quality and to meet required specifications. Hard lump ore with a low proportion of fine sized particles is still a premium product. It is not available in quantity in North America, and most of the ore now used consists of mineral concentrates that are sintered or pelletized to provide the desired physical and chemical properties and grade required for a particular smelting process.

The most desirable iron ore will have the highest iron and lowest silica and alumina contents possible; low contents of magnesium and calcium carbonate, depending on the amount of fluxing agents needed in a particular furnace burden; less than 2% manganese; and low contents (<0.05%) of phosphorus, sulphur, arsenic, sodium, potassium, titanium, and other ferride elements. A high proportion of the iron ore used throughout the world is processed and marketed as pellets, concentrate, or sinter feed and products.

The compositions of typical high quality iron ore pellets produced in Canada in 1985 and sintered ore blended in special processes are given in Table 3-2. Iron ore containing more than 2% manganese is usually sold separately as manganiferous iron ore and used for blending with iron ore from other sources, or in the production of ferromanganese.

Table 3-2. Typical composition of iron ore pellets and sinter produced in Canada in 1985.

%	Pellets	Sinter
Fe	64.76	47.81
P	0.008	0.016
SiO_2	4.90	8.31
Mn	0.10	2.24
Al_2O_3	0.31	1.31
Ca	0.51	12.69
Mg	0.38	6.46
S	0.003	0.083
LOI	0.19	0.87
Moisture	1.15	–
LOI = Loss on ignition		

Iron ore containing as much as 1% titanium has been used, but because titanium causes slags to become viscous, it is generally not accepted for blast furnace burden.

ACKNOWLEDGMENTS

This introduction was edited by C.W. Jefferson and R.I. Thorpe. Word processing assistance was provided by C.M. Plant and L.C. O'Neill.

SELECTED BIBIOGRAPHY

References marked with asterisks (*) are considered to be the best sources of information on the deposit subtypes in this group.

Anderson, F.D.
1986: Woodstock, Millville, and Coldstream map-areas, Carleton map-areas, Carleton and York counties, New Brunswick; Geological Survey of Canada, Memoir 353, p. 45-55.

*Anonymous
1979: La genèse des nodules de manganèse; Colloques internationaux du Centre national de la recherche scientifique, Éditions du Centre national de la recherche scientifique, Paris, no. 289, 405 p.

Appel, P.W.U.
1979: Stratabound copper sulfides in a banded iron-formation and basaltic tuffs in the Early Precambrian Isua Supracrustal Belt, west Greenland; Economic Geology, v. 74, p. 45-52.
1980: On the early Archaean Isua iron-formation, west Greenland; Precambrian Research, v. 11, p. 73-87.
*1986: Strata bound scheelite in the Archean Malene supracrustal belt, west Greenland; Mineralium Deposita, v. 21, p. 207-215.

Barrett, T.J. and Jambor, J.L. (ed.)
1988: Seafloor hydrothermal mineralization; Canadian Mineralogist, v. 26, pt. 3, p. 429-888.

Barrett, T.J., Wares, R.P., and Fox, J.S.
1988: Two stage hydrothermal formation of a Lower Proterozoic sediment-hosted massive sulfide deposit, northern Labrador Trough, Quebec; Canadian Mineralogist, v. 26, p. 871-888.

Bell, R.
1879: Report on an exploration of the east coast of Hudson's Bay, Nastapoka Islands; Geological Survey Canada, Report of Progress for 1877-78, p. 15c-18c.

Beukes, N.J.
*1973: Precambrian iron-formations of Southern Africa; Economic Geology, v. 68, p. 960-1004.
1983: Paleoenvironmental setting of iron-formation in the depositional basin of the Transvaal Supergroup; in Iron-formation: Facts and Problems, (ed.) A.F. Trendall and R.C. Morris; Elsevier, Amsterdam, p. 131-198.

Bischoff, J.L.
1969: Red Sea geothermal brine deposits, their mineralogy, chemistry, and genesis; in Hot Brines and Recent Heavy Metal Deposits in the Red Sea, (ed.) E.T. Degens and D.A. Ross; Springer-Verlag Inc., New York, p. 368-402.

Bischoff, J.L., Rosenbouer, R.J., Aruscavage, P.J., Baedecker, P.A., and Crock, J.G.
1983: Sea-floor massive sulfide deposits from the 21°N, East Pacific Rise, Juan de Fuca Ridge, and Galapagos Rift: bulk chemical composition and economic implications; Economic Geology, v. 78, p. 1711-1720.

*Bolis, J.L. and Bekkala, J.A.
1986: Iron ore availability-market economy countries, a minerals availability appraisal; Bureau of Mines, United States Department of the Interior, Information Circular 9128, 56 p.

Bornhold, B.D., Gross, G.A., McLeod, C.R., and Pasho, D.W.
1982: Polymetallic sulphide deposits on ocean ridges; The Canadian Mining and Metallurgical Bulletin, v. 75, no. 841, p. 24-28.

Boström, K. and Widenfalk, L.
1984: The origin of iron-rich muds at the Kameni Islands, Santorini, Greece; Chemical Geology, v. 42, p. 203-218.

Brandt, R.T., Gross, G.A., Gruss, H., Semenenko, N.P., and Dorr, J.V.N., II
1972: Problems of nomenclature for banded ferruginous-cherty sedimentary rocks and their metamorphic equivalents; Economic Geology, v. 67, p. 682-684.

Brodtkorb, M.K. and Brodtkorb, A.
1977: Strata-bound scheelite deposits in the Precambrian basement of San Luis (Argentina); in Time- and Strata-bound Ore Deposits, (ed.) D.D. Klemm and H.-J. Schneider; Springer-Verlag, Berlin, p. 141-152.

Burchard, U.
1977: Genesis of the King Island (Tasmania) scheelite mine; in Time- and Strata-bound Ore Deposits, (ed.) D.D. Klemm and H.-J. Schneider; Springer-Verlag, Berlin, Heidelberg, New York, p. 199-204.

Cronan, D.S.
1980: Underwater Minerals; Academic Press, London, 351 p.

Degens, E.T. and Ross, D.A. (ed.)
1969: Hot Brines and Recent Heavy Metal Deposits in the Red Sea; Springer-Verlag New York Inc., 600 p.

De Villiers, J.
1956: The manganese deposits of the Union of South Africa; Twentieth International Geological Congress, Mexico, 1956, (ed.) J.R. Reyna; Symposium sobre Yacimentos de Manganeso, v. 2, Africa, p. 39-71.

Devaraju, T.C. and Laajoki, K.
1986: Mineralogy and mineral chemistry of the manganese-poor and manganiferous iron-formations from the high-grade metamorphic terrain of Southern Karnataka, India; Journal of the Geological Society of India, v. 28, p. 134-164.

*Dimroth, E. and Chauvel, J.J.
1973: Petrography of the Sokoman iron formation in part of the central Labrador trough, Quebec, Canada; Geological Society of America, Bulletin, v. 84, p. 111-134.

Dorf, E.
1959: Cretaceous flora from beds associated with rubble iron-ore deposits in the Labrador trough; Geological Society of America, Bulletin, v. 70, p. 1591.

*Dorr, J.V.N., II
1973: Iron-formation in South America; Economic Geology, v. 68, p. 1005-1022.

Dorr, J.V.N., II, Coelho, I.S., and Horen, A.
1956: The manganese deposits of Minas Gerais, Brazil; in Twentieth International Geological Congress, Mexico, 1956, (ed.) J.R. Reyna, Symposium sobre Yacimentos de Manganeso, v. 3, America, p. 277-346.

*Edwards, R. and Atkinson, K.
1986: Sediment-hosted copper-lead-zinc deposits; Chapter 6 in Ore Deposit Geology; Chapman and Hall, London, New York, 445 p.

*El Shazly, E.M.
1990: Red Sea deposits; in Ancient Banded Iron Formations (Regional Presentations), (ed.) J.J. Chauvel, Y. Cheng, E.M. El Shazly, G.A. Gross, K. Laajoki, M.S. Markov, K.L. Rai, V.A. Stulchikov, and S.S. Augustithis; Theophrastus Publications, S.A., Athens, Greece, p. 157-222.

Eugster, H.P. and Chou, I.M.
1973: The depositional environments of Precambrian banded iron-formation; in Precambrian Iron-formations of the World, (ed.) H.L. James and P.K. Sims; Economic Geology, v. 68, p. 1144-1168.

Franklin, J.M. and Thorpe, R.I.
1982: Comparative metallogeny of the Superior, Slave and Churchill Provinces; in Precambrian Sulphide Deposits, (ed.) R.W. Hutchinson, C.D. Spence, and J.M. Franklin; Geological Association of Canada, Special Paper 25, p. 3-90.

Franklin, J.M., Lydon, J.W., and Sangster, D.F.
1981: Volcanic-associated massive sulfide deposits; Economic Geology, Seventy-fifth Anniversary Volume, 1905-1980, (ed.) B.J. Skinner; p. 485-627.

*Frietsch, R.
1974: The occurrence and composition of apatite with special reference to iron ores and rocks in Northern Sweden; Sveriges Geologiska Undersökning, Serie C, NR 694, Arsbok 68, NR 1, 49 p.

1977: The iron ore deposits in Sweden; in Iron Ore Deposits of Europe and Adjacent Areas, Volume I, (ed.) A. Zitzmann; Bundesanstalt für Geowissenschaften und Rohstoffe, Hannover, p. 279-293.

1980a: Metallogeny of the copper deposits of Sweden; in European Copper Deposits, (ed.) S. Janković and R.H. Sillitoe; Department of Economic Geology, Belgrade University, Belgrade, p. 166-179.

1980b: Precambrian ores of the northern part of Norbotten county, northern Sweden, Guide to excursions 078A + C, Part 1 (Sweden), Twenty-sixth International Geological Congress, Paris, 1980; Geological Survey of Finland, Espoo, 1980, 35 p.

1980c: The ore deposits of Sweden; Geological Survey of Finland, Bulletin 306, 20 p.

*Frietsch, R. (cont.)
1982a: Chemical composition of magnetite and sphalerite in the iron and sulphide ores of central Sweden; Geologiska Föreningens i Stockholm Förhandlingar, v. 104, pt. 1, p. 43-47.

1982b: A model for the formation of the non-apatitic iron ores, manganese ores and sulphide ores of Central Sweden; Sveriges Geologiska Undersökning, Serie C, NR 795, Arsbok 76, NR 8, 43 p.

Fryer, B.J.
1983: Rare earth elements in iron-formation; in Iron-formation: Facts and Problems: Developments in Precambrian Geology, (ed.) A.F. Trendall and R.C. Morris; Elsevier, Amsterdam, p. 345-358.

Gale, G.H.
1983: Proterozoic exhalative massive sulphide deposits; Geological Society of America, Memoir 161, p. 191-207.

*Garrels, R.M.
1987: A model for the deposition of the microbanded Precambrian iron-formations; American Journal of Science, v. 287, p. 81-106.

*Gauthier, M., Brown, A.C., and Morin, G.
1987: Small iron-formations as a guide to base- and precious-metal deposits in the Grenville Province of southern Quebec; in Precambrian Iron-formations, (ed.) P.W.U. Appel and G.L. Laberge; Theophrastus Publications, S.A., Athens, Greece, p. 297-328.

Geijer, P. and Magnusson, N.H.
1952: Geological history of the iron ores of Central Sweden; in Report of the Eighteenth Session, Part 13, Proceedings of Section M, Other Subjects, (ed.) R.M. Shackleton; International Geological Congress, London, 1948, p. 84-89.

*Gilmour, P.
1965: The origin of the massive sulphide mineralization in the Noranda district, northwestern Quebec; Proceedings of the Geological Association of Canada, v. 16, p. 63-81.

Gole, M.J. and Klein, C.
1981: Banded iron-formations through much of Precambrian time; Journal of Geology, v. 89, p. 169-183.

Goodwin, A.M.
*1956: Facies relations in the Gunflint iron-formation; Economic Geology, v. 51, p. 565-595.

1961: Genetic aspects of Michipicoten iron-formations; The Canadian Institute of Mining and Metallurgy, Transactions, v. 64, p. 24-28.

1962: Structure, stratigraphy and origin of iron formation, Michipicoten area, Algoma district, Ontario, Canada; Geological Society of America, Bulletin, v. 73, p. 561-585.

1965: Mineralized volcanic complexes in the Porcupine-Kirkland Lake-Noranda region, Canada; Economic Geology, v. 60, p. 995-971.

*1973: Archean iron-formation and tectonic basins of the Canadian Shield; Economic Geology, v. 68, p. 915-933.

1982: Distribution and origin of Precambrian banded iron formation; Revista Brasileira de Geociências, v. 12, no. 1-3, p. 457-462.

Gross, G.A.
1959a: A classification for iron deposits in Canada; Canadian Mining Journal, v. 80, p. 87-91.

1959b: Metallogenic Map, Iron in Canada; Geological Survey of Canada, Map 1045A-M4, scale 1 inch to 120 miles.

1964a: Iron-formation, Snake River area, Yukon and Northwest Territories; in Report of Activities, Field, 1964; Geological Survey of Canada, Paper 65-1, p. 143.

*1964b: Primary features in cherty iron-formations; in Genetic Problems of Ores, (ed.) R.K. Sundaram; Proceedings of Section 5, Twenty-second International Geological Congress, India, p. 102–117.

*1965: General geology and evaluation of iron deposits; Volume 1 in Geology of Iron Deposits in Canada; Geological Survey of Canada, Economic Geology Report 22, 181 p.

*1967: Iron deposits in the Appalachian and Grenville regions of Canada; Volume II in Geology of Iron Deposits in Canada; Geological Survey of Canada, Economic Geology Report 22, 111 p.

*1968: Iron ranges of the Labrador geosyncline; Volume III in Geology of Iron Deposits in Canada; Geological Survey of Canada, Economic Geology Report 22, 179 p.

1970a: Geological concepts leading to mineral discovery; Canadian Mining Journal, April, 1970, p. 51-53.

*1970b: Nature and occurrence of iron ore deposits: iron ore deposits of Canada and the West Indies; in United Nations Survey of World Iron Ore Resources; United Nations Publication, Sales No. E.69, II.C.4, New York, p. 13-31, 237-269.

*1973: The depositional environment of principal types of Precambrian iron-formation; in Genesis of Precambrian Iron and Manganese Deposits, Proceedings of Kiev Symposium, 1970, UN Educational, Scientific and Cultural Organization, Earth Sciences 9, p. 15-21.

Gross, G.A. (cont.)

*1980: A classification of iron-formation based on depositional environments; Canadian Mineralogist, v. 18, p. 215-222.

*1983a: Tectonic systems and the deposition of iron-formation; Precambrian Research, v. 20, p. 171-187.

*1983b: Low grade manganese deposits – a facies approach; in Unconventional Mineral Deposits, (ed.) W.C. Shanks; Society of Mining Engineers, American Institute of Mining, Metallurgical and Petroleum Engineers Inc., New York, N.Y., p. 35-47.

*1986: The metallogenetic significance of iron-formation and related stratafer rocks; Journal of the Geological Society of India, v. 28, no. 2 and 3, p. 92-108.

1987: Mineral deposits on the deep seabed; Marine Mining, v. 6, p. 109-119.

*1988a: Gold content and geochemistry of iron-formation in Canada; Geological Survey of Canada, Paper 86-19, 54 p.

1988b: A comparison of metalliferous sediments, Precambrian to Recent; Krystalinikum, v. 19, p. 59-74.

*1990: Geochemistry of iron-formation in Canada; in Ancient Banded Iron Formations (Regional Presentations), (ed.) J.J. Chauvel, Y. Cheng, E.M. El Shazly, G.A. Gross, K. Laajoki, M.S. Markov, K.L. Rai, V.A. Stulchikov, and S.S. Augustithis; Theophrastus Publications, S.A., Athens, Greece, p. 3-26.

*1991: Genetic concepts for iron-formation and associated metalliferous sediments; in Historical Perspectives of Genetic Concepts and Case Histories of Famous Discoveries, (ed.) R.W. Hutchinson and R.I. Grauch; Economic Geology, Monograph 8, p. 51-81.

*1993a: Iron-formation metallogeny and facies relationships in stratafer sediments; in Proceedings of the Eighth Quadrennial International Association on the Genesis of Ore Deposits Symosium, (ed.) Y. Maurice; E. Schweizerbart'sche Verlagsbuchhandlung (Nägele u. Obermiller), Stuttgart, p. 541-550.

*1993b: Rare earth elements and niobium in iron-formation at Bayan Obo, Inner Mongolia, China; in Proceedings of the Eighth Quadrennial International Association on the Genesis of Ore Deposits Symposium, (ed.) Y. Maurice; E. Schweizerbart'sche Verlagsbuchhandlung (Nägele u. Obermiller), Stuttgart, p. 477-490.

Gross, G.A. and Donaldson, J.A. (ed.)

1990: Iron-formation and metalliferous sediments in central Canada; Field Trip 8 Guidebook, Eighth International Association on the Genesis of Ore Deposits Symposium; Geological Survey of Canada, Open File 2163, 66 p.

Gross, G.A. and McLeod, C.R.

1980: A preliminary assessment of the chemical composition of iron formations in Canada; Canadian Mineralogist, v. 18, p. 223-229.

1983: Worldwide distribution of ocean bed metallic minerals; GEOS, v. 12, no. 3, p. 3-9.

*1987: Metallic minerals on the deep seabed; Geological Survey of Canada, Paper 86-21, 65 p.

*Gross, G.A. and Zajac, I.S.

1983: Iron-formation in fold belts marginal to the Ungava craton; in Iron-formation: Facts and Problems: Developments in Precambrian Geology, (ed.) A.F. Trendall and R.C. Morris; Elsevier, Amsterdam, p. 253-294.

Gross, G.A., Glazier, W., Kruechl, G., Nichols, L., and O'Leary, J.

*1972: Iron ranges of Labrador and northern Quebec; Guidebook, Field Excursion A55, Twenty-fourth International Geological Congress, (ed.) D.J. Glass; Geological Survey of Canada, Ottawa, 58 p.

Grout, F.F.

1919: The nature and origin of the Biwabik iron-bearing formations of the Mesabi Range, Minnesota; Economic Geology, v. 14, no. 6, p. 452-464.

Gruner, J.W.

1922: Organic matter and the origin of the Biwabik iron-bearing formations of the Mesabi Range; Economic Geology, v. 17, p. 407-460.

Han, Tsu-Ming

1968: Ore mineral relations in the Cuyuna sulfide deposit, Minnesota; Mineralium Deposita, v. 3, p. 109-134.

Hannington, M.D. and Scott, S.D.

1988: Mineralogy and geochemistry of a hydrothermal silica-sulfide-sulfate spire in the caldera of Axial Seamount, Juan de Fuca Ridge; Canadian Mineralogist, v. 26, p. 603-627.

Hekinian, R.

*1982: Petrology of the Ocean Floor; Elsevier, Amsterdam, 393 p.

*Hodder, R.W. and Petruk, W. (ed.)

1982: Geology of Canadian gold deposits; The Canadian Institute of Mining and Metallurgy, Special Volume 24, 286 p.

Hodgson, C.J. and MacGeehan, P.J.

1982: Geological characteristics of gold deposits in the Superior Province of the Canadian Shield; in Geology of Canadian Gold Deposits, (ed.) R.W. Hodder and W. Petruk; The Canadian Institute of Mining and Metallurgy, Special Volume 24, p. 211-232.

Höll, R.

1977: Early Paleozoic ore deposits of the Sb-W-Hg Formation in the eastern Alps and their genetic interpretation; in Time- and Strata-bound Ore Deposits, (ed.) D.D. Klemm and H.-J. Schneider; Springer-Verlag, Berlin, p. 169-198.

*Holland, H.D.

1973: The oceans: a possible source of iron in iron-formations; Economic Geology, v. 68, p. 1169-1172.

Honnorez, J.B., Honnorez-Guerstein, J., Valette, J., and Wauschkuhn, A.

1973: Present day formation of an exhalative sulfide deposit at Vulcano (Tyrrhenian Sea), part II: active crystallization of fumarolic sulfides in the volcanic sediments of the Baia di Levante; in Ores in Sediments, (ed.) G.C. Amstutz and A.J. Bernard; Springer-Verlag, New York, p. 139-166.

Hutchinson, R.W.

1965: Genesis of Canadian massive sulphides reconsidered by comparison to Cyprus deposits; The Canadian Institute of Mining and Metallurgy, Transactions, v. 68, p. 286-300.

*1973: Volcanogenic sulfide deposits and their metallogenic significance; Economic Geology, v. 68, p. 1223-1246.

Hutchinson, R.W., Ridler, R.H., and Suffel, G.G.

1971: Metallogenic relationships in the Abitibi Belt, Canada: a model for Archean metallogeny; The Canadian Institute of Mining and Metallurgy, Bulletin, v. 64, no. 708, p. 48-57.

*James, H.L.

1954: Sedimentary facies in iron-formation; Economic Geology, v. 49, p. 235-293.

*James, H.L. and Sims, P.K. (ed.)

1973: Precambrian iron-formations of the world; Economic Geology, v. 68, no. 7, p. 913-1173.

*James, H.L. and Trendall, A.F.

1982: Banded iron formation: distribution in time and paleo-environmental significance; in Mineral Deposits and the Evolution of the Biosphere, (ed.) H.D. Holland and M. Schidlowski; Dahlem Konferenzen, Springer-Verlag, Berlin, p. 199-218.

Janardhan, A.S., Shadakshara, S.N., and Capdevila, R.

1986: Banded iron-formation and associated manganiferous horizons of the Sargur Supracrustals, southern Karnataka; Journal of the Geological Society of India, v. 28, p. 179-188.

*Jenkyns, H.C.

1970: Submarine volcanism and the Toarcian iron pisolites of western Sicily; Eclogae Geologicae Helvetiae, Basle, v. 63, no. 2, p. 549-572.

Kalinin, V.V.

1965: The iron-manganese ores of the Karadzhal deposit; Institute of the Geology of Ore Deposits, Petrography, Mineralogy and Geochemistry, USSR, Academy of Sciences, Moscow (translation of Russian), 123 p.

Kerswill, J., Woollett, G.N., Strachan, D.M., and Moffett, R.

1983: The Lupin gold deposit: some observations regarding geological setting of gold distribution (abstract); The Canadian Institute of Mining and Metallurgy, Bulletin, v. 76, no. 851, p. 81.

1986: Gold deposits hosted by iron formation in the Contwoyto Lake area, Northwest Territories (extended abstract); in Gold '86 (Poster Volume), (ed.) A.M. Charter, Konsult International Inc., Willowdale, Ontario, Canada, p. 82-85.

Koski, R.A., Clague, D.A., and Oudin, E.

1984: Mineralogy and chemistry of massive sulfide deposits from the Juan de Fuca Ridge; Geological Society of America, Bulletin, v. 95, p. 930-945.

Krauskopf, K.B.

1956: Separation of manganese from iron in the formation of manganese deposits in volcanic association; in Symposium sobre Yacimentos de Manganeso, v. 1, El Manganeso en General, Twentieth International Geological Congress, Mexico, 1956, (ed.) J.R. Reyna; p. 119-131.

Krumbein, W.C. and Garrels, R.M.

1952: Origin and classification of chemical sediments in terms of pH and oxidation-reduction potentials; Journal of Geology, v. 60, p. 1-33.

Laajoki, K.

1986: Main features of the Precambrian banded iron-formations of Finland; Journal of the Geological Society of India, v. 28, p. 251-270.

Laberge, G.L.
1973:   Possible biological origin of Precambrian iron-formations; Economic Geology, v. 68, p. 1098-1109.

*Lang, A.H., Goodwin, A.M., Mulligan, R., Whitmore, D.R.E., Gross, G.A., Boyle, R.W., Johnston, A.G., Chamberlain, J.A., and Rose, E.R.
1970:   Economic minerals of the Canadian Shield; Chapter V in Geology and Economic Minerals of Canada; Geological Survey of Canada, Economic Geology Report 1, fifth edition, p. 152-223.

Lepp, H.
1963:   The relation of iron and manganese in sedimentary iron formations; Economic Geology, v. 58, p. 515-526.
1968:   The distribution of manganese in the Animikian iron formation of Minnesota; Economic Geology, v. 63, p. 61-75.

Macdonald, A.J.
1990:   Banded oxide facies iron formation as a host for gold mineralization; in Ancient Banded Iron Formations (Regional Presentations), (ed.) J.J. Chauvel, Y. Cheng, E.M. El Shazly, G.A. Gross, K. Laajoki, M.S. Markov, K.L. Rai, V.A. Stulchikov, and S.S. Augustithis; Theophrastus Publications, S.A., Athens, Greece, p. 63-81.

McAllister, A.L.
1959:   Massive sulphide deposits in New Brunswick; The Canadian Institute of Mining and Metallurgy, Transactions, v. 63, p. 50-60.

Mellon, G.B.
1962:   Petrology of Upper Cretaceous oolitic iron-rich rocks from northern Alberta; Economic Geology, v. 57, p. 921-940.

*Moore, E.S. and Maynard, J.E.
1929:   Solution, transportation, and precipitation of iron and silica; Economic Geology, v. 24, no. 3, p. 272-303; no. 4, p. 365-402; no. 5, p. 506-527.

Natarajan, W.K. and Mukherjee, M.M.
1986:   A note on the auiferous banded iron-formation of Kolar schist belt; Journal of the Geological Society of India, v. 28, p. 218-222.

*Nealson, K.H.
1982:   Microbiological oxidation and reduction of iron; in Mineral Deposits and the Development of the Biosphere, (ed.) H.D. Holland and M. Schidlowski; Springer-Verlag, Berlin, p. 51-66.

*Oftedahl, C.
1958:   A theory of exhalative-sedimentary ores; Geologiska Föreningens, i Stockholm Förhandlingar, No. 492, Band 80, Hafte 1, 19 p.

*Ohmoto, H. and Skinner, B.J. (ed.)
1983:   The Kuroko and related volcanogenic massive sulfide deposits; Economic Geology, Monograph 5, 570 p.

Park, C.F., Jr., Dorr, J.V.N., II, Guild, P.W., and Barbosa, A.J.M.
1951:   Notes on manganese ores of Brazil; Economic Geology, v. 46, p. 1-22.

Roscoe, S.M.
1965:   Geochemical and isotopic studies, Noranda and Matagami areas; The Canadian Institute of Mining and Metallurgy, Transactions, v. 68, p. 279-285.

Roy, S.
*1981: Manganese Deposits; Academic Press, London, 458 p.

Rozendaal, A.
*1980: The Gamsberg zinc deposit, South Africa: a banded stratiform base-metal sulfide ore deposit; in Proceedings of the Fifth Quadrennial International Association on the Genesis of Ore Deposits Symposium, E.Schweizerbart'sche Verlagsbuchhandlung (Nägele u. Obermiller) Germany, Stuttgart, p. 619-633.

Ruckmick, J.C.
1961:   Tropical weathering and the origin of the Cerro Bolivar iron ores; in Program of the 1961 Annual Meeting, Geological Society of America, Cincinnati, 1961, p. 134A.

Rye, D.M. and Rye, R.O.
*1974: Homestake gold mine, South Dakota: I. Stable isotope studies; Economic Geology, v. 69, p. 293-317.

Saager, R., Oberthur, T., and Tomschi, H.-P.
1987:   Geochemistry and mineralogy of banded iron-formation-hosted gold mineralization in the Gwanda greenstone belt, Zimbabwe; Economic Geology, v. 69, p. 293-317.

Sangster, D.F.
*1972: Precambrian volcanogenic massive sulphide deposits in Canada: a review; Geological Survey of Canada, Paper 72-22, 44 p.

Sangster, D.F. and Scott, S.D.
1976:   Precambrian strata-bound massive Cu-Zn-Pb sulfide ores of North America; in Handbook of Strata-bound and Stratiform Ore Deposits, (ed.) K.H. Wolf; Elsevier, New York, p. 129-214.

Sapozhnikov, D.G.
1963:   The Karadzhal iron-manganese deposits; Transactions of the Institute of the Geology of Ore Deposits, Petrography, Mineralogy and Geochemistry, No. 89, USSR Academy of Sciences, Moscow (translation of Russian), 193 p.

Schmidt, R.G.
1963:   Geology and ore deposits of the Cuyuna North Range, Minnesota; United States Geological Survey, Professional Paper 407, 96 p.

Schneider, H.-J. and Lehmann, B.
1977:   Contribution to a new genetical concept on the Bolivian tin province; in Time- and Strata-bound Ore Deposits, (ed.) D.D. Klemm and H.-J. Schneider; Springer-Verlag, Berlin, p. 153-168.

Schultz, R.W.
1966:   Lower Carboniferous cherty ironstones at Tynagh, Ireland; Economic Geology, v. 61, p. 311-342.

Shegelski, R.J.
*1987: The depositional environment of Archean iron formations, Sturgeon-Savant greenstone belt, Ontario, Canada, in Precambrian Iron-formations, (ed.) P.W.U. Appel and G.L. Laberge; Theophrastus Publications, S.A., Athens, Greece, p. 329-344.

Smith, P.A. and Cronan, D.S.
*1983: The geochemistry of metalliferous sediments and waters associated with shallow submarine hydrothermal activity (Santorini, Aegean Sea); Chemical Geology, v. 39, p. 241-262.

Sohnge, P.G.
1977:   Timing aspects of the manganese deposits of the northern Cape Province (South Africa); in Time- and Strata-bound Ore Deposits, (ed.) D.D. Klemm and H.-J. Schneider; Springer-Verlag, Berlin, p. 115-122.

Sozen, A.
1977:   Geological investigations on the genesis of the cinnabar deposit of Kalecik/Karaburun (Turkey); in Time- and Strata-bound Ore Deposits, (ed.) D.D. Klemm and H.-J. Schneider; Springer-Verlag, Berlin, p. 205-219.

*Stanton, R.L.
1959:   Mineralogical features and possible mode of emplacement of the Brunswick Mining and Smelting orebodies, Gloucester County, New Brunswick; The Canadian Institute of Mining and Metallurgy, Bulletin, v. 52, no. 570, p. 631-643.
1960:   General features of the conformable "pyritic" orebodies – Part 1, field associations; The Canadian Institute of Mining and Metallurgy, Bulletin, v. 53, no. 573, p. 24-29.
1976:   Petrochemical studies of the ore environment at Broken Hill, New South Wales; 3, Banded iron formations and sulphide orebodies: constitutional and genetic ties; Institution of Mining and Metallurgy, Transactions, v. 85, p. B132-B141.
1991:   Understanding volcanic massive sulfides – past, present and future; Economic Geology, Monograph 8, p. 82-95.

Tatsumi, T. (ed.)
1970:   Volcanism and Ore Genesis; University of Tokyo Press, Tokyo, 448 p.

Trendall, A.F.
1973:   Iron-formations of the Hamersley Group of Western Australia: type examples of varved Precambrian evaporites; in Genesis of Precambrian Iron and Manganese Deposits; Proceedings of the Kiev Symposium, 1970, UN Educational, Scientific and Cultural Organization, Earth Sciences, v. 9, Paris, p. 257-270.

*Trendall, A.F. and Blockley, J.G.
1970:   The iron-formations of the Precambrian Hamersley Group, Western Australia; Geological Survey of western Australia, Bulletin 119, 346 p.

Tu Guangzhi, Zhao Zhenhua, and Qiu Yuzhuo
1985:   Evolution of Precambrian REE mineralization; Precambrian Research, v. 27, p. 131-151.

UNESCO
*1973: Genesis of Precambrian iron and manganese deposits; Proceedings of the Kiev Symposium, UN Educational, Scientific and Cultural Organization, Earth Sciences, v. 9, 382 p.

United Nations
1955:   Survey of world iron ore resources, occurrence, appraisal and use; United Nations Department of Economic and Social Affairs, New York, 345 p.
*1970: Survey of world iron ore resources, occurrence and appraisal; United Nations, New York, 479 p.

Van Staal, C.R. and Williams, P.F.
1984:   Structure, origin, and concentration of the Brunswick 12 and 6 orebodies; Economic Geology, v. 79, p. 1669-1692.

Varentsov, I.M. and Grasselly, G. (ed.)
*1980a: Geology and Geochemistry of Manganese; Volume I, General problems, mineralogy, geochemistry, methods; E.Schweizerbart'sche Verlagsbuchhandlung (Nägele u. Obermiller), Stuttgart, Germany, 463 p.
1980b: Geology and Geochemistry of Manganese; Volume II, Manganese deposits on continents; E.Schweizerbart'sche Verlagsbuchhandlung (Nägele u. Obermiller), Stuttgart, Germany, 510 p.
1980c: Geology and Geochemistry of Manganese; Volume III, Manganese on the bottom of Recent basins; E.Schweizerbart'sche Verlagsbuchhandlung (Nägele u. Obermiller), Stuttgart, Germany, 357 p.

Winn, R.D. and Bailes, R.J.
1987: Stratiform lead-zinc sulfides, mudflows, turbidites: Devonian sediments along a submarine fault scarp of extensional origin, Jason deposit, Yukon Territory, Canada; Geological Society of America, Bulletin, v. 98, p. 528-539.

Yeo, G.M.
*1978: Iron-formation in the Rapitan Group, Mackenzie Mountains, Yukon and Northwest Territories; in Mineral Industry Report, 1975, Northwest Territories: Department of Indian and Northern Affairs, Report EGS 1978-5, p. 170-175.

Yeo, G.M. (cont.)
1986: Iron-formations in the late Proterozoic Rapitan Group, Yukon and Northwest Territories; in Mineral Deposits of Northern Cordillera, (ed.) J.A. Morin; The Canadian Institute of Mining and Metallurgy, Special Volume 37, p. 142-153.

*Zajac, I.S.
1974: The stratigraphy and mineralogy of the Sokoman Formation in the Knob Lake area, Quebec and Newfoundland; Geological Survey of Canada, Bulletin 220, 159 p.

Zantop, H.
1978: Geologic setting and genesis of iron oxides and manganese oxides in the San Francisco manganese deposit, Jalisco, Mexico; Economic Geology, v. 73, p. 1137-1149.

Zelenov, K.K.
1958: On the discharge of iron in solution into the Okhotsk Sea by thermal springs of the Ebeko volcano (Paramushir Island); Proceedings of the Academy of Sciences, USSR, v. 120, p. 1089-1092. (In Russian; English translation published by Consultants Bureau, Inc., New York, 1959, p. 497-500).

3.1 LAKE SUPERIOR-TYPE IRON-FORMATION

G.A. Gross

INTRODUCTION

Lake Superior-type iron-formations are chemically precipitated (hydrolithic) banded sedimentary rocks composed of iron oxide, quartz (chert), silicate, carbonate, and sulphide lithofacies that were deposited along Paleoproterozoic craton margins, on marine continental shelves, and in shallow rift basins. They are typically associated with texturally and compositionally mature sedimentary rocks such as quartz arenite, dolostone, black shale, and argillite, with additional tuffaceous strata and other volcanic rocks. Lithofacies that are not highly metamorphosed or altered by weathering processes have been referred to as taconite and the more highly metamorphosed as metataconite, itabirite, or banded iron-formation (BIF). The iron oxide lithofacies selected for iron ore usually contain at least 30% iron and a minimum amount of carbonate and silicate lithofacies that must be separated from the iron oxide minerals in the processing of the crude ore and production of high grade ore concentrates. The highly metamorphosed iron-formations are coarser grained and are generally more amenable to processing and concentration of the iron oxide minerals than the fine grained lithofacies. Iron ore consisting of mineral concentrates is usually processed further by pelletizing and sintering, and blended with other types of ore to improve the overall grade and structural quality of the furnace burden, the blended mixture of iron ore, coke, coal or carbon fuel, and carbonate rock charged to blast furnaces.

IMPORTANCE, SIZE, AND GRADE OF DEPOSITS

Lake Superior-type iron-formations include the thickest and most extensive stratigraphic units of iron-formation, and provide the largest iron deposits and the source rocks for more than 60% of the iron ore resources of the world. About 35 million long tons (35.6 Mt) of iron ore concentrate have been produced in Canada annually from 82 million long tons (83.3 Mt) of highly metamorphosed hematite and magnetite lithofacies of iron-formation from the Humphrey, Smallwood, and Wabush mines in southwest Labrador, and the Mount Wright mine in northeast Quebec (Fig. 3.1-1). In 1985, the Humphrey and Smallwood mines produced 34 million long tons (34.6 Mt) crude ore containing 38.9% iron, which yielded 15 million long tons (15.2 Mt) of concentrate grading 63.9% iron; and the Mount Wright mine produced 37 million long tons (37.6 Mt) of crude ore containing 39.9% iron, to provide 15 million long tons (15.2 Mt) of concentrate grading 63.9% iron.

Taconite deposits are not mined in Canada, but a major part of the 40 million or more long tons (40.7 Mt) of ore concentrate produced annually from taconite in the United States has come from the Minnetac, Hibbing, Hoyt Lakes

Gross, G.A.
1996: Lake Superior-type iron-formation; in Geology of Canadian Mineral Deposit Types, (ed.) O.R. Eckstrand, W.D. Sinclair, and R.I. Thorpe; Geological Survey of Canada, Geology of Canada, no. 8, p. 54-66 (also Geological Society of America, The Geology of North America, v. P-1).

Plant, Eveleth (Thunderbird), Minorca, and National Steel Pellet Project, Erie, and Reserve mine properties on the Mesabi Range in Minnesota, and the Tilden, Empire, and Republic mines in Michigan. Taconite mines in the Great Lakes Region produced 118 million long tons (119.9 Mt) of crude ore containing 32% iron in 1986, to provide 40 million long tons (40.7 Mt) of concentrate grading 63.9% iron.

Potential resources of magnetite-rich taconite have been outlined in large deposits in the Schefferville area of Quebec and Labrador, and near Lac Albanel in Quebec. Other large taconite deposits occur on the Belcher Islands, in northeastern Quebec, and in the Gunflint iron-formation, northwestern Ontario. Many iron-formation units in Canada are host rocks for more than 1000 Mt of measured iron resources and some have even greater resource potential.

GEOLOGICAL FEATURES

Tectonosedimentary setting

Lake Superior-type iron-formations form a major part of the successions of folded Proterozoic sedimentary and volcanic rocks that were deposited within extensive basins, some interconnected, along the northeastern and southwestern craton margins of the Superior Province of the Canadian Shield (see Introduction; Fig. 3-3). The Labrador-Quebec fold belt, consisting of sequences of sedimentary and volcanic rocks and mafic intrusions deposited in smaller interconnected subbasins, is the largest continuous stratigraphic-tectonic unit that extends along the eastern margin of the Superior-Ungava craton for more than 1200 km (Fig. 3.1-2). The principal iron-formation unit, the Sokoman Formation, forms a continuous stratigraphic unit that thickens and thins from subbasin to subbasin throughout this fold belt.

As the tectonic systems and volcanic arcs developed along the margins of the Superior-Ungava craton, iron-formation units were deposited in the marginal basins in association with thick sequences of shale, argillite, dolostone, stromatolitic dolostone, chert, chert breccia, black carbon-rich shale, quartz arenite, conglomerate, siltstone, redbeds, tuff, and other volcanic rocks. Lateral transitions occur in many basins from stratigraphic sequences of iron-formation and typical mature clastic sediments that were deposited in nearshore environments on the platform, to sequences of greywacke, turbidites, and shale interbedded with iron-formation, tuff, and volcanic rocks deposited near volcanic centres offshore from the craton margins (Fig. 3.1-3). The thickest sections of iron-formation in the Labrador-Quebec and Belcher-Nastapoka fold belts were deposited in basins adjacent to thick accumulations of volcanic rocks, gabbro, and ultramafic intrusions in the volcanic belts offshore.

The Knob Lake Group in the Labrador-Quebec fold belt (Table 3.1-1) typifies a stratigraphic section of Lake Superior-type iron-formation.

Characteristic structural settings for iron-formations in the fold belts along the craton margins include low-dipping homoclines of quartz arenite and iron-formation that lie unconformably on Archean gneisses, granulites, and granitoid cratonic rocks near the original basin shorelines. Outward from the craton to its faulted margins, these homocline structures are succeeded by broad, open folds that are intensely deformed by complex isoclinal folds and faults, imbricate thrust sheets, and nappe structures that

developed by tectonic transport directed toward the craton. Structural deformation of the marginal basins appears to be related to the initial stages of plate subduction that took place along the craton margins.

The southern part of the Labrador-Quebec fold belt is truncated by the Grenville Front along the northern margin of this orogenic belt (Fig. 3.1-2), but iron-formations and associated shelf metasediments extend southwest into the Grenville Province for more than 200 km. Iron-formations and associated rocks within the Grenville orogen are highly metamorphosed and complexly folded, forming numerous isolated structural segments that have been mapped in detail in mines at Lac Jeannine, Fire Lake, and Mount Wright, and at the Smallwood, Humphrey, and Scully mines in the Wabush Lake area. Iron-formations north of the Grenville belt in the central part of the Labrador-Quebec fold belt are not metamorphosed beyond greenschist facies, but metamorphism increases to lower amphibolite facies in a large area west of Ungava Bay (Gross, 1962, 1967, 1968). Iron-formation in the basins marginal to the Superior Province was deposited between 2100 and 1850 Ma ago, with major deposition about 2000 Ma ago (Morey, 1983; Barrett et al., 1988). Because basin development may have migrated along the craton margin, it is unlikely that major episodes for the deposition of iron and silica were contemporaneous in all areas. Folding and metamorphism related to the Grenville orogeny between 1200 Ma and 800 Ma ago played an essential role in the development of the large iron deposits in the southern Labrador-Quebec fold belt (Fig. 3.1-2) (Gross, 1968; Gross and Zajac, 1983).

Typical successions of Proterozoic rocks within the deeper basins along the Quebec-Labrador belt include basal argillite and shale overlain in sequence by dolostone, quartz arenite, black shale, iron-formation, and black shale. Stratigraphic units become thinner or pinch out near the margins of the depositional basins, and only thin members of quartz arenite, iron-formation, and shale are present in some areas (Fig. 3.1-3). Thin units of iron-formation

Table 3.1-1. Stratigraphy of the Knob Lake Group, Schefferville area, Labrador and Quebec.

Menihek Formation – Carbonaceous slate, shale, quartzite, greywacke, mafic volcanic rocks; minor dolomite and chert.

Purdy Formation – dolomite, developed locally.

Sokoman Iron-formation – oxide, silicate, and carbonate lithofacies, minor sulphide lithofacies, interbedded mafic volcanic rocks, ferruginous slate, slaty iron-formation, black and brown slate, and carbonaceous shale.

Wishart Formation – feldspathic quartz arenite, arkose; minor chert, greywacke, slate, and mafic volcanic rocks.

Fleming Formation – chert breccia, thin-bedded chert, limestone; minor lenses of shale and slate.

Denault Formation – dolomite and minor chert.

Attikamagen Formation – green, red, grey, and black shale, and argillite, interbedded with mafic volcanic rocks.

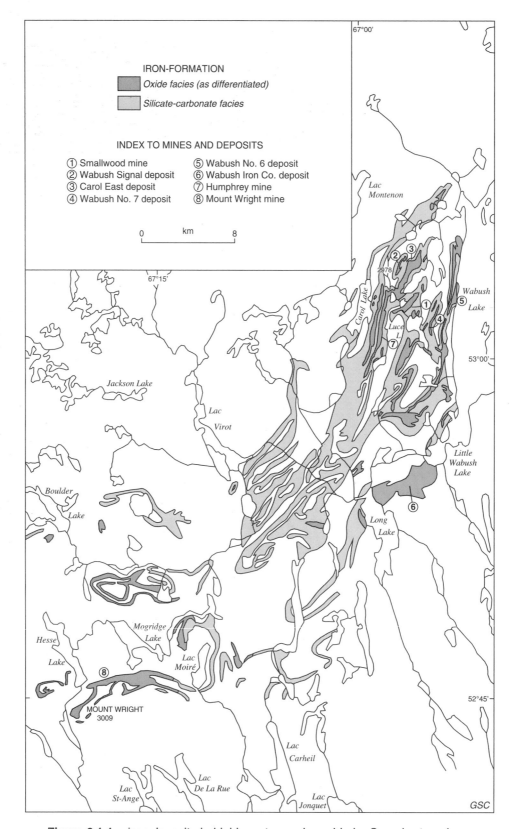

Figure 3.1-1. Iron deposits in highly metamorphosed Lake Superior-type iron-formation in the Wabush Lake and Mount Wright areas, Labrador and Quebec (from Gross, 1968).

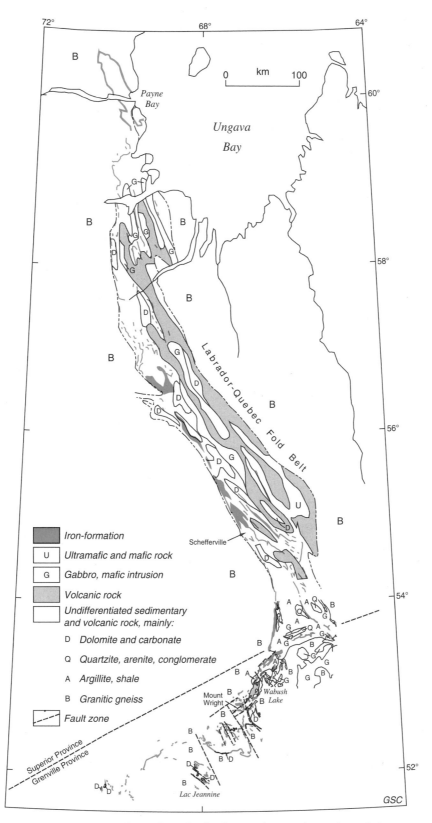

Figure 3.1-2. Iron-formation distribution and tectonic setting of the Labrador-Quebec fold belt (after Gross, 1961, 1968).

unconformably overlie gneissic rocks along the western margin of the Labrador-Quebec fold belt west of Knob Lake (Fig. 3.1-2, 3.1-3).

Lithology

The principal ore zones throughout the Sokoman Formation consist of discrete stratigraphic units of oxide lithofacies of iron-formation (Fig. 3.1-4). Silicate and carbonate lithofacies are consistently developed in the lower parts of the iron-formation and locally in the middle and upper parts, where they interfinger on a microscopic and macroscopic scale (Fig. 3.1-5). Sulphide lithofacies iron-formation occurs near the base of the upper black shale member, and with greywacke and volcanic rocks in some areas that have been mapped in detail. Macroscopic bands commonly range in thickness from 1 to 10 cm, and exceed 1 m in a few places; microscopic bands (<1 mm thick) are rare.

Primary sedimentary features such as granules and oolites in a chert or carbonate matrix, interlayered beds of chert or quartz and iron oxide minerals, crossbedding, intraformational breccia, slump folds, compaction and desiccation structures, and stromatolite-like forms are well preserved and widely distributed in the Sokoman Formation (Gross, 1964, 1968; Zajac, 1974).

Metamorphism was mainly isochemical; lithofacies of metamorphosed iron-formation reflect the distribution of primary sedimentary facies. Element mobility was minimal in oxide lithofacies iron-formation, even under high-rank metamorphism, but was maximized in silicate, carbonate, and sulphide lithofacies.

Form of deposits

Mineable deposits consist of selected oxide lithofacies of iron-formation with cumulative stratigraphic thicknesses ranging from 30 to 300 m and strike lengths of several

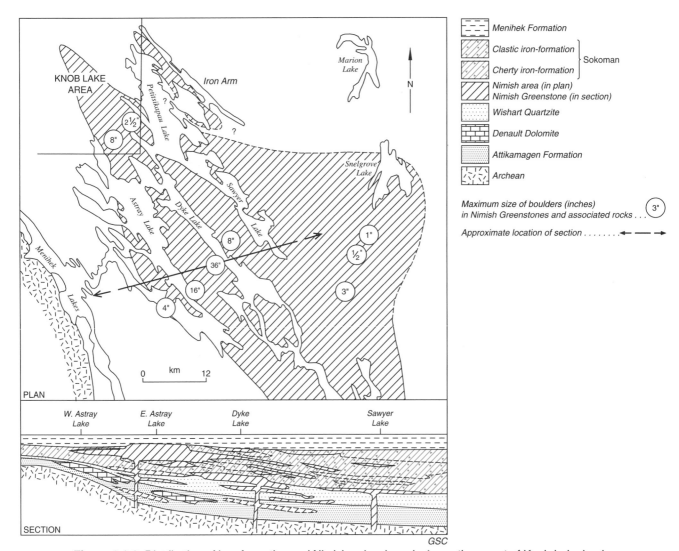

Figure 3.1-3. Distribution of iron-formation and Nimish volcanic rocks in southern part of Knob Lake basin, Quebec-Labrador fold belt (from Zajac, 1974).

kilometres. In the Wabush Lake and Mount Wright areas of Labrador and Quebec, repetition of iron-formations by complex folds and thrust faults has produced extensive thicknesses of ore. Here, too, amphibolite-grade metamorphism has enlarged the grain size of magnetite, hematite, and quartz, greatly improving its amenability to concentration and beneficiation (Fig. 3.1-6). Difficulties commonly are encountered in achieving uniform grade and quality in crude ore mined from deposits in which iron silicate or carbonate lithofacies are infolded and interlayered with oxide lithofacies.

Mineralogy

Taconite ores in iron ranges of the Lake Superior region have not been affected greatly by metamorphism; mineral distribution reflects the composition of the layers and beds formed during their sedimentation and diagenesis. The iron and silica remain distributed as very fine grained and intimately intermixed mineral aggregates of quartz, magnetite, and hematite, typically in discrete beds, layers, granules, and oolites

of the crude taconite ore. Consequently, mineral assemblages in taconite ore tend to be very complex in comparison with the highly metamorphosed iron oxide lithofacies.

Hematite and magnetite, the principal ore minerals in Lake Superior-type iron-formation, are associated with minor amounts of goethite, and with pyrolusite, manganite, and hollandite in manganiferous oxide lithofacies. Other minerals in the finely laminated crude ore are quartz, in granular and chert form, iron silicates, iron carbonates, and iron sulphides as primary mineral assemblages or their metamorphic derivatives.

Metamorphism

Typical sequential changes in grain size, texture, and mineralogy associated with increasing rank of metamorphism in the oxide lithofacies of iron-formation are illustrated in Figure 3.1-6. As grain size of both quartz and iron oxide

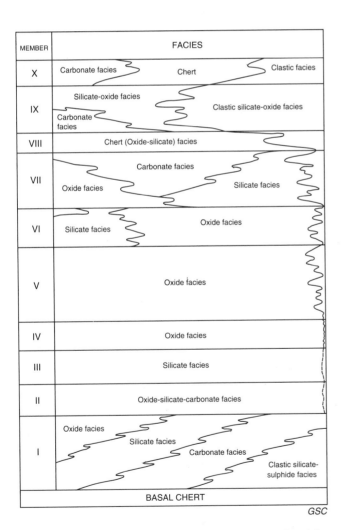

Figure 3.1-4. Subdivisions of the Sokoman Formation (after Zajac, 1974).

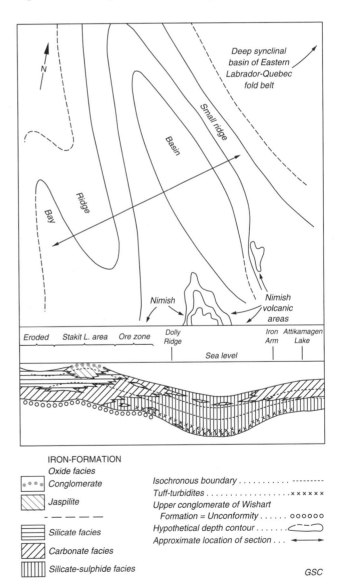

Figure 3.1-5. Interpretation of depositional environment, Member 1 of the Sokoman Formation (after Zajac, 1974).

minerals increases, primary chert is transformed to coarse granular quartz, and discrete grains of iron oxide are distributed in a granular quartz matrix. Iron silicate minerals are recrystallized to form higher rank metamorphic minerals such as minnesotaite, grunerite, cummingtonite, and hypersthene or forsterite. Iron silicate minerals also form under special conditions by the reaction of quartz and iron oxide. As the rank of metamorphism increases, siderite, ankerite, and dolomite break down and magnetite, actinolite, cummingtonite, grunerite, minnesotaite, and various other silicate minerals are formed. Foliation developed during metamorphism is parallel or subparallel to primary bedding. The evidence for mobilization of elements during metamorphism and their transfer to fracture and cleavage zones is minimal in oxide facies iron-formation, but increases significantly in silicate, sulphide, and carbonate lithofacies.

Composition of ore

The typical composition of Lake Superior-type iron-formations mined in Labrador-Newfoundland, Quebec, Minnesota, and Michigan, and of the iron ore concentrate or pellets produced from them, is given in Table 3.1-2.

DEFINITIVE CHARACTERISTICS OF ORE

Ore deposits in Lake Superior-type iron-formation are characterized by the following:

1. Iron content is 30% or greater;

2. Discrete units of oxide lithofacies iron-formation are clearly segregated from silicate, carbonate, or sulphide facies and other barren rock, and are amenable to concentration and beneficiation of the iron to meet required chemical and physical specifications;

3. Iron is uniformly distributed in discrete grains or grain-clusters of hematite, magnetite, and goethite in a cherty or granular quartz matrix;

4. Iron-formations, repeated by folds and faults, provide thick sections for mining; and

5. Metamorphic enlargement of mineral grain size has improved the quality of the crude ore for concentration and processing.

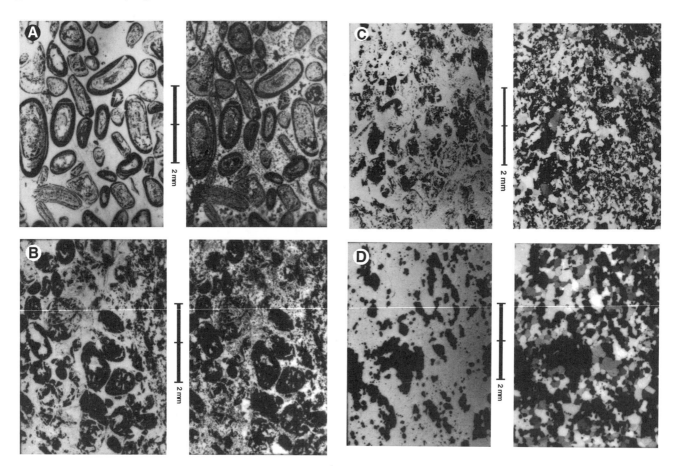

Figure 3.1-6. Textures of Lake Superior-type magnetite-hematite-chert iron-formation, illustrating effect of metamorphism on grain size (Gross, 1968). In each pair of photomicrographs, the left-hand view is in plane light, and the right-hand view is with crossed nicols. **A)** Gunflint Range near Thunder Bay, Ontario; very little metamorphism. GSC 204578X **B)** Near Schefferville, Quebec; greenschist facies. GSC 204578Y **C)** Southwest of Ungava Bay, Quebec; epidote-amphibolite facies. GSC 204538T **D)** Mount Wright area, Quebec; upper epidote-amphibolite facies. GSC 204538R

GENESIS OF LAKE SUPERIOR-TYPE IRON-FORMATION

Two principal genetic models have been considered for the origin of iron-formations and the associated stratafer lithofacies (see Introduction, Type 3; Gross, 1965, 1990, 1991). One model has emphasized volcanogenic and hydrothermal effusive or exhalative processes and the other hydrogenous-sedimentary processes with derivation of the iron, silica, and other constituents by deep weathering of a landmass. Controversy on the origin of the Lake Superior-type of iron-formation is here attributed to inadequate consideration of the diversity in lithofacies, and other types of iron-formation that have been documented throughout the geological record (Gross, 1965, 1991). The history of genetic concepts is summarized as follows.

Prior to the early part of this century it was generally believed that the iron and silica in iron-formations were derived from deeply weathered rocks and deposited in restricted basins where chemical precipitation predominanted over clastic sedimentation. Important contributions to understanding the solution, transportation, and deposition of iron and silica in thinly laminated beds by inorganic processes or with the assistance of bacteria were made by Harder (1919), Grout (1919), Gruner (1922), Moore and Maynard (1929), Woolnough (1941), Sakamoto (1950), and Spencer and Percival (1952).

Deposition of metalliferous sediments by volcanogenic and hydrothermal effusive or exhalative processes was recognized as early as 1650 and seriously considered in Europe in the nineteenth century. More recently, volcanogenic processes were considered by Van Hise and Leith (1911), Collins et al. (1926), Landergren (1948), Oftedahl (1958), Zelenov (1958), Gross (1959a, 1965), Goodwin (1962), Eugster and Chou (1973), and many others.

Controversy regarding the origin of iron-formation has continued in recent years, and many theories still fail to account for important features of stratafer sediments. Trendall (1973), Button (1976), and Garrels (1987) for instance, have suggested that some iron-formations were formed as evaporites in restricted, brine-filled basins. The evaporite suggestions do not account for observed facies (sulphide and carbonate as well as oxide) and thickness variations in the basins discussed by these writers.

Table 3.1-2. Typical composition of Lake Superior-type iron-formations, concentrates, and pellets (weight per cent)

1/ Mesabi taconite, Biwabik iron-formation, Minnesota 2/ Hibbing taconite pellets 3/ Sokoman iron-formation, oxide facies, Knob Lake, Quebec and Labrador 4/ Sokoman iron-formation, oxide facies, French mine, Knob Lake Ranges					5/ Wabush Lake area iron-formation, oxide facies, Labrador-Newfoundland 6/ Carol Lake concentrate, Labrador-Newfoundland 7/ Wabush Mines pellets 8/ Mount Wright concentrate, northeast Quebec 9/ Negaunee iron-formation, Empire mine, Michigan 10/ Empire pellets (acidic)				
	1	**2**	**3**	**4**	**5**	**6**	**7**	**8**	**9**	**10**
Fe (total)	28.36	63.61	26.41	33.97	33.46	64.08	63.62	64.15	30.20	63.67
Fe_2O_3	18.70	nd	45.76	45.98	39.93	nd	nd	nd	6.72	nd
FeO	19.71	nd	nd	2.89	2.33	nd	7.11	nd	32.81	nd
P	0.035	0.010	0.020	0.017	0.009	0.008	0.008	0.010	0.017	0.017
SiO_2	46.40	4.66	49.11	48.35	45.82	4.46	3.06	4.74	33.63	5.43
Mn	0.49	0.07	0.08	0.019	0.25	0.11	1.90	0.02	0.86	0.08
TiO_2	0.04	nd	nd	0.00	nd	nd	nd	nd	0.11	nd
Al_2O_3	0.90	0.18	0.43	0.48	0.63	0.16	0.30	0.32	0.50	0.42
CaO	1.60	0.21	0.09	0.002	1.62	0.44	0.06	0.02	0.03	0.23
MgO	2.98	0.31	0.65	0.31	0.89	0.29	0.04	0.02	2.22	0.27
Na_2O	0.004	0.02	0.15	0.033	0.04	nd	0.024	nd	nd	0.29
K_2O	0.13	0.02	0.10	0.01	0.009	nd	0.03	nd	nd	0.047
S	nd	0.001	0.007	0.013	0.006	0.004	0.004	0.007	nd	0.003
CO_2	6.90	nd	0.50	0.033	3.22	nd	nd	nd	nd	nd
C	0.17	nd	0.136	0.083	0.88	nd	nd	nd	nd	nd
$H_2O+/-$	1.92	3.65	1.04	2.045	0.31	2.77	2.88	3.02	18.32	2.50
LOI	1.72	nd	nd	nd	nd	0.25	nd	nd	18.32	nd

DATA SOURCES

Note: Analyses for natural material. nd = no data

1. Bayley and James, 1973. 2. American Iron Ore Association, 1987, p. 65; Cliffs Mining Company, 1987, p. 18. 3. Gross, 1990. 4. Gross and Zajac, 1983. 5. Gross, 1990. 6. American Iron Ore Association, 1987, p. 70. 7. American Iron Ore Association, 1987 p. 70; Cliffs Mining Company, 1987, p. 19. 8. American Iron Ore Association, 1987, p. 70. 9. Davy, 1983. 10. American Iron Ore Association, 1987, p. 64; Cliffs Mining Company, 1987, p. 18.

MacGregor (1927), Lepp and Goldich (1964), and Cloud (1973) considered that Lake Superior-type iron-formations were the product of atmospheric evolution. In this model, iron was derived from weathered rocks and through an extensive time period became concentrated in seawater as ferrous carbonate. The ferrous iron was then rapidly and widely deposited as ferric iron because of a sudden increase of atmospheric oxygen that was generated by the wide-spread action of photosynthesizing organisms which evolved in the early Proterozoic. This theory has received wide attention, but faces the following problems:

1. No clear evidence has been established for saturation of ferrous carbonate in Archean seas, nor for the specific role of organisms in the deposition of iron.

2. Some evidence has been presented to show that significant concentrations of oxygen existed in the Archean atmosphere (Dimroth and Kimberly, 1976).

3. According to the "atmospheric change" model, deposition of oxide-facies iron-formation should be restricted to a relatively brief period of geological time. Geochronological data, however, indicate that Lake Superior-type iron-formations were deposited over several hundred million years (Gole and Klein, 1981; James and Trendall, 1982; Gross, 1983a).

4. Other iron-formation and stratafer lithofacies (oxide, carbonate, silicate, and sulphide) are also widely distributed in Archean, Phanerozoic, and Recent marine basins (Gross, 1964, 1967, 1968, 1970, 1983a, 1987).

Lake Superior-type iron-formations throughout the world are directly related to, and form a part of, the tectonic belts developed along the margins of cratons or plates in the early Proterozoic. Deposition of the successions of clastic sediments, iron-formation, stratafer lithofacies, and volcanic rocks in these marginal basins was related to basin architecture and the many factors controlling their depositional environments. Direct evidence of hydrothermal effusive or exhalative processes active during basin development is registered in the composition and trace element content of the iron-formations, and in volcanic centres which have been mapped (Fig. 3.1-3). Evidence of many of the volcanic and effusive centres has been lost through extensive overthrusting and deformation of basin areas related to subduction at the plate margins. Unlike many areas with Algoma-type iron-formation, the spatial relationship of the sedimentary facies developed and volcanic centres is not well preserved.

Many have suggested that iron-oxidizing micro-organisms played a key role in concentrating and precipitating iron. Evidence demonstrating that organisms played an essential part in the deposition and accumulation of large quantities of iron appears to be lacking, but bacteria, morphologically similar to fossils in Lake Superior-type iron-formations, are present in modern iron and manganese sediments (Nealson, 1982). Tunnicliffe and Fontaine (1987) have demonstrated a microbial origin for iron-rich particles associated with abundant inorganic deposition of iron in modern metalliferous sediments.

Borchert (1960) suggested that diagenetic porewater from clastic sediments could transport iron in a ferrous state from anoxic to oxidizing environments where ferric oxides would be precipitated. Holland (1973) and Button et al. (1982) suggested that very low concentrations of ferrous iron in anoxic deep basins were transported landward by upwelling seawater and precipitated with silica as ferric hydroxide in shallow water oxygenated by photosynthesizing organisms.

Gross (1965, 1968, 1983a, 1986) emphasized the extensive evidence for hydrothermal effusive or exhalative processes in the origin of iron-formation, and showed that the distribution and correlation patterns for the major and minor elements are consistent with volcanogenic processes. Analytical data indicate that the bulk composition, correlation, and distribution patterns for the minor and rare-earth elements in more than fifty iron-formation units, that range in age from early Archean to Recent, are similar to those in modern marine deposits of metalliferous sediments which are forming by hydrothermal effusive processes (Gross and McLeod, 1980, 1987; Gross, 1988, 1990, 1991, 1993c). With few exceptions, tuffaceous beds and volcanic rocks are now known to be present in, or closely associated with, iron-formations of all types.

Gross (1965) suggested that hot springs along a volcanic arc could have been an adequate source of iron and silica to form the iron-formations in the 1200 km long Labrador-Quebec fold belt in about 50 000 years. Previous objections to this model had hinged on (1) the apparent lack of evidence in some areas for contemporary volcanism (evidence now available), and (2) failure to recognize other primary stratafer lithofacies (carbonate and sulphide), which are now known to be associated with iron-formations, and are generally accepted as having formed by hydrothermal effusive or exhalative processes.

A convincing body of data thus indicates that iron-formations have formed by volcanogenic or hydrothermal effusive processes, and that the depositional environment, tectonic setting, and the composition of the exhalative hydrothermal fluids were the principal factors that controlled the composition of the lithofacies developed from them. Modern submarine metalliferous sediments, which have formed by hydrothermal effusive processes that are widely distributed on the seafloor, are the final proof and considered as analogous to facies protolithic to iron-formation.

RELATED TYPES OF DEPOSITS

Lake Superior-type iron-formations extend continuously through transitions in their depositional environments from sites where they are associated with mature clastic sediments in typical continental shelf basins to distal off-shore sites where they are associated with greywacke, tuff, volcanic, and other members of the stratafer group (such as base-metal-bearing sulphide lithofacies) that are more characteristic of depositional environments for Algoma-type iron-formations, as reported in the Krivoy Rog basin (Alexandrov, 1973; Belevtsev, 1973); Damara Supergroup, South Africa (Beukes, 1973, 1983); Cuyuna Range, Lake

Superior Region (Bayley and James, 1973); and in parts of the Sokoman Formation in the Quebec-Labrador belt (Gross, 1968; Zajac, 1974). Examples of related deposits are provided in the following.

Lake Superior-type iron-formations are the source rocks or protore for the large high grade direct-shipping iron ore deposits formed by secondary enrichment processes that have provided the extensive resources of iron ore now being mined in Australia, Brazil, Venezuela, India, South Africa, present day Russia and Ukraine, and in the Lake Superior and Labrador-Quebec regions in North America in the past.

Manganese facies of iron-formation in the Lake Superior, Labrador-Quebec, and other areas have provided manganiferous iron ore containing more than 3% manganese. Large, rich manganese deposits are associated with iron-formations in the Transvaal Supergroup in South Africa (Beukes, 1983), and with iron-formation in the Nova Lima Group in Minas Gerais, Brazil (Dorr, 1973).

Manganese-bearing facies of Lake Superior-type iron-formation in North America are relatively uncommon. Small occurrences in Canada include: beds rich in manganese (<1 m thick) in carbonate facies iron-formation that have been traced for several kilometres on Belanger and Flint islands, on the east side of Hudson Bay (Bell, 1879; Chandler, 1988); thin beds of oxide facies containing up to 2% manganese at Mount Reed, Quebec; and manganiferous facies iron-formation in the Wabush Lake area, Labrador, and at Sutton Lake, Ontario. Manganiferous iron-formations have been investigated in the Cuyuna Range of Minnesota. Most iron-formations of this type were deposited in relatively shallow basins in platform and continental shelf environments where conditions fluctuated and lithofacies range from predominantly banded manganese- and hematite-rich strata developed under highly oxidizing conditions, to predominantly siderite, chert-siderite, sulphide, and manganese carbonate lithofacies, deposited under reducing conditions.

Highly metamorphosed iron-formation containing 5 to 6% rare-earth elements is part of a Middle Proterozoic sequence of dolostone, potassium-rich black schist, shale, quartz arenite, and arkose that is intruded in places by dykes of alkaline rock in the Bayan Obo mine area in Inner Mongolia. The iron-formation sequence extends for a distance of 20 km, and more than 1500 Mt of iron ore containing 0.10% niobium and 5 to 6% rare-earth elements has been defined in 16 deposits. The rare-earth elements are distributed throughout the iron-formation and in some of the dolostones closely associated with it. Although highly deformed and altered (see Fig. 3.1-6 for typical progression of textural changes in Labrador iron-formations with metamorphism), in many places the iron-formation preserves granular and oolitic textures, microbanding, and other relict sedimentary features; it is typical in scale for Lake Superior-type iron-formation, and is interpreted to have formed in a marginal basin along a major fault zone at the edge of the North China Platform. A syngenetic origin for the niobium, rare-earth elements, and fluorite in the iron-formation and dolostone has been clearly demonstrated (Gross, 1986, 1993b). The highly anomalous niobium, rare-earth elements, and fluorite in the iron-formation and dolostone has been ascribed to hydrothermal metasomatic processes by

Chao et al. (1993), and others have considered its possible affinity to carbonatite deposits. The writer is satisfied that the range of isotope dates mark different stages or periods in metamorphism of the iron-formation.

Sulphide lithofacies in Lake Superior-type iron-formations are less abundant than in Algoma-type, but still common. Sediment-hosted sulphide lithofacies in the northern Labrador-Quebec fold belt form lenses 40 m thick and 400 m long, and have an average content of 2% copper and zinc, and less than 0.2% lead (Barrett et al., 1988). Extensive sulphide lithofacies have been described in the Lake Superior iron ranges by James (1954), and Han (1968), and in many other parts of the world.

Anomalous amounts of lead and zinc occur in many of the carbonate sequences associated with Lake Superior-type iron-formations, and small lead-zinc deposits in the Nastapoka Group were mentioned by Chandler (1988). Likewise, copper deposits hosted by black shale and by redbeds may be genetically related to hydrothermal systems that produced iron-formations in their vent areas.

The origin and distribution of gold in Lake Superior-type iron-formations has not been widely documented. The background content of gold in Sokoman oxide lithofacies at Schefferville, Quebec, deposited in a shallow water shelf environment, is about half (0.02 ppm) the average content in oxide lithofacies of Algoma-type iron-formation (Gross, 1988).

Small amounts of riebeckite, arfvedsonite, crocidolite, and other sodium-rich amphibole and silicate minerals are associated with stilpnomelane in many of the metamorphosed Lake Superior-type iron-formations, and "tiger's eye" quartz is genetically related to these mineral occurrences. In most cases, the sodium-rich silicate minerals appear to have developed where tuffaceous material was intermixed with the iron-formation. Blue asbestos, crocidolite, and riebeckite deposits in iron-formation in South Africa and Australia have been mined for many years (Trendall and Blockley, 1970).

Graphite-iron-sulphide- and kyanite-bearing gneisses and schist associated with highly metamorphosed iron-formation in the northeastern Grenville orogenic belt, noted by Gross (1968), appear to be the metamorphosed equivalents of carbonaceous and aluminous Menihek slate rocks which overlie the Sokoman iron-formation. The Lac Knife graphite deposit located about 35 km south of Fermont in the Mount Wright and Wabush Lake area of Quebec and Labrador, now under development, is a highly metamorphosed carbonaceous sedimentary deposit (Bonneau and Raby, 1990).

EXPLORATION GUIDELINES

Lithofacies of Lake Superior-type iron-formation selected for iron ore have the following characteristics:

1. The composition, texture, and grade of the crude ore must be such that ore concentrates can be produced that meet required chemical and physical specifications in a particular industrial area.

2. Discrete, well defined magnetite and hematite lithofacies of iron-formation are preferred with a minimum of other lithofacies and clastic sediment interbedded in the crude ore.

3. Granular, medium- to coarse-grained textures with well defined, sharp grain boundaries are desirable, to enable liberation and clean separation of mineral grains in the concentration and beneficiation of the crude ore. These features are achieved in highly metamorphosed iron-formation where there is maximum grain enlargement, and magnetite lithofacies are usually preferred over hematite or mixed facies.

4. Mineable thicknesses of ore, usually 30 to 100 m or greater, may vary depending on location and economic factors in the market area. Ore deposits may consist of thick primary stratigraphic units or a succession of ore beds that were repeated by folding and faulting.

5. Uniform granularity and mineral composition, with a minimum of infolding of barren sediment or marginal facies of iron-formation, are advantageous factors for grade control and in processing of the crude ore.

6. Oxide facies iron-formation deposited in highly oxidizing environments normally has a low content of minor elements, especially sodium, potassium, sulphur, and arsenic, which have deleterious effects in the processing of the ore and quality of steel produced from it.

7. The content of minor elements varies significantly in some different iron-formations, depending on the kinds of associated volcanic, sedimentary, intrusive, and other stratafer lithofacies.

8. Most iron-formations are regional-scale stratigraphic units that are relatively easy to define by mapping or with the aid of aeromagnetic and gravity surveys. Detailed stratigraphic information is an essential part of the data base required to define grade, physical and chemical quality, and beneficiation and concentration characteristics of the ore.

9. Basin analysis and sedimentation modelling enable definition of factors that controlled the development, location, and distribution of different iron-formation lithofacies.

10. Metamorphic mineral assemblages reflect the composition and mineralogy of the primary sedimentary facies.

SELECTED BIBLIOGRAPHY

References marked with asterisks (*) are considered to be the best sources of information on the deposit subtypes in this group. For additional references, see "Introduction", Type 3.

*Alexandrov, A.A.
1973: The Precambrian banded iron-formations of the Soviet Union; in Precambrian Iron-formations of the World, (ed.) H.L. James and P.K. Sims; Economic Geology, v. 68, no. 7, p. 1035-1063.

American Iron Ore Association
1987: Iron Ore – 1986; American Iron Ore Association, Cleveland, Ohio, 100 p.

*Barrett, T.J., Wares, R.P., and Fox, J.S.
1988: Two-stage hydrothermal formation of a Lower Proterozoic sediment-hosted massive sulphide deposit, northern Labrador Trough, Quebec; Canadian Mineralogist, v. 26, p. 871-888.

*Bayley, R.W. and James, H.L.
1973: Precambrian iron-formations of the United States; Economic Geology, v. 68, no. 7, p. 942.

*Belevtsev, Y.N.
1973: Genesis of high-grade iron ores of Krivoyrog type; in Genesis of Precambrian Iron and Manganese Deposits, Proceedings of the Kiev Symposium 20-25 August, 1970, UN Educational, Scientific and Cultural Organization, Paris, p. 167-180.

Bell, R.
1879: Nastapoka Islands, Northwest Territories; Report of Progress, 1877-78, Geological Survey of Canada, p. 15C-18C.

*Beukes, N.J.
1973: Precambrian iron-formations of Southern Africa; in Precambrian Iron-formations of the World; (ed.) H.L. James and P.K. Sims; Economic Geology, v. 68, no. 7, p. 960-1005.
1983: Paleoenvironmental setting of iron-formations in the depositional basin of the Transvaal Supergroup, South Africa; in Iron-formation: Facts and Problems, (ed.) A.F. Trendall and R.C. Morris; Developments in Precambrian Geology 6, Elsevier, Amsterdam, p. 131-209.

Bonneau, J. and Raby, R.
1990: The Lac Knife graphite deposit; Mining Magazine, July 1990, p. 12-18.

*Borchert, H.
1960: Genesis of marine sedimentary ores; Institution of Mining and Metallurgy, Bulletin, v. 640, p. 261-279.

Button, A.
1976: Iron-formation as an end-member in carbonate sedimentary cycles in the Transvaal Supergroup, South Africa; Economic Geology, v. 71, p. 193-201.

*Button, A., Brock, T.D., Cook, P.J., Eugster, H.P., Goodwin, A.M., James, H.L., Margulis, L., Nealson, K.H., Nriagu, J.O., Trendall, A.F., and Walter, M.R.
1982: Sedimentary iron deposits, evaporites and phosphorites – state of the art report; in Mineral Deposits and Evolution of the Biosphere, (ed.) H.D. Holland and M. Schidlowski; Springer-Verlag, Berlin, p. 259-273.

Chandler, F.W.
1988: The Early Proterozoic Richmond Gulf Graben, east coast of Hudson Bay, Quebec; Geological Survey of Canada, Bulletin 362, 76 p.

Chao, E.C.T., Mitsunobu, Tatsumoto, Minkin, J.A., Back, J.M., McKee, E.H., and Ren, Yingchen
1993: Multiple lines of evidence for establishing the mineral paragenetic sequence of the Bayan Obo rare earth ore deposits of Inner Mongolia, China; in Proceedings of the Eighth Quadrennial International Association on the Genesis of Ore Deposits Symposium, (ed.) Y.T. Maurice; E. Schweizerbart'sche Verlagsbuchhandlung (Nagele u. Obermiller) Stuttgart, 1993, p. 55-73.

Cliffs Mining Company
1987: Iron Ore Analyses – 1987; Cliffs Mining Company, Cleveland, Ohio, 86 p.

*Cloud, P.
1973: Paleoecological significance of the banded iron-formation; in Precambrian Iron-formations of the World, (ed.) H.L. James and P.K. Sims; Economic Geology, v. 68, no. 7, p. 1135-1143.

*Collins, W.H., Quirke, T.T., and Thomson, E.
1926: Michipicoten iron ranges; Geological Survey of Canada, Memoir 147, 175 p.

Davy, R.
1983: A contribution on the chemical composition of Precambrian iron-formations; in Iron-formation: Facts and Problems, (ed.) A.F. Trendall and R.C. Morris; Developments in Precambrian Geology 6, Elsevier, Amsterdam, p. 325-343.

*Dimroth, E. and Kimberley, M.M.
1976: Precambrian atmospheric oxygen: evidence in the sedimentary distribution of carbon, sulphur, uranium and iron; Canadian Journal of Earth Sciences, v. 13, p. 1161-1185.

*Dorr, J.V.N., II
1973: Iron-formations in South America; in Precambrian Iron-formations of the World, (ed.) H.L. James and P.K. Sims; Economic Geology, v. 68, no. 7, p. 1005-1022.

*Eugster, H.P. and Chou, I-M.
1973: The depositional environments of Precambrian banded iron-formation, in Precambrian Iron-formations of the World, (ed.) H.L. James and P.K. Sims; Economic Geology, v. 68, no. 7, p. 1144-1168.

Garrels, R.M.
1987: A model for the deposition of the microbanded Precambrian iron-formations; American Journal of Science, v. 287, p. 81-106.

*Gole, M.J. and Klein, C.
1981: Banded iron-formations through much of Precambrian time; Journal of Geology, v. 89, p. 169-183.

*Goodwin, A.M.
1956: Facies relations in the Gunflint iron-formation; Economic Geology, v. 51, no. 6, p. 565-595.

*Goodwin, A.M. (cont.)

1962: Structure, stratigraphy, and origin of iron-formation, Michipicoten area, Algoma district, Ontario, Canada; Geological Society of America, Bulletin, v. 73, p. 561-586.

Gross, G.A.

1959a: A classification of iron deposits in Canada; Canadian Mining Journal, v. 80, no. 10, p. 87-91.

1959b: Metallogenic Map, Iron in Canada; Geological Survey of Canada, Map 1045A-M4, scale 1 inch to 120 miles.

1961: Iron-formations and the Labrador Geosyncline; Geological Survey of Canada, Paper 60-30, 7 p.

1962: Iron deposits near Ungava Bay, Quebec; Geological Survey of Canada, Bulletin 82, 54 p.

*1964: Primary features in cherty iron-formations; in Genetic Problems of Ores, (ed.) R.K. Sundaram; Part V, Proceedings of Section 5, XXII International Geological Congress, India, p. 102-117.

*1965: General geology and evaluation of iron deposits; in Volume I, Geology of Iron Deposits in Canada; Geological Survey of Canada, Economic Geology Report 22, 181 p.

*1967: Iron deposits in the Appalachian and Grenville regions of Canada; in Volume II, Geology of Iron Deposits in Canada; Geological Survey of Canada, Economic Geology Report 22, 111 p.

*1968: Iron ranges of the Labrador geosyncline; in Volume III, Geology of Iron Deposits in Canada; Geological Survey of Canada, Economic Geology Report 22, 179 p.

*1970: Nature and occurrence of iron ore deposits; iron ore deposits of Canada and the West Indies; in United Nations Survey of World Iron Ore Resources, United Nations Publication Sales No. E.69.II.C.4, New York, p. 13-31, p. 237-269.

*1973: The depositional environment of principal types of Precambrian iron-formation; in Genesis of Precambrian Iron and Manganese Deposits: Proceedings of Kiev Symposium, 1970, UN Educational, Scientific and Cultural Organization, Earth Sciences 9, Paris, p. 15-21.

*1983a: Tectonic systems and the deposition of iron-formation; Precambrian Research, v. 20, p. 171-187.

*1983b: Low grade manganese deposits – a facies approach; in Cameron Volume on Unconventional Mineral Deposits, (ed.) W.C. Shanks; American Institute of Mining, Metallurgical, and Petroleum Engineers Inc., New York, New York, p. 35-46.

1986: The metallogenetic significance of iron-formation and related stratafer rocks; Journal of the Geological Society of India, v. 28, no. 2 and 3, p. 92-108.

1987: Mineral deposits on the deep seabed; Marine Mining, v. 6, p. 109-119.

*1988: Gold content and geochemistry of iron-formation in Canada; Geological Survey of Canada, Paper 86-19, 54 p.

*1990: Geochemistry of iron-formation in Canada; in Ancient Banded Iron-formations (Regional Presentations), (ed.) J.-J. Chauvel; Theophrastus Publications, S.A., Athens, Greece, p. 3-26.

*1991: Genetic concepts for iron-formation and associated metalliferous sediments; in Historical Perspectives of Genetic Concepts and Case Histories of Famous Discoveries, (ed.) R.W. Hutchinson and R.I. Grauch; Economic Geology Monograph 8, Economic Geology, p. 51-81.

1993a: Iron-formation metallogeny and facies relationships in stratafer sediments; in Proceedings of the Eighth Quadrennial International Association on the Genesis of Ore Deposits Symosium, (ed.) Y. Maurice; E. Schweizerbart'sche Verlagsbuchhandlung (Nagele u. Obermiller), D-7000 Stuttgart 1, p. 541-550.

*1993b: Rare earth elements and niobium in iron-formation at Bayan Obo, Inner Mongolia, China; in Proceedings of the Eighth Quadrennial International Association on the Genesis of Ore Deposits Symposium, (ed.) Y. Maurice; E. Schweizerbart'sche Verlagsbuchhandlung (Nagele u. Obermiller), D-7000 Stuttgart 1, p. 477-490.

*1993c: Element distribution patterns as metallogenetic indicators in siliceous metalliferous sediments; in Proceedings of Symposium II-16-5: Present and Past Seafloor Hydrothermal Mineralization, Twenty-ninth International Geological Congress, 1992; Resource Geology Special Issue, no. 17, p. 96-107.

Gross, G.A. and McLeod, C.R.

1980: A preliminary assessment of the chemical composition of iron-formations in Canada; Canadian Mineralogist, v. 18, p. 215-222.

*1987: Metallic minerals on the deep seabed; Geological Survey of Canada, Paper 86-21, 65 p., and Map 1659A, scale 1:40 000 000.

*Gross, G.A. and Zajac, I.S.

1983: Iron-formations in fold belts marginal to the Ungava Craton; in Iron-formation: Facts and Problems; (ed.) A.F. Trendall and R.C. Morris; Developments in Precambrian Geology 6, Elsevier, Amsterdam, p. 253-294.

Grout, F.F.

1919: The nature and origin of the Biwabik iron-bearing formations of the Mesabi Range, Minnesota; Economic Geology, v. 14, no. 6, p. 452-464.

Gruner, J.W.

1922: Organic matter and the origin of the Biwabik iron-bearing formations of the Mesabi Range; Economic Geology, v. 17, p. 407-460.

Han, Tsu-Ming

1968: Ore mineral relations in the Cuyuna sulfide deposit, Minnesota; Mineralium Deposits, v. 3, p. 109-134.

Harder, E.C.

1919: Iron-depositing bacteria and their geologic relations; United States Geological Survey, Professional Paper 113, p. 89.

*Holland, H.D.

1973: The oceans: a possible source of iron in iron-formations; Economic Geology, v. 68, no. 7, p. 1169-1172.

*James, H.L.

1954: Sedimentary facies in iron-formation; Economic Geology, v. 49, p. 235-293.

*James, H.L. and Sims, P.K. (ed.)

1973: Precambrian iron-formations of the world; Economic Geology, v. 68, no. 7, p. 913-1173.

*James, H.L. and Trendall, A.F.

1982: Banded iron-formation: distribution in time and paleo-environmental significance; in Mineral Deposits and the Evolution of the Biosphere, (ed.) H.D. Holland, and M. Schidlowski; Springer-Verlag, Berlin, p. 199-218.

Landergren, S.

1948: On the geochemistry of Swedish iron ores and associated rocks: a study of iron-ore formation; Sveriges Geologiska Undersökning, Serie C, NR. 496, Arsbok 42, NR. 5, Stockholm, 182 p.

*Lang, A.H., Goodwin, A.M., Mulligan, R., Whitmore, D.R.E., Gross, G.A., Boyle, R.W., Johnston, A.G., Chamberlain, J.A., and Rose, E.R.

1970: Economic minerals of the Canadian Shield; in Chapter V, Geology and Economic Minerals of Canada, Geological Survey of Canada, Economic Geology Report 1, fifth edition, p. 152-223.

Lepp, H. and Goldich, S.S.

1964: Origin of Precambrian banded iron-formations; Economic Geology, v. 59, p. 1025-1060.

*MacGregor, A.M.

1927: Problem of pre-Cambrian atmosphere: South Africa Journal of Science, v. 24, p. 155-172.

Moore, E.S. and Maynard, J.E.

1929: Solution, transportation, and precipitation of iron and silica; Economic Geology, v. 24, no. 3, p. 272-303; no. 4, p. 365-402; no. 5, p. 506-527.

*Morey, G.B.

1983: Animikie Basin, Lake Superior Region, U.S.A.; in Iron-formation: Facts and Problems, (ed.) A.F. Trendall and R.C. Morris; Elsevier, Amsterdam-Oxford-New York-Tokyo, p. 13-68.

Nealson, K.H.

1982: Microbiological oxidation and reduction of iron; in Mineral Deposits and the Development of the Biosphere, (ed.) H.D. Holland and M. Schidlowski; Springer-Verlag, Berlin, p. 51-66.

*Oftedahl, C.

1958: A theory of exhalative-sedimentary ores; Geologiska Föreningens i Stockholm Förhandlingar no. 492, Band 80, Hafte 1, 19 p.

Sakamoto, T.

1950: The origin of the pre-Cambrian banded iron ores; American Journal of Science, v. 248, no. 7, p. 449-474.

Spencer, E. and Percival, F.G.

1952: The structure and origin of the banded hematite jasper of Singbhum, India; Economic Geology, v. 47, no. 4, p. 365-383.

Trendall, A.F.

1973: Iron-formations of the Hamersley Group of Western Australia: type examples of varved Precambrian evaporites; in Genesis of Precambrian Iron and Manganese Deposits; Proceedings of the Kiev Symposium, 1970, UN Educational, Scientific and Cultural Organization, Earth Sciences 9, Paris, p. 257-270.

*Trendall, A.F. and Blockley, J.G.

1970: The iron-formations of the Precambrian Hamersley Group, Western Australia; Geological Survey of Western Australia, Bulletin 119, 346 p.

*Tunnicliffe, V. and Fontaine, A.R.
1987:   Faunal composition and organic surface encrustations at hydrothermal vents on the Southern Juan de Fuca Ridge; Journal of Geophysical Research, v. 92, no. B11, p. 11, 303-11, 314.

Van Hise, C.R. and Leith, C.K.
1911:   The geology of the Lake Superior region; United States Geological Survey, Monograph 52, 641 p.

Woolnough, W.G.
1941:   Origin of banded iron deposits – a suggestion; Economic Geology, v. 36, no. 5, p. 465-489.

*Zajac, I.S.
1974:   The stratigraphy and mineralogy of the Sokoman Formation in the Knob Lake area, Quebec and Newfoundland; Geological Survey of Canada, Bulletin 220, 159 p.

Zelenov, K.K.
1958:   On the discharge of iron in solution into the Okhotsk Sea by thermal springs of the Ebeko volcano (Paramushir Island); Proceedings of Academy of Sciences, U.S.S.R., v. 120, p. 1089-1092. (In Russian; English translation published by Consultants Bureau, Inc., New York, 1959, p. 497-500).

3.2 ALGOMA-TYPE IRON-FORMATION

G.A. Gross

INTRODUCTION

Algoma-type iron-formations consist primarily of microscopic to macroscopic alternating layers and beds of chert or quartz, and iron-rich minerals, magnetite, hematite, pyrite, pyrrhotite, iron carbonates, and iron silicates (see "Introduction", Type 3). They are composed of a variety of interbedded oxide, carbonate, sulphide, and silicate lithofacies, and include extensive manganese-rich lithofacies; sulphide lithofacies rich in copper, zinc, lead, tin, and gold; oxide and carbonate lithofacies bearing rare-earth elements; tungsten-bearing lithofacies; and various lithofacies of iron-formation that host syngenetic and epigenetic gold deposits (Gross, 1970, 1988a, b, 1990a, 1993a, b). These iron-rich lithofacies form a major part of the large assemblage of siliceous hydrolithic sediments referred to as the stratafer group (Gross, 1986, 1991, 1993a; see "Introduction", Type 3).

Algoma-type iron-formations were deposited with volcanic rocks and greywacke, turbidite, and pelitic sediments in volcanic arc and spreading ridge tectonic settings (Gross, 1965, 1970, 1983a; Lang et al., 1970). They range in age from 3.2 Ga to modern, possibly protolithic facies on the seafloor. Iron ore is recovered from Algoma-type iron-formation, from naturally enriched deposits, and from selected zones of oxide lithofacies which are the principal kind discussed here.

Gross, G.A.
1996:   Algoma-type iron-formation; in Geology of Canadian Mineral Deposit Types, (ed.) O.R. Eckstrand, W.D. Sinclair, and R.I. Thorpe; Geological Survey of Canada, Geology of Canada, no. 8, p. 66-73 (also Geological Society of America, The Geology of North America, v. P-1).

World class examples include Kudremuk in India, Cerro Bolivar in Venezuela (enriched), Carajas in Brazil (enriched), large highly metamorphosed oxide lithofacies in Sweden, oxide lithofacies in northern China such as Qinan, and numerous deposits in Precambrian terrane in Canada, the Scandinavian Shield, Australia, Africa, India, and the former U.S.S.R.

IMPORTANCE, SIZE, AND GRADE

Metamorphosed magnetite oxide and carbonate lithofacies of Algoma-type iron-formation are the second most important source of iron ore after the taconite and enriched deposits in Lake Superior-type iron-formations. They range in size from about 1000 to 100 Mt or smaller.

In 1986 production from oxide lithofacies at the Adams, Griffith, and Sherman mines in Ontario amounted to more than 8.1 Mt of crude ore grading 19 to 27% iron for the recovery of 2.1 Mt of ore-concentrate and pellets. In 1986, Algoma Ore Division at Wawa, Ontario, produced more than 1.7 Mt of siderite crude ore grading 34.15% iron that provided 1.2 Mt of sinter and agglomerate.

The ore-concentrate, pellets, sinter, and agglomerate produced from these mines provided about 10% of the total iron ore produced in Canada in 1986, and the compositions of the ore products recovered from Algoma-type iron-formation are given in Table 3.2-1.

Iron ore deposits in Precambrian Algoma-type iron-formations usually contain less than 2% manganese, but many Paleozoic iron-formations such as those near Woodstock, New Brunswick, contain 10 to 40% manganese and have Fe/Mn ratios of 40:1 to 1:50 (Gross, 1967).

GEOLOGICAL FEATURES

Algoma-type iron-formations and associated stratafer sediments are interbedded with volcanic rocks of mafic to felsic composition, and with greywacke, turbidites, argillite, and shale of all ages. They formed both near to and distal from extrusive centres along volcanic belts, deep fault systems, and rift zones and may be present at any stage in a volcanic succession. Algoma-type iron-formations include silicate, oxide, carbonate, and sulphide facies and base metal lithofacies of the stratafer group (Type 3, "Introduction"), and are interbedded in assemblages of black shale, carbonate, turbidite, and siltstone sediments and volcanic rocks, or their metamorphosed equivalents (Stanton, 1959, 1976, 1991; Gross, 1965, 1986, 1991; Gauthier et al., 1987; Beeson, 1990).

Depositional environments for Algoma-type iron-formations are varied and diverse. The proportions of volcanic and clastic sedimentary rocks vary greatly from one fold belt to another and are rarely mutually exclusive. In Ontario, volcanic rocks are predominant in areas around Adams mine in Boston township; Sherman mine near Temagami Lake; Moose Mountain mine and at Bending Lake, and metasediments are most abundant at Griffith mine, Nakina, and at Lake St. Joseph, where deltaic fan sequences have been described (Gross, 1965, 1983a; Lang et al., 1970; Meyn and Palonen, 1980). A series of preliminary maps compiled by Gross (1963) show the extensive distribution of Algoma-type iron-formations in "greenstone belts" of the Superior Structural Province and outline their rock associations and geological settings. Algoma-type iron-formations and associated stratafer sediments are consistently present in the "greenstone belts" of the Canadian Shield and are marker beds for locating other sulphide and stratafer lithofacies, as well as indicating ancient sites of mineral deposition by syngenetic and epigenetic hydrothermal processes (see Fig. 3-4, Introduction).

Oxide, silicate, carbonate, and sulphide lithofacies are commonly interbedded in iron-formations composed of microscopic to macroscopic alternating layers or beds of silica (chert or quartz) and iron-rich minerals. Thin beds of iron-formation are usually interbedded with clastic sediments and volcanic strata at the margins of stratigraphic units. Rocks associated with Algoma-type iron-formations vary greatly in composition, even within local basins, and range from felsic to mafic and ultramafic volcanic rocks, and from greywacke to black shale, argillite, and chert interlayered with pyroclastic and other volcanoclastic beds or their metamorphic equivalents.

Primary sedimentary features, including those related to compaction, contraction, and desiccation of the beds, microscopic banding, bedding, and penecontemporaneous deformation features of the hydroplastic sediment such as slump folds and faults, are abundant, and can be recognized in many cases in highly metamorphosed oxide lithofacies of iron-formation and even in some granulite metamorphic lithofacies (Gross, 1964). Mineral distribution in highly metamorphosed iron-formations closely reflects the composition and distribution of primary sedimentary facies. Metamorphism is mainly isochemical and produces grain enlargement and the segregation of quartz and gangue from the iron oxide and carbonate minerals, thus enhancing the beneficiation qualities of crude ore (Fig. 3.2-1). Element mobility during metamorphism is minimal in oxide lithofacies even under high rank conditions, but increases in silicate and carbonate lithofacies to

a maximum in sulphide lithofacies, in which many of the primary sedimentary features are destroyed during diagenesis and metamorphism.

Most of the iron ore resources in Algoma-type iron-formation consist of specially selected oxide lithofacies composed mainly of magnetite, hematite, and quartz. Oxide facies usually contain variable amounts of magnetite, hematite, siderite or ferruginous ankerite, and dolomite, manganoan siderite, and silicate minerals. Silicate lithofacies are characterized by iron-silicate minerals including grunerite, minnesotaite, hypersthene, reibeckite, and stilpnomelane, associated with chlorite, sericite, amphibole, and garnet. Siderite ore containing pyrite and pyrrhotite beds and lenses has been mined in the Wawa area of Ontario for more than 50 years.

Economic iron ore deposits have been developed mainly where metamorphosed sequences of oxide lithofacies of iron-formation are 30 to 100 m thick and several kilometres in strike length. Thick sequences of ore beds have been developed in many areas where strata have been repeated by isoclinal folding and thrust faulting and economic feasibility for mining them has been greatly enhanced.

Algoma-type iron-formations are most widely distributed and achieve their greatest thicknesses in Archean terrane. They range in age from 3.2 Ga in India, China, Canada, and Greenland, to the large number deposited between 2.5 and 2.9 Ga in the volcanic belts of the Canadian Shield and in Scandinavia, to the smaller and frequently manganiferous facies in Paleozoic rocks, to Mesozoic facies in the Cordillera, and to potentially protolithic facies of iron-formation that are being deposited at the present time near volcanic centres located along island arcs and spreading ridges (Gross, 1973, 1983a, 1988a, b; Gross and McLeod, 1987).

Table 3.2-1. Composition of ore-concentrate, pellets, and sinter produced from Algoma-type iron-formation in Ontario in 1986.

	Range in composition (%)	
	pellets and ore-concentrates	sinter and agglomerates
Fe	66.2 - 66.7	48.32
SiO_2	3.5 - 5.5	8.0
P	0.03 - 0.02	0.018
Mn	0.03 - 0.12	2.36
Al_2O_3	0.29 - 0.4	1.06
CaO	4.7 - 4.9	12.14
MgO	1.7 - 1.8	7.19
S	<0.027	0.09

DEFINITIVE CHARACTERISTICS OF IRON ORE IN ALGOMA-TYPE IRON-FORMATION

Iron ore deposits in Algoma-type iron-formation consist mainly of metamorphosed oxide and carbonate lithofacies that contain 20 to 40% iron. The quality of oxide facies crude ore is greatly enhanced by metamorphism which leads to the development of coarse granular textures and discrete grain enlargement (Gross, 1961). Mining feasibility has been improved greatly where the ore beds have been repeated and thickened by folding and faulting. Carbonate lithofacies containing a minimum of interbedded chert and sulphide minerals provide suitable crude ore for processing and beneficiation, and the production of sinter and agglomerated ore products.

Coarse, granular, magnetite-quartz lithofacies are most easily concentrated and beneficiated to remove gangue quartz, silicate, and alkali-bearing minerals, and significant amounts of minor elements that are commonly present in Algoma-type iron-formations (Landergren, 1948; Gross, 1965, 1980, 1990b; Frietsch, 1974).

The compositions of Algoma-type iron-formations mined in Canada and the ore concentrate, pellets, and sinter produced from them at five mines are given in Table 3.2-2.

GENESIS OF ALGOMA-TYPE IRON-FORMATION

It is generally accepted that Algoma-type iron-formations are hydrolithic sediments formed by volcanogenic and hydrothermal-effusive (exhalative) processes. Genetic models invoking deep weathering of a landmass and transport of the iron and silica to restricted basins along with the depletion of other major rock forming constituents do not account for the broad range in depositional environments, facies development, composition, and the diversity in content and distribution of minor elements found in Algoma-type iron-formations. The bulk composition, distribution and correlation patterns for major and minor elements in Algoma-type iron-formation are strikingly similar to those in marine siliceous metalliferous sediments that are forming on the modern seafloor in many parts of the

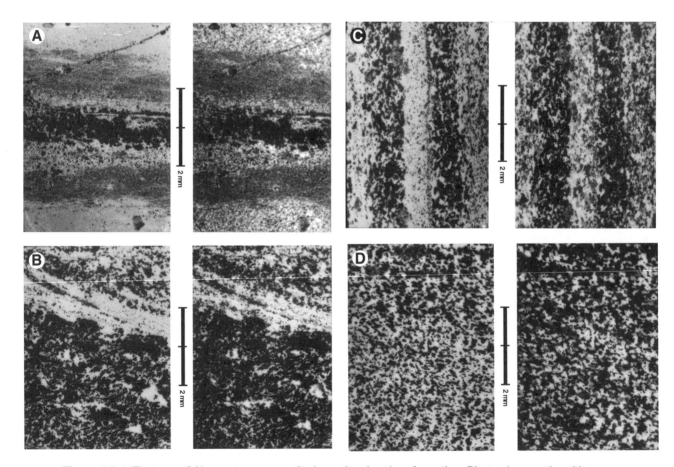

Figure 3.2-1. Textures of Algoma-type magnetite-hematite-chert iron-formation. Photomicrographs with transmitted light on left and polarized light on right. **A)** Lower greenschist metamorphic facies from Temagami Lake, Ontario. GSC 204578T **B)** Greenschist metamorphic facies from Kirkland Lake, Ontario. GSC 204578U **C)** Upper greenschist metamorphic facies from Capreol area, Ontario. GSC 204578S **D)** Associated with garnet-biotite-hornblende schist from Zealand township, Ontario. GSC 204578W (from Gross, 1965).

world by hydrothermal-effusive processes, and considered to be protolithic facies of iron-formation and other stratafer sediments (Gross, 1987, 1988a, 1990a, b, 1991, 1993c).

Iron-formations were formed by the deposition of iron and silica in colloidal size particles by chemical and biogenic precipitation processes. Their main constituents evidently came from hydrothermal-effusive sources and were deposited in euxinic to oxidizing basin environments, in association with clastic and pelagic sediment, tuff, volcanic rocks, and a variety of clay minerals. The variety of metal constituents consistently present as minor or trace elements evidently were derived from the hydrothermal plumes and basin water and adsorbed by amorphous iron

and manganese oxides and smectite clay components in the protolithic sediment (Gross, 1993c). Their development and distribution along volcanic belts and deep-seated faults and rift systems was controlled mainly by tectonic rather than by biogenic or atmospheric factors.

Algoma-type iron-formations were deposited relatively close to volcanic centres and include a broad range of lithofacies. Study of available data on recent metalliferous sediments on the seafloor (Gross and McLeod, 1987; Gross, 1987) shows that sulphide facies were deposited close to the higher temperature effusive centres, iron oxide and silicate facies were intermediate, and manganese-iron facies were deposited from cooler hydrothermal vents and in areas

Table 3.2-2. Compositions of Algoma-type iron-formation concentrates, pellets, and sinters.

	Average composition %									
	1	2	3	4	5	6	7	8	9	10
Fe total	29.680	60.800	27.900	60.240	31.930	66.750	38.830	63.35	36.60	48.32
Fe_2O_3	28.170	*nd	24.840	nd	31.050	nd	37.420	37.42	nd	nd
FeO	12.810	nd	13.520	nd	13.110	nd	16.260	16.26	nd	nd
P	0.07	0.02	0.05	0.018	0.090	0.020	0.050	0.05	0.017	0.029
SiO_2	53.640	5.410	55.220	5.270	46.460	3.550	42.150	42.15	9.10	8.00
Mn	0.040	0.060	0.090	0.030	0.080	0.110	0.050	0.05	2.19	2.36
Ti	0.010	nd	0.050	nd	0.070	nd	0.010	0.01	0.042	nd
Al_2O_3	0.440	0.270	1.200	0.400	3.540	0.430	0.590	0.59	1.145	1.06
CaO	2.130	4.950	0.770	4.730	1.270	nd	1.580	1.58	1.065	12.14
MgO	1.620	1.850	1.560	1.700	1.520	nd	1.890	1.89	4.33	7.17
Na_2O	0.100	0.052	0.040	0.035	0.450	nd	0.100	0.10	0.055	nd
K_2O	0.150	0.062	0.100	0.056	1.450	nd	0.110	0.11	0.05	nd
S	0.170	0.030	0.300	0.027	0.080	nd	0.213	0.213	5.62	0.099
CO_2	0.480	nd	0.760	nd	0.200	nd	0.270	0.27	nd	nd
C	0.130	nd	0.210	nd	0.050	nd	0.070	0.07	nd	nd
H_2O+/-	0.220	0.980	1.270	0.830	0.470	0.680	0.570	1.30	nd	nd

*nd = no data
Sources of data:
1. Gross, 1990b;   **2.** American Iron Ore Association, 1987, p. 69; Cliffs Mining Company, 1987, p. 19; Cliffs Mining Company, 1987, p. 13;
3. Gross, 1990b;   **4.** American Iron Ore Association, 1987, p. 69; Cliffs Mining Company, 1987, p. 19;   **5.** Gross, 1990b;
6. Cajka and Cadieux, 1985, p. 13;   **7.** Gross, 1990b;   **8.** Boucher, 1979, p. 16;   **9.** Goodwin et al., 1985;
10. American Iron Ore Association, 1987, p. 68.

1. Adams mine iron-formation, oxide facies
2. Adams mine, fluxed pellets
3. Temagami iron-formation, oxide facies
4. Sherman mine, fluxed pellets
5. Griffith mine iron-formation, oxide facies
6. Griffith mine, fluxed pellets
7. Moose Mountain, oxide facies
8. Moose Mountain, fluxed pellets
9. Algoma iron-formation, oxide facies
10. Algoma sinter, super-fluxed

distal from active hydrothermal discharge. Overlapping and lateral transitions of one kind of lithofacies to another appear to be common and are to be expected (Gross, 1988a, 1991, 1993a).

RELATED TYPES OF DEPOSITS

Algoma-type iron-formations are protore for high grade, direct shipping types of residual-enriched iron ore deposits that are mined in many parts of the world. Residual-enriched iron ore was mined in the Old Helen mine in the Michipicoten district, and at Steep Rock Lake near Atikokan, Ontario, that was derived from siderite and sulphide lithofacies of iron-formation. Other enriched iron ore deposits of this type derived mainly from oxide lithofacies include the Cerro Bolivar deposits in Venezuela (Dorr, 1973), the Carajas deposits in Brazil (Hoppe et al., 1987; Gibbs and Worth, 1990), and deposits in Swaziland, southern Africa (Beukes, 1973).

Algoma-type iron-formations and associated stratafer sediments commonly show a prolific development of different facies types within a single stratigraphic sequence of rocks or metallogene. Oxide lithofacies are usually the thickest and most widely distributed units of iron-formation in a region and serve as excellent metallogenetic markers (Frietsch, 1977, 1980a, c, 1982a, b; Gross, 1986, 1988a, 1991, 1993a, c; Gauthier and Brown, 1986; Gauthier et al., 1987).

Transitions from Lake Superior- to Algoma-type iron-formations occur in areas where iron-formations extend from continental shelf to deep water environments along craton margins as reported in the Krivoy Rog iron ranges (Belevtsev, 1973).

Manganese-iron lithofacies

Significant amounts of manganese are associated with Algoma-type iron-formation in several types of deposits (Gross, 1983b, 1990a). Manganese is commonly associated with carbonate lithofacies which range from predominantly siderite ($FeCO_3$) to rhodochrosite ($MnCO_3$). Siderite does not contain more than 2% MnO and the distribution of manganese in the dolomite-ankerite-kutnahorite group of minerals is more complex. Dolomite and ankerite may have ratios of Mg:(Fe, Mn) less than 1:2.6, and Mn-rich carbonate may have ratios of Mn:(Mg, Fe) less than 1.32. Some dolomitic lithofacies of iron-formation contain less than 26% FeO and 23.4% MnO, but the MnO content in most of the Precambrian Algoma-type iron-formations is usually less than 15% (Gross, 1965). Sinter products from the siderite ore in the Wawa area, Ontario, contain 2 to 2.5% manganese which has enhanced their value for blending with other iron ores that have a low content of manganese.

The world-wide distribution of manganese-rich facies in oxide and carbonate facies of iron-formation was studied by Varentsov and Grasselly (1980a, b, c), Roy (1981), and Gross (1983b). Important deposits of this type occur in India, Japan, China, Brazil, Africa, and Australia. Typical examples of transitions from cherty manganese to iron oxide lithofacies in Devonian and Tertiary iron-formations are found at the Karadzhal deposit in Kazakhstan (Sapozhnikov, 1963; Kalinin, 1965) and at Jalisco, Mexico (Zantop, 1978), in which Fe/Mn ratios range from 10:1 to

1:1.5 and the manganese content is 40% or less. The manganiferous jasper-hematite lithofacies at Woodstock, New Brunswick, are less than 30 m thick and the iron content ranges from 11 to 30% and the manganese content from 12 to 25%. The overall iron to manganese ratio is about 1.5 (Gross, 1967, 1983b, 1990a, b).

Sulphide deposits in Algoma-type iron-formation

The common association of sulphide and oxide lithofacies of Algoma-type iron-formation has been overlooked or not mentioned in many cases in the descriptions and documentation of sulphide lithofacies that contain large stratiform "massive" sulphide deposits. The following large deposits illustrate the important association of oxide and sulphide lithofacies: Manitouwadge, Bathurst-Newcastle, Matagami, Sturgeon Lake (Shegelski, 1987), Sherridon, and Michipicoten (Goodwin, 1962) in Canada; Gamsberg, South Africa (Rozendal, 1980); Broken Hill, Australia (Edwards and Atkinson, 1986); Tynagh, Ireland (Schultz, 1966); Cyprus; Kuroko-type deposits, Japan (Ohmoto and Skinner, 1983); deposits in Sweden (Frietsch, 1980a, b, c, 1982a, b); and Atlantis II deep in the Red Sea (El Shazly, 1990).

Gauthier and Brown (1986) described the association and relationship of various zinc-rich stratigraphic zones in pyrrhotite, pyrite, graphite, and magnetite lithofacies of iron-formation in the Maniwaki-Gracefield district of Quebec and concluded that these iron-formations formed by submarine exhalative processes (Gauthier et al., 1987). Occurrences of sphalerite in magnetite iron-formations in Sweden have been described by Frietsch (1982a, b).

Gold in Algoma-type iron-formation

Algoma-type iron-formations are considered to be prime metallogenetic markers in exploration for gold because of their syngenetic gold content, as at Homestake mine in South Dakota, and the Lupin mine, Northwest Territories; remobilization of gold in iron-formations during metamorphism; and the favorable chemical-structural environment provided by iron-formations for deposition of gold in veins, fractures, and shear zones (Hodder and Petruk, 1982; Macdonald, 1990). Gross (1988b) found that the average content of gold in oxide facies, based on random sampling of 40 Algoma-type iron-formations in the Canadian Shield, was about 0.04 ppm.

Minor elements in Algoma-type iron-formation

The average content of most of the minor elements in Algoma-type iron-formations is usually about double that in similar facies of Lake Superior-type iron-formation. This reflects the nature of the hydrothermal systems that produced them, and the prominent types of associated igneous rocks (Gross, 1990b, 1991, 1993a). Algoma-type iron-formations are therefore regarded as good metallogenetic marker beds in prospecting for other metals. For example, the tungsten-bearing iron-formations in Greenland were probably related genetically to a granitic igneous system (Appel, 1986) and are part of a sequence of stratafer lithofacies.

Furthermore, many Algoma-type iron-formations have a significantly high content of rare-earth elements. Lithofacies that are rich in rare-earth elements (analogous to the Bayan Obo deposits in China in Lake Superior-type iron-formations) would be expected where Algoma-type iron-formations are associated with alkaline rocks (Gross, 1993b).

EXPLORATION GUIDELINES

Because of their smaller size, more complex facies development, and the higher content and erratic distribution of minor elements, large iron deposits are not as numerous or as easily defined in Algoma-type iron-formations as in Lake Superior-type. Electromagnetic, magnetic, and electrical conductance and resistivity survey methods are used effectively in tracing and defining the distribution of Algoma-type beds, either in exploring for iron and manganese ore, or for using these stratafer beds as metallogenetic markers. The use of iron-formation geochemistry for understanding the metallogeny of an area and as an exploration guideline is considered to be in its initial stages of development.

Guidelines and recommendations for exploring for iron ore outlined for Lake Superior-type iron-formations also apply for Algoma-type:

1. The composition, texture, and grade of the crude ore must be such that ore concentrates can be produced that meet required chemical and physical specifications in a particular industrial area.

2. Discrete, well defined magnetite and hematite lithofacies of iron-formation are preferred with a minimum of other lithofacies and clastic sediment interbedded in the crude ore.

3. Granular, medium- to coarse-grained textures with well defined, sharp grain boundaries are desirable, to enable liberation and a clean separation of mineral grains in the concentration and beneficiation of the crude ore. These features are usually achieved in highly metamorphosed iron-formation where there is maximum grain enlargement, and magnetite lithofacies are usually preferred over hematite or mixed facies.

4. Mineable thicknesses of ore, usually 30 to 100 m or greater, may vary depending on location and economic factors in the market area. Ore deposits may consist of thick primary stratigraphic units or a succession of ore beds that have been repeated by folding and faulting.

5. Uniform granularity and mineral composition with a minimum of infolding of barren sediment or marginal facies of iron-formation, are advantageous factors for grade control and in processing of the crude ore.

6. Oxide facies iron-formation normally has a low content of minor elements, especially sodium, potassium, sulphur, and arsenic, which have deleterious effects in the processing of the ore and quality of steel produced from it.

7. The content of minor elements may vary significantly in different iron-formations depending on the kinds of volcanic, igneous, and stratafer rocks associated with them.

8. Iron-formations are usually large regional geological features that are relatively easy to define by mapping or with the aid of aeromagnetic or gravity surveys. Detailed stratigraphic information is an essential part of the database required for defining grade, physical and chemical quality, and beneficiation and concentration characteristics of the ore.

9. Basin analysis and sedimentation modelling enable definition of factors that controlled the development, location, and distribution of different iron-formation lithofacies.

10. Metamorphic mineral assemblages reflect the composition and mineralogy of primary sedimentary facies.

ACKNOWLEDGMENTS

This paper was critically read by D.F. Sangster, R.I. Thorpe, and C.W. Jefferson. Carol Plant and Lara O'Neill assisted with word processing.

SELECTED BIBLIOGRAPHY

References marked with asterisks (*) are considered to be the best sources of general information on this deposit subtype. For additional references, see "Introduction", Type 3.

American Iron Ore Association
1987: Iron Ore - 1986; American Iron Ore Association, Cleveland, Ohio, 100 p.
*Appel, P.W.U.
1986: Strata bound scheelite in the Archean Malene supracrustal belt, West Greenland; Mineralium Deposita, v. 21, p. 207-215.
*Beeson, R.
1990: Broken Hill-type lead-zinc deposits - an overview of their occurrence and geological setting; Transactions of the Institute of Mining and Metallurgy (Section B), v. 99, p. 163-175.
*Belevtsev, Y.N.
1973: Genesis of high-grade iron ores of Krivoy Rog type; in Genesis of Precambrian Iron and Manganese Deposits, Proceedings of the Kiev Symposium, 20-25 August, 1970, UN Educational, Scientific and Cultural Organization, Paris, p. 167-180.
Beukes, N.J.
1973: Precambrian iron-formations of Southern Africa; in Precambrian Iron-formations of the World, (ed.) H.L. James and P.K. Sims; Economic Geology, v. 68, no. 7, p. 960-1005.
*Bolis, J.L. and Bekkala, J.A.
1986: Iron ore availability – market economy countries, a minerals availability appraisal; Bureau of Mines, United States Department of the Interior, Information Circular 9128, 56 p.
Boucher, M.A.
1979: Canadian iron ore industry statistics, 1977-78; Energy, Mines and Resources, Canada, Mineral Policy Sector, Internal Report, MRI 79/2, 30 p.
Cajka, C.J. and Cadieux, A.
1985: Canadian iron ore industry statistics, 1984; Energy, Mines and Resources, Canada, Mineral Policy Sector, Internal Report MRI 85/2, p. 12-13.
Cliffs Mining Company
1987: Iron Ore Analyses - 1987; Cliffs Mining Company, Cleveland, Ohio, 86 p.
*Collins, W.H., Quirke, T.T., and Thomson, E.
1926: Michipicoten iron ranges; Geological Survey of Canada, Memoir 147, 175 p.
*Dorr, J.V.N., II
1973: Iron-formations in South America; in Precambrian Iron-formations of the World, (ed.) H.L. James and P.K. Sims; Economic Geology, v. 68, no. 7, p. 1005-1022.
*Edwards, R. and Atkinson, K.
1986: Sediment-hosted copper-lead-zinc deposits; Chapter 6 in Ore Deposit Geology, Chapman and Hall, London, New York, 445 p.
*El Shazly, E.M.
1990: Red Sea deposits; in Ancient Banded Iron Formations (Regional Presentations), (ed.) J.-J. Chauvel, Cheng Yuqi, E.M. El Shazly, G.A. Gross, K. Laajoki, M.S. Markov, K.L. Rai, V.A. Stulchikov, and S.S. Augustithis; Theophrastus Publications, S.A., Athens, Greece, p. 157-222.
*Frietsch, R.
1974: The occurrence and composition of apatite with special reference to iron ores and rocks in northern Sweden; Sveriges Geologiska Undersökning, Ser. C, NR 694, Arsbok 68, NR 1, 49 p.

*Frietsch, R. (cont.)

1977: The iron ore deposits in Sweden; in Iron Ore Deposits of Europe and Adjacent Areas, Volume I, (ed.) A. Zitzmann; Bundesanstalt für Geowissenschaften und Rohstoffe, Hannover, p. 279-293.

1980a: Metallogeny of the copper deposits of Sweden; in European Copper Deposits, (ed.) S. Janković and R.H. Sillitoe; Department of Geology, Belgrade University, Belgrade, p. 166-179.

1980b: Precambrian ores of the northern part of Norbotten county, northern Sweden; Guide to Excursions 078A+C, Part 1 (Sweden), Twenty-sixth International Geological Congress, Paris, 1980; Geological Survey of Finland, Espoo, 1980, 35 p.

1980c: The ore deposits of Sweden; Geological Survey of Finland, Bulletin 306, 20 p.

1982a: Chemical composition of magnetite and sphalerite in the iron and sulphide ores of central Sweden; Geologiska Föreningens i Stockholm Förhandlingar, v. 104, pt. 1, p. 43-47.

1982b: A model for the formation of the non-apatitic iron ores, manganese ores and sulphide ores of central Sweden; Sveriges Geologiska Undersökning, Serie C, NR 795, Arsbok 76, NR 8, 43 p.

*Gauthier, M. and Brown, A.C.

1986: Zinc and iron metallogeny in the Maniwaki-Gracefield district, southwestern Quebec; Economic Geology, v. 81, no. 1, p. 89-112.

*Gauthier, M., Brown, A.C., and Morin, G.

1987: Small iron-formations as a guide to base- and precious-metal deposits in the Grenville Province of southern Quebec; in Precambrian Iron-Formations, (ed.) P.W.U. Appel and G.L. LaBerge; Theophrastus Publications, S.A., Athens, Greece, p. 297-327.

*Gibbs, A.K. and Worth, K.R.

1990: Geologic setting of the Serra dos Carajas iron deposits Brazil; in Ancient Banded Iron Formations (Regional Presentations), (ed.) J.-J. Chauvel, Cheng Yuqi, E.M. El Shazly, G.A. Gross, K. Laajoki, M.S. Markov, K.L. Rai, V.A. Stulchikov, and S.S. Augustithis; Theophrastus Publications, S.A., Athens, Greece, p. 83-102.

*Goodwin, A.M.

1962: Structure, stratigraphy, and origin of iron-formation, Michipicoten area, Algoma district, Ontario, Canada; Geological Society of America, Bulletin, v. 73, p. 561-586.

1973: Archean iron-formations and tectonic basins of the Canadian Shield; in Precambrian Iron-formations of the World, (ed.) H.L. James and P.K. Sims; Economic Geology, v. 68, no. 7, p. 915-933.

*Goodwin, A.M., Thode, H.G., Chou, C.-L., and Karkhansis, S.N.

1985: Chemostratigraphy and origin of the Late Archean siderite-pyrite-rich Helen Iron-Formation, Michipicoten belt, Canada; Canadian Journal of Earth Sciences, v. 22, no. 1, p. 72-84.

Gross, G.A.

*1961: Metamorphism of iron-formations and its bearing on their beneficiation; The Canadian Mining and Metallurgical Bulletin, Transactions, v. 64, p. 24-31.

Gross, G.A. (comp.)

1963a: Distribution of iron deposits, Rivière Gatineau, Superior Structural Province, Quebec-Ontario; Geological Survey of Canada, Preliminary Map 13-1963, scale 1:1 000 000.

1963b: Distribution of iron deposits, Broadback River, Superior Structural Province, Quebec-Ontario; Geological Survey of Canada, Preliminary Map 14-1963, scale 1:1 000 000.

1963c: Distribution of iron deposits, Fort George River, Superior Structural Province, Quebec-Ontario; Geological Survey of Canada, Preliminary Map 15-1963, scale 1:1 000 000.

1963d: Distribution of iron deposits, Montreal River, Superior Structural Province, Ontario; Geological Survey of Canada, Preliminary Map 16-1963, scale 1:1 000 000.

1963e: Distribution of iron deposits, Albany River, Superior Structural Province, Ontario; Geological Survey of Canada, Preliminary Map 17-1963, scale 1:1 000 000.

1963f: Distribution of iron deposits, Ekwan River, Superior Structural Province, Ontario; Geological Survey of Canada, Preliminary Map 18-1963, scale 1:1 000 000.

1963g: Distribution of iron deposits, Ogoki River, Superior Structural Province, Ontario and Manitoba; Geological Survey of Canada, Preliminary Map 19-1963, scale 1:1 000 000.

1963h: Distribution of iron deposits, Sachigo River, Superior Structural Province, Ontario and Manitoba; Geological Survey of Canada, Preliminary Map 20-1963, scale 1:1 000 000.

Gross, G.A.

*1964: Primary features in cherty iron-formations; in Genetic Problems of Ores, (ed.) R.K. Sundaram, Part V, Proceedings of Section 5, XXII International Geological Congress, India, p. 102-117.

*1965: General geology and evaluation of iron deposits; in Volume I, Geology of Iron Deposits in Canada, Geological Survey of Canada, Economic Geology Report 22, 181 p.

Gross, G.A. (cont.)

*1967: Iron deposits in the Appalachian and Grenville regions of Canada; in Volume II, Geology of iron deposits in Canada, Geological Survey of Canada, Economic Geology Report 22, 111 p.

*1968: Iron ranges of the Labrador geosyncline; in Volume III, Geology of iron deposits in Canada, Geological Survey of Canada, Economic Geology Report 22, 179 p.

*1970: Nature and occurrence of iron ore deposits; iron ore deposits of Canada and the West Indies; in United Nations Survey of World Iron Ore Resources; United Nations Publication Sales No. E.69.II.C.4, New York, p. 13-31, p. 237-269.

*1973: The depositional environment of principal types of Precambrian iron-formation; in Genesis of Precambrian Iron and Manganese Deposits, Proceedings of Kiev Symposium, 1970, UN Educational, Scientific and Cultural Organization, Earth Sciences 9, Paris, p. 15-21.

*1980: A classification of iron-formation based on depositional environments; Canadian Mineralogist, v. 18, p. 215-222.

*1983a: Tectonic systems and the deposition of iron-formation; Precambrian Research, v. 20, p. 171-187.

*1983b: Low grade manganese deposits - a facies approach; in Cameron Volume on Unconventional Mineral Deposits, (ed.) W.C. Shanks; American Institute of Mining, Metallurgical, and Petroleum Engineers Inc., New York, New York, p. 35-46.

*1986: The metallogenetic significance of iron-formation and related stratafer rocks; Journal of the Geological Society of India, v. 28, no. 2, 3, p. 92-108.

*1987: Mineral deposits on the deep seabed; Marine Mining, v. 6, p. 109-119.

*1988a: A comparison of metalliferous sediments, Precambrian to Recent; Kristalinikum, v. 19, p. 59-74.

*1988b: Gold content and geochemistry of iron-formation in Canada; Geological Survey of Canada, Paper 86-19, 54 p.

*1990a: Manganese and iron facies in hydrolithic sediments; Special Publication of the International Association of Sedimentologists, no. 11, p. 31-38.

1990b: Geochemistry of iron-formation in Canada; in Ancient Banded Iron-formations (Regional Presentations); (ed.) J.-J. Chauvel, Cheng Yuqi, E.M. El Shazly, G.A. Gross, K. Laajoki, M.S. Markov, K.L. Rai, V.A. Stulchikov, and S.S. Augustithis; Theophrastus Publications, S.A., Athens, Greece, p. 3-26.

*1991: Genetic concepts for iron-formation and associated metalliferous sediments; in Historical Perspectives of Genetic Concepts and Case Histories of Famous Discoveries; (ed.) R.W. Hutchinson and R.I. Grauch; Economic Geology Monograph 8, Economic Geology, p. 51-81.

1993a: Iron-formation metallogeny and facies relationships in stratafer sediments; in Proceedings of the Eighth Quadrennial International Association on the Genesis of Ore Deposits Symposium, (ed.) Y.T. Maurice; E. Schweizerbart'sche Verlagsbuchhandlung (Nagele u. Obermiller), D-7000 Stuttgart 1, p. 541-550.

1993b: Rare earth elements and niobium in iron-formation at Bayan Obo, Inner Mongolia, China; in Proceedings of the Eighth Quadrennial International Association on the Genesis of Ore Deposits Symposium, (ed.) Y.T. Maurice; E. Schweizerbart'sche Verlagsbuchhandlung (Nagele u. Obermiller), D-7000 Stuttgart 1, p. 477-490.

*1993c: Element distribution patterns as metallogenetic indicators in siliceous metalliferous sediments; Proceedings of the Twenty-ninth International Geological Congress, Resource Geology Special Issue, no. 17, p. 96-107.

*Gross, G.A. and McLeod, C.R.

1980: A preliminary assessment of the chemical composition of iron-formations in Canada; Canadian Mineralogist, v. 18, p. 215-222.

1987: Metallic minerals on the deep seabed; Geological Survey of Canada, Paper 86-21, 65 p., and Map 1659A, scale 1:40 000 000.

*Hodder, R.W. and Petruk, W. (ed.)

1982: Geology of Canadian Gold Deposits; The Canadian Institute of Mining and Metallurgy, Special Volume 24, 286 p.

*Hoppe, A., Schobbenhaus, C., and Walde, D.H.G.

1987: Precambrian iron formations in Brazil; in Precambrian Iron-Formations, (ed.) P.W.U. Appel and G.L. LaBerge; Theophrastus Publications, S.A., Athens, Greece, p. 347-392.

Kalinin, V.V.

1965: The iron-manganese ores of the Karadzhal deposit; Institute of the Geology of Ore Deposits, Petrography, Mineralogy and Geochemistry, USSR Academy of Sciences, Moscow, (translation of Russian, 123 p.).

*Landergren, S.

1948: On the geochemistry of Swedish iron ores and associated rocks: a study of iron-ore formation; Sveriges Geologiska Undersökning, Serie C, NR. 496, Arsbok 42, NR 5, Stockholm, 182 p.

*Lang, A.H., Goodwin, A.M., Mulligan, R., Whitmore, D.R.E.,
Gross, G.A., Boyle, R.W., Johnston, A.G., Chamberlain, J.A.,
and Rose, E.R.
1970:   Economic minerals of the Canadian Shield; Chapter 5 in Geology
         and Economic Minerals of Canada; Geological Survey of Canada,
         Economic Geology Report 1, fifth edition, p. 152-223.

*Macdonald, A.J.
1990:   Banded oxide facies iron-formation as a host for gold mineralization;
         in Ancient Banded Iron Formations (Regional Presentations), (ed.)
         J.-J. Chauvel, Cheng Yuqi, E.M. El Shazly, G.A. Gross, K. Laajoki,
         M.S. Markov, K.L. Rai, V.A. Stulchikov, and S.S. Augustithis;
         Theophrastus Publications, S.A., Athens, Greece, p. 63-82.

*Meyn, H.D. and Palonen, P.A.
1980:   Stratigraphy of an Archean submarine fan; Precambrian Research,
         v. 12, p. 257-285.

*Ohmoto, H. and Skinner, B.J. (ed.)
1983:   The Kuroko and related volcanogenic massive sulfide deposits;
         Economic Geology, Monograph 5, 604 p.

*Roy, S.
1981:   Manganese Deposits; Academic Press, London-New York-Toronto-
         Sydney-San Francisco, 458 p.

Sapozhnikov, D.G.
1963:   The Karadzhal iron-manganese deposit; Transactions, Institute of
         the Geology of Ore Deposits, Petrography, Mineralogy and
         Geochemistry, No. 89, USSR Academy of Sciences, Moscow, chap. 3,
         no. 89 (translation of Russian, 193 p.).

*Schultz, R.W.
1966:   Lower Carboniferous cherty ironstones at Tynagh, Ireland:
         Economic Geology, v. 61, p. 311-342.

*Shegelski, R.J.
1987:   The depositional environment of Archean iron formations,
         Sturgeon-Savant greenstone belt, Ontario, Canada; in Precambrian
         Iron-Formations, (ed.) P.W.U. Appel and G.L. LaBerge; Theophrastus
         Publications, S.A., Athens, Greece, p. 329-344.

Stanton, R.L.
*1959:   Mineralogical features and possible mode of emplacement of the
         Brunswick Mining and Smelting orebodies, Gloucester County,
         New Brunswick; The Canadian Mining and Metallurgy Bulletin,
         v. 52, p. 631-643.
*1960:   General features of the conformable "pyritic" orebodies; The
         Canadian Mining and Metallurgy Bulletin, v. 53, p. 24-29.
*1976:   Petrochemical studies of the ore environment at Broken Hill, New
         South Wales: 3 – banded iron formations and sulphide orebodies:
         constitutional and genetic ties; Institution of Mining and
         Metallurgy Transactions, v. 85, p. B132-141.
1991:   Understanding volcanic massive sulfides – past, present and
         future; Economic Geology, Monograph 8, p. 82-95.

Varentsov, I.M. and Grasselly, G. (ed.)
1980a: Geology and Geochemistry of Manganese: Volume I, General Problems,
         Mineralogy, Geochemistry, Methods; E. Schweizerbart'sche
         Verlagsbuchhandlung (Nägele u. Obermiller), Stuttgart, Germany, 463 p.
1980b: Geology and Geochemistry of Manganese: Volume II, Manganese
         Deposits on Continents; E. Schweizerbart'sche Verlagsbuchhandlung
         (Nägele u. Obermiller), Stuttgart, Germany, 510 p.
1980c: Geology and Geochemistry of Manganese: Volume III, Manganese on the
         Bottom of Recent Basins; E. Schweizerbart'sche Verlagsbuchhandlung
         (Nägele u. Obermiller), Stuttgart, Germany, 357 p.

Zantop, H.
1978:   Geologic setting and genesis of iron oxides and manganese oxides
         in the San Francisco manganese deposit, Jalisco, Mexico; Economic
         Geology, v. 73, p. 1137-1149.

3.3 IRONSTONE

G.A. Gross

INTRODUCTION

Following established precedent (Gross, 1965), the term ironstone refers to lithofacies of hydrolithic sediments composed of yellow to brown or blood-red iron oxide, clay minerals, and fine grained clastic and fossil detritus that commonly have well developed oolitic and granular textures. Selected lithofacies of ironstone requiring a minimum of processing are used as iron ore and because of their relatively high content of calcium and magnesium provide suitable fluxing components in blast furnace burden. Iron ore should have the highest content of iron and the lowest possible content of slag-forming constituents (alumina and silica), and meet grade and quality specifications of a particular iron and steel industry. The use of ironstone for iron ore is restricted because of its high content of phosphorus, but it is accepted for use in areas where the phosphorus-rich slag produced from it can be used in the production of fertilizers or for other industrial purposes. The content of deleterious constituents, such as sulphur and arsenic, must be kept low and controlled to very rigid specifications.

Goethite is the principal iron oxide mineral in the brown ironstones, e.g., **Minette-type** Jurassic Lorraine Basin in France and Luxembourg (Bubenicek, 1961, 1971; Teyssen, 1984), in the Midlands of England (Taylor, 1949), and Peace River area of Alberta (Mellon, 1962). Hematite is the most abundant iron oxide in the purple to red **Clinton-type** ironstones, e.g., Lower Ordovician in Wabana mine of Newfoundland, and in Alabama and Appalachian area of United States. Both types have significantly high contents of alumina, phosphorus, calcium, and magnesium.

Descriptive data and interpretation of Ordovician ironstone in Wabana mine, Newfoundland, are used here to illustrate this deposit type and its genesis.

Gross, G.A.
1996:   Ironstone; in Geology of Canadian Mineral Deposit Types, (ed.)
         O.R. Eckstrand, W.D. Sinclair, and R.I. Thorpe; Geological
         Survey of Canada, Geology of Canada, no. 8, p. 73-80 (also
         Geological Society of America, The Geology of North America,
         v. P-1).

IMPORTANCE

Ironstone has provided a relatively small amount of iron ore in Canada. About 80 million long tons (81.3 Mt) of Clinton-type ironstone, mainly from underground stopes, were shipped from Wabana mine on Bell Island, Newfoundland, during its operation from 1892 to 1965. Nearly 35 million long tons (35.6 Mt) of this ore were used in Canada and the remainder was shipped to England and Germany. The highest rate of production was achieved in 1960 when 2.8 million long tons (2.84 Mt) of ore were mined.

Large resources of Minette-type ironstone have been explored in the Peace River area of Alberta, and about 350 000 long tons (355 600 t) of ore were produced from the Nictaux-Torbrook ironstone in Nova Scotia between 1825 and 1916.

A significant amount of iron ore has been mined from ironstone deposits in Alabama in the past. The Clinton Formation extends through the Appalachian fold belt for more than 1000 km into New York State (Cotter and Link, 1993). The Minette ironstone deposits in England, France, Luxembourg, and Germany were one of the principal sources of iron ore in western Europe until production was phased out in recent years.

Ironstone has provided a source of ferrous iron in the preparation of fertilizer products for use in tropical areas where iron is oxidized by weathering processes and the soils become depleted in essential ferrous iron compounds.

SIZE AND GRADE OF DEPOSITS

The iron content in Clinton-type ironstone in the Lower, Middle, and Upper ore beds in Wabana mine, Newfoundland, ranged from 45.00 to 59.60%, and the silica content from 6.5 to 20%. The average iron content in ore mined in 1965 was 48% and after treatment in the heavy media plant to remove shale the ore shipped contained 50.18% iron, 12.92% silica, and 1.77% moisture.

The Jurassic Minette-type ironstones of the Lorraine Basin in France and Luxembourg, and in Northhamptonshire in England contain from 25 to 35% iron; silica ranges from 5 to 20%; and the amounts of other constituents vary greatly.

Grade and quality are of special concern in the use of ironstone iron ore. The total thickness of stratigraphic units of ironstone rarely exceeds 10 m but estimates of iron resources in single basins often exceed one billion tonnes.

GEOLOGICAL FEATURES

Ironstone lithofacies range in age from Late Precambrian to Recent. They are associated with black shale, siltstone, sandstone, shale, limestone, and manganiferous, phosphatic, or pyritic shale, and were deposited in neritic oxygenated and euxinic environments in continental shelf and estuarine basins. Ironstones are included in the stratafer group of hydrolithic sediments (see Type 3 "Introduction").

Stratigraphic and sedimentary features typical of many Clinton-type ironstones are exposed in a sequence of Cambro-Ordovician sedimentary rocks at Wabana mine on Bell Island, Newfoundland (Table 3.3-1). The ironstone ore zones and enclosing strata dip gently to the north (Fig. 3.3-1) and extend under Conception Bay.

The description of the stratigraphy of the Wabana ironstone within the Lower Ordovician Wabana and Bell Island groups (Table 3.3-2) is based mainly on data from Hayes (1915) and illustrated in Figure 3.3-1. Oolitic ironstone and ferruginous rocks occur in six zones on Bell Island and are described in order from the upper to the lower parts of the stratigraphic column in Table 3.3-2.

Sedimentary features of ore

Beds of ironstone in the ore zones that are less than one metre thick are separated by thin beds of ferruginous shale that contain isolated iron-rich granules or oolites. Transitional zones between ironstone and shale consist of interbedded shale and oolitic hematite. Sandy beds containing chamosite granules in a matrix of siderite and fine clastic mud are commonly interlayered with shale and ferruginous beds. The oolitic hematite beds show many sedimentary features that formed in shallow water such as crossbedding, ripple-marks, scour-and-fill structures, worm burrows, and abraded fossil fragments.

Oolites and ooids in the ironstone consist of alternating concentric rings of siderite or mixtures of siderite, hematite, and chamosite surrounding nuclei of fossil fragments, sand grains, or granules, and are distributed in a matrix of hematite or siderite. The average size of the oolites is about 0.5 mm and their outer rings usually consist of hematite. Hayes (1915) described delicate, well-preserved algal borings that cut across some of the oolite rings and showed that there has been little alteration of the oolites, ooids, or spherules since they were formed.

Stratigraphic features of ore

The lenticular nature of the ironstone beds is shown in Figure 3.3-1. Iron-rich beds, less than 10 m thick, extend along strike for more than 1000 m and appear to have been sand bars that were parallel to the ancient shoreline (Fig. 3.3-2).

Ironstone ore was mined from three stratigraphic units, which were less than 12 m thick, typically massive, deep red to purplish red, and composed of oolites or spherules consisting of hematite, chamosite, and siderite. The ore broke into rectangular blocks along well developed sets of joints and fractures.

The "Lower" ironstone ore zone, the Dominion bed, maximum thickness about 12 m, was the thickest of the three zones mined, and provided most of the ore. It consisted of several hematite-rich lenses separated by lenses of leaner shale or sandstone. The upper part of this zone was rich in siderite and similar to the Upper bed, probably recording a change in sedimentation with deeper submergence of the depositional basin. Good quality hematite-chamosite ore, present mainly in the upper and lower parts of this ore zone, graded laterally over a distance of 30 to 100 m into leaner siliceous material in the main bar or lens of iron-rich sediment (Fig. 3.3-2). The top of the Lower bed was defined by a disconformity, and by a persistent bed of sandy conglomerate less than 2 m thick that contains nodules of black shale and pyrite in a cherty matrix. A pyrite bed (Zone 3, 0.5 m thick), composed of oolites and spherules of pyrite in a cherty matrix, lies above the sandy bed.

The "Middle" ironstone ore zone, Scotia bed, was mineralogically and texturally like the Lower bed but richer in iron. The average thickness mined was about 2.5 m. It contained 59.6 to 51.5% iron and 6.4 to 12.0% SiO_2. The lenses of oolitic hematite, 1-2 m thick, strike northwest and are parallel to the present shoreline. They thin to less than one metre in areas one to two kilometres from the shore.

The "Upper" ironstone ore zone (5), marks the upper limits of iron deposition in the basin. It is located 10 to 15 m above the Middle zone, and is characterized by the lensy, erratic distribution of hematite-chamosite-siderite facies interbedded with sand and shale, a lower iron content, and siderite present in wavy bands and as matrix to the oolites.

Structure in ore zones

Two prominent sets of faults with parallel sets of joints are developed in the east limb of the Conception Bay Syncline. One set strikes 030° and dips 85°SE and the other set strikes 290° and dips 85°SW. The ore breaks into rectangular fragments along joints spaced 10 to 30 cm apart. Both strike-slip and dip-slip movement took place along most faults; right-hand displacement was along northwest-trending faults; and left-hand displacement was along the northeast-trending faults. The northeast-trending faults appear to have developed somewhat later than the northwesterly faults, but movement along the faults caused very little brecciation or contraction of adjacent beds. Vertical movement of 30 m has taken place on some of the major faults that pass through the three ore zones (Fig. 3.3-3).

DEFINITIVE CHARACTERISTICS OF IRONSTONE IRON ORE

The most desirable ironstone beds at Wabana had a minimum of shale and clay intermixed with the iron-rich oolites and spherules, a high proportion of iron as hematite and hydrated iron oxide minerals relative to siderite and chamosite, and iron distributed in discrete mineral grains, oolites, and spherules, rather than intermixed in clay-sized or amorphous aggregates. Facies rich in siderite frequently contained an undesirably high content of manganese. Beds of well-sorted hematite, siderite, and chamosite oolites and granules, of a mineable thickness, 2 to 10 m, provided the best ore in Wabana mine. Sharp boundaries between the

Table 3.3-1. Regional stratigraphy of the Wabana mine area (based on data from Rose, 1952; Hutchinson, 1953; and Gross, 1967a).

WABANA GROUP – Lower Ordovician, >300 m thick. This group overlies the Dominion or "Lower ore bed". Lithofacies are similar to those in the Bell Island Group, except that black shale units containing beds of oolitic pyrite and pyrite-bearing shale associated with oolitic hematite-chamosite-siderite ironstone are more abundant in basal part of unit.
BELL ISLAND GROUP – Lower Ordovician, estimated thickness >1220 m. Composed of massive red oolitic hematite-chamosite beds, ferruginous sandstone, and shale; thin-bedded, grey, grey-brown, and greenish sandstone; grey, brown, and black shale; light micaceous sandy shale and sandstone. Sandstone beds consist of subangular quartz grains, glauconite, chamosite, altered feldspar, ferromagnesian minerals, and accessory zircon, sphene, and magnetite, which are enclosed in a matrix of siderite in some of the beds. Deposition of the ironstone and clastic sediments in shallow water is indicated by numerous sedimentary features – crossbedding, ripple-marks, rain-drop impressions, worm burrows, algae tubes, and fossil fragments.
ELLIOT COVE GROUP – Upper Cambrian, >150 m thick. Composed of thin-bedded, dark grey and black shale with nodules of pyrite, and lenses of limestone and sandstone.
ACADIAN GROUP – Middle Cambrian, 150 m thick. Composed of black and green shale, slate, and siltstone with nodules, lenses, and thin beds of red and grey limestone, pyritic slate, and manganiferous and phosphatic beds.
LOWER CAMBRIAN GROUP – 150 to 245 m thick. Red to pink to green wavy banded limestone interbedded with red or green slate. Quartz-pebble conglomerate cemented by red limestone at the base of the section is in angular unconformity with Precambrian rocks.

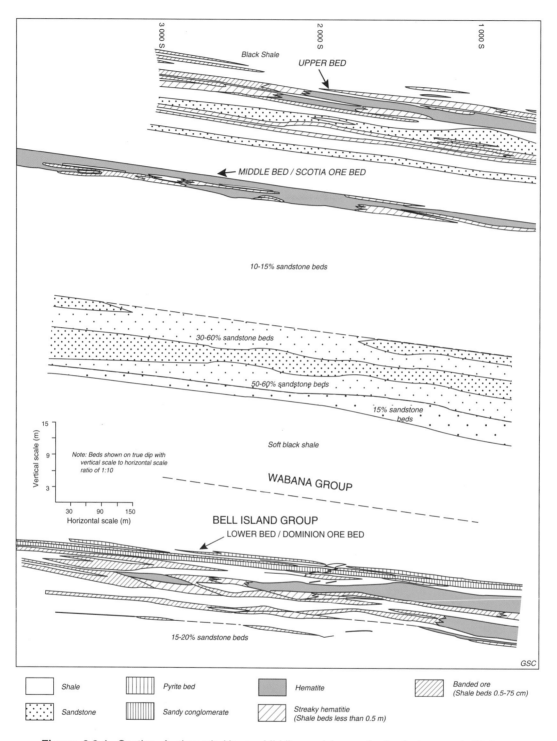

3 000 S

2 000 S

1 000 S

Black Shale

UPPER BED

◄ **MIDDLE BED / SCOTIA ORE BED**

10-15% sandstone beds

30-60% sandstone beds

50-60% sandstone beds

15% sandstone beds

Note: Beds shown on true dip with vertical scale to horizontal scale ratio of 1:10

Vertical scale (m)

15

9

3

30   90   150
Horizontal scale (m)

Soft black shale

WABANA GROUP

BELL ISLAND GROUP

LOWER BED / DOMINION ORE BED

15-20% sandstone beds

GSC

☐ Shale		⦀ Pyrite bed		▨ Hematite		⧄ Banded ore (Shale beds 0.5-75 cm)	
⠿ Sandstone		⦀ Sandy conglomerate		⧄ Streaky hematitie (Shale beds less than 0.5 m)			

Figure 3.3-1. Section A, through Upper, Middle, and Lower beds, facing west, Wabana, Newfoundland (after Lyons, 1957).

Table 3.3-2. Stratigraphy of the Wabana ironstone within the Lower Ordovician Wabana and Bell Island groups.

Zone 5, Upper bed. The lowest ironstone bed of this zone is separated from the Scotia bed by about 12 m of sandstone and shales, and oolitic ironstone beds are prominant throughout the upper 15 m of this stratigraphic zone.
Zone 4, Middle "Scotia" bed ironstone, ore zone. This zone is about 4.5 m thick and is 60 m stratigraphically above zone 3.
Zone 3, Pyrite beds. Three beds of oolitic pyrite, ranging in thickness from 5 cm to 1.2 m, are interbedded with black shale in 3 m of the stratigraphic succession immediately above the oolitic hematite beds of zone 2.
Zone 2, Dominion bed, the main ore zone. This is about 30 m thick and consists of a series of oolitic hematite ironstone beds interbedded with shale and crossbedded, fine grained sandstone. The main Dominion ore bed in the upper 10.5 m of zone 2 is composed of thick beds of oolitic hematite separated by thin layers of ferruginous sandstone and shale. The lowest oolitic hematite beds in the Dominion zone are 180 m stratigraphically above zone 1.
Zone 1. This zone is partially exposed by shallow cuttings along the tramways at the mine. It contains bands of oolitic hematite, two of which appear to be continuous, one attaining a thickness of about 60 cm.
Zone 0. This consists of a thin ferruginous band located in the extreme southwestern part of the island.

ironstone and associated sediments, shallow-dipping to flat-lying strata, and a minimum of disruption of the ore strata by faults, were favourable factors in defining ore zones.

GENETIC MODEL FOR IRONSTONE

The oolitic iron ore, ferruginous shale, and sandstone form part of a series of sedimentary rocks of Lower Ordovician age. Three stages in the deposition of the Lower ironstone bed at Wabana were described by Lyons (1957):

1. deposition of a thin bed of oolitic hematite, 30 to 60 cm thick, was followed by deposition of ferruginous shale beds, 30 cm to 3 m thick;

2. development in favourable areas of a complex zone of hematite- and chamosite-rich beds, 1 to 5 m thick, that were relatively free from shale partings, except at the base and top of the zone; and

3. the iron-rich beds were succeeded by another disconformity zone (reactivation surface) in which iron deposition was much more limited and a bed of mud 10 cm to 2 m thick accumulated. Hematite and chamosite beds deposited above this disconformity zone are 2.5 to 5 m thick.

Precipitation of iron took place separately or during the normal deposition of mud, silt, shale, sand, and some limestone. Observations by Hayes (1915) on the Wabana beds outlined features of special genetic significance which are summarized below.

The ore beds are characterized by ripple-marked surfaces and crossbedded layers, and contain remains of organisms which lived in shallow water. The spherules, oolites, and granules in the ore range in size from 0.1 to 0.5 mm and are composed of alternating concentric layers of hematite and chamosite. Some of the oolites were pierced by living, boring algae that flourished on the sea bottom, which suggests that precipitation of iron and the development of the oolites took place near the sediment-water interface.

Chamosite appears to have formed in advance of the other iron-rich minerals. The conspicuous increase in hematite in the outer parts of oolites suggests that part of

it may have formed by oxidation of chamosite or during a later stage of precipitation. Ferrous iron in solution reacted with complexes of clay and colloidal clay particles to form chamosite on the surface of convenient nuclei.

Much of the siderite appears to have formed during diagenesis of the beds. Siderite is less abundant than hematite and chamosite and replaces these minerals and some of the detrital quartz in the ironstone. Fossilized algae is found in all parts of the ore beds and evidently algae were very abundant in the marine plant life growing on the sea bottom. Tubules of the algae preserved in the siderite are usually coated on the exterior with hematite. The siderite was probably precipitated below the sediment-water interface where concentrations of ammonia and carbon dioxide were produced from decaying organic matter. Evidently hematite and chamosite formed on the surface of the sediment whereas siderite formed contemporaneously during early diagenesis of the underlying sediments.

The iron ore beds occur as primary sedimentary deposits that are essentially in the same condition today as when they were deposited, except for induration, faulting, and the addition of small amounts of secondary calcite and quartz in joints and fractures. Limestone is not associated with the ore beds and all of the original calcium in the ore, average content about 2.5%, is present in the fossils, which are composed largely of calcium phosphate, or calcium phosphate derived from organic matter. The phosphorus in the iron ore is also distributed in the fossil material. No evidence of diagenetic transformation from an original oolitic limestone to an oolitic iron ore has been found and there is no evidence that addition or concentration of iron has occurred since the deposition of these ferruginous sediments.

Oolitic pyrite beds in the same sedimentary succession as the hematite ore beds are characterized by a planktonic fauna indicative of open ocean currents and deeper water. The layers of pyrite show distinct stratification and are probably similar in origin to modern deposits of pyrite now forming in the Black Sea. The pyrite oolites and spherules are composed of fine concentric laminae of pyrite, and some oolites have alternate fine laminae of phosphatic material and pyrite. Pyritized and unpyritized graptolites and brachiopod remains occur together in contact with the spherules, indicating that some mechanical mixing took place on the depositional surfaces.

It is evident that complex and delicate adjustments of the composition and chemistry of the basin water, water depth, distribution of clastic material, and of Eh and pH of the seawater were required for the formation of this ironstone. The oolites are evidence of agitation and disturbance of the water, probably by wave or tidal action, in a shallow bay or shelf.

The source of the iron in the Wabana ironstone has not been established with any degree of certainty. Many have inferred that the iron was derived by weathering processes and transported in solution as inorganic and organic acid compounds. The imposition of chemically precipitated iron at specific intervals and locations in the sequence of ordinary shelf sediments, and the abundance of shales and coarse grained clastic material intimately associated with these hydrolithic ironstones, raises some doubt as to whether sedimentation factors in a basin and weathering of terrestrial rocks can solely account for, or are compatible with, the concentration of iron as found in ironstones.

The iron in some of the ironstones may have come from submarine volcanic emanations which could have been transported by currents to the sedimentary basin where it was oxidized and precipitated in shallow water near the shore (Gross, 1965; see also Kimberley, 1979, 1994). The large deposits of oolitic brown hematite, goethite, and silicate ironstones deposited on river beds and flats in Oligocene time at Lisakov in the southern Ural Mountains (Gross, 1967b), and the large Oligocene ironstone deposits around Kerch near the Black Sea and Sea of Azov, clearly show that the iron in large ironstone deposits can be derived by sedimentary processes, since evidence of hot springs and effusive sources of iron at these sites has not been reported. On the other hand, iron from hydrothermal sources is being deposited in the hydrolithic sediments forming along the shore of Santorini Island, Greece, and

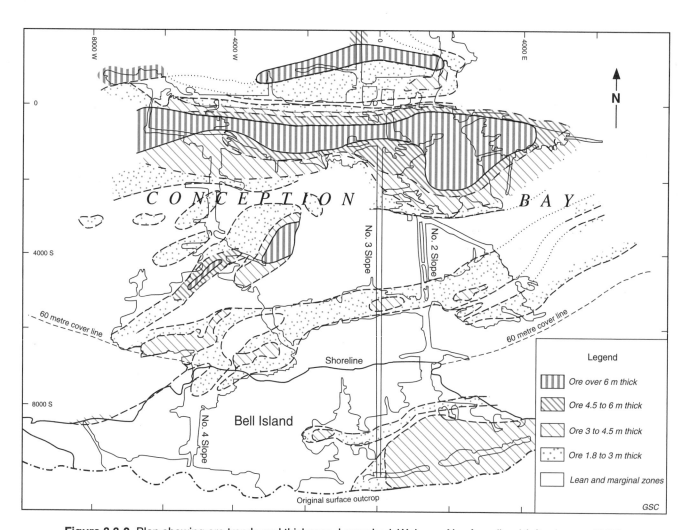

Figure 3.3-2. Plan showing ore trends and thickness, Lower bed, Wabana, Newfoundland (after Lyons, 1957).

Volcano Island in the Tyrrhenian Sea. Jenkyns (1970) concluded that exhalations associated with volcanism were the most likely source of iron in Jurassic ferruginous pisolitic sediments in western Sicily.

A fundamental difference between ironstones and iron-formation may be a land source for some of the iron and most of the other constituents in ironstones, as opposed to hydrothermal-effusive sources for the iron and silica in iron-formations and other stratafer lithofacies. The bulk chemistry of most ironstones is compatible with a land source for their components but the content of minor elements is not well known. Iron may have been derived from either one of these sources, but environmental factors controlled the composition and the kinds of lithofacies that developed.

The genetic relationship between ironstones and iron-formations or other stratafer rocks is not clearly understood, and a distinction between them is maintained because of marked differences in their composition, lithology, and

mineralogy, and their qualities as iron ore. Because of the high contents of alumina and clay material in the iron-stones and the high content of silica in the form of chert and quartz in iron-formation, it has generally been accepted that these two main groups of iron-rich hydrolithic sediments were formed by different processes in different environments and that deposition of silica and alumina in large quantities in the same basin environment was chemically incompatible. Banded chert is found occasionally in sequences of ironstone, but an exclusive hydrogenous-sedimentary or hydrothermal genesis has not been demonstrated.

RELATED TYPES OF DEPOSITS

Possible genetic relationships between ironstones and manganese-, copper-, and zinc-rich metalliferous sediments require further investigation.

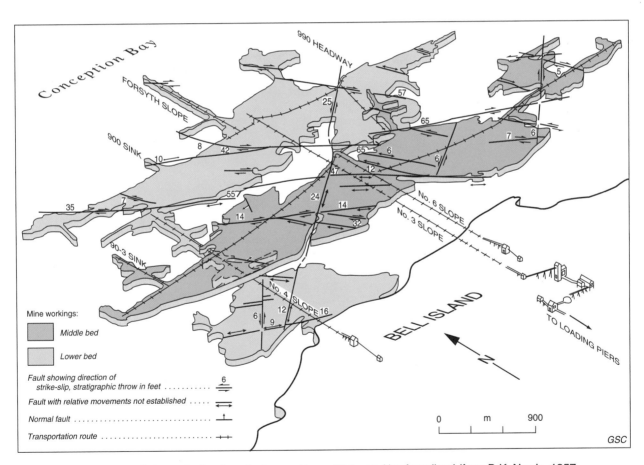

Figure 3.3-3. Schematic drawing of mine workings, Wabana, Newfoundland (from D.K. Norris, 1957, GSC; courtesy Dominion Wabana Ore Limited).

The typically high content of phosphorus in ironstones suggests that some facies of ironstone might be used as metallogenetic markers for locating beds and nodules rich in phosphate.

Manganese-rich facies stratigraphically equivalent to ironstone have not been described as such and apparently are not an important source of manganese. This suggests further differences between ironstone and iron-formations (see iron-formations and residual-enriched iron ore subtypes 3.1, 3.2, and 4.1).

Ironstones provide a source of ferrous iron in the preparation of fertilizer products for application in tropical areas where the iron is oxidized by weathering processes and the soils become depleted in essential ferrous iron compounds.

EXPLORATION GUIDELINES

1. Most of the ironstone facies selected for ore should be sought in strata which record neritic to estuarian environments, have a predominance of chemically precipitated material, a minimum of clastic material, and distinctive mineral assemblages and textures that are conducive to beneficiation and smelting.

2. Ironstone facies suitable for iron ore usually contain 30% or more iron, and the acceptable ranges of contents of other constituents have been indicated above.

3. The bulk of the iron should be in discrete coarse granular aggregates of iron oxide and carbonate minerals, with a minimum of associated fine grained admixtures of clay and complex iron-silicate or carbonate minerals.

4. Ironstone selected for ore should consist of well defined, uniform lithofacies that are relatively free from interbedded shale and other kinds of sediment, and have an adequate thickness to meet requirements for mining.

5. Intense folding, steep dips, and excessive structural disruption adversely affect mining feasibility.

6. Location of the ore with respect to the smelter site or shipping routes is important, as a large proportion of the cost of bulk commodities is usually attributable to transportation.

7. Phosphate minerals usually associated with organic debris, and other constituents, such as sulphide minerals, which necessitate a high slag volume in the furnaces, should be kept to a minimum in the ore.

ACKNOWLEDGMENTS

This paper was reviewed by R.T. Bell, R.I. Thorpe, and C.W. Jefferson. Assistance in word processing was provided by Lara O'Neill and Carol Plant. Figures were expedited by R. Lancaster.

REFERENCES

References marked with asterisks (*) are considered to be the best sources of general information on this deposit subtype. For additional references, see "Introduction", Type 3.

Bubenicek, L.
1961: Recherches sur la constitution et la repartition des minerais de fer dans l'Aalenien de Lorraine; Sciences de la Terre, Tome VIII, Nancy, France, 204 p.
*1971: Géologie du gisement de fer de Lorraine; Bulletin du Centre de Recherche Pau-SNPA, v. 5, p. 223-320.

*Cotter, E. and Link, J.E.
1993: Deposition and diagenesis of Clinton ironstones (Silurian) in the Appalachian Foreland Basin of Pennsylvania; Geological Society of America, Bulletin, v. 105, p. 911-922.

Gross, G.A.
1965: General geology and evaluation of iron deposits; in Geology of Iron Deposits in Canada: Volume I, Geological Survey of Canada, Economic Geology Report 22, 181 p.
*1967a: Iron deposits in the Appalachian and Grenville regions of Canada; in Geology of Iron Deposits in Canada: Volume II, Geological Survey of Canada, Economic Geology Report 22, 111 p.
1967b: Iron deposits of the Soviet Union; The Canadian Institute of Mining and Metallurgical Bulletin, December, 1967, 6 p.

*Hayes, A.O.
1915: Wabana iron ore of Newfoundland; Geological Survey of Canada, Memoir 78, 163 p.

Hutchinson, R.D.
1953: Geology of Harbour Grace map-area, Newfoundland; Geological Survey of Canada, Memoir 275, p. 35.

Jenkyns, H.C.
1970: Submarine volcanism and the Toarcian iron pisolites of western Sicily; Eclogae Geologicae Helvetiae, Basle, v. 63, no. 2, p. 549-572.

Kimberley, M.M.
1979: Origin of oolitic iron formations: Journal of Sedimentary Petrology, v. 49, p. 111-132.
*1994: Debate about ironstone: has solute supply been surficial weathering, hydrothermal convection, or exhalation of deep fluids?; Terra Nova, v. 6, no. 2, p. 116-132.

*Lyons, J.C.
1957: Wabana iron ore deposits; in Structural Geology of Canadian Ore Deposits (Congress Volume), The Canadian Institute of Mining and Metallurgy, Montreal, Quebec, p. 503-516.

Mellon, G.B.
1962: Petrology of Upper Cretaceous oolitic iron-rich rocks from northern Alberta; Economic Geology, v. 57, p. 921-940.

Rose, E.R.
1952: Torbay map-area, Newfoundland; Geological Survey of Canada, Memoir 265, 64 p.

*Taylor, J.H.
1949: Petrology of the Northhampton Sand Ironstone Formation; Memoirs of the Geological Survey of Great Britain, London, 111 p.

*Teyssen, T.A.L.
1984: Sedimentology of the Minette oolitic ironstones of Luxembourg and Lorraine: a Jurassic subtidal sandwave complex; Sedimentology, v. 31, p. 195-211.

Author's address

G.A. Gross
Geological Survey of Canada
601 Booth Street
Ottawa, Ontario
K1A 0E8

Printed in Canada

4. RESIDUALLY ENRICHED DEPOSITS

4.1 Enriched iron-formation

4.2 Supergene base metals and precious metals

4.2a Supergene zones developed over massive sulphide deposits

4.2b Oxidation zones developed in the upper parts of vein, shear/fault-related, and replacement deposits

4.2c Supergene oxide and sulphide zones formed over porphyry Cu-Mo-(Au, Ag) deposits

4.3 Residual carbonatite-associated deposits

4. RESIDUALLY ENRICHED DEPOSITS

INTRODUCTION

Residually enriched deposits are sedimentary concentrations of commodities achieved by in situ weathering of a suitable precursor rock. By the mechanism of uplift, surficial exposure, and weathering of the precursor rock, the original low concentrations of the commodities of interest become enriched to economic or improved economic grades in residual deposits that overlie, or are adjacent to, the parent rocks. The enrichment is achieved by two separate processes, depending on the solution characteristics of the protore minerals. If the protore minerals are resistant to chemical weathering, these minerals become concentrated through dissolution and removal of the surrounding matrix; this can be termed "concentration of resistates". If the protore minerals are relatively soluble in oxygenated surficial waters, but insoluble in reduced solutions, the enrichment proceeds by dissolution of the protore minerals in the oxidizing zone and reprecipitation, generally as new ore minerals, in the lowest part of the oxidizing zone and/or in the underlying reduced environment; this process is commonly referred to as "supergene enrichment".

Three principal types of residually enriched deposits, reflecting different types of precursor rocks and processes of enrichment, are distinguishable in Canada. Residually enriched ore deposits of iron (subtype 4.1, "Enriched iron-formation") have resulted from the actions of both concentration of resistates and supergene enrichment operating on Lake Superior-type and Algoma-types of iron-formation (subtypes 3.1, 3.2) as protore. Secondary deposits of base metals and precious metals (subtype 4.2, "Supergene base metals and precious metals") are the product mainly of supergene enrichment acting on several sulphidic protore types including massive sulphides (subtype 6.4), various vein and fault related deposits, and porphyry Cu-Mo-Au-Ag deposits (Type 19). Deposits of niobium, phosphate, and rare-earth elements (subtype 4.3, "Residual carbonatite-associated deposits") are the result of weathering of carbonatite, and enrichment of the ore elements by both concentration of resistates and supergene enrichment.

4.1 ENRICHED IRON-FORMATION

G.A. Gross

INTRODUCTION

Main geological characteristics and commodities

Residually enriched iron deposits are large, tabular and irregular in situ masses of high grade direct-shipping iron and manganese ore hosted in iron-formation protore. The iron ore consists of hard lumpy fragments and earthy masses of friable and porous, fine grained, granular to dusty iron oxide minerals. The ore consists mainly of residual and secondary fine grained mineral aggregates of hematite and goethite which were derived from iron-formation protore and iron-rich rocks by oxidation, leaching, and deep weathering processes. They are associated with ancient regoliths and were formed by the action of circulating groundwater that penetrated along fractures, faults, and porous permeable zones in the iron-formation. Silica, carbonate, and gangue minerals were removed leaving residual masses and rich concentrations of ferric iron oxides.

Zones of disseminated manganese oxide in the residual iron deposits provide manganiferous iron ore, and pockets of rich manganese ore occur within, or in association with, the residually enriched iron deposits. Most of the deposits of high grade manganese ore throughout the world are hosted in manganese-bearing lithofacies of iron-formation and formed by secondary enrichment processes.

Some of the common characteristic features of residually enriched iron deposits that have been derived from Lake Superior-type iron-formation protore in the Knob Lake iron ranges of Quebec and Labrador are described in this paper.

Examples

The most important residual iron deposits in Canada occur in Lake Superior-type iron-formation in the Knob Lake iron ranges in the Labrador-Quebec fold belt near Schefferville, Quebec (Fig. 4.1-1). Other large well known deposits of this type occur in the Mesabi Range of Minnesota, in the Marquette Range in Michigan, in the Belo Horizonte area of Brazil, at Mount Whaleback in northwest Australia, in India, near Krivoy Rog in the Ukraine and Kursk in Russia, and in South Africa.

Gross, G.A.
1996:   Enriched iron-formation; in Geology of Canadian Mineral Deposit Types, (ed.) O.R. Eckstrand, W.D. Sinclair, and R.I. Thorpe; Geological Survey of Canada, Geology of Canada, no. 8, p. 82-92 (also Geological Society of America, The Geology of North America, v. P-1).

Other world-class examples derived from oxide lithofacies of Algoma-type iron-formation include Cerro Bolivar in Venezuela (Ruckmick, 1962), Carajas in northern Brazil, Nimba in Liberia, and large deposits in Swaziland. The Steep Rock Lake deposits in Canada appear to have formed by leaching and enrichment of a carbonate-sulphide lithofacies of Algoma-type iron-formation (Table 4.1-1).

IMPORTANCE

Large high-grade residually enriched iron ore deposits are the principal resource base for modern iron and steel industries in North and South America, Eastern Europe, Japan, India, and Africa, and have become an important source of ore in recent years for Western Europe and China. Approximately half of the world iron ore production in recent years, including lump ore and sinter fines, have come from residually enriched iron-formation.

Mine development in the Knob Lake-Schefferville area in the Labrador-Quebec fold belt of Canada was based on residually enriched deposits. The maximum annual production during the operation of the mines in this area from 1954 to 1984 was achieved in 1959 when more than 12 million tonnes of direct shipping ore was produced.

SIZE AND GRADE OF DEPOSITS

Individual deposits in the Knob Lake iron range of Labrador and Quebec are relatively small and the largest contain about 50 Mt (million tonnes) of ore. Deposits of this type in Australia and Brazil range from 100 Mt to more than 1000 Mt. Residually enriched deposits contain from 45 to 69% iron, overall grades average from 50 to 60% iron, and many deposits have large mineable sections that contain from 65 to 69% iron (Table 4.1-1.)

GEOLOGICAL FEATURES

Geological setting

Residually enriched iron ore deposits occur in lithofacies of Lake Superior- and Algoma-type iron-formation in most of the major iron ranges of the world (see Lake Superior- and Algoma-type iron-formation, deposit subtypes 3.1 and 3.2). Prominent geological features of residually enriched deposits in Lake Superior-type iron-formation are illustrated in the Labrador-Quebec fold belt (Fig. 4.1-1), and have been described in detail by Gross (1968) and Gross et al. (1972).

Stratigraphy of protore iron-formation

More than 50 residually enriched deposits of iron ore are hosted in the Sokoman Formation which has a maximum thickness of about 150 m in the Knob Lake basin in the

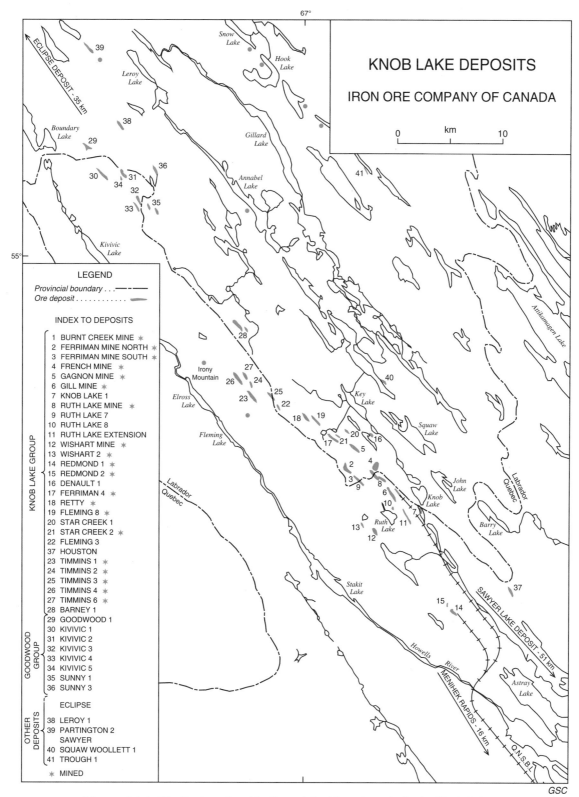

Figure 4.1-1. Distribution of residually enriched iron deposits in the Knob Lake iron range, Quebec and Labrador (after Gross et al., 1972).

central part of the belt (Fig. 4.1-1). It is part of an Early Proterozoic sequence of continental shelf sediments outlined in Table 4.1-2.

Typical lithofacies of protore iron-formation near the French mine, which hosts residual ore, are described in Table 4.1-3, and their composition is given in Table 4.1-4. Typical oolitic textures in protore oxide lithofacies of iron-formation from the northwestern part of the Knob Lake Basin are shown in Figure 3.1-6 (see subtype 3.1, Lake Superior-type iron-formation).

Table 4.1-1. Iron content in iron ore resources in major deposits of enriched iron-formation (Bolis and Bekkala, 1986).

COUNTRY/DEPOSIT	% IRON	RESOURCES IN SITU (Mt)
AUSTRALIA		
Mount Tom Price	61.9	650
Mount Whaleback	61.4	1676
Paraburdoo	63.4	450
Robe River	57.0	140
Rhodes Ridge	61.8	1000
Wittenoom	54.9	1000
Yandicoogina	58.5	1243
BRAZIL		
Aguas Claras	62.2	260
Carajas	66.1	1340
Caue	56.1	620
Conceicao-Dos Corregos	66.6	1050
Timbopeba	66.6	170
GABON		
Belinga	63.9	514
INDIA		
Bailadila #5	64.4	209
Bolani	58.9	480
LIBERIA		
Nimba	59.1	62
Western area	52.2	411
SENEGAL		
Faleme area	63.6	340
SOUTH AFRICA		
Sishen	64.0	1314
VENEZUELA		
Altamira	63.1	130
Cerro Bolivar	63.1	187
San Isidro	64.1	391

Structure in main ore zone, Knob Lake Basin

The central part of the Knob Lake Basin area has been highly deformed by doubly plunging isoclinal folds, with axes trending northwest, that are cut by steep easterly dipping thrust faults which give rise to a complex imbricate regional structure. Cross faults strike east to northeast and have vertical displacements of 20 m or less. Many faults of both sets were reactivated during Cretaceous or later time, resulting in further thrusting from the northeast and causing brecciation in some pockets of ore and overlapping of Cretaceous clay and gravel or rubble ore by unaltered strata (Fig. 4.1-2).

Form and relationship of residual ore and host iron-formation

Deep residual enrichment of the iron-formation took place along the height of land that now forms the watershed between streams flowing to the Atlantic coast and northward or westward to Ungava and Hudson Bay. Unlike the surrounding area, this highland area probably was not covered by Paleozoic rocks and may have been exposed to deep chemical weathering and erosion from middle Proterozoic to relatively recent times.

Residually enriched ore was developed in lenticular, tabular, or irregular masses in the cherty iron-formation protore under stratigraphic and structural control, where ground water could circulate in permeable fold, fault, and fracture zones. The distribution of ore bodies along the height of land in the Knob Lake area is shown in Figure 4.1-1. Typical relationships of ore to stratigraphy and to the folded and faulted structures are shown in a cross-section of the Ferriman mine in Figure 4.1-3.

Ore types, mineralogy, and textures

Residually enriched ore in the Knob Lake deposits consists of red to blue, fine grained hematite, yellow to brown or black goethite, and mixtures of these main mineral aggregates. Porosity, density, and the relative proportions of dense lumpy material and dusty granular fine grained ore of clay-like consistency may vary greatly in the different types of ore. The physical and chemical properties of the prominent types of ore in the Knob Lake Range are shown in Table 4.1-5. The mineralogy and physical properties of the main types of residual ore reflect the lithofacies of iron-formation from which they were derived. Relict bedding and evidence of primary sedimentary features are preserved in some of the ore zones (Gross, 1964a, b, 1968).

Blue ore was derived from cherty oxide lithofacies of iron-formation and is composed of fine grained, blue to dark grey-black hematite with lesser amounts of red hematite, martite, and brown goethite. It has a distinctly porous granular texture and contains primary ovoid granules or oolites of iron oxide replaced by later hematite or goethite, or enlarged by overgrowths of iron oxide. Blue ore may be friable to crumbly or nodular, or hard coherent and lumpy where a high proportion of secondary iron oxide cements relict fragments. Silica remaining in the ore after leaching and oxidation is mostly very fine grained, less than 100 mesh in size, with a sugary to equigranular texture.

Table 4.1-2. Stratigraphy of the Knob Lake Group, Schefferville area, Labrador and Quebec.

Menihek Formation	carbonaceous slate, shale, quartzite, greywacke, mafic volcanic rocks; minor dolomite and chert.
Purdy Formation	dolomite, developed locally.
Sokoman Formation	oxide, silicate, and carbonate lithofacies iron-formation, minor sulphide lithofacies iron-formation, interbedded mafic volcanic rocks, ferruginous slate, slaty iron-formation, black and brown slate, and carbonaceous shale.
Wishart Formation	feldspathic quartz arenite, arkose, minor chert, greywacke, slate and intercalated mafic volcanic rocks.
Fleming Formation	chert breccia, thin-bedded chert, limestone; minor lenses of shale and slate.
Denault Formation	dolomite and minor chert.
Attikamagen Formation	green, red, grey, and black shale, and argillite, interbedded with mafic volcanic rocks.

Table 4.1-3. Iron-formation lithofacies, French mine, Knob Lake range.

Lithofacies	Description of lithofacies	
Thickness of section sampled	Megascopic	Microscopic
B523 --- 15 m Silicate-carbonate	Olive-green to brown with orange-red to brown weathered surface, fine grained magnetite, iron-silicates and chert, beds 1-5 cm thick with laminae <5 mm thick.	Microbanding <1 mm thick, dense felty mass of minnesotaite with ovoid granules, granules are sheared or distorted in some microbeds, minor quartz and carbonate.
B524 --- 12 m Lower red chert	Bright red jasper and grey-blue hematite chert in bands, stubby lenses, laminae, nodules, 5 to 20 mm thick, granules and oolites of chert and hematite in chert matrix, dense blue-black hematite beds and fine grained specular hematite.	Hematite and chert in ovoid granular to oolitic texture, chert granules rimmed by coarser hematite, coarser quartz patches in matrix and cores of granules and oolites.
B525 --- 8 m Pink chert	Thin (5-10 mm) banded pink chert and hematite interbedded with blue-grey hematite-rich beds, 5-20 mm thick, some wavy laminated beds with coarse granules, some brown chert beds.	Coarse granular, oolitic and nodular chert with patches of coarse grained quartz and hematite, minor brown iron-oxide.
B526 --- 15 m Grey chert	Mainly pink to grey or brown thin banded <5 mm, to crudely laminated or lenticular beds, some massive beds 5-15 cm thick. Chert beds and iron-rich beds are well differentiated.	Medium grained quartz with fine 0.5 mm granules, granular to oolitic textures in some beds, hematite in discrete grains.
B527 --- 6 m Brown chert	Brownish grey to pink jasper facies in lenticular beds 5 to 10 cm thick, blue to brown iron-rich beds, coarse granular texture, nodes and lenses of pink or brown jasper 10 mm thick.	Pure hematite, goethite and chert, oxide facies, jasper nodules, coarse hematite and secondary goethite.
B528 --- 20 m Upper red chert	Thick massive beds <30 cm thick with patches of blue to grey-pink, iron-rich beds interbedded with banded, lenticular, and nodular jasper, 1 to 2 cm thick, magnetite- and hematite-rich lenses in granular jasper.	Granular textured hematite, magnetite, and chert, minor goethite and minnesotaite, coarse patches of quartz in matrix and cores of granules.
B529 --- 25 m Grey upper chert	Grey-green magnetite-carbonate-chert with blue to brown hematite-goethite-rich beds, spotty carbonate nodes in grey chert and iron oxide beds, granular texture in jasper beds, metallic lustre on hematite beds.	Hematite, magnetite, and goethite in coarse chert, fine chert in cores of granules, and coarse grained chert in matrix, hematite grains on rims of granules, brown iron oxide replaces hematite, siderite, and iron-silicate in green chert.

Table 4.1-4. Chemical analyses of samples of iron-formation lithofacies, French mine, Knob Lake range, Labrador and Quebec (from Gross, 1968).

B523 - Silicate-carbonate facies	B527 - Brown cherty facies
B524 - Lower red cherty facies	B528 - Upper red cherty facies
B525 - Pink cherty facies	B529 - Grey upper cherty facies
B526 - Grey cherty facies	

Sample	B523	B524	B525	B526	B527	B528	B529
SiO_2	49.41	41.42	48.16	51.24	43.77	49.01	56.49
Al_2O_3	0.68	0.79	0.53	0.42	0.42	0.37	0.37
Fe_2O_3	16.34	54.49	46.96	41.97	49.85	44.50	38.10
FeO	24.19	1.35	1.50	3.25	2.27	3.65	1.99
CaO	0.02	0.00	0.01	0.00	0.00	0.00	0.00
MgO	2.95	0.37	0.31	0.62	0.37	0.19	0.00
Na_2O	0.03	0.08	0.03	0.02	0.02	0.03	0.02
K_2O	0.07	0.01	0.01	0.01	0.01	0.01	0.01
H_2O+	5.20	0.98	2.04	2.10	2.54	1.94	2.42
H_2O-	0.38	0.06	0.04	0.05	0.05	0.02	0.03
TiO_2	0.00	0.00	0.00	0.00	0.00	0.00	0.00
P_2O_5	0.08	0.04	0.04	0.03	0.04	0.05	0.04
MnO	0.65	0.02	0.02	0.02	0.03	0.03	0.03
CO_2	0.22	0.02	0.02	0.06	0.02	0.04	0.04
S	0.05	0.05	0.03	0.00	0.00	0.00	0.00
C	0.15	0.12	0.10	0.08	0.13	0.04	0.03
Total	100.42	99.80	99.80	99.87	99.52	99.88	99.57

Analysts: G.A. Bender and W.F. White, Geological Survey of Canada;
Spectrographic analyses, composition of samples within these ranges: Co- 0.1-1.0%, Ni- <0.01%, Ti- <0.01%, Cr-0.01-0.1%, Cu- <0.01%, Ba- <0.01%, B- not detected

Beds originally rich in iron oxide minerals usually remain hard and relatively unaltered and have a distinct blue metallic lustre. Ore derived from oxide lithofacies commonly has a red to pink or grey cast, depending on the predominant colour of the cherty protore beds. Much of the blue hematite, martite, and goethite form vermicular to irregular microscopic intergrowths, and some are euhedral and fine grained. Partial leaching of the equigranular blue ore usually results in a sandy friable mass of well segregated silica and iron-oxide grains, and many lean or partially leached lithofacies of iron-formation can be beneficiated by simple washing and gravity processes.

Yellow to brown ore, referred to as "SCIF" ore, was derived from silicate-carbonate iron-formation lithofacies and is composed mainly of goethite or hydrous iron oxide minerals and dark brown martite. It is earthy to ochreous and has a high proportion of clay-size particles. Textures vary from fine colloform intergrowths of brown or yellow goethite, to crudely banded fragments with patchy needle-like or radial intergrowths of pseudomorphous iron oxide developed over the textural features of the protore iron-silicate and carbonate-chert lithofacies. Silica from the chert or iron-silicate minerals is intimately mixed with iron oxide in yellow ore and has a spongy clay-like cohesion when squeezed in the hand. Porosity is high in this ore and it has a high moisture content due to adsorption of water on the earthy clay particles. Silica is readily removed from minnesotaite-carbonate-chert lithofacies of iron-formation by weathering processes, and felty textured iron-silicate and siderite mineral aggregates are the first to show signs of oxidation and leaching. The yellow ore is much more difficult to beneficiate than the blue.

Red ore is composed of earthy red hematite, goethite, soft aluminous silicates, and very fine grained chert or quartz. It is derived from the upper iron-rich part of the Ruth slate at the base of the iron-formation and from black slaty facies within it. Individual layers consist of dense earthy hematite or nodular to colloform hematite intergrowths, and most of the red ore is soft, spongy, and clay-like and has high porosity and moisture content. Red, blue, and blood-red laminae give the ore a distinct banded appearance. When silica, calcium, magnesium, and iron are leached from fine grained black shale and slate, the primary beds and banding stand out in the red ore as punky, soft saprolitic alumina-rich clay layers that contain no appreciable amount of iron. These leached slaty beds vary

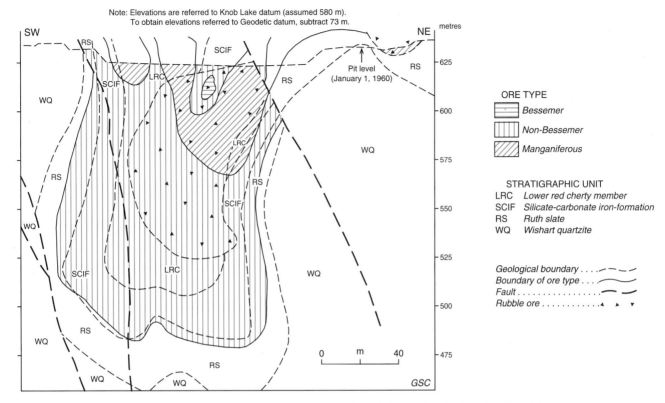

Figure 4.1-2. Typical cross-section of Ruth Lake 3N deposit, Burnt Creek mine, showing distribution of ore (from Iron Ore Company of Canada).

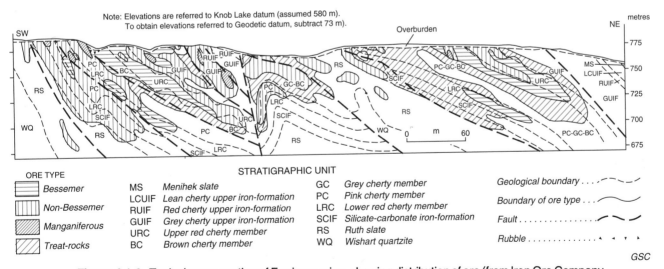

Figure 4.1-3. Typical cross-section of Ferriman mine, showing distribution of ore (from Iron Ore Company of Canada; Gross, 1968).

Table 4.1-5. Physical and chemical properties of typical iron ores, Knob Lake range (data from Stubbins et al., 1961).

Type of ore Relative abundance in mines	Blue 66%	Yellow 20%	Red 14%
Average composition of natural material (analyses in %)			
Fe	54.5	50.0	48.0
P	0.038	0.103	0.123
Mn	0.56	1.13	2.41
SiO_2	8.2	5.6	6.2
Al_2O_3	0.72	1.0	2.1
Moisture	8.9	14.2	13.7
Critical moisture	9-11	15-16	15-18
Porosity in % of volume	10-44	19-49	7-47
Average porosity	31	36	30
Permeability	Medium	Low	Low
Specific gravity	3.2	2.6	2.8
Tonnage factor, m^3/t	0.096	0.120	0.108
Typical sieve analyses, dry screening (%)			
+1 inch	19	22	20
+6 mesh	26	35	14
+35 mesh	20	25	26
+100 mesh	15	9	9
-100 mesh	20	9	11
Ore minerals	Blue hematite goethite magnetite	Geothite limonite	Earthy red hematite goethite limonite
Amenability to concentration by washing			
	Medium to high	Very low	Very low

Table 4.1-6. Iron Ore Company of Canada, mine production, 1959, Knob Lake Iron Range (from Elver, 1960; Gross, 1968).

Kind of ore Long tons	Analyses of unprocessed ore from 5 mines					
	Fe Iron	P_2O_5	Mn	Silica	Al_2O_3 Alum	Moisture
Bessemer 1 991 104	55.7	0.026	0.319	6.64	0.84	8.73
Non-Bessemer 9 179 134	51.4	0.083	1.104	6.55	1.46	11.8
Manganiferous 1 451 019	47.1	0.108	4.69	6.44	1.5	11.5
Total all mines 12 621 256	51.6	0.077	1.39	6.56	1.36	11.29

from white to pink or deep red and are known as "paint rock". The intimate intergrowths of iron oxide minerals and aluminous clay or silica make the red ore difficult to beneficiate.

Rubble ores (solid triangle pattern in Fig. 4.1-2), the fourth main type in the Knob Lake iron range, are similar to the canga and talus deposits developed in tropical areas. They overlie some of the banded residual ore deposits and consist of irregular masses and pockets of clastic ore fragments, bedded clay, and rock clasts and gravel, and in some deposits contain an abundance of fossil wood and plant debris.

Three main kinds of rubble ore, which consist of sand and gravel, talus, and breccia ores, were contemporaneous in their development and grade from one type to another in some of the mine pits. They represent reworked parts of ore bodies that formed at an earlier stage and contain fossil wood, leaves, and insects considered to be early Late Cretaceous (Dorf, 1959). Rubble ore is usually uniform, high grade material mostly of non-Bessemer quality (0.045 to 0.18% P).

Sand and gravel ores are mainly well sorted, unconsolidated sediments that occur in the upper parts of ore deposits. They are composed of subangular ore fragments, mostly 5 cm to 20 cm in size, which are distributed in a matrix of sandy ore, chert, and quartz grains. Layers and crossbeds dip toward the centre of the rubble pockets and rubble ore deposits may be more than 100 m thick and have no topographic expression.

Sand and gravel rubble ores grade downward to talus ores, which are poorly sorted and consist of angular to subangular blocks of bedded ore from all parts of the iron-formation sequence that have slumped into fissures and crevices along fault scarps. Breccia ores are developed along fault zones and are composed of broken layered ore fragments less than a centimetre to several metres in size cemented by sandy material. Breccia ore contains vugs filled with colloform secondary iron and manganese oxide minerals or quartz.

The margins of enriched ore bodies are usually irregular, defined by cut-off grades, composition, and textural criteria, and extend transitionally into altered iron-formation from which only part of the silica has been leached. Some beds and laminae are more extensively oxidized and leached than others and broad marginal zones around enriched ore bodies are composed of friable, sugary-textured iron-formation, silica, and iron oxide sand. The marginal, partially leached oxide facies of the iron-formation are usually amenable to beneficiation by removal of the silica grains by washing, and concentration of the iron oxide mineral grains and oolites by flotation and gravity processes.

This partly leached iron-formation has been termed "treat rock" or "wash ore" and provides an important fifth type of ore derived from residually enriched iron-formation. Several million tonnes of iron ore concentrate were produced annually during the later years of mining on the Knob Lake range from treat rock containing 46 to 56% iron.

The various types of deposits reflect the different lithofacies of iron-formation from which they were derived by oxidation, leaching of silica, and enrichment of iron and manganese by the accumulation of residual and secondary oxide minerals.

Table 4.1-7. Compositions of typical residually enriched iron ore mined in Canada and the United States.

Composition in per cent			
	1	2	3
Fe=metal equivalent	54.0	57.45	54.0
P	0.064	0.02	0.02
SiO_2	7.94	5.13	8.45
Mn	0.17	0.17	0.52
Al_2O_3	1.01	1.01	1.5
H_2O	3.17	3.17	8.5

1/ Average for 3 Mt of direct shipping non-Bessemer quality ore mined in the Schefferville area in the Knob Lake range in 1981.
2/ Residually enriched ore, including both coarse and fine material, probably derived from carbonate-sulphide lithofacies protore, from the Caland mine in the Steep Rock iron range in Ontario.
3/ Average for 35 Mt of typical residually enriched iron ore mined in 1962 in the Lake Superior area of United States, mostly from the Mesabi range.

Residually enriched ore deposits are related to present and paleotopographic features. They were developed in permeable zones in the iron-formation protore in fold, fault, and fracture zones where circulating groundwater caused oxidation, leaching, and enrichment of the host rocks in iron and manganese. The compositions of typical residually enriched iron ore mined in Canada in the past are given in Table 4.1-7. Table 4.1-6 shows the average composition and proportions of Bessemer (<0.045% P), non-Bessemer (0.045 to 0.18% P), and manganiferous ore produced from the Gagnon, French, Ferriman, Ruth Lake, and Burnt Creek mines in the Knob Lake range in 1959, prior to the processing of treat rock ore.

Definitive characteristics of residually enriched iron ore in iron-formation

Residually enriched iron ore deposits hosted in iron-formation protore are irregular, lenticular or tabular masses of hematite and goethite. The compositions and proportions of hard lump and fine grained dusty ore vary from deposit to deposit and are directly related to the kind and quality of protore lithofacies from which they have been derived (Gross, 1968; Gross et al., 1972) (Table 4.1-2, 4.1-3, 4.1-4).

Blue ores are lumpy to granular aggregates of blue to dark grey and red hematite, martite, and brown goethite that have been derived mainly from oxide lithofacies of the protore iron-formation. In some cases, they retain relict oolitic textures pseudomorphous after the textures of the protore. Yellow to brown ores are earthy to ochreous to clay-like goethite and hydrous iron oxide minerals, that were derived from silicate-carbonate lithofacies of protore. Red ore consists of earthy red hematite, goethite, soft aluminous silicate minerals, clay minerals, and fine grained quartz derived from iron-rich slate.

Rubble ores are accumulations of alluvial and residual fragments and fine clastic material composed of enriched and weathered iron-formation, fossil carbon and plant debris, sand, silt, and gravel. They may represent brecciated, reworked, or residual accumulations of the various types of residually enriched iron ore and commonly have a higher content of manganese than the iron-formation hosting them.

The compositions of typical residually enriched iron ores mined in North America in the past are shown in Table 4.1-7. Most of the iron ore resources of this type in countries outside of North America contain significantly higher amounts of iron with less silica and other constituents as shown in Table 4.1-1.

GENESIS OF RESIDUALLY ENRICHED IRON ORE IN IRON-FORMATION

The soft "direct-shipping" type of hematite-goethite residually enriched iron ore was derived from iron-formation as a result of two principal processes: 1) leaching of silica, carbonate, and gangue minerals, and 2) enrichment of iron by oxidation, residual accumulation, and deposition of goethite, hematite, and other iron oxide minerals. Leaching and enrichment processes were carried out by groundwater circulating in deep porous zones along faults, fractures, and folds, and by weathering processes under tropical climatic conditions. As a result of these processes the protore iron-formation lithofacies were reconstituted. The structures, primary textures, and sedimentary features of the host iron-formation were almost completely destroyed and secondary iron and manganese oxides were deposited with the residual iron oxide minerals in the weathered and leached protore.

Consistent relationships between the blue, yellow, and red types of ore and protore lithofacies from which they were derived were clearly defined in the Knob Lake range by geological mapping of the mine areas and zones of oxidation and enrichment in the protore iron-formation. Removal of silica by leaching took place in advance of secondary iron enrichment, as shown by a lateral transition along beds from protore through zones of partially leached friable protore to dense, hard, enriched masses of iron oxide minerals and iron ore. Goethite forms a matrix around hematite, chert, and iron-oxide granules in the hard, lumpy, dense ore, and is present in veins and discordant masses. Selective alteration of carbonate and iron-silicate minerals to goethite, and oxidation of magnetite to martite, are usually the first changes detected in the conversion of protore iron-formation to ore and indicate that leaching took place in strongly oxidizing environments.

Controversy in the past has centred on whether the residually enriched iron ores were formed by the action of descending groundwater, hydrothermal solutions, or meteoric waters mixed with ascending emanations from deep seated, probably magmatic sources. Several different lines of evidence lead to the conclusion that the residual iron deposits on the Knob Lake range formed near a surface of considerable topographic relief by the action of descending groundwater, and the environmental conditions at the time of their genesis appear to have been similar to those observed in deposits forming today in temperate, warm, or tropical climates.

The solubility of silica increases directly with increasing temperatures and with an increase in pH above 8.9, but ferric iron is relatively insoluble and is particularly stable under conditions of high pH and Eh. Weathering processes in most tropical environments remove silica and leave residual iron. When humic acids are present in the groundwater, the ratio of iron to silica in the solutions increases at temperatures below 20°C and decreases above this temperature. Higher temperatures and an increase in the activity of bacteria cause the decomposition of humic acids. The organic constituents formed by the decomposition of the humic acids are effective in leaching silica from soils, but ferric iron is not affected significantly. Evidently, temperature fluctuations have been a major factor in controlling the content of humic acids and the leaching and transportation of silica and iron.

Humic acids appear to have been an important media for solution and deposition of the secondary goethite and enrichment of iron in this type of ore deposit. Ferric oxides and hydroxides are relatively soluble in humic acids and in many other organic compounds, and with increasing pH or saturation in these acids they coagulate and are precipitated as amorphous hydrous iron oxide.

Interpretation of climatic conditions in the Knob Lake area from fossil plants (Dorf, 1959) indicate a warm, humid environment with abundant vegetation in the Late Cretaceous. These conditions would have been suitable for both organic and inorganic solution and transportation of iron and silica and may have existed throughout the Mesozoic. Ruckmick (1962) suggested that the Cerro Bolivar deposit in Venezuela formed in about 26 million years, or since the Oligocene, if climatic conditions were similar to those at present and had prevailed during that period in the past. Springs emerging from the lower flanks of the ore bodies have a pH of 6.1 and contain an average of 7 ppm silica and 0.05 ppm iron, indicating that the rate of removal of silica is about 80 times greater than that for ferric iron.

Study of the rubble ore pockets in the Knob Lake range show that much of the enrichment and leaching in the development of the ore may have taken place prior to Cretaceous time but relatively little definitive work has been done on the timing of formation of the ore in the Knob Lake iron range. It is highly possible that major enrichment of the iron-formation was related to deep weathering and regolith development at the close of the Precambrian as observed in many other parts of Canada.

Morris (1980) referred to work on several major deposits which indicated that deep weathering and/or enrichment processes may have started in the Precambrian, probably as early as 1800 Ma for the Tom Price and Whaleback deposits in Australia, that the hematite ores of Nimba Range, Liberia, formed in the Precambrian, and that deposits at Sishen and in the Manganore iron-formation in the Postmasburg Group in South Africa probably were formed in the early to middle Proterozoic. Ore zones in the Krivoy Rog iron range related to Precambrian paleotopography extend to depths greater than 2000 m (Belevtsev, 1973) and evidence from microspores show that deep weathering processes were active in the late Precambrian.

The sequence of events leading to the development of the residually enriched iron ore deposits on the Knob Lake range in Labrador and Quebec are summarized as follows:

1. Deposition of the Sokoman Formation in the Knob Lake Basin on a continental shelf along the margin of the Ungava Craton coincided with deep faulting and deformation in the offshore tectonic belt, and probably took place during the initial stages of subduction of the craton under the eastern crustal block.

2. Folding and faulting of the Labrador-Quebec fold belt, probably associated with subduction and orogenic processes between 1400 and 1900 Ma, produced permeable structures in the iron-formation.

3. Erosion of the fold belt, accompanied by deep weathering and ore forming processes, was followed by submergence and possibly deposition of Paleozoic marine sediments.

4. Rugged highland topography was developed during the Permian and Mesozoic when further uplift and sufficient erosion exposed the Knob Lake range and most of the shelf strata of the fold belt to weathering in a warm and arid climate.

5. Deep chemical weathering developed in this highland belt, probably before and during the Mesozoic. Evidently groundwater penetrated from surface along permeable structures and produced extensive oxidized zones that were associated with leaching of silica and carbonate minerals, hydration and martitization of iron oxides, deposition of secondary goethite, and redistribution of iron oxide in residual zones near the surface.

6. During the Late Cretaceous an abundance of vegetation developed in a warm, humid, tropical climate; about the same time, recurrence of reverse movement on easterly dipping thrust faults, reactivation of cross-faults, and downfaulting of some fault blocks resulted in brecciation of residual iron ore along fault zones.

7. Rubble ore derived by sloughing and erosion from the rising blocks and along the reactivated fault scarps accumulated in the downfaulted blocks, and extensive secondary enrichment and redistribution of goethite and manganese oxide minerals took place in the rubble ore and upper permeable parts of the ore deposits.

8. Further enrichment of manganese oxides has taken place in porous and permeable zones in the residual iron deposits since the Cretaceous.

9. Glacial erosion during the Pleistocene, the last major event identified, appears to have removed most of the weathered rock, plucked off most of the hard surface cappings and canga from the orebodies, and scoured the ridges to expose fresh, unaltered rock around the ore deposits. The ore bodies remain as parts of the deeply leached and altered zones that were protected by topographic and structural features and subsequently covered by glacial till and soil.

RELATED TYPES OF DEPOSITS

Residually enriched iron deposits are associated with unaltered iron-formation and taconite deposits, with partially leached and enriched iron-formation which provides wash ore and treat rock, and with manganiferous iron ore and manganese deposits of various grades and quality.

The friable and sandy, partially leached and enriched iron-formation on the margins of ore zones has provided wash ore or treat rock for concentration of the iron oxide mineral aggregates and removal of silica and clay constituents through processing by washing, jigging, gravity spirals and cones, and by flotation. Extensive amounts of treat rock were processed from the mines in the Lake Superior and Knob Lake ranges by utilizing crude ore containing less than 45% iron and upgrading it to 55 to 65% iron with 5 to 8% silica for blending as sinter fines or in pellets.

Manganiferous iron ore was recovered from many of the deposits in the Lake Superior and Knob Lake ranges. At the beginning of mining operations in the Knob Lake range in 1954, estimates of proven ore reserves in 44 deposits included 12 to 13% manganiferous ore in the total 42.4 Mt of reserves. Of the 55.8 Mt of ore shipped in the first five years of production from the Knob Lake range, about 12% was manganiferous grade and contained Fe – 46.9%, Mn – 5.26%, P – 0.118%, SiO_2 – 6.15%, Al_2O_3 – 1.52%, CaO – 0.12%, MgO – 0.02%, S – 0.005%, Moisture – 10.44%, and Loss on ignition – 5.97%. The manganese was derived mainly from carbonate facies iron-formation. Possibly some manganese-bearing oxide facies was oxidized and mobilized by enrichment processes, and redistributed within the ore zones. Highly aluminous or shaly protore facies provided a favoured site for manganese deposition. Most of the high grade manganese deposits throughout the world have developed by enrichment of manganese-bearing lithofacies associated directly with iron-formation or stratafer lithofacies. Iron and manganese are strongly differentiated under highly oxidizing conditions in the depositional basins, and manganese-rich lithofacies are usually distal from the cherty iron-rich lithofacies and have been deposited near the margins of the sedimentary basins. Iron and manganese are separated further by oxidation and enrichment processes and secondary manganese-rich lithofacies of ore may be deposited some distance from their protore facies.

Important examples of enriched manganese deposits developed in iron-formation are Moanda in Gabon; Serra do Navio in Amapa, Brazil; deposits in the Dharwar, Iron Ore, and Aravalli supergroups of India; and in the Khondalite, Sausar, and Gangpur Groups of India; in the Transvaal and Damara supergroups of southern Africa and Namibia; and in Ghana.

The distribution of gold in residually enriched iron ore deposits is being investigated in different parts of the world.

EXPLORATION GUIDELINES

1. Areas should be selected in which:

   a) iron-formations are 30 to 100 m thick, and contain 30% or more iron;

   b) silica in the iron-formation has cherty fine grained textures in preference to coarse granular textures, enabling a maximum exposure of grain surfaces to the leaching solutions;

   c) lithofacies consist of a high proportion of minerals that are readily amenable to leaching and oxidation, such as siderite, iron silicates, and fine grained cherty iron oxide, and a minimum of sulphide lithofacies;

   d) iron-formation has relatively thin banding and primary sedimentary features that enhance permeability;

   e) structural deformation through faulting, fracturing, and shearing has developed high permeability in the iron-formation, and has repeated stratigraphic units to provide large structural blocks of protore lithofacies of the iron-formation; and

   f) there is evidence of paleoenvironments suitable for deep penetration and circulation of groundwater to promote oxidation, leaching, and enrichment of the iron-formations over extended periods of geological time.

2. The stratigraphy, structure, and distribution of the iron-formation protore should be clearly defined through geological mapping and drilling to provide a basis for defining deposit settings and characteristics, and to enable assessment of resource potential. Drilling and test pitting are usually required to define the borders and distribution of the iron-formation units.

3. Aeromagnetic surveys are used effectively for tracing the regional distribution of iron-formations. Combinations of gravity and magnetic surveys have been used successfully for locating residually enriched zones within units of iron-formation.

4. Residually enriched ore deposits within the protore iron-formation may be detected directly through observation and study of topographic and related features such as:

   a) the presence of gossans, red and brown stained soil, the presence of enriched fragments of iron oxide and iron-formation in the soil (a number of ore deposits were found in the Knob Lake range by examining the frost boils and searching for enriched fragments of iron-formation); rock fragments in the soil are commonly used in deeply weathered terrain to trace the distribution of iron-formation; iron-formation fragments serve as important lithological markers for tracing the direction of transport and source areas for till, eskers, and moraines, and the glacial deposits may be followed to the iron-formation source areas;

   b) outcrops and cores of iron-formation may show evidence of leaching of silica and enrichment of iron, and secondary manganese distributed in fractures, veins, pods, or stain, may indicate significant mobilization of highly oxidized iron and manganese in the vicinity of enriched zones;

   c) synclinal fold structures and traces of fault zones, in iron-formation; and

   d) ridges of iron-formation broken by areas of low relief, which may mark areas of leaching, enrichment, and ore development.

5. Reconstruction of paleotopography to trace regoliths and deeply weathered zones which may be capped by hard rock overthrust by late faults, as in the case of the deep pockets of rubble ore in the Ruth Lake deposits of the Knob Lake range (Fig. 4.1-2).

ACKNOWLEDGMENTS

This manuscript was reviewed and edited by C.W. Jefferson and R.I. Thorpe. Assistance with word processing by C.M. Plant and L.C. O'Neill was appreciated.

SELECTED BIBLIOGRAPHY

References with asterisks (*) are considered to be the best source of general information on this deposit subtype. For additional references, see Type 3.

*Belevtsev, Y.N.
1973:   Genesis of high-grade iron ores of Krivoyrog type; in Genesis of Precambrian Iron and Manganese Deposits, Proceedings of the Kiev Symposium, 20-25 August 1970, UN Educational Scientific and Cultural Organization, Paris, p. 167-180.

*Bolis, J.L. and Bekkala, J.A.
1986:   Iron ore availability–market economy countries, a minerals availability appraisal; Bureau of Mines, United States Department of the Interior, Information Circular 9128, 56 p.

*Dorf, E.
1959:   Cretaceous flora from beds associated with rubble iron-ore deposits in the Labrador Trough; Geological Society of America, Bulletin, v. 70, p. 1591.

Elver, R.B.
1960:   Survey of the Canadian iron ore industry during 1959; Department of Mines and Technical Surveys, Ottawa, Mineral Resources Division, Mineral Information Bulletin MR 45, 121 p.

*Garrels, R.M.
1960:   Mineral Equilibria at Low Temperature and Pressure; Harper and Brothers, New York, 254 p.

Gross, G.A.
*1964a: Primary features in cherty iron-formations; in Genetic Problems of Ores, (ed.) R.K. Sundaram; Part V, Proceedings of Section 5, XXII International Geological Congress, India.
1964b: Mineralogy and beneficiation of Quebec iron ores; The Canadian Institute of Mining and Metallurgy Bulletin, v. 67, p. 17-24.
*1965:   General geology and evaluation of iron deposits; Volume I in Geology of Iron Deposits in Canada; Geological Survey of Canada, Economic Geology Report No. 22, 181 p.
*1968:   Iron ranges of the Labrador geosyncline; Volume III in Geology of Iron Deposits in Canada; Geological Survey of Canada, Economic Geology Report 22, 179 p.

*Gross, G.A. and Zajac, I.S.
1983:   Iron-formations in fold belts marginal to the Ungava Craton; in Iron-formation: Facts and Problems, (ed.) A.F. Trendall and R.C. Morris; Developments in Precambrian Geology, v. 6, Elsevier, Amsterdam-Oxford-New York-Tokyo, p. 253-294.

*Gross, G.A., Glazier, W., Kruechl, G., Nichols, L., and O'Leary, J.
1972:   Iron Ranges of Labrador and northern Quebec; Guidebook, Field Excursion A55, Twenty-forth International Geological Congress, Ottawa, Ontario, Canada,1972, 58 p.

*Morris, R.C.
1980:   A textural and mineralogical study of the relationship of iron ore to banded iron-formation in the Hamersley iron province of Western Australia; Economic Geology, v. 75, no. 2, p. 184-209.

*Ruckmick, J.C.
1962:   Tropical weathering and the origin of the Cerro Bolivar iron ores (abstract); Geological Society of America, Special Paper No. 68, p. 258.

*Stubbins, J.B., Blais, R.A., and Zajac, S.I.
1961:   Origin of the soft iron ores of the Knob Lake range; Canadian Institute of Mining and Metallurgy, Transactions,v. 64, p. 37-52.

4.2 SUPERGENE BASE METALS AND PRECIOUS METALS

D.R. Boyle

INTRODUCTION

Prior to glaciation, the Canadian geological landscape was subjected to a long period of physico-chemical weathering during Cambrian to late Tertiary with considerable variations in regional climates. Sulphide-bearing deposits at or near the land surface at this time, especially during the

Boyle, D.R.
1996:   Supergene base metals and precious metals; in Geology of Canadian Mineral Deposit Types, (ed.) O.R. Eckstrand, W.D. Sinclair, and R.I. Thorpe; Geological Survey of Canada, Geology of Canada, no. 8, p. 92-108 (also Geological Society of America, The Geology of North America, v. P-1).

Tertiary, underwent extensive oxidation leading to formation of supergene precious metal and base metal deposits. Many of these deposits experienced considerable enrichments in these metals. The various supergene mineral deposits associated with primary sulphide-bearing deposits in Canada (Fig. 4.2-1) can be divided into three subtypes 4.2a: supergene zones developed over massive sulphide deposits; 4.2b: oxidation zones developed in the upper parts of vein, shear/fault-related, and replacement deposits; and 4.2c: supergene oxide and sulphide zones formed over porphyry Cu-Mo-(Au-Ag) deposits.

The term oxidation as used here does not simply denote formation of oxide, carbonate, or sulphate minerals *sensu stricto*, but rather a state in which primary sulphide minerals are progressively oxidized, or precipitated from oxidizing solutions, through attainment of higher oxidation states represented by formation of both sulphide (e.g., chalcocite-covellite-digenite) and oxide-carbonate-sulphate-silicate-phosphate secondary minerals. In this context, four processes are important in the formation of supergene mineral deposits: 1) alteration of an existing primary mineral to form a secondary mineral of similar composition but higher oxidation state (e.g., chalcocite after chalcopyrite), 2) coating of primary grains (reduction sites) by secondary sulphides (e.g., sooty chalcocite on chalcopyrite and pyrite), 3) replacement of primary sulphides by secondary minerals of different composition (e.g., chalcocite and covellite replacement of sphalerite and galena), and 4) precipitation of secondary minerals in primary and dissolution voids of the host or country rocks (e.g., precipitation of most oxides, sulphates, and native gold, silver, and copper).

The economic commodities associated with the three types of supergene deposits mentioned above are: subtype 4.2a – Au, Ag, Cu, Ni, and Hg; subtype 4.2b – Cu, Au, Ag, Zn, Pb, and Mn; and subtype 4.2c – Cu, Mo, Au, and Ag.

In Canada, the best examples of subtype 4.2a deposits are found in the Bathurst mining camp of New Brunswick (e.g., Murray Brook, Caribou, Heath Steele deposits) and at the Windy Craggy deposit in British Columbia. For subtype 4.2b deposits the best examples are the Copper Rand, Henderson II, Selbaie (Detour) deposits in Quebec, the Ross and Keeley mines in Ontario, the Bayonne mine in southern British Columbia, and the Keno Hill, Mount Nansen, Ketza River, Brewery Creek, and Williams Creek deposits in Yukon Territory. For subtype 4.2c deposits the best examples are the Casino deposit in Yukon Territory and the Afton, South Kemess, Gibraltar, Mount Polley, Bell, Berg, and Krain deposits in British Columbia (see Fig. 4.2-1 and Table 4.2-1).

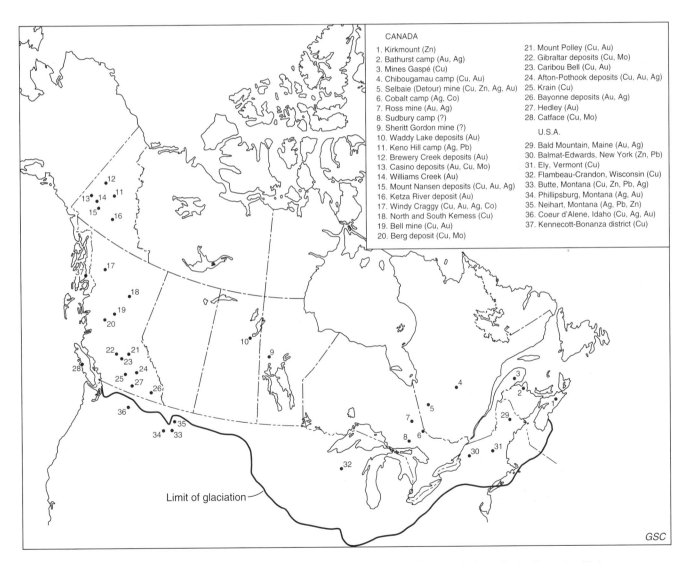

Figure 4.2-1. Location of supergene precious and base metal mineral deposits in Canada. Major supergene mineral camps in northern U.S.A. are also shown.

Table 4.2-1. Tonnages, grades, and supergene characteristics of Canadian precious metal and base metal supergene mineral deposits. For large mining camps in which supergene mineral deposits occur, only representative deposits are given. See Figure 4.2-1 for deposit locations.

Deposit		Tonnage (Mt)	Grades*	% of Deposit ‡ Supergene	Depth of Oxidation (m)	Ref.
Massive sulphide deposits (subtype 4.2a)						
Bathurst camp						
- Murray Brook	- gossan	1.02	Au(1.52), Ag(65.9)	15	60	1
	- primary	21.50	Cu(0.20), Zn(4.23), Pb(1.62) Au(0.6), Ag(54.3)	---	---	1,2
- Caribou	- gossan	0.06	Au(6.02), Ag(242.3)	2	20	2
	- secondary S₂	0.83	Cu(3.40), Zn(2.60), Pb(1.10)	20	20	2
	- primary	5.40+	Cu(<0.4), Zn(8.30) Pb(3.7), Au(1.5), Ag(102)	---	---	2
- Heath Steele B	- gossan	0.18	Au(5.03), Ag(171.6)	2	30	2
	- primary	15.80	Cu(1.0), Zn(4.48), Pb(2.88) Au(1.5), Ag(69.7)	---	---	2
Windy Craggy	- gossan	4.0+	Au(1.69), Ag(13.0), Cu(0.28)	0.2	5	3
	- secondary S₂	1.6+	Au(0.41), Ag(11.0), Cu(4.21)	0.1	2	3
	- primary	200+	Au(0.20), Ag(3.83), Cu(1.44) Co(0.07)	---	---	3
Vein, shear/fault-related, and replacement deposits (subtype 4.2b)						
Chibougamau camp						
- Copper Rand		9.5	Cu(1.81), Au(2.05)	20-30	350	4
- Henderson II		N/A	N/A	15-20	300	5
Selbaie (Detour)						
	- A1 zone	32.1	Cu(0.39), Zn(2.30), Au(0.31) Ag(36)	<20	150	6,7
	- B zone	0.9	Cu(3.51), Au(0.96), Ag(30.8)	50-60	200	6,7
Mount Nansen	- Webber lode	N/A	Au(12.0), Ag(836)	50	70	8
Bayonne mine	- oxide zone	N/A	Au(35 to 70)	>80	150	9
	- primary	N/A	Au(<13)	---	---	9
Ross mine		N/A	oxide ore considerably enriched in Au and Ag compared to primary ore	?	120	10
Cobalt Camp - Keeley mine		N/A	Ag(3300; up to 102 000) Au(19.9)	?	170	11
Keno Hill	- oxide ore	1.0	Ag(2300-16 600), Pb(25-60)	>80	200	12
	- primary ore	2.0	Ag(340-3400), Pb(4-22)	---	---	12
Ketza River	- oxide ore	0.5	Au(18.0)	100	150	13
Brewery Creek	- oxide zone	10.0	Au(1.99)	100	30	14
Williams Creek	- No. 1 zone	13.6	Cu(1.07), Au(0.5)	90	250	15
Porphyry Cu-Mo (Au, Ag) deposits (subtype 4.2c)						
Casino	- leached capping	31.0	Cu(0.11), Mo(0.024), Au(0.8)	5	60	16
	- secondary	95.0	Cu(0.43), Mo(0.031), Au(0.4)	15	130	16
	- primary	489.0	Cu(0.23), Mo(0.024), Au(0.3)	---	---	16
South Kemess	- all zones	200.4	Cu(0.22), Au(0.63)	20	70	17
Mount Polley	- all zones	155.0	Cu(0.29), Au(0.39)	25	250	18
Gibraltar	- four deposits	360.0	Cu(0.371), Mo(0.016)	30	100	19
Krain	- secondary	4.9	Cu(0.64)	50	100	20
	- primary	9.1	Cu(0.53)	---	---	20
Bell - all zones		116.0	Cu(0.48), Au(0.35)	25	70	21
Berg - all zones		400.0	Cu(0.4), Mo(0.05)	60	130	22
Afton - all zones		30.8	Cu(1.0), Au(0.58), Ag(4.19)	>80	500	23
Gaspe mines						
Mines Gaspé- Copper Mountain -secondary		30.4	Cu(0.45)	15	100	24
- primary		208.6	Cu(0.40)	---	---	24

N/A - not available   * Au, Ag grades in g/t; base metals in %   S₂ - sulphides   ? - unknown   ‡ - by volume

1. Rennick and Burton (1992) and author's data
2. McCutcheon (1992)
3. B. Downing (pers. comm., 1994)
4. Mining Magazine (1983)
5. G. Allard (pers. comm., 1993)
6. Bouillon (1990)
7. Sinclair and Gasparini (1980)
8. Saager and Bianconi (1971)
9. Rice (1941)
10. Moore (1938)
11. Bell (1923)
12. Boyle (1965)
13. Abercrombie (1990)
14. R. Dimet (pers. comm., 1994)
15. K. McNaughton (pers. comm., 1994)
16. Payne et al. (in press)
17. Rebagliati et al. (in press)
18. Nikic (in press)
19. Drummond et al. (1976)
20. Christie (1976)
21. Carson et al. (1976)
22. Panteleyev et al. (1976)
23. Kwong et al. (1982)
24. Canadian Mines Handbook (1971-1972).

In other parts of the world the best examples of subtype 4.2a deposits can be found in the famous Rio Tinto massive sulphide belt of Spain (Williams, 1934; Palermo et al., 1986), and the Mt. Lyell, Mt. Morgan, and Elura deposits of Australia (Solomon, 1967; Hughes, 1990). Supergene deposits associated with subtype 4.2b deposits are numerous throughout all base metal metallogenic regions in the world. Good examples of this group can be found in the U.S.A. in such mining camps as Butte and Granite-Bimetallic, Montana; Creede and Cripple Creek, Colorado; Tonopah, Nevada; and Tintic, Utah (Lindgren, 1923), and in practically all of the base and precious metal mining camps of Australia (Hughes, 1990). Much of the silver production from the famous Potosi tin-silver veins of Bolivia came from supergene ore zones. Good examples of subtype 4.2c deposits outside Canada can be found in Chile (e.g., Chuquicamata, El Salvador, La Escondida), U.S.A. (e.g., Bingham, Utah; Butte, Montana; Ray, Morenci, Santa Rita, Twin Buttes, and Sierrita, Arizona), Iran (Sar Cheshmeh), and many other porphyry copper metallogenic provinces.

For the descriptions of supergene mineral deposits that follow, it is assumed that the reader is familiar with the various primary mineralization features of the main deposit groups mentioned above (see subtypes 6.3, 6.4 and Types 16, 17, 19).

IMPORTANCE

Supergene weathering processes are economically important since they often lead to the enrichment of precious and base metals to the point that even very low grade primary sulphide or precious metal-bearing mineralized zones can evolve into economic mineral deposits. This can be accomplished by two main processes: secondary sulphide or precious metal enrichment contained within a smaller volume of rock than the primary ore, and conversion of original, less extractable, forms of primary sulphides or refractory precious metals into mineral forms which are more economical to extract, the primary and supergene grades largely remaining the same.

In unglaciated regions of the world, supergene precious and base metal deposits have made a major contribution to the mineral economy. Thus, if it were not for supergene enrichment processes many of the well known porphyry Cu deposits in the southwestern U.S.A., Chile, Middle East, and elsewhere would not be economic (Sillitoe and Clark, 1969; Titley, 1975). For many deposits in these regions, all of the ore is supergene (with enrichment). In Australia about 70% of Au reserves today are in supergene ores (calculated from Hughes, 1990). Indirectly, precious metal supergene ores were the source for the large bonanza Au, Pt placer deposits in Yukon Territory; Ballarat, Australia; Urals, Russia; and smaller placer regions such as those at Barkerville, British Columbia, and the Eastern Townships of Quebec.

In Canada, in situ supergene precious and base metal deposits have made only a modest contribution to overall mineral production due to the scouring effects of glaciation and the difficulties in detecting such ores under glacial cover. However, in some individual metallogenic provinces contributions have been quite significant. Thus, in the Bathurst base metal camp of New Brunswick, practically all of the Au production has come from supergene gossans,

and in the western Cordillera porphyry Cu-Mo province, supergene ores have contributed significantly to the metal endowment. According to Ney et al. (1976), the widespread belief by explorationists that glaciation had removed the important supergene zones that make porphyry deposits such important sources of Cu was largely responsible for the late (mid-1950s) recognition and exploration of the porphyry Cu-Mo province of the Canadian Cordillera.

SIZE AND GRADE OF DEPOSITS

Tonnages for supergene ore deposits associated with massive sulphide deposits (subtype 4.2a) are typically in the 0.5 to 10 Mt range with grades for Au and Ag ranging from 1.0 to 10.0 g/t and 50 to 300 g/t, respectively. Typically Au is enriched in these gossans by factors of 2 to 4 (Table 4.2-1).

Tonnages for supergene ores associated with vein, shear/fault-related, and replacement deposits (subtype 4.2b) range from 2 to 15 Mt with grades for Au, Ag, Cu, and Pb typically ranging from 0.5-20.0 g/t, 30-1000 g/t, 1-5%, and 20-60%, respectively. Zinc is enriched only in willemite and smithsonite deposits such as those at Balmat-Edwards, New York (Brown, 1936) and Kirkmount, Nova Scotia (Sangster, 1986; Fig. 4.2-1), otherwise this element is severely depleted from these zones. Occasionally these deposits represent bonanza 'free Au' and 'native Ag' deposits (e.g., Keno Hill, Yukon Territory; Cobalt silver deposits, Ontario, and Bayonne and other mines in the Nelson area of British Columbia). Gold and silver grades in these deposits in many cases reach levels of 20-100 g/t and 3000-100 000 g/t, respectively. These represent considerable enrichments over protore grades.

Tonnages for supergene ore zones developed over porphyry Cu-Mo deposits (subtype 4.2c) can be as small as 30 Mt (e.g., Afton mine, British Columbia) to as large as the super giant porphyries, which are primarily supergene enriched deposits (e.g., Chuquicamata; 1 Bt). Grades for supergene Cu-Au-Ag porphyries are typically 0.2-3.0% Cu, 0.2-1.0 g/t Au, and 1-10 g/t Ag. Supergene enrichment factors for Cu over primary mineralization are in the order of 1.25 to 5.0 (Ney et al., 1976; Titley, 1982). Some deposits have been considered not to be enriched, but simply to represent a conversion of Cu-sulphides to a secondary form such as native Cu (e.g., Afton, South Kemess). True enrichment must be calculated on a mass balance basis, not on the basis of ore reserve calculations. From this viewpoint, practically all supergene zones associated with porphyry systems represent some overall metal enrichment.

Tonnages and grades for Canadian supergene precious and base metal deposits for which data are available are given in Table 4.2-1. For some deposits a comparison of supergene and primary ore grades and tonnages are also given together with the approximate percentage of the ore deposit that is supergene.

A number of supergene precious and base metal mineral deposits have failed at the mining stage due to improper ore reserve calculations. Unlike primary deposits, supergene ores are highly porous and simple use of "specific gravity x volume" methods are not applicable. Also, the use of the water balance method for calculating specific gravity is ineffective because the total porosity factor is not included in this measurement. Calculations may be further hampered by the fact that porosities will vary considerably between supergene zones within a given deposit. Statistically representative analyses of porosity and use of powder pycnometry for specific gravity should be used for the calculation of supergene ore reserves.

GEOLOGICAL FEATURES

General zonation and mineralogy

Irrespective of the type of sulphide-bearing or precious metal deposit over which supergene ores form, there is a general zonation scheme that is common to all. Specific deviations from this scheme and descriptions of the behaviour of elements within each of the main supergene zones developed over the three primary deposit types mentioned above are given in detailed descriptions below and in Table 4.2-1. Generally, supergene precious metal and base-metal ore deposits can be divided into three main zones (see Fig. 4.2-2 to 4.2-6). The uppermost zone has been commonly termed "leached capping". It can vary in composition from highly siliceous-kaolinized and slightly ferruginized material to massive goethite-hematite-jarosite gossan. Below the leached capping an "oxide zone" generally forms which is characterized by the formation of base metal oxide, carbonate, sulphate, phosphate, silicate, halide, and native minerals formed after secondary sulphides as precipitates in rock voids. A "supergene sulphide zone", varying considerably in thickness, forms between the oxide zone and the primary ore. This zone is characterized by dissolution of carbonate minerals to create secondary voids, the coating of primary sulphides with 'sooty' secondary sulphides, and the replacement of primary sulphides by secondary

sulphides. Depending on conditions of formation, this zone may or may not be characterized by strong enrichments of Cu and Ni. Zinc and lead are generally depleted in this zone.

The principal minerals found in each of these three zones are presented in Table 4.2-2. It should be noted that adjacent zones, and thus their mineral assemblages, overlap.

The behaviour and disposition of Au during formation of the above zones is important to the economic viability of some of these deposits. Generally Au accumulates only in the leached capping and oxide zones in which it has two main forms of occurrence; as in situ freed grains initially occluded in primary sulphides, and as precipitated micrometre-size native Au and electrum grains formed in rock voids during descent of oxidizing solutions. Generally, only the latter process will lead to enrichment of Au in the upper supergene zones. This enrichment is generally in the order of 2 to 5 times primary Au concentrations. Both of the above mentioned processes lead to Au reserves that can be recovered using methods that are generally cheaper than those used in primary sulphide ore extraction.

Depth of oxidation

The depths to which supergene mineral deposits extend will be dependent largely on the position of the water table during the predominant time of formation. Emmons (1917) and Lindgren (1923) have described the formation of significant supergene sulphide and oxide zones down to depths of 500-700 m in some of the mining camps in the U.S.A. In Canada the various recorded depths of supergene ore development are presented in Table 4.2-1. For massive sulphide deposits the depths of oxidation are relatively shallow (7-60 m), but greater depths have been recorded elsewhere (Palermo et al., 1986; Hughes, 1990). For vein, shear/fault-related, and replacement deposits, the depth of oxidation in Canada varies from 30-350 m; depths which are similar to those recorded in unglaciated terranes. For porphyry Cu deposits in the Canadian Cordillera the depths of oxidation vary from 60 to 500 m, which makes them comparable in supergene development to the porphyry Cu provinces of the southwestern U.S.A. and the Andes Mountains of South America.

Although oxidation above the water table is the most dominant supergene process in these deposits, numerous examples have been given where significant oxidation has continued for considerable depths below the water table (Emmons, 1917; Lindgren, 1923; Guilbert and Park, 1986). This process is considered to be driven by differential electrochemical oxidation potentials involving anode-cathode cells (Sato and Mooney, 1960; Blain and Brotherton, 1975) and/or lateral influx into ore zones of groundwaters still containing considerable amounts of dissolved O_2.

Age of mineralization

All of the deposits outlined in Figure 4.2-1 and Table 4.2-1 are preglacial, but exact ages are very difficult to determine due to the lack of good geochronological dating methods for supergene deposits and the general absence of datable stratigraphic marker horizons. Some deposits, such

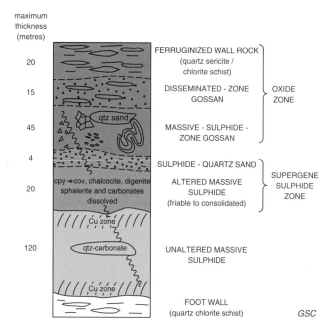

Figure 4.2-2. Representative stratigraphic profile of the supergene zones developed over massive sulphide deposits in the Bathurst camp, New Brunswick (after Boyle, 1993). Abbreviations: qtz = quartz; cpy = chalcopyrite; cov = covellite.

as Afton (late Paleocene-middle Paleocene), Krain (early Tertiary), and Bell Copper (post-Eocene to pre-Pleistocene), can be dated reasonably well due to the presence of overlying Tertiary volcanic and sedimentary rocks. Most of the supergene zones associated with porphyry Cu deposits in the Canadian Cordillera can be bracketed in age between late Cretaceous and late Tertiary since the host rocks for primary ore span this period. For the Selbaie (Detour) mine in Quebec (Sinclair and Gasparini, 1980) and the Flambeau deposit in Wisconsin (May, 1977) it has been suggested that oxidation took place in late Precambrian to Cambrian times, but in the opinion of the author evidence for this conclusion is weak. Recent paleomagnetic research on the supergene deposits overlying the massive sulphide deposits of the Bathurst camp places the period of weathering at 1-3 million years ago, just before onset of Pleistocene glaciation (Symons and Boyle, unpub. data, 1994).

It is the opinion of the author that most of the supergene deposits in Canada and the northern United States (see Fig. 4.2-1 and Table 4.2-1) were formed during late Cretaceous to late Tertiary times under regional climates that varied from tropical through semiarid to arid.

Subtype 4.2a
Supergene zones developed over massive sulphide deposits
Importance of structure and stratigraphy

Compared to other types of primary sulphide deposits, structure plays a minimal role in the formation of supergene zones over massive sulphide deposits. Where significant postmineralization faulting has occurred, these faults may act as channelways for localized oxidation within the deposit. Stratigraphy, however, does play a major role in sulphide oxidation processes. Massive sulphide deposits generally exhibit cyclical sulphide-gangue layering. Sphalerite-, pyrrhotite-, and carbonate-rich layers oxidize at a much faster rate than galena-, chalcopyrite-, pyrite-, and magnetite-rich layers. Deposits with a high proportion of the former mineral assemblages will oxidize faster, more completely, and to greater depths over a given time period than those with a predominance of the latter mineral sequence. Oxidation of sphalerite-pyrrhotite-carbonate zones will lead to significant secondary porosities since

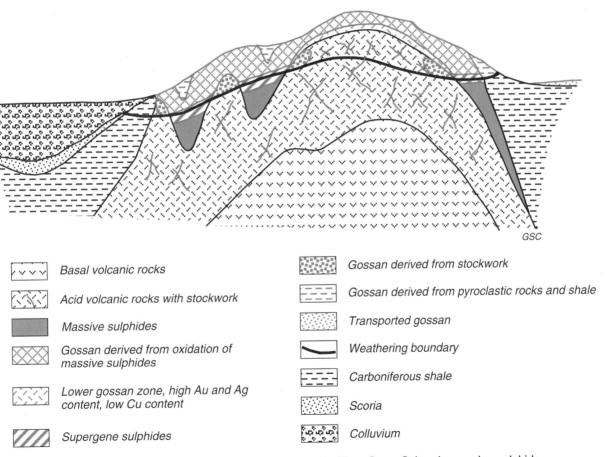

	Basal volcanic rocks		Gossan derived from stockwork
	Acid volcanic rocks with stockwork		Gossan derived from pyroclastic rocks and shale
	Massive sulphides		Transported gossan
	Gossan derived from oxidation of massive sulphides		Weathering boundary
	Lower gossan zone, high Au and Ag content, low Cu content		Carboniferous shale
	Supergene sulphides		Scoria
			Colluvium

Figure 4.2-3. Schematic cross-section of the Rio Tinto Cerro Colorado massive sulphide district of Spain showing secondary and primary zones (after Palermo et al., 1986).

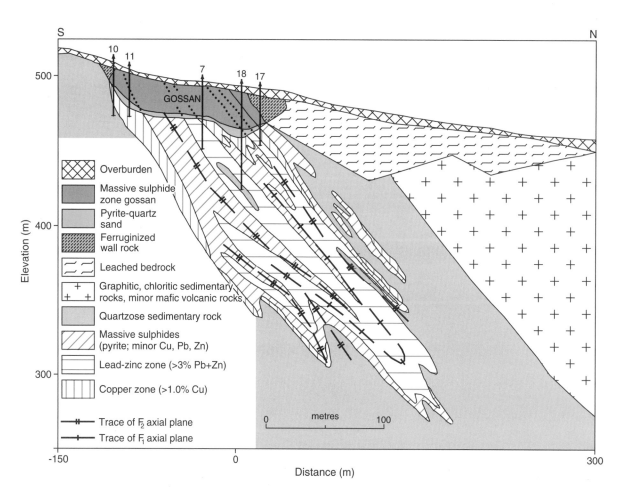

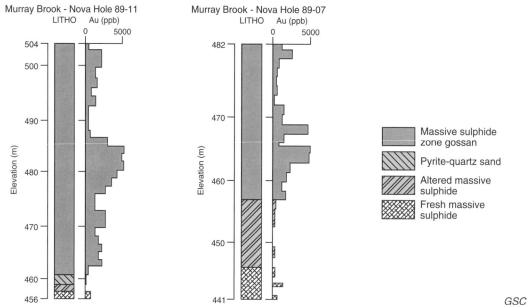

Figure 4.2-4. Cross-section and representative Au profiles of the Murray Brook gossan deposit, Bathurst mining camp, New Brunswick (modified after Boyle, 1993).

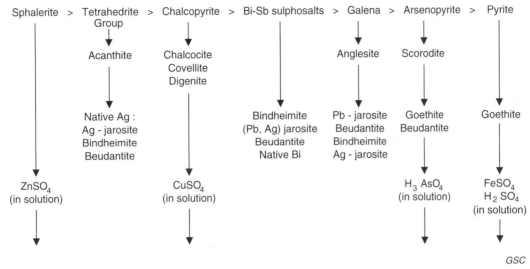

Figure 4.2-5. Typical alteration sequence for the oxidation of low-pyrrhotite massive sulphide deposits (after Boyle, 1993).

their weathering products are largely leached out of the sulphide zone. Greater porosity leads to accelerated oxidation and the formation of a larger reservoir for the precipitation of secondary phases that remain within the zone.

Alteration and texture

A typical stratigraphic section through an oxidized massive sulphide deposit is shown in Figure 4.2-2. The surrounding host rocks to these deposits are generally strongly ferruginized and argillized. If exposed, the stringer zone will be altered to a "disseminated-zone gossan" formed predominantly of goethite which can be very rich in Au.

When altered, the main body of the massive sulphide deposit forms a "massive-sulphide-zone gossan" consisting mainly of massive goethite and hematite showing highly vesicular, cellular-boxwork and pseudomorphic-replacement textural features of completely oxidized massive sulphide. Principal minerals in this zone are goethite, hematite, silica (anhedral and amorphous), hydroxy-sulphates and oxides (beudantite, plumbojarosite, jarosite, binheimite, scorodite), and traces of cinnabar, primary cassiterite, and native Au, Ag, and Bi. Primary carbonates (siderite, ankerite, dolomite, calcite) are completely dissolved by acid solutions, often leaving small lenticular cavities in the gossan.

Beneath the massive-sulphide-zone gossan, oxidation is dominated by almost total dissolution of carbonate-gangue phases, dissolution of sphalerite, alteration of chalcopyrite to supergene Cu sulphides (chalcocite, covellite, digenite), and partial oxidation of galena and arsenopyrite to anglesite and scorodite, respectively.

A unique feature of Canadian supergene deposits overlying massive sulphides is the presence between the gossan and supergene sulphide zones of a thin, 1-3 m, layer of

pyrite-quartz sandy-mud which is almost totally devoid of base metals and precious metals. It has been suggested that this is a postglacial feature (Boyle, 1993).

The relational aspects of the various zones mentioned above are exemplified by the schematic cross-section in Figure 4.2-3 for the famous Rio Tinto massive sulphide belt of Spain and by the cross-section of the Murray Brook deposit in the Bathurst mining camp of New Brunswick shown in Figure 4.2-4.

Figure 4.2-5 shows the typical alteration sequence for low grade pyrrhotite-bearing massive sulphide deposits. When pyrrhotite is present in concentrations generally above 10-15%, this sequence is greatly disrupted, and pyrrhotite accelerates the oxidation of sphalerite and galena, and especially of pyrite.

Enrichment processes and metal zoning

For gossan deposits overlying massive sulphides, Cu and Zn are almost totally leached from the oxide zone. Zinc leaves the system as a whole, whereas Cu may be strongly enriched or follow a history similar to Zn. For example, in the Bathurst mining camp, both the Murray Brook and Caribou deposits have similar primary ore compositions, yet Cu is strongly enriched in the supergene sulphide zone at Caribou whereas at Murray Brook the element simply takes part in a 'constant grade' conversion of chalcopyrite to chalcocite-covellite-digenite. These two deposits are only 10 km apart and occur at the same elevation. The reasons for this disparity for Cu are not clearly understood, but it would appear that the difference between the enrichment and constant Cu grade scenarios is exemplified by strong replacement of sphalerite and galena by chalcocite within the enriched zones in addition to the conversion of chalcopyrite to secondary sulphides.

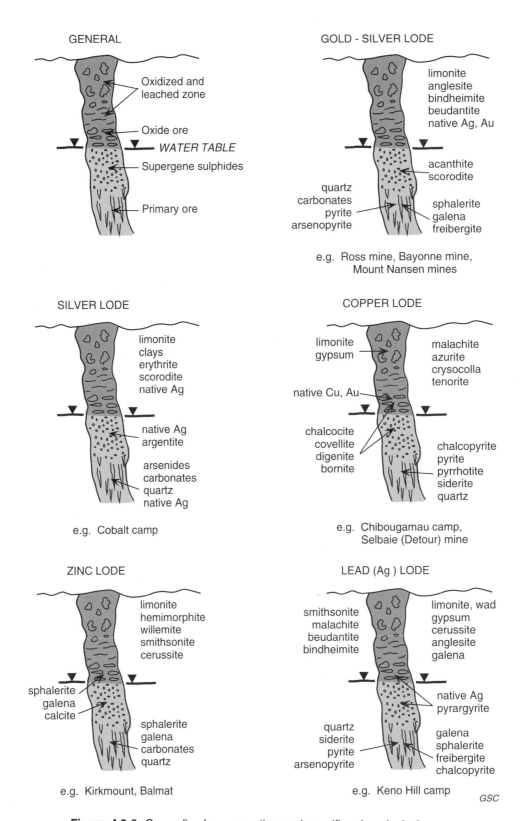

GENERAL

Oxidized and leached zone

Oxide ore

WATER TABLE

Supergene sulphides

Primary ore

GOLD - SILVER LODE

limonite
anglesite
bindheimite
beudantite
native Ag, Au

acanthite
scorodite

quartz
carbonates
pyrite
arsenopyrite

sphalerite
galena
freibergite

e.g. Ross mine, Bayonne mine,
Mount Nansen mines

SILVER LODE

limonite
clays
erythrite
scorodite
native Ag

native Ag
argentite

arsenides
carbonates
quartz
native Ag

e.g. Cobalt camp

COPPER LODE

limonite
gypsum

malachite
azurite
crysocolla
tenorite

native Cu, Au

chalcocite
covellite
digenite
bornite

chalcopyrite
pyrite
pyrrhotite
siderite
quartz

e.g. Chibougamau camp,
Selbaie (Detour) mine

ZINC LODE

limonite
hemimorphite
willemite
smithsonite
cerussite

sphalerite
galena
calcite

sphalerite
galena
carbonates
quartz

e.g. Kirkmount, Balmat

LEAD (Ag) LODE

smithsonite
malachite
beudantite
bindheimite

limonite, wad
gypsum
cerussite
anglesite
galena

native Ag
pyrargyrite

quartz
siderite
pyrite
arsenopyrite

galena
sphalerite
freibergite
chalcopyrite

e.g. Keno Hill camp

GSC

Figure 4.2-6. Generalized cross-section and specific mineralogical cross-sections for precious metal and base metal lode deposits.

Table 4.2-2. Principal minerals in supergene zones associated with sulphide-bearing ore deposits. Main zones are presented in descending lithostratigraphic order. In some deposits transition zones may occur between these main zones and thus contain minerals of each zone. To the author's knowledge all of the minerals listed below have been found in Canadian supergene mineral deposits.

Leached cappings

Goethite - $FeO(OH)$
Hematite - Fe_2O_3
Jarosite - $KFe_3(SO_4)_2(OH)_6$
Pitch limonite - $H(Fe,Cu)O_2$
Native - Au, Cu
Kaolinite - $Al_2Si_2O_5(OH)_4$
Chlorite - $(Mg,Fe)_5Al(Si_3Al)O_{10}(OH)_8$
Smectites

Antlerite - $Cu_3(SO_4)(OH)_4$
Brochantite - $Cu_4(SO_4)(OH)_6$
Chrysocolla - $(Cu,Al)_2H_2Si_2O_5(OH)_4 \cdot nH_2O$
Cuprite - Cu_2O
Tenorite - CuO
Turquoise - $CuAl_6(PO_4)_4(OH)_8 \cdot 4H_2O$
Neotocite - $(Mn,Fe)SiO_3 \cdot H_2O$
Ferrimolybdite - $Fe_2(MoO_4)_3 \cdot 7H_2O$

Oxide zones

Goethite - $FeO(OH)$
Silica - SiO_2
Hematite - Fe_2O_3
Alunite - $KAl_3(SO_4)_2(OH)$
Jarosite - $KFe_3(SO_4)_2(OH)_6$
Plumbojarosite - $PbFe_6(SO_4)_4(OH)_{12}$
Argentojarosite - $AgFe_3(SO_4)_2(OH)_6$
Cuprite - Cu_2O
Tenorite - CuO
Malachite - $Cu_2(CO_3)(OH)_2$
Azurite - $Cu_3(CO_3)_2(OH)_2$
Chrysocolla - $(Cu,Al)_2H_2Si_2O_5(OH)_4 \cdot nH_2O$
Brochantite - $Cu_4(SO_4)(OH)_6$
Chalcanthite - $CuSO_4 \cdot 5H_2O$
Antlerite - $Cu_3(SO_4)(OH)_4$
Turquoise - $CuAl_6(PO_4)_4(OH)_8 \cdot 4H_2O$
Hemimorphite - $Zn_4Si_2O_7(OH)_2 \cdot H_2O$
Smithsonite - $ZnCO_3$
Hydrozincite - $Zn_5(CO_3)_2(OH)_6$
Goslarite - $ZnSO_4 \cdot 7H_2O$
Willemite - Zn_2SiO_4
Gypsum - $CaSO_4 \cdot 2H_2O$
Calcite - $CaCO_3$
Dolomite - $CaMg(CO_3)_2$
Kaolinite - $Al_2Si_2O_5(OH)_4$
Chlorite - $(Mg,Fe)_5Al(Si_3Al)O_{10}(OH)_8$
Smectites

Native - Au, Ag, Cu, Bi, Zn, S
Anglesite - $PbSO_4$
Cerussite - $PbCO_3$
Scorodite - $FeAsO_4 \cdot 2H_2O$
Beudantite - $PbFe_3(AsO_4)(SO_4)(OH)_6$
Bindheimite - $Pb_2Sb_2O_6(O,OH)$
Minium - Pb_3O_4
Litharge/Massicot - PbO
Ferrimolybdite - $Fe_2(MoO_4)_3 \cdot 8H_2O$
Wulfenite - $PbMoO_4$
Cinnabar - HgS
Valentinite - Sb_2O_3
Stibiconite - $Sb_3O_6(OH)$
Pyrolusite MnO_2
Manganite - $MnO(OH)$
Rhodochrosite - $MnCO_3$
Rhodonite - $(Mn,Fe,Mg,Ca)SiO_3$
Neotocite - $(Mn,Fe)SiO_3 \cdot H_2O$
Chlorargyrite - $AgCl$
Iodargyrite - AgI
Marshite - CuI
Miersite - $(Ag,Cu)I$
Bromargyrite - $AgBr$
Atacamite - $Cu_2Cl(OH)_3$
Mimetite - $Pb_5(AsO_4)_3Cl$
Pyromorphite - $Pb_5(PO_4)_3Cl$
Vanadinite - $Pb_5(VO_4)_3Cl$

Supergene sulphide zones[*]

Chalcocite - Cu_2S
Covellite - CuS
Digenite - Cu_9S_5
Bornite - Cu_5FeS_4
Djurleite - $Cu_{31}S_{16}$
Acanthite - Ag_2S
Native - Au, Ag, Cu, Bi, S
Sphalerite - $(Zn,Fe)S$ (rare)
Kaolinite - $Al_2Si_2O_5(OH)_4$
Chlorite - $(Mg,Fe)_5Al(Si_3Al)O_{10}(OH)_8$

Pyrite - FeS_2
Marcasite - FeS_2
Violarite - $FeNi_2S_4$
Bravoite - $(Ni,Fe)S_2$
Krennerite - $AuTe_2$
Pyrargyrite - Ag_3SbS_3
Sylvanite - $(Au,Ag)_2Te_4$
Stromeyerite - $AuCuS$
Galena - PbS (rare)

[*] secondary sulphides only

Gold is greatly enriched within the gossan portion of these deposits. For the Murray Brook, Caribou, Heath Steele, and Windy Craggy deposits the Au enrichment factors (on a weight/weight basis) are 2.5, 4.0, 3.5, and 8.5, respectively. Practically all of the Au occurs as fine submicrometre grain precipitates in voids within the gossan. For the same deposits mentioned above, Ag enrichment factors are 1.2, 2.4, 2.5, and 3.4, respectively. The Ag is partitioned between three main phases: 1) native Ag in voids, 2) argentojarosite, and 3) sequestration in Fe-oxyhydroxide and hydrated-sulphate minerals. As shown in the Au profiles for the Murray Brook deposit (Fig. 4.2-4), this element (and Ag, not shown) is distinctly zoned within the gossan. Zonation appears to be strongly controlled by the pyrite concentration in primary ore. During oxidation of massive sulphides, Pb and As are immobile and thus become concentrated in the gossan (see minerals, Table 4.2-2).

Subtype 4.2b

Oxidation zones developed in the upper parts of vein, shear/fault-related, and replacement deposits

Importance of structure

Structural elements play an important role in the development and localization of vein, shear/fault-related, and replacement precious metal and base metal deposits. These same structures also play a strong role in the development of supergene ore zones in the upper parts of these deposits. Primary structures become the channelways for descending oxidizing solutions which first attack the carbonate, sphalerite, and chalcopyrite phases. Because these deposits in many cases contain high concentrations of carbonates, a significant secondary porosity for precipitation of enriched supergene base metal and precious metal minerals can develop and oxidation zones can extend for considerable depths (see Table 4.2-1).

One of the most interesting features of these deposits is that in a given mining camp containing a number of vein or shear zone-related deposits, only a few will show extensive oxidation, and in some cases (e.g., Chibougamau and Cobalt camps) down to the deepest depths recorded from throughout the world. In the opinion of the author, this selective oxidation may be due to the relative locations of the deposits to groundwater discharge and recharge zones. Deposits in recharge zones would be more susceptible to deeper oxidation.

Alteration and texture

A general oxidation profile for this deposit type and specific mineralogical profiles for Au-Ag, Ag, Cu, Zn, and Pb-Ag lodes are presented in Figure 4.2-6. The general oxidation scheme is very similar to that described above for massive sulphide deposits, but the gossan zone may not be as developed in lode deposits. These deposits are known for their spectacular development of botryoidal, reniform, vermiform, and massive crystal aggregates of secondary base metals and precious metals. Many of these deposits have generated considerable income as sources of rare or exceptional quality minerals for the mineral collection industry.

Enrichment process and metal zoning

For the precious metal-bearing deposits, Au is generally enriched in the upper oxide zones (e.g., Ross and Bayonne mines) wheras Ag is generally enriched in the lower oxide and supergene sulphide zones (e.g., Cobalt and Keno Hill camps). For some deposits, enrichments of Au and Ag can reach spectacular levels. Thus in the Keeley mine in the Cobalt camp, metres of almost solid native Ag and pyrargyrite were mined from the lower oxidized levels, whereas at the Bayonne and Ross gold mines, Au in the 30-100 g/t range was mined from the upper oxide zones.

The mobilization and enrichment processes for lode supergene deposits are very similar to those of the massive sulphide deposits, except that a much larger proportion of the enriched Au comes simply from concentration of primary free gold grains by volume wastage. Silver migrates down the lode structure to become concentrated in both the lower oxide and upper supergene sulphide zones as native silver, pyrargyrite, and acanthite. Lead in lodes such as the rich Pb-Ag Keno Hill veins of Yukon Territory is generally shielded by anglesite or cerussite coatings and thus becomes concentrated by volume wastage. Zinc rarely accumulates in lode deposits in which it is a minor constituent, although in Zn lode deposits, especially those with high carbonate contents, this element may become enriched through the formation of secondary willemite-sphalerite-galena deposits or secondary smithsonite oxide deposits (Fig. 4.2-6).

Subtype 4.2c

Supergene oxide and sulphide zones formed over porphyry Cu-Mo-(Au, Ag) deposits

Importance of structure and stratigraphy

Porphyry Cu-Mo systems are generally concentrically zoned, and have a potassic low grade, pyrite-poor core surrounded by a quartz-sericite-pyritic (1-3%) ore shell (Cu, Mo), which in turn is surrounded by a strongly pyritized (10-30%) low grade propylitic zone (Fig. 4.2-7). The types of supergene alteration associated with porphyry Cu-Mo deposits are therefore controlled largely by the styles of primary mineralization and alteration, the compositions of rocks hosting the porphyry system, and fracture densities of the various zones. Highly siliceous and kaolinized cappings containing Cu-silicates and -oxides form over the potassic core of these deposits. These cappings generally contain low concentrations of Au. Supergene alteration developed over the ore shell and outer propylitic zone consists of a goethite-hematite-jarosite-clay caprock underlain by well developed oxide-supergene sulphide zones. Fracture densities are usually greatest over these two zones (Titley, 1982)

Porphyry Cu-Mo systems may be hosted in rocks varying from highly felsic intrusive-volcanic to intermediate-mafic intrusive-volcanic as well as volcanic-sedimentary stratigraphies. Occasionally skarn zones develop in calcareous host rocks. The types of supergene ores that develop are therefore strongly affected by host rock lithologies. Mineralogically, the main controlling parameters are pyrite, magnetite, hornblende, biotite, and carbonate contents. Strongly pyritized rocks favour formation of rich chalcocite blankets, whereas rocks rich in magnetite, biotite-hornblende, and carbonates favour formation of secondary native Cu deposits after chalcocite.

Alteration and texture

The supergene zones developed over potassic core zones consist of vesicular, highly siliceous and kaolinized, and slightly limonitized caprocks that contain crustations of Cu-oxides, Cu-carbonates, and Cu-silicates lining voids. These zones are generally not very thick due to minimal fracturing. For some porphyry Cu provinces this is an important source of ore, but in Canada this type of mineralization is not well developed.

Leached cappings developed over phyllic and propylitic zones consist of highly vesicular, spongy, and often pulverulent rocks composed of Fe-Mn oxyhydroxides and sulphates in a clay-rich matrix. Replacement boxwork textures are generally evident and the proportions of goethite, hematite, jarosite, and alunite are controlled by the original composition of the mineralized host rock (Anderson, 1982).

The alteration and textures in the supergene sulphide zone are dominated by sooty coatings of chalcocite-covellite-digenite on primary sulphides, pseudomorphic replacement textures of primary sulphides by these minerals, kaolinization and chloritization of silicates, and dissolution of calcite and gypsum, if present. In some deposits, the upper portion, and occasionally all, of the chalcocite blanket has been further oxidized leading to void fillings of Cu-oxides and native Cu (e.g., Afton and South Kemess).

Enrichment process and zonation

During formation of supergene mineralization over porphyry Cu-Mo deposits, Cu is strongly leached from the capping zone and transported down to the water table to form an enriched supergene sulphide blanket. If sometime during the formation of this enrichment blanket the water table is lowered, either through epeirogenic uplift or a climatic shift to more arid conditions, the chalcocite blanket will be oxidized to form Cu oxide-sulphate-carbonate, native Cu, hematite mineralization. For some deposits in the Canadian Cordillera, native Cu ore predominates in the

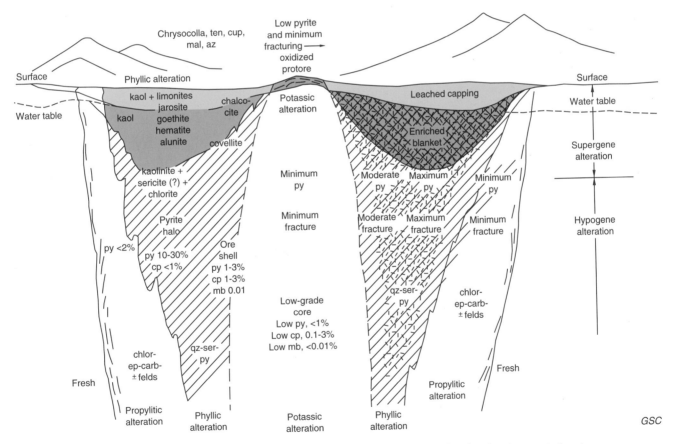

Figure 4.2-7. Schematic cross-section of a porphyry copper system showing the development of various types of supergene ore associated with underlying stratigraphy, ore and gangue tenor, and structure (after Guilbert and Park, 1986). Abbreviations: py = pyrite; ten = tenorite; cup = cuprite; mal = malachite; az = azurite; kaol = kaolinite; cp = chalcopyrite; chlor = chlorite; ep = epidote; carb = carbonates; qz = quartz; ser = sericite; felds = feldspars; mb = molybdenite.

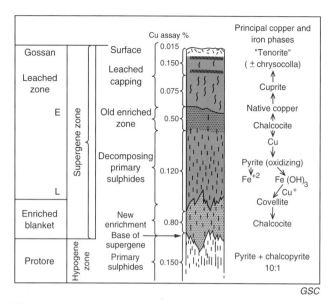

Figure 4.2-8. Profile of supergene zones developed over Plesyumi porphyry copper deposit, Papua New Guinea showing effects of multicycle weathering due to lowering of water table (modified from Guilbert and Park, 1986; after Titley, 1978).

supergene zone (e.g., Afton, South Kemess, Mt. Polley), whereas others show a more typical oxide-sulphate-carbonate and supergene sulphide zonation. Multicycle oxidation is a characteristic feature of most porphyry Cu-Mo provinces (Titley, 1982; Fig. 4.2-8). The formation of native Cu in these deposits is still not clearly understood. The deposits in which this metal predominates are generally characterized by low pyrite concentrations and high magnetite, carbonate, and biotite or hornblende contents. These minerals may have a controlling influence on the redox conditions under which chalcopyrite-chalcocite mineralization is oxidized and native Cu precipitated (Carr and Reed, 1976; Kwong et al., 1982). Afton, South Kemess, and Mt. Polley, which contain significant concentrations of native Cu, are hosted by alkaline suite intrusions and/or volcanic rocks, as opposed to the more dominant calc-alkaline intrusive hosts which are typified by oxide-carbonate-sulphate and chalcocite zones.

Molybdenum is relatively immobile during the oxidation of porphyry Cu-Mo deposits and therefore generally shows little or no enrichment. In the leached capping and oxide zones it may be partially converted to ferrimolybdite.

During the oxidation process, Au may be enriched in the leached capping of some deposits (e.g., Casino), but generally this metal is merely freed from its occluded habit in sulphides with little remobilization. Silver is generally leached from the upper zones and may occasionally be concentrated in the supergene sulphide zone.

Deposits that are rich in anhydrite-gypsum and carbonates may produce a gypsum layer below the chalcocite blanket that effectively seals the primary ore from further oxidation (e.g., Berg deposit).

GLACIATION

The perception in the early part of this century that glaciation in Canada destroyed preglacial weathered landscapes created serious doubts in the minds of explorationists that viable supergene ore deposits could be discovered in Canada. However, in the Canadian Cordillera the continual discovery of preserved, well developed supergene zones over porphyry Cu-Mo deposits demonstrates the often minimal effects glaciation has had. Today, most porphyry systems in the Canadian Cordillera, unless very large, would not be economic without rich supergene Cu and Au zones.

In the Appalachians, the discovery over the years of numerous economic gossan deposits in the massive sulphide camps of New Brunswick and Maine, and the important supergene enrichments over the Gaspésie Cu deposits (Fig. 4.2-1), indicate that glaciation has not been as destructive towards supergene ore deposits as earlier believed.

With respect to the Canadian Shield the above-mentioned perception still remains the norm for most explorationists, but compilation of existing data by the author indicates that important supergene mineral deposits (e.g., Au-Ag, base metals, iron, phosphate, rare-earth elements, kaolin, silica) have been discovered throughout the present century at an average rate of one every five years. No doubt more are to be discovered under thick glacial cover.

DEFINITIVE CHARACTERISTICS

The definitive characteristics for supergene precious metal and base metal deposits relate entirely to the recognition of the effects of secondary chemical weathering on primary metal-bearing zones. Thus these deposits are characterized by three main zonational features: 1) a leached capping composed of variable concentrations of Fe-Mn oxyhydroxides and sulphates (goethite, hematite, wad, jarosite), clays (kaolinite, chlorite, smectites), and occasionally Cu-silicates; this zone may contain important reserves of Au and Cu (rarely Ag and Mo), 2) an oxide zone which is characterized by a large array of Cu, Pb, Zn, Ag, and Sb oxides, sulphates, carbonates, phosphates, and halides, native Au, Ag, Cu, Bi, Zn, and S, as well as the same mineralogy for the capping zone; these zones constitute important reserves of Cu, Au, Ag, Zn, and Pb; and 3) a supergene sulphide zone typified largely by the formation, through precipitation or replacement of secondary Cu, Ni, and Ag sulphides (see Table 4.2-2). This zone often displays supergene enrichments of base metals and precious metals.

Transitional zones may occur between these three main zones. The host rocks surrounding these zones are in many cases strongly ferruginized and/or altered to clay-silica rock. At surface, the most definitive feature of these deposits is the presence of various types of gossanous-siliceous-argillized rocks.

GENETIC MODEL

The sizes, grades, mineralogical compositions, and preservation of supergene mineral deposits will be controlled to a large extent by the following physicochemical factors and processes: 1) climatic history, 2) tectonic history, 3) geomorphological development, 4) position of the sulphide body in the hydrological regime, 5) size, geometry, and structure of the sulphide body, 6) primary sulphide composition of the ore body, 7) nature and distribution of acid-neutralizing gangue minerals, 8) primary permeability (porosity) and degree to which secondary permeability can be developed during oxidation, 9) strength and sustainability of electrochemical oxidation cells within the sulphide body, 10) diversity and strength of microbiological activity within the ore zone, and 11) diffusion rate of atmospheric O_2 into the sulphide body.

If there has not been a sufficient period of tectonic quiescence, or if the tectonic history after formation has been largely destructive, these deposits will either not form or not be preserved, regardless of whether all other factors mentioned above were optimal. For this reason a detailed understanding of recent tectonic history, largely early Cenozoic to present, is paramount to the recognition of supergene base metal and precious metal metallogenic provinces. In Canada these provinces are most apparent in the northern Appalachian regions of New Brunswick (Bathurst camp) and Quebec (Gaspé copper camp), in a large central region of the Canadian Shield stretching from the Labrador Trough to the Flin Flon base metal camp in Manitoba, and throughout the entire Intermontane region of the Yukon Territory-British Columbia Cordillera (see Fig. 4.2-1).

Climatic conditions will control the rate of oxidation and ultimate composition of secondary phases. Temperature controls the rate of chemical reaction, whereas precipitation controls the rate of leaching and depth of oxidation. Under humid conditions shallow oxidation prevails. Under semiarid conditions deep oxidation, leaching, and enrichment will occur, and under arid conditions deep oxidation may occur, but there will be little leaching and enrichment. It is therefore important to know the historical characteristics of climate during ore formation and the durations of each climatic period.

High primary sulphide contents will produce low pH capillary water conditions and thus greater mobility of some elements out of the oxidation zone. Some elements (e.g., Zn) will not be concentrated in sulphide-rich environments, whereas others (e.g., Cu, Ag) move freely downward to become concentrated under more reducing conditions in the supergene sulphide zone.

High silica and carbonate contents in primary ore, together with moderate to low pyrite contents, will lead to the formation of metal-oxide-silicate-carbonate-sulphate deposits.

It is not difficult to see from the diversity of factors mentioned above that a given mineralized province can produce supergene deposits with widely different tonnages, grades, and mineral assemblages.

RELATED DEPOSIT TYPES

Base metal and precious metal supergene deposits represent just two groups of a large variety of weathered regolith deposits which have been preserved from the effects of glaciation in various regions throughout Canada.

The entire class of preglacial regolith deposits includes residually enriched iron ore deposits (see "Enriched iron-formation", subtype 4.1), residual enrichments over REE- and phosphate-bearing carbonatites (e.g., Cargill and Martinson, Ontario; Lac Shortt, Quebec; see subtype 4.3), kaolin deposits (e.g., Moose River basin, Ontario), silica sand deposits (e.g., Sussex, New Brunswick; north shore of St. Lawrence River, Quebec), gold and platinum placers remobilized from Cenozoic residual profiles (e.g., central Yukon Territory; Similkameen area, British Columbia; Lower St. Lawrence, Quebec), and various saprolitic and lateritic profiles located throughout all of the Canadian geological provinces. To date no lateritic Au, Ni, or Al deposits have been discovered beneath glacial cover or in unglaciated regions of the country, although evidence of lateritization does exist.

EXPLORATION GUIDES

Gossan, ironstone, and leached capping evaluation

The occurrence of gossan zones and gossanous/siliceous cappings at surface is not always indicative of underlying supergene or primary ore. Ironstone formations (false gossans) may occur in many different forms (fault, drainage, stratigraphic, leakage, chemically transported, and lateritic ironstones) and many intrusive bodies, especially those rich in magnetite and/or biotite can have associated leached cappings that are very similar to those overlying porphyry Cu-Mo-Au systems. For this reason a number of mineralogical (Emmons, 1917; Blanchard, 1939; Anderson, 1982), textural (Locke, 1926; Blanchard, 1939), colour (Blanchard, 1939; Anderson, 1982), and lithogeochemical (Clema and Stevens-Hoare, 1973; Bull and Mazzucchelli, 1975, Gulson and Mizon, 1979; Andrew, 1984) techniques have been proposed for differentiating fertile from barren gossans and leached cappings. Today, the most promising lithogeochemical exploration methods involve multi-element geochemical evaluations in concert with sound mineralogical and textural analyses. The creation of a database on known fertile and barren gossans and leached cappings is important. Lithogeochemical interpretations may be further complicated by the fact that chemically transported gossans may form from emergent groundwaters passing through primary ore extant from the gossan, and ferruginized wall rock gossans may be the only exposed portion of an economic supergene zone. Both are, in their own right, indicative of nearby mineralization and their identification requires a good geochemical database of known indicative and false occurrences.

Hydrogeochemistry

Because many indicator elements in gossans and leached cappings are strongly depleted and other elements are converted to stable secondary mineral forms, surficial hydrogeochemical methods are often ineffective in outlining this type of mineralization. For supergene deposits that are still undergoing development or alteration by a lowering water table, groundwater geochemical surveys may be effective in outlining hydrogeochemical haloes around deposits.

Pedogeochemistry

Although many supergene deposits have been well pre-served in the Canadian landscape (Fig. 4.2-1), most have been at least partially affected by glacial scouring. Some deposits have been affected by glacial erosion to the point that only their supergene roots remain. The use of till geochemistry for the detection of this type of mineralization requires the application of different indicator element associations than might be traditionally expected from primary sulphide occurrences. For example, in the Bathurst massive sulphide camp, the gossans overlying mineralization have had more than 95% of their Cu and Zn leached out and Pb and As concentrations are greatly enhanced (Boyle, 1993). Analysis of tills dispersed from these zones would generally give strong anomalies for Pb and As, and occasionally Hg and Bi, but these anomalies would not be supported by coincident anomalies for Cu and Zn. Anomalies unsupported by Cu and Zn in overburden and stream sediments have traditionally been given low exploration priority. Further caution is required when using Au as a tracer of gossan or leached cappings, since this metal is generally present in grain sizes <10 μm and specialized concentration methods must therefore be used. If glaciation has scoured the metal oxide or supergene sulphide zones, special attention must be paid to the fact that the sulphide minerals are generally much finer grained (e.g., sooty chalcocite) or of different composition (e.g., native Cu, Cu-oxides, carbonates) than primary sulphides. This is especially important if the heavy mineral fraction of till is to be used as a sampling medium.

Biogeochemistry

Supergene base metal and precious metal deposits may in some climatic zones be characterized by certain base metal indicator plants, thus permitting the use of geobotanical surveys. Occasionally sulphide oxidation zones may also be outlined as biological 'kill zones' where the overlying soils are incapable of supporting the vegetation typical of that climate. In this respect remote sensing methods can be useful in detecting potential mineralization. Biogeochemical methods, where a specific genus of plant is analyzed for specific indicator elements, can be used in areas of residual and shallow glacial overburden.

Geophysics

Most oxidation zones developed over sulphide mineralization will display strong negative self potentials in the order of -100 to -500 mV (Blain and Brotherton, 1975) and surface self potential surveys are therefore not only useful in detecting supergene zones, but are also helpful in differentiating between fertile and barren (ironstone) gossan zones. The self potential method is not effective where the depth of oxidation is deep (>200 m). Because of differences in acoustic characteristics between oxidized and primary mineral zones, ground penetrating radar can be useful in mapping the thicknesses and topology of supergene zones.

ACKNOWLEDGMENTS

I would like to thank M. Rebagliati, Rebagliati Geological Consulting, Ltd. (South Kemess deposit); the staff of Pacific Sentinel Gold Corp. (Casino deposit); B. Downing, Teck Corporation (Windy Craggy deposit); G. Allard, University of Georgia and J. Guha, Université du Québec à Chicoutami (Chibougamau camp); R. Dimet, Loki Gold Ltd. (Brewery Creek deposits); K. McNaughton, Western Copper Holdings Ltd. (Williams Creek deposit); and Z. Nikic, Imperial Metals Corp. (Mt. Polley deposit) for supplying the author with valuable unpublished deposit data. A review by S.B. Ballantyne of the Geological Survey of Canada is greatly appreciated.

SELECTED BIBLIOGRAPHY

References with asterisks (*) are considered to be the best source of general information on this deposit subtype.

Abercrombie, S.M.
1990: Geology of the Ketza River gold mine; in Mineral Deposits of the Northern Canadian Cordillera, Yukon-Northeastern British Columbia, (ed.) J.G. Abbott and R.J.W. Turner; Geological Survey of Canada, Open File 2169, p. 259-267.

*Alpers, C.N. and Brimhall, G.H.
1989: Paleohydrologic evolution and geochemical dynamics of cumulative supergene metal enrichment at La Escondida, Atacama Desert, northern Chile; Economic Geology, v. 84, p. 229-254.

*Anderson, J.A.
1982: Characteristics of leached capping and techniques of appraisal; in Advances in Geology of the Porphyry Copper Deposits, Southwestern North America, (ed.) S.R. Titley; University of Arizona Press, Tucson, Arizona, p. 275-295.

*Andrew, R.L.
1980: Supergene alteration and gossan textures of base-metal ores in southern Africa; Minerals Science and Engineering, v. 12, p. 193-215.

1984: The geochemistry of selected base-metal gossans, southern Africa; Journal of Geochemical Exploration, v. 22, p. 161-192.

Bell, J.M.
1923: Deep-seated oxidation and secondary enrichment at the Keeley Silver Mine; Economic Geology, v. 18, p. 684-694.

*Blain, C.F. and Andrew, R.L.
1977: Sulphide weathering and the evaluation of gossans in mineral exploration; Minerals Science and Engineering, v. 9, no. 3, p. 119-150.

Blain, C.F. and Brotherton, R.L.
1975: Self potentials in relation to oxidation of nickel sulphide bodies within semi-arid climatic terrains; Institution of Mining and Metallurgy Transaction, v. 84, p. B123-B127.

*Blanchard, R.
1939: Interpretation of leached outcrops; Nevada Bureau of Mines and Geology, Bulletin 66, 196 p.

Bouillon, J.J.
1990: Les Mines Selbaie story - Geology; the Canadian Institute of Mining and Metallurgy, v. 83, no. 936, p. 79-87.

*Boyle, D.R.
1993: Oxidation of massive sulphide deposits in the Bathurst mining camp, New Brunswick - natural analogues for acid drainage in temperate climates; in Environmental Geochemistry of Sulphide Oxidation (ed.) C.N. Alpers, and D.W. Blowes; American Chemical Society Symposium Series, no. 550, p. 535-550.

*Boyle, R.W.
1965: Geology, geochemistry and origin of the lead-zinc-silver deposits of the Keno Hill-Galena Hill area, Yukon Territory; Geological Survey of Canada, Bulletin 111, 302 p.

*Boyle, R.W. and Dass, A.S.
1971: The geochemistry of the supergene processes in the native silver veins of the Cobalt-South Lorraine area, Ontario; in The Silver-Arsenide Deposits of the Cobalt-Gowganda Region, Ontario; Canadian Mineralogist, v. 11, pt. 1, p. 358-390.

*Brimhall, G.H., Alpers, C.N., and Cunningham, A.C.
1985: Analysis of supergene ore-forming processes and groundwater solute transport using mass balance principles; Economic Geology, v. 80, p. 1227.

Brown, J.S.
1936: Supergene sphalerite, galena and willemite at Balmat, New York; Economic Geology, v. 31, p. 331-354.

Bull, A.J. and Mazzucchelli, R.H.
1975: Application of discriminant analysis to the geochemical evaluation of gossans; in Geochemical Exploration 1974, (ed.) I.L. Elliott and W.L. Fletcher; Elsevier, Amsterdam, p. 219-226.

*Butt, C.R.M.
1988: Genesis of lateritic and supergene gold deposits in the Yilgarn Block, Western Australia; in Bicentennial Gold 88, Geological Society of Australia, Abstracts Series (Extended), no. 22, p. 359-364.

*Butt, C.R.M. and Nickel, E.H.
1981: Mineralogy and geochemistry of the weathering of the disseminated nickel sulphide deposit at Mt. Keith, Western Australia; Economic Geology, v. 76, p. 1736-1751.

Carr, J.M. and Reed, A.J.
1976: Afton: a supergene copper deposit; in Porphyry Deposits of the Canadian Cordillera; The Canadian Institute of Mining and Metallurgy, Special Volume 15, p. 376-387.

Carson, D.J.T., Jambor, J.L., Ogryzlo, P., and Richards, T.A.
1976: Bell Copper: geology, geochemistry and genesis of a supergene-enriched biotized porphyry copper deposit with a superimposed phyllic zone; in Porphyry Deposits of the Canadian Cordillera, The Canadian Institute of Mining and Metallergy Special Volume 15, p. 245-263.

Christie, J.S.
1976: Krain; in Porphyry Deposits of the Canadian Cordillera; The Canadian Institute of Mining and Metallurgy, Special Volume 15, p. 182-185.

*Clark, A.H., Cooke, R.V., Mortimer, C., and Sillitoe, R.H.
1967: Relationships between supergene mineral alteration and geomorphology, Southern Atacama Desert, Chile – an interim report; Institution of Mining and Metallurgy Transactions, v. 76, p. 89-96.

Clema, J.M. and Stevens-Hoare, N.P.
1973: A method of distinguishing Ni gossans from other ironstones on the Yilgarn Shield; Journal of Geochemical Exploration, v. 2, p. 393-402.

Drummond, A.D., Sutherland-Brown, A., Young, R.J., and Tennant, S.J.
1976: Gibraltar: regional metamorphism, mineralization, hydrothermal alteration and structural development; in Porphyry Deposits of the Canadian Cordillera; The Canadian Institute of Mining and Metallurgy, Special Volume 15, p. 195-205.

*Emmons, W.H.
1917: The enrichment of ore deposits; United States Geological Survey, Bulletin 625.

*Gilbert, G.
1924: Oxidation and enrichment at Ducktown, Tennessee; Transactions of the American Institute of Mining, Metallurgy and Petroleum Geology Engineers, v. 70, p. 998-1023.

*Guilbert, J.M. and Park, C.F. Jr.
1986: Deposits related to weathering; in The Geology of Ore Deposits, W.H. Freeman and Co., New York, p. 774-831.

Gulson, B.L. and Mizon, K.J.
1979: Lead isotopes as a tool for gossan assessment in base metal exploration; in Geochemical Exploration in Deeply Weathered Terrain, (ed.) R.E. Smith; Commonwealth Scientific and Industrial Research Organization, Perth, Australia, p. 195-201.

*Gustafson, L.B. and Hunt, J.P.
1975: The porphyry copper deposits at El Salvador, Chile; Economic Geology, v. 70, p. 857-912.

*Habashi, F.
1966: The mechanism of oxidation of sulphide ores in nature; Economic Geology, v. 61, p. 587-591.

*Heyl, A.V. and Bozion, C.N.
1962: Oxidized zinc deposits of the United States: Pt 1. General geology; United States Geological Survey, Bulletin 1135-A, p. A1-A52.

*Hughes, F.E. (ed.)
1990: Geology of the mineral deposits of Australia and Papua New Guinea; Australasian Institute of Mining and Metallurgy Publication, Monograph 14, v. 1 and 2, 1828 p.

*Jarrel, O.W.
1944: Oxidation at Chuquicamata, Chile; Economic Geology, v. 39, p. 251-285.

Kwong, Y.T.J., Brown, T.H., and Greenwood, H.J.
1982: A thermodynamic approach to the understanding of the supergene alteration at the Afton copper-mine, south-central British Columbia; Canadian Journal of Earth Sciences, v. 19, p. 2378-2386.

Lindgren, W.
1923: Mineral Deposits (4th edition); McGraw-Hill, New York, 930 p.

*Locke, A.
1926: Leached Ooutcrops as Guides to Copper Ore; Williams and Wilkins Co., Baltimore, 166 p.

*Locke, A., Hall, D.A., and Short, M.N.
1924: Role of secondary enrichment in genesis of the Butte Chalcocite; Transaction of the American Institute of Mining and Metallurgical Engineers, v. 70, p. 933-963.

*Mann, A.W.
1984: Mobility of gold and silver in lateritic weathering profiles: some observations from Western Australia; Economic Geology, v. 79, p. 38-49.

May, E.
1977: Flambeau - a Precambrian supergene enriched massive sulphide deposit; Geoscience Wisconsin, v. 1, p. 1-24.

McCutcheon, S.R.
1992: Base-metal deposits of the Bathurst-Newcastle district: characteristics and depositional models; Exploration and Mining Geology, v. 1, no. 2, p. 105-120.

*Moore, E.S.
1938: Deep oxidation in the Canadian Shield; Canadian Institute of Mining and Metallorgy, Bulletin, v. 41, p. 172-182.

*Muller, D.W.
1972: Geology of the Beltana willemite deposits; Economic Geology, v. 67, no. 8, p. 1146-1167.

*Ney, C.S., Cathro, R.J., Panteleyev, A., and Rotherham, D.C.
1976: Supergene copper mineralization; in Porphyry Deposits of the Canadian Cordillera, The Canadian Institute of Mining and Metallurgy, Special Volume 15, p. 72-78.

*Nickel, E.H.
1984: The mineralogy and geochemistry of the weathering profile of the Teutonic Bore Cu-Pb-Zn-Ag sulphide deposit; Journal of Geochemical Exploration, v. 22, p. 239-264.

*Nickel, E.H., Allchurch, P.D., Mason, M.G., and Wilmhurst, J.R.
1977: Supergene alteration at the Perseverance nickel deposit, Agnew, Western Australia; Economic Geology, v. 72, p. 184-203.

*Nickel, E.H. and Daniels, J.L.
1985: Gossans; in Handbook of Strata-Bound and Stratiform Ore Deposits, Part IV, (ed.) K.H. Wolf; v. 13, p. 261-390.

*Nickel, E.H., Ross, J.R., and Thornber, M.R.
1974: The supergene alteration of pyrrhotite-pentlandite ore at Kambalda, Western Australia; Economic Geology, v. 69, p. 93-107.

Nikic, Z.
in press: Mount Polley; in Porphyry Deposits of the Northwestern Cordillera of North America, The Canadian Institute of Mining and Metallurgy, Special Volume 46.

Palermo, F.G., Fernandez, J.L.B., Magarino, M.G., and Sides, E.J.
1986: Recent investigations and assessment of gossan reserves at Rio Tinto mines; (in Spanish), Boletin Geologico y Minero, v. 97, no. 5, p. 622-642, 31 p. (Geological Survey of Canada Translation No. 3251).

Panteleyev, A., Drummond, A.D., and Beaudoin, P.G.
1976: Berg; in Porphyry Deposits of the Canadian Cordillera, The Canadian Institute of Mining and Metallurgy, Special Volume 15, p. 274-283.

Payne, J., Bower, B., and Delong, C.
in press: Casino deposit; in Porphyry Deposits of the Northwestern Cordillera of North America, The Canadian Institute of Mining and Metallurgy, Special Volume 46.

Rebagliati, C.M., Bowen, B.K., Copeland, D., and Niosi, D.
in press: Kemess Souht and Kemess North porphyry gold-copper deposits, northern British Columbia; in Porphyry Deposits of the Northwestern Cordillera of North America, The Canadian Institute of Mining and Metallurgy, Special Volume 46.

Rennick, M.P. and Burton, D.M.
1992: The Murray Brook deposit, Bathurst Camp, New Brunswick: geological setting and recent developments; Exploration and Mining Geology, v. 1, no. 2, p. 137-142.

Rice, H.M.A.
1941: Nelson map-area, east half, British Columbia; Geological Survey of Canada, Memoir 228, p. 62-83.

Saager, R. and Bianconi, F.
1971: The Mount Nansen gold-silver deposit, Yukon Territory, Canada; Mineralium Deposita, v. 6, p. 209-224.

Sangster, A.L.
1986: Willemite and native silver occurrences, Kirkmount, Pictou County, Nova Scotia; in Current Research, Part A; Geological Survey of Canada, Paper 86-1A, p. 151-158.

Sato, M.
*1950:   Oxidation of sulphide ore bodies: Pt 1. Geochemical environments in terms of Eh and pH; Economic Geology, v. 5, p. 928-961.
*1960:   Oxidation of sulphide ore bodies: Pt 2. Mechanisms of oxidation of sulphide minerals at 25°C; Economic Geology, v. 55, p. 1202-1231.

Sato, M. and Mooney, H.M.
1960:   The electrochemical mechanism of sulphide self-potentials; Geophysica, v. 25, p. 226-249.

*Sillitoe, R.H. and Clark, A.H.
1969:   Copper and copper-iron sulphides as the initial products of supergene oxidation, Copiapo Mining District, Northern Chile;. American Mineralogist, v. 54, p. 1684-1710.

*Sillitoe, R.H., Mortimer, C., and Clark, A.H.
1968:   A chronology of landform evolution and supergene mineral alteration, southern Atacama Desert, Chile; Transactions, Institution of Mining and Metallurgy, v. 77, sec. B, p. B166-B169.

Sinclair, I.G.L. and Gasparrini, E.
1980:   Textural features and age of supergene mineralization in the Detour copper-zinc-silver deposit, Quebec; Economic Geology, v. 75, no. 3, p. 470-477.

Solomon, M.
1967:   Fossil gossans at Mt. Lyell, Tasmania; Economic Geology, v. 62, p. 757-772.

*Stoffregen, R.
1986:   Observations on the behavior of gold during supergene oxidation at Summitville, Colorado, U.S.A. and implications for electrum stability in the weathering environment; Applied Geochemistry, v. 1, p. 549-558.

*Takahashi, T.
1960:   Supergene alteration of zinc and lead deposits in limestone; Economic Geology, v. 55, p. 1083-1115.

*Thornber, M.R.
1975a: Supergene alteration of sulphides I. A chemical model based on massive nickel sulphide deposits at Kambalda, Western Australia; Chemical Geology, v. 15, p. 1-14.
1975b: Supergene alteration of sulphides. II. A chemical study of the Kambalda nickel deposits; Chemical Geology, v. 15, p. 117-144.

Titley, S.R.
1975:   Geological characteristics and environment of some porphyry copper occurrences in the southwestern Pacific; Economic Geology, v. 70, p. 499-514.
1978:   Geologic history, hypogene features, and processes of secondary sulphide enrichment at the Plesyumi copper prospect, New Britain, Papua-New Guinea; Economic Geology, v. 73, no. 5, p. 768-784.
*1982:   The style and progress of mineralization and alteration in porphyry copper systems - American Southwest; in Advances in Geology of the Porphyry Copper Deposits, Southwestern North America, (ed.) S.R. Titley; University of Arizona Press, Tucson, Arizona, p. 93-116.

*Webster, J.G. and Mann, A.W.
1984:   The influence of climate, geomorphology and primary geology on the supergene migration of gold and silver; Journal of Geochemical Exploration, v. 22, p. 21-42.

Williams, D.
1934:   The geology of the Rio Tinto Mines, Spain; Transactions, Institution of Mining and Metallurgy, v. 43, p. 594-640.

4.3 RESIDUAL CARBONATITE-ASSOCIATED DEPOSITS

D.G. Richardson and T.C. Birkett

INTRODUCTION

Weathering and erosion of carbonatites can produce both mechanical accumulations of inert minerals, such as apatite, and secondary, chemically-enriched deposits of relatively immobile elements such as niobium, titanium, yttrium, and rare-earth oxides (REOs). In some cases, vermiculite deposits have also formed during weathering of carbonatite.

IMPORTANCE

Weathered carbonatites are major sources of phosphates in Brazil (Araxá and Catalão) and Finland (Sokli). They dominate the world niobium supply (Araxá and Catalão l) and contain enormous exploitable titanium reserves in the form

of anatase in Brazilian carbonatites (e.g., Minas Gerais State – Salitre I and II, Serra Negra, Tapira; Goiás State – Catalão I; Pará State – Serra de Maicuru). In Canada, residual accumulations of apatite have been formed on carbonatites within Cargill Township and at Martison Lake, Ontario (Fig. 4.3-1) that are potential sources of phosphate. Although no tonnage or grade figures have been published, residuum developed on the carbonatites of the Tomtor alkaline-carbonatite complex, northern Russia (latitude 71°N and longitude 117°E), contains significant concentrations of rare-earth elements, including scandium and yttrium (predominantly monazite, REE-carbonates). According to Epstein et al. (1994), the residuum at Tomtor has potentially economic reserves comparable to those of Araxá (i.e., several hundred million tonnes), covers an area of 8 km by 6 km, ranges in thickness from 20 to 350 m, and has average grades of 3.9% $(REE)_2O_3$, 12% P_2O_5, and 0.74% Nb_2O_5.

SIZE AND GRADE OF DEPOSITS

Significant production of phosphate (apatite, francolite $[Ca_5(PO_4,CO_3)_3(F,OH)]$), niobium (pyrochlore/bariopyrochlore), and titanium (anatase) is derived from very large

Richardson, D.G. and Birkett, T.C.
1996:   Residual carbonatite-associated deposits; in Geology of Canadian Mineral Deposit Types, (ed.) O.R. Eckstrand, W.D. Sinclair, and R.I. Thorpe; Geological Survey of Canada, Geology of Canada, no. 8, p. 108-119 (also Geological Society of America, The Geology of North America, v. P-1).

(>150 Mt of ore), high grade (>8% P_2O_5; >1.2 % Nb_2O_5; >15% TiO_2), enriched carbonatite-associated residual deposits in Brazil and elsewhere. Residual carbonatite-associated phosphate/niobium/titanium deposits range in size from 4 to 1000 Mt grading 7% to >30% P_2O_5; 0.4% to >2.5% Nb_2O_5; and 13% to >27% TiO_2. Published grades and tonnages of Canadian and world deposits are presented in Tables 4.3-1 and 4.3-2, respectively. The grade-tonnage relationships of these deposits are shown in Figure 4.3-2.

GEOLOGICAL FEATURES

Geological setting

Carbonatite deposits, and associated residual enriched deposits, are typically located in relatively stable, anorogenic settings, but some are found near plate margins and may be linked with orogenic activity or rifting. Carbonatites tend to form clusters or provinces (e.g., Brazil, East African rift system), and are commonly located either on

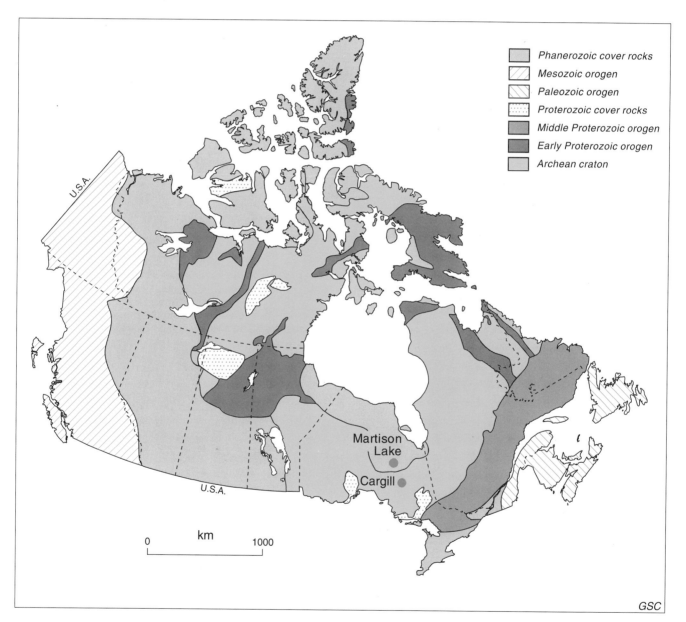

Figure 4.3-1. Location of Canadian residual carbonatite-associated phosphate-niobium deposits and occurrences.

Table 4.3-1. Production/reserves of selected Canadian residual carbonatite-associated deposits and occurrences.

DEPOSIT	PRODUCTION/RESERVES/GRADE	COMMENTS/REFERENCES
Cargill, Ontario Lat. - 49°19'N Long. - 82°49'W	South subcomplex carbonatite (to a depth of ~170 m) diamond drill indicated reserves: 62.5 Mt grading 19.5% P_2O_5. Crandallite-rich zone between residuum and periglacial sediments contains 1 Mt grading 0.015% U_3O_8, 0.07-0.4% Nb_2O_5, and 1.0-4.0% REO (Sage, 1988)	Undeveloped preglacial residual-type phosphate deposit developed over Precambrian (1906 ± 30 Ma) carbonatite-alkalic complex. Of the three subcomplexes present at Cargill, only the southern one has been drilled in detail; the north subcomplex has been explored in a superficial manner; and the small west subcomplex has not been drilled to date. Karstic weathering of the carbonatite in Cretaceous time produced an apatite-rich residuum that has variable thickness. In the preglacial troughs, thickness locally exceeds 170 m and thins to a few metres or disappears altogether on ridges. The Cargill carbonatite complex has undergone post-emplacement deformation and recrystallization as evident by the presence of cataclastic rock textures (Sandvik and Erdosh, 1977; Vos, 1981; Sage, 1988, 1991a; Erdosh, 1989).
Martison Lake, Ontario Lat. - 50°20'N Long. - 83°25'W	Carbonatite A (to a depth of ~150 m) proven + probable + possible diamond drill indicated reserves: 145 Mt grading 20.1% P_2O_5, 0.35% Nb_2O_5. Within a portion of the unconsolidated residuum, a zone of 57 Mt grading 0.4% REO has been delineated (Potapoff, 1989; Mariano, 1989b; Sage, 1991b).	Undeveloped preglacial residual-type phosphate-niobium deposit, covered by 30 to 90 m of glacial till. Economic mineralization is contained in both cemented and unconsolidated residuum. The residuum averages 11 m in thickness, but reaches 50 to 100 m in several locations. The cemented residuum appears to be composed of the same minerals and grain sizes as the unconsolidated residuum. (Woolley, 1987; Potapoff, 1989).

major lithospheric domes or along major lineaments (Woolley, 1989). The Cargill deposit appears to be associated with the Lepage fault, which is contained in the larger regional northeast-trending Kapuskasing magnetic-gravity high/tectonic zone (Percival and West, 1994).

Age of deposits

Because the development of residual enriched carbonatite deposits is dependent on the formation of residuum in response to prolonged periods of weathering of carbonatitic protore, the ages of these deposits are limited by the ages of the associated carbonatites. The Mt. Weld carbonatite in Australia has been dated at 2064 Ma (Middlemost, 1990), and the overlying residual deposit, which formed during the Mesozoic, represents a laterite that has developed over one of the oldest known carbonatites (Mariano, 1989a). The known carbonatites of the Amazon Basin, including Morro do Seis Lagos, in the State of Amazonas, and Serra de Maicuru, in the state of Pará, are thought to be Precambrian. In contrast, the carbonatites that rim the periphery of the Paraná basalt flow basin (i.e., Catalão I and II, Serra Negra, Salitre I and II, Araxá and Tapira) were emplaced during the Cretaceous (Eby and Mariano, 1992). Development of extensive residuum over Brazilian carbonatites began at the start of the Cenozoic, and, in the case of carbonatites in the Amazon Basin, is ongoing (CBMM, 1984). The Cargill and Martison Lake residual carbonatites are thought to have formed during Cretaceous time (Sage, 1988).

Form of deposit

The differential leaching of carbonatitic material, notably along joints, related fractures, and particularly faults, resulted in the development in many carbonatite complexes of a highly irregular, karstic topography characterized by pinnacles of unweathered carbonatite bedrock separating pits and depressions filled with residual weathered material. The faults not only determine the sites of deeper weathering, but also control the development of internal drainage and sinkholes. Therefore, depending on the nature of the topography on the underlying primary carbonatite, enriched carbonatite-associated deposits commonly have blanket-like to highly irregular forms that overlie carbonatite protore, and phosphate-rich residuum produced in this manner usually contains both primary apatite derived from the carbonatite and supergene crandallite-group minerals (Notholt, 1980). The well-developed, high relief, buried preglacial topography of the Cargill deposit is a good example of the development of a karstic system. The south subcomplex is dominated by the development of three subparallel northeast-trending troughs in the carbonatite (Fig. 4.3-3A, B). Two major troughs, both filled with residuum greater than 170 m thick, wrap around the outer edge of the carbonatite and converge in a large buried topographic low of undetermined depth, which has been interpreted by Sandvik and Erdosh (1977, 1984) and Erdosh (1989) to be a large 'master sinkhole' solution collapse structure. It is through this structure that leached carbonate solution products were removed, leaving behind the highest concentration of apatite-rich residuum. The troughs are commonly steep-sided, and at several locations, depth exceeds width. The Martison Lake residual carbonatite deposits (Fig. 4.3-4, 4.3-5) lack such karstic features, and instead consist of nearly horizontal layers that range from 1 to 55 m thick and have flat tops and irregular keels.

Mineralogy and ore zonation

The mineralogy of residual carbonatite-associated deposits is highly variable. Chemical weathering of carbonatites and the dissolution of Ca and Mg contained in carbonate minerals (calcite, dolomite, and siderite) results in concentration of resistant primary minerals (e.g., apatite, magnetite, pyrochlore, ilmentite, rutile, zircon, and quartz), as well as formation of supergene phosphates, carbonates, and sulphates containing less mobile elements released during carbonate dissolution. Potentially economic supergene mineralization associated with enriched residual carbonatites includes the following:

1) In situ lateritic decalcification of perovskite ($CaTiO_3$) in pyroxenites of the carbonatite complexes of Goiás, Minas Gerais, and Pará states of Brazil produces polycrystalline aggregates of micrometre-size anatase (TiO_2) platelets (Mariano, 1989a; Kurtanjek and Tandy, 1989; Mariano and Mitchell, 1990).

2) The most common supergene REE mineral formed is monazite [(REE)PO_4]. Other REE minerals commonly found in carbonatite laterites include bastnaesite [(Ce,La)(CO$_3$)F], parisite [(Ce,La)$_2$Ca(CO$_3$)$_3$F], synchysite [(Ce,La)Ca(CO$_3$)$_2$F], cerianite [(Ce^{+4},Th)O$_2$], and the crandallite-group minerals florencite [(REE)Al$_3$(PO$_4$)$_2$(OH)$_6$], gorceixite [(Ba,REE)Al$_3$(PO$_4$)$_2$(OH$_5$·H$_2$O)], and goyazite [(Sr,REE)Al$_3$(PO$_4$)$_2$(OH)$_5$·H$_2$O)]. Rhabdophane [(REE)PO$_4$·H$_2$O] is less common. Light REE-bearing supergene minerals generally predominate over heavy REE minerals. At Mt. Weld, supergene light REE-bearing monazite and rhabdophane are contained in the upper part of the residuum, whereas heavy REEs and Y are selectively concentrated at depth as xenotime [YPO$_4$] and churchite [YPO$_4$·2H$_2$O].

Vertical mineralogical zonation displayed by carbonatites that have undergone extreme lateritic weathering has been summarized by Mariano (1989a, b). In the upper parts, lateritic residuum consists primarily of insoluble ferric iron oxides, aluminum oxides, clays, supergene monazite, crandallite-group minerals rich in Ba, Nb, light REEs and Sr, and cerianite. In this zone all primary minerals have disappeared with the exception of zircon and

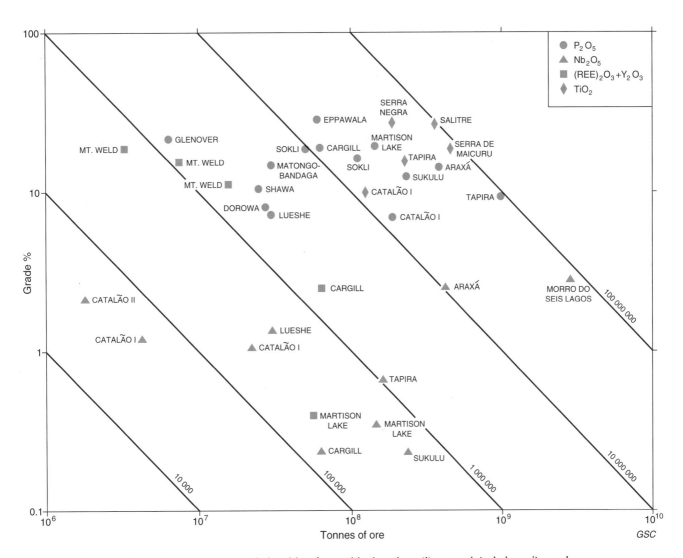

Figure 4.3-2. Grade-tonnage relationships for residual carbonatite-associated deposits and occurrences. P$_2$O$_5$, Nb$_2$O$_5$, (REE)$_2$O$_3$ + Y$_2$O$_3$, and TiO$_2$ are expressed in weight per cent. The diagonal lines indicate the quantity, in tonnes, of the contained commodity in the deposits.

Table 4.3-2. Production/reserves of selected foreign residual carbonatite-associated deposits and occurences.

DEPOSIT	PRODUCTION/RESERVES/GRADE	COMMENTS/REFERENCES
Mt. Weld, Western Australia	Indicated reserves: 15.4 Mt; 11.2% (REO + Y_2O_3), or 7.43 Mt; 15.7% (REO + Y_2O_3), or 3.47 Mt; 19.6% (REO + Y_2O_3)	Proterozoic (2021 ± 13 Ma) circular carbonatite structure having a diameter of about 4 km. A residuum rich in primary igneous apatite overlies the carbonatite and in turn is overlain by a supergene zone that contains abundant secondary phosphates and aluminophosphates with elevated concentrations of REEs, U, Th, Nb, Ta, Zr, Ti, Ba, Sr, and Fe. Supergene monazite is the principal REE hostmineral;lesser quantities are in apatite, crandallite group minerals, cerianite, churchite [$YPO_4 \cdot 2H_2O$] and rhabdophane [$(Nd,Ce,La)PO_4 \cdot H_2O$] (Duncan and Willett, 1989; Mariano, 1989a; Lottermoser, 1990).
Sokli, Finland	Proven reserves: 50 Mt; 19% P_2O_5 Indicated reserves: 110 Mt; 16.5% P_2O_5	The circular Sokli residual carbonatite has an aerial extent of 18 km^2 and is part of the Kola Peninsula igneous alkaline province. Carbonatite-apatite-francolite regolith forms the phosphate deposit at Sokli. The thickness of the regolith averages 25 m but locally attains a maximum thickness of 72 m. Residual primary minerals derived from carbonatite include apatite, magnetite, amphiboles, micas, dolomite, ilmenite, pyrochlore, zircon, baddeleyite, and rutile; secondary minerals of supergene origin include francolite, goethite, siderite, Mn-oxides, hematite, and rhabdophane (Notholt, 1979; Vartiainen, 1989; Harben and Bates, 1990b).
Eppawala, Anuradhapura, Sri Lanka	Proven & inferred reserves (Block A to F): 60 Mt; 20-38% P_2O_5, 0.24% Nb_2O_5	Residuum/lateritic soil developed to a depth of 75 m over a Proterozoic (?) apatite-magnetite-bearing carbonatite. Phosphate deposit is composed of chlor-apatite, francolite, martite, and goethite. Six deposits (Block A to F) have been delineated by drilling (Notholt, 1980; Jayawardena, 1989).
African deposits Sukulu, Uganda	Reserves in North, South, and West valleys: 230.7 Mt; 12.8% P_2O_5, 0.24% Nb_2O_5	Apatite-rich residuum, ranging in thickness from 15 m to a maximum of 67 m, developed on a circular carbonatite body having a diameter of approximately 4 km. The apatite-rich soil consists of 20-25% apatite, 20-25% crandallite, 15-20% quartz, 20% magnetite, and 10% hematite (Kabagambe-Kaliisa, 1989).
Glenover, western Transvaal, Republic of South Africa	Drill indicated reserves: 6.3 Mt; 21.8% P_2O_5	Ferruginous carbonate apatite residuum formed by karstic weathering of carbonatite. Economic deposit is restricted to a central pear-shaped high grade zone (maximum of 32% P_2O_5 and 10% Fe_2O_3) (Deans, 1978).
Dorowa and Shawa, Zimbabwe	Indicated reserves - Dorowa south orebody to a depth of 120 m: 27 Mt; 8% P_2O_5 Indicated reserves at Shawa to a depth of 23 m: 25 Mt; 10.8% P_2O_5	Residuum developed over late Paleozoic to Mesozoic decalcified Dorowa and Shawa ijolite-syenite-carbonatite pipe-like bodies that have intruded Archean granite gneiss (Deans, 1978; Fernandes, 1989).
Lueshe, Zaire	Estimated reserves: 30 Mt; 1.34% Nb_2O_5, 4.2-10% P_2O_5	Mantle of niobium-rich ferruginous residuum developed over an elliptical (3.0 km x 1.9 km) plug of pyrochlore-aegirine-carbonatite. Moderate degree of weathering has resulted in pyrochlore becoming friable (Gittins, 1966; Mariano, 1989b).
Matongo-Bandaga, Burundi	Drill indicated reserves: 30 Mt; 11% P_2O_5 (high grade zone = 13.2 Mt; 13% P_2O_5)	Phosphate deposits lie in subhorizontal lateritic masses directly overlying fresh carbonatite. Thicknesses of individual apatite-rich bands are highly variable, and the maximum thickness observed to date is 55 m. Troughs formed in the irregular subcrop of the carbonatite are the sites of deepest weathering and are marked by thickening of phosphate-rich zones. Karstic weathering is evident and residuum minerals are apatite, niobian rutile, ilmenite, quartz, K-feldspar, and goethite. Fluorapatite, cacoxenite [$Fe_4(PO_4)_3(OH)_3 \cdot 12H_2O$], goyazite [$SrAl_3(PO_4)_2(OH)_5 \cdot H_2O$] and crandallite/pseudo-wavellite [$CaAl_3(PO_4)_2(OH)_5 \cdot H_2O$] may also be present in the phosphate-rich zones (Kurtanjek and Tandy, 1989; Mariano, pers. comm., 1994).

Table 4.3-2. (cont.)

DEPOSIT	PRODUCTION/RESERVES/GRADE	COMMENTS/REFERENCES
Brazilian deposits		
Barreiro, Araxá, Minas Gerais	Proven and indicated reserves: 375 Mt; 14.5% P_2O_5 418 Mt; 2.5% Nb_2O_5 495 000 t; 10-11% REO	Deposit is in Late Cretaceous circular, 16 km², carbonatite (minor pyroxenite and glimmerite). Main carbonatite complex is capped by 150 to 230 m thick, deeply weathered residuum. Three distinct zones of the residuum have been delineated: 1) niobium as bariopyrochlore $[(Ba,Sr)_2(Nb,Ti)_2(O,OH)_7]$ in a 1 km² area, in the centre of the complex with associated REEs; 2) secondary carbonate apatite in the northwest and southeast part of the complex; and 3) REEs contained in earthy monazite and goyazite in the northeast quadrant of the complex (800 000 tonnes grading 13% REO). Only niobium and phosphate are presently being mined (Woolley, 1987; Mariano, 1989b; Gomes et al., 1990)
Catalão I & II, Goias	Proven + inferred reserves, Catalão I: 188 Mt; 7.04% P_2O_5 4.2 Mt; 1.20% Nb_2O_5 22 Mt; 1.07% Nb_2O_5 125 Mt; 10% TiO_2	Catalão I is hosted in an approximately 27 km² circular stock of pyroxenite-serpentinized peridotite-glimmerite-carbonatite. The lateritic residuum developed on the central core carbonatite has an average thickness of 30 m. The residuum contains supergene monazite, rhapdophane, florencite, and gorceixite/goyazite. Anatase, derived from the weathering of perovskite, overlies the phosphate-rich ore (CBMM, 1984; Woolley, 1987; Mariano, 1989a, b; Gomes et al., 1990).
	Estimated reserves, Catalão II: 1.8 Mt; 2.19% Nb_2O_5	Catalão II, located 15 km north of the Catalão I deposit, is hosted in a 14 km² circular calcitic carbonatite-phoscorite-pyroxenite intrusive dome. Bariopyrochlore is contained in lateritic soil that is 3 to 10 m thick (CBMM, 1984; Woolley, 1987; Gomes et al., 1990).
Tapira, Minas Gerais	976 Mt; 8.11% P_2O_5 162 Mt; 0.67% Nb_2O_5 229 Mt; 15.98% TiO_2	Late Cretaceous, oval-shaped, 35 km² intrusion of different generations of carbonatite emplaced in Precambrian quartzites. The complex is deeply weathered and the average thickness of the residuum is 30 m, but locally attains 250 m. The residual deposit consists of 1) apatite ores; 2) titanium (anatase) ore, which overlies the phosphate ores; and 3) niobium (bariopyrochlore and pyrochlore) mineralization (CBMM, 1984; Woolley, 1987; Harben and Bates, 1990c; Gomes et al., 1990).
Patrocinio area (Bananeira deposit), Minas Gerais	Salitre: 353 Mt; 27.50% TiO_2	The Bananeira deposit is actually composed of the Salitre and Serra Negra deposits. The 500 m² Salitre Late Cretaceous carbonatite plug is associated with a larger 35 km² syenite-nepheline syenite intrusive complex. The residuum is generally 37 m to 60 m thick, but locally exceeds 100 m. Residuum contains anatase, francolite, perovskite, and magnetite with high values of U, Th, and REEs (CBMM, 1984; Woolley, 1987; Gomes et al., 1990).
	Serra Negra: 181 Mt; 27.68% TiO_2	Circular carbonatite-pyroxenite stock, having a diameter of 9 km, contained in 65 km² oval complex of nepheline syenites and major peripheral bebedourite pyroxenites containing significant perovskite and magnetite. The residuum overlying the carbonatite averages 150 m in thickness and contains a significant concentration of anatase (Woolley, 1987; Gomes et al., 1990).
Morro do Seis Lagos, Amazonas	Indicated resources: 2800 Mt; 2.81% Nb_2O_5	Reported to be the world's largest niobium deposit; consists of three north-south aligned circular structures with approximate diameters of 5.5, 0.75, and 0.5 km. Laterite covered hills developed over ferrocarbonatite, carbonatite breccias, and syenite which intrude Precambrian gneisses and migmatites. Laterites have an average thickness of 230 m, are radioactive and consist of siderite, ankerite, dolomite, barite, goethite, pyrite, and niobium rutile (Woolley, 1987; Verwoerd, 1989; Gomes et al., 1990).
Serra de Maicuru, Pará	Potential reserves: 450 Mt; 19% TiO_2	Deeply weathered laterite developed over circular topographic high consisting of pyroxenites, alkali syenites, and carbonatites that are intrusive into rocks of the Guyana Shield. Economic mineralization consists primarily of anatase derived from decalcification of perovskite (Woolley, 1987; Mariano, 1989b; Gomes et al.,1990).

rutile. Vadose water or low-temperature hydrothermal solutions can introduce quartz and supergene apatite (francolite) into the laterite. As lateritic weathering diminishes with depth, magnetite is encountered, followed by partly decomposed pyrochlore rimmed by crandallite-type minerals rich in Nb. At deeper levels, fresh pyrochlore is present, and primary apatite is prominent. With increasing depth, partly decomposed calcite/dolomite or siderite is encountered, below which is fresh unweathered carbonatite.

The mineralogy of the residuum of the Cargill south subcomplex has been described by Sandvik and Erdosh (1977, 1984), Vos (1981), and Erdosh (1989). The unconsolidated sand-sized residuum consists of primary white or colourless apatite, with minor goethite, siderite, magnetite, and crandallite [$CaAl_3(PO_4)_2(OH)_5 \cdot H_2O$]. Calcite and dolomite are absent. In many places, the apatite residuum is diluted by other weathering products, such as clay, vermiculite, iron oxide, quartz, and chlorite. A thin supergene REE-bearing crandallite-rich blanket, which has not been investigated in detail, is present in many places at the top of the apatite residuum. Secondary pyrite occurs locally near the top of the residuum (Vos, 1981; Sandvik and Erdosh, 1984).

At Martison Lake, the cemented residuum constitutes about 10% of the total mass of residuum, except in those areas where the thickness exceeds 20 m, where it comprises 25% of the total residuum (Fig. 4.3-5). The cemented residuum appears to be composed of the same minerals and grain sizes (i.e., predominantly silt size – 0.002-0.06 mm; with subordinate amounts of clay size – <0.002 mm, and sand size – 0.06-1.1 mm) as the unconsolidated residuum, but contains greater amounts of secondary phosphate minerals and lesser amounts of lanthanide (light REEs), clay, and iron minerals (Potapoff, 1989). The main primary minerals of the residuum are apatite, magnetite, phlogopite, biotite, and pyrochlore; the dominant secondary minerals are francolite, florencite [$(La,Ce)Al_3(PO_4)_2(OH)_6$], goethite, limonite, hematite, and clay minerals. Primary and secondary accessory minerals, which occur in minor quantities and only in restricted areas of the residuum, include: columbite, chlorite, alkali feldspar, perovskite [$CaTiO_3$], zirkelite [$(Ca,Th,Ce)Zr(Ti,Nb)_2O_7$], monazite [$(La,Ce,Nd)PO_4$], bastnaesite [$(La,Ce)(CO_3)F$], siderite, crandallite,

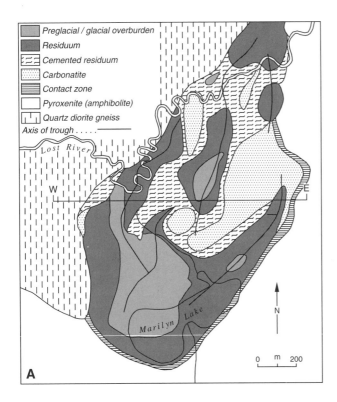

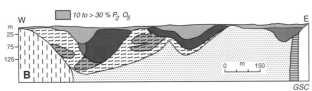

Figure 4.3-3. A) Geology of the South Cargill Subcomplex - 50 m level (after Erdosh, 1989), **B)** Typical west-east section across the central part of the South Subcomplex; areas in red denote zones of high grade (10% to >30%) P_2O_5 mineralization (after Sandvik and Erdosh, 1977).

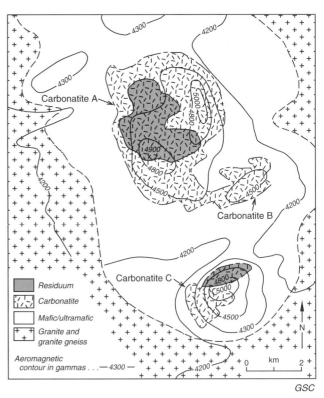

Figure 4.3-4. Aeromagnetic expression of the Martison Lake area, showing location of carbonatite bodies (after Potapoff, 1989).

woodhousite [$CaAl_3(PO_4)(SO_4)(OH)_6$], waylandite [$(Bi,Ca)Al_3(PO_4,SiO_4)_2(OH)_6$], barite, pyrite, ilmentite, rutile, cerianite [$(Ce^{+4},Th)O_2$], zircon, quartz, and sphene (Potapoff, 1989; Sage, 1991b). As shown in Figures 4.3-5B and C, the cemented, and to a lesser extent, the unconsolidated residuum contain the most significant mineralized zones ($P_2O_5>30\%$; $Nb_2O_5>2\%$). Potapoff (1989) noted that anomalous (>0.12% rare-earth oxides) lanthanide concentrations only occur in unconsolidated residuum that is also enriched in niobium (>0.5% Nb_2O_5). This correlation is likely attributed to the substitution of REEs in residual apatite and pyrochlore, rather than the presence of independent REE-bearing minerals in the residuum.

DEFINITIVE CHARACTERISTICS

Residual enriched carbonatite deposits form irregular to blanket-like layers draped over eroded weathered carbonatitic intrusions (Dawson and Currie, 1984). Although the geometry and mineralogy of these deposits can be highly variable, these deposits all formed by supergene processes and/or the residual concentration of primary materials during prolonged periods of weathering. Weathering develops residuum that can contain concentrations of economically important elements (P, Nb, Ti, REEs) that are as much as ten times greater than those found in the underlying unweathered carbonatite (Sage, 1991a).

GENETIC MODEL

Carbonatites weather relatively easily under a variety of climatic conditions and in some cases contain mineral deposits formed by surficial processes. The climate during weathering determines which rock is most resistant to chemical erosion, and consequently which particular types of deposits are developed. According to Mariano (1989b), the best conditions for the development of abundant supergene REE mineralization in carbonatites exist in humid tropical climates with moderate to high rainfall conditions (e.g., Amazon basin of Brazil), and in complexes in which a karst system is absent and interior drainage and a basin-type topography allows for the entrapment of both eluvial and supergene residuum from decalcified carbonatites. Sandvik and Erdosh (1977) and Erdosh (1979) noted that in dry, hot climates (e.g., South Africa and east Africa) carbonatites in general tend to form topographic highs; under more temperate cool, moist climates (e.g., Cargill and Martison Lake), a greater degree of leaching results in the development of karstic features. The development of topographic lows associated with karstic terrane enhances the potential for larger eluvial accumulations of primary apatite and/or pyrochlore, but not for the development of extensive accumulations of secondary REE minerals. This is because karst weathering effectively removes and flushes the elements dissolved from carbonates and apatite (i.e., REEs, Ca, Mg, Sr, and Ba) out of the carbonatite. As noted by Mariano (1989b), in carbonatites in which karstic weathering has occurred to an advanced level (e.g., Matongo-Bandaga, Burundi, and Cargill, Canada), the accumulation of significant concentrations of secondary REE minerals is negligible.

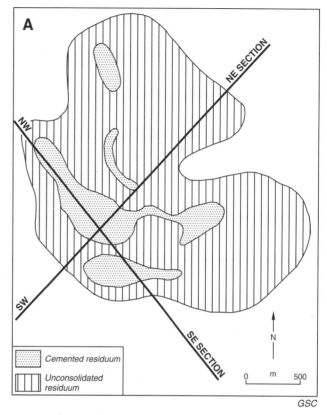

Figure 4.3-5A. Geology of the Martison Lake 'A' carbonatite, showing location of unconsolidated and cemented residuum and traces of NW-SE and SW-NE sections (after Potapoff, 1989).

The preservation of resistant phases is probably linked to a number of factors which, according to Mariano (1989b), include:

1. the concentration of H_2CO_3 in meteoric waters;

2. the development of H_2SO_4 from accessory sulphides; and

3. the development of internal drainage systems within carbonatites in response to carbonatite lithology.

Alternatively, supergene phosphate, carbonate, and sulphate minerals can form subsequent to the release of less mobile elements during chemical dissolution of calcite, dolomite, and siderite (e.g., REE mineralization at Mt. Weld, and crandallite-rich portions of the Cargill deposit of Ontario). Mariano (1989a, b) and Möller (1989a) have suggested that REEs derived from the decalcification of primary carbonates, apatites, and perovskite in the presence of oxygen, PO_4^{3-}, and F^-, commonly form residual supergene REE minerals that remain stable under conditions of lateritic weathering.

115

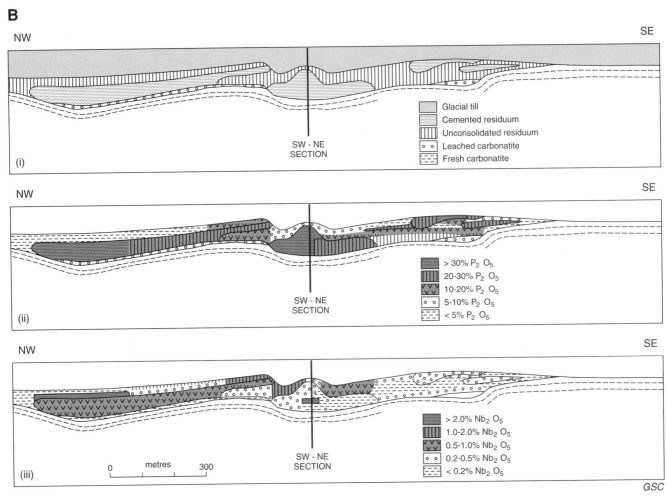

B

NW

SE

(i)

SW - NE
SECTION

Glacial till
Cemented residuum
Unconsolidated residuum
Leached carbonatite
Fresh carbonatite

NW

SE

(ii)

SW - NE
SECTION

> 30% P_2O_5
20-30% P_2O_5
10-20% P_2O_5
5-10% P_2O_5
< 5% P_2O_5

NW

SE

(iii)

SW - NE
SECTION

> 2.0% Nb_2O_5
1.0-2.0% Nb_2O_5
0.5-1.0% Nb_2O_5
0.2-0.5% Nb_2O_5
< 0.2% Nb_2O_5

0   metres   300

GSC

Figure 4.3-5B. NW-SE section of the Martison Lake A carbonatite including: i) geology, ii) P_2O_5 grade, and iii) Nb_2O_5 grade (after Potapoff, 1989). Areas in red denote potential ore zones,

RELATED DEPOSIT TYPES

Deep chemical weathering of other types of mineral deposits and lithologies can result in the formation of enriched residual deposits of other types. Significant examples include:

1. Oxidized and chemically enriched zones of porous, friable iron-formation and earthy hydrated iron oxide developed on Lake Superior-type and Algoma-type iron-formation protore (see subtype 4.1 – "Enriched iron-formation").

2. Physico-chemical oxidation processes which led to the formation of supergene precious and base metal zones that overlie massive sulphide deposits; the upper portion of sulphide vein, shear/fault related, and replacement deposits; and porphyry Cu-Mo-Au-Ag deposits (see subtype 4.2 - "Supergene base metals and precious metals").

3. Lateritic or silicate aluminum bauxites associated with, and derived from, the weathering of rocks rich in aluminum silicates (e.g., nepheline syenites, kaolinitic sandstones and arkoses and clays) (Harben and Bates, 1990a).

4. Lateritic nickeliferous ores (garnierite [$(Ni,Mg)_3$ $Si_2O_5(OH)_4$]) that have resulted from the weathering of ultramafic rocks, peridotites, and serpentinites in tropical to subtropical climates (e.g., New Caledonia, Indonesia). In São Paulo state, Brazil, which contains numerous residual carbonatite-associated deposits, part of the residuum overlying the peridotites-pyroxenites of the Jacupiranga alkaline-carbonatite complex contains in excess of 2.7 Mt of garnierite-rich ore with an average grade of 1.39% Ni (Woolley, 1987; Gomes et al., 1990).

5. Residual manganese oxide (pyrolusite) deposits that usually overlie manganiferous carbonaceous black shale or manganese-bearing carbonate rocks (e.g., Moanda mine, Gabon; Nsuta mine, Ghana; Serra do Navio deposit, Brazil) (DeYoung et al., 1984; Machamer, 1987).

C

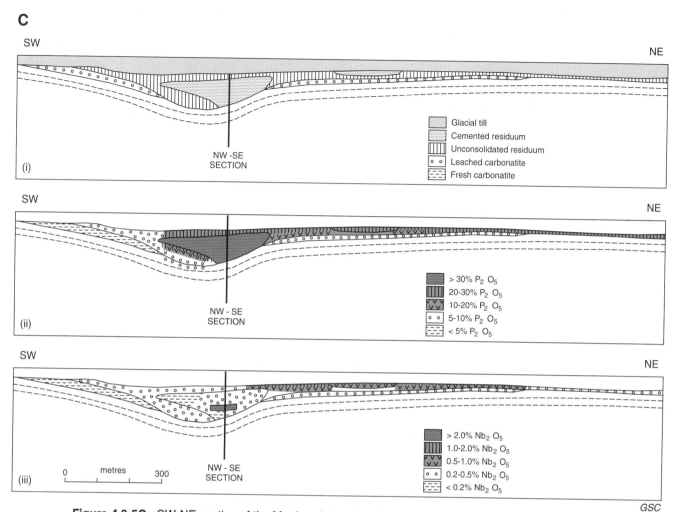

Figure 4.3-5C. SW-NE section of the Martison Lake A carbonatite showing: i) geology, ii) P₂O₅ grade, and iii) Nb₂O₅ grade (after Potapoff, 1989). Areas in red denote potential ore zones.

6. Vermiculite deposits produced by the supergene alteration of mafic/ultramafic minerals (biotite, phlogopite, diopside, hornblende, serpentine) by the combined effects of weathering and circulating meteoric water.

EXPLORATION GUIDES

Exploration guidelines for residual enriched carbonatite deposits include the following:

1. Broad scale features: when the carbonatites are exposed, topography is usually the best exploration guide. Because the protoliths of these deposits usually consist of nepheline syenite-carbonatite bodies that form rounded plugs and curved bodies centrally located within relatively small (generally <50 km²), steep-sided alkaline complexes, recognition of discrete annular features can be particularly useful in identifying sites that may contain mineralized carbonatites (Notholt, 1980). Potentially economic minerals can be found in thin surface crustifications in cool temperate climates and to depths exceeding 300 m in tropical regions (Mariano, 1989b). Usually, fresh rock in outcrop is absent or extremely rare. In the Brazilian residual deposits, concentric/radial faults and fractures have resulted in the development of distinctive ring-like, and radial drainage which can control the distribution of the deposits (Gomes et al., 1990).

2. Geochemical approaches: stream sediment and soil sampling, and panning of heavy mineral concentrates (anatase, pyrochlore, monazite) from the residuum developed over certain ultramafic-carbonatite complexes in Brazil has been successful in delineating zones of both primary and secondary mineralization (Costa et al., 1991). However, in areas where the background levels of crucial indicative trace elements (REEs, Nb, Ti, P, etc.) are high (i.e., small carbonatite(s) that have intruded areally extensive alkaline plutons), geochemical sampling may be inadequate to localize potential mineralized zones (Möller, 1989b). According to Mariano (pers. comm., 1994) in Minas Gerais and Goiás

117

states, Brazil, once a carbonatite has been discovered, the standard exploration procedure includes detailed soil sampling and analysis for Nb_2O_5, REOs, PO_4, and TiO_2.

3. Geophysical approaches: most residual enriched carbonatite deposits, including Cargill and Martison Lake, are characterized by pronounced, closed, positive magnetic anomalies and coincident elevated radiometric responses (Fig. 4.3-4) (Notholt, 1979; Sage, 1979; CBMM, 1984; Mariano, 1989a; Potapoff, 1989). However, because strongly alkali conditions typically associated with late-stage carbonatites and fenite aureoles can, in many cases, stabilize iron in the ferric state and prevent the development of magnetite, deposits may also have a negative magnetic response. Depending on U and Th contents in the residuum, airborne gamma ray spectrometry results can be as much as 10 to 20 times regional background. Large amounts of magnetite (martite) are typically found in residual enriched carbonatites and result in well-defined circular aeromagnetic anomalies. If the host rocks contrast sufficiently in density from surrounding phases of the complexes, a marked gravity response in conjunction with a positive magnetic anomaly may reflect the presence of residual-carbonatite deposits in alkaline-carbonatite complexes (Erdosh, 1979).

4. Specific geochemical and geophysical exploration approaches are required for each carbonatite. The techniques applied should be appropriate to the characterisitics displayed by the deposit (i.e., petrology, mineralogy, and geochemistry of both protolith and laterite).

ACKNOWLEDGMENTS

A.N. Mariano is thanked for reviewing the manuscript and for providing insightful comments and significant new information regarding Brazilian residual carbonatite deposits. R.P. Sage and W.D. Sinclair are also thanked for critically reading the manuscript.

SELECTED BIBLIOGRAPHY

References with asterisks (*) are considered to be the best source of general information on this deposit subtype. The references listed below that have not been specifically cited in the text are provided for the reader's interest.

*Companhia Brasileira de Metalurgia e Mineração (CBMM)
1984: Carbonatitic complexes of Brazil: geology; Companhia Brasileira de Metalurgia e Mineração (CBMM), São Paulo, 44 p.

Costa, M.L., Fonseca, L.R., Angélica, R.S., Lemos, V.P., and Lemos, R.L.
1991: Geochemical exploration of the Maicuru alkaline-ultramafic-carbonatite complex, northern Brazil; Journal of Geochemical Exploration, v. 40, p. 193-204.

*Dawson, K.R. and Currie, K.L.
1984: Carbonatite-hosted deposits; in Canadian Mineral Deposit Types: A Geological Synopsis, (ed.) O.R. Eckstrand; Geological Survey of Canada, Economic Geology Report 36, p. 48-49.

Deans, T.
1966: Economic mineralogy of African carbonatites; in Carbonatites, (ed.) O.F. Tuttle and J.C. Gittins; John Wiley & Sons, New York, p. 385-413.
1978: Mineral production from carbonatite complexes: a world review; in Proceedings of the First International Symposium on Carbonatites, Poços de Caldas, Brazil, 1976; Departamento Nacional da Produçao, Rio de Janeiro, p. 123-133.

DeYoung, J.H., Sutphin, D.M., and Cannon, W.F.
1984: International strategic minerals inventory summary report - manganese; United States Geological Survey, Circular 930-A, 22 p.

Duncan, R.K. and Willett, G.C.
1989: High grade lanthanide and yttrium mineralization in the palaeoregolith of the Mt. Weld carbonatite, Western Australia; in Program with Abstracts, Geological Asssociation of Canada/Mineralogical Association of Canada, v. 14, p. A20.

Eby, G.N. and Mariano, A.N.
1992: Geology and geochemistry of carbonatites and associated alkaline rocks peripheral to the Paraná Basin, Brazil-Paraguay; Journal of South American Earth Sciences, v. 6, no. 3, p. 207-216.

Epstein, E.M., Danil'chenko, N.A., and Postnikov, S.A.
1994: Geology of the unique Tomtor deposit of rare metals (north of the Siberian Platform); Geology of Ore Deposits, v. 36, no. 2, p. 75-100.

Erdosh, G.
1979: The Ontario carbonatite province and its phosphate potential; Economic Geology, v. 74, p. 331-338.
1989: Cargill carbonatite complex, Canadian Precambrian Shield; in Phosphate Deposits of the World: Volume 2 - Phosphate Rock Resources, (ed.) A.J.G. Notholt, R.P. Sheldon, and D.F. Davidson; International Geological Correlation Programme, Project 156: Phosphorites, Cambridge University Press, p. 36-41.

Fernandes, T.R.C.
1989: Dorowa and Shawa: Late Paleozoic to Mesozoic carbonatite complexes in Zimbabwe; in Phosphate Deposits of the World: Volume 2 - Phosphate Rock Resources, (ed.) A.J.G. Notholt, R.P. Sheldon, and D.F. Davidson; International Geological Correlation Programme, Project 156: Phosphorites, Cambridge University Press, p. 171-175.

Gittins, J.C.
1966: Summaries and bibliographies of carbonatite complexes; in Carbonatites, (ed.) O.F. Tuttle and J.C. Gittins; John Wiley & Sons, New York, p. 417-570.

Gomes, C.B., Ruberti, E., and Morbidelli, L.
1990: Carbonatite complexes from Brazil: a review; Journal of South American Earth Sciences, v. 3, no. 1, p. 51-63.

Harben, P.W. and Bates, R.L.
1990a: Bauxite; in Industrial Minerals: Geology and World Deposits, Industrial Minerals Division, Metal Bulletin Plc., London, p. 19-26.
1990b: Phosphate rock; in Industrial Minerals, Geology and World Deposits, Industrial Minerals Division, Metal Bulletin Plc., London, p. 190-204.
1990c: Titanium and zirconium minerals; in Industrial Minerals: Geology and World Deposits, Industrial Minerals Division, Metal Bulletin Plc., London, p. 282-294.

Heinrich, E.W.
1966: The Geology of Carbonatites; Rand McNally & Company, Chicago, 555 p.

Jayawardena, D.E. de S.
1989: The phosphate resources of the Eppawala carbonatite complex, northern Sri Lanka; in Phosphate Deposits of the World: Volume 2 - Phosphate Rock Resources, (ed.) A.J.G. Notholt, R.P. Sheldon, and D.F. Davidson; International Geological Correlation Programme, Project 156: Phosphorites, Cambridge University Press, p. 458-460.

Kabagambe-Kaliisa, F.A.
1989: The Sukulu phosphate deposits, south-eastern Uganda; in Phosphate Deposits of the World: Volume 2 - Phosphate Rock Resources, (ed.) A.J.G. Notholt, R.P. Sheldon, and D.F. Davidson; International Geological Correlation Programme, Project 156: Phosphorites, Cambridge University Press, p. 184-186.

Kurtanjek, M.P. and Tandy, B.C.
1989: The igneous phosphate deposits of Matongo-Bandaga, Burundi; in Phosphate Deposits of the World: Volume 2 - Phosphate Rock Resources, (ed.) A.J.G. Notholt, R.P. Sheldon, and D.F. Davidson; International Geological Correlation Programme, Project 156: Phosphorites, Cambridge University Press, p. 262-266.

Lottermoser, B.G.
1990: Rare-earth element mineralization within the Mt. Weld carbonatite laterite, Western Australia; Lithos, v. 24, p. 151-167.

Machamer, J.F.
1987: A working classification of manganese deposits; Mining Magazine, October 1987, p. 348-351.

Mariano, A.N.
*1989a: Economic geology of rare earth elements; in Geochemistry and Mineralogy of Rare Earth Elements, (ed.) B.R. Lipin and G.A. McKay; Reviews in Mineralogy, v. 21, p. 309-336.
*1989b: Nature of economic mineralization in carbonatites and related rocks; in Carbonatites: Genesis and Evolution, (ed.) K. Bell; Unwin Hyman, London, p. 149-176.

Mariano, A.N. and Mitchell, R.H.

1990: Mineralogy and geochemistry of perovskite-rich pyroxenites; in Program with Abstracts, Geological Association of Canada/ Mineralogical Association of Canada, v. 15, p. A83.

Middlemost, E.

1990: Mineralogy and petrology of the rauhaugites of the Mt Weld cabonatite complex of western Australia; Mineralogy and Petrology, v. 41, p. 145-161.

Möller, P.

1989a: Prospecting for rare-earth element deposits; in Lanthanides, Tantalum and Niobium, (ed.) P. Möller, P. Černý, and F. Saupe; Society for Geology Applied to Mineral Deposits, Special Publication No. 7, p. 263-265.

1989b: REE(Y), Nb, and Ta enrichment in pegmatities and carbonatite-alkalic rock complexes; in Lanthanides, Tantalum and Niobium, (ed.) P. Möller, P. Černý, and F. Saupe; Society for Geology Applied to Mineral Deposits, Special Publication No. 7, p. 103-144.

Notholt, A.J.G.

1979: The economic geology and development of igneous phosphate deposits in Europe and the USSR; Economic Geology, v. 74, p. 339-350.

1980: Igneous apatite deposits: mode of occurrence, economic development and world resources; in Fertilizer Mineral Potential in Asia and the Pacific, (ed.) R.P. Sheldon and W.C. Burnett; Proceedings of the Fertilizer Raw Materials Resources Workshop, August 20-24, 1979, p. 263-285.

Percival, J.A. and West, G.

1994: The Kapuskasing uplift: a geological and geophysical synthesis; Canadian Journal of Earth Sciences, v. 31, p. 1256-1286.

Potapoff, P.

1989: The Martison carbonatite deposit, Ontario, Canada; in Phosphate Deposits of the World: Volume 2 - Phosphate Rock Resources, (ed.) A.J.G. Notholt, R.P. Sheldon, and D.F. Davidson; International Geological Correlation Programme, Project 156: Phosphorites, Cambridge University Press, p. 71-78.

Sage, R.P.

1979: No. 16 - alkalic rocks carbonatite complexes; in Summary of Field Work, 1979, (ed.) V.G. Milne, O.L. White, R.B. Barlow, and C.R. Kustra; Ontario Geological Survey, Miscellaneous Paper 90, p. 70-75.

1983: Literature review of alkalic rocks - carbonatites; Ontario Geological Survey, Open File Report 5436, 277 p.

1986: Alkalic rock complexes - carbonatites of northern Ontario and their economic potential; PhD. thesis, Carleton University, Ottawa, Ontario, 355 p.

1988: Geology of carbonatite-alkalic rock complexes in Ontario: Cargill Township Carbonatite Complex, District of Cochrane; Ontario Geological Survey, Study 36, 92 p.

1991a: Chapter 18: alkalic rock, carbonatite and kimberlite complexes of Ontario, Superior Province; in Geology of Ontario, (ed.) P.C. Thurston, H.R. Williams, R.H. Sufcliffe, and G.M. Stott; Ontario Geological Survey, Special Volume 4, pt. 1, p. 683-709.

1991b: Geology of the Martison carbonatite complex; Ontario Geological Survey, Open File report 5420, 74 p.

Sandvik, P.O. and Erdosh, G.

*1977: Geology of the Cargill phosphate deposit in northern Ontario; The Canadian Mining and Metallurgical Bulletin, v. 69, no. 777, p. 90-96.

*1984: Geology of the Cargill phosphate deposit in northern Ontario; in The Geology of Industrial Minerals in Canada, (ed.) G.R. Guillet and G. Martin; The Canadian Institute of Mining and Metallurgy, Special Volume 29, p. 129-133.

Sinclair, W.D., Jambor, J.L., and Birkett, T.C.

1992: Rare earths and the potential for rare-earth deposits in Canada; Mining and Exploration Geology, v. 1, p. 265-281.

Vartiainen, H.

1989: The phosphate deposits of the Sokli carbonatite complex, Finland; in Phosphate Deposits of the World: Volume 2 - Phosphate Rock Resources, (ed.) A.J.G. Notholt, R.P. Sheldon, and D.F. Davidson; International Geological Correlation Programme, Project 156: Phosphorites, Cambridge University Press, p. 398-402.

Verwoerd, W.J.

1989: Genetic types of ore deposits associated with carbonatites; in Abstracts Volume 3 of 3, Twenty-eighth International Geological Congress, Washington, District of Columbia, p. 3-295.

Vos, M.A.

1981: Industrial minerals of the Cargill Complex; in Summary of Field Work by the Ontario Geological Survey, (ed.) J. Wood, O.L. White, R.B. Barlow, and A.C. Colvine; Ontario Geological Survey, Miscellaneous Report 100, p. 224-229.

Woolley, A.R.

*1987: Alkaline rocks and carbonatites of the world: part 1. North and South America; British Museum (Natural History), London, 216 p.

1989: The spatial and temporal distribution of carbonatite; in Carbonatites: Genesis and Evolution, (ed.) K. Bell; London, Unwin Hyman, p. 15-37.

Authors' addresses

T.C. Birkett
SOQUEM
2600, boul. Laurier
Tour Belle Cour, bureau 2500
Sainte-Foy, Quebec
G1V 4M6

D.R. Boyle
Geological Survey of Canada
601 Booth Street
Ottawa, Ontario
K1A 0E8

G.A. Gross
Geological Survey of Canada
601 Booth Street
Ottawa, Ontario
K1A 0E8

D.G. Richardson
Geological Survey of Canada
601 Booth Street
Ottawa, Ontario
K1A 0E8

Printed in Canada

5. EVAPORITES

5. EVAPORITES

R.T. Bell

IDENTIFICATION

The distribution of the main evaporite deposits in Canada is shown in Figure 5-1. Three subtypes of these deposits can be recognized: marine evaporite deposits, nonmarine evaporite deposits, and duricrust deposits.

In **marine environments**, evaporite deposits comprise essentially pure halite (NaCl), and any or all of: gypsum, anhydrite ($CaSO_4 \cdot 2H_2O$, $CaSO_4$), sulphur, sylvite (KCl), and various halides (Cl, Br, I) of Ca, K, Na, Mg, and Sr.

In Canada the best examples of this type are: in the Salina Formation in Ontario (halite and gypsum; Fig. 5-2); in the Windsor Group in the Appalachian region (halite, sylvite, gypsum, celestite; Fig. 5-3); in the Prairie Formation in Saskatchewan (sylvite, halite, brine); at Gypsumville, Manitoba (gypsum); and at Windermere, British Columbia (gypsum).

Foreign examples are numerous, of which the most important are: in the Michigan, Delaware, and Paradox basins, U.S.A. (halite, sylvite, gypsum); Gulf Coast salt domes, U.S.A. and Mexico (halite, native sulphur, gypsum); in the Zechstein and Messinian basins in Europe; Nepa Basin in Siberian Russia; Khorat Basin in Thailand; Qaidam depression in China; and various Triassic basins in western Europe and north Africa; Devonian of the Donbass in Ukraine; Eocambrian-Cambrian of the Persian Gulf through to the Salt Ranges in Pakistan; and Cenozoic deposits in the Red Sea and Horn of Africa.

In **nonmarine environments (lacustrine)** evaporite deposits comprise large amounts of trona, borates, Na, Mg, Ca, halite, Na-sulphates, together with variably large amounts of sylvite, nitrates, phosphates, Li-, Mg-, and W-rich brines, zeolites, and clay minerals deposited in lake basins.

In Canada these occur in alkaline lakes in Saskatchewan, Alberta, and British Columbia (mainly for sodium sulphates). Important foreign examples include the following: in the Green River Formation, Wyoming, U.S.A and Lake Magadi, Kenya (trona); in Searles Lake, California, U.S.A., and Bigadiuz, Emet, and Kirka, Turkey (borates, halite, sylvite, Li, Br, Mg, and W); Great Salt Lake, Utah, U.S.A. and Dead Sea area in Israel and Jordan (halite, sylvite, Mg, and Br); Atacama desert, Altiplano of Chile (nitrates); Basin and Range playa lake sediments of southwestern U.S.A. (zeolites).

Duricrust sediments formed in arid climates are also included in the evaporite class. Examples are at Dusa Mareb, Somalia, and Yeelirrie, Western Australia (calcrete that contains U), and in Mali (phoscrete that contains U; Hirono et al., 1987).

IMPORTANCE

Marine evaporites in Canada are the most important sources for salt (11 to 12 x 10^6 t/a) (tonnes/year), potash (40% of world's production, at about 7 million t/a), and gypsum (6 x 10^6 to 8 x 10^6 t/a) in Canada. Waste brines from some of the mines in Saskatchewan (normally considered a deleterious feature) have been collected, and used directly and advantageously as a dust deterrent on roads rather than solid $CaCl_2$. Evaporite deposits are also considered as potential sites for petroleum storage and nuclear waste storage (Johnson and Gonzales, 1978). In the U.S.A., salt domes in the Gulf of Mexico region were formerly important sources of elemental sulphur.

Bell, R.T.
1996:   Evaporites; in Geology of Canadian Mineral Deposit Types, (ed.) O.R. Eckstrand, W.D. Sinclair, and R.I. Thorpe; Geological Survey of Canada, Geology of Canada, no. 8, p. 121-127 (also Geological Society of America, The Geology of North America, v. P-1).

Lacustrine evaporites in Canada are the main source for sodium sulphates (in the order of 300 000 t/a), otherwise they are of minor importance.

Worldwide these types of deposits are important sources for borates, trona, and lithium, especially in the U.S.A. There, Searles Lake also contains in brines about 50% of the known tungsten reserves in the U.S.A. (Smith, 1979). Nitrate deposits in Chile were the most important for a century, but are of minor importance now.

Duricrust deposits may become significant sources for uranium in the future, but are not being mined at present. There are no significant deposits of this class in Canada.

SIZE AND GRADE OF DEPOSITS
Marine evaporite deposits

Deposit grades are typically 90 to 100% for halite; 15 to 30% for K_2O-equivalent in sylvite; 90 to 100% for gypsum. Deposits range in size from tens of millions to greater than one billion tonnes.

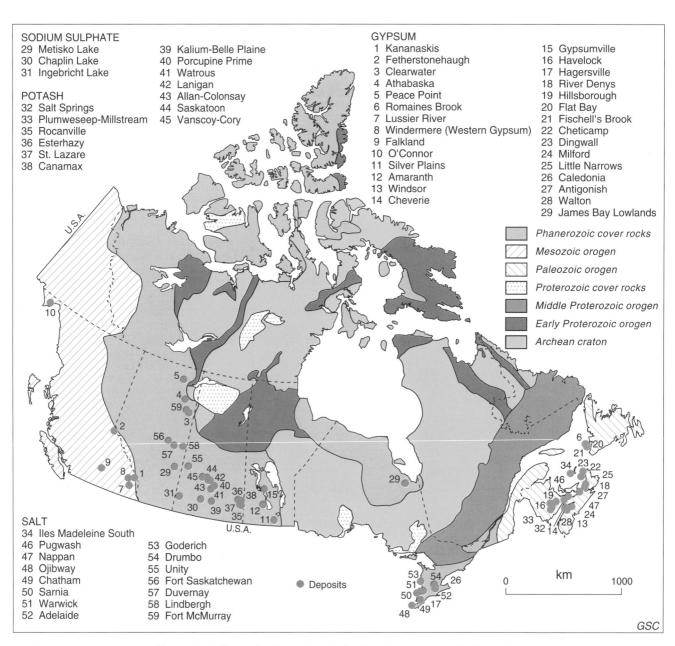

SODIUM SULPHATE
29 Metisko Lake
30 Chaplin Lake
31 Ingebricht Lake

POTASH
32 Salt Springs
33 Plumweseep-Millstream
35 Rocanville
36 Esterhazy
37 St. Lazare
38 Canamax

39 Kalium-Belle Plaine
40 Porcupine Prime
41 Watrous
42 Lanigan
43 Allan-Colonsay
44 Saskatoon
45 Vanscoy-Cory

GYPSUM
1 Kananaskis
2 Fetherstonehaugh
3 Clearwater
4 Athabaska
5 Peace Point
6 Romaines Brook
7 Lussier River
8 Windermere (Western Gypsum)
9 Falkland
10 O'Connor
11 Silver Plains
12 Amaranth
13 Windsor
14 Cheverie

15 Gypsumville
16 Havelock
17 Hagersville
18 River Denys
19 Hillsborough
20 Flat Bay
21 Fischell's Brook
22 Cheticamp
23 Dingwall
24 Milford
25 Little Narrows
26 Caledonia
27 Antigonish
28 Walton
29 James Bay Lowlands

Phanerozoic cover rocks
Mesozoic orogen
Paleozoic orogen
Proterozoic cover rocks
Middle Proterozoic orogen
Early Proterozoic orogen
Archean craton

SALT
34 Iles Madeleine South
46 Pugwash
47 Nappan
48 Ojibway
49 Chatham
50 Sarnia
51 Warwick
52 Adelaide

53 Goderich
54 Drumbo
55 Unity
56 Fort Saskatchewan
57 Duvernay
58 Lindbergh
59 Fort McMurray

● Deposits

km   0   1000

GSC

Figure 5-1. Evaporite deposits and significant occurrences in Canada.

Figure 5-2. Relatively undeformed bedded halite, A-2 unit of the Upper Silurian Salina Formation, in the Sifto Salt Company mine, Goderich, Ontario. Scale is indicated by the vehicle in the foreground. Photo courtesy of Sifto Salt Division, DOMTAR Ltd. GSC 203368

Figure 5-3. Canadian Rock Salt Company Ltd. mine, Pugwash, Nova Scotia: highly contorted anhydrite (darker grey) inter-layered with bedded halite, Lower Mississippian Windsor Group. Photo by H. Wiele, Bedford Institute of Oceanography. B.I.O. 4392-12

The potash deposits in Saskatchewan are estimated to contain greater that 14 billion tonnes of recoverable K_2O-equivalent by conventional mining at an average grade of 25%. The deposits in New Brunswick contain several million tonnes of ore averaging 20 to 30% K_2O-equivalent.

Elemental sulphur deposits in some salt domes around the Gulf of Mexico contain individually greater than 50 Mt of extractable sulphur.

Nonmarine evaporite deposits

Alkaline lake deposits in Saskatchewan contain total estimated reserves of 55 to 65 Mt of sodium sulphate. Deposits in Wyoming contain greater than 30 Gt of trona. The Searles Lake deposits and brines contain more than 200 Mt of extractable soluble salts. The Emet and Kirka areas in Turkey are estimated to contain, respectively, 45 Mt averaging 45% B_2O_3 and 500 Mt averaging 27% B_2O_3. Nitrate reserves in Chile are greater than 2.5 Gt of ore at a cutoff grade of 7% $NaNO_3$. The calcrete deposits at Yeelirrie in Australia contain an estimated 41 000 t of U_3O_8 at an average grade of 0.15% U_3O_3; similar deposits occur in Namibia and Somalia; none have been identified in Canada. Phoscrete occurrences have been identified and described in Mali (Hirono et al., 1987); these contain 0.02% U on average in phoscrete that contains 30-36% P_2O_5. In the Baker Lake area in the Northwest Territories slightly uraniferous apatite is contained in Mesoproterozoic phoscretes at the base of the Thelon Formation; these do not comprise a resource.

GEOLOGICAL FEATURES

General

In Canada (Gould and De Mille, 1968; Baar, 1972; Sanford and Norris, 1975; Fuzesy, 1982; Boehner, 1986; Meijer-Drees, 1986; Howie, 1988; Bezys, 1990) evaporites have been found in rocks of Meso- to Neoproterozoic and Cambrian ages in the Mackenzie Mountains and Amundsen Basin; of Ordovician age in the Sverdrup Basin; of Silurian age in the Williston and Michigan basins; and of Devonian and Early Carboniferous age throughout the western provinces, Hudson Platform, and the Michigan Basin. Early Carboniferous evaporites are notable in southern Saskatchewan and adjacent Manitoba, the Sverdrup Basin of the Arctic Islands, and in the Gulf of St. Lawrence and adjacent Maritime Provinces. Triassic and Jurassic evaporites are also present in the Prairie Provinces and along the margins of the Atlantic Ocean. Quaternary continental evaporites are preserved and currently forming in semiarid areas of southern Saskatchewan and Alberta, and the southern interior of British Columbia.

All evaporitic deposits (Mattox, 1968; Dellwig and Evans, 1969; Kirkham, 1984; Sonnenfeld, 1984, 1992; Light et al., 1987) are formed in areas, marine or nonmarine, in which evaporation exceeds precipitation and other water inflow, essentially in arid areas. In the marine setting, marginal and restricted basins with extensive coastal sabkhas are important. Present environments are in the Red Sea and Persian Gulf. Restricted basins like these are often a consequence of continental rifting (Red Sea) or continental collision (Persian Gulf). Some modern lagoonal settings, such as the Coorong region of South Australia where magnesite is being formed, may have ancient analogues in bedded magnesites such as Neoproterozoic Skillogalee "Dolomite" of Australia (von der Borch and Lock, 1979), and in the Mesoproterozoic rocks of the Grenville Province in the Otish Mountains, Quebec.

In the nonmarine setting, large desert salt lakes and inland sabkhas or playas are important in desert and semiarid areas. Modern environments are around the Great Salt Lake in Utah, in lakes such as Lake Frome in South Australia, Lake Chad in central Africa, in the Dead Sea, and in the Gulf of Kara Bug just southeast of the Caspian Sea. These closed drainage systems are in many cases the result of continental rifting, as in the Afar Rift and the East African Rift Valley (i.e., Lake Magadi).

Age of host rocks and mineralization

Traces of evaporite deposits of these types are clearly evident as far back as the Archean (Sighinolfi et al., 1980), and become more common around 2000 Ma (Badham and Stanworth, 1977). Significant accumulations occur in the Neoproterozoic at Bitter Springs, and Mt. Isa, Australia (McClay and Carlile, 1978); and in the Coates Lake Group, Mackenzie Mountains, Northwest Territories. The first very large amounts of preserved evaporites occur in Eocambrian-Cambrian of the Hormuz Formation and equivalents throughout southwest Asia; these have been mined in Iran through to the Salt Ranges in Pakistan. On a worldwide basis bedded evaporite deposits are known in every interval of the Phanerozoic except the Middle Ordovician and at the Cretaceous-Tertiary boundary.

Initial mineralization of these deposits is contemporaneous with the deposition of the adjacent host rock, which may be composed of almost any sort of sediment. The only restriction is that clastic sedimentation be (somewhat) suppressed while salts are being deposited by precipitation from the brine under desiccating conditions. After burial, the evaporite beds may undergo partial dewatering as lithification proceeds, and with further burial the evaporites, usually dominated by salt, may develop pinch-and-swell structures and rolls ("salt anticlines") from the original tabular beds. With further dewatering these rolls grow into diapirs which rise through the overlying sedimentary column (Heroy, 1968; Mattox, 1968; Rios, 1968; Stöcklin, 1968; Tortochaux, 1968; Kent, 1979; Jenyon, 1986; Lerche and O'Brien, 1987; Light et al., 1987). Some diapirs even pierce through the cover and flow out on the surface as salt glaciers. This is happening at present in the southern Zagros Mountains of Iran (this salt is of Eocambrian age, but is now being mechanically emplaced, intruded, and extruded). All of this salt eventually will be dissolved and carried away; some is being reprecipitated in new sites. Similar processes have occurred in the Maghreb in northern Africa as well as in northern Spain (Triassic salt, Cretaceous-Tertiary diapirism and erosion).

The high solubility of the evaporite minerals means that fresh meteoric water will dissolve evaporites and remove them from their deposits either by groundwater dissolution or by direct erosion. For these reasons evaporite deposits have poor preservation potential. Deposits of significant size are increasingly rare with age (see comment 8 under "Exploration guides", regarding metamorphism).

Associated structures and form of deposits

Initially, evaporites are in tabular bodies conformable with and/or interbedded with dolomites, fetid dolomites, shales and oil shales, reefal limestones, varicoloured siltstones and redbed sandstones in paralic and sabkha or lacustrine environments (Perthuisot, 1975). Regionally the deposits are associated with rifts, e.g., Red Sea deposits, or with large shallow continental basins that are commonly bounded by reefs, e.g., Michigan Basin.

Under deep burial, the low specific gravity, the flowage properties, and the high thermal conductivity of salt in the solid state, may lead salt and anhydrite to be displaced upwards as rolls and diapirs, some of which pierce through the entire cover sequence and flow out as salt glaciers (described above), or stand out as prominent pinnacles of gypsum as in the Sverdrup Basin, in the Canadian Arctic. Even in nondiapiric settings, abundant small folds have formed due to intrastratal flowage.

Nature of the ore, distribution of ore minerals, ore textures

In marine deposits, salt, potash, gypsum, and anhydrite beds intergrade with one another. In diapirs, the structures are usually capped and encased by a 'cap rock' of secondary calcite, dolomite, and anhydrite; the salt and the anhydrite are intensely folded.

Nonmarine deposits comprise trona, salt, borate, potash, gypsum, and glauberite, as well as variable proportions of interbedded shale and/or siltstone of the host sequence.

In duricrusts, carnotite and uraniferous phosphate are the main minerals of economic interest.

Ore composition and mineralogy

Most minerals are syngenetic, and constitute the host rock itself. In addition, multiple, extensive diagenetic modification (recrystallization, ion exchange, hydration and dehydration and rehydration, metasomatism) has taken place. Some minerals, such as the celestite in the deposits at Lake Enon, Nova Scotia are believed to be diagenetic. Native sulphur in cap rocks (itself a replacement and displacement deposit formed at the top of a rising salt diapir) is believed to be formed by bacterial reduction of sulphates.

Marine deposits

Gypsum, anhydrite, halite, sylvite, carnallite, polyhalite, bischofite, rinneite, kieserite, tachyhydrite, kainite, langbeinite, celestite, strontianite, and native sulphur are the ore minerals. The associated minerals are dolomite, calcite, ankerite, clay minerals, quartz, pyrite, marcasite, and barite.

Nonmarine deposits

Variable amounts of Na, Ca, and Mg minerals are characteristic. Gypsum, anhydrite, halite, glauberite, trona, shortite, nahcolite, borax, kernite, hanksite, burkeite, ulexite, colemanite, epsomite, mirabilite, and thenardite may be present, as well as associated dolomite, calcite, magnesite, siderite, ankerite, clay minerals, quartz, talc, pyrite, marcasite, and iron oxides. An example of the extreme variability is provided by the Chilean nitrate ores, which consist of soda niter, niter, darapskite, lautarite, humberstonite, brueggenite, and dietzeite within associated gypsum, anhydrite, halite, glauberite, lopezite, tarapacaite, and ulexite.

Duricrust sediments

The ore mineral in the Yeelirrie uranium deposit is carnotite, and associated minerals are dolomite, calcite, chalcedony, gypsum, anhydrite, and sepiolite.

DEFINITIVE CHARACTERISTICS

General

1. These deposits form in arid and superarid environments.
2. a) They are associated with continental rifting, either during an early (or aborted) phase, producing lacustrine environments with closed drainage, or a later phase when there was restricted circulation with the world ocean (e.g., Miocene Mediterranean; present day Red Sea).
   b) The deposits may also form under conditions of closed or restricted circulation within epicontinental basins.
3. They are commonly associated with redbeds.
4. They are commonly associated with fetid dolomites.
5. The sequences are generally of Phanerozoic age.

GENETIC MODEL

The general genetic model (Liechti, 1968; Mattox, 1968; Moine et al., 1981; Sonnenfeld, 1984) hinges on the deposition of soluble minerals *by evaporation* in salt lakes (salinas) and low-lying salt flats (sabkhas) and in generally shallow marine basins restricted from circulation with the world ocean and suffering from a deficit of input of fresh water, and by precipitation from subsurface brines in both marginal marine and inland arid basins.

Precipitation from seawater follows a pattern: aragonite/dolomite(carbonates) – gypsum (sulphates) – halite(salt) – sylvite – tachyhydrite (bitterns), and the density of the brine increases through evaporation. In deeper basins, beneath the photic zone, anaerobic bacterial action may transform the sulphate ions to sulphide, and these bacteria will reduce already precipitated gypsum if sufficient organic matter is present (deposits of fetid dolostones are potential source beds for petroleum).

In lakes, the initial calcium carbonate is usually calcite, not aragonite. At sulphate saturation these continental brines start to precipitate out minerals corresponding to the major cations leached out of the surrounding terrain. For example, in the interior of British Columbia, basic igneous rocks dominating in that terrane lead to Mg-sulphate brines, and produce epsomite. In many continental settings sodium bicarbonate is high; in the presence of sodium bicarbonate gypsum does not precipitate, but instead calcite, calcium-sodium sulphate, and sodium sulphate (i.e., glauberite and mirabilite, respectively) are deposited. Because chloride ions are low in groundwaters, halite is a minor phase in continental evaporites, except where previously formed marine evaporites are being recycled (e.g., Permian marine salt recycled into Tertiary evaporites in basins in the Basin and Range terrane of Arizona).

Most ore minerals are syngenetic, and comprise the host rock itself, but there is multiple, extensive diagenetic modification (recrystallization, ion exchange, hydration and dehydration and rehydration, metasomatism; Dellwig and Evans, 1969; Dunsmore, 1977a, b; Jonasson and Dunsmore, 1979; Behr et al., 1983; Schmidt-Mumm et al., 1987). Abundant volcanic ash beds in salt lakes are favourable for the formation of zeolites. Nitrates are concentrated only under extremely arid conditions. Economic deposits of gypsum normally result from near-surface rehydration of anhydrite.

Under deep burial, the low specific gravity, the flowage properties, and the high thermal conductivity of salt (mainly) in the solid state, commonly result in the upward displacement of salt and anhydrite as rolls and diapirs, some of which may even pierce through the entire cover sequence as at present in southern Iran (Kent, 1979); the Maghreb and northern Spain (Rouvier et al., 1985); and in the Sverdrup Basin (Gould and DeMille, 1968). Many salt diapirs are tens of kilometres across, and have risen 10 km or more from the source beds. Most diapirs are zoned downward from cap rock, through anhydrite, to salt. Many contain rafted fragments and blocks, as much as 3 km across, of rocks which have originated above, or immediately below, the original or "mother salt" beds.

Duricrusts require evaporitic environments in semiarid to arid continental conditions. In many cases they form in discharge areas of briny groundwaters saturated with Ca^{+2}, CO_3^{-2} and/or PO_4^{-3}. In the case of uranium-bearing calcrete, which normally contains vanadates, or phoscrete, the uranium probably is derived in large catchment areas from a separate source from that for the vanadium and phosphate. In some areas iron-rich, silica-rich, apatite-rich, or gypsum-rich duricrusts form due to the abundance of Fe, Si, phosphate, or sulphate in the host sequence: they are known as ferricrete, silicrete, phoscrete, or gypcrete, respectively.

RELATED DEPOSIT TYPES

In large evaporite basins, deposits are, in general, laterally associated with sandstone-hosted copper and uranium deposits, and in some places with sedimentary exhalative sulphide deposits (e.g., African Copperbelt – Zaire to Zambia; for a discussion of relationships here, see Bell, 1989). Evaporites may thus share definitive characteristics with some sandstone-hosted Cu, V, and Pb(-Zn, Ag) deposits (Renfro, 1974; Kirkham, 1989), and in many cases occur in the same districts (for example, Uravan district in Utah, U.S.A.).

Other types of mineralization are associated with some evaporite diapirs (Rouvier et al., 1985; De Magnée and François, 1988; Bell, 1989), namely vein and replacement deposits of: barite (Flinders Range, South Australia); Zn-Pb-Sr-Ba (Mississippi salt basin, U.S.A.; Saunders and Swann, 1994); Pb-Zn-Ag (Gulf Coast, U.S.A.: Price et al., 1983; Light et al., 1987), (Tunisia: Rouvier et al., 1985); Pb-Zn-Hg (Tunisia: Rouvier et al., 1985; Hatira et al., 1990); (Donbass Basin, Ukraine: Belous et al., 1984); siderite (weathered mainly to limonite/hematite as in Bilbao-type Fe deposits of Spain: Rouvier et al., 1985; and of Tunisia and Algeria: Hatira et al., 1990); elemental sulphur

(Gulf Coast, U.S.A.); and, perhaps, copper (Corocoro, Bolivia: Pellisonnier, 1964; Cox et al., 1991; Dongchuan, China: Ruan et al., 1991).

Large marine evaporite basins in many cases contain associated fetid dolomites, and the latter may provide source beds for some large petroleum fields (Eugster, 1985). Also, structures associated with the emplacement of diapirs are important for petroleum and gas accumulation (Gulf Coast, U.S.A. and Mexico, Romania, Iran, northern Germany, and Spain).

EXPLORATION GUIDES

1. See Definitive characteristics for geological features.

2. Target areas are evaporitic basins, preferably of a large size, or small basins with unusual concentrations of economically significant elements (e.g., borates, lithium, nitrates).

3. Identify, outline, and verify the distribution of valuable components within the evaporite sequence.

4. Geochemistry of springs and seeps (Na^+, K^+, Li^+, Br^-, Cl^-, B, etc.) has proved useful in discovery of the New Brunswick deposits.

5. Gravity and seismic surveys are useful in outlining salt domes and rolls.

6. Mining sites are usually near or at surface (open pit); salt domes and rolls bring salt, anhydrite, and potash beds near the surface. For example, "sylvinite" (mixture of halite and sylvite) in southwestern Ontario is too deep for commercial exploitation. Mining by in situ solution methods is also used in Saskatchewan.

7. As preservation potential is poor for some, if not most, lacustrine evaporites, and extremely poor for calcretes, sequences of Quaternary and late Tertiary age are most favourable.

8. Metamorphism of evaporite sequences does lead to unusual suites of minerals, themselves useful. For example, lapis lazuli, sodalite, and cancrinite may be useful as ornamental stone or gems (Hogarth and Griffin, 1978). Some of the magnesite deposits in the Grenville Province may be of evaporitic origin. Deposits that have been extensively metamorphosed are certainly difficult and controversial to clearly identify as metaevaporites (Moine et al., 1981; Rozen, 1982; Behr et al., 1983; Schmidt-Mumm et al., 1987; Bone, 1988). Sequences made up of mainly tourmaline ("tourmalinites": Bone, 1988) or magnesite are worth a closer look in exploration for tungsten. At Broken Hill, Australia, Plimer (1994) recognized scheelite mineralization in tourmaline-bearing metaevaporites.

ACKNOWLEDGMENTS

Much of the inspiration for this paper came from long discussions with the late H.E. Dunsmore. This paper was built upon the framework of a synopsis account by Kirkham (1984). In addition to literature sources, some information was obtained from an unpublished account of the distribution and general character of Canadian evaporites by P. Sonnenfeld.

REFERENCES

References with asterisks (*) are considered to be the best source of general information on this deposit type.

Baar, C.A.
1972: Actual geological problems in Saskatchewan potash mining; Saskatchewan Research Council, Report E72-18, 50 p.

Badham, J.P.N. and Stanworth, C.W.
1977: Evaporites from the lower Proterozoic of the East Arm, Great Slave Lake; Nature (London), v. 268, p. 516-517.

Behr, H.J., Ahrendt, H., Martin, H., Porada, H., Röhrs, J., and Weber, K.
1983: Sedimentology and mineralogy of Upper Proterozoic playa-lake deposits in the Damara Orogen; in Intracontinental Fold Belts, (ed.) H. Martin and F.W. Eder; Springer-Verlag, Berlin, p. 576-611.

Bell, R.T.
1989: A conceptual model for development of megabreccias and associated mineral deposits in Wernecke Mountains, Canada, Copperbelt, Zaire, and Flinders Range, Australia; in Uranium Resources and Geology of North America, Proceedings of IAEA Workshop, Saskatoon, Saskatchewan, August, 1987, International Atomic Energy Agency, Vienna, IAEA-TECDOC-500, p. 149-169.

Belous, I.R., Kirikilitsa, S.I., Levenshteyn, M.L., Rodina, E.K., and Florinskaya, V.N.
1984: Occurrences of mercury in northeastern Donbass salt domes; International Geological Review, v. 26, p. 573-582.

Bezys, R.K.
1990: Geology of gypsum deposits in the James Bay Lowlands, Ontario Geological Survey, Open File Report 5728, 109 p.

*Boehner, R.C.
1986: Salt and potash resources in Nova Scotia; Nova Scotia Department of Mines and Energy, Bulletin 5, 346 p.

Bone, Y.
1988: The geological setting of tourmaline at Rum Jungle, N.T., Australia - genetic and economic implications; Mineralium Deposita, v. 23, p. 34-41.

Cox, D., Carrasco, R., Orlando, A., Hinojosa, A., and Long, K.R.
1991: Copper deposits in Tertiary red beds in Bolivia (abstract); in Seventh Annual McKelvey Forum on Mineral and Energy Resources, Reno, February 11-14, 1991; United States Geological Survey, Circular 1062, p. 11.

Dellwig, L.F. and Evans, R.
1969: Depositional processes in the Salina salt of Michigan, Ohio and New York; American Association of Petroleum Geologists, Bulletin, v. 53, p. 949-956.

De Magnée, I. and François, A.
1988: The origin of the Kipushi (Cu, Zn, Pb) deposit in direct relation with a Proterozoic salt diapir, Copperbelt of Central Africa, Shaba, Republic of Zaire; in Base Metal Sulfide Deposits in Sedimentary and Volcanic Environments, (ed.) G.H. Friedrich and P.M. Herzig; Society of Geology Applied to Mineral Deposits, Special Publication 5, p. 74-93.

Dunsmore, H.E.
1977a: A new genetic model for uranium-copper mineralization, Permo-Carboniferous Basin, northern Nova Scotia; in Report of Activities, Part B, Geological Survey of Canada, Paper 77-1B, p. 247-253.
1977b: Uranium resources of the Permo-Carboniferous basin, Atlantic Canada; in Report of Activities, Part B, Geological Survey of Canada, Paper 77-1B, p. 341-344.

Eckstrand, O.R. (ed.)
1984: Canadian Mineral Deposit Types: a Geological Synopsis; Geological Survey of Canada, Economic Geology Report 36, 86 p.

Eugster, H.P.
1985: Oil shales, evaporites and ore deposits; Geochimica et Cosmochimica Acta, v. 49, p. 619-635.

*Fuzesy, A.
1982: Potash in Saskatchewan; Saskatchewan Geological Survey, Report 181, 44 p.

Gould, D.B. and De Mille, G.
1968: Piercement structures in Canadian Arctic Islands; in Diapirism and Diapirs, (ed.) J. Braumstein and G.D. O'Brien; American Association of Petroleum Geologists, Memoir 8, p. 183-214.

Hatira, N., Perthuisot, V., and Rouvier, H.
1990: Les minéraux à Cu, Sb, Ag, Hg des minerais de Pb-Zn de Sakiet Koucha (diapir de Sakiet Sidi Youssef, Tunisie septentrionale); Mineralium Deposita, v. 25, p. 112-117.

Heroy, W.B.
1968:   Thermicity of salt as a geologic function; in Saline Deposits, (ed.) R.B. Mattox; Geological Society of America, Special Paper 88, p. 619-629.

Hirono, S., Hirakawa, K., and Hanada, K.
1987:   Uranium-bearing phoscrete from Mali, West Africa; Chemical Geology, v. 60, p. 281-286.

Hogarth, D.D. and Griffin, W.L.
1978:   Lapis lazuli from Baffin Island - a Precambrian meta-evaporite; Lithos, v. 11, p. 37-60.

*Howie, R.D.
1988:   Upper Paleozoic evaporites of southeastern Canada; Geological Survey of Canada, Bulletin 380, 120 p.

*Jenyon, M.K.
1986:   Salt Tectonics; Elsevier, Amsterdam, 191 p.

Johnson, K.S. and Gonzales, S.
1978:   Salt deposits in the United States and regional geological characteristics important for storage of radioactive waste; United States Department of Energy, Office of Waste Isolation, Report No. Y/OWI/SUB-7414/1/W -7405 - eng - 26, Washington, District of Columbia.

Jonasson, I.R. and Dunsmore, H.E.
1979:   Low grade uranium mineralization in carbonate rocks from some salt domes in the Queen Elizabeth Islands, District of Franklin; in Current Research, Part A; Geological Survey of Canada, Paper 79-1A, p. 61-70.

Kent, P.E.
1979:   The emergent Hormuz salt plugs of southern Iran; Journal of Petroleum Geology, v. 2, no. 2, p. 117-144.

Kirkham, R.V.
1984:   Evaporites and brines; in Canadian Mineral Deposit Types: a Geological Synopsis, (ed.) O.R. Eckstrand; Geological Survey of Canada, Economic Geology Report 36, p. 13-15.
1989:   Distribution, settings, and genesis of sediment-hosted stratiform copper deposits; in Sediment-hosted Stratiform Copper Deposits, (ed.) R.W. Boyle, A.C. Brown, C.W. Jefferson, E.T. Jowett, and R.V. Kirkham; Geological Association of Canada, Special Paper 36, p. 3-38.

Lerche, I. and O'Brien, J. (ed.)
1987:   Dynamical Geology of Salt and Related Structures; Academic Press, Orlando, Florida, 798 p.

Liechti, P.
1968:   Salt features of France; in Saline Deposits, (ed.) R.B. Mattox; Geological Society of America, Special Paper 88, p. 83-106.

Light, M.P., Posey, H.H., Kyle, J.R., and Price, P.E.
1987:   Model for the origins of geopressured brines, hydrocarbons, cap rocks and metallic mineral deposits, Gulf Coast, U.S.A.; in Dynamical Geology of Salt and Related Structures, (ed.) I. Lerche and J.J. O'Brien; Academic Press, Inc., p. 789-830.

*Mattox, R.B. (ed.)
1968:   Saline Deposits; Geological Society of America, Special Paper 88, 701 p.

McClay, K.R. and Carlile, D.G.
1978:   Mid-Proterozoic sulphate evaporites at Mount Isa mine, Queensland, Australia; Nature (London), v. 274, p. 240-241.

*Meijer-Drees, N.C.
1986:   Evaporitic deposits of western Canada; Geological Survey of Canada, Paper 85-20, 120 p.

Moine, B., Sauvan, P., and Jarousse, J.
1981:   Geochemistry of evaporite-bearing series: a tentative guide for the identification of metaevaporites; Contributions to Mineralogy and Petrology, v. 76, p. 401-412.

Pellisonnier, H.
1964:   Structure géologique et genèse du gisement de cuivre de Corocoro (Bolivie); Bulletin Société Géologique, France, série 7, v. 7, p. 502-514.

Perthuisot, V.
1975:   Le sabkha El Melah de Zarsis. Genèse et évolution d'un bassin salin paralique; Travaux du Laboratoire de Géologie; École Normandie Supérieure, v. 9, 252 p.

Plimer, J.R.
1994:   Stratabound scheelite in meta-evaporites, Broken Hill, Australia; Economic Geology, v. 89, p. 423-437.

Price, P.E., Kyle, J.R., and Wessel, G.R.
1983:   Salt dome related zinc-lead deposits; in Proceedings, International Conference on Mississippi Valley Type Lead-Zinc deposits, (ed.) G. Kisvarsanyi, S.K. Grant, W.P. Pratt, and J.W. Koenig; University of Missouri at Rolla, p. 558-571.

Renfro, A.R.
1974:   Genesis of evaporite-associated stratiform metalliferous deposits - a sabkha process; Economic Geology, v. 69, p. 33-45.

Rios, J.M.
1968:   Saline deposits of Spain; in Saline Deposits, (ed.) R.B. Mattox; Geological Society of America, Special Paper 88, p. 59-74.

Rouvier, H., Perthuisot, V., and Mansouri, A.
1985:   Pb-Zn deposits and salt-bearing diapirs in southern Europe and North Africa; Economic Geology, v. 80, p. 666-687.

Rozen, O.M.
1982:   Geochemical variations in sulphate-bearing sediments and their metamorphic products (the problem of Precambrian evaporites); Lithology and Mineral Resources, Plenum Publishing Co., no. 2, p. 175-183 (translated from Litologia i Poleznye Iskopaemye, March-April, 1982, no. 2, p. 94-103).

Ruan, H., Hua, R., and Cox, D.P.
1991:   Copper deposition by fluid mixing in deformed strata adjacent to a salt diapir, Dongchuan area, Yunnan Province, China; Economic Geology, v. 86, p. 1539-1545.

Sanford, B.V. and Norris, A.W.
1975:   Devonian stratigraphy of the Hudson Platform. Part I: Stratigraphy and economic geology; Geological Survey of Canada, Memoir 379, 124 p.

Saunders, J.A. and Swann, C.T.
1994:   Mineralogy and geochemistry of a cap-rock Zn-Pb-Sr-Ba occurrence at the Hazlehurst salt dome, Mississippi; Economic Geology, v. 89, p. 381-390.

Schmidt-Mumm, A., Behr, H.J., and Horn, E.E.
1987:   Fluid systems in metaplaya sequences in the Damara Orogen (Namibia): evidence for sulphur-rich brines – general evolution and first results; Chemical Geology, v. 61, p. 135-145.

Sighinolfi, G.P., Kronberg, B.I., Gordoni, C., and Fyfe, W.S.
1980:   Geochemistry and genesis of sulphide-anhydrite-bearing Archean carbonate rocks from Bahia (Brazil); Chemical Geology, v. 29, p. 323-331.

Smith, G.I.
1979:   Subsurface stratigraphy and geochemistry of Late Quaternary evaporites, Searles Lake, California; United States Geological Survey, Professional Paper 1043, 130 p.

Sonnenfeld, P.
*1984:   Brines and Evaporites; Orlando, Florida, Academic Press Inc., 683 p.
1992:   Sulphatization and desulphatization of marine evaporite facies; in Seventh International Salt Symposium, Proceedings, Volume 1, Amsterdam: Elsevier, p. 211-219.

Stöcklin, J.
1968:   Salt deposits of the Middle East; in Saline Deposits, (ed.) R.B. Mattox; Geological Society of America, Special Paper 88, p. 157-181.

Tortochaux, F.
1968:   Occurrence and structure of evaporites in North Africa; in Saline Deposits, (ed.) R.B. Mattox; Geological Society of America, Special Paper 88, p. 107-138.

von der Borch, C.C. and Lock, D.
1979:   Geological significance of Coorong dolomites; Sedimentology, v. 26, p. 813-824.

Author's address

R.T. Bell
Box 287
North Gower, Ontario
K0A 2T0

Printed in Canada

6. EXHALATIVE BASE METAL SULPHIDES

6.1 Sedimentary exhalative sulphides (Sedex)

6.2 Sedimentary nickel sulphides

6.3 Volcanic-associated massive sulphide base metals

6.4 Volcanic-associated massive sulphide gold

6. EXHALATIVE BASE METAL SULPHIDES

Introduction

Stratiform exhalative sulphide deposits are generally concordant, massive to semi-massive accumulations of sulphide (primarily iron sulphide) – sulphate minerals (barite, anhydrite) that formed on or immediately below the seafloor penecontemporaneously with their host rocks. They range in age from those that are actively forming within modern oceanic spreading ridges and back-arc basins, to those preserved in ca. 2.0 Ga sedimentary basins, and in ca. 3.4 Ga oceanic crust. Canada is particularly well endowed with classical examples of all subtypes, as well as with deposits that may be regarded as hybrid or of mixed character between the different subtypes. Four subtypes are identified:

Subtype 6.1, Sedimentary exhalative sulphides (Sedex) occur in terranes dominated by sedimentary strata. Volcanic rocks (lava, tuff) may be a minor component of the associated strata, and penecontemporaneous intrusions (mafic sills, dykes) may be present regionally. Zinc, lead, and silver are the primary metals recovered from the subtype; copper is generally a minor product. Barite may be abundant (deposits of mainly Phanerozoic age) or minor (deposits of mainly Proterozoic age); many deposits have none at all. Iron sulphide (pyrite, pyrrhotite) content is also highly variable.

Subtype 6.2, Sedimentary nickel sulphides occur in basinal sedimentary terranes similar to those that host Sedex deposits, and in fact have much in common with them in terms of physical characteristics and genesis. However the primary metals available from this subtype are nickel, zinc, molybdenum, and platinum group elements (PGEs).

Subtype 6.3, Volcanic-associated massive sulphide base metals. All volcanic-associated deposits occur in terranes dominated by volcanic rocks. The deposits may occur in volcanic or sedimentary strata that are integral parts of a volcanic complex. Volcanic-associated massive sulphide deposits may be divided into two compositional groups: base metal enriched (subtype 6.3) and auriferous (subtype 6.4). Deposits of the first group contain highly variable amounts of economically-recoverable copper, lead, zinc, silver, and gold. Minor elements such as tin, cadmium, bismuth, and selenium may also be important smelter byproducts; however base metals are the *primary* commodities recovered. Some deposits are dominated by iron sulphide (up to 90% pyrite) and have been mined primarily for iron and sulphur.

Subtype 6.4, Volcanic-associated massive sulphide gold. In deposits of this group, gold is the *primary* commodity, with copper, zinc, silver, and lead being of lesser economic importance. Mineralization may be massive, disseminated, or in stockworks.

6.1. SEDIMENTARY EXHALATIVE SULPHIDES (SEDEX)

John W. Lydon

INTRODUCTION

The term "Sedex" is an acronym for "sedimentary exhalative" that was proposed by Carne and Cathro (1982) as a short and convenient name for a class of deposits that was referred to by a variety of terms, including "sediment-hosted stratiform Zn-Pb", "shale-hosted", and "sedimentary-exhalative" deposits. The deposit class includes important producers of zinc and lead ores, such as Broken Hill and Mount Isa in Australia, Sullivan in Canada, Red Dog in Alaska, and Navan in Ireland.

A working definition of the deposit class is *a sulphide deposit formed in a sedimentary basin by the submarine venting of hydrothermal fluids and whose principal ore minerals are sphalerite and galena*. This definition is deliberately loose because more explicit definitions tend to exclude examples that should be included in the class. However, the definition serves to distinguish Sedex deposits from other seafloor metalliferous deposits related to hydrothermal venting, of which volcanogenic massive sulphides (VMS), Besshi-type, iron-formations, manganese formations, and bedded baritite deposits are the most important examples (see "Definitive characteristics", below).

This definition requires a knowledge of the processes by which a deposit has been formed. In highly metamorphosed or intensely deformed terranes, primary textural evidence for genetic processes may not be preserved. Under these conditions, it may be difficult to distinguish Mississippi Valley-type (MVT) deposits from Sedex-type deposits, because both have similar hydrothermal mineralogy and bulk composition (Sangster, 1990), and a MVT deposit that has been structurally transposed into geometric conformity with its host rocks may assume an overall morphology comparable to a Sedex deposit. Even for well preserved deposits, the distinction between the two types may be unclear. Considering that the subsurface deposition of sulphides (with its resultant epigenetic textures) is an important component of the ore-forming process of most Sedex deposits, those hosted in carbonate rocks may only be distinguished from MVT deposits by evidence that the ores were emplaced below a submarine hydrothermal vent field. Similarly, there is a continuum between Sedex deposits and VMS deposits in both depositional environment and deposit characteristics, and again it is more a matter of prejudice than of verifiable fact, that certain deposits are classified as one type instead of the other.

Lydon, J.W.
1996: Sedimentary exhalative sulphides (Sedex); in Geology of Canadian Mineral Deposit Types, (ed.) O.R. Eckstrand, W.D. Sinclair, and R.I. Thorpe; Geological Survey of Canada, Geology of Canada, no. 8, p. 130-152 (also Geological Society of America, The Geology of North America, v. P-1).

IMPORTANCE

Sedex deposits are a major source of zinc and lead, and an important source of silver. A compilation by Tikkannen (1986) indicates that, on the global scale, Sedex deposits accounted for about 40% of zinc production and about 60% of lead production. For known resources that are not being mined, the proportions are considerably greater, though no accurate statistics are available. In Canada, most production of zinc and lead from Sedex-type deposits has been obtained from the Sullivan mine, British Columbia, and the Faro and Vangorda deposits of the Anvil district in Yukon Territory (Fig. 6.1-1). Over the last decade, this has accounted for about 15% and 20% of primary Canadian zinc and lead production respectively (Fig. 6.1-2).

Despite their importance in terms of proportion of metal production and reserves, Sedex deposits are a comparatively rare type of deposit. A compilation of known Sedex deposits (Table 6.1-1) contains only about 70 examples, of which only 24 have been, or are being mined. The list is not exhaustive, especially for deposits in Asia, South America, and Eastern Europe, for which the literature is not sufficiently informative for accurate compilation. However, even if the list contained twice the number, it would not change the basic facts that: i) Sedex deposits are comparatively rare; ii) the majority of Sedex deposits are uneconomic because the ores are of too low a grade or because they are too fine grained for high beneficiation recoveries; and iii) a minority of Sedex deposits constitutes the greatest individual concentrations of zinc and lead ores known.

The positive aspect of this third fact outweighs the negative considerations of the other two, and makes Sedex deposits the most attractive targets for zinc and lead exploration.

SIZE AND GRADE OF DEPOSITS

The average size and grade of 62 deposits classified here as of the Sedex type (exclusive of Howards Pass deposits) is 41.3 Mt grading 6.8% Zn, 3.5% Pb, and 50 g/t Ag. The size of Sedex deposits range up to 120 Mt, with a few, the giant Sedex deposits, containing reserves in excess of this figure (Fig. 6.1-3C). Grades of zinc and lead range up to 18% and 9% respectively (Fig. 6.1-3A, B). Exclusive of the giant Sedex deposits, the amount of contained metal ranges up to 20 Mt of combined zinc and lead (Fig. 6.1-3F) and up to 10 million kilograms silver (Fig. 6.1-3E).

The Zn:Pb ratios of Sedex deposits, expressed as (Zn x 100)/(Zn+Pb), range from 15 to 100 (Fig. 6.1-3F), a seemingly nonsystematic spread. However, there seems to be some geological significance to the ratios.

i) The (Zn x 100)/(Zn+Pb) ratio of aqueous chloride solutions saturated with respect to sphalerite and galena varies from 75 to 85, depending on temperature and chlorinity. This would be the ratio expected of a Sedex deposit if the upflowing ore fluids (see "Genetic models", below) were saturated with metals and all the metal were precipitated and preserved in the deposit. About 30% of Sedex deposits are in this range of ratios.

ii) Sedex deposits are typically zoned, with the bulk of the galena being concentrated in the highest grade ores close to the upflow zone. Most of the deposits plotting between 40 and 70 are those for which mining, and not geological reserves, are reported. Thus, to varying degrees, the grades of the ore reserves do not include the zinc-dominant lower grade outer margins of the geological sulphide body. For example, drill indicated

1975 reserves for the Tom deposit were calculated at 15.7 Mt at 7.0% Zn and 4.61% Pb (Zn x 100)/(Zn+Pb)=60 but these figures do not include the known 600 m northward low grade extension to the West zone. Using a cut-off of Zn+Pb=7% for the "geological reserves", mining reserves are reported as 9.2 Mt at 7.49% Zn and 6.19% Pb (Zn x 100)/(Zn+Pb)=55 (McClay and Bidwell, 1986).

iii) Those deposits with (Zn x 100)/(Zn+Pb) values of less than 40 include those that have been subjected to oxidative weathering during their history. The oxidative weathering of a sphalerite-galena body may lead to the preferential removal of zinc; sphalerite is solubilized as the sulphate but galena develops a protective rind of insoluble cerussite. The effect can be seen by comparing the compositions of the sulphide protore and secondary

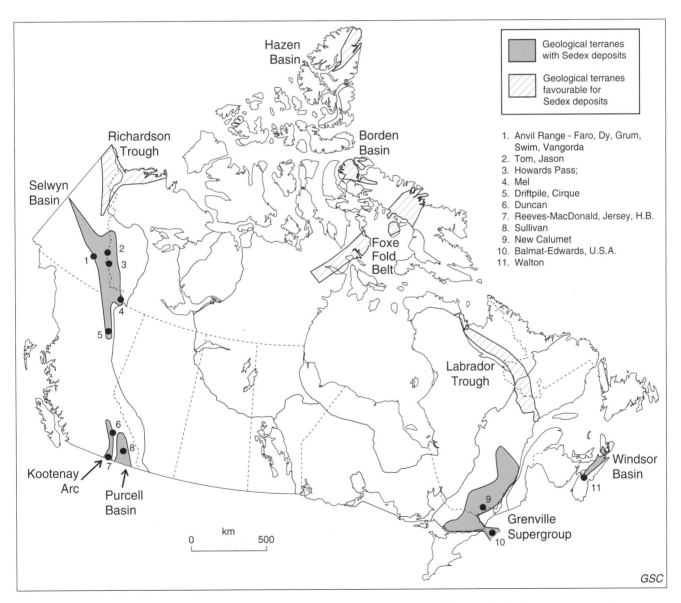

Figure 6.1-1. Distribution of Sedex deposits and geological terranes with potential for the occurrence of Sedex deposits in Canada.

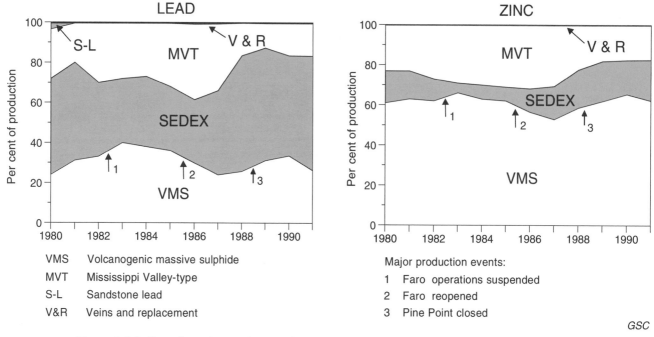

Figure 6.1-2. Canadian mine production of zinc and lead by deposit type for the period 1980-1991.

ore at Dariba (Table 6.1-1). The Tynagh ores also largely consist of the insoluble residue of subaerial weathering. It is interesting to note that the Broken Hill and Big Syn deposits of the Aggeneys area, other lead-dominant examples, are associated with magnetite iron-formation and baritite, both products of an oxidative environment, and it may well be that these deposits represent the remains of sulphide deposits that were partially oxidized while on the seafloor.

GEOLOGICAL FEATURES

Morphology and architecture of deposits

The idealized characteristics of a Sedex deposit are schematically illustrated in Figure 6.1-4. Sedex deposits usually comprise a conformable to semiconformable stratiform lens or lenses of sulphide and associated hydrothermal products (Fig. 6.1-4). The stratiform lens typically has an aspect ratio (the ratio of maximum lateral extent to maximum thickness) of 20, and maximum thicknesses are most commonly in the range 5 to 20 m.

Typically the stratiform lens can be divided into three distinct facies: i) vent complex; ii) bedded ores; and iii) distal hydrothermal products. In some deposits, discordant vein and replacement mineralization stratigraphically below the vent complex is significant enough to warrant a fourth facies type being distinguished: iv) feeder zone.

Vent complex

The vent complex part of the deposit lies within the upflow zone of the hydrothermal fluids, and consists of hydrothermal products that formed immediately below and within the submarine vent field. Ore textures (see Fig. 6.1-5A, D, G) typically reflect recurrent cycles of hydrofracturing, hydrothermal cementation, and replacement. The vent complex usually contains the highest grade mineralization, and for some deposits is the only part that is of ore grade.

Bedded ores

The bedded ores facies is a compositionally layered apron of hydrothermal products, which may or may not be interlayered with host rock lithologies, that is usually asymmetrically distributed around the vent complex. Examples are illustrated in Figure 6.1-5B, C, F. The thickness of individual compositional layers of hydrothermal products ranges from the scale of millimetres to the scale of metres. Individual layers are commonly very persistent laterally, and they decrease in thickness only very gradually away from the vent complex. The layers of hydrothermal products in most deposits are internally laminated. Their appearance, which is varve-like, has led to their interpretation as products of brine pool sedimentation or fallout from a hydrothermal plume that has accumulated in a low energy environment. In some deposits, the bedded ores contain fragmental, graded, or massive layers of hydrothermal products that have been interpreted to be debris flows, turbidites, and mudflows, respectively, and whose source area was a topographically elevated vent complex. Other schools of thought interpret sedimentary textures of the bedded ores to represent pseudomorphs of precursor lithologies formed by selective replacement of the host sediments by hydrothermal metasomatism just below the seafloor. In most deposits, the sphalerite and galena contents of the bedded ores gradually decrease with increasing distance from the vent complex.

Table 6.1-1. List of Sedex deposits for which there are sufficient published data for their classification.

Deposit	Location	Zn %	Pb %	Cu %	Ag (g/t)	Au (g/t)	BaSO$_4$	Size (Mt)	Age (Ma)	References
Santa Lucia	Cuba	5.7	1.8	0.0	30.0	0.0	Yes	19.4	150	Valdes-Nodarse et al., 1993; Mining Journal, Sept. 3, 1993
Gunga	Pakistan	5.0	1.0	0.0	0.0	0.0	Yes	10.0	150	Anderson and Lydon, 1990
Duddar	Pakistan	16.4	3.9	0.0	0.0	0.0	Yes	1.0	150	Anderson and Lydon, 1990
Filizchai	Azerbaijan	7.0	1.0	2.1	0.0	0.0	?	5.0	175	Laznicka, 1981
Lik	Alaska	8.8	3.0	0.0	28.0	0.0	Yes	25.0	340	Forrest and Sawkins, 1984
Galmoy	Ireland	10.9	1.0	0.0	0.0	0.0	?	6.7	340	Mining Journal, Sept. 1990, p. 231
Lisheen	Ireland	12.0	1.5	0.0	32.0	0.0	?	20.0	340	Shearley et al., 1992
Navan	Ireland	10.1	2.6	0.0	0.0	0.0	No	69.9	340	Andrew and Ashton, 1985
Silvermines	Ireland	6.4	2.5	0.0	23.0	0.0	Yes	17.7	340	Hitzman and Large, 1986
Tynagh	Ireland	4.5	4.8	0.4	46.5	0.0	Yes	13.6	340	Morrissey et al., 1971
Walton	Canada	1.4	4.3	0.6	305.0	0.0	Yes	0.7	340	Patterson, 1988
Red Dog	U.S.A.	17.1	5.0	0.0	82.3	0.0	Yes	77.0	350	Moore et al., 1986
Tom	Canada	7.0	4.6	0.0	49.1	0.0	Yes	15.7	350	McClay and Bidwell, 1986; Goodfellow and Rhodes, 1990
Elura	Australia	8.4	5.6	0.0	139.0	0.0	Yes	27.0	370	Schmidt, 1990
Cirque	Canada	8.0	2.0	0.0	0.0	0.0	Yes	52.2	370	Jefferson et al., 1983; Canadian Mines Handbook 1991-2, p. 127
Driftpile	Canada	11.9	3.1	0.0	0.0	0.0	Yes	2.4	370	MacIntyre, 1983; Teck Corporation, 1994 (circular)
Jason	Canada	7.4	6.5	0.0	79.9	0.0	Yes	10.1	370	Turner, 1990; Northern Miner, May 28, 1990, p. 21
Meggen	Germany	7.0	1.0	0.1	14.0	0.0	Yes	50.0	370	Krebs, 1981
Rammelsberg	Germany	16.4	7.8	1.0	103.0	0.0	Yes	27.2	370	Hannak, 1981; Krebs, 1981
Zhairem	Khazakhstan	5.0	2.0	0.5	0.0	0.0	Yes	N/A	370	Smirnov and Gorzhersky, 1977; Laznicka, 1981
Tekeli	Khazakhstan	6.0	5.0	1.0	0.0	0.0	Yes	6.0	380	Smirnov and Gorzhersky, 1977; Laznicka, 1981
Howards Pass	Canada	5.0	2.0	0.0	9.0	0.0	No	476.0	435	Goodfellow and Jonasson, 1986; Placer Dev., Annual Report 1982
Peary Land	Greenland	8.0	1.0	0.0	?	?		12.0	435	Northern Miner, Oct. 10, 1994, p. 3
El Aguilar	Argentina	8.5	6.5	0.1	100.0	0.0	No	30.0	450	Sureda and Martin, 1990; Gemmell et al., 1992
Bleikvassli	Norway	7.5	3.0	0.4	0.0	0.0	No	6.0	460	Skauli et al., 1992
Mofjellet	Norway	4.7	1.0	0.4	0.0	0.0	No	4.2	460	Laznicka, 1981
Dy	Canada	6.7	5.5	0.1	84.0	1.0	Yes	21.1	510	Jennings and Jilson, 1986
Faro	Canada	5.7	3.4	0.0	36.0	0.0	Yes	57.6	510	Jennings and Jilson, 1986
Grum	Canada	4.9	3.1	0.0	49.0	0.0	Yes	31.0	510	Jennings and Jilson, 1986
Swim	Canada	4.7	3.8	0.0	42.0	0.0	Yes	4.3	510	Jennings and Jilson, 1986
Vangorda	Canada	4.9	3.8	0.3	54.0	0.8	Yes	7.5	510	Jennings and Jilson, 1986; Canadian Mines Handbook 1988-89, p. 135
Koushk	Iran	15.0	3.0	0.0	0.0	0.0	?	5.0	510	Mining Journal, June 14, 1991, p. 454
Portel	Portugal	1.4	0.2	0.0	0.0	0.0	Yes	35.0	510	J.W. Lydon, unpub. data, 1968
Fuenteheridos	Spain	2.0	0.2	0.0	0.0	0.0	Yes	100.0	510	J.W. Lydon, unpub. data, 1968
Duncan	Canada	3.1	3.3	0.0	0.0	0.0	?	2.8	550	Höy, 1982
H.B.	Canada	4.1	0.8	0.0	4.8	0.0	?	6.5	550	Höy, 1982
Jersey	Canada	3.5	1.7	0.0	3.1	0.0	?	7.7	550	Höy, 1982
Mel	Canada	5.6	2.1	0.0	0.0	0.0	Yes	4.8	550	Miller and Wright, 1983
Reeves-MacDonald	Canada	3.5	1.0	0.0	3.4	0.0	?	5.8	550	Höy, 1982
Aberfeldy	Scotland	1.2	0.4	0.0	0.0	0.0	Yes	N/A	600	Coates et al., 1980; Willan and Coleman, 1983
Kholodnina	Russia	7.0	1.5	0.0	0.0	0.0	?	N/A	1075	Smirnov and Gorzhersky, 1977; Laznicka, 1981
Rosh Pinah	Namibia	7.0	2.0	0.1	0.0	0.0	Yes	17.4	1100	Page and Watson, 1976; VanVuuren, 1986
Jiashengpan	China	3.8	1.3	0.0	0.0	0.0	No	N/A	1300	Lang and Xingjun, 1987
Balmat-Edwards	U.S.A.	10.1	0.3	0.0	0.0	0.0	No	17.4	1300	Lea and Dill, 1968
New Calumet	Canada	8.8	2.8	0.2	0.0	0.1	No	1.3	1300	McLarsn, 1946
Sullivan	Canada	5.5	5.8	0.0	59.0	0.0	No	170.0	>1468	Hamilton et al., 1982; H.E. Anderson, D. Davis, and R.R. Parrish,
Sargipali	India	0.4	2.0	0.0	65.0	0.0	?	N/A	1600	Sarkar, 1974
Big Syn	South Africa	2.5	1.0	0.1	12.9	0.0	Yes	101.0	1650	Ryan et al., 1986; Reid et al., 1987
Black Mtn.	South Africa	0.6	2.7	0.8	29.8	0.0	Yes	81.6	1650	Reid et al., 1987; Ryan et al., 1986
Broken Hill	South Africa	1.8	3.6	0.3	48.1	0.0	Yes	85.0	1650	Reid et al., 1987; Ryan et al., 1986
Gamsberg	South Africa	7.1	0.5	0.0	0.0	0.0	Yes	150.0	1650	Reid et al., 1987; Ryan et al., 1986
Century	Australia	10.0	1.5	0.0	30.0	0.0	?	120.0	1670	Mining Magazine, Oct. 1991, p. 233-237
Hilton	Australia	10.2	6.5	0.0	137.9	0.0	No	72.0	1670	Forrestal, 1990
Lady Loretta	Australia	14.0	8.0	0.0	110.0	0.0	Yes	9.0	1670	Australian Mining Yearbook, 1988, p. 20; Louden et al., 1975
Mt. Isa	Australia	6.0	7.0	0.1	160.0	0.0	No	125.0	1670	Mathias and Clark, 1975; Forrestal, 1990
Dugald River	Australia	15.0	2.0	0.0	62.0	0.0	No	12.0	1670	Whitcher, 1975; Mining Journal, May 4 1990, p. 352
H.Y.C.	Australia	9.5	4.1	0.0	40.0	0.0	No	227.0	1690	Lambert, 1976; Logan et al., 1990
Broken Hill	Australia	12.0	13.0	0.2	175.0	0.0	No	300.0	1690	Both and Rutland, 1976; Wright et al., 1987
Balaria(Zawar)	India	5.6	1.1	0.0	0.0	0.0	No	19.0	1700	Deb et al., 1989
Baroi magra	India	1.2	4.3	0.0	0.0	0.0	No	11.0	1700	Deb et al., 1989
Mochia(Zawar)	India	3.8	1.7	0.0	0.0	0.0	No	27.0	1700	Deb et al., 1989
Zarwarmala(Zawar)	India	3.7	2.1	0.0	0.0	0.0	No	18.0	1700	Deb et al., 1989
Dariba	India	7.3	2.0	1.0	5.0	0.0	No	13.0	1800	Nandan et al., 1981; Deb and Bhattacharya, 1980
Dariba(secondary)	India	1.2	4.0	0.0	200.0	0.0	No	31.6	1800	Nandan et al., 1981; Deb and Bhattacharya, 1980
Dariba-Rajpura	India	5.1	1.2	1.1	122.0	0.0	No	51.0	1800	Nandan et al., 1981; Deb and Bhattacharya, 1980
Rampura-Agucha	India	13.5	1.6	0.0	45.0	0.0	No	61.1	1800	Nandan et al., 1981; Deb and Bhattacharya, 1980
Saladipura	India	1.0	0.0	0.0	0.0	0.0	No	115.0	1800	Deb et al., 1989
Sindesar	India	2.1	0.5	0.0	0.0	0.0	No	70.0	1800	Deb, 1982; Deb et al., 1989

Distal hydrothermal products

The lateral lithological equivalents of the bedded ores, outside the economic limits of the Sedex orebody, are termed here "distal hydrothermal products", reflecting their greater distance from the vent complex than the bedded ores themselves. Although the boundary between the bedded ores and distal hydrothermal products is primarily an economic one, in some deposits it may also have geological significance, separating two distinct sedimentological or metasomatic facies.

Feeder zone

Though not documented for most deposits, the vent complex is rooted in a feeder zone of discordant vein and/or replacement-type mineralization. The fracture systems of the feeder zone may either be due to tectonism, and related to movement on synsedimentary faults with which many Sedex deposits are spatially associated, and/or may be due to hydrofracturing accompanying hydrothermal eruption. In contrast to VMS deposits, the feeder zone of Sedex deposits rarely contributes any significant ore reserves. There are two main reasons for this.

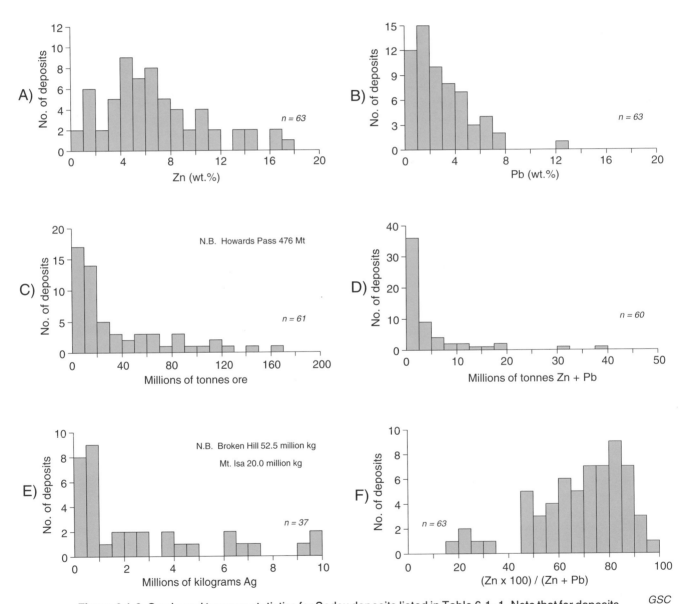

GSC

Figure 6.1-3. Grade and tonnage statistics for Sedex deposits listed in Table 6.1-1. Note that for deposits that have not been mined, the figures reported are usually geological reserves; for deposits that have been or are being mined, the figures are mining reserves plus production. **A)** Grade of zinc in ore; **B)** grade of lead in ore; **C)** tonnage of ore (production plus reserves); **D)** tonnes of contained zinc plus lead per deposit; **E)** kilograms of contained silver per deposit (note that silver grades are not reported for a majority of deposits); **F)** Average Zn:Pb ratio of deposit expressed as (Zn x 100)/(Zn + Pb).

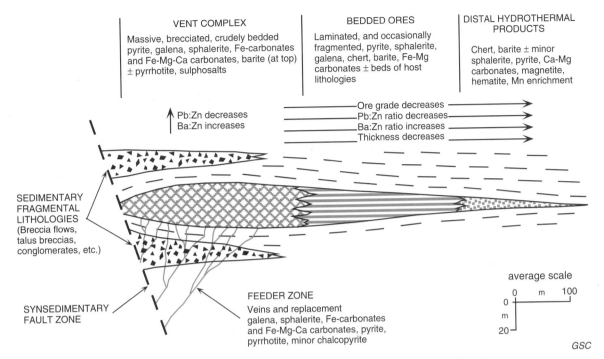

Figure 6.1-4. Schematic illustration of the characteristic features of the idealized Sedex deposit.

1. The formation of high grade vein networks and mineralized breccias of feeder zone ores requires a rock with sufficient mechanical strength to brecciate and form rigid clasts that can support a high fracture porosity. In contrast to volcanic rocks, which attain a high mechanical strength immediately upon cooling on the seafloor, unconsolidated sediments have such low mechanical strength that they tend to flow rather than brecciate when subjected to differential pore fluid pressures. Fault breccias, hydrothermal vent breccias, and hydrothermal eruption breccias may readily form in a substrate of volcanic rocks to provide the host for typical feeder zone mineralization. In unconsolidated sediments, however, the equivalent expressions of tectonic movement or hydrothermal eruption are zones of disrupted sediment, discordant columns of turbated sediment and concordant mud flows, none of which have a significantly higher porosity than surrounding undisturbed sediments, and hence do not provide a host for preferential subsurface hydrothermal precipitation.

2. The main economic mineral in feeder zone ores of VMS deposits is chalcopyrite which, because of its higher economic value compared to sphalerite and galena, lowers the concentration of ore minerals required to achieve ore grade. The average relative proportion of copper in Sedex deposits (Table 6.1-1) is very much less than in VMS deposits, so that even though in Sedex deposits chalcopyrite tends to be concentrated in feeder zone mineralization, it rarely attains absolute concentrations of economic significance.

Mineralogy
Primary hydrothermal minerals

Typical hydrothermal products of Sedex deposits include sulphides, carbonates, quartz, and barite. Pyrite is usually the most abundant sulphide, and commonly is the only iron sulphide in the deposit. However, in some Proterozoic deposits, such as Mt. Isa (Mathias and Clark, 1975), pyrrhotite is common, and in the Sullivan deposit is more abundant than pyrite (Hamilton et al., 1982). At the Duddar deposit, Pakistan, which is one of the youngest fossil Sedex deposit known, marcasite is the dominant primary iron sulphide.

Sphalerite and galena are invariably the main economic minerals. Chalcopyrite is usually a very minor constituent, although it was abundant enough in the Rammelsberg deposit to be economically recoverable (Hannak, 1981). Antimony, arsenic, and bismuth, particularly in the sulphosalts tetrahedrite, freibergite, and boulangerite, are typically concentrated in and around the vent complex, and in cases are the host minerals for much of the silver content of the ores.

Carbonates are much more important constituents of Sedex deposits than VMS deposits. Siderite and ankerite may occur as veins in the feeder zone, as replacive masses in the vent complex and as beds or laminations in the bedded ores, as, for example, in the Tom deposit, Yukon Territory (Goodfellow and Rhodes, 1990), and at Silvermines, Ireland (Taylor, 1984; Andrew, 1986). Calcite is relatively common as veins, interstitial cement, and laminations in

both the vent complex and bedded ores of the Sullivan deposit (Hamilton et al., 1982). Abundant bedded calcium carbonate is restricted to the ore zone of the XY deposit at Howards Pass (Goodfellow and Jonasson, 1986). Calcite nodules are a common hydrothermally induced feature of unconsolidated sediment surrounding the modern metalliferous hydrothermal vents of Middle Valley (Goodfellow et al., 1994). Dolomite, either as a precipitate or hydrothermal alteration product, is common around Sedex deposits in carbonate rocks (e.g., Tynagh and Silvermines, Ireland: Hitzman and Large, 1986; Fuenteheridos, Spain). Secondary dolomite is co-extensive with the stratiform sulphide mineralization of the McArthur River deposits, Australia (Williams, 1978a), but there is debate whether the widespread silica dolomite alteration at Mt. Isa is genetically related to the Zn-Pb mineralization. Quartz, as chert, is abundant in many Sedex deposits (e.g. Tom, Silvermines), but is conspicuously absent in others (e.g. Sullivan). Where it is present, the proportion of quartz, relative to other hydrothermal products, increases with increasing distance from the vent complex. Barite, notably as bedded baritite in the bedded ores and distal hydrothermal facies, is a common constituent of most Paleozoic and younger Sedex deposits, but is much less common in Proterozoic deposits. Notable exceptions are the Howards Pass and Navan deposits of Paleozoic age which do not contain significant barite, and the Proterozoic deposits of the Aggeneys area, South Africa, which do. Tourmaline is spatially associated with several Sedex deposits, including Sullivan, where it is the major component in the footwall feeder zone (Hamilton et al., 1982), Broken Hill, Australia (Slack et al., 1993), and the Rampura-Agucha deposits, India (Ranawat and Sharma, 1990). Because tourmaline is not readily recognizable, especially in a fine grained form, it may be more common in Sedex deposits than has been recognized.

Metamorphic minerals

The effect of high temperature and pressure metamorphism on mineral assemblages typical of Sedex deposits does not generally obscure recognition of the original mineralogy of the deposit, even though some significant mineralogical changes may be produced. Pyrite may be converted to pyrrhotite by reaction with ferrous silicate minerals or by the degassing of sulphur from the sulphide body during metamorphism. At least some of the pyrrhotite at Broken Hill, Australia, in the Aggeneys district, South Africa, and in the Rampura-Agucha (Ranawat and Sharma, 1990) and Rajpura-Dariba (Deb, 1990) deposits of India, is probably of metamorphic origin. During metamorphic retrogression, pyrrhotite may react to form pyrite and magnetite, although the quantities of pyrite and magnetite so formed would be minor unless a source of oxygen were available. At least some of the pyrite-magnetite assemblage at the Sullivan deposit appears to have formed by this mechanism.

Sphalerite may react with aluminosilicates to form gahnite ($ZnAl_2O_4$) at amphibolite facies and higher metamorphic grades. It is a common though minor mineral at Broken Hill, Australia (Mackenzie and Davies, 1990; van de Hayden and Edgecombe, 1990) and in the deposits of the Aggeneys area, South Africa (Spry, 1987).

Barite may react with aluminosilicates at greenschist facies to form cymrite, celsian, or hyalophane. At the Aberfeldy deposit, Scotland, a cherty layer of rock several metres thick, and composed largely of celsian, forms an envelope around the distal parts of the deposit (Coates et al., 1980).

Zonation

Sedex deposits exhibit compositional zonation in both lateral and vertical directions.

Lateral zonation

Characteristic of most Sedex deposits is a lateral mineralogical, chemical, and thickness zonation with respect to the vent complex. Fundamentally this zonation can be attributed to two main effects:

1. A decrease in the relative proportion of hydrothermal products with respect to intercalated or admixed indigenous sediment with increasing lateral distance from the vent complex. This effect can be ascribed to mechanisms of hydrothermal dispersal as a function of distance from the local hydrothermal upflow zone.

2. A zonation due to systematic changes in the relative proportion of different hydrothermal products with lateral distance from the vent complex. This effect can be ascribed to the different chemical behaviour of different components during dispersal from their local source in the hydrothermal upflow zone.

Figure 6.1-5. Photographs of some macroscopic textures and relationships in Sedex deposits

A) Vent complex ore, Sullivan deposit. Polished drill core (dark grey with white flecks – galena; light grey – pyrrhotite; black – silicates). GSC 1994-640C

B) Bedded ore, Sullivan deposit. Interlayered sulphides (light grey) and argillite (dark grey), "A" ore band. Height of photograph about 2 m. GSC 1994-640H

C) Bedded ore, Sullivan deposit. Interlaminated light grey sulphides (pyrrhotite, sphalerite, galena) and dark grey to black argillite, "B-Triplet" ore band. Light grey flecks in argillite are pyrrhotite laths. Note both tectonic low strain (well laminated) and high strain (turbated) sulphide layers; polished slab. GSC 1994-640F

D) Vent complex mineralization, Tom deposit. Polished slab; dark grey – pyrite; medium grey – ankerite; white – calcite. GSC 1994-640A

E) Bedded ore, Tom deposit. Polished slab; dark grey - chert; medium grey – sphalerite and siderite; light grey – barite. GSC 1994-640I

F) Bedded ore, Tom deposit. Height of photo about 1 m. Laminated barite and sphalerite (light grey) interlayered with siliceous argillite (black). GSC 1994-640D

G) Vent complex mineralization, Tom deposit. Polished slab, black – brecciated siliceous argillite; medium grey – pyrite; white – ankerite. GSC 1994-640G

The first effect largely determines the morphology of the deposit, in the sense that the deposit is defined as that volume of rock containing hydrothermal products. The second effect largely determines the compositional architecture of the deposit, especially its mineralogical zonation. Grades of individual hydrothermal ore components are a function of both effects, because the concentration of any one hydrothermal component may be diluted both by indigenous sediment and by other hydrothermal products.

However, such a simple and consistent zonation pattern reflecting steady state dispersal and accumulation processes is rarely exhibited by most Sedex deposits. The theoretically simple zonation patterns that should result are usually disrupted by various unpredictable events, such as:

1. Dilution of hydrothermal products by locally derived terrigenous sediments. Sedimentary fragmental rocks are a common feature of most Sedex deposits and include talus breccias and debris flows from the wasting of synsedimentary fault scarps, hydrothermal eruption breccias, and mud volcano extrusions.

2. Mass transportation of hydrothermal products, particularly as slide sheets and debris flows, from topographically elevated portions of the deposit.

3. Mass removal of hydrothermal and terrigenous accumulations from the hydrothermal upflow zone by hydrothermal eruption.

4. Preferential chemical removal of previously accumulated hydrothermal components from the upflow zone during the zone refinement process. The removal of barite from the vent complex is perhaps the most outstanding example. Barite, which is formed by hydrothermal barium combining with water column sulphate, is easily solubilized by reduced hydrothermal fluids in the upflow zone and may be recycled to be reprecipitated in more oxidized parts of the deposit.

Notwithstanding these complications, which produce zonation patterns whose details are unique for each Sedex deposit, there are trends that are common to a high proportion of the class as a whole:

1. There is a zonation from reduced mineral facies (e.g., sulphides, ferroan carbonates) within the hydrothermal upflow zone to more oxidized facies (e.g., barite, iron oxides, calcic carbonates) at the periphery of the deposit. Chemically, this trend may be reflected, for example, in decreasing ratios of Zn:Ba and Zn:Mn.

2. Amongst the sulphides, there is a general zonation outwards from the core of the upflow zone in the sequence chalcopyrite, pyrrhotite, galena, sphalerite, and pyrite. The zonation from chalcopyrite via galena to sphalerite, as in VMS deposits, largely reflects a thermal gradient. Chalcopyrite rarely attains significant concentrations in most Sedex deposits because the temperature of the hydrothermal fluids ($<300°C$) is too low to transport copper in reduced sulphidic fluids. The Rammelsberg deposit (Hannak, 1981), which contains 1.0% Cu, is the notable exception. Primary pyrrhotite is common in only a few Sedex deposits, notably of Proterozoic age (e.g. Sullivan, Mt. Isa). If primary iron monosulphide was a common quenching product in Sedex deposits, as it is in modern black smoker buoyant plumes, then most of it has been sulphidized to pyrite. In terms of metal ratios the most consistent zonation is an increase in Zn:Pb ratios outwards from the vent complex. In deposits in which most of the iron is contained in pyrite, ratios of Fe:Pb and Fe:Zn also generally increase with increasing distance from the vent complex. The polarity of the normal upward and outward increase in Fe:Pb and Fe:Zn ratios is reversed in the core of the Sullivan deposit by the replacement of sphalerite and galena by pyrrhotite, upward and outward from the base of the vent complex.

3. Tin, bismuth, arsenic, and mercury tend to be concentrated in the vent complex in sulphosalt minerals and arsenopyrite. Silver, an important economic constituent in most Sedex deposits, is frequently hosted by sulphosalts as well as being held in solid solution in galena. Consequently, the highest grades of silver are typically in the vent complex. A significant proportion of the trace copper content of Sedex deposits may also be contained in sulphosalts (e.g., Tynagh, Tom).

4. Carbonates are probably an integral hydrothermal component of most Sedex deposits, but their importance may have been overlooked in many cases. It might be expected that carbonates are more abundant in Phanerozoic deposits than in Proterozoic deposits, because of the increased abundance of marine calcareous faunal remains in the younger terrigenous source rocks for the hydrothermal fluids. For example, the Tom deposits contains 15-20% CO_2 in the vent complex (Goodfellow and Rhodes, 1990), whereas the Sullivan deposits contains 0.2-1.0% CO_2 (Hamilton et al., 1982). The carbonates tend to be zoned from ferruginous, in the core of the upflow zone, to calcic at the periphery of the deposit, and occur both as subsurface infillings and replacements in the upflow zone, as well as layers in the stratified part of the deposit. For example, siderite, occurs as vein fillings and as an interstitial cement of discordant mineralization at Jason deposit (Bailes et al., 1986; Turner, 1990), and as a basal layer to the massive bedded sulphides in the Mogul B deposit, Silvermines, Ireland (Taylor, 1984; Andrew, 1986).

5. Sulphide grades, especially for Zn and Pb, generally systematically decrease towards the periphery of the deposit from maxima in the vent complex. Elemental ratios diagnostic of the gradual dilution of hydrothermal sulphide by indigenous sediment (e.g. $Zn:Al_2O_3$; $S(sulphide):Al_2O_3$) or by other hydrothermal products (e.g., Zn:Ba) gradually decrease towards the periphery of the deposit.

6. Manganese tends to be concentrated at the margins of the deposit in carbonates, iron oxides or, in the case of more highly metamorphosed deposits, in garnet. A manganese enrichment in carbonates surrounding the Meggen deposit may be detected several kilometres away from the deposit (Gwosdz and Krebs, 1977). At Tynagh, Ireland, manganese is concentrated in the hematite iron-formation which constitutes a distal hydrothermal products facies, but also forms a halo of manganese enrichment in the host carbonates for a distance of 7 km around the deposit (Russell, 1974, 1975). At Sullivan, manganese garnets are concentrated in the bedded ores, particularly at the margins of the deposit, and in the footwall conglomerate.

7. The thickness of the deposit generally decreases towards its periphery, though its maximum thickness may not be coincident with the central part of the vent complex.

8. Maximal Zn or Pb grade may or may not coincide with the thickest part of the deposit. This is particularly true where there has been dichotomy of processes that produced maximum thicknesses of hydrothermal products on the one hand (e.g., sedimentation, tectonic deformation) and maximum zinc and lead concentrations on the other (e.g., zone refinement). The Tom deposit (McClay and Bidwell, 1986; Goodfellow and Rhodes, 1990) and the Cirque (Stronsay) deposit (Jefferson et al., 1983; Pigage, 1986; MacIntyre, 1992) are examples in which the thickest part of the deposit occurs in the bedded ores, but the highest Zn and Pb grades occur in the vent complex. The Sullivan deposit is an example where both the highest grades and the thickest part of the deposit occur in the vent complex.

Vertical zonation

In most deposits, especially in the vent complex, the vertical zonation of mineral assemblages and chemistry mimics the lateral zonation. This upward and outward zonation is similar to that observed above the feeder zone in VMS deposits, and would seem to be most logically interpreted to be the result of a subsurface zone refinement process.

The sequence of vertical zonation within the bedded part of the deposit need not necessarily duplicate that observed in the vent complex portion, nor is there a consistency in zonal order of mineral facies from deposit to deposit. For example, baritite, which usually is most abundant towards the periphery of the deposit e.g., Silvermines deposit (Taylor, 1984; Andrew, 1986), Tom deposit (McClay and Bidwell, 1986), and Jason deposit (Bailes et al., 1986), may be concentrated near the stratigraphic top of the bedded part of the deposit, for example, Rammelsberg (Hannak, 1981) or at the stratigraphic base, for example, Duddar deposit. The variation of metal ratios as a function of stratigraphic position in the bedded portion of Sedex deposits has been examined by Lydon (1983). Lead:zinc ratio generally decreases stratigraphically upwards in some deposits (e.g., Tom, Sullivan), decreases downwards in others (Rammelsberg, Broken Hill) whereas in others there is no discernible trend (H.Y.C. and Howards Pass [according to Lydon, 1983]). Silver:lead ratios generally decrease stratigraphically upward in most deposits (e.g., Tom, Sullivan, Broken Hill), although a Ag enrichment in the stratigraphically highest ores is apparent in some deposits. Stratigraphically upward Zn:Fe ratios may increase overall (e.g., Sullivan), decrease overall (e.g., Rammelsberg), or not show any discernible trend (e.g., H.Y.C.).

Fluid inclusions

There are few published data on fluid inclusions from Sedex deposits. At the Tom (Gardner and Hutcheon, 1985) and nearby Jason (Ansdell et al., 1989) deposits of the Selwyn Basin, homogenization temperatures average about 260°C and salinities average about 9 wt.% NaCl equivalent (i.e. 2 to 3 times sea water salinity). At the Silvermines deposit, Ireland, homogenization temperatures range between 50°C and 260°C and salinities between 8 and 28 wt.% NaCl equivalent (Samson and Russell, 1987). A negative correlation between homogenization temperature and salinity in quartz was interpreted by Samson and Russell (1987) to indicate the mixing of higher temperature, lower salinity, hydrothermal fluids with either lower temperature high salinity brines of a seafloor brine pool or shallow pore fluids formed by contemporaneous evaporitic processes. At the Sullivan deposit, Leitch (1992) reported homogenization temperatures that range from 150°C to 320°C and salinities that range from 8 to 36 wt.% NaCl equivalent but no prima facie evidence to link the values to the ore fluids.

GEOLOGICAL SETTING

Sedex deposits occur in sedimentary basins that are controlled by tectonic subsidence associated with major intracratonic or epicratonic rift systems. Most commonly, Sedex deposits occur within rift-cover sequences as opposed to the rift-fill sequences (Fig. 6.1-6). That is, they occur in that sequence of sediments that covers the coarse clastic sediments, turbidites, and/or volcanic rocks which were deposited within the rift during its most active stages of extension and subsidence. The rift-cover sequence, which commonly consists of shallow water sedimentary facies, is deposited during the thermal subsidence or rift sag stage (of rifting) and covers both the buried site of the rift and its adjacent platformal shoulders.

The time lapse between initiation of rifting and deposition of Sedex deposits may be as much as 400 million years. For example, in the Mount Isa area, the initiation of rifting and infilling of the Leichhardt River Fault Trough was at about 1800 Ma. Infilling of the rapidly subsiding rift zone with about 16 km of fluviatile to shallow marine clastics and mafic volcanics was completed by about 1740 Ma. A rift-cover sequence, 2.5 km thick, dominated by quartzites and carbonates, but with subaerial felsic volcanic rocks at it base, is overlain by a marine transgressive sequence that is 6 km thick. The transgressive sequence consists of basal arenites, siltites, and shales and overlying fine grained terrigenous and dolomitic sediments that form the host rocks to Sedex deposits. Tuff beds in the ore host have been dated at about 1670 Ma (Page, 1981), indicating for the Mount Isa, Hilton, and H.Y.C. deposits a time lapse of about 130 Ma between initiation of rift-fill and the formation of Sedex deposits.

Similarly, the Cambrian to Devonian basinal rocks hosting Sedex deposits in the Selwyn Basin of Canada can be viewed as a rift-cover sequence to Upper Proterozoic rift-fill clastics of the Windermere Supergroup. Windermere deposition began along a newly rifted margin of western North America at about 770 Ma (Abbott et al., 1986; Gabrielse and Campbell, 1991). Although only the eastern margin of the westward thickening rift-fill sequence is exposed, at least 3 km of coarse feldspathic turbidites occur beneath the lowest calcareous rift-cover strata of Cambrian age (Eisbacher, 1981). Sedex deposits in Cambrian to Mississippian host rocks therefore postdate rift initiation by between 200 and 400 Ma.

In Ireland, extensional basins were initiated during the Lower Devonian, either in response to crustal extension or as pull-apart basins along transcurrent faults. More than 6 km of conglomerate, sandstone, and mudstone fill the Munster Basin (Phillips and Sevastopulo, 1986). The Irish deposits of middle Dinantian age were thus formed only about 40 Ma after initiation of rift-controlled subsidence.

The major exception to the generalization that Sedex deposits occur in a rift-cover sequence is the Sullivan deposit of the middle Proterozoic Belt-Purcell Supergroup.

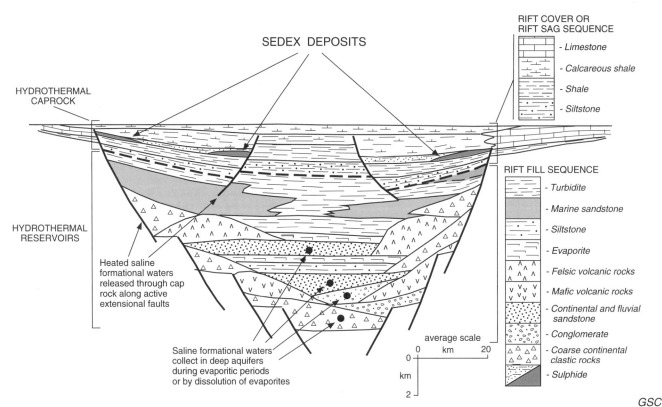

SEDEX DEPOSITS

RIFT COVER OR
RIFT SAG SEQUENCE

- Limestone
- Calcareous shale
- Shale
- Siltstone

HYDROTHERMAL
CAPROCK

HYDROTHERMAL
RESERVOIRS

RIFT FILL SEQUENCE

- Turbidite
- Marine sandstone
- Siltstone
- Evaporite
- Felsic volcanic rocks
- Mafic volcanic rocks
- Continental and fluvial
  sandstone
- Conglomerate
- Coarse continental
  clastic rocks
- Sulphide

Heated saline
formational waters
released through cap
rock along active
extensional faults

Saline formational waters
collect in deep aquifers
during evaporitic periods
or by dissolution of evaporites

average scale

0   km   20

0

km

2

GSC

Figure 6.1-6. Schematic representation of the geological setting of Sedex deposits. Sedex deposits are hosted by the cover sequence to an intracontinental rift system that has been filled by continental clastics, volcanics, and/or marine clastics. Chloride brines, formed during an evaporitic period of rift filling or by the later subsurface dissolution of the evaporites, collect in the deep part of the rift fill sequence. The rift cover sequence acts as a hydrothermal caprock (base marked by bold dashed line) to the brines during heating by burial or deep magmatism. The heated brines flow to the contemporaneous surface of the cover sequence when the caprock is ruptured by renewed extensional tectonism.

The deposit occurs in a rift-fill sequence of turbidites and tholeiitic sills of the Aldridge Formation in the lower part of the exposed stratigraphy. However, some exploration potential can be inferred from the presence of a few relatively small baritic Zn-Pb deposits, possibly of Sedex type, (Mineral King, Paradise, Leg) which are located near the top of the preserved rift-cover sequence (Dutch Creek and Mount Nelson formations).

AGE AND PALEOGEOGRAPHIC DISTRIBUTION

Sedex deposits span the range from the Middle Proterozoic to the present. There are two frequency peaks to the distribution of deposit ages, namely in the Middle Proterozoic and in the Paleozoic (Fig. 6.1-7A). These peaks are accentuated in terms of contained metal (Fig. 6.1-7B), emphasizing that the bulk of known reserves of the Sedex type occur in the Middle Proterozoic of Australia, and in the middle Paleozoic of Western Canada, Alaska, and Western Europe (Table 6.1-1).

The oldest Sedex deposits may be those of the highly metamorphosed Bhilwara and Aravalli supergroups of northwestern India. Model Pb dating indicates an age of

about 1800 Ma for the deposits themselves, and Rb-Sr dating of an intrusive granite constrains their host rocks to a minimum age of 1500 Ma (Deb et al., 1989). The Black Angel deposit of Greenland, which is hosted by Lower Proterozoic (i.e. >1800 Ma) carbonate rocks of the Marmorilik Formation (Garde, 1978; Thomassen, 1991), has been metamorphosed and structurally deformed beyond unequivocal classification. Though considered by Sangster (1990) to be of the Sedex type and by Pedersen (1981) as a syndiagenetic (i.e., Irish) type , it could also be of the Mississippi Valley type (Carmichael, 1988). If the last, it could belong to the same metallogenetic province as the MVT deposits of the Lower Proterozoic Ramah Group in Labrador (Wilton et al., 1993) or the Middle Proterozoic Society Cliffs Formation on Baffin Island (Clayton and Thorpe, 1982). Because of its uncertain genesis, it has been excluded from the present classification. A deposit interpreted to be of Sedex type, with a Pb-Pb model age of 1350 Ma, is reported for the Lower Proterozoic of China (Hou and Zhao, 1993), but because it occurs in a Pb-Zn skarn metallogenetic province, and is itself situated within the metamorphic aureole of a granite and contains elevated molybdenum concentrations, it is excluded from the present classification because of its dubious genesis. The Boquira deposit of Brazil, containing 5.6 Mt at 1.43% Zn

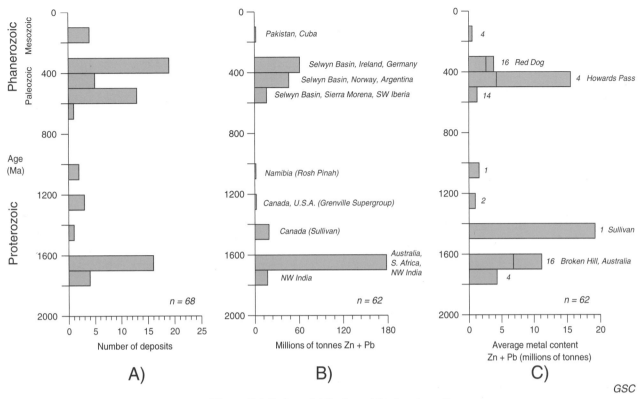

GSC

Figure 6.1-7. Age distribution of Sedex deposits.

A) Number of deposits per 100 Ma interval. Note the two frequency peaks during the Middle Proterozoic (1600-1800 Ma) and the Cambrian-Mississippian span of the Phanerozoic (300-600 Ma).

B) Tonnes of Zn+Pb per 100 Ma interval. Note that the great majority of Sedex resources are about equally divided between the Middle Proterozoic deposits of Australia and South Africa on the one hand and the Paleozoic deposits of Canada and Western Europe on the other.

C) Average metal content of deposits per 100 Ma interval. The number of deposits in each group is indicated. The vertical bar indicates the average metal content of deposits exclusive of the "giant" in each group, the name of which is indicated.

and 8.85% Pb in an amphibolite-magnetite iron-formation of attributed late Archean age (Espourteille and Fleischer, 1988), and suggested to be of sedimentary exhalative origin (Carvalho et al., 1982), is also excluded from the classification because of its suspect genesis.

There does not appear to be any compelling reason, based on current understanding, why Sedex deposits should not occur in rocks older than Middle Proterozoic. The onset of the Middle Proterozoic does not seem to coincide with any major permanent change in global climate, ocean water composition, atmosphere composition, or geotectonic processes. For example, the >3.2 Ga (Kroner et al., 1991) sedimentary baritite of the Fig Tree Group, South Africa (Heinrichs and Reimer, 1977) indicates that a deposit type most closely related to Sedex deposits (see below) was being formed very much earlier in the Earth's history than Middle Proterozoic.

The majority of Proterozoic Sedex ore deposits were formed in the time range 1650-1700 Ma and, together with the Sullivan deposit (1467 Ma – H.E. Anderson, D. Davis, and R.R. Parrish, unpub. data, Geological Survey of Canada, Minerals Colloquium, January, 1994, Ottawa, Ontario),

seem to be spatially associated with the suture systems that separate the Australian, Antarctican, and South African cratons from the North American craton on a reconstruction of the Upper Proterozoic Supercontinent (see Fig. 6.1-8A). This remarkable correlation might be fortuitous, because the mineralization and the rifting with which it is associated predate the orogenic belts of Grenville age, along which assembly of the Upper Proterozoic supercontinent is inferred to have taken place. If, as suggested by Hoffman (1991), the Laurentian, Australian, and Antarctican cratons were fellow travellers from 1900 Ma to 600 Ma, then the rift systems along which the Sedex deposits are located did not lead to oceanic spreading at that time but remained intracratonic. They did, however, provide the cratonic perforations along which the cratons ultimately separated during the latest Proterozoic.

The bulk of Paleozoic Sedex resources are located in the Cambro-Silurian Road River Group of Selwyn Basin and the Devono-Carboniferous Earn Group of northwestern Canada and their continuation into Alaska. The deposits are clustered at Cambrian, Silurian, and Devonian ages. Selwyn Basin is filled by a condensed sequence of shales and carbonaceous cherts representing a semistarved basin, is bounded by a carbonate shelf, and is interpreted to cover the site of an Upper Proterozoic rift zone (Abbott et al., 1986). The unconformably overlying Earn Group comprises shales to conglomerates of coarsening upward clastic wedges of a western provenance (Gordey et al., 1982). Mississippian carbonate rocks of Ireland and Devonian shales of the Rhenish Basin in Germany are the two other most important host successions for Paleozoic Sedex deposits (Fig. 6.1-8B).

The youngest fossilized Sedex deposits reported in the literature are of Jurassic age (Table 6.1-1). Those hosted by Jurassic carbonate rocks of the Lasbela-Khuzdar Belt of Pakistan (Anderson and Lydon, 1990) were formed on the flanks of the rift system that separated Madagascar from the Indian craton. Lead-zinc-barium mineralization near the top of a thick deltaic sequence that comprises the Jurassic San Cayetano Formation in Cuba, has been interpreted to be of seafloor origin (Zhidkov and Jalturin, 1976; Simon et al., 1990; Valdes-Nodarse et al. 1993).

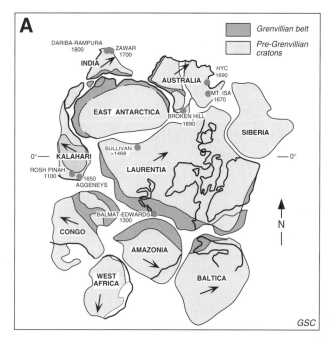

Figure 6.1-8. **A)** Geographical distribution of Proterozoic Sedex deposits shown on the continental reconstruction for the Upper Proterozoic by Hoffman (1991). Numbers refer to the most reliable radiometric age of the host rocks and bold lines indicate known outlines of modern day margins of continents. Arrows indicate present day north. **B)** Geographic distribution of Paleozoic Sedex deposits shown on the continental reconstruction for the Upper Devonian by Scotese (1984).

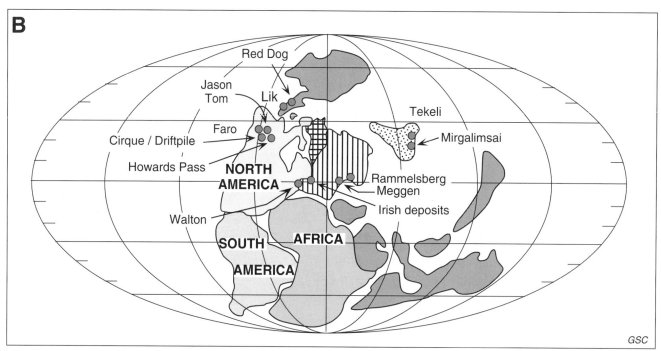

It is debatable whether there are any modern analogues of Sedex deposits. The metalliferous sediments of the Red Sea brine pools (Degens and Ross, 1969), which are often cited as the analogue for Sedex deposit genesis in connection with the brine pool model, have a Zn:Pb ratio of about 50:1 and Zn:Cu ratio of 4:1 (Bignell et al., 1976), which is not typical for Sedex deposits. The sulphide deposits of Middle Valley (Goodfellow and Franklin, 1993) and the Escanaba Trough (Zierenberg et al., 1993) are composed dominantly of pyrrhotite, and have elevated copper concentrations. The Bent Hill sulphide deposit at Middle Valley (Juan de Fuca Ridge), which is the only seafloor sulphide deposit investigated by deep drilling, has the morphology of a discordant pipe (Davis et al., 1992b), which is in contrast to typically stratiform Sedex deposits. In the geological record these modern sulphide deposits of sediment-covered oceanic ridges would be preserved in, or just above, ophiolite sequences. The brines of the Salton Sea geothermal system, which occur within deltaic sediments that infill an intracontinental rhombochasm, have both the chemical composition and geological setting typical of most Sedex deposits. If situated in a submarine, rather than a subaerial environment, brines of this hydrothermal plume would have the potential to form Sedex-type deposits.

DISTRIBUTION IN CANADA

The distribution of the major Canadian Sedex deposits is shown in Figure 6.1-1. The Sullivan deposit hosted by turbidites in the bottom part of the Middle Proterozoic Purcell Supergroup has been the major producer. Cambro-Silurian shales and carbonaceous cherts of the Selwyn Basin and the overlying Devono-Mississippian Earn Group in Yukon Territory and northern British Columbia, contain the great bulk of known Canadian Sedex resources. Only the Faro and Vangorda deposits of the Anvil district have been mined. The Howards Pass deposits (XY, Anniv, OP) are one of the world's greatest concentrations of sphalerite and galena. The Windsor Basin in Nova Scotia has similar age and geological attributes to the Irish metallogenetic province, of which it may be considered to be a pre-Atlantic opening extension. The major Zn-Pb deposit in this area, Gays River, is classified as a MVT deposit. The Walton deposit, which has Irish-type Sedex characteristics, was mined only for barite. The Grenville Supergroup contains many minor sphalerite-rich deposits in carbonates and shales that accumulated on the flanks of a rift, and is very similar in geological attributes to the Middle Cambrian Sedex metallogenetic province of the Iberian Peninsula. Intense deformation of the major deposit at Balmat-Edwards, New York, precludes an unequivocal classification as a Sedex deposit, (for example it may be a salt-dome related sulphide deposit).

Terranes considered favourable for the occurrence of Sedex deposits, but in which no examples of Sedex deposits are known, are also shown on Figure 6.1-1. The Richardson Trough has similar geological characteristics to the Selwyn Basin. The Hazen Basin, of similar age to the Selwyn Basin, is considered favourable because its extension in northern Greenland contains Sedex-type mineralization. Lower Proterozoic sediments of the Labrador Trough have all the positive regional and local indicators for the occurrence of Sedex deposits (H.S. Swinden and F. Santaguida, Geological

Survey of Canada, Minerals Colloquium, January, 1994, Ottawa, Ontario), including initiation of sedimentation as redbed intracontinental rift-fill, with subsequent marine sedimentation in a volcanically active spreading environment (see "Exploration guidelines"). The Proterozoic Borden Basin and Foxe Fold Belt are considered as having Sedex potential because of their thick shale accumulations, abundant evidence for synsedimentary faulting, and scattered centres of contemporaneous magmatic activity (Sangster, 1981).

HOST ROCKS AND RELATED LITHOLOGIES

Sedex deposits occur in sedimentary rock types considered typical of most epicontinental marine environments or playa lakes of inland drainage systems. Their depositional settings range from deep water distal turbidite fans (e.g. Sullivan) through relatively deep water, off-shelf to slope shales (e.g. Rammelsberg), and shelf carbonates (e.g. Irish and Pakistani deposits) to back-reef or playa lake calcareous siltstones and shales (e.g. H.Y.C. and Mt. Isa, Australia). The siliceous carbonaceous shales of the Selwyn basin deposits reflect a euxinic, starved, semirestricted marine basin, and their depositional setting may have been comparable to the Black Sea.

Whatever the dominant lithology of the host rocks, locally derived fragmental sedimentary rocks are a typical feature of most Sedex deposits. These fragmental rocks comprise slide sheets, debris flows, stratiform and discordant breccias, conglomerates, and mud flows, and are usually composed of rock types which occur at the same or deeper stratigraphic levels as the Sedex deposit itself. The fragmental rocks may occur in the near footwall, within, or in the near hanging wall to the deposit, or as local lithological facies some distance from the locus of mineralization. The fragmental rocks may be divided into two categories:

1. Those associated with tectonic activity, and include fault scarp talus, slide sheets, slumps, and debris flows; and

2. Those associated with fluid upflow and include hydrothermal vent and eruption breccias; discordant zones of hydraulically fractured or turbated sediment; and extrusion breccias and mudflows resulting from mud volcano activity.

These sedimentary fragmental rocks are considered to be most positive indicators for favourable environmental conditions (i.e., tectonic activity and fluid upflow) for the formation of Sedex deposits.

HOST ROCK ALTERATION

Hydrothermal alteration of the host rocks to Sedex deposits has not been documented at the same level of detail as it has for VMS deposits for two main reasons:

1. Feeder zone mineralization associated with Sedex deposits usually is not, in contrast to VMS deposits, of economic grade. Hence the hydrothermal upflow zone where the most intense hydrothermal alteration takes place is not made accessible for study via underground workings or systematic drill cores.

2. The most common alteration minerals of fossilized seafloor hydrothermal systems (chlorite, muscovite, quartz, and accessory sulphides, carbonates, and oxides) is similar to the rock-forming mineralogy of terrigenous sedimentary rocks that have been metamorphosed to greenschist facies (essentially quartz, muscovite, and chlorite, and accessory carbonates, and iron sulphides). A hydrothermal alteration pipe in metamorphosed terrigenous sediments may therefore be difficult to recognize on mineralogical grounds alone, and requires significant metasomatic or textural signatures for it to be readily distinguished.

The lack of a sufficient number of case examples of host rock alteration associated with Sedex deposits precludes generalizations to the same extent as is possible for VMS deposits, and so only a range of examples is presented here.

One of the most remarkable Sedex alteration pipes is that associated with the Sullivan deposit, which is underlain by a tourmalinite pipe about 1000 m in diameter (Hamilton et al., 1982). Whether this tourmalinite is directly related to hydrothermal upflow that formed the sulphide deposit is debatable, because tourmalinite pipes without any base metal sulphide associations occur elsewhere within the Belt-Purcell Supergroup (e.g. Beaty et al., 1988). At Sullivan, the tourmalinite is cut by veins and networks of sulphides that appear to be feeder mineralization to the overlying massive ore, which suggests that the bulk of the tourmalinite predates the sulphide mineralization. Significant silicification, either by replacement or as bedded chert, has not been recognized at the Sullivan deposit. Alteration patterns at Sullivan are complicated by a postore overprinting by an albite-chlorite-pyrite alteration that is most extensively developed in the hanging wall of the deposit (Hamilton et al., 1982) and may be related to the emplacement of a thick tholeiitic sill and its offshoots (Jardine, 1966; Ethier et al., 1976) about 200 m below the ore horizon.

Silicification may have been an integral part of hydrothermal alteration at both the Tom and nearby Jason deposits, but the host siliceous carbonaceous shales make it difficult to distinguish hydrothermal silicification from normal diagenetic effects. The most obvious and extensive footwall alteration at the Tom deposit is ankerite/pyrite alteration (Goodfellow and Rhodes, 1990). Carbonatization, notably by ferroan ankerite, is strongest within the feeder zone and vent complex, where there appears to have been a concomitant removal of silica. At the Jason deposit, footwall alteration consists of variably silicified and carbonatized rocks (Bailes et al., 1986). The feeder zone and vent complex are characterized by a lower siderite zone and upper ferroan dolomite zone (Turner, 1990).

The Rammelsberg deposit is stratigraphically underlain by a zone of silica-rich lenses termed "Kniest" (Hannak, 1981). Large (1983) stated that the Kniest has a pipe-like geometry that distinctly crosscuts the bedding and contains veins of chalcopyrite. At Mount Isa, an extensive zone of quartz and dolomite with associated chalcopyrite ores occurs down dip from the Sedex-type Zn-Pb ores, but there is debate whether it is contemporaneous with the formation of the Zn-Pb ores (e.g., Stanton, 1963; Mathias and Clark, 1975) or whether it is a much later superimposed event (e.g., Perkins, 1984; Bell et al., 1988).

Dolomitization, with or without silicification, is associated with several Sedex deposits whose host rocks have a calcareous component. Generally throughout Ireland, a pervasive dolomitization is interpreted to have preceded the main sulphide-depositing event (Hitzman and Large, 1986; Hitzman, 1986). At the Silvermines deposit, pervasive iron-free dolomitization of the footwall rocks preceded seafloor sulphide and siderite mineralization, whereas iron-bearing dolomite, as fracture fillings and secondary cements, was precipitated both during and after accumulation of the stratiform ores (Andrew, 1986). Similarly at Tynagh, replacement and cementation by dolomite and ferroan calcite preceded sulphide mineralization whereas precipitation of iron-bearing dolomite took place both during and after the main sulphide event (Boast et al., 1981; Clifford et al., 1986). At the Navan deposit, replacement of micritic host rock by low-iron dolomite and minor chalcedony preceded and accompanied sulphide mineralization (Andrew and Ashton, 1985). In the H.Y.C. and Ridge II deposits of the McArthur River area, Australia, multiple generations of dolomite and the development of nodular dolomite distinguish mineralized calcareous siltstones from nonmineralized lithological equivalents (Williams, 1978a).

If the above examples are representative of the deposit type as a whole, then iron- and magnesium-bearing carbonatization is probably the most common easily recognizable hydrothermal alteration type associated with Sedex deposits. Silicification may also be significant, though more difficult to recognize. Argillic alteration, including the prominent chloritization and sericitization typical of VMS deposits, may be too cryptic to recognize, unless the host rock originally contained significant concentrations of feldspar.

ASSOCIATED STRUCTURES

As discussed above, Sedex deposits occur in structurally controlled sedimentary basins, and most of these basins are related to major rift zones. Large (1983) classified the sedimentary basins that host the deposits according to their lateral dimensions. First-order basins have lateral dimensions of the order of 100 km and correspond to the major rift structure. Second-order basins generally have lateral dimensions in excess of 10 km and third-order basins range from hundreds of metres to ten kilometres, both commonly having the form of a half-graben. Sedex deposits are spatially associated with the synsedimentary faults that control these second- and third-order basins. As noted above, Sedex deposits, and the second- and third-order basins with which they are associated, occur in the rift-cover sequence and hence postdate the main phase of rifting. It is not clear whether the localized extensional faults, which control the second- and third-order basins, form an integral part of the tectonic evolution of a sediment-covered rift, and hence are predictable, or whether they represent a separate tectonic event, and hence are unpredictable features of sediment-covered rifts. Most authors agree that a rift-cover sequence represents basinal development due to thermal relaxation after extension or the rift sag stage. For example, the Mount Isa Group of Australia is considered to represent a rift sag sequence following renewed extension of the Leichardt River Fault Trough at

1678 Ma. Large (1983) considered Sedex deposits to be associated with a tectonic extensional pulse superimposed upon postrift subsidence.

Not all Sedex deposits are directly related to a tectonic rifting process per se. In some cases at least, the second- and third-order basins are interpreted as rhombochasms or pull-apart basins associated with transcurrent faulting in a sedimentary basin. For example, the Macmillan Pass Central Block, hosting the Tom and Jason Sedex deposits, is thought to represent a Devonian pull-apart basin (Abbott et al., 1986). The clastic-filled Devonian extensional basins that control the siting of the Sedex deposits in the Mississippian carbonate rift-cover sequence in Ireland are good candidates for rhombochasms developed along transcurrent faults that are controlled by the Caledonian structural grain of the basement. It is significant to note that the Salton Sea geothermal system, the best modern analogue for a hydrothermal system capable of forming Sedex-type deposits, is located in a rhombochasm along the San Andreas transcurrent fault system (McKibben et al., 1988; McKibben and Eldridge, 1989).

DEFINITIVE CHARACTERISTICS

Sedex deposits were loosely defined in the "Introduction" as *a sulphide deposit formed in a sedimentary basin by the submarine venting of hydrothermal fluids and whose principal ore minerals are sphalerite and galena.* Irrespective of the definition used, examples of mineral deposits can invariably be cited that are intermediate in characteristics between the Sedex type on the one hand and the VMS, MVT, bedded barite, or iron-formation types on the other. The key concepts in the definition used here that distinguish Sedex deposits from these other deposit types are:

1. *Sedex deposits occur in a sedimentary basin.* The scale of sedimentary basin referred to is in the order of 10^2 to 10^3 km^2. In other words Sedex deposits occur in a geological terrane that is defined by the presence of a specified sequence of sedimentary rocks (i.e., a sedimentary basin). This criterion helps to distinguish Sedex deposits from VMS deposits, which typically occur in a geological terrane distinguished by the presence of volcanic rocks (i.e., a volcanic belt). The mere presence of sedimentary rocks in the immediate footwall of a conformable sulphide deposit is in itself not diagnostic of a Sedex deposit – the immediate host rocks of many VMS deposits are sedimentary rocks. (see subtype 6.3 "Volcanic-associated massive sulphide base metals").

2. *The principal economic minerals are sphalerite and galena.* The importance of this concept is that it not only distinguishes Sedex deposits from other nonsulphide seafloor metalliferous sediments such as baritite deposits, iron-formations, and manganese formations, but also is another criterion to distinguish them from VMS deposits, in which chalcopyrite, is almost invariably a principal ore mineral. The presence or absence of chalcopyrite as a primary hydrothermal product in a seafloor sulphide deposit directly reflects the temperature of the hydrothermal fluids. The solubility of chalcopyrite in a reduced (i.e. $H_2S \gg SO_4^{2-}$) aqueous fluid in which metals are present mainly as chloride complexes is insignificant (<1 ppm) below about 300°C. Fluid inclusions and other geothermometers indicate that the temperature of the hydrothermal fluids

responsible for Sedex deposits are in the range 150°C-300°C, whereas for VMS deposits the temperatures of the ore fluids are generally >300°C.

3. *Sedex deposits were formed during the submarine venting of hydrothermal fluids.* This concept emphasizes that Sedex deposits formed at or just below the seafloor and are essentially a quasi-synsedimentary phenomenon, although there is debate whether the deposits are dominantly seafloor metalliferous sediments around hydrothermal vent fields or are dominantly the products of subsurface replacement of sediments around the hydrothermal upflow conduits. This depositional environment distinguishes them from MVT deposits, which are essentially epigenetic deposits formed in lithified carbonate rocks, and for which there is a significant difference between the age of the mineralization and the age of the host rocks.

GENETIC MODELS

There are two distinct sets of problems concerning the genesis of Sedex deposits. One is with regard to the origins of the ore fluids and the reasons for their upflow to the seafloor. The other pertains to the processes of sulphide precipitation and accumulation to form the ore deposit.

Generation and upflow of hydrothermal fluids

The majority of opinion is that the ore fluids for Sedex deposits are formational waters of the sedimentary basin that for whatever reason have become unusually hot, saline, and metalliferous (e.g., Walker et al., 1981; Badham, 1981; Carne and Cathro, 1982; Lydon, 1983; Sawkins, 1984). The salient points of this model are illustrated in Figure 6.1-6. The reason the formational waters become unusually hot is thought to be unusually high geothermal gradient of an area undergoing extensional tectonism. Although this high heat flow can be ultimately linked to contemporaneous, usually deep, magmatic activity of a spreading geotectonic environment, there is little or no evidence to support direct heating by a magmatic body of the hydrothermal systems responsible for Sedex deposits. Heating by an elevated geothermal gradient, even as highly anomalous as 70°C/km, requires that the hydrothermal fluids originate from a depth of several kilometres. Lydon (1983) suggested that the most plausible geological environment which would allow the heating of large volumes of formational water to greater than 200°C was one in which a sequence of porous rocks, forming the aquifer or hydrothermal reservoir, was overlain by a sequence of argillaceous rocks. The argillaceous sequence performs the double function of an impermeable caprock, which prevents the dissipation of heat by the upward convective mass flow of heated fluids, and also a thermal insulator, which prevents the dissipation of heat by conduction, and thus allows the buildup of high temperatures in the reservoir. This is the configuration of most sediment-covered rift systems that host Sedex deposits, in which coarse clastic and volcanic rocks of the rift-fill sequence are overlain by argillaceous rocks of the rift-cover sequence, deposited during the marine transgression associated with the thermal subsidence stage of the rifting cycle.

Although the sparse fluid inclusion data (see above) suggest that the ultra high salinities of the Salton Sea, for example, are not necessary to form an ore fluid, they do suggest that the ore fluids must have salinities at least two to three times that of seawater. The origin of these high salinities in the ore fluids, like their modern analogues, is most likely to be a result of the dissolution of evaporites in the sedimentary sequence stratigraphically below the deposit. In terms of Lydon's (1983) model outlined above, these evaporites could be within or at the base of the rift-fill sequence. Candidates for such a role in the formation of Sedex deposits in the Selwyn Basin would be the evaporites of the Little Dal Group and the Coates Lake Group. These may extend completely beneath Selwyn Basin at the base of the Windermere Supergroup, which is the Upper Proterozoic rift-fill sequence to the overlying Paleozoic rift-cover sequence of the Selwyn Basin.

Processes of sulphide precipitation and accumulation

A wide spectrum of hypotheses have been suggested in attempts to explain the precipitation, accumulation, and preservation of sulphides and other hydrothermal products in Sedex deposits. The common theme in most cases is an explanation of the bedded or laminated textures typical of the bedded ores and distal hydrothermal products facies of the idealized Sedex deposit (Fig. 6.1-4). The ideas expressed can be conveniently grouped into four models.

Brine pool / bottom-hugging brine model

The brine pool model advocates that the bedded ores and distal hydrothermal products facies represent sedimentation from a stagnant brine pool, formed by the collection of hydrothermal effluent in a topographic depression adjacent to the vent area or from a migrating bottom-hugging brine. Advocates of a dense brine model for Sedex deposits include Solomon and Walshe (1979), Finlow-Bates, (1980), Russell et al. (1981), Carne and Cathro (1982), Lydon (1983), and Samson and Russell (1987). The brine pool variant of the model has a modern analogy in the brine pools and metalliferous sediments of the Red Sea (e.g., Degens and Ross, 1969; Bäcker, 1975). The merits of this model are that it explains:

i) The finely laminated nature of the bedded ores, which is consistent with the very low energy depositional environment provided by a stagnant brine pool.

ii) The sheet-like morphology of the bedded part of a Sedex deposit, suggesting uniform depositional rates of compositionally unique sediments of hydrothermal origin within the confines of a limited area.

iii) The lateral continuity of individual lamina of hydrothermal products, many of which are monomineralic, suggests synchronous precipitation and settling from an overlying water column that is distinctly different from normal seawater and whose chemical characteristics undergo repetitive cyclic changes.

iv) The lack of bioturbation of the laminated hydrothermal products, indicating a bottom environment hostile to burrowing organisms, even though, in the Phanerozoic at least, such organisms are common.

v) The preferential siting of Sedex deposits in third- or second-order basins, indicating that a topographic depression on the seafloor is required for their formation. Although a topographic depression is a requisite for the brine pool model, it is not an integral part of alternative models.

vi) The large average tonnage of metals contained in a Sedex deposit. A brine pool confines the upward dispersal of hydrothermal products to below the pycnocline which defines its upper surface, and limits lateral dispersal to the brine pool margins. As a consequence, a high proportion of the metals that are carried upward by the hydrothermal fluids are precipitated within the limits of the brine pool and not dispersed in the open ocean.

The bottom-hugging brine variant of the model was inspired by the experiments of Turner and Gustafson (1978) and Solomon and Walshe (1979). A difference between the bottom-hugging brine model and the brine pool model is that there is only a single pass of each pulse of hydrothermal fluid over the proximal site of hydrothermal precipitation (which is used to explain mineralogical zonation). Another difference is that the migrating brines may collect in a basin distal from the upflow zone and result in the hydrothermal sediments being disconnected from a hydrothermal vent zone.

Buoyant plume model

Calculations by Sato (1972) have shown that in order for hydrothermal effluent cooled to below 100°C by mixing with seawater to be denser than seawater and hence to form a bottom-hugging brine or brine pool, the vented hydrothermal solutions must have a salinity greater than four times seawater if their vent temperature is above 250°C. The basis of this model is that the only fluid inclusion data available (see above) suggests that ore fluids for Sedex deposits have salinites in the range of only two to three times seawater. Therefore, the plume formed by mixing seawater with hydrothermal fluids of these salinities and temperatures around 250°C, would remain buoyant for most of its cooling and precipitational history.

The model has only been explicitly applied to Sedex deposits (Goodfellow and Jonasson, 1986; Goodfellow and Rhodes, 1990) in conjunction with a postulated stratified oceanic water column during times of oceanic anoxia (Goodfellow, 1987). The stratified water column is required to provide sulphide for ore deposition over a long time period and to prevent the rapid oxidation of sulphides nucleated in the water column or accumulated on the normally oxic ocean floor.

Clastic apron model

Clastic sedimentary textures in Sedex deposits, such as debris flows of fragmental sulphides (e.g. Sullivan; Hamilton et al., 1983) and graded bedding (e.g. Rammelsberg; Large, 1983) have been cited as evidence for their synsedimentary seafloor origin. The origin of the clastic sulphides is by contemporaneous erosion or collapse of a topographically elevated sulphide mound formed above the hydrothermal upflow zone, with the resulting talus breccia, debris flows, and turbidites collecting as a clastic apron around the base of the mound. Modern analogies for this process have been described for Middle Valley (Goodfellow and Franklin, 1993). Although no one has advocated that the bedded ores of Sedex deposits have formed entirely by a process of clastic reworking of vent complex, the process is evidently of some significance in certain deposits.

Subsurface replacement model

This model is gaining increasing advocacy, especially by those working on the Australian Proterozoic deposits of the McArthur Basin and Mount Isa Inlier. The essence of the model is that the bedded ores of Sedex deposits represent the very early diagenetic, precompaction replacement of fine grained bedded sediments in the subsurface around seafloor hydrothermal vents. The sedimentary textures of Sedex deposits are therefore not primary but result from pseudomorphism of those of the host sediment. Prominent advocates of the model, based on studies of the McArthur River deposits (H.Y.C., Ridge II), have been Williams and Rye (1974), Williams (1978a, b), and Eldridge et al. (1993), who showed that texturally the ore sulphides postdate early diagenetic pyrite and that sulphur isotope data require interpretations involving different sulphur sources for the diagenetic and ore sulphides. It has been argued that the geochemistry of the Mount Isa ores indicates that the Cu-silica dolomite and Pb-Zn ores are cogenetic during diagenesis, and that the finely layered Pb-Zn ores inherited their textures from the original clastic sediments. Andrew and Ashton (1985) have shown that the bulk of the Zn-Pb ore of the Navan deposit postdates the earliest carbonate cements.

The merits of the subsurface replacement model are that:

i) It gives the most satisfactory explanation for the characteristic mineralogical and chemical zonation of the ores around the upflow zone. Zone boundaries form concentric shells, both in vertical and lateral sections, around the upflow zone. This suggests a zone refinement process that simultaneously affected the entire stratigraphic sequence of the ore body. Zonation patterns of the vent complex ores, which are generally agreed to be the products of subsurface replacement and infilling, continue with the same polarity and gradients into the bedded ores (e.g., Sullivan, Tom, and Jason deposits). This suggests that mineralogical and chemical distribution patterns for the bedded ores are also a product of subsurface dispersal or zone refinement.

ii) Subsurface replacement is consistent with the observations at the two modern analogues of metalliferous hydrothermal systems in unconsolidated sediments (Salton Sea and Middle Valley), that the bulk of subsurface hydrothermal fluid is contained in a plume that pervasively occupies the porosity of the host sediments and is not confined to channels.

Source of sulphide sulphur

Sedex deposits not only contain on average large quantities of metal, but also large quantities of sulphur. Experimental investigations have shown that aqueous chloride solutions with five times the salinity of seawater can carry hundreds of parts per million of zinc, lead, and hydrogen sulphide in stoichiometric proportions to form sphalerite and galena at temperatures above 200°C and pH<4 (Barrett and Anderson, 1982). However, these concentrations are reduced by orders of magnitude at the pH of hydrothermal fluids emanating from sedimentary rocks which are in the range 5.5 to 7.0 (e.g., Salton Sea pH=5.5 (McKibben et al., 1988) and Guaymas Basin pH=5.9 (Von Damm et al., 1985). For the ore fluids to carry sufficient concentrations of metal (tens or hundreds parts per million) to form a large Sedex deposit, they must be sulphur deficient (i.e., the metal:sulphide ratio in the fluid greater than stoichiometric proportions for the metal sulphide). In other words, at least some of the sulphide sulphur must be supplied at the depositional site of the Sedex deposit. The only viable source of local seafloor sulphur in the sulphide form is that produced by the bacteriogenic reduction of seawater sulphate, and stored at the depositional site either in the subsurface as hydrogen sulphide dissolved in porewater and/or as diagenetic iron sulphides (e.g., Williams, 1978a; Samson and Russell, 1987), or in the lower part of a stratified anoxic water column (e.g., Goodfellow, 1987; Turner, 1992).

In Sedex deposits that contain barite, the sulphur isotope ratios of the barite are usually close to those of coeval seawater sulphate, indicating that much of the barite was precipitated by hydrothermal barium fixing ambient marine sulphate. The sulphur isotope ratios of sulphides in the same deposit usually range from the barite values to values with an increasing proportion of ^{32}S, consistent with the explanation that most of the sulphide in Sedex deposits is derived by the bacterial reduction of coeval marine sulphate. Goodfellow and Jonasson (1986) have shown that the average sulphur isotope ratios of Sedex deposits in the Selwyn Basin closely track the stratigraphic variation in sulphur isotope ratios of diagenetic pyrite and thus supports this conclusion. Studies by Shanks et al. (1987) of the Anvil Range deposits suggested a mixed source involving reduced seawater sulphate and hydrothermal hydrogen sulphide. Interpretations of sulphur isotope analysis of bulk mineral separates from the H.Y.C. deposit led Smith and Croxford (1973) to interpret a hydrothermal source of sulphide for sphalerite and galena but a biogenic reduction of ambient sulphate for pyrite. Williams and Rye (1974) and Williams (1978a, b) interpreted the same data to indicate that all sulphur was supplied as sulphate in the ore fluid. The reduction of the sulphate supplied sulphide to form early diagenetic pyrite, which in turn was partially redissolved to form a hybrid sulphide supply, from which the texturally later ore metal sulphides were formed. Interpretation of more recent SHRIMP analyses concluded that both diagenetically early and diagenetically late pyrite were formed from the same bacterially reduced sulphate batch but that the ore sulphides were formed from an unknown independent supply of sulphide (Eldridge et al., 1993). A similar two-stage model has been advocated for the Rammelsberg deposit (Eldridge et al., 1988), although the interpretation has been disputed and a single marine sulphate source suggested (Goodfellow and Turner, 1989).

RELATED DEPOSIT TYPES

Although all seafloor metalliferous deposits that formed directly or indirectly by the submarine venting of hydrothermal fluids, such as VMS deposits, Besshi-type deposits, iron-formations and manganese formations, can be considered to be genetically related to Sedex deposits, only stratiform barite deposits appear to be consistently related to the same geological environments in which Sedex deposits occur. Barite is an important accessory mineral in the bedded ores facies and the most important constituent of the distal hydrothermal products facies of many Phanerozoic Sedex deposits. In some cases, as at Silvermines, Ireland and Walton, Nova Scotia, baritite has been mined as an ore in its own right. The Selwyn Basin is the prime example where barite deposits are directly related to Sedex deposits. Here, especially in Middle Devonian to Lower Mississippian strata, there are numerous "barren barite" deposits. In the Macmillan Pass area alone, site of the Tom and Jason Sedex deposits, there are thirteen known baritite occurrences at approximately the same stratigraphic level. The barren baritites consist of various proportions of bedded barite, limestone, and chert, and form lens-shaped bodies that usually contain less than a million tonnes of hydrothermal products.

It has been proposed by Lydon et al. (1979, 1985) that the barren baritite deposits of the Selwyn Basin were formed from hydrothermal systems that were contemporaneous with, but derived from different reservoirs than those that formed the Sedex deposits. The hydrothermal fluids that formed the baritite deposits were cooler, less saline, and derived from shallower reservoirs within the carbonaceous sediments of the Paleozoic rift-cover sequence, whereas the hydrothermal fluids for the Sedex deposits were derived from deeper reservoirs within the feldspar-rich clastic rocks of the Upper Proterozoic rift-fill sequence.

EXPLORATION GUIDELINES

Geological attributes and interpretations listed below that serve as guides for exploration for Sedex deposits are grouped according to scale, in the sequence: regional, local, and deposit scale.

Regional scale

1. A sedimentary basin that accumulated in a tectonically active environment. Extensional tectonic regimes, in which magmatic activity is contemporaneous with sedimentation, are the most favourable.

2. The deeper parts of the sedimentary basin, or precursor basin, contain, or did contain, evaporites. Low latitude intracontinental rift-fill sequences are the most favourable.

3. The most productive stratigraphic intervals for Sedex deposits are those in the rift-cover sequence, that accumulated during the thermal subsidence stage of the rifting cycle.

Local scale

1. Evidence of synsedimentary faulting. The presence of synsedimentary fragmental rocks representing fault scarp talus and debris flows are the most easily recognized criteria during reconnaissance mapping.

2. Evidence for synsedimentary hydrothermal sulphide mineralization. The presence of sulphides or barite, either as epigenetic veins along the synsedimentary

fault or as clasts in synsedimentary fragmental rocks, are strong indicators that the synsedimentary fault was a conduit for mineralizing fluids.

3. Evidence of hydrothermal upflow. Discordant zones of disrupted sediments, especially those indicating hydrothermal metasomatism, and concordant mud flows and debris flows, indicating mud volcano activity, are among the most easily recognized criteria.

4. Evidence of hydrothermal sediments. Local lenses of chert, baritite, carbonate, and magnetite/hematite iron-formation are the most useful nonsulphide indicators.

Deposit scale

Increasing Pb:Zn ratios and Ag content are the best indicators for increasing proximity to the vent complex, which contains the highest grade ores.

ACKNOWLEDGMENTS

Ian Jonasson is especially thanked for freely sharing of his data and for his outstanding editorial job in reducing the weight of the original manuscript. Jari Paakki tracked down references and helped with compilations. Janice Nunney retrieved references and constructed CAD reproductions of diagrams. Ralph Thorpe, Ian Jonasson, and Charlie Jefferson made many suggestions that improved the quality and content of the manuscript.

SELECTED BIBLIOGRAPHY

References marked with asterisks (*) are considered to be the best sources of general information on this deposit subtype.

*Abbott, J.G., Gordey, S.P., and Tempelman-Kluit, D.J.
1986: Setting of sediment-hosted stratiform lead-zinc deposits in Yukon and northeastern British Columbia; in Mineral Deposits of Northern Cordillera (ed.) J.A. Morin; The Canadian Institute of Mining and Metallurgy, Special Volume 37, p. 1-18.

Anderson, W.L. and Lydon, J.W.
1990: SEDEX and MVT deposits in Jurassic carbonates of Pakistan; in Program with Abstracts, International Association on the Genesis of Ore Deposits, Eighth Symposium, Ottawa, p. A257.

*Andrew, C.J.
1986: The tectono-stratigraphic controls to mineralization in the Silvermines area, County Tipperary, Ireland; in The Geology and Genesis of Mineral Deposits in Ireland (ed.) C.J. Andrew, R.W.A. Crowe, S. Finlay, W.M. Pennell, and J.F. Pyne; Irish Association of Economic Geology and Geological Survey of Ireland, p. 377-418.

Andrew, C.J. and Ashton, J.H.
1985: Regional setting, geology and metal distribution patterns of Navan orebody, Ireland; Institution of Mining and Metallurgy Transactions, v. 94, p. B66-B93.

Ansdell, K.M., Nesbitt, B.E., and Longstaffe, F.J.
1989: A fluid inclusion and stable isotope study of the Tom Ba-Pb-Zn deposit, Yukon Territory, Canada; Economic Geology, v. 84, p. 841-856.

Bäcker, H.,
1975: Exploration of the Red Sea and Gulf of Aden during the M.S. VALDIVIA cruises "Erzschlämme A" and "Erzschlämme B"; Geologisches Jarbuch, D.17, p. 3-78.

*Badham, J.P.N.
1981: Shale-hosted Pb-Zn deposits: products of exhalation of formation waters?; Institution of Mining and Metallurgy Transactions, Section B, v. 90, p. B70-B76.

*Bailes, R.J., Smee, B.W., Blackadar, D.W., and Gardner, H.D.
1986: Geology of the Jason lead-zinc-silver deposits, Macmillan Pass, eastern Yukon; in Mineral Deposits of Northern Cordillera (ed.) J.A. Morin; The Canadian Institute of Mining and Metallurgy, Special Volume 37, p. 87-99.

Barrett, T.J. and Anderson, G.M.
1982: The solubility of sphalerite and galena in NaCl brines; Economic Geology, v. 77, p. 1923-1933.

Beaty, D.W., Hahn, G.A., and Threlkeld, W.E.
1988: Field, isotopic and chemical studies of tourmaline-bearing rocks in the Belt-Purcell Supergroup: genetic constraints and exploration significance for Sullivan type ore deposits; Canadian Journal of Earth Sciences, v. 25, p. 392-402.

Bell, T.H., Perkins, W.G., and Swager, C.P.
1988: Structural controls on the development and localization of syntectonic copper mineralization at Mount Isa, Queensland; Economic Geology, v. 83, p. 69-85.

Bignell, R.D., Cronan, D.S., and Tooms, J.S.
1976: Red Sea metalliferous brine precipitates; in Metallogeny and Plate Tectonics (ed.) D.F. Strong; Geological Association of Canada, Special Paper 14, p. 147-179.

Boast, A.M., Coleman, M.L., and Halls, C.
1981: Textural and stable isotopic evidence for the genesis of the Tynagh base metal deposit, Ireland; Economic Geology, v. 76, p. 27-55.

Both, R.A. and Rutland, R.W.R.
1976: The problem of identifying and interpreting stratiform orebodies in highly metamorphosed terrains: the Broken Hill example; in Handbook of Stratabound and Stratiform Ore Deposits (ed.) K.H. Wolfe; Elsevier, Amsterdam, v. 4, p. 261-325.

Carmichael, A.J.
1988: The tectonics and mineralization of the Black Angel Pb-Zn deposits, central West Greenland; PhD. thesis, Goldsmith's College, University of London, 371 p.

Carne, R.C. and Cathro, R.J.
1982: Sedimentary exhalative (SEDEX) zinc-lead deposits, northern Canadian Cordillera; The Canadian Mining and Metallurgy Bulletin, v. 75, p. 66-78.

Carvalho, I.G., Zantop, H., and Torquato, J.R.F.
1982: Geologic setting and genetic interpretation of the Boquira Pb-Zn deposits, Bahia State, Brazil; Revista Brasiliera de Geociencias, v. 12, p. 414-425.

Clayton, R.H. and Thorpe, L.
1982: Geology of the Nanisivik zinc-lead deposit; in Precambrian Sulphide Deposits (ed.) R.W. Hutchinson, C.D. Spence, and J.M. Franklin; Geological Association of Canada, Special Paper 25, p. 739-760.

Clifford, J.A., Ryan, P., and Kucha, H.
1986: A review of the geological setting of the Tynagh orebody, Co. Galway; in The Geology and Genesis of Mineral Deposits in Ireland (ed.) C.J. Andrew, R.W.A. Crowe, S. Finlay, W.M. Pennell, and J.F. Pyne; Irish Association of Economic Geology and Geological Survey of Ireland, p. 419-440.

Coates, J.S., Smith, C.G., Fortney, N.J., Gallager, M.J., May, F., and McCourt, W.J.
1980: Stratabound barium-zinc mineralization in Dalradian schist near Aberfeldy, Scotland; Institution of Mining and Metallurgy, Transactions, v. B.89, p. 110-122.

Deb, M.
1990: Regional metamorphism of sediment-hosted, conformable base-metal sulfide deposits in the Aravalli-Delhi Orogenic Belt, NW India; in Regional Metamorphism of Ore Deposits and Genetic Implications, (ed.) P.G. Spry and L.T. Bryndzia; (Proceedings of the Twenty-eighth International Geology Congress), VSP Utrecht, The Netherlands, p. 117-140.
1982: Crustal evolution and Precambrian metallogenesis in western India; Revista Brasiliera de Geociencias, v. 12, p. 94-104.

Deb, M. and Bhattacharya, A.K.
1980: Geological setting and conditions of metamorphism of Rajpura-Dariba polymetallic ore deposit, Rajasthan, India; in Proceedings of the Fifth International Association on the Genesis of Ore Deposits Symposium, (ed.) J.D. Ridge; E. Schweizerbart'sche Verlagsbuchhandlung, Stuttgart, v. 1, p. 679-697.

Deb, M., Thorpe, R.I., Cumming, G.L., and Wagner, P.A.
1989: Age, source and stratigraphic implications of Pb isotope data for conformable, sediment-hosted, base metal deposits in the Proterozoic Aravalli-Delhi orogeneic belt, northwestern India; Precambrian Research, v. 43, p. 1-22.

Degens, E.T. and Ross, D.A. (ed.)
1969: Hot Brines and Recent Heavy Metal Deposits in the Red Sea; Springer-Verlag, New York, 600 p.

Eisbacher, G.H.
1981: Sedimentary tectonics and glacial record in the Windermere Supergroup, Mackenzie Mountains, northwestern Canada; Geological Survey of Canada, Paper 80-27, 40 p.

Eldridge, C.S., Compston, W., Williams, I.S., Both, R.A., Walshe, J.L., and Ohmoto, H.
1993: Sulfur isotope variability in sediment-hosted massive sulfide deposits as determined using the ion microprobe SHRIMP: II. A study of the H.Y.C. deposit at McArthur River, Northern Terrotory, Australia; Economic Geology, v. 88, p. 1-26.

Eldridge, C.S., Williams, N., and Walshe, J.L.
1988: Sulfur isotope variability in sediment-hosted massive sulfide deposits as determined using the ion microprobe SHRIMP: 1. An example from the Rammelsberg orebody - a discussion; Economic Geology, v. 83, p. 443-449.

Espourteille, F. and Fleischer, R.
1988: Mina de chumbo de Boquira, Bahia; in Principais Depositos Minerais do Brasil, Volume III, (co-ordinators) C. Schobbenhaus and C.E. Silva Coelho; Ministerio das Minas e Energia, Republica Federativa do Brasil, Brasilia, chap. X, p. 91-99.

Ethier, V.G., Campbell, F.A., Both, R.A., and Krouse, H.R.
1976: Geological setting of the Sullivan Orebody and estimates of temperatures and pressures of metamorphism; Economic Geology, v. 71, p. 1570-1588.

Finlow-Bates, T.
1980: The chemical and physical controls on the genesis of submarine exhalative orebodies and their implications for formulating exploration concepts; Geologisches Jahrbuch, D. 40, p. 131-168.

Forrest, K. and Sawkins, F.J.
1984: The Lik deposit, western Brooks: Sedex mineralization along axial vent sites in a structural basin; in Abstracts with Program, Geological Society of America, Ninty-seventh Annual Meeting, Reno, Nevada, p. 511.

Forrestal, P.J.
1990: Mount Isa and Hilton silver-lead-zinc deposits; in Geology of the Mineral Deposits of Australia and Papua New Guinea (ed.) F.E. Hughes; The Australasian Institute of Mining and Metallurgy, Melbourne, Monograph 14, v. 1, p. 927-934.

Gabrielse, H. and Campbell, R.B.
1991: Upper Proterozoic assemblages, Chapter 6 in Geology of the Cordilleran Cordilleran Orogen in Canada (ed.) H. Gabrielse and C.J. Yorath; Geological Survey of Canada, Geology of Canada, no. 4, p. 125-150 (also Geological Society of America, The Geology of North America, v. G-2).

Garde, A.A.
1978: The Lower Proterozoic Marmorilik Formation, east of Marmorilik, West Greenland; Meddelelser om Grønland, v. 200, no. 3, 71 p.

Gardner, H.D. and Hutcheon, I.
1985: Geochemistry, mineralogy and geology of the Jason Pb-Zn deposits, Macmillan Pass, Yukon, Canada; Economic Geology, v. 80, p. 1257-1276.

Gemmell, J.B., Zantop, H., and Meinert, L.D.
1992: Genesis of the Aguilar zinc-lead-silver deposit, Argentina: contact metasomatic vs. sedimentary exhalative; Economic Geology, v. 87, p. 2085-2112.

Goodfellow, W.D.
1987: Anoxic stratified oceans as a source of sulphur in sediment-hosted stratiform Zn-Pb deposits (Selwyn Basin, Yukon, Canada); Chemical Geology, v. 65, p. 359-382.

Goodfellow, W.D. and Franklin, J.M.
1993: Geology, mineralogy and chemistry of sediment-hosted clastic massive sulfides in shallow cores, Middle Valley, northern Juan de Fuca Ridge; Economic Geology, v. 88, p. 2033-2064.

***Goodfellow, W.D. and Jonasson, I.R.**
1986: Environment of formation of the Howards Pass (XY) Zn-Pb deposit, Selwyn Basin, Yukon; in Mineral Deposits of Northern Cordillera (ed.) J.A. Morin; The Canadian Institute of Mining and Metallurgy, Special Volume 37, p. 19-50.

Goodfellow, W.D., and Rhodes, D.
1990: Geological setting, geochemistry and origin of the Tom stratiform Zn-Pb-Ag-barite deposits; in Mineral Deposits of the Northern Canadian Cordillera (ed.) J.G. Abbott and R.J.W. Turner; International Association on the Genesis of Ore Deposits, Eighth Symposium, Ottawa, Field Trip 14: Guidebook, Geological Survey of Canada, Open File 2169, p. 177-244.

Goodfellow, W.D. and Turner, R.J.
1989: Sulfur isotope variability in sediment-hosted massive sulfide deposits as determined using the ion microprobe SHRIMP: 1. An example from the Rammelsberg orebody - a discussion; Economic Geology, v. 84, p. 451-452.

Goodfellow, W.D., Grapes, K., Cameron, B., and Franklin, J.M.
1994: Hydrothermal alteration associated with massive sulphide deposits, Middle Valley, Northern Juan de Fuca Ridge; Canadian Mineralogist, v. 31, p. 1025-1060.

Gordey, S.P., Abbott, J.G., and Orchard, M.J.
1982: Devono-Mississippian (Earn Group) and younger strata in east-central Yukon; in Current Research, Part B; Geological Survey of Canada, Paper 82-1B, p. 93-100.

Gwosdz, W. and Krebs, W.
1977: Manganese halo surrounding Meggen ore deposit, Germany; Institution of Mining and Metallurgy Transactions, Section B, v. 86, p. B73- B77.

*Hamilton, J.M., Bishop, D.T., Morris, H.C., and Owens, O.E.
1982: Geology of the Sullivan orebody, Kimberley, B.C., Canada; in Precambrian Sulphide Deposits (ed.) R.W. Hutchinson, C.D. Spence, and J.M. Franklin; Geological Association of Canada, Special Paper 25, H.S. Robinson Memorial Volume, p. 597-665.

Hamilton, J.M., Delaney, G.D., Hauser, R.L., and Ransom, P.W.
1983: Geology of the Sullivan deposit, Kimberley, B.C., Canada; in Sediment-hosted Stratiform Lead-zinc Deposits (ed.) D.F. Sangster; Mineralogical Association of Canada, Short Course Handbook, v. 8, p. 31-83.

Hannak, W.W.
1981: Genesis of the Rammelsberg ore deposit near Goslar/Upper Hartz, Federal Republic of Germany; in Handbook of Stratabound and Stratiform Ore Deposits, (ed.) K.H. Wolfe; Elsevier, Amsterdam, v. 9, p. 551-642.

Heinrichs, T.K. and Reimer, T.O.
1977: A sedimentary barite deposit from the Archean Fig Tree Group of the Barberton Mountain Land (South Africa); Economic Geology, v. 72, p. 1426-1441.

Hitzman, M.W.
1986: Geology of the Abbeytown mine, Co. Sligo, Ireland; in The Geology and Genesis of Mineral Deposits in Ireland (ed.) C.J. Andrew, R.W.A. Crowe, S. Finlay, W.M. Pennell, and J.F. Pyne; Irish Association of Economic Geology and Geological Survey of Ireland, p. 341-354.

Hitzman, M.W. and Large, D.E.
1986: A review and classification of the Irish carbonate-hosted base metal deposits; in Geology and Genesis of the Mineral Deposits in Ireland (ed.) C.J. Andrew, R.W.A. Crowe, S. Finlay, W.A. Pennell, and J.F. Pyne; Irish Association for Economic Geology and Geological Survey of Ireland, p. 217-237.

Hoffman, P.F.
1991: Did the breakout of Laurentia turn Gondwanaland inside-out?; Science, v. 252, p. 1409-1412.

Hou, B. and Zhao, D.
1993: Geology and genesis of the Bajiazi polymetallic sufide deposits, Liaoning, China; International Geology Review, v. 35, p. 920-943.

Höy, T.
1982: Stratigraphic and structural setting of stratabound lead-zinc deposits in southeastern B.C.; The Canadian of Mining and Metallurgical Bulletin, v. 75, p. 114-134.

Jardine, D.E.
1966: An investigation of brecciation associated with the Sullivan mine ore body at Kimberley, B.C.; MSc. thesis, University of Manitoba, Winnipeg, Manitoba, 121 p.

Jefferson, C.W., Kilby, D.B., Pigage, L.C., and Roberts, W.J.
1983: The Cirque barite-zinc-lead deposits, northeastern British Columbia; in Sediment-hosted Stratiform Lead-zinc Deposits (ed.) D.F. Sangster; Mineralogical Association of Canada, Short Course Handbook, v. 8, p. 121-139.

*Jennings, D.S. and Jilson, G.A.
1986: Geology and sulphide deposits of the Anvil Range, Yukon; in Mineral Deposits of Northern Cordillera (ed.) J.A. Morin; The Canadian Institute of Mining and Metallurgy, Special Volume 37, p. 339-361.

Krebs, W.
1981: The geology of the Meggen ore deposit; in Handbook of Stratabound and Stratiform Ore Deposits (ed.) K.H. Wolfe; Elsevier, Amsterdam, v. 9, p. 510-549.

Kröner, A., Byerly, G.R., and Lowe, D.R.
1991: Chronology of early Archean granite-greenstone evolution in the Barberton Mountain Land, South Africa, based on precise dating by single zircon evaporation; Earth and Planetary Science Letters, v. 103, p. 41-54.

Lambert, I.B.
1976: The McArthur zinc-lead-silver deposits: features, metallogenesis, and comparisons with some other stratiform ores; in Handbook of Stratabound and Stratiform Ore Deposits, (ed.) K.H. Wolfe; Elsevier, Amsterdam, v. 6, p. 535-585.

Lang, D.Y. and Znang Xingjun
1987: Geological setting and genesis of the Jiashengpan Pb-Zn-S ore belt, Inner Mongolia; People's Republic of China, v. 6, p. 39-54 (Geological Survey of Canada Translation no. 3124).

*Large, D.E.
1983: Sediment-hosted massive sulphide lead-zinc deposits: an empirical model; in Sediment-hosted Stratiform Lead-zinc Deposits (ed.) D.F. Sangster; Mineralogical Association of Canada, Short Course Handbook, v. 8, p. 1-29.

Laznicka, P.
1981: Data on the worldwide distribution stratiform and stratabound ore deposits; in Handbook of Stratiform and Stratabound Ore Deposits, (ed.) K.H. Wolfe; Elsevier, Amsterdam, v. 9, p. 479-576.

Lea, E.R. and Dill, D.B.
1968: Zinc deposits of the Balmat-Edwards District; in Ore Deposits of the United States, 1933-1967: Graton-Sales Volume 1, (ed.) J.D. Ridge; American Institute of Mining, Metallurgical and Petroleum Engineers, p. 20-48.

Leitch, C.H.B.
1992: A progress report of fluid inclusion studies of veins from the vent zone, Sullivan stratiform sediment-hosted Zn-Pb deposit, B.C.; in Current Research, Part E; Geological Survey of Canada, Paper 92-1E, p. 71-82.

Logan, R.G., Murray, W.J., and Williams, N.
1990: HYC Silver-lead-zinc deposit, McArthur River; in Geology of the Mineral Deposits of Australia and Papua New Guinea (ed.) F.E. Hughes; The Australasian Institute of Mining and Metallurgy, Melbourne, Monograph 14, v. 1, p. 907-911.

Louden, A.G., Lee, M.K., Dawling, J.F., and Bourn, R.
1975: Lady Loretta silver-lead-zinc deposit, Northern Territory; in Economic Geology of Australia and Paupa New Guinea, 1. Metals, (ed.) C.L. Knight; Australasian Institute of Mining and Metallurgy, Monograph 5, p. 377-382.

*Lydon, J.W.
1983: Chemical parameters controlling the origin and deposition of sediment-hosted stratiform lead-zinc deposits; in Sediment-hosted Stratiform Lead-zinc Deposits, (ed.) D.F. Sangster; Mineralogical Association of Canada, Short Course Handbook, v. 8, p. 175-250.

Lydon, J.W., Goodfellow, W.D., and Jonasson, I.R.
1985: A general genetic model for stratiform baritic deposits of the Selwyn Basin, Yukon Territory and District of Mackenzie; in Current Research, Part A; Geological Survey of Canada, Paper 85-1A, p. 651-660.

Lydon, J.W., Lancaster, R.D., and Karkkainen, P.
1979: Genetic controls of Selwyn Basin stratiform barite/sphalerite/galena deposits: an investigation of the dominant barium mineraloly of the TEA deposit, Yukon; in Current Research, Part B; Geological Survey of Canada, Paper 79-1B, p. 223-229.

MacIntyre, D.G.
1983: Geologic setting of recently discovered stratiform barite-sulphide deposits in northeastern British Columbia; Canadian Institute Mining and Metallurgy Bulletin, v. 75, p. 99-113.
1992: Geological setting and genesis of sedimentary exhalative barite and barite-sulfide deposits, Gataga District, northeastern British Columbia; Exploration and Mining Geology, v. 1, no. 1, p. 1-20.

Mackenzie, D.H. and Davies, R.H.
1990: Broken Hill lead-silver-zinc deposit at Z.C. mines; in Geology of the Mineral Deposits of Australia and Papua New Guinea (ed.) F.E. Hughes; The Australasian Institute of Mining and Metallurgy, Melbourne, Monograph 14, v. 2, p. 1079-1084.

Mathias, B.V. and Clark, G.J.
1975: Mount Isa copper and silver-lead-zinc orebodies - Isa and Hilton mines; in Economic Geology of Australia and Paupa New Guinea, 1. Metals (ed.) C.L. Knight; Australasian Institute of Mining and Metallurgy, Monograph 5, p. 351-372.

*McClay, K.R. and Bidwell, G.E.
1986: Geology of the Tom deposit, Macmillan Pass, Yukon; in Mineral Deposits of Northern Cordillera, (ed.) J.A. Morin; The Canadian Institute of Mining and Metallurgy, Special Volume 37, p. 100-114.

McKibben, M.A. and Eldridge, C.S.
1989: Sulfur isotopic variations among minerals and aqueous species in the Salton Sea geothermal system: a SHRIMP ion microprobe and conventional study of active ore genesis in a sediment-hosted environment; American Journal of Science, v. 289, p. 661-707.

McKibben, M.A., Andes, J.P. Jr., and Williams, A.E.
1988: Active ore-formation at a brine interface in metamorphosed deltaic-lacustrine sediments: the Salton Sea geothermal system, California; Economic Geology, v. 83, p. 511-523.

McLarsn, D.C.
1946:   The New Calumet mines; Canadian Mining Journal, v. 67, p. 233-241.

Miller, D. and Wright, J.
1983:   Mel barite-zinc-lead deposit, Yukon - an exploration case history: in Mineral Deposits of Northern Cordillera (ed.) J.A. Morin; The Canadian Institute of Mining and Metallurgy, Special Volume 37, p. 129-141.

Moore, D.W., Young, L.E., Modene, J.S., and Plahuta, J.T.
1986:   Geological setting and genesis of the Red Dog zinc-lead-silver deposit, Western Brooks Range, Alaska; Economic Geology, v. 81, p. 1696-1727.

Morrissey, C.J., Davis, G.R., and Steed, G.M.
1971:   Mineralization in the Lower Carboniferous of central Ireland; Institution of Mining and Metallurgy Transactions, Section B, v. 80, p. 174-185.

Nandan Raghu, K.R., Dhruva Rao, B.K., and Singhal, M.L.
1981:   Exploration for copper, lead and zinc ores in India (incorporating the compilation made by late S. Narayanaswamy); Bulletin of the Geological Survey of India, Series A, no. 47, 222 p.

Page, D.C. and Watson, M.D.
1976:   The Pb-Zn deposit of Rosh Pinah, South West Africa; Economic Geology, v. 75, p. 1022-1041.

Page, R.W.
1981:   Depositional ages of the stratiform base metal deposits at Mount Isa and McArthur River, Australia, based on U-Pb zircon dating of concordant tuff horizons; Economic Geology, v. 76, p. 648-658.

Patterson, J.M.
1988:   Exploration potential for argentiferous base metals at the Walton deposit, Hand County, Nova Scotia; in Report of Activities 1987, Nova Scotia Department of Mines and Energy, Report 88-1, p. 129-134.

Pedersen, F.D.
1981:   Polyphase deformation of the massive sulphide ore of the Black Angel mine, central West Greenland; Mineralium Deposita, v. 16, p. 157-176.

Perkins, W.G.
1984:   Mount Isa silica-dolomite and copper orebodies: the result of a syntectonic hydrothermal alteration system; Economic Geology, v. 79, p. 601-637.

Phillips, W.E.A. and Sevastopulo, G.D.
1986:   The stratigraphic and structural setting of Irish mineral deposits; in The Geology and Genesis of Mineral Deposits in Ireland (ed.) C.J. Andrew, R.W.A. Crowe, S. Finlay, W.M. Pennell, and J.F. Pyne; Irish Association of Economic Geology and Geological Survey of Ireland, p. 1-30.

*Pigage, L.C.
1986:   Geology of the Cirque barite-zinc-lead-silver deposits, northeastern British Columbia; in Mineral Deposits of Northern Cordillera, (ed.) J.A. Morin; The Canadian Institute of Mining and Metallurgy, Special Volume 37, p. 71-86.

Ranawat, P.S. and Sharma, N.K.
1990:   Petrology and geochemistry of the Precambrian lead-zinc deposit, Rampura-Agucha, India; in Regional Metamorphism of Ore Deposits and Genetic Implications, (ed.) P.G. Spry and L.T. Bryndzia; (Proceedings of the Twenty-eighth International Geological Congress), VSP Utrecht the Netherlands, p. 197-227.

Reid, D.L., Welke, H.J., Erlank, A.J., and Betton, P.J.
1987:   Composition, age and tectonic setting of amphibolites in the central Bushmanland Group, western Namaqua Province, southern Africa; Precambrian Research, v. 36, p. 99-126.

Russell, M.J.
1974:   Manganese halo surrounding the Tynagh ore deposit, Ireland: a preliminary note; Institution of Mining and Metallurgy Transactions, Section B, v. 83, p. B65-B66.
1975:   Lithogeochemical environment of the Tynagh base metal deposit, Ireland, and its bearing on ore deposition; Institution of Mining and Metallurgy Transactions, Section B, v. 84, p. B128-B133.

*Russell, M.J., Solomon, M., and Walshe, J.L.
1981:   The genesis of sediment-hosted, exhalative zinc + lead deposits; Mineralium Deposita, v. 16, p. 113-127.

Ryan, P.J., Lawrence, A.L., Lipson, A.L., Moore, J.M., Paterson, A., Stedman, D.P., and Van Zyl, D.
1986:   The Aggeneys base metal sulphide deposits, Namaqualand District; in Mineral Deposits of Southern Africa (ed.) C.R. Anhaeusser and S. Maske; Geological Society of South Africa, Johannesburg, p. 1447-1474.

Sangster, D.F.
1981:   Three potential sites for the occurrence of stratiform, shale-hosted lead-zinc deposits in the Canadian Arctic; in Current Research, Part A; Geological Survey of Canada, Paper 81-1A, p. 1-8.
*1990:   Mississippi Valley-type and Sedex lead-zinc deposits: a comparative examination; Institution of Mining and Metallurgy Transactions, Section B, v. 99, p. B21-B42.

Samson, I.M. and Russell, M.J.
1987:   Genesis of the Silvermines zinc-lead-barite deposit, Ireland: fluid inclusion and stable isotope evidence: Economic Geology, v. 82, p. 371-394.

Sarkar, S.C.
1974:   Sulphide mineralization at Sargipali, Orissa, India; Economic Geology, v. 69, p. 206-217.

Sato, T.
1972:   Behaviours of ore-forming solutions in sea water; Mining Geology, v. 22, p. 31-42.

*Sawkins, F.J.
1984:   Ore genesis by episodic dewatering of sedimentary basins: application to giant Proterozoic lead-zinc deposits; Geology, v. 12, p. 451-454.

Schmidt, B.L.
1990:   Elura zinc-lead deposit, Cobar; in Geology of the Mineral Deposits of Australia and Papua New Guinea (ed.) F.E. Hughes; The Australasian Institute of Mining and Metallurgy, Melbourne, Monograph 14, v. 2, p. 1329-1336.

Scotese, C.R.
1984:   An introduction to this volume: Paleozoic paleomagnetism and the assembly of Pangea; in Plate Reconstruction from the Paleozoic Paleomagnetism (ed.) R. Van der Voo, C.R. Scotese, and N. Bonhommet; American Geophysical Union, Geodynamic Series, v. 12, p. 1-10.

Shanks, W.C., Woodruff, L.G., Jilson, G.A., Jennings, D.S., Modene, J.S., and Ryan, B.D.
1987:   Sulfur and lead isotope studies of stratiform Zn-Pb-Ag deposits, Anvil Range, Yukon: basinal brine exhalation and anoxic bottom-water mixing; Economic Geology, v. 82, p. 600-634.

Shearley, E., Hitzman, M.W., Walton, G., Redmond, P., Davis, R., King, M., Duffy, L., and Goodman, R.
1992:   Structural controls of mineralization, Lisheen Zn-Pb-Ag deposit, County Tipperary, Ireland; in Abstracts with Program, Geological Society of America, Annual Meeting, Cincinnati, p. A.354.

Shipboard Scientific Party
1992a: Site 858; in Proceedings of the Ocean Drilling Program, Initial Reports, Leg 139, College Station, TX(Ocean Drilling Program), p. 431-569.
1992b: Site 856; in Proceedings of the Ocean Drilling Program, Initial Reports, Leg 139, College Station, TX(Ocean Drilling Program), p. 161-281.

Simon, A.A., Garcia, L.A., and Barzana, J.A.
1990:   Jurassic metallogenesis in the greater Antilles: the Matahambre-Santa Lucia ore district, western Cuba; in Program with Abstracts, International Association on the Genesis of Ore Deposits, Eighth Symposium, Ottawa, p. A184-185.

Skauli, H., Boyce, A.J., and Fallick, A.E.
1992:   A sulphur isotope study of the Bleikvassli Zn-Pb-Cu deposit, Nordland, northern Norway; Mineralium Deposita, v. 27, p. 284-294.

Slack, J.F., Palmer, M.R., Stevens, B.P., and Barnes, R.G.
1993:   Origin and significance of tourmaline-rich rocks in the Broken Hill District, Australia; Economic Geology, v. 88, p. 505-541.

Smirnov, V.I. and Gorzhersky, D.I.
1977:   Deposits of lead and zinc; in Ore Deposits of the USSR, (ed.) V.I. Smirnov; v. 2, Pitman, London, p. 182-256.

Smith, J.W. and Croxford, N.J.W.
1973:   Sulphur isotope ratios in McArthur Pb-Zn-Ag deposit; Nature, v. 245, p. 10-12.

Solomon, M. and Walshe, J.L.
1979:   The formation of massive sulphide deposits on the sea floor; Economic Geology, v. 74, p. 797-813.

Spry, P.G.
1987:   The chemistry and origin of zincian spinel associated with the Aggeneys Cu-Pb-Zn-Ag deposits, Namaqualand, South Africa; Mineralium Deposita, v. 22, p. 262-268.

Stanton, R.L.
1963:   Constitutional features of the Mount Isa sulphide ores and their interpretation; Proceedings of the Australasian Institute of Mining and Metallurgy, v. 205, p. 131-153.

Sureda, R.J. and Martin, J.L.

1990: El Aguilar mine: an Ordovician sediment-hosted stratiform lead-zinc deposit in the Central Andes; in Stratabound Ore Deposits in the Andes, (ed.) L. Fontbote, G.C. Amstutz, M. Cardozo, E. Cedillo, and J. Frutos; Special Publication of the Society for Geology Applied to Ore Deposits, No. 8, Springer Verlag, Berlin, p. 161-174.

Sverjensky, D.A.

1984: Oil field brines as ore-forming solutions; Economic Geology, v. 79, p. 23-37.

Taylor, S.

1984: Structural and paleotopographic controls of lead-zinc mineralization in the Silvermines orebodies, Republic of Ireland; Economic Geology, v. 79, p. 529-548.

Thomassen, B.

1991: The Black Angel lead-zinc mine 1973-90; in Current Research, Grønlands Geologiske Undersøgelse, (ed.) A.K. Higgins and M. Sonderholm; Rapport 152, p. 46-50.

Tikkanen, G.D.

1986: World resources and supply of lead and zinc; in Economics of Internationally Traded Minerals, (ed.) W.R. Bush; Society of Mining Engineers, Inc., p. 242-250.

Turner, J.S. and Gustafson, L.B.

1978: The flow of hot saline solutions from vents in the sea floor - some implications for exhalative massive sulfide and other ore deposits; Economic Geology, v. 73, p. 1082-1100.

Turner, R.J.W.

1990: Jason stratiform Zn-Pb-barite deposit, Selwyn Basin, Canada (NTS 105-O-1): Geological setting, hydrothermal facies and genesis; in Mineral Deposits of the Northern Canadian Cordillera, (ed.) J.G. Abbott and R.J.W. Turner, International Association on the Genesis of Ore Deposits, Field Trip 14: Guidebook, Geological Survey of Canada, Open File 2169, p. 137-175.

1992: Formation of Phanerozoic stratiform sediment-hosted zinc-lead deposits: evidence for the critical role of oceanic anoxic events; Chemical Geology, v. 99, p. 165-188.

Valdes-Nodarse, E.L., Diaz-Carmona, A., Davies, J.F., Whitehead, R.E., and Fonseca, L.

1993: Cogenetic sedex Zn-Pb and stockwork ores, western Cuba; Exploration and Mining Geology, v. 2, no. 4, p. 297-306.

van den Heyden, A. and Edgecombe, D.R.

1990: Silver-lead-zinc deposit at South Mine, Broken Hill; in Geology of the Mineral Deposits of Australia and Papua New Guinea, (ed.) F.E. Hughes; The Australasian Institute of Mining and Metallurgy, Melbourne, Monograph 14, v. 2, p. 1073-1077.

van Vuuren, C.J.J.

1986: Regional setting and structure of the Rosh Pinah zinc-lead deposit, South West Africa / Namibia; in Mineral Deposits of Southern Africa, (ed.) C.R. Anhaeusser and S. Maske; Geological Society of South Africa, Johannesburg, p. 1593-1607.

Von Damm, K.L., Edmond, J.M., Measures, C.I., and Grant, B.

1985: Chemistry of submarine hydrothermal solutions at Guaymas Basin, Gulf of California; Geochimica et Cosmochimica Acta, v. 49, p. 2221-2237.

Whitcher, I.G.

1975: Dugald River zinc-lead lode; in Economic Geology of Australia and Papua New Guinea, 1. Metals, (ed.) C.L. Knight; Australasian Institute of Mining and Metallurgy, Monograph 5, p. 372-376.

Willan, R.C.R. and Coleman, M.L.

1983: Sulphur isotope study of the Aberfeldy barite, zinc, lead deposit and minor sulphide mineralization in the Dalradian Metamorphic Terrain, Scotland; Economic Geology, v. 78, p. 1619-1656.

Williams, N.

*1978a: Studies of base metal sulphide deposits at McArthur River, Northern Territory, Australia: I. The Cooley and Ridge deposits; Economic Geology, v. 73, p. 1005-1035.

*1978b: Studies of base metal sulfide deposits at McArthur River, Northern Territory, Australia: II. The sulfide-S and organic-C relationships of the concordant deposits and their significance; Economic Geology, v. 73, p. 1036-1056.

Williams, N. and Rye, D.M.

1974: Alternative interpretation of sulphur isotope ratios in the McArthur lead-zinc-silver deposit; Nature, v. 247, p. 535-537.

Wilton, D., Archibald, S., Hussey, A., and Butler, R.

1993: Report on metallogenetic investigations on the north Labrador coast during 1993; Newfoundland Department of Mines and Energy, Open File 996, 17 p.

Wright, J.V., Haydon, R.C., and McConachy, G.W.

1987: Sedimentary model for the giant Broken Hill Pb-Zn deposit, Australia; Geology, v. 15, p. 598-602.

Zhidkov, A. and Jalturin, N.L.

1976: Mineralizacion estratiforme piritico-polimetallica, Zona La Oriental-Baritina; Revista La Mineria en Cuba, no. 3, p. 28-39.

Zierenberg, R.A., Koski, R.A., Morton, J.L., and Bouse, R.M.

1993: Genesis of massive sulfide deposits on a sediment-covered spreading center, Escanaba Trough, southern Gorda Ridge; Economic Geology, v. 88, p. 2065-2094.

6.2 SEDIMENTARY NICKEL SULPHIDES

Larry J. Hulbert

INTRODUCTION

The significant nickel occurrences of this type that have been recognized to date are typically thin, sheet-like, nickel-enriched pyritic sulphide layers of great lateral extent in phosphoritic marine shale basins. In the Nick

Hulbert, L.J.

1996: Sedimentary nickel sulphides; in Geology of Canadian Mineral Deposit Types, (ed.) O.R. Eckstrand, W.D. Sinclair, and R.I. Thorpe; Geological Survey of Canada, Geology of Canada, no. 8, p. 152-158 (also Geological Society of America, The Geology of North America, v. P-1).

basin, Yukon Territory, the main associated metals are Zn and platinum-group elements (PGEs), whereas those in several deposits in Lower Cambrian strata of southern China are mainly Mo, but also include PGEs, Cu, and Zn. These two districts contain the only presently known, near- or subeconomic examples, and the following account is based almost entirely on findings from the Nick property (Hulbert et al., 1992).

The Nick property is centred on 64°43′N latitude and 135°13′W longitude in the Yukon Territory, Canada (Fig. 6.2-1). At this locality a thin, sheet-like, Ni-Zn-PGE-enriched, pyritic massive sulphide layer was deposited over the entire expanse of a small Middle to Upper Devonian shale subbasin known informally as the "Nick basin" that represents an outlier of (eroded) Selwyn Basin sediments. A concretionary Limestone Ball member represents an important stratigraphic marker in the immediate footwall of the mineralized horizon.

In the discussion to follow, this new style of Ni-Zn-PGE mineralization is considered from the viewpoint of (i) regional and local geology; (ii) stratigraphic, structural, and tectonic setting; (iii) base metal, metalloids, noble metal, and stable isotope characteristics; and (iv) mineralogy. A model for the origin of this unusual style of mineralization, with its extraordinary assemblage of ore-forming and ore-associated elements will be presented.

IMPORTANCE

The Nick mineralization is believed to represent a new geological environment and potentially economic deposit type for Ni and PGEs. Globally, similar mineralization is only known from southern China (Fan, 1983; Coveney and Chen Nansheng, 1991). Now that it is fully appreciated that high grade Ni mineralization can be hosted in black shale environments, there is reason to believe that additional occurrences and exploitable deposits will be discovered. One deposit near Zunyi, Guizhou province, southern China has been mined for molybdenum and oil shale since 1985 (Coveney et al., 1992).

SIZE AND GRADE OF DEPOSIT

The mineralization in the Nick subbasin has been traced around the circumference of two major synclines, and constitutes a potentially mineralized area greater than 80 km^2 (Fig. 6.2-2). Assays of sulphide mineralization from this horizon indicate average grades of 5.3% Ni, 0.73% Zn, and 776 ppb PGEs+Au based on 9 samples (Table 6.2-1). Anomalous levels of Re, U, Mo, Ba, Se, As, V, and P are also present. Conservative estimates of the amount of Ni deposited at this mineralized horizon is about 0.90x10^6 t of Ni metal (based on an average thickness of 3 cm), clearly indicating a major North American Ni-metallogenic event. This is large by comparison with the amount of contained Ni in various major Ni camps in the world (Naldrett, 1973; Hulbert et al., 1992).

GEOLOGICAL FEATURES

Geological setting

The Nick property is located within the Mackenzie Platform tectonic province (Fig. 6.2-1). The associated stratiform Ni-Zn-PGE mineralization is hosted by two synclinal outliers of a late Paleozoic shale sequence (Fig. 6.2-2) that belong to the Road River Group and Earn Group strata typically associated with the contiguous Selwyn Basin to the south. The shale outliers overlie Cambrian to Ordovician carbonate rocks of the Mackenzie Platform and therefore the Nick basin is considered to be the erosional remnant of a local trough or embayment on the northeastern margin of the Selwyn Basin.

The main syncline is approximately 16 km by 2 km. A second syncline of about equal strike length, but of considerably narrower width, lies to the north (Fig. 6.2-2). These regional north-northwest-trending folds were generated in response to Cretaceous Laramide compression. Prominent regional-scale normal faults (Green, 1972) occur to the north, south, and east of the Paleozoic shale and are believed to represent the reactivated margins of a graben that was the site of Ordovician to Devonian deep water sedimentation within the platform (Fig. 6.2-2).

The Ni-Zn-PGE mineralization in the Nick basin occurs as a thin conformable massive sulphide horizon located near the stratigraphic contact between lower Earn Group strata, which consist of siliceous shale, mudstone, phosphatic chert, and concretionary limestone, and older calcareous rocks of the Road River Group. All strata are correlative with similar rocks of the Selwyn Basin. Preliminary age determinations based on conodonts extracted from the concretionary limestone in the immediate footwall of the mineralized horizon, suggest a Givetian-Frasnian age bracket (Middle-Upper Devonian boundary). The lower Earn Group is overlain by the Devonian-Mississippian upper Earn Group comprising noncalcareous siliceous and fine grained clastic rocks. Field relationships are illustrated in Figure 6.2-2 and stratigraphic relationships between the units and their approximate thickness are depicted in the stratigraphic column presented in Figure 6.2-3.

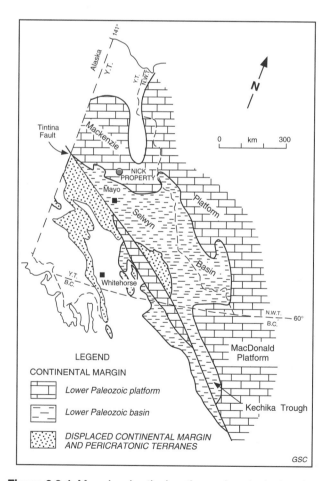

Figure 6.2-1. Map showing the location and geological setting of the Nick property and tectonic elements controlling the Lower Paleozoic facies distribution in the Northern Cordillera (modified after Tipper et al., 1978).

Stratigraphy

The oldest rocks in the vicinity of the Nick basin are part of an unnamed sequence of basinal Cambrian-Ordovician limestones (perhaps equivalent to the Rabbit Kettle Formation of the Selwyn Basin) formed during subsidence along a major east-trending, westerly deepening graben. The basin margins are now represented by regional faults. These dark calcareous rocks consist of platey dolomitic limestones that grade upward into calcareous shales. This sequence is at least 300 m thick but the base has not been observed in the map area. Following deposition of this sequence, Road River Group fetid calcareous graptolitic shale dominated sedimentation for the duration of the Ordovician to Lower Devonian. These dark grey to black shales attain a thickness of at least 100 m in the study area. The lower Earn Group rocks were deposited unconformably on Road River Group shales. Sedimentation had changed from that of a relatively quiescent and euxinic starved basinal sequence environment to an environment with increased circulation and ventilation and accompanying clastic sedimentation (Gordey et al., 1982). Locally, the

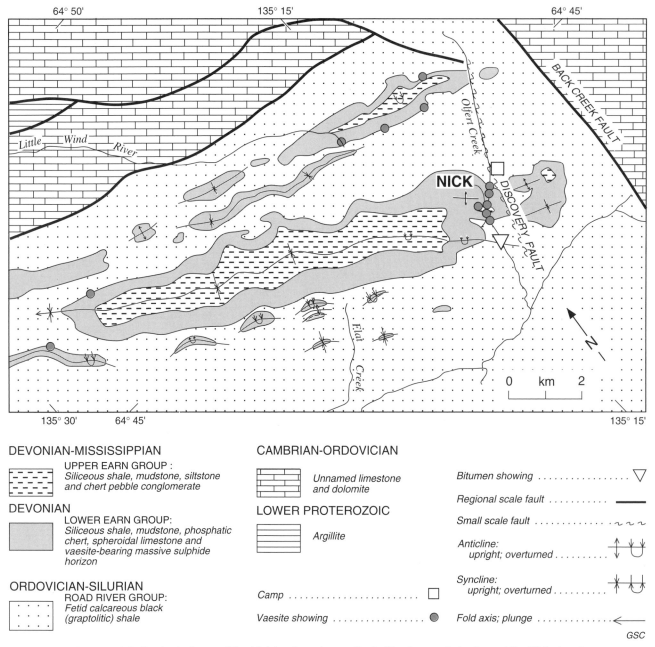

Figure 6.2-2. Geological map of the Nick basin; note reactivated basin margin faults and the Nick showing (discovery site) (after Hulbert et al., 1992).

Table 6.2-1. Metal content of 9 representative samples from the Nick horizon: C, S in wt.%; Pt, Pd, Au, Ir, Ru, Rh, Os, Re in ppb; all others in ppm (after Hulbert et al., 1992); " – " signifies no data.

Sample	1035	NICK-2	NICK-3	NICK-4	NICK-5	NICK-7A	NICK-7B	NICK-8X	DDH-1
C	2.2	2.20	2.50	1.40	1.30	1.80	2.20	1.70	1.90
S	29.10	32.40	20.70	32.70	31.70	27.80	31.70	26.10	20.20
F	349	353	360	293	336	275	533	225	229
Cl	240	<100	<100	123	<100	112	<100	<100	<100
V	880	590	740	370	720	530	570	730	560
Cr	240	160	140	160	140	160	150	180	190
Co	250	350	240	290	250	290	390	170	130
Cu	350	390	250	400	360	340	410	230	170
Ni	48000	78000	51000	62000	59000	56000	76000	38000	23000
Pb	82	90	64	100	95	84	100	57	67
Zn	10000	8700	3500	12000	2900	11000	13000	1400	6100
Se	941	2000	1400	2000	1800	1600	2400	1100	610
Sb	65	83	57	105	95	69	94	64	44
As	2434	3500	2170	3600	3500	3000	4200	3300	1900
Bi	0.42	1.7	1.3	1.5	1.5	1.3	1.5	1.3	1.1
Mo	3920	2467	2372	1907	1704	2363	2968	2472	1411
U	40	60.5	59	44.7	42.7	51.4	107.7	15.8	15.8
Y	74	170	160	110	87	87	200	44	27
Zr	29	31	53	18	24	27	33	25	31
Ba	4300	3800	4200	2300	4900	2300	2900	3900	1900
Ag	4	6	3	4	5	4	4	3	3
Pd	—	308	228	264	247	91	319	158	99
Pt	—	618	427	510	446	149	609	314	208
Au	—	103	67	138	141	29	82	69	57
Os	—	60	45	42	<15	70	60	<15	17
Ir	—	2	1.4	2.2	1.8	2.4	3	1.2	0.8
Ru	—	<30	<30	<30	<30	<30	<30	<30	<30
Rh	—	12	13	5	8	11	14	8	5
Re	—	18000	23000	10000	9600	61000	40000	34000	11400

Middle to Upper Devonian lower Earn Group is composed of four members (Fig. 6.2-3). The Transition member (20-80 m) marks the base of the Earn Group and preserves thin bedded (10 to 25 cm), black calcareous and cherty shales deposited in a restricted euxinic environment. The Limestone Ball member (3 to 20 m) is unique because of unusual texture, composition, stratigraphic confinement, and enigmatic genesis. The member consists of black- to grey-weathering, moderately phosphatic, siliceous shale containing 35 to 40% limestone spheroids. The spheroids range in size from 5 cm to 1.5 m and are interpreted to be concretions. These were probably formed under dystrophic conditions. The Phosphatic Chert member (5 to 8 m) represents a return to euxinic conditions and is dark grey, thin- to medium-bedded, and grades into the overlying siliceous shale over a distance of several metres. The base is marked by the recessive Ni-Zn-PGE sulphide horizon which characteristically occurs 20 to 120 cm above the top of the Limestone Ball member. The uppermost member of the lower Earn Group is an unnamed, 175 to 225 m sequence of rhythmically banded, thin, dark grey to black siliceous shale beds. Limy beds are common near the base of the member, whereas ochrous weathering pyritic beds are more common up-section. The pyrite occurs as thin (5 mm to 1.5 cm) interbeds and ubiquitous disseminations.

Upper Earn Group strata are composed of two lithofacies: Sequence I consists of a fine grained, black carbonaceous muddy siltstone that coarsens upwards into chert pebble conglomerate approximately 80 to 120 m from the base; Sequence II is a fine- to medium-grained, similarly coarsening upwards, siltstone sequence which ranges in thickness from 60 to 100 m. The top of the group is not exposed in the map area.

Stratiform Ni-Zn-PGE mineralization

The Nick mineralization consists of a thin nickeliferous massive sulphide layer ("vaesite horizon") that lies at the base of the Phosphatic Chert member, approximately 20 to 120 cm above the top of the Limestone Ball member (Fig. 6.2-2 and 6.2-3). The thickness of the sulphide layer ranges from 0.4 to 10 cm. The horizon is readily oxidized and highly recessive, and for this reason its recognition on the surface has been restricted to creek beds and cliff faces. However, accurate prediction of the position of the mineralized horizon can be made based on location of the underlying Limestone Ball member marker horizon. Evidence to date suggests that this mineralized horizon is a continuous sheet of sulphide that formed over the entire floor of the Nick subbasin during the Middle Devonian.

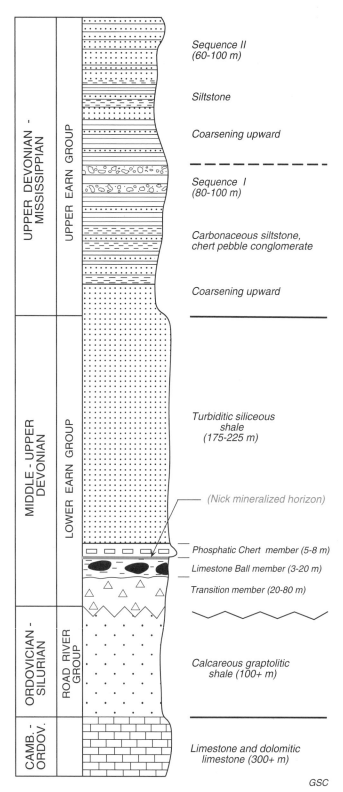

Figure 6.2-3 column content:

UPPER DEVONIAN - MISSISSIPPIAN	UPPER EARN GROUP		Sequence II (60-100 m)
			Siltstone
			Coarsening upward
			Sequence I (80-100 m)
			Carbonaceous siltstone, chert pebble conglomerate
			Coarsening upward
MIDDLE - UPPER DEVONIAN	LOWER EARN GROUP		Turbiditic siliceous shale (175-225 m)
			(Nick mineralized horizon)
			Phosphatic Chert member (5-8 m)
			Limestone Ball member (3-20 m)
			Transition member (20-80 m)
ORDOVICIAN - SILURIAN	ROAD RIVER GROUP		Calcareous graptolitic shale (100+ m)
CAMB. - ORDOV.			Limestone and dolomitic limestone (300+ m)

GSC

Figure 6.2-3. Stratigraphic column for the Nick basin showing the position of the stratiform Ni-Zn-PGE (Nick) mineralization and the footwall Limestone Ball member (after Hulbert et al., 1992).

Fresh mineralized specimens commonly contain 40 to 65% sulphides and display a variety of sedimentary features, the most notable being a soft sediment deformation of laminated metalliferous sediments (Fig. 6.2-4). The sulphides form thin (1 to 3 mm) convolute and discontinuous laminae ≤5 cm in length. The sulphide laminae are enclosed in a dark grey siliceous matrix that contains fine disseminations of pyrite. Some of the sulphide laminae have a vermiform structure (Fig. 6.2-4) and may have formed by replacement of organic matter or precipitation in pore space.

The stratiform Ni-Zn-PGE mineralization in the Nick basin is mineralogically unique. It consists of pyrite (46%), vaesite (NiS_2; 10%), melnikovite (2%), sphalerite and wurtzite (2%), with a gangue (39%) of phosphatic-carbonaceous chert, amorphous silica, and intergrown bitumen (1%). Although the thickness of the horizon varies, the grade and mineralogical composition are extremely consistent.

Sulphur-isotope values ($\delta^{34}S$) in sulphides of the Nick mineralization and overlying lower Earn Group pyritic intervals exhibit a compositional range of 34.6‰. The heaviest sulphur (+11.5 to +19.9‰) occurs in the pyritic bands in the overlying siliceous shale, whereas the lightest sulphur (-14.7 to -10.0‰) was found in the Nick horizon.

Bitumen veins

A number of float and outcrop occurrences of bituminous vein material have been discovered in the area and all are confined to the Road River Group. The largest is known as the Bitumen showing and is located in a fault zone on Olfert Creek (Fig. 6.2-2). It is a tabular body, 3 m wide, that is exposed in a canyon wall for a vertical extent of 9 m and for a strike length of 21 m. Internal siliceous pipe-like features are present within the bitumen infilled vein structures, and localized silicified bituminous wall rocks have also been

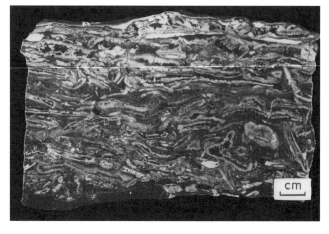

Figure 6.2-4. Typical stratiform Ni-Zn-PGE (Nick) mineralization illustrating soft sediment deformational fabrics that include load casts and slump fragmentation of sulphide-shale laminae. Note local colloform sulphides and sulphide rims surrounding some fragments. GSC 1995-026

noted. Consolidated and ashed bitumen samples contain anomalous concentrations of Ni, Zn, As, Mo, V, and Re and have a chemical signature similar to that of the Nick mineralization. These veins have been interpreted to be possible feeder structures to the stratiform Nick sulphide mineralization (Hulbert et al., 1992).

STRATIFORM Mo-Ni MINERALIZATION, SOUTHERN CHINA

The only occurrences of mineralization similar to the Nick deposit are those of the Mo-Ni sulphide mineralization of southern China (Fan Delian, 1983; Coveney and Chen Nansheng, 1991). This mineralization occurs in lowermost Cambrian black shales and is consistently associated with the widespread distribution of sapropelic organic-rich rocks that occur in a linear belt for a distance of more than 2000 km. The Mo-Ni sulphide ores consist of a mixture of sulphide clasts, phosphorite pellets and low-grade anthracitic coal referred to as "stone coal".

DEFINITIVE CHARACTERISTICS

The most definitive characteristics of the Nick mineralization are (a) the basin-wide distribution of the Ni-Zn-PGE massive sulphide layer and the ubiquitous presence of the footwall Limestone Ball member 20-120 cm below the mineralization; (b) the lateral consistency of the grade and mineralogy; (c) the laminated nature of the sulphides and associated soft sediment textural features; (d) affiliation with the most carbonaceous and phosphatic stratigraphic intervals; (e) the association of anomalous concentrations of Ba, U, V, Se, As, and Re with the Ni-Zn-PGE-rich sulphide mineralization; (f) presence of bitumen veins with chemical signatures similar to those of the stratiform Ni-Zn-PGE mineralization; and (g) the presence of sulphides isotopically enriched in ^{32}S. Most of these characteristics also apply generally to the occurrences of southern China.

GENETIC MODEL

Evidence for a sedimentary-diagenetic process of deposition for the Nick mineralization includes (a) the basin-wide stratiform distribution of the Ni-Zn-PGE massive sulphide layer and associated footwall Limestone Ball member, (b) the fine rhythmically laminated character of the mineralization, (c) the characteristic basin-wide, soft sediment deformation of the mineralization, and (d) the stratigraphic distribution of chemical sediments, i.e., phosphatic cherts and barite-rich shales within the mineralized sequence and enclosing rocks.

It is envisaged that hot basinal brines, perhaps already carrying considerable dissolved sulphate and organic matter, migrated through the organic-rich Upper Silurian and Lower Devonian sediments and extracted Ni, Zn, and other metals associated with the organic material in these sediments. The migrating hot fluids, which were localized by faults that acted as conduits, were discharged towards surface. The well-laminated nature of the mineralization suggests that periodic influxes of this fluid gave rise to the Nick mineralized horizon. The introduction of this nutrient-rich fluid into ooze-like, carbonaceous bottom sediments stimulated biogenic activity and led to sulphate reduction and sulphide precipitation. The resulting sulphides are strongly depleted in ^{34}S suggesting that they were generated by bacterial reduction of sulphate in the restricted reservoir of pore fluids in the bottom sediments, rather than from the oceanic water column. The sulphides consist mainly of a pyrite-vaesite assemblage without bravoite, which implies a minimum temperature of formation of about 137°C (Kullerud, 1962).

The presence of pyrobitumens and the organic compound-rich nature of the host rocks invites speculation on the possible role of hydrophobic hydrocarbons. Petroliferous fluid inclusions were recently discovered in hydrothermal minerals of chimneys and mounds on the seafloor in the southern trough of the Guaymas Basin, central Gulf of California (Peter et al., 1990). They contain a wide range of hydrocarbon and aqueous fluid components, indicating that the hydrothermal fluid and hydrocarbons were never a homogeneous solution, but that the hydrocarbons were transported as immiscible and, possibly, solvated forms. The presence of such hydrocarbons associated with the Nick mineralizing fluids may have facilitated generation of the ooze layer and its discharge may be likened to an oil seep. The pyrobitumen veins scattered through the Nick basin may represent degradation products of trapped petroliferous material.

EXPLORATION GUIDES

Although the presently disclosed thickness and areal extent of the Nick mineralized horizon is limited, it seems clear that this mineralizing process has been operative on an extensive scale. Future exploration should focus on identification of restricted embayments or troughs with indications of bitumen occurrences and phosphatic sediments. Known mineralization in North America and southern China appears to be associated with deep fissures developed along the margins of large epicratonic and foreland basins. Hydrothermal activity may be related to episodes of thermal subsidence following periods of extension and deep rifting

Detailed geochemical profiles through the lower Earn Group, hosting the Nick mineralization, reveal that these black shales are chemically similar to other North American black shales. However, the unusual Ni, Zn, PGE, P, U, V, and Ba element associations and their anomalous concentrations in stream sediments draining areas of "Nick"-type mineralization can be used as an exploration tool.

REFERENCES

References with asterisks (*) are considered to be the best source of general information on this deposit subtype.

Coveney, R.M., Jr. and Chen Nansheng
1991: Ni-Mo-PGE-Au-rich ores in Chinese black shales and speculations on possible analogues in the United States; Mineralium Deposita, v. 26, p. 83-88.
Coveney, R.M. Jr., Murowchick, J.B., Grauch, R.I., Chen Nansheng, and Glascock, M.D.
1992: Field relations, origins and resource implications for platiniferous molydenum-nickel ores in black shales of South China; Exploration and Mining Geology, v. 1, no. 1, p. 21-28.

*Fan Delian
1983: Polyelements in the Lower Cambrian black shale series in southern China; in The Significance of Trace Metals in Solving Petrogenetic Problems and Controversies, (ed.) S.S. Augustithis; Theophrastus Publications S.A., Athens, p. 447-474.

Gordey, S.P., Abbott, J.G., and Orchard, M.J.
1982: Devono-Mississippian Earn Group and younger strata in east central Yukon; in Current Research, Part B; Geological Survey of Canada, Paper 82-1B, p. 93-100.

Green, L.H.
1972: Geology of Nash Creek, Larsen Creek and Dawson map areas, Yukon; Geological Survey of Canada, Memoir 364, 155 p.

*Hulbert, L.J., Carne, R.C., Grégoire, D.C., and Paktunc, D.
1992: Sedimentary nickel, zinc and platinum-group element mineralization in Devonian black shales at the Nick Property, Yukon, Canada: a new deposit type; Exploration and Mining Geology, v. 1, no. 1, p. 39-62.

Kullerud, G.
1962: The Fe-Ni-S system; in Carnegie Institute of Washington Year Book, v. 61, p. 144-150.

Naldrett, A.J.
1973: Nickel sulphide deposits - their classification and genesis, with special emphasis on deposits of volcanic association; The Canadian Institute of Mining and Metallurgy, Transactions, v. 76, p. 183-201.

Peter, J.M., Simoneit, B.R.T., Kawka, O.E., and Scott, S.D.
1990: Liquid hydrocarbon-bearing inclusions in modern hydrothermal chimneys and mounds from the southern trough of Guaymas Basin, Gulf of California; Applied Geochemistry, v. 5, p. 51-63.

Tipper, H.W., Woods, G.J., and Gabrielse, H.
1978: Tectonic Assemblage Map of the Canadian Cordilleran; Geological Survey of Canada, Map 1505A, scale 1:2 000 000.

6.3 VOLCANIC-ASSOCIATED MASSIVE SULPHIDE BASE METALS

J.M. Franklin

INTRODUCTION

All volcanic-associated massive sulphide deposits occur in terranes dominated by volcanic rocks. However, the individual deposits may be hosted predominantly by volcanic or sedimentary strata, all of which form integral parts of a volcanic complex. Such deposits are also commonly referred to as volcanogenic massive sulphides, or simply as VMS.

These deposits occur in two distinct compositional groups, the **copper-zinc group** and the **zinc-lead-copper group,** according to their total contained copper, lead, and zinc (Fig. 6.3-1; Franklin et al., 1981). Using the Zn/Zn+Pb ratio, the division between these two groups is established at 0.90. All are within sequences dominated by submarine volcanic rocks, and contain about 90% iron sulphide (pyrite dominant). They consist of two parts: massive sulphide ore that formed either on or immediately below the seafloor, and generally less important vein and disseminated ore (stringer zone) that immediately underlies the massive sulphide ore. The stringer ore is usually within an intensely metasomatically altered "alteration pipe". Deposits of the volcanic-associated massive sulphide type are important sources of copper, zinc, and lead; many deposits contain economically recoverable silver and gold. Cadmium, tin, indium, bismuth, and selenium are also recovered as smelter byproducts.

Deposits of the copper-zinc group are concordant to semiconcordant massive iron sulphide bodies, commonly underlain by stringer ore, within volcanic sequences that are dominated by mafic volcanic rocks, with locally important felsic and/or sedimentary rocks. Examples in Canada (Fig. 6.3-2) are the deposits near Noranda, Quebec; Flin Flon-Snow Lake, Manitoba; and the Juan de Fuca and Explorer ridges in the northeastern Pacific. Other deposits are those of the Cyprus and Oman ophiolite sequences, and the Besshi-type deposits of the Shikoku district, Japan.

Deposits of the zinc-lead-copper group are tabular, concordant massive pyritic bodies, typically underlain by less prominent stringer ore, in felsic volcanic sequences; sedimentary rocks may form a significant portion of the footwall. Canadian examples are the Buttle Lake, British Columbia; Bathurst, New Brunswick; and Buchans, Newfoundland deposits. Other deposits are those of the Hokuroku basin, Japan; the Iberian Pyrite Belt, Spain; Neves-Corvo, Portugal; and the Tasman Geosyncline, Australia.

Volcanic-associated massive sulphide deposits have been extensively reviewed by Klau and Large (1980), Franklin et al. (1981), Lydon (1984, 1988), and Franklin (1986). These accounts provide descriptive, experimental and theoretical data that are beyond the scope of this review which will focus on Canadian examples (Table 6.3-1), and draw on other areas only where the Canadian deposits do not provide a complete perspective.

Franklin, J.M.
1996: Volcanic-associated massive sulphide base metals; in Geology of Canadian Mineral Deposit Types, (ed.) O.R. Eckstrand, W.D. Sinclair, and R.I. Thorpe; Geological Survey of Canada, Geology of Canada, no. 8, p. 158-183 (also Geological Society of America, The Geology of North America, v. P-1).

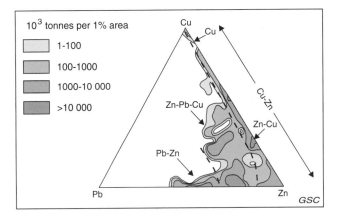

Figure 6.3-1. Contours of tonnes of contained copper, lead, and zinc, in approximately 800 massive sulphide deposits of both the volcanic-associated and sediment-associated types from Canada, U.S.A., Japan, Scandinavia, Spain, and Portugal (after Franklin et al., 1981).

IMPORTANCE

Volcanic-associated massive sulphide deposits are important sources of base metals and precious metals in Canada; in 1988 they produced 32.8% of Canada's copper, 29.4% of its lead, 56.3% of its zinc, 3.6% of its gold, and 30.4% of its silver.

SIZE AND GRADE OF DEPOSITS

Copper-zinc group

The mean and median (in brackets) grades and sizes of 142 Canadian deposits are: 5 300 000 (1 241 000) t containing 1.95% (1.60%) Cu, 4.23% (3.07%) Zn, 0.09% (0.01%) Pb, 0.8 (0.6) g/t Au, and 19.0 (8.0) g/t Ag. The lead content of most deposits in this group is rarely determined; its statistical values are approximate. The largest Canadian deposit is the Kidd Creek mine, which contains 12 000 000 t combined copper, zinc, and lead. The Horne mine at Noranda, Quebec, has a very large but sub-ore grade pyrite-sphalerite zone (No. 5 zone: 170 Mt, 0.1% Cu, 0.5% Zn, Kerr and Mason, 1990) that is not included in the calculations; with the No. 5 zone included, the Horne mine may contain more base metal than Kidd Creek. The deposit sizes of the combined Cu-Zn and Zn-Pb-Cu groups are log-normally distributed (Fig. 6.3-3).

Zinc-lead-copper group

The mean and median (in brackets) grades and sizes of 92 Canadian deposits are 5 600 000 (1 177 000) t containing 1.23% (1.01%) Cu, 3.60% (2.80%) Zn, 1.46% (0.97%) Pb, 2.0 (0.5) g/t Au, and 79.0 (57.0) g/t Ag. Brunswick #12 is the largest Canadian deposit, containing at least 10 500 000 t of combined copper, lead, and zinc.

GEOLOGICAL FEATURES

Volcanic-associated massive sulphide deposits occur in submarine volcanic rocks of all ages, from the presently-forming deposits in modern, actively-spreading ridges to deposits in the pre-3400 Ma volcanic strata of the Pilbara Block in Australia. They occur in a wide variety of tectonic regimes. Almost all deposits have a close association with at least minor amounts of sedimentary rock.

Copper-zinc group

Geological setting

These deposits occur in two principal geological settings; 1) in mafic-volcanic dominated areas, such as Archean and Proterozoic greenstone belts (Fig. 6.3-2) and modern and Phanerozoic spreading ridges and seamounts; 2) in areas containing subequal amounts of both mafic volcanic rocks and sedimentary strata, such as are in Phanerozoic arc sequences.

Volcanic rock-dominated areas

Significant variation in the composition of these deposits, and the alteration associated with them, has been related to the depth of water under which the deposits formed. Morton and Franklin (1987) defined two groups. 1) Deposits typified by the Noranda and Matagami Lake districts, Quebec (Fig. 6.3-4A), were formed at depths of considerably more than 500 m. These are associated with sequences composed primarily of massive to pillowed mafic flows. Felsic ash-flow tuff beds are usually prominent immediately below the deposits, and felsic domes may immediately underlie or enclose the ore. However, the amount of felsic rock in the footwall sequence may be only minor (Flin Flon, Manitoba), or comprise as much as 30% (e.g. Noranda). 2) A second group of deposits, typified by those near Sturgeon Lake, Ontario, Hackett River, Northwest Territories, and possibly the Kidd Creek mine near Timmins, Ontario, are associated with volcanic rocks deposited in subaerial to shallow marine environments (<500 m). These include mafic and felsic amygdaloidal and scoriaceous flows and pyroclastic rocks, volcanic breccia, and epiclastic strata (Fig. 6.3-4B). Felsic rocks typically comprise 30% of the footwall sequence.

Both groups of deposits occur in volcanic sequences that have prominent subvolcanic intrusions near their base. Trondhjemitic intrusions predominate (Noranda, Sturgeon Lake, Flin Flon, Snow Lake), but a layered mafic intrusion forms the base of the Matagami Lake sequence.

Actively forming deposits within modern mid-ocean spreading ridges occur as two types, those in sediment-free, basalt-dominated terranes (primarily axial grabens), and those in sediments. Some of the first type are associated with off-axis seamounts. The deposits in grabens within basalt-dominated active ridges may be further divided into two groups (Kappel and Franklin, 1989): small deposits forming in ridge axes which are in the active and early phases of volcanic construction, such as the Cleft segment of the southern Juan de Fuca Ridge (and the East Pacific Rise deposits at 11°N and 21°N), and larger deposits in more volcanically evolved ridge crests, such as those of the Endeavour and Explorer segments of the Juan de Fuca Ridge, and the TAG area of the Mid-Atlantic Ridge.

Deposits associated with areas of recent prolific volcanism lie along the bounding faults of the narrow central graben developed on top of a prominently elongate, but locally inactive volcano, typically 500 m high, and composed of inflated (over-filled) pillows. They are comparatively large (10^5-10^6 t), are composed of coalesced mounds, and are associated with highly fissured pillowed basalts.

The deposits in the Cyprus, Oman, U.S.A., and Canadian ophiolite sequences are the forerunners of the spreading-ridge association. In Canada, the best examples of preserved ophiolite-associated deposits are in the Ordovician sequences of Newfoundland; a few small deposits are also in Ordovician ophiolites of southeastern Quebec. These deposits occur within basaltic to andesitic pillow sequences, typically a few kilometres thick, that overlie the sheeted dyke and gabbroic portions of ophiolitic sequences.

Sediment-dominated areas

Terranes commonly ascribed to arc-related basins, composed of relatively monotonous, regular sequences of volcanic and sedimentary strata, contain many massive sulphide deposits. Deposits formed close to a tectonic boundary between ocean floor and island arcs, ocean floor and cratons, or ocean floor and continental crust are included. The volcanic component is usually dominant, and composed predominantly of mafic volcanic rocks. However, some areas also have minor quantities of felsic volcanic strata. The sedimentary rocks are dominantly pelitic. The ratio of volcanic to sedimentary strata associated with the deposits is highly variable. These terranes are typically highly deformed, making identification of primary tectonic relationships difficult. Terms such as "Besshi-type" or "Keislager-type" may be used.

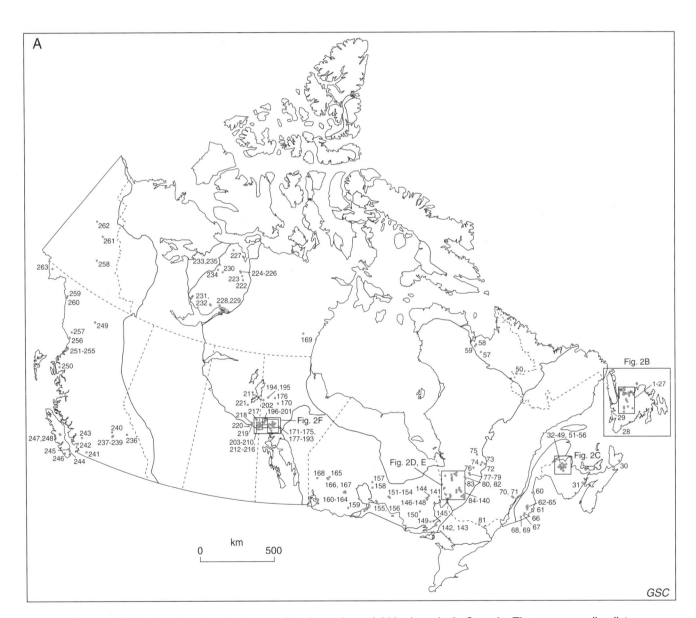

Figure 6.3-2A. Location of volcanic-associated massive sulphide deposits in Canada. The corresponding list of deposits is in Table 6.3-1. Major districts within the boxes are shown as separate maps 6.3-2B to 2E.

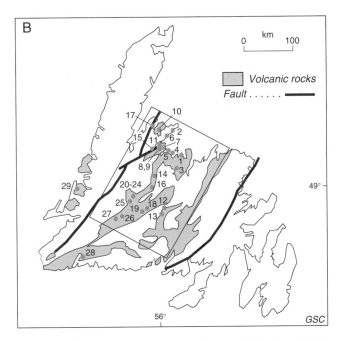

Figure 6.3-2B. Volcanogenic massive sulphide deposits in Newfoundland.

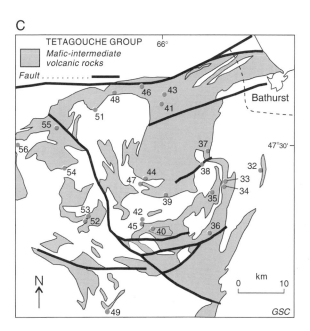

Figure 6.3-2C. Distribution of volcanogenic massive sulphide deposits in the Bathurst district, New Brunswick.

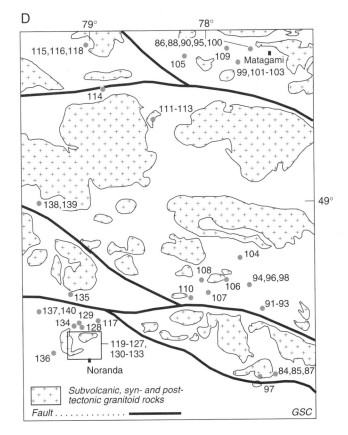

Figure 6.3-2D. Distribution of volcanogenic massive sulphide deposits in the Noranda and Val d'Or districts, Quebec. Location of deposits in box are shown in Figure 6.3-2E.

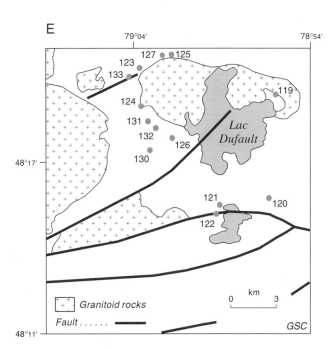

Figure 6.3-2E. Distribution of volcanogenic massive sulphide deposits in the Noranda district, Quebec.

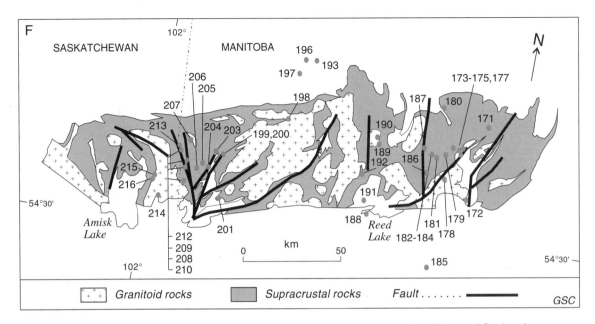

Figure 6.3-2F. Distribution of deposits in the Flin Flon-Snow Lake districts, Manitoba and Saskatchewan. Amisk volcanic rocks are shown with a light stipple.

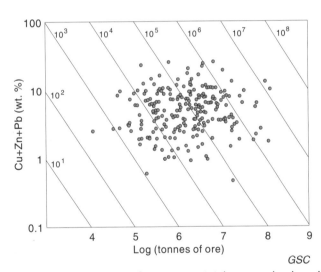

Figure 6.3-3. Tonnes of ore versus total copper, lead, and zinc (per cent) in Canadian volcanogenic massive sulphide deposits (see Table 6.3-1 for list). The size of deposits is log-normally distributed. Diagonal lines indicate tonnes of contained Cu + Zu + Pb.

Examples of copper-zinc deposits in sediment-dominated settings (Fig. 6.3-4C) are not common in Canada, but include the Granduc deposit, deposits in the Anyox and Kutcho Creek areas, and the Windy Craggy deposit, all in northern British Columbia. Possibly the Britannia mine in southwestern British Columbia (Payne et al., 1980), the Sherridon deposit in northern Manitoba, and deposits such as the Soucy #1 in the Labrador Trough, are also in this group. The Windy Craggy deposit, although low grade, is possibly the largest massive sulphide deposit in Canada; the Granduc and Hidden Creek (Anyox area) deposits are also in the upper quartile of deposit tonnages in Canada (43 and 40 million tonnes, respectively).

Deposits in the sediment-covered portions of actively spreading ridges are much larger than those in modern volcanic ridges. The Middle Valley Bent Hill deposit, one example of this type, is in the complex northern Endeavour segment of the Juan de Fuca Ridge. It has formed within the upper portion of a sequence of pelite and turbidite, about 300 m thick, that overlies a volcanic basement (Goodfellow and Franklin, 1993). The brine pools and metalliferous muds in the Red Sea are the product of a highly saline, high-temperature hydrothermal system and contribute a large sediment-hosted copper-zinc deposit in the process of formation.

Form and composition

The typical form of the Noranda-type deposits (Fig. 6.3-5) is a generally conformable bulbous or conical mound consisting of at least 60% of pyrite, sphalerite, chalcopyrite, and pyrrhotite. The length-to-thickness aspect ratio of a typical (1 million tonne) undeformed deposit is usually between 3 and 10 to 1. The largest deposits have a much greater ratio. The upper contact is sharp, but the lower is transitional into the stringer zone, composed of chalcopyrite, pyrrhotite, and pyrite. Deposits of the Mattabi type are more stratiform in shape, with contact relationships similar to those of the Noranda-type deposits. Stringer zones of deposits of the latter type commonly contain economically recoverable ore for several tens of metres below the deposit. In the Mattabi-type deposits, although alteration pipes may be extensive, the amount of ore contained in them is relatively small.

Table 6.3-1. Canadian volcanic-associated massive sulphide deposits.

NO.	NAME	LATITUDE / LONGITUDE		NTS	Cu %	Pb %	Zn %	Ag (g/t)	Au (g/t)	TONNES	REFERENCE
1	LOCKPORT	49°27'18"	55°29'53"	2E/6	0.75		1.21			555 934	1
2	TILT COVE	49°53'24"	55°37'12"	2E/13	6.00					8 165 000	2
3	POINT LEAMINGTON	49°16'36"	55°37'53"	2E/5	0.50		2.00	18.00	0.90	13 800 000	3, 1
4	PILLEYS ISLAND	49°30'31"	55°43'07"	2E/12	1.23					1 025 000	1, 3
5	MILES COVE	49°32'22"	55°47'26"	2E/12	1.55			12.00	0.34	200 000	1, 2
6	BETT'S COVE	49°48'46"	55°48'23"	2E/13	6.00					119 000	1, 2
7	LITTLE BAY	49°36'52"	55°56'20"	2E/12	2.10					2 934 464	1, 2
8	WHALESBACK	49°35'45"	56°00'22"	12H/9	0.95					3 793 561	1, 2, 3
9	LITTLE DEER	49°35'20"	56°01'08"	12H/9	1.53					285 200	3
10	RAMBLER MAIN	49°54'02"	56°03'32"	12H/16	1.30		2.16	23.24	2.4	572 000	1, 3
11	COLCHESTER	49°38'20"	56°04'51"	12H/9	1.30					1 001 000	3, 2
12	SOUTH POND	48°25'36"	56°08'23"	12A/8	1.33			12		293 000	3, 2
13	GREAT BURNT LAKE	48°20'19"	56°09'06"	12A/8	2.40					796 500	3, 2
14	GULLBRIDGE	49°11'54"	56°09'18"	12H/1	1.02					4 072 430	3
15	RENDELL JACK MAN	49°33'47"	56°10'52"	12H/9	2.50				1.00	13 081	3, 2
16	LAKE BOND	49°01'39"	56°11'11"	12H/1	0.31		2.10			1,200,000	1
17	TERRA NOVA	49°55'17"	56°13'44"	12H/16	2.41			9.94	1.68	257 417	3, 2
18	BOUNDARY	48°39'22"	56°26'47"	12A/9	3.50	1.00	4.00	34		500 000	3
19	DUCK(TALLY) POND	48°38'04"	56°29'16"	12A/9	3.60	7.00		70.4		3 860 000	3, 2
20	ORIENTAL	48°49'00"	56°53'00"	12A/15	1.48	7.90	14.18	154.63	1.95	3 326 876	4
21	MACLEAN	48°49'00"	56°53'00"	12A/15	1.16	7.32	13.20	128.91	1.03	3 313 271	4
22	BUCHANS	48°49'00"	56°53'00"	12A/15	1.37	7.73	14.97	127.89	1.58	567 782	4
23	ROTHERMERE	48°49'00"	56°53'00"	12A/15	1.16	7.72	12.74	135.09	1.03	3 263 386	4
24	LUCKY STRIKE	48°49'00"	56°53'00"	12A/15	1.51	8.20	15.20	112.46	1.6	5 899 128	1, 4
25	SKIDDER	48°42'23"	56°56'07"	12A/10	2.00		2.00			900 000	3, 2
26	TULK'S EAST	48°32'15"	57°07'54"	12A/11	0.24	0.12	1.50	8.5		6 230 000	3, 2
27	TULK'S HILL	48°30'53"	57°12'07"	12A/11	1.30	2.00	5.60	41	0.40	720 000	3, 2
28	STRICKLAND	47°48'29"	58°18'00"	11O/16		1.00	1.00	195.00		1 010 000	3
29	YORK HARBOUR	49°03'00"	58°18'32"	12G/1	1.92		4.67			338 039	3
30	STIRLING	45°43'40"	60°26'15"	11F/9	0.66	1.30	5.50	68.57	0.96	995 990	3
31	TEAHAN	45°42'50"	64°59'20"	21H/10	0.46		1.46	29.83		122 445	3
32	KEY ANACON	47°26'12"	65°42'20"	21P/5	0.22	3.47	8.41	111.43		1 519 722	3
33	BRUNSWICK #6	47°24'30"	65°49'00"	21P/5	0.39	2.16	5.43	67.00		12 100 000	5
34	AUSTIN BROOK	47°23'48"	65°49'21"	21P/5	0.10	1.86	2.93	34.29		907 000	3
35	FLAT LANDING	47°22'55"	65°52'36"	21P/5		0.94	4.90	19.54		1 750 510	3
36	CAPTAIN DEPOSIT	47°17'03"	65°52'58"	21P/5	1.99			9.60	0.58	311 101	3
37	BRUNSWICK #12	47°28'00"	65°53'00"	21P/5	0.30	3.60	8.87	100.00		134 100 000	5
38	PABINEAU RIVER	47°26'58"	65°55'07"	21P/5		0.87	2.65			136 050	3
39	NINE MILE BROOK	47°23'25"	65°55'59"	21P/5	0.42	1.00	1.07	90.86	1.03	133 329	3
40	NEPISIGUIT	47°22'00"	66°02'00"	21O/8	0.40	0.60	2.80	10.29		3 537 300	3
41	ARMSTRONG	47°36'06"	66°02'33"	21O/9	0.29	0.42	2.29	2.40	0.68	3 100 000	3
42	HEATH STEELE	47°17'00"	66°05'00"	21O/8	0.70	2.50	6.30	60.00	0.62	33 777 000	3, 6
43	ROCKY TURN	47°38'00"	66°04'15"	21O/9	0.30	1.50	7.00	78.86		180 000	3
44	CANOE LANDING	47°24'39"	66°06'24"	21O/8	0.56	0.64	1.82	29.50	1.07	20 380 000	3
45	STRATMAT WEST	47°18'30"	66°07'31"	21O/8	0.59	2.44	6.29	31.89		4 353 600	3
46	ORVAN BROOK	47°37'37"	66°08'07"	21O/9	0.00	3.25	6.30	34.29		181 400	3
47	WEDGE	47°23'50"	66°08'30"	21O/8	2.17	0.24	0.46	0.70	0.03	1 500 000	5

Table 6.3-1. (cont.)

NO.	NAME	LATITUDE / LONGITUDE		NTS	Cu %	Pb %	Zn %	Ag (g/t)	Au (g/t)	TONNES	REFERENCE
48	MCMASTER	47°36′26″	66°13′59″	21O/9	0.60					181 400	3
49	CHESTER	47°05′54″	66°14′50″	21O/1	0.44					16 235 300	3
50	FREDRICKSON LAKE	55°03′02″	66°15′09″	23O/1	0.77		4.38	42.17	0.69	279 356	3
51	CARIBOU	47°34′50″	66°17′56″	21O/9		3.84	8.46	109.60	1.70	5 694 000	3
52	HALF MILE NORTH	47°18′00″	66°19′00″	21O/8	0.39	0.87	4.95	8.23	0.17	789 250	3
53	HALF MILE	47°17′00″	66°19′00″	21O/8	0.39	2.49	6.60			7 847 142	3
54	DEVILS ELBOW	47°25′51″	66°23′57″	21O/8	1.00					471 640	3
55	MURRAY BROOK	47°31′30″	66°26′00″	21O/9	0.44	0.86	1.95	31.20	0.34	20 861 000	3
56	RESTIGOUCHE	47°30′20″	66°33′50″	21O/10	0.32	5.99	7.72	124.11	1.20	997 700	7 (26/11/90)
57	BOYLEN(KOKE)	57°39′42″	69°27′00″	24F/11	0.70	1.03	6.86	54.51	1.03	1 060 433	3
58	SOUCY NO.1 ZONE	58°19′12″	69°52′12″	24K/5	1.40	0.10	1.09	19.20	2.06	4 324 032	3
59	PRUDHOMME NO.1	58°15′36″	69°54′30″	24K/5	1.50	0.20	1.90	26.06	1.37	5 729 161	3
60	PANET METALS	46°35′18″	70°13′00″	21L/9	0.40	0.43	2.67	19.89	2.06	272 100	3
61	CLINTON COPPER	45°27′36″	70°54′30″	21E/7	1.84		1.58	13.71		1 380 809	3
62	CUPRA DESTRIE	45°46′24″	71°18′48″	21E/14	2.75	0.47	1.48	37.71	0.51	3 007 451	3
63	SOLBEC COPPER	45°49′00″	71°18′30″	21E/14	1.60	0.70	4.60	59.66	0.79	1 860 000	8, 9
64	LINGWICK	45°39′18″	71°23′00″	21E/11	0.60		6.00	17.14		317 450	3
65	WEEDON	45°42′00″	71°23′00″	21E/11	2.50	0.08	0.40	6.86		1 700 000	8, 9
66	MOULTON HILL	45°24′30″	71°49′30″	21E/5	1.00					331 055	3
67	SUFFIELD	45°19′12″	71°57′36″	21E/5	0.83		4.25	63.09	4.94	205 889	3
68	IVES	45°17′36″	72°19′24″	31H/8	3.50					771	3
69	HUNTINGDON	45°15′48″	72°19′54″	31H/8	0.90			1.03	0.10	1 814 000	3
70	TETREAULT	46°49′30″	72°20′00″	31I/16		1.40	4.00	100.46	1.23	2 670 208	8
71	MONTAUBAN	46°50′06″	72°20′48″	31I/16		1.07	3.46	38.74	0.55	3 339 514	3
72	LEMOINE(PATINO)	49°45′44″	74°06′10″	32G/16	4.20		9.50	92	4.4	750 000	5
73	ANTOINETTE(TACHE)	49°56′30″	74°24′18″	32G/16	0.33		3.31	11.31	1.89	2 305 594	3
74	SCOTT TWP	49°51′52″	74°37′47″	32G/15	0.14		6.91	11.66	0.34	704 739	3, 7 (21/01/91)
75	DOMERGUE(LESSARD)	50°38′30″	74°38′30″	32J/10	1.82		3.30	38.40	0.72	952 350	3
76	LA-RIBOURDE	49°49′00″	75°31′24″	32G/13	1.35		2.73	42.51		408 150	3
77	EMPIRE OIL	49°25′04″	76°08′09″	32F/8	0.23		2.45	5.49		300 217	8
78	SOMA ALTA	49°26′48″	76°09′30″	32F/8			2.70	0.34		31 745	3, 8
79	CONIAGAS S	49°29′54″	76°09′48″	32F/8			10.70	182.00		700 000	3, 5, 10
80	GREVET (M ZONE)	49°14′06″	76°40′02″	32F/2	0.48	0.10	8.65	34.00		6 200 000	10
81	NEW CALUMET	45°42′12″	76°40′24″	31F/10		1.77	6.00	80.23	0.48	3 651 618	3
82	GREVET (B ZONE)	49°14′00″	76°45′00″	32F/2	0.58		9.67	24.00		477 000	7 (10/90)
83	TONNANCOURT-3	48°51′48″	76°59′24″	32C/15	2.07		3.17	25.03	0.38	54 420	3
84	LOUVEM(ZN ORE)	48°05′54″	77°31′00″	32C/4	0.21		5.59	34.29	0.69	2 358 200	3
85	LOUVICOURT	48°05′54″	77°31′00″	32C/4	3.40		2.20	31.00	0.90	15 700 000	11(22/04/94)
86	RADIORE	49°45′03″	77°33′03″	32F/13	1.57		1.34	6.86	0.34	150 000	8, 5
87	QUE MANITOU	48°05′30″	77°35′06″	32C/4	1.26			3.43		692 041	3
88	GARON LAKE	49°46′24″	77°34′06″	32F/13	1.45		2.22	4.53		512 455	3, 8
89	MANITOU-BARVUE	48°05′12″	77°36′36″	32C/4	0.07	0.07	0.36	11.66	0.34	13 750 120	3
90	BELL CHANNEL#1	49°46′12″	77°37′24″	32F/13	1.95		0.57	29.14	0.34	82 084	3, 8
91	BELFORT (ROYMONT)	48°25′24″	77°39′42″	32C/5	0.21	0.12	7.00	23.04	0.38	226 750	3
92	BARVALLEE	48°25′24″	77°39′42″	32C/5	1.23		5.71	48.69		181 400	3
93	MOGADOR	48°25′24″	77°39′42″	32C/5	0.47	0.34	7.30	55.89	1.17	1 107 447	3
94	BARVUE	48°31′03″	77°40′05″	32C/12	0.05	0.00	3.50	41.14		8 341 679	8

Table 6.3-1. (cont.)

NO.	NAME	LATITUDE / LONGITUDE		NTS	Cu %	Pb %	Zn %	Ag (g/t)	Au (g/t)	TONNES	REFERENCE
95	NORITA (RADIORE A)	49°46'00"	77°41'00"	32F/13	1.80		3.80	27.43	0.69	4 000 000	8, 5
96	CONS. PERSHCOURT	48°31'24"	77°41'06"	32C/12			2.42	91.89		4 824 333	8
97	EAST SULLIVAN	48°04'18"	77°42'30"	32C/4	1.07		4.0	7.92	0.27	15 186 638	3, 10
98	FREBERT	48°31'00"	77°43'00"	32C/12			2.50	50.40		2 721 000	8
99	MATTAGAMI LAKE	49°43'18"	77°43'00"	32F/12	0.42		5.10	21.60	0.30	25 600 000	5
100	LYNX (OBASKA)	49°38'48"	77°43'18"	32F/12	1.60		0.35			204 075	3
101	BELL ALLARD	49°41'00"	77°44'00"	32F/12	1.14		9.30	41.14		234 000	10
102	ORCHAN S	49°42'00"	77°44'00"	32F/12	1.20		8.70	37	0.5	4 500 000	10
103	ISLE DIEU	49°43'05"	77°44'06"	32F/12	1.03		17.80	82.00	0.45	2 060 000	10
104	TRINITY PROPERTY	48°42'30"	77°45'36"	32C/12	1.18		0.74			133 329	3
105	NEW HOSCO	49°47'24"	77°50'06"	32F/13	1.41		1.11	4.11	0.03	2 040 750	3
106	MONPAS (ALBAR)	48°36'54"	77°52'12"	32C/12	2.00		0.75	20.57		45 350	3
107	CONIGO	48°35'48"	78°03'30"	32D/9	1.26			17.14		2 735 512	3
108	JAY COPPER	48°36'45"	78°03'30"	32D/9	1.26			6.86		1 959 120	7 (22/04/91)
109	LA GAUCHETIERE	49°45'54"	78°10'00"	32E/16	1.10		4.90			1 541 900	3
110	NEWCONEX FIGUERY	48°28'43"	78°10'15"	32D/8			5.00	68.57		453 500	3
111	CONS. NORTH EXPL.	49°26'06"	78°19'54"	32E/8	0.73		6.95	34.29		996 893	3
112	JOUTEL COPPER	49°27'12"	78°21'12"	32E/8	2.16		3.90			1 702 439	3
113	POIRIER	49°26'36"	78°23'18"	32E/8	2.22		5.08	7.60		7 082 143	3, 5
114	ESTRADES	49°35'12"	78°51'24"	32E/10	1.02	0.90	9.90	218.40	6.72	2 000 000	3, 10, 12
115	DETOUR (A2 ZONE)	49°49'00"	78°55'55"	32E/15	2.26		1.24	16.80	0.79	8 852 320	8
116	DETOUR (A1 ZONE)	49°49'00"	78°55'58"	32E/15	0.39		2.30	35.66	0.31	32 107 800	8
117	MOBRUN	48°24'00"	78°56'00"	32D/7	.63		4.66	31.4	1.55	8 640 000	5
118	DETOUR (B ZONE)	49°49'04"	78°56'16"	32E/15	4.49		0.80	39.43	1.23	3 061 125	8
119	GALLEN	48°19'30"	78°57'12"	32D/7	0.08		3.36	2.40	0.06	8 100 000	3, 10
120	DELBRIDGE	48°15'53"	78°57'52"	32D/7	0.55		8.60	68.60	2.40	360 000	5
121	QUEMONT	48°15'00"	78°58'00"	32D/6	1.32	0.02	2.44	30.90	5.50	13 800 000	10
122	HORNE	48°15'18"	79°00'42"	32D/6	2.20			13.00	6.10	54 000 000	10
123	EAST WAITE	48°21'00"	79°02'00"	32D/6	4.10		3.25	31.00	1.80	1 500 000	5
124	AMULET (F ZONE)	48°20'00"	79°03'00"	32D/6	3.40		8.60	46.30	0.30	270 000	5
125	NORBEC	48°21'12"	79°03'05"	32D/6	2.77		4.50	48.00	0.70	3 950 000	5
126	MILLENBACH	48°18'04"	79°03'15"	32D/6	3.46		4.33	56.20	1.00	3 560 000	5
127	LD-75	48°18'14"	79°03'39"	32D/6	2.8		4.2	60.00	1.71	90 000	8
128	LAC DUFAULT 2	48°18'27"	79°03'45"	32D/6	0.50		8.30	113.49	1.51	90 700	8
129	VAUZE	48°21'35"	79°04'50"	32D/6	2.90		0.94	24.00	0.70	354 000	5
130	CORBET	48°18'00"	79°04'55"	32D/6	3.0		1.96	21.00	1.00	2 780 000	5
131	AMULET (C ZONE)	48°19'00"	79°05'00"	32D/6	2.20		8.50	86.70	0.60	570 000	5
132	AMULET (LOWER A)	48°19'00"	79°05'00"	32D/6	5.14		5.30	44.20	1.40	4 690 000	5
133	OLD WAITE	48°20'24"	79°05'20"	32D/6	4.70	0.04	2.98	22.00	1.10	1 120 000	5
134	ANSIL	48°21'15"	79°07'00"	32D/6	7.22		0.94	26.50	1.60	1 580 000	5, 10
135	HUNTER	48°33'06"	79°08'24"	32D/11	1.06					438 988	3
136	ALDERMAC	48°13'12"	79°14'00"	32D/3	1.40		4.12	7.0	0.3	1 880 000	5
137	NEW INSCO	48°26'27"	79°21'06"	32D/6	2.59			20.57	0.9	886 139	3, 10
138	NORMETAL	49°00'18"	79°22'18"	32D/14	0.79	0.18	5.30	65.0	0.8	10 100 000	10
139	NORMETMAR	49°00'18"	79°22'18"	32E/3			11.72	7.06		571 410	3
140	MAGUSI RIVER	48°26'30"	79°22'24"	32D/6	1.20		3.55	31.20	1.10	3 727 770	3
141	POTTER	48°29'27"	80°11'50"	42A/9	15.20		4.15	92.5	1.54	1 758	13

Table 6.3-1. (cont.)

NO.	NAME	LATITUDE / LONGITUDE		NTS	Cu %	Pb %	Zn %	Ag (g/t)	Au (g/t)	TONNES	REFERENCE
142	ERRINGTON	46°32′15″	81°15′30″	41I/11	1.20	0.99	3.82	54.51	0.82	12 391 541	3
143	VERMILION LAKE	46°31′11″	81°21′20″	41I/11	1.26	0.90	3.92	48.00	0.86	5 410 003	3
144	KIDD CREEK	48°41′30″	81°22′00″	42A/11	2.20	0.28	7.25	147.43		117 547 200	13
145	LAKE GENEVA	46°47′35″	81°30′57″	41I/13		3.34	9.21	754.29	2.74	227 102	3
146	JAMELAND	48°34′00″	81°34′00″	42A/12	.99		.88	3.12	0.03	461 805	13, 8
147	CAN JAMIESON	48°30′55″	81°34′07″	42A/12	2.39		4.05	30.17	0.31	800 600	13, 8
148	KAM KOTIA	48°36′09″	81°36′45″	42A/5	1.09		1.03	3.39	0.03	6 007 194	13, 8
149	STRALAK	46°48′09″	81°41′50″	41I/13	0.50	0.50	3.18	68.57		680 250	3
150	SHUNSBY	47°42′48″	82°39′30″	41O/10	0.40		2.40			2 176 800	3
151	GECO	49°09′15″	85°47′40″	42F/4	1.86	0.15	3.45	50.06		58 400 000	8, 13
152	WILLROY	49°09′26″	85°48′30″	42F/4	1.64		2.84	27.77		3 949 985	8, 13
153	BIG NAMA CREEK	49°09′51″	85°50′46″	42F/4	0.83	0.02	4.16	35.66		180 493	8, 13
154	WILLECHO	49°10′30″	85°53′00″	42F/4	0.50	0.18	4.43	67.89		1 961 841	8, 13
155	ZENITH	48°58′40″	87°21′48″	42D/14	0.95		17.81	25.37	0.86	2 840 724	8, 13
156	WINSTON LAKE	49°00′10″	87°24′30″	42E/3	1.00		15.60	30.87	1.02	3 077 000	13
157	KENDON COPPER	50°25′20″	87°35′13″	42L/5	1.22		4.20	84.00	0.79	2 237 569	3
158	HEADVUE	50°01′14″	87°39′38″	42L/4		0	4.60	49.71		272 100	3
159	NORTH COLDSTREAM	48°36′05″	90°35′05″	52B/10	2.51					1 814 000	8
160	CREEK ZONE	49°52′54″	90°52′00″	52G/15	1.66	0.76	8.80	141.50		908 000	5,13
161	STURGEON LAKE	49°52′30″	90°52′00″	52G/15	2.55	1.21	9.17	164.20		2 070 000	5, 13
162	LYON LAKE	49°53′00″	90°52′00″	52G/15	1.24	0.63	6.53	141.50		3 945 000	5,13
163	F GROUP	49°52′30″	90°58′30″	52G/15	0.64	0.64	9.51	60.40		340 000	5,13
164	MATTABI S	49°52′30″	90°58′30″	52G/15	0.74	0.85	8.28	104.00		11 400 000	5,13
165	SOUTH BAY	51°06′30″	92°40′45″	52N/2	2.30		14.50	82.29		1 534 644	8
166	COPPER LODE (E ZONE)	50°57′54″	92°52′48″	52K/15	0.60		4.36	13.71		306 799	3
167	COPPER LODE (MAIN ZONE)	50°58′48″	92°57′18″	52K/15	1.01			19.54		774 584	3
168	TROUT BAY CU	51°00′10″	94°12′20″	52M/1	1.50	0.24	7.80	58.29	0.24	125 643	3
169	HENINGA GEMEX	61°46′25″	96°12′10″	65H/16	1.30		9.00	68.57	1.03	5 442 000	3
170	RUTTAN LAKE	56°28′00″	99°28′00″	65B/5	1.28		1.40	7.54	0.38	69 839 000	14 (90)
171	OSBORNE LAKE	54°57′48″	99°43′42″	63J/13	3.14		1.50			3 380 000	3, 13
172	COPPER MAN	54°39′00″	99°52′24″	63J/12	2.63		4.46			221 308	3, 13
173	ROD 2	54°51′38″	99°55′12″	63J/13	7.23		3.08			688 000	13
174	LINDA	54°50′00″	99°56′00″	63J/13	0.30		0.80	10.29	1.71	11 791 000	3
175	STALL LAKE	54°51′34″	99°56′50″	63J/13	4.42		0.50	11.66	1.47	6 513 000	8, 13
176	KNOBBY(MCBRIDE)	56°53′06″	99°55′00″	64B/13	0.35		8.77	11.66	1.47	1 819 666	3
177	ANDERSON LAKE	54°51′00″	100°00′30″	63K/16	3.46		0.10	11.66	1.47	3 354 000	15, 13
178	MORGAN LAKE	54°45′42″	100°13′06″	63K/16			16.00		3.43	362 800	3
179	JOANNIE	54°49′42″	100°02′00″	63K/16	1.28					394 545	8, 3
180	WIM	55°01′30″	100°02′36″	63N/1	2.91			8.23	1.71	989 000	3
181	GHOST LAKE	54°49′10″	100°04′00″	63K/16	1.38	0.75	8.50	44.91	1.65	646 000	15, 13
182	CHISEL LAKE NORTH	54°50′05″	100°06′05″	63K/16	0.30	0.40	9.00			2 457 000	13
183	LOST LAKE	54°49′47″	100°06′36″	63K/16	1.45	1.00	4.90	89.14	3.22	76 188	8
184	CHISEL LAKE	54°49′43″	100°07′00″	63K/16	0.50	1.40	10.90	56.23	2.40	7 490 000	15, 13
185	DYCE SIDING	54°24′24″	100°09′00″	63K/8	2.08		1.60			1 197 240	3
186	POT LAKE	54°46′48″	100°10′54″	63K/16	1.43		4.50	18.86	3.77	101 584	3
187	BOMBER(COOK L)	54°51′18″	100°11′00″	63K/16	0.04		1.00	8.57	0.10	620 388	3
188	SPRUCE POINT	54°34′30″	100°24′00″	63K/9	2.70		4.30	32.57	1.92	1 763 000	13

Table 6.3-1. (cont.)

NO.	NAME	LATITUDE / LONGITUDE		NTS	Cu %	Pb %	Zn %	Ag (g/t)	Au (g/t)	TONNES	REFERENCE
189	DICKSTONE	54°51′18″	100°29′24″	63K/16	2.42		3.17	12.69	0.69	1 083 000	3, 13
190	NORRIS LAKE	54°53′00″	100°30′18″	63K/15	2.51		4.82	20.9	0.55	226 750	3
191	REED LAKE	54°38′12″	100°32′54″	63K/10	1.30					1 361 000	3, 13
192	RAIL LAKE	54°44′54″	100°35′30″	63K/10	3.00		0.70			295 000	3, 13
193	JUNGLE LAKE	55°09′48″	100°58′18″	63N/2	1.42		1.10			3 355 900	3
194	DH FL GROUPS	56°49′00″	101°01′30″	64C/14	0.91		1.86			513 362	3
195	Z DEPOSIT	56°49′42″	101°01′30″	64C/14	1.11		2.49		0.55	138 771	3
196	BOB LAKE	55°09′30″	101°02′30″	63N/3	1.33		1.18	9.26	0.34	2 158 660	3
197	SHERRIDON	55°06′39″	101°05′00″	63N/3	2.75		3.39	30.86	0.69	7 018 366	8
198	VAMP LAKE	54°56′18″	101°10′05″	63K/14	1.34		1.90	13.03	3.98	739 205	3
199	NORTH STAR	54°46′00″	101°35′00″	63K/13	6.10			8.57	0.34	242 000	15
200	DON JON	54°46′00″	101°35′00″	63K/13	3.09			15.09	0.93	79 000	15
201	COPPER REEF	54°36′45″	101°36′24″	63K/12	1.50		0.50			453 500	16
202	FOX LAKE	56°38′00″	101°37′00″	64C/12	1.81		1.77	4.46	0.17	10 799 649	8
203	PINEBAY CU	54°45′48″	101°37′18″	63K/13	1.3					1 361 000	13, 15
204	CENTENNIAL	54°42′03″	101°39′59″	63K/12	1.56		2.20	26.40	1.51	2 366 000	13, 15
205	CUPRUS	54°43′45″	101°42′13″	63K/12	3.24		6.40	31.89	1.51	462 000	13, 15
206	WHITE LAKE	54°42′40″	101°43′30″	63K/12	1.97	0.50	4.63	36.00	0.69	850 000	13, 15
207	TROUT(EMBURY)	54°49′45″	101°49′18″	63K/13	1.80		5.80	11.20	1.45	6 600 000	7 (19/11/90)
208	SCHIST LAKE	54°43′00″	101°49′50″	63K/12	4.30	0.03	7.25	39.43	1.41	1 871 000	13, 15
209	MANDY	54°43′45″	101°50′00″	63K/12	8.03		15.05	61.71	3.09	150 000	13
210	WESTARM	54°38′36″	101°50′12″	63K/12	3.70		1.50			1 702 000	13, 15
211	LAR(LAURIE)	56°38′06″	101°52′48″	64C/12	0.80		2.15			1 360 500	3
212	FLIN FLON	54°45′28″	101°53′00″	63K/13	2.20		4.10	43.20	2.85	62 927 000	13, 15
213	CALLINAN	54°47′03″	101°53′08″	63K/13	1.5		4.1	11.20	1.45	2 619 000	13, 15
214	CORONATION	54°35′00″	102°00′00″	63K/12	4.25		0.24	5.66	2.26	1 282 498	8
215	FLEXAR	54°40′37″	102°01′43″	63L/9	4.08		0.44	5.49	1.17	305 659	8
216	BIRCH LAKE	54°39′41″	102°01′58″	63L/9	6.17			4.53		278 449	8
217	SCHOTTS LAKE	55°05′45″	102°13′35″	63M/1	0.61		1.35			1 982 702	3
218	RAMSAY SHOWING	54°44′17″	102°45′05″	63L/10	2.16		1.77	6.96		738 884	3
219	HANSON(MCILVENNA)	54°37′30″	102°51′00″	63L/10	0.95		5.76	30.17	0.69	8 888 600	12
220	BIGSTONE	54°34′58″	103°11′40″	63L/11	1.87		1.11	9.94	0.45	3 628 000	3
221	MCKENZIE(PEG)	56°07′20″	103°42′04″	64D/4	0.55		4.84	0.18	18.63	3 695 000	3
222	MUSK	65°19′20″	107°36′00″	76G/5	1.20	1.40	10.00	343.00		340 000	3
223	YAVA	65°36′10″	107°56′00″	76G/12	0.50	0.50	3.00	102.80	2.09	1 500 000	3
224	HACKETT (A ZONE)	65°55′00″	108°22′00″	76F/16	0.25	1.40	8.50	240.00	1.89	4 535 000	3
225	BOOT LAKE	65°54′55″	108°25′50″	76F/16	0.29	4.97	0.99	200.91	0.48	4 535 000	3
226	EAST CLEAVER	65°55′55″	108°27′30″	76F/16	0.47	0.94	4.08	108.34	0.48	7 256 000	3
227	HIGH LAKE	67°22′45″	110°51′19″	76M/7	3.53	0.20	2.46	37.71	0.79	4 722 749	3
228	KENNEDY LAKE	63°01′57″	110°56′57″	75M/2		1.10	7.30	137.49		39 001	3
229	INDIAN MTN(BB)	63°01′57″	110°56′57″	75M/2	0.20	0.70	9.54	116.57		879 790	3
230	GONDOR	65°33′43″	111°48′00″	76E/12	0.50	0.50	6.00	50.00		7 500 000	3
231	SUNRISE LAKE	62°54′06″	112°22′30″	85I/16	0.02	4.20	8.90	404.57	0.96	1 865 699	3, 13
232	BEAR	62°53′36″	112°23′30″	85I/16		2.32	6.11	219.77	0.89	809 700	3, 13
233	HOOD RIVER #10	66°03′35″	112°45′15″	86I/2	5.00		3.50	34.29		453 500	3
234	IZOK LAKE	65°37′50″	112°47′50″	86H/10	2.85	1.46	14.40	75.09		9 795 600	3
235	HOOD RIVER #41	66°02′30″	112°48′00″	86I/2	1.57	0.20	4.12	17.83		272 100	3
236	GOLDSTREAM RIVER	51°37′00″	118°13′00″	82M/9	4.81		3.08	20.60		2 287 886	17
237	REA GOLD	51°09′00″	119°49′00″	82M/4	0.57	2.14	2.25	73.37	6.51	242 822	3

Table 6.3-1. (cont.)

NO.	NAME	LATITUDE / LONGITUDE		NTS	Cu %	Pb %	Zn %	Ag (g/t)	Au (g/t)	TONNES	REFERENCE
238	SAMATOSUM	51°09'40"	119°49'00"	82M/4	1.10	1.50	2.50	685.00	1.40	483 000	7 (01/04/91)
239	HOMESTAKE	51°06'27"	119°49'45"	82M/4	0.55	2.50	4.00	240.00		919 296	17
240	CHU CHUA	51°22'10"	120°03'30"	92P/8	2.00		0.50	8.00	0.50	3 000 000	3
241	SENECA	49°19'00"	121°56'37"	92H/5	0.63	0.15	3.60	41.14	0.82	1 506 334	3
242	BRITANNIA	49°36'40"	123°08'30"	92G/11	1.90		0.65	6.86	0.69	48 815 534	3, 17
243	TEDI (BRANDY)	50°05'00"	123°08'30"	92J/3	0.22	2.07	2.04	126.17	0.34	132 255	3
244	TWIN J	48°52'00"	123°47'00"	92B/13	1.60	0.65	6.60	140.57	4.11	594 992	3
245	LARA (HOPE)	48°52'50"	123°54'20"	92B/13	1.01	1.22	5.87	100.11	4.73	529 000	17
246	SUNRO	48°27'00"	124°01'00"	92C/8	1.47			1.37	0.14	2 032 587	3
247	WESTMIN (HW)	49°34'20"	125°35'10"	92F/12	2.20	0.30	5.30	37.71	2.40	13 815 424	18
248	WESTMIN(L,M,P)	49°34'32"	125°36'07"	92F/12	1.50	1.10	7.60	109.71	2.06	5 646 075	18
249	KUTCHO CREEK	58°12'10"	128°21'40"	104I/7	1.76	0.06	2.54	35.00	0.37	14 300 000	3, 7 (25/03/91)
250	ECSTALL RIVER	53°52'25"	129°30'40"	103H/13	0.86	0.20	2.30	24.34	0.69	4 535 000	3, 17
251	HIDDEN CREEK(ANYOX)	55°26'33"	129°49'35"	103P/5	0.90			8.23		23 721 524	3, 17
252	BONANZA(ANYOX)	55°23'45"	129°51'15"	103P/5	1.88					882 456	3, 17
253	DOUBLE ED(ANYOX)	55°24'45"	129°52'35"	103P/5	1.00		0.60			3 628 800	3
254	RED WING	55°22'50"	129°53'00"	103P/5	1.84			29.49		181 400	3
255	EDEN	55°25'40"	129°53'30"	103P/5	2.00					226 800	3
256	GRANDUC	56°12'45"	130°20'30"	104B/1	1.79	0.02	0.10	10.63	0.17	25 063 140	8, 17
257	ESKAY CREEK	56°37'01"	30°27'55"	104B/9		2.20	5.40	998.40	26.40	3 954 520	7 (31/12/90)
258	SWIM LAKE	62°12'50"	133°01'50"	105K/3		3.80	4.70	50.4		4 308 250	7 (25/03/91)
259	TULSEQUAH CHIEF	58°44'20"	133°34'30"	104K/12	1.60	1.31	7.02	100.50	2.74	6 193 609	3, 8, 17
260	BIG BULL	58°38'24"	133°35'24"	104K/12	1.32	1.31	6.06	113.14	3.15	4 107 803	8
261	MARG	64°00'40"	134°28'20"	106D/1	1.90	2.60	4.99	64.11	0.89	2 857 050	3, 7 (3/12/90)
262	HART RIVER	64°38'00"	136°51'00"	116A/10	1.45	0.87	3.65	49.71	1.41	1 067 539	3
263	WINDY CRAGGY	59°44'00"	137°45'00"	114P/12	1.59			3.60	0.20	113 000 000	3, 17

[1]**Evans, D.T.W., Swinden, H.S., Kean, B.S., and Hogan, A.**
1992: Metallogeny of the vestiges of Iapetus, Island of Newfoundland; Newfoundland Department of Mines and Energy, Map 92-19, scale 1:500 000.

[2]**Swinden, H.S. and Kean, B.F. (ed.)**
1988: The Volcanogenic Sulphide Districts of Central Newfoundland; Geological Association of Canada Guidebook, 250 p.

[3]**Energy, Mines and Resources Canada**
1989: Canadian Mineral Deposits Not Being Mined in 1989; Mineral Policy Sector, Energy Mines and Resources Mineral Bulletin, MR223.

[4]**Swanson, E.A., Strong, D.F., and Thurlow, J.G. (ed.)**
1981: The Buchans Orebodies: Fifty Years of Geology and Mining; Geological Association of Canada, Special Paper 22, 350 p.

[5]**Gibson, H.L. and Kerr, D.J.**
1993: Giant volcanic-associated massive sulphide deposits: with emphasis on Archean examples; in Giant Ore Deposits (ed.) B.H. Whiting, R. Mason, and C.J. Hodgson; Society of Economic Geologists Special Publication No. 2, p. 319-348.

[6]**Anon**
1993: Canadian Mines Handbook 1992-93; (ed.) D. Giancola; Northern Miner Press, Toronto, Ontario.

[7]**Northern Miner newspaper** (issue in brackets)

[8]**National Mineral Index**, Geological Survey of Canada, Room 652, 601 Booth Street, Ottawa, Ontario.

[9]**Fyffe, L.R., van Staal, C.R., and Winchester, J.A.**
1990: Late Precambrian-early Paleozoic volcanic regimes and associated massive sulfide deposits in northeast mainland Appalacians; The Canadian Mining and Metallurgical Bulletin, v. 83, no. 938, p. 70-78.

[10]**Rive, M., Verpaelst, P., Gaganon, Y., Lulin, J.M., Riverin, G. and Simard, A. (ed.)**
1990: The Northwestern Quebec Polymetallic Belt; The Canadian Institute of Mining and Metallurgy, Special Volume 43, 423 p.

[11]**Mining Journal magazine** (London) (issue in brackets).

[12]**Mining Annual Review**, 1990: Mining Journal Press, London

[13]**International Association for the Genesis of Ore Deposits, Eighth Symposium, Ottawa,**
1990: Guidebooks:
2158 - Lithotectonic and associated mineralization of the eastern extremity of the Abitibi Greenstone belt
2159 - Mineral deposits of Noranda, Quebec, and Cobalt, Ontario
2161 - Geology and ore deposits of the Timmins district, Ontario
2164 - Mineral deposits in the western Superior Province, Ontario
2165 - Geology and mineral deposits of the the early Proterozoic Flin Flon Belt and Thompson Belt, Manitoba
2168 - Mineral deposits of the Slave Province, N.W.T.

[14]**Manitoba Energy and Mines Report of Activities** (issue in brackets)

[15]**Bailes, A.H., Syme, E.C., Galley, A., Price, D.P., Skirrow, R., Ziehlke, D.J. (ed.)**
1987: Early Proterzoic Volcanism, Hydrothermal Activity, and Associated Ore Deposits at Flin Flon and Snow Lake, Manitoba; Geological Association of Canada/Mineralogical Association of Canada guidebook, 95 p.

[16]**Gale, G.H., Baldwin, D.A., and Koo, J.**
1980: Geological evaluation of Precambrian massive sulphide deposit potential in Manitoba; Manitoba Energy and Mines, Economic Geology Report 79-1, 137 p., 23 maps.

[17]**McMillan, W.J., Howy, T., MacIntyre, D.G., Nelson, J.L., Nixon-Graham, T., Hammack, J.L., Panteleyev, A., Ray, G.E., and Webster, I.C.L.**
1991: Ore deposits, teconics and metallogeny in the Canadian Cordillera; British Columbia, Ministry of Energy, Mines and Petroleum Resources, Paper 1991-4, 276 p.

[18]**Fleming, J., Walker, R., and Wilton, P. (ed.)**
1983: Mineral Deposits of Vancouver Island: Westmin Resources (Au-Ag-Cu-Pb-Zn), Island Copper (Cu-Au-Mo), Argonaut (Fe); Geological Association of Canada/Mineralogical Association of Canada guidebook.

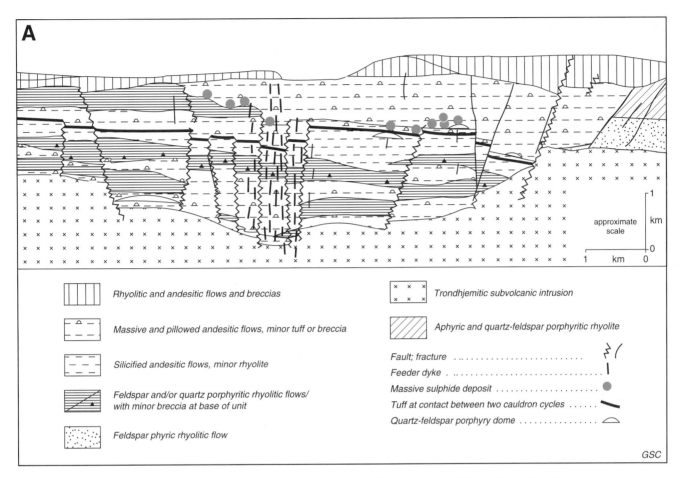

Figure 6.3-4A. Cross-section of the main caldera, Noranda massive sulphide district (after Gibson and Watkinson, 1990). This section is oriented approximately north-south, and is based on information from mines and drill holes.

In contrast, deposits in sediment-covered areas are tabular, circular to elliptical in plan, and commonly display distinctive layering or bedding within the sulphide zones. The sulphides are locally interbanded with silicate layers. Some parts of the Besshi-type deposits are composed of massive ore. Lateral dimensions of more than a kilometre are common. Ore thickness is typically a few tens of metres. Iron-formations (both oxide and silicate facies) are common at the ore horizon (e.g., the Vasskis of Norway). Alteration zones are not as pronounced as for the deposits in more volcanic-dominated areas, but iron- and magnesium-enriched zones, containing disseminated sulphides, underlie many of the Caledonide deposits, such as those near Bathurst, New Brunswick.

Volcanic rock-dominated deposits are commonly overlain by sedimentary strata, composed of both clastic and chemical components. These strata may have considerable lateral continuity; the Key Tuffite horizon, coincident with or immediately overlying all of the deposits on both limbs of the Mattagami Lake "anticline", is a laminated chert-pyrite unit containing some volcaniclastic fragments

(Piche et al., 1990). Graphitic shale overlies the Kidd Creek orebody. This shale may represent the residues of vent-specific bacterial colonies, but does not extend far beyond the limits of the deposit. Tuff horizons are associated with some of the deposits in the Noranda district (e.g. Ansil, Corbet, and the Amulet bodies). Iron-formation is not common, but is prominently associated with the hanging wall of the Manitouwadge, Ontario deposits, and the footwall to the Lyon Lake and Creek Zone deposits near Sturgeon Lake, Ontario. Iron oxide zones (ochre) occur at the top of some of the ophiolite-associated deposits, and in some cases may be the product of seafloor oxidation of sulphides, as observed on the Mir mound (TAG), Mid-Atlantic Ridge (Hannington and Jonasson, 1992).

Mineralogical composition, textures, and structures within orebodies

Virtually all massive sulphide deposits are mineralogically simple. They contain at least 50%, and commonly more than 80% sulphides by volume (Sangster and Scott, 1976).

B

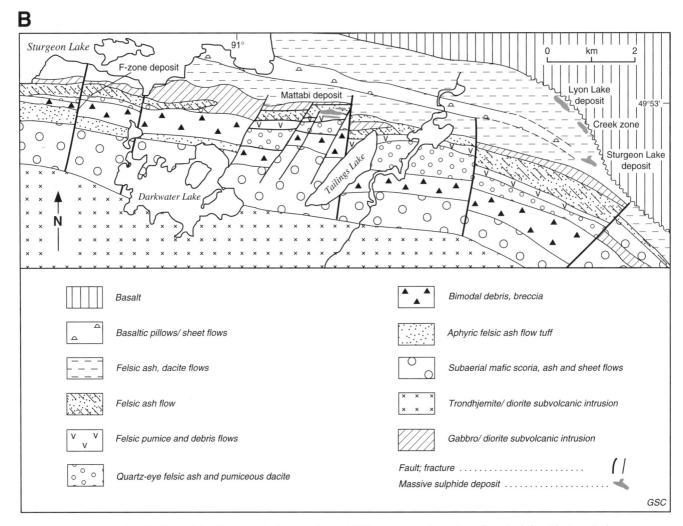

Figure 6.3-4B. Map of the Sturgeon Lake massive sulphide camp, northwestern Ontario (after Morton et al. 1990). This map represents closely a cross-section through the Sturgeon Lake caldera, and is at about the same scale as Figure 6.3-4A. Both diagrams closely reflect a right-section through their respective districts. Note that both districts have subvolcanic intrusions at their base, and subequal amounts of felsic and mafic volcanic rock. However, the Noranda district is dominated by felsic flows, whereas Sturgeon Lake is dominated by subaerial mafic scoria deposits, and subaqueous pyroclastic flows and breccia.

Pyrite typically constitutes 50-90% of the massive ore, with sphalerite, chalcopyrite, and galena forming about 10%. Some metamorphosed deposits, such as those at Manitouwadge, Ontario, contain abundant pyrrhotite (Kissin, 1974). Deposits formed in deep water (Noranda type) contain only sphalerite and chalcopyrite as their principal ore minerals, but those that formed in shallow water typically also contain recoverable galena. Curiously, barite occurs in the oldest deposits in Australia and South Africa (pre-3.0 Ga) and in some Phanerozoic deposits, but is much less common in late Archean and Proterozoic deposits. Massive magnetite constitutes about one-third of the Ansil deposit, Quebec (Riverin et al., 1990).

Gangue minerals are poorly documented, and consist primarily of quartz, with chlorite, sericite, and alumino-silicate minerals (or their metamorphic equivalents)

predominant. Gahnite is an accessory mineral at virtually all deposits that have attained amphibolite grade of metamorphism.

The mineral assemblage of the stringer ores is usually also very simple, with chalcopyrite, pyrite, pyrrhotite, sphalerite, and magnetite present. One orebody and the stringer zone at the Kidd Creek mine have a local bornite zone with an unusual proliferation of selenium minerals (Thorpe et al., 1976). Tellurides occur beneath the Mattagami Lake deposit (Thorpe and Harris, 1973).

Within the deposits in volcanic-dominated areas, the well-known pattern of high $Cu/(Cu+Zn)$ ratios in the lower zones and lower ratios in the upper zones is so common that Sangster (1972) suggested that it can be used reliably as an indicator of stratigraphic facing. The massive zones are

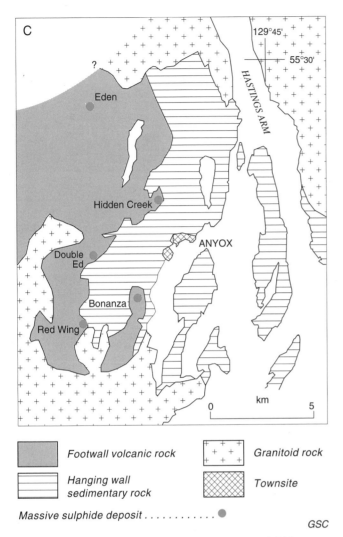

Figure 6.3-4C. Geology of the Anyox massive sulphide area, northern British Columbia. Note that most of the deposits are at or near the contact between hanging wall sedimentary and footwall volcanic rocks (after Alldrick, 1986).

iron-rich (pyritic) at the tops and flanks of the deposit. Some stringer zones exhibit a narrow zinc-rich fringe zone (Purdie, 1967; Simmons, 1973) and a copper-rich core. Copper-zinc ratios within some of the less deformed orebodies (e.g., Millenbach deposit) indicate the presence of a distinct copper-rich spine (Knuckey, 1975) that is transgressive to bedding. This "spine" is present even where the pyrite-sphalerite zone is brecciated and locally transported, indicating that it "grew" after emplacement of the zinc ore. The spine is composed of massive chalcopyrite and pyrite; pyrrhotite is more abundant in the laterally adjacent pyrite-sphalerite ore. The ophiolite deposits, and some of the deposits in which sedimentary rocks are important, are usually copper-rich (Cu/Zn = 3/1 to 4/1). Some of the latter group of deposits have high (~1000 ppm) cobalt contents.

Layered (and possibly bedded) sulphide is present at the tops and peripheral parts of many of the massive sulphide mounds. The central parts of the orebodies are paragenetically complex, with zoned "veins" or pseudo-beds of sulphide following fractures and joints within massive sulphides, as at Mine Gallen (Watkinson et al., 1990). The margins of the massive bodies may consist of sulphide breccia, possibly formed from fallen chimney structures.

Alteration beneath Cu-Zn massive sulphide deposits

Alteration has been studied more extensively than most attributes of these deposits. Alteration mineral assemblages and associated chemical changes have been very useful exploration guides. Alteration occurs in two distinct zones beneath these deposits (Fig. 6.3-6 and 6.3-4A). **Alteration pipes** occur immediately below the massive sulphide zones; here a complex interaction has occurred between the immediate substrata to the deposits and both ore-forming (hydrothermal) fluids and locally-advecting seawater. **Lower, semiconformable alteration zones** (Franklin et al., 1981) occur several hundreds of metres or more below the massive sulphide deposits, and may represent in part the "reservoir zone" (Hodgson and Lydon, 1977) where the metals and sulphur were leached (Spooner and Fyfe, 1973) prior to their ascent to and expulsion onto the seafloor.

Beneath Precambrian deposits formed in deep water (Noranda type), alteration pipes typically have a chloritic core, surrounded by a sericitic rim. Some, such as at Mattagami Lake, contain talc, magnetite, and phlogopite (Costa et al., 1983). The pipes usually taper downwards within a few tens of metres to hundreds of metres below the deposits, to a fault-controlled zone less than a metre in diameter. Beneath deposits formed in shallow water (Mattabi type; Fig. 6.3-7), the pipes are silicified and sericitized; chlorite is subordinate and is most abundant on the periphery of the pipes. Aluminosilicate minerals are prominent.

Alteration pipes under Phanerozoic Cu-Zn deposits are similar to, but more variable than those under their Precambrian counterparts. For example, Aggarwal and Nesbitt (1984) described a talc-enriched alteration core, surrounded by a silica-pyrite alteration halo, beneath the Chu Chua, British Columbia deposit. The Newfoundland, Cyprus, Oman, and East Galapagos Ridge deposits have Mg-chlorite in the peripheral parts of their pipes, together with illite. The presence of abundant Fe-chlorite, quartz, and pyrite typify the central parts of the pipes.

The lower semiconformable alteration zones have been recognized under deposits in several massive sulphide districts (Galley and Jonasson, 1992). These include laterally extensive (several kilometres of strike length) quartz-epidote zones that are several hundred metres thick, and extend downwards from a few hundred metres stratigraphically below the Noranda, Matagami, and Snow Lake deposits of the Canadian Shield. Zones containing epidote, actinolite, and quartz in the lower pillow lavas and sheeted dykes of the ophiolite sequences at Cyprus (Gass and Smewing, 1973) and in East Liguria, Italy, were explained by Spooner and Fyfe (1973) as being due to increased heat flow as a result of convective heat transfer away from the cooling intrusions at the base of these sequences. All of the

171

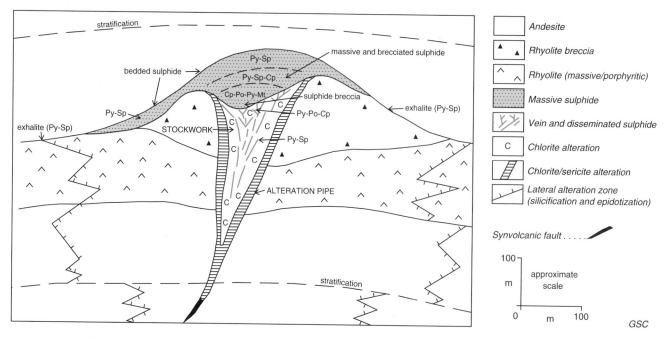

Figure 6.3-5. Summary cross-section of a typical Noranda type deposit. Py: pyrite; Sp: sphalerite; Cp: chalcopyrite; Po: pyrrhotite; Mt: magnetite.

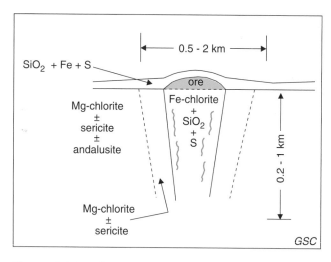

Figure 6.3-6. Synthesis of alteration zones immediately beneath Cu-Zn deposits, such as at Noranda, Cyprus, and the Proterozoic deposits of the Canadian Shield. Note that the "Mg-chlorite±sericite" zone is gradational into both the central Fe-chlorite+SiO₂+S and the outermost andalusite zones.

epidote-quartz zones are metal depleted. They represent the zone of high temperature hydrothermal reaction (ca. 400°C), under low water-rock ratio conditions, where the metals and sulphur entered into the ore-forming solution (Spooner, 1977; Spooner et al., 1977a, b; Richards et al., 1989).

Zones of alkali-depleted, variably carbonatized strata occur immediately beneath some deposits in sedimentary rocks, and also in volcanic rocks under the Mattabi-type

deposits. These may extend for as much as tens of kilometres along strike, and occur in the upper few hundred metres of the footwall. These zones probably represent a sealed cap to the hydrothermal reservoir, and formed through progressive heating of downward percolating seawater, with some possible input of CO_2 from an underlying magma chamber, or by thermal breakdown of organic compounds in the footwall.

Zinc-lead-copper group

These deposits, most commonly Phanerozoic in age, occur primarily in arc-related terranes where felsic volcanic rocks, with or without associated sedimentary strata, are dominant (Fig. 6.3-8). The deposits range from those in felsic volcanic-dominated terranes, such as those in the Buttle Lake area of Vancouver Island and the Buchans area of Newfoundland, to those with thick sedimentary sequences in their footwall section, such as the deposits of the Bathurst district, New Brunswick. Deposits of the latter group are similar in many of the characteristics of their geological settings to the sediment-associated deposits in the Cu-Zn group. Notably, however, the deposits in the Zn-Pb-Cu group have little or no mafic volcanic rock in their footwall sequences. Modern examples of this group have been discovered in the Lau Basin (Von Stackelberg and Brett, 1990) and the Okinawa Trough (Urabe et al., 1990).

Geological setting

Volcanic rock-dominated areas

The best known major district of this type is the Hokuroku district of Japan. This area of the Green Tuff Belt is a 13 Ma old basin containing a bimodal suite of island arc-related volcanic rocks and mudstone. Extensive reviews of the

"Kuroko deposits" of this district (Tatsumi, 1970; Tatsumi and Watanabe, 1971; Ishihara and Terashima, 1974; Ohmoto and Skinner, 1983) serve as a comparative model for similar deposits in Canada, at Buchans, Newfoundland and Buttle Lake, British Columbia (Juras and Pearson, 1990), as well as for deposits in the Tasman Geosyncline (Lambert, 1979). The Canadian and Australian deposits of this group are in much more deformed terranes, however, and some characteristics of the orebodies are best exemplified in the Hokuroku district.

The footwall sequences to these deposits are composed primarily of calc-alkaline felsic porphyritic ash-flow tuff, rhyolite domes and flows, and some felsic epiclastic rocks. Basalt may occur near the base of the sequence (Buchans and Buttle Lake). Most of the host strata probably were deposited in submarine calderas. The deposits formed during periods of extension following island arc construction, and may be temporally related to caldera resurgence.

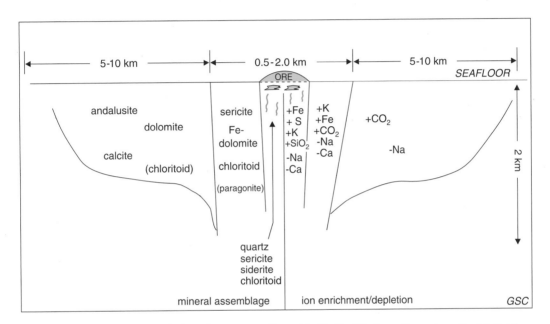

Figure 6.3-7. Synthesis of alteration zones adjacent to Zn-Cu-Pb deposits such as the Mattabi deposit, Ontario and others that formed in sequences dominated by felsic tuff, breccia, and mafic scoria and sheet flows. Adjacent to the main alteration pipe, the alteration zones are not well defined. Sodium depletion is laterally extensive but confined to a few hundred metres vertically. Metasomatic carbonate alteration is pervasive in the footwall. Minerals in brackets are present in minor amounts.

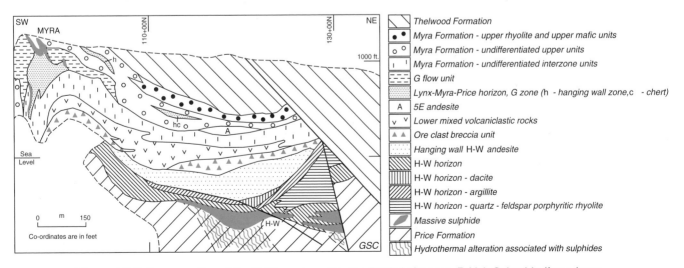

Figure 6.3-8. Cross-section of the H-W ore deposit in the Buttle Lake area, British Columbia (from Juras and Pearson 1990). This is a typical Kuroko-like deposit.

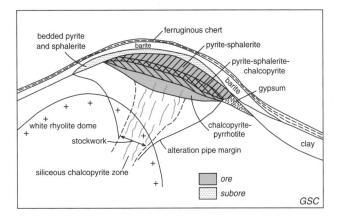

Figure 6.3-9. Typical cross-section of Pb-Zn-Cu ore zone illustrating the principal ore zones. Compiled from Eldridge et al. (1983), Lambert and Sato (1974), and others.

Volcano-sedimentary areas

The largest deposits of this group are in lower Carboniferous rocks of the Iberian pyrite belt in southern Portugal and southwestern Spain. The best Canadian examples are in the Bathurst district, New Brunswick; the Sudbury Basin, Ontario; the Omineca crystalline belt, British Columbia; and in several enclaves in the Coast Batholith in British Columbia. The great lateral continuity of ore, lack of extensive alteration, and association with sedimentary rocks make deposits of this group similar to arc-related deposits of the Cu-Zn group. However, the Zn-Pb-Cu deposits are usually in sequences in which felsic volcanic rocks predominate near the ore horizon. Deposits in the Bathurst camp, for example, have a few hundred metres of felsic ash flow tuff as their most common immediate footwall rock, underlain by thousands of metres of greywacke and pelite. Iron-formation is the immediate hanging wall to 7 of the 30 deposits in this camp. Basalt is restricted to the hanging wall sequence of most of the deposits. Van Staal (1987) has established that the Tetagouche rocks are part of a series of thrust slices, which include ophiolite-associated slices near the base, and arc-like volcanic rocks in the uppermost slice. Van Staal interpreted the strata containing the massive sulphide deposits to have formed as part of a back-arc basin sequence, immediately southeast of a magmatic arc. The Iapetus ocean floor separated this arc from the North America craton, further to the northwest.

Form and composition

Deposits in volcanic-dominated areas are typically well zoned bulbous massive bodies, underlain by variably developed stringer ore (Fig. 6.3-9). Lambert and Sato (1974) described the following zones for the Kuroko deposits: 1) Siliceous ore (keiko): pyrite-chalcopyrite-quartz stockwork ore; 2) "Gypsum ore (sekkoko): gypsum-anhydrite-(pyrite-chalcopyrite-sphalerite-galena-quartz-clays) stratabound ore; 3) "Pyrite ore (ryukako): pyrite-(chalcopyrite-quartz) ore"; 4) "Yellow ore (oko): pyrite-chalcopyrite-(sphalerite-barite-quartz) stratiform ore"; 5) "Black ore (kuroko): sphalerite-galena-chalcopyrite-pyrite-barite stratiform ore." Towards the top

of the zone, tetrahedrite-tennantite is common. Bornite occurs in a few deposits. 6) "Barite ore: thin well-stratified bedded ore consisting almost entirely of barite, but sometimes containing minor amounts of calcite, dolomite, and siderite"; 7) "Ferruginous chert (tetsusekiei bed): a thin bed of cryptocrystalline quartz and hematite." Similar zoning is present at many Canadian examples, including the Buttle Lake deposits (Juras and Pearson, 1990).

Mechanically-transported breccias are commonly associated with this type of deposit, and are particularly prominent in the Buchans, Newfoundland deposits (Binney, 1987), where they constitute about 50% of the ore. In most deposits, the breccia formed as a sulphide talus at the base of the endogenous mounds that were presumably primary vent sites. Lateral transport of the Buchans sulphide breccias for hundreds of metres or more indicates that these deposits formed on steep slopes, or were rapidly uplifted immediately after deposition.

Orebodies in sediment-dominated terranes, such as those in the Bathurst district, New Brunswick, are tabular and laterally extensive. These deposits exhibit primary metal zoning, although less prominent than their volcanic-hosted counterparts. Luff (1977) described three zones at Brunswick # 12. Unit 1, at the bottom and south of the ore, consists of massive pyrite with minor sphalerite and galena, and variable but locally significant amounts of chalcopyrite and pyrrhotite. Unit 2, the main zone, is distinctly layered sphalerite-galena-pyrite ore with minor chalcopyrite and pyrrhotite. Unit 3, the upper pyrite unit, consists of massive, fine grained pyrite with minor sphalerite, galena, and chalcopyrite.

Iron-rich rocks are common in the immediate hanging wall sequence of deposits in both the volcanic- and sediment-associated deposits of this compositional group. These are of two distinctly different genetic types. Some ferruginous strata have formed through seafloor weathering of sulphides, forming poorly bedded, massive oxide zones. Others are ferruginous (and usually cherty) precipitates which formed from low-temperature hydrothermal fluids. The tetsusekiei zone that caps many of the Kuroko deposits, and the manganiferous iron-formation that overlies the Bathurst deposits is "Algoma type" (Gross, 1965). These are probably not a product of oxidation of sulphides, but may have precipitated from low-temperature fluids that were associated with the terminal stage of their respective hydrothermal events.

Mineralogical composition, textures, and structures within orebodies

Deposits of this group are more mineralogically complex than those of the copper-zinc group. In addition to pyrite, sphalerite, galena, and chalcopyrite, barite is common, particularly in the volcanic-associated deposits such as Buchans and Buttle Lake. In the Bathurst district, barite is absent; sulphosalts, arsenopyrite, and stannite occur in minor amounts (Chen and Petruk, 1980). Nonsulphide minerals in the deposits in the Bathurst district include quartz (major mineral); chlorite, siderite, calcite, and dolomite are all present (Jambor, 1979). In the Bousquet district, Quebec, Au-rich sulphide deposits have similar mineralogical characteristics to this group, and may be highly deformed volcanic-associated massive sulphide

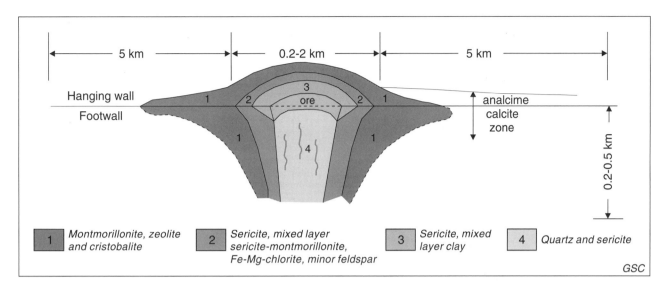

Figure 6.3-10. Alteration immediately associated with Zn-Pb-Cu deposits (after Shirozu, 1974).

deposits (Marquis et al., 1990). Most of the zinc-lead-copper deposits are fine grained, which causes recovery problems during processing. Most copper-zinc deposits, in contrast, are coarse grained and easily milled.

Alteration beneath zinc-lead-copper massive sulphide deposits

Alteration associated with Zn-Pb-Cu deposits, including Canadian deposits, is typified by that in the Hokuroku district of Japan (Fig. 6.3-10). The "lower semiconformable" alteration zones, such as those which underlie the Cu-Zn deposits, are unknown under these deposits. However, the lower strata are difficult to access in Japan, and structural complexities at Buchans and Buttle Lake, and in the Tasman geosyncline, may have removed these parts of their sequences.

Four alteration zones (Fig. 6.3-10) in the Hokuroko district have been described by Shirozu (1974), Iijima (1974), and Date et al. (1983). The most intense zone of alteration, zone 4, is immediately below the deposits, and consists of silicified, sericitized rock, with a small amount of chlorite. Zone 3 contains sericite, Mg-chlorite, and montmorillonite, and is not silicified. Feldspar is absent from zones 3 and 4. Zone 2 consists of sericite, mixed-layer smectite minerals, and feldspar. Zone 1 contains zeolite (typically analcime) as an essential mineral, as well as montmorillonite. Outside these four zones, the volcanic rocks have been affected by deuteric alteration, which formed clinoptilolite and mordenite. Under many older deposits, metamorphism and deformation obscure the alteration minerals associated with zones 1 to 3. At Buttle Lake, for example, zone 4 alteration is most prominent; carbonate is also present, in contrast to the Hokuroko deposits. Although chlorite is much less abundant under Zn-Pb-Cu deposits than under Cu-Zn deposits, the alteration

pipe under the Woodlawn deposit (New South Wales, Australia) is highly chloritic (Petersen and Lambert, 1979). Alteration under the sediment-associated deposits of this group consists of locally distributed sericite-quartz; many deposits do not have obvious alteration zones.

GENETIC MODEL

The basic process of formation of massive sulphide deposits, as syngenetic accumulations on or near the seafloor, of sulphide and sulphate minerals from hydrothermal fluids has been well established since the late 1950s (Oftedahl, 1958). Their stratiform nature, sharp upper contact, close association with immediately overlying, well-bedded chemical sedimentary rocks that contain abundant transported products from the hydrothermal vents, beds of locally transported sulphide breccia, and extensive alteration and stringer zones confined almost exclusively to the stratigraphic footwall of the massive sulphides, all gave credibility to this model. Discovery of active seafloor vent systems from which massive sulphide deposits are forming has since ended any doubt about the applicability of the general model. Many of the genetic aspects have been considered in depth by Lydon (1988), and are summarized here. The basic tenets of the genetic model apply to both the Cu-Zn and Zn-Pb-Cu groups of deposits (Fig. 6.3-11).

Source of fluid

Two possible sources of fluid are a) circulating seawater and b) magmatic water. The hydrothermal fluids emanating from modern seafloor vents are considered to be seawater, that has been modified by progressive heating and water-rock interactions during its descent into heated crust, leading to loss of Mg, Sr, and some Ca (due to formation of magnesium hydroxysulphate, anhydrite, zeolites, and clay minerals), and eventual reaction with deeper parts of the crust under high temperature conditions. Excluding the hypersaline brines of the Red Sea (7x sea-water), where the fluid has interacted with Miocene evaporites, salinities of mid-ocean ridge hydrothermal fluids vary from 20% to 150% of seawater for the cooled vapour-phase and brine fluids respectively, at Axial Seamount, to close to seawater values for fluids at the majority of the East Pacific Rise and central and northern Juan de Fuca vent areas, and to almost twice seawater at the southern Juan de Fuca site (Von Damm and Bischoff, 1987). Experimental data (e.g. Bischoff and Seyfried, 1978; Seyfried and Janecky, 1985) together with investigations of the silicified and epidotized lower semiconformable alteration zones (e.g. Spooner and Fyfe, 1973; Gibson et al., 1983), indicate that the reaction between modified seawater and basalt at about 385°C causes sufficient depression of pH to mobilize metals and sulphur from the rocks into the fluid. Virtually any volcanic rock or immature sediment contains sufficient metals to produce a viable ore-forming hydrothermal fluid. Lower semiconformable silicified alteration zones are greatly depleted in metals (MacGeehan and MacLean, 1980; Skirrow, 1987), and a few hundred cubic kilometres of this type of alteration, commensurate with observations at Noranda, Mattagami Lake, and Snow Lake, could provide the metals for a major mining district. The metal contents between districts are highly variable, and some of this variation, particularly between the two principal groups of massive sulphide deposits, could be explained by the difference in metal contents of the source of the rocks (Lydon, 1988). Differences between camps that have closely similar geological attributes, such as Noranda and Sturgeon Lake, are more probably related to thermal and mixing conditions affecting the fluids in the ascent and precipitation regimes. Metals and sulphur could be derived directly from sediments or felsic rocks at lower temperature, as they buffer the fluid to a low pH at lower temperatures than does basalt.

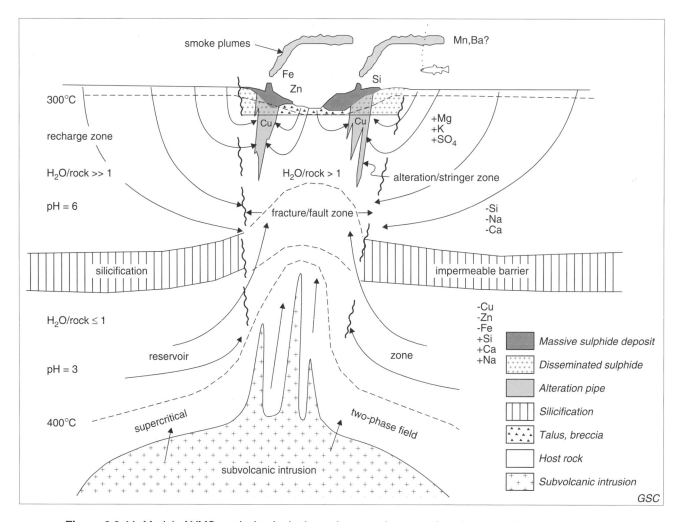

Figure 6.3-11. Model of VMS-producing hydrothermal system, incorporating elements of all subtypes of VMS deposits. The subvolcanic intrusion may be contributing some metals and gases to the hydrothermal fluid. Copper precipitates within the alteration pipes, as well as in the core of the massive sulphide mounds. This diagram is drawn perpendicular to the fracture system which controls hydrothermal discharge.

Hydrothermal fluids of seawater salinity are very buoyant at 385°C, and would rise quickly to the seafloor (Franklin, 1986). Thus identification of high-temperature reaction zones within footwall strata may be an important guide to areas with massive sulphide potential.

Some contribution from magmatic sources is possible. This is indicated by anomalous salinities, the carbon and oxygen isotope compositions (Taylor, 1990) and high CO_2 and ^{3}He contents of some hydrothermal fluids, together with the loss of sulphur in highly fractionated magmas associated with the Galapagos deposits (Perfit et al., 1983). This contribution, although a small percentage of the total fluid, might be responsible for a significant proportion of its metals.

The source of the heat for a hydrothermal system has been examined by Cathles (1981, 1983) and Cann et al. (1985). Sufficient heat is required to produce fluid temperatures in the reaction zone that were high enough for these fluids to dissolve sufficient metals (at least several parts per million) to ultimately form deposits. Obviously, the ability of this zone (called the "reservoir" by Hodgson and Lydon, 1977) to attain these temperatures also requires sufficient hydrological "insulation" to prevent excessive cooling through mixing with downward advecting cold seawater. Cathles (1983) has calculated that sufficient heat to form all the deposits in the Hokuroko basin could only be attained from a magma chamber containing at least 169 km^3 of felsic melt. The Noranda deposits would require a magma chamber larger than 78 km^3. Using the parameters suggested by Cann et al. (1985), a district of moderate size containing 30 million tonnes of sulphide would require a chamber of 300 km^3. Intrusions of an appropriate size are associated with most of the Archean and Proterozoic Cu-Zn districts in Canada. The provision of heat for the deposits in the back-arc, inter-arc, and off-ridge areas where sedimentary rocks form a significant footwall component is more problematic. The sediments at Middle Valley, which are 300 to 600 m thick, provide natural hydrological and thermal insulation, greatly reducing heat loss from the underlying cooling oceanic crust (Davis et al., 1987). The necessity of contemporaneous intrusions as a heat source is unresolved. In the Guaymas Basin and Escanaba Trough, small sills, themselves insufficient in size to provide all the heat needed for a major hydrothermal system, may be indicative of larger intrusions near the base of the sedimentary sequence. At Middle Valley, carbon and oxygen isotopic data (Taylor, 1990) indicate that no actively degassing magma chamber presently underlies the area, but at Escanaba Trough such a chamber probably is present.

An impermeable "cap" to the hydrothermal reaction zone is most important, as suggested by Hodgson and Lydon (1977) and demonstrated in mid-ocean ridge areas by Kappel and Franklin (1989). This cap can take many forms. The aforementioned sediments in the back-arc and off-ridge areas for both Cu-Zn and Pb-Zn-Cu deposits probably provide this. Massive pyroclastic ash flows which overlie more permeable mafic sequences, as at Mattagami Lake, Sturgeon Lake, and Snow Lake, are similar impermeable caps. The hydrothermal system itself may have formed an impermeable zone within the footwall strata. For example, the lower semiconformable alteration zones that are composed of silica, epidote, albite, and tremolite, and which are excellent candidates as the sources of constituents for the hydrothermal fluids, contain uppermost zones of pervasive silicification (Gibson et al., 1983; Skirrow, 1987) that probably sealed the rocks very effectively, thus preventing the escape of buoyant hydrothermal fluid through unfocused discharge. The broad-scale carbonate alteration associated with copper-zinc deposits in shallow water regimes such as Sturgeon Lake, Ontario and the Hackett River district in Slave Province would similarly reduce cross-stratal permeability.

Fluid composition

Analyses of fluids from seafloor vents in volcanic-dominated ridge crests are available for most of the East Pacific Rise and Juan de Fuca hydrothermal areas, as well as for some vents along the Mid-Atlantic Ridge. Fluids in sediment-covered ridges are less well studied, although data are available for the Escanaba Trough and Guaymas Basin vents. All of the end-member compositions are calculated for zero Mg content, on the assumption the Mg is totally removed from the descending, progressively heated seawater, through retrograde solubility of Mg-hydroxysulphates such as caminite, and loss through metasomatic alteration of basalt (Humphris and Thompson, 1978). The maximum temperature is near 400°C for fluids from Endeavour Ridge (Delaney et al., 1984); maxima for many other sites are about 350°C, but are lower at southern Juan de Fuca, Axial, Explorer, and the sediment-covered ridge sites.

Summary

Combined information from modern and ancient hydrothermal systems, lead to the following important points.

Chemical composition of the fluids: a) the salinities of the fluids range widely, from less than 20% of seawater for one of the fluids of Axial Seamount, to slightly less than seawater for some East Pacific Rise and Axial Seamount fluids, and to almost twice seawater for the southern Juan de Fuca fluids; b) the zinc contents range to as much as 59 mg/kg, and Fe to more than 1000 mg/kg (both at Southern Juan de Fuca); c) SiO$_2$ contents are about 1000 mg/kg; d) pH is about 3.5; e) both the vapour-dominated and residual fluids at Axial Seamount contain subequal amounts of gold, H$_2$S, and arsenic, but the vapour-dominated phase has no base metals; f) virtually all measured fluids have been partially cooled prior to venting and many of their constituents may have precipitated from them during cooling.

Mechanisms of focusing fluid discharge: the faults or fracture zones that form the conduits for discharging hydrothermal fluids are all tectonically generated. Specific patterns of fractures have been established in some districts; Knuckey and Watkins (1982) have identified specific faults in the Noranda district, and Scott (1978) has identified two linear elements in the Hokuroku district which may have provided the fluid channelways. Identification of subvolcanic faults requires careful mapping in each district. Discovery of such faults usually occurs after initial discovery of deposits, and may best be applied in well explored districts.

Alteration and stringer zone formation: stringer zones beneath all types of massive sulphide deposits are copper-rich. Experimental and thermodynamic data indicate that copper solubility in most highly reduced hydrothermal fluids decreases relatively quickly with decreasing temperature, compared with zinc and lead (Franklin et al., 1981). Thus conductive cooling of a hydrothermal fluid near the seafloor would promote precipitation of chalcopyrite, provided that no mixing with ambient oxidizing seawater took place, and a copper-stringer zone would form. Mixing with oxidizing seawater seems unlikely (Solomon et al., 1988) in the stringer zone, although it is more possible in the massive ore zone (Ohmoto and Skinner, 1983).

The alteration assemblages associated with these stringer zones are variable. That each alteration assemblage is closely related to a specific group of deposits, with silicification associated primarily with the Zn-Pb-Cu group (Franklin et al., 1981), is clearly an over-simplification. Silicification occurs beneath copper-zinc deposits of the "Mattabi type", as well as beneath many ophiolite-associated deposits in Cyprus and Oman, and 5000 year old lavas of the Galapagos Rift (Embley et al., 1988). Chloritization occurs beneath both the Noranda type deposits, and those representing the Zn-Pb-Cu group (Solomon et al., 1988).

Seafloor hydrothermal fluids generally contain abundant silica, which is highly over-saturated with respect to cold seawater; its precipitation as amorphous silica and cristobalite (as observed in the Galapagos alteration zone; Embley et al., 1988) through cooling may be slow, as the kinetic effect on silica precipitation seems to be important. Thus in areas of the pipe where the fluid flow was restricted, silica precipitation through cooling was likely more effective. Boiling may also promote silica precipitation (Lydon and Galley, 1986), explaining the abundant silicification beneath deposits formed in relatively shallow water, such as the Mattabi type. The abundance of Mg-chlorite may be related either to the rate of advection of cold seawater into the area of the sub-seafloor hydrothermal fluid conduit, or to the Mg content of the hydrothermal fluid. All data on the composition of modern seafloor hydrothermal fluids is calculated to zero magnesium content, assuming that magnesium was completely stripped from progressively heated seawater during its descent into the reaction zone; such an assumption is consistent with experimental data (Seyfried and Janecky, 1985). These data are inconsistent with the Mg loss observed in the major silicified areas in most lower semiconformable alteration zones. In the advection model, cold seawater was probably drawn down rapidly, particularly in highly permeable areas, and Mg lost through reaction with wall rocks. This mechanism would have been particularly effective beneath Archean deposits, as seawater at that time probably contained much less sulphate than did Proterozoic and younger seawater, due to the lack of an early oxidizing atmosphere. Thus during the Archean (or at any time or place where bottom water was reducing), Mg might not have been lost to Mg-hydroxysulphate, but would have been available to form Mg-smectites and chlorite. As noted by Lydon (1988), the amount of magnesium in a pipe zone may increase as the pipe collapses upwards; thus the composition of alteration pipes may be largely established during the waning stages of formation of the deposits. Even

in a modern example at Galapagos (Embley et al., 1988), Mg-silicates formed later, or peripheral to, the core Fe-S-SiO_2 alteration zone.

Sulphides are precipitated on or near the seafloor primarily by rapid cooling of the hydrothermal fluid. Reduction of seawater sulphate is unimportant as a principal source of sulphur (Janecky and Shanks, 1988). Precipitation processes for seafloor sulphides in individual chimneys in ridge crests have been examined by Haymon (1983) and Goldfarb et al. (1983), and have been summarized by Franklin (1986) and Lydon (1988). The initial precipitate at a vent orifice is anhydrite, derived entirely from ambient seawater. As the chimney grows, the porous accumulation of coarse anhydrite crystals is infiltrated by hydrothermal fluid, which mixes with seawater; a combination of overgrowth and replacement of anhydrite results in formation of a well zoned chimney, typically with an isocubanite- or chalcopyrite-rich lining of the orifice, a sphalerite (and/or wurtzite)-pyrite-chalcopyrite-anhydrite inner ring, and an outer wall of marcasite-pyrite-sphalerite-anhydrite-silica with surficial organic material. Individual vents have very short lives (Johnson and Tunnicliffe, 1986), and fluid flow is tortuous through most chimneys. Individual chimneys with multiple orifices commonly vent grey to black smoke (from high-temperature fluids) and much more rarely, clear water (from depleted fluids) from separate orifices within a few centimetres of each other, attesting to a very complex "plumbing system" within their uppermost parts.

Less is known about the internal composition of the mounds beneath the chimneys. On their outer edges, pyrite is abundant. In the inner parts of the oldest seafloor deposits that have been studied, the inactive deposits at Galapagos (Embley et al., 1988) and on a seamount at 12°N (Hekinian and Fouquet, 1985), chalcopyrite has replaced much of the pre-existing sulphide. Initial growth of a mound probably occurred as chimneys collapsed, and formed sulphide talus through which hydrothermal fluid continued to pass. The talus became infilled, the sulphide blocks coalesced, and a mound probably began to grow by "inflation" or overgrowth, through a very complex and irregular sequence of precipitation and dissolution mechanisms (Watkinson et al., 1990). Metal zonation formed through incremental precipitation of metals from high temperature fluids across a large thermal gradient within the chimneys or mounds. Silicification throughout the outer portions of mounds and chimneys partially sealed them from massive ingress of cold seawater, and also provided some strength (Tivey and Delaney, 1986).

Ancient deposits have undoubtedly experienced the same complex precipitation history. The "copper-spine" observed at Millenbach (Knuckey et al., 1982) is, in modified form, present in many massive sulphide deposits. Progressive heating of a sulphide mound could induce replacement of earlier-formed pyrite by sphalerite and then chalcopyrite. Thus, sulphide mounds that have been subject to heating by the passage of high temperature fluid for an extensive period of time may become progressively enriched in copper. This process could have formed the "Yellow-ore" in the Kuroko deposits (Eldridge et al., 1983).

Barite is forming in modern seafloor hydrothermal discharge areas regardless of source rock or depositional environment. Its presence is more dependent on the temperature of discharge, with low temperature (about 250°C or less) vent fluids, which are depleted in most base metals and sulphur, forming barite-rich chimneys. Most of the barite-rich chimneys are isolated from the metalliferous (higher temperature) vent sites by at least a few tens of metres. Thus near ancient deposits, barite formed where the hydrothermal system continued to vent after (or away from) its peak (or geographic centre) of high temperature activity, and where the seawater contained abundant sulphate. Hydrothermal systems which essentially were terminated while venting high temperature fluids would not have formed significant barite accumulations, most barium being lost to smoke plumes.

Coincidentally, gold-rich massive sulphides are commonly associated with barite-rich assemblages. Hannington and Scott (1989) showed that gold is effectively preserved in hydrothermal fluids by bisulphide transport, and efficiently precipitated from such fluids by ligand oxidation. Bisulphide complexing in a typical seafloor hydrothermal fluid is enhanced by cooling, raising of pH, and increasing P_{O_2}. Some of these conditions are promoted by boiling. Boiling also causes precipitation of copper, iron, and zinc as sulphides (through cooling) in the subsurface, leaving the remaining fluid relatively enriched in gold, barium, and lead. Thus gold-enriched massive sulphide deposits might form preferentially in depositional environments that are relatively shallow (<1900 m), and in which ambient seawater is oxidizing. The high gold content of the Axial Seamount and Explorer Ridge sulphides is evidence that shallow water depth (i.e. boiling can occur) may be a contributing factor to gold enrichment (Kappel and Franklin, 1989).

Significantly, however, gold-rich samples from the TAG hydrothermal field on the Mid-Atlantic Ridge (Hannington et al., 1990), which formed at about 3700 m depth, are attributed, in part, to a "diagenetic" redeposition process that involves dissolution of gold from paragenetically early (about 0.5 ppm) sulphides by later fluids that are drawn through the sulphide mound, and deposition of this gold in the cooler, well oxidized outer edges of the mound. In addition, zones of seafloor-oxidized sulphides (i.e. "supergene" altered zones) are prospective areas for gold enrichment, through a scavenging and redeposition mechanism.

Finally, the strata of presumed hydrothermal origin that cap many of the deposits appear to have highly diverse origins. Particulate sulphides are abundant in black smoke, but in the modern oceans these become widely dispersed (Converse et al., 1984; McConachy, 1988); oxidation destroys the primary minerals, and sulphide "fallout" is minimal, even immediately adjacent to the deposits. If the near-bottom waters were reducing, however, sulphide fallout could have accumulated. The abundant sulphides in the tuff horizons associated with many of the Noranda deposits and in the Key Tuffite horizon at Mattagami Lake may have been plume particulates. Clay minerals have been noted in hydrothermal plumes and probably contribute to the thin layers of pelagic sediment commonly associated with modern vent sites. Such layers, once they reach a metre or more in thickness, could be important in maintaining well-focused discharge at vent sites, as well as preventing cooling of the rising fluid by inhibiting local advection of cold seawater.

The ferruginous chert zone overlying many deposits may be the product of low temperature discharge. Considerable dissolved silica is discharged through the high temperature vents, but does not appear to form an important particulate constituent of the "smoke". Slow precipitation rates for silica from vent fluids mixing with seawater may cause it to be diluted prior to precipitation. Silica is precipitated in and around low temperature vents, particularly where the discharging fluid has no H_2S. Jonasson and Walker (1987) and Juniper and Fouquet (1988) have shown that opal-A is precipitated on filamentous bacteria, and in open spaces in chimneys, as a silica gel. These bacteria require an H_2S supply, but if this supply is stopped, they die and are replaced by iron oxide, on which opal-A precipitates. Some ferruginous caps probably are the products of syndepositional oxidation of seafloor sulphide mounds. Such oxidation in the modern ocean can destroy a small deposit in about 100 000 years, unless it is protected by sediments or volcanic strata, or by an oxidized cap, especially if it is silicified (i.e., iron-formation).

EXPLORATION GUIDES

1. Presence of submarine volcanic strata. Paleo-water depth probably controlled some variations in volcanic morphology, as well as alteration assemblages and ore composition. Volcanological studies will provide useful information to help determine which assemblages and compositions to expect.

2. Presence of a subvolcanic magma chamber at shallow crustal levels (about 2 km). These can be any composition represented in the overlying volcanic rocks, and are sill-like, but locally transect stratigraphy; are texturally variable, multiple intrusions; are highly fractionated, with "reverse zonation" common in felsic intrusions; have little or no metamorphic halo relative to intrusions emplaced at deeper, drier crustal levels; and are potential hosts to very low grade porphyry copper zones that are superimposed on all rock types. In some districts, synvolcanic dyke swarms occupy faults and fractures, and may be spatially and temporally coincident with hydrothermal conduits.

3. Presence of high-temperature reaction zones (one form of semiconformable alteration) within about 1.5 km of the subvolcanic intrusions. Quartz-epidote-albite alteration, commonly mistakenly mapped as intermediate to felsic rocks, are prevalent under many copper-zinc deposits.

4. Near deposits that formed in relatively shallow water (accompanied by explosion breccia, debris flows, some subaerial volcanic products), laterally extensive carbonatized (and more locally silicified) volcanic strata which are depleted in sodium, are prevalent. These possibly represent the zone in which ambient seawater reacted with the upper part of the hydrothermal reservoir. Some massive sulphide deposits that formed under relatively shallow water may have the alteration and compositional characteristics of epithermal deposits.

5. Synvolcanic faults are recognizable because: they do not extend far into the hanging wall of most deposits; they are commonly altered, with pipe-like assemblages, in their stratigraphically highest portions; they may have asymmetric zones of growth-fault-induced talus; and

they may be locally occupied by synvolcanic dykes. Virtually all of these formed in tensional tectonic regimes, and may be listric. Some may be related to caldera margins, and thus curvilinear; others may be margins of elongate axial summit depressions (grabens), and subparallel to the axis of spreading.

6. Alteration pipes may have sufficient vertical stratigraphic extent to be mappable. Virtually all are sodium depleted, but mineralogical characteristics vary. Most commonly, these are silicified near the deposits, sericitized less locally, and have variable amounts of both Mg- and Fe-rich chlorite or smectite. Less commonly, but important in many Cu-Zn districts, these may have intensely chloritized cores, with sericitized rims. Peripheral to the distinctive pipes, commonly there is a broad zone of more subtly altered rock; smectite and zeolite minerals may be important. Chemical changes in these latter alteration zones may be very subtle, requiring mineralogical or isotopic studies to detect them.

   Metamorphosed pipe assemblages are usually relatively easy to recognize. Magnesium-enriched pipes will be recrystallized to anthophyllite and cordierite. Adjacent, less intensely altered rocks may contain staurolite. Gahnite and Mn-rich garnets may be important accessories. The relatively high ductility of altered volcanic rocks compared with their hosts has resulted in them being exceptionally deformed in some districts. They may have become detached completely from their attendant orebodies. Structural mapping and synthesis are essential to finding additional resources in established massive sulphide districts.

7. The immediately overlying strata may contain indications of mineralization. Hanging wall volcanic rocks may contain alteration pipe assemblages, or at least zeolite-smectite assemblages similar to the peripheral alteration associated with the pipes.

   More importantly, hydrothermally-related precipitates, such as ferruginous chert, sulphidic tuff, and products of oxidation of sulphide mounds may be sufficiently laterally extensive to be detected. Base metal contents within these may increase towards the deposits, whereas Ni, V, Mn, and Ba may increase away from deposits.

RELATED DEPOSIT TYPES

Although some Algoma-type iron-formations are spatially related to massive sulphide deposits (Gross, 1965), this is rather rare. Both Algoma-type and Superior-type oxide-facies iron deposits originated as precipitated products on the seafloor, but almost no evidence indicates that they are the product of focused discharge. At Axial Seamount on the Juan de Fuca Ridge, a laterally-extensive bed of ferruginous silica surrounds the area of high temperature venting. Silica is being fixed by bacterial remains, presumably from low temperature, silica-enriched, but metal- and sulphur-depleted hydrothermal fluid that is discharging throughout the area of the caldera floor at Axial Seamount. Some iron-formation may thus be indirectly related to the high temperature massive sulphide-forming process. As Shegelski (1978) has shown, however, much other Algoma-type iron-formation is not related to this process.

A few epithermal vein deposits may be the product of high temperature hydrothermal systems, similar to those which formed massive sulphide deposits, that discharged into subaerial environments. The Headway-Coulee prospect in northwestern Ontario (Osterberg et al., 1987; Anglin et al., 1988) is an example of a syndepositional epithermal occurrence of zinc-lead-copper-silver mineralization.

Synvolcanic intrusions underlying many of the copper-zinc districts contain extensive low grade porphyry copper occurrences (Franklin et al., 1977). These may be the product of collapsed hydrothermal systems, that penetrated the predominantly crystallized magma chambers following termination of volcanism.

DEFINITIVE CHARACTERISTICS

Volcanic-associated massive sulphide deposits have these distinctive geological characteristics:

1. Spatial association with submarine volcanic rocks and commonly with associated sedimentary sequences.

2. Bulbous to tabular stratiform accumulations of massive pyrite and subordinate pyrrhotite, and contain about 8 to 10% total combined zinc, lead, and copper.

3. Underlain by discordant alteration zones, which may occupy synvolcanic faults.

4. Regionally characterized by manifestations of high heat flow; subvolcanic intrusions, laterally extensive semiconformable alteration zones, and anomalous petrogenetic trends in the volcanic sequence, all of which indicate proximity to massive sulphide deposits.

SELECTED BIBLIOGRAPHY

References noted with an asterisk (*) are summary papers, providing a thorough review of volcanic-associated massive sulphide deposits.

Aggarwal, P.K. and Nesbitt, B.T.
1984: Geology and geochemistry of the Chu Chua massive sulphide deposit, British Columbia; Economic Geology, v. 79, p. 815-825.
Alldrick, D.J.
1986: Stratigraphy and structure in the Anyox area (103 P/5); British Columbia Ministry of Energy, Mines and Petroleum Resources, Geological Fieldwork, 1985, British Columbia Department of Energy and Mines, Paper 1986-1, p. 211-216.
Anglin, C.D.A., Franklin, J.M., Loveridge, W.D., Hunt, P.A., and Osterberg, S.A.
1988: Use of zircon U-Pb ages of felsic intrusive and extrusive rocks in eastern Wabigoon Subprovince, Ontario, to place constraints on base metal and gold mineralization; in Radiogenic Age and Isotopic Studies: Report 2, Geological Survey of Canada, Paper 88-2, p. 109-115.
Binney, W.P.
1987: A sedimentological investigation of MacLean channel transported sulphide ores; in Buchans Geology, Newfoundland, (ed.) R.V. Kirkham; Geological Survey of Canada, Paper 86-24, p. 107-147.
Bischoff, J.L. and Seyfried, W.E.
1978: Hydrothermal chemistry of seawater from 25° to 350°C; American Journal of Earth Science, v. 278, p. 838-860.
Cann, J.R., Strens, M.R., and Rice, A.
1985: A simple magma-driven thermal balance model for the formation of volcanogenic massive sulfides; Earth and Planetary Science Letters, v. 76, p. 123-134.
Cathles, L.M.
*1981: Fluid flow and genesis of hydrothermal ore deposits; in Economic Geology, Seventy-Fifth Anniversary Volume 1905-1980, (ed.) B.J. Skinner; p. 424-457.

Cathles, L.M., (cont.)
1983: An analysis of the hydrothermal system responsible for massive sulphide deposition in the Hokuroku basin of Japan; in Kuroko and Related Volcanogenic Massive Sulphide Deposits, (ed.) H. Ohmoto and B.J. Skinner; Economic Geology Monograph 5, p. 439-487.

Chen, T.T. and Petruk, W.
1980: Mineralogy and characteristics that affect recoveries of metals and trace elements from the ore at Heath Steele mines, New Brunswick; The Canadian Mining and Metallurgical Bulletin, v. 73, no. 823, p. 167-179.

Converse, D.R., Holland, H.D., and Edmond, J.M.
1984: Flow rates in the axial hot springs of East Pacific Rise (21°N): implications for the heat budget and the formation of massive sulphide deposits; Earth and Planetary Science Letters, v. 69, p. 159-175.

Costa, U.R., Barnett, R.L., and Kerrich, R.
1983: The Mattagami Lake mine Archean Zn-Cu sulphide deposit, Quebec: hydrothermal coprecipitation of talc and sulphides in a seafloor brine pool - evidence from geochemistry, $^{18}O/^{16}O$, and mineral chemistry; Economic Geology, v. 78, no. 6, p. 1144-1203.

Date, J., Watanabe, Y., and Saeki, Y.
1983: Zonal alteration around the Fukazawa Kuroko deposits, Akita prefecture, northern Japan; in Kuroko and Related Volcanogenic Massive Sulphide Deposits, (ed.) H. Ohmoto and B.J. Skinner; Economic Geology, Monograph 5, p. 365-386.

Davidson, A.J.
1977: Petrography and chemistry of the Key Tuffite at Bell Allard, Matagami, Quebec; MSc. thesis, McGill University, Montreal, Quebec, 131 p.

Davis, E.E., Goodfellow, W.D., Bornhold, B.D., Adshead, J., Blaise, B., Villinger, H., and Lecheminant, G.M.
1987: Massive sulphides in a sedimented rift valley, northern Juan de Fuca Ridge; Earth and Planetary Science Letters, v. 82, p. 49-61.

Delaney, J.R., McDuff, R.E., and Lupton, J.E.
1984: Hydrothermal fluid temperatures of 400°C on the Endeavour segment, northern Juan de Fuca Ridge; EOS transactions, American Geophysical Union, v. 65, p. 973.

Eldridge, C.S., Barton, P.B., Jr., and Ohmoto, H.
1983: Mineral textures and their bearing on formation of the Kuroko orebodies; in Kuroko and Related Volcanogenic Massive Sulphide Deposits, (ed.) H. Ohmoto and B.J. Skinner; Economic Geology, Monograph 5, p. 241-281.

Embley, R.W., Jonasson, I.R., Perfit, M.R., Tivey, M., Malahoff, A., Franklin, J.M., Smith, M.F., and Francis, T.J.G.
1988: Submersible investigation of an extinct hydrothermal system on the eastern Galapagos Ridge: sulfide mounds, stockwork zone, and differentiated lavas; Canadian Mineralogist, v. 26, p. 517-540.

*Franklin, J.M.
1986: Volcanic-associated massive sulphide deposits - an update; in Geology and Genesis of Mineral Deposits in Ireland (ed.) C.J. Andrew, R.W.A. Crowe, S. Finlay, W.M. Pennell, and J.F. Pyne; Irish Association for Economic Geology, Dublin, p. 49-69.

Franklin, J.M., Gibb, W., Poulsen, K.H., and Severin, P.
1977: Archean metallogeny and stratigraphy of the south Sturgeon Lake area; Institute on Lake Superior Geology, 23rd Annual Meeting, Thunder Bay, Ontario, Guidebook, 73 p.

*Franklin, J.M., Lydon, J.W., and Sangster, D.F.
1981: Volcanic-associated massive sulfide deposits; in Economic Geology Seventy-fifth Anniversary Volume 1905-1980, (ed.) B.J. Skinner; p. 485-627.

Galley, A.G. and Jonasson, I.R.
1992: Semi-conformable alteration and volcanogenic massive sulphide deposits; Proceedings, 14th New Zealand Geothermal Workshop, (ed.) S.F. Simmons, J. Newson, and K.C. Lee; p. 279-284.

Gass, I.G. and Smewing, J.D.
1973: Intrusion, extrusion and metamorphism at constructive plate margins: evidence from the Troodos massif, Cyprus; Nature, v. 242, p. 26-29.

Gibson, H.L. and Watkinson, D.H.
1990: Volcanogenic massive sulphide deposits of the Noranda Cauldron and Shield volcano, Quebec; in The Northwestern Quebec Polymetallic Belt, (ed.) M. Rive, P. Verpaelst, Y. Gagnon, J.M. Lulin, G. Riverin, and A. Simard; The Canadian Institute of Mining and Metallurgy, Special Volume 43, p. 119-133.

Gibson, H.L., Watkinson, D.H., and Comba, C.D.A.
1983: Silicification: hydrothermal alteration in an Archean geothermal system within the Amulet Rhyolite Formation, Noranda, Quebec; Economic Geology, v. 78, p. 954-971.

Goldfarb, M.S., Converse, D.R., Holland, H.D., and Edmond, J.M.
1983: The genesis of hot spring deposits on the East Pacific Rise, 21°N. in Kuroko and Related Volcanogenic Massive Sulphide Deposits, (ed.) H. Ohmoto and B.J. Skinner; Economic Geology, Monograph 5, p. 507-522.

Goodfellow, W.D. and Franklin, J.M.
1993: Geology, mineralogy and chemistry of sediment-hosted clastic massive sulphides in shallow cores, Middle Valley, northern Juan de Fuca Ridge; Economic Geology, v. 88, no. 8, p. 2037-2068.

Gross, G.A.
1965: General geology and evaluation of ore deposits; Volume 1 in Geology of Iron Deposits in Canada; Geological Survey of Canada, Economic Geology Report 22, 181 p.

Hannington, M.D and Jonasson, I.R.
1992: Fe and Mn oxides at seafloor hydrothermal vents; Catena Supplement 21, p. 351-370.

Hannington, M.D. and Scott, S.D.
1989: Gold mineralization in volcanogenic massive sulphide deposits: implications of data from active hydrothermal vents on the modern sea floor; in The Geology of Gold Deposits, (ed.) R.R. Keays, W.R.H. Ramsay, and D.I. Groves; Economic Geology, Monograph 6, p. 491-507.

Hannington, M.D., Herzig, P.M., and Scott, S.D.
1990: Auriferous hydrothermal precipitates on the seafloor; in Gold Metallogeny and Exploration, (ed.) R.P. Foster; Blackie and Son, Glasgow, p. 250-282.

Haymon, R.
1983: The growth history of hydrothermal "black smoker" chimneys; Nature, v. 301, p. 695-698.

Hekinian, R. and Fouquet, Y.
1985: Volcanism and metallogenesis of axial and off-axial structures on the East Pacific Rise near 13°N. Economic Geology, v. 80, p. 221-249.

*Hodgson, C.J. and Lydon, J.W.
1977: The geological setting of volcanogenic massive sulfide deposits and active hydrothermal systems: some implications for exploration; The Canadian Institute of Mining and Metallurgy, Bulletin, v. 70, p. 95-106.

Humphris, S.E. and Thompson, G.
1978: Hydrothermal alteration of oceanic basalts by seawater; Geochimica et Cosmochimica Acta, v. 42, p. 107-125.

Iijima, A.
1974: Clay and zeolitic alteration zones surrounding Kuroko deposits in the Hokuroko district, Northern Akita, as submarine hydrothermal-diagenetic alteration products; Society of Mining Geologists of Japan, Special Issue 6, p. 267-290.

Ishihara, S. and Terashima, S.
1974: Base metal contents of the basement rocks of Kuroko deposits - an overall view to examine their effect on the Kuroko mineralization; Society of Mining Geologists of Japan, Special Issue 6, p. 421-431.

Jambor, J.L.
1979: Mineralogical evaluation of proximal-distal features in New Brunswick massive sulfide deposits; Canadian Mineralogist, v. 17, p. 649-664.

Janecky, D.R. and Shanks, W.C., III
1988: Computational modelling of chemical and sulphur isotopic reaction processes in seafloor hydrothermal systems: chimneys, massive sulfides, and subjacent alteration zones; Canadian Mineralogist, v. 26, pt. 3, p. 805-826.

Johnson, H.P. and Tunnicliffe, V.
1986: Time-lapse camera measurements of a high temperature hydrothermal system on Axial Seamount, Juan de Fuca Ridge; EOS, Transactions of the American Geophysical Union, v. 67, no. 44, p. 1283.

Jonasson, I.R. and Walker, D.A.
1987: Micro-organisms and their debris as substrates for base metal sulfide nucleation and accumulation in some mid-ocean ridge deposits; (abstract) EOS, Transactions of the American Geophysics Union, v. 68, no. 44, p. 1546.

Juniper, S.K. and Fouquet, Y.
1988: Filamentous iron-silica deposits from modern and ancient hydrothermal sites; Canadian Mineralogist, v. 26, pt. 3, p. 859-870.

Juras, S.J. and Pearson, C.A.
1990: The Buttle Lake Camp, Central Vancouver Island, B.C.; Chapter 9 in Geology and Regional Setting of the Major Mineral Deposits in Southern British Columbia, Guidebook, International Association on the Genesis of Ore Deposits, Field Trip 12, Geological Survey of Canada, Open File 2167, p. 145-161.

Kappel, E.S. and Franklin, J.M.

1989: Relationships between geologic development of ridge crests and sulfide deposits in the northeast Pacific Ocean; Economic Geology and the Bulletin of the Society of Economic Geologists, v. 84, no. 3, p. 485-505.

Kerr, D.J. and Mason, R.

1990: A re-appraisal of the geology and ore deposits of the Horne mine complex at Rouyn-Noranda, Quebec; in The Northwestern Quebec Polymetallic Belt, (ed.) M. Rive, P. Verpaelst, Y. Gagnon, J.M. Lulin, G. Riverin, and A. Simard; The Canadian Institute of Mining and Metallurgy, Special Volume 43, p. 153-166.

Kissin, S.A.

1974: Phase relations in a portion of the Fe-S system; PhD. thesis, University of Toronto, Toronto, Ontario, 234 p.

Klau, W. and Large, D.E.

1980: Submarine exhalative Cu-Pb-Zn deposits, a discussion of their classification and metallogenesis; Geologische Jahrbuch, sec. D, no. 40, p. 13-58.

Knuckey, M.J.

1975: Geology of the Millenbach copper-zinc orebody; (abstract) Economic Geology, v. 70, no. 1, p. 247.

Knuckey, M.J. and Watkins, J.J.

1982: The geology of the Corbet massive sulphide deposit, Noranda, Quebec, Canada; in Precambrian Sulphide Deposits, (ed.) R.W. Hutchinson, C.D. Spence, and J.M. Franklin; Geological Association of Canada, Special Paper 25, p. 296-317.

Knuckey, M.J., Comba, C.D.A., and Riverin, G.

1982: Structure, metal zoning, and alteration at the Millenbach deposit, Noranda, Quebec; in Precambrian Sulphide Deposits, (ed.) R.W. Hutchinson, C.D. Spence, and J.M. Franklin;, Geological Association of Canada, Special Paper 25, p. 255-295.

Lambert, I.B.

1979: Massive copper-lead-zinc deposits in felsic volcanic sequences of Japan and Australia; comparative notes; Mining Geology, v. 29, p. 11-20.

Lambert, I.B. and Sato, T.

1974: The Kuroko and associated ore deposits of Japan: a review of their features and metallogenesis; Economic Geology, v. 69, p. 1215-1236.

Luff, W.M.

1977: Geology of the Brunswick no. 12 mine. The Canadian Institute of Mining and Metallurgy, Bulletin, v. 70, no. 782, p. 109-119.

*Lydon, J.W.

1984: Volcanogenic massive sulphide deposits. Part 1: a descriptive model; Geoscience Canada, v. 11, p. 195-202.

1988: Ore deposit models #14. Volcanogenic massive sulphide deposits. Part 2: genetic models; Geoscience Canada, v. 15, no. 1, p. 43-65.

Lydon, J.W. and Galley, A.

1986: The chemical and mineralogical zonation of Mathiati alteration pipe, Cyprus, and genetic significance; in Metallogeny of Basic and Ultrabasic Rocks, (ed.) M.J. Gallagher, R.A. Ixer, C.R. Neary, and H.M. Prichard; Institution of Mining and Metallurgy, London, U.K., p. 49-68.

MacGeehan, P.J. and MacLean, W.H.

1980: Tholeiitic basalt-rhyolite magmatism and massive sulphide deposits at Mattagami, Quebec; Nature, v. 283, p. 153-157.

Marquis, P., Hubert, C., Brown, A.C., and Rigg, D.M.

1990: An evaluation of genetic models for gold deposits of the Bousquet district, Quebec, based on their mineralogic, geochemical and structural characteristics; in The Northwestern Quebec Polymetallic Belt, (ed.) M. Rive, P. Verpaelst, Y. Gagnon, J.M. Lulin, G. Riverin, and A. Simard; The Canadian Institute of Mining and Metallurgy, Special Volume 43, p. 383-399.

McConachy, T.F.

1988: Hydrothermal plumes over spreading ridges and related deposits in the northeast Pacific Ocean: the East Pacific Rise near 11°N and 21°N, Explorer Ridge and J. Tuzo Wilson seamounts; PhD. thesis, University of Toronto, Toronto, Ontario, 403 p.

Morton, R.L. and Franklin, J.M.

1987: Two-fold classification of Archean volcanic-associated massive sulfide deposits; Economic Geology, v. 82, p. 1057-1063.

Morton, R.L., Hudak, G., Walker, J., and Franklin, J.M.

1990: Physical volcanology and hydrothermal alteration of the Sturgeon Lake caldera complex; in Mineral Deposits in the Western Superior Province, Ontario; International Association on the Genesis of Ore Deposits, Field Trip Guidebook No. 9, (ed.) J.M. Franklin, B.R. Schneiders, and E.R. Koopman; Geological Survey of Canada, Open File 2164, p. 74-94.

Oftedahl, C.

1958: On exhalative-sedimentary ores; Geologiska Föreningen i Stockholm Förhandlingar, v. 80, p. 1-19.

Ohmoto, H. and Skinner, B.J.

1983: The Kuroko and related volcanogenic massive sulphide deposits: introduction and summary of new findings; in Kuroko and Related Volcanogenic Massive Sulphide Deposits, (ed.) H. Ohmoto and B.J. Skinner; Economic Geology, Monograph 5, p. 1-8.

Osterberg, S.A., Morton, R.L., and Franklin, J.M.

1987: Hydrothermal alteration and physical volcanology of Archean rocks in the vicinity of the Headway-Coulee massive sulfide occurrence, Onaman area, northwestern Ontario; Economic Geology, v. 82, no. 6, p. 1505-1520.

Payne, J.G., Bratt, J.A., and Stone, B.G.

1980: Deformed Mesozoic volcanogenic Cu-Zn sulphide deposits in the Britannia district, British Columbia; Economic Geology, v. 75, p. 700-721.

Perfit, M.R., Fornari, D.J., Malahoff, A., and Embley, R.W.

1983: Geochemical studies of abyssal lavas recovered by DSRV ALVIN from eastern Galapagos rift, Inca transform and Ecuador rift, 3. Trace element abundances and petrogenesis; Journal of Geophysical Research, v. 88, p. 10551-10572.

Petersen, M.D. and Lambert, I.B.

1979: Mineralogical and chemical zonation around the Woodlawn Cu-Pb-Zn ore deposit, southeastern New South Wales; Journal of the Geological Society of Australia, v. 26, p. 169-186.

Piche, M., Guha, J., Sullivan, J., Bouchard, G., and Daigneault, R.

1990: Structure, stratigraphie et implication metallogenique - les gisements volcanogenes du camp minier de Matagami; in The Northwestern Quebec Polymetallic Belt, (ed.) M. Rive, P. Verpaelst, Y. Gagnon, J.M. Lulin, G. Riverin, and A. Simard; The Canadian Institute of Mining and Metallurgy, Special Volume 43, p. 327-336.

Purdie, J.J.

1967: Lake Dufault Mines, Ltd.; The Canadian Institute of Mining and Metallurgy, Annual Meeting, Montreal, Centennial Field Excursion, Northwestern Quebec and Northern Ontario, p. 52-57.

Richards, H.G., Cann, J.R., and Jensenius, J.

1989: Mineralogical zonation and metasomatism of the alteration pipes of Cyprus sulfide deposits; Economic Geology, v. 84, p. 91-115.

Riverin, G., Labrie, M., Salmon, B., Cazavant, A., Asselin, R., and Gagnon, M.

1990: The geology of the Ansil deposit, Rouyn-Noranda, Quebec; in The Northwestern Quebec Polymetallic Belt; (ed.) M. Rive, P. Verpaelst, Y. Gagnon, J.-M. Lulin, G. Riverin, and A. Simard; The Canadian Institute of Mining and Metallurgy, Special Volume 43, p. 2.

Sangster, D.F.

1972: Precambrian volcanogenic massive sulphide deposits in Canada: a review; Geological Survey of Canada, Paper 72-22, 44 p.

Sangster, D.F. and Scott, S.D.

1976: Precambrian stratabound massive Cu-Zn-Pb sufide ores of North America; in Handbook of Strata-bound and Stratiform Ore Deposits, (ed.) K.H. Wolf; Elsevier, Amsterdam, p. 129-222.

Scott, S.D.

1978: Structural control of the Kuroko deposits of the Hokuroku district, Japan; Mining Geology, v. 28, p. 301-311.

*Seyfried, W.E., Jr. and Janecky, D.R.

1985: Heavy metal and sulfur transport during subcritical and supercritical hydrothermal alteration of basalt: influence of fluid pressure and basalt composition and crystallinity; Geochimica et Cosmochimica Acta, v. 49, p. 2545-2560.

Shegelski, R.I.

1978: Stratigraphy and geochemistry of Archean iron formation in the Sturgeon Lake and Savant Lake greenstone terrains, northwestern Ontario; PhD. thesis, University of Toronto, Toronto, Ontario, 251 p.

Shirozu, H.

1974: Clay minerals in altered wall rocks of the Kuroko-type deposits. Society of Mining Geologists of Japan, Special Issue 6, p. 303-311.

Simmons, B.D.

1973: Geology of the Millenbach massive sulphide deposit, Noranda, Quebec; The Canadian Institute of Mining and Metallurgy, Bulletin, v. 166, no. 739, p. 67-78.

Skirrow, R.G.

1987: Silicification in a semiconformable alteration zone below the Chisel Lake massive sulphide deposit, Manitoba; MSc. thesis, Carleton University, Ottawa, Ontario, 171 p.

Solomon, M., Eastoe, C.J., Walshe, J.L., and Green, G.R.

1988: Mineral deposits and sulphur isotope abundances in the Mount Read volcanics between Que River and Mount Darwin, Tasmania; Economic Geology, v. 83, no. 6, p. 1307-1328.

Spooner, E.T.C.
1977:   Hydrodynamic model for the origin of the ophiolitic cupriferous pyrite ore deposits of Cyprus; in Volcanic Processes in Ore Genesis; Geological Society of London, Special Publication 7, p. 58-71.

Spooner, E.T.C. and Fyfe, W.S.
1973:   Sub-sea floor metamorphism, heat and mass transfer; Contributions to Mineralogy and Petrology, v. 42, p. 287-304.

Spooner, E.T.C., Bechinsdale, R.D., England, P.C., and Senior, A.
1977a: Hydration ^{18}O enrichment and oxidation during ocean floor hydrothermal metamorphism of ophiolitic metabasic rocks from E. Liguria, Italy; Geochimica et Cosmochimica Acta, v. 41, p. 857-872.

Spooner, E.T.C., Chapman, H.J., and Smewing, J.D.
1977b: Strontium isotopic contamination and oxidation during ocean floor hydrothermal metamorphism of the ophiolitic rocks of the Troodos Masiff, Cyprus; Geochimica et Cosmochimica Acta, v. 41, p. 873-890.

Tatsumi, T. (ed.)
1970:   Volcanism and Ore Genesis; University of Tokyo Press, Tokyo, 488 p.

Tatsumi, T. and Watanabe, T.
1971:   Geological environment of formation of the Kuroko-type deposits; Society of Mining of Geologists Japan, Special Issue 3, p. 216-220.

Taylor, B.E.
1990:   Carbon dioxide and methane in hydrothermal vent fluids from Middle Valley, a sediment-covered ridge segment; EOS, Transactions, American Geophysical Union, v. 71, no. 43, p. 1569.

Thorpe, R.I. and Harris, D.C.
1973:   Mattagamite and tellurantimony, two new telluride minerals from Mattagami Lake mine, Matagami area, Quebec; Canadian Mineralogist, v. 12, pt. 1, p. 55-60.

Thorpe, R.I., Pringle, G.J., and Plant, A.G.
1976:   Occurrence of selenide and sulphide mineral in bornite ore of the Kidd Creek massive suphide deposit, Timmins, Ontario; in Report of Activities, Part A; Geological Survey of Canada, Paper 76-1A, p. 311-317.

Tivey, M.K. and Delaney, J.R.
1986:   Growth of large sulphide structures on the Endeavour segment of the Juan de Fuca Ridge; Earth and Planetary Science Letters, v. 77, p. 303-317.

Urabe, T., Marumo, K., and Nakamura, K.
1990:   Mineralization and related hydrothermal alteration in Izena Cauldron (Jade Site), Okinawa Trough, Japan; Geological Society of America, Abstracts with Programs, v. 22, no. 7, p. A9.

van Staal, C.R.
1987:   Tectonic setting of the Tetagouche Group in northern New Brunswick: implications for plate tectonic models of the northern Appalachians; Canadian Journal of Earth Sciences, v. 24, no. 7, p. 1329-1351.

von Damm, K.L. and Bischoff, J.L
1987:   Chemistry of hydrothermal solutions from the southern Juan de Fuca Ridge; Journal of Geophysical Research, v. 92, p. 11334-11346.

von Stackelberg, U. and Brett, R.
1990:   Extended hydrothermal activity in the Lau Basin (southwest Pacific): first results of R.V. Sonne cruise 1990; EOS, Transactions, American Geophysical Union, v. 71, no. 43, p. 1680.

Watkinson, D.H., McEwen, J.H., and Jonasson, I.R.
1990:   Mine Gallen, Noranda, Quebec: geology of an Archean massive sulphide mound; in The Northwestern Quebec Polymetallic Belt: a Summary of 60 Years of Mining Exploration; (ed.) M. Rive, P. Verpaeist, Y. Gagnon, J.-M. Lulin, G. Riverin, and A. Simard; The Canadian Institute of Mining and Metallurgy, Special Volume 43, p. 167-174.

6.4   VOLCANIC-ASSOCIATED MASSIVE SULPHIDE GOLD

K.H. Poulsen and M.D. Hannington

INTRODUCTION

Volcanic-associated massive sulphide deposits contain variable abundances of gold (Fig. 6.4-1). There are three ways in which massive sulphide deposits may be considered "gold-rich" (Table 6.4-1). First are those deposits that contain high absolute concentrations of gold: volcanic-associated massive sulphide deposits typically contain 1 to 2 ppm, although a few attain anomalous values of 10-15 ppm or more; some of these are not primary gold producers, but nevertheless have significantly higher than average gold grades (e.g., D'Eldona, Lemoine-Patino; Fig. 6.4-1). Second are those deposits (e.g., Flin Flon, Buchans, Britannia; Fig. 6.4-1) that, by virtue of modest to

high gold concentration and large tonnages, have a relatively large total amount of contained gold. Third, a consideration of the relative proportions of contained gold, silver, and base metals (Fig. 6.4-2) shows that volcanic-associated sulphide deposits can be divided into two main compositional groups. The first and largest corresponds to base-metal massive sulphide deposits, including some like Flin Flon and Lemoine with anomalous gold concentrations and significant byproduct gold. The second group corresponds to auriferous sulphide deposits, mainly massive but also as stockworks and disseminations, in which gold is a primary commodity and base metals are of lesser economic importance. They are gold deposits in a strict economic sense (Fig. 6.4-2) and include several important Canadian and overseas deposits (Table 6.4-2).

Auriferous volcanic-associated sulphide deposits in Canada are exemplified by three main varieties: (1) Archean copper-gold deposits such as Horne, Quebec; (2) Archean pyritic gold deposits such as Bousquet No. 1, Quebec; and (3) auriferous polymetallic sulphide deposits such as the Jurassic Eskay Creek deposit, British

Poulsen, K.H. and Hannington, M.D.
1996:   Volcanic-associated massive sulphide gold; in Geology of Canadian Mineral Deposit Types, (ed.) O.R. Eckstrand, W.D. Sinclair, and R.I. Thorpe; Geological Survey of Canada, Geology of Canada, no. 8, p. 183-196 (also Geological Society of America, The Geology of North America, v. P-1).

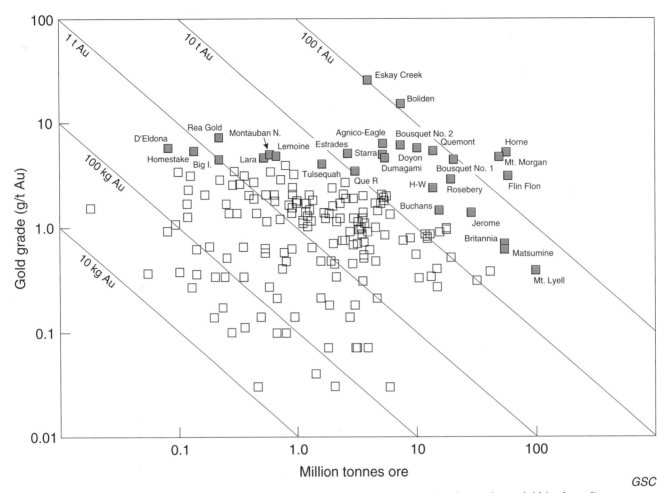

Figure 6.4-1. Plot of tonnage versus gold grade for volcanic-associated massive sulphide deposits, including 184 deposits in Canada and selected gold-rich deposits world-wide. Diagonal lines indicate quantities of contained gold. Open symbols: conventional "gold-poor" base metal-rich massive sulphides. Solid symbol: auriferous volcanic-associated massive sulphide deposits and selected "gold-rich" base metal massive sulphides (see Table 6.4-1 and text for discussion).

Columbia. Other examples that fit into one or more of these categories include the Bousquet No. 2 and Dumagami deposits in the southern Abitibi belt, Quebec; the Agnico-Eagle deposit near Joutel, Quebec; the Proterozoic Montauban North deposit in the Grenville Supergroup; and a variety of Devonian and younger, high-grade polymetallic sulphide deposits in British Columbia. Examples of major gold-producing deposits of this type elsewhere in the world include the early Proterozoic Boliden deposit in the Skellefte district of Sweden and the Paleozoic Mt. Morgan and Mt. Chalmers deposits in metamorphosed volcanic rocks of the Tasman geosyncline, Australia.

Although the auriferous massive sulphide deposits possess many of the geological characteristics of submarine base-metal massive sulphide deposits, many of them are unique in their bulk composition and mineralogy and may have formed under somewhat different geological conditions (Hannington, 1993). The fact that gold deposits of this type co-exist regionally with other base-metal-rich massive sulphides that do not contain anomalous gold is one of their most notable characteristics and implies that their auriferous nature may be due to an enriched source or to a particularly efficient local means of gold precipitation and concentration (e.g., boiling).

IMPORTANCE

Auriferous volcanic-associated massive sulphide deposits account for approximately 5% of Canada's historical gold production and reserves. The giant Horne mine, with a total past production of approximately 330 t Au, has been the largest Canadian producer of this deposit type. Together, the Horne and adjacent Quemont deposits accounted for more than 400 t Au (13 million ounces) and represent fully one-quarter of the total gold production and reserves from deposits of this kind in Canada. Gold-rich pyritic deposits, such as those of the Bousquet district, are also an important type of gold mineralization, and these deposits account for approximately 10% of current Canadian gold output.

Table 6.4-1. Tonnage and grade statistics for selected "gold-rich" volcanic-associated massive sulphide deposits (see Fig. 6.4-1).

Deposit	Location	Type	Age	Tonnes Ore	Au ppm	Ag ppm	Cu %	Zn %	Pb %	Tonnes Au[1]
Bousquet No. 1	Bousquet district, Quebec	Pyritic gold	Archean	20 737 000	4.5	-	-	-	-	93
Bousquet No. 2	Bousquet district, Quebec	Copper-gold	Archean	7 400 000	6.1	16	0.6	-	-	45
Dumagami	Bousquet district, Quebec	VMS (Cu-Zn-Au)	Archean	5 500 000	4.6	9	-	-	-	25
Doyon	Bousquet district, Quebec	Intrusion-hosted Au	Archean	10 243 000	5.7	-	-	-	-	58
Agnico-Eagle	Joutel, Quebec	Pyritic-gold (I.F.)	Archean	5 279 000	6.4	<10	-	-	-	34
Estrades	Joutel, Quebec	VMS (Cu-Zn-Au)	Archean	2 670 000	5.1	110	0.8	9.6	0.9	14
Horne	Noranda, Quebec	Copper-gold	Archean	54 300 000	6.1	13	2.2	-	-	330
Quemont	Noranda, Quebec	Copper-gold	Archean	13 925 000	5.4	20	1.3	2.4	0.02	75
D'Eldona	Noranda, Quebec	VMS (Cu-Zn-Au)	Archean	90 000	4.1	30	0.3	5.0	-	<1
Louvicourt	Val d'Or, Quebec	VMS (Cu-Zn)	Archean	15 700 000	0.9	30	3.4	2.2	-	14
Montauban North	Grenville, Quebec	Pyritic gold	Proterozoic	600 000	5.0	-	-	-	-	3
Flin Flon	Flin Flon, Manitoba	VMS (Cu-Zn-Au)	Proterozoic	58 416 000	3.1	50	2.3	4.3	-	181
Big Island	Flin Flon, Manitoba	VMS (Cu-Zn-Au)	Proterozoic	220 000	4.5	80	1.0	14.0	-	<1
Eskay Creek[2]	Northeastern B.C.	Auriferous polymetallic	Jurassic	3 968 000	26.0	1000	<1.0	5.0	2.0	103
Tulsequah	Northeastern B.C.	VMS (Zn-Cu-Pb-Au)	Permian	1 620 000	4.0	140	1.3	6.9	1.3	6
Lara	Vancouver Island	VMS (Zn-Cu-Pb-Au)	Devonian	529 000	4.7	100	1.0	5.9	1.2	2
H-W mine	Vancouver Island	VMS (Zn-Cu-Pb)	Devonian	13 818 000	2.4	35	2.2	5.3	0.3	33
Britannia	South central B.C.	VMS (Zn-Cu-Pb)	Cretaceous	55 000 000	0.7	<10	1.1	0.7	0.1	39
Rea Gold	South central B.C.	Auriferous polymetallic	Devonian	134 000	5.4	60	0.7	2.4	2.4	<1
Homestake	South central B.C.	Auriferous polymetallic	Devonian	220 000	7.4	70	0.5	7.3	6.2	1
Buchans	Central Newfoundland	VMS (Zn-Cu-Pb)	Ordovician	15 809 000	1.5	130	1.3	14.6	7.6	24
Rambler Cons.	Central Newfoundland	VMS (Zn-Cu)	Ordovician	399 000	5.1	30	1.3	2.2	-	2
Lemoine-Patino	Chibougamau, Quebec	VMS (Cu-Zn-Au)	Archean	750 000	4.4	90	4.8	10.6	-	3
Boliden	Skellefte district, Sweden	Copper-gold	Proterozoic	8 340 000	15.2	50	1.4	-	-	126
Mt. Morgan	East-central Queensland	Copper-gold	Devonian	50 000 000	4.8	<10	0.7	0.1	0.05	240
Mt. Chalmers	East-central Queensland	Copper-gold	Permian	3 600 000	2.0	15	1.8	1.0	0.2	7
Starra	Mt. Isa Inlier, Queensland	Pyritic-gold (I.F.)	Cambrian	5 300 000	5.0	-	2.0	-	-	26
Mt. Lyell	Mt. Read, Tasmania	Copper-gold	Cambrian	98 574 000	0.4	<10	1.2	<0.1	<0.1	39
Mt. Lyell (Blow)	Mt. Read, Tasmania	Copper-gold	Cambrian	5 600 000	2.0	60	1.3	-	-	11
Rosebery	Mt. Read, Tasmania	VMS (Zn-Cu-Pb-Au)	Cambrian	19 400 000	2.9	160	0.7	16.2	5.0	56
Que River	Mt. Read, Tasmania	VMS (Zn-Cu-Pb-Au)	Cambrian	3 100 000	3.4	200	0.6	13.5	7.5	10

Notes: [1] deposits with greater than 30 tonnes Au are considered to be major gold deposits (i.e., more than 1 million ounces of contained gold)
    [2] base metal values quoted for Eskay Creek are approximate grades for the 21B zone
    - signifies no data available
   I.F. = iron formation
  VMS = volcanic-associated massive sulphide

SIZE AND GRADE

Auriferous volcanic-associated massive sulphide deposits are characterized by grades typical of other gold deposits and by tonnages similar to those of other base-metal massive sulphide deposits (Fig. 6.4-1). Their base metal contents and gold/silver ratios are distinct from those of typical massive sulphide deposits (Fig. 6.4-2) and also differ somewhat from other types of gold deposits that occur in greenstone belts (see Fig. 15.4-4). The giant Horne deposit (54.3 Mt grading 6.1 g/t Au) and the much smaller, but high grade Eskay Creek deposit (3.95 Mt grading 26.4 g/t Au) illustrate the range of sizes and gold grades of the individual orebodies.

Many auriferous massive sulphide deposits occur in close proximity to base metal massive sulphides, but few camps contain more than one distinctly auriferous deposit. Of the 22 sulphide deposits at Noranda, the giant Horne and Quemont gold deposits have produced more than 90%

of the gold and also account for more than 65% of the total massive sulphide tonnage (Kerr and Mason, 1990). In the Skellefte district of Sweden, the Boliden deposit (7.6 million tonnes grading 15.2 g/t Au) is among the largest of 20 different massive sulphide orebodies, but also has ten times the average gold grade for the district. In contrast, all 14 deposits within the Bousquet district have high gold grades ranging from 4.3 to 9.7 g/t Au (Marquis et al., 1990b). Collectively these deposits contain nearly 60 million tonnes of ore averaging 5 g/t Au.

GEOLOGICAL FEATURES

Setting

Auriferous massive sulphide deposits occur in rocks of dominantly volcanic derivation, typical of host rocks for other volcanic-associated base metal massive sulphide deposits of all ages. In nearly all cases the paleotectonic

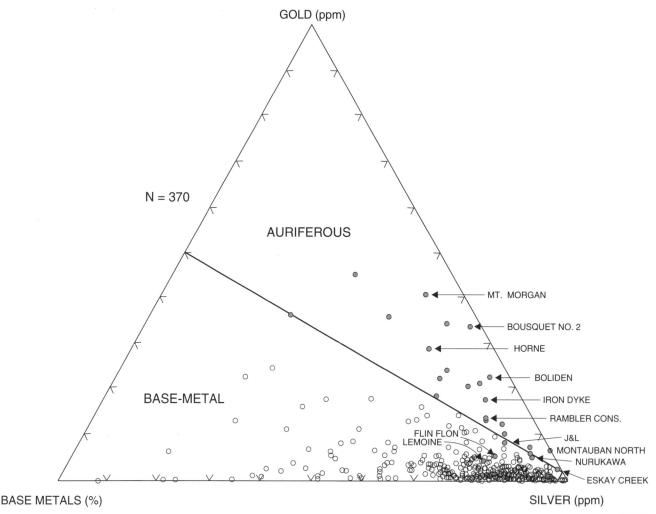

Figure 6.4-2. Ternary diagram portraying the relative abundance of gold (ppm), silver (ppm), and base metals (total per cent combined) for selected volcanic-associated massive sulphide deposits world-wide. The centrally located diagonal line can be used to make an approximate compositional separation of auriferous and base metal massive deposits. Red dots portray volcanic-associated massive sulphide deposits with high absolute abundances of gold.

GSC

settings are inferred to be those of island arcs, rifted arcs, or back-arc basins and, at regional scales, the gold deposits occur within the same lithological units as base metal deposits. The host sequences typically comprise two contrasting components: a mafic one, in the form of basalt, andesite, or amphibolite, and a felsic one, in the form of tuffs and volcanic breccias.

Archean deposits of this type in Canada include the Horne, Bousquet (No. 1 and No. 2), and Agnico-Eagle deposits, which occur in the 2.7 Ga greenstone belts of the Abitibi Subprovince. The Horne deposit occurs in a sequence of felsic fragmental rocks and flows of the Blake River Group with virtually no mafic volcanic rocks (Kerr and Gibson, 1993). The Bousquet district (Fig. 6.4-3), also within the Blake River Group, hosts important gold

Table 6.4-2. Geological characteristics of selected auriferous volcanic-associated massive sulphide deposits.

Deposit	Host rocks	Metamorphism	Structure	Intrusions	Nature of orebodies	Ore mineralogy/ chemistry	Gangue/alteration
Canadian examples:							
Horne	felsic volcaniclastics and rhyolite flows	lower greenschist	relatively undeformed, E-W cleavage	mafic, felsic dykes subvolc. tonalite	pipe-like mass. and stringer py-cp, mass. py	pyrite-pyrrhotite-chalcopyrite-tellurides-sphalerite	quartz-sericite-chlorite
Quemont	rhyolite breccias	lower greenschist	relatively undeformed, E-W cleavage	mafic, felsic dykes subvolc. tonalite	mass. and dissem. sulphides, multiple lenses	pyrite-pyrrhotite-chalcopyrite-tellurides-sphalerite	quartz-sericite-chlorite
Bousquet No.1	felsic and mafic tuffs	middle-upper greenschist	intense E-W foliation and shearing	none known	pyritic qtz-ser. schists, dissem. sulphides and foliation-parallel stringers	pyrite-chalcopyrite-sphalerite-arsenopyrite-tellurides-gudmundite	quartz-muscovite kyanite-andalusite-chloritoid
Bousquet No. 2	felsic tuffs	middle-upper greenschist	intense E-W foliation and shearing	none known	foliation-parallel stringers mass. py, remob. cp-bn	pyrite-chalcopyrite-bornite-tellurides	quartz-muscovite andalusite-kyanite-paragonite
Dumagami	felsic tuffs	middle-upper greenschist	intense E-W foliation and shearing	none known	mass. pyritic sulphides	pyrite-sphalerite-chalcopyrite-bornite	quartz-muscovite andalusite-kyanite-paragonite
Doyon	felsic tuffs, intrusion	middle-upper greenschist	intense E-W foliation	adjacent tonalite	quartz-pyrite veins	pyrite-chalcopyrite-tellurides	quartz-muscovite andalusite-kyanite-chloritoid
Agnico-Eagle	felsic tuffs and cherty sulphide-carb. iron-fm.	middle-upper greenschist	NW foliation near regional shear zone	postore diabase	chemical and clastic seds., mass. sulphides	pyrite, minor chalcopyrite and arsenopyrite	chert, Fe-carbonate, Fe-silicates
Montauban North	felsic volcanics (quartz-biotite gneiss)	amphibolite	folded N-S foliation	none known	dissem. sulphides	pyrite-sphalerite-chalcopyrite-pyrrhotite±glena	quartz-biotite garnet-cordierite-anthophyllite
Eskay Creek	felsic tuffs, breccias, mudstone at rhyolite-basalt contact	weak	N-S folds	diorite sheets	stockwork and stratiform massive sulphides	sphalerite-tetrahedrite-pyrite-galena, stibnite-realgar-cinnabar-arsenopyrite	quartz-chlorite-sericite
Non-Canadian deposits with similar characteristics:							
Boliden	dacite-rhyolite	greenschist-amphibolite	intense E-W foliation and isoclinal folds	subvolc. tonalite intrusion	multiple mass. py-cp and mass. py-aspy lenses	pyrite-pyrrhotite-arsenopyrite-chalcopyrite-tetrahedrite-tellurides	quartz-chlorite-sericite-rutile-andalusite-corundum
Mt. Morgan	felsic tuffs, siltstones, lavas and porphyries	weak	moderately deformed, intense E-W foliation	subvolc. tonalite intrusion	pipe-like dissem.-stringer py-cp, mass. py-cp	pyrite-pyrrhotite-arsenopyrite-galena-sphalerite	quartz-biotite andalusite-staurolite-sillimanite
Mt. Chalmers	rhyolite-dacite dome, volcaniclastics, siltstone	weak	relatively undeformed	qtz.-feld. porphyry, andesite sills	dissem.-stringer py-cp, minor mass. sulphides	pyrite-chalcopyrite-sphalerite-galena-barite	quartz-sericite-chlorite-dolomite
Mt. Lyell (The Blow)	felsic pyroclastics and flows	lower greenschist	moderately deformed, folded	minor int. and felsic intrusives	dissem.-stringer py-cp-bn, mass. py-cp, mass. sp-gn	pyrite-pyrrhotite-chalcopyrite-bornite± sphalerite±galena	quartz-sericite-chlorite-siderite-hematite-barite
Starra; Trough Tank	oxide facies iron-fm., mafic-felsic tuffs	middle-upper greenschist	multiphase, intense foliation and shearing	none known	iron-formation	quartz-magnetite-pyrite-chalcopyrite±scheelite	chlorite-magnetite-hematite±sericite
Nurukawa, Japan	rhyolite-dacite flows and pyroclastics	unmetamorphosed	relatively undeformed	none known	stockwork and dissem. sulphides	pyrite-chalcopyrite	quartz-sericite± chlorite- kaolinite-pyrophyllite

Abbreviations: carb., carbonate; fm., formation; qtz., quartz; feld., feldspar; mass., massive; remob., remobilized; dissem., disseminated; seds., sedimentary rocks; subvolc., subvolcanic; int., intermediate; py, pyrite; cp, chalcopyrite; bn, bornite; sp, sphalerite; gn, galena; aspy, arsenopyrite; ser., sericite.

deposits of three different types, including Bousquet No. 1 (pyritic gold), Bousquet No. 2 (copper-gold; Fig. 6.4-4A), and the Dumagami mine (auriferous polymetallic). A number of intrusion-related pyritic gold deposits also occur in the vicinity of the Bousquet mines (e.g., Doyon-Silverstack, Ellison) and these may be related to the near-surface exhalative deposits. Although in such a deformed terrane it is difficult to distinguish between volcanogenic deposits and deep-seated intrusion-related deposits, the presence of an extensive Mn-garnet "exhalite" unit and the local stratiform base metal mineralization (e.g., Warrenmac) near Doyon argues for a volcanogenic origin for the Bousquet deposits. The Agnico-Eagle pyritic gold deposit, north of Joutel, is interpreted to sit near the top of the Joutel volcanic complex, which hosts a number of gold-poor base metal deposits (Poirier, Joutel Cu, Consolidated Northern). The host rocks are interpreted to consist of a stratiform carbonate-sulphide-silicate-oxide iron-formation, intercalated with cherty sedimentary rocks and tuffs, and an apparent chloritic footwall (Barnett et al., 1982). Although the nearby base metal massive sulphides at Joutel are gold-poor, the small Estrades Cu-Zn deposit, northeast of Joutel, is anomalously gold-rich (2.67 Mt grading 5.1 g/t Au).

Montauban North is the only notable Canadian example of a Proterozoic deposit of this type, but deposits of this age elsewhere in the world include Boliden, Sweden and Yavapai, U.S.A. Gold-rich pyritic sulphides occur along strike from the Montauban Zn-Pb orebody in Middle Proterozoic gneisses of the Grenville Supergroup in Quebec. The North and South zones of the Montauban deposit produced subequal amounts of gold (0.7 t Au) and silver (0.9 t Ag) from a zone of pyrite-sphalerite-chalcopyrite mineralization associated with cordierite-anthophyllite and quartz-biotite-garnet assemblages within quartz-biotite and quartz-sillimanite gneisses (Morin, 1987). The gold zones contained as much as 30% disseminated suphides, but high gold values appear to have been independent of base metals. Although of high metamorphic grade, the quartz-plagioclase-biotite gneiss adjacent to the Montauban deposit has been interpreted to be derived mainly from felsic volcaniclastic rocks that contained local sedimentary intercalations (Morin, 1987). The Boliden auriferous massive sulphide deposit occurs in the circa 1.8 Ga greenstone belt of the Skellefte district in Sweden. The Boliden orebody consists of two large pyrite-chalcopyrite orebodies (6 Mt), which envelop several smaller arsenopyrite-rich lenses (2 Mt), and crosscutting quartz-tourmaline veins. The deposit is hosted by pyritic quartz-sericite±chlorite and andalusite-rich schists of the Skellefte Volcanics (Fig. 6.4-4B). Likewise, the stratiform pyritic gold occurrences in Yavapai County, Arizona, are of Early Proterozoic age. The Yavapai deposits (e.g., Iron Dyke) consist of gold-bearing disseminated zones and local thin massive sulphide lenses with minor base metals in cherty, quartz-sericite and quartz-chlorite schists (Swan et al., 1981).

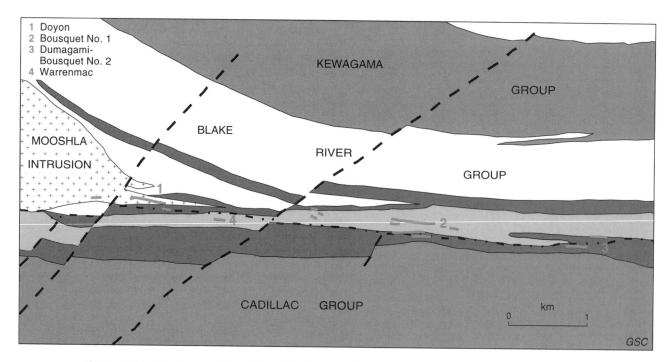

Figure 6.4-3. Regional setting of the gold-rich volcanic-associated massive sulphide deposits in the Bousquet district, Quebec (adapted after Marquis et al., 1990b). The rocks of the Blake River Group include felsic volcanics (light grey); intermediate composition tuffs and epiclastic rocks (dark grey); basalt (uncoloured). Heavy dashed and dash-dot lines represent faults and shear zones respectively, and areas in red represent the surface traces of gold orebodies and showings.

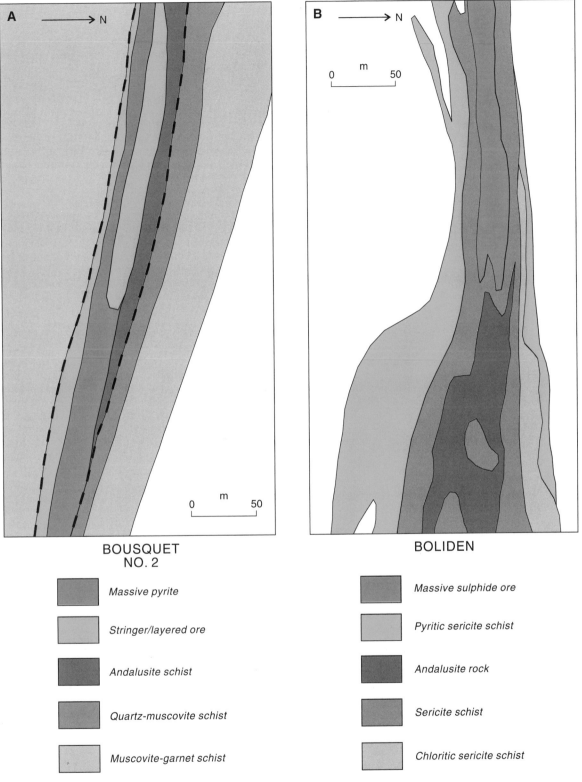

BOUSQUET
NO. 2

BOLIDEN

Massive pyrite

Stringer/layered ore

Andalusite schist

Quartz-muscovite schist

Muscovite-garnet schist

Massive sulphide ore

Pyritic sericite schist

Andalusite rock

Sericite schist

Chloritic sericite schist

GSC

Figure 6.4-4. Distribution of alteration minerals in and adjacent to auriferous massive sulphide orebodies which are shown in red: **A)** cross-section of the Bousquet No. 2 deposit, Bousquet district; dashed lines correspond to faults (adapted after Tourigny et al., 1993); **B)** comparative section through Boliden deposit, Sweden (adapted after Grip and Wirstam, 1970).

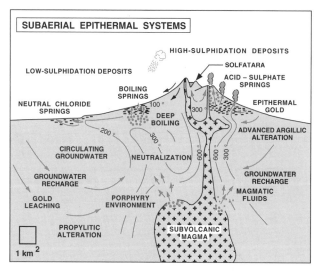

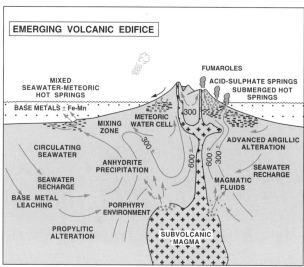

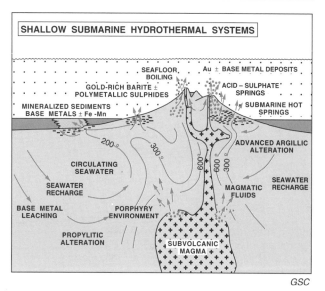

GSC

Volcanic-associated massive sulphide deposits are well known in contemporaneous felsic volcanic rocks of the Jerome area, and the Yavapai Volcanics also host the nearby Iron King base metal massive sulphide deposit (4.97 Mt grading 7.3% Zn, 2.5% Pb, 0.2% Cu, and 4.2 g/t Au).

The only Paleozoic examples of this deposit type in Canada include small, high grade auriferous polymetallic sulphide deposits that are found in volcanic rocks of Devonian age in central British Columbia (e.g., Höy, 1991). Of these, Rea Gold, Homestake, and J&L are the most gold-rich examples, but they are compositionally transitional between typical base metal massive sulphide deposits and truly auriferous deposits (Fig. 6.4-2). Important Paleozoic auriferous massive sulphide deposits do occur, however, in the Tasman Geosyncline. The large Mt. Morgan deposit in the Mt. Windsor volcanic rocks of eastern Australia has been one of the country's largest single lode-gold deposits, having produced nearly 238 t of gold. Mt. Morgan and the nearby Reward deposit are copper-gold deposits hosted by pyritic quartz-sericite schists (Large, 1992). In addition, the Mt. Lyell copper-gold deposit in Tasmania consists of massive, gold-rich pyrite ore (e.g., The Blow), siliceous barite-chalcopyrite-bornite ore (e.g., North Lyell), and disseminated pyrite-chalcopyrite ore (e.g., Prince Lyell and others), all of which are hosted by quartz-sericite and quartz-sericite-chlorite schists. These orebodies were originally worked for gold and remain among the largest gold producers in Tasmania (Large et al., 1990). A number of other uncommonly gold-rich base metal deposits are also known in the Mt. Read volcanic rocks of Tasmania (e.g., Rosebery, Hellyer, Que River, South Hercules, Mt. Charter) and resemble Canadian gold-rich base-metal massive sulphides such as Flin Flon and H.W.

Among Mesozoic and Cenozoic deposits of this age, the best example of this type occurs at Eskay Creek in British Columbia in Jurassic volcanic rocks of the Upper Hazelton Group. The ore zones at Eskay Creek are situated near the contact between rhyolite breccia, tuff, and mudstone and overlying pillowed andesite (Britton et al., 1990). These are distinctive deposits, typically consisting of small, high-grade massive sulphides which resemble the Miocene Kuroko-type, polymetallic Zn-Pb-Cu deposits of Japan. Although these gold-rich deposits are modelled after Kuroko-type massive sulphides, most Kuroko deposits have rather restricted zones of gold enrichment (e.g., Shakanai No. 1-3). A particularly high-grade, Kuroko-type copper stringer zone has recently been documented in the Nurukawa deposit in Japan (Yamada et al., 1987), and this may be analogous to some older, metamorphosed, pipe-like copper-gold deposits.

Intrusions constitute a large proportion of the rocks in most of the above districts and include dykes, subvolcanic sills, and porphyry stocks that range from mafic to felsic in composition and from pre- to post-tectonic in timing.

Figure 6.4-5. Schematic diagram of a hypothetical shallow marine to epithermal transition (after Hannington, 1993). Note the expected progression from marine to emergent epithermal conditions and the possible involvement of shallow subvolcanic intrusions.

Tonalitic subvolcanic intrusions are present at Quemont (Powell pluton) and Horne, Bousquet (Mooshla pluton), Boliden (Jarn granitoids), and Mt. Morgan (Mt. Morgan tonalite). In many cases the intrusions are hydrothermally altered and contain vein- and stockwork-style mineralization that locally constitutes significant gold ore in itself (e.g., Doyon deposit in the Mooshla intrusion).

Regional dynamothermal metamorphism of low to medium grade has affected the rocks in most of the older greenstone belts that contain gold deposits of this type. Middle to upper greenschist metamorphic conditions are inferred at Bousquet and Agnico-Eagle, and metamorphism at Montauban fully attained the amphibolite facies. The Boliden deposit in Sweden occurs in rocks at the transition from greenschist to amphibolite facies. The deposits at Noranda and Eskay Creek are virtually unmetamorphosed.

Structure

Regional metamorphism of the older greenstone belts has resulted in penetrative deformation of many of the deposits and the development of at least one generation of tectonic fabric that overprints the mineralization. In most cases, the degree of deformation and transposition has obscured many of the primary relationships between the ore deposits and their host rocks, and significant remobilization of gold has complicated the paragenetic relationships between gold and sulphide mineralization, leading to considerable debate about possible syntectonic versus synvolcanic origins for much of the gold. The existence of well preserved, relatively unmetamorphosed gold deposits of this type in younger volcanic sequences, and detailed mapping at a number of metamorphosed examples, indicate that a synvolcanic origin for gold in many cases is valid (see "Genetic models" below).

A common feature of older, metamorphosed deposits is a strong foliation, amplified within discrete shear zones, that strikes subparallel to the regional lithological trend. In the Bousquet district, pyritic gold and copper-gold deposits are contained within a zone of intense brittle-ductile deformation known as the Dumagami structural zone (Tourigny, 1991). Here the local structure is dominated by metre-scale anastomosing shear zones in which the transposition of bedding into parallelism with the inherited schistosity is common (Tourigny et al., 1989; Marquis et al., 1990a; Tourigny et al., 1993). Such transposition accounts for the "straightness" of the rock units (e.g., Fig. 6.4-3) and usually obscures any original relationships between the ore deposits and their host rocks. Strong linear fabrics, such as the axes of asymmetric minor folds and mineral and shape lineations, are also characteristic of these areas and are of consistent orientation within a district. Moderate to steep plunges have been noted at Bousquet, whereas fold hinges and lineations of shallow plunge are common at Montauban. The relationship of gold mineralization to the mixed volcanic, chemical, and clastic sedimentary host rocks at Agnico-Eagle is also complicated by the fact that the rocks have a moderate tectonic fabric, and the presence of gold within late carbonate veins and fractured pyrite is evidence for introduction or remobilization of gold during brittle deformation (Wyman et al., 1986). The Boliden deposit is tightly isoclinally folded, and a well developed axial planar schistosity gives the appearance of a shear zone (Grip and Wirstam, 1970; Rickard, 1986).

Orebodies

Auriferous massive sulphide orebodies may have formed directly on the seafloor as massive sulphide accumulations or in the immediate subseafloor as stratiform replacements. However, many of the auriferous volcanic-associated pyritic and copper-gold deposits consist largely of disseminated and stockwork-like vein systems and are, strictly speaking, not massive sulphide ores. These deposits are commonly pipe-like bodies with relatively minor massive stratiform sulphide accumulations (e.g., Mt. Morgan, Boliden). Nevertheless, the presence of minor amounts of exhalative sulphides in most cases argues strongly for a submarine setting. Such deposits are generally considered to be stratabound at the scale of a district (e.g., within or along strike from well-defined lithotectonic packages of rocks), even where transposition has resulted in structural and lithological trends that are parallel to one another. Furthermore, the deposits commonly occur at contacts between distinctive lithological units or solely within a particular unit. The lenticular to tabular shapes of most orebodies is such that they are geometrically concordant with their host rocks, even though the main sources of ore-grade massive sulphides may be discordant (e.g., Bousquet No. 1 and No. 2); where deformation has been extreme, the long axes of orebodies are commonly parallel to other linear fabrics in the district. Where deformation is less severe (e.g., Horne, Eskay Creek) the concordance of orebodies and host rocks is more clearly related to pretectonic processes. Orebody thicknesses, especially for the massive sulphide type deposits, range from 2 m to 60 m, a factor which, along with consistent ore grades, has made many of them amenable to open pit mining methods (Bousquet, Mt. Morgan, Boliden).

Ore composition

Specific deposits in this category display considerable variation in both the distribution and composition of ore types.

The Horne mine was a pipe-like copper-gold deposit (Upper H and Lower H orebodies) capped by a large, lower-grade pyritic massive sulphide lens (No. 5 zone). Most of the ore produced from the mine consisted of massive pyrite-pyrrhotite-chalcopyrite and associated stringer mineralization from the Upper H and Lower H orebodies. Gold was present mainly as Au-Ag tellurides and as native metal.

At Bousquet most ores contain 5-20% pyrite, locally substantial amounts of chalcopyrite, and lesser quantities of other sulphide and telluride minerals. The auriferous suphides are interpreted to be synvolcanic in origin, and subsequent deformation and remobilization of the ore constituents are responsible for many of the veins present (Tourigny et al., 1989). The Bousquet No. 1 mine consists of gold-bearing, pyritic quartz-muscovite schists with multiple, foliation-parallel and foliation-oblique, disseminated and vein-type ore lenses. Bousquet No. 2 has characteristics of both pyritic gold and base-metal-rich deposits. The orebodies consist of foliation-parallel, pyritic stringers, massive pyritic sulphides, sulphide breccias, chalcopyrite-rich layered stringer ore with minor sphalerite and galena, and semimassive remobilized bornite-chalcopyrite veins (Tourigny et al., 1993). This latter ore type accounts for the major part of the gold mineralization in Bousquet No. 2. As in Bousquet No. 1, the mineralization is largely structurally-controlled, but mainly as the result of metamorphic remobilization of a synvolcanic copper-gold protore. The ore mineralogy is also unusually complex, consisting of pyrite, chalcopyrite, bornite, digenite, chalcocite, sphalerite, galena, tennantite, mawsonite, colusite, native gold, and Au-, Ag-, and Pb-tellurides. The Dumagami orebodies consist of massive pyritic suphides, which include abundant sphalerite, chalcopyrite, and bornite, and most closely resemble exhalative massive sulphides (Marquis et al., 1990a, b).

The Agnico-Eagle gold deposit consists of auriferous chemical and volcaniclastic sediments, which contain about 20% disseminated pyrite and local massive to semimassive stratiform pyrite (Barnett et al., 1982; Wyman et al., 1986). Minor amounts of base-metals are also present. Fine grained gold occurs within distinct pyritic laminae and within late carbonate veins and fractures.

The mineralization at Eskay Creek is exceptionally high grade (e.g., 26 g/t Au and 100 g/t Ag) and consists of stockwork and disseminated sulphides in an epithermal-style vein system (21A zone), as well as stratiform, bedded sulphides that include sphalerite, galena, tetrahedrite, stibnite, realgar, cinnabar, and arsenopyrite (21B zone) (Britton et al., 1990; Roth, 1993). The mineralization is hosted in a quartz-sericite-chlorite assemblage derived from altered rhyolite breccias and tuffs.

Mt. Morgan consists of a large pyritic massive sulphide lens capped by stratiform pyritic and cherty horizons (Main Pipe orebody) and a siliceous zone of disseminated and stringer mineralization (Sugarloaf orebody), which together form a pipe-like body, similar to the Horne deposit (Taube, 1986). The ore mineral assemblage is pyrite, pyrrhotite, chalcopyrite, magnetite, and minor sphalerite, together with native gold and gold tellurides. In addition, nearly 72 t of native gold (30% of the total production) was recovered from a 2.4 million tonne gossan cap with an average grade of 30.6 g/t Au. In the Mt. Morgan area, there are also several conventional volcanogenic massive sulphides (Ajax, Upper Nine Mile Creek), as well as porphyry-related mineralization and alteration (Taube, 1990). The nearby Mt. Chalmers copper-gold deposit has a somewhat higher base-metal-to-gold ratio, but also possesses a gold-rich pyritic massive sulphide body and associated copper-stringer zone (Large and Both, 1980).

Boliden ore averaged 15.2 g/t Au from 8.34 million tonnes mined over a 43 year period. Although pyrite is the dominant sulphide, the major ore minerals are chalcopyrite, arsenopyrite, and sphalerite; minor galena, pyrrhotite, and locally abundant sulphosalts are present. Arsenic, silver, cobalt, selenium, and mercury were also enriched in the Boliden ores; the average arsenic grade for the mine was 6.9 wt.% As, making it one of the largest arsenic deposits in the world. Gold occurs mainly as native metal and electrum and shows a positive correlation with arsenic. Bonanza-type gold mineralization also occurred locally in large crosscutting quartz-tourmaline veins, as well as in arsenopyrite-rich ore lenses (e.g., 600 g/t Au over 1 m in the Gold Rise and 200 g/t Au over 2 m in the root zone of the Eastern Ore arsenopyrite lenses).

As a group, deposits of this type are distinguishable mainly by their high gold content relative to base metals (Fig. 6.4-2) and in some cases by their unique mineralogy. However, their variability in bulk composition is broadly similar to that of massive base metal sulphide deposits of similar age and host rock lithology (e.g. subtype 6.3). The pyritic gold (e.g., Bousquet No. 1) and copper-gold (e.g., Horne) deposits typically have much higher Au/Ag ratios than auriferous polymetallic sulphides, the latter sometimes having exceptional silver concentrations (e.g., Eskay Creek). The nature of the ore-bearing material differs from deposit to deposit, with highly variable proportions of sulphide-to-silicate host rock. The ore mineral assemblages commonly show greater complexity than in similar gold-poor massive sulphides and may include a variety of minor minerals such as bornite, arsenopyrite, tellurides, and high-sulphidation minerals (e.g., enargite-tennantite) which are less common in conventional base-metal massive sulphides (e.g., Table 6.4-1). Notable concentrations of massive to semimassive bornite ore, together with abundant and complex assemblages of sulphosalt minerals, are common in many pyritic and copper-gold deposits of this type (e.g., Bousquet No. 2, North Lyell, Boliden) and in some auriferous polymetallic sulphides. High gold grades are also commonly associated with those parts of conventional base-metal massive sulphides with similar high-sulphidation mineral assemblages (e.g., H.W., typical Kuroko-type deposits: Hannington and Scott, 1989b). Auriferous polymetallic sulphides commonly display a wide range of minor and trace minerals, dominated by sulphosalts of silver, arsenic, antimony, lead, and mercury which are typical of epithermal gold deposits. Tetrahedrite, stibnite, realgar, cinnabar, and arsenopyrite are present in both ore types at Eskay Creek, and the 21B zone contains massive, bedded stibnite ore. These deposits are commonly capped by or associated with abundant barite mineralization, and in some cases barite may occur within auriferous stockworks (e.g., Mt. Charter and North Lyell).

Alteration

Aluminous mineral assemblages and distinctly acid alteration (e.g., alunite or pyrophyllite common) are common features of the rocks adjacent to many gold deposits of this type. Pyritic, quartz-sericite schists are the most common hosts, although in some unmetamorphosed and weakly metamorphosed deposits, advanced argillic alteration (quartz, kaolinite, pyrophyllite, and other clay minerals) is well preserved. Numerous Kuroko-type deposits in Japan are characterized by clay-rich alteration minerals, and these assemblages are recognized in a number of the younger auriferous polymetallic sulphide deposits (e.g., Marumo, 1990). The aluminous nature of these rocks is similar, in many respects, to that ascribed to alunite-kaolinite alteration associated with certain porphyry deposits and high-sulphidation epithermal gold deposits (see subtype 15.1). However, the significance of this alteration in relation to gold mineralization in auriferous sulphide deposits is not fully understood, and similar alteration is in some cases found in association with gold-poor massive sulphides (e.g., Mattabi, Bathurst-Norsemines). Typical base-metal sulphide deposits are noted for their distinctive chloritic, footwall alteration "pipes", but many gold-only deposits possess an enveloping alteration halo consisting dominantly of sericite and silica. Nevertheless, some gold-only deposits also have discordant alteration "pipes" that are not noticeably different from those of conventional base-metal massive sulphides and reflect ordinary seafloor hydrothermal activity. For example, the main mine formation at the Horne deposit consists of rhyolite flows and felsic volcaniclastic rocks, and the No. 5 pyritic lens is contained within a well bedded tuffaceous unit. All of the felsic rocks have been altered to quartz-sericite and quartz-chlorite-sericite assemblages.

The alteration in the Bousquet district (Fig. 6.4-4A) is typical of metamorphosed deposits of this type. The altered rocks are strongly foliated quartz-muscovite±andalusite±kyanite schists that locally contain pyrophyllite and diaspore, and are interpreted to be the metamorphosed equivalents of advanced argillic alteration (Valliant et al., 1983; Tourigny et al., 1989). At Dumagami, peraluminous alteration also hosts the massive pyrite and massive sphalerite-galena bodies, and is surrounded by a sericitic envelope (Marquis et al., 1990a). The aluminous nature of the alteration associated with many pyritic gold deposits of this type is also obvious in other metamorphosed Archean and Proterozoic examples. For example, at Boliden, alteration is dominated by quartz-sericite-andalusite in close proximity to the ore, and this core is surrounded by a chloritic outer envelope (Fig. 6.4-4B). The laminated quartz-sericite schists containing abundant andalusite, and local kyanite, that occur on both sides of certain of the Bousquet deposits (Valliant et al., 1983; Marquis et al., 1990a) are remarkably similar in disposition to those at Boliden (Fig. 6.4-4B).

In addition to aluminosilicates, ferromagnesian aluminous minerals, such as chloritoid, staurolite, cordierite, and garnet, are noteworthy in many of these deposits (Fig. 6.4-4). Chloritoid and manganiferous garnets are also notable at Bousquet (Valliant and Barnett, 1982). At Montauban, a unit of sillimanite gneiss envelops the deposit, and cordierite, anthophyllite, and manganiferous garnets are locally abundant (Morin, 1987; Jourdain et al., 1987).

DEFINITIVE CHARACTERISTICS

As a group, auriferous sulphide deposits possess a number of definitive characteristics.

1. The deposits have low contents of base metals relative to gold (i.e., less than one per cent combined base metals for each part per million gold).

2. Layered, stratiform massive sulphides typically contribute to at least some of the ore, although the most gold-rich zones may be restricted to disseminated and stockwork-like feeders.

3. Sulphide ores commonly include a complex assemblage of minor and trace minerals such as bornite, sulphosalts, arsenopyrite, tellurides, and other high-sulphidation minerals, locally with high concentrations of the epithermal suite of elements (e.g., Ag, As, Sb, Hg).

4. Individual orebodies are commonly associated with zones of sericitic alteration and silicification and, in some cases, are enclosed by aluminous, acid alteration zones; the presence of magnesium-rich and manganiferous alteration minerals may be indicative of seafloor hydrothermal alteration, and abundant carbonate may indicate boiling.

5. The deposits typically occur together with conventional base metal massive sulphide deposits and share many of their geological characteristics; they are typically stratabound at the district and deposit scale and commonly occur at or near interfaces between felsic volcanic rocks and either mafic volcanic or clastic sedimentary rocks.

GENETIC MODEL

Currently two genetic models are applied to these deposits:

1. They are viewed to be variants of conventional massive sulphide deposits that are distinguished by inherently anomalous fluid chemistry and/or deposition within a shallow marine to subaerial volcanic setting in which boiling may have had a significant impact on the chemistry of the ore fluids; or

2. They are viewed to be syntectonic sulphide replacement deposits in shear zones or, at least, as massive sulphide deposits that have been overprinted by gold-bearing fluids during regional deformation and metamorphism.

In the first case, gold deposits of this type are seen as transitional between submarine base metal sulphide deposits and terrestrial epithermal gold deposits (Fig. 6.4-5). The Eskay Creek deposit, which is relatively undeformed and possesses unequivocal exhalative affinities, is an example of a distinctly gold-rich massive sulphide which likely formed in very shallow water (Britton et al., 1990). For deposits of this kind that have escaped regional metamorphism and significant deformation, the relationship between gold enrichment and massive sulphide mineralization is usually obvious and provides strong evidence for a synvolcanic origin for the gold. The genetic models for these deposits are essentially variations on those for other volcanogenic sulphide deposits, and recent discoveries of gold-rich massive sulphides actively forming on the ocean floor have provided unique opportunities to study the processes of gold mineralization in this environment.

The concept that some pyritic gold deposits are not inherently gold-rich, but were overprinted by auriferous fluids during deformation and metamorphism, is founded in the fact that many of the type localities for these deposits are strongly deformed and that gold and its associated minerals, at the mesoscopic and microscopic scales, exhibit a late textural paragenesis. Features such as sulphide veins that are discordant to regional foliation and ore zones that locally are parallel to foliation, but at a high angle to transposed bedding, are particularly common in the Bousquet district and similar structural complexities have also been identified at Agnico-Eagle. The important structural controls in these districts have led to alternate hypotheses concerning the emplacement of the ore: (1) the deposits are wholly syntectonic in origin – i.e., sulphide replacements in shear zones (note that this same interpretation was applied to the Boliden and Horne deposits earlier in this century); (2) the sulphide ores are pretectonic and synvolcanic, but some or all of the gold has been superimposed on them during regional tectonism (e.g., Wyman et al., 1986; Tourigny et al., 1989; Marquis et al., 1990a); (3) the sulphide ores are pretectonic and inherently auriferous, and the regional metamorphism and deformation served only to modify the deposits and locally remobilize some of the ore constituents into structurally controlled sites (Barnett et al., 1982; Valliant and Barnett, 1982; Valliant et al., 1983; Tourigny et al., 1993). Although the third hypothesis is compelling, it is unlikely that the alternatives can be conclusively excluded in the case of Bousquet and Agnico-Eagle.

In older metamorphosed terranes, the apparent stratigraphic controls on the location of the deposits at a district scale is the strongest point in favour of a volcanic exhalative model. In most districts, pyritic gold deposits that have all of the attributes of volcanogenic massive sulphide deposits occur within the same stratigraphic sequences and adjacent to conventional base metal deposits. The Bousquet gold orebodies occur within the same sequence as the Dumagami Zn-Cu-Au deposit, and the Montauban gold orebodies occur immediately along strike from the Montauban Zn-Pb deposit. In both of these cases the base metal deposits are considered by most geologists to be of volcanogenic massive sulphide type. In a re-examination of the geology of the Horne deposit, Kerr and Mason (1990) offered numerous reasons why late tectonic superposition of gold is unlikely. Foremost among these are the observations that, locally, an unmineralized debris flow deposit has unconformably cut down into the gold-copper mineralization and that otherwise barren pyroclastic tuffs in the stratigraphic-hanging wall contain blocks of the underlying gold-rich massive sulphides. It also has been argued that, because inherently auriferous chemical and volcaniclastic sediments are common host rock for some gold deposits of this type and because high gold grades occur in regional chemical sediments away from the deposits, a volcanogenic origin for at least some of the gold is likely. If one accepts a volcanic exhalative origin for some of these deposits, the most relevant inquiry pertains to the controls on mineralization that distinguish them from other volcanic-associated deposits containing relatively little gold.

Some insight on gold in volcanogenic massive sulphide systems has also been gained from an evaluation of data from hydrothermal vents on the modern seafloor (Hannington and Scott, 1989a; Hannington et al., 1991). Analysis of vent fluids and precipitates from the seafloor shows that submarine hydrothermal systems have enormous capacities to transport gold. Large differences in the gold contents of seafloor sulphide deposits can be explained either in terms of gold-enriched sources or in terms of a favourable mechanism by which to effectively concentrate gold from solution. Gold may be precipitated at low temperatures by oxidation of aqueous sulphur complexes during mixing with cold seawater or at higher temperatures from chloride complexes in fluids ascending through the volcanic pile. The locus of mixing, extent of sulphide-sulphate equilibrium, and degree of interaction between the fluid and wall rocks will be important controls on the efficiency and site of gold deposition, and such factors have been used to explain gold enrichment in both modern and ancient seafloor sulphide deposits (e.g., Hannington and Scott, 1989a, b; Large et al., 1989). Boiling is an additional factor that may have a major impact on the nature and efficiency of gold precipitation, but, although the efficiency of gold deposition largely determines whether or not a deposit will be gold-rich, the possibility of gold-enriched source fluids for some deposits cannot be excluded. Some gold-rich seafloor precipitates are characterized by high-sulphidation mineral assemblages in association with advanced argillic alteration and resemble epithermal systems on land. This style of mineralization and alteration is thought to reflect direct input of magmatic volatiles to the hydrothermal fluids, and this input may also be responsible for significant contributions of gold (see "Related deposit types" below).

The above observations suggest that there may be a number of end-member explanations for primary gold enrichment in volcanic-associated sulphide deposits. In the first case, gold enrichment is a consequence of the efficient precipitation of gold from cooling hydrothermal fluids and the continuous hydrothermal reworking of gold into high-grade zones within the sulphide deposits. In the second case, sustained boiling allows for the effective separation of gold from base metals and the deposition of a significantly gold-enriched massive sulphide at the seafloor. The third case is one in which a uniquely gold-rich fluid may be produced by direct contributions from a high-level degassing magma (e.g., a porphyritic subvolcanic intrusion) into the submarine hydrothermal system. When abundant volcanic gases are introduced, a strongly acidic fluid chemistry may evolve which, combined with conditions that are shallow enough for boiling, may promote alteration and mineralization similar to that of subaerial epithermal systems. The proposed conditions that fit this scenario might be those of an emerging (or submerging) volcanic arc (e.g., Fig. 6.4-5). This model is attractive in that it allows for co-existence of gold-rich sulphides with gold-poor varieties, perhaps on the same volcanic edifice but at different water depths.

RELATED DEPOSIT TYPES

Three other distinct classes of gold deposits appear to share many of the attributes of auriferous volcanic-associated sulphide deposits:

1. High sulphidation epithermal Cu-Au and porphyry Cu-Au deposits.

A number of the auriferous volcanic-associated sulphide deposits described above have similarities to high-sulphidation epithermal gold deposits (e.g., acid alteration, a history of boiling, separation of gold and base metals, and possible contributions from magmatic fluids). High-sulphidation epithermal copper-gold deposits, also commonly referred to as being of acid-sulphate-type, alunite-kaolinite-type or enargite-type, are characterized by "high sulphur" mineral assemblages (e.g., enargite-tennantite) and by advanced argillic or acid alteration (e.g., alunite or pyrophyllite common). These conditions are thought to arise from acidic fluids in which SO_2 is contributed directly as a volcanic gas. The associated mineralization is similar to auriferous sulphide deposits in their relative contents of gold, silver, and base metals, although these deposits typically form as subsurface replacements (e.g., in tuffaceous rocks) within subaerial andesitic volcanic complexes. Subaerial examples have also commonly developed above high-level porphyry stocks in a geometric configuration that is virtually identical to that of submarine volcanic-associated massive sulphide deposits and their subvolcanic intrusions (Fig. 6.4-5). In some emerging (or submerging) arc settings, there may be a continuum between subaerial epithermal gold deposits and auriferous submarine massive sulphides, the main variables being water depth and the degree of interaction between evolved seawater and rising magmatic vapours (Hannington, 1993). Recently, a number of high-grade, gold-rich base metal deposits with distinctive epithermal characteristics have been recognized in volcanic arcs of the Pacific Rim. As well, a number of modern analogues of high-sulphidation and Eskay-type deposits may be forming on the modern seafloor in several back-arc basins and submerged volcanic arcs (e.g., Hine Hina vent field, Lau Basin: Herzig et al., 1993; Jade deposit, Okinawa Trough: Halbach et al., 1993).

In some volcanic-associated auriferous sulphide districts, there is a close spatial, and perhaps genetic, relationship between the gold deposits and high-level porphyry systems. The Doyon deposit in the Bousquet district may be an example. Unlike other gold deposits in the area, Doyon possesses most of the features of subvolcanic porphyry gold deposits (e.g., pretectonic stockworks, veins and disseminations in trondhjemite-diorite host). The relationship between subvolcanic intrusions and gold deposits of the type described above is also well illustrated at Mt. Morgan, Australia which has been interpreted to be both a massive sulphide deposit (Taube, 1986 and other authors) and an intrusion-related sulphide replacement deposit (Arnold and Sillitoe, 1989).

2. Disseminated and replacement gold deposits.

Auriferous volcanic-associated sulphide deposits of the type described above also share some characteristics with other disseminated, stockwork, and replacement gold deposits, particularly those that are volcanic-hosted. Deposits such as Hope Brook, Hemlo, and Equity Silver (see subtype 15.4) share moderately aluminous mineral assemblages and comprise mainly disseminated to locally massive sulphides. These deposits occur in moderately to strongly deformed terranes and it may be difficult to distinguish them definitively from other types of sulphide deposits. The absence of distinctive exhalative units and the local primary discordance of the ore zones with respect to stratigraphic units may be important distinguishing criteria.

3. Vein Cu-Au deposits.

The copper-gold deposits of the Chibougamau and Chapais camps are dominantly sulphide-rich veins, many of which are hosted by shear zones in mafic subvolcanic intrusions (i.e., the Dore Lake anorthosite complex and Cummings mafic-ultramafic complex). Although these deposits occur in late tectonic shear zones, the mineralization appears to be close in age to that of the mafic sills and may be partly synvolcanic. In a morphological and compositional sense there is considerable similarity between the sulphide-rich veins at Chapais and Bousquet No. 2. A number of Cu-Zn massive sulphides also occur in the hanging wall volcanic rocks of the Dore Lake anorthosite at Chibougamau (e.g., the Lemoine-Patino that contained 0.75 Mt grading 4.4 g/t Au) and in volcanic rocks adjacent to the Chapais deposit (No. "8-5" zone that contains 50 000 t grading 0.4 g/t Au at the Cooke mine).

EXPLORATION GUIDES

On a regional scale, gold-rich volcanic-associated massive sulphide deposits occur at major lithological contacts that mark distinctive changes in volcanic and sedimentary facies, and that typically host other massive sulphide deposits that may not be particularly gold-rich. The close spatial relationship between deposits of both types at Noranda, Joutel, Bousquet, and Montauban suggest that auriferous deposits may exist in any base metal camp. At the mine scale, the presence of acid alteration and aluminous mineral assemblages in rocks in which they are not normally expected may be a useful exploration guide, as these features are well developed at Bousquet and Boliden. The sulphide contents of many of these deposits are sufficient to produce geophysical responses and, owing to the disseminated to massive nature of the sulphides, induced polarization methods should be the most effective geophysical tools.

SELECTED BIBLIOGRAPHY

References marked with asterisks (*) are considered to be the best sources of general information on this deposit subtype.

Arnold, G.O. and Sillitoe, R.H.
1989:   Mount Morgan copper-gold deposit, Queensland, Australia: evidence for an intrusion-related replacement origin; Economic Geology, v. 84, p. 1805-1816.

Barnett, E.S., Hutchinson, R.W., Adamcik, A., and Barnett, R.
1982:   Geology of the Agnico-Eagle deposit, Quebec; in Precambrian Sulphide Deposits, (ed.) R.W. Hutchinson, C.D. Spence, and J.M. Franklin; Geological Association of Canada, Special Paper 25, p. 403-426.

Britton, J.M., Blackwell, J.D., and Schroeter, T.G.
1990:   #21 Zone deposits, Eskay Creek, northwestern British Columbia; in Exploration in British Columbia 1989, (ed.) B. Grant and J. Newell; British Columbia Geological Survey Branch, p. 197-223.

Grip, E. and Wirstam, A.
1970:   The Boliden sulphide deposit; Sveriges Geologiska Undersökning, v. C651, 68 p.

Halbach, P., Pracejus, B., and Marten, A.
1993:   Geology and mineralogy of massive sulfide ores from the Central Okinawa Trough, Japan; Economic Geology, v. 88, p. 2210-2225.

Hannington, M.D.
1993:   Shallow submarine hydrothermal systems in modern island arc settings; The Gangue, no. 43, p. 6-8.

Hannington, M.D. and Scott, S.D.
*1989a: Gold mineralization in volcanogenic massive sulfides: implications of data from active hydrothermal vents on the modern sea floor; Economic Geology, Monograph 6, p. 491-507.
1989b: Sulfidation equilibria as guides to gold mineralization in volcanogenic massive sulfides: evidence from sulfide mineralogy and the composition of sphalerite; Economic Geology, v. 84, p. 1978-1995.

Hannington, M.D., Herzig, P.M., and Scott, S.D.
1991:   Auriferous hydrothermal precipitates on the modern sea floor; in Gold Metallogeny and Exploration, (ed.) R.P. Foster; Blackie, Glasgow and London, p. 249-282.

Herzig, P.M., Hannington, M.D., Fouquet, Y., von Stackelberg, U., and Petersen, S.
1993:   Gold-rich polymetallic sulfides from the Lau back arc and implications for the geochemistry of gold in sea-floor hydrothermal systems of the southwest Pacific; Economic Geology, v. 88, p. 2182-2209.

Höy, T.
1991:   Volcanic massive sulphide (VMS) deposits in British Columbia; Chapter 5 in Ore Deposits, Tectonics and Metallogeny in the Canadian Cordillera; Ministry of Energy Mines and Petroleum Resources, Paper 1991-4, p. 89-123.

Jourdain, V., Roy, D.W., and Simard, J-M.

1987: Stratigraphy and structural analysis of the North Gold Zone at Montauban-les-mines, Quebec; Canadian Institute of Mining and Metallurgy Bulletin, v. 80, p. 61-66.

Kerr, D. and Gibson, H.L.

1993: A comparison of the Horne volcanogenic massive sulfide deposit and intracauldron deposits of the Mine Sequence, Noranda, Quebec; Economic Geology, v. 88, p. 1419-1442.

Kerr, D.J. and Mason, R.

1990: A reappraisal of the geology and ore deposits of the Horne mine complex at Rouyn-Noranda, Quebec; in The Northwestern Quebec Polymetallic Belt, (ed.) M. Rive, P. Verpaelst, Y. Gagnon, J.-M. Lulin, G. Riverin, and A. Simard; The Canadian Institute of Mining and Metallurgy, Special Volume 43, p. 153-166.

Large, R.R.

1992: Australian volcanic-hosted massive sulfide deposits: features, styles, and genetic models; Economic Geology, v. 87, p. 571-500.

Large, R.R. and Both, R.A.

1980: The volcanogenic sulfide ores at Mt. Chalmers, eastern Queensland; Economic Geology, v. 75, p. 992-1009.

Large, R.R., Huston, D.L., McGoldrick, P.J., McArthur, G., Wallace, D., Carswell, J., Purvis, G., Creelman, R., and Ramsden, A.

1990: Gold in western Tasmania: Australasian Institute of Mining and Metallurgy, Monograph 17, p. 71-82.

Large, R.R., Huston, D.L., McGoldrick, P.J., Ruxton, P.A., and McArthur, G.

1989: Gold distribution and genesis in Australian volcanogenic massive sulfide deposits and their significance for gold transport models; Economic Geology, Monograph 6, p. 520-536.

*Marquis, P., Hubert, C., Brown, A.C., and Rigg, D.M.

1990a: Overprinting of early, redistributed Fe and Pb-Zn mineralization by late-stage Au-Ag-Cu deposition at the Dumagami mine, Bousquet district, Abitibi greenstone belt, Quebec; Canadian Journal of Earth Sciences, v. 27, p. 1651-1657.

1990b: An evaluation of genetic models for gold deposits of the Bousquet district, Quebec, based on their mineralogic, geochemical, and structural characteristics; Canadian Institute of Mining and Metallurgy, Special Volume 43, p. 383-399.

Marumo, K.

1990: Genesis of kaolin minerals and pyrophyllite in Kuroko deposits of Japan: implications for the origins of the hydrothermal fluids from mineralogical and stable isotope data; Geochimica et Cosmochimica Acta, v. 53, p. 2915-2924.

Morin, G.

1987: Gîtologie de la region Montauban; Ministère de l'Energie et des Ressources, Québec, Report MM-86-02, 59 p.

Rickard, D.

1986: The Skellefte Field; 7th International Association on the Genesis of Ore Deposits Symposium Excursion No. 4, Sveriges Geologiska Undersökning, ser. Ca, v. 62, p. 54.

Roth, T.

1993: Geology and alteration in the 21A zone, Eskay Creek, northwestern British Columbia; MSc. thesis, University of British Columbia, Vancouver, British Columbia, 186 p.

Swan, M.M., Hausen, D.M., and Newell, R.A.

1981: Lithological, structural, chemical and mineralogical patterns in a Precambrian stratiform gold occurrence, Yavapai County, Arizona; in Process Mineralogy, Extractive Mineralogy, Mineral Exploration, and Energy Resources, (ed.) D.M. Hausen and W.C. Park; Proceedings 110th American Institute of Mining, Metallurgical and Petroleum EngineersAnnual Meeting, p. 143-157.

Taube, A.

1986: The Mount Morgan gold-copper mine and environment, Queensland: a volcanogenic massive sulfide deposit associated with penecontemporaneous faulting; Economic Geology, v. 81, p. 1322-1340.

1990: Mount Morgan gold-copper deposit, Queensland, Australia: evidence for an intrusion-related replacement origin - a discussion; Economic Geology, v. 85, p. 1947-1955.

Tourigny, G.

1991: Archean volcanism and sedimentation in the Bousquet gold district, Abitibi greenstone belt, Quebec: implications for stratigraphy and gold concentration - alternative interpretation and reply; Geological Society of America Bulletin, v. 103, p. 1253-1257.

Tourigny, G., Brown, A.C., Hubert, C., and Crepeau, R.

1989: Syn-volcanic and syntectonic gold mineralization at the Bousquet Mine, Abitibi greenstone belt, Quebec; Economic Geology, v. 84, p. 1875-1890.

*Tourigny, G., Doucet, D., and Bourget, A.

1993: Geology of the Bousquet 2 Mine: an example of a deformed, gold-bearing, polymetallic sulfide deposit; Economic Geology, v. 88, p. 1578-1597.

Valliant, R.L. and Barnett, R.L.

1982: Manganiferous garnet underlying the Bousquet gold orebody, Quebec: metamorphosed manganese sediment as a guide to gold ore; Canadian Journal of Earth Sciences, v. 19, p. 993-1010.

*Valliant, R.L., Barnett, R.L., and Hodder, R.W.

1983: Aluminous rock and its relation to gold mineralization, Bousquet Mine, Quebec; Canadian Institute of Mining and Metallurgy Bulletin, v. 76, p. 811-819.

Wyman, D.A., Kerrich, R., and Fryer, B.J.

1986: Gold mineralization overprinting iron-formation at the Agnico-Eagle deposit, Quebec, Canada: mineralogical, microstructural and geochemical evidence; in Proceedings of Gold '86, an International Symposium on the Geology of Gold, (ed.) A.J. Macdonald; Toronto, 1986, p. 108-123.

Yamada, R., Suyama, T., and Ogushi, O.

1987: Gold-bearing siliceous ore of the Nurukawa kuroko deposit, Akita Prefecture, Japan; Mining Geology, v. 37, p. 109-118 (in Japanese).

Authors' addresses

J.M. Franklin
Geological Survey of Canada
601 Booth Street
Ottawa, Ontario
K1A 0E8

M.D. Hannington
Geological Survey of Canada
601 Booth Street
Ottawa, Ontario
K1A 0E8

L.J. Hulbert
Geological Survey of Canada
601 Booth Street
Ottawa, Ontario
K1A 0E8

J.W. Lydon
Geological Survey of Canada
601 Booth Street
Ottawa, Ontario
K1A 0E8

K.L. Poulsen
Geological Survey of Canada
601 Booth Street
Ottawa, Ontario
K1A 0E8

Printed in Canada

7. UNCONFORMITY-ASSOCIATED URANIUM

7. UNCONFORMITY-ASSOCIATED URANIUM

V. Ruzicka

INTRODUCTION

Unconformity-associated uranium deposits typically consist of uranium concentrations at the base of a Proterozoic sandstone sequence where it unconformably overlies pre-Middle Proterozoic metamorphic basement rocks, which commonly include graphitic pelitic units. The deposits are associated with faults or fracture zones.

The principal commodity is uranium. It is commonly accompanied by other metals, particularly Ni, Co, and As, but none of these constitute significant recoverable byproducts at present. Examples of important deposits of this type in Canada are Cigar Lake, Key Lake, Rabbit Lake, McArthur River (also known as P2 North), and Eagle Point, all in Saskatchewan (Fig. 7-1). The most notable foreign examples are the Australian deposits Ranger I and III, and Jabiluka I and II in the Pine Creek Geosyncline, Northern Territory (Ruzicka, 1993).

IMPORTANCE

In 1993 about one third of the world's (excluding the former Soviet Union and China) Reasonably Assured Resources of uranium recoverable at prices up to US$130 /kg U was of the unconformity-associated type. This proportion is increasing as a result of new discoveries, and the diminishing viability of other types of lower grade uranium resources. Deposits of this type (Table 7-1) account for a major portion of

Ruzicka, V.
1996: Unconformity-associated uranium; in Geology of Canadian Mineral Deposit Types, (ed.) O.R. Eckstrand, W.D. Sinclair, and R.I. Thorpe; Geological Survey of Canada, Geology of Canada, no. 8, p. 197-210 (also Geological Society of America, The Geology of North America, v. P-1).

Canadian uranium resources. In 1993 the annual output from unconformity-associated deposits in Canada and Australia represented about 30% of the world's total production of uranium. Canadian output in 1993 from deposits of this type was more than ten times the output from deposits of the paleoplacer (quartz-pebble conglomerate) type (see subtype 1.1) at Elliot Lake, Ontario.

SIZE AND GRADE OF DEPOSITS

Unconformity-associated uranium deposits display a wide range of size and grade (Table 7-1, Fig. 7-2). In general, these deposits are much smaller than those of the quartz-pebble conglomerate type. For example, the Cigar Lake deposit is less than one hundredth the size of the Quirke zone at Elliot Lake. The average grades range from a few tenths of one per cent U (e.g., Ranger III, 0.17% U) to as much as 12.2% U (the main pod of the Cigar Lake deposit). As a result, the amounts of contained uranium metal range from a few thousand tonnes (e.g., Cluff Lake) to more than one hundred thousand tonnes (e.g., the main pod of the Cigar Lake deposit).

Australian deposits of the unconformity-associated type (Table 7-1) are of lower grade than the Canadian, but exhibit a larger range of ore tonnage (Battey et al., 1987). Because of the relatively sharp decline in grade at the fringes of mineralization, the sizes of orebodies are rather insensitive to lowering of cutoff grades.

GEOLOGICAL FEATURES
Geological setting

The majority of Canadian uranium deposits associated with pre-Middle Proterozoic unconformities occur in the Athabasca Basin, Saskatchewan. Some deposits of this type have been discovered in the Thelon Basin, Northwest Territories, and some occurrences are also known from the Otish Basin, Quebec (Ruzicka, 1984; Fig. 7-1).

Basement under the Athabasca Basin comprises Archean and Lower Proterozoic rocks. These include granulites of the Western Craton (2.6-2.9 Ga, Sm-Nd model ages; Bickford et al., 1990) and granitoid rocks of the Wollaston Domain (2.5-2.6 Ga, U-Pb zircon method; Ray and Wanless, 1980). The latter commonly form elongate domes, which are flanked by Lower Proterozoic folded strata that include graphitic, pyritic, and aluminous pelites and semipelites, calc-silicate rocks, banded iron-formation, volcanic rocks, and greywackes. These basement rocks

experienced at least three main deformation events and various grades of metamorphism (Lewry and Sibbald, 1980) during early Proterozoic time.

The crystalline Archean and Aphebian basement rocks were subjected to peneplanation and development of regolith (Macdonald, 1985) prior to deposition of the Middle Proterozoic cover rocks of the Athabasca Group. These rocks, which compose the Athabasca Basin, consist of fluviatile and marine or lacustrine redbed sequences of unmetamorphosed, flat-lying and little disturbed sandstone, siltstone, and conglomerate. The sediments were deposited upon an intensely weathered surface or, where the regolith had been eroded, on the unaltered basement.

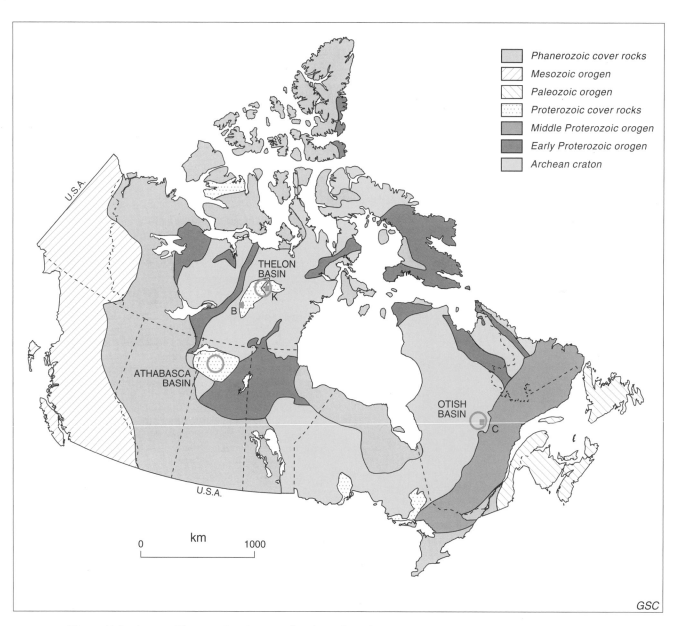

Figure 7-1. Areas with unconformity-associated uranium deposits in Canada. For location of deposits in Athabasca Basin see Figure 7-3; K = Kiggavik, formerly Lone Gull, deposit; B = Boomerang Lake deposit; C = Camie River occurrence.

There were initially three northeasterly trending tectonic depressions, the western Jackfish, central Mirror, and eastern Cree subbasins, which later coalesced into a single Athabasca Basin. Sedimentation was locally accompanied by volcanism. The clastic sedimentary rocks underwent diagenetic changes, such as silicification, hematization, clay alteration, and cementation by carbonates, and in some areas by phosphates. Layers of sulphides and organic substances, such as kerogen, occur locally (Ramaekers, 1990). Both basement rocks and cover rocks were intruded during the late Proterozoic by diabase dykes.

The deposits display two main types of metal association. Some deposits are polymetallic (U-Ni-Co-As) and occur immediately at the unconformity. Others are **monometallic** (U) and generally occur either below or (rarely) above the unconformity.

Deposits with mainly monometallic (U) mineralization generally occur either in basement rocks or in the upper parts of the Middle Proterozoic sedimentary sequence. For instance, the Rabbit Lake, Eagle Point, Raven, Horseshoe, and Dominique-Peter deposits (Fig. 7-1 and 7-3; Table 7-2) are within altered basement rocks beneath the unconformity; they are confined to various horizons of the Lower Proterozoic sequences, such as the Wollaston Group and Peter River Group. The Fond-du-Lac deposit is within sandstone in the cover rocks, some distance above the unconformity. The more recently discovered McArthur River (P2 North) deposit, which consists of monometallic mineralization directly at the unconformity, represents the only known exception from this rule. Several monometallic uranium deposits associated with the sub-Thelon unconformity occur in the Kiggavik Trend, which approximately parallels the southeastern margin of the Thelon Basin, Northwest Territories.

Table 7-1. Reasonably Assured and Estimated Additional Resources of uranium (including past production) in selected unconformity-associated deposits (data from Battey et al., 1987, and Geological Survey of Canada database).

Deposit	Ore (kt)	Grade (% U)	U (t)	Status as of 1992
MONOMETALLIC				
Canada				
Claude	583	0.36	2097	Depleted
Cluff Lake 'N'	505	0.34	1729	Dormant
Cluff Lake 'OP'	60	0.28	150	Depleted
Dominique-Janine	23	3.8	874	Producing
South Dominique-Janine	95	5.8	5510	Under development
Dominique-Peter	1756	0.66	11 587	Producing
Eagle Point	3300	1.55	51 152	Under development
Kiggavik	3022	0.51	15 384	Dormant
Rabbit Lake	5840	0.27	15 769	Depleted
McArthur River (P2 North)	2230	3.4	76 000	Advanced exploration
Australia				
Jabiluka I	1373	0.21	2883	Dormant
Jabiluka II	52 422	0.33	172 992	Dormant
Koongarra	4946	0.228	11 278	Dormant
Nabarlek	558	1.56	8700	Depleted
Ranger I	12 057	0.273	32 915	Dormant
Ranger III	42 425	0.17	72 123	Producing
POLYMETALLIC				
Canada				
Cigar Lake	902	12.2	110 000	Advanced exploration
Cluff Lake 'D'	128	3.41	4370	Depleted
Collins Bay 'A'	140	4.83	6500	Dormant
Collins Bay 'B'	3000	0.38	11 400	Depleted
Collins Bay 'D'	120	1.86	2500	Dormant
Key Lake	3518	1.99	70 000	Producing
McClean	352	1.53	5385	Dormant
Midwest	1200	1.6	19 300	Advanced exploration
Australia				
Kintyre	5936	0.5	29 680	Exploration

The principal host rocks of the Rabbit Lake deposit (Fig. 7-4 and 7-5) are albite-rich rocks, derived apparently from arkosic to semipelitic rocks, which were subjected to sodic metasomatism (producing rocks termed "plagioclasites" by Sibbald, 1976; Appleyard, 1984); meta-arkose; calc-silicate; and graphitic granulites. The plagioclasites form part of the footwall complex of the deposit. The host rocks also include a unit of partly graphitic semipelite and a layer of dolomite. The metasedimentary sequence has been intruded by granitic rocks.

The Eagle Point deposit is hosted by the lower pelitic unit of the Wollaston Group, which consists of quartzofeldspathic gneiss that is locally graphitic, quartzite, and granite pegmatite. This suite unconformably overlies folded Archean granitoid rocks. In the Eagle Point deposit Andrade (1989) identified two generations of euhedral uraninite, which belong to the oldest phases of mineralization, three forms of pitchblende (veinlets, coatings, and inclusions), which are younger than uraninite and represent the bulk of the mineralization, and minor amounts of boltwoodite and coffinite, which represent the youngest members of the uranium mineral assemblage.

The Raven and Horseshoe deposits occur within the quartz-amphibolite unit of the Wollaston Group, which consists of sillimanite meta-arkose, amphibolite, graphitic metapelite and quartzite, calc-silicate rocks, phosphates, and sillimanitic quartzite. The metasedimentary sequence has been folded into a syncline and intruded by dykes of granite pegmatite. The mineralization is confined mainly to the graphitic quartzite horizon, which is fractured and altered by sericitization, chloritization, and argillization.

The Dominique-Peter deposit, which is located in the Carswell Structure, is confined to a mylonite zone. This zone is entirely within basement gneisses at a contact between the Peter River gneiss and the Earl River gneiss complex. Most of the mineralization occurs in the mylonitized Peter River gneiss.

The Fond-du-Lac deposit occurs in hematitized, carbonatized, and silicified sandstone of the Athabasca Group, about 30 m above the unconformity. The mineralization is composed of a stockwork of steeply dipping fractures and disseminations in the adjacent porous, coarse grained facies of the sandstone.

The McArthur River (P2 North) deposit, which is located about 70 km northeast of the Key Lake deposit, contains prevailingly monometallic uranium (pitchblende) mineralization just above the sub-Athabasca unconformity and in the footwall of a thrust fault. The mineralized zone has been traced for 1850 m along strike by vertical drillholes. It averages 30 m wide and 7 m thick, but is locally more than 50 m wide and its vertical thickness is as much as 46 m. The main orebody is located from about 500 to about 600 m below the surface. As of 1992 it was estimated that the deposit contained in excess of 76 000 t of uranium metal in ores grading 3.4% U (Marlatt et al., 1992). The orebody consists of massive pitchblende and trace amounts of galena, pyrite, and chalcopyrite. The basement rocks in the footwall of the orebody consist of quartzite interbedded with garnetiferous and cordieritic gneisses, and are capped by a few metres of chloritic and hematitic regolith. The overthrust basement rocks consist of Aphebian graphitic and sericitic schists, quartzites, and minor amounts of pegmatites and calc-silicate rocks. The basement rocks are

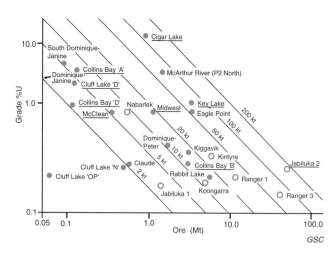

Figure 7-2. Grade/tonnage relationships in selected unconformity-associated uranium deposits (see Table 7-1). Dots = Canadian deposits; circles = Australian deposits. Deposits that are primarily polymetallic are underlined.

unconformably overlain by conglomerate and sandstone of the Helikian Athabasca Group. The host rocks are strongly silicified, but otherwise only relatively weakly altered by illite, chlorite, kaolinite, hematite, limonite, siderite, and dravite. Except for the silicification, the alteration of the deposit is restricted to a narrow aureole around the orebody. The pitchblende has yielded two main U-Pb ages: an older and prevailing age of 1514 ± 18 Ma and a younger age of 1327 ± 8 Ma (Cumming and Krstic, 1992). The older date represents the oldest known mineralization among the deposits associated with the sub-Athabasca unconformity.

The Kiggavik deposit is a large uranium concentration associated with the sub-Thelon unconformity in the Northwest Territories. The deposit occurs in Lower Proterozoic basement rocks, mica-rich nongraphitic quartzofeldspathic metasedimentary rocks, and unmetamorphosed fluorite-bearing granite (Miller et al., 1984; Ashton, 1988; Fuchs and Hilger, 1989; Henderson et al., 1991; LeCheminant and Roddick, 1991; Dudas et al., 1991). The mineralization lies at an undefined distance below the assumed sub-Thelon unconformity (Fuchs and Hilger, 1989).

Deposits of **polymetallic** (U-Ni-Co-As) character in the Athabasca Basin occur immediately at the sub-Athabasca unconformity. Examples include the Key Lake, Cigar Lake, Collins Bay 'A', Collins Bay 'B', McClean, Midwest, Sue, and Cluff Lake 'D' (Fig. 7-3; Table 7-2) deposits, which occur in the basal part of the Middle Proterozoic Athabasca Group clastic sedimentary sequence and/or the uppermost part of the Lower Proterozoic basement rocks.

The Key Lake deposit consists of two orebodies (Gärtner and Deilmann), which occur at the unconformity between the Athabasca Group rocks and the rocks of the underlying Wollaston Group. The deposition of the orebodies was controlled structurally by the intersection of the sub-Athabasca unconformity and a major reverse fault zone. The orebodies occur in proximity to graphitic metapelite layers of the Wollaston Group, which also contains biotite-plagioclase-quartz-cordierite gneiss, garnet-quartz-feldspar-cordierite gneiss, amphibolite, calc-silicate rocks, migmatite, and

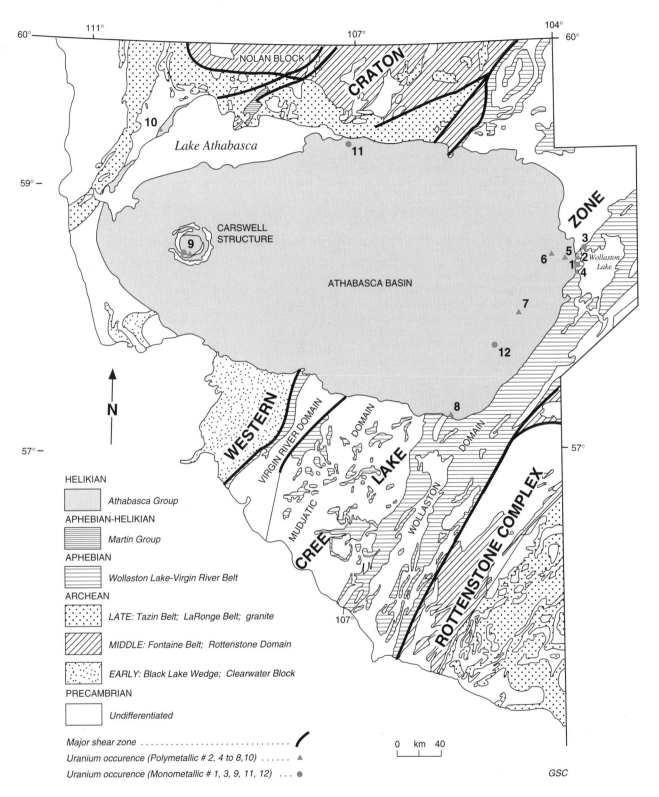

Figure 7-3. Unconformity-associated uranium deposits in the Athabasca Basin region, Saskatchewan. (Geology after Lewry and Sibbald, 1979.) 1 – Rabbit Lake; 2 – Collins Bay 'A' and 'B' zones; 3 – Eagle Point; 4 – Raven and Horseshoe; 5 – McClean Lake; 6 – Midwest and Dawn Lake; 7 – Cigar Lake; 8 – Key Lake; 9 – deposits in the Carswell Structure – (Cluff Lake 'D', Dominique-Peter, Claude, Cluff Lake 'OP', and Dominique-Janine); 10 – Maurice Bay; 11 – Fond-du-Lac; 12 – McArthur River (P2 North).

granite pegmatite. The Wollaston Group rocks unconformably overlie Archean granitic rocks, which are exposed in northeasterly elongated domal structures. The sedimentary rocks of the Athabasca Group have been subjected to alteration by diagenetic and mineralization processes. The diagenetic alteration, which is preserved outside the mineralized zone, is characterized by clay alteration of feldspars, corrosion of quartz grains by kaolinite and chlorite, partial bleaching (removal) of the original hematite, development of several generations of secondary hematite, and dravitization and carbonatization of the kaolinite matrix. In the immediate vicinity of ore, the Athabasca Group and the basement rocks have been altered to illite, chlorite, and kaolinite.

The world's largest high-grade uranium deposit (with ores of the world's highest average grade), Cigar Lake, contains not only polymetallic, but also some monometallic mineralization. Most of the mineralization occurs in clay-altered rocks at the base of the Athabasca Group, i.e., immediately at the unconformity (Fig. 7-6; Fouques et al., 1986). Small amounts of mineralization are contained within altered basement rocks just beneath the unconformity and up to 200 m above the unconformity in fractured Athabasca Group sediments. The mineralization is present in three assemblages of elements: (i) uranium, nickel, cobalt, and arsenic; (ii) uranium and copper; and (iii) uranium alone (mainly coffinite).

Table 7-2. Structural and lithological controls of selected unconformity-associated uranium deposits.

Deposit (ore type)	Relation of orebodies to Unconformity	Mineralized structures	Principal host rock	Main age of mineralization	Principal alteration
Cigar Lake (polymetallic)	Along	Two sets of faults	Athabasca sandstone, cordierite-feldspar augen gneiss	1.33 Ga (U-Pb)	Cover: Fe-Mg-illite; Basement: chlorite, Mg-illite
Cluff Lake 'D' (polymetallic)	Along	Mylonite zone and faults	Athabasca sandstone, garnet-rich aluminous gneiss	1.2 Ga (U-Pb)	Hematitization, bleaching
Collins Bay 'A' (polymetallic)	Along	Collins Bay Fault	Athabasca sandstone, quartzofeldspathic gneiss	N/A	Illitization, kaolinization, bleaching, hematitization
Collins Bay 'B' (polymetallic)	Along	Collins Bay Fault	Athabasca sandstone, paragneiss	1.38 Ga (U-Pb)	Illitization, kaolinization, bleaching, hematitization
Dominique-Peter (monometallic)	Below	Mylonite zone and 2 sets of faults	Quartzofeldspathic gneiss	1.05 Ga (U-Pb)	Chloritization, sericitization, illitization
Eagle Point (monometallic)	Below	Collins Bay and Eagle Point faults, wrench faults	Quartzofeldspathic gneiss, locally graphitic, quartzite, pegmatite	1.4 Ga (U-Pb)	Chloritization, illitization, hematitization, bleaching
Key Lake (polymetallic)	Along	Key Lake Fault	Graphitic metapelite, Athabasca sandstone	1.39 Ga (U-Pb)	illitization, kaolinization
Maurice Bay 'A' (monometallic)	Below	Two faults	Mylonitic gneiss	1.3 Ga (U-Pb)	Chloritization, illitization
McClean (polymetallic)	Along	Fracture zones	Athabasca sandstone, regolith	1.3-1.17 Ga (Ar-Ar, U-Pb)	Illitization, chloritization, kaolinization
Midwest (polymetallic)	Along	Fracture zone	Athabasca sandstone, pelitic gneiss	1.3 Ga (U-Pb)	Sericitization, chloritization, kaolinization
Rabbit Lake (monometallic)	Below	Rabbit Lake Fault	Plagioclasite, meta-arkose, calc-silicate rock, granulite	1.3 Ga (U-Pb)	Mg-chloritization, carbonatization, tourmalinization
N/A = Not available					

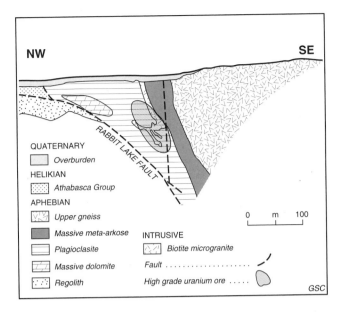

Figure 7-4. Geology of the Rabbit Lake deposit; cross-section (after Sibbald in Heine, 1981).

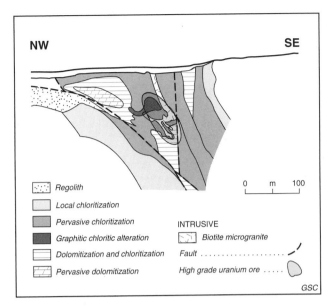

Figure 7-5. Rock alteration at the Rabbit Lake deposit; cross-section (after Sibbald in Heine, 1981).

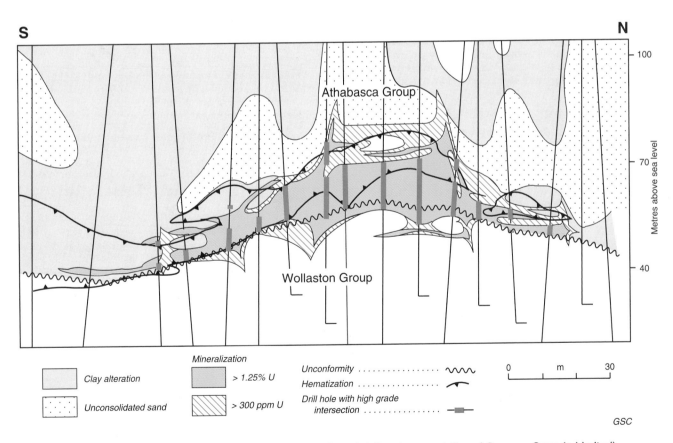

Figure 7-6. Cross-section through the Cigar Lake deposit (after documentation of Cogema Canada Limited).

203

The Collins Bay 'A' and 'B' zones occur at the unconformity, partly in clay-altered sedimentary rocks of the Athabasca Group and partly in altered metamorphic rocks of the Wollaston Group, along the Collins Bay Fault.

The Sue deposits occur within the submeridional part of the Sue structural trend, which consists of a series of faults adjacent to the southwestern margin of the Collins Bay Granitic Dome. The trend includes a layer of graphitic gneiss within the Aphebian sequence. The mineralization is predominantly polymetallic (U, Ni, Co, As, V, Cu, and Pb) and hosted by sandstone, but is in part monometallic (U) and hosted by basement rocks. The mineralization occurs in several zones, such as the Sue 'B', Sue 'A', Sue 'C', Sue 'CQ', Sue 'D', and Sue 'E'. The Sue deposits are excellent examples of the consanguinity of the sandstone- and basement-hosted mineralization (Ruzicka, 1992).

In the Thelon Basin region the only known polymetallic uranium occurrence is at Boomerang Lake, at the southwestern rim of the basin (Fig. 7-1). The mineralization occurs at the unconformity in altered sandstone of the flat-lying Helikian Thelon Formation and in the underlying graphitic metapelites of the Elk River belt, of inferred Early Proterozoic age (Davidson and Gandhi, 1989).

In the Otish Basin, Quebec (Fig. 7-1) polymetallic uranium occurrences are associated with an unconformity between the Archean volcanic rocks and the overlying Lower Proterozoic unmetamorphosed Otish Group. Sedimentation in the Otish Group varied from fluvial in the basal Indicator Formation to marginal marine in the overlying Peribonca Formation. The basement and the sedimentary cover rocks have been intruded by gabbroic sills and dykes. Age dating of the uranium mineralization and the associated rocks indicates Hudsonian events in the basin.

Age of host rocks and deposits

Unconformity-associated uranium deposits are in general hosted by Proterozoic rocks (about 2.5 to 0.6 Ga old). The mineralization is diagenetic and epigenetic, and formed during several stages.

Mineralization in the Athabasca Basin region is hosted by Lower Proterozoic and Middle Proterozoic (Athabasca Group) rocks, whose age is bracketed by Archean granitoid units (about 2500 Ma) and by intrusion of diabase dykes of the Mackenzie swarm (U-Pb age of 1267 ± 2 Ma on baddeleyite, LeCheminant and Heaman, 1989). Uranium-lead analyses of fluorapatites from the Upper Wolverine Point and Fair Point formations of the Athabasca Group indicated at least two distinct ages in the range 1650-1700 Ma (Cumming et al., 1987).

An important source of uranium apparently was older granitic and metasedimentary rocks. Archean granitic plutons containing above normal contents of uranium occur in the vicinity of most deposits. Uraninite-bearing pegmatites and metasedimentary rocks (U-Pb age >2.2 Ga; Robinson, 1955) are present in the Beaverlodge area, a short distance to the north of the Athabasca Basin. Hudsonian felsic intrusive rocks, and particularly their pegmatitic derivatives, are abundant, for instance, in the source area for the Manitou Falls Formation that surrounds the Key Lake deposits. Concerning metasedimentary sources, Ray (1977) speculated that "the initial Aphebian sedimentation under

anaerobic conditions could have produced suitable conditions for syngenetic concentration of uranium within the basal pelites; these may have formed a source for some uranium deposits in northern Saskatchewan". Ramaekers (pers. comm., 1981) considered the Athabasca Group sedimentary rocks as the "immediate source for uranium and base metals found in the unconformity deposits at their base...".

Numerous age determinations have indicated that the ores in the majority of the deposits in this region were formed and/or remobilized and isotopically reset during the period between 1.4 and 0.8 Ga ago (Cumming and Rimsaite, 1979; Worden et al., 1985). However, more recently, Cumming and Krstic (1992) presented results of geochronological studies on a number of the uranium deposits of the Athabasca Basin, including the major deposits at Collins Bay, Cigar Lake, Dawn Lake, Eagle Point, Midwest, Rabbit Lake, and McArthur River, and concluded that "almost all the deposits formed in a restricted time interval between about 1330 and 1360 Ma. The one major exception is, however, the recently discovered NiAs-free deposit at McArthur River where a well determined age of 1521 ± 8 Ma (2σ) has been obtained" (pitchblende U-Pb). Remobilization and redeposition of pitchblende in the deposits took place at about 1070 Ma, 550 Ma, and 225 Ma ago (Cumming and Krstic, 1992). The main 1360-1330 Ma stage of mineralization corresponds to the age of 1326 ± 10 Ma published earlier by Ruzicka and LeCheminant (1986) for ore from the main pod of the Cigar Lake deposit. Other age determinations also fall close to the main mineralization stage established by Cumming and Krstic (1992). For example, U-Pb isotope analyses for 26 anisotropic uraninites from the Key Lake deposit, done at the Institut für Geowissenschaften und Rohstoffe in Germany, yielded a slightly older age of crystallization for uraninite of 1386 ± 4 Ma (Federal Institute for Geosciences and Natural Resources, 1989).

A Rb-Sr isochron age of 1477 ± 57 Ma was obtained for illites from various deposits associated with the sub-Athabasca unconformity (Kotzer and Kyser, 1990b). This age apparently reflects the beginning of the diagenetic-hydrothermal ore-forming process that led to accumulation of uranium and associated metals.

In the Thelon Basin, which is lithostratigraphically correlative with the Athabasca Basin (Miller et al., 1989), U-Pb isotope dating on ores in the Kiggavik deposit suggests three mineralization events; the oldest at 1400 Ma, a later one at about 1000 Ma, and the youngest indicating rejuvenation of mineralization at 10 Ma (Fuchs and Hilger, 1989). The mineralization thus postdates the deposition of the Thelon Formation, for which a minimum U-Pb age of 1720 ± 6 Ma was obtained by Miller et al. (1989) by dating uraniferous phosphate minerals that cement sedimentary units within the Thelon Basin. The mineralization associated with the sub-Thelon unconformity has been described by Miller (1983) and Miller et al. (1984).

Structural features

The most important structures controlling localization of unconformity-associated deposits are the unconformity itself, and faults and fracture zones that intersect this surface. In the Athabasca Basin, the mineralization is structurally controlled by the pre-Middle Proterozoic (sub-Athabasca) unconformity and by intersecting northeasterly and easterly trending faults. For instance, the Rabbit Lake deposit is localized at the intersection of the sub-Athabasca unconformity and the northeast-trending Rabbit Lake thrust fault. The Eagle Point deposit occurs in the hanging wall of the Collins Bay Fault at its intersection with the sub-Athabasca unconformity. Similarly the Key Lake deposit is associated with a regional northeast-trending steep fault where it intersects the sub-Athabasca unconformity. The Cigar Lake deposit is located where the sub-Athabasca unconformity is intersected by an east-trending fracture zone that coincides with graphitic pelite layers in the Lower Proterozoic basement. The Dominique-Peter deposit in the Cluff Lake area is confined to a mylonite zone which occurs at the contact between two gneissic lithostratigraphic units, presumably not far below the unconformity. Distribution of the orebodies of the Kiggavik deposit is controlled by several intersecting fault zones. Structural controls of selected deposits are summarized in Table 7-2.

Form of deposits

The forms of the orebodies are controlled by generally subvertical faults, shear zones, and fracture zones, and by the subhorizontal plane of the unconformity. Orebodies of the monometallic subtype consist typically of lenses in veins and thin veinlets in stockworks. Orebodies of the polymetallic subtype form pods and lenses aligned along the controlling structures and, to a lesser extent, veinlets and impregnations in the host rocks. A typical shape for such orebodies is plume-like lobes that formed from ascending fluids. The orebodies are commonly surrounded by clay, chlorite, or carbonate alteration zones.

Ores

Ores of the monometallic deposits, such as Rabbit Lake and Eagle Point, consist of pitchblende (in massive, globular, and sooty forms), coffinite, and, locally, secondary uranium minerals such as boltwoodite, sklodowskite, and kasolite. Carbonates (calcite, dolomite, siderite), sericite, chlorite, clay minerals (illite, kaolinite), celadonite, and tourmaline (dravite) are common gangue minerals.

The polymetallic ores, such as the Key Lake, Cigar Lake, Collins Bay 'B', and Midwest consist of several generations of pitchblende and coffinite; arsenides and sulpharsenides of nickel and cobalt; sulphides of nickel, copper, lead, molybdenum, iron, and zinc; and oxides and hydroxides of iron. Silver, gold, and platinum group minerals occur locally. Chlorite, illite, kaolinite, and siderite are the most common gangue minerals.

Some deposits of the polymetallic subtype (e.g., Cigar Lake) have vertically zoned mineral assemblages. At the unconformity, U-Ni-Co-Ag-As assemblages grade locally upward into a zone with the U-Cu assemblage, whereas monometallic uranium is found in upper and lower extremities of the orebodies. The zonal arrangement of these assemblages suggests that they are contemporaneous and are apparently related to the geochemical mobilities of individual elements and stabilities of the minerals.

Proportions of metals in ores of the polymetallic subtype differ from one deposit to another. In the Key Lake ores, the contents of the principal constituents, uranium and nickel, are 1:0.55 (A. de Carle, verbal comm., 1986); whereas in the ores of the main pod of the Cigar Lake deposit the contents of these metals are 1:0.078 (Fouques et al., 1986).

Alteration

Both the monometallic and polymetallic deposits have associated zones of host rock alteration that appear to result from three distinct processes.

1. Paleoweathering of the metamorphic basement rocks prior to deposition of the Middle Proterozoic clastic sedimentary rocks led to formation of regolith. The regolith persists throughout the basin and shows many features compatible with present-day lateritic soil profiles formed in subtropical to tropical climates (Macdonald, 1985). Development of regolith was characterized by chloritization and hematitic alteration of ferromagnesian minerals, sericitization, illitization, or kaolinization of K-feldspars, and saussuritization of plagioclase (de Carle, 1986). The weathered material was an important source of metals for the Athabasca Group rocks. Regolith is also host for mineralization of some deposits (Tremblay, 1982; Table 7-2).

2. Diagenetic and epigenetic alteration was coeval with the mineralization, and affected not only the rocks of the Atahabasca Group, but also the basement rocks, particularly in the vicinity of the deposits. Oxygen- and hydrogen-isotope analyses of illite, kaolinite, and chlorite associated with uranium mineralization indicated (Kyser et al., 1989; Kotzer and Kyser, 1990a, b; Kotzer, 1992) that (i) the basement fluids produced clinochlore with $\delta^{18}O = +2$ to $+4‰$ and $\delta D = -45$ to $-15‰$ and sudoite with $\delta^{18}O = -25$ to $-60‰$ and $\delta D = 7$ to $9‰$; (ii) the basinal fluids produced illite and kaolinite with $\delta^{18}O = +2$ to $+4‰$ and $\delta D = -60 \pm 20‰$; and (iii) the retrograde fluids (i.e., meteoric waters that circulated along fault zones) produced a late stage kaolinite with $\delta^{18}O = -16‰$ and $\delta D = -130 \pm 10‰$. The alterations have various forms and intensities depending on the character of the host rocks and the nature of the fluids. For example, at Key Lake kaolinization of the Athabasca Group rocks was superimposed on illitization and extends for several hundred metres laterally from the mineralization. At Cigar Lake the orebody is surrounded by an alteration halo, which contains hematite, illite, ferromagnesian illite, chlorite and its Al-Mg variety – sudoite, kaolinite, iron-rich kaolinite, locally unconsolidated sand (quicksand), and a quartz-cemented cap (Fouques et al., 1986; Percival and Kodama, 1989). At Rabbit Lake, where mineralization is entirely hosted by the basement rocks, chloritization, graphitic chloritic alteration, and dolomitization are the main forms of alteration (Fig. 7-5). At the Midwest deposit, uranium and boron, and at Key Lake, boron and lead, are enhanced in the host rocks around mineralization (Sopuck et al., 1983). The orebodies are commonly surrounded by clay envelopes. Quartz grains in rocks at the unconformity are corroded or even totally replaced by clay. Silicification of the sandstone in the form of vein systems and pervasive cements has occurred in places in the overlying sandstone and is a manifestation of intense illitization and desilicification at depth. Partial destruction of graphite and carbon from metapelitic rocks and formation of limonite and hematite, or bleaching of the host rocks, are other

characteristics of the alteration zones. Brecciation and development of collapse structures in the immediate vicinity of the mineralization were associated with the alteration processes. The diagenetic and epigenetic alterations were apparently enhanced by ionization effects of the radionuclides, particularly by the radiolysis of water and by reactions of hydrogen and oxygen with the rocks. Effects of water radiolysis were observed at the Cigar Lake deposit (Cramer, 1986).

3. Postore alteration (about 1.2 to 0.8 Ga) succeeded the main episode of uranium mineralization. Tectonic uplift of the Athabasca Basin about 300 Ma ago (Hoeve and Quirt, 1984) triggered circulation of basinal fluids, which caused corrosion of the ores; formation of new alteration minerals, particularly chlorite, smectite, and mixed-layer clays; and kaolinization of illite and quartz (Ruhrmann and von Pechmann, 1989).

Mineralogy

Pitchblende is the principal uranium mineral in deposits of both the monometallic and polymetallic types. A crystalline variety of pitchblende (alpha-triuranium heptaoxide-U_3O_7-tetrauraninite) has been identified in some deposits (e.g., Key Lake, Cigar Lake, and Eagle Point). Coffinite is another common uranium mineral. Locally thucholite and uranoan carbon are present as veinlets, globules, and lenses. Thorium-bearing uraninite, brannerite, and U-Ti mineral aggregates are rare. Secondary uranium minerals are present in some deposits, even at depths exceeding 100 m (e.g., in the Eagle Point and Rabbit Lake deposits). They include uranophane, kasolite, boltwoodite, sklodowskite, becquerelite, vandendriesscheite, woelsendorfite, tyuyamunite, zippeite, masuyite, bayleyite, and yttrialite (Ruzicka, 1989). Minerals of nonradioactive metals occur in relatively large quantities in the polymetallic subtype, but minor amounts are present also in the monometallic subtype. Nickeline and rammelsbergite are the most common arsenides. Skutterudite, pararammelsbergite, safflorite, maucherite, and modderite occur locally. Gersdorffite is the most common representative of the sulpharsenides. Cobaltite, glaucodot, and tennantite are relatively rare. Chalcopyrite, pyrite, and galena are the most common sulphides; others include bornite, chalcocite, sphalerite, marcasite, bravoite, millerite, jordisite, covellite, and digenite. Some deposits contain selenides such as clausthalite, freboldite, trogtalite, and guanajuatite. Tellurides, such as altaite and calaverite, occur in some deposits in the Carswell Structure. Locally native metals, such as gold, copper, and arsenic, accompany the uranium minerals. A detailed list of ore-forming minerals in individual deposits has been given by Ruzicka (1989).

DEFINITIVE CHARACTERISTICS

1. Unconformity-associated uranium deposits typically occur in close spatial association with unconformities that separate crystalline (Archean and Lower Proterozoic) basement rocks from overlying Middle Proterozoic clastic sedimentary rocks, which are generally unmetamorphosed and flat-lying.

2. Granitic rocks, which commonly form domal structures, and metamorphosed graphitic pelitic rocks are the most distinctive members of the basement complexes. Most of the deposits are located along the flanks of the domal structures and in proximity to the metapelitic rocks.

3. The clastic sedimentary cover rocks were deposited on weathered basement rocks, in large intracratonic basins.

4. The deposits are commonly associated with the unconformity where it is intersected by faults, shear zones, or fracture zones.

5. The uranium mineralization is either monometallic (containing predominantly uranium minerals), or polymetallic (i.e., accompanied by arsenides, sulpharsenides, and sulphides of nickel and cobalt, and sulphides of copper, iron, lead, zinc, bismuth, and molybdenum). Locally a transitional phase of mineralization, consisting of uranium and only one or a few other base metal minerals (e.g., at Cigar Lake) may occur.

6. The polymetallic mineralization generally occurs at the unconformity, either in the overlying sedimentary cover (e.g., Athabasca Group) or in subjacent crystalline basement rocks, whereas the monometallic mineralization is farther from the unconformity, usually localized in the basement rocks or, less commonly, in the cover rocks.

7. The mineralization occurs only in areas of alteration, which comprises illite, kaolinite, and chlorite.

GENETIC MODEL

Three conceptual genetic models have been proposed for Canadian uranium deposits associated with a Middle Proterozoic (particularly sub-Athabasca) unconformity (Hoeve et al., 1980):

1. a near-surface supergene origin, which involves derivation of uranium and other ore constituents from basement rocks by supergene processes, their transport by surface and ground waters, and their deposition in host rocks under reducing conditions;

2. a magmatic or metamorphic hydrothermal origin, whereby the uranium is derived from deep-seated sources, and transported by and deposited from ascending solutions; and

3. a diagenetic-hydrothermal origin, which relates uranium mineralization to diagenetic processes active under elevated temperatures in Athabasca Group sediments after their deposition, and precipitation of uranium from diagenetic fluids by local reductants. Ruzicka (1993) analyzed the geological features of unconformity deposits of the Athabasca Basin and Pine Creek Geosyncline metallogenic provinces and established two sets of models for each province: regional models, which summarized the geological histories, and deposit models, which reflected the ore-forming processes and environments within these provinces. A modified version of these models, applied to Athabasca Basin deposits, is presented here (Fig. 7-7):

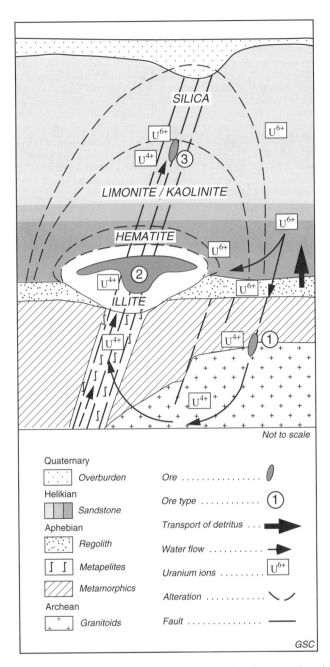

Figure 7-7. Conceptual model of unconformity-associated uranium deposits. A generalized vertical cross-section. Arrows indicate flow paths of oxidized and reduced convective waters. Circled numbers indicate locations of various styles of mineralization: (1) high grade polymetallic mineralization at the unconformity, (2) medium grade monometallic mineralization below the unconformity, (3) low grade monometallic mineralization in sedimentary cover rocks above the unconformity.

i) The deposits are part of a uranium-enriched sub-province within the Churchill Structural Province. The uranium may originally have been introduced into the geochemical cycle in the form of granitic magmatism, during the late Archean or Early Proterozoic. Uraniferous monazite and uraninite yielding U-Pb ages >2.2 Ga are present in pegmatites.

ii) In a subsequent stage uranium was concentrated in sedimentary rocks (e.g., certain Lower Proterozoic metasedimentary rocks contain up to 50 ppm uranium).

iii) Further concentration of uranium took place during the waning phase of the Trans-Hudson Orogeny, at which time the pitchblende-brannerite deposits (Type 13; "Vein uranium") in the Beaverlodge area were formed.

iv) Subsequent peneplanation and lateritic weathering of the uranium-enriched rocks resulted in liberation of uranium- and associated metal-bearing minerals and their incorporation in the detritus. The detritus was deposited in Middle Proterozoic intracratonic basins, which thus became reservoirs for the metals. The basinal sediments underwent profound diagenesis from about 1700 to 1400 Ma. During that time portions of the sequence evolved in chemical maturity, accompanied by breakdown of minerals and release of metals.

v) Tectonic events associated with rapid subsidence and rifting of the basin activated hydrological systems and thus caused convective cycling of fluids and mobilization of the metals from the reservoirs. The events are reflected in the Rb-Sr isochron age of 1477 ± 57 Ma for illites from deposits associated with the sub-Athabasca unconformity; the date thus marks a diagenetic ore-forming process that led to mineralization with uranium, nickel, cobalt, and other associated metals. Three types of fluids took part in the hydrological system: (a) oxidized basinal fluids, which also included metalliferous connate waters brought into the basins along with detritus; (b) reduced basement fluids; and (c) retrograde fluids, derived from meteoric waters. The oxidized metalliferous fluids moved laterally and downwards (their salinity and metal contents caused their high density); when these fluids encountered the hydrological barrier at the unconformity, they continued flowing laterally along the unconformity. However, part of the fluids circulated through fault and fracture zones in the basement rocks, and became reduced. These reduced basement fluids then re-entered the sedimentary rocks in ascending flows along faults and fracture zones, where they mingled with the oxidized metalliferous basinal waters at the unconformity and in the upper parts of the cover rocks. Whether or not the ascending fluids also contained water from deep-seated sources is not known.

vi) Deposition of the metals and associated gangue minerals, took place at the interface between the oxidizing and reducing fluids, i.e., at the redox front, during the diagenesis and epigenesis of the sedimentary cover rocks. Depending upon the location of the redox front, mineralization took place in diverse parts of the basin and the basement: (a) high grade polymetallic mineralization directly at the unconformity (see location 1 in Fig. 7-7; examples are at the Key Lake, Cigar Lake, and Midwest deposits); (b) medium grade monometallic mineralization in fractures and faults below the unconformity (see location 2 in Fig. 7-7; examples are the Rabbit Lake and Eagle Point deposits); and (c) low grade monometallic mineralization in the sedimentary cover sequence at some distance above the unconformity (see location 3 in Fig. 7-7; examples are the Fond-du-Lac deposit and the "perched" mineralization at Cigar Lake).

vii) The mineralization was accompanied by alteration of the host rocks, primarily argillization (illitization, kaolinization) and chloritization. The argillic alteration was superimposed on the earlier lateritization. The hydrothermal fluids introduced carbon dioxide, hydrogen sulphide, and methane, and caused dissolution of quartz in the area of mineralization, and silicification of the sandstone (including crystallization of euhedral quartz in vugs) more distant in an aureole around the orebodies. Local tourmalinization (e.g., at Key Lake) and magnesian metasomatism (e.g., at Rabbit Lake) accompanied the ore-forming process. The alteration was enhanced by the ionization effects of radiation and by partial hydrolysis of the waters.

viii) Deposition of uranium and associated metals was structurally controlled by the unconformity and intersecting faults and fracture zones. It was lithologically controlled by altered graphitic pelites in the basement, and altered and porous clastic sediments above the unconformity. Geochemical processes (e.g., Eh and pH changes, adsorption, and complexing) gave rise to the deposition of the ore-forming minerals, the specific mineral assemblages that formed, and their zonal arrangement (Wallis et al., 1986).

ix) The mineralization was remobilized and redeposited in at least two later periods, but this had little effect on the main concentrations or economic viability of the uranium deposits.

RELATED DEPOSIT TYPES

The unconformity-associated deposits exhibit some geological features that are also typical of other types of uranium deposits, such as Phanerozoic sediment-hosted and vein (subtype 8.1, "Sandstone uranium"; Types 13, "Vein uranium"; and 14, "Arsenide vein silver, uranium") deposits.

Phanerozoic, sediment-hosted uranium ("Sandstone uranium"; subtype 8.1) deposits, also are associated with coarse clastic sedimentary rocks in continental intracratonic basins; however, they contain orebodies of lower grade and usually of different morphology. Their mineralization is commonly disseminated to semimassive and occurs in 'C-shape', 'blanket', or 'stack' orebodies. Their principal uranium minerals are carnotite, tyuyamunite, coffinite, and urano-organic complexes; pitchblende is less abundant than in the unconformity deposits. Vanadium, molybdenum, and selenium are commonly associated with uranium in the sandstone deposits, whereas nickel, cobalt, and arsenic are the elements that typically accompany uranium in the unconformity-associated deposits. The sandstone uranium deposits formed from flowing intrastratal oxidizing fluids at their entry into reducing parts of the sandstone (i.e., at a moving redox front), whereas the unconformity-associated deposits may have formed at the confluence of flows of oxidized basinal and reduced basement waters (i.e., at a stationary redox front). The difference in grades of these two deposit types seems to be related to the different forms of the redox processes. Effects of hydrolysis are less intense in formation of the Phanerozoic sediment-hosted deposits and their clay envelope is small or absent.

The U-Ni-Co-As mineral assemblages that are typical for the polymetallic unconformity-associated deposits are also found in uranium vein deposits (subtype 14.2, "Arsenide vein uranium-silver"), such as those at Port Radium; the Jáchymov deposit, Czech Republic; and the Shinkolobwe deposit, Zaire. However, polymetallic vein deposits consist of relatively irregular discontinuous orebodies, in contrast to the massive concentrations that comprise unconformity-associated deposits. Monometallic uranium vein deposits contain, as do the monometallic unconformity-associated deposits, only a few principal ore-forming minerals, namely pitchblende and coffinite.

Mineralization of both the vein and unconformity deposits is structurally controlled. For example, the structural locus of the Eldorado (Ace-Fay-Verna) vein deposit (Type 13, "Vein uranium") in the Beaverlodge area is the St. Louis Fault; the unconformity-associated Eagle Point deposit is spatially related to the Eagle Point and Collins Bay faults. The Ace-Fay-Verna deposit is also spatially related to the sub-Martin Formation unconformity.

EXPLORATION GUIDES

Exploration criteria for uranium and uranium-polymetallic deposits associated with unconformities include the following:

1. Regional scale guides are:

   i) Middle Proterozoic intracratonic basins containing clastic sediments, which rest unconformably on Lower Proterozoic or Archean supracrustal and granitoid rocks. The Middle Proterozoic basinal cover rocks must be thoroughly oxidized.

   ii) Presence of a regolith at the top of the basement rocks.

   iii) Presence in the basement terrane of reductants such as carbonaceous/graphitic or pyrite-bearing pelitic units.

2. District and local scale guides are:

   i) Features indicating the presence of an unconformity, a regolith, steep faults or fracture zones; the bedrock surface below the overburden may exhibit effects of glacial scouring in areas occupied by these faults or fracture zones.

ii) Alteration of the rocks and mineralization; zones of alteration may have a low density and be detectable by gravity surveys.

iii) Reducing horizons, particularly graphitic/pyritic pelites, may comprise geophysically detectable conductive layers.

iv) Radiometric anomalies, although weak, and particularly those with a high U/Th ratio. However, weak or even negative results of ground radiometric surveys (using Geiger-Müller counters, scintillometers, or gamma-ray spectrometers) do not preclude the possible existence of uranium mineralization at depth.

v) Geochemical anomalies of U and/or associated elements, such as As, Ni, Co, Pb, Cu, Bi, Li, and B.

vi) During geological and radiometric logging of drill core particular attention should be paid to: argillized (illitized/kaolinized) and chloritized rocks; recognition of the unconformity, and intersecting faults and fracture zones along the unconformity; limonite/hematite alteration; dilational and hydrothermal effects associated with the mineralization process; presence of collapse structures; vugs filled with euhedral quartz and silicification aureoles within the clastic sedimentary cover rocks; depletion of graphite in pelitic layers in the basement; sampling for chemical analyses for U, Ni, Co, and for other elements (discretional).

ACKNOWLEDGMENT

Critical reading of the manuscript and helpful recommendations by A.R. Miller and R.I. Thorpe of the Geological Survey of Canada are gratefully acknowledged.

SELECTED BIBLIOGRAPHY

References marked with asterisks (*) are considered to be the best sources of general information on this deposit type.

Andrade, N.
1989:   The Eagle Point uranium deposit, northern Saskatchewan, Canada; in Uranium Resources and Geology of North America, Proceedings of a Technical Committee Meeting, Saskatoon, 1987, International Atomic Energy Agency, IAEA-TECDOC-500, Vienna, p. 455-490.

Appleyard, E.C.
1984:   The origin of plagioclasite in the vicinity of the Rabbit Lake uranium deposit; in Summary of Investigations 1984, Saskatchewan Geological Survey, Saskatchewan Energy and Mines, Miscellaneous Report 84-4, p. 68-71.

Ashton, K.E.
1988:   Precambrian Geology of the southeastern Amer Lake area (66H/1), near Baker Lake, N.W.T.; PhD. thesis, Queen's University, Kingston, Ontario, 335 p.

Battey, G.C., Miezitis, Y., and McKay, A.D.
1987:   Australian uranium resources; Australia, Bureau of Mineral Resources and Geophysics, Resource Report 1, Canberra, Australia, 67 p.

Bickford, M.E., Collerson, K.D., Lewry, J.F., van Schmus, W.R., and Chiarenzelli, J.R.
1990:   Proterozoic collisional tectonism in the Trans-Hudson orogen, Saskatchewan; Geology, v. 18, p. 14-18.

*Cameron, E.M. (ed.)
1983:   Uranium Exploration in Athabasca Basin, Saskatchewan, Canada; Geological Survey of Canada, Paper 82-11, 310 p.

Cramer, J.J.
1986:   A natural analog for a fuel waste disposal vault; in Proceedings of the Second International Conference on Radioactive Waste Management; Winnipeg, Manitoba, September 7-11, 1986; Canadian Nuclear Association, p. 679-702.

Cumming, G.L. and Krstic, D.
1992:   The age of unconformity-related uranium mineralization in the Athabasca Basin, northern Saskatchewan; Canadian Journal of Earth Sciences, v. 29, p. 1623-1639.

Cumming, G.L. and Rimsaite, J.
1979:   Isotopic studies of lead-depleted pitchblende, secondary radioactive minerals and sulphides from the Rabbit Lake uranium deposit, Saskatchewan; Canadian Journal of Earth Sciences, v. 16, p. 1702-1715.

Cumming, G.L., Krstic, D., and Wilson J.A.
1987:   Age of the Athabasca Group, northern Alberta; in Geological Association of Canada/Mineralogical Association of Canada, Program with Abstracts, v. 12, p. 35.

Davidson, G.I. and Gandhi, S.S.
1989:   Unconformity-related U-Au mineralization in the Middle Proterozoic Thelon Sandstone, Boomerang Lake Prospect, Northwest Territories, Canada; Economic Geology, v. 84, p. 143-157.

de Carle, A.
1986:   Geology of the Key Lake deposits; in Uranium Deposits of Canada, (ed.) E.L. Evans; The Canadian Institute of Mining and Metallurgy, Special Volume 33, p. 170-177.

Dudás, F.Ö., LeCheminant, A.N., and Sullivan, R.W.
1991:   Reconnaissance Nd isotopic study of granitoid rocks from the Baker Lake region, District of Keewatin, N.W.T., and observations on analytical procedures; in Radiogenic Age and Isotopic Studies: Report 4, Geological Survey of Canada, Paper 90-2, p. 101-112.

*Evans, E.L. (ed.)
1986:   Uranium Deposits of Canada; The Canadian Institute of Mining and Metallurgy, Special Volume 33, 324 p.

Federal Institute for Geosciences and Natural Resources
1989:   Activity Report 1987/88; Federal Institute for Geosciences and Natural Resources, Hannover, p. 43.

*Ferguson, J. (ed.)
1984:   Proterozoic unconformity and stratabound uranium deposits; International Atomic Energy Agency, IAEA-TECDOC-315, Vienna, 338 p.

*Fouques, J.P., Fowler, M., Knipping, H.D., and Schimann, K.
1986:   The Cigar Lake uranium deposit: discovery and general characteristics; in Uranium Deposits of Canada, (ed.) E.L. Evans; The Canadian Institute of Mining and Metallurgy, Special Volume 33, p. 218-229.

Fuchs, H. and Hilger, W.
1989:   Kiggavik (Lone Gull): an unconformity related uranium deposit in the Thelon Basin, Northwest Territories, Canada; in Uranium Resources and Geology of North America, Proceedings of a Technical Committee Meeting, Saskatoon, 1987, International Atomic Energy Agency, IAEA-TECDOC-500, Vienna, p. 429-454.

Heine, T.
1981:   The Rabbit Lake deposit and the Collins Bay deposits; in Uranium Guidebook; The Canadian Institute of Mining and Metallurgy, Geology Division, September 8-13, Saskatchewan, p. 87-102.

Henderson, J.R., Henderson, M.N., Pryer, L.L., and Cresswell, R.G.
1991:   Geology of the Whitehills-Tehek area, District of Keewatin: an Archean supracrustal belt with iron-formation-hosted gold mineralization in the central Churchill Province; in Current Research, Part C; Geological Survey of Canada, Paper 91-1C, p. 149-156.

Hoeve, J. and Quirt, D.H.
1984:   Mineralization and host-rock alteration in relation to clay mineral diagenesis and evolution of the Middle Proterozoic Athabasca Basin, northern Saskatchewan, Canada; Saskatchewan Research Council, Technical Report No. 187, 187 p.

*Hoeve, J., Sibbald, T.I.I., Ramaekers, P., and Lewry, J.F.
1980:   Athabasca Basin unconformity-type uranium deposits: a special class of sandstone-type deposits?; in Uranium in the Pine Creek Geosyncline, (ed.) J. Ferguson and A.B. Goleby; International Atomic Energy Agency, Vienna, p. 575-594.

Kotzer, T.
1992:   Origin and geochemistry of fluid events in the Proterozoic Athabasca Basin; PhD. thesis, University of Saskatchewan, Saskatoon, Saskatchewan, 329 p.

Kotzer, T.G. and Kyser, T.K.
1990a: Fluid history of the Athabasca Basin and its relation to uranium deposits; in Summary of Investigations 1990; Saskatchewan Geological Survey, Saskatchewan Energy and Mines, Miscellaneous Report 90-4, p. 153-157.

Kotzer, T.G. and Kyser, T.K. (cont.)

1990b: The use of stable and radiogenic isotopes in the identification of fluids and processes associated with unconformity-type uranium deposits; in Modern Exploration Techniques, (ed.) L.S. Beck and T.I.I. Sibbald; Saskatchewan Geological Society, Special Publication 9, p. 115-131.

Kyser, T.K., Kotzer, T.G., and Wilson, M.R.

1989:   Isotopic constraints on the genesis of unconformity uranium deposits; in Geological Association of Canada/Mineralogical Association of Canada, Program with Abstracts, v. 14, p. A120.

*Laine, R., Alonso, D., and Svab, M. (ed.)

1985:   The Carswell Structure uranium deposits, Saskatchewan; Geological Association of Canada, Special Paper 29, 230 p.

LeCheminant, A.N. and Heaman, L.M.

1989:   Mackenzie igneous events, Canada: Middle Proterozoic hotspot magmatism associated with ocean opening; Earth and Planetary Science Letters, v. 96, p. 38-49.

LeCheminant, A.N. and Roddick, J.C.

1991:   U-Pb zircon evidence for widespread 2.6 Ga felsic magmatism in the central District of Keewatin, N.W.T.; in Radiogenic Age and Isotopic Studies: Report 4, Geological Survey of Canada, Paper 90-2, p. 91-99.

*Lewry, J.F. and Sibbald, T.I.I.

1979:   A review of pre-Athabasca basement geology in northern Saskatchewan; in Uranium Exploration Techniques, Proceedings of a Symposium, Regina, 1978, (ed.) G.R. Parslow; Saskatchewan Geological Society, Special Publication No. 4, p. 19-58.

1980:   Thermotectonic evolution of the Churchill Province in northern Saskatchewan; Tectonophysics, v. 68, p. 45-82.

Macdonald, C.

1985:   Mineralogy and geochemistry of the sub-Athabasca regolith near Wollaston Lake; in Geology of Uranium Deposits, (ed.) T.I.I. Sibbald and W. Petruk; The Canadian Institute of Mining and Metallurgy, Special Volume 32, p. 155-158.

Marlatt, J., McGill, B., Matthews, R., Sopuck, V., and Pollock, G.

1992:   The discovery of the McArthur River uranium deposit, Saskatchewan, Canada; in New Developments in Uranium Exploration, Resources, Production and Demand, Proceedings of a Technical Committee Meeting jointly organized by the International Atomic Energy Agency and the Nuclear Energy Agency of the Organization for Economic Cooperation Development, Vienna, 26-29 August, 1991, IAEA-TECDOC-650, p. 118-127.

Miller, A.R.

1983:   A progress report: uranium-phosphorous association in the Helikian Thelon Formation and sub-Thelon saprolite, central District of Keewatin; in Current Research, Part A; Geological Survey of Canada, Paper 83-1A, p. 449-456.

Miller, A.R., Blackwell, J.D., Curtis, L., Hilger, W., McMillan, R.H., and Nutter, E.

1984:   Geology and discovery of Proterozoic uranium deposits, central District of Keewatin, Northwest Territories, Canada; in Proterozoic Unconformity and Stratabound Uranium Deposits, (ed.) J. Ferguson; International Atomic Energy Agency, Vienna, IAEA-TECDOC-315, p. 285-312.

Miller, A.R., Cumming, G.L., and Krstic, D.

1989:   U-Pb, Pb-Pb and K-Ar isotopic study and petrography of uraniferous phosphate-bearing rocks in the Thelon Formation, Dubawnt Group, Northwest Territories, Canada; Canadian Journal of Earth Sciences, v. 26, p. 867-880.

Percival, J.B. and Kodama, H.

1989:   Sudoite from Cigar Lake, Saskatchewan; Canadian Mineralogist, v. 27, p. 633-641.

Ramaekers, P.

1990:   Geology of the Atahabasca Group (Helikian) in northern Saskatchewan; Saskatchewan Geological Survey, Report 195, 49 p.

Ray, G.E.

1977:   The geology of Highrock Lake-Key Lake vicinity, Saskatchewan; Saskatchewan Geological Survey, Report 197, 29 p.

Ray, G.E. and Wanless, R.K.

1980:   The ages of the Wathaman Batholith, Johnson River Granite and Peter Lake Complex and their geological relationships to the Wollaston, Peter Lake and Rottenstone Domains of northern Saskatchewan; Canadian Journal of Earth Sciences, v. 17, p. 333-347.

Robinson, S.C.

1955:   Mineralogy of uranium deposits, Goldfields, Saskatchewan; Geological Survey of Canada, Bulletin 31, 128 p.

Ruhrmann, G. and von Pechmann, E.

1989:   Structural and hydrothermal modification of the Gaertner uranium deposit, Key Lake, Saskatchewan, Canada; in Uranium Resources and Geology of North America, Proceedings of a Technical Committee Meeting, Saskatoon, 1987; International Atomic Energy Agency, IAEA-TECDOC-500, Vienna, p. 363-377.

*Ruzicka, V.

1984:   Unconformity-related uranium deposits in the Athabasca Basin region, Saskatchewan; in Proterozoic Unconformity and Stratabound Uranium Deposits, (ed.) J. Ferguson; International Atomic Energy Agency, Vienna, IAEA-TECDOC-315, p. 219-267.

1989:   Monometallic and polymetallic deposits associated with the sub-Athabasca unconformity in Saskatchewan; in Current Research, Part C; Geological Survey of Canada, Paper 89-1C, p. 67-79.

1992:   Uranium in Canada, 1991; in Current Research, Part D; Geological Survey of Canada, Paper 92-1D, p. 49-57.

1993:   Unconformity-type uranium deposits; in Mineral Deposit Modelling, (ed.) R.V. Kirkham, W.D. Sinclair, R.I. Thorpe, and J.M. Duke; Geological Association of Canada, Special Paper 40, p. 125-149.

Ruzicka, V. and LeCheminant, G.M.

1986:   Developments in uranium geology in Canada, 1985; in Current Research, Part A; Geological Survey of Canada, Paper 86-1A, p. 531-540.

Sibbald, T.I.I.

1976:   Uranium metallogenic studies: Rabbit Lake; in Summary of Investigations 1976 by the Saskatchewan Geological Survey, (ed.) J.E. Christopher and R. Macdonald; Province of Saskatchewan, Department of Mineral Rersources, p. 115-123.

*Sibbald, T.I.I. and Petruk, W. (ed.)

1985:   Geology of uranium deposits; The Canadian Institute of Mining and Metallurgy, Special Volume 32, 268 p.

Sopuck, V.J., de Carle, A., Wray, E.M., and Cooper, B.

1983:   Application of lithogeochemistry to the search for unconformity-type uranium deposits in the Athabasca Basin; in Uranium Exploration in Atahabasca Basin, (ed.) E.M. Cameron; Geological Survey of Canada, Paper 82-11, p. 192-205.

Tremblay, L.P.

1982:   Geology of the uranium deposits related to the sub-Athabasca unconformity, Saskatchewan; Geological Survey of Canada, Paper 81-20, 56 p.

Wallis, R.H., Saracoglu, N., Brummer, J.J., and Golightly, J.P.

1986:   The geology of the McClean uranium deposits, northern Saskatchewan; in Uranium Deposits of Canada, (ed.) E.L. Evans; The Canadian Institute of Mining and Metallurgy, Special Volume 33, p. 193-217.

Worden, J.M., Cumming, G.L., and Baadsgaard, H.

1985:   Geochronology of host rocks and mineralization of the Midwest uranium deposit, northern Saskatchewan; in Geology of Uranium Deposits, (ed.) T.I.I. Sibbald and W. Petruk; The Canadian Institute of Mining and Metallurgy, Special Volume 32, p. 67-72.

Author's address

V. Ruzicka
Geological Survey of Canada
601 Booth Street
Ottawa, Ontario
K1A 0E8

Printed in Canada

8. STRATABOUND CLASTIC-HOSTED URANIUM, LEAD, COPPER

INTRODUCTION

Disseminations of the ore minerals in clastic sedimentary rocks characterize this deposit type. Deposits occur in marine, paralic, lacustrine, and continental host rocks at or near a redox boundary. Three subtypes are distinguished on the basis of the major, economically-viable commodity and on the relation to the redox boundary. Sandstone uranium (8.1) and sandstone lead (8.2) deposits are associated with reductants within porous oxidizing sandstones; sediment-hosted stratiform copper (8.3) deposits occur in reduced host rocks adjacent to redbeds.

8.1 SANDSTONE URANIUM

R.T. Bell

IDENTIFICATION

The deposits occur as disseminated (mainly pore fillings) uranium minerals in arenaceous rocks, predominantly of continental origin and setting, in successor or foreland basins, and are mainly of post-Silurian age. Vanadium, molybdenum, selenium, copper, phosphorus, manganese, and chromium enrichments are common but at present are of negligible economic importance. The first three elements are commonly used as pathfinder elements for uranium.

The historically most important examples are in the U.S.A. There are several subtypes: roll, peneconcordant (blanket or trend), stacked, coal or lignite, and basal channel. The economically significant examples in Canada are restricted to the basal channel subtype in the Beaverdell area in southern British Columbia and a hybrid (vein modified trend type) at Mountain Lake in the northwestern part of District of Mackenzie, Northwest Territories.

IMPORTANCE

Canada

In 1990, the economic deposits, all in British Columbia, accounted for less than 2% of Canada's Reasonably Assured Resources of uranium. The deposits have yet to be exploited and remain dormant.

Foreign

In the western world prior to 1982, sandstone uranium deposits accounted for 40% of Reasonably Assured Resources and 54% of Estimated Additional Resources, but are steadily declining in importance. These deposits are mainly in the U.S.A. where they are the dominant source.

Significant deposits are (or have been) exploited in Niger, Gabon (of Precambrian age), Argentina, former Czechoslovakia, China, the former U.S.S.R. (Russia, Kazakhstan, Uzbekistan; Ruzicka, 1992), the former Yugoslavia (Slovenia), and France. Significant resources occur in Australia, Brazil, Mexico, Turkey, South Africa, and Japan.

Bell, R.T.
1996: Sandstone uranium; in Geology of Canadian Mineral Deposit Types, (ed.) O.R. Eckstrand, W.D. Sinclair, and R.I. Thorpe; Geological Survey of Canada, Geology of Canada, no. 8, p. 212-219 (also Geological Society of America, The Geology of North America, v. P-1).

SIZE AND GRADES

Canada

The deposits contain as much as 4000 t U in ore bodies grading from 0.1 to 0.2% U. Blizzard is the largest deposit with about 4000 t U at an average grade of about 0.18% U.

Foreign

The common range is 1000 to 10 000 t of contained U in ores grading from 0.03 to 2%. Many are smaller and a few contain as much as 30 000 t U. Clusters of small, very low grade (averaging as little as 0.04% U) deposits at shallow depths in very friable sandstones and in surficial deposits are economically recovered by relatively inexpensive in situ leaching processes (examples: Crow Butte, Nebraska and others in Texas, U.S.A.).

GEOLOGICAL FEATURES

Geological setting

Most uranium sandstone deposits are in continental sandstones and associated conglomerates that compositionally are immature to submature (i.e., arkosic, lithic, and/or tuffaceous), and texturally are mature (well sorted). Predominantly the sandstone sequences, as successor or foreland basins, lie on, or adjacent to, uplifted, deformed, and metamorphosed basement of felsic rocks such as the foreland basins to Nevadan-Laramide orogenies in western North America or as successor basins to the Hercynian orogeny in Europe.

Relation of ore to host rocks

Generally, deposits are confined to porous arenaceous units commonly bound by relatively impermeable mudstones. The deposits may follow organic-rich, high porosity trends roughly concordant with the strata (peneconcordant subtypes) in the sequence; if they are equidimensional in plan, they are referred to as blanket subtype (classical area: Lisbon Valley, Utah) or, if elongate in plan, trend subtypes (classical area: Ambrosia Lake, New Mexico). Basal channel subtypes (classical area: Tono, Japan) occur near the base of fluvial sediments in paleovalleys; coal or lignite subtype (classical area: South Dakota) is found in very organic-rich layers adjacent to porous sand strata. The deposits may occur at discordant oxidation-reduction (redox) boundaries (roll subtype, commonly C-shaped in section, (classical areas: Wyoming, Texas Gulf Coast, and Uravan Belt in Colorado) within individual porous sandstone units. In these roll subtype deposits, geochemical zoning (for example, Se⇒U⇒U+Mo) across the roll front is common. In some oxidized sandstones the deposits follow vague discordant and concordant bleached zones and occur as irregular stacked peneconcordant tabular bodies at or near faults or fracture systems (stacked subtype, linear in plan and rectangular in section; classical area: Ambrosia Lake, New Mexico). Some writers include the polymetallic deposits in collapse structures in sandstones (collapse breccia pipe type; Organization for Economic Cooperation Development/International Atomic Energy Agency, 1992) as in the Grand Canyon area, Arizona (Finch, 1967) as a sandstone type deposit. Some consider coal or lignite subtypes as a separate type.

Nature of the ore

Vertical and horizontal (facies trends) changes, resulting in porosity differences, control deposits by acting as conduits for, or as barriers to, mineralizing fluids. Faults may also serve as barriers or as conduits; fluids moving along faults are most likely to be those that act as reductants.

Dominance of uranium-bearing minerals in pore spaces and/or as partial to complete replacement of particles (clasts, matrix, cement, organic debris) in contrast to veins and gashes in arenaceous rocks is the most definitive feature of this deposit class. In the reverse situation the deposit is classified as vein type. Hybrid cases as at Mountain Lake, Northwest Territories occur (see "Mountain Lake", below).

Mineralogy

The ore occurs dominantly as pitchblende and/or coffinite in pore spaces or strongly absorbed by organic material, or weakly absorbed by clays and zeolitic material. In strongly oxidized portions of ore bodies, uranium occurs as a large variety of secondary (U in hexavalent form) minerals, of which autunite, uranophane, carnotite, and torbernite are common in pore spaces, in clays, limonite, opaline wood, and phosphates.

Canadian examples

(See Fig. 8.1-1 for locations)

Beaverdell area

The Beaverdell area deposits lie in the Okanagan Highlands of south-central British Columbia (Fig. 8.1-2). The basement rocks are felsic gneisses and granites formed during Mesozoic compressive orogenies. This basement has been "loosened" by pervasive faulting during east-west extension and uplift during the Eocene, which accompanied emplacement of high level granitic and volcanic rocks. Ductile and brittle low-angle normal and listric faulting and tectonic denudation also occurred in the Okanagan metamorphic core complex. This Eocene environment has been compared with that of the later Tertiary of the Basin and Range in the southwestern U.S.A. Rate of uplift in the Okanagan region has gradually decreased since the Eocene.

The host sequences for the uranium deposits are in paleovalleys produced in these uplands during the Miocene (Fig. 8.1-2, 8.1-3, and 8.1-4). The sediments comprise a basal sequence of very coarse conglomeratic sediments of torrential and braided stream environment with a few lenses of mud-rich, debris-flow or slump deposits. These give way abruptly to an overlying fine sequence of sandstone, siltstone, and mudstone of lacustrine and meandering stream environments. The onset of the fine sequence coincided with local onset of olivine basaltic volcanism which produced flows that dammed the stream valleys and with which the finer sediments were interstratified in the lower reaches of the valleys.

In the Intermontane region to the west and to the north the main 'plateau basalts' were formed during numerous extrusive events peaking at 11 and 6 Ma. The basalts in the Okanagan Highlands are the distal aspects of these and range from earlier 'valley basalts' to a later blanket or 'plateau' phase. Locally the earlier valley basalts caused damming and diversion of the streams and resulted in widespread development of the fine facies. The later blanketing basalts are about 5 Ma in age and effectively ended the sedimentary phase, and for a time protected much of the basement terrane of this area from erosion while regional uplift proceeded.

Late Pliocene and Quaternary erosion has removed most of these basalts in the Okanagan Highlands, and particularly so in the lower reaches of the Miocene paleodrainage. The catch-up of erosion into the basement, after prolonged protection by the 'plateau' basalt cover, gives the appearance of renewed uplift. There remain only small basalt caps in tributary paleovalleys and on the flanks of the main or trunk paleovalleys.

The uranium deposits lie in both coarse and fine sedimentary facies immediately beneath and adjacent to basalt caps in the uppermost tributary paleovalleys (Fig. 8.1-3, 8.1-4). Sediment thickness rarely exceeds 45 m. The individual deposits are roughly lenticular, tabular bodies usually only a metre or two in thickness. In the coarse facies ore boundaries are fairly gradational, whereas in the fine facies boundaries are sharp. Carbonaceous material and minor pyrite or marcasite accompany most ore but with two important exceptions.

The first exception is in the Blizzard deposit where virtually all the ore at the top of the coarse facies (Fig. 8.1-3, 8.1-4: zones B and C) is in the form of saléeite, autunite, and uranophane in light brown sandstone. Here, and in all but a very few places (usually restricted to the top) in the coarse sandstones, virtually all pyrite has been oxidized to limonite and only plant imprints and a few silicified wood fragments remain. Much of the ore in the lowest part of the fine facies (Fig. 8.1-3: zone A) is likewise oxidized in the northernmost part of the deposit. In the highest levels (fine facies, zone A) uranium, in association with black organic material and minor pyrite, occurs as pitchblende and coffinite(?). Between the oxidized and reduced facies both saléeite and ningyoite occur. Boyle (1982) has documented the sequence of ningyoite replacing saléeite and suggests that saléeite and autunite are primary and ningyoite and pitchblende are secondary. The predominance of reduced facies ore elsewhere suggests the reverse and that the phenomenon at Blizzard is due to fluctuating redox boundaries in the presence of a high phosphate background.

The other exception is in the Tyee deposit (Fig. 8.1-2) in which the core of the lower part of the orebody is heavily cemented by marcasite. The uranium mineralization is associated with organic material and was identified as ningyoite.

Minor ore grade zones occur along the baked mudstone contacts of basalt flows (Fig. 8.1-3, 8.1-4: zone F) and rarely in fractures in feeder dykes. A very minor part occurs in the regolith beneath the sediments (Fig. 8.1-3, 8.1-4: zone D) where the base of zone B reaches the base of the channel of the coarse facies (elsewhere the base of zone B is several metres above the base of the coarse facies). Another interesting potential ore zone is in the fine matrix of the uppermost (100 m) oxidized (rusty) part of a breccia pipe at the northern part of the Blizzard deposit (Fig. 8.1-3, 8.1-4: zone E).

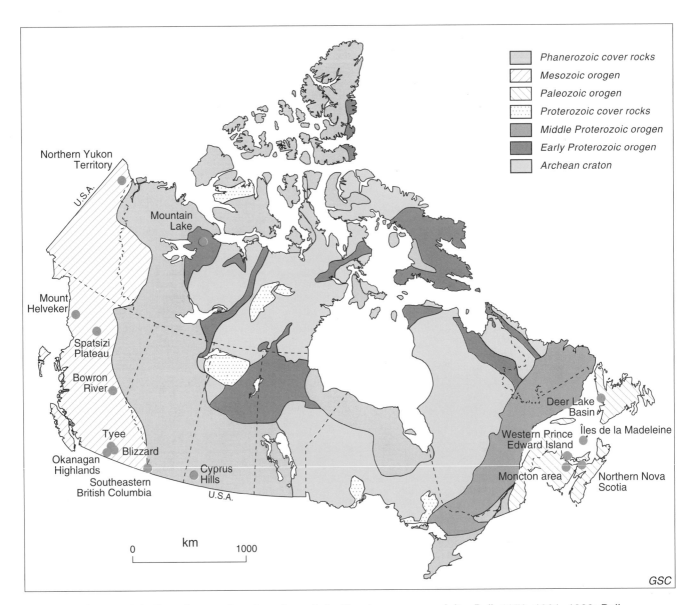

Figure 8.1-1. Canadian uranium deposits and significant occurrences (after Bell, 1976, 1981, 1982; Bell et al.,1976; Dunsmore, 1977a, b; Bell and Jones, 1979; Geological Survey of Canada, 1980, 1981; Boyle, 1982; and Jones, 1990). 1 – Blizzard (Cup lake, Lassie, and Fuki-Donan nearby); 2 – Tyee (a.k.a. Hydraulic Lake, with Venus and Haynes Lake nearby); 3 – Okanagan surficial (22 clusters of deposits between Oliver and Kelowna); 4 – Mountain Lake (PEC-YUK); <u>Significant groups of occurrences:</u> 5 – southeastern British Columbia (Lin, Commerce); 6 – Bowron River; 7 – Spatsizi Plateau (Edozadelly); 8 – Mount Helveker; 9 – northern Yukon Territory (Bou, Bon); 10 – Cypress Hills; 11 – Moncton area; 12 – western Prince Edward Island; 13 – northern Nova Scotia (Tatamagouche); 14 – Îles de la Madeleine; 15 – Deer Lake Basin.

Content of P (but not Th, Mo, V, or Se) is slightly elevated in the Beaverdell deposits, although Mo is strongly elevated in the Recent surficial occurrences west of the Okanagan valley.

Age of mineralization is still poorly defined. A single U-Pb date of 2.8 Ma in the transition from oxidized to reduced facies in the Blizzard deposit suggests that the original accumulation was at least of the same age and it is likely that remobilization is still going on in that deposit. Elsewhere uranium concentrations in the basal baked zones of the earliest basalt flows suggest a (re)mobilization at the time of the earliest basalt in this area, likely at least before 5 Ma.

Others

Mountain Lake

A Precambrian deposit occurs near Mountain Lake, Northwest Territories in the middle part of the lower white and light brown sandstone unit of the Neohelikian Dismal Lakes

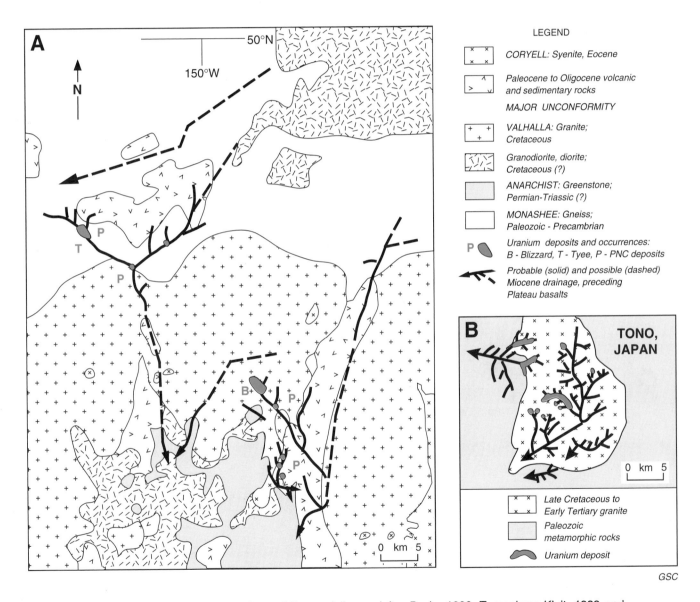

Figure 8.1-2. A) Simplified geology of Beaverdell area (after Boyle, 1982; Tempelman-Kluit, 1989 and references therein) showing uranium deposits and late Miocene paleodrainage. Plateau basalts (late Miocene-Pliocene) not shown. **B)** Paleodrainage and deposits in Tono area, Japan (after Katayama et al., 1974).

Group. The unit unconformably overlies the Paleohelikian Hornby Bay Group and, locally, Aphebian basement gneisses and felsic volcanic rocks. The unit is conformably overlain by grey siltstone and shale.

The uranium occurs as disseminated sooty pitchblende and coffinite(?) in irregular stratiform lenses, 2 m thick, as much as 400 m long, and 200 m wide. Grades are of the order of 0.1 to 0.3% U but actual size and overall grade have not been reported. Yellow and green secondary minerals occur in fractures and in weathered zones. Some chalcopyrite and pyrite are present, as well as traces of cobalt-nickel arsenides. Trigg (1986) reported up to 0.5% Cu, and associated high Co, Ni, and Ag values. Red hematite staining is common along the margins.

Trigg (1986) also reported grades up to about 5% U (as both pitchblende and secondaries) that occur in a zone characterized by veins occupying steeply dipping fractures at the north end of the deposit in association with a major fault. This part of the deposit may be regarded as vein-modified sandstone uranium.

The host sequence is between 1500 and 1275 Ma old and a single U-Pb isotope analysis suggests an age of 794 Ma (Gandhi, 1986) for latest (re)mobilization of uranium.

Surficial deposits of the Okanagan valley

West of the Okanagan valley a large number of small, 'surficial' or 'young' low grade deposits of uranium occur to depths of 10 m in Recent sediments (Culbert et al., 1984) in mainly fine grained fluvial, lacustrine, and bog deposits.

They are accompanied by high Mo content and have grades the order of 0.01% U and are young in that there are negligible daughter elements present (hence normal gamma-ray geophysical methods are useless for exploration and delineation). At present these 'young' deposits are uneconomic. Despite being small and of low grade these occurrences could eventually prove economic, as indicated by the attempt at development of a similar deposit just south of there at Flodelle Creek in the state of Washington.

Miscellaneous occurrences

Occurrences of uranium with vanadium (Prince Edward Island), with copper (northern Nova Scotia), and with silver (Newfoundland) are found in Carboniferous redbed successions, commonly associated with carbonaceous debris (Dunsmore, 1977a, b). Dunsmore (1977a) pointed to an association with evaporites for some of these occurrences.

In northern Yukon Territory, latest Devonian or Early Carboniferous continental, basal sandstones and conglomerates south and east of the Barn Mountains contain phosphatic uraniferous occurrences. These are deeply weathered and the uranium occurs in part as uranophane.

Uranium in lignites and in organic-rich siltstones occurs in Upper Cretaceous sequences in Cypress Hills, Saskatchewan, and Bowren River Basin, British Columbia. Large mammal bones in the gravels of the Cypress Hills Formation (Oligocene) of southwestern Saskatchewan and in the Echo Lake aquifer (Pleistocene) near Fort Qu'Appelle, Saskatchewan, are uraniferous. Ash-tuffs containing organic fragments and

A LONGITUDINAL SECTION

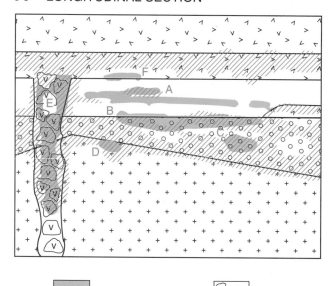

B TRANSVERSE SECTION

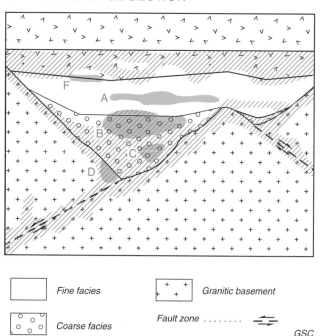

| | Ore zone | | Breccia | | Fine facies | | Granitic basement |
| | Oxidized zone | | Basalt | | Coarse facies | | Fault zone .... |

GSC

Figure 8.1-3. Schematic diagram of host sequence of the Blizzard deposit with sections **A)** longitudinal and **B)** transverse to the paleovalleys illustrating location of ore zones (in red): A – fine facies; B – transecting upper border of coarse facies; C – entirely in coarse facies; D – in regolith; E – in upper most part of breccia pipe; and F – at basal margins of flows and in dykes.

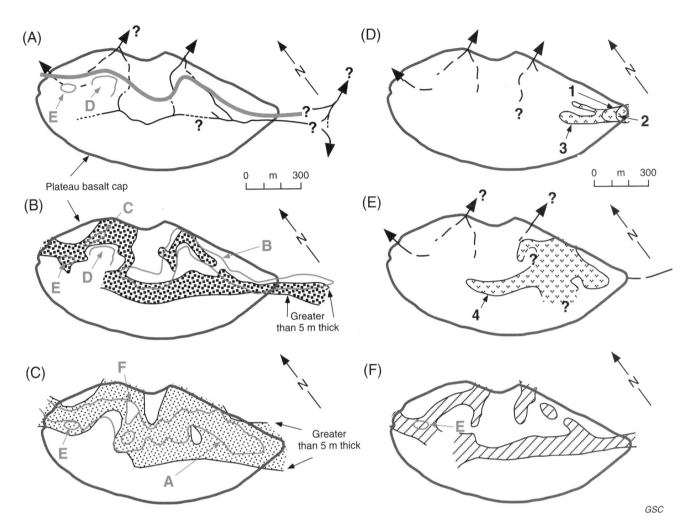

Figure 8.1-4. Plan views of Blizzard deposit showing ore zones (A to E) listed in Figure 8.1-3 and described in text. Plateau basalt cap shown as heavy outline. Based in part from Norcen (1979) and their submissions to the Bates Commission (Bates et al., 1980).

A) Initial paleodrainage at the base of the sedimentary sequence. Note the offset from the initial paleodrainage, of the trend of the axis of maximum ore grade-thickness (heavy red line) which coincides roughly with the initial drainage divide.

B) Distribution of coarse facies thicker than 5 m.

C) Distribution of fine facies thicker than 5 m.

D) Distribution of first (1), second (2), and third (3) basaltic flow units (valley basalts) and suggested stream diversion (dashed lines).

E) Distribution of fourth (4) flow. Succeeding flows (true 'Plateau' basalts) cover the area. A sixth flow (not shown) is very heavily oxidized (weathered) and in turn covered by two very thick and massive flows which constitute the main Plateau flows in the area.

F) Distribution of continuous and thick oxidized facies in sediments below basalt cap. Almost entirely in the coarse facies (see Fig. 8.1-4B).

some sandstones are uraniferous in the Upper Cretaceous of the Spatsizi Plateau and Mount Helveker areas in British Columbia. Radioactive anomalies occur in association with small copper showings in Grinnell Formation Purcell (Precambrian) rocks in the Sage Creek area in southeastern British Columbia.

Most of these occurrences are unimportant, but the sequences bearing them are all favourable targets for further uranium exploration and, in the case of the eastern Canadian Carboniferous basins, the environments are favourable for copper, silver, and lead deposits as well.

DEFINITIVE CHARACTERISTICS
General

1. Continental sandstones adjacent/superjacent to felsic basement rocks;
2. Sandstones with permeability contrasts (shale/mudstone beds);
3. Submature sandstone, especially with felsic clasts;
4. Poorly cemented (at least during mineralization);
5. Presence of reductants (coaly material, sulphides);
6. Presence of secondary oxidation/reduction features;
7. Sequence generally younger than Silurian.

Specific (Beaverdell area)

8. Upper reaches of paleodrainage;
9. Protective cap of volcanics and mudstone;
10. Structurally "loosened" basement (extensional tectonics).

Most of these features (1 to 7) are brought together in continental successor and foreland basins adjacent to, or in terrains dominated by, felsic rocks. Younger deposits are favoured by the present preservation potential (poor in the long term) and by the presence of abundant reductants (advent of land plants in the Silurian). Most importantly, continental conditions favour input of oxidized and oxidizing surface and groundwaters.

GENETIC MODEL

The general genetic model hinges on the divalent nature of uranium – being strongly soluble in oxidizing conditions (uranic: U^{+6}) and relatively insoluble in reducing conditions (uranous: U^{+4}). The paleohydrology of meteoric groundwaters comprise the most important aspect for the primary mineralizing solutions. The model involves three stages: i) oxidative leaching of uranium from slightly enriched felsic rocks (in basement, clasts within host sequence); ii) transportation by oxidizing surface waters and groundwaters into and through porous rocks; and iii) precipitation on entering reducing environments. Active bacterial biochemical effects in organic and sulphide-rich sediments enhance this trap. However reductant fluids may be introduced through fault zones or other aquifers or from brines from rising salt domes to mix and react with the uranium-bearing meteoric waters. Phosphates and vanadates also effect precipitation.

Semiarid (wet-dry climatic) conditions during sedimentation and especially during diagenesis and mineralization, afford favourable conditions. If the climate is too wet, heavy and rapid flow of oxidizing groundwaters will eventually flush out the deposits. If too dry, sufficient indigenous organic material and sulphides may not accumulate in the sediments. Semiarid conditions favour the development of more mature groundwaters (bicarbonate-rich) at shallow depths, which in turn favour complexing with uranyl ions in the upstream end of the aquifer.

Shale (mudstone) layers and lateral facies changes help constrain the plumbing system for groundwater flow and may, along with local closed basins, produce stagnation zones. Lower rates and volume of groundwater flow in first- and second-order (paleo)stream channels are likely critical to affording the proper conditions for deposition as significant deposits of the basal channel subtype are restricted to the first- and second-order paleochannels in both Japan and the Beaverdell area (Fig. 8.1-2). Organic-rich fine facies, as well as providing physical restraints to the plumbing system, may also provide reductant buffers. Impermeable mudstones or volcanic rocks provide the deposits a degree of protection from severe destruction.

The mineralization occurs during (and is a part of) the process of diagenesis of the sandstones. Commonly this process starts shortly after deposition of the sands, but may become dormant and then restart several times and much later, and continues until porosity and permeability is greatly restricted by more or less complete lithification of the sandstone.

RELATED DEPOSIT TYPES

In general sandstone-hosted uranium deposits share definitive characteristics with other sandstone-hosted Cu, V, and Pb(-Zn±Ag) deposits and commonly occur in the same district (for example, Uravan district in Utah, U.S.A.).

Some sandstone-hosted uranium deposits contain juvenile and eroded felsic volcanic material. The host sequence may grade laterally and/or vertically to fully volcanic rocks, and all gradations may be present from volcanic- to sandstone-hosted deposits (for example, those in northern Italy).

In some unconformity-type uranium deposits, part or most of the deposit may be in overlying continental sandstones, such as in the Athabasca Basin in Saskatchewan. In such cases the reductants appear to have originated in the basement and the uranium appears to have been entrapped as a stationary plume in the overlying sandstones and regoliths. Conversely, in basal channel subtype deposits there are some minor zones of mineralization within the regolith, as in deposits in the Beaverdell area (Fig. 8.1-3 and 8.1-4, zone D).

Some, if not all, paleoplacer-uranium deposits show minor remobilization (modified placer model).

EXPLORATION GUIDES

1. Attention should be paid to the geological characteristics as described previously, in particular, elevated amounts of uranium in basement rocks and abundance of uranium occurrences (even if small) in associated felsic

volcanics, in late stage and post-tectonic granites and pegmatites. Basement metamorphic complexes should also be regarded as favourable geological environments.

2. Some caution must be exercised with geochemical methods. For example, in south-central British Columbia the most arid areas give high responses for uranium in surface (stream) waters and low responses in stream sediments, whereas in the damper areas, as in the nearby Okanagan Highlands, the reverse is true.

3. Similarly, caution must be used with radiometric methods. Younger (some Tertiary and most Quaternary) deposits are commonly out of equilibrium or have not yet reached equilibrium with daughter elements. This is especially true for surficial deposits (see Culbert et al., 1984).

SELECTED BIBLIOGRAPHY

An immense amount has been written in English with regard to sandstone-hosted uranium deposits; most publications are based on studies in the U.S.A. The Finch and Davis (1995) and Grutt (1972) references provide an introduction to this literature. The asterisk (*) in the references below denotes important publications for this deposit-type.

Bates, D.V., Murray, J.W., and Raudsepp, V.
1980: British Columbia - Royal Commission of Inquiry, Health and Environmental Protection, Uranium Mining; Queens Printer, Victoria, British Columbia, 3 volumes.
[Excellent source for documentary material and references for Beaverdell area.]

Bell, R.T.
1976: Geology of uranium in sandstones in Canada; in Report of Activities, Part A; Geological Survey of Canada, Paper 76-1A, p. 337-338.
1981: Preliminary evaluation of uranium in Sustut and Bowser successor basins, British Columbia; in Current Research, Part A; Geological Survey of Canada, Paper 81-1A, p. 241-246.
1982: Notes on uranium investigations in the Canadian Cordillera, 1981; in Current Research, Part A; Geological Survey of Canada, Paper 82-1A, p. 438-440.

Bell, R.T. and Jones, L.D.
1979: Geology of some uranium occurrences in western Canada; in Current Research, Part A; Geological Survey of Canada, Paper 79-1A, p. 397-399.

Bell, R.T., Steacy, H.R., and Zimmerman, J.B.
1976: Uranium-bearing bone occurrences; in Report of Activities, Part A; Geological Survey of Canada, Paper 76-1A, p. 339-340.

*Boyle, D.R.
1982: The formation of basal-type uranium deposits in south central British Columbia; Economic Geology, v. 77, p. 1176-1209.

*Culbert, R.R. and Leighton, D.G.
1988: Young uranium; Ore Geology Reviews, v. 3, p. 313-332.

*Culbert, R.R., Boyle, D.R., and Levinson, A.A.
1984: Surficial uranium deposits in Canada; in Surficial Uranium Deposits, (ed.) P.D. Toens; International Atomic Energy Agency, Vienna, Austria, IAEA-TECDOC-322, p. 179-191.

*Dunsmore, H.E.
1977a: A new genetic model for uranium-copper mineralization, Permo-Carboniferous Basin, northern Nova Scotia, in Report of Activities, Part B; Geological Survey of Canada, Paper 77-1B, p. 247-253.
1977b: Uranium resources of the Permo-Carboniferous basin, Atlantic Canada; in Report of Activities, Part B; Geological Survey of Canada, Paper 77-1B, p. 341-344.

*Finch, W.I.
1967: Geology of epigenetic uranium deposits in sandstone in the United States; United States Geological Survey, Professional Paper 538, 121 p. plus 2 plates.
[An excellent and detailed review of the geological studies and exploration histories of uranium deposits in sandstone in the U.S.A. from 1943 to 1959.]

*Finch, W.I. and Davis, J.F. (ed.)
1985: Geological environments of sandstone-type uranium deposits; International Atomic Energy Agency, Vienna, Austria, IAEA-TECDOC-328, 408 p.

Gandhi, S.S.
1986: Mountain Lake deposit, Northwest Territories; in Uranium Deposits of Canada, (ed.) E.L. Evans; The Canadian Institute of Mining and Metallurgy, Special Volume 33, p. 293-302.

Geological Survey of Canada
1980: Non-hydrocarbon mineral resource potential of parts of northern Canada; Geological Survey of Canada, Open File 716, 376 p.
1981: Assessment of mineral and fuel resource potential of the proposed Northern Yukon National Park and adjacent areas (Phase I); Geological Survey of Canada, Open File 760, 31 p., plus appendices.

*Grutt, E.W., Jr.
1972: Prospecting criteria for sandstone-type uranium deposits; in Uranium Prospecting Handbook, Proceedings of a NATO-sponsored Advanced Study Institute on Methods of Prospecting for Uranium Minerals, (ed.) S.H.U. Bowie, M. Davis, and D. Ostle; London, 21 September - 2 October, 1971, The Institute of Mining and Metallurgy 1972, p. 47-76, Discussion p. 77-78.
[A superior short discussion of this deposit type (and subtypes) in the U.S.A. including detailed local criteria regarded as favourable for exploration. Remainder of this book is also a superior reference on subject of uranium prospecting.]

*International Atomic Energy Agency
1985: Geological environments of sandstone-type uranium deposits, report of the working group on uranium geology; International Atomic Energy Agency, Vienna, Austria, IAEA-TECDOC-328, 408 p.
[Important comprehensive update on the subject.]

*Jones, L.D.
1990: Uranium and thorium occurrences in British Columbia; British Columbia Ministry of Energy, Mines and Petroleum Resources, Geological Survey Branch, Open File 1990-32, 78 p., 1 map, and 1 MINFILE dataset disc.
[The definitive documentation for uranium and thorium deposits and occurrences in British Columbia and important reference for property assessment reports available to the public.]

*Katayama, N., Kubo, K., and Hirono, S.
1974: Genesis of uranium deposits of the Tono mine, Japan; in Formation of Uranium Ore Deposits, International Atomic Energy Agency, Vienna, Austria, IAEA-SM-183, p. 437-452.

*Langford, F.F.
1977: Surficial origin of North American pitchblende and related uranium deposits; American Association of Petroleum Geologists, Bulletin, v. 61, p. 28-42.
[Treatment of both sandstone and unconformity-type deposits.]

Norcen
1979: Statement of evidence relating to summary of the geology of the Blizzard deposit, Phase 1, overview; presented September 1979 to British Columbia Royal Commission of Inquiry into Uranium Mining, Document 20255, 11 p. plus figures.
[Excellent summary of the framework of the Blizzard deposit.]

Organization for Economic Cooperation Development/ International Atomic Energy Agency
1992: Uranium 1991 resources, production and demand; A joint report by the OECD Nuclear Energy Agency and the International Atomic Energy Agency, Paris, 255 p.

Ruzicka, V.
1992: Types of uranium deposits in the former U.S.S.R.; Colorado School of Mines Quarterly Review, v. 92, no. 1, p. 19-27.

Tempelman-Kluit, D.J.
1989: Geology, Penticton, British Columbia; Geological Survey of Canada, Map 1736A, scale 1:250 000.

Trigg, C.M.
1986: PEC uranium deposit, Hornby Bay Basin, Northwest Territories; in Uranium Deposits of Canada, (ed.) E.L. Evans; The Canadian Institute of Mining and Metallurgy, Special Volume 33, p. 295-302.

Yeo, G.M.
1981: Sandstone stratigraphy and uranium potential of the eastern Hornby Bay Basin, in Mineral Industry Report 1977, Northwest Territories, (ed.) C. Lord, P.J. Laporte, W.A. Gibbins, J.B. Seaton, J.A. Goodwin, and W.A. Padgham; Indian and Northern Affairs, Canada, Exploration and Geological Services Unit, EGS 1981-11, p. 132-147.

8.2 SANDSTONE LEAD

D.F. Sangster

INTRODUCTION

Disseminated galena and minor sphalerite, in transgressive basal quartzite or quartzofeldspathic sandstones resting on sialic basement are the definitive geological features of sandstone-lead deposits. The commodities produced from this type of deposit are mainly lead with lesser zinc; silver is rarely recovered. Important examples include: Yava, Nova Scotia; George Lake, Saskatchewan; Laisvall, Sweden; Maubach and Mechernich, Germany; Largentière, France; and Zeida, Morocco.

IMPORTANCE

Only two examples of this deposit type are known in Canada (Fig. 8.2-1); neither is in production at the present time. Compared with other deposit types, sandstone-lead deposits are a relatively minor type, although in some countries they constitute a major source of metal (e.g. Sweden).

SIZE OF DEPOSIT

Deposits range in grade from 2 to 5% Pb, 0.2 to 0.8% Zn, and 1 to 20 g/t Ag; most are less than 10 million tonnes in size. Because of the disseminated nature of the ore, tonnages and grades can be markedly affected by changes in cut-off grades. At Yava, for example, at cut-off grades of 1, 2, and 3%, tonnages and grades are as follows: 71.2 million at 2.09% Pb, 30.3 million at 3.01%, and 12.6 million at 3.95%, respectively.

GEOLOGICAL FEATURES

Geological setting

Host rock sandstones, quartzitic or quartzofeldspathic in composition, were deposited in environments ranging from continental fluvial (Yava) to shallow marine or tidal beach (Laisvall). The most common environment is one of mixed continental and marine character (i.e. paralic). Host rocks in most districts are succeeded by marine sediments, suggestive of marine transgression onto the craton.

Without exception, basement rocks underlying sandstone-lead deposits are of sialic composition and most are granites or granitic gneisses (Bjørlykke and Sangster, 1981). In several instances, basement rocks are demonstrably anomalous in lead content compared to the world average for granitic rocks (~22 ppm).

Sandstone-lead deposits occur in marine sandstone or in fluvial sandstone with terrestrial organic material. Host rock sandstones are grey or white but never red.

Paleomagnetic data available in several districts indicate a low paleolatitude position (0-30°). The presence of evaporites in some deposits indicates that paleoclimatic conditions during deposition of host rocks ranged from warm arid to semiarid. Alternatively, abundant organic debris in other districts may suggest a somewhat more humid climate prevailed in those areas. Taken together, paleoclimatic conditions for sandstone-lead deposits indicates host rock depositional conditions varied from area to area but with a majority of them being semiarid and warm (Bjørlykke and Sangster, 1981).

Age of host rocks

Deposits are found in rocks ranging from Middle Proterozoic to Cretaceous; those in Canada are Middle Proterozoic (George Lake) and Pennsylvanian (Yava). Rocks of Late Proterozoic-Early Cambrian (Norway, Sweden) and Triassic (France, Germany, Morocco) ages contain a majority of deposits of this type.

Relations of ore to host rocks

The orebodies are commonly conformable to bedding in the sandstone, especially on a mine scale. In detail, however, the ore zones may actually transgress bedding at a low angle. Sedimentary channels in the sandstone are preferentially mineralized.

Form of deposits

Because sedimentary channels are the preferred sites for sandstone-lead deposits (Bjørlykke and Sangster, 1981; Sangster and Vaillancourt, 1990a), most of them have a generally lensoid, broadly conformable form. In plan, ore zones tend to be sinuous, again illustrating the sedimentary control, and laterally discontinuous. At Largentière (France), higher grade zones occur in, and adjacent to, steep faults; consequently, in this deposit, many ore zones are narrow, lenticular bodies oriented at high angles to bedding.

Distribution of ore minerals

The preferred site of ore minerals is as cement between sand grains resulting in disseminated sulphide blebs or spots in massive sandstones or concentrations of sulphides along the lower, more porous, portions of graded beds. Where carbonaceous material is present, sulphides fill wood cells or replace cell walls. Concretionary-like sulphide concentrations are abundant in most deposits.

Sangster, D.F.
1996: Sandstone lead; in Geology of Canadian Mineral Deposit Types, (ed.) O.R. Eckstrand, W.D. Sinclair, and R.I. Thorpe; Geological Survey of Canada, Geology of Canada, no. 8, p. 220-223 (also Geological Society of America, The Geology of North America, v. P-1).

Ore zones tend to be delimited by assay, rather than by geological boundaries. Characteristically, a higher grade core is surrounded by material that progressively decreases in grade outward.

Ore compositions; zoning of ore

Deposits are normally lead dominant; $Pb/(Pb+Zn)$ values are all greater than 0.85 and most are greater than 0.95. George Lake (Canada), in contrast, is zinc-dominant but all other geological features are typical of this deposit type. Few data are available on metal zoning but there is evidence of an upward, and basinward, increase in zinc relative to lead in some deposits (e.g. Yava, Laisvall).

Alteration

Hosted as they are in siliciclastic rocks, it is not surprising, perhaps, that evidence of strong wall rock alteration is characteristically lacking in sandstone-lead deposits. In fact, alteration is generally restricted to weathering of basement rocks underlying the ore-bearing sandstones.

Mineralogy

The most common sulphide minerals are galena, sphalerite, pyrite, and chalcopyrite although they differ markedly in abundance both within, and between, deposits. Replacement of these minerals by secondary analogues have been reported in several deposits (e.g. Zeida, Maubach, Mechernich).

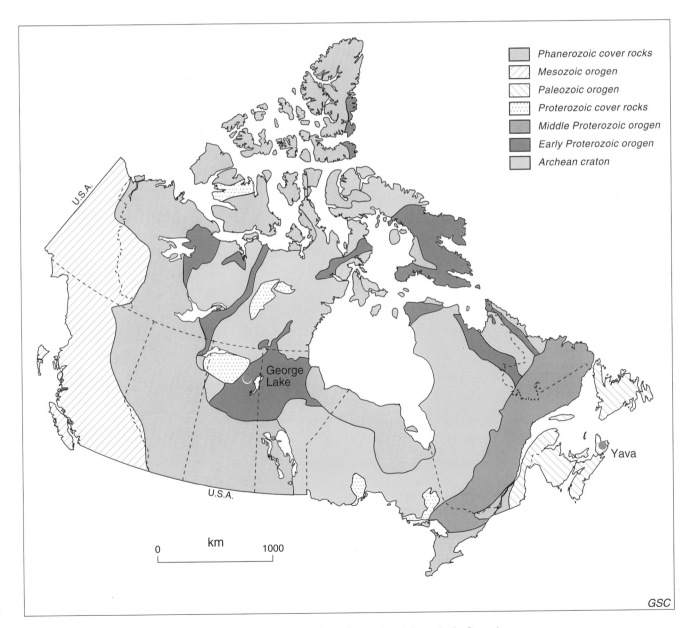

Figure 8.2-1. Location of sandstone-lead deposits in Canada.

Quartz and various carbonate minerals constitute the most abundant nonsulphide cement; barite and fluorite are minor cements in some deposits. Silica, usually chalcedonic, is characteristically more abundant than carbonate.

Ore textures

Sulphide minerals in sandstone-lead deposits, as a result of their typical occurrence as cement between detrital sand grains, characteristically display a disseminated texture. However, the disseminated sulphides are not normally homogeneously dispersed throughout the sandstone. Two very common textures are: i) poikiloblastic "spots" as much as 2 cm in diameter, representing local accumulations of galena; spots may be randomly distributed in the sandstone or may show a slight preferential alignment parallel to bedding; ii) discontinuous galena-rich streaks distributed parallel to bedding, including crossbedding.

Another characteristic texture is the abundant epitaxial quartz overgrowths on detrital quartz grains (Fig. 8.2-2). This texture, in some deposits at least, is more abundant within or near ore zones than regionally. Paragenetic studies indicate the epitaxial quartz predates galena.

DEFINITIVE CHARACTERISTICS

Sandstone-lead deposits can be recognized by the following characteristic features:

1. Host rock is a relatively clean, reasonably mature, quartz sandstone which is transgressive onto a sialic platform.

2. Galena is the dominant sulphide; pyrite and sphalerite are present in decidedly subordinate amounts. Copper minerals occur in trace quantities only.

3. Sulphide minerals are mainly disseminated, either as random aggregates or distributed along bedding planes in the host sandstone.

4. Epitaxial quartz overgrowths are abundant, especially in the ore zones.

5. Ore zones tend to follow sedimentary channels in the sandstone; higher grade zones are surrounded by lower grade material.

GENETIC MODEL

As discussed by Bjørlykke and Sangster (1981), two main genetic models have been proposed: 1) the hydrothermal or basin brine model, and 2) the groundwater or meteoric model. Inasmuch as deposits are found in both marine and continental depositional environments, it is unlikely that either of these models applies to all sandstone-lead deposits.

Based on research in the marine sandstone-hosted Laisvall deposit (Sweden), Rickard et al. (1979) proposed a basin brine model, relating migration of the ore fluid with nappe movement in the Caledonides. They suggested that compression by the overriding nappes was the main force driving ore fluids eastward. Marine shales, lying westward (i.e. basinward) of the Laisvall area were suggested as the source of metals; seawater sulphate was identified as the source of sulphur. This model has many points in common with that of the "squeegee" model for Mississippi Valley-type (MVT) deposits (Oliver, 1986), also hosted in marine sedimentary rocks. A variation on the basin brine model was proposed by Bjørlykke

et al. (1991) who drew attention to the presence of high heat-producing (HHP) granites beneath certain MVT and sandstone-lead deposits and suggested that basinal brines might have participated in vertically-directed convection cells generated by high heat-producing basement granites.

For sandstone-lead deposits in a continental environment, the author prefers a meteoric groundwater model involving the following: i) in situ breakdown, by passage of groundwaters, of potassium feldspar in arkosic sandstone; ii) transport of the lead released in this way, through porous channels in the sandstone, to an environment having a sufficiently high reduced sulphur content to precipitate sulphides. Detailed lead isotopic studies at Yava (Sangster and Vaillancourt, 1990b) have shown lead in the deposit was derived from underlying basement. During weathering, lead, released from the basement-derived feldspar in the sandstone, was carried in groundwater to the site of ore formation. Within the oxidizing environment of meteoric and shallow groundwater, lead will remain in solution provided the dissolved carbonate and sulphate content of

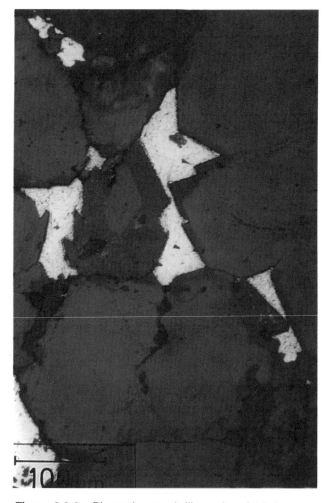

Figure 8.2-2. Photomicrograph illustrating detrital quartz grains (light grey) partially surrounded by silica overgrowths (dark grey) in characteristic straight-edge contact with galena cement (white), Yava (from Sangster and Vaillancourt, 1990b). GSC 204077-F

the groundwater is low. Upon reaching a reducing environment provided by the accumulation of organic material in the sediments, lead combined with biogenically reduced sulphur and precipitated as galena. Source of sulphur in the continental sandstone deposits is conjectural although, at Yava, gypsum nodules in underlying marine shale was suggested as a sulphur source (Sangster and Vaillancourt, 1990b).

RELATED DEPOSIT TYPES

Sandstone-uranium and sedimentary copper deposit types share the following genetic features with sandstone-lead deposits, particularly those hosted in continental sandstones: i) metals were derived from basement rocks (felsic in the case of Pb and U; mafic in the case of Cu); ii) transport was effected by oxidized subsurface waters through permeable clastic rocks; iii) precipitation occurred upon reaching a reducing environment. Sandstone-lead deposits in marine or paralic sandstones may share a basin brine-related genetic model with Mississippi Valley-type lead-zinc deposits, especially those situated immediately above basement containing high heat-producing granites.

EXPLORATION GUIDES

The following parameters may be useful in the search for sandstone-lead deposits:

1. Sialic basement; those with average lead content greater than ~30 ppm are particularly significant.

2. Basal portion of grey or white (not red) quartzitic sandstone of a transgressive sequence on sialic basement.

3. Channels in sandstone, especially on the periphery of the sedimentary basin.

4. Permeable zones in sandstone, i.e. the "cleaner" portions with minimum intergranular material.

5. Epitaxial quartz overgrowths on detrital quartz grains.

SELECTED BIBLIOGRAPHY

References marked with asterisks (*) are considered to be the best sources of general information on the deposit subtype.

*Bjørlykke, A. and Sangster, D.F.
1981: An overview of sandstone-lead deposits and their relation to red-bed copper and carbonate-hosted lead-zinc deposits; in Economic Geology, Seventy-fifth Anniversary Volume 1905-1980, (ed.) B.J. Skinner; p. 179-213.

Bjørlykke, A. and Thorpe, R.I.
1983: The source of lead in the Osen sandstone-lead deposit on the Baltic Shield, Norway; Economic Geology, v. 76, p. 1205-1210.

Bjørlykke, A., Sangster, D.F., and Fehn, U.
1991: Relationships between high heat-producing (HHP) granites and stratabound lead-zinc deposits; in Source, Transport and Deposition of Metals, Proceedings of the 25th Anniversary Meeting of the Society for Geology Applied to Mineral Deposits, Nancy, France, (ed.) M. Pagel and J.L. Leroy; p. 257-260.

Oliver, J.
1986: Fluids expelled tectonically from orogenic belts: their role in hydrocarbon migration and other geologic phenomena; Geology, v. 14, p. 99-102.

*Rickard, D.T., Wildén, M.Y., Marinder, N.E., and Donnelly, T.H.
1979: Studies on the genesis of the Laisvall sandstone lead-zinc deposits; Economic Geology, v. 74, p. 1255-1285.

Sangster, D.F. and Vaillancourt, P.D.
1990a: Paleo-geomorphology in the exploration for undiscovered sandstone-lead deposits, Salmon River Basin, Nova Scotia; The Canadian Institute of Mining and Metallurgy, Bulletin, v. 83, p. 62-68.

*1990b: Geology of the Yava sandstone-lead deposit, Cape Breton Island, Nova Scotia, Canada; in Mineral Deposit Studies in Nova Scotia, Volume I, (ed.) A.L. Sangster; Geological Survey of Canada, Paper 90-8, p. 203-244.

8.3 SEDIMENT-HOSTED STRATIFORM COPPER

R.V. Kirkham

Sediment-hosted stratiform copper (SSC) or 'diagenetic sedimentary' copper deposits belong to a large family of diverse, more or less concordant and discordant copper deposits and occurrences that show a close relationship to their sedimentary host rocks. In this sense they are 'sedimentary', even though available evidence supports diagenetic precipitation of metals at redox boundaries rather than syngenetic metal deposition. Nomenclature for deposits of this type is problematic and no general term referring to them is entirely satisfactory. Individual deposits have been variously referred to as syngenetic, sedimentary, syndiagenetic, diagenetic, stratiform, concordant, peneconcordant, stratabound, sediment-hosted, shale-hosted, sandstone-hosted, and carbonate-hosted. Because of the diverse lithologies of host rocks and different

morphologies, close relationship of these deposits to the environment of formation of their host rocks and inferred diagenetic origin, the term 'diagenetic sedimentary' copper deposits is a reasonable designation for these deposits, but as the term 'sediment-hosted stratiform' copper deposits is

Kirkham, R.V.
1996: Sediment-hosted stratiform copper; in Geology of Canadian Mineral Deposit Types, (ed.) O.R. Eckstrand, W.D. Sinclair, and R.I. Thorpe; Geological Survey of Canada, Geology of Canada, no. 8, p. 223-240 (also Geological Society of America, The Geology of North America, v. P-1).

widely accepted for such deposits (e.g., Boyle et al., 1989) it will be used in this paper. Nevertheless, the deposits under consideration are not truly 'stratiform' and this nongenetic term also permits inclusion of deposits of entirely unrelated genesis, such as syngenetic exhalative (Sedex) and high-temperature replacement deposits. However, in this paper only 'diagenetic sedimentary sediment-hosted stratiform' copper deposits will be considered. 'Diagenetic sedimentary' copper deposits can be divided into two subtypes: 'Kupferschiefer-type' which occur in rocks deposited in paralic marine (or large-scale saline lacustrine) environments, and 'redbed-type' which occur in rocks deposited in continental environments. Much of the data presented here is from recent compilations by Kirkham (1989) and Kirkham et al. (1994). Kirkham (1989) and sources in Boyle et al. (1989) should be consulted for further information and discussion.

8.3a Kupferschiefer-type

IDENTIFICATION

Kupferschiefer-type deposits occur typically as zonally distributed, disseminated sulphides at oxidation-reduction boundaries in anoxic rocks at the base of a marine or large-scale saline lacustrine transgressive cycle overlying or interbedded with continental redbeds. Redbeds and evaporites are characteristic associated rock types.

Copper is the most important metal found in these deposits and silver and cobalt are the most important byproduct or, in a few localities, coproduct metals. Other characteristic associated base metals, such as lead and zinc, are either of little or no economic importance. Platinum group elements have been reported to occur in some deposits in Zaire and the Lubin deposit in Poland.

The Redstone (Coates Lake) deposit in the Northwest Territories (Fig. 8.3-1) (Kirkham, 1974; Ruelle, 1982; Chartrand and Brown, 1985; Jefferson and Ruelle, 1986; Chartrand et al., 1989; Lustwerk and Wasserman, 1989) is the most completely documented deposit of this type in Canada. Most Kupferschiefer-type occurrences shown in Figure 8.3-1, apart from the Redstone area, are thin, low grade, and of minor importance. Nevertheless, their distribution indicates areas that might be prospective for this deposit type. Kupferschiefer deposits in Germany and Poland (Rentzsch, 1974; Jung and Knitzschke, 1976; Oszczepalski and Rydzewski, 1983; Jowett et al., 1987; Oszczepalski, 1989; Peryt, 1989), most of the main deposits in the central African Copperbelt of Zambia and Zaire (Darnley, 1960; Garlick, 1961, 1969; Mendelsohn, 1961; Lombard and Nicolini, 1962, 1963; Bartholomé et al., 1972; Bartholomé, 1974; Fleischer et al., 1976; Annels, 1979a, b, 1989; Lefebvre, 1989a, b) and the White Pine deposit in Michigan (White and Wright, 1966; Ensign et al., 1968; Brown, 1971; White, 1971) are typical examples of this type in other parts of the world.

IMPORTANCE

In Canada, no significant production has come from Kupferschiefer-type deposits and none is anticipated in the near future. Nevertheless, they are the world's second most important source of copper (after porphyry deposits) and areas favourable for their occurrence, such as the Redstone copperbelt in the Northwest Territories, have been identified in Canada.

Most world production from deposits of this type comes from a few very large deposits and districts, such as the Lubin deposit in Poland and the central African Copperbelt of Zambia and Zaire. The White Pine deposit in Michigan is the only current producer of this type in North America. The Dongchuan district in southern China and deposits in the Aynak area near Kabul, Afghanistan could be of Kupferchiefer-type.

Kupferschiefer-type deposits in the central African Copperbelt are also the world's most important source of cobalt, and some Kupferschiefer-type deposits, such as Lubin in Poland, produce significant amounts of silver. Small amounts of platinum group elements, gold, lead, uranium, and rhenium(?) have also been recovered from deposits of this type.

SIZE AND GRADE OF DEPOSITS

Figure 8.3-2 shows the grade and tonnage distribution for 74 sediment-hosted stratiform copper deposits. Kupferschiefer- and redbed-type deposits are indicated by separate symbols. This plot shows that Kupferschiefer-type deposits tend to be larger and richer than most redbed-type deposits and that some of them contain as much copper as some of the world's largest porphyry copper deposits. An average grade and tonnage for Kupferschiefer-type deposits is 44 million tonnes at 1.8% Cu, amounting to about 0.8 million

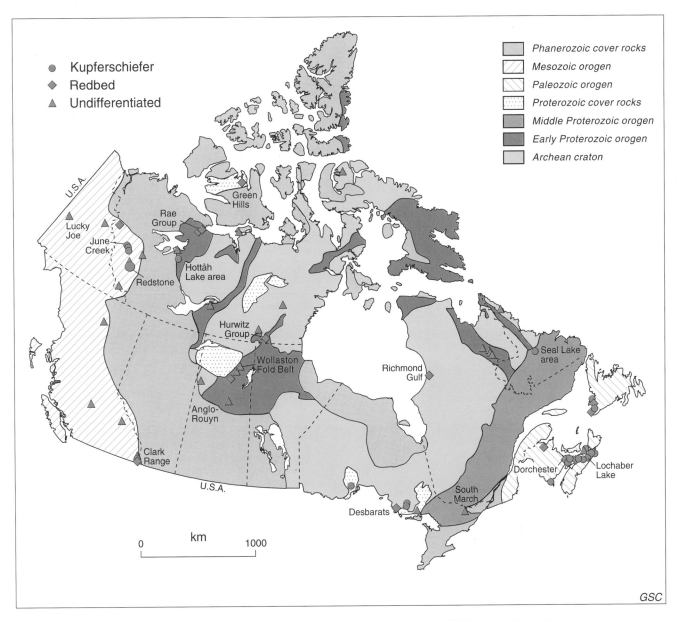

Figure 8.3-1. Distribution of selected sediment-hosted stratiform copper (SSC) deposits and occurrences in Canada. Beside the Redstone deposit in the Northwest Territories, most of the localities shown are of minor importance. Nevertheless, they outline areas and rock sequences favourable for the occurrence of such deposits. See Kirkham et al. (1994) for deposit listings.

tonnes of copper metal. The Redstone deposit, even though it is of significant size and grade (drill-indicated 37 million tonnes averaging 3.9% Cu and 11.3 g/t Ag), is not attractive for mining because it occurs in a remote, undeveloped region with a severe climate, dips about 35° to 45°, and only averages about 1 m thick (Ruelle, 1982).

The tabular nature of most Kupferschiefer-type deposits is not conducive to open-pit or block-caving, low-cost mining methods, and therefore higher grades are necessary in these deposits to support relatively selective underground mining methods. Cut-off grades generally define sharp ore limits and mining of barren adjacent rocks can lead to significant dilution.

GEOLOGICAL FEATURES

The classic setting of the very bituminous, anoxic Permian Kupferschiefer (copper shale) of Europe at the base of a marine transgressive sequence overlying continental redbeds is illustrated in Figure 8.3-3. However, in some areas, such as the central African Copperbelt of Zambia (Clemmey, 1978); White Pine, Michigan (Daniels, 1982); and Redstone, Northwest Territories (Jefferson and Ruelle, 1986), it has been suggested that ore host rocks were deposited in large-scale saline lacustrine, rather than marine, environments. Some lacustrine evaporite environments are similar in many respects to marine evaporite environments (e.g., Dead Sea). Both Kupferschiefer- and redbed-type sedimentary copper deposits typically occur in rocks that were deposited in arid and semiarid areas within 20 to 30 degrees of the paleo-equator (e.g., Fig. 8.3-4). This consistent feature of deposits of these types is an important factor bearing on their genesis. For further discussion of this subject and for plots for other geological periods see Kirkham (1989). They also only occur in rocks that postdate oxygenation of the Earth's atmosphere or "oxyatmoversion" at about 2.4 Ga (Roscoe, 1973; Kirkham and Roscoe, 1993).

Many Kupferschiefer-type deposits occur in rifts, in areas where basalts, redbeds, and evaporites form part of the stratigraphic succession. Typical features of a Kupferschiefer-type deposit occurring in a rift are shown in Figure 8.3-5, modelled after the Redstone deposit. In some areas, such as Creta in Oklahoma (Johnson and Croy, 1976; Dingess, 1976; Smith, 1976) and in upper Windsor Group rocks in Nova Scotia and New Brunswick (Kirkham, 1985), Kupferschiefer-type deposits are interbedded with, rather than overlying, redbeds.

Sulphides in Kupferschiefer-type deposits are characteristically disseminated and occur in a series of overlapping mineral zones (hematite, native copper, chalcocite, bornite, chalcopyrite, galena, to sphalerite and/or pyrite), outward and upward, from the oxidized to the reduced side (Fig. 8.3-3, 8.3-5). Not all of these minerals are present in every deposit. In a given deposit not all beds are equally mineralized. Very bituminous, reduced beds (such as the Kupferschiefer ore unit) are high grade, whereas less reduced or oxidized beds are typically lower grade. In a sequence containing many reduced beds, the lowest one typically has the highest grade and reduced beds above the deposit can be entirely barren. In some deposits carbonaceous shale and microbial carbonates (e.g., Redstone) contain higher metal concentrations than adjacent arenaceous and evaporitic units. Other deposits, such as that in the Kona

Dolomite (Taylor, 1972); Mufulira (Annels, 1979a, b); and White Pine (Hamilton, 1967; Kelly and Niskioka, 1985) show that arenaceous aquifers have been important in controlling fluid flow, mineral zoning, and possibly, also in containing fluid hydrocarbons that controlled metal deposition.

Veins, ranging from small discontinuous gash structures 1 to 2 cm long to more through-going structures, are common, but in most deposits have complex histories and have not been studied in detail. Some veins, such as ones at White Pine, Michigan (Carpenter, 1963) and at Spar Lake, Montana (Hayes and Balla, 1986) have copper depletion halos around them and clearly postdate the main disseminated stratiform sulphides. In some deposits, such as those in the Kupferschiefer and White Pine, veins with addition halos are also common (Gregory, 1930; Jowett, 1987; Mauk et al., 1992), indicating at least local addition of metals from the fractures. Nevertheless, in all Kupferschiefer-type deposits stratigraphic controls are more important than structural controls and the deposits show no relationship to igneous rocks. In many deposits, the disseminated sulphides fill original porosity and, in the case of microbial carbonate host rocks in Zaire and at Redstone, which should lose their permeability readily, support an early diagenetic age for mineralization (Bartholomé et al., 1972; Chartrand and Brown, 1985; Chartrand et al., 1989). Jowett (1987) and Jowett et al. (1987), however, have argued for a late diagenetic origin for deposits in the Kupferschiefer.

Kupferschiefer-type deposits are characterized by a red, hematitic, oxidative alteration (Rote Fäule) beneath and adjacent to the copper deposit, indicative of infiltration of metalliferous, oxic fluids into the anoxic host rocks (Rentzsch, 1974; Jung and Knitzschke, 1976; Oszczepalski and Rydzewski, 1983; Jowett et al., 1987).

In some Kupferschiefer-type deposits, particularly favourable host rocks are uniformly mineralized over large areas and only at the outer limits of the deposit do the sulphide zones cut the stratigraphy (e.g., White Pine). In some deposits they cut 10 m to as much as 100 m of section. In deposits containing other base metals, copper sulphide zones are flanked and/or overlain by lead and zinc zones. In the central African Copperbelt cobalt is generally concentrated in pyritic rocks near the outer edge of copper zones.

DEFINITIVE CHARACTERISTICS

1. The deposits are peneconcordant concentrations of copper at oxidation-reduction boundaries in anoxic marine (or lacustrine) rocks in contact with continental redbeds.

2. Deposits in many cases occur in association with redbeds and evaporites.

3. Sulphides are generally disseminated and occur in overlapping mineral zones upward and outward from the base of the deposit.

4. The economic deposits are dominated by copper, with variable amounts of silver as the most diagnostic associated ore element. In the central African Copperbelt and a few other areas, cobalt is also an important associated element. In many deposits lead and zinc are present in upper and outer mineral zones in subeconomic concentrations.

GENETIC MODEL

Genesis of Kupferschiefer-type deposits has been debated widely since their recognition. For perhaps 70 years or more views have tended to be polarized, supporting either a syngenetic or late hydrothermal epigenetic origin. As can be seen in a number of papers in Boyle et al. (1989), several different views of genesis are currently held. If any consensus exists, an intermediate diagenetic origin has the greatest support (e.g., White and Wright, 1966; Ensign et al., 1968; Brown, 1971, 1992; Bartholomé et al., 1972; Rentzsch, 1974; Gustafson and Williams, 1981; Chartrand and Brown, 1985; Haynes, 1986; Jowett et al., 1987; and many papers cited in Boyle et al., 1989). This writer favours a diagenetic origin(s) and recognizes no convincing evidence supporting syngenesis or late hydrothermal epigenetic origins. On the other hand, the consistent relationship of these deposits with redbeds and evaporites and low-latitude arid and semiarid areas of deposition, despite diagenetic origins, indicates an essential control by the environment of sedimentation. This important relationship between redbeds, evaporites, and the environment of sedimentation has been known for some time and was emphasized by Strakhov (1962) and other workers.

Kirkham (1989) has considered briefly some of the main factors in the genesis of sediment-hosted stratiform copper deposits and has discussed some of their variability. Some of these factors are timing of ore emplacement, nature and source of ore fluids, sources of metals, sources of sulphur and reductants, and controls on fluid migration. For elaboration on these and other aspects of genesis the reader should consult Kirkham (1989) and other papers in Boyle et al. (1989).

As mentioned previously, evidence from little deformed and well studied deposits can be interpreted to support a diagenetic origin, but the precise timing and duration of ore formation, even for well-documented deposits, is difficult to determine. Probably a spectrum of time of formation ranging from early to late diagenesis, but definitely predating folding and metamorphism of the rocks (e.g., Garlick, 1965; Hayes and Einaudi, 1986), is most reasonable. The most likely ore fluids are brines derived from evaporites. Experimental work (Rose, 1976, 1989) indicated that the solubilities of Cu and Ag at low temperatures are dependent on chloride ion concentration, which supports the concept of evaporitic brines as the ore fluids and is consistent with the observed spatial association with evaporites. Labile constituents, such as mafic rock fragments and minerals (e.g., hornblende, pyroxene, biotite, oxides, and sulphides) in first-cycle redbeds, are viewed as the most likely source of the copper. The presence of lead in the deposits probably indicates a mixed provenance of the redbeds and that they contained silicate components such as potassium feldspar. The copper might be released directly into the pore fluid or converted to a form that could be stripped easily by the later passage of brines during the oxidative diagenetic formation of redbeds (Walker, 1967, 1989; Zielinski et al., 1983). Experimental studies by Rose and Bianchi-Mosquera (1993) indicate that metal associations (e.g., Cu-Ag, Cu-Co, Cu-Pb-Zn-Ag) can be explained by differences in pH, Eh, temperature, chemical composition of ore fluids, and Fe-oxide character of the diagenetic environment, but that in some circumstances copper and other metals are strongly absorbed on goethite and hematite and would be relatively immobile in diagenetic redbed environments. Older ore deposits in basement rocks are not viewed as likely sources of metals, as cupriferous formations hosting some Kupferschiefer-type deposits contain an enormous amount of copper, far exceeding that found in most economic deposits.

The sources of sulphur and reductants are important aspects of the oxidation-reduction processes controlling ore deposition. Reductants could be such materials as pyrite, solid carbonaceous matter, H_2S, CH_4, or liquid hydrocarbons indigenous to, or added to, the host rocks. Work by Hoy et al. (1986) and Hoy and Ohmoto (1989) indicated that significant sulphur, as sulphate in the ore fluid, might have been added to deposits during ore deposition. The nature and distribution of sulphur and reductants probably differed considerably from deposit to deposit and had profound effects on metal grades and distribution.

The paleohydrology of the ore systems is not well understood and is only now receiving considerable attention. The lack of definitive data on the timing and on constraints on the nature of the ore-forming systems has precluded rigorous analysis of the paleohydrology. Many possibilities exist (see Kirkham, 1989), but a body of evidence is appearing for many deposits supporting relatively early diagenetic formation dewatering, while coarse grained sediments were still permeable and fine grained sediments were compacting (White, 1971; Ruelle, 1982; Kirkham, 1989; Lustwerk and Wasserman, 1989). Magara (1976) and Bredehoeft et al. (1988) indicated that in an active, compacting sedimentary basin with mixed sand and shale, the sand will act as drains for compactional fluids and most of the fluids will migrate laterally within the sandstone aquifers towards the edge of the basin. In thick shale sequences without sand, fluid flow will be primarily vertically upward. Galloway (1982) summarized basic fluid genesis and migration in dynamic compacting sedimentary basins. Jowett (1986) and Jowett et al. (1987), however, have proposed a late diagenetic, thermally driven fluid flow model for the Kupferschiefer deposit in Germany and Poland.

The general aspects of evaporite-derived ore fluid, redbed metal source rocks and aquifers, and an overlying anoxic transgressive marine (or lacustrine) succession of ore host rocks in a rift environment are illustrated schematically in Figure 8.3-5. The processes that controlled ore formation probably differed considerably from one area to another, but, whatever the details of processes were, available information indicates that large amounts of metals (Cu, Ag, Co) were moved and concentrated into economic deposits in such sedimentary sequences of many different ages in several parts of the world. So much copper (and cobalt) has been concentrated into so many large, rich deposits in the central African Copperbelt that no copper source or concentrating process seem adequate. Possibly in this exceptional area, compressional forces related to collision events to the south and west in the Zambesi and/or Damaran belts at a critical stage during diagenesis formed large gravity-driven fluid-flow systems that moved vast quantities of cupriferous fluids that were responsible for the formation of these extraordinary deposits (Fig. 8.3-6). Higher temperatures of ore deposition could be expected of deep basinal brines and could account for the high cobalt contents of these ores (Annels, 1989; Rose and Bianchi-Mosquera, 1993).

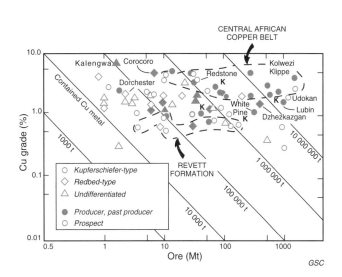

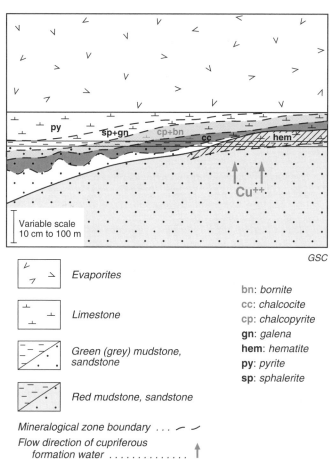

Figure 8.3-2. Size and grade of 74 sedimentary copper deposits. Producers and past producers are shown in solid and undeveloped deposits are shown as open symbols. "K" denotes deposits in the Kupferschiefer in Europe.

Figure 8.3-3. Typical setting, rock units, and metal zonation for the Kupferschiefer-type deposits. The zonal pattern of sulphides (Cu sulphide zones in red) is from White and Wright, 1966; Jung, 1968; Brown, 1971; Rentzsch, 1974; Jung and Knitzschke, 1976; Oszczepalski and Rydzewski, 1983; and several other studies (after Kirkham, 1989).

Figure 8.3-4 A) Distribution of copper deposits and occurrences and evaporite deposits (modified from Kozary et al., 1968 and Zharkov, 1984) in Carboniferous rocks and paleolatitudes at about 280 Ma ago (latest Carboniferous-earliest Permian) (after Irving, 1983). 1) Firth of Forth, Drumshantie, and Larkfield; 2) Mallow and Ballyvergin; 3) W Araba and Sarabit Al Khadmin; 4) southern end of Teniz Basin (many deposits and occurrences); 5) Dzhezkazgan district (many deposits and occurrences); 6) Chu River; 7) Mirgalim-Sai and Zhanatas; 8) northern Kirgizia and Przheval; 9) Kurpandzha (Devono-Carboniferous); 10) Tung-Chuan; 11) Kengir (Devono-Carboniferous); 12) Pajarito Azule and Coyote Creek; 13) Mogollon Rim; 14) Bronze Lake and Ridenour; 15) Western Star and Cotopaxi Cordova; 16) Watercress Canyon; 17) Hot Brook Canyon; 18) numerous occurrences (e.g., Dorchester, Midway, and Goshen); 19) numerous occurrences in Windsor, Pictou, and other groups (e.g., Canfield, Oliver, Limerock, McLellan Brook, Rights River, Yankee Line Road, and Frenchvale); 20) Searston, Bald Mountain, Boswarlos, etc. (after Kirkham, 1989). **B)** Reconstructed Earth diagram for Late Carboniferous (circa 280 Ma) (after Irving, 1983) showing the distribution of copper deposits and occurrences and evaporite deposits in Carboniferous rocks. See Irving (1983) for explanation of patterns and abbreviations on base map (after Kirkham, 1989).

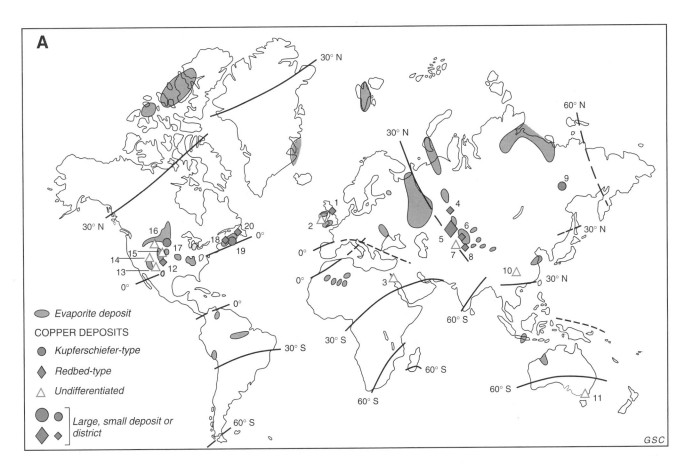

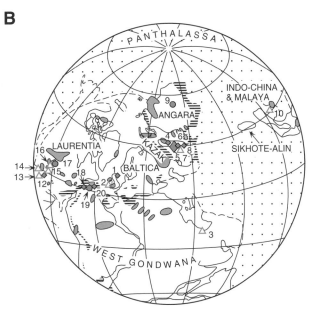

RELATED DEPOSIT TYPES

In some interlayered marine-continental sequences Kupferschiefer- and redbed-type copper deposits occur together, supporting a close genetic relationship for these deposit subtypes. Volcanic redbed copper (VRC) deposits, to a large degree, are probably the analogues in volcanic sequences of diagenetic sedimentary copper deposits in sedimentary sequences. In many areas, such as the Keweenaw Peninsula in Michigan, the Coppermine River area in the Northwest Territories, the Seal Lake area in Labrador, and in the Andes of Chile, both of these types of deposits occur together. Sediment-hosted stratiform copper deposits have some features in common with sandstone-lead and uranium, Mississippi Valley-type (carbonate-hosted) lead-zinc, and unconformity-type uranium deposits

but, as concluded by Bjørlykke and Sangster (1981), these various types of deposits were probably formed by somewhat different processes, by different fluids, at different times, and in different places. Metalliferous brines trapped in deeply buried, pressurized redbed reservoirs, an important component in the genesis of diagenetic sedimentary copper deposits, may have been responsible for the formation of other sediment-hosted copper deposits, such as Mount Isa and Gunpowder in Queensland; Kapunda and Kanmantoo in South Australia; Nifty in Western Australia; Kipushi, Zaire; Tsumeb and Kombat in Namibia; Apex, Utah; Sheep Creek, Montana; and Ruby Creek and Kennecott, Alaska, and even for some exhalative lead-zinc deposits in sedimentary sequences. However, little evidence is available to support such connections. More work, especially isotopic tracer studies, is required to evaluate such possibilities.

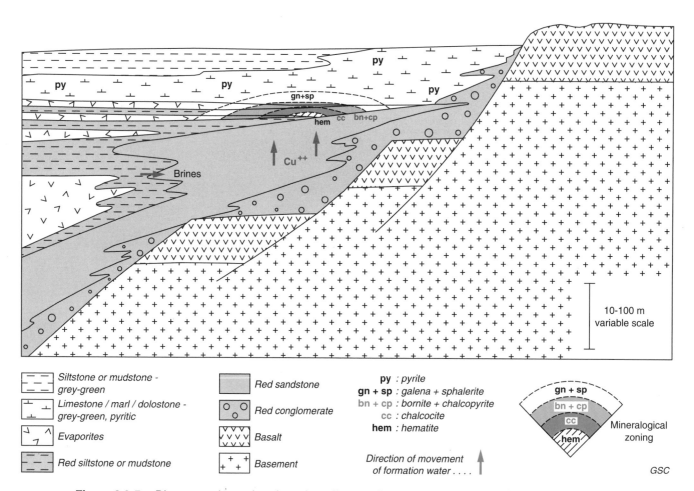

Figure 8.3-5. Diagrammatic section through a rift-controlled sedimentary basin showing many of the essential features in the formation of Kupferschiefer-type copper deposits. The Cu sulphide zones are shown in red. Modelled after Redstone, Northwest Territories. 'Basement' means any pre-existing rock type, at Redstone referring to platformal sedimentary rocks and tholeiitic continental basalts (after Kirkham, 1989).

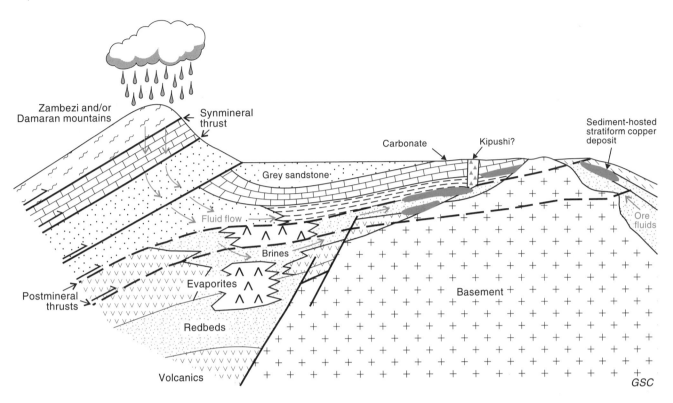

Figure 8.3-6. Schematic model for a gravity-driven fluid flow for genesis of sediment-hosted stratiform copper deposits in the central African Copperbelt. Diagram is not to scale but distance from source mountains to sites of copper deposition could have been several hundred kilometres (Daly, 1986).

EXPLORATION GUIDES

Markedly reduced marine or extensive, saline lacustrine units in contact with redbeds and associated with evaporites constitute the prime settings in which to search for Kupferschiefer-type deposits. Most major Kupferschiefer-type deposits occur where the anoxic rocks overlie thick redbed sequences. A good geochemical source of copper is required, such as immature redbeds with labile detritus derived from rift-related basalts or other copper-bearing rocks or minerals, but in some districts, such as the central African Copperbelt, potential source rocks might be far removed from the deposits. Fetid microbial carbonate rocks and bituminous shales and/or sandstones in many deposits contain higher copper contents than mildly reduced units. The presence of broad-scale, well defined mineral zones covering several kilometres or tens of kilometres is a positive feature indicating large mineralized systems as opposed to small, erratically mineralized deposits. In some deposits, such as that in the Kona Dolomite and possibly Mufulira, prominent aquifers within the deposit could have controlled metal distribution, zoning, and concentration. No deposits have been found in rocks older than 2.4 Ga that predated oxygenation of the Earth's atmosphere. Significant differences between areas indicate that each deposit and area should be evaluated carefully on its own merits.

SELECTED BIBLIOGRAPHY

References marked with asterisks (*) are considered to be the best sources of general information on the deposit subtype.

Annels, A.E.

1979a: The genetic relevance of recent studies at Mufulira mine, Zambia; Annales de la Société Géologique de Belgique, v. 102, p. 431-449.

1979b: Mufulira greywackes and their associated sulphides; Institution of Mining and Metallurgy, Transactions, v. 88, p. B15-B23.

1989: Ore genesis in the Zambian Copperbelt, with particular reference to the northern sector of the Chambishi Basin; in Sediment-hosted Stratiform Copper Deposits, (ed.) R.W. Boyle, A.C. Brown, C.W. Jefferson, E.C. Jowett, and R.V. Kirkham; Geological Association of Canada, Special Paper 36, p. 427-452.

*Bartholomé, P. (ed.)

1974: Gisements Stratiformes et Provinces Cuprifères; Centenaire de la Société Géologique de Belgique, Liège, 427 p.

*Bartholomé, P., Evrard, P., Katekesha, P., Lopez-Ruiz, J., and Ngongo, M.

1972: Diagenetic ore-forming processes at Kamoto, Katanga, Republic of the Congo; in Ores in Sediments, (ed.) G.C. Amstutz and A.J. Bernard; Springer, Berlin, p. 21-41.

Binda, P.L., Koopman, H.T., and Koopman, E.R.

1989: A stratiform copper occurrence in the Helikian Siyeh Formation of Alberta and British Columbia; in Sediment-hosted Stratiform Copper Deposits, (ed.) R.W. Boyle, A.C. Brown, C.W. Jefferson, E.C. Jowett, and R.V. Kirkham; Geological Association of Canada, Special Paper 36, p. 269-285.

Bjørlykke, A. and Sangster, D.F.

1981: An overview of sandstone-lead deposits and their relation to red-bed copper and carbonate-hosted lead-zinc deposits; in Economic Geology, Seventy-fifth Anniversary Volume, 1905-1980, (ed.) B.J. Skinner, p. 179-213.

*Boyle, R.W., Brown, A.C., Jefferson, C.W., Jowett, E.C., and Kirkham, R.V. (ed.)

1989: Sediment-hosted Stratiform Copper Deposits; Geological Association of Canada, Special Paper 36, 710 p.

Bredehoeft, J.D., Djevanshir, R.D., and Belitz, K.R.

1988: Lateral flow in a compacting sand-shale sequence: South Caspian basin; The American Association of Petroleum Geologists, Bulletin, v. 72, p. 416-424.

*Brown, A.C.

1971: Zoning in the White Pine copper deposit, Ontonagon County, Michigan; Economic Geology, v. 66, p. 543-573.

1992: Sediment-hosted stratiform copper deposits; Geoscience Canada, v. 19, p. 125-141.

Carpenter, R.H.

1963: Some vein-wall rock relationships in the White Pine mine, Ontonagon Co., Michigan; Economic Geology, v. 58, p. 643-666.

Chandler, F.W.

1989: Lower Proterozoic sabkha-related copper mineralization, paleoenvironment and diagenesis, Cobre Lake, Ontario; in Sediment-hosted Stratiform Copper Deposits, (ed.) R.W. Boyle, A.C. Brown, C.W. Jefferson, E.C. Jowett, and R.V. Kirkham; Geological Association of Canada, Special Paper 36, p. 225-244.

Chartrand, F.M. and Brown, A.C.

1985: The diagenetic origin of stratiform copper mineralization, Coates Lake, Redstone copper belt, N.W.T., Canada; Economic Geology, v. 80, p. 325-343.

Chartrand, F.M., Brown, A.C., and Kirkham, R.V.

1989: Diagenesis, sulphides and metal zoning in the Redstone Copper deposit, Northwest Territories; in Sediment-hosted Stratiform Copper Deposits, (ed.) R.W. Boyle, A.C. Brown, C.W. Jefferson, E.C. Jowett, and R.V. Kirkham; Geological Association of Canada, Special Paper 36, p. 189-206.

Clemmey, H.

1978: A Proterozoic lacustrine interlude from the Zambian Copperbelt; International Association of Sedimentologists, Special Publication 2, p. 257-278.

Daly, M.C.

1986: Crustal shear zones and thrust belts: their geometry and continuity in Central Africa; Philosophical Transactions of the Royal Society of London, v. A317, p. 111-128.

Daniels, P.A.

1982: Upper Precambrian sedimentary rocks: Oronto Group, Michigan-Wisconsin; in Geology and Tectonics of the Lake Superior Basin, (ed.) R.J. Wold and W.J. Hinze; Geological Society of America, Memoir 158, p. 107-133.

Darnley, A.G.

1960: Petrology of some Rhodesian Copperbelt orebodies and associated rocks; The Institution of Mining and Metallurgy, Transactions, v. 69, p. 137-173, 371-398, 540-569.

Dingess, P.R.

1976: Geology and mining operations at Creta copper deposit of Eagle-Picher Industries, Inc.; in Stratiform Copper Deposits of the Midcontinent Region, a Symposium, (ed.) K.S. Johnson and R.L. Croy; Oklahoma Geological Survey, Circular 77, p. 15-24.

*Ensign, C.O., White, W.S., Wright, J.C., Patrick, J.L., Leone, R.J., Hathaway, D.J., Trammell, J.W., Fritts, J.J., and Wright, T.L.

1968: Copper deposits in the Nonesuch Shale, White Pine, Michigan; in Ore Deposits of the United States, 1933-1967; The Graton-Sales Volume, (ed.) J.D. Ridge; The American Institute of Mining, Metallurgical and Petroleum Engineers, Inc., New York, p. 460-488.

*Fleischer, V.D., Garlick, W.G., and Haldane, R.

1976: Geology of the Zambian Copperbelt; in Handbook of Strata-bound and Stratiform Ore Deposits, (ed.) K.H. Wolf; Elsevier, Amsterdam, v. 8, p. 223-352.

Galloway, W.E.

1982: Epigenetic zonation and fluid flow history of uranium-bearing fluvial aquifer systems, south Texas uranium province; Texas, Bureau of Economic Geology, Report of Investigations, no. 119, 31 p.

Garlick, W.G.

1961: The syngenetic theory; in The Geology of the Northern Rhodesian Copperbelt, (ed.) P. Mendelsohn; MacDonald, London, p. 146-162.

1965: Criteria for the recognition of syngenetic sedimentary mineral deposits and veins formed by their remobilization; Australasian Institute of Mining and Metallurgy, Proceedings, v. 6, p. 1393-1418.

1969: Special features and sedimentary facies of stratiform sulphide deposits in arenites; in Sedimentary Ores - Ancient and Modern, (ed.) C.H. James; University of Leicester, Department of Geology, Special Publication No. 1, p. 107-169.

Gregory, J.W.

1930: The copper-shale (Kupferschiefer) of Mansfeld; Institution of Mining and Metallurgy, Transactions, v. 40, p. 3-55.

*Gustafson, L.B. and Williams, N.

1981: Sediment-hosted stratiform deposits of copper, lead, and zinc; Economic Geology, Seventy-fifth Anniversary Volume, 1905-1980, (ed.) B.J. Skinner; p. 139-178.

Hamilton, S.K.

1967: Copper mineralization in the upper part of the Copper Harbor Conglomerate at White Pine, Michigan; Economic Geology, v. 82, p. 885-904.

Hayes, T.S. and Balla, J.C.

1986: Troy mine: in Proterozoic Sediment-hosted Stratiform Copper Deposits of Upper Michigan and Belt Supergroup of Idaho and Montana, (ed.) A.C. Brown and R.V. Kirkham; Geological Association of Canada, Mineralogical Association of Canada, Canadian Geophysical Union, Joint Annual Meeting, Ottawa '86, Field Trip 1: Guidebook, p. 54-84.

Hayes, T.S. and Einaudi, M.T.

1986: Genesis of the Spar Lake strata-bound copper-silver deposit, Montana: Part I. Controls inherited from sedimentation and preore diagenesis; Economic Geology, v. 81, p. 1899-1931.

Haynes, D.W.

1986: Stratiform copper deposits hosted by low-energy sediments: I. Timing of sulfide precipitation - an hypothesis; Economic Geology, v. 81, p. 250-265.

Hoy, L.D. and Ohmoto, H.

1989: Constraints for the genesis of redbed-associated stratiform Cu deposits from sulphur and carbon mass-balance relations; in Sediment-hosted Stratiform Copper Deposits, (ed.) R.W. Boyle, A.C. Brown, C.W. Jefferson, E.C. Jowett, and R.V. Kirkham; Geological Association of Canada, Special Paper 36, p. 135-149.

Hoy, L.D., Ohmoto, H., Rose, A.W., Dimanche, F., and Coipel, J.

1986: Constraints for the genesis of red-bed-associated stratiform Cu deposits from S and C mass-balance relations (abstract); Canadian Mineralogist, v. 24, p. 189.

Irving, E.

1983: Fragmentation and assembly of the continents, Mid-Carboniferous to present; Geophysical Surveys, v. 5, p. 299-333.

*Jefferson, C.W. and Ruelle, J.C.

1986: The Late Proterozoic Redstone copper belt, Northwest Territories; in Mineral Deposits of Northern Cordillera, (ed.) J.A. Morin; The Canadian Institute of Mining and Metallurgy, Special Volume 37, p. 154-168.

Johnson, K.S. and Croy, R.L.

1976: Stratiform copper deposits of the Midcontinent region, a symposium; Oklahoma Geological Survey, Circular 77, 99 p.

Jowett, E.C.

1986: Genesis of Kupferschiefer Cu-Ag deposits by convective flow of Rotliegendes brines during Triassic rifting; Economic Geology, v. 81, p. 1823-1837.

1987: Formation of sulfide-calcite veinlets in the Kupferschiefer Cu-Ag deposits in Poland by natural hydrofracturing during basin subsidence; Journal of Geology, v. 95, p. 513-526.

*Jowett, E.C., Rydzewski, A., and Jowett, R.J.

1987: The Kupferschiefer Cu-Ag ore deposits in Poland: a re-appraisal of the evidence of their origin and presentation of a new genetic model; Canadian Journal of Earth Sciences, v. 24, p. 2016-2037.

*Jung, W.

1968: Sedimentary rocks and deposits in Saxony and Thuringia; Guide to Excursion 38AC, XXIII International Geological Congress, Prague, 33 p.

Jung, W. and Knitzschke, G.

1976: Kupferschiefer in the German Democratic Republic (GDR) with special reference to the Kupferschiefer deposit in the southeastern Harz foreland; in Handbook of Strata-bound and Stratiform Ore Deposits, (ed.) K.H. Wolf; Elsevier, Amsterdam, v. 6, p. 353-406.

Kelly, W.C. and Niskioka, G.K.
1985: Precambrian oil inclusions in late veins and the role of hydrocarbons in copper mineralization at White Pine, Michigan; Geology, v. 13, p. 334-337.

Kirkham, R.V.
1974: A synopsis of Canadian stratiform copper deposits in sedimentary sequences; Centenaire de la Société Géologique de Belgique, Gisements Stratiformes et Provinces Cuprifères, Liège, p. 367-382.
1985: Base metals in upper Windsor (Codroy) Group oolitic and stromatolitic limestones in the Atlantic Provinces; in Current Research, Part A; Geological Survey of Canada, Paper 85-1A, p. 573-585.
*1989: The distribution, settings and genesis of sediment-hosted, stratiform copper deposits; in Sediment-hosted Stratiform Copper Deposits, (ed.) R.W. Boyle, A.C. Brown, C.W. Jefferson, E.C. Jowett, and R.V. Kirkham; Geological Association of Canada, Special Paper 36, p. 3-38.

Kirkham, R.V. and Roscoe, S.M.
1993: Atmospheric evolution and ore deposit formation; Resource Geology, Special Issue, no. 15, p. 1-17.

Kirkham, R.V., Carrière, J.J., Laramée, R.M., and Garson, D.F.
1994: Global distribution of sediment-hosted stratiform copper deposits and occurrences; Geological Survey of Canada, Open File 2915, scale 1:35 000 000.

Kozary, M.T., Dunlap, J.C., and Humphrey, W.E.
1968: Incidence of saline deposits in geologic time; in Saline Deposits, (ed.) R.B. Mattox; Geological Society of America, Special Paper 88, p. 43-57.

Lefebvre, J.-J.
1989a: Depositional environment of copper-cobalt mineralization in the Katangan sediments of southeast Shaba, Zaire; in Sediment-hosted Stratiform Copper Deposits, (ed.) R.W. Boyle, A.C. Brown, C.W. Jefferson, E.C. Jowett, and R.V. Kirkham; Geological Association of Canada, Special Paper 36, p. 401-426.
1989b: Les gisements stratiformes en roche sédimentaire d'Europe centrale (Kupferschiefer) et de la ceinture cuprifère du Zaïre et de Zambie; Annales de la Société Géologique de Belgique, v. 112, p. 121-135.

Lefebvre, J.-J. and Tshiauka, T.
1986: Le groupe des mines à Lubembe, Shaba, Zaïre; Annales de la Société Géologique de Belgique, v. 109, p. 557-571.

Lombard, J. and Nicolini, P. (ed.)
1962: Stratiform Copper Deposits in Africa; First Volume: Lithology, sedimentology; Association of African Geological Surveys, Paris, 212 p.
1963: Stratiform Copper Deposits in Africa; Second Volume: Tectonics; Association of African Geological Surveys, Paris, 265 p.

Lustwerk, R.W. and Wasserman, M.D.
1989: Water escape structures in the Coates Lake Group, Northwest Territories, Canada and their relationship to mineralization at the Redstone stratiform copper deposit; in Sediment-hosted Stratiform Copper Deposits, (ed.) R.W. Boyle, A.C. Brown, C.W. Jefferson, E.C. Jowett, and R.V. Kirkham; Geological Association of Canada, Special Paper 36, p. 207-224.

Magara, K.
1976: Water expulsion from clastic sediments during compaction – directions and volumes; American Association of Petroleum Geologists, Bulletin, v. 80, p. 543-553.

Mauk, J.L., Kelly, W.C., van der Pluijm, B.A., and Seasor, R.W.
1992: Relations between deformation and sediment-hosted copper mineralization: evidence from the White Pine part of the Midcontinent rift system; Geology, v. 20, p. 427-430.

*Mendelsohn, F. (ed.)
1961: The Geology of the Northern Rhodesian Copperbelt; MacDonald, London, 523 p.

*Oszczepalski, S.
1989: Kupferschiefer in southwestern Poland; sedimentary environments, metal zoning, and ore controls; in Sediment-hosted Stratiform Copper Deposits, (ed.) R.W. Boyle, A.C. Brown, G.W. Jefferson, E.C. Jowett, and R.V. Kirkham; Geological Association of Canada, Special Paper 36, p. 571-600.

Oszczepalski, S. and Rydzewski, A.
1983: Copper distribution in Permian rocks in the area adjoining the Lubin-Sieroszowice deposits; Przeglad Geologiczny, no. 7, p. 437-443 (in Polish with English summary and captions).

Peryt, T.M.
1989: Basal Zechstein in southwestern Poland: sedimentation, diagenesis, and gas accumulations; in Sediment-hosted Stratiform Copper Deposits, (ed.) R.W. Boyle, A.C. Brown, C.W. Jefferson, E.C. Jowett, and R.V. Kirkham; Geological Association of Canada, Special Paper 36, p. 601-625.

*Rentzsch, J.
1974: The Kupferschiefer in comparison with the deposits of the Zambian Copperbelt; in Gisements Stratiformes et Provinces Cuprifères, (ed.) P. Bartholomé; Centenaire de la Société Géologique de Belgique, Liège, p. 395-418.

Richards, J.P., Krogh, T.E., and Spooner, E.T.C.
1988: Fluid inclusion characteristics and U-Pb rutile age of late hydrothermal alteration and veining at the Musoshi stratiform copper deposits, Central African copper belt, Zaire; Economic Geology, v. 83, p. 118-139.

Roscoe, S.M.
1973: The Huronian Supergroup, a Paleoaphebian succession showing evidence of atmospheric evolution; in Huronian Stratigraphy and Sedimentation, (ed.) G.M. Young; Geological Association of Canada, Special Paper 12, p. 32-47.

Rose, A.W.
1976: The effect of cuprous chloride complexes in the origin of red bed copper and related deposits; Economic Geology, v. 71, p. 1036-1048.
1989: Mobility of copper and other heavy metals in sedimentary environments; in Sediment-hosted Stratiform Copper Deposits, (ed.) R.W. Boyle, A.C. Brown, C.W. Jefferson, E.C. Jowett, and R.V. Kirkham; Geological Association of Canada, Special Paper 36, p. 97-110.

Rose, A.W. and Bianchi-Mosquera, G.C.
1993: Adsorption of Cu, Pb, Zn, Co, Ni, and Ag on goethite and hematite: a control on metal mobilization from red beds into stratiform copper deposits; Economic Geology, v. 88, p. 1226-1236.

Ruelle, J.C.
1982: Depositional environments and genesis of stratiform copper deposits of the Redstone Copper Belt, Mackenzie Mountains, N.W.T.; in Precambrian Sulphide Deposits, H.S. Robinson Memorial Volume, (ed.) R.W. Hutchinson, C.D. Spence, and J.M. Franklin; Geological Association of Canada, Special Paper 25, p. 701-737.

Smith, G.E.
1976: Sabkha and tidal-flat facies control of stratiform copper deposits in north Texas; Stratiform Copper Deposits of the Midcontinent Region, a Symposium, (ed.) K.S. Johnson and R.L. Croy; Oklahoma Geological Survey, Circular 77, p. 25-39.

*Strakhov, N.M.
1962: Principles of Lithogenesis; translated by J.P. Fitzsimmons, 1970, (ed.) S.I. Tomkeieff and J.E. Hemingway; Plenum Publishing Corporation, New York, and Oliver and Boyd, Edinburgh, v. 3, 577 p.

Taylor, G.L.
1972: Stratigraphy, sedimentology, and sulfide mineralization of the Kona Dolomite; PhD. thesis, Michigan Technological University, Houghton, Michigan, 112 p.

Walker, T.R.
1967: Formation of red beds in modern and ancient deserts; Geological Society of America, Bulletin, v. 78, p. 353-368.
1989: Application of diagenetic alterations in red beds to the origin of copper in stratiform copper deposits; in Sediment-hosted Stratiform Copper Deposits, (ed.) R.W. Boyle, A.C. Brown, C.W. Jefferson, E.C. Jowett, and R.V. Kirkham; Geological Association of Canada, Special Paper 36, p. 85-96.

*White, W.S.
1971: A paleohydrologic model for mineralization of the White Pine copper deposit, northern Michigan; Economic Geology, v. 66, p. 1-13.

White, W.S. and Wright, J.C.
1966: Sulfide-mineral zoning in the basal Nonesuch Shale, northern Michigan; Economic Geology, v. 61, p. 1171-1190.

Zharkov, M.A.
1984: Paleozoic Salt-Bearing Formations of the World; Springer-Verlag, New York, 316 p.

Zielinski, R.A., Bloch, S., and Walker, T.R.
1983: The mobility and distribution of heavy metals during the formation of first cycle red beds; Economic Geology, v. 78, p. 1574-1589.

8.3b Redbed-type

IDENTIFICATION

Redbed-type copper deposits typically comprise disseminated sulphides at oxidation-reduction boundaries in anoxic rocks within or at the top of continental redbed sequences.

Copper is the most important metal found in these deposits and silver is the most important byproduct or, in some localities, coproduct metal. Lead and zinc are typical associated metals, but they have been of little or no economic value.

Reasonably typical redbed-type copper deposits are Dorchester, New Brunswick; several minor occurrences in the Carboniferous Pictou Group of Nova Scotia (Fig. 8.3-1); the Nacimiento deposit, New Mexico (Woodward et al., 1974); and other deposits in New Mexico (LaPoint, 1976). The Dzhezkazgan district in Kazakhstan, and possibly deposits in the Revett Formation in Montana and in the Lisbon Valley area, Utah are economically more important, but less typical, deposits formed in sediments deposited in continental redbed environments.

IMPORTANCE

Redbed-type copper deposits are unimportant in Canada and, with the exception of the Dzhezkazgan and Revett Formation deposits, have been unimportant in other parts of the world. These deposits are relatively poorly understood and the economic incentives have been insufficient to justify extensive documentation and exploration. Nevertheless, deposits at Dzhezkazgan have been a major source of copper in the former Soviet Union and the Revett Formation contains some significant copper deposits with important coproduct levels of silver. The Paoli deposit in the Permian Wellington Formation in Oklahoma with 158 g/t Ag (Thomas et al., 1991) and the Silver Reef deposit in Jurassic rocks of the Moenave Formation in southwestern Utah, which contains approximately 155 g/t Ag (Proctor and Brimhall, 1986; James and Newman, 1986; L.P. James, pers. comm., 1987), are silver-rich redbed-type copper deposits.

SIZE AND GRADE OF DEPOSITS

The size and grade of some redbed-type copper deposits are shown in Figure 8.3-2, however, data are limited. Corocoro and Charcarilla in Bolivia are the only deposits, in addition to Dzhezkazgan and the Revett deposits, that have sufficient grade and tonnage to be attractive exploration targets in Canada. Deposits with relatively low copper grades in the Revett Formation can be mined underground only because of the coproduct levels of silver (50-90 g/t Ag) and efficient trackless mining methods. A total of forty-six, 30 cm long channel samples in a roll-front type concentration in the Paoli deposit, Oklahoma averaged 158 g/t Ag and 0.68% Cu, indicating that some redbed-type copper deposits may contain substantial amounts of silver (Shockey et al., 1974; Thomas et al., 1991).

The subeconomic Dorchester deposit in New Brunswick evidently has a higher grade linear or curvilinear zone in contact with barren pyrite near an oxidation-reduction 'front' and a lower grade zone away from the 'front'. Although incompletely documented, this morphology is similar to that of uranium roll-front deposits.

GEOLOGICAL FEATURES

In some areas, such as in rocks of the Pictou Group in Nova Scotia and Mesozoic formations in the Colorado Plateau region of the western United States, redbed-type copper occurrences are abundant. In many areas examined by the writer and based on published descriptions, mixed red and grey, fining-upward meandering stream deposits are the dominant host rocks for this deposit type (Fig. 8.3-7). As illustrated in Figure 8.3-7, in many such sequences caliche (calcite) nodules are found within the upper overbank parts of the fining-upward cycles and the cycles are reddened diagenetically from their tops toward their bases. The lowermost channel lag and point bar deposits, because of a concentration of reductants, typically are the last parts of the cycles to be oxidized and also generally contain the highest concentrations of copper. Entirely grey, reduced, in places coal-bearing, fining-upward meandering stream sequences are typically barren. Furthermore, entirely oxidized, red sequences contain only trace amounts of copper, even in copper-bearing sequences (e.g., Pictou Group, Nova Scotia). An anoxic grey unit, either overlying or within a thick sequence of redbeds, is a particularly favourable site for this type of deposit.

The consistent stratigraphic setting for many redbed-type copper occurrences indicates a relatively specialized environment of sedimentation consisting of a well developed, mature system of relatively low-gradient meandering streams that periodically became desiccated, especially in the upper overbank parts of the cycles (Fig. 8.3-7). This probably occurred in arid or semiarid areas with limited vegetation. The caliche zones (calcrete in some localities) signify soil formation. The lower channel lag and point bar deposits remained in reduced states provided that they stayed below the water table. Figure 8.3-8 is a schematic illustration of the environment of sedimentation. Occurrences in lacustrine rocks are less abundant than those in fluviatile rocks.

The Nacimiento and Eureka deposits in the Triassic Chinle Formation in New Mexico also occur in crudely fining-upward fluvial units (Woodward et al., 1974; LaPoint, 1976, 1989). However, the streams that deposited these sediments probably had steeper gradients and were more braided in nature than the more typical sinuous meandering streams that deposited host rocks of many redbed copper deposits. These large-volume stream deposits overlie lower energy redbed deposits of the saline Permian Abo Formation.

Prior to Siluro-Devonian time, fining-upward fluvial sequences were considerably different, without the stabilizing influence of vascular land plants. In the Grinnell Formation of the Middle Proterozoic Purcell Supergroup in southwestern Alberta and southeastern British Columbia, several copper occurrences are known in relatively clean, white quartzite units within a redbed sequence (Fig. 8.3-9). Collins and Smith (1977) suggested that these might be fluvial floodplain deposits. These might be Precambrian analogues of the very abundant redbed-type copper occurrences in Phanerozoic fluvial sequences.

Copper sulphides, dominantly chalcocite, are typically disseminated and replace early diagenetic pyrite and wood debris. As the early diagenetic pyrite and wood debris are commonly concentrated in the lower parts of the fluvial cycles, this is also where copper is concentrated.

Most deposits have not been studied sufficiently to document mineral and metal zoning. Nevertheless, preliminary work at Dorchester suggests that, in the lower fining-upward fluvial cycle, copper sulphides may occur in contact with barren pyrite and that in the overlying cycle, anomalous lead and zinc occur above the copper zone (Fig. 8.3-10).

In many deposits the cell structure of wood has been well preserved by sulphides. However, in most areas, insufficient information is available to indicate if copper sulphides fill-in or replace the plant structure directly or if the carbonaceous material was first filled in or replaced by pyrite. In any event, the excellent preservation of cell structures indicates preservation of plant debris by some sulphide before compaction and coalification. Isolated disseminated sulphide grains and small gash veins are also common.

At Dzhezkazgan and in the case of deposits in the Revett Formation, the redbeds might have been reduced on a regional scale by hydrocarbon-bearing formation waters emanating from underlying marine beds (Lur'ye and Gablina, 1978; Gablina, 1981; Kirkham, 1989) (Fig. 8.3-11, 8.3-12). The reduction of redbeds by mobile reductants, if such a mechanism can be substantiated, is an important variation in the genetic model for redbed-type copper deposits and may account for the deposition of much larger concentrations of metals than those in typical redbed-type deposits in which copper minerals were precipitated by immobile reductants. Ryan (1993) suggested that the location of deposits in the Revett Formation might be controlled by syndepositional faults. Recently discovered copper occurrences in the Rae Group on Victoria Island in northern Canada have some characteristics of Dzhezkazgan-type deposits (Rainbird et al., 1992, 1994).

DEFINITIVE CHARACTERISTICS

1. Redbed-type copper deposits characteristically are disseminated deposits in reduced rocks in continental redbed sequences.

2. Evaporites, caliche, calcrete, mudcracks, and other features record arid and semiarid continental environments of sedimentation.

3. Copper is generally the dominant metal, although some deposits contain byproduct or coproduct silver.

GENETIC MODEL

Most redbed-type copper deposits are accepted as being diagenetic replacements of early diagenetic pyrite and wood debris. They were probably formed from the infiltration of cupriferous fluids into permeable continental sediments. As illustrated in Figure 8.3-10, deposits show evidence of infiltration of oxic cupriferous fluids into reduced sandstone aquifers in otherwise less permeable redbeds. The

sands probably acted as drains for the less permeable, compacting red shale and siltstone. As discussed for Kupferschiefer-type deposits, the most probable ore fluids were oxic brines derived from evaporites that extracted metals from the redbeds and deposited them by reaction with any anoxic rocks and/or fluids that they encountered. For many occurrences fluid flow was mainly in sandstone aquifers and, in deposits such as Dorchester, deposition occurred at redox boundaries or 'fronts' where the fluids came in contact with early diagenetic pyrite and wood debris.

Important variations of this diagenetic model are illustrated in Figures 8.3-11 and 8.3-12, whereby more concentrated metal precipitation occurred in a series of stacked aquifers at regional redox boundaries between redbeds and diagenetically reduced redbeds. Such environments have, with both mobile oxic ore fluids and reductants, evidently resulted in much greater concentrations of metals than in more typical redbed environments with more localized, immobile reductants.

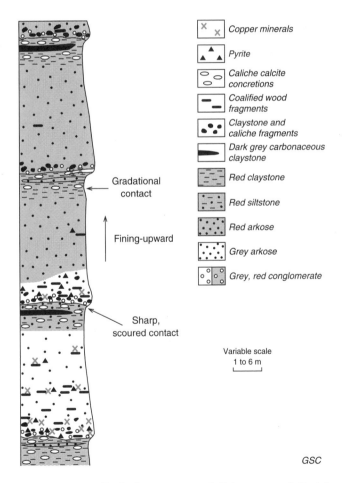

Copper minerals

Pyrite

Caliche calcite concretions

Coalified wood fragments

Claystone and caliche fragments

Dark grey carbonaceous claystone

Red claystone

Red siltstone

Red arkose

Grey arkose

Grey, red conglomerate

Gradational contact

Fining-upward

Sharp, scoured contact

Variable scale
1 to 6 m

GSC

Figure 8.3-7. Typical sequence of fining-upward fluvial cycles. Increased oxidization (reddening) and caliche nodules are in the upper parts of the cycles. Wood debris and early diagenetic pyrite reductants and copper sulphides are in the grey, reduced lower parts of the units.

235

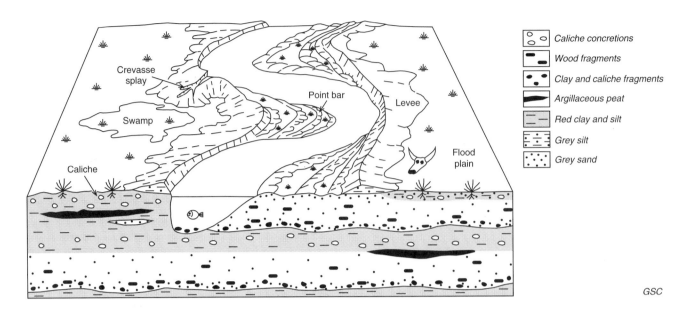

Figure 8.3-8. Schematic block diagram illustrating a meandering stream in a semiarid environment with near-surface oxidation and caliche formation in soils. The fining-upward fluvial cycles, typical host rocks of many redbed copper occurrences, probably formed in such an environment.

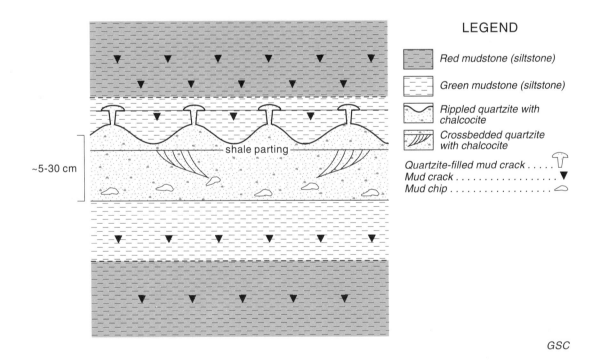

Figure 8.3-9. Typical copper-bearing cyclic unit in the Middle Proterozoic Grinnell Formation in Alberta and British Columbia (adapted from Collins and Smith, 1977). These cyclic quartzite-siltstone units could be Proterozoic analogues of younger fluvial host rocks of redbed copper deposits postdating vascular land plants in the Siluro-Devonian.

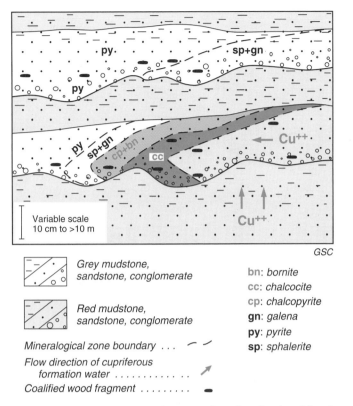

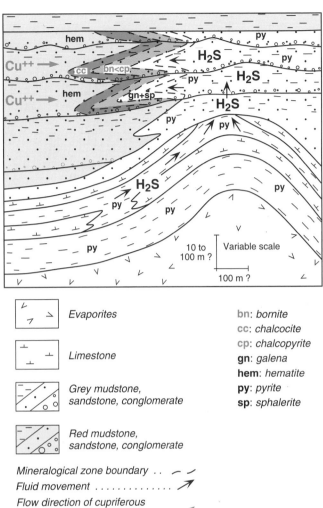

Figure 8.3-10. Interlayered grey, reduced and red, oxidized fining-upward fluvial host rocks typical of many redbed copper deposits. The Cu sulphide zones are shown in red. Metalliferous brines migrating in sandstone aquifers precipitated base-metal sulphides upon reaction with carbonaceous wood debris and early diagenetic pyrite reductants (after Kirkham, 1989).

Figure 8.3-11. Model for the Dzhezkazgan region, Kazakhstan, involving both mobile reductants and oxic metalliferous brines (after Lur'ye and Gablina, 1978; Gablina, 1981; Kirkham, 1989). The Cu sulphide zones are shown in red.

Figure 8.3-13, for the Lisbon Valley area in Utah and Colorado, shows another variation on the general oxidation-reduction processes responsible for the formation of Kupferschiefer- and redbed-type copper deposits. In this area, metalliferous brines trapped in redbed aquifers were evidently released upward along faults during later tectonism or a reversal of the topographic gravity gradient (Schmitt, 1968; Morrison and Parry, 1986; Breit et al., 1987; G.N. Breit, pers. comm., 1987). In the Lisbon Valley area these heated and pressurized fluids evidently rose as much as a kilometre or more along the faults until they encountered reduced aquifers, such as the Triassic Wingate Sandstone and Cretaceous Morrison Formation. Copper sulphides were precipitated as veins and disseminations near the faults and their abundance decreases rapidly away from the faults. Price et al. (1985, 1988) proposed an analogous thermally-driven model for some vein and fault-controlled Cu-Ag deposits in Precambrian, Permian, and Cretaceous redbeds in the Trans-Pecos region of Texas. Evidence supports diagenetic fluid flow of oxic metalliferous brines in aquifers and along permeable structures, such as faults, and metal precipitation where the brines encountered anoxic conditions. The basic oxidation-reduction processes of ore formation are similar for Kupferschiefer-, redbed-, and volcanic redbed-type copper deposits, but they evidently varied considerably from one area to another.

RELATED DEPOSIT TYPES

Kupferschiefer-, redbed-, and volcanic redbed-type copper deposits, as mentioned above for Kupferschiefer-type deposits, are all related to some degree and in many areas they occur together. Other deposit types, such as sandstone lead and uranium deposits, Mississippi Valley-type lead-zinc, and unconformity uranium deposits, have some features in common with redbed-type copper deposits, but they were probably formed at different times, in different areas, and from different ore fluids. The association of lead with copper in some redbed-type copper deposits probably indicates that the redbeds had mixed felsic and mafic provenance.

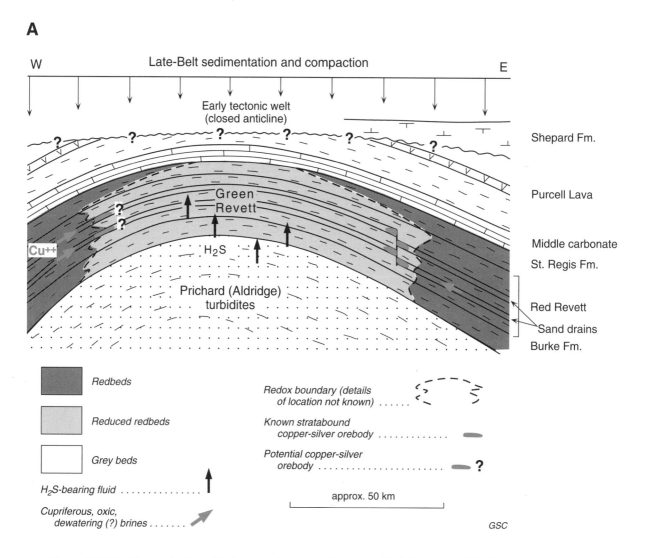

A

W   Late-Belt sedimentation and compaction   E

Early tectonic welt
(closed anticline)

Shepard Fm.

Purcell Lava

Green
Revett

Middle carbonate

St. Regis Fm.

H₂S

Cu⁺⁺

Prichard (Aldridge)
turbidites

Red Revett

Sand drains

Burke Fm.

Redbeds

Reduced redbeds

Grey beds

H₂S-bearing fluid . . . .

Cupriferous, oxic,
dewatering (?) brines . . . .

Redox boundary (details
of location not known) . . . . . .

Known stratabound
copper-silver orebody . . . .

Potential copper-silver
orebody . . . .   ?

approx. 50 km

GSC

Figure 8.3-12. Conceptual models for zoned copper-silver deposits (shown in red) in the Revett Formation in Montana. Mobile reductants (e.g., H₂S, CH₄, hydrocarbons) are modelled as derived from the underlying Prichard Formation. Oxic metalliferous brines migrated in permeable sandstone units (mineral zoning after Hayes and Einaudi, 1986). **A)** Model for reductants trapped in early anticlinal structure. **B)** Model for reductants entering Revett aquifers along faults (after comment by Wodzicki, 1990, quoted in Adkins, 1993).

EXPLORATION GUIDES

Reduced units within continental redbed requences deposited in low-latitude arid and semiarid areas, are suitable sites for the occurrence of redbed-type copper deposits. Fluvial rocks, especially in more permeable lower parts of fining-upward cycles, are the most typical host rocks. Higher copper grades can occur at reaction fronts between base-metal zones and barren pyritic host rocks.

Although more work is required to confirm relationships, sulphide deposition at the redox boundaries between redbeds and diagnetically reduced redbeds has resulted in larger concentrations of metals than in areas with immobile reductants. Areas that contain such diagenetically reduced redbeds might occur where redbeds overlie anoxic marine rocks and also possibly around salt domes and anticlines. Deposits might also occur in anoxic units along fault zones above redbeds and evaporites (e.g., Fig. 8.3-13) or in reduced redbed units where reductants migrating up fault zones encountered metalliferous brines in aquifers (e.g., Fig. 8.3-12).

B

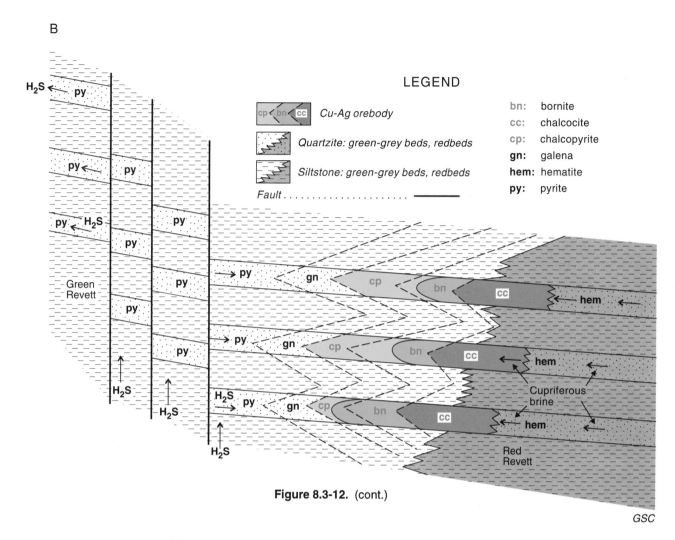

LEGEND

cp ◄ bn ◄ cc	*Cu-Ag orebody*
	Quartzite: green-grey beds, redbeds
	Siltstone: green-grey beds, redbeds
Fault	━━

bn: bornite
cc: chalcocite
cp: chalcopyrite
gn: galena
hem: hematite
py: pyrite

Green
Revett

Cupriferous
brine

Red
Revett

Figure 8.3-12. (cont.)

GSC

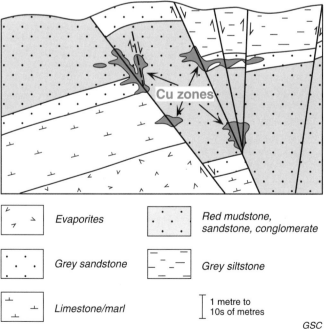

Cu zones

	Evaporites
	Grey sandstone
	Limestone/marl

	Red mudstone, sandstone, conglomerate
	Grey siltstone

1 metre to
10s of metres

Figure 8.3-13. A relatively late diagenetic model for the Lisbon Valley area of Utah and Colorado, where heated oxic metalliferous brines migrated up fault-zone aquifers (after Schmitt, 1968; D.R. Shawe, pers. comm., 1980; Morrison and Parry, 1986; Breit et al., 1987; G.N. Breit, pers. comm., 1987; Kirkham, 1989).

GSC

SELECTED BIBLIOGRAPHY

References marked with asterisks (*) are considered to be the best sources of general information on this deposit subtype.

Adkins, A.R.
1993:   Geology of the Montanore stratabound Cu-Ag deposit, Lincoln and Sanders counties, Montana (extended abstract); in Belt Symposium III, Program and Abstracts, Whitefish, Montana, August 14-21, 1993, 3 p.

Avila-Salinas, W.
1990:   Origin of the copper ores at Corocoro, Bolivia; in Stratabound Ore Deposits in the Andes, (ed.) L. Fontboté, G.C. Amstutz, M. Cardozo, E. Cedillo, and J. Frutos; Springer-Verlag, Berlin, p. 559-670.

*Boyle, R.W., Brown, A.C., Jefferson, C.W., Jowett, E.C., and Kirkham, R.V. (ed.)
1989:   Sediment-hosted Stratiform Copper Deposits; Geological Association of Canada, Special Paper 36, 710 p.

Breit, G.N., Meunier, J.D., Rowan, E.L., and Goldhaber, M.B.
1987:   Alteration related to red bed copper mineralizing brines and other fault-controlled solutions in Lisbon Valley, Utah, and the Slick Rock district, Colorado (abstract); in United States Geological Survey Research on Mineral Resources-1987, Program and Abstracts, (ed.) J. Somerville Sach; United States Geological Survey, Circular 995, p. 7-8.

Collins, J.A. and Smith, L.
1977:   Genesis of cupriferous quartz arenite cycles in the Grinnell Formation (Spokane equivalent), Middle Proterozoic (Helikian) Belt-Purcell Supergroup, eastern Rocky Mountains, Canada; Bulletin of Canadian Petroleum Geology, v. 25, p. 713-735.

*Gablina, I.F.
1981:   New data on formation conditions of the Dzhezkazgan copper deposit; International Geology Review, v. 23, p. 1303-1311.

*Hayes, T.S. and Einaudi, M.T.
1986:   Genesis of the Spar Lake strata-bound copper-silver deposit, Montana: Part I. Controls inherited from sedimentation and preore diagenesis; Economic Geology, v. 81, p. 1899-1931.

James, L.P. and Newman, E.W.
1986:   Subsurface character of mineralization at Silver Reef, Utah, and a possible model for ore genesis; in Thrusting and Extensional Structures and Mineralization in the Beaver Dam Mountains, Southwestern Utah, (ed.) D.T. Griffen and W.R. Phillips; Utah Geological Association, Publication 15, p. 149-158.

Kirkham, R.V.
*1989:   The distribution, settings and genesis of sediment-hosted, stratiform copper deposits: in Sediment-hosted Stratiform Copper Deposits, (ed.) R.W. Boyle, A.C. Brown, C.W. Jefferson, E.C. Jowett, and R.V. Kirkham; Geological Association of Canada, Special Paper 36, p. 3-38.

LaPoint, D.J.
1976:   A comparison of selected sandstone copper deposits in New Mexico; in Stratiform Copper Deposits of the Midcontinent Region, a Symposium, (ed.) K.S. Johnson and R.L. Croy; Oklahoma Geological Survey, Circular 77, p. 80-96.
*1989:   A model for the diagenetic formation of sandstone copper deposits in sedimentary rocks of Permian and Triassic age, in New Mexico, U.S.A.; in Sediment-hosted Stratiform Copper Deposits, (ed.) R.W. Boyle, A.C. Brown, C.W. Jefferson, E.C. Jowett, and R.V. Kirkham; Geological Association of Canada, Special Paper 36, p. 357-370.

*Lur'ye, A.M. and Gablina, I.F.
1978:   Basic scheme of formation of supergene copper deposits; Doklady Akademii Nauk SSSR, v. 241, p. 126-129.

*Morrison, S.J. and Parry, W.T.
1986:   Formation of carbonate-sulfate veins associated with copper ore deposits from saline basin brines, Lisbon Valley, Utah: fluid inclusion and isotopic evidence; Economic Geology, v. 81, p. 1853-1866.

Narkelyun, L.F. and Fatikov, R.F.
1989:   Sedimentary control of mineralization at the Dzhezkazgan copper deposit; International Geology Review, v. 31, p. 196-203.

*Popov, V.M.
1962:   Geologic regularities in the distribution of cupriferous sandstones in central Kazakhstan; International Geology Review, v. 4, p. 393-411.

Price, J.G., Henry, C.D., Standen, A.R., and Posey, J.S.
1985:   Origin of silver-copper-lead deposits in red-bed sequences of Trans-Pecos Texas: Tertiary mineralization in Precambrian, Permian, and Cretaceous sandstones; Texas Bureau of Economic Geology, Report of Investigations, no. 145, 65 p.

Price, J.G., Rubin, J.N., and Tweedy, S.W.
1988:   Geochemistry of the Vigas red-bed copper deposit, Chihuahua, Mexico; Economic Geology, v. 83, p. 1993-2001.

Proctor, P.D. and Brimhall, W.H.
1986:   Silver Reef mining district revisited, Washington County, Utah; in Thrusting and Extensional Structures and Mineralization in the Beaver Dam Mountains, Southwestern Utah, (ed.) D.T. Griffen and W.R. Phillips; Utah Geological Association, Publication 15, p. 159-177.

Rainbird, R.H., Darch, W., Jefferson, C.W., Lustwerk, R., Reese, M., Telmer, K., and Jones, T.A.
1992:   Preliminary stratigraphy and sedimentology of the Glenelg Formation, lower Shaler Group, and correlatives in the Amundsen Basin, Northwest Territories: relevance to sediment-hosted copper; in Current Research, Part C; Geological Survey of Canada, Paper 92-1C, p. 111-119.

Rainbird, R.H., Jefferson, C.W., Hildebrand, R.S., and Worth, J.K.
1994:   The Shaler Supergroup and revision of Neoproterozoic stratigraphy in the Amundsen Basin, Northwest Territories; in Current Research 1994-C; Geological Survey of Canada, p. 61-70.

Ryan, P.C.
1993:   Stratabound Cu-Ag mineralization and implications for Revett Formation deposition, Flathead Indian Reservation, western Montana (extended abstract); in Belt Symposium III, Program and Abstracts, Whitefish, Montana, August 14-21, 1993, 3 p.

Schmitt, L.J.
1968:   Uranium and copper mineralization in the Big Indian Wash-Lisbon Valley mining district, southeastern Utah; PhD. thesis, Columbia University, New York, New York, 173 p.

Shockey, P.N., Renfro, A.R., and Peterson, R.J.
1974:   Copper-silver solution fronts at Paoli, Oklahoma; Economic Geology, v. 69, p. 266-268.

Thomas, C.A., Hagni, R.D., and Berendsen, P.
1991:   Ore microscopy of the Paoli silver-copper deposit, Oklahoma; Ore Geology Reviews, v. 6, p. 229-244.

Winston, D.
1990:   Evidence for intracratonic, fluvial and lacustrine settings of Middle to Late Proterozoic Basins of western U.S.A.; in Mid-Proterozoic Laurentia-Baltica, (ed.) C.F. Gower, T. Rivers, and B. Ryan; Geological Association of Canada, Special Paper 38, p. 535-564.

*Woodward, L.A., Kaufman, W.H., Schumacher, O.L., and Talbott, L.W.
1974:   Strata-bound copper deposits in Triassic sandstone of Sierra Nacimiento, New Mexico; Economic Geology, v. 69, p. 108-120.

Authors' addresses

R.T. Bell
Geological Survey of Canada
601 Booth Street
Ottawa, Ontario
K1A 0E8

R.V. Kirkham
Geological Survey of Canada
100 West Pender Street
Vancouver, B.C.
V6B 1R8

D.F. Sangster
Geological Survey of Canada
601 Booth Street,
Ottawa, Ontario
K1A 0E8

Printed in Canada

9. VOLCANIC REDBED COPPER

9. VOLCANIC REDBED COPPER

R.V. Kirkham

INTRODUCTION

Volcanic redbed copper (VRC) deposits occur as concordant and peneconcordant disseminated and crosscutting vein and fault-controlled copper sulphide and/or native copper deposits in predominantly subaerial volcanic sequences. They are characterized by relatively simple copper sulphide and/or native copper mineral assemblages, contain variable amounts of silver, and are distinct from submarine, polymetallic, volcanic-associated massive sulphide base metal deposits. The name 'volcanic redbed copper' is used to draw attention to the similarity with redbed- and Kupferschiefer-type deposits in sedimentary sequences and to the fact that oxidized flow tops and some red interlayered sedimentary rocks are probably essential in the genesis of these deposits. In mixed volcanic and sedimentary sequences, sedimentary and volcanic redbed copper deposits occur together and, in places, no clear distinction can be drawn between the two deposit types.

Areas in Canada with volcanic redbed copper deposits and occurrences include the following: Coppermine River area (Kindle, 1972; Carrière and Kirkham, 1993) and Natkusiak Formation, Victoria Island, Northwest Territories (Jefferson et al., 1985); Seal Lake area, Labrador (Brummer and Mann, 1961; Gandhi and Brown, 1975); Mamainse Point, Ontario (Heslop, 1970; Hak et al., 1977; Pearson et al., 1985); Sustut, British Columbia (Harper, 1977; Wilton and Sinclair, 1988); Triassic and Lower Jurassic volcanic rocks in central British Columbia (Kirkham, 1970); Karmutsen Formation, Vancouver Island, British Columbia (Surdam, 1968; Lincoln, 1981), and White River

area, Yukon Territory (Kirkham, 1970; Sinclair et al., 1979; Carrière et al., 1981; Fig. 9-1, 9-2). Deposits in other countries that are probably of this type include the famous native copper deposits in the Keweenaw Peninsula, Michigan (White, 1968; Weege and Pollack, 1972), and many stratabound copper deposits in Jurassic, Cretaceous, and Tertiary volcanic sequences in Chile (Ruiz et al., 1971; Losert, 1973; Lortie and Clark, 1974, 1987; Sato, 1984; Camus, 1986). The Bleïda deposit in Late Proterozoic rocks of the Anti-Atlas area of Morocco also shows features indicative of volcanic redbed copper deposits (Leblanc and Billaud, 1978, 1990; R.V. Kirkham, pers. observations, 1992).

IMPORTANCE

These deposits have not been important in Canada and are unlikely to be of major importance in the future. Nevertheless, they are widely distributed in Canada (Fig. 9-1) and could result in some significant copper mines.

Native copper deposits on the Keweenaw Peninsula were the most important source of copper in the United States from 1845 to 1885, but now this deposit type is generally of minor importance throughout the world. From 1845 to 1968 the native copper district of northern Michigan produced in excess of 5 Mt of copper metal, making it a major copper district (White, 1968). Deposits are currently being re-evaluated in this district, and a number of small- and medium-sized mines in Chile produce significant amounts of copper (Ruiz et al., 1971; Sato, 1984; Camus, 1986).

SIZE AND GRADE OF DEPOSITS

Production and reserve data for volcanic redbed copper deposits are shown in Table 9-1 and Figure 9-3. Some of the deposits have sufficient grade to support underground mining and four of the deposits (Calumet-Hecla, El Soldado, Mantos Blanco, and Kearsarge) contained more than one million tonnes of copper metal.

Kirkham, R.V.
1996: Volcanic redbed copper; in Geology of Canadian Mineral Deposit Types, (ed.) O.R. Eckstrand, W.D. Sinclair, and R.I. Thorpe; Geological Survey of Canada, Geology of Canada, no. 8, p. 241-252 (also Geological Society of America, The Geology of North America, v. P-1).

The Coppercorp mine, Mamainse Point, produced about 850 000 t of ore containing 1.15% Cu, 8.3 g/t Ag, and 0.06 g/t Au, and has about 1.4 Mt grading 1.7% Cu remaining (Pearson et al., 1985). The 47 zone in the Coppermine River area, Northwest Territories contains about 3.2 Mt grading 3.44% Cu (Northern Miner, October 17, 1968), and the Sustut deposit in British Columbia contains a total in situ, drill-indicated geological inventory of 43.5 Mt grading 0.81% Cu (Falconbridge Limited, written comm., 1988).

The Calumet and Kearsarge deposits on the Keweenaw Peninsula are among the largest known deposits of this type. White (1968) estimated that the Calumet Conglomerate lode produced about 72.4 Mt grading 2.64% Cu and R.J. Weege (pers. comm., 1988) indicated that it still contains subeconomic indicated and inferred resources of 38.3 Mt grading 1.92% Cu. The Kearsarge Amygdaloid lode produced about 89.1 Mt of ore grading 1.05% Cu (White, 1968) and contains inferred and potential subeconomic resources of 72.1 million tonnes grading 1.0% Cu (R.J. Weege, pers. comm., 1988). Other substantial deposits were also mined in the Keweenaw Peninsula (White, 1968).

Most volcanic redbed copper deposits are narrow and tabular, requiring expensive underground mining techniques, and are of limited size and grade. Nevertheless, White (1968) indicated that in the Keweenaw Peninsula the average grade tends to increase with the size of the deposit; therefore, large, rich deposits, such as the Calumet Conglomerate, would still be attractive exploration targets today.

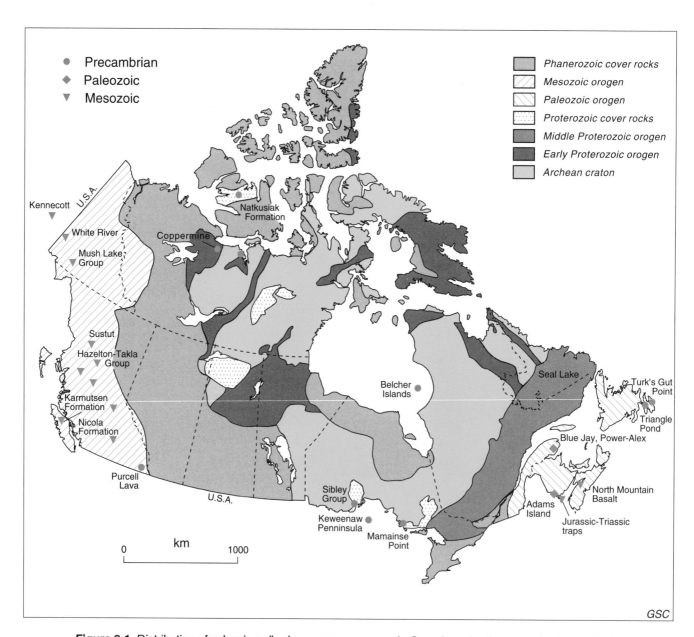

Figure 9-1. Distribution of volcanic redbed copper occurrences in Canada and adjacent ports of the U.S.A.

GEOLOGICAL FEATURES

Geological setting

This deposit type is widely distributed in subaerial flood basalt sequences, and some authors have referred to them as 'basalt-hosted' copper (native copper) deposits (e.g., Sutherland Brown et al., 1971; Cox, 1986). However, these deposits are also widespread in differentiated continental arc and island-arc sequences, and occur in rocks which are as felsic as rhyolite (e.g., Jardin, Chile (Lortie and Clark, 1974, 1987) and NH Group, British Columbia (Kirkham, 1969)). Peneconcordant deposits occur in permeable amygdaloidal flow top breccias (e.g., Kearsarge and Osceola lodes of the Keweenaw Peninsula; Butler and Burbank, 1929; Weege and Pollack, 1972) and Natkusiak basalt, Victoria Island (Jefferson et al., 1985) and Karmutsen Formation, Vancouver Island (Surdam, 1968; Lincoln, 1981), and they are also common in interlayered sedimentary units

within volcanic sequences (e.g., in the Calumet Conglomerate on the Keweenaw Peninsula (Weege and Pollack, 1972); Talcuna, Chile (Ruiz et al., 1971); Abbas-Abad, Iran (Khadem, 1964); Bleïda, Morocco; and Northstar, British Columbia (Sutherland Brown, 1968)). In the Keweenaw Peninsula and at Buena Esperanza in Chile (Ruiz et al., 1971; Losert, 1973), peneconcordant en echelon orebodies are stacked through the volcanic sequence. Discordant deposits occur as zoned, disseminated and stringer copper sulphide deposits in shear (fault) zones (e.g., 47 zone, Coppermine River area; Mount Bohemia, Keweenaw Peninsula (Robertson, 1975), and Coppercorp mine (Heslop, 1970)) and as well defined, layered quartz-carbonate-copper sulphide and/or native copper veins (e.g., native copper-arsenic veins, Keweenaw Peninsula (Butler and Burbank, 1929; Broderick et al., 1946); Copper Lamb and Jack Lake, Coppermine River area (Kindle, 1972); and the Main showing, Seal Lake area, Labrador (Brummer and Mann, 1961)).

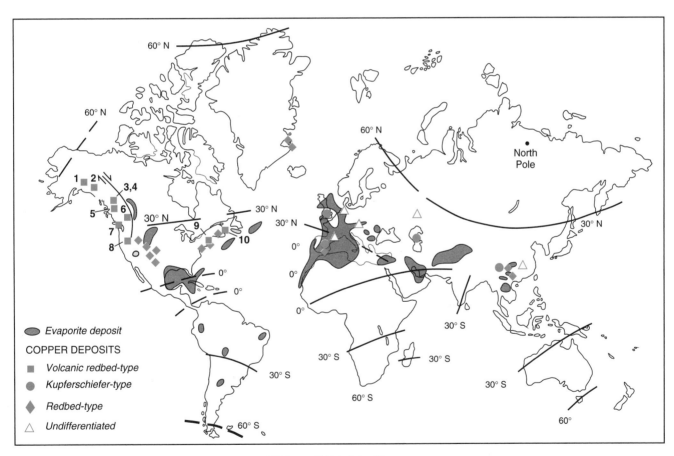

Evaporite deposit

COPPER DEPOSITS

■ *Volcanic redbed-type*

● *Kupferschiefer-type*

◆ *Redbed-type*

△ *Undifferentiated*

1	*Kennecott, Alaska*	5	*NH Group, British Columbia*	8	*Seven Devils, Oregon, U.S.A.*
2	*White River, Yukon Territory*	6	*Shamrock, British Columbia*	9	*Newark Basin, eastern U.S.A.*
3	*Sustut, British Columbia*	7	*Quadra Island, British Columbia*	10	*North Mountain Basalt, Nova Scotia*
4	*Northstar, British Columbia*				

GSC

Figure 9-2. Distribution of volcanic redbed copper occurrences in Triassic and Jurassic rocks in North America and their relationship to paleolatitudes, evaporites, and diagenetic sedimentary copper occurrences.

Table 9-1. Production/reserves of selected Canadian and foreign volcanic redbed copper deposits.

Deposit	Production/Reserves	Comments/References
Canadian deposits:		
Coppercorp (Mamainse Point), Ontario	0.85 Mt; 1.15% Cu 8.3 g/t Ag 0.06 g/t Au	Production 1965-1972; Pearson et al. (1985)
47 Zone, Coppermine River area, Northwest Territories	3.2 Mt; 3.44% Cu	Estimated reserves; Northern Miner, October 17, 1968, p. 1
Sustut, British Columbia	43.5 Mt; 0.81% Cu	Drill-indicated subeconomic resource; Falconbridge Limited personnel, pers. comm., 1988
Foreign deposits:		
Bleïda, Morocco	5.2 Mt; 4.1% Cu	Production and reserves; P. Billaud, pers. comm., 1994
Carolina de Michilla district, Chile	~40 Mt; 2% Cu	Approximate district reserves; H. Alfaro Huente, pers. comm., 1990
Ivan, Chile	5.3 Mt; 2.3% Cu	Estimated reserves; Northern Miner, July 2, 1990, p. 12
Sierra Valenzuela, Chile	10-15 Mt; 1.75% Cu	Reserve potential; Northern Miner, November 23, 1993, p. 6
San José de Tuina, Chile	4.4 Mt; 1.5% Cu	Production and reserves; H. Ramirez, Carrasco, pers. comm., 1990
Mantos Blancos, Chile	220 Mt; 1.2% Cu	Production and reserves, sulphide and oxide ore; R. Ramirez Rodriguez, pers. comm., 1990
Caleta del Cobre, Chile	11 Mt; 2.2% Cu	Production and reserves; Camus (1986)
Jardin, Chile	3.29 Mt; 1.4% Cu 80 g/t Ag	Production 1960-1989 plus reserves; N. Perez, pers. comm., 1990
Amolanas, Chile	10 Mt; ~1.8% Cu ~20 g/t Ag	Estimated possible and potential reserves; company official, pers. comm., 1990
Talcuna, Chile	4.3 Mt; 1.47% Cu 38 g/t Ag	Camus (1986)
El Soldado, Chile	136 Mt; 1.6% Cu 6-10 g/t Ag	Production plus reserves; H. Ruge R., pers. comm., 1990
Cerro Negro, Chile	7.2 Mt; 2% Cu 20 g/t Ag	Estimated past production, plus reserves; J. Oyarzun Bustos, pers. comm., 1990
Mantos de Catemu, Chile	2.0 Mt; 1.75% Cu 20 g/t Ag	Camus (1986)
El Salado, Chile	1.0 Mt; 1.25% Cu	Camus (1986)
Lo Aguirre, Chile	11 Mt; 2.24% Cu	Camus (1986)
Altamira, Chile	12.1 Mt; 1.7% Cu 41 g/t Ag	Reserves; Mining Journal, December 3, 1993, p. 381
Calumet-Hecla, Michigan	114 Mt; 2.4% Cu	Production plus reserves; White (1968) and R. Weege (pers. comm.,1988)
Kingston, Michigan	5.4 Mt; 1.2% Cu	Production and reserves; Weege and Pollack (1972)
Kearsarge, Michigan	161 Mt; 1.0% Cu	Production plus reserves; White (1968) and R. Weege (pers. comm.,1968)
Pewabic, Michigan	38.9 Mt; 1.26% Cu	Production; White (1968)
Allouez, Michigan	41.3 Mt; 0.77% Cu	Production plus reserves; White (1968) and R. Weege (pers. comm.,1968)
Isle Royale, Michigan	17.1 Mt; 0.90% Cu	Production; White (1968)
Baltic, Michigan	55.6 Mt; 1.5% Cu	Production; White (1968)
Osceola, Michigan	37.7 Mt; 0.92% Cu	Production plus reserves; White (1968) and R. Weege (pers. comm.,1988)
Kennecott, Alaska	4.2 Mt; 12.8% Cu 55 g/t Ag	Production 1911-1938; Maloney and Bottge (1973)

Similar to diagenetic sedimentary copper deposits, these deposits evidently occur only in subaerial to shallow marine rocks that postdate oxygenation of the Earth's atmosphere, i.e., those that were formed from about 2.4 Ga to Present (Kirkham and Roscoe, 1993). They do not occur in deep water, pillow lava sequences. Emplacement of ore minerals clearly postdated deposition of the host rocks, but in many areas evidence indicates that the host rocks had not completely lost their permeability at the time of ore deposition.

Form of deposits

In many of the economically more important districts evidence supports migration of ore fluids up dip along more permeable aquifers in the sequence (e.g., Keweenaw Peninsula; stratabound deposits in Chile, and Sustut, British Columbia; Fig. 9-4). However, in other areas fluid flow was dominantly cross-stratal along faults and fractures (e.g., Coppermine River area). In such areas, copper occurrences show a prominent structural control (Fig. 9-5). As a result of the highly variable nature of their structural and stratigraphic settings, volcanic redbed copper deposits have a great diversity of morphologies, including simple veins, through discordant and partly concordant disseminated and complex stringer, fault breccia deposits (e.g., 47 zone, Fig. 9-5), and to relatively conformable-deposits (e.g., Keweenaw Peninsula, Chile, and Sustut, British Columbia).

Mineralogy and mineral zoning

Some deposits, such as the 47 zone (Fig. 9-5), Sustut (Fig. 9-4; Harper, 1977; Wilton and Sinclair, 1988), Coppercorp (Heslop, 1970), Mount Bohemia (Robertson, 1975), and Mantos Blancos (Chavez, 1983) show prominent mineral zoning with native copper (only identified at Sustut) and chalcocite in the core of the deposits and bornite, chalcopyrite, and pyrite in successive, overlapping, somewhat irregular zones outward. Native copper is regionally distributed in amygdaloidal basalts away from the 47 zone and Mount Bohemia deposits. Although not well documented, this pattern of zoning probably occurs in other volcanic redbed copper deposits (e.g., stringer zones at Northstar, British Columbia and White River, Yukon Territory). This sequence of mineral zones is similar to that found in diagenetic sedimentary copper (SSC) deposits.

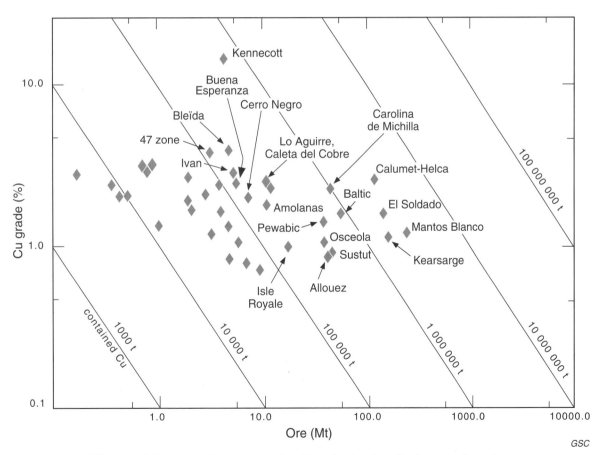

GSC

Figure 9-3. Grade and tonnage relationships of volcanic redbed copper deposits.

Many volcanic redbed-type copper deposits are characterized by low grade, regional metamorphic minerals, such as quartz, epidote, albite, chlorite, calcite, prehnite, pumpellyite, and laumontite, that occur as vesicle and fracture fillings and as patchy "metadomain" replacement bodies (Stoiber and Davidson, 1959; Jolly and Smith, 1972; Losert, 1973; Jolly, 1974; Lincoln, 1981; Jefferson et al., 1985).

Native copper, chalcocite, digenite, djurleite, bornite, sulphur-rich bornite, hematite, minor pyrite, and trace amounts of native silver are typical minerals found in volcanic redbed copper deposits. Galena, sphalerite, and other sulphides and sulphosalts are typically minor or absent, although minor galena and sphalerite are present in the outer mineral zones at Mount Bohemia, Michigan (Robertson, 1975). Fissure vein deposits elsewhere on the Keweenaw Peninsula also contain copper-arsenic minerals such as domeykite, algodonite, and whitneyite. Specular hematite, and less commonly magnetite, are present in some deposits.

Nature of ore

In many of the concordant or peneconcordant deposits the copper minerals fill intergranular spaces in clastic sedimentary host rocks, and spaces in the matrices of flow-top breccias and in amygdules in amygdaloidal flow tops. Small

gash veins are also typical in such deposits. Some fissure- and fault-controlled discordant deposits are relatively simple or layered, multiple-opening, tension vein fillings with minor or major amounts of gangue minerals, such as quartz and carbonates. In some more significant discordant deposits, such as the 47 zone and JUNE deposits in the Coppermine River area, deposits in the Quill Creek and White River areas, Yukon Territory (most significant showings in the Kluane Ranges are truncated by major, complex, postore faults), Coppercorp, Ontario, and possibly Mount Bohemia, Keweenaw Peninsula (Robertson, 1975), the copper sulphides occur, with little or no gangue, as minute sulphide stringers and disseminations. Copper in these deposits is in most cases concentrated in wide zones of fault brecciated rocks, at the intersections with favourable lithologies, such as amydaloidal flow-top breccia, interlayered sedimentary units, and felsic ash-flow tuff (Fig. 9-6).

DEFINITIVE CHARACTERISTICS

1. Relatively simple copper sulphide and/or native copper mineral assemblages in volcanic sequences with or without minor amounts of silver. They are generally not polymetallic, and do not contain large amounts of iron sulphides.

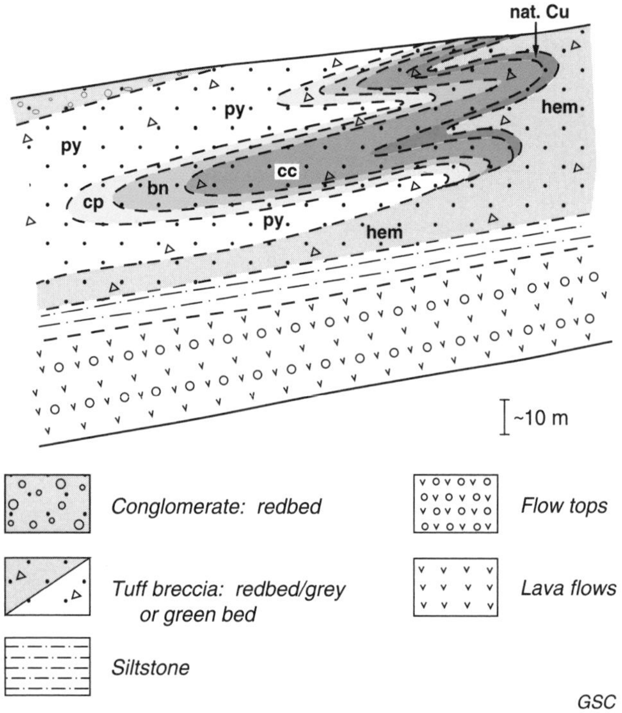

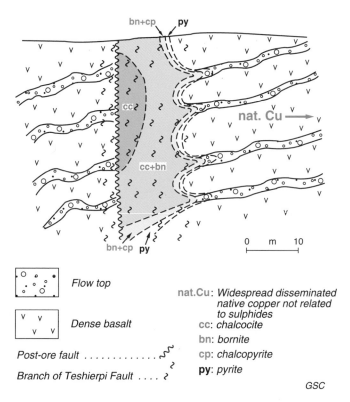

Figure 9-4. Typical mineral zoning in the peneconcordant Sustut deposit, British Columbia. Minerals: py – pyrite, cp – chalcopyrite, bn – bornite, cc – chalcocite, nat. Cu – native copper, hem – hematite.

Conglomerate: redbed		Flow tops	
Tuff breccia: redbed/grey or green bed		Lava flows	
Siltstone			

GSC

Figure 9-5. Typical mineral zoning in the discordant DOT 47 zone, Coppermine River area, Northwest Territories.

Flow top

Dense basalt

Post-ore fault . . . .

Branch of Teshierpi Fault . . . .

nat.Cu: *Widespread disseminated native copper not related to sulphides*
cc: *chalcocite*
bn: *bornite*
cp: *chalcopyrite*
py: *pyrite*

GSC

2. Deposits consist of copper minerals typically disseminated or in veins, rather than as massive sulphide bodies.

3. Wall rock alteration tends to be insignificant or absent, but many deposits are accompanied by low-grade regional metamorphic minerals such as quartz, epidote, albite, chlorite, calcite, prehnite, pumpellyite, and laumontite.

4. A significant part of the rock sequence in which these deposits occur was deposited in subaerial environments and was in an oxidized state.

5. As for sediment-hosted stratiform, diagenetic sedimentary copper deposits, paleomagnetic and sedimentological evidence indicates that the host rocks were deposited in arid and semiarid, low-latitude areas.

6. In many districts volcanic redbed and diagenetic sedimentary copper deposits occur in the same sequences.

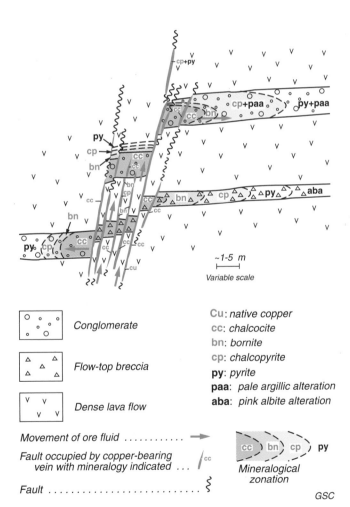

Cu: *native copper*
cc: *chalcocite*
bn: *bornite*
cp: *chalcopyrite*
py: *pyrite*
paa: *pale argillic alteration*
aba: *pink albite alteration*

Conglomerate

Flow-top breccia

Dense lava flow

Movement of ore fluid . . . .

Fault occupied by copper-bearing vein with mineralogy indicated . . .

Fault . . . .

Mineralogical zonation

GSC

Figure 9-6. Schematic diagram illustrating typical mineral zoning of volcanic redbed copper deposits formed in favourable stratigraphic units and along fault and fracture zones.

GENETIC MODEL

After the decline and closure of the Keweenawan native copper district, economic incentives, in most areas, have been insufficient to justify extensive documentation of, and exploration for, this deposit type. Chile is the only country that currently contains significant producing mines of this type. Hence, the genesis of volcanic redbed copper deposits is relatively poorly understood. A common view regarding their genesis is that they are metamorphogenic; i.e., that copper was released elsewhere at higher metamorphic grades during dehydration reactions, and was deposited in oxidized subaerial host rocks under lower metamorphic grade conditions associated with hydration reactions (Stoiber and Davidson, 1959; Surdam, 1968; Losert, 1973; Jolly, 1974; Harper, 1977; Lincoln, 1981; Wilton and Sinclair, 1988). However, the environment of formation was probably much more specialized; a significant portion of the pile had to be in an oxidized state and some original permeability probably remained at the time of ore formation. The metamorphism was probably of a relatively early burial type not accompanied by significant regional, penetrative deformation and, in some areas, the conditions of ore deposition were probably 'diagenetic' rather than 'metamorphic'. Chloride (±sulphate) brines, possibly derived from evaporites, were the probable ore fluids.

Many aspects of these deposits suggest that they are the analogues in volcanic sequences of diagenetic sedimentary copper deposits in sedimentary sequences. Figure 9-2 demonstrates the spatial relationship of volcanic redbed copper deposits to diagenetic sedimentary copper deposits, evaporites, and paleolatitude in Triassic and Jurassic time. Nevertheless, many volcanic redbed copper deposits in volcanic sequences are accompanied by low grade, regional metamorphic minerals. What is not clear is whether copper in some areas was deposited under submetamorphic diagenetic conditions preceding low-grade, prograde metamorphism. Of course, metamorphogenic fluids that rise to the unconsolidated upper parts of a volcanic-sedimentary pile become diagenetic fluids, not necessarily accompanied by significant drops in temperature or changes in composition. The importance of original permeability controls in many concordant and peneconcordant deposits indicates that a delicate balance must have been maintained between metamorphic and/or diagenetic losses of permeability and ore deposition. In volcanically active areas, with rapid accumulation rates and high heat flows, the lower parts of the piles could have been subjected to burial metamorphism during (or closely following) volcanism in the upper parts of the sequences. In such environments, burial metamorphism could have had direct links with shallower diagenetic processes (Fig. 9-7).

For volcanic redbed copper deposits to form, a significant part of the pile must have been deposited in subaerial environments because volcanic flows and most red sedimentary rocks must have been in an oxidized state and the lavas must have had an opportunity to desulphurize. The oxidized rocks permitted later passage of oxic metalliferous fluids (brines?). Associated pillow lavas could not have been erupted under significant water depths, because such lavas retain some of their sulphur, and this would inhibit later extraction and migration of metals. In the Triassic Karmutsen Formation on Vancouver Island, British Columbia, volcanic redbed copper occurrences are associated with

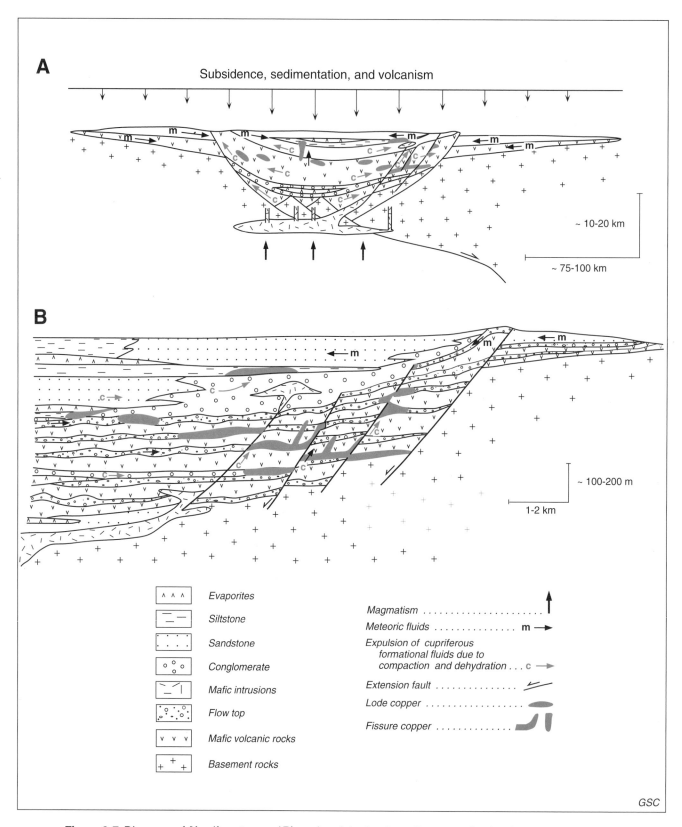

Figure 9-7. Diagrams of **A)** a rift system and **B)** a subaerial volcanic-sedimentary pile showing possible migration routes of cupiferous brines through flow top and clastic sedimentary aquifers, and along discordant faults. Possible metal deposition sites at redox boundaries, stratigraphic pinchouts, in veins, and fault zones are shown in red (modelled after the Keweenawan rift).

subaerial flows and interlayered redbeds in the upper part of the pile, but not with pillow lavas, pillow breccias, and aquagene tuffs that comprise most of the lower part of the pile (Surdam, 1968; Lincoln, 1981).

The zoned distribution patterns of ore minerals in the 47 zone (Fig. 9-5), Mount Bohemia, and possibly Sustut (Fig. 9-4), are poorly defined, because the zones do not occur at redox boundaries and are not obviously related to host rock sources of sulphur. In these deposits sulphur was probably carried in the ore fluid as sulphate, which was reduced to sulphide at the sites of ore deposition. The nature of this reduction process is unknown, but the presence of anthraxolite in amygdules in the Coppermine River

basalts suggests the possibility that liquid hydrocarbons might have been present in the volcanic pile at the time of ore deposition and could have fuelled bacterial reduction of sulphate to sulphide. More research is required in this area to refine ore transport and depositional models.

Figure 9-8 shows a comparison of sulphur isotope values for a variety of volcanic redbed copper occurrences in Canada and for redbed copper and sandstone uranium deposits. The wide spread of values in Canadian volcanic redbed-type, sediment-hosted redbed-type copper, and sandstone uranium deposits suggests that similar processes could have been involved in their genesis. Large ranges of values in individual occurrences and strongly

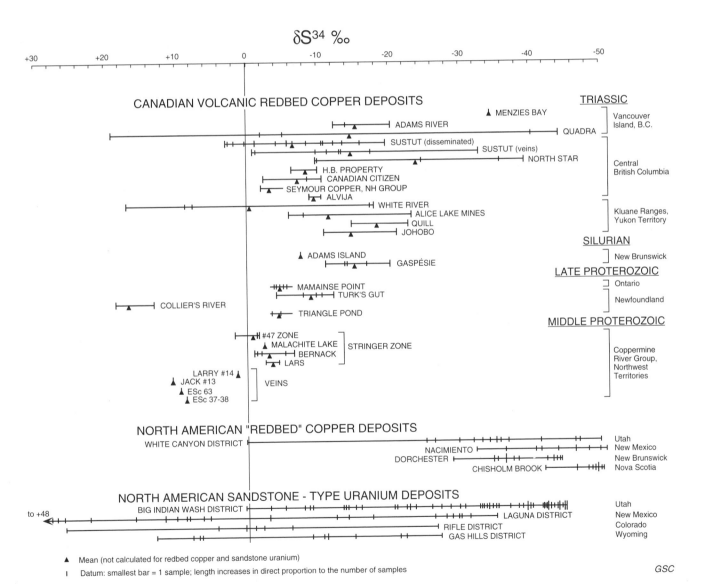

Figure 9-8. Diagram comparing sulphur-isotope compositions of volcanic redbed copper occurrences in Canada with other sediment-hosted redbed-type copper and sandstone-type uranium deposits. Most of the data for Canadian volcanic redbed copper occurrences are unpublished data of the author. Some examples for Sustut were kindly contributed by D.H.C. Wilton; and for White River by A.J. Sinclair. The sediment-hosted redbed copper and sandstone uranium data are from Jensen (1967).

negative [34]S values also suggest possible bacterial reduction of sulphate to produce sulphide in some deposits. The general patterns of sulphur isotope distribution support the concept that volcanic redbed-type copper deposits are related to redbed-type copper and sandstone uranium deposits, and had similar sources of sulphur and depositional mechanisms. Similar mineral and metal contents, zonal patterns of minerals, association with redbeds, low paleolatitudes of volcanic and sedimentary rocks, permeability and porosity controls, and, in some deposits, metal deposition at redox boundaries also support this conclusion. On the other hand, the characteristic location of this type of deposit at or near the metamorphic facies transitions from the prehnite-pumpellyite to epidote zones indicates that, in general, volcanic redbed copper deposits may have formed at somewhat higher temperatures and pressures than diagenetic sedimentary copper deposits.

RELATED DEPOSIT TYPES

As indicated above, a close relationship exists between volcanic redbed and diagenetic sedimentary copper deposits. In many localities the two deposit types occur together in the same rock sequence. A characteristic site for volcanic redbed copper deposits is in interlayered sedimentary units within favourable volcanic sequences (e.g., Fig. 9-7, for the White Pine district and for native copper deposits in the Portage Lake Volcanics, Michigan). In areas where volcanic sequences contain extensive interlayered sedimentary units or where copper-bearing beds occur at or near the base of overlying sedimentary formations, distinguishing between the two deposit types is difficult.

Other close relatives of the volcanic redbed copper deposit type are the unusual, carbonate-hosted, rich copper deposits at Kennecott, Alaska near the base of the Triassic Chitistone Limestone (Bateman and McLaughlin, 1920; Armstrong and Mackevett, 1975, 1982). These deposits formed about 30 m above the Nikolai Greenstone in small dissolution cavities in limestone, beneath a dolomite unit, and along fracture systems that extend down into the Nikolai Greenstone. Volcanic redbed-type copper occurrences are characteristic of the Nikolai Greenstone immediately below the Kennecott orebodies and elsewhere in the district.

The important synmetamorphic and syndeformational copper deposits at Mount Isa, Australia (Bell et al., 1988) represent another unusual type of copper deposit that might be related to volcanic redbed copper deposits. At Mount Isa, copper-bearing fluids contained in oxidized Eastern Creek Volcanics might have migrated along active fault systems and have been released into the anoxic Urquhart Shale during deformation to form the large, economic copper deposits at an unusual type of redox boundary for ore deposition (Swager et al., 1987; Hannan et al., 1993). However, more work is required to demonstrate whether ore-forming fluids derived from an oxidized volcanic pile were important at Mount Isa.

EXPLORATION GUIDES

Effective exploration for this deposit type requires a broad knowledge of the environment of ore formation, of the processes involved, and a knowledge of possible local sites favourable for the concentrations of metals. Copper-bearing fluids, probably basinal brines, extracted their metals from the enclosing oxidized volcanic or volcaniclastic rocks, followed permeable concordant aquifers and discordant structures up dip, and deposited their contained metals under a variety of conditions conducive to metal precipitation. Within this broad framework, undoubtedly much diversity existed. Therefore, each area and deposit should be evaluated on its own merits. Subaerial to shallow marine volcanic-sedimentary sequences deposited in low-latitude, arid and semiarid areas with favourable diagenetic and/or metamorphic depositional sites should be sought. More work is required to evaluate depositional mechanisms, such as redox gradients at interlayered anoxic sedimentary and hydrocarbon-bearing units within the volcanic pile.

The native copper deposits in the Keweenaw Peninsula show a strong control by through-going zones of original permeability. Broderick et al. (1946) suggested that a knowledge of "inlets" and "outlets" of fluid flow, in the area of investigation, can be very useful in exploration. Much of the native copper was concentrated in the main fluid flow channel, such that only one major orebody formed in a favourable, permeable lithological unit (Broderick et al., 1946).

White (1978) developed a perceptive and innovative model for the exploration of Keweenaw-type native copper deposits, based largely on a hydrological model for metamorphogenic hydrothermal ore fluids that moved updip through permeable aquifers. He suggested that the boundary between epidote and pumpellyite metamorphic zones is the most favourable target, particularly if it is characterized by numerous copper occurrences, which are considered to indicate the passage of copper-bearing fluids through rocks of the area. White (1968) observed that the average grade of deposits in the Keweenaw Peninsula seems to correlate with deposit size, and suggested that very large deposits, therefore, could be valid exploration targets.

Many of the volcanic redbed copper occurrences in Canada are in orogenic belts, and therefore have been subjected to postmineral faulting and penetrative deformation. Hence, a knowledge of fault offsets and fold geometries are essential for tracing disrupted mineral zones.

Typically, volcanic redbed copper occurrences are deceptively inconspicuous, principally because they lack associated pyrite and obvious alteration and gossans. Finely disseminated chalcocite in unaltered, dark maroon volcanic host rocks can be difficult to recognize unless copper stains occur on the outcrop. The Sustut deposit in British Columbia probably would not have been discovered had not copper stains on cliff faces been clearly visible from a helicopter (Harper, 1977).

SELECTED BIBLIOGRAPHY

References with asterisks (*) are considered to be the best source of general information on this deposit type.

Armstrong, A.K. and MacKevett, E.M., Jr.
1975: Relations between Triassic carbonate sabkhas and Kennecott-type copper deposits, Wrangell Mountains, Alaska; Economic Geology, v. 70, p. 1316-1317.
1982: Stratigraphy and diagenetic history of the lower part of the Triassic Chitistone Limestone, Alaska; United States Geological Survey, Professional Paper 1212-A, 26 p.

Bateman, A.M. and McLaughlin, D.H.
1920: Geology of the ore deposits of Kennecott, Alaska; Economic Geology, v. 15, p. 1-80.

Bell, T.H., Perkins, W.G., and Swager, C.P.
1988: Structural controls on development and localization of syntectonic copper mineralization at Mount Isa, Queensland; Economic Geology, v. 83, p. 69-85.

Bornhorst, T.J., Paces, J.B., Grant, N.H., Obradovich, J.D., and Huber, N.H.
1988: Age of native copper mineralization, Keweenaw Peninsula, Michigan; Economic Geology, v. 83, p. 619-625.

Broderick, T.M., Hohl, C.D., and Eidenmiller, H.N.
1946: Recent contributions to the geology of the Michigan copper district; Economic Geology, v. 41, p. 675-725.

Brummer, J.J. and Mann, E.L.
1961: Geology of the Seal Lake area, Labrador; Geological Society of America, Bulletin, v. 72, p. 1361-1382.

*Butler, B.S. and Burbank, W.S.
1929: The copper deposits of Michigan; United States Geological Survey, Professional Paper 144, 238 p.

*Camus, F.
1986: Los yacimientos estratoligados de Cu, Pb-Zn y Ag de Chile; in Geologia y Recursos Minerales de Chile, Tomo II, (ed.) J. Frutos, R. Oyarzun, and M. Pincheira; Universidad de Concepcion, Chile, p. 549-635.

Carrière, J.J. and Kirkham, R.V.
1993: Copper deposits and occurrences in Northwest Territories with sheet 1 (N.W.T.) and sheet 2 (NTS 86N and 86O); Geological Survey of Canada, Open File 2575, scale 1:500 000.

Carrière, J.J., Sinclair, W.D., and Kirkham, R.V.
1981: Copper deposits and occurrences in Yukon Territory; Geological Survey of Canada, Paper 81-12, 62 p.

Chavez, W.X., Jr.
1983: The geologic setting of disseminated copper sulfide mineralization of the Mantos Blancos copper-silver district, Antofagasta Province, Chile; Society of Mining Engineers of American Institute of Mining, Metallurgical and Petroleum Engineers, Annual Meeting, Atlanta, Georgia, preprint, 83-193, 20 p.

Church, B.N.
1975: Geology of the Sustut area; in Geology, Exploration and Mining in British Columbia 1974; British Columbia Department of Mines and Petroleum Resources, p. 305-309.

Cox, D.P.
1986: Descriptive model of basaltic Cu; in Mineral Deposit Models, (ed.) D.P. Cox and D.A. Singer; United States Geological Survey, Bulletin 1693, p. 130.

Gandhi, S.S. and Brown, A.C.
1975: Cupiferous shales of the Adeline Island Formation, Seal Lake Group, Labrador; Economic Geology, v. 70, p. 145-163.

Hak, J., Tupper, W.M., and Heslop, J.B.
1977: The mineralogy of the Coppercorp deposit, Mamainse Point, Ontario, Canada; Acta Univeristatis Carolinae-Geologica, Slavik volume, no. 3-4, p. 267-238.

Hannan, K.W., Golding, S.D., Herbert, H.K., and Krouse, H.R.
1993: Contrasting alteration assemblages in metabasites from Mount Isa, Queensland: implications for copper ore genesis; Economic Geology, v. 88, p. 1135-1175.

Harper, G.
1977: Geology of the Sustut copper deposit in B.C.; The Canadian Institute of Mining and Metallurgy, Bulletin, v. 70, no. 777, p. 97-104.

Heslop, J.B.
1970: Geology, mineralogy and textural relationships of the Coppercorp deposit, Mamainse Point area, Ontario; MSc. thesis, Carleton University, Ottawa, Ontario, 103 p.

Jefferson, C.W., Nelson, W.E., Kirkham, R.V., Reedman, I.H., and Scoates, R.F.J.
1985: Geology and copper occurrences of the Natkusiak basalts, Victoria Island, District of Franklin; in Current Research, Part A; Geological Survey of Canada, Paper 85-1A, p. 203-214.

Jensen, M.L.
1967: Sulphur isotopes and mineral genesis; in Geochemistry of Hydrothermal Ore Deposits (ed.) H.L. Barnes; Holt, Rinehart and Winston, New York, p. 143-165.

Jolly, W.T.
1974: Behaviour of Cu, Zn, and Ni during prehnite-pumpellyite rank metamorphism of the Keweenawan basalts, northern Michigan; Economic Geology, v. 69, p. 1118-1125.

Jolly, W.T. and Smith, R.E.
1972: Degradation and metamorphic differentiation of the Keweenawan tholeiitic lavas of northern Michigan, U.S.A.; Journal of Petrology, v. 13, p. 273-309.

Khadem, N.
1964: Types of copper ore deposits in Iran; in Symposium on Mining Geology and the Base Metals, Central Treaty Organization, Ankara, p. 101-115.

Kindle, E.D.
1972: Classification and description of copper deposits, Coppermine River area, District of Mackenzie; Geological Survey of Canada, Bulletin 214, 109 p.

Kirkham, R.V.
1969: NH; in Lode Metals; British Columbia Ministry of Mines and Petroleum Resources, Annual Report for the year ended December 31, 1968, p. 121-124.
1970: Certain copper deposits in Jurassic volcanic rocks of central British Columbia (93L, M, 94D, 103L); in Report of Activities, Part A: April to October 1969; Geological Survey of Canada, Paper 70-1, pt. A, p. 62-63.
1971: Geology of copper and molybdenum deposits; in Report of Activities, Part A, April to October, 1970; Geological Survey of Canada, Paper 71-1, pt. A, p. 85-88.
1982: Volcanic red bed copper deposits - environments of formation and distribution in accreted terranes of western North American (abstract); in Rocks and Ores of the Middle Ages, Programme and Abstracts; Cordilleran Section, The Geological Association of Canada, p. 14-16.
1984: Volcanic redbed copper; in Canadian Mineral Deposit Types: a Geological Synopsis, (ed.) O.R. Eckstrand; Geological Survey of Canada, Economic Geology Report 36, p. 37.
*1989: The distribution, settings and genesis of sediment-hosted, stratiform copper deposits; in Sediment-hosted, Stratiform Copper Deposits, (ed.) R.W. Boyle, A.C. Brown, C.W. Jefferson, E.C. Jowett, and R.V. Kirkham, Geological Association of Canada, Special Paper 36, p. 3-38.

Kirkham, R.V. and Roscoe, S.M.
1993: Atmospheric evolution and ore deposit formation; Resource Geology, Special Issue, no. 15, p. 1-17.

Klohn, E., Holmgren, C., and Ruge, H.
1990: El Soldado, a stratabound copper deposit associated with alkaline volcanism in the central Chilean Coastal Range; in Stratabound Ore Deposits in the Andes, (ed.) L. Fontboté, C.G. Amstutz, M. Cardozo, E. Cedillo, and J. Frutos; Springer-Verlag, Berlin, p. 435-448.

Leblanc, M. and Billaud, P.
1978: A volcano-sedimentary copper deposit on a continental margin of upper Proterozoic age: Bleïda (Anti-Atlas, Morocco); Economic Geology, v. 73, p. 1101-1111.
1990: Zoned and recurrent deposition of Na-Mg-Fe-Si exhalites and Cu-Fe sulfides along synsedimentary faults (Bleïda, Morocco); Economic Geology, v. 85, p. 1759-1769.

Lincoln, T.N.
1981: The redistribution of copper during low-grade metamorphism of the Karmutsen volcanics, Vancouver Island, British Columbia; Economic Geology, v. 76, p. 2147-2161.

Lortie, R.B. and Clark, A.H.
1974: Stratabound fumarolic copper deposits in rhyolitic lavas and ash-flow tuffs, Copiapó District, Atacama, Chile; in Problems of Ore Deposition, Fourth International Association on the Genesis of Ore Deposits Symposium, Varna, v. 1, p. 256-264.
1987: Strata-bound cupriferous sulphide mineralization associated with continental rhyolitic volcanic rocks, northern Chile: I. The Jardin copper-silver deposit; Economic Geology, v. 82, p. 546-570.

Losert, J.
1973: Genesis of copper mineralizations and associated alterations in the Jurassic volcanic rocks of the Buena Esperanza mining area (Antofagasta Province, northern Chile); Universidad de Chile, Facultad de Ciencias Fisicas y Matematicas, departmento de Geologia, Publicacion 40, 64 p.

Maloney, R.P. and Bottge, R.G.
1973: Estimated costs to produce copper at Kennicott, Alaska; United States Bureau of Mines, Information Circular 8602, 35 p.

Mayer, C.K. and Fontboté, L.
1990: The stratiform Ag-Cu deposit El Jardin, northern Chile; in Stratabound Ore Deposits in the Andes, (ed.) L. Fontboté, G.C. Amstutz, M. Cardozo, E. Cedillo, and J. Frutos; Springer-Verlag, Berlin, p. 637-646.

Muñoz, J.O.
1975: On stratiform copper deposits of Chile; Annales de la Société Geologique de Belgique, v. 98, p. 17-21.

Palacios, C.M., Hein, U.F., and Dulski, P.
1986: Behaviour of rare earth elements during hydrothermal alteration at the Buena Esperanza copper-silver deposit, northern Chile; Earth and Planetary Science Letters, v. 80, p. 208-216.

Pearson, W.N., Bretzlaff, R.E., and Carrière, J.J.
1985: Copper deposits and occurrences in the north shore region of Lake Huron, Ontario; Geological Survey of Canada, Paper 83-28, 34 p.

Robertson, J.M.
1975: Geology and mineralogy of some copper sulfide deposits near Mount Bohemia, Keweenaw County, Michigan; Economic Geology, v. 70, p. 1202-1224.

*Ruiz, C., Aguilar, A., Egert, E., Espinosa, W., Peebles, F., Quezada, R., and Serrano, M.
1971: Strata-bound copper sulphide deposits of Chile; The Society of Mining Geologists of Japan, Special Issue 3, p. 252-260.

*Sato, T.
1984: Manto type copper deposits in Chile - a review; Bulletin of the Geological Survey of Japan, v. 35, p. 565-582.

Sinclair, A.J., Bentzen, A., and McLeod, J.A.
1979: Geology of the White River Native Copper Deposit, Yukon Territory; Indian and Northern Affairs, Canada, QS-Y001-000-EE-A1, 27 p.

*Stoiber, R.E. and Davidson, E.S.
1959: Amygdule mineral zoning in the Portage Lake Lava Series, Michigan copper district; Economic Geology, v. 54, p. 1250-1277, 1444-1460.

Surdam, R.C.
1968: Origin of native copper and hematite in the Karmutsen Group, Vancouver Island, B.C.; Economic Geology, v. 63, p. 961-966.

Sutherland Brown, A.
1968: Northstar; in British Columbia Minister of Mines and Petroleum Resources Report, 1967, p. 86-88.

Sutherland Brown, A., Cathro, R.J., Panteleyev, A., and Ney, C.S.
1971: Metallogeny of the Canadian Metallurgy, Cordillera; in Canadian Institute of Mining and Metallurgy Transactions, v. 74, p. 121-145.

Swager, C.P., Perkins, W.G., and Knights, J.G.
1987: Stratabound phyllosilicate zones associated with syntectonic copper orebodies at Mt. Isa, Queensland; Australian Journal of Earth Sciences, v. 34, p. 463-476.

Weege, R.J. and Pollack, J.P.
1971: Recent developments in the native-copper district of Michigan; in Guidebook for Field Conference, Michigan Copper District, September 30-October 2, 1971, Society of Economic Geologists, p. 18-43.
1972: The geology of two new mines in the native copper district of Michigan; Economic Geology, v. 67, p. 622-633.

*White, W.S.
1968: The native-copper deposits of northern Michigan; in Ore Deposits of the United States, 1933-1967; The Graton-Sales Volume, (ed.) J.D. Ridge; American Institute of Mining, Metallurgical and Petroleum Engineers, Inc., New York, p. 303-325.
1978: A theoretical basis for exploration for native copper in northern Wisconsin; United States Geological Survey, Circular 769, 19 p.

Wilton, D.H.C. and Sinclair, A.J.
1988: Ore petrology and genesis of a strata-bound disseminated copper deposit at Sustut, British Columbia; Economic Geology, v. 83, p. 30-45.

Author address

R.V. Kirkham
Geological Survey of Canada
100 West Pender Street
Vancouver, B.C.
V6B 1R8

Printed in Canada

10. MISSISSIPPI VALLEY-TYPE LEAD-ZINC

MISSISSIPPI VALLEY-TYPE LEAD-ZINC

D.F. Sangster

INTRODUCTION

Mississippi Valley-type (MVT) lead-zinc deposits typically are composed of galena and sphalerite occurring as open-space fillings in carbonate breccias; replacement of host rocks is relatively minor except in high grade zones. The deposits owe their commonly accepted name to the fact that several classical districts occur within the drainage basin of the Mississippi River, central U.S.A. Important Canadian examples include Pine Point, Polaris, Nanisivik, Newfoundland Zinc, and Monarch-Kicking Horse deposits; of lesser importance are the deposits at Gays River, Nova Scotia and Robb Lake, British Columbia (Fig. 10-1).

IMPORTANCE

Prior to the opening of the Pine Point mine in 1965, Mississippi Valley-type deposits contributed only negligibly to Canadian lead-zinc production. Since that time, and until recently, about 30% of annual production has been derived from deposits of this type. With the closures of Pine Point and Newfoundland Zinc mines, however, the proportion of zinc and, to a lesser extent, lead derived from MVT deposits has decreased dramatically.

SIZE AND GRADE OF DEPOSITS

Data for individual deposits are difficult to obtain, mainly because of company reporting policies or because the deposits themselves tend to be interconnected. Most deposits, however, are small and fall between 1 and 10 Mt (Table 10-1). Grades generally range between 5 and 10% combined lead-zinc with a majority of deposits being decidedly zinc-rich relative to lead. The tendency for MVT deposits to occur in clusters, up to as many as 400 in the Upper Mississippi Valley district, for example, renders them amenable to exploitation through establishment of common infrastructures. Some of these districts can be as large, in terms of contained lead and zinc, as certain of the world's giant stratiform, sediment-hosted (Sedex) deposits.

GEOLOGICAL FEATURES

The geological characteristics of these deposits have been summarized many times, most recently by Anderson and Macqueen (1988). A review by Sverjensky (1986) concisely lists the more important attributes of MVT deposits and Leach and Sangster (1993) present a comprehensive review of MVT deposits, their characteristics, and genetic models.

Geological setting

The deposits are hosted in carbonate rocks, usually platformal in origin, peripheral to cratonic sedimentary basins and characterized by tectonically stable depositional conditions (Anderson and Macqueen, 1988; Leach and Sangster, 1993). Only a few have been significantly affected by postore deformational events. In Canada, for example, the Polaris deposit has been tilted roughly 20° eastward by postore folding. Most deposits/districts are found below unconformities or nonconformities, presumably produced by minor uplift or warping of the host strata.

Sangster, D.F.
1996: Mississippi Valley-type lead-zinc; in Geology of Canadian Mineral Deposit Types, (ed.) O.R. Eckstrand, W.D. Sinclair, and R.I. Thorpe; Geological Survey of Canada, Geology of Canada, no. 8, p. 253-261 (also Geological Society of America, The Geology of North America, v. P-1).

Age of host rocks and mineralization

Host rocks to Canadian deposits range from Upper Proterozoic (Nanisivik, Gayna River) to Lower Carboniferous (Gays River) but most are in lower to mid-Paleozoic strata. Pine Point, the largest Canadian Mississippi Valley-type district, is found in rocks of Middle Devonian age.

Age of mineralization in MVT deposits in general, and in Canadian ones in particular, is known for only a few districts. The ore-forming process itself does not consistently yield material, either within the deposit or by alteration of adjacent host rocks, amenable to dating by traditional radiometric methods (<u>see</u> discussion <u>in</u> Sangster, 1986). Furthermore, the characteristic relatively stable tectonic regime of MVT districts generally precludes bracketing of mineralization age by field observations of postore events. Some success has recently been achieved through Rb-Sr dating of carefully selected sphalerite (e.g. Brannon et al., 1992) or through Ar-Ar methods on alteration minerals (e.g. Kontak et al., 1994). The most widespread technique, however, has been paleomagnetic dating of ore and host rocks, resulting in successful dating of several North American MVT deposits and districts (Pan et al., 1990, 1993; Symons and Sangster, 1991, 1992; Sangster and Symons, 1991; Symons et al., 1993). Results from all these recent methods have yielded mineralization dates which coincide with periods of orogenic uplift in the regions adjacent to the respective MVT deposits/districts. This correspondence in ages has been interpreted as support for the "gravity-driven fluid flow" model for MVT genesis (<u>see</u> discussion under "Genetic model" below).

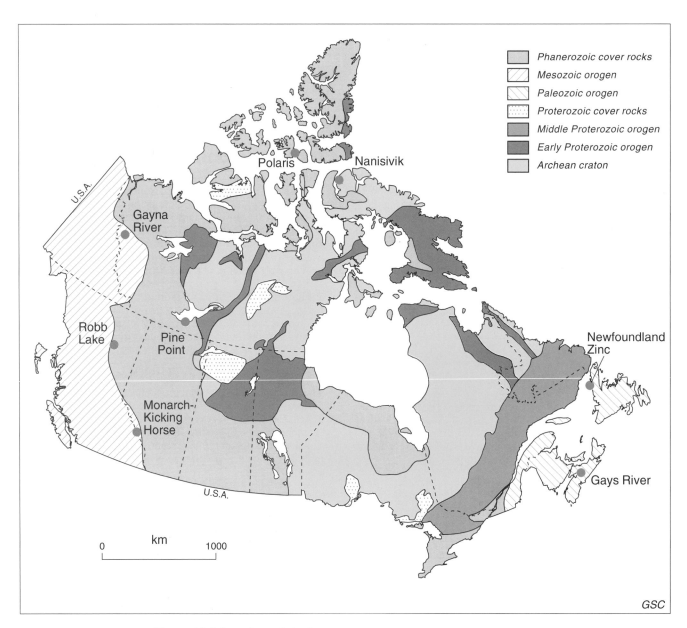

Figure 10-1. Locations of significant Canadian MVT deposits and districts.

Table 10-1. Grade-tonnage data for selected MVT deposits/districts.

DEPOSIT/DISTRICT	Mt	%Zn	%Pb	COMMENTS	REFERENCE
Canada					
Newfoundland Zinc deposit	7.2	8.0	-	Production 1975-1990	Lane,1990
Polaris deposit	22	14.0	4.0	Reserves + production, 1994	Randell, 1994
Nanisivik deposit	>10	10.0	1.0	Production 1976-1993	McNeil et al., 1993
Gays River deposit	2.4	8.6	6.3	Reserves 1992	Kontak, 1993
Pine Point district	76.1	6.5	2.9	Reserves + production, 1983	Rhodes et al., 1984
Robb Lake deposit	5.53	5.0	2.3	Reserves, 1975	Macqueen and Thompson, 1978
U.S.A.					
Tri-State district	>500	2.4	0.6	Production 1907-1964	Brockie et al., 1968
East Tenn. district	250	3.0	-	Production 1912-1986	Briskey et al.,1986
Old Lead Belt district	206.8	-	2.8	Production 1915-1972	Hagni,1989
Viburnum district	111.6	0.8	5.8	Production 1960-1984	Hagni,1989
Morocco-Algeria					
Bou Beker-El Abed district	62.1	3.8	3.1	Reserves + production, 1993	Bouabdellah,1993
- = not reported					

Relations of ore to host rock

Deposits are discordant on a deposit scale but stratabound on a district scale. Ore-hosting structures are most commonly zones of highly brecciated dolomite; in some instances (e.g. Pine Point, Newfoundland Zinc) these zones are arranged in linear patterns suggesting a structural control, although faults of large displacement are never present. The breccia zones may range from more-or-less concordant, occurring in one or two beds, to highly discordant cylindrical structures which penetrate tens of metres of sedimentary succession (Fig. 10-2, 10-3).

Although the ore-hosting breccia bodies are commonly interpreted to be collapse breccias brought about by dissolution of underlying beds, there is little consensus on whether the dissolution and brecciation is due to meteoric waters (meteoric karst) or the ore-bearing fluids themselves (hydrothermal karst). A large majority of MVT deposits are overlain by a disconformity or unconformity and this has been cited in support of the meteoric hypothesis (e.g. Kyle, 1981). However, as pointed out by Sangster (1988, p. 114) "little or no direct evidence of a preore meteoric component....has yet been recognized" in MVT deposits. The hydrothermal karst origin for the collapse breccias has been most strongly supported for the Polish Silesian district (e.g. Sass-Gustkiewicz et al., 1982), the Upper Mississippi Valley district (Heyl et al., 1959), and the Pine Point district (Qing, 1991).

Distribution of ore minerals

Characteristically, sulphide minerals in MVT deposits are found as open-space fillings between angular dolomite fragments in the ore-hosting breccias. In some instances, sulphides fill primary porosity cavities such as fossil molds or voids between the carbonate grains. Replacement of carbonate host rocks is generally minor except in those deposits, or portions of deposits, containing high-grade mineralization.

Ore composition and zoning

A majority of MVT deposits are of simple composition and consist of lead, zinc, and iron sulphides. Copper is not normally a constituent of MVT ores and is important only in the Southeast Missouri district, U.S.A. Cadmium, germanium, barite, and fluorite are, or have been, recovered from some districts. In the Southeast Missouri district and, to a lesser degree, the Upper Mississippi Valley district, Cu, Ni, and Co are diagnostic accessory elements but are atypical of the deposit type as a whole. Silver contents are generally very low; in most deposits silver is not reported in reserve statements. In those few deposits where it is reported, silver grades generally average in the 30-40 g/t range. Nanisivik contains the highest silver grade in Canadian MVT ores with an average around 60 g/t.

With the exception of the Southeast Missouri (U.S.A.) and Touissit-Beddiane (Morocco) districts, all major MVT districts are zinc-rich relative to lead and possess Zn/(Zn+Pb) ratios greater than 0.5. In the Southeast Missouri district, metal ratios are less than 0.1. Some ores, such as those of East Tennessee and Newfoundland Zinc, are essentially lead-free and have Zn/(Zn+Pb) ratios approaching 1.0.

Consistent, well developed metal or mineralogical zoning is not a characteristic feature of MVT deposits. The simplicity of the ores all but precludes systematic zoning in most districts. Again, however, the unusually mineralogically complex Southeast Missouri district is an exception and exhibits its own unique zoning patterns of Pb, Zn, Cu, Ni, and Co.

Of the other world-class MVT districts, metal zoning has been described only for the Upper Mississippi Valley and Pine Point districts. In the former, Heyl et al. (1959, p. 143) have defined the district boundaries in terms of the distribution of lead-bearing deposits. Within the mainly east-trending lead district, zinc and copper occur mainly in a north-south belt through the east-central portion. An elongate north-south barite zone coincides with the western edge of the copper zone. A nickel zone, within which all lead-zinc orebodies contain millerite, is centred within the northern quarter of the barite zone.

Compositional zoning in individual deposits at Pine Point has been described by Kyle (1981). Prismatic, discordant orebodies in this district feature an outward increase in the Fe/(Fe+Zn+Pb) ratio and a decrease in the Pb/(Pb+Zn) ratio; tabular, concordant orebodies in the same district are not zoned.

Alteration

Dissolution, recrystallization, and brecciation of the host carbonate rocks within, and peripheral to, mineralization is common to virtually all MVT deposits. These effects, which occur in conjunction with silicification and dolomitization, constitute the major form of wall rock alteration in MVT districts.

Silicification is most characteristic of the Tri-State district in south-central U.S.A. (Brockie et al., 1968). The typical pattern there is a central dolomitic core surrounded progressively outward by the main ore zone, then by the silicified zone (consisting of microcrystalline quartz referred to as "jasperoid"), and finally by unaltered limestone.

Dolomitization, of pre-, syn-, and postore age, is a widespread and characteristic feature of MVT districts. In spite of this, however, its significance and genetic history relative to burial diagenesis and the mineralizing event(s) is not well understood. In some districts it displays a complex relationship to host rocks and ore paragenesis. At Pine Point, for example, Krebs and Macqueen (1984) reported eight major paragenetic stages within which are at least seven periods of dolomite or calcite deposition.

Alteration of rocks other than carbonates has only been reported in two districts. In the Upper Mississippi Valley district, Heyl et al. (1964) showed that minor changes occur in illite crystal structure outward from the ore zones. In basement granitic fragments occurring within the Southeast Missouri district, Stormo and Sverjensky (1983) observed alteration of K-feldspar and albite to kaolinite, quartz, and minor sericite.

Mineralogy

Simple mineralogy has long been identified as one of the main characteristics of MVT deposits but, as Ohle (1959, p. 775) pointed out, simplicity is apparent only on a district scale. When several MVT districts are considered together, an impressive list of more than three dozen minerals can be compiled. When, however, the mineralogically unique and complex Southeast Missouri district is omitted, together with those minerals reported in only one or two districts, the most common minerals are: pyrite, marcasite, sphalerite, and galena, although relative mineral abundances vary markedly between deposits and/or districts. Barite and/or fluorite are accessory minerals in

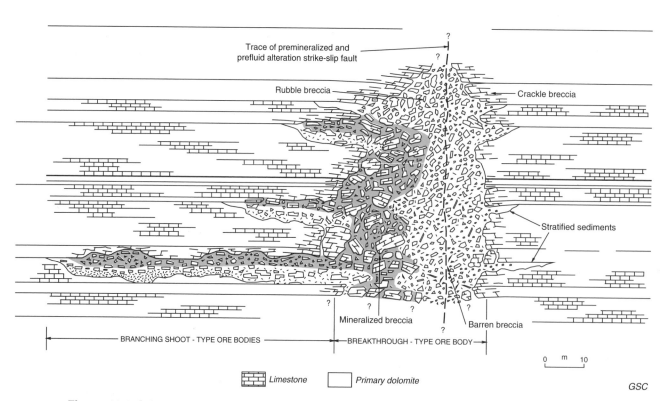

Figure 10-2. Schematic representation of a columnar ore-bearing breccia body in the East Tennessee MVT district (modified from Ohle, 1985).

some lead-zinc districts (e.g. Central Tennessee) but, in a majority of districts, these minerals are entirely lacking. Lead and zinc are the most common elements that determine economic viability. In some districts, silver is an important byproduct but, in most deposits, it is a negligible factor. Cadmium, germanium, barite, and fluorite are, or have been, recovered from some deposits.

Organic material

Organic material, usually in the form of mature, tarry hydrocarbon ("pyrobitumen") is common in some, but not all, MVT districts and occurs as a paragenetically late phase (e.g., Macqueen and Powell, 1983). Methane and/or liquid petroleum have been identified in fluid inclusions in sphalerite in a few instances.

Ore textures

The most diagnostic textures of MVT deposits are those produced by open-space filling. Replacement is relatively minor except in high grade zones. The open spaces may be primary (e.g. fossil molds) but are usually secondary. Secondary porosity is dominantly in the form of collapse (solution) breccias; sulphides and carbonates constitute the cement between breccia fragments.

Ore textures are, to a considerable degree, controlled by ore-bearing breccias. The latter vary considerably, from one MVT district to the other, in the nature and abundance of their internal features. Three main breccia types or textures have been recognized: crackle breccia, mineralized breccia (sometimes referred to as "ore-matrix breccia"), and barren breccia ("rock-matrix breccia") (Sangster, 1988). As suggested by the name, sulphides are most common and abundant in the mineralized breccia which is characterized by dolostone fragments cemented by white or pink crystalline sparry dolomite and sulphides (Fig. 10-4). Dolostone fragments are typically rotated and can usually be demonstrated to have dropped from their original stratigraphic position.

Both ore and gangue minerals are commonly, but not invariably, coarse grained. The Tri-State district, U.S.A., for example, is world-renowned for its exceptionally large, euhedral sphalerite and galena crystals yet the ore-related chert and jasperoid consists of microcrystalline quartz. Some districts, in contrast, are characterized by their extremely fine grained sulphides as exemplified by the banded, colloform sphalerite containing skeletal galena crystals as at Pine Point and Polaris in Canada, the Silesia district in Poland, and the Cadjebut deposit in Australia.

Three sulphide textures in particular are characteristic of MVT deposits although it is uncommon for all three to occur in a single deposit: trash zones, internal sediments, and "snow-on-roof". During dissolution of the host carbonate, resulting in formation of the ore-bearing collapse breccias, much insoluble material accumulated at the bottom of the breccia bodies. These trash zone accumulations can range up to a few metres in thickness, consist of poorly sorted grains of a variety of minerals, most of which are dolomite clay, chert, quartz, sulphides, and black, carbonaceous material (Hill et al., 1971). Internal sediments, well stratified material which partly or entirely fills the space between breccia fragments (Fig. 10-5), consist mainly of

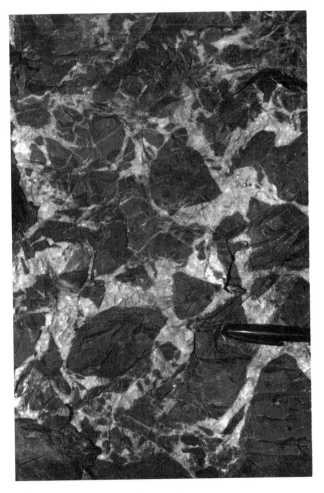

Figure 10-4. Ore-matrix breccia, Jefferson City mine, East Tennessee district. White material is sparry dolomite cement. Length of pen visible is 6 cm. GSC 202229-D

Figure 10-3. Prismatic ore breccia, O-28 deposit, Pine Point district. Note beds on left bend downward as they approach the breccia body (outlined by short dashes). GSC 203510-M

Figure 10-5. Internal sediments in K-57 orebody, Pine Point district. Sediment consists of detrital dolomite sand grains. Knife is 8 cm. GSC 203510-F

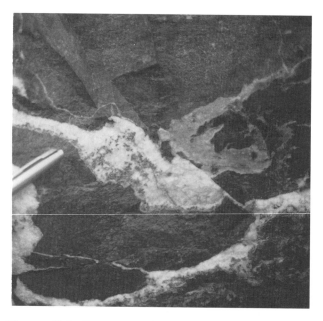

Figure 10-6. Snow-on-roof texture in ore-matrix breccia, K zone, Newfoundland Zinc mine. Sphalerite is the grey material within the white sparry dolomite cement and immediately on top of the dark fragment under the pen. Note the sphalerite is absent underneath the same fragment but is present again on top of the next fragment, lower right corner. Pen is 6 cm. GSC 203510-H

dolomite grains but commonly contain a large component of detrital sand-size sulphide grains and composite sulphide-carbonate fragments (Kendall, 1960; Sass-Gustkiewicz et al., 1982; Rhodes et al., 1984; Randell and Anderson, 1990; Bouabdellah, 1993). Graded bedding is not uncommon in internal sediments. First described by Oder and Hook (1950), "snow-on-roof" texture is, in the author's opinion, the most diagnostic texture of MVT deposits because it has not been reported in any other type of mineral deposit. Typically, sphalerite occurs as coarse crystals preferentially coating the tops of dolomite breccia fragments in ore-matrix breccias (Fig. 10-6). Undersides of the same fragments are coated either not at all or to a much lesser degree. A variety of snow-on-roof texture is sphalerite lining the bottoms of cavities (Fig. 10-7). In either case, the feature can be regarded as a geopetal indicator.

DEFINITIVE CHARACTERISTICS

1. Deposits occur principally in limestone or dolomite in platform carbonate sequences;
2. Ore is strongly controlled by individual strata (i.e. stratabound);
3. The most common minerals are pyrite, sphalerite, galena, dolomite, calcite, and quartz;
4. Deposits are not associated with igneous rocks;
5. They may occur in areas of mild deformation expressed as brittle fracture, uplift of broad domes, and subsidence of related basins;
6. There is always evidence of dissolution of carbonate host rock expressed by slumping, collapse, brecciation, or some combination of these;
7. Fluid inclusions in the constituent minerals always contain dense, saline, aqueous fluids. Total dissolved salts range from 10 to 30 wt.% and are predominantly sodium and calcium chlorides.

Figure 10-7. Coarse grained sphalerite (dark) lying on bottom of cavity; remainder of void filled with carbonate. Elmwood mine, Central Tennessee district. Pen is 16 cm. GSC 203510-E

GENETIC MODEL

The deposits are obviously epigenetic, having been emplaced after host rock lithification. The ore-hosting breccias are considered to have resulted from dissolution of more soluble sedimentary layers with subsequent collapse of overlying beds (see discussion in Sangster, 1988). The major mineralizing processes appear to have been open-space filling between breccia fragments and replacement of fragments or wall rock, although the relative importance of these two processes varies widely among deposits.

Based on fluid inclusion data, MVT deposits clearly formed from hot, saline, aqueous solutions which, in many respects, are similar to modern oil-field brines (e.g. Sverjensky, 1984). Thus, some form of basinal brine migration out of sedimentary basins, through aquifers, to the basin periphery is the most widely accepted general mode of formation of MVT ores (Sverjensky, 1986).

To move the ore solutions out of the basin clastics and into the platform carbonates, two main models have been proposed:

1. Basin compaction-driven fluid flow

In this proposal, two subtypes may be considered: (i) continuous outward flow which assumes that simple compaction of sediments in a subsiding basin will drive pore fluids laterally along aquifers; and (ii) episodic outward flow. The latter subtype was proposed to overcome some of the objections raised in the concept of continuous flow, notably the problem of maintaining high initial fluid temperatures during transport of up to hundreds of kilometres from basin source to platform depositional site. To address this difficulty, a process involving overpressuring of subsurface aquifers by rapid sedimentation, followed by rapid and episodic release of basinal fluids, has been proposed (Sharp, 1978; Cathles and Smith, 1983).

2. Gravity-driven fluid flow

The gravity-driven fluid flow model advocates flushing of subsurface brines out of a sedimentary basin by groundwater flow from recharge areas, in elevated regions of a foreland basin, to discharge areas in regions of lower elevation (Garven, 1985; Bethke, 1986; Bethke and Marshak, 1990).

The pros and cons of each of these mechanisms have been debated in the scientific literature (e.g. Sverjensky, 1986) and, as pointed out by Bethke and Marshak (1990), considerable evidence is being accumulated in support of the gravity-driven fluid concept, particularly with respect to the MVT districts of mid-continent U.S.A. "Increasingly it is clear that the migrating brines originated in the forelands of North American tectonic belts, and that the migrations coincided in time with intervals during which the belts were deformed" (Bethke and Marshak, 1990, p. 287). Evidence in support of this concept includes: i) a steady decrease in MVT ore-stage fluid inclusion homogenization temperatures away from the uplifted orogenic belt and toward the continental interior; ii) the development of alteration potassium feldspar and clay minerals in the host sedimentary rocks, interpreted as due to K-metasomatism associated with brine movement;

iii) epigenetic dolomite cement with chemical and isotopic compositions indicating flow directions away from the uplifted orogen; iv) correlation of cathodoluminescence banding in dolomite cements over very large areas, indicating widespread brine migration; v) paleomagnetic ages in MVT ores which coincide with periods of orogenic uplift; and vi) radiometric ages of authigenic K-feldspar in platform carbonate rocks, also coinciding with orogenic uplift. Results of these studies are still being evaluated relative to other evidence from MVT deposits but seem to indicate that MVT deposits may have formed in response to brine migration through carbonate successions in the forelands of tectonic belts.

In addition to the problem of transport of fluid from the basin to carbonate platforms, three geochemical models have been proposed to account for chemical transport and deposition of the ore constituents:

i) mixing model – base metals are transported by fluids of low sulphur content. Precipitation is effected by a) mixing with fluids containing hydrogen sulphide, or b) replacement of diagenetic iron sulphide, or c) thermal degradation of organic compounds and resultant release of sulphur;

ii) sulphate reduction model – base metals are transported together with sulphate in the same solution. Precipitation is the result of reduction of sulphate by reaction with organic matter or methane;

iii) reduced sulphur model – base metals are transported together with reduced sulphur in the same solution. Precipitation is brought about by any or all of: a) change of pH, or b) dilution, or c) cooling.

In attempting to explain the genesis of MVT deposits, it has been traditional to emphasize the common features of several MVT districts. In recent years, however, several writers have begun to question this approach and to suggest, instead, that a re-examination of the differences might be a more fruitful line of approach. Thus, for example, Ohle (1980, p. 163) reflected on the fact that the "individualities raise the question as to whether each ore had a different source, a different plumbing system, and a different timing from all the others". Similarly, Sangster (1983, p. 7) concluded that the differences between MVT districts outweighed the similarities, both numerically and in significance, and that, therefore, a common genetic model was perhaps precluded. In a review of MVT deposit models, Sverjensky (1986) concluded that significant differences exist between MVT districts and that, as a consequence, all districts may not have formed in exactly the same way. Evidence from Pine Point, for example, suggests a two-fluid mixing model whereas Southeast Missouri ores appear to have been precipitated from a single fluid containing base metals and reduced sulphur.

RELATED DEPOSIT TYPES

Fracture-controlled, F-dominant deposits (with subordinate Ba, Pb, and Zn) such as those of Illinois-Kentucky and the English Pennines, the central Missouri Ba(Pb) district, and the Tennessee Sweetwater F-Ba(Pb-Zn) district may represent MVT F-Ba deposits rather than MVT Pb-Zn deposits.

Inasmuch as the weight of current evidence favours the formation of MVT Pb-Zn ores from sedimentary basin waters, these deposits have been genetically linked to sedimentary exhalative deposits (Sedex) Pb-Zn which are also considered to have formed from basinal fluids (Sangster, 1990, 1993).

EXPLORATION GUIDES

Because of the significant differences between districts noted above, universal exploration guides for MVT deposits are virtually precluded. Only broad guidelines such as platformal carbonates adjacent to large sedimentary basins, the presence of unconformities in the carbonate platform, and the abundance of dolomite in and around MVT districts can be regarded as common to all districts. The presence of basement highs underneath the carbonate cover, if detectable by geophysical means, might be useful in some instances. Extension of exploration outward from known districts should focus on the ore-containing strata of the known district. Beyond this rather self-evident truism, however, there is little to guide the explorationist. This is borne out by the "random walk" method used to site exploration holes used in the discovery of the Central Tennessee district situated in similar strata to, but 200 km distant from, the well-known East Tennessee district (Callahan, 1977).

SELECTED BIBLIOGRAPHY

References with asterisks (*) are considered to be the best source of general information on this deposit type.

Anderson, G.M. and Macqueen, R.W.
1988:   Ore deposit models - 6. Mississippi Valley-type lead-zinc deposits: in Ore Deposit Models, (ed.) R.G. Roberts and P.A. Sheahan; Geoscience Canada, Reprint Series 3, p. 79-90.

Bethke, C.M.
1986:   Hydrologic constraints on the genesis of the Upper Mississippi Valley mineral district from Illinois Basin brines; Economic Geology, v. 81, p. 233-249.

*Bethke, C.M. and Marshak, S.
1990:   Brine migrations across North America - the plate tectonics of groundwater; Annual Review and Earth Planetary Science, v. 18, p. 287-315.

Bouabdellah, M.
1993:   Métallogenèse d'un district de type Mississippi Valley, cas de Beddiane, District de Toussit-Bou Becker, Maroc; PhD. thesis, École Polytechnique, Montreal, Canada, 338 p.

Brannon, J.C., Podosek, F.A., and McLimans, R.K.
1992:   Alleghanian age of the Upper Mississippi Valley zinc-lead depposit determined by Rb-Sr dating of sphalerite; Nature, v. 356, p. 509-511.

Briskey, J.A., Dingess, P.R., Smith, F., Gilbert, R.C., Armstrong, A.K., and Cole, G.P.
1986:   Localization and source of Mississippi Valley-type zinc deposits in Tennessee, U.S.A., and comparisons with Lower Carboniferous rocks of Ireland; p. 635-661, in Geology and Genesis of Mineral Deposits in Ireland, (ed.) C.J. Andrew, R.W.A. Crowe, S. Findlay, W.M. Pennel, and J.F. Pyne; Irish Association for Economic Geology, Cahill Printers Ltd., Dublin, 699 p.

Brockie, D.C., Hare, E.H., Jr., and Dingess, P.R.
1968:   The geology and ore deposits of the Tri-State District of Missouri, Kansas, and Oklahoma; Ore Deposits of the United States, (ed.) J.D. Ridge; American Institute of Mining, Metallurgical, and Petroleum Engineers, Inc., New York, v. l, p. 400-430.

Callahan, W.H.
1977:   The history of the discovery of the zinc deposit at Elmwood, Tennessee: concept and consequence; Economic Geology, v. 72, p. 1382-1392.

Cathles, L.M. and Smith, A.T.
1983:   Thermal constraints on the formation of Mississippi Valley-type lead-zinc deposits and their implications for episodic basin dewatering and deposit genesis; Economic Geology, v. 78, p. 983-1002.

*Garven, G.
1985:   The role of regional fluid flow in the genesis of the Pine Point deposit, Western Canada sedimentary basin; Economic Geology, v. 80, p. 307-204.

Garven, G. and Freeze, R.A.
1984a: Theoretical analysis of the role of groundwater flow in the genesis of stratabound ore deposits: 1. Mathematical and numerical model; American Journal of Science, v. 284, p. 1085-1124.
1984b: Theoretical analysis of the role of groundwater flow in the genesis of stratabound ore deposits: 2. Quantitative results; American Journal of Science, v. 284, p. 1125-1174.

Hagni, R.D.
1989:   The southeast Missouri lead district: a review; in Mississippi Valley-type Mineralization of the Viburnum Trend, Missouri, (ed.) R.D. Hagni and R.M. Coveney, Jr.; Guidebook Series Volume 5, Society of Economic Geologists, p. 12-57.

Heyl, A.V., Agnew, A.F., Lyons, E.J., and Behre, C.H., Jr.
1959:   Geology of the Upper Mississippi Valley zinc-lead district; United States Geological Survey, Professional Paper 309, 310 p.

Heyl, A.V., Hosterman, J.W., and Brock, M.R.
1964:   Clay mineral alteration in the Upper Mississippi Valley zinc-lead district; Twelfth National Conference on Clays and Clay Minerals, 1963, Proceedings, New York, MacMillan Co., p. 445-453.

Hill, W.T., Morris, R.G., and Hagegeorge, C.G.
1971:   Ore controls and related sedimentary features at the Flat Gap mine, Treadway, Tennessee; Economic Geology, v. 66, p. 748-756.

Jackson, S.A. and Beales, F.W.
1967:   An aspect of sedimentary basin evolution: the concentration of Mississippi Valley-type ores during late stages of diagenesis; Canadian Petroleum Geology Bulletin, v. 15, p. 383-433.

Kendall, D.L.
1960:   Ore deposits and sedimentary features, Jefferson City mine, Tennessee; Economic Geology, v. 55, p. 985-1003.

Kontak, D.J.
1993:   A preliminary report on geological, geochemical, fluid inclusion, and isotopic studies of the Gays River Zn-Pb deposit, Nova Scotia; Nova Scotia Department of Natural Resources, Open File Report 92-014, 223 p.

Kontak, D.J., Farrar, E., and McBride, S.L.
1994: $^{40}Ar/^{39}Ar$ dating of fluid migration in a Mississippi Valley-type deposit: the Gays River Zn-Pb deposit, Nova Scotia, Canada; Economic Geology, v. 89, p. 1501-1517.

Krebs, W. and Macqueen, R.
1984:   Sequence of diagenetic and mineralization events, Pine Point lead-zinc property, Northwest Territories, Canada; Canadian Petroleum Society Bulletin, v. 32, p. 434-464.

Kyle, J.R.
1981:   Geology of the Pine Point lead-zinc district; Handbook of strata-bound and stratiform ore deposits, (ed.) K.H. Wolf; Elsevier Publishing Company, New York, v. 9, p. 643-741.

Lane, T.E.
1990:   Dolomitization, brecciation, and zinc mineralization and their paragenetic, stratigraphic, and structural relationships in the Upper St. George Group (Ordovician) at Daniel's Harbour, western Newfoundland; PhD. thesis, Memorial University of Newfoundland, St. John's, Newfoundland, 565 p.

*Leach, D.L. and Sangster, D.F.
1993:   Mississippi Valley-type lead-zinc deposits; in Mineral Deposit Modeling (ed.) R.V. Kirkham, J.M. Duke, W.D. Sinclair, and R.I. Thorpe; Geological Association of Canada, Special Paper 40, p. 289-314.

Macqueen, R.W. and Powell, T.G.
1983:   Organic geochemistry of the Pine Point lead-zinc ore field and region, Northwest Territories, Canada; Economic Geology, v. 78, p. 1-25.

McNeil, W.H., Rawling, K.R., and Sutherland, R.A.
1993:   Nanisivik Mine - operations and innovations in an Arctic environment; in World Zinc '93, Proceedings of the International Symposium on Zinc, (ed.) I.G. Mathews; Australasian Institute of Mining and Metallurgy, Publication Series No. 7/93, p. 41-52,

Oder, C.R.L. and Hook, J.W.
1950:   Zinc deposits in the southeastern states; in Symposium on Mineral Resources of the South-eastern United States, (ed.) F.G. Snyder; Knoxville, Tennessee, University of Tennessee Press, p. 72-87.

Ohle, E.L.
1959: Some considerations in determining the origin of ores of the Mississippi Valley type; Economic Geology, v. 54, p. 769-789.
1980: Some considerations in determining the origin of ores of the Mississippi Valley type, Part II; Economic Geology, v. 75, p. 161-172.
1985: Breccias in Mississippi Valley-type deposits; Economic Geology, v. 80, p. 1736-1752.

Oliver, J.
1986: Fluids expelled tectonically from orogenic belts: their role in hydrocarbon migration and other geologic phenomena; Geology, v. 14, p. 99-102.

Pan, H., Symons, D.T.A., and Sangster, D.F.
1990: Paleomagnetism of the Mississippi Valley-type ores and host rocks in the northern Arkansas and Tri-State districts; Canadian Journal of Earth Sciences, v. 27, p. 923-931.
1993: Paleomagnetism of the Gays River zinc-lead deposit, Nova Scotia: Pennsylvanian ore genesis; Geophysical Research Letters, v. 20, p. 1159-1162.

Qing, H.
1991: Diagenesis of Middle Devonian Presqu'ile dolomite, Pine Point, N.W.T. and adjacent subsurface; PhD. thesis, McGill University, Montreal, Quebec, 287 p.

Randell, R.N.
1994: Geology of the Polaris Zn-Pb Mississippi Valley-type deposit, Canadian Arctic Archipelago; PhD. thesis, University of Toronto, Toronto, Ontario.

Randell, R.N. and Anderson, G.M.
1990: The geology of the Polaris carbonate-hosted Zn-Pb deposit, Canadian Arctic Archipelago; in Current Research, Part D; Geological Survey of Canada, Paper 90-1D, p. 47-53.

Rhodes, D.A., Lantos, E.A., Lantos, J.A., Webb, R.J., and Owens, D.C.
1984: Pine Point orebodies and their relationship to the stratigraphy, structure, dolomitization, and karstification of the Middle Devonian barrier complex; Economic Geology, v. 79, p. 991-1055.

Sangster, D.F.
1983: Mississippi Valley-type deposits: a geological mélange; in Proceedings, International Conference on Mississippi Valley-type Lead-zinc Deposits, (ed.) K. Kisvarsanyi, S.K. Grant, W.P. Pratt, and J.W. Koenig; University of Missouri-Rolla, Missouri, p. 7-19.
1986: Age of mineralization in Mississippi Valley-type (MVT) deposits: a critical requirement for genetic modelling; in Geology and Genesis of Mineral Deposits in Ireland, (ed.) C.J. Andrew, R.W.A. Crowe, S. Findlay, W.M. Pennel, and J.F. Pyne; Irish Association for Economic Geology, Cahill Printers Ltd. Dublin, p. 625-634.

Sangster, D.F. (cont.)
1990: Mississippi Valley-type and sedex lead-zinc deposits: a comparative examination; Institution of Mining and Metallurgy, Transactions, (Sect. B: Applied earth science), v. 99, p. B21-B42.
1988: Breccia-hosted lead-zinc deposits in carbonate rocks; in Paleokarst, (ed.) N.P. James and P.W. Choquette; Society of Economic Paleontologists and Mineralogists, p. 102-116.
1993: Evidence for, and implications of, a genetic relationship between MVT and SEDEX zinc-lead deposits; in World Zinc '93, Proceedings of the International Symposium on Zinc, (ed.) I.G. Mathews; Australasian Institute of Mining and Metallurgy, Publication Series No. 7/93, p. 85-94.

Sangster, D.F. and Symons, D.T.A.
1991: Methodology and genetic implications of paleomagnetic dating of Mississippi Valley-type lead-zinc deposits in the midcontinental region of the U.S.A.; in Source, Transport and Deposition of Metals, Proceedings of the 25th Anniversary Meeting of the Society for Geology Applied to Mineral Deposits, (ed.) M. Pagel and J.L. Leroy; p. 413-416.

Sass-Gustkiewicz, M., Dzulynski, S., and Ridge, J.D.
1982: The emplacement of zinc-lead sulfide ores in the Upper Silesian district - a contribution to the understanding of Mississippi Valley-type deposits; Economic Geology, v. 77, p. 392-412.

Sharp, J.M., Jr.
1978: Energy and momentum transport model of the Ouachita basin and its possible impact on formation of economic mineral deposits; Economic Geology, v. 73, p. 1057-1068.

Stormo, S. and Sverjensky, D.A.
1983: Silicate hydrothermal alteration in a Mississippi Valley-type deposit, Viburnum, southeast Missouri lead district; Geological Society of America, Abstracts with Programs, v. 15, p. 699.

Sverjensky, D.A.
1984: Oil field brines as ore-forming solutions; Economic Geology, v. 79, p. 23-37.
*1986: Genesis of Mississippi Valley-type lead-zinc deposits; Annual Review, Earth and Planetary Science, v. 14, p. 177-199.

Symons, D.T.A. and Sangster, D.F.
1991: Paleomagnetic age of the central Missouri barite deposits and its genetic implications; Economic Geology, v. 86, p. 1-12.
1992: Late Devonian paleomagnetic age for the Polaris Mississippi Valley-type Zn-Pb deposit, Canadian Arctic Archipelago; Canadian Journal of Earth Sciences, v. 29, p. 15-25.

Symons, D.T.A., Pan, H., Sangster, D.F., and Jowett, E.C.
1993: Paleomagnetism of the Pine Point Zn-Pb deposits; Canadian Journal of Earth Sciences, v. 30, p. 1028-1036.

Author's address

D.F. Sangster
Geological Survey of Canada
601 Booth Street
Ottawa, Ontario
K1A 0E8

Printed in Canada

11. ULTRAMAFIC-HOSTED ASBESTOS

11. ULTRAMAFIC-HOSTED ASBESTOS

J.M. Duke

INTRODUCTION

"Asbestos" is the general term applied to fibrous silicate minerals which are commercially valuable because they are resistant to heat and chemical attack and exhibit high tensile strength. Chrysotile is by far the most important asbestos mineral, accounting for more than 95% of consumption. The balance is made up by asbestiform varieties of amphibole, including riebeckite (variety crocidolite), cummingtonite (variety amosite), anthophyllite, tremolite, and actinolite. Almost all chrysotile is mined from vein deposits in serpentinized ultramafic rocks. A small amount is derived from either mass fibre deposits in serpentinites or from serpentinized dolomites.

This section deals only with deposits of chrysotile asbestos in serpentinized ultramafic rocks. The most important deposits in Canada include the Jeffrey mine at Asbestos, Quebec; the Bell-King-Beaver deposit at Thetford Mines, Quebec; the British Canadian and Black Lake mines at Black Lake, Quebec; the Asbestos Hill deposit in the Ungava region of Quebec; the Advocate mine at Baie Verte, Newfoundland; the Cassiar mine in British Columbia; and the Clinton Creek mine in Yukon Territory (Fig. 11-1). Important foreign deposits include Dzhetygara in Kazakhstan; Bazhenovo in Russia; Msauli in the Republic of South Africa; Havelock in Swaziland; Mang'ai in Qinghai, China; and Cana Brava in Goias, Brazil.

Duke, J.M.
1996: Ultramafic-hosted asbestos; in Geology of Canadian Mineral Deposit Types, (ed.) O.R. Eckstrand, W.D. Sinclair, and R.I. Thorpe; Geological Survey of Canada, Geology of Canada, no. 8, p. 263-268 (also Geological Society of America, The Geology of North America, v. P-1).

IMPORTANCE

Chrysotile asbestos accounts for about 1.5% of the value of Canadian nonfuel mineral production. Canada accounts for about 16% of world asbestos production, ranking second behind Russia, and ahead of Kazakhstan, Brazil, China, Zimbabwe, and the Republic of South Africa.

SIZE AND GRADE OF DEPOSITS

Chrysotile asbestos deposits typically contain on the order of 10 to 1000 Mt averaging 3 to 10% recoverable fibre. Past production plus reserves of the Jeffrey, Bell-King-Beaver, and British Canadian mines amount to 800 Mt, 250 Mt, and 150 Mt, respectively, averaging 6% fibre. The Advocate mine contained about 60 Mt with 3% recoverable fibre, whereas the Cassiar deposit amounted to 23 Mt with fibre recoveries in the range of 7 to 10%. The reserves of the McDame deposit in the Cassiar district are 16 Mt with 5.6% recoverable fibre. The Clinton Creek mine yielded about 17 Mt of ore averaging from 4.4% to 9.2% recoverable fibre. (Note that in the asbestos industry, "grade" refers to fibre length rather than to the asbestos content of the rock.)

GEOLOGICAL FEATURES
Geological setting

The largest deposits occur in allochthonous bodies of serpentinized ophiolitic or Alpine-type ultramafic rocks. The deposits of the Appalachians, Cordillera, and the Urals fall into this category. Within the Northern Appalachians, the major ophiolites and associated asbestos deposits occur along a lineament known as the Baie Verte-Brompton Line, which extends from the Burlington Peninsula, across Newfoundland, through Gaspésie and southern Quebec, and through Vermont to northern Massachusetts. In the

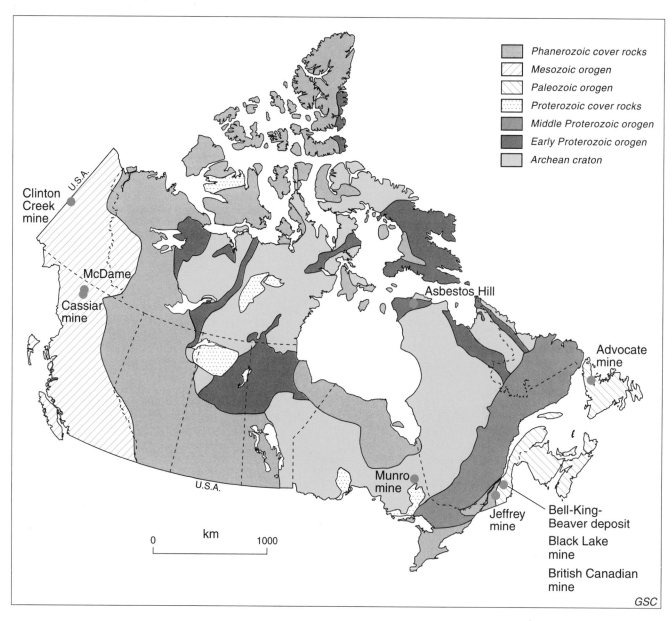

Figure 11-1. Distribution of major asbestos deposits in Canada.

Cordillera, the Cassiar and McDame deposits occur in Alpine-peridotite within the Sylvester allochthon (Burgoyne, 1986), whereas the Clinton Creek deposit is in the Anvil allochthon (Abbott, 1983). The Asbestos Hill deposit in the Ungava Peninsula occurs in the Proterozoic Purtuniq ophiolite.

The ideal ophiolite succession comprises, from the top downwards, marine sediments, pillowed basaltic lavas, sheeted diabase dykes, noncumulate and cumulate mafic rocks, ultramafic cumulates, and ultramafic tectonites. Chrysotile asbestos deposits are most commonly developed in the tectonite member of the succession, although the Asbestos Hill deposit occurs in the cumulate sequence.

Chrysotile asbestos deposits also occur in serpentinized ultramafic rocks of synvolcanic intrusions of komatiitic affinity in Archean greenstone belts. Examples include the Munro mine near Matheson, Ontario (Kretchmar and Kretchmar, 1986); the Msauli mine, which is the most important deposit in the Republic of South Africa; the Havelock mine in Swaziland; and the King mine in Zimbabwe (Anhaeusser, 1976).

Age of host rocks and mineralization

Ultramafic-hosted asbestos deposits are generally considered to be syntectonic. The deposits in southeastern Quebec occur within the Ordovician ophiolitic succession which was emplaced during the Taconic orogeny. Chrysotile vein formation took place at a relatively late stage in this deformational event (O'Hanley, 1987). The Cassiar ultramafic-mafic sheet that hosts the Cassiar and McDame deposits in north-central British Columbia occurs within the Sylvester allochthon, which is dominated by volcanic-sedimentary sequences of Early Mississippian to Early Permian age (Nelson and Bradford, 1989). The allochthon was emplaced in the Jurassic, but the asbestos stockworks are probably Early Cretaceous. The rocks of the Purtuniq ophiolite, which host the Asbestos Hill deposit are 1998 Ma old (Scott et al., 1989), and asbestos vein formation occurred some time after peak thermal metamorphism (1888 Ma). The age of the synvolcanic intrusions that host the Munro mine and other deposits near Matheson, Ontario is approximately 2715 Ma, whereas orogenesis occurred between 2700 and 2670 Ma (Corfu, 1993).

Form of deposits

The shapes of chrysotile asbestos orebodies are variously described as equidimensional, ellipsoidal, lenticular, and tabular, and are of the order of 100 to 1000 m on a side (Fig. 11-2). The ore zone at the Advocate mine in Newfoundland and the Bazhenovo deposit in Russia (Petrov and Znamensky, 1981) are annular in plan and synformal in cross-section. The ore may be characterized as a stockwork of chrysotile veins within masses of partly or completely serpentinized ultramafic rocks as seen at

Normandie mine (Fig. 11-3). Individual veins are for the most part less than 1 cm thick, but range from less than a centimetre to several metres in length. The veins are commonly randomly oriented, although preferred orientations are developed in some deposits. For example, the majority of veins in the Vimy Ridge deposit near Black Lake, Quebec have a single orientation, which gives the ore a ribboned or laminated appearance (Riordon, 1975), and in the Asbestos Hill orebody the veins define two predominant directions at right angles (Stewart, 1981).

The host rock of the asbestos deposits in ophiolites is commonly harzburgite that has been partially or completely serpentinized: asbestos ore is unusual in the dunites. In the deposits of southeastern Quebec, Riordon (1975) noted that the best commercial grades of asbestos are associated with peridotite masses that have been partially (30 to 95%) serpentinized. The rock commonly exhibits "kernal pattern" in which a rim of completely serpentinized peridotite surrounds a core of partially or unserpentinized peridotite (O'Hanley, 1992). The harzburgite commonly contains small lenses or layers of dunite, pyroxenite, and chromite. Abundant, irregularly-shaped bodies of granitic and intermediate rocks, from a few centimetres to several hundred metres in size, are common in the deposits of southeastern Quebec and also in those of the Urals (Lamarche and Riordon, 1981). Rodingite is observed in those deposits that contain granitic and intermediate masses. Talc and carbonate are common, especially in fault or shear zones.

The most common host rock of asbestos deposits in Archean greenstone belts is serpentinized dunite, although some ore also occurs in serpentinized harzburgite and wehrlite. These ultramafic rocks occur in the lower parts of differentiated sills.

Ore mineralogy and textures

Chrysotile is the only ore mineral (although nephrite jade was occasionally recovered as a byproduct at the Cassiar mine). Magnetite and brucite are commonly associated with chrysotile in the veins. Lizardite, antigorite, magnetite, brucite, chlorite, tremolite, talc, magnesite or dolomite, olivine, orthopyroxene, and chromite occur in the wall rocks.

Asbestos veins are normally characterized as being either "cross fibre", in which the chrysotile fibres are oriented at a high angle to the plane of the vein, or "slip fibre", in which the fibres generally parallel the vein. Cross fibre veins constitute most of the ore in Canadian deposits. Chrysotile also occurs as "mass fibre" in which up to 80% of the rock consists of a tangled mass of asbestos fibre.

DEFINITIVE CHARACTERISTICS

This type is characterized by stockwork of chrysotile veins in serpentinized ultramafic rocks in ophiolitic terranes or Precambrian greenstone belts.

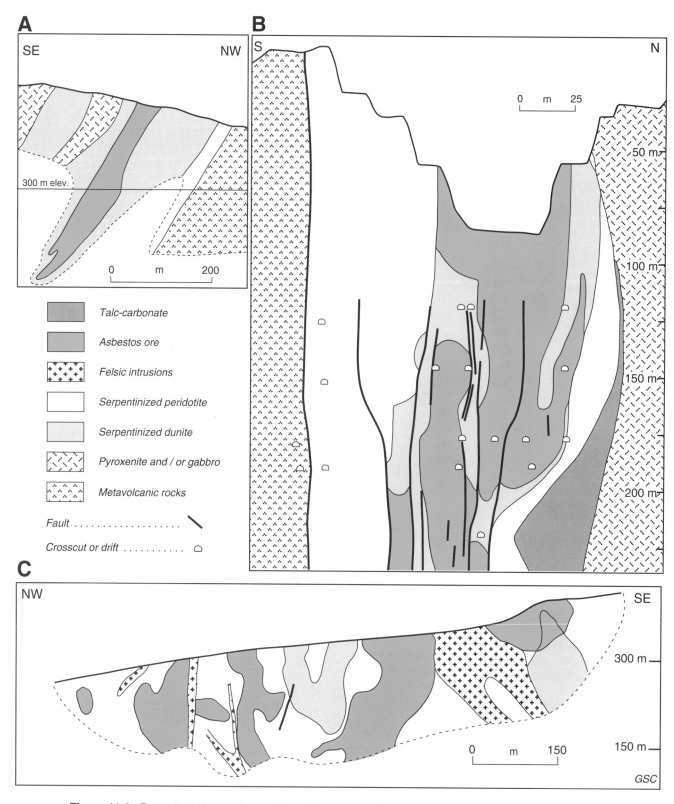

Figure 11-2. Representative sections of Canadian asbestos deposits: **A)** Asbestos Hill, Ungava Peninsula (after Stewart, 1981); **B)** Munro A orebody, Matheson, Ontario (after Vos, 1971); **C)** British Canadian mine, Black Lake, Quebec (after Riordon, 1975).

GENETIC MODEL

Chrysotile asbestos deposits develop during deformation and alteration of ultramafic rocks under relatively low grade metamorphic conditions, with temperatures of $300 \pm 50°C$ and water pressures less than 1 kbar (O'Hanley, 1991). Wicks and Whittaker (1977) concluded that chrysotile vein formation requires specific conditions of prograde metamorphism. The preponderance of cross fibre veins in the stockworks indicates that chrysotile grew during fracturing under tensional stress or dilation of pre-existing fractures. O'Hanley (1988) has proposed a model for Alpine peridotite-hosted deposits based on common elements observed at Thetford Mines, Cassiar, and Woodsreef (Australia). In each case, chrysotile developed during a transition in regional stress regime from convergent to transcurrent motion. The tension fractures which were filled by cross fibre chrysotile formed in massive serpentinite. If the rock had developed a schistose fabric prior to the change in stress regime, it would have continued to deform by ductile shear rather than brittle fracture. It is necessary that the temperature, pressure, and fluid composition are in the stability field of chrysotile during tensional fracturing and, furthermore, that deformation and metamorphism not proceed to such an extent following vein formation that chrysotile is destroyed. O'Hanley (1988) has observed that the presence of a major fault that was active during the transition in stress regime was an important factor at both Cassiar and Thetford Mines.

Figure 11-3. Asbestos ore from the Normadie mine, Thetford Mines, Quebec, showing 2 to 4 mm wide veins of cross fiber chrysotile asbestos in serpentinized peridotite. Length of specimen is 10.5 cm. GSC 1995-173

RELATED DEPOSIT TYPES

Podiform chromite deposits (see Type 28, "Mafic/ultramafic-hosted chromite") are often spatially associated with asbestos deposits in ophiolitic sequences, although there is no genetic relationship. Some talc and magnesite deposits are genetically related to asbestos deposits, having been formed by the alteration of ultramafic rocks by relatively CO_2-rich fluids.

EXPLORATION GUIDES

1. On the regional scale, exploration should be directed at serpentinized ultramafic rocks in either ophiolite sequences or Archean greenstone terranes.

2. Within ophiolite sequences, asbestos deposits most commonly occur in the harzburgite tectonite. In greenstone terranes, asbestos deposits occur in serpentinized dunite or peridotite members in the lower parts of layered sills of komatiitic affinity.

3. The most prospective serpentinized ultramafic rocks are massive and of low metamorphic grade.

4. The occurrence of a major fault in proximity to the serpentinite mass appears to be an important feature of most deposits.

ACKNOWLEDGMENTS

I am grateful to David O'Hanley and Tucker Barrie for their critical reviews of the manuscript.

SELECTED BIBLIOGRAPHY

References with asterisks (*) are considered to be the best source of general information on this deposit type.

Abbott, G.
1983:   Origin of the Clinton Creek asbestos deposit; in Yukon Exploration and Geology 1982, Geology Section, Department of Indian and Northern Affairs, p. 18-25.

*Anhaeusser, C.R.
1976:   The nature of chrysotile asbestos occurrences in Southern Africa; Economic Geology, v. 71, p. 96-116.

Burgoyne, A.A.
1986:   Geology and exploration, McDame asbestos deposit, Cassiar, B.C.; The Canadian Mining and Metallurgical Bulletin, v. 79, no. 889, p. 31-37.

Corfu, F.
1993:   The evolution of the southern Abitibi greenstone belt in light of precise U-Pb geochronology; Economic Geology, v. 88, p. 1323-1340.

Kretschmar, U. and Kretschmar, D.
1986:   Talc, magnesite and asbestos deposits in the Timmins-Kirkland Lake area, districts of Timiskaming and Cochrane; Ontario Geological Survey, Study 28, 100 p.

*Lamarche, R.Y. and Riordan, P.H.
1981:   Geology and genesis of chrysotile asbestos deposits of northern Appalachia; in Geology of Asbestos Deposits, (ed.) P.H. Riordon; American Institute of Mining and Metallurgical Engineers, New York, p. 11-23.

*Laurent, R.

1979: Paragenesis of serpentine assemblages in harzburgite tectonite and dunite cumulate from the Quebec Appalachians; The Canadian Mineralogist, v. 17, p. 857-869.

Nelson, J.L. and Bradford, J.A.

1989: Geology and mineral deposits of the Cassiar and McDame map areas, British Columbia (104P3/5); in British Columbia Ministry of Energy, Mines and Petroleum Resources, Geological Fieldwork, 1988, Paper 1989-1, p. 323-338.

O'Hanley, D.S.

1987: The origin of the chrysotile asbestos veins in southeastern Quebec; Canadian Journal of Earth Sciences, v. 24, p. 1-9.

*1988: The origin of alpine peridotite-hosted, cross fiber, chrysotile asbestos deposits; Economic Geology, v. 83, p. 256-265.

1991: Fault-related phenomena associated with hydration and serpentine recrystallization during serpentinization; The Canadian Mineralogist, v. 29, p. 21-35.

1992: Solution to the volume problem in serpentinization; Geology, v. 20, p. 705-708.

Petrov, V.P. and Znamensky, V.S.

1981: Asbestos deposits of the USSR; in Geology of Asbestos Deposits, (ed.) P.H. Riordon; American Institute of Mining and Metallurgical Engineers, New York, p. 45-52.

*Riordon, P.H.

1975: Geology of the asbestos deposits of southeastern Quebec; Ministere des Richesses naturelles du Quebec, Etude speciale 18, 100 p.

Scott, D.J., St-Onge, M.R., Lucas, S.B., and Helmstaedt, H.

1989: The 1998 Ma Purtuniq ophiolite: imbricated and metamorphosed oceanic crust in the Cape Smith Thrust Belt, northern Quebec; Geoscience Canada, v. 16, p. 144-147.

Stewart, R.V.

1981: Geology and evaluation of the Asbestos Hill ore body; in Geology of Asbestos Deposits, (ed.) P.H. Riordon; American Institute of Mining and Metallurgical Engineers, New York, p. 53-62.

Vos, M.A.

1971: Asbestos in Ontario; Ontario Department of Mines, Industrial Minerals Report 36, 69 p.

*Wicks, F.J. and Whittaker, E.J.W.

1977: Serpentine textures and serpentinization; The Canadian Mineralogist, v. 15, p. 459-488.

Yu, S.

1989: Geological features and mineralization-controlling conditions of asbestos deposit in Mang'ai, Qinghai Province, China; Progress in Geosciences of China (1985-1988) - Papers to Twenty-eighth International Geological Congress, v. 1, p. 67-71.

Author's address

J.M. Duke
Geological Survey of Canada
601 Booth Street
Ottawa, Ontario
K1A 0E8

Printed in Canada

12. VOLCANIC-ASSOCIATED URANIUM

12. VOLCANIC-ASSOCIATED URANIUM

S.S. Gandhi and R.T. Bell

IDENTIFICATION

Deposits of this type consist of disseminations and veins of uraninite/pitchblende in volcanic rocks that are commonly felsic and subalkaline to peralkaline felsic in composition, but some potassic alkaline mafic to intermediate rocks also host uranium occurrences. The volcanic rocks are formed mainly in late- or postorogenic settings and continental extensional tectonic environments. They occur in all volcanic lithofacies, from proximal to distal in setting, including volcaniclastic sediments. Molybdenum, flourine, thorium, rare-earth elements, and lead-zinc, are commonly enriched in the deposits and may be of economic importance.

The best examples in Canada are the trachyte-hosted Rexspar deposit in British Columbia and the rhyolite-hosted Michelin deposit in Labrador (Fig. 12-1). The important foreign examples include the ignimbrite-hosted Sierra Pena Blanca deposits in Mexico, rhyolite-hosted Pleutajokk deposit in Sweden, Anderson deposit in moat sediments in southwestern United States, and deposits in a collapsed caldera complex in the Streltsov region in eastern Russia.

Felsic volcano-plutonic complexes also host genetically related, uraniferous polymetallic deposits characterized by abundant iron oxides that form the matrix of breccias, e.g. the giant Olympic Dam deposit in South Australia and the Sue-Dianne deposit in the Northwest Territories. These are treated separately under the Kiruna/Olympic Dam-type deposits (see Type 22), which include deposits that contain little or no uranium.

IMPORTANCE

In Canada, mineable deposits of the volcanic-associated type account for less than 1% of the uranium resources in the 'reasonably assured' category. At present in the western world, they probably account for less than 5% of the 'reasonably assured' resources. However, in the Russian Republic and China they represent a significant proportion of the uranium resources.

No uranium has yet been produced from the Canadian deposits of this type, unless the U-Ag-Cu-Co-Ni veins (described as "Arsenide vein uranium-silver", subtype 14.2) of mined out deposits in the Great Bear Lake area in Northwest Territories are included. These veins are hosted by volcanic and volcaniclastic rocks, but are much younger (isotopic ages of 1870-1840 Ma for the host rocks versus 1775-1500 Ma for the veins), and hence apparently have no direct genetic relation to those discussed here.

SIZE AND GRADE OF DEPOSIT

The Michelin deposit in Labrador contains about 7000 t U at an average grade of 0.1% U. Several smaller deposits occur in this volcanic district, and together they contain an additional uranium resource of comparable magnitude. The Rexspar deposit in British Columbia contains about 700 t at an average grade of 0.077% U.

Foreign deposits commonly range between 500 and 10 000 t of contained uranium in ores grading from 0.04 to 4% U (Cox and Singer, 1986, p. 162-164). The Anderson deposit in U.S.A. contains 20 000 t. A cluster of deposits in the Streltsov district in the Russian Republic contains as much as 100 000 t of uranium.

Gandhi, S.S. and Bell, R.T.
1996: Volcanic-associated uranium; *in* Geology of Canadian Mineral Deposit Types, (ed.) O.R. Eckstrand, W.D. Sinclair, and R.I. Thorpe; Geological Survey of Canada, Geology of Canada, no. 8, p. 269-276 (*also* Geological Society of America, The Geology of North America, v. P-1).

GEOLOGICAL FEATURES

Geological setting

The deposits are associated with felsic volcanic complexes that developed in late- and postorogenic settings and extensional tectonic environments. Their local settings are varied: calderas, lava and ash flow fields, taphrogenic volcanic-sedimentary basins, domes, and breccias (including diatremes).

The deposits in Labrador and Sweden are hosted by areally extensive sequences of flows, tuffs, and volcaniclastic sediments that developed during the late- and posttectonic stages of the Makkovikian-Svecofennian orogeny 1900-1700 Ma (Gandhi, 1978, 1986; Lindroos and Smellie, 1979; Gustafsson, 1981; Hålenius et al., 1986; Schärer et al., 1988). A number of uranium occurrences are also found in felsic volcanic rocks of the Great Bear magmatic zone, which was developed on the west side of the Wopmay

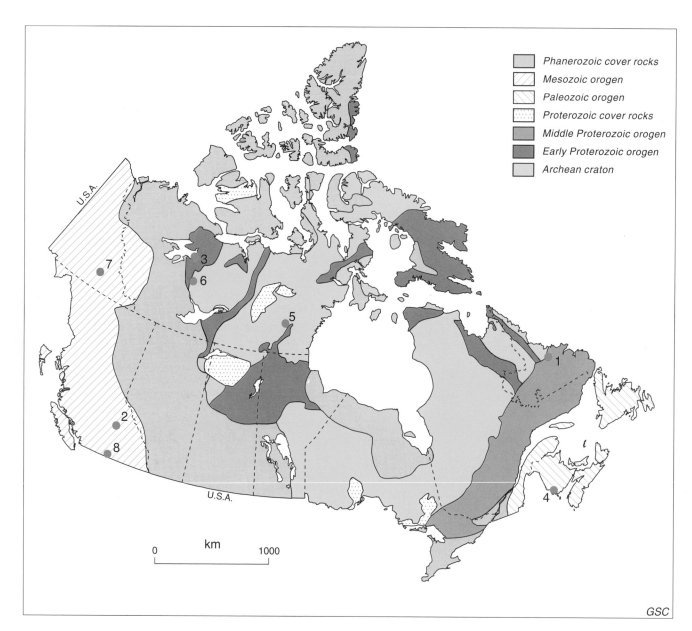

Figure 12-1. Volcanic-associated uranium deposits and main occurrences/districts in Canada.
Deposits: 1 – Michelin, Labrador; 2- Rexspar, British Columbia; 3 – Eldorado, Great Bear Lake area, Northwest Territories (U-Ag-Cu-Co-Ni-As-bearing veins hosted by volcanic rocks but genetically unrelated to them).
Occurrences/districts: 4 – Mount Pleasant, New Brunswick (Curtis, 1981); 5 – Lower Dubawnt Group volcanic rocks, Baker Lake area, Northwest Territories (Curtis, 1981; Miller and LeCheminant, 1985; LeCheminant et al., 1987); 6 – UGI-DV, Devries Lake, Northwest Territories (Gandhi and Prasad, 1993); 7 – Nokluit, Yukon Territory (Bell, 1985); 8 – Marron volcanic rocks, British Columbia (Bell, 1985).

orogen 30 to 60 million years after the peak of the orogeny about 1900 Ma (Hoffman, 1980; Hildebrand et al., 1987). Most of these are vein-type and much younger than the host rocks, but a few at DeVries Lake are of disseminated type such as the Michelin deposit, and are approximately of the same age as the host rocks (Miller, 1982; Gandhi and Prasad, 1993). Another Canadian uranium district, about 1840 Ma in age, is the intracratonic extensional Baker Lake basin in Northwest Territories which contains an areally extensive and voluminous assemblage of subaerial potassic to ultrapotassic trachyandesite and trachyte flows, related bostonite and other intrusions, continental clastic sediments, and a 1760 Ma, high silica, fluorine-rich, rhyolite-granite assemblage (Miller, 1980; Miller and LeCheminant, 1985; Miller et al., 1986; LeCheminant et al., 1987). Syngenetic U-Th mineralization occurs in bostonite dykes, which

represent a highly fractionated phase of alkalic potassic magma, and epigenetic uranium and uranium-polymetallic (Cu+Pb±Zn±Au±Ag±Se±Mo) occurrences are associated with the mafic to intermediate volcanic rocks. Anomalous radioactivity is also noted in a high silica, topaz-bearing rhyolite.

The Rexspar deposit occurs in an areally restricted volcanic centre in Paleozoic rocks of the Canadian Cordillera (Preto, 1978; Morton et al., 1978). This deposit and those in the Labrador district have been affected by postmineralization deformation that has imparted schistosity to their host sequences. A collapsed caldera of Jurassic age, developed at the margin of a Paleozoic granitic dome, hosts deposits in the Streltsov district, eastern Russia (Ruzicka, 1992). Another Paleozoic uranium district is in northern Italy, where the Permo-Carboniferous Collio basin or half-graben, has abundant

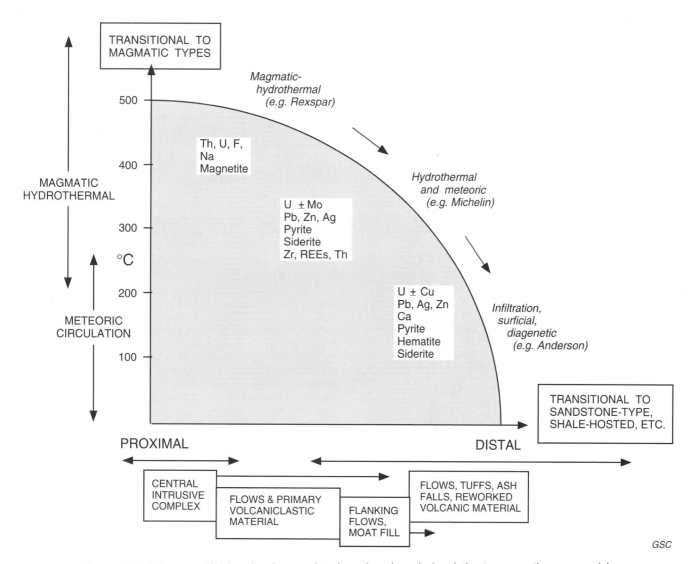

Figure 12-2. Diagram of felsic volcanic-associated uranium deposits in relation to magmatic source, mixing of magmatic fluids and meteoric waters, and temperature of mineralizing solutions.

271

felsic volcanic rocks which host stratabound and fracture-controlled polymetallic uranium deposits (Cadel, 1986; Cadel et al., 1987; Fuchs, 1989). In southeast China, numerous epigenetic uranium occurrences are hosted by rhyolites and granites that were formed at late stages of subduction of the Pacific plate under mainland China during Jurassic-early Cretaceous time (Xu, 1990). The uranium mineralization is believed to have occurred during the extensional tectonic stage that followed the end of subduction, and the mineralizing solutions may have been derived from deep crustal anatexis (Chen and Fang, 1985; Yao et al., 1989).

The Anderson deposit in organic-rich tuffaceous shales, and other felsic volcanic-associated deposits in the southwestern U.S.A. (e.g., those in the McDermitt caldera), and the ignimbrite-hosted Pena Blanca deposits in Mexico, are in the extensional Basin and Range Province-Rio Grande Rift Zone of Tertiary age (Sherborne et al., 1979; Rytuba and Glazman, 1979; Cárdenas-Flores, 1985; Dayvault et al., 1985; Goodell, 1985; Wenrich, 1985). Quaternary volcaniclastic rocks and vitric tuffs of the potassic volcanic Roman Province, controlled by a graben structure, contain exhalative-supergene concentrations of uranium, thorium, and iron (Locardi and Mittemperger, 1971).

Age of host rocks and mineralization

For this type of uranium deposits in general, the host rocks range in age from early Proterozoic to Cenozoic. No Archean examples have been documented to date.

The mineralization may be synvolcanic in the case of magmatic-hydrothermal deposits in volcanic vents and proximal exhalative settings, or it may be epigenetic resulting from the action of circulating heated meteoric waters, and/or by groundwaters during weathering processes. The epigenetic mineralization ranges in age from close to that of the host rocks to significantly younger. It may occur anywhere in the volcanic region, but appears to be more important in the distal facies.

Most of the deposits in the Labrador uranium district have yielded isotopic ages in the range 1800-1750 Ma, and their felsic volcanic host rocks, which dominate the bimodal upper part of the Aillik Group, are between 1860 and 1805 Ma (Gandhi, 1978, 1986; Schärer et al., 1988). In Sweden, the Pleutajokk and other deposits in its vicinity, and their host rocks are essentially coeval with those in Labrador (Gustafsson, 1981; Hålenius et al., 1986). They are in an early Proterozoic terrane which was conterminous with Greenland and Labrador prior to the birth of Atlantic Ocean (Gower et al., 1990, Gower, 1992; Gandhi and Bell, 1993). Some younger occurrences are also found in this terrane, a notable one being the Duobblon deposit in about 1725 Ma ignimbrite in northern Sweden (Lindroos and Smellie, 1979). Fracture-controlled occurrences in the Baker Lake basin have yielded discordant pitchblende ages between 1800 and 1700 Ma (Miller and LeCheminant, 1985). The Rexspar deposit is Paleozoic, and no younger than Permian (Morton et al., 1978; Bell, 1985). Most of the other foreign examples are Phanerozoic, as mentioned above.

Associated structure

Block faulting, breccia development, weak to moderately intensive folding and shearing related to volcanic activity, and late tectonic structural adjustments are common prior to, during, and after mineralization. No major orogenic deformation, however, occurred immediately before or after the mineralization.

Relation of ore to host rocks

Ore zones are stratabound away from the volcanic vent, as in the case of the Michelin and Rexspar deposits, as well as the foreign deposits in lava and ash flow fields mentioned above. In proximal deposits, ore zones are primarily discordant veins, although breccia fillings and disseminations also occur.

The Michelin deposit is hosted by rhyolitic, nonwelded ash-flow and air-fall crystal tuffs, which are about 250 m thick, have a strike length of more than 4 km, and dip 55° to the south-southeast. The host rocks are part of the much thicker and extensive, rhyolite-dominated, upper Aillik Group, deposited about 1860 Ma and deformed during the Makkovikian 'orogeny' about 1800 Ma (Gandhi, 1978, 1986; Bailey, 1979; Schärer et al., 1988). They are layered, and the layering is defined by abrupt textural changes across strike from coarse feldspar-porphyritic to subporphyritic (with relatively smaller and less abundant phenocrysts) to nonporphyritic rhyolites. The mineralized zones are as much as a few metres thick and several hundred metres long, and are concordant with the layering in the host sequence. Four main mineralized zones and a few subsidiary zones occur in, but are not restricted to, coarse feldspar porphyritic units within the 40 m thick lower part of the host crystal tuffs, and within a strike length of 1 km. The host rocks are variably foliated and sheared, and the mineralization predates at least the final stages of deformation. A number of other smaller occurrences similar to the Michelin deposit occur in the Aillik Group (Gandhi, 1978, 1986).

The Rexspar mineralized zones occur in metamorphosed potassium-rich, feldspar-porphyritic trachyte and tuff units, whose stratigraphic tops are not known with certainty, although they are most likely overturned (Bell, 1985). The three main mineralized zones are widely separated, and occur in pyritic, tuffaceous, argillaceous lenses containing pyroclastic fragments. Mid-Mesozoic deformation has rendered the host rocks schistose. The mineralized zones are as much as 20 m thick and 200 m long, and dip less than 30°.

Form of deposits

The stratabound disseminated-type deposits are lenticular to tabular. Discordant veins and fracture-fillings are common in volcanic vents and proximal deposits. They also occur in the districts that contain stratabound deposits, but are relatively subordinate in importance; they may be primary fracture fillings or may have resulted from remobilization of uranium during geological events postdating the primary mineralization.

Ore composition

The ores may be either compositionally simple or complex. The Michelin deposit is of simple type, and contains only traces of other elements. Some complex deposits with high Th/U ratios include significant amounts of other elements, in particular REEs, Mo, F, Pb, and Zn, as in the Rexspar deposit. In some cases proximal vein deposits with low Th/U ratios also contain significant amounts of Mo, F, Pb, and Zn.

Alteration

Hematitization, carbonatization, and albitization are common. Hematitization is distinctive in the Michelin ore, but not in the sulphide-rich ore of the Rexspar deposit. Hematite occurs as very fine grains or interstitial films in the Michelin ore, but does not represent a significant addition of iron to the host rock. Magnetite and ilmeno-magnetite in the rhyolitic host rock have been hematitized to a variable degree.

Carbonatization is reflected in widely distributed calcite in and around ore zones as interstitial grains, fracture-coatings, and veins. Pink calcite is typical of the uranium deposits. Purple fluorite is a common associate of the calcite.

Albitization of the host rock is an important feature of many of these deposits. It is strongly developed at the Michelin deposit, where mineralized rhyolites contain 7% to 11% Na_2O and less than 0.5% K_2O, in contrast to the unmineralized rhyolites which contain 4% to 6% Na_2O and 5% to 7% K_2O. The strongly metasomatized rocks also show, in addition to alkali exchange and alkali enrichment, some desilication (White and Martin, 1980). In the Rexspar deposit, some albite-rich rocks have been noted, but albitization is apparently not extensive.

Mineralogy

Uranium occurs mainly as pitchblende, and to a lesser extent as coffinite in the thorium-poor occurrences, such as the Michelin deposit and many high grade veins and disseminated deposits. In the thorium-rich Rexspar deposit and the Nokluit occurrence in Yukon Territory (Fig. 12-1), the dominant uranium minerals are uraninite, uranoan thorite, and uranothorite (Bell, 1985).

Associated minerals at the Michelin deposit are nearly pure albite (99% Ab), sodic pyroxene, sodic amphibole, zircon, sphene, calcite, and fluorite (Gandhi, 1978, 1986). In the Rexspar deposit, a great variety of minerals are present: pyrite, fluorophlogopite, apatite, fluorite, celestite, galena, sphalerite, molybdenite, scheelite, siderite, calcite, barite, quartz, albitic plagioclase, bastnaesite, and monazite (Preto, 1978). Fluorite, celestite, and rare-earth elements are abundant enough to be of economic interest.

Texture

Uranium-bearing minerals are commonly finely disseminated. In the Michelin deposit, they occur as grains less than 10 μm in diameter that are interstitial to other minerals, as inclusions in sphene and in mafic silicate minerals, and as small aggregates with mafic silicate grains,

zircon, and sphene. In the Rexspar deposit, uranium-bearing minerals are tiny, discrete grains, from less than a micrometre to a few tens of micrometres in diameter, and commonly occur as inclusions in fluorphlogopite.

DEFINITIVE CHARACTERISTICS

1. The host rocks are commonly felsic volcanic and associated clastic sediments of post-Archean age.
2. The host rocks were deposited in an extensional tectonic environment, during late- and post-tectonic stages in orogenic zones or in continental anorogenic settings.
3. The hosting volcanic suites are generally felsic dominated or bimodal, are meta- or peraluminous, and have a mildly to highly alkaline character.
4. Fluorine and molybdenum are commonly associated with uranium.
5. Hematitization, carbonatization, and albitization of the host rocks are associated features of the deposits.

GENETIC MODEL

Two stages are involved in the model:

a) an initial enrichment of uranium in the felsic magma source, and

b) a later concentration of the metal at the deposit site, during and after extrusion of the felsic magma, by magmatic hydrothermal fluids and/or meteoric waters circulating through the volcanic pile.

The main debate is whether or not the primary magmatic enrichment in uranium and associated incompatible elements is of deep-seated mantle origin (Locardi, 1985; Treuil, 1985) or of metamorphic and anatectic origin from the middle or lower regions of a thick sialic crust (Chen and Fang, 1985). Adherents to both schools of thought agree that a dilational (viz., taphrogenic) geological regime is important regionally in the formation of volcanogenic uranium deposits, and that the volcanic rocks are enriched in uranium substantially above normal crustal levels.

An important dilatant geological regime is that of the rising metamorphic core complex, e.g. the Basin and Range Province in the southwestern U.S.A. and its extension in the Rio Grande Rift (Wenrich, 1985), and the Shuswap terrane in British Columbia (Bell, 1985). Rise of the complexes allow for a regional structural ground preparation through progressive mylonitization, development of detachment zones, listric faults, brittle deformation, and brecciation. These are coupled with intrusion of deep-seated alkali-rich melts, possibly mantle-derived, with or without underplating and delamination as visualized by Wyborn et al. (1987) and Wyborn (1988). The surface expressions of the phenomenon are fault-bounded, terrestrial clastic-volcanic basins.

A lower crustal derivation of the felsic magma is favoured here, but it is possible that this felsic magma is derived by remelting of previously underplated igneous bodies in the manner visualized by Wyborn et al. (1987) and Wyborn (1988). The silicate melts and the accompanying hydrous fluids were enriched in the incompatible elements through partial melting of the source rocks. Further enrichment occurred through differentiation during crustal residence and during ascent of the magma. Although this process is considered adequate by most workers, input of additional uranium in these melts and fluids by mixing with meteoric waters is suggested by Chen and Fang (1985) in their "double mixing" model. This model involves deep circulation of oxygenated uraniferous meteoric waters in volcanic terranes.

The concentration of uranium at the deposit site may be brought about by magmatic hydrothermal fluids, or oxidized meteoric waters near surface or at shallow depths, with or without direct input of the metal from magmatic fluids. The volcanic-associated uranium deposits vary considerably in their characters, reflecting a variety of mechanisms for uranium transport and deposition in the complex volcanic-hypabyssal environment (Fig. 12-2). Some features suggest relatively higher temperatures of deposition, e.g. intense soda metasomatism at the Michelin deposit (Gandhi, 1978, 1986; White and Martin, 1980); other features, e.g., fluid inclusions and metal associations, indicate boiling off of volatiles, as in the case of the Rexspar deposit (Morton et al., 1978; Preto, 1978; Bell, 1985), and of the Moonlight mine in the McDermitt caldera complex in the southwestern U.S.A. (Rytuba and Glazman, 1979). Posteffusive, circulating oxidized meteoric waters may, if the volcanic heat supply persists, leach uranium from devitrifying glass (Zielinski, 1985) and from zones mineralized by magmatic fluids, and deposit it at favourable sites within the volcanic pile or in nearby sedimentary basins or lakes where reducing conditions are encountered. White and Martin (1980) and Gower et al. (1982) proposed such convection cells for the deposits in Labrador, and Marten (1977) and Evans (1980) suggested similar processes, but with mineralizing solutions of metamorphic and meteoric origin respectively, without input from a magmatic source, and with deposition controlled by permeable shear zones produced by premineralization deformation. Gandhi (1986) has, however, emphasized features indicating predeformation mineralization, and the role of a magmatic source at depth in generation of mineralizing solutions. In addition to synvolcanic hydrothermal mineralization, subaerial volcanism may contribute uraniferous tuffaceous sediments to the sedimentary basins in, or adjacent to, the volcanic field, and uranium from them may be concentrated during sedimentation and diagenesis. Examples of deposits formed by surface and groundwater transport of uranium in the volcaniclastic sedimentary environment are several occurrences in the tuffaceous beds in the upper part of the Aillik Group (Gandhi, 1978, 1986) and the Anderson deposit in organic-rich tuffaceous shales (Sherborne et al., 1979).

An interesting aspect of most felsic volcanic-associated uranium deposits is that they are in distinctly bimodal or felsic igneous suites that are late- and postorogenic and anorogenic, rather than in classical andesite-dominated volcanic suites associated with continental margin orogenies.

RELATED DEPOSIT TYPES

The upper Aillik Group underlies an area more than 130 km by 40 km and hosts, in addition to a number of volcanic-associated uranium occurrences which resemble the Michelin deposit, a few cogenetic molybdenite-pyrite-fluorite occurrences. The largest molybdenite deposit in the district is the Cape Makkovik deposit, and this contains mainly stratabound disseminated mineralization in a granulated and sheared zone (Gandhi, 1978, 1986). Veins carrying these minerals, with or without pitchblende, galena, and sphalerite, occur elsewhere in the Aillik Group in the vicinity of granite intrusions and hence are regarded as granite-related. The lower Aillik Group includes some argillaceous-mafic tuffaceous beds near the base of its rhyolite-dominated upper part, and these beds host disseminated and associated vein-type uranium occurrences, of which the best known is the Kitts deposit in pyrrhotite-rich graphitic argillite (Gandhi, 1978, 1986).

The boundaries of the host volcanic rocks with related plutonic suites, and with associated volcaniclastic sediments, which in some cases host uranium deposits (e.g., the tuffaceous sediments of caldera-fill at the Anderson mine, U.S.A.), are for the most part ill-defined and arbitrary. It is apparent, however, that the volcanic-associated uranium deposits encompass a wide spectrum of volcano-plutonic environments. They are most closely related to the continental sandstone-type deposits, and indeed many workers have regarded felsic volcanic rocks as a fertile source of uranium in the sandstone-type deposits, insofaras they are among the most uraniferous rocks and are associated with sandstones in many places. Among other related deposits are precious metal-bearing veins in caldera complexes, which contain little or no uranium. The granite-related polymetallic deposits, particularly those carrying uranium, appear in many cases to be deep-seated equivalents of the felsic volcanic-associated uranium deposits, e.g., the Mount Pleasant Bi-Mo-Sn-W deposits in New Brunswick (Curtis, 1981).

Within the felsic volcanic domain, the deposit classification by type seems somewhat arbitrary. Thus felsic volcanic-hosted uranium deposits have been separated from Kiruna/Olympic Dam-type deposits, although a broader category of deposits genetically related to felsic volcano-plutonic complexes would include both. Furthermore, the polymetallic veins of the Great Bear Lake area, although they are hosted by volcanic rocks, are distinguished on the basis of isotopic age constraints. These ages call for a genetic model different from that for the volcanic-associated deposits.

EXPLORATION GUIDES

Terranes of profuse post-Archean felsic volcanic rocks are favourable for volcanic-associated uranium deposits. Of particular interest are those that developed in continental, extensional tectonic environments. Such felsic or bimodal (basalt-rhyolite) suites were apparently deposited during the interval 2000 to 1500 Ma over large parts of the Canadian, Baltic, and Australian shields. Detailed stratigraphic mapping of felsic volcanic sequences will be helpful in the recognition of favourable horizons for stratabound deposits, and of the vent facies favourable for discordant deposits. Alkali metasomatism is a useful guide as it extends for large areas surrounding the deposits.

REFERENCES

References with asterisks (*) are considered to be the best source of general information on this deposit type.

Bailey, D.G.
1979: Geology of the Walker-McLean Lake area, (13K/9E, 13J/12), Central Mineral Belt, Labrador; Newfoundland Department of Mines and Energy, Mineral Development Division, Report 78-3, 17 p.

*Bell, R.T.
1985: Overview of uranium in volcanic rocks of the Canadian Cordillera; in Uranium Deposits in Volcanic Rocks, Proceedings of a Technical Committee Meeting, El Paso, Texas, April 1984; International Atomic Energy Agency, Vienna, IAEA-TECDOC-490, p. 319-335.

Cadel, G.
1986: Geology and uranium mineralization of the Collio basin (central southern Alps, Italy); Uranium, v. 2, no. 3, p. 215-240.

Cadel, G., Fuchs, Y., and Meneghel, L.
1987: Uranium mineralization associated with the evolution of a Permo-Carboniferous volcanic field - examples from Novazza and Val Vedello (northern Italy); Uranium, v. 3, no. 2/4, p. 407-421.

Cárdenas-Flores, D.
1985: Volcanic stratigraphy and U-Mo mineralization of the Sierra de Pena Blanca District, Chihuahua, Mexico; in Uranium Deposits in Volcanic Rocks, Proceedings of a Technical Committee Meeting, El Paso, Texas, April 1984; International Atomic Energy Agency, Vienna, IAEA-TECDOC-490, p. 125-136.

Chen, Zhaobao and Fang, Xiheng
1985: Main characteristics and genesis of Phanerozoic vein-type uranium deposits; in Uranium Deposits in Volcanic Rocks, Proceedings of a Technical Committee Meeting, El Paso, Texas, April 1984; International Atomic Energy Agency, Vienna, IAEA-TECDOC-490, p. 69-82.

*Cox, D.P. and Singer, D.M. (ed.)
1986: Mineral deposit models; United States Geological Survey, Bulletin 1693, p. 162-164.

Curtis, L.
1981: Uranium in volcanic and volcaniclastic rocks - examples from Canada, Australia and Italy; American Association of Petroleum Geologists, Studies in Geology, no. 13, p. 55-62.

Dayvault, R.D., Castor, S.B., and Berry, M.R.
1985: Uranium associated with volcanic rocks of the McDermitt caldera, Nevada and Oregon; in Uranium Deposits in Volcanic Rocks, Proceedings of a Technical Committee Meeting, El Paso, Texas, April 1984; International Atomic Energy Agency, Vienna, IAEA-TECDOC-490, p. 379-409.

Evans, D.
1980: Geology and petrochemistry of the Kitts and Michelin uranium deposits and related prospects, Central Mineral Belt, Labrador; PhD. thesis, Queen's University, Kingston, Ontario, 312 p.

Fuchs, Y.
1989: Hydrothermal alteration at the Novazza volcanic field and its relation to the U-Mo-Zn Novazza deposit, northern Italy; in Metallogenesis of Uranium Deposits, Proceedings of a Technical Committee Meeting, Vienna, March 1987; International Atomic Energy Agency, Vienna, p. 137-152.

*Gandhi, S.S.
1978: Geological setting and genetic aspects of uranium occurrences in the Kaipokok Bay-Big River area, Labrador; Economic Geology, v. 73, p. 1492-1522.
1986: Uranium in Early Proterozoic Aillik Group, Labrador; in Uranium Deposits of Canada, (ed.) E.L. Evans; The Canadian Institute of Mining and Metallurgy, Special Volume 33, p. 70-82.

Gandhi, S.S. and Bell, R.T.
1993: Metallogenic concepts to aid exploration for the giant Olympic Dam-type deposits and their derivatives; Proceedings of the Eighth Quadrennial IAGOD Symposium, Ottawa, Canada, August 1990; International Association on the Genesis of Ore Deposits; (ed.) Y.T. Maurice; E. Schweizerbart'sche Verlagsbuchhandlung, Stuttgart, p. 787-802.

Gandhi, S.S. and Prasad, N.
1993: Regional metallogenic significance of Cu, Mo, and U occurrences at DeVries Lake, southern Great Bear magmatic zone, Northwest Territories; in Current Research, Part C; Geological Survey of Canada, Paper 93-1C, p. 29-39.

Goodell, P.C.
1985: Chihuahua city uranium province, Chihuahua, Mexico; in Uranium Deposits in Volcanic Rocks, Proceedings of a Technical Committee Meeting, El Paso, Texas, April 1984; International Atomic Energy Agency, Vienna, IAEA-TECDOC-490, p. 97-124.

Gower, C.F.
1992: The relevance of Baltic Shield metallogeny to mineral exploration in Labrador; in Current Research, Newfoundland Department of Mines and Energy, Report 92-1, p. 331-366.

Gower, C.F., Flanagan, M.J., Kerr, A., and Bailey, D.G.
1982: Geology of the Kaipokok Bay-Big River area, Central Mineral Belt, Labrador; Newfoundland Department of Mines and Energy, Report 82-7, 77 p.

Gower, C.F., Ryan, B., and Rivers, T.
1990: Mid-Proterozoic Laurentia-Baltica: an overview of its geological evolution and a summary of contributions made in this volume; in Mid-Proterozoic Laurentia-Baltica, (ed.) C.F. Gower, B. Ryan, and T. Rivers; Geological Association of Canada, Special Paper 32, p. 1-20.

Gustafsson, B.
1981: Uranium exploration in the N. Vasterbotten - S. Norbotten Province, northern Sweden; in Uranium Exploration Case Histories; Proceedings of an Advisory Group Meeting, November 1979; International Atomic Energy Agency, Vienna, p. 333-352.

Hålenius, U., Smellie, J.A.T., and Wilson, M.R.
1986: Uranium genesis within the Arjeplog-Arvidsjaur-Sorcelle uranium province, northern Sweden; in Vein Type Uranium Deposits, International Atomic Energy Agency, Vienna, IAEA-TECDOC-361, p. 21-42.

Hildebrand, R.S., Hoffman, P.F., and Bowring, S.A.
1987: Tectono-magmatic evolution of the 1.9-Ga Great Bear magmatic zone, Wopmay orogen, northwestern Canada; Journal of Volcanology and Geothermal Research, v. 32, p. 99-118.

Hoffman, P.F.
1980: Wopmay Orogen: a Wilson cycle of Early Proterozoic age in the northwest of the Canadian Shield; in Continental Crust and its Mineral Deposits; (ed.) D.W. Strangway; Geological Association of Canada, Special Paper 20, p. 523-549.

LeCheminant, A.N., Miller, A.R., and LeCheminant, G.M.
1987: Early Proterozoic alkaline igneous rocks, District of Keewatin, Canada: petrogenesis and mineralization; in Geochemistry and Mineralization of Proterozoic Volcanic Suites, (ed.) T.C. Pharoah, R.D. Beckinsale, and D. Rickard; Geological Society Special Publication, no. 33, p. 219-240.

Lindroos, H. and Smellie, J.
1979: A stratabound uranium occurrence within Middle Precambrian ignimbrites at Duobblon, Northern Sweden; Economic Geology, v. 74, no. 5, p. 1131-1152.

Locardi, E.
1985: Uranium in acidic volcanic environments; in Uranium Deposits in Volcanic Rocks, Proceedings of a Technical Committee Meeting, El Paso, Texas, April 1984; International Atomic Energy Agency, IAEA-TECDOC-490, p. 17-27.

Locardi, E. and Mittemperger, M.
1971: Exhalative supergenic uranium, thorium and marcasite occurrences in Quarternary volcanites of central Italy; Bulletin Volcanologique, v. 35, p. 173-184.

Marten, B.E.
1977: The relationship between the Aillik Group and the Hopedale Complex, Kaipokok Bay, Labrador; PhD. thesis, Memorial University of Newfoundland, St. John's, Newfoundland, 244 p.

Miller, A.R.
1980: Uranium geology of the eastern Baker Lake basin, District of Keewatin, Northwest Territories; Geological Survey of Canada, Bulletin 330, 63 p.

Miller, A.R. and LeCheminant, A.N.
1985: Geology and uranium metallogeny of Proterozoic supracrustal successions central District of Keewatin, N.W.T. with comparisons to northern Saskatchewan; in Geology of Uranium Deposits, (ed.) T.I.I. Sibbald and W. Petruk; The Canadian Institute of Mining and Metallurgy, Special Volume 32, p. 167-185.

Miller, A.R., Stanton, R.A., Cluff, G.R., and Male, M.J.
1986: Uranium deposits and prospects of the Baker Lake Basin and sub-basins Central District of Keewatin, N.W.T.; in Uranium Deposits of Canada, (ed.) E.L. Evans; The Canadian Institute of Mining and Metallurgy, Special Volume 33, p. 263-285.

Miller, R.G.
1982: The geochronology of uranium deposits in the Great Bear batholith, Northwest Territories; Canadian Journal of Earth Sciences, v. 19, no. 7, p. 1428-1448.

Morton, R.D., Aubut, A., and Gandhi, S.S.
1978: Fluid inclusion studies and genesis of the Rexspar uranium-fluorite deposit, Birch Island, British Columbia; in Current Research, Part B; Geological Survey of Canada, Paper 78-1B; p. 137-140.

Preto, V.A.
1978: Setting and genesis of uranium mineralization at Rexspar; The Canadian Institute of Mining and Metallurgy, Bulletin, v. 71, no. 800, p. 82-88.

*Ruzicka, V.
1992: Types of uranium deposits in the former U.S.S.R.; Colorado School of Mines, Quarterly Review, v. 92, no. 1, p. 19-27.

Rytuba, J.J. and Glazman, R.K.
1979: Relation of mercury, uranium and lithium deposits of the McDermitt caldera complex, Nevada-Oregon; in Papers on Mineral Deposits of Western North America, Nevada Bureau of Mines and Geology, Report 33, p. 109-118.

Schärer, U., Krogh, T.E., Wardle, R.J., Ryan, A.B., and Gandhi, S.S.
1988: U-Pb ages of Lower and Middle Proterozoic volcanism and metamorphism in the Makkovik Orogen, Labrador; Canadian Journal of Earth Sciences, v. 25, no. 7, p. 1987-1107.

*Sherborne, J.E., Jr., Buckovic, W.A., DeWitt, D.B., Hellinger, T.S., and Pavlak, S.J.
1979: Major uranium discovery in volcaniclastic sediments, Basin and Range Province, Yavapai County, Arizona; American Association of Petroleum Geologists, v. 63, no. 4, p. 621-646.

Treuil, M.
1985: A global geochemical model of uranium distribution and concentration in volcanic rock series; in Uranium Deposits in Volcanic Rocks, Proceedings of a Technical Committee Meeting, El Paso, Texas, April 1984, International Atomic Energy Agency, Vienna, IAEA-TECDOC-490, p. 53-68.

Wenrich, K.J.
1985: Geochemical characters of uranium enriched rocks; in Uranium Deposits in Volcanic Rocks, Proceedings of a Technical Committee Meeting, El Paso, Texas, April 1984; International Atomic Energy Agency, Vienna, IAEA-TECDOC-490, p. 29-51.

White, M.V.W. and Martin, R.F.
1980: The metasomatic changes that accompany uranium mineralization in the non-orogenic rhyolites of the Upper Aillik Group, Labrador; Canadian Mineralogist, v. 18, p. 459-479.

Wyborn, L.A.I.
1988: Petrology, geochemistry and origin of a major Australian 1880-1840 Ma felsic volcano-plutonic suite: a model for continental felsic magma generation; Precambrian Research, v. 40/41, p. 37-60.

Wyborn, L.A.I., Page, R.W., and Parker, A.J.
1987: Geochemical and geochronological signatures in Australian Proterozoic igneous rocks; in Geochemistry and Mineralization of Proterozoic Volcanic Suites, (ed.) T.C. Pharoah, R.D. Beckinsale, and D. Rickard; Geological Society of London, Special Publication No. 33, p. 377-394.

Xu, Zhigang
1990: Mesozoic volcanism and volcanogenic iron-ore deposits in eastern China; Geological Society of America, Special Paper 237, 46 p.

Yao, Z., Liu, Y., Zhu, R., Yang, S., and Li, D.
1989: Uranium metallogenesis in southeast China; in Metallogenesis of Uranium Deposits, Proceedings of a Technical Committee Meeting, Vienna, March 1987; International Atomic Energy Agency, Vienna, p. 345-357.

*Zielinski, R.A.
1985: Volcanic rocks as sources of uranium: current perspectives and future directions; in Uranium Deposits in Volcanic Rocks, Proceedings of a Technical Committee Meeting, El Paso, Texas, April 1984; International Atomic Energy Agency, Vienna, IAEA-TECDOC-490, p. 83-96.

Authors' addresses

R.T. Bell
P.O. Box 287
2430 Garmil Cr.
North Gower, Ontario
K0A 2T0

S.S. Gandhi
Geological Survey of Canada
601 Booth Street
Ottawa, Ontario
K1A 0E8

13. VEIN URANIUM

13.1 Veins in shear zones
13.2 Granitoid-associated veins

13. VEIN URANIUM

V. Ruzicka

INTRODUCTION

Uranium vein deposits are concentrations of uranium minerals, such as pitchblende, coffinite, and brannerite, in fractures, shear zones, and stockworks. The uranium minerals are commonly accompanied by quartz and/or carbonate gangue. The mineralization is, as a rule, lithologically and structurally controlled. The vein deposits occur either (a) in shear or mylonite zones in various geological environments (veins in shear zones), or (b) within granitic or syenitic plutons or in rocks mantling granitic batholiths (granitoid-associated veins). The host rocks are parts of metasedimentary, metavolcanic, or plutonic complexes. Classical veins in mylonite zones occur in the Beaverlodge area, Saskatchewan; classical veins in shear zones are those in the Bohemian Massif; classical granitoid-associated vein deposits are those of the Massif Central in France and the Gunnar property in Saskatchewan; typical peribatholithic veins occur in rocks mantling the Central Bohemian Pluton, in the Czech Republic.

Ruzicka, V.
1996: Vein uranium; in Geology of Canadian Mineral Deposit Types, (ed.) O.R. Eckstrand, W.D. Sinclair, and R.I. Thorpe; Geological Survey of Canada, Geology of Canada, no. 8, p. 277 (also Geological Society of America, The Geology of North America, v. P-1).

13.1 VEINS IN SHEAR ZONES

V. Ruzicka

IMPORTANCE

In Canada, uranium vein deposits in mylonite zones have been important sources of uranium in the past. The mines in the Beaverlodge area, Saskatchewan (Fig. 13.1-1), produced in excess of 25 000 t of uranium metal between 1950 and 1982 (Ward, 1984). They were the major producers of uranium in Canada before the uranium mines in the Elliot Lake area, Ontario, commenced operations in 1957. In Europe, mineralized shear zones which extend vertically for more than 1000 m have been exploited in the eastern and western parts of the Bohemian Massif, at Rožná – Olší and Zadní Chodov, the Czech Republic, respectively (Ruzicka, 1971, 1993).

SIZE AND GRADE OF DEPOSITS

The vein deposits associated with mylonite or shear zones contain mineralization, as a rule, in lenses and in narrow veinlets. Individual orebodies consist of irregularly distributed lenses. The largest uranium vein deposits of this subtype in Canada were exploited in the Beaverlodge area in northern Saskatchewan (Tremblay, 1972). Numerous orebodies contained from a few tonnes to as much as several thousand tonnes of uranium metal. The mining grades (i.e., grades of ore diluted to mining width) were on average 0.2% U, and the maximum average grade was as great as 0.4% U. Metal tonnage and pertinent mining grades of individual orebodies mined by Eldorado Mines Limited in the Beaverlodge area are shown in Table 13.1-1 (see also Fig. 13.1-2 and Figure 13.1-3). The largest metal tonnage (in the 09 Fay orebody) was in excess of 8000 t of U in ore grading 0.25% U. The highest grade of ore was 0.42% U (the 38 Hab orebody, containing in excess of 500 t U; Ward, 1984).

The Rožná - Olší deposits produced in excess of 20 kt of uranium metal from ore containing on average several kilograms of uranium per tonne.

Ruzicka, V.
1996: Veins in shear zones; in Geology of Canadian Mineral Deposit Types, (ed.) O.R. Eckstrand, W.D. Sinclair, and R.I. Thorpe; Geological Survey of Canada, Geology of Canada, no. 8, p. 278-283 (also Geological Society of America, The Geology of North America, v. P-1).

GEOLOGICAL FEATURES

The characteristic geological features of Canadian uranium vein deposits associated with mylonite zones are best demonstrated by deposits in the Beaverlodge area, specifically the deposit exploited by the Ace, Fay, and Verna mines (Fig. 13.1-3). This deposit consists of numerous orebodies hosted by mylonitized feldspathic quartzite, brecciated and mylonitized granitic gneiss, altered argillite, and brecciated feldspar-carbonate rocks.

The host rocks are part of the Lower Proterozoic metasedimentary sequence of the Tazin Group, and to a lesser extent the Middle Proterozoic sedimentary and volcanic suite of the Martin Formation (Fig. 13.1-4). The rocks of the area have been deformed during at least two orogenies: Kenoran (about 2.6 Ga; Stockwell, 1982) and Hudsonian (about 1.7 Ga; Stockwell, 1982). During the Kenoran Orogeny, gneiss domes and uranium-bearing pegmatite dykes were formed. The Hudsonian Orogeny caused mylonitization of Tazin Group rocks and reactivation of major fault systems, such as the St. Louis, which provided the main structural control for the Ace-Fay-Verna deposit. Deposition of the rocks of the Martin Formation during the late stages of this orogeny was accompanied by volcanism.

Most of the orebodies are spatially related to the St. Louis Fault and to fracture systems associated with crossfaults, such as the Larum, South Radiore, and George Lake faults (Fig. 13.1-4). Stratigraphically the orebodies occur near the Tazin-Martin unconformity. The rocks in the footwall of the St. Louis Fault represent units of the Tazin domain and may form part of a subduction zone assemblage; the fault itself apparently represents a suture zone in the region.

The ore is epigenetic. The main episode of mineralization was associated with the Hudsonian Orogeny, but pitchblende was subsequently remobilized. Ward (1984) reported the results of a statistical analysis of a total of approximately 120 age determinations (U-Pb) on pitchblendes from the Beaverlodge deposit. This analysis shows clustering of isotopic ages in three periods, namely at 1.8 to 1.7 Ga, which corresponds with Hudsonian events, at 1.1 to 0.9 Ga, which corresponds with Grenvillian events, and at about 0.2 Ga.

The mineralization exhibits vertical zoning (Ruzicka, 1989). The upper parts of the main orebodies consist essentially of pitchblende, whereas the lower (deeper) parts contain substantial amounts of brannerite. Locally the

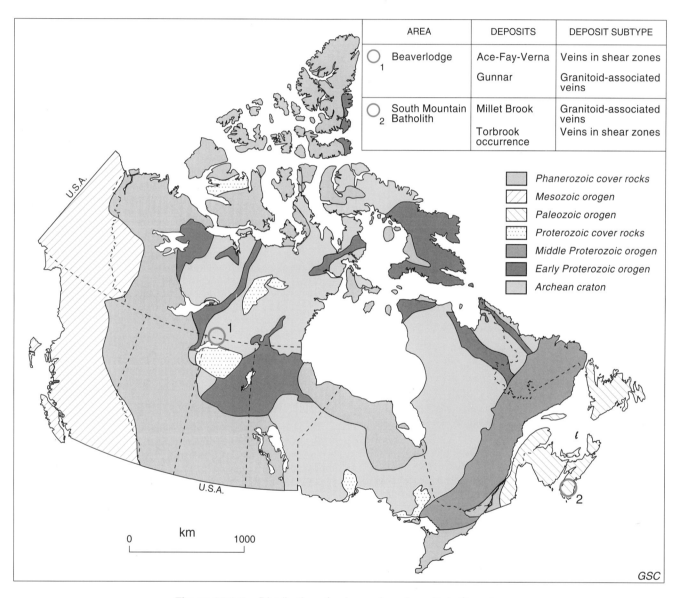

	AREA	DEPOSITS	DEPOSIT SUBTYPE
1	Beaverlodge	Ace-Fay-Verna	Veins in shear zones
		Gunnar	Granitoid-associated veins
2	South Mountain Batholith	Millet Brook	Granitoid-associated veins
		Torbrook occurrence	Veins in shear zones

Phanerozoic cover rocks
Mesozoic orogen
Paleozoic orogen
Proterozoic cover rocks
Middle Proterozoic orogen
Early Proterozoic orogen
Archean craton

Figure 13.1-1. Distribution of vein uranium deposits in Canada.

Table 13.1-1. Tonnages of ores, their grades, and contained uranium metal in orebodies in mines of former Eldor Resources Limited Beaverlodge operations. (Cumulative production and reserves as of 1975.)

Symbol	Orebody	Ore (kt)	U (t)	Grade (U%)
1	01 Fay	1347	3033	0.23
2	04 Center	98	175	0.18
3	09 Fay	3247	8362	0.25
4	16 Center	274	615	0.23
5	38 Hab	128	532	0.42
6	39 Hab	55	130	0.24
7	43 West Fay	113	202	0.18
8	44 Verna	315	476	0.15
9	55 West Fay	151	244	0.16
10	64 Center	327	389	0.12
11	71 Verna	76	111	0.14
12	73 Verna	1355	2104	0.15
13	76 Verna	738	1436	0.2
14	79 Verna	893	1186	0.14
15	91 Zone	178	249	0.14
16	93 Verna	990	1762	0.18
17	Other (<100 t U each)	258	353	0.14

metastable form of pitchblende, may correspond to the euhedral alpha-triuranium-heptaoxide form in the Key Lake, Eagle Point, and Cigar Lake unconformity-associated uranium deposits (Dahlkamp and Tan, 1977; Ruzicka and Littlejohn, 1982; Ruzicka, 1984). The botryoidal pitchblende is locally cut by selenides. The massive pitchblende occurs as veins and breccia fragments. Sooty pitchblende represents a finely disseminated variety that has been derived from massive or botryoidal pitchblende by hydration shattering, as indicated by the preserved forms and structures.

Fluid inclusion studies of quartz and carbonate gangue minerals from the Ace-Fay-Verna deposit indicate (Sassano, 1972) that the mineralizing fluids were derived from pore fluids in the host rocks and that the exchange reactions took place initially at a temperature of 500°C, and that the temperature gradually dropped to 80°C. Oxygen isotopic studies show that no additional fluids were later introduced and that the system was essentially closed, i.e. that only redistribution and redeposition of uranium and associated metals took place during subsequent reactivation of the vein system. The presence of albite in the deposits indicates that sodic metasomatism occurred during the early stages of the mineralization process. Wall rocks adjacent to the orebodies are commonly hematitized, feldspathized, chloritized, and carbonatized. Oligoclase and quartz locally cement wall rock breccias.

The pitchblende-coffinite mineralization of the Rožná Olší and Zadní Chodov deposits, in the Czech Republic, is associated with shear and fault zones containing chlorite, carbonate and, locally, graphite, quartz, and albite gangue. The mineralized zones of the Rožná – Olší deposits, several tens of kilometres long, are parts of the Labe Lineament fault system, which also contains several other smaller deposits. The Zadní Chodov deposit is spatially related to the Bohemian Quartz Lode, a major fault, which is part of a 60 km long regional fault system (Ruzicka, 1971).

DEFINITIVE CHARACTERISTICS

Vein deposits associated with mylonite zones in the Beaverlodge area are lithologically controlled by Aphebian rocks of the Tazin Group and Helikian rocks of the Martin Formation. They are structurally controlled by major faults and apparently by the Aphebian-Helikian unconformity. The principal ore-forming minerals are pitchblende, which is abundant in upper portions of the deposits, and brannerite, which is associated with pitchblende in the deeper portions. The deposits in the Bohemian Massif are lithologically controlled by Precambrian metasedimentary and metavolcanic mylonitized rocks as well as by younger granitoid intrusions, and structurally by major regional fault systems (Ruzicka, 1971, 1993).

veins contain coffinite, uranoan carbon (thucholite), and secondary uranium minerals, such as metauranocircite, liebigite, and becquerelite, which are alteration products after pitchblende. Also, locally, sulphides of iron, copper, lead, and zinc, and selenides of lead, are present.

The principal gangue minerals in the Ace-Fay-Verna deposit are carbonates and the minor ones are quartz, chlorite, and albite. Sassano (1972) distinguished five generations of carbonates, four of calcite, and one of dolomite. The quartz gangue occurs in two generations: older comb quartz and a younger clear variety.

The main ore-forming mineral, pitchblende, occurs in a number of forms: euhedral (Robinson, 1955), botryoidal, massive, and sooty. The euhedral pitchblende, which is a

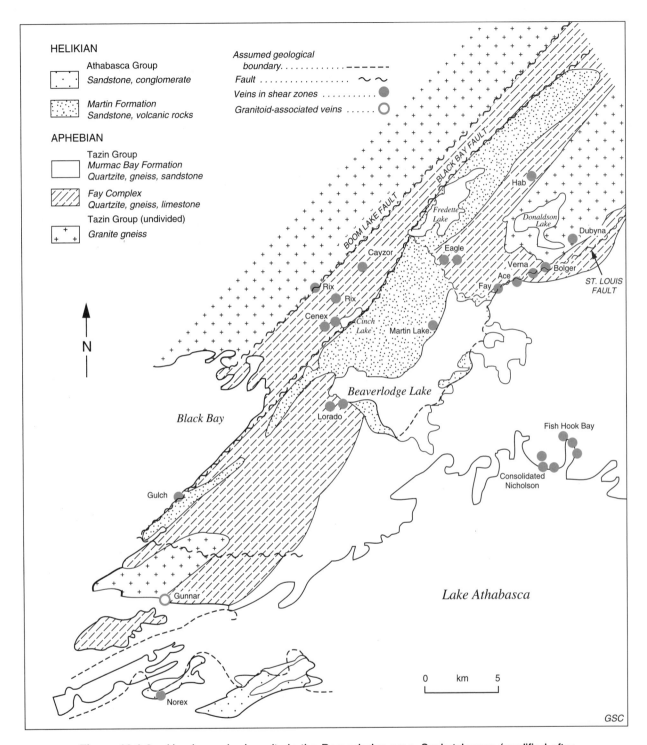

Figure 13.1-2. Uranium vein deposits in the Beaverlodge area, Saskatchewan (modified after L.P. Tremblay, unpub. report, 1978).

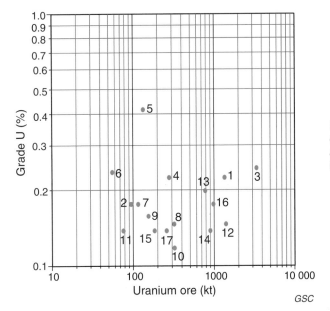

Figure 13.1-3. Uranium metal/grade relationships of orebodies, Ace-Fay-Verna mining field, Beaverlodge area, Saskatchewan. Numbers correspond to deposits in Table 13.1-1.

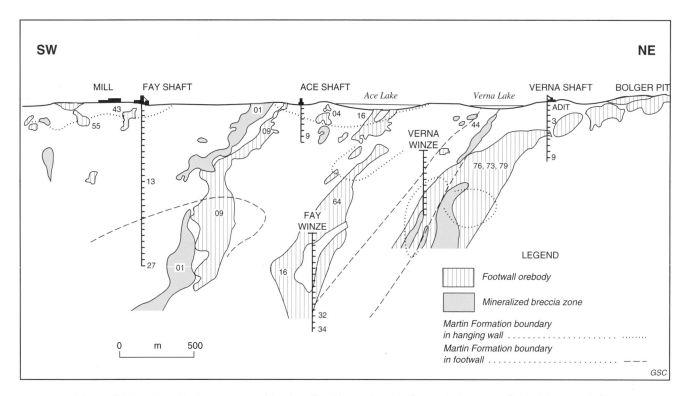

Figure 13.1-4. Longitudinal section of the Ace-Fay-Verna deposit, Beaverlodge area, Saskatchewan (after unpublished documentation of the former operator, Eldor Resources, Limited, 1980). Red vertical line pattern indicates orebodies in the footwall of the St. Louis Fault. Red contoured areas indicate mineralized breccia zones. Rocks of the Martin Formation are shown by dashed (in the footwall of the St. Louis Fault) or dotted (in the hanging wall of the St. Louis Fault) lines. Numbers designate orebodies keyed to Table 13.1-1.

GENETIC MODEL

In the Canadian Shield the vein deposits in shear and mylonite zones are part of a major metallogenic cycle, which started with introduction of uranium that was incorporated within granitoid plutons during the Kenoran Orogeny. Uranium and associated elements were further concentrated in sedimentary rocks of the Tazin Group, which were mylonitized and faulted during the Hudsonian Orogeny. The metals from the sediments were remobilized and redeposited by hydrothermal processes to form fracture fillings, stockworks, and disseminations in the host rocks. The mineralization was controlled lithologically and structurally. Sodic metasomatism (albitization) took part in the early stages of the mineralization process. Hematitization, chloritization, feldspathization, and carbonatization are the main types of alteration associated with the mineralization.

In the Bohemian Massif, the uranium in the original sedimentary and volcanic rocks has been mobilized by metamorphic, tectonic, and igneous events and redistributed in mylonitic and fault zones (Ruzicka, 1971, 1993). The mineralization processes were accompanied by chloritization, carbonatization, and locally, by albitization of the host rocks.

EXPLORATION GUIDES

Vein uranium deposits similar to those in the Beaverlodge area, Saskatchewan, should be sought in areas underlain by granitoid plutons of the Kenoran Orogeny containing elevated radionuclide contents. Aphebian metasedimentary and metavolcanic rocks should flank these granitoid bodies, and clastic sedimentary rocks should unconformably overlie the metasedimentary-metavolcanic sequence. The complex should be faulted, mylonitized, and subjected to retrograde metamorphism. Hematitization, feldspathization, and chloritization of wall rocks, adjacent to carbonate or quartz-carbonate veins would indicate that hydrothermal processes, a prerequisite for mineralization, were operative in the area.

13.2 GRANITOID-ASSOCIATED VEINS

V. Ruzicka

INTRODUCTION

Granitoid-associated uranium-bearing veins are concentrations of uranium minerals that fill fractures within granitic or syenitic plutons (intragranitic veins) or in rocks mantling granitic batholiths (peribatholithic veins).

Intragranitic uranium veins are typically developed in highly differentiated granitic rocks, e.g., in two-mica leucocratic granites that were subjected to preceding alteration, such as albitization and desilicification (episyenitization). The deposits are spatially related to regional faults or lineaments. Their principal uranium minerals, pitchblende and coffinite, are commonly associated with sulphides and gangue minerals, such as carbonates, quartz, chalcedony, fluorite, and barite.

Ruzicka, V.
1996: Granitoid-associated veins; in Geology of Canadian Mineral Deposit Types, (ed.) O.R. Eckstrand, W.D. Sinclair, and R.I. Thorpe; Geological Survey of Canada, Geology of Canada, no. 8, p. 283-285 (also Geological Society of America, The Geology of North America, v. P-1).

In Canada the best representatives of the intragranitic veins are the Gunnar deposit, which occurs on the north shore of Lake Athabasca, about 30 km southwest from the Ace-Fay-Verna deposit (Evoy, 1960; Gandhi, 1983), and the Millet Brook deposit in the South Mountain Batholith, Nova Scotia (Chatterjee and Strong, 1984). The host rocks of the Gunnar deposit are essentially 'episyenitic' in the terminology used for the altered granitoid plutons in France, pervasively albitized and carbonatized. The Millet Brook deposit occurs in sheared, metasomatically altered granitic rocks. Typical intragranitic vein deposits are associated with Hercynian granitoid plutons in France, Portugal, and China (Ruzicka, 1993).

Peribatholithic veins have been discovered in Canada in Cambro-Ordovician rocks of the Meguma Group and their analogues in Nova Scotia, in the proximity of their contacts with the South Mountain Batholith (e.g., occurrences at Torbrook, Lamb's Lake, and Inglisville along the southern margin of the Annapolis valley). The occurrences consist of massive pitchblende associated mainly with carbonate gangue in short tension fractures in pelitic and psammitic rocks. Important deposits of this type occur in Europe in the Bohemian Massif (e.g., the Příbram deposit, Czech Republic).

IMPORTANCE AND SIZE OF DEPOSITS

In Canada uranium resources in granitoid-associated veins are much less significant than those in veins in shear zones. The Gunnar deposit yielded, during nine years of production, little more than 6000 t of uranium metal from ores grading 0.13% U (Evoy, 1960). The Millet Brook deposit contains dormant resources amounting to less than 500 t U (Chatterjee and Strong, 1984). In Europe, particularly in France and Portugal, however, intragranitic vein deposits represent a substantial source of uranium production. In China intragranitic vein deposits contain a major part of the country's uranium resources. The most important deposits occur in the Xiazhuang mining district, Guandong Province, southern China.

Peribatholithic veins account for only a small amount of Canada's subeconomic uranium resources. However, the Příbram deposit, Czech Republic, produced in the period 1950 to 1990 in excess of 55 000 t of U from ores grading about 0.64% U (NUEXCO, 1990).

GEOLOGICAL FEATURES

The Gunnar uranium deposit occurs in rocks of the Aphebian Tazin Group, at a contact between granite gneiss and syenite, which in turn are in contact with paragneiss. The pipe-like orebody consists of pitchblende and uranophane. The host rock, which originally had a granitic composition, is highly albitized, irregularly carbonatized, and locally silicified. The albitization of the granitic rock produced albite syenite and albite granite by complete or partial metasomatic replacement of the quartz, microcline, and perthite in the original rocks. Carbonate (calcite) was introduced during the subsequent uranium-bearing stage and replaced some albite and some remaining quartz in the albite granite, and in this way produced calcite syenite (Evoy, 1960).

The ore, consisting of pitchblende, coffinite, uranophane, and associated gangue minerals such as calcite, quartz, chlorite, and hematite, was formed from epigenetic hydrothermal solutions in the metasomatic syenite (or "sponge rock" of Lang et al., 1962). The mineralization process was succeeded by local kaolinization of the host rocks.

Metasomatic replacement of quartz and some other rock-forming minerals by sodic feldspars and micas, and subsequent carbonatization, is associated with some intragranitic vein deposits that are related to Phanerozoic Hercynian granites in France. This process, similar to that at the Gunnar deposit, has been called "episyenitization" by French geologists (Poty et al., 1974).

The Millet Brook deposit, Nova Scotia, occurs in alteration zones in granodiorite of the South Mountain Batholith. The alteration zones, as much as 30 m wide, were developed along fractures in four stages: potassic, sodic, ferruginous, and calcic, by replacement of quartz in the biotite granodiorite, which is 370 Ma old. The uranium-bearing solutions, which produced the mineralization and alteration, were apparently derived from a leucomonzogranite body that intruded the granodiorite (Chatterjee and Strong, 1984).

In China some deposits, for example, Zhushanxia, Shijiaowei, and Xiwang, are associated with granitic rocks of the Yanshanian structural unit. The host granites were, however, emplaced during a preceding orogeny. The host rocks formed 185 to 135 Ma ago, but uranium was deposited 85 to 70 Ma ago (Li Tiangang and Huang Zhizhang, 1986; Du Letian, 1986; Chen Zuyi and Huang Shijie, 1986).

CONCEPTUAL MODEL AND EXPLORATION GUIDES

Intragranitic vein uranium deposits occur, as a rule, in alteration zones which have developed along structures of high permeability in granitic bodies. Some of the fluids that caused the alteration (sodic, potassic, carbonatic) also introduced uranium mineralization into fractures or other open spaces, developed by dissolution of rock-forming minerals. Therefore, areas underlain by granitoid rocks that have been affected by sodic, potassic, or calcic metasomatic processes should be considered favourable for the occurrence of intragranitic vein uranium deposits.

Favourable environments for peribatholithic veins exist in areas containing sedimentary and metamorphic rocks that mantle granitic plutons, particularly those in which granitic bodies have intruded black shales. The black shales are considered by some authors (e.g., Ruzicka, 1971) not only as a metal source, but also as favourable host rocks.

RELATED DEPOSIT TYPES

These vein uranium deposits (13.1 and 13.2) are similar in some features to other deposit types. They are closely similar in morphology and mineral composition, in particular, to some unconformity associated deposits (Tremblay, 1982). For example, the shapes and monometallic compositions of the orebodies of the Eagle Point deposit, Saskatchewan, which is associated with the unconformity underlying the Athabasca group rocks, strongly resemble those of the Ace, Fay, and Verna deposits. Furthermore, some of the vein deposits are, at least spatially, related to unconformities. For example, the Ace-Fay-Verna orebodies occur mainly below, but also just above, the unconformity underlying the Martin Formation rocks. The Proterozoic rocks of the Bohemian Massif host the veins of the Příbram deposit, which extend from the unconformity below the Cambrian rocks to a depth of more than 2000 m.

ACKNOWLEDGMENT

Critical reading of the manuscript by R.I. Thorpe of the Geological Survey of Canada is gratefully acknowledged.

SELECTED BIBLIOGRAPHY

References with asterisks (*) are considered to be the best source of general information on this deposit type.

Chatterjee, A.K. and Strong, D.F.
1984: Discriminant and factor analysis of geochemical data from granitoid rocks hosting the Millet Brook uranium mineralization, South Mountain Batholith, Nova Scotia; Uranium, Elsevier Science Publishers B.V. Amsterdam, v. 1, p. 289-305.

Chen Zuyi and Huang Shijie
1986: Summary of uranium geology of China; Beijing Uranium Geology Research Institute, Beijing, unpublished report, 21 p.

Dahlkamp, F.J. and Tan, B.
1977: Geology and mineralogy of the Key Lake U-Ni deposits, northern Saskatchewan, Canada; in Geology, Mining and Extraction Processing of Uranium, (ed.) M.J. Jones; Institute of Mining and Metallurgy, London, p. 145-158.

Du, Letian
1986: Granite-type uranium deposits of China; in Vein Type Uranium Deposits, (ed.) H. Fuchs; International Atomic Energy Agency, Vienna, IAEA-TECDOC-361, p. 377-393.

Evoy, E.F.
1960: Geology of the Gunnar uranium deposit, Beaverlodge area, Saskatchewan; PhD. thesis, University of Wisconsin, Madison, Wisconsin, 62 p.

Gandhi, S.S.
1983: Age and origin of pitchblende from the Gunnar deposit, Saskatchewan; in Current Research, Part B; Geological Survey of Canada, Paper 83-1B, p. 291-297.

*Lang, A.H., Griffith, J.W., and Steacy, H.R.
1962: Canadian deposits of uranium and thorium; Geological Survey of Canada, Economic Geology Report 16 (second edition), 324 p.

Li Tiangang and Huang Zhizhang
1986: Vein uranium deposits in granites of Xiazhuang ore field; in Vein Type Uranium Deposits, (ed.) H. Fuchs; International Atomic Energy Agency, Vienna, IAEA-TECDOC-361, p. 359-376.

NUEXCO
1990: Pribram Uranium District; Raw Materials, no. 267, p. 27-31.

Poty, B.P., Leroy, J., and Cuney, M.
1974: Les inclusions fluides dans les minerais des gisements d'uranium intragranitiques du Limousin et du Forez (Massif Central, France); in Formation of Uranium Ore Deposits, International Atomic Energy Agency, Vienna, SM-183/17, p. 569-582.

Robinson, S.C.
1955: Mineralogy of uranium deposits, Goldfields, Saskatchewan; Geological Survey of Canada, Bulletin 31, 128 p.

Ruzicka, V.
*1971: Geological comparison between East European and Canadian uranium deposits; Geological Survey of Canada, Paper 70-48, 195 p.
1984: Unconformity-related uranium deposits in the Athabasca Basin Region, Saskatchewan; in Proterozoic Unconformity and Stratabound Uranium Deposits, International Atomic Energy Agency, Vienna, IAEA-TECDOC-315, p. 219-267.
1989: Conceptual genetic models for important types of uranium deposits and areas favourable for their occurrence in Canada; in Proterozoic Unconformity and Stratabound Uranium Deposits, (ed.) J. Ferguson; International Atomic Energy Agency, Vienna, IAEA-TECDOC-315, p. 49-79.
*1993: Vein uranium deposits; in Vein-type Ore Deposits, (ed.) S.J. Haynes; Ore Geology Reviews, v. 8, p. 247-256.

Ruzicka, V. and Littlejohn, A.L.
1982: Notes on mineralogy of various types of uranium deposits and genetic implications; in Current Research, Part A; Geological Survey of Canada, Paper 82-1A, p. 341-349.

*Sassano, G.
1972: The nature and origin of the uranium mineralization at the Fay mine, Eldorado, Saskatchewan, Canada; PhD. thesis, University of Alberta, Edmonton, Alberta, 272 p.

Stockwell, C.H.
1982: Proposals for time classification and correlation of Precambrian rocks and events in Canada and adjacent areas of the Canadian Shield; Geological Survey of Canada, Paper 80-19, 135 p.

Tremblay, L.P.
*1972: Geology of the Beaverlodge Mining area, Saskatchewan; Geological Survey of Canada, Memoir 367, 265 p.
1982: Geology of the uranium deposits related to the sub-Athabasca unconformity, Saskatchewan; Geological Survey of Canada, Paper 81-20, 56 p.

Ward, D.M.
1984: Uranium geology, Beaverlodge area; in Proterozoic Unconformity and Stratabound Uranium Deposits, (ed.) J. Ferguson, International Atomic Energy Agency, Vienna, IAEA-TECDOC-315, p. 269-284.

Author's address

V. Ruzicka
Geological Survey of Canada
601 Booth Street
Ottawa, Ontario
K1A 0E8

14. ARSENIDE VEIN SILVER, URANIUM

14.1 Arsenide vein silver-cobalt
14.2 Arsenide vein uranium-silver

14. ARSENIDE VEIN SILVER, URANIUM

V. Ruzicka and R.I. Thorpe

INTRODUCTION

Arsenide-silver-uranium vein deposits are epigenetic concentrations of silver minerals, arsenides, sulpharsenides, sulphides, and accompanying chalcophile and siderophile elements, and of uranium oxides and silicates and accompanying lithophile (oxyphile) elements. Proportions of the ore constituents in these deposits differ substantially in different areas. For instance, the deposits in the Great Bear Lake area, Northwest Territories, contain a whole spectrum of uranium, silver, cobalt, copper, and associated minerals, whereas in the deposits in the Cobalt area, Ontario, uranium minerals are absent.

As demonstrated by analyses of paragenetic mineral sequences in the Jáchymov deposit, Czech Republic, and the Eldorado deposit at Great Bear Lake, Canada, the introduction of uranium into these deposits took place at distinct stages in the mineralization process. Although the uranium and arsenide-silver assemblages may locally occur together, their deposition took place in separate mineralization stages (Ruzicka, 1971).

Because of the above-mentioned features, deposits of the arsenide silver-uranium vein type have been subdivided. The subtypes are arsenide vein silver-cobalt and arsenide vein uranium-silver.

Ruzicka, V. and Thorpe, R.I.
1996: Arsenide vein silver, uranium; in Geology of Canadian Mineral Deposit Types, (ed.) O.R. Eckstrand, W.D. Sinclair, and R.I. Thorpe; Geological Survey of Canada, Geology of Canada, no. 8, p. 287 (also Geological Society of America, The Geology of North America, v. P-1).

14.1 ARSENIDE SILVER-COBALT VEINS

INTRODUCTION

Vein deposits of the arsenide silver-cobalt subtype are epigenetic and contain important amounts of silver, nickel, cobalt, and bismuth in predominantly arsenide, sulpharsenide, and native forms. The host rocks include Precambrian metasedimentary and metavolcanic rocks, which are, as a rule, intruded by dykes and sills of diabase. The metallic minerals occur in (a) lenses or veinlets that fill fractures, (b) in stockworks, or (c) as impregnations in wall rocks. These minerals are usually associated with carbonate and/or quartz gangue. The wall rocks adjacent to the veins are commonly hydrothermally altered.

IMPORTANCE

The arsenide vein deposits in Canada were important sources of silver during the first half of this century. In addition to silver they yielded cobalt, copper, nickel, arsenic, and bismuth.

The largest silver-producing area in Canada was, for a long period of time, the Cobalt district in Ontario (Fig. 14.1-1). Large amounts of silver from arsenide veins were also produced from the Great Bear Lake mining district, Northwest Territories. Veins in the Thunder Bay area have been of much less economic significance.

SIZE AND GRADE OF DEPOSITS

The arsenide silver-cobalt vein deposits vary from single veins to sets of numerous veins, and range greatly in size and grade. For instance, in the Cobalt district, Ontario, more than 70 mines produced, from 1904 until the end of 1985, 14 545 t of silver and about 25 000 t of cobalt (Thorpe, 1984; Ontario Ministry of Natural Resources, 1984, 1985, 1986; Mohide, 1985). In 1993 there was no production of silver and cobalt in the Cobalt district and the remaining reserves were small; Agnico-Eagle Mines Limited, however, was considering the possibility of recovery of cobalt from old tailings and waste dumps.

In the Northwest Territories the operations of Terra Mines Limited, which included the Silver Bear, Norex, and Smallwood mines, produced about 250 t of silver from 1976 to 1983. More than 70% of the silver was produced from the Silver Bear mine (Brophy, 1985). In 1993 there was no silver production from arsenide veins in the Northwest Territories.

Ruzicka, V. and Thorpe, R.I.
1996:   Arsenide silver-cobalt veins; in Geology of Canadian Mineral Deposit Types, (ed.) O.R. Eckstrand, W.D. Sinclair, and R.I. Thorpe; Geological Survey of Canada, Geology of Canada, no. 8, p. 288-296 (also Geological Society of America, The Geology of North America, v. P-1).

GEOLOGICAL FEATURES

Silver-bearing veins of the Silver Islet mine, within the Island Belt of veins in the Thunder Bay district, Ontario, are arsenide-rich. However, the more important and most typical Canadian representatives of the arsenide-silver-cobalt veins are deposits in the Cobalt district, Ontario. The deposits in this district are associated with Aphebian rocks of the Cobalt Group (Coleman Member of the Gowganda Formation), which consist of conglomerate, quartzite, and greywacke, as well as with major sill-like bodies of Nipissing diabase and with Archean mafic and intermediate lavas and intercalated pyroclastic and sedimentary rocks. Locally the Archean rocks have been intruded by felsite dykes and quartz and feldspar porphyry bodies. The Archean and Aphebian sequences have been intruded by minor quartz diabase dykes, which are younger than the diabase sills (Fig. 14.1-2 and Table 14.1-1). The rocks of the Cobalt district are part of the Superior and Southern structural provinces.

The deposits in the Cobalt district contain three principal mineral assemblages: (i) a relatively minor base metal sulphide assemblage, which is confined to Archean metasedimentary and metavolcanic rocks; (ii) the arsenide silver-cobalt assemblage, which occurs prevailingly at and near the contacts between the Nipissing diabase and the sedimentary rocks of the Cobalt Group, and is present to a lesser extent along contacts between the diabase and the Archean rocks; and (iii) a late stage sulphide assemblage, which is in part distributed along the margins of arsenide-rich veins, where these have apparently been reopened (Table 14.1-2).

The age of the arsenide silver-cobalt veins has been established from geological evidence and from dating of the associated diabase sheets. In the Cobalt area the arsenide silver-cobalt veins cut the Nipissing diabase, but are displaced by postore reverse faults, which are contemporaneous with the intrusion of the quartz diabase dykes. Therefore the deposition of the ore must have taken place after intrusion of the Nipissing diabase sheets, but before intrusion of the quartz diabase dykes, i.e., between 2.22 and 1.45 Ga. The bulk of the ore apparently formed shortly after intrusion of the Nipissing diabase sheets, which took place about 2.22 Ga (Jambor, 1971a; Corfu and Andrews, 1986).

Distribution of the silver-cobalt veins in the Cobalt district is controlled by the contact between the Nipissing diabase sheets and the rocks of the Cobalt Group (Gowganda Formation). The veins occur in the diabase and in the sedimentary rocks within about 200 m of their contact with the diabase. They dip steeply, extend horizontally as much as 1000 m and vertically as much as 120 m, and are as wide as 1.2 m (Fig. 14.1-3). A typical deposit consists of a few short anastomosing veins of variable thickness from a few centimetres to two or three decimetres. The metallic minerals occur in irregular lenses of high grade ore surrounded by aureoles of low grade material (Fig. 14.1-4). Arsenides, sulpharsenides, and antimonides of nickel, cobalt, and iron in various proportions, as well as large amounts of native silver, are the principal metallic constituents of the ore. Carbonates (dolomite, calcite), quartz, and chlorite are typical gangue minerals (Fig. 14.1-5).

The host rocks of the deposits in the Cobalt district were affected by several phases of alteration. Intrusion of the diabase sheets was accompanied by contact metasomatic alteration of the country rocks and by deuteric alteration of the diabase itself. A specific kind of contact alteration is the spotted chloritic alteration, which developed in the vicinity of the Nipissing diabase prior to ore formation. It is characterized by the occurrence of chlorite-rich spots, which are surrounded by chlorite-deficient aureoles, and affected many of the rocks intruded by the diabase.

The most prominent alteration was, however, associated with formation of the ore veins. Its effects depended upon the composition of the rocks involved. For instance, the alteration of diabase resulted in: (i) replacement of pyroxene by actinolite and some chlorite; (ii) retrogression of plagioclase to muscovite, epidote, and albite; and (iii) replacement of ilmenite and magnetite by leucoxene and titanite (Andrews et al., 1986). The hydrothermal wall rock alteration along the ore veins is developed in narrow zones, typically a few centimetres wide. The most distinct alteration zones are developed in the diabase and consist of two or three layers. The first (inner) layer, immediately adjacent to the veins, contains albite, chlorite, and anatase; the

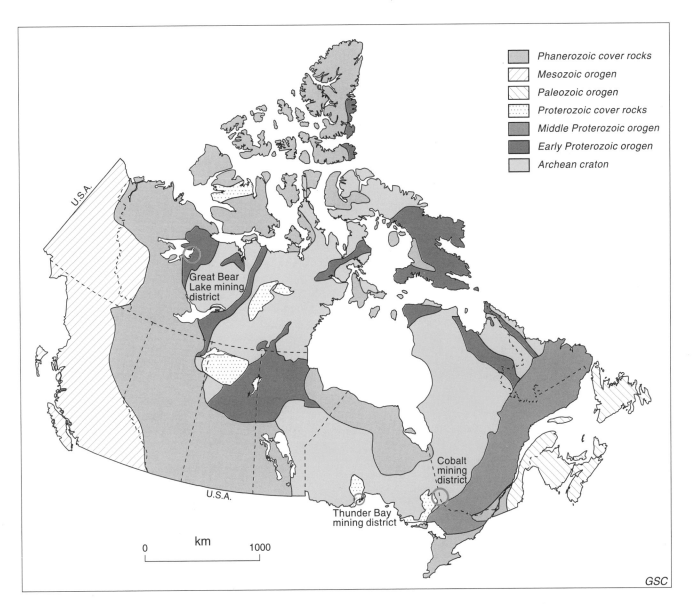

Figure 14.1-1. Locations of the Cobalt, Thunder Bay, and Great Bear Lake districts.

second layer has calcite, epidote, and small amounts of muscovite; and the third (outer) layer comprises increased amounts of muscovite (Fig. 14.1-6; Jambor, 1971b; Andrews et al., 1986).

The ore veins in the Cobalt area contain native silver, dyscrasite, acanthite, rammelsbergite, skutterudite, arsenopyrite, gersdorffite, cobaltite, glaucodot, nickeline, breithauptite, chalcopyrite, tetrahedrite, and native bismuth. Native silver and the cobalt-nickel arsenides are the most abundant ore minerals. Quartz, chlorite, calcite, and dolomite are the most common gangue minerals (Lang et al., 1970; Petruk et al., 1971a, b, c, d; Jambor, 1971c).

The ore minerals occur in masses, lenses, veinlets, and disseminations with or without associated gangue minerals. They are present in distinct mineral assemblages, such as nickel-arsenide, nickel-cobalt-arsenide, cobalt-arsenide, cobalt-iron-arsenide, iron-arsenide, sulphide, and oxide (Petruk, 1971). The nickel-arsenide assemblage is localized in many cases at the periphery of major veins, but also occurs in various places in small veins. The nickel-cobalt-arsenide assemblage occupies a transitional position between the nickel-arsenide and cobalt-arsenide assemblages. Much of the best silver ore is associated with this assemblage. The cobalt-arsenide assemblage occurs generally in the main parts of the veins. The cobalt-iron-arsenide assemblage occurs in various textural forms, such as intergrowths, disseminations, dendrites, rosettes, and monocrystals; however, this assemblage is less common than the preceding ones. Minerals of the iron-arsenide assemblage tend to be concentrated at the ends of the veins. They are commonly accompanied by native bismuth, galena, and marcasite.

Table 14.1-1. Generalized stratigraphy of the Cobalt mining district, Ontario (modified after Jambor, 1971a; Russell, 1983; Owsiacki, 1984).

Eon	Era	Units	Lithology
Phanerozoic	Holocene and Pleistocene		Till, sand, gravel, clay
		--------------- Unconformity --------------- Wabi Group Liskeard Group	Dolomite, limestone, shale
	Silurian and Ordovician		
		--------------- Unconformity --------------- Diabase dykes	Olivine and quartz diabase
Proterozoic	Keweenawan	--------------- Intrusive contact --------------- Nipissing sill	Quartz diabase
	Huronian or equivalent	--------------- Intrusive contact --------------- Cobalt Group: - Lorrain Formation - Gowganda Formation	Conglomerate, quartzite, arkose, greywacke
Archean	Post-Algoman	--------------- Unconformity --------------- Matachewan	Diabase and lamphryphyre
	Algoman	--------------- Intrusive contact ---------------	Granite
	Pre-Algoman	--------------- Intrusive contact --------------- (Haileyburian)	Mafic rocks, lamprophyre, serpentinite
		--------------- Intrusive contact --------------- (Timiskaming)	Greywacke and conglomerate
		--------------- Unconformity --------------- (Keewatin)	Volcanic rocks, iron-formation

The sulphide assemblages typically contain chalcopyrite and tetrahedrite, although more than thirty sulphide minerals have been reported (Petruk, 1971). They occur in some of the main carbonate veins, usually in the peripheral portions of highly mineralized ore sections.

Oxide minerals, hematite, magnetite, rutile, anatase, ilmenite, and wolframite, occur in the veins only in small amounts. They are typically associated with the carbonate gangue.

Analyses for selected elements in samples from individual arsenide assemblages are shown in Table 14.1-3. Twenty-seven samples of the Ni-Co-As assemblage exhibit a large content (1.15%) of antimony. An even larger amount of antimony (5.5%) was found in two samples from the Ni-As assemblage. On the other hand, the lead contents in all of the assemblages are relatively low (from traces of Pb in the Co-As and Co-Fe-As assemblages to a maximum of 0.14% Pb in the Fe-As assemblage).

DEFINITIVE CHARACTERISTICS

The arsenide silver-cobalt vein deposits occur in their classical form in the Cobalt area, Ontario. They are localized in areas affected by basinal subsidence and rifting and are spatially related to regional fault systems and closely associated with intrusions of mafic rocks. Distribution of the veins is structurally controlled by regional fault systems and by the contact zones between the sill-like bodies of Nipissing diabase and the Huronian sedimentary rocks, and less commonly between the Nipissing diabase and Archean rocks.

The silver-nickel-cobalt-arsenide mineral concentrations occur in short (generally less than 100 m), steeply dipping veins at the contacts between diabase and Huronian sedimentary and Archean metasedimentary and metavolcanic rocks. The vein gangue is calcite and dolomite. Alteration haloes are developed in the wall rocks along the veins as narrow (less than 10 cm) zones of calcite, chlorite, epidote, K-feldspar, muscovite, and anatase. Chlorite occurs locally in spots, 1 to 5 mm in diameter.

GENETIC MODEL

The solutions that deposited silver-arsenide ores were initially as hot as 400°C in some cases, although wide ranges of fluid inclusion temperatures (mostly 100°-250°C) and salinities have been recorded (Franklin et al., 1986; Kerrich et al., 1986; Jennings, 1987; Kissin, 1988). In the Thunder Bay district, fluid inclusion temperatures increase with depth (Kissin, 1989; Kissin and Sherlock, 1989). The fluids may have been variable mixtures of basinal brines and meteoric waters. The absence of Ni-Co arsenides in the silver veins of the Mainland Belt in the Thunder Bay district was attributed by Kissin (1989) to dilution of the hydrothermal fluids by meteoric water. There is abundant evidence for repetitive opening of fractures and ore deposition. Kissin (1988) has suggested that the deposits were formed in an environment characterized by incipient rifting of continental crust. Granitic or felsic tuffaceous sequences may be necessary in the metal source areas in the case of U-bearing arsenide vein systems (Kissin, 1988). Kissin also pointed out that the relationship between ores and rifting is strongest at Thunder Bay (Keweenawan rift system), but is also evident in the case

Table 14.1-2. Generalized mineral succession in the main ore-bearing stage of the arsenide silver-cobalt deposits, Cobalt mining district, Ontario (modified after Andrews et al., 1986 and Goodz et al., 1986).

Mineral assemblages	Stage 1 (Early)	Stage 2 (Main)	Stage 3 (Late)
Silicate assemblage: quartz albite epidote K-feldspar chlorite			
Arsenide assemblage: Arsenides Sulpharsenides Sulphantimonides Sulphides Native silver Native bismuth Dolomite Actinolite			
Carbonate assemblage: Calcite			

of the Cobalt and Gowganda districts (Timiskaming rift); Black Hawk, New Mexico (Rio Grande rift); Wickenburg, Arizona (Basin and Range extensional faults); Kongsberg-Modum, Norway (Oslo Graben); and Wittichen and Nieder-Ramstadt, Germany (Rhine Graben). Extensive discussion of alternative genetic models for arsenide silver-cobalt deposits in general is beyond the scope of this summary account, but the reader may find the treatment of genesis by Kissin (1992, 1993) to be helpful.

Many deposits are associated spatially, and perhaps temporally, with granitic intrusions, although no possible granitic source is evident for the important Cobalt-Gowganda and Thunder Bay districts of Ontario (Kissin, 1988). These districts, in fact, are spatially associated with diabasic and gabbroic intrusions. The U-bearing veins in the Port Radium area, Great Bear Lake, are also in close spatial association with gabbroic sheets.

In the case of the arsenide silver-cobalt veins in the Cobalt area, genetic models have been postulated that involve derivation of the Ag, Ni, Co, As, Sb, Bi, Cu, and Hg either from the Archean sedimentary beds, with minor

contributions from certain volcanic flows (Boyle and Dass, 1971), or, more recently, from the formational brines of the Archean carbonaceous, pyritic tuffs or their clastic derivatives in the Proterozoic sedimentary sequence (Watkinson, 1986). The latter hypothesis is supported by fluid inclusion and oxygen isotopic data. A conceptual model, which incorporates sulphidic tuffaceous units as source rocks, was suggested by Watkinson (1986). He inferred from the relatively homogeneous Pb isotopic ratios, as demonstrated by Thorpe et al. (1986), that the metalliferous brines had a long residence time in the sulphide-bearing rocks, but were released into tensional fractures upon intrusion of the Nipissing diabase sheet. The sudden release of pressure caused rapid precipitation of the ore-forming minerals in fractures at the diabase contacts (Watkinson, 1986). According to sulphur isotope studies, the mineralization took place under temperatures between 130°C and 254°C (Goodz et al., 1986). The ore components, principally native silver, arsenic, and cobalt, were introduced into the fractures along with carbonate gangue by hydrothermal solutions of high pH and low Eh.

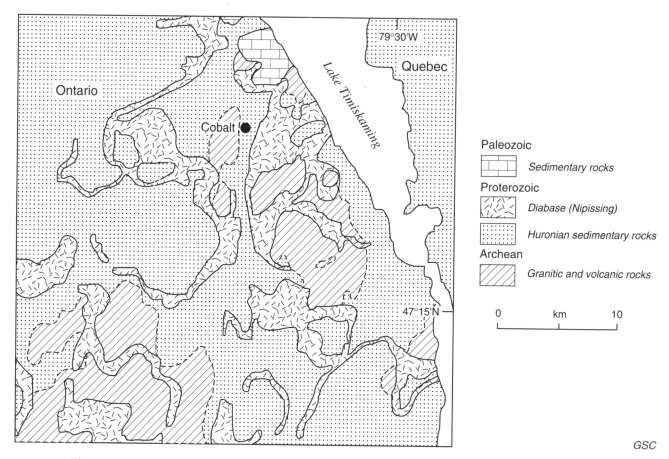

GSC

Figure 14.1-2. Distribution of the principal lithological units in the Cobalt mining district and adjacent area in northeastern Ontario (after Ministry of Natural Resources, 1977).

RELATED DEPOSIT TYPES

The most closely related deposits are those of the arsenide uranium-silver vein subtype. In some deposits, such as those in the Jáchymov area, Czech Republic, almost all the veins with arsenide-silver mineralization contain minor amounts of uranium minerals. In the Great Bear Lake area, a major part of the Silver Bear deposit contains arsenide-silver mineralization without uranium. In the case of both the Silver Bear and Jáchymov area deposits, however, monomineralic uranium veins are also present.

The Bou Azzer deposit, Morocco, which is associated with serpentinites of a late Proterozoic ophiolite, consists of veins, lodes, and stockworks that contain cobalt arsenides and accessory sulpharsenides, copper, and molybdenum sulphides in a quartz-carbonate gangue (Leblanc and Billaud, 1982; Leblanc, 1986). The cobalt arsenide orebodies are controlled by tectonic structures along the borders and top of the serpentinite bodies. Silver is randomly distributed in the cobalt arsenide ore, and is more enriched in association with löllingite in a peripheral zone (Leblanc, 1986). However, silver is a minor constituent in these ores in comparison to the ores of the Cobalt and Great Bear Lake districts.

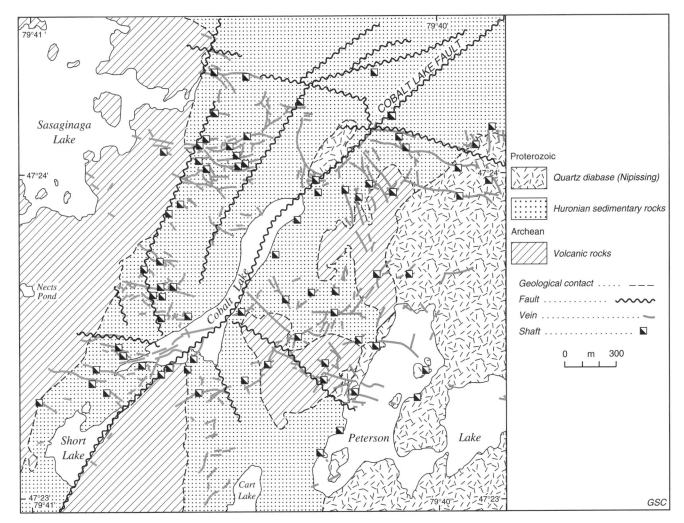

Figure 14.1-3. Arsenide silver-cobalt veins in the central part of the Cobalt mining district, Ontario (modified after Ontario Department of Mines, 1964).

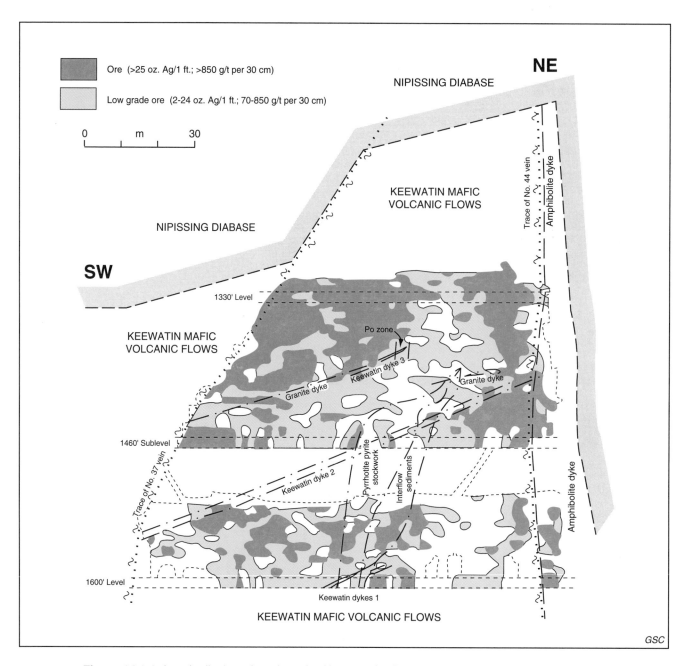

Figure 14.1-4. Longitudinal section along the No. 41 vein, Beaver-Timiskaming mine, Cobalt mining district, Ontario, showing distribution of silver ore (after Robinson, 1984).

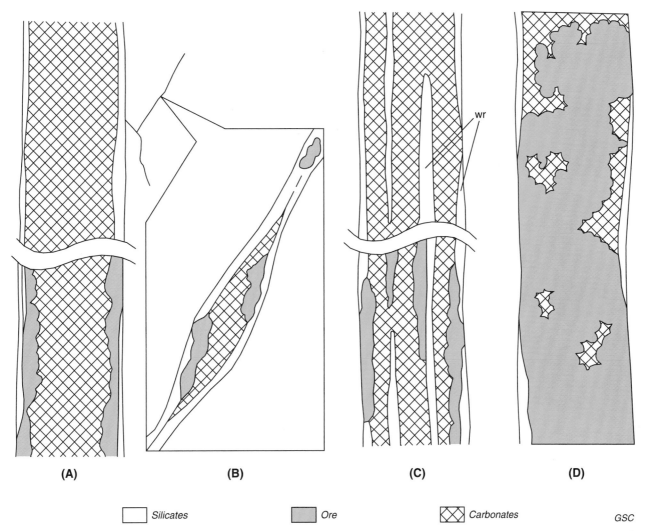

(A)   (B)   (C)   (D)

☐ Silicates   ▨ Ore   ⊠ Carbonates   *GSC*

Figure 14.1-5. Generalized distribution of ore and gangue mineral assemblages in arsenide silver veins, Cobalt mining district, Ontario. A = simple ore vein; B = mineralized apophysis; C = ore vein with chloritized fragments of wall rock (wr); D = rich ore vein (after Andrews et al., 1986).

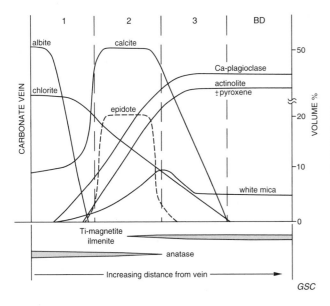

Figure 14.1-6. Distribution of diagnostic minerals in alteration zones (usually each 1-2 cm wide) in diabase adjacent to ore veins; generalized from arsenide silver deposits, Cobalt mining district, Ontario. BD = background diabase (after Andrews et al., 1986).

Table 14.1-3. Elemental and oxide constituents in ore samples from arsenide silver-cobalt veins, Cobalt mining district, Ontario (from Petruk et al., 1971a).

Element/oxide	Assemblage				
	Ni-Co-As	Co-As	Fe-As	Ni-As	Co-Fe-As
Number of samples	27	49	9	2	19
CaO	11.08	13.91	17.01	NA	12.05
MgO	4.64	2.88	3.85	NA	4.11
MnO	0.26	0.18	0.19	NA	0.15
Fe	3.03	5.88	8.83	1.6	9.96
Co	8.52	7.25	2.04	1.6	5.49
Ni	7.74	2.24	0.39	24.6	0.71
Cu	0.06	0.15	0.07	0.03	0.07
Zn	tr	NA	NA	ND	NA
Pb	0.09	tr	0.14	ND	tr
Mo	0.02	0.02	0.01	0.04	0.04
Sb	1.15	0.1	ND	5.5	ND
Bi	0.13	0.07	0.05	0.19	0.1
Cr	0.06	0.03	0.07	0.04	0.01
In	0.02	0.02	0.01	ND	0.02
Ti	0.03	0.04	0.04	ND	0.04
V	0.03	0.03	NA	NA	0.03
Si	0.47	0.46	0.41	0.53	0.44
Al	0.15	0.17	0.23	0.14	0.15

NA = not analyzed; ND = not detected: tr = trace

EXPLORATION GUIDES

Selection of exploration targets for arsenide-silver vein deposits should be based on the occurrence of sulphide-bearing carbonaceous tuffaceous horizons in the Archean and/or Proterozoic complexes located beneath diabase sills. In addition, target areas should contain, in the overlying sequence, permeable rocks capable of yielding formational metalliferous brines. The target area should also have favourable structural features, which include broad dome-like arches of the base of a diabase sill and possible associated structural traps in the form of fracture systems favourable for deposition of metallic minerals from hydrothermal solutions. Identification of the above-mentioned features, which are considered as factors controlling arsenide silver-cobalt mineralization, should be based upon detailed geological mapping combined with electromagnetic and selective multi-element (Ni-Co-As-Ag) geochemical surveys.

14.2 ARSENIDE VEIN URANIUM-SILVER

INTRODUCTION

Deposits of the arsenide uranium vein subtype represent concentrations of uranium minerals, typically pitchblende and coffinite; of arsenides, sulphides, and sulpharsenides of nickel and cobalt; and of native metals. The metallic and associated gangue minerals fill fractures, faults, and shear zones in crystalline rocks and form veins, lodes, and stockworks. The ore deposits usually occur in marginal facies and/or in rocks mantling granitic plutons or batholiths. The common gangue minerals are carbonates (dolomite, calcite, ankerite, rhodochrosite), fluorite, and quartz. The host rocks immediately adjacent to the veins have, as a rule, been affected by hematitization, argillization, and chloritization (Ruzicka, 1989). Commodities produced from deposits of this subtype are uranium, silver, and copper, but locally bismuth, cobalt, and selenium are recovered as byproducts.

Ruzicka, V. and Thorpe, R.I.
1996: Arsenide vein uranium-silver; in Geology of Canadian Mineral Deposit Types, (ed.) O.R. Eckstrand, W.D. Sinclair, and R.I. Thorpe; Geological Survey of Canada, Geology of Canada, no. 8, p. 296-306 (also Geological Society of America, The Geology of North America, v. P-1).

Classical examples of deposits of this category are the former Eldorado mine at Port Radium, Great Bear Lake, Northwest Territories (Stockwell et al., 1970; Thorpe, 1984); deposits in the Jáchymov mining district, northwestern Bohemia (Czech Republic); in the Aue mining district, Saxony, Germany; and the Shinkolobwe deposit in Zaïre. All the above-mentioned deposits have been mined out (Ruzicka, 1971, 1993).

IMPORTANCE

In Europe arsenide uranium vein deposits were, for a long period of time, an important source of silver and, at the end of the nineteenth and first half of the twentieth centuries, they were the sole source of uranium, radium, and cobalt. For example, silver was produced from the Jáchymov deposit in Bohemia from the beginning of the sixteenth century, and it was pitchblende from this deposit that was used by Klaproth when he discovered elemental uranium in 1797. The same deposit has also been the source of cobalt and uranium for a local glass industry since the nineteenth century and, after discovery of radium in tailings of a uranium-dye factory in Jáchymov, also a source of this element. The Eldorado mine at Port Radium was the sole Canadian producer of radium from 1934 until market demand essentially ended, and the sole producer of uranium in Canada from 1938 to 1955. The Eldorado mine also produced large amounts of silver, copper, and cobalt during its lifetime.

SIZE AND GRADE OF DEPOSITS

The above-mentioned deposits and districts contained uranium resources suitable for underground mining in the range from several thousands to several tens of thousands of tonnes from ores that averaged 0.2 to 1.0% U. For instance, the Eldorado mine at Port Radium produced, from 1938 to 1960, about 6000 t of uranium metal in concentrates (Cranstone and Whillans, 1986) from ores grading between 0.3 and 0.6% U. The Jáchymov deposits produced, between 1946 and 1964, 6850 t of uranium metal (Hrádek, 1993). The Shinkolobwe mine, Zaïre produced an estimated total 27 000 to 33 000 t of uranium metal from oxide, silicate, and phosphate ores (Heinrich, 1958) that graded 0.4 to 0.8% uranium (International Atomic Energy Agency, 1977). Numerous mines in the Aue mining district of Germany produced, from the commencement of mining in the fifteenth century, large amounts of silver, some cobalt, selenium, and radium, and several tens of thousands of tonnes of uranium metal in ores that graded between 0.4% and several per cent of uranium metal.

GEOLOGICAL FEATURES

Arsenide uranium-silver vein deposits occur in areas underlain by metasedimentary and metavolcanic rocks that have been intruded by felsic igneous rocks. A typical representative of this subtype in Canada is the Eldorado deposit at Port Radium, Great Bear Lake area, Northwest Territories (Fig. 14.1-1). Other deposits of similar geological setting and morphology, but of somewhat different mineralogical composition and lesser uranium contents, also occur

elsewhere in the Great Bear Lake uranium metallogenic domain, which occupies part of the Great Bear Batholith and some of its roof pendants (Fig. 14.2-1).

Geological setting

The oldest rocks of the Great Bear Lake uranium metallogenic domain belong to the Aphebian Echo Bay Group, which consists of a lower, predominantly metasedimentary sequence, and an upper unit of prevalently volcanic rocks (Mursky, 1973). The rocks of the Echo Bay Group are

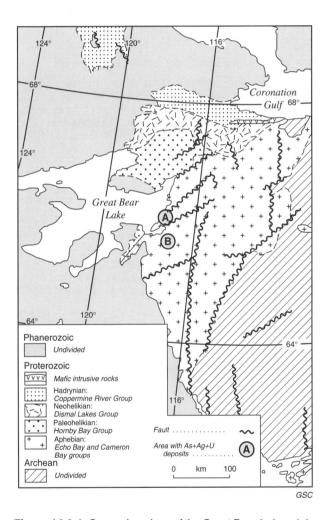

Figure 14.2-1. General geology of the Great Bear Lake mining district and adjacent area, Northwest Territories (modified after Fraser et al., 1978). Area 'A' includes the following deposits: Eldorado (118°02'00"W, 66°05'00"N); Echo Bay (118°01'00"W, 65°05'20"N); Contact Lake (117°47'00"W, 66°00'00"N); El Bonanza (118°04'25"W, 66°00'00"N); and Bonanza (118°05'30"W, 66°00'38"N). Area 'B' includes the following deposits: Terra (118°07'00"W, 65°35'20"N); Northrim (117°59'00"W, 65°35'15"N); Norex (117°58'00"W, 65°35'10"N); and Smallwood (117°57'00"W, 65°35'00"N).

unconformably(?) overlain mainly by sedimentary, and to a lesser extent volcanic rocks of the Cameron Bay Group. Explosive volcanism occurred intermittently during deposition of rocks of the Echo Bay and Cameron Bay groups. The sedimentary-volcanic complexes of the Echo Bay and Cameron Bay groups have been intruded by a suite of igneous rocks that includes feldspar-hornblende porphyry, rhyolite porphyry, granite, quartz monzonite, and granodiorite (Fig. 14.2-1 and Table 14.2-1). Diabase dykes and giant quartz veins cut both the stratified complexes and the intrusive rocks.

The general regional structure of the domain is homoclinal, but locally the rocks are very complexly faulted and folded, particularly at the contacts with intrusive bodies. A northeast trend predominates in the regional structural pattern.

Age of mineralization and host rocks

The polymetallic mineralization of the deposits in the Great Bear Lake area formed during several mineralization stages. Deposition of uranium followed after initial deposition of quartz and hematite, but preceded deposition of nickel and cobalt arsenides (Jory, 1964; Ruzicka, 1971).

The ore-forming processes were related to thermal events within the broader Bear Structural Province. However, the age of this mineralization is not well constrained.

The host rocks belong to a suite of calc-alkaline volcanic rocks that are widespread within the Great Bear magmatic zone. They are dominated by intermediate tuffs and lavas that generally have ages in the range 1875 to 1860 Ma (Hildebrand and Bowring, 1987; Gandhi and Mortensen, 1992), and include the LaBine Group within the Port Radium-Camsell River mining district. These volcanic rocks and their related subvolcanic phases overlie and have intruded an older sialic basement that is apparently represented along the western margin of the Great Bear magmatic zone by the continental-arc-type Hottah Terrane, in which 1936 to 1914 Ma granitoid intrusions have been emplaced (Hildebrand et al., 1983; Housh et al., 1989). These rocks were deformed by the Calderian Orogeny, which culminated at about 1885 Ma (Hoffman and Bowring, 1984). In association with this orogeny, granitic plutons of the Hepburn intrusive suite were emplaced at 1896 to 1878 Ma (Hildebrand et al., 1987; Lalonde, 1989) within the sedimentary back-arc basin succession in the Hepburn belt to the east of the Great Bear magmatic zone. Hornblende-biotite-bearing plutons penecontemporaneous with volcanism and a younger (1860-1840 Ma) series of

Table 14.2-1. Generalized stratigraphy of the Great Bear Lake mining district, Northwest Territories (modified after Mursky, 1973).

Eon	Era	Units	Lithology
Pleistocene and Holocene		Glacial deposits	Sand, gravel, silt, erratics
		------- Unconformity ------- Mineralization Quartz stockworks Diabase dykes and sills	Polymetallic quartz-carbonate veins Giant quartz veins Diabase
		------- Intrusive contact ------- Granitic intrusions	Granite, granodiorite, dirorite, aplite, quartz monzonite
		------- Intrusive contacts ------- Porphyry intrusions	Feldspar-hornblende porphyry, rhyolite porphyry
		------- Intrusive contact -------	
Proterozoic	Cameron Bay Group	------- Unconformity -------	Andesite Argillite Tuff Arkose, sandstone, greywacke Conglomerate
	Echo Bay Group	Upper unit	Andesite flows, quartzite Argillite, tuff, conglomerate, breccia
		(?) (May be younger)	Crystalline tuff
		Lower unit	Chert, argillite, quartzite, tuff, arkose, conglomerate, breccia, andesite, trachyte

epizonal, postfolding biotite syenogranitic plutons, with no known eruptive equivalents, have intruded the suite of widespread volcanic rocks (Hildebrand and Bowring, 1987). Jory (1964) obtained an apparent zircon U-Pb age of 1770 ± 30 Ma for the granitic intrusion off LaBine Point, at Port Radium.

The sedimentary sequence of the Hornby Bay Group, within the upper part of which the Narakay suite of volcanic rocks has yielded a zircon U-Pb age of 1663 ± 8 Ma (Bowring and Ross, 1985), unconformably overlies rocks of the Great Bear magmatic zone within the Coppermine homocline to the north of Great Bear Lake. Cook and MacLean (1993) have suggested that the intracratonic Forward Orogeny in the Colville Hills and Anderson Plain regions to the north and west of Great Bear Lake may be close in age to the Hornby Bay Group. Badham (1973) suggested that a set of vertical, northeast-trending faults (Hildebrand, 1986) that cut the volcanic and plutonic rocks were most active between 1700 and 1400 Ma. A whole-rock Rb-Sr isochron age of 1392 ± 48 Ma has been obtained for the prominent Western Channel Diabase sill or sheet in the vicinity of the Eldorado mine (Wanless and Loveridge, 1978) and Fahrig obtained a K-Ar age of 1400 ± 75 Ma for a biotite-hornblende mixture from another diabase sheet on Hogarth Island, about 26 km northeast of the Eldorado mine (Wanless et al., 1970). Yet another diabase sheet in the area, the Port Radium or Cameron Bay sill, is older. Magnetite-actinolite-apatite veins and pods in the vicinity of the pitchblende-arsenide-silver veins in the Eldorado mine have yielded a K-Ar age of 1408 ± 60 Ma (Robinson, 1971).

Lead isotope data for galena from the arsenide veins (Jory, 1964; Thorpe, 1974; Changkakoti et al., 1986a) yield model ages, based on Pb evolution mixing models recently derived by one of the authors (R.I. Thorpe), of about 1755 to 1600 Ma. The basic parameters for these models are included in Warren et al. (1995). Based on the assumption of continuous in situ radiogenic addition, the moderately radiogenic Pb isotopic compositions for two dolomite specimens (Changkakoti et al., 1986a) from different properties yield a calculated mineralization age of 1855 to 1665 Ma. This range of age is in agreement with the range of possible model ages. However, many of the vein carbonates define a general trend of Pb isotope compositions that suggest a resetting(?) age of roughly 1500 Ma (Changkakoti et al., 1986a).

The Pb isotope data presented by Housh et al. (1989) for feldspars from plutons in the different arc terranes of the Wopmay Orogen also provide useful evidence regarding the age of the arsenide veins. The authors presented the least radiogenic Pb isotopic compositions they obtained from multiple leaches of feldspar separates, and these likely correspond to, or are near initial compositions. The composition for a feldspar from a pluton dated at 1843 Ma is in agreement with the galena data for the arsenide veins (Jory, 1964; Thorpe, 1974; Changkakoti et al., 1986a) when it is forward calculated to 1770 to 1760 Ma. The same is true for a cluster of slightly more primitive feldspars from plutons of the Great Bear magmatic zone at about 1770 Ma, assuming these plutons were emplaced at about 1870 Ma;

likewise for feldspars from the Hepburn intrusive suite at 1780 to 1770 Ma when an age of 1885 Ma is assumed for intrusion of the plutons. The veins could be somewhat younger than this calculated 1780 to 1760 Ma range if a major component of the vein lead was leached from feldspars rather than being representative of the bulk country rock lead.

The ages based on U-Pb data for pitchblende from the veins are significantly younger than the ages based on the Pb isotope data. A moderately radiogenic isotopic composition ($^{206}Pb/^{204}Pb = 99.95$, $^{207}Pb/^{204}Pb = 23.54$) obtained by Jory (1964) for a uranium-bearing tetrahedrite-chalcopyrite specimen can be interpreted, assuming continuous accumulation of radiogenic lead, to indicate primary mineralization was at 1580 to 1550 Ma. However, the more common age result for pitchblende from the Great Bear Lake area is about 1425 to 1420 Ma. Three of six analyses by Jory (1964) for pitchblende from a single specimen yield, on a plot of $^{204}Pb/^{206}Pb$ versus $^{207}Pb/^{206}Pb$, a precise line that indicates a radiogenic $^{207}Pb/^{206}Pb$ ratio of about 0.0898, corresponding to an age of 1421 Ma. Older analyses, as noted by Thorpe (1974) and Miller (1982) are in general accord with this result. Miller (1982) reported a concordia intercept age of 1424 ± 29 Ma based on nine samples, including four of Jory's analyses and three new analyses for specimens from Achook Island, about 31 km north of Eldorado mine. Miller (1982) suggested that an age of about 1419 Ma, based on the two most concordant analyses by Jory, was the most accurate estimate of age. It must be noted, however, that three analyses by Miller (1982) for uranium-bearing specimens from Achook Island indicated a mineralization age of 1500 ± 10 Ma.

The relationship of the pitchblende-arsenide-silver veins at Port Radium to the diabase sheets and dykes in the area is not entirely clear. Jory (1964) concluded that the diabase sheets were intruded after formation of the veins and were fractured during late stage movements on these veins. He noted that the No.1 vein at the Eldorado mine pinched to a few discontinuous veinlets of quartz and carbonate where it entered the diabase. Robinson and Morton (1972) reported that a veinlet of native bismuth and arsenides occurred in the diabase about 6 m from the contact. In the case of such a diabase-hosted vein (possibly the same vein) in the Echo Bay mine, however, projection of the generally regular contact of the sheet of Western Channel Diabase suggested that the vein was likely in an older diabase intrusion.

In summary, we tentatively conclude that the uranium-bearing arsenide veins formed at about 1775 to 1665 Ma. Vein formation may have been related to Narakay volcanism at about 1665 Ma, or perhaps to mafic magmatism represented by the Port Radium-Cameron Bay sill. Based on U-Pb data for pitchblendes and Pb isotope data for vein carbonates, isotopic resetting seems to have occurred at about 1500 Ma and again at about 1420 Ma. It is possible, however, that primary deposition of some pitchblende in the area was at about 1500 Ma. In the light of the available geochronological data, a pitchblende U-Pb age of about 1420 Ma can be provisionally correlated with emplacement of the sheet of Western Channel Diabase and with formation of magnetite-apatite veins and pods.

Associated structure and host rocks

The regional structural pattern of the Great Bear Lake domain is reflected in the Eldorado Port Radium deposit (Fig. 14.2-2). The most prominent northeasterly-trending regional fault system dominates the structural pattern of the deposit.

The structural backbone of the deposit is the northeasterly trending Bear Bay shear, a shear zone 1.5 to 7.0 m wide that is filled with brecciated rock, clay, hematitic and chloritic material, and metallic mineral assemblages. It is a first order dislocation, to which the other ore veins are spatially related.

The other veins, which have been mined out, occurred in the footwall of the Bear Bay shear. They contained the highest concentrations of ore minerals in segments where the veins were deflected, branched, or were intersected by other veins, faults, or dykes. The character of the veins was closely related to the physical properties of the enclosing rocks. The mineralization was commonly greater in rocks favourable for fracturing and maintenance of open spaces, such as tuffs, metamorphosed cherty sedimentary rocks, andesite, and other competent rocks of the Echo Bay Group.

Form of deposits

The ore minerals were deposited as lensoid concentrations in veins, breccia zones, and shear zones which together formed several elongate orebodies that generally trended northeasterly and dipped steeply (Fig. 14.2-3). The length of some veins exceeded 1000 m. The orebodies extended from the surface down to about 500 m. Individual pitchblende lenses locally exceeded one metre in thickness. The ore concentrations were either monomineralic or polymetallic. A typical feature of the mineralization was superposition or telescoping of various mineral assemblages in one vein.

Distribution of ore minerals

The minerals in the Eldorado Port Radium deposit crystallized in several distinct stages that were separated by tectonic movements, which caused brecciation of the vein material. Individual mineralization stages were represented by specific elemental assemblages rather than by individual mineral species (Campbell, 1957; Ruzicka, 1971; Robinson and Ohmoto, 1973; Table 14.2-2). The first (quartz) mineralization stage consisted predominantly of

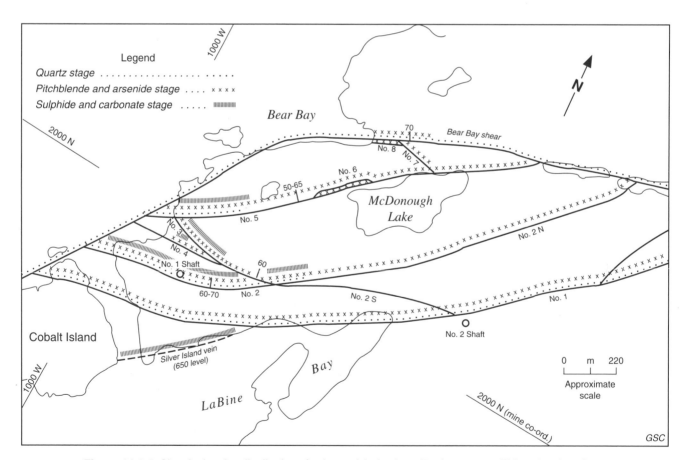

Figure 14.2-2. Sketch showing distribution of veins and their mineralization stages; Eldorado mine, Great Bear Lake mining district, Northwest Territories (after Ruzicka, 1971).

quartz and hematite and occupied extensively the peripheral structures of the deposit, i.e., the Bear Bay shear and the No. 1 vein, and parts of the No. 2 and No. 5 veins adjacent to the Bear Bay shear (Fig. 14.2-2). Mineral assemblages of the second and third (pitchblende and arsenide) stages represented the principal components of the ores and consisted of pitchblende, nickel-, cobalt-, and iron-arsenides, sulpharsenides, and Ni-sulphides associated with carbonates, chlorite, and fluorite. These assemblages occupied all veins in various places, except along the Bear Bay shear. The fourth (sulphide) stage included Cu-, Fe-, Pb-, Zn-, Sb-, As-, Bi-, and Ag-sulphides, and, very rarely, Ag-telluride, associated with carbonate gangue. Minerals of this stage formed in all vein systems in various places, but to a much lesser extent than those of the second and third stages. Mineral assemblages of the fifth (carbonate) stage comprised predominantly calcite, with some associated quartz, and some native silver and bismuth. They occupied the latest fractures and locally the central portions of the main veins.

Ore composition

The proportional amounts of uranium, arsenic, copper, cobalt, nickel, and bismuth in gravity concentrates from the Eldorado mine ores were, according to four-year records, approximately 20:13:7:7:4:1, respectively (Table 14.2-3). Principal constituents of the ores were uranium oxides

(pitchblende), arsenides and sulpharsenides of nickel and cobalt, copper sulphides, native silver, quartz, and carbonates.

Randomly collected specimens of the vein material from the Echo Bay and Contact Lake mines and from a trench near the Echo Bay mine (Ruzicka, 1971; Table 14.2-3)

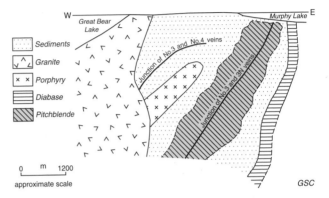

Figure 14.2-3. Longitudinal section along No. 3 vein, Eldorado mine, Great Bear Lake mining district, Northwest Territories (from Ruzicka, 1971).

Table 14.2-2. Generalized paragenesis of constituents in the Eldorado and Echo Bay deposits, Great Bear Lake mining district, Northwest Territories. (Compiled after Jory, 1964; Mursky, 1973; Changkakoti et al., 1986b. The oxygen and carbon isotopic values were calculated from data in Changkakoti et al., 1986b.)

Mineral Assemblages/minerals	Stage				
	1 Quartz	2 Pitchblende	3 Arsenide	4 Sulphide	5 Carbonate
Quartz					
Hematite					
Pitchblende					
Ni, Co, Fe, Ag arsenides					
Ni, Co, Fe, Sb Sulpharsenides					
Ni sulphides					
Cu, Fe, Pb, Zn, Sb					
Mo, As, Bi, Ag					
Sulphides					
Ag tellurides					
Native Ag					
Native Bi					
Chlorite					
Fluorite					
Carbonates					
Temperature (°C)	150-250	150-250	220-480	150-250	90-250
$\delta^{18}O$ ‰ (n)	16.5(1)	15.15(2)	22.2(12)	23.7(7)	13.97(13)
$\delta^{13}C$ ‰ (n)	-73.5(1)	-4.05(2)	-3.64(12)	-2.7(7)	-4.9(13)

contained, in addition to U, Co, Cu, and Ni, relatively high amounts of rare-earth elements, such as up to 2% Y, and up to 0.2% each of Yb, La, and Ce. These values compare with clarkes (crustal/in granites/in diabase) as follows: for Y = 33/13/25 ppm, Yb = 3.4/1.1/2.1 ppm, La = 30/101/9.8 ppm, and Ce = 60/170/23 ppm (Mason and Moore, 1982). Mason and Moore (1982) observed that the rare-earth elements are sensitive indicators to different igneous processes: for example, elevated contents of lanthanum and cerium above their respective chondrite-normalized concentrations are characteristic for felsic igneous rocks, whereas elevated contents of rare-earth elements of higher atomic numbers, such as yttrium and ytterbium, indicate relationship to mafic igneous rocks. Elevated contents of lanthanum, cerium, yttrium, and ytterbium in the mineralized samples from the Great Bear Lake district (Table 14.2-3) apparently indicate affiliation of the mineralization both to felsic and mafic magmatism.

Alteration associated with ore

The wall rocks of the veins have been altered during various stages of the ore-forming process: pervasive hematite staining developed during the earliest stages of the hydrothermal activity. Argillization and chloritization were associated with the main (pitchblende and arsenide) mineralization stages. Chloritization preferentially affected rocks containing ferromagnesian minerals. Carbonatization was associated with the late stages (sulphide and carbonate) of mineralization. It affected all types of rocks, but its distribution was irregular and related only to those parts of the veins in which the carbonate-bearing hydrothermal solutions were in direct contact with the wall rocks. Locally the wall rocks were affected by silicification, sericitization, and development of authigenic sulphides and apatite (Campbell, 1957; Ruzicka, 1971).

Mineralogy

The uranium minerals in the Eldorado deposit at Port Radium were represented almost exclusively by pitchblende. Only in the oxidation zone of the deposit were gummite, uranophane, and zippeite present. Nickeline, safflorite-löllingite, gersdorffite, and skutterudite were the principal minerals of the arsenide stage; polydymite, glaucodot, and cobaltite occurred in smaller amounts. Erythrite and annabergite were typical oxidation products of the arsenide assemblages in the deposit. Chalcopyrite, chalcocite, bornite, tetrahedrite, and acanthite were principal minerals of the sulphide stage. Galena, sphalerite, stibnite, tennantite, aikinite, and covellite were present in the sulphide assemblages in lesser amounts. Malachite and azurite were present in the oxidation zone. Native silver and native bismuth constituted components of the arsenide and carbonate mineralization stages. Quartz, dolomite, montmorillonite, chlorite, rhodochrosite, and hematite were the main gangue minerals. In addition, manganese oxides, such as pyrolusite and psilomelane, were present in the oxidation zone of the deposit (Mursky, 1973).

Ore textures

The mineral fillings of the No. 1 and No. 2 veins, which contained the richest metallic mineralization, had typical banded textures: quartz and hematite were directly deposited on the wall rocks; pitchblende occupying adjacent bands occurred in several forms: botryoidal, spherulitic, brecciated, dendritic, massive, and as veinlets. Veinlets of pitchblende alternating with carbonate layers, and accompanied by lenses of arsenides and sulphides, were typical of the No. 2 and No. 1 veins. The remaining veins of the deposit contained mainly disseminated or irregularly distributed pitchblende, arsenides, and sulphides (Lang et al., 1962). In order of tonnages of contained uranium ores the most important resources of the Eldorado deposit were in the No. 1, No. 3, No. 2, and No. 5 veins.

Table 14.2-3. Elemental constituents in samples from selected polymetallic occurrences in the Great Bear Lake mining district, Northwest Territories (from Ruzicka, 1971).

Constituent	Sample			
	1	2	3	4
Si	1	0.5	10	ND
Al	0.7	0.2	1	ND
Fe	5	5	7	10.8
Ca	10	3	3	3.4
Mg	3	10	0.07	3.2
Ti	0.02	NF	0.05	0.14
Mn	2	1.5	0.15	0.7
Sr	0.01	NF	NF	ND
Ba	0.05	0.003	0.05	ND
Cr	<.002	0.003	<.002	tr
Zr	0.07	NF	0.01	tr
V	0.03	0.01	0.03	0.09
Ni	0.015	0.02	NF	2.9
Ce	0.1	0.2	0.2	ND
Cu	0.5	0.5	0.2	4.8
Y	0.07	2	0.1	0.11
Nb	NF	NF	NF	ND
Co	0.02	0.3	0.08	4.6
La	0.02	0.02	0.15	ND
Pb	0.1	0.1	1	2.56
Th	NF	NF	0.13	ND
B	<0.008	NF	<0.008	0.02
Yb	0.005	0.2	0.005	ND
Be	NF	NF	NF	tr
U	0.7	1	1	13.81
Mo	NF	NF	NF	0.05
Bi	NF	NF	NF	0.68
Ag	0.1	0.007	0.015	0.09

Sample 1: Echo Bay mine adit No. 2 (Ruzicka, 1971)
Sample 2: Contact Lake mine (Ruzicka, 1971)
Sample 3: Trench East of Echo Bay mine (Ruzicka, 1971)
Sample 4: Four-year production record (Mursky, 1973)
Spectrochemical analyses by the Geological Survey of Canada
NF = Not found
ND = No data available
tr = traces
Data in weight per cent

DEFINITIVE CHARACTERISTICS

The arsenide uranium vein deposits represent concentrations of uranium oxides associated with arsenides and sulpharsenides of nickel and cobalt, sulphides of base metals, particularly copper, and native metals, such as silver and bismuth. The metallic minerals fill fractures, stockworks, breccia, and shear zones in metasedimentary and metavolcanic rocks, which have, as a rule, been intruded by granitic or syenitic plutons and by younger suites of dykes. The ore minerals were deposited in several stages. Deposition of the uranium-bearing minerals took place during the earlier stages, followed by deposition of nickel-cobalt arsenides, sulpharsenides, and native silver and bismuth; sulphides and some native metals were deposited during the latest stages. Deposition of quartz, hematite, and carbonates accompanied deposition of the metallic minerals. The host rocks were pervasively altered by hematitization, argillization, chloritization, and carbonatization and, to a lesser extent, by silicification, sericitization, and formation of sulphides and apatite.

GENETIC MODEL

The arsenide uranium vein deposits are epigenetic concentrations of uranium, arsenides, and sulpharsenides of nickel and cobalt, sulphides of silver and base metals, and native silver and bismuth (Ruzicka, 1993). The deposits were formed by hydrothermal processes, in which the ore constituents were mobilized from formational brines and redeposited in open fractures and cavities. This model involves the further concentration of uranium in a reservoir by a continuing process of dissolution or leaching from the country rocks. In the Great Bear Lake area the highest concentration of uranium apparently took place within the clastic sedimentary rocks of the Hornby Bay Group, which were overlying the basement rocks at the time of the mineralization. Based on fluid inclusion, mineralogical, and S isotope evidence, Robinson and Ohmoto (1973) concluded that in the Echo Bay mine, deposition of early mineralization stages was at about 120°C, and that the sulphide and late silver stages formed at about 200°C. The fluid inclusion evidence suggested a salinity of approximately 30 wt.% NaCl equivalent.

The mineral succession in the veins indicates that the mineralizing fluids changed with time from lower to higher pH values and in chemical composition from silicic to carbonatic and finally to sulphidic. Changkakoti et al. (1986b) interpreted O-, H-, and C isotopic data to indicate that magmatic water and carbon from a magmatic source were predominant in the early stages of mineralization, and that meteoric water had a more pronounced influence during the late stages. After the initial stage of hydrothermal activity the metal-saturated formational brines of low pH and high Eh descended from the reservoir and released uranium at the interface with reducing ascending hydrothermal fluids. Nickel-cobalt arsenides and most of the native silver and bismuth were deposited after re-opening of the fractures. The sulphides and some native metals were deposited during the last stages of the hydrothermal process under high pH and low Eh conditions. According to oxygen and carbon isotope studies (Table 14.2-2) the mineralizing solutions changed from endogenous hydrothermal to a mixture of endogenous hydrothermal and exogenous waters. Temperature of the fluids reached the highest (culmination) point (at about 480°C) during the third (arsenide) mineralization stage.

It appears that the physical properties of the enclosing rocks, rather than their petrochemical composition, controlled the amount of mineral deposition (Jory, 1964; Ruzicka, 1971). The general tectonic settings and geological environments for deposits of this type, and the factors that have been considered of greatest genetic significance, have been discussed above with regard to the arsenide cobalt-silver subtype of veins.

RELATED DEPOSIT TYPES

The geological setting and morphology of vein orebodies of the arsenide uranium-silver subtype resemble those of uranium vein deposits with simple uranium mineralogy (i.e., monometallic uranium vein deposits such as the Eldorado mine at Uranium City, Saskatchewan). These deposits are treated separately in this volume (see Type 13, "Vein uranium").

The mineral compositions of arsenide uranium-silver vein deposits resemble those of unconformity-associated uranium deposits (see Type 7), such as the Key Lake, Cigar Lake, and 'D' orebody at Cluff Lake, all in the Athabasca Basin region, Canada. The 'polymetallic subtype' of unconformity deposits contains elemental assemblages, including uranium, nickel, cobalt, silver, copper, and other associated metals, which are typical for the arsenide uranium-silver veins.

EXPLORATION GUIDES

Target areas for exploration should be selected using a conceptual genetic model such as that mentioned above. Attention should be paid in particular to the geological environment, which should consist of: a crystalline basement including metasedimentary, metavolcanic, and felsic plutonic rocks; a clastic sedimentary cover sequence; and intrusive dykes. Such environments should exhibit prominent regional faults and should contain indications of hydrothermal activity.

Airborne and ground radiometric surveys can assist in pinpointing areas with anomalous concentrations of radionuclides, and electromagnetic surveys should help in definition of regional structural and lithological patterns and the locations of dykes. Geochemical surveys can be used to indicate the presence of dispersion haloes in residual soils, and anomalous contents of elements (U, Ni, Co, As, Cu, Bi, and Ag) in stream and lake-bottom sediments, indicative of the polymetallic elemental assemblages.

Prospecting for diagnostic minerals, such as uranophane, erythrite, annabergite, dark purple fluorite, and for quartz-carbonate veins should complement the geophysical and geochemical surveys. In addition, attention should be paid to alteration phenomena within the rocks adjacent to veins, particularly to hematitization, argillization, and chloritization, and the presence of radioactive gossans. However, it should also be kept in mind that the metallic minerals occur in lenses and that the distribution of the orebodies is, as a rule, irregular.

SELECTED BIBLIOGRAPHY

References marked with asterisks (*) are considered to be the best sources of general information on this deposit type.

*Andrews, A.J., Owsiacki, L., Kerrich, R., and Strong, D.F.
1986: The silver deposits at Cobalt and Gowganda, Ontario I: Geology, petrography and whole-rock geochemistry; Canadian Journal of Earth Sciences, v. 23, p. 1480-1506.

Badham, J.P.N.
1973: Volcanogenesis, orogenesis and metallogenesis, Camsell River, N.W.T., Canada; PhD. thesis, University of Alberta, Edmonton, Alberta, 363 p.

Bowring, S.A. and Ross, M.G.
1985: Geochronology of the Narakay volcanic complex: implications for the age of the Coppermine Homocline and Mackenzie igneous events; Canadian Journal of Earth Sciences, v. 22, no. 5, p. 774-781.

*Boyle, R.W. and Dass, A.S.
1971: Origin of the native silver veins at Cobalt, Ontario; in The Silver-arsenide Deposits of the Cobalt-Gowganda Region, Ontario, (ed.) L.G. Berry; The Canadian Mineralogist, v. 11, pt. 1, p. 414-417.

Brophy, J.A.
1985: Operating mines; in Mineral Industry Report 1982-83, Northwest Territories, Indian and Northern Affairs, Canada, p. 13-74.

*Campbell, D.D.
1957: Port Radium mine; in Structural Geology of Canadian Ore Deposits, v. II; The Canadian Institute of Mining and Metallurgy, Congress Volume, p. 177-189.

Changkakoti, A., Ghosh, D.K., Krstic, D., Gray, J., and Morton, R.D.
1986a: Pb and Sr isotopic compositions of hydrothermal minerals from the Great Bear Lake silver deposits, N.W.T., Canada; Economic Geology, v. 81, p. 739-743.

Changkakoti, A., Morton, R.D., Gray, J., and Yonge, J.
1986b: Oxygen, hydrogen and carbon isotopic studies of the Great Bear Lake silver deposits, Northwest Territories; Canadian Journal of Earth Sciences, v. 23, p. 1463-1469.

Cook, D.G. and MacLean, B.C.
1993: The intra-cratonic Paleoproterozoic Forward Orogeny, and implications for regional correlations, Northwest Territories, Canada; Geological Society of America, Abstracts with Programs, v. 25, no. 6, p. A388.

Corfu, F. and Andrews, A.J.
1986: A U-Pb age for mineralized Nipissing diabase, Gowganda, Ontario; Canadian Journal of Earth Sciences, v. 23, p. 107-109.

Cranstone, D.A. and Whillans, R.T.
1986: Costs and rates of uranium discovery in Canada; unpublished report, Energy, Mines and Resources, Canada, 10 p.

Franklin, J.M., Kissin, S.A., Smyk, M.C., and Scott, S.D.
1986: Silver deposits associated with the Proterozoic rocks of the Thunder Bay district, Ontario; Canadian Journal of Earth Sciences, v. 23, p. 1576-1591.

Fraser, J.A., Heywood, W.W., and Mazurski, M.A. (comp)
1978: Metamorphic map of the Canadian Shield; Geological Survey of Canada, Map 1475A, scale 1:3 500 000.

Gandhi, S.S. and Mortensen, J.K.
1992: 1.87-1.86 Ga old felsic volcano-plutonic activity in southern Great Bear magmatic zone, N.W.T.; in Geological Association of Canada/Mineralogical Association of Canada, Abstracts Volume, v. 17, p. A37.

Goodz, M.D., Watkinson, D.H., Smejkal, V., and Pertold, Z.
1986: Sulphur-isotope geochemistry of silver sulpharsenide vein mineralization, Cobalt, Ontario; Canadian Journal of Earth Sciences, v. 23, p. 1551-1567.

*Heinrich, E.W.
1958: Mineralogy and geology of radioactive raw materials; McGraw-Hill Book (Publishing) Company Inc., New York, Toronto, London, 654 p.

Hildebrand, R.S.
1986: Kiruna-type deposits: their origin and relationship to intermediate subvolcanic plutons in the Great Bear magmatic zone, northwest Canada; Economic Geology, v. 81, p. 640-659.

Hildebrand, R.S. and Bowring, S.A.
1987: Continental arc magmatism in the early Proterozoic Wopmay Orogen, northwestern Canadian Shield (abstract); EOS (Transactions, American Geophysical Union), v. 68, no. 44, p. 1517.

Hildebrand, R.S., Bowring, S.A., Steer, M.E., and Van Schmus, W.R.
1983: Geology and U-Pb geochronology of parts of the Leith Peninsula and Riviere Grandin map areas, District of Mackenzie; in Current Research, Part A; Geological Survey of Canada, Paper 83-1A, p. 329-342.

Hildebrand, R.S., Hoffman, P.F., and Bowring, S.A.
1987: Tectono-magmatic evolution of the 1.9 Ga Great Bear magmatic zone, Wopmay orogen, northwestern Canada; Journal of Volcanology and Geothermal Research, v. 32, p. 99-118.

Hoffman, P.F. and Bowring, S.A.
1984: Short-lived 1.9 Ga continental margin and its destruction, Wopmay orogen, northwest Canada; Geology, v. 12, p. 68-72.

Housh, T., Bowring, S.A., and Villeneuve, M.
1989: Lead isotope study of Early Proterozoic Wopmay Orogen, NW Canada: role of continental crust in arc magmatism; Journal of Geology, v. 97, p. 735-747.

Hrádek, J.
1993: Uranium deposits of the Czech Republic; International Atomic Energy Agency, Technical Committee Meeting on Recent Developments in Uranium Resources and Supply, Vienna, Austria, 24-28 May, 1993, preprint of a paper, 18 p.

International Atomic Energy Agency
1977: National favourability studies no. 119: Zaïre; International Uranium Resource Evaluation Project (IUREP), Publication No. 77-10265, 14 p.

Jambor, J.L.
*1971a: General geology; in The Silver-arsenide Deposits of the Cobalt-Gowganda Region, Ontario, (ed.) L.G. Berry; The Canadian Mineralogist, v. 11, pt. 1, p. 12-33.
1971b: Wall rock alteration; in The Silver-arsenide Deposits of the Cobalt-Gowganda Region, Ontario, (ed.) L.G. Berry; The Canadian Mineralogist, v. 11, pt. 1, p. 272-304.
*1971c: Gangue mineralogy; in The Silver-arsenide Deposits of the Cobalt-Gowganda Region, Ontario, (ed.) L.G. Berry; The Canadian Mineralogist, v. 11, pt. 1, p. 232-261.

Jennings, E.A.
1987: A survey of the Mainland and Island belts, Thunder Bay silver district, Ontario: fluid inclusions, mineralogy and sulfur isotopes; MSc. thesis, Lakehead University, Thunder Bay, Ontario, 159 p.

*Jory, L.T.
1964: Mineralogical and isotopic relations in the Port Radium pitchblende deposit, Great Bear Lake, Canada; PhD. thesis, California Institute of Technology, Pasadena, California, 275 p.

Kerrich, R., Strong, D.F., Andrews, A.J., and Owsiacki, L.
1986: The silver deposits at Cobalt and Gowganda, Ontario: III. Hydrothermal regimes and source reservoirs – evidence from H, O, D, and Sr isotopes and fluid inclusions; Canadian Journal of Earth Sciences, v. 23, p. 1519-1550.

Kissin, S.A.

*1988: Nickel-cobalt-native silver (five-element) veins: a rift-related ore type; in North American Conference on Tectonic Control of Ore Deposits and the Vertical and Horizontal Extent of Ore Systems, Proceedings Volume, (ed.) G. Kisvarsanyi and S.K. Grant; University Missouri, Department of Geology and Geophysics, Rolla, Missouri, p. 268-279.

1989: Genesis of silver vein deposits on the north shore of Lake Superior, Thunder Bay district, Ontario; Geological Society of America, Abstracts with Programs, v. 21, no. 6, p. A130.

*1992: Five-element (Ni-Co-As-Ag-Bi) veins; Geoscience Canada, v. 19, no. 3, p. 113-124.

1993: The geochemistry of transport and deposition in the formation of five-element (Ag-Ni-Co-As-Bi) veins; in Proceedings of the Eighth Quadrennial International Association on the Genesis of Ore Deposits Symposium, (ed.) Y.T. Maurice; E. Schweizerbart'sche Verlagsbuchhandlung, Stuttgart, p. 773-786.

Kissin, S.A. and Sherlock, R.L.

1989: The genesis of silver vein deposits in the Thunder Bay area, northwestern Ontario; in Geoscience Research Grant Program Summary of Research 1988-1989, (ed.) V.G. Milne; Ontario Geological Survey, Miscellaneous Paper 143, p. 33-41.

Lalonde, A.E.

1989: Hepburn intrusive suite: peraluminous plutonism within a closing back-arc basin, Wopmay orogen, Canada; Geology, v. 17, p. 261-264.

*Lang, A.H., Goodwin, A.M., Mulligan, R., Whitmore, D.R.E., Gross, G.A., Boyle, R.W., Johnston, A.G., Chamberlain, J.A., and Rose, E.R.

1970: Economic minerals of the Canadian Shield; in Geology and Economic Minerals of Canada, (ed.) R.J.W. Douglas; Geological Survey of Canada, Economic Geology Report 1, p. 153-226.

*Lang, A.H., Griffith, J.W., and Steacy, H.R.

1962: Canadian deposits of uranium and thorium; Geological Survey of Canada, Economic Geology Report 16, 324 p.

Leblanc, M.

1986: Co-Ni arsenide deposits, with accessory gold, in ultramafic rocks from Morocco; Canadian Journal of Earth Sciences, v. 23, p. 1592-1602.

Leblanc, M. and Billaud, P.

1982: Cobalt arsenide orebodies related to an Upper Proterozoic ophiolite: Bou Azzer (Morocco); Economic Geology, v. 77, p. 162-175.

Mason, B. and Moore, C.B.

1982: Principles of Geochemistry; John Wiley and Sons, New York, 344 p.

Miller, R.G.

1982: The geochronology of uranium deposits in the Great Bear batholith, Northwest Territories; Canadian Journal of Earth Sciences, v. 19, no. 7, p. 1428-1448.

Ministry of Natural Resources

1977: Sudbury-Cobalt; Ontario Geological Survey, Map 2361, Geological compilation series, scale 1:253 440.

Mohide, T.P.

1985: Silver; Ontario Ministry of Natural Resources, Mineral Policy Background Paper No. 20, 406 p.

*Mursky, G.

1973: Geology of the Port Radium map-area, District of Mackenzie; Geological Survey of Canada, Memoir 374, 40 p.

Ontario Department of Mines

1964: Cobalt silver area, northern sheet; Ontario Department of Mines, Map 2050, scale 1 inch: 1000 feet.

Ontario Ministry of Natural Resources

1984: Ontario Mineral Score 1983; Video Census Series No. 4.

1985: Ontario Mineral Score 1984; Video Census Series No. 4.

1986: Ontario Mineral Score 1985; Video Census Series No. 4.

*Owsiacki, L.

1984: Geology and silver deposits of the Cobalt area; in Geology, Silver and Gold Deposits: Cobalt and Kirkland Lake, (ed.) L. Owsiacki and H. Lovell; Geological Association of Canada, Field Trip Guidebook, Field Trip 4, p. 1-16.

*Petruk, W.

1971: Mineralogical characteristics of the deposits and textures of the ore minerals; in The Silver-arsenide Deposits of the Cobalt-Gowganda Region, Ontario, (ed.) L.G. Berry; The Canadian Mineralogist, v. 11, pt. 1, p. 108-139.

*Petruk, W. and staff, Mineral Sciences Division, Mines Branch, Department of Energy, Mines and Resources

1971c: Geochemistry of the ores; in The Silver-arsenide Deposits of the Cobalt-Gowganda Region, Ontario, (ed.) L.G. Berry; The Canadian Mineralogist, v. 11, pt. 1, p. 140-149.

1971d: Characteristics of the sulphides; in The Silver-arsenide Deposits of the Cobalt-Gowganda Region, Ontario, (ed.) L.G. Berry; The Canadian Mineralogist, v. 11, pt. 1, p. 196-227.

*Petruk, W., Harris, D.C., and Stewart, J.M.

1971b: Characteristics of the arsenides, sulpharsenides and antimonides; in The Silver-arsenide Deposits of the Cobalt-Gowganda region, Ontario, (ed.) L.G. Berry; The Canadian Mineralogist, v. 11, pt. 1, p. 150-186.

*Petruk, W., Harris, D.C., Cabri, L.J., and Stewart, J.M.

1971a: Characteristics of the silver-antimony minerals; in The Silver-arsenide Deposits of the Cobalt-Gowganda Region, Ontario, (ed.) L.G. Berry; The Canadian Mineralogist, v. 11, pt. 1, p. 187-194.

*Robinson, B.W.

1971: Studies on the Echo Bay silver deposit, N.W.T., Canada; PhD. thesis, University of Alberta, Edmonton, Alberta, 229 p.

*Robinson, B.W. and Morton, R.D.

1972: The geology and geochronology of the Echo Bay area, Northwest Territories, Canada; Canadian Journal of Earth Sciences, v. 9, p. 158-171.

Robinson, B.W. and Ohmoto, H.

1973: Mineralogy, fluid inclusions, and stable isotopes of the Echo Bay U-Ni-Ag-Cu deposits, Northwest Territories, Canada; Economic Geology, v. 68, p. 635-656.

Robinson, D.

1984: General geology of the Beaver-Temiskaming Mine, Cobalt, Ontario; in Geology, Silver and Gold Deposits: Cobalt and Kirkland Lake, (ed.) L. Owsiacki and H. Lovell; Geological Association of Canada, Field Trip Guidebook, Field Trip 4, p. 1-16.

Russell, D.J.

1983: Geology of the Paleozoic outliers of the Canadian Shield; in Summary of Field Work, (ed.) J. Wood, O.L. White, R.B. Barlow, and A.C. Colvine; Ontario Geological Survey, Miscellaneous Paper 116, p. 104-106.

Ruzicka, V.

1971: Geological comparison between East European and Canadian uranium deposits; Geological Survey of Canada, Paper 70-48, 195 p.

1989: Conceptual genetic models for important types of uranium deposits and areas favourable for their occurrence in Canada; in Uranium Resources and Geology of North America, Proceedings of a Technical Committee Meeting, organized by the International Atomic Energy Agency, Saskatoon, Canada, 1-3 September 1987, International Atomic Energy Agency, IAEA-TECDOC-500, p. 49-79.

*1993: Vein uranium deposits; in Vein-type Ore Deposits, (ed.) S.J. Haynes; Ore Geology Reviews, v. 8, p. 247-256.

Ruzicka, V. and LeCheminant, G.M.

1986: Developments in uranium geology in Canada, 1985; in Current Research, Part A; Geological Survey of Canada, Paper 86-1A, p. 531-540.

1987: Uranium investigations in Canada, 1986; in Current Research, Part A; Geological Survey of Canada, Paper 87-1A, p. 249-262.

Stockwell, C.H., McGlynn, J.C., Emslie, R.F., Sanford, B.V., Norris, A.W., Donaldson, J.A., Fahrig, W.F., and Currie, K.L.

1970: Geology of the Canadian Shield; in Geology and Economic Minerals of Canada, (ed.) R.J.W. Douglas; Geological Survey of Canada, Economic Geology Report 1, p. 43-150.

Thorpe, R.I.

1974: Lead isotope evidence on the genesis of the silver-arsenide vein deposits of the Cobalt and Great Bear Lake areas, Canada; Economic Geology, v. 69, no. 6, p. 777-791.

*1984: Arsenide vein silver, uranium; in Canadian Mineral Deposit Types: a Geological Synopsis, (ed.) O.R. Eckstrand; Geological Survey of Canada, Economic Geology Report 36, p. 63.

Thorpe, R.I., Goodz, M.D., Jonasson, I.R., and Blenkinsop, J.

1986: Lead-isotope study of mineralization in the Cobalt district, Ontario; Canadian Journal of Earth Sciences, v. 23, p. 1568-1575.

Wanless, R.K. and Loveridge, W.D.

1978:   Rubidium-strontium isotopic age studies, Report 2 (Canadian Shield); Geological Survey of Canada, Paper 77-14, 70 p.

Wanless, R.K., Stevens, R.D., Lachance, G.R., and Delabio, R.N.

1970:   Age determinations and geological studies, K-Ar isotopic age Report 9; Geological Survey of Canada, Paper 69-2A, p. 45.

Wanless, R.K., Stevens, R.D., Lachance, G.R., and Edmonds, C.M.

1968:   Age determinations and geological studies, K-Ar isotopic ages, Report 8; Geological Survey of Canada, Paper 67-2, Part A, 141 p.

Warren, R.G., Thorpe, R.I., Dean, J.A., and Mortensen, J.K.

1995:   Pb-isotope data from base-metal deposits in Central Australia: implications for Proterozoic stratigraphic correlations; AGSO Journal of Australian Geology and Geophysics, v. 15, no. 4, p. 501-509.h

Watkinson, D.H.

1986:   Mobilization of Archean elements into Proterozoic veins; an example from Cobalt, Canada; in Proceedings of the Conference on the Metallogeny of the Precambrian (IGCP Project 91), Geological Survey of Czechoslovakia (UUG), Prague, p. 133-138.

Authors' addresses

V. Ruzicka
Geological Survey of Canada
601 Booth Street
Ottawa, Ontario
K1A 0E8

R.I. Thorpe
Geological Survey of Canada
601 Booth Street
Ottawa, Ontario
K1A 0E8

Printed in Canada

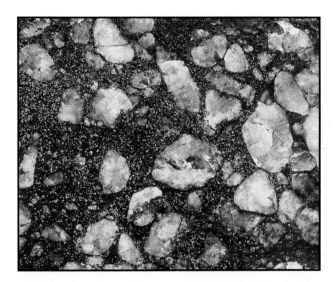

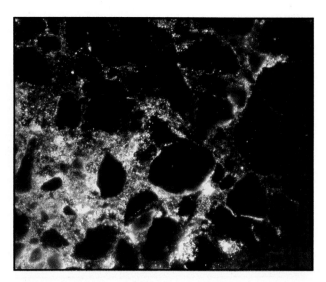

Plate 1. *Deposit subtype 1.1,* Paleoplacer uranium, gold; Uraniferous, pyritic quartz-pebble conglomerate. The uranium minerals (mainly uraninite and a uranium-titanium phase; see Pl. 2) occur in the matrix, associated with pyrite (white) and other heavy minerals. AB reef, Denison mine, Elliot Lake, Ontario. Maximum dimension of photo is 14 cm. GSC 1995-200A

Plate 2. *Deposit subtype 1.1,* Paleoplacer uranium, gold; Autoradiograph of the sample shown in Plate 1. The white areas are caused by radiation from uranium- and thorium-bearing minerals. The distribution of radioactive minerals and pyrite is similar, but different in detail. Elliot Lake, Ontario.

Plate 3. *Deposit subtype 1.2,* Placer gold, platinum. Gold nuggets from the Klondike area, Yukon Territory. NRCan Photo 8397

Plate 4. *Deposit subtype 1.2,* Placer gold, platinum. Basal part of gold-bearing high-level White Channel bench gravels (white) with remnants of Klondike gravels (brown). Early placer miners (ca. 1902) followed the lower bedrock surface with a timbered drift (about 1.2 m wide). Later bench mining was done hydraulically and with bulldozer in 1960. Paradise Hill on Hunker Creek, Klondike area, Yukon Territory. GSC 202282V

Plate 5. *Deposit type 3,* Stratiform iron. Dolomite drop-stone of glacial origin in jasper-hematite lithofacies of Rapitan type iron-formation, Snake River area, Yukon Territory. GSC 1995-202

Plate 6. *Deposit subtype 3.1,* Lake Superior-type iron-formation. Jasper-hematite lithofacies of Lake Superior type iron-formation, Knob Lake-Schefferville area, Quebec-Labrador. GSC 1995-203A

Plate 7. *Deposit subtype 3.1,* Lake Superior-type iron-formation. Jasper-hematite-magnetite facies of iron-formation, showing typical, relatively thick, well preserved sedimentary bedding. Magnet is 3.5 cm. Schefferville area, Quebec-Labrador. GSC 1995-203B

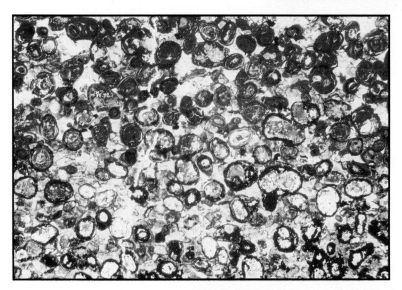

Plate 8. *Deposit subtype 3.1,* Lake Superior-type iron-formation. Photomicrograph of typical hematite-magnetite-chert oxide lithofacies in the Lower Red Chert member of the Sokoma Formation. Typical granular and oolitic texture; ooides of hematite (red), partly recrystallized to magnetite (black), and fine grained quartz (chert) in the granular matrix. Maximum dimension of photo is 13 mm. Near Wishart mine, Schefferville area, Quebec-Labrador. GSC 1995-203C

Plate 9. *Deposit subtype 3.2,* Algoma-type iron-formation. Quartz-feldspar layers in highly contorted magnetite-quartz lithofacies of iron-formation. Moose Mountain mine, Ontario. GSC 1995-204A

Plate 11. *Deposit subtype 4.1,* Enriched iron-formation. Early stages of mining the G-orebody at Steep Rock Lake, Ontario (1958). Massive hematite-goethite ore (red) derived from carbonate and sulphide lithofacies of Algoma-type iron-formation protore. GSC 1995-204C

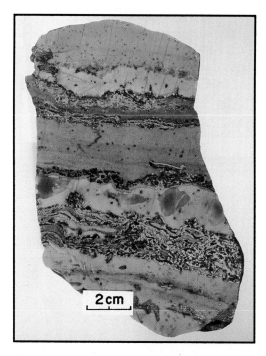

2 cm

Plate 10. *Deposit subtype 3.2,* Algoma-type iron-formation. Pyrite-pyrrhotite, siderite, and quartz layers in carbonate-sulphide lithofacies of iron-formation. Wawa area, Michipicoten district, Ontario. GSC 1995-204B

Plate 12. *Deposit type 5,* Evaporites. Folded halite beds. Rock Salt mine, Pugwash, Nova Scotia. NRCan Photo 2950

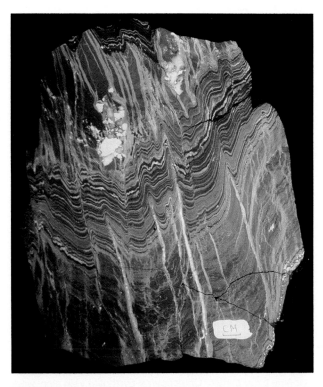

Plate 14. *Deposit subtype 6.3,* Volcanic-associated massive sulphide base metals. Well-bedded massive pyrite (brass-yellow), chalcopyrite (bright yellow), and sphalerite (dark metallic blue-grey). Lac Dufault mine, Rouyn-Noranda, Quebec. GSC 1995-209

Plate 13. *Deposit subtype 6.1,* Sedimentary exhalative sulphides (Sedex). Sphalerite, pyrite, and galena form finely laminated sedimentary beds within a mudstone that also contains nodular pyrite in the upper, carbonaceous cherty mudstone unit of the Active Member (Lower Silurian; Road River Group). Buckle folding and dewatering during early diagenesis resulted in pressure solution and diffusive mass transfer (40% shortening) that led to loss of silica and accumulation of sphalerite and garnet in striped cleavages. Howards Pass XY deposit, Selwyn Basin, Yukon-Northwest Territories. GSC 1995-205

Plate 15. *Deposit subtype 6.3,* Volcanic-associated massive sulphide base metals. Alteration zone and mineralization developed within talus breccia, 10 m below a massive sulphide lens. The breccia matrix is infilled with coarse grained sphalerite. Pillow basalt fragments exhibit zoned alteration from core to rim; chlorite, quartz, sericite, and sphalerite, respectively. Corbet mine, No. 3 lens, 15 level, Rouyn-Noranda, Quebec. GSC 1995-206A

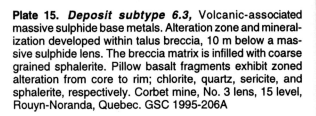

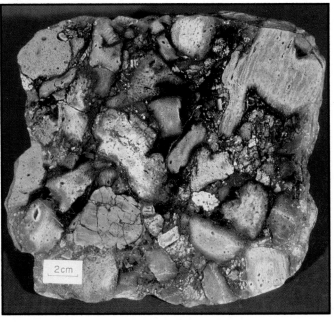

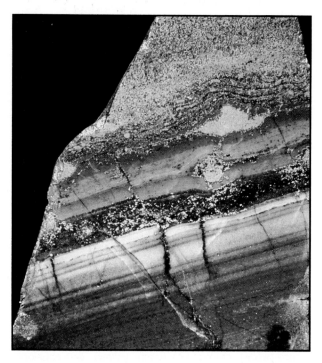

Plate 16. *Deposit subtype 6.3,* Volcanic-associated massive sulphide base metals. Delicately laminated aquagene quartz porphyry tuff; coarser parts of normally graded beds have been impregnated and replaced, first by sphalerite/pyrite (brown and black) then by chalcopyrite/pyrrhotite (yellow/grey). This process is a microcosm of the mineralizing events that led to the formation of the replacement-type deposit. Ansil mine, level 9C, Rouyn-Noranda, Quebec. Minimum dimension of photo is 5 cm. GSC 1995-211A

Plate 17. *Deposit subtype 6.3,* Volcanic-associated massive sulphide base metals. Chalcopyrite-pyrrhotite vein stockwork directly underlying a massive sulphide lens. Ansil mine, sublevel 8A, crosscut no. 2, Rouyn-Noranda, Quebec. Roof bolt plates are 20 cm wide. GSC 1995-211B

Plate 18. *Deposit subtype 6.4,* Volcanic-associated massive sulphide gold. Primary ore textures; dark grey quartz, light grey sphalerite, and pale yellow pyrite. Note primary botryoidal texture of pyrite. Mattabi mine, Sturgeon Lake district, Ontario. Photomicrograph, reflected light, maximum dimension is 2.5 mm. GSC 1995-212

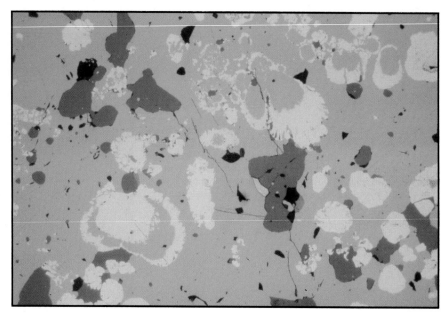

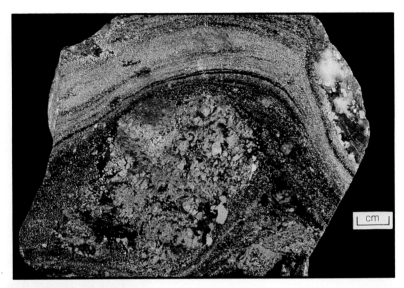

Plate 19. *Deposit subtype 6.4,* Volcanic-associated massive sulphide gold. In this strongly deformed deposit, recrystallized chalcopyrite and pyrite comprise boudins, around which sphalerite, minor associated galena, and incorporated granular pyrite have flowed. Estrades mine, Joutel, northwestern Quebec. GSC 1995-206B

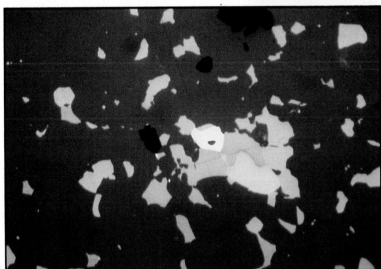

Plate 20. *Deposit subtype 6.4,* Volcanic-associated massive sulphide gold. Electrum (bright yellow), intergrown with chalcopyrite (pale yellow), galena (white), and pyrrhotite (pale grey) in a matrix of massive sphalerite (grey). Photomicrograph reflected light; the largest grain of electrum is 20 μm in greatest dimension. Estrades mine, Joutel, Quebec. GSC 1995-213

Plate 21. *Deposit type 10,* Mississippi Valley-type lead-zinc. Dolomite (black) breccia formed by solution-collapse processes has been infilled with sparry calcite (white) and coarsely crystalline sphalerite (orange). The latter forms a "snow-on-roof" texture due to its precipitation from percolating solutions inside open cavities. Zone B, Gayna River deposit, Gayna River, Northwest Territories. GSC 1995-086B

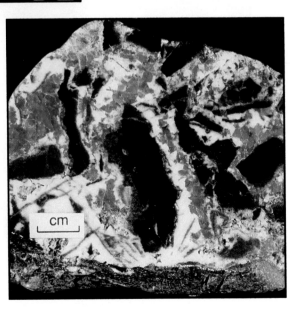

Plate 22. *Deposit type 10,* Mississippi Valley-type lead-zinc. Typical botryoidal textured sulphides; layered sphalerite (dark brown and white) is overlain by coarse grained galena (steel blue). Sulphides were precipitated in a cavity, beginning at the lower right, with successive layers added upward into open space; the last sulphide deposited was galena. Minimum dimension of the photo is 5.5 cm. Pine Point mine, Northwest Territories. GSC 1995-214

Plate 23. *Deposit type 10,* Mississippi Valley-type lead-zinc. Cathodoluminescence photomicrograph of dolomite-cement showing typical saddle-shaped compositional zoning (red and black) associated with lead-zinc mineralization. The dolomite was partially dissolved prior to the deposition of sphalerite (blue, yellow, and brown). Luminescent orange calcite (extreme right) filled remaining cavities. The width (short dimension) of the dolomite grain is 1 mm. Gayna River deposit, Gayna River, Northwest Territories. GSC 1995-215

Plate 24. *Deposit type 11,* Ultramafic-hosted asbestos. Veins of cross-fibre chrysotile asbestos in serpentinite, Cassiar mine, British Columbia. GSC 1995-216

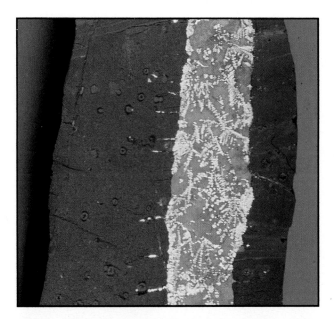

Plate 25. *Deposit subtype 14.1,* Arsenide vein silver-cobalt. Vein of safflorite botryoids and dendrites, cored by native silver. Host rock is Huronian Coleman Formation conglomerate (tillite) that is weakly bleached and has chlorite alteration spots close to the vein. Native silver has been precipitated in small tension gashes subparallel to bedding. Maximum width of vein is 3.5 cm. 235 level, Langis mine, New Liskeard, Ontario. GSC 1995-207A

Plate 27. *Deposit subtype 15.2,* Quartz-carbonate vein gold. Gold-bearing quartz-carbonate-tourmaline (black) vein in sheared mafic volcanic rocks. The vein is fringed by small extensional veins. Sigma mine, Val d'Or, Quebec. GSC 1995-222

1 cm

Plate 26. *Deposit subtype 15.2,* Quartz-carbonate vein gold. High-grade quartz-calcite vein with coarse gold, associated with minor chlorite, pyrite, chalcopyrite, and tennantite. Gold lies in tiny tension gashes orthogonal to localized shear planes that contain the sulphides. Hearne-Taurus mine, Cassiar, British Columbia. GSC 1995-208

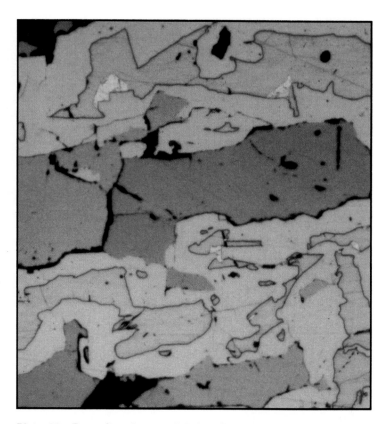

Plate 28. *Deposit subtype 15.3,* Iron-formation-hosted stratabound gold. Arsenic-rich gold-bearing sulphide iron-formation, immediately adjacent to a late quartz vein showing sulphide-arsenide megacrysts distributed along bedding. Lupin mine, Northwest Territories. Scale bar is 1 cm. GSC 1995-201A

Plate 29. *Deposit subtype 15.3,* Iron-formation-hosted stratabound gold. Gold grains (bright yellow) occur along boundaries between arsenopyrite (white rims) and loellingite (bluish cores), within arsenopyrite-loellingite-pyrrhotite (brown) megacrysts (see Pl. 28). Maximum dimension is approximately 0.5 cm. Photomicrograph, reflected light, oil immerson, after dilute nitric acid etch. GSC 1995-201B

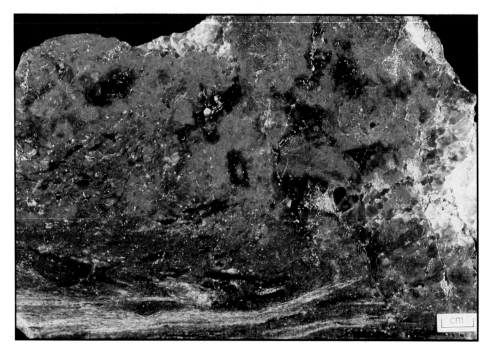

Plate 30. *Deposit subtype 15.4,* Disseminated and replacement gold. Boudinaged, recrystallized quartz vein that contains realgar (orange), cinnabar (red), and stibnite (black). Wall rock consists of highly sheared, barian mica schist. Discovery zone ("A" Pit), Williams mine, Hemlo, Ontario. GSC 1995-207B

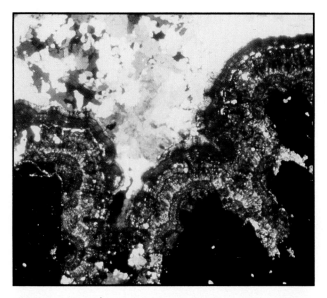

Plate 31. *Deposit type 18,* Vein stockwork tin, tungsten. Photomicrograph in transmitted light (crossed nicols) of banded, colloform cassiterite ("wood tin"; various shales of yellow, red, green, and brown) that is overgrown on fluorite (black). Quartz (bright white to grey) is later. Endozone tin deposit, Mount Pleasant, New Brunswick. Maximum dimension of photo is 2.7 mm. GSC 1995-217A

Plate 32. *Deposit type 18,* Vein stockwork tin, tungsten. Angular, silicified granite fragments in a dark matrix of fine grained sphalerite and cassiterite in a specimen from a tin-bearing breccia pipe. Fire Tower zone, Mount Pleasant, New Brunswick. GSC 1995-217B

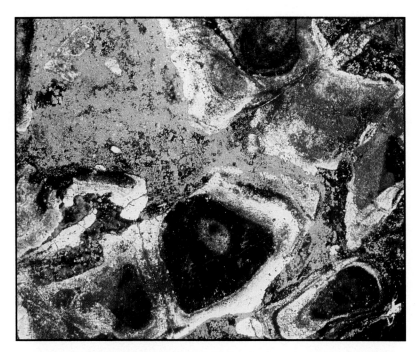

Plate 33. *Deposit type 18,* Vein stockwork tin, tungsten. Chloritized breccia fragments of granite (dark green to black) rimmed by fine grained quartz-topaz alteration (white) and cassiterite (brown); breccia matrix consists of arsenopyrite (silver-grey) and fluorite (purple, green). Endozone tin deposit, Mount Pleasant, New Brunswick. Maximum dimension of photo is 18 cm. GSC 1995-217C

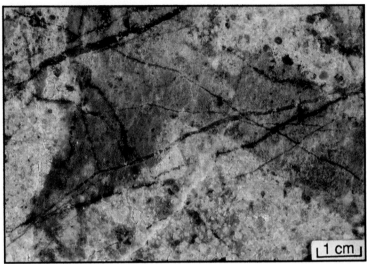

Plate 35. *Deposit type 19*, Porphyry copper, molybdenum, gold, tungsten, tin, silver. Molybdenite-bearing quartz veinlets cutting sericitized granodiorite, Red Mountain molybdenum deposit, Yukon Territory. GSC 1995-218

Plate 34. *Deposit type 19*, Porphyry copper-molybdenum, gold, tungsten, tin, silver. Wolframite-bearing quartz veins in granite, truncated by a slightly later stage of unmineralized granite. Fire Tower zone tungsten-molybdenum deposit, Mount Pleasant, New Brunswick. Minimum dimension of photo is 12 cm. GSC 1995-217D

Plate 36. *Deposit subtype 20.1,* Skarn zinc-lead-silver. Massive to layereed pyrite-marcasite, sphalerite, and galena. Midway deposit, British Columbia. Scale is in centimetres. GSC 1995-219

Plate 38. *Deposit subtype 20.5,* Skarn tungsten. Pyrrhotite (bronze-brown) with scheelite (not visible) in fractured limestone. E-zone orebody, Cantung mine, Northwest Territories. GSC 1995-220A

Plate 37. *Deposit subtype 20.2,* Skarn copper. Massive pods and coarse disseminated grains of chalcopyrite (bright yellow) with associated magnetite (grey), potassium feldspar (pink), calcite (white to light grey), and chlorite (black). Craigmont mine, British Columbia. GSC 1995-210

Plate 39. *Deposit subtype 20.5,* Skarn tungsten. High-grade tungsten ore; scheelite crystals (white) in biotite-diopside-pyrrhotite skarn. E-zone orebody, Cantung mine, Northwest Territories. GSC 1995-220B

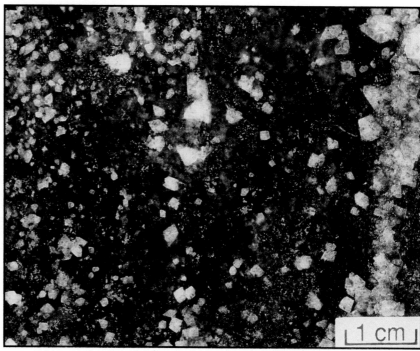

Plate 40. *Deposit type 21,* Granitic pegmatites. Zone of beryl crystals (white) between the quartz zone (dark grey) and layered aplites (shades of grey); beryl crystals are generally oriented perpendicular to the contact. Tanco mine, southeastern Manitoba. GSC 1995-223A

Plate 41. *Deposit type 21,* Granitic pegmatites. Saccharoidal albite (centre) rimmed by layered aplite that contains disseminated tantalum oxide minerals (black). From the eastern flank of the albitic aplite zone. Tanco mine, southeastern Manitoba. GSC 1995-223B

Plate 42. *Deposit type 23,* Peralkaline rock-associated rare metals. Coarse grained, pegmatitic eudialyte (red), sodic amphibole (black), and albite (white) in amphibole-albite-eudialyte gneiss. Kipawa yttrium-zirconium deposit, Quebec. GSC 1995-221A

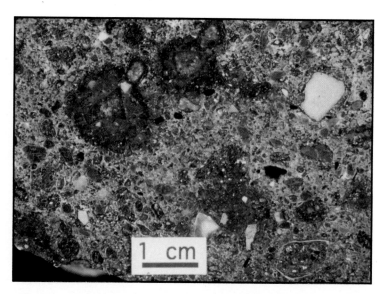

Plate 43. *Deposit type 24,* Carbonatite-associated deposits. Magmatic flow layering in niobium-bearing carbonatite; dark layers are composed of biotite, apatite, and fine grained pyrochlore. Niobec mine, Quebec. GSC 1995-221B

Plate 44. *Deposit subtype 25.1,* Kimberlite-hosted diamond. Tuffisitic kimberlite (diatreme facies), composed dominantly of pelletal lapilli, altered olivine (yellow-green), crustal limestone xenoliths (white), and garnet xenocrysts (red). Guigues pipe, Kirkland Lake-Timiskaming kimberlite field, Guigues Township, Quebec. GSC 1995-224

Plate 45. *Deposit subtype 27.1,* Nickel-copper sulphide. Typical disseminated nickel-copper-bearing sulphides in inclusion-rich noritic sublayer of the Sudbury Igneous Complex. Most of the larger, sulphide-free areas are inclusions. Longest dimension of sample is approximately 20 cm. Clarabelle mine, Sudbury, Ontario. GSC 1995-225A

321

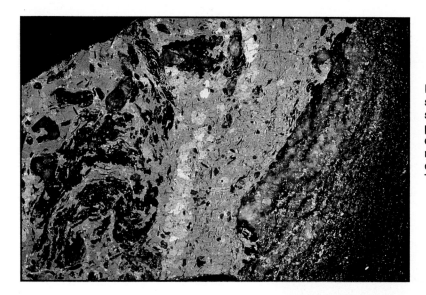

Plate 46. *Deposit subtype 27.1,* Nickel-copper sulphide. Typically deformed massive nickel sulphide ore, showing partially disaggregated pelitic schist inclusions, and a central layer of coarse pentlandite (lighter yellow) in a dominantly pyrrhotitic (yellow) matrix. Longest dimension of photo is 13 cm. Thompson mine, Thompson, Manitoba. GSC 1995-226

Plate 47. *Deposit subtype 27.1,* Nickel-copper sulphide. Matrix nickel sulphides (yellow) in serpentinized peridotitic host (black), typical of komatiitic nickel sulphide ores. The richer sulphide layer containing skeletal black crystals probably represents downward invasion of sulphide liquid that has replaced the matrix of the underlying spinifex-textured flow top. Longest dimension of photo is 11 cm. Alexo mine, Timmins, Ontario. GSC 1995-225B

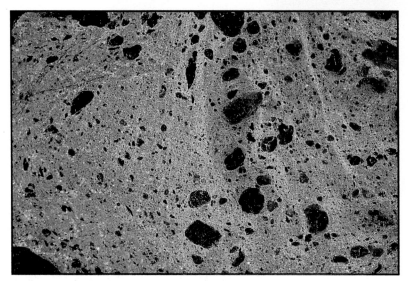

Plate 48. *Deposit subtype 27.1,* Nickel-copper sulphide. Massive, fine grained nickel-copper sulphide ore (pyrrhotite-pentlandite-chalcopyrite) that contains small abraded inclusions of wallrock. Longest dimension of photo is 13 cm. Falconbridge Main mine, Sudbury, Ontario. GSC 1995-225C

322

15. LODE GOLD

K.H. Poulsen

INTRODUCTION

Gold is a commodity that occurs in Canada in a wide variety of both geological settings and ore deposit types. **Byproduct** gold from volcanogenic massive sulphide deposits, nickel-copper deposits, porphyry copper-molybdenum deposits, and the Chibougamau copper deposits accounts for approximately one-third of Canadian resources. The remainder occurs in **gold-only** deposits which comprise **placers** (5%) and bedrock sources (60%), termed **lode gold** deposits (e.g. Cooke, 1946). Lode gold deposits are present in all of the major tectonic subdivisions of the Canadian landmass but occur dominantly in terranes with an abundance of volcanic and clastic sedimentary rocks of low to medium metamorphic grade. Economically viable deposits are concentrated primarily in the Archean greenstone terranes of Superior and Slave provinces, with lesser numbers in the Mesozoic-Cenozoic rocks of the Cordillera, the Proterozoic greenstone sequences of Trans-Hudson Orogen and Grenville Province, and the Paleozoic sequences of the Appalachians (Fig. 15-1).

The geological classification of lode gold deposits is problematical owing to the diversity of their host rocks. There are four aspects that historically have been important in arriving at a coherent classification of these deposits.

First is the temperature-depth concept. Classification is still largely influenced by the scheme of Lindgren (1933) who divided hydrothermal ore deposits, including those of gold and silver, into thermal types such as *epithermal*, *mesothermal*, and *hypothermal*. Ore deposit geologists now fully appreciate that thermal conditions for gold and silver deposition are rather similar (200-400°C) for all of these types so that Lindgren's choice of terms was unfortunate. Nonetheless, Lindgren fully recognized that his scheme also applied in a qualitative way to the depths in the Earth's crust at which various types of deposits form

Poulsen, K.H.
1996: Lode gold: in Geology of Canadian Mineral Deposit Types, (ed.) O.R. Eckstrand, W.D. Sinclair, and R.I. Thorpe; Geological Survey of Canada, Geology of Canada, no. 8, p. 323-328 (also Geological Society of America, The Geology of North America, v. P-1).

323

(Fig. 15-2) and it is this aspect of his classification scheme which has persisted to the present day. Thus, *epithermal* gold deposits are those for which there is evidence of a shallow crustal origin (less than 1 or 2 km), *mesothermal* deposits are those inferred to have formed at 1 to 3 km, and *hypothermal* deposits at 3 km to more than 5 km. The depth ranges implied for each of the three types are not firmly fixed, but are guidelines that reflect variations in lithostatic pressure, fluid pressure, crustal temperature and metamorphic facies transitions, availability of meteoric fluids, and the vertical extent of brittle and ductile fields of deformation and seismicity. For example, in areas of high heat flow such that both the brittle-ductile transition and metamorphic facies boundaries occupy elevated positions in the crust, and where a reduced permeability permits establishment of high fluid pressures, it is theoretically possible

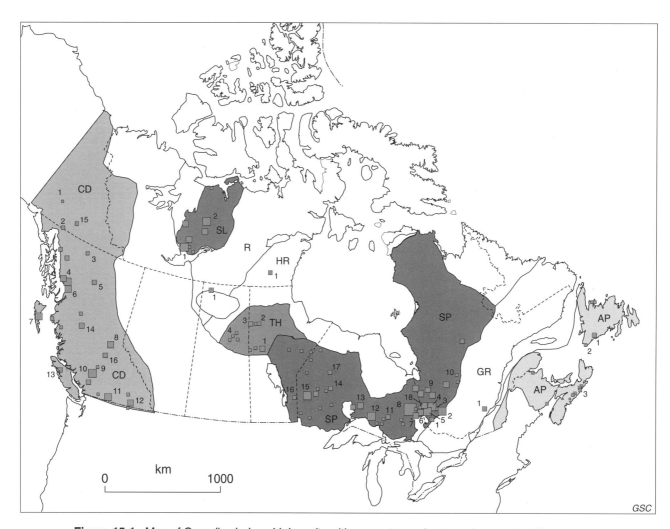

Figure 15-1. Map of Canadian lode gold deposits with respect to major tectonic domains. The smallest symbol size represents deposits that are estimated to contain less than 10 t Au; the next largest symbols, deposits containing 10 to 100 t; the second largest symbols, deposits and districts containing 100 to 1000 t; and the largest symbol size, the giant Timmins district containing more than 1000 t Au. Selected deposits and districts include: Appalachians Orogen (AP) – Hope Brook (1), Cape Ray (2), Goldenville (3); Grenville (GR) – Montauban (1); Superior Province (SP) – Belleterre (1), Val d'Or (2), Cadillac-Malartic (3), Bousquet (4), Noranda (5), Kirkland Lake-Larder Lake (6), Matachewan (7), Timmins (8), Agnico-Eagle (9), Chibougamau district (Norbeau) (10), Renabie (11), Hemlo (12), Geraldton (13), Pickle Lake (14), Red Lake (15), Bissett (16), North Cariboo Lake (17), Beattie (18); Trans-Hudson Orogen (TH) – Snow Lake (1), Farley Lake (2), MacLellan (3), LaRonge (4); Hearne Province (HR) – Cullaton Lake (1); Rae Province (R) – Box (1): Slave Province (SL) – Yellowknife (1), Lupin (2); Cordilleran Orogen (CD) – Brewery Creek (1), Mount Skukum (2), Cassiar (3), Iskut River (4), Toodoggone (5), Premier- Stewart (6), Cinola (7), Cariboo (8), Blackdome (9), Bridge River (10), Hedley (11), Rossland (12), Zeballos (13), Equity Silver (14), Ketza River (15), QR (16).

that the depth interval encompassing both "epithermal" and "mesothermal" conditions could be substantially compressed to 2 or 3 km. Burial and uplift histories are also important factors in assigning gold deposits to a particular depth zone because they can result in the superposition of a style of mineralization that characterizes one zone onto a style that characterizes deeper or shallower conditions.

Second is the vein-replacement distinction. There has been a historical distinction between vein deposits and those in which gold is disseminated, along with sulphide minerals, throughout the matrix of a particular host rock. Some lode gold deposits are particularly rich in sulphide minerals (10-70%) and these sulphides are not distributed in any particular relationship to associated veins. Such deposits have been termed "replacement" deposits in the past (e.g. Cooke, 1946). This term is no longer widely used because it was also used historically to describe magmatic nickel sulphide deposits and volcanic exhalative massive sulphide deposits which are not now believed to have formed entirely by replacement processes. Nonetheless, replacement is still a relevant process in the formation of many hydrothermal ore deposits (e.g. mantos and "zone refinement" in massive sulphide deposits) and is of particular importance in many gold deposit types.

Third is the concordance-discordance distinction. In the past two decades, increasing emphasis has been placed on the geometric relationships between gold orebodies and their host rocks. Thus concordance or discordance of orebodies is an important parameter because many other ore deposit types, such as volcanogenic massive sulphide and magmatic nickel deposits, once portrayed as being of "mesothermal" and "replacement" origins, are now regarded to have formed syngenetically. Many gold deposits are stratabound at a large scale but composed of discordant orebodies at a smaller scale.

Fourth is the compositional aspect. For many decades, geologists have noted the fact that gold deposits differ from one another in their relative contents of gold and silver and that, in the case of many epithermal deposits, the ratio of these two elements changes with depth. A corollary is that some gold deposits contain significant base metals,

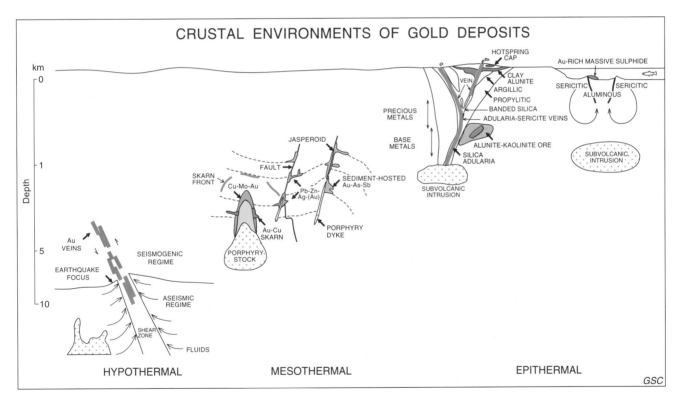

Figure 15-2. Schematic models of the crustal settings of gold deposits. For the deeper "hypothermal" environment a steep shear zone is illustrated to transect the boundary between seismogenic and aseismic crust and as a control on fluid (curved arrows) movement (after Sibson et al., 1988); for a shallower "mesothermal" environment the relative positions of porphyry Cu-Mo-Au, Au-skarn, and distal "Carlin-type" Au-As-Sb mineralization are illustrated (after Sillitoe and Bonham, 1990); for a shallow "epithermal" environment the relative position of subaerial hotspring mineralization is illustrated with respect to deeper epithermal veins (after Buchanan as reproduced in Panteleyev, 1986) as well as a hypothetical shallow marine environment corresponding to the formation of gold-rich volcanogenic massive sulphides.

whereas many base metal deposits yield significant byproduct gold. The definition of what constitutes a "gold deposit" in economic terms depends both on relative abundance of gold, silver, and base metals and on the prevailing prices of these commodities. Figure 15.3 shows the compositional ranges for common geologically-defined types of hydrothermal ore deposits as well as for individual examples. Note the wide variation in gold:silver ratios, lower values favouring epithermal, porphyry, and massive sulphide deposits, as well as the overlap of some deposit types (massive sulphide, porphyry, skarn) into the economically defined fields of both gold and base metal deposits.

Although there are many classification schemes that embody combinations of the above concepts, as well as parameters such as types of host rocks and alteration, lode gold deposits possess such a diversity of characteristics that there is little consensus among geologists as to their division into unique geological types. Most classifications (e.g. Boyle, 1979) emphasize aspects of the structure and geological setting of deposits with a decided emphasis on the nature of host rocks, whereas others, such as the classical scheme of Lindgren (1933), rely on inferred genetic variables such as depth and temperature. The most common current practice is to identify gold deposits with a "typical" deposit or geological setting (Table 15-1). This approach is practical and useful for some purposes but suffers from several weaknesses: a) many important deposits differ sufficiently from the type example to require constant definition of new types; b) the use of this approach also tends to obscure common processes that may link more than one deposit type; and, c) objective classification of deposits, particularly those that have attributes of more than one typical example, is difficult in highly deformed and metamorphosed terranes. The scheme is nonetheless satisfactory for the broad recognition of groups of deposits having like characteristics.

The summaries of Canadian lode gold deposits in this volume are organized less on the basis of a unified classification scheme than on broad groupings of deposits that illustrate common geological problems. The treatment

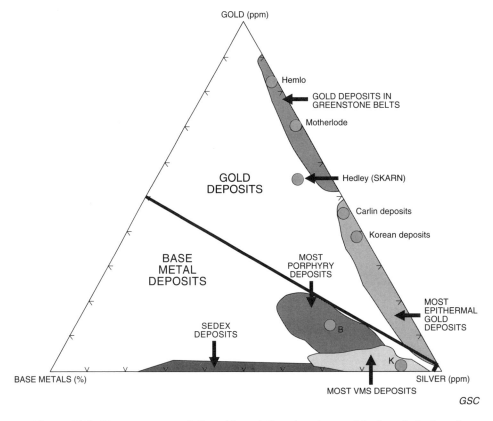

Figure 15-3. Ternary representation of the relative abundance of Au (ppm), Ag (ppm), and base metals (%) in a variety of ore deposit types world-wide: specific examples of individual gold-bearing deposits are shown for comparison. Note that the boundaries of the field defining "gold deposits" in an economic sense are elastic (a function of prevailing metal prices) but the plotted positions for specific deposits are fixed only by their bulk composition. "B" identifies the Bingham Canyon, U.S.A. porphyry deposit and "K" the Kidd Creek, Ontario massive sulphide deposit. "Gold deposits in greenstone belts" refers to deposits discussed below as subtypes 15.2, 15.3, and 15.4.

emphasizes the geological setting and the nature of the ore constituting the deposits; where possible, widely used and accepted nomenclature has been retained. Clearly, the gold deposits, as defined in economic terms, are of diverse geological types. Some are treated in separate sections of this volume, together with their related deposit types: examples include placers (see subtype 1.2), paleoplacers (see subtype 1.1), volcanic-associated massive sulphide

gold deposits (see subtype 6.4), gold-bearing porphyry deposits (see Type 19), and skarn gold (see subtype 20.3). Others, as summarized below, are described as separate subtypes of "Lode gold" (15) deposits.

"Epithermal gold" (subtype 15.1) deposits account for approximately 5% of Canada's lode gold production and reserves. Most contain more silver than gold (Fig. 15-3). Deposits of this type occur mainly in extensional settings

Table 15-1. Common type of lode gold deposits with Canadian examples.

Type	Characteristics	Typical example	World examples	Canadian examples	References
SHALLOW ENVIRONMENTS					
a) Witwatersrand type	quartz arenite-conglomerate; paleoplacers	Witwatersrand, South Africa	Jacobina, Brazil; Tarkwa, Ghana	minor occurrences in Huronian rocks, Ont.	Minter, 1991
b) Hotspring type	gold disseminated in sinters	McLaughlin, California	Round Mountain, Nevada	Cinola, B.C.	Bonham, 1989
c) Submarine exhalative type	stratabound sulphides; shallow(?) felsic volcanic association; Cu, Pb, Zn, Ag common	Boliden, Sweden	Mount Lyell; Mount Morgan, Australia	Eskay Creek, B.C.; Montauban, Bousquet, Agnico-Eagle, Horne, Que.	Hannington and Scott, 1989
d) Alunite-kaolinite (Acid-Sulphate) type	advanced argillic alteration; vein and disseminated; high-sulphur mineral assemblages (i.e. enargite)	Goldfield, Nevada	El Indio, Chile	rare - minor development at Island Copper, B.C.	Heald et al., 1987; Berger and Henley, 1989
e) Adularia-sericite (Bonanza) type)	crustiform veins; chalcedonic quartz; vertical metal zonation; calderas	Creede, Colorado	Thames, New Zealand; Hishikari, Japan	Blackdome, Lawyers, B.C.; Grew Creek, Y.T.	Heald et al., 1987; Berger and Henley, 1989
MODERATELY DEEP ENVIRIONMENTS					
a) Porphyry type	mainly intrusion-hosted; stockwork and disseminated; K-silicate alteration, Cu-Bi association	Lepanto, Philippines	Yu Erya, China	Fish Lake, Kemess, B.C.; Young-Davidson, Ross, Ont.; Doyon, Douay, Que.	Sillitoe, 1991
b) Breccia pipe type	magmatic-hydrothermal and phreatomagmatic breccias; pipes and cylindrical sheeted fractures	Kidston, Australia	Golden Sunlight, Montana	Sunbeam Kirkland, Man.; Chadbourne, Que.	Sillitoe, 1991
c) Skarn type	Al-rich skarn assemblages adjacent to diorite-granodiorite; As, Bi, Te association	Fortitude, Nevada	Red Dome, Australia; Suan, Korea	Hedley, Tillicum, B.C.; Akasaba, Que.	Meinert, 1989
d) Manto type	i) carbonate replacement ii) noncarbonate replacement	i) Cove, Nevada ii) Porgera, Papua New Guinea	i) Foley Ridge, South Dakota; ii) Andacollo, Chile	i) Ketza River, Y.T.; Mosquito Creek, B.C. ii) Equity Silver, QR, B.C.; Brewery Creek, Y.T.; Hemlo, Ont.; Beattie, Que.; Hope Brook, Nfld.	Sillitoe, 1991
e) Carlin type	sediment-hosted disseminated micrometre-size Au; As-Sb-Hg association	Carlin, Nevada	Mercur, Utah; Guizhou, China	rare	Berger and Bagby, 1991
f) Korean type	intrusion-related, fault-controlled quartz veins and disseminated zones	Shandong Province, China	Charters Towers, Australia	Venus, Y.T.; Zeballos, B.C.	Shelton et al., 1988
DEEP ENVIRONMENTS					
a) Motherlode type	shear zone-related, volcanic-hosted ribboned quartz veins; carbonatization	Mother Lode district, California	Kalgoorlie-Norseman, Australia	Bralorne, B.C.; Giant-Con, N.W.T.; Noracme, Man.; Kerr Addison, McIntyre-Hollinger, Ont.; Sigma-Lamaque, Que.; Deer Cove, Nfld.	Knopf, 1929
b) Grass Valley type	shear zone-related, plutonic-hosted ribboned quartz veins; sericitization and carbonatization	Grass Valley district, California	Alleghaney district, California	Star Lake, Contact Lake, Sask.; Renabie, Ont.; Silidor, Ferderber, Que.	Johnston, 1944
c) Bendigo type	fold and fault-controlled, turbidite-hosted quartz veins; arsenopyrite common	Victoria Goldfields, Australia	Otago, New Zealand; Ashanti, Ghana; Murantau, Uzbekistan	Camlaren, N.W.T.; Pamour, Little Long Lac, Ont.; Meguma, N.S.	Boyle, 1986
d) Homestake type	vein-related, iron-formation-hosted; sulphidic alteration; arsenopyrite common	Homestake, South Dakota	Jardine, Montana; Cuiaba, Brazil; Hill 50, Australia	Lupin, N.W.T.; Farley, Man.; Central Patricia, McLeod-Cockshutt, Ont.	Phillips et al., 1984; Caddy et al., 1991

in Mesozoic and Tertiary rocks in the Canadian Cordillera. Important examples include Mount Skukum in Yukon Territory, and the Blackdome, Cinola, and Toodoggone deposits in British Columbia (Fig. 15-1). The discussion covers those epithermal deposits that are generally regarded to be of alunite-kaolinite, adularia-sericite, and hotspring types, as well as "transitional" or "deep epithermal" veins and Carlin-type sediment-hosted gold-arsenic-antimony deposits, although the latter are not particularly important in Canada.

"Quartz-carbonate vein gold" (subtype 15.2) deposits typify metamorphic terranes of all ages and account for approximately 80% of the production from Canadian lode gold deposits. By contrast with epithermal deposits, they contain little silver relative to gold (Fig. 15-3). The Canadian Shield, and Superior Province in particular, contains the most significant producers (Fig. 15-1). Typical examples of these deposits include Goldenville, Nova Scotia; Sigma-Lamaque at Val d'Or, Quebec; Dome and Campbell Red Lake, Ontario; San Antonio at Bissett, Manitoba; Star Lake, Saskatchewan; and Bralorne-Pioneer and Cariboo-Island Mountain, British Columbia. The deposits consist of simple to complex vein systems with significant vertical extent (in some cases greater than 2 km), hosted by deformed and metamorphosed volcanic, plutonic, and clastic sedimentary rocks in compressional tectonic settings. Sulphide-rich vein deposits such as at Chibougamau, Quebec and Rossland, British Columbia overlap in some attributes with these deposits.

"Iron-formation-hosted stratabound gold" (subtype 15.3) deposits represent approximately 5% of Canada's total lode gold production and reserves. These occur only within the Canadian Shield (Fig. 15-1) and important Canadian examples include the McLeod Cockshutt-Hardrock deposit at Geraldton and Central Patricia deposit at Pickle Lake in Ontario, the Farley deposit in Manitoba, and the Lupin deposit in the Northwest Territories. All of these deposits occur in complexly folded iron-formation containing quartz veins, and ore commonly consists of disseminated to massive sulphides adjacent to veins. They are a special subtype of the quartz-carbonate veins but are treated separately because they embody aspects of two ore deposit classes, the host iron-formations and their contained gold mineralization. Due to the inherent exhalative nature of the iron formations, there is a possibility that some deposits in this category owe their gold enrichment to prevein processes.

"Disseminated and replacement gold" (subtype 15.4) deposits represent approximately 10% of Canada's historical production and reserves. Important examples include the Hope Brook deposit in Newfoundland, the Hemlo and Madsen (Red Lake) deposits in Ontario, the MacLellan deposit in Manitoba, and the Equity Silver deposit and the sulphide lodes of the Island Mountain mine in the Cariboo district of British Columbia (Fig. 15-1). The Ketza River manto deposits, as well as several intrusion-related disseminated deposits in Superior Province (e.g. Beattie and Lac Shortt in Quebec), also belong to this general category. All of these deposits comprise auriferous bodies of disseminated to massive sulphides, typically pyrite or pyrrhotite, in which ore distribution is not dictated by the presence of vein quartz and, with a few exceptions, they have low contents of base metals and a gold content exceeding that of silver (Fig. 15-3).

REFERENCES

Berger, B.R. and Bagby, W.C.
1991:   The geology and origin of Carlin-type deposits; in Gold Metallogeny and Exploration, (ed.) R.P. Foster; Blackie and Son, Ltd., Glasgow, p. 210-248.

Berger, B.R. and Henley, R.W.
1989:   Advances in the understanding of epithermal gold-silver deposits, with special reference to the Western United States; in The Geology of Gold Deposits: the Perspective in 1988, (ed.) R.R. Keays, W.R.H. Ramsay, and D.I. Groves; Economic Geology, Monograph 6, p. 405-423.

Bonham, H.F., Jr.
1989:   Bulk mineable gold deposits of the western United States; in The Geology of Gold Deposits: the Perspective in 1988, (ed.) R.R. Keays, W.R.H. Ramsay, and D.I. Groves; Economic Geology, Monograph 6, p. 193-207.

Boyle, R.W.
1979:   The geochemistry of gold and its deposits; Geological Survey of Canada, Bulletin 280, 584 p.
1986:   Gold deposits in turbidite sequences: their geology, geochemistry and history of theories of their origin; in Turbidite-hosted Gold Deposits, (ed.) J.D. Keppie, R.W. Boyle, and S.J. Haynes; Geological Association of Canada, Special Paper 32, p. 1-13.

Caddy, S.W., Bachman, R.L., Campbell, T.J., Reid, R.R., and Otto, R.P.
1991:   The Homestake gold mine, an Early Proterozoic iron-formation-hosted gold deposit, Lawrence County, South Dakota; United States Geological Survey, Bulletin 1857-J, 67 p.

Cooke, H.C.
1946:   Canadian lode gold areas (summary account); Canadian Department of Mines and Resources, Economic Geology Series, v. 15, 86 p.

Hannington, M.D. and Scott, S.D.
1989:   Gold mineralization in volcanogenic massive sulfides: implications of data from active hydrothermal vents on the modern sea floor; in The Geology of Gold Deposits: the Perspective in 1988, (ed.) R.R. Keays, W.R.H. Ramsay, and D.I. Groves; Economic Geology, Monograph 6, p.491-507.

Heald, P., Foley, N.K., and Hayba, D.O.
1987:   Comparative anatomy of volcanic-hosted epithermal deposits: acid-sulfate and adularia-sericite types; Economic Geology, v. 82, p. 1-26.

Johnston, W.D., Jr.
1944:   The gold-quartz veins of Grass Valley, California; United States Geological Survey, Professional Paper 194, 101 p.

Knopf, A.
1929:   The Mother Lode system of California; United States Geological Survey, Professionl Paper 157, 88 p.

Lindgren, W.
1933:   Mineral Deposits; McGraw-Hill, New York and London, 930 p.

Meinert, L.
1989:   Gold skarn deposits - geology and exploration criteria; in The Geology of Gold Deposits: the Perspective in 1988, (ed.) R.R. Keays, W.R.H. Ramsay, and D.I. Groves; Economic Geology, Monograph 6, p. 537-552.

Minter, W.E.L.
1991:   Ancient placer gold deposits; in Gold Metallogeny and Exploration (ed.) R.P. Foster; Blackie and Son, Ltd., Glasgow, p. 283-308.

Panteleyev, A.
1986:   A Canadian Cordilleran model for epithermal gold-silver deposits; Geoscience Canada, v. 13, p. 101-111.

Phillips, G.N., Groves, D.I., and Martyn, J.E.
1984:   An epigenetic origin for Archean banded iron-formation-hosted gold deposits; Economic Geology, v. 79, p. 162-171.

Shelton, K.L., So, C.-S., and Chang, J.-S.
1988:   Gold-rich mesothermal vein deposits of the Republic of Korea: geochemical studies of the Jungwon gold area; Economic Geology, v. 83, p. 1221-1237.

Sibson, R.H., Robert, F., and Poulsen, K.H.
1988:   High-angle reverse faults, fluid pressure cycling, and mesothermal gold-quartz deposits; Geology, v. 16, p. 551-555.

Sillitoe, R.H.
1991:   Intrusion-related gold deposits; in Gold Metallogeny and Exploration, (ed.) R.P. Foster; Blackie and Son, Ltd., Glasgow, p. 165-209.

Sillitoe, R.H. and Bonham, H.F., Jr.
1990:   Sediment-hosted gold deposits: distal products of magmatic hydrothermal systems; Geology, v. 18, p. 157-161.

15.1 EPITHERMAL GOLD DEPOSITS

15.1a Quartz-(kaolinite)-alunite deposits
15.1b Adularia-sericite deposits

Bruce E. Taylor

INTRODUCTION

Current usage of the term "epithermal (gold) deposit" encompasses a somewhat broader range of physico-chemical conditions than originally envisioned by Lindgren (1922, 1933) in his depth-temperature zoning concept of deposit classification according to environment. Although still generally accepted as "shallow" (e.g., <1500 m), the depth of formation is perhaps the most difficult characteristic of epithermal deposits to quantify. Mesothermal deposits ("transitional" deposits of Panteleyev, 1986), which may form as deep as perhaps 3000 m, and in close association with intrusions, have many characteristics in common with epithermal deposits, including aspects of their origin. Indicated temperatures of formation for epithermal deposits range from about 100°C for hot spring or steam-heated deposits to about 350-400°C for deeper vein and replacement deposits. However, it is the marked variations in temperature and pressure which best characterize the ore-forming epithermal environment. Pronounced changes in the physical, thermal, and chemical properties of the hydrothermal solutions occur over short distances, promoting ore deposition.

During the past decade, classification and terminology of epithermal deposits have evolved substantially, as has our understanding of these deposits. This is evidenced by a rapidly-growing body of literature covering geological aspects and genetic hypotheses, such as discussed and further cited in recent papers by Hayba et al. (1985), Berger and Henley (1989), White and Hedenquist (1990), Panteleyev (1991), and Sillitoe (1993); see also the accompanying list of references.

Alteration and ore mineral assemblages are used here to distinguish two principal subtypes of deposits: quartz-(kaolinite)-alunite (QAL; subtype 15.1a) and adularia-sericite (ADS; subtype 15.1b; Hayba et al., 1985; Heald et al., 1987). The descriptive mineralogical terms quartz-(kaolinite)-alunite (QAL) and adularia-sericite (ADL) are preferred here, although the former deposits have been variously termed "high-sulphur" (Bonham, 1988), "acid-sulphate" (Heald et al., 1987), "high-sulphidation"

(Hedenquist, 1987), or kaolinite-alunite (Berger and Henley, 1989) deposits. Adularia-sericite-subtype deposits have also been named "low-sulphur" (Bonham, 1988), or "low-sulphidation" (Hedenquist, 1987). Differing levels of acidity and oxidation state (adularia-sericite: near-neutral, intermediate to reduced; quartz-(kaolinite)-alunite: acidic, oxidized), largely distinguish the ("end-member") hydrothermal environments of these two general deposit types. In this context, the "high/low sulphur" or "high/low sulphidation" terminology (not to be confused with high/low sulphide contents) can be misleading, as discussed later.

Adularia-sericite and quartz-(kaolinite)-alunite deposits can be further subdivided. Two subenvironments of quartz-(kaolinite)-alunite alteration are recognized: magmatic-hydrothermal and steam-heated; either may be mineralized, or barren. Subdivision of adularia-sericite-subtype deposits by volcanic host rock (alkalic; subakalic, rhyolite; and subakalic, andesite-rhyolite) has recently been suggested by Sillitoe (1993; cf. Bonham, 1988; Mutschler and Mooney, 1993). The subakalic, andesite-rhyolite association includes sulphide-rich adularia-sericite-subtype deposits and some deep epithermal Canadian gold deposits described here.

Hot spring deposits, which may be associated with either acid-sulphate (quartz-(kaolinite)-alunite) alteration and/or argillic alteration (adularia-sericite-subtype), can constitute significant surface expressions of some geothermal systems, and form a subset of epithermal deposits. These deposits may or may not contain economic abundances of gold (±silver) in associated near-surface silicified zones, sinters, or phreatic breccias (e.g., Bonham, 1989). However, unless sinters can be positively identified, or evidence for near 100°C (boiling-limited) temperatures is available, it is difficult to distinguish a hot-spring setting per se from the general near-surface, steam-heated environment. Hot spring environments have been inferred for gold deposits in Canada hosted by both volcanic (e.g., Silver Pond; Toodoggone River, British Columbia; Diakow et al., 1993) and sedimentary (e.g., Cinola, British Columbia; Tolbert and Froc, 1988; Christie, 1989) rocks; sinters have not been identified, however.

Gold is the principal commodity of epithermal gold (±silver) deposits, and it usually occurs as free gold and alloyed with silver, but may also occur in tellurides or be included in sulphides. Copper and the other base metals, lead and zinc, may also occur with gold, especially in deposits with high silver grades. Indeed, quartz-(kaolinite)-alunite-subtype deposits are sometimes referred to as enargite-gold deposits (Ashley, 1982). Hot spring deposits,

Taylor, B.E.
1996: Epithermal gold deposits; in Geology of Canadian Mineral Deposit Types, (ed.) O.R. Eckstrand, W.D. Sinclair, and R.I. Thorpe; Geological Survey of Canada, Geology of Canada, no. 8, p. 329-350 (also Geological Society of America, The Geology of North America, v. P-1).

Table 15.1-1. Comparative mineralogical, geological, and production data for selected epithermal Au deposits in Canada and several non-Canadian, "type" examples.

District and/or deposit	Age[1] Host	Age[1] [Min.]	Size[2] (R+P) Ore[5]	Size[2] (R+P) Au[6]	Grade[3]	Ag/Au	Base metal	%S*	Ad	Al	Cpy	En	Ss	Ags	Sp	Gn	Ba	Fl	Rc	Cc	Ank	Host rock	Alt'n[7] vn → w.r.	Form[8]	Selected Refs.
QUARTZ-KAOLINITE-ALUNITE (QAL) TYPE:[9] Volcanic host rocks																									
Toodoggone River, B.C.	189-198;182																								
Al (Bonanza; Thesis)		[196]	0.348	3.210	9.6					x	x		x	x	x	x	x					and./dacite	Si/A	vn	10,11,30,31,35
BV		[190-197]	0.053	0.55	10.4		x										x					andesite		vn,bx	30
Summitville, Colorado	20.2-22.0	[22.3]	83.51	3.5		1.2	x	5?		XX	XX	x	XX	x	x	x	x					qtz. latite	Vgy-Si/Qtz-Al/A	repl,vn	28,32
Nansatsu, Japan	3.4-7.6	[2.7-5.5]		>18	3-6	0.1-1.0	x	≤10		x	x	x	x		[x]	[x]	x			x		andesite	Vgy-Si/Al-A/Ph/P	repl.	33
El Indio, Chile[10]	13.7	[8.6]	8.7	108	1.7-218	0.5-10	XX	≤30[11]	x	[X]	[X]	[XX]	x	x	[x]	[x]	x					rhyodacite	Si/Ph-A (Al-A)	vn,bx	36,37
ADULARIA-SERICITE (ADS) TYPE: Volcanic and plutonic host rocks																									
Mt. Skukum, Y.T.	53.2	[50.7]	0.200	2.49	25.0	0.9		<1	x	x							x	x	x		x	andesite	Si/±K/Ph/A/P	vn,bx,st	7,8,9
Mt. Nansen, Y.T.	Tertiary		0.288	3.15	11.1	39.0									x	x	x					andesite	Ph	vn	14
Laforma, Y.T.	>140	[78?]	0.191	2.13	11.2		x				x		x		x	x	x			x	x	granodiorite	Ph	vn	34
Venus, Y.T.	L. Jur.		>0.07	>0.66	9.3	26.5	XX	15-60						x	x	x	x					and./dacite	Si/A	vn	5,6,16
Toodoggone River, B.C.	189-198;182																								
Lawyers		[180]	0.880	6.73	7.4	46.7	x		x	x	x	x	x	x	x	x	x	x				andesite	Si/A/P	st,bx	1,2,3,24,31,35
Baker (Chapelle)			0.055	1.05	19.5	9.1		3-15		x	x		x		x	x				x		and./basalt	Si/Ph/A/P	vn	13
Blackdome, B.C.	Eocene [>24,<51.5]		0.368	7.35	20.6	3.1	x	≤5		x	x		x	x	x	x	x			x	x	and./dacite	Si/K/A/P	vn,bx	4,23,26,35
Stewart-Iskut region, B.C.	210																								
Silbak-Premier	[194.8?]		9.622	66.24	7.0	22.6	XX	≤5[12]	X[13]		x	x	x	x	x	x	x	x	x	x	x	and./dacite	Si/K/Ph/P	vn,st,bx	21
Sulphurets (Snowfield)	[192.7]		25	0.78	2.4	0.6	x						x		x	x	x	x	x	x		bslt.-and./andesite	K/Ph/A/P	st,vn,diss.	41
Creede, Colorado	Tertiary		1.4	21.0	1.5	400	x		x	x			x	x	x	x	x	x	x	x	x	and./dacite	Si/A	vn,st,diss.	22,28
ADULARIA-SERICITE (ADS) TYPE: Sedimentary and/or mixed host rocks																									
Cinola, B.C.	Tert./Cret. [14]		23.80	58.31	2.45	2	x	≤10		x			x		x	x	x			x		congl./s.s./shale	Si/A	diss.,bx,vn	25,27,35
Equity Silver, B.C.[14]	57.2 [>48; 57.2]		31.42	24.41	4.2	128.2	x				x		x		x	x						dacite/tuff/congl.	A	vn,st,diss.	17,18,19
Dusty Mac, B.C.	Eocene		0.093	0.60	7.2	21.5	x	≤15			x		x		x	x		x		x		s.s./sh./and. pyroclastic	Si/A	bx,st	35,40
Carlin, Nevada	Paleozoic [Tertiary]		10	109.7	11.0		x			X[15]			x	x	x	x	x			x		silty lms.	Si	diss.,vn	29
Hishikari, Japan	0.51-1.78/Cret. [0.8-1.0]			121.7	70	1.27	x		XX		x		x		x	x	x		x	x	x	shale-s.s./and./dacite	Si/A	vn	38,39

* Principal deposits plus several others selected to represent part of the spectrum of variation in type and setting.

1. Based on reported mineral ages; Ma., exclusive of uncertainty limits. Host= age of host rocks; [Min.] = age of mineralization. 2. P = cumulative production; R = reserves; 3. Average grade in g/t; 4. characteristic; in addition to quartz (+pyrite±sericite±clays); 5. tonnes of ore x10[6]. 6. grams of gold (Au) x10[6]; 7. Alteration facies, vein (vn) to wall rock (w.r.): Vgy-Si; vuggy silica; Qtz, quartz; Al, alunite (advanced argillic); Si, silicification; K, potassic; Ph, phyllic (sericitic); A, argillic/advanced argillic; P, propylitic (sequence from vein to wall rock); 8. Form of deposit (in order of importance) vn, vein; bx, breccia; st, stockwork; diss., disseminated, repl., replacement; 9. Classification is based on available data; uncertain in some cases; 10. main gold deposition probably not from alunite-kaolinite type system, *sensu stricto*, see text; 11. older, alunite-associated (Cu) veins contain 30-90% sulphide. 12. base metal-rich veins and breccias contain 20 to 45% sulphide; 13. potassium feldspar; species not confirmed; 14. contains metamorphosed advanced argillic mineral assemblage; low-pH conditions approached those of magmatic-hydrothermal QAL subtype deposits, 15. in oxidized ore.

Abbreviations: %S*, per cent sulphide; Ad, adularia; Al, alunite; Cpy, chalcopyrite; En, enargite; Ss, sulphosalts (e.g., tennantite-tetrahedrite); Ags, silver sulphides; Sp, sphalerite; Gn, galena; Ba, barite; Fl, fluorite; CO₃*, carbonate: Rc, rhodochrosite; Cc, calcite; Ank, ankerite; X = present; x = minor to rare; XX = abundant; [ ] = absent or unknown; and. = andesite; bstl. = basalt; congl. = conglomerate; s.s. = sandstone; lms. = limestone; sh. = shale; Tert. = Tertiary; Cret. = Cretaceous; L. Jur. = Lower Jurassic; NB: *[ ] = not in paragenetic association with Au.*

References: 1) Schroeter, 1986; 2) Schroeter, 1985; 3) Schroeter, 1982; 4) D. Rennie, 1986, written commun.; 5) Walton, 1986; 6) Walton and Nesbitt, 1986; 7) McDonald et al., 1986; 8) McDonald and Godwin, 1986; 9) Pride and Clark, 1985; 10) Clark and Williams-Jones, 1985; 11) Schroeter, 1986; 12) Andrew et al., 1986; 13) Barr et al., 1986; 14) Morin and Downing, 1984; 15) Duke and Godwin, 1986; 16) McFall, 1981; 17) Shen and Sinclair, 1982; 18) Cyr et al., 1984; 19) Wojdak and Sinclair, 1984; 20) Love, 1989; 21) McDonald, 1990; 22) Mosier et al., 1985; 23) Faulkner, 1986; 24) Vulimiri et al., 1986; 25) Champigny and Sinclair, 1982; 26) Vivian et al., 1987; 27) Christie, 1989; 28) Heald et al., 1987; 29) Bagby and Berger, 1986; 30) Diakow et al., 1991; 31) Clark and Williams-Jones, 1993; 33) Hedenquist et al., 1994; 34) McInnes et al., 1990; 35) Anon., 1992, B.C. Geol. Survey (MINFILE/pc), 1992; 36) Jannas et al., 1990; 37) Siddeley and Araneda, 1986; 38) Izawa et al., 1990; 39) Bakken and Einaudi, 1986; 40) Church, 1973; 41) Margolis, 1993.

some other epithermal deposits, and the well-known Carlin-type deposits, exhibit a characteristic association of Hg, As, Sb, and Tl with gold.

The best examples in Canada of low-sulphide, volcanic-hosted adularia-sericite epithermal gold deposits include the Blackdome deposit and deposits in the Toodoggone River camp in British Columbia, and the Mt. Skukum deposit in Yukon Territory (Table 15.1-1). The largest adularia-sericite epithermal deposit (Silbak-Premier) is sulphide-rich, but has a low-sulphide precious metal stage (McDonald, 1990). Quartz-(kaolinite)-alunite deposits are not prominent in Canada. The Al deposit (Toodoggone River camp: Diakow et al., 1993) is the single example chosen (Table 15.1-1). A quartz-alunite bearing alteration mineral assemblage also occurs in a small area at Mt. Skukum (McDonald, 1987), and advanced argillic mineral assemblages (or metamorphic equivalents) are found in several other deposits (e.g., Snowfield: Margolis, 1993; Equity Silver: Cyr et al., 1984; Chetwynd: McKenzie, 1986), but the quartz-muscovite-pyrite±chlorite (adularia-sericite) alteration mineral assemblage is typical of the majority of deposits in Canada.

Cinola is the principal (clastic) sediment-hosted adularia-sericite epithermal gold deposit in Canada. According to Champigny and Sinclair (1982), it shares some features with Carlin-type deposits in the western United States, including close association with felsic dykes, fine grained gold in silicified rocks, Tertiary age, and high Hg contents. The deposits differ in that mudstone was silicified and replaced at Cinola, whereas decarbonated, silicified impure carbonate rocks are the host to ore at the Carlin mine (Bakken and Einaudi, 1986; Christie, 1989). The depth of formation of these deposits is controversial and not well established, however. Tolbert and Froc (1988) and Christie (1989) interpreted Cinola to be a hot spring deposit based on classic "epithermal" characteristics, such as chalcedonic silica and hydrothermal breccias, whereas Shen et al. (1982) suggested a depth of formation of 1.8 km based on fluid inclusion data. Isotopic data (discussed below) require deep circulation of meteoric waters. The commonly accepted shallow emplacement level of the Carlin-type deposits has been questioned on the basis of geological arguments (Bakken and Einaudi, 1986; Sillitoe and Bonham, 1990) and isotopic data (e.g., Taylor, 1987; Holland et al., 1988).

IMPORTANCE

Epithermal gold deposits represent a minor proportion of the gold reserves and production in Canada, where meso- and hypothermal gold deposits are the principal producers. For example, the average annual production from epithermal gold deposits in Canada during 1985-1987 was 2725 kg/a, or about 2.7% of the total annual gold produced. In British Columbia and Yukon Territory, epithermal gold deposits contributed relatively more (24%) of the total gold produced than elsewhere in Canada.

SIZE AND GRADE OF DEPOSITS

The sizes of principal Canadian epithermal gold vein deposits and selected "type" deposits elsewhere, are given in Table 15.1-1 in millions of tonnes (Mt) of ore (geological reserves plus past production) and kilograms (kg) of gold.

The size estimates, which range from about 0.05 to 42 Mt of ore, give an order of magnitude basis for comparison as they depend on cut-off grades and economics. Grades and tonnages of these Canadian deposits are compared to other selected gold vein deposits and to gold-producing deposits of other types in Figure 15.1-1. The grades (grams/tonne; g/t) of Canadian epithermal gold vein deposits (most about 2.5-25 g/t), and of several deep epithermal or mesothermal vein deposits, are similar to those of most hypothermal quartz-carbonate gold vein deposits, but the epithermal deposits tend to be smaller in size. Epithermal vein deposits are distinguished from hypothermal quartz-carbonate vein deposits by higher silver:gold ratios (>1:1; see Fig. 15-3). Although variable, silver:gold ratios tend to be higher in adularia-sericite-subtype deposits than in quartz-(kaolinite)-alunite-subtype deposits (Table 15.1-1). The Canadian deposits are comparable to the smaller deposits in major epithermal terranes (e.g., western Pacific: Sillitoe, 1989; central Andes: Erickson and Cunningham, 1993).

At Cinola and in areas of the Sulphurets district of British Columbia, ore comprises disseminated gold in silicified and/or finely veined rocks; grades are typically lower, but tonnage larger, than in other vein-type epithermal deposits (Table 15.1-1). The Cinola deposit contains 58 310 kg of Au, based on a reported grade of 2.45 g/t and 23.80 Mt of ore, which compares to the median gold grade of 2.5 g/t and 5.1 Mt size of Carlin-type deposits in Nevada (Bagby et al., 1986). The Cinola deposit is potentially the second largest epithermal gold deposit in Canada (Table 15.1-1).

Variable Ag:Au ratios characterize deposits as shown in Table 15.1-1. Lower values of Ag:Au are shown in Table 15.1-1 for several quartz-(kaolinite)-alunite deposits, but this ratio varies widely on a world-wide basis (e.g., 0.5, Kasuga, Japan: Hedenquist et al., 1994; >>500, Cerro Rico de Potosi: Erickson and Cunningham, 1993). Both differing magmatic metal budgets (Sillitoe, 1993) and depth of formation (Hayba et al., 1985) have been suggested to influence this ratio. Very silver-rich quartz-(kaolinite)-alunite deposits like those found in Bolivia and Peru (e.g., Erickson and Cunningham, 1993) are not presently known in Canada. The deep epithermal (mesothermal) Equity Silver deposit (e.g., Cyr et al., 1984; Wojdak and Sinclair, 1984; Table 15.1-1) represents the closest analogue to a magmatic-hydrothermal quartz-(kaolinite)-alunite deposit in Canada (silver-rich; contact-metamorphosed); alunite has not been reported. At the Lawyers deposit, the ratio varies northward, from less than 20 to more than 80, and higher Ag:Au ratios are also found at deeper levels of the deposit (Vulimiri et al., 1986; average = 46.7).

Quartz-(kaolinite)-alunite deposits of magmatic-hydrothermal origin (see Rye et al., 1992; discussed below) are restricted to areas in close proximity to (above) a related source of magmatic heat and volatiles. Altered rocks of the Summitville, Colorado deposit outcrop over an area of 1.5 by 1.0 km, and produced about 3500 kg of gold (Heald et al., 1987). Shallow, steam-heated environments may produce wide-spread altered areas, typically (but not always) barren; bulk-tonnage mining of these zones may be possible if they are mineralized. For example, mineralized areas altered to quartz+clay+alunite(+barite+dickite) at the Al deposit, Toodoggone River area, measure about 250 m by as much as 1.5 km (Diakow et al., 1993). Fault-controlled, quartz-(kaolinite)-alunite alteration zones occur topographically above the Mt. Skukum deposit, in an area measuring roughly 200 by 250 m (McDonald, 1987).

Adularia-sericite-subtype deposits in some cases cover large areas, even though alteration mineral assemblages are restricted to generally narrow zones enclosing veins and breccias. At the Blackdome mine, British Columbia, quartz veins as much as 0.7 m thick and 2200 m long, within an area of about 2 by 5 km, are estimated to contain 8860 kg of gold. Veins comprising the Lawyers deposit and the Baker mine in the Toodoggone district, British Columbia, are commonly 2-7 m wide and as much as several hundred metres in length. The Silbak-Premier deposit in British Columbia has yielded thus far about 56 440 kg of Au from veins and breccia zones as wide as 40 m and as long as 1200 m (Main Zone; McDonald, 1990). Elsewhere,

mineralized veins have been mined for a strike length of more than 5 km at Creede, Colorado (Heald et al., 1987), and occur for a distance of about 2 km at the Hishikari mine, Japan (Izawa et al., 1990). Alteration zones around the veins at Hishikari have been mapped in an area measuring as much as 2 km wide by more than 3 km long (Izawa et al., 1990).

GEOLOGICAL FEATURES

Epithermal gold deposits are in many cases fault controlled, and occur in igneous (generally volcanic), sedimentary, or, less commonly, metamorphic rocks. They may be of similar age to their host rocks where these are volcanic, or much younger. A magmatic heat source is commonly associated. The deposits comprise veins and/or related

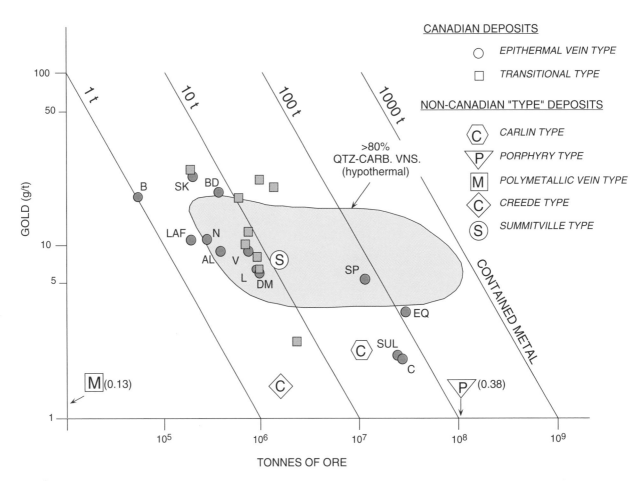

Figure 15.1-1. Plot of gold grade versus total tonnes (reserves + production) for Canadian deposits in Tables 15.1-1 and 15.1-2 (solid circles); SK, Mt. Skukum; B, Baker; BD, Blackdome; LAF, Laforma; N, Mt. Nansen; AL, Al; V, Venus; L, Lawyers; DM, Dusty Mac; SP, Silbak Premier; EQ, Equity Silver; SUL, Sulphurets; and C, Cinola. Also plotted (small squares) are hydrothermal vein deposits of a possible "transitional" or deep epithermal nature, and median grades and tonnages for several comparable "types" of deposits (open symbols; see Cox and Singer,1986); M, polymetallic veins associated with felsic intrusions; C (diamond), Creede-type; C̄ (octagon), Carlin-type; P, porphyry Cu-Au; S, Summitville-, or quartz-(kaolinite)-alunite-subtype; position of L (Lawyers) corresponds closely with that of Cox and Singer's (1986) Comstock-type (no symbol). Qtz-Carb. Vns. = quartz-carbonate vein gold deposits (subtype 15.2, this volume).

Table 15.1-2. Summary of geological setting, definitive characterisitics[1] and several examples of typical Au-bearing epithermal systems.

	QUARTZ-KAOLINITE-ALUNITE subtype Hosted in volcanic rocks	ADULARIA-SERICITE subtype Hosted in volcanic and plutonic rocks	ADULARIA-SERICITE subtype Hosted in sedimentary and mixed host rocks
Geological setting	volcanic terrane, often in caldera-filling volcaniclastic rocks; hot spring deposits and acid lakes may be associated.	Spatially related to intrusive centre; veins in major faults, locally ring fracture type faults; hot springs may be present.	In calcareous to clastic sedimentary rocks; may be intruded at depth by magma; can form at variety of depths.
Ore mineralogy	native gold, electrum, tellurides; *magmatic-hydrothermal:* py (+bn), en, tennantite, cv, sp, gn; Cu typically > Zn, Pb; Au-stage may be distinct, base metal poor; *steam-heated:* base-metal poor; gangue: quartz ('vuggy' silica), barite	electrum (lower Au/Ag with depth); gold; sulphides include: py, sp, gn, cpy, ss); gangue: quartz, adularia, sericite, calcite, chlorite; variable base metal content, high sulphide veins closer to intrusions.	gold (micrometre): within or on sulphides (e.g., pyrite unoxidized ore), native (in oxidized ore), electrum, Hg-Sb-As sulphides, pyrite, minor base metals; gangue: quartz, calcite.
Alteration mineralogy	advanced argillic + alunite, kaolinite, pyrophyllite (deeper); ± sericite (illite); adularia, carbonate absent; chlorite and Mn-minerals rare; no selenides; barite with Au; *steam-heated:* vertical zoning.	sericitic replaces argillic facies (adularia ± sericite ± kaolinite); Fe-chlorite, Mn-minerals, selenides present; carbonate (calcite and/or rhodochrosite) may be abundant, lamellar if boiling occurred; quartz-kaolinite-alunite-subtype minerals possible in steam-heated zone.	silicification, decalcification, sericitization, sulphidation; alteration zones may be controlled by stratigraphic permeability rather than by faults and fractures; quartz (may be chalcedonic)-sericite (illite)-montmorillonite.
Host rocks	silicic to intermediate (andesite)	intermediate to silicic intrusive/extrusive rocks.	felsic intrusions; most sedimentary rocks except massive carbonates (hosts to mantos and skarns).
$^{18}O/^{16}O$ - shift in wall rocks	may be less pronounced, or superposed on earlier high-^{18}O alteration.	moderate to large; pronounced in and immediately adjacent to veins.	very limited ^{18}O-shift of altered rocks, if present at all.
C-H-S isotopes	magmatic fluids indicated ($\delta^{13}C_{CO2}$ = -5±2; $\delta^{18}O_{H2O}$ = +7±2; $\delta^{34}S_{\Sigma S}$ = 0; *magmatic-hydrothermal* alunite: δD_{H2O} = -35±10; *steam-heated* alunite: $\delta^{34}S$ > sulphide minerals; $\delta^{18}O$ data indicate hydrothermal origin.	magmatic water (H_2O) may be obscured by mixing; surface waters dominate; C, S typically indicate a magmatic source, but mixtures with wall rock derived C, S possible	hydrogen isotope data (sericite, clays, fluid inclusions) in some cases indicate presence of evolved surface waters; organic carbon ($\delta^{13}C$ = -26±2) may be derived from wall rocks.
Ore fluids (examples from-fluid inclusion studies)	160-240 °C; ≤ 1 wt.% NaCl (late fluids); possibly to 30 wt.% NaCl in early fluids; boiling common; (Nansatsu district, Japan; Hedenquist et al., 1994).	sulphide-poor: 180-310°C, ≤ 1 wt.% NaCl, about 1.0 molal CO_2 (Mt. Skukum: McDonald, 1987). sulphide-rich: ave. 250 °C, < 1 to 4 wt.% NaCl (Silbak-Premier: McDonald, 1990)	bimodal: 150-160 (most); 270-280°C, ≤ 1.5 wt.% NaCl; nonboiling: (Cinola: Shen et al., 1982); 230-250°C, ≤ 1 wt.% NaCl; nonboiling (Dusty Mac: Zhang et al., 1989)
Age of mineralization and host rocks	host rocks and mineralization of similar age.	mineralization variably younger (> 1 Ma) than host rocks.	mineralization variably younger (> 1 Ma) than host rocks.
Deposit size	small areal extent (e.g., ca. 1 km²) and size (e.g. 2500-3500 kg Au)	may occur over large area (e.g., several tens of km²); may be large (e.g. 100 000 kg Au).	may have large areal extent (e.g. >> 1 km²), large size (e.g., 58 000 kg Au), low grades (e.g., 2.5 g/t).
Examples Canadian	Mt. Skukum, Y.T. (alunite 'cap') Al deposit, Toodoggone River, B.C.	Blackdome, B.C.; Mt. Skukum, Y.T. Silbak-Premier, B.C.	Cinola, B.C.
Foreign	Summitville, Colorado Kasuga, Japan	Hishikari, Japan Creede, Colorado	Carlin, Nevada
Modern analogues:	Matsukawa, Japan[2]	Broadlands, New Zealand[3]	Salton Sea geothermal field, California[4]

1) based, in part, on Heald et al., 1987; Taylor, 1987; Berger and Henley, 1989; Panteleyev, 1991; Rye et al., 1992; Sillitoe, 1993, and data reported for Canadian deposits and other examples cited in the text;
2) Nakamura et al., 1970;   3) Browne in Henley et al., 1986;   4) Williams and McKibben, 1989; but analogy not complete.
Abbreviations: py, pyrite; bn, bornite; en, enargite; sp, sphalerite; gn, galena; ss, sulphosalts; cv, covellite; cpy, chalcopyrite.

mineralized breccia and wall rock (e.g., Mt. Skukum), or replacement bodies associated with zones of silicification (e.g., Cinola). Principal geological and other characteristics of each subtype of epithermal gold (±silver) deposit are listed in Table 15.1-2 (see Table 15.1-1 for data on individual examples). As may be seen from Table 15.1-2, both subtypes share many features.

The deposit subtype is distinguished primarily on the basis of associated alteration mineral assemblage, largely reflecting differences in the ore-forming fluids (i.e., oxidized, low pH: quartz-(kaolinite)-alunite; reduced, near-neutral: adularia-sericite). Alteration mineral assemblages and associated minor and trace elements are also influenced by

the composition of the host rocks. Identification of the epithermal nature of a gold deposit principally entails recognition of evidence for a shallow origin; this may be based on, among other things: geological (stratigraphic) reconstruction of the depth of formation, nature and zoning of alteration and ore minerals, presence of hydrothermal breccia, form and structure of deposit, and mineralogical and textural characterisitics (noted later). Corroborative features such as high-temperature/low-pressure conditions indicated by primary fluid inclusions, or oxygen isotope depletion of wall rocks indicating high (meteoric) water:rock ratios (except in some quartz-(kaolinite)-alunite-subtype deposits with large amounts of magmatic

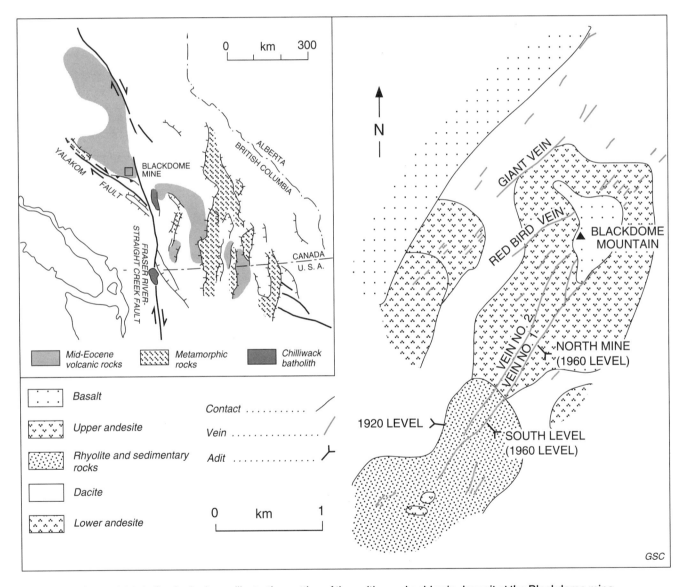

Figure 15.1-2. Geological map illustrating setting of the epithermal gold vein deposit at the Blackdome mine, British Columbia (data from: D. Rennie, unpub. rep., 1987). Inset map illustrates the regional tectonic setting of the Blackdome mine area (red square), between the Yalakom (55-45 Ma) and younger Fraser River-Straight Creek (40-35 Ma) strike-slip faults (Coleman and Parrish, 1990; R.R. Parrish, pers. comm., 1991).

volatiles) may also be helpful. Well-documented modern geothermal fields have many features in common with epithermal deposits, as discussed in following sections.

Geological setting

The tectonic settings of volcanic-associated epithermal gold deposits are numerous, and include island-arc volcanoes (e.g., Papua New Guinea; Sillitoe, 1989), continental-based arcs, and volcanic centres (e.g., Silverton caldera, Colorado). Extensional tectonism, whether on a local or regional scale, is often a common factor, as indicated by the steep normal faults which in some deposits (e.g., Blackdome, British Columbia) control the emplacement of dykes, veins, and breccia zones. The deposits of the Toodoggone River area are thought to have formed in an elongate, tectonically controlled graben in the medial portion of an island arc (Diakow et al., 1993). That this site was one of active deposition of younger rocks, rather than one of constructional volcanism and uplift, in a climate providing high erosion rates such as found today in Melanesia (e.g., Chivas et al., 1984), probably explains the preservation of these Early Jurassic epithermal deposits.

Volcanic structures, such as caldera ring fractures (e.g., Summitville, Colorado; Lipman, 1975) and radial fractures (e.g., Lake City, Colorado; Slack, 1980) may control magmatic and hydrothermal activity, and tension-producing resurgent doming may create vein-hosting extensional faults (e.g., Creede, Colorado). In British Columbia, however, regional tectonic stress fields related to Eocene strike-slip faulting appear to have been particularly important. For example, northeast-trending synvolcanic faults at Blackdome controlled the emplacement of dykes and gold-bearing quartz veins (Fig. 15.1-2). These faults have a similar orientation to normal faults found southeast of the Blackdome area and east of the Fraser River-Straight Creek fault, which are attributed to northwest extension along the Yalakom fault (e.g., Ewing, 1980; Parrish and Coleman, 1990).

The Mt. Skukum deposit, Yukon Teritory, is situated between two northwest-trending, dextral strike-slip faults, the Tintina and Denali-Shakwak. North-northeast- and northeast-trending veins are located in an eroded caldera within an andesitic stratovolcano (Pride, 1986). The mineralized veins are interpreted to have been controlled by Riedel shears, whose orientations are consistent with northeastward compression during the Eocene (Love, 1989). The more northerly trending vein orientation is consistent with extension, whereas Love (1989) suggested that the largest ore body developed in a dilational jog localized by the intersection of a strike-slip fault and a felsic dyke. Synchronous tectonic and hydrothermal activity is indicated in some deposits by the fact that many of the vein-bearing faults were active during and after vein-filling (e.g., Blackdome, Mt. Skukum, and Toodoggone deposits); tectonic vein breccias and displaced mineralized and altered rocks resulted.

Small volcanic- and volcaniclastic-hosted deposits in Canada are also found in other structural-tectonic settings. These include Dusty Mac deposit, British Columbia (e.g., Church, 1973; Zhang et al., 1989; Table 15.1-1), located in breccia and stockwork zones along reverse faults at the margin of the White Lake basin. Eocene rhyolitic dykes are associated with areas of adularia-sericite and gold-mineralized zones of silicification and argillic alteration along faults in the Tintina Trench, characterized on the surface by superimposed steam-heated or (probably) supergene quartz-(kaolinite)-alunite alteration mineral assemblages (cf. Duke and Godwin, 1986).

Sediment-hosted gold (±silver) deposits occur in a variety of settings in which sedimentary sequences have been intruded by magmas, and also in sedimentary rocks not obviously closely associated with intrusions. The deposits are located in some cases in the outer zones of hydrothermally altered rocks adjacent to intrusions (e.g., Cinola; Equity Silver, British Columbia). Carlin-type deposits are typically referred to as the classic (carbonate) sediment-hosted precious metal deposits, although they have a restricted distribution. Thirty-five of 39 such deposits tabulated by Bagby and Berger (1986) occur in Nevada. These deposits occur in "belts" in Paleozoic rocks which appear to be underlain by major structural discontinuities (e.g., thrust faults; deep fault zones in the basement, or along basement margins; cf. Cunningham, 1985) that have guided the emplacement of magma and controlled the distribution of hydrothermal fluids. For example, a granitic pluton about 120 m below gold ore at the Carlin-type Gold Acres deposit is thought to have been responsible for the overlying, fault-controlled mineralization (Wrucke and Armbrustmacher, 1975).

Age

Epithermal gold deposits, especially in volcanic terranes, are commonly Tertiary in age or younger; older deposits are more likely to have been removed from the geological record through erosion. However, gold deposits and their host rocks in the Toodoggone district, British Columbia are Jurassic in age (see Table 15.1-1). A Lower Paleozoic quartz-(kaolinite)-alunite-subtype gold deposit (Gidginbung) has been described by Lindhorst and Cook (1990) from the Lachlan fold belt, New South Wales, Australia, and Early Devonian hot spring sinter deposits have been recognized in Scotland (Nicholson, 1989). Other examples of pre-Tertiary epithermal gold deposits include those of Paleozoic age in the (former) Soviet Union (Y. Sofonov, pers. comm., 1990) and Queensland, Australia (Wood et al., 1990), and the Late Proterozoic Mahd adh Dhahab deposit, Arabian Shield (Huckerby et al., 1983).

The age of quartz-(kaolinite)-alunite-subtype deposits is typically within 0.5 Ma of their volcanic host rocks, whereas adularia-sericite-subtype deposits may also occur in considerably older volcanic and/or sedimentary (or other) host rocks (see Table 15.1-1). Essentially coincident ages of host rock alteration (and probably mineralization; 50.7 Ma) and volcanic host rocks (53.2 Ma) characterize the adularia-sericite-subtype deposits at Mt. Skukum, Yukon Territory. Quartz-(kaolinite)-alunite-subtype alteration at the Al deposit, Toodoggone River area coincided with the first of two periods of volcanism, whereas the principal adularia-sericite-subtype deposits may have formed several million years later (Clark and Williams-Jones, 1991). The relationship of the barren alunite-silica-pyrophyllite-kaolinite-bearing zone ("alunite cap") to the adularia-sericite gold deposit of the Cirque vein (and others) at Mt. Skukum is unclear. Love (1989) suggested that the acid-sulphate alteration formed on a different fault than that hosting the vein; the presence of pyrophyllite suggests a deeper (higher temperature) rather than shallower setting. Steam-heated quartz-(kaolinite)-alunite alteration also occurred above adularia-sericite-subtype alteration and mineralization of the same age at Sulphurets (Margolis, 1993). The relative positions of quartz-(kaolinite)-alunite and adularia-sericite alteration is similar in the Toodoggone and Mt. Skukum occurrences, but the exact age relationships are not clear; magmatic-hydrothermal origins, though less likely, cannot be excluded.

The sediment-hosted Cinola deposit is spatially associated with altered porphyritic rhyolite, which evidently invaded the same fault that focused hydrothermal fluids. Christie (1989) suggested that the intrusion is unrelated to the genesis of the deposit. The age of mineralization has not yet been determined directly, but it is likely Miocene, based on the 14 Ma K-Ar age for the altered rhyolite (Champigny and Sinclair, 1982).

Form and structure

In volcanic terranes, epithermal gold deposits typically occur in or adjacent to steeply-dipping faults, breccia zones, and fractures, and, therefore, comprise tabular mineralized zones within which there are higher-grade "pipes" or vertical zones (e.g., Cirque vein, Mt. Skukum; Fig. 15.1-3; Table 15.1-1). Mineralized faults may have also earlier served to localize intrusions.

Active faulting has in some cases occurred during and after mineralization, resulting in brecciation of previously emplaced veins (e.g., Mt. Skukum). Permeable zones can form along irregularities in fault planes: vertically-plunging ore zones in faults with strike-slip motion, and horizontal ore zones in dip-slip faults. Topographic (i.e., paleosurface) control of boiling by hydrostatic pressure can also result in horizontal or subhorizontal mineralized zones, limiting the vertical distribution of ore. The distribution of quartz-(kaolinite)-alunite alteration in steam-heated settings (possibly in the Toodoggone River camp, British Columbia) may also reflect a topographic control of the paleowatertable.

Silicified rocks are common in epithermal deposits. For example, irregularly silicified and mineralized wall rocks occur adjacent to faults and fractures in both volcanic (e.g., Blackdome) and sedimentary (e.g., Cinola) host rocks. Silicified and decarbonated rocks comprise the principal ore host of Carlin-type gold deposits (e.g., Bagby and Berger, 1986). At the Carlin mine, the ore zone comprises irregular, partially oxidized, pod-like lenses in the footwall of the Roberts Mountain thrust (Bakken and Einaudi, 1986). The silicification of wall rocks (and the distribution of ore) can be controlled by available primary permeability caused by, for example, bedding planes or rock fabric (e.g., Cinola) in addition to differences in host rock composition. Secondary permeability can also be produced by the hydrothermal fluids themselves. Sudden release of pressure on hydrothermal fluid (e.g., by faulting) can produce

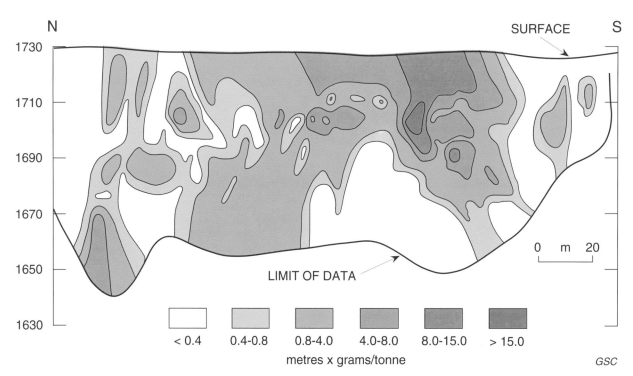

Figure 15.1-3. Longitudinal section of the Cirque vein, Mt. Skukum (modified after McDonald, 1987) illustrating distribution of gold (thickness x grade).

brecciation which forms slightly dismembered to completely milled zones in brittle rocks (e.g., Cinola, British Columbia). This can occur in geothermal systems within several hundred metres of the Earth's surface (e.g., Hedenquist and Henley, 1985). Hydrothermal fluids can also produce secondary permeability in carbonate rocks by dissolution.

Ore mineralogy, composition, texture, and zonation

Distinctive ore minerals and textures characterize each subtype of epithermal gold deposit. Mineralogical zonation around veins or zones of replacement, although different in mineralogical composition, records chemical and/or thermal gradients in both adularia-sericite- and quartz-(kaolinite)-alunite-subtypes of deposits. Both subtypes of deposits can contain very fine grained gold and gangue mineral assemblages in hot-spring and steam-heated environments.

Quartz-(kaolinite)-alunite volcanic-hosted deposits

Native gold and electrum are associated with the characteristic mineral assemblage pyrite+enargite±covellite± bornite±chalcocite. In addition to sulphosalts and base metal sulphides, tellurides and bismuthinite are present in some deposits. The total sulphide content is typically higher than in ores of many adularia-sericite-subtype deposits, although high sulphide contents may also characterize polymetallic adularia-sericite deposits (e.g., Silbak-Premier, British Columbia). Copper contents can vary enormously in quartz-(kaolinite)-alunite deposits (<0.1% to 5%: Sillitoe, 1993), but in quartz-(kaolinite)-alunite deposits in which base metals are present, copper typically predominates over zinc.

Adularia-sericite deposits

Native gold and electrum occur in typically sulphide-poor veins. Typical sulphide contents are usually a few per cent or less (e.g., Blackdome, British Columbia); pyrite is dominant. In deposits in which sulphide minerals are abundant, these sulphides commonly, in addition to pyrite, include chalcopyrite, tetrahedrite, galena, sphalerite, and arsenopyrite (e.g., Venus; Silbak-Premier: sulphide-rich stage).

Gold grains are mostly micrometre-size in Carlin-type deposits, although visible gold occurs in oxidized portions of some deposits. Gold is found in some cases coating sulphides and/or encapsulated in quartz in silicified rocks. Mercury, antimony, and arsenic sulphides accompany gold in these deposits, and are also found in deposits hosted by volcaniclastic rocks (e.g., McDermitt, Nevada). At Cinola, gold is most abundant in silicified sediments and hydrothermal breccia. Unique to deposits hosted by sedimentary rocks, or deposits which formed in hydrothermal systems encompassing sedimentary rocks, is the possibility for inclusion of sediment-derived hydrocarbons (e.g., Owen Lake, British Columbia; Thomson et al., 1992) during vein formation.

Hot spring deposits

Hot spring deposits may form as surface expressions of epithermal vein systems. These deposits (and steam-heated zones in general) are thought to have characteristically high precious metal/base metal ratios (Buchanan, 1981). Buchanan (1981) suggested that this results from deposition of gold in an upper, gas-rich, or boiling portion of the geothermal system. Base metals are deposited in deeper, more saline, liquid-dominated, portions of the system. Siliceous sinter deposits, which contain sulphate minerals, clays, and minor pyrite, typically form at the surface; broadly vertical zonation of alteration mineral assemblages is characteristic.

Gangue mineralogy and zonation

Quartz is the predominant gangue mineral in all epithermal gold deposits. Quartz-(kaolinite)-alunite deposits in the magmatic-hydrothermal environment characteristically contain quartz as a "vuggy silica" facies, formed by an increase in porosity associated with base leaching (particularly of feldspar) by very acidic fluids, and by concentration of residual silica (e.g., Summitville, Colorado; Stoffregen, 1987). This has occurred in quartz-(kaolinite)-alunite-subtype alteration zones at Mt. Skukum, Yukon Territory (alunite cap zone; Love, 1989) and at the Al deposit (Toodoggone River, British Columbia; Diakow et al., 1993). In quartz-(kaolinite)-alunite deposits, alunite is characteristic; barite (especially associated with gold), and sulphur (in some deposits) may be common. Manganese minerals and fluorite are rare.

In both adularia-sericite- and (steam-heated) quartz-(kaolinite)-alunite-subtype deposits, silica may be chalcedonic and form laminated veins (e.g., Cinola, British Columbia), occur as massive, sugary white quartz veins and breccia cements (e.g., Mt. Skukum), or form cockade or comb structures, as zones of inward-pointing crystals in laminated veins (e.g., Venus, Blackdome). Such textures are consistent with, but do not necessarily prove, a relatively shallow origin.

Gangue minerals in some adularia-sericite deposits include calcite, chlorite, adularia, barite, rhodochrosite, fluorite, and sericite. Anhydrite has been noted in minor amounts in hypogene veins at Mt. Skukum (McDonald, 1987). Lamellar or platy ("angel wing") calcite, in some cases pseudomorphically replaced by silica (e.g., Mt. Skukum, Yukon Territory), is of particular significance because it forms in boiling zones in adularia-sericite-subtype systems (e.g., Simmons and Christenson, 1994; see de Ronde and Blattner, 1988). Calcite is not characteristic of quartz-(kaolinite)-alunite-subtype deposits due to the high acidity of the hydrothermal fluids.

In sediment-hosted (adularia-sericite) deposits, especially those of Carlin type, gangue minerals constitute a characteristic assemblage and commonly include cinnabar, orpiment-realgar, and stibnite, in addition to jasperoid, quartz, dolomite, and calcite. Quartz veins (chalcedonic) and jasperoid are typically associated with ore, whereas calcite veins are in many cases more common further from ore, or paragenetically late.

Alteration mineralogy and zoning

Hydrothermal alteration mineral assemblages associated with both quartz-(kaolinite)-alunite- and adularia-sericite-subtype deposits are commonly regularly zoned about vein- or breccia-filled fluid conduits. In near-surface environments, or where permeable rocks have been replaced, zoning is in many cases less well defined. Characteristic alteration mineral assemblages in both deposit subtypes can give way to propylitically altered rocks containing quartz+chlorite+albite+carbonate±sericite±epidote±pyrite. The distribution and formation of the propylitic assemblage generally bear no obvious direct relationship to ore-related alteration mineral assemblages; propylitic alteration typically predates mineralization.

Quartz-(kaolinite)-alunite deposits are characterized by advanced argillic alteration mineral assemblages that include: quartz+kaolinite+alunite+dickite+pyrite in and adjacent to veins or zones of replacement in the magmatic-hydrothermal environment. Pyrophyllite is found in place of kaolinite in deeper (higher temperature and pressure) deposits. Argillic (smectite)±sericite mineral assemblages may occur in some outer zones (e.g., alunite "cap"; Mt. Skukum). Minerals such as topaz and tourmaline in high temperature zones indicate the presence of F and B in the hydrothermal fluids. The alteration minerals indicate a very low pH hydrothermal environment (occasionally below even that for alunite stability: Stoffregen, 1987), and one of high oxidation state (for hematite and sulphate stability). Consequently, carbonates are absent. Zones of silica replacement and "vuggy silica" are characteristic.

Acid-sulphate (quartz-(kaolinite)-alunite) alteration can also form by reaction of host rocks with steam-heated meteoric waters acidified by oxidation of H_2S (probably of magmatic origin: e.g., Rye et al., 1992), or by dissolution of CO_2. The distinction of the steam-heated environments, so-called because they are localized above deeper, boiling hydrothermal systems (Henley and Ellis, 1983), from shallow magmatic hydrothermal environments is less straightforward. Steam-heated quartz-(kaolinite)-alunite alteration zones may occur as "blankets" above adularia-sericite deposits, truncating older adularia-sericite

alteration zones (e.g., Sulphurets, British Columbia; Margolis, 1993), and also at the top of quartz-(kaolinite)-alunite systems; such occurrences may or may not overlie mineral deposits (Henley, 1985). Alunite occurs in the steam-heated environment, and stable isotope studies (discussed below) may be required to distinguish between alunite of supergene, steam-heated, or magmatic-hydrothermal hypogene origin.

Supergene alteration of sulphide-bearing deposits can also produce quartz-(kaolinite)-alunite alteration minerals. Alunite of fine grained, poorly crystallized nature, evidence of sulphide oxidation, and presence of halloysite,

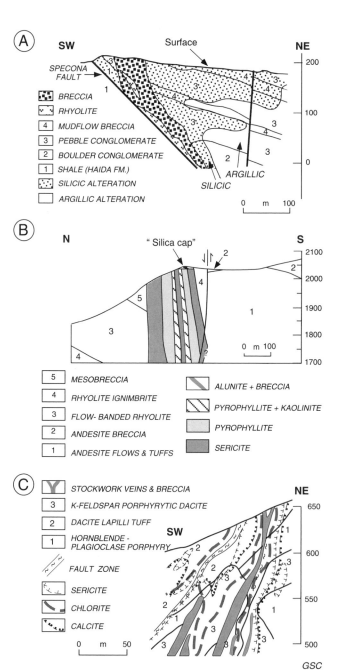

Figure 15.1-4. A) Geological cross-section through the Cinola sedimentary-hosted gold deposit (after Christie, 1989; adularia-sericite-subtype) illustrates localization of both magma (interpreted from faulted dyke) and hydrothermal fluids by the Specona fault, and the control exerted by primary lithological permeability on the distribution of zones of silicification and argillic alteration. **B)** Geological cross-section through a portion of the Mt. Skukum area (after Love,1989; adularia-sericite-subtype, volcanic host) illustrating the distribution of alunite+silica, and advanced argillic alteration mineral assemblages near a principal normal fault. Supergene alteration has been superposed on quartz-(kaolinite)-alunite alteration zones. **C)** Geological cross-section through a portion of the Silbak-Premier deposit (after McDonald,1990) illustrating the distribution of hydrothermal alteration minerals indicative of propylitic, sericitic, and potassic alteration assemblages in relation to a fault-controlled vein stockwork and breccia, and to porphyritic dacite.

iron oxides, jarosite, possible supergene enrichment, and subhorizontal mineral zonation are some of the features that characterize supergene quartz-(kaolinite)-alunite deposits. Additional, definitive evidence comes from the stable isotope composition of alunite (Rye et al., 1992).

Altered rocks in adularia-sericite deposits generally comprise two mineralogical zones: (1) an inner zone of silicification (characterized by replacement of wall rocks by quartz or chalcedonic silica); and (2) an outer zone of potassic-sericitic (phyllic) alteration (characterized by quartz+K-feldspar and/or sericite, or sericite and illite-smectite). Adularia is the typical K-feldspar in these deposits, but its prominence varies greatly; it may be absent altogether. Chlorite and carbonate are present in many deposits, especially in wall rocks of intermediate composition, and in some cases (e.g., Shasta deposit, Toodoggone River: Thiersch and Williams-Jones, 1990; Silbak-Premier: McDonald, 1990) chloritic alteration accompanies the potassic alteration and silicification. Argillic alteration (forming kaolinite and smectite clay minerals) occurs still farther from the vein. In some deposits (e.g., Cinola; Christie, 1989), argillic alteration predates silicification, giving evidence of the waxing and waning of hydrothermal systems. More commonly, argillic mineral assemblages are superposed on the above, or form higher level alteration zones (e.g., Toodoggone River area; Diakow et al., 1993), in which adularia is replaced by kaolinite. Kaolinite may occur closer to veins than smectite.

Bagby and Berger (1986) distinguished two types of sediment-hosted deposits based on the nature of silicification: jasperoidal- and Carlin-type deposits. Jasperoidal deposits occur in clastic sedimentary rocks, whereas those of Carlin type occur in carbonate or calcareous host rocks. These types of silicification are gradational and differ primarily in the manner in which silica occurs: quartz veins are common, and accompany replacement silicification in the jasperoidal type (e.g., Cinola, British Columbia), whereas replacement silicification is relatively more common in Carlin-type deposits. The effects of alteration are otherwise similar in the two deposit types, and include decarbonation and argillization. Alteration minerals include quartz, calcite, illite, cinnabar, orpiment, realgar, stibnite, pyrite, pyrrhotite, marcasite, and arsenopyrite. Subsequent weathering has markedly changed the appearance and mineralogy of much of the carbonaceous Carlin-type ore through bleaching and oxidation. Supergene minerals include calcite, iron oxides, sulphates, and supergene alunite (e.g., Arehart et al., 1992), as well as clay minerals.

The zonation of alteration minerals and their relationship to lithology are illustrated for portions of three Canadian deposits in Figures 15.1-4A-C. The examples chosen represent a sediment-hosted, adularia-sericite-subtype (Cinola, British Columbia), a sulphide-rich, volcanic-hosted adularia-sericite subtype (Silbak-Premier, British Columbia), and rhyolite/andesite-hosted quartz-(kaolinite)-alunite alteration of wall rocks topographically above the adularia-sericite subtype (Mt. Skukum deposit, Yukon Territory). In each case, the zones of alteration minerals are structurally controlled, and crosscut the host rocks. Zoning is broadly symmetrical about some veins (e.g., Mt. Skukum), but markedly asymmetrical in other cases (e.g., Cinola).

Stable isotopes and fluid inclusions

In addition to determining the source(s) of H, C, O, and S in ores and altered rocks (Table 15.1-2), stable isotope and fluid inclusion data may be utilized to map paleogeothermal systems and, in part, deduce their time-space hydrological evolution. Oxygen and hydrogen isotope and fluid inclusion investigations thus far indicate that gold-precipitating hydrothermal fluids in epithermal deposits comprised mixtures of low-salinity, meteoric waters and more saline waters. In some cases (magmatic-hydrothermal quartz-(kaolinite)-alunite deposits), the saline waters are magmatic fluids, in others (adularia-sericite deposits) these are largely evolved (reacted, boiled) surface waters with some magmatic components.

Hydrothermal alteration involving meteoric (or marine) waters results in a lowering of the $^{18}O/^{16}O$ ratios of the wall rocks and a concomitant increase of the $^{18}O/^{16}O$ ratio of the hydrothermal fluid (Fig. 15.1-4B). The extent of ^{18}O depletion of the host rocks in geothermal systems depends principally on the amount and isotopic composition of the recharged geothermal fluid, the isotopic composition of the wall rocks, and the temperature and lifetime (i.e., amount of heat supplied) of the system. At all but extremely small water:rock ratios, the D/H ratio of the altered rock is completely determined by the hydrogen isotope composition of the hydrothermal fluid. The hydrothermal fluids responsible for alteration in specific deposits, and in groups of deposits, plotted in Figure 15.1-5 predominantly represent altered or "evolved" meteoric waters whose compositions have been shifted to the right (i.e., to higher $\delta^{18}O$) of the present day meteoric water (PDMW) line during hydrothermal alteration of the host rocks. Involvement of seawater or low-latitude meteoric water is indicated for the Sulphurets area (Margolis, 1993).

Altered sedimentary wall rocks typically are less depleted in ^{18}O, but the hydrothermal fluids more enriched in ^{18}O, than in volcanic terranes. For example, the markedly higher $^{18}O/^{16}O$ ratios of hydrothermal fluids accompanying alteration and mineralization at Cinola, in comparison to those at Mt. Skukum and Blackdome (see Fig. 15.1-5), can be attributed largely to greater relative $\overline{^{18}O}$ enrichment of deeply circulating hydrothermal fluids by reaction with sedimentary wall rocks with high $^{18}O/^{16}O$ ratios.

Mixtures of either marine, meteoric, or magmatic waters would plot along straight lines in Figure 15.1-5, whereas reacted, or evolved (i.e., isotopically altered) meteoric waters would plot along curved lines. Detailed paragenetic and isotopic studies are required to elucidate such processes. Evidence of the mixing of distinct fluids with distinct isotopic ratios and salinities has been reported for some vein deposits of the deep epithermal or "transitional" category (e.g., Finlandia vein, Peru; Kamilli and Ohmoto, 1977). Variations of δD or $\delta^{18}O$ of the fluids are usually accompanied by variations in wt.% NaCl$_{equiv.,}$ suggesting the presence of a (deeper) saline fluid, and a (shallower) dilute fluid (summarized in Taylor, 1987). In some cases (e.g., Creede, Colorado), incorporation of the dilute fluids occurred abruptly, and late in the paragenesis (e.g., Foley et al., 1989). An unusual range in δD (-151 to -54) and $\delta^{18}O$ (recalculated: -7.6 to -2.6) for vein-depositing fluids in the Laforma vein was attributed by McInnes et al. (1990) to extensive boiling; fluid inclusion data and carbon, sulphur, oxygen, and hydrogen isotope data are also consistent, however with a magmatic-meteoric water mixing scenario. Meteoric waters formed the major component of the ore-forming fluids at the Blackdome (Vivian et al., 1987), Dusty Mac (Zhang et al., 1989), and Mt. Skukum (McDonald, 1987) deposits. Data reported in Diakow et al. (1993) indicates a broadly similar scenario in adularia-sericite deposits of the Toodoggone River area. Progressive mixing of magmatic water and seawater during potassic to sericitic to advanced argillic alteration at Sulphurets, British Columbia was inferred by Margolis (1993) from isotopic data and water-rock reaction modelling.

There is isotopic evidence from alunite (Rye et al., 1992) for a major component of magmatic water in some magmatic-hydrothermal quartz-(kaolinite)-alunite deposits. On the one hand, magmatic-hydrothermal alunite is characterized by $\delta^{34}S$ greater (by perhaps 8 permil or more) than that of associated sulphides due to disproportionation of magmatic SO_2 gas during cooling and reaction with hydrothermal water below about 400°C, according to (Holland, 1965): $4SO_2 + 4H_2O \rightarrow 3H_2SO_4 + H_2S$.

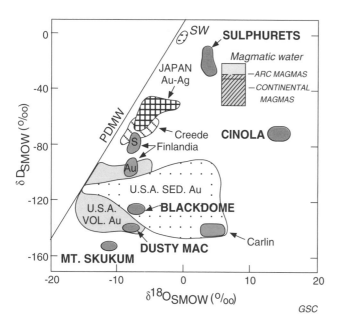

Figure 15.1-5. Plot of $\delta^{18}O$ versus δD for present day meteoric waters and for waters in equilibrium with gangue minerals in selected epithermal gold deposits; Canadian examples are shown in red. This diagram illustrates the origin and oxygen isotope enrichment of meteoric waters in many epithermal vein systems. Abbreviations are, for the Finlandia vein, Colqui district, Peru (S = sulphide stage; Au = precious metal stage); SW, sea water; PDMW, present day meteoric water; U.S.A. SED. Au= U.S.A. sedimentary rock-hosted Au; U.S.A. VOL. Au = U.S.A. Volcanic rock-hosted Au. Data for magmatic water are from Taylor (1987,1992). Sources of data for non-Canadian deposits: Taylor (1987); for Blackdome: Vivian et al. (1987); Dusty Mac: Zhang et al. (1989); Mt. Skukum: McDonald (1987) and B. Taylor (unpub. data,1991); and Cinola: B. Taylor and A. Christie (unpub. data,1991).

On the other hand, alunites of steam-heated and supergene origin have $\delta^{34}S$ values similar to those of accompanying sulphides, although oxygen isotope data indicate the hydrothermal deposition of the former (Rye et al., 1992).

Recognition of the source of sulphur depends on being able to establish the relative mass balance for the contributing sources. Host rock sulphur may comprise a significant component in some adularia-sericite deposits, whereas in other deposits, particularly quartz-(kaolinite)-alunite deposits, magmatic sulphur (as SO_2; $\delta^{34}S$ about 0-4 for felsic magmas; Taylor, 1987) dominates.

Carbon isotope data for calcite or CO_2 from fluid inclusions typically reveals its magmatic (ultimately mantle) origin, even in systems dominated by meteoric water (e.g., Laforma, Yukon Territory: McInnes et al., 1990; Mt. Skukum, Yukon Territory: McDonald, 1987). Admixture with terrestrial carbon sources may also occur (e.g., organic carbon in the Owen Lake deposit: Thomson et al., 1992).

Studies of fluid inclusions typically have shown the preponderance of fluids of low salinity (less than about 5 wt.% $NaCl_{equiv.}$) and filling temperatures of 150 to 300°C. Histograms of homogenization temperatures commonly have maxima in the neighbourhood of 260-280°C (e.g., Equity Silver; Shen and Sinclair, 1982; Blackdome, Vivian et al., 1987). This is not surprising, since a dilute vapour-dominated system at or near a boiling water table tends to evolve toward a rather uniform temperature of about 240°C due to the limitation imposed by a maximum in enthalpy of steam+liquid (e.g., White et al., 1971). Higher temperatures and salinities characterize some deep epithermal (transitional) environments close to genetically-related intrusions. Formation of fluid inclusions present at different times in a dynamic system complicates interpretation of the evolution of the system. Temporal changes within the Creede hydrothermal system, as identified by marked changes in the chemical and isotopic compositions of fluid inclusions associated with growth zones or fracture planes in crystals, demonstrates that the identity of ore-transporting fluids can be obscured by inappropriate sampling and analysis (Foley et al., 1989).

DEFINITIVE CHARACTERISTICS

Mineral assemblages, geological reconstruction demonstrating shallow emplacement, presence of sinters, fluid inclusion or textural evidence (e.g., lamellar calcite) for boiling, hydrothermal breccias and eruption deposits (e.g., Izawa et al., 1993), open-space crustiform veins, and marked ^{18}O-depletion of wall rocks comprise independent, definitive (but not always present) characteristics of epithermal deposits. Vertical zoning of alteration minerals, Au:Ag ratios in electrum, and spatial and temporal separation of gold and abundant base metals are also characteristic. Mineralogical characteristics which distinguish quartz-(kaolinite)-alunite (proximal) from adularia-sericite (distal) subtypes of gold-bearing systems are listed in Table 15.1-2. Quartz-(kaolinite)-alunite-subtype deposits are characterized uniquely by enargite+pyrite±covellite, and by advanced argillic alteration minerals, including hypogene alunite (Heald et al., 1987).

Most known epithermal gold deposits are Tertiary in age, and, therefore, age is a semi-"definitive characteristic", especially in volcanic terranes. However, preservation of epithermal vein gold deposits in pre-Tertiary geological terranes has occurred, probably due to special circumstances which prevented their erosion. In cases in which deformation and/or metamorphic recrystallization may have destroyed the mineralogical and geological characteristics noted above, $^{18}O/^{16}O$ ratios of silicate rocks can provide a unique record of alteration by meteoric waters. For example, five gold deposits in the Carolina Slate Belt are hosted by rocks containing pyrophyllite, andalusite, topaz, and traces of diaspore. These minerals suggest metamorphism of rocks subjected to the base leaching (and silica enrichment) associated with advanced argillic alteration in epithermal systems, and oxygen-isotope depletion of these and adjacent rocks (Klein and Criss, 1988) provides evidence for alteration by paleometeoric waters.

Metamorphosed magmatic-hydrothermally altered rocks can be discerned from those altered in supergene or steam-heated environments. McKenzie (1986) suggested the auriferous and pyrite-bearing quartz-andalusite-sericite schists at the Chetwynd disseminated gold deposits, Newfoundland, were the metamorphosed equivalents of an advanced argillic assemblage. Rocks of the Hope Brook zone at Chetwynd lack evidence of ^{18}O depletion characteristic of meteoric alteration fluids (B. Taylor and P. Stewart, unpub. data, 1990), but the data are consistent with a magmatically-dominated alteration system. The geochemical association of Hg-Sb-As-Tl (e.g., Harris, 1989) and the range in the $\delta^{34}S$ values of pyrite (Cameron and Hattori, 1985) at the large Hemlo (Ontario) disseminated gold deposit in the Precambrian Shield might suggest a metamorphosed epithermal deposit. However, mineralized rocks from the Hemlo deposit are not depleted in ^{18}O (Kuhns, 1988). A deep level of emplacement from dominantly magmatic fluids is suggested for Hemlo, and is supported by sulphate-sulphide sulphur isotope fractionations (Hattori and Cameron, 1986).

GENETIC MODELS

Recent documentation of epithermal gold deposits worldwide has helped to clarify their geological settings and characteristic mineral assemblages. Stable isotope and fluid inclusion studies have contributed to our knowledge of the origins and temperatures of the hydrothermal fluids. Studies of modern geothermal systems, volcanic gases, mineral solubility experiments and phase relations, and numerical water-rock reaction simulations have especially contributed to our knowledge of the chemical and physical nature of hydrothermal fluids, and also to our understanding of the processes which lead to the transport and deposition of gold, silver, and base metals. Large deposits appear to require a sustained (magmatic) heat source, and efficient, localized processes leading to supersaturation in one or more ore minerals (e.g., cooling and degassing by boiling and fluid mixing, and/or reaction with wall rocks). The site of deposition and the formation of an epithermal deposit are also influenced by many other factors, both local and regional (e.g., structural setting, paleohydrology, and climate; White and Hedenquist, 1990).

Lindgren (1922, 1933) originally suggested that ore-forming constituents were derived from degassing magmas. This supposition appears to be essentially correct for magmatic-hydrothermal quartz-(kaolinite)-alunite deposits (Stoffregen, 1987; Rye et al., 1992). However, for many deposits (e.g., the majority of adularia-sericite subtypes) stable isotope data permit only a very small fraction (i.e., <10%) of the hydrothermal water to be of magmatic origin, despite the close association of some deposits with cooling magmatic rocks. A complex origin, involving a more distant link to magmatic degassing, is indicated.

Schematic cross-sections illustrating the principal environments of adularia-sericite and quartz-(kaolinite)-alunite epithermal vein and hot spring deposits and their related geothermal systems, as discussed above, are shown in Figure 15.1-6. The figure was drawn to emphasize features found in at least some of the Canadian deposits noted previously.

Two fundamentally different hypotheses for the source of gold in epithermal vein gold deposits are: (1) the metals are supplied by the magma that is also the heat source, or (2) the metals are leached from the rocks which host the geothermal system. Proof of the involvement of meteoric waters has encouraged proponents of the second hypothesis. Stable isotope data indicate that sulphur and carbon are of magmatic origin in certain deposits (e.g., Summitville, Colorado: Rye et al., 1992) and magmatically-heated geothermal systems (e.g., Taylor, 1987), even when the dominant source of water was meteoric. The association of precious and base metals with magmatic Cl-, C-, and S-bearing gases has been demonstrated by Symonds et al. (1987) from analyses of (very high temperature) volcanic gases. Although both magmatic and nonmagmatic wall rocks can be sources for various chemical components (e.g., major elements and some metals) in hydrothermal fluids, and these can be released during alteration of the wall rocks, the introduction of metals with C- and S-rich magmatic volatiles into adjacent or overlying meteoric geothermal systems is also potentially significant. It is perhaps too simplistic to insist upon a (direct) sole source of all metals in epithermal deposits.

The alteration mineral assemblages described earlier indicate two broadly different chemical environments of alteration and mineralization: low to very low pH, oxidized fluids (quartz-(kaolinite)-alunite alteration) and near-neutral, more reduced fluids (adularia-sericite alteration). These two environments are contrasted in Figure 15.1-7, along with selected mineral stability fields and isopleths of gold solubility (after Giggenbach, 1992).

Because of sulphide abundances and the fact that enargite (rather than tetrahedrite) may dominate sulphide minerals in these deposits (e.g., Goldfield: Ashley, 1982), terms such as "high sulphur" (Bonham, 1988) and "high sulphidation" (Hedenquist, 1987) have been proposed to convey the fact that these features imply higher activities of sulphur in the hydrothermal fluids. Such terms can be misleading, despite their increasing usage, and do not convey the essential differences of the adularia-sericite and quartz-(kaolinite)-alunite geochemical environments. It is true that, from a thermodynamic point of view, higher activities of sulphur promote enargite in place of tennantite.

However, equally high (or even higher) activities of sulphur may have prevailed in Cu- or As-poor hydrothermal fluids (liquids) in adularia-sericite systems. For example, chalcopyrite (found in deposits of both subtypes) can be stable at higher activities of sulphur than those indicated by the presence of enargite, and over broad ranges of temperature (e.g., 200-300°C) and oxidation state (e.g., Henley et al., 1984, p. 109; Heald et al., 1987, p. 20). Some authors have used the term "high sulphidation" to also imply "high oxidation states of sulphur"; the terms are not synonymous, although magmatic hydrothermal quartz-(kaolinite)-alunite environments are highly oxidized owing to chemical buffering by magmatic SO_2 (e.g., Giggenbach, 1992).

The two principal geochemical environments of epithermal mineralization and alteration are determined largely by the source and relative abundance of two different fluids, and by water-rock reaction. On the one hand, magmatic-hydrothermal environments that are dominated by acidic, magmatic fluids (epithermal environment: saline liquids with dissolved CO_2, HCl, and S-species; volcanic environment: CO_2-, HCl-, and SO_2-rich vapour) produce quartz-(kaolinite)-alunite mineral assemblages characterized by oxidized forms of iron (e.g., hematite) and sulphur (e.g., alunite) and by base leaching of wall rocks. This environment may overlie porphyry systems (Sillitoe and Bonham, 1984). On the other hand, near-neutral, more

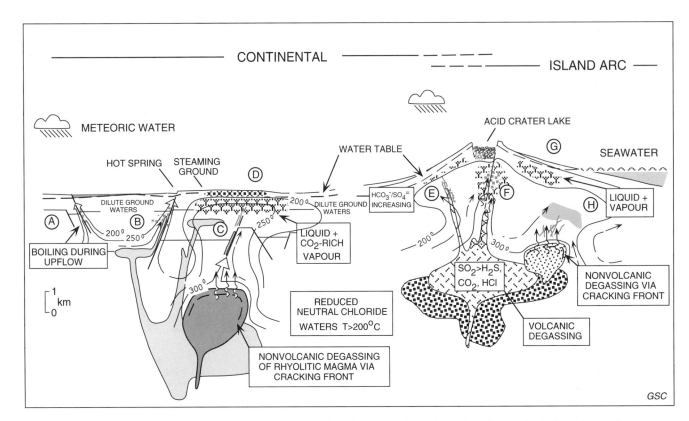

Figure 15.1-6. Schematic cross-section illustrating the general geological and hydrological settings of quartz-(kaolinite)-alunite and adularia-sericite deposits (includes concepts partially adapted from Henley and Ellis,1983; Rye et al.,1992). Characteristics shown evolve with time; all features illustrated are not implied to be synchronous. Interpreted settings are indicated for several Canadian deposits discussed in the text; see also Table 15.1-1. Local environments and examples of adularia-sericite deposits include: (A) basin margin faults: Dusty Mac; (B) disseminated ore in sedimentary rocks: Cinola; (C) veins in degassing, CO_2-rich, low sulphide content systems: Blackdome, Mt. Skukum; (E) porphyry-associated vein-stockwork, sulphide-rich and sulphide-poor stages: Silbak-Premier; and (H) disseminated replacement associated with porphyry-type and stockwork deposits, involving seawater: Sulphurets. Examples of quartz-(kaolinite)-alunite environments include: (D and G) steam-heated advanced argillic alteration (quartz-(kaolinite)-alunite) zone: Toodoggone River district; (F) magmatic-hydrothermal quartz-(kaolinite)-alunite replacement deposit: Summitville, Colorado, or Nansatsu district, Japan. Fluid flow parallels isotherms. Upflow zones shown schematically by arrowhead-shaped isotherm. Volcanic degassing refers to magmatic degassing driven by depressurization during emplacement ("first boiling"). Nonvolcanic degassing refers to vapour exsolution during crystallization ("second boiling"). The SO_2 disproportionates to H_2S and H_2SO_4 during ascent beneath environment (F). Note that free circulation occurs only in crust above about 400°C. All shown temperatures are in Celsius degrees.

reduced, meteoric-dominated waters containing Cl, H_2S, and CO_2, yield adularia-sericite mineral assemblages through hydrolysis reactions involving feldspar in the wall rocks. The chemical state of magmatic-hydrothermal quartz-(kaolinite)-alunite fluids may be said to be buffered by magmatic volatiles (especially HCl and SO_2), whereas the chemical state of adularia-sericite fluids is buffered largely by water-rock reaction (e.g., feldspar hydrolysis).

The upwardly welling, highly acidic, magmatic-hydrothermal plume may produce a quartz-(kaolinite)-alunite deposit (magmatic-hydrothermal environment). The hydrothermal-mineralization event is likely to be short-lived, limited by shallow degassing of the magma in response to depressurization during its ascent (so-called "first boiling") and by the eventual neutralization of the fluids due to reaction with wall rocks and/or dilution by meteoric fluids. In contrast, meteoric fluids, heated by cooling magmatic rocks, can provide potential fluids for mineralization and alteration over somewhat longer periods of time, and at sites further removed from the magmatic heat source. With time, the meteoric water dominated "environment" may encroach upon the earlier, hotter, hydrothermal-magmatic environment.

Slowly cooling epizonal plutons which undergo "subvolcanic" rather than "volcanic" degassing as they crystallize (i.e., "second boiling"), probably provide mineralizing constituents to overlying or adjacent meteoric hydrothermal systems via protracted leakage of magmatic volatiles across cracking fronts at the margins of the crystallizing magma. Variations on this theme derive also from differences in the sulphur content of rhyolitic (lower) to andesitic (higher) magmatic volatiles, from differences in crustal level at which magmatic degassing occurs, and from the relative proportions of magmatic and meteoric fluids involved through time. Additional discussion can be found, for example, in Henley (1985), Stoffregen (1987), White and Hedenquist (1990), Rye et al. (1992), Giggenbach (1992), and Sillitoe (1993).

Active geothermal systems provide instructive analogues to adularia-sericite-producing hydrothermal systems. Geochemical studies of dominantly volcanic-hosted geothermal systems in the Taupo Volcanic Zone, New Zealand (see Henley and Hedenquist, 1986) have demonstrated the existence of two principal types of fluids: (1) a deep chloride water, generally 200° to about 300°C, and (2) a shallower, less than 100° to 200°C steam-heated, low chlorinity, acidic water. The interface between waters with markedly different salinities has been described in the Salton Sea geothermal system by Williams and McKibben (1989). These deep chloride waters produce adularia-sericite-subtype alteration (e.g., Henley, 1985), and where they are rapidly depressurized, degassed of CO_2 and H_2S, and cooled, precious and base metals are deposited (Clark and Williams-Jones, 1990). The well scales studied by Clark and Williams-Jones (1990) revealed a vertical separation of precious metals (higher) and base metals (lower) analogous to that described by Ewers and Keays (1977) for the Broadlands geothermal field (New Zealand), and by Buchanan (1981) for a number of deposits.

Simple conductive cooling is, itself, sufficient to cause gold precipitation (see Fig. 15.1-7). Boiling also causes cooling, chemical fractionation, and an increase in pH. This leads to saturation and precipitation of chloride-complexed metals (e.g., Cu, Pb, Zn; Drummond and Ohmoto, 1985; Spycher and Reed, 1989). Also, degassing of initially CO_2-rich fluids in gas-rich systems depletes the liquid in H_2S which is carried off in a CO_2-rich vapour. The loss of H_2S eventually leads to precipitation of sulphur-complexed metals (e.g., gold; Drummond and Ohmoto, 1985; Henley, 1985; Hayashi and Ohmoto, 1991). Carbon dioxide and hydrogen sulphide are well correlated in some geothermal fluids (reviewed in Taylor, 1987). Boiling and chemical fractionation of the hydrothermal fluid provides an explanation for the separation of precious and base metals. This separation results in a vertical zonation where fluids are upwardly flowing (Clark and Williams-Jones, 1990), or in relative temporal stages, such as at Silbak-Premier and El Indio. As a corollary, larger vein deposits require the movement of larger amounts of fluid through localized zones of boiling, and thus the importance of structural analysis in exploration is obvious. Neutralization and cooling of ore fluids may also occur (1) by mixing with dilute groundwaters, and (2) by water-rock reaction (e.g., sulphidation of ferrous iron-bearing minerals), especially during formation of disseminated and replacement-type ore bodies.

The steam-heated acid waters, formed by the oxidation and condensation of H_2S (boiled off deeper geothermal reservoirs) in groundwater, produces quartz-(kaolinite)-alunite-subtype alteration of the volcanic rocks (Henley and Hedenquist, 1986). The Champagne pool, in the CO_2-rich Waitapu geothermal field (quartz-(kaolinite)-alunite), New Zealand, is a hydrothermal eruption feature below which gold and silver are being deposited in response to boiling and loss of H_2S over the approximate temperature interval 250-175°C (Hedenquist, 1986). Ore-grade, gold-bearing amorphous sulphides precipitate in the pool at 75°C, and base metal sulphides occur below the zone of boiling. Acidic waters produce advanced argillic alteration, and, with variation in P_{CO_2}, evolve to cause the replacement of adularia and albite by sericite. Thus, by chemical evolution, a geothermal field, initially boiling and producing quartz-(kaolinite)-alunite-subtype alteration, may eventually produce minerals characteristic of adularia-sericite-subtype alteration.

The precious metal content of steam-heated alteration zones may also be related to the rate of fluid ascent versus the extent of boiling and H_2S loss: faster moving fluids and/or those less depleted in H_2S may produce higher grades of precious metals in steam-heated alteration zones. This might apply to the ascension of boiling magmatic hydrothermal plumes as well as to boiling meteoric and marine geothermal fluids.

RELATED DEPOSIT TYPES

Lindgren (1933) included in his scheme of classification a qualitative assessment of the depth of formation which is, in a relative sense, largely still valid, despite the sometimes broader application today of his term "epithermal deposit". Deep epithermal (or shallow mesothermal) veins ("transitional" deposits of Panteleyev, 1986) provide an example of the extended depth of formation currently included in the broad sense of epithermal. Intrusion-related vein deposits in the Sulphurets, Mt. Washington, and Zeballos camps, all in British Columbia, are possible examples (Anon., 1992 – B.C. MINFILE; Margolis, 1993). Other hydrothermal deposits are also broadly related to epithermal vein deposits by

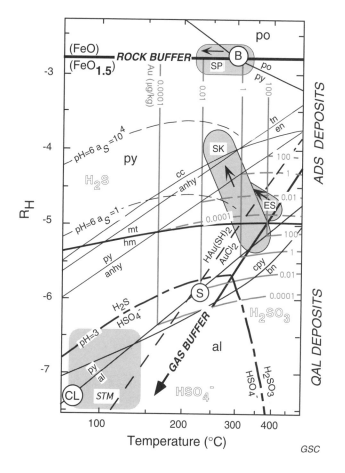

Figure 15.1-7. Diagram of redox potential (R_H = log f_{H_2}/f_{H_2O}) versus temperature (modifed after Giggenbach, 1992; Hedenquist et al., 1994). Calculated isopleths of gold in µg/kg solubility (as the disolved species HAu(SH)$_2$; (Giggenbach, 1992) are shown in red. An equimolar isopleth for HAu(SH)$_2$ and AuCl$_2$ (after Hedenquist et al., 1994) is shown for pH = 3 and 1.0 wt. % Cl (or, pH = 5 and 10 wt.% Cl). The thiogold complex HAu(SH)$_2$ probably dominates as the gold-transporting agent in much of the epithermal environment at pH<5 (Giggenbach, 1992). Redox conditions for mineral deposition in the Broadlands-Ohaaki geothermal system (B), Summitville deposit (S), and in Crater lakes (CL) (e.g., Ruapehu, New Zealand) are from Giggenbach (1992). Fields showing approximate conditions of formation, and their variation with time shown by arrows, for Mt. Skukum (SK), Equity Silver (ES), and Silbak-Premier (SP; also Blackdome and others) are based on data in references cited in Table 15.1-1. Also shown are approximate conditions for steam-heated deposits (STM) of quartz-(kaolinite)-alunite subtype. The diagram shows stability limits and reactions for several minerals discussed in the text. Diagram represents a large variation in system composition, and not all mineral reactions are implied to occur in all deposits. The "gas buffer" curve represents redox control by a magmatic SO$_2$-H$_2$S gas mixture; the Fe^{+2}/Fe^{+3} couple provides redox control in "rock dominated" systems (rock buffer). Quartz-(kaolinite)-alunite deposits form under oxidizing, acidic conditions of the lower one-third of the diagram. Conditions of formation of adularia-sericite deposits are represented by the upper half of diagram, with most forming near the "rock buffer" curve. Isopleths of gold solubility (in µg/kg) are shown in red for two sets of buffered conditions, one set buffered by the coexistence of pyrite and alunite, the other by pyrite and anhydrite. Dashed isopleths apply to the pyrite-anhydrite buffered conditions. Abbreviations are: $a_S = a_{H_2S}/a_{SO_4^{\bar{}}}$; po, pyrrhotite; py, pyrite; anhy, anhydrite; tn, tennantite; en, enargite; cpy, chalcopyrite; bn, bornite; mt, magnetite; hm, hematite; cc, calcite; al, alunite.

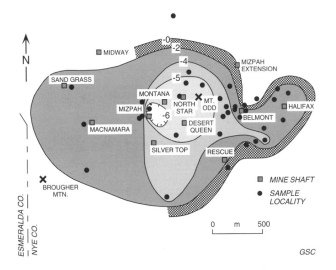

Figure 15.1-8. A δ^{18}O isopleth map for whole-rock samples (filled circles) in the vicinity of epithermal gold deposits at Tonopah, Nevada (after H.P. Taylor, 1973) illustrates oxygen isotope depletion of wall rocks by reaction with a meteoric water hydrothermal system. Greatest depletions (i.e., most negative δ^{18}O values) are closely associated with higher density of deposits as indicated by mine shafts (square symbols).

virtue of similar magmatic-hydrothermal and meteoric-hydrothermal processes. These deposits include gold-bearing skarns (high-temperature, silica-replacement deposits; e.g., Hedley, British Columbia) and manto deposits (sulphide-rich replacement; e.g., Ketza River, British Columbia) in carbonate rocks (see "Skarn gold", subtype 20.3).

Figure 15-2 (see Introduction above) represents a schematic cross-section which, although not meant to imply a continuum, graphically represents the relative geological environments of hot spring and epithermal vein deposits, skarns, mantos, Carlin-type deposits, and porphyry systems. Indeed, the geothermal systems which formed epithermal veins may have been situated well above zones in which porphyry-type deposits formed, and separated from them by a "barren gap". The geothermal systems would have acted as "traps" for the magmatic hydrothermal plumes. In addition, Figure 15-2 compares the environments of these deposits with that of "hypothermal" quartz-carbonate gold veins (see "Quartz-carbonate vein gold", subtype 15.2).

The occurrence of epithermal and hot spring environments in close spatial (but variable temporal) proximity is well documented in the Taupo Volcanic Zone, New Zealand (Henley et al., 1986). A genetic link with high-level intrusions is suggested by several studies. Vein quartz-(kaolinite)-alunite deposits at Butte, Montana are superposed upon earlier, disseminated, or porphyry-type mineralized rocks (Brimhall and Ghiorso, 1983). This suggests that juxtaposition of epithermal vein and porphyry deposits may develop largely as a consequence of a change in the relative position of the meteoric geothermal system and magmatic heat source. Epithermal gold-bearing veins are superposed on slightly older Cu-Mo-Au porphyry deposits in the Coromandel Peninsula, New Zealand (Merchant in Henley et al., 1986), and superposition of an adularia-sericite-subtype deposit on a porphyry copper-type deposit has been documented in the Philippines (Acupan: Cooke and Bloom, 1990). The extent of superposition of related epithermal and porphyry environments is tectonically and climatically controlled, facilitated by rapid uplift, high erosion rates, and volcanic sector collapse (Sillitoe, 1993). Such an association has provided a genetic framework useful in the exploration for epithermal gold deposits in the United States (Berger and Eimon, 1983) and in British Columbia (Panteleyev, 1986).

The term "epithermal" is commonly reserved for non-marine deposits. However, volcanogenic massive sulphide (VMS) deposits are also epithermal deposits in the broadest sense, having formed from submarine hot spring and geothermal systems, although they are sulphide rich and are mined principally for base metals. Gold-bearing VMS deposits do occur (e.g., Horne mine, Noranda; Eskay Creek 21B, British Columbia: Britton et al., 1989), although sub-aerial oxidation is essential to render some VMS deposits economically mineable for gold.

EXPLORATION GUIDES

A young age of intermediate to felsic volcanism (Toodoggone River camp excepted) is particularly significant in selecting a prospective area. Specific exploration guides for epithermal vein gold deposits depend on the search scale. At a regional scale (within a volcanic province) documentation of structural setting, and identification of areas which have experienced high heat flow and geothermal activity (e.g., in or adjacent to volcanic structures or rift zones), is of foremost importance. Although Tertiary and Recent terranes are preferable, older extensional terranes that have not been eroded deeply should not be overlooked. Geochemical analysis of stream sediments for precious and base metals (especially Pb and Zn), Hg, As, Tl, and Sb may be useful.

At the district scale, mapping (including structural analysis) and geochemical and stable isotopic analyses are important to define the size of the hydrothermal system, and the prospective locations of potential deposits. These studies may entail documentation of (1) mineralogical evidence of wall rock alteration (type and zoning; recognition of deposit subtype); (2) open-space filling vein textures; (3) oxygen isotope depletion and zoning in wall rocks; (4) evidence for boiling (e.g., presence of platy calcite; variable vapour/liquid ratios in primary fluid inclusions); (5) evidence regarding the origin of alunite; and (6) the relative distribution of gold, silver, and base metals, plus the volatile elements, Hg, Sb, As, and Tl, by mapping and obtaining analytical data.

Oxygen isotope mapping (e.g., Fig. 15.1-8 and 15.1-9) may also prove useful for recognizing hydrothermal alteration associated with hidden vein systems. Because the oxygen isotope composition of surface-derived, marine or meteoric waters ($\delta^{18}O < 0$; Fig. 15.1-5) differs markedly from that of typical rocks ($\delta^{18}O > 6$), and because surface-derived waters dominate in epithermal systems, mappable oxygen isotope alteration of rocks occurs during reaction with hydrothermal fluids. The magnitude of the change in the rock's initial isotopic composition is proportional to temperature and fluid flux, as previously noted, and oxygen isotope mapping thus provides a potential tool for delineating zones of upflow and potential mineralization in the paleohydrothermal system. For example, Figure 15.1-8 shows that mines in a classic district of adularia-sericite deposits are concentrated within the area of greatest ^{18}O depletion of volcanic host rocks, an area of presumable hydrothermal upflow. Areas of upflow can be associated with boiling and cooling which lead to precipitation of ore minerals.

Oxygen isotope zonation is readily detected in the andesitic wall rocks of the veins at Hishikari, Japan. A general correlation between patterns of ^{18}O depletion and alteration mineral assemblages around the veins can also be noted in Figure 15.1-9. It can be seen from the cross-section in this figure, that oxygen isotope evidence for hydrothermal circulation extends several hundreds of metres into the wall rocks, above the buried veins. In some cases, oxygen isotope alteration may be more readily detected than alteration mineral assemblages.

Prospective areas for hidden deposits may also be targeted by various geophysical techniques. These may include gravity surveys to map subsurface structure, and electrical resistivity and electromagnetic surveys to map hydrothermally altered rocks.

At the deposit or mine scale, the mapping of hydrothermal conduits and their structural expression, the identification of faults and their sense and magnitude of displacement, and determination of the distribution of gold (and silver) become most important. All of the above have as their goals the location of permeable zones. Documentation of mineralogical and structural guides to ore distribution and zones of boiling can also guide mine development.

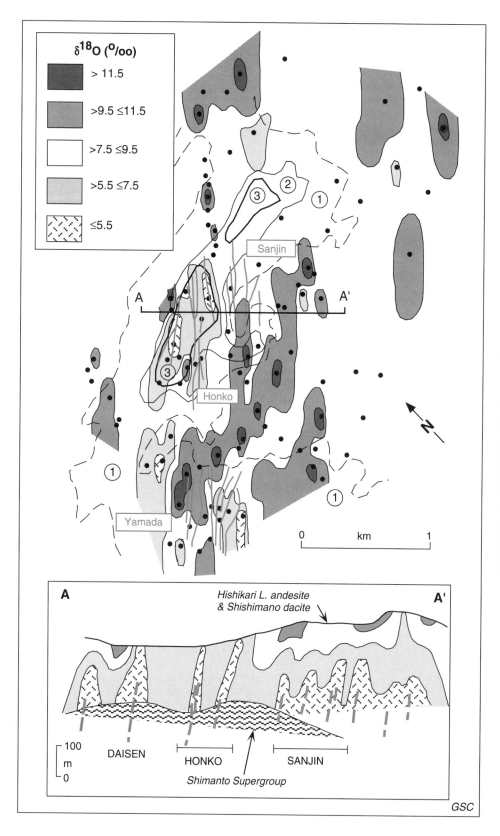

Figure 15.1-9. The $\delta^{18}O$ isopleth map and cross-section of the Hishikari area, Japan, modified after Naito et al. (1993). Positions of Daisen quartz vein, Honko vein system, and Sanjin vein system (shown in red) are projected to surface and into plane of section A-A'. Oxygen isotope data of Naito et al. (1993) were recontoured to emphasize structural control of oxygen isotope zonation. Fresh andesite and dacite are presumed to have $\delta^{18}O$ of 8.5 (cf., Naito et al.,1993). Locations of whole-rock oxygen isotope samples are shown by filled circles, but the locations of subsurface samples, which provide the basis for isotopic zonation in the cross-section, are omitted for simplicity. Alteration mineral zones (after Izawa et al.,1990) are: (1) chlorite- smectite and/or sericite-smectite; (2) quartz- smectite and/or kaolin minerals; and (3) cristobalite-smectite+kaolinite or halloysite. Chlorite-sericite dominate at depth.

SELECTED BIBLIOGRAPHY

References with asterisks (*) are considered to be the best source of general information on this deposit subtype.

Andrew, K.P.E., Gordon, C.I., and Cann, R.M.
1986: Wolf epithermal precious metal vein prospect central British Columbia (93F/3); Geological Fieldwork 1985, British Columbia Ministry of Energy, Mines, and Petroleum Resources, Paper 1986-1, p. 317-320.

Anon.
1992: Minfile/pc V. 3.0; British Columbia Ministry of Energy, Mines and Petroleum Resources, Information Circular 1992-28.

Arehart, G.B., Kesler, S.E., O'Neil, J.R., and Foland, K.A.
1992: Evidence for the supergene origin of alunite in sediment-hosted micron gold deposits, Nevada; Economic Geology, v. 87, p. 263-270.

Ashley, R.P.
1982: Occurrence model for enargite-gold deposits; United States Geological Survey, Open File Report 82-795, p. 144-147.

*Bagby, W.C. and Berger, B.R.
1986: Geologic characteristics of sediment-hosted, disseminated precious-metal deposits in the western United States; Reviews in Economic Geology, v. 2, p. 169-199.

*Bagby, W.C., Menzie, W.D., Mosier, D.L., and Singer, D.A.
1986: Grade and tonnage model of carbonate-hosted Au-Ag; in Mineral Deposit Models, (ed.) D.P. Cox and D.A. Singer; United States Geological Survey, Bulletin 1693, p. 175-177.

Bakken, B.M. and Einaudi, M.T.
1986: Spatial and temporal relations between wall rock alteration and gold mineralization, Main pit, Carlin gold mine, Nevada, U.S.A.; in Proceedings Gold '86, an International Symposium on the Geology of Gold, (ed.) A.J. Macdonald; Konsult International, Toronto, Ontario, p. 388-403.

Barr, D.A., Drown, T.J., Law, T.W., McCreary, G.A., Muir, W.W., Paxton, J.M., and Roscoe, R.L.
1986: The Baker mine operation; The Canadian Institute of Mining and Metallurgy Bulletin, v. 79, no. 893, p. 61-68.

*Berger, B.R. and Bethke, P.M. (ed.)
1986: Geology and geochemistry of epithermal systems; Reviews in Economic Geology, v. 2, Society of Economic Geologists, El Paso, Texas, 298 p.

Berger, B.R. and Eimon, P.I.
1983: Conceptual models of epithermal precious-metals deposits; in Cameron Volume on Unconventional Mineral Deposits, (ed.) W.C. Shanks, III; Society of Mining Engineering, p. 191-205.

*Berger, B.R. and Henley, R.W.
1989: Advances in the understanding of epithermal gold-silver deposits, with special reference to the western United States; in the Geology of Gold Deposits: the Perspective in 1988, (ed.) R.R. Keays, W.R.H. Ramsay, and D.I. Groves; Economic Geology, Monograph 6, p. 405-423.

*Bethke, P.M.
1984: Controls on base- and precious-metal mineralization in deeper epithermal environments; United States Geological Survey, Open File Report 84-890, 40 p.

*Bonham, J.F., Jr.
1988: Models for volcanic-hosted precious metal deposits; a review; in Bulk Mineable Precious Metal Deposits of the Western United States, (ed.) R.W. Schafer, J.J. Cooper, and P.G. Vikre; Geological Society of Nevada, p. 259-271.
1989: Bulk mineable gold deposits of the western United States; in The Geology of Gold Deposits; the Perspective in 1988, (ed.) R.R. Keays, W.R.H. Ramsay, and D.I. Groves; Economic Geology, Monograph 6, p. 193-207.

*Bonham, J.F., Jr. and Giles, D.L.
1983: Epithermal gold/silver deposits: the geothermal connection; Geothermal Resources Council, Special Report 13, p. 257-262.

Brimhall, G.H., Jr. and Ghiorso, M.S.
1983: Origin and ore-forming consequences of the advanced-argillic alteration process in hypogene environments by magmatic gas contamination of meteoric fluids; Economic Geology, v. 78, p. 73-90.

Britton, J.M., Blackwell, J.D., and Schroeter, T.G.
1989: #21 zone deposits, Eskay Creek, northwestern British Columbia; in Exploration in British Columbia 1989; British Columbia Ministry of Energy, Mines, and Petroleum Resources, p. 197-223.

*Buchanan, L.J.
1981: Precious metal deposits associated with volcanic environments in the southwest; in Relations of Tectonics to Ore Deposits in the South Cordillera, (ed.) W.R. Dickinson and W.D. Payne; Arizona Geological Society Digest, v. XIV, p. 237-261.

Cameron, E.M. and Hattori, K.
1985: The Hemlo gold deposit, Ontario: a geochemical and isotopic study; Geochimica et Cosmochimica Acta, v. 49, p. 2041-2050.

Champigny, N. and Sinclair, A.J.
1982: The Cinola gold deposit, Queen Charlotte Islands, British Columbia; in Proceedings of the Gold Symposium, (ed.) R.W. Hodder and W. Petruk; The Canadian Institute of Mining and Metallurgy, Special Volume 24, p. 243-254.

Chivas, A.R., O'Neil, J.R., and Katchan, G.
1984: Uplift and submarine formation of some Melanesian porphyry copper deposits: stable isotope evidence; Earth and Planetary Science Letters, v. 68, p. 326-334.

Christie, A.B.
1989: Cinola deposit, Queen Charlotte Islands; Geological Fieldwork 1988, British Columbia Ministry of Energy, Mines, and Petroleum Resources, Paper 1989-1, p. 423-428.

*Christie, A.B. and Braithwaite, R.L.
1986: Epithermal gold-silver and porphyry copper deposits of the Hauraki goldfield – a review; Chapter 13 in Guide to Active Epithermal (Geothermal) Systems and Precious Metal Deposits, (ed.) R.W. Henley, J.W. Hedenquist, and P.J. Roberts; Bureau of Mineral Resources, Australia, Mineral Deposits, Monograph 26, 211 p.

Church, N.B.
1973: Geology of the White Lake basin; British Columbia Department of Mines and Petroleum Reources, Bulletin 61, 120 p.

Clark, J.R. and Williams-Jones, A.E.
1986: Geology and genesis of epithermal gold-barite mineralization, Verrenas deposit, Toodoggone district, B.C. (abstract); Program with Abstracts, Geological Association Canada-Mineralogical Association Canada, Annual Meeting, Ottawa, 1986, v. 11, p. 57.
1990: Analogues of epithermal gold-silver deposition in geothermal well scales; Nature, v. 346, p. 644-645.
1991: New K-Ar isotopic ages of epithermal alteration from the Toodoggone River area, British Columbia (94E); Geological Fieldwork 1990, British Columbia Ministry of Energy, Mines, and Petroleum Resources, Paper 1991-1, p. 207-216.

Coleman, M.E. and Parrish, R.R.
1990: Eocene deformation in the Bridge River terrane, B.C.: constraints on dextral movement on the Fraser fault (abstract); Geological Association Canada, Program with Abstracts, v. 15, p. A26.

Cooke, D.R. and Bloom, M.S.
1990: Epithermal and subjacent porphyry mineralization, Acupan, Baguio district, Philippines: a fluid-inclusion and paragenetic study; in Epithermal Gold Mineralization of the Circum-Pacific; Geology, Geochemistry, Origin and Exploration, II, (ed.) J. Hedenquist, N.C. White, and G. Siddeley; Journal of Geochemical Exploration, v. 35, p. 297-340.

*Cox, D.P. and Singer, D.A. (ed.)
1986: Mineral deposit models; United States Geological Survey, Bulletin 1693, United States Government Printing Office, Washington, D.C., 379 p.

Cunningham, C.G.
1985: Relationship between disseminated gold deposits and a regional paleothermal anomaly in Nevada; United States Geological Survey, Open File Report 85-722, 20 p.

Cyr, J.B., Pease, R.B., and Schroeter, R.G.
1984: Geology and mineralization at Equity Silver mine; Economic Geology, v. 79, p. 947-968.

de Ronde, C.E.J. and Blattner, P.
1988: Hydrothermal alteration, stable isotopes, and fluid inclusions of the Golden Cross epithermal gold-silver deposit, Waihi, New Zealand; Economic Geology, v. 83, p. 895-917.

Diakow, L.J., Panteleyev, A., and Schroeter, T.G.
1993: Geology of the Early Jurassic Toodoggone Formation and gold-silver deposits in the Toodoggone River map area, northern British Columbia; British Columbia Geological Survey, Bulletin 86, 72 p.

Drummond, S.E. and Ohmoto, H.
1985: Chemical evolution and mineral deposition in boiling hydrothermal systems; Economic Geology, v. 80, p. 126-147.

Duke, J.L. and Godwin, C.I.
1986: Geology and alteration of the Grew Creek epithermal gold-silver prospect, south-central Yukon; in Yukon Geology, v. 1; Exploration and Geological Services Division, Yukon, Indian and Northern Affairs Canada, p. 72-82.

Erickson, G.E. and Cunningham, C.G.

1993: Epithermal precious-metal deposits hosted by the Neogene and Quaternary volcanic complex of the central Andes; in Mineral Deposit Modeling, (ed.) R.V. Kirkham, W.D. Sinclair, R.I. Thorpe, and J.M. Duke; Geological Association of Canada, Special Paper 40, p. 419-431.

Ewers, G.R. and Keays, R.R.

1977: Volatile and precious metal zoning in the Broadlands geothermal field, New Zealand; Economic Geology, v. 72, p. 1337-1354.

Ewing, T.E.

1980: Paleogene tectonic evolution of the Pacific Northwest; Journal of Geology, v. 88, p. 619-638.

Faulkner, E.L.

1986: Blackdome deposit (92O/7E, 8W); Geological Fieldwork 1985, British Columbia Ministry of Energy, Mines and Petroleum Resources, Paper 1986-1, p. 106-109.

Foley, N.K., Bethke, P.M., and Rye, R.O.

1989: A reinterpretation of the δD_{H2O} of inclusion fluids in contemporaneous quartz, sphalerite, Creede mining district, Colorado: a generic problem for shallow orebodies?; Economic Geology, v. 84, p. 1966-1977.

Giggenbach, W.

1992: Magma degassing and mineral deposition in hydrothermal systems along convergent plate boundaries; Economic Geology, v. 87, p. 1927-1944.

Harris, D.C.

1989: The mineralogy and geochemistry of the Hemlo gold deposit, Ontario; Geological Survey of Canada, Economic Geology Report 38, 88 p.

Hattori, K. and Cameron, E.M.

1986: Archean magmatic sulphate; Nature, v. 319, p. 45-47.

Hayashi, K.-I. and Ohmoto, H.

1991: Solubility of gold in NaCl- and H_2S-bearing aqueous solution at 259-350°C; Geochimica et Cosmochimica Acta, v. 55, p. 211-216.

Hayba, D.O., Bethke, P.M., Heald, P.M., and Foley, N.K.

1985: Geologic, mineralogic, and geochemical characteristics of volcanic-hosted epithermal precious-metal deposits; Reviews in Economic Geology, v. 2, p. 129-167.

*Heald, P., Foley, N.K., and Hayba, D.O.

1987: Comparative anatomy of volcanic-hosted epithermal deposits: acid-sulphate and adularia-sericite types; Economic Geology, v. 82, p. 1-26.

Hedenquist, J.W.

1986: Waiotapu geothermal field; Chapter 6 in Guide to Active Epithermal (Geothermal) Systems and Precious Metal Deposits, (ed.) R.W. Henley, J.W. Hedenquist, and P.J. Roberts; Mineral Deposits, Monograph 26, 211 p.

1987: Mineralization associated with volcanic-related hydrothermal systems in the circum-Pacific Basin; in Transactions of the Fourth Circum-Pacific Energy and Mineral Resources Conference, Singapore, (ed.) M.K. Horn; American Association of Petroleum Geologists, p. 513-524.

Hedenquist, J.W. and Henley, R.W.

1985: Hydrothermal eruptions in the Waiotapu geothermal system, New Zealand: their origin, associated breccias, and relation to precious metal mineralization; Economic Geology, v. 80, p. 1640-1668.

Hedenquist, J.W., Matsuhisa, Y., Izawa, E., White, N.C., Giggenbach, W.F., and Aoki, M.

1994: Geology, geochemistry, and origin of high sulfidation Cu-Au mineralization in the Nansatsu district, Japan; Economic Geology, v. 89, p. 1-30.

Hedenquist, J., White, N.C., and Siddeley, G. (ed.)

1990a: Epithermal gold mineralization of the Circum-Pacific; Geology, Geochemistry, Origin and Exploration, I; Journal of Geochemical Exploration, v. 35, 447 p.

1990b: Epithermal gold mineralization of the Circum-Pacific; Geology, Geochemistry, Origin and Exploration, II; Journal of Geochemical Exploration, v. 36, 474 p.

*Henley, R.W.

1985: The geothermal framework of epithermal deposits; in Geology and Geochemistry of Epithermal Systems, (ed.) B.R. Berger and P.M. Bethke; Reviews in Economic Geology, Volume 2, Society of Economic Geologists, p. 1-24.

*Henley, R.W. and Ellis, A.J.

1983: Geothermal systems, ancient and modern; Earth Science Reviews, v. 19, p. 1-50.

Henley, R.W. and Hedenquist, J.W.

1986: Introduction to the geochemistry of active and fossil geothermal systems; Chapter 1 in 1986, Guide to Active Epithermal (Geothermal) Systems and Precious Metal Deposits, (ed.) R.W. Henley, J.W. Hedenquist, and P.J. Roberts; Mineral Deposits, Monograph 26, 211 p.

*Henley, R.W., Hedenquist, J.W., and Roberts, P.J. (ed.)

1986: Guide to Active Epithermal (Geothermal) Systems and Precious Metal Deposits; Mineral Deposits, Monograph 26, 211 p.

Henley, R.W., Truesdell, A.H., and Barton, P.B., Jr.

1984: Fluid-mineral equilibria in hydrothermal systems; Reviews in Economic Geology, Volume 1; Society of Economic Geologists, 267 p.

Holland, H.D.

1965: Some applications of thermochemical data to problems of ore deposits. II. Mineral assemblages and the composition of ore-forming fluids; Economic Geology, v. 60, p. 1101-1166.

Holland, P.T., Beaty, D.W., and Snow, G.G.

1988: Comparative elemental and oxygen isotope geochemistry of jasperoid in the northern Great Basin: evidence for distinctive fluid evolution in gold-producing hydrothermal systems; Economic Geology, v. 83, p. 1401-1423.

Huckerby, J.A., Moore, J.M., and Davis, G.R.

1983: Tectonic control of mineralization at Mahd adh Dhahab gold mine, western Saudia Arabia; The Institute of Mining and Metallurgy Transactions, v. 92, p. B171-B182.

Izawa, E., Naito, K., Ibaraki, K., and Suzuki, R.

1993: Mudstones in a hydrothermal eruption crater above the gold-bearing vein system of the Yamada deposit at Hishikari, Japan; Resource Geology Special Issue, no. 14, p. 85-92.

Izawa, E., Urashima, Y., Ibaraki, K., Suzuki, R., Yokoyama, T., Kawasaki, K., Koga, A., and Taguchi, S.

1990: The Hishikari gold deposit: high-grade epithermal veins in Quaternary volcanics of southern Kyushu, Japan; Journal of Geochemical Exploration, v. 36, p. 1-56.

Jannas, R.R., Beane, R.E., Ahler, B.A., and Brosnahan, D.R.

1990: Gold and copper mineralization at the El Indio deposit, Chile; in Epithermal Gold Mineralization of the Circum-Pacific; Geology, Geochemistry, Origin and Exploration, II, (ed.) J. Hedenquist, N.C. White, and G. Siddeley; Journal of Geochemical Exploration, v. 36, p. 233-266.

Kamilli, R.J. and Ohmoto, H.

1977: Paragenesis, zoning, fluid inclusion, and isotopic studies of the Finlandia vein, Colqui District, central Peru; Economic Geology, v. 72, p. 950-982.

Klein, T.L. and Criss, R.E.

1988: An oxygen isotope and geochemical study of meteoric-hydrothermal systems at Pilot Mountain and selected other localities, Carolina Slate Belt; Economic Geology, v. 83, p. 801-821.

Kuhns, R.J.

1988: The Golden Giant deposit, Hemlo, Ontario: geologic and geochemical relationships between mineralization, alteration, metamorphism, magmatism and tectonism; PhD. dissertation, University of Minnesota, Minneapolis, Minnesota, 254 p.

Lindgren, W.

1922: A suggestion for the terminology of certain minerals deposits; Economic Geology, v. 17, p. 202-294.

1933: Mineral Deposits; 4th edition, New York, McGraw-Hill, 930 p.

Lindhorst, J.W. and Cook, W.G.

1990: Gidginbung gold-silver deposit, Temora; in Geology of the Mineral Deposits of Australia and Papua New Guinea, (ed.) F.E. Hughes; The Australasian Institute of Mining and Metallurgy; Melbourne, p. 1365-1370.

Lipman, P.W.

1975: Evolution of the Platoro caldera complex and related volcanic rocks, southeastern San Juan Mountains, Colorado; United States Geological Survey, Professional Paper 852, 128 p.

Love, D.S.

1989: Geology of the epithermal Mount Skukum gold deposit, Yukon Territory; Geological Survey of Canada, Open File 2123, 46 p.

Margolis, J.

1993: Geology and intrusion-related copper-gold mineralization, Sulphurets, British Columbia; PhD. dissertation, University of Oregon, Eugene, Oregon, 289 p., 5 maps.

McDonald, B.W.R.

1987: Geology and genesis of the Mount Skukum Tertiary epithermal gold-silver vein deposit, southwestern Yukon Territory (NTS 105D SW); MSc. thesis, University of British Columbia, Vancouver, British Columbia, 177 p.

McDonald, B.W.R. and Godwin, C.I.
1986: Geology of main zone at Mt. Skukum, Wheaton River area, southern Yukon; in Yukon Geology, v. 1; Exploration and Geological Services Division, Yukon, Indian and Northern Affairs Canada, p. 6-10.

McDonald, B.W.R., Stewart, E.B., and Godwin, C.I.
1986: Exploration geology of the Mt. Skukum epithermal gold deposit, southwestern Yukon; in Yukon Geology, v. 1; Exploration and Geological Services Division, Yukon, Indian and Northern Affairs Canada, p. 11-18.

McDonald, D.W.A.
1990: The Silbak Premier silver-gold deposit: a structurally controlled, base metal-rich Cordilleran epithermal deposit, Stewart, B.C.; PhD. dissertation, University of Western Ontario, London, Ontario, 411 p.

McFall, J.
1981: The geology and mineralization of the Venus Pb-Zn-Ag mine Yukon Territory (abstract); The Canadian Institute of Mining and Metallurgy Bulletin, v. 74, p. 64.

McInnes, B.I.A., Crocket, J.H., and Goodfellow, W.D.
1990: The Laforma deposit, an atypical epithermal-Au system at Freegold Mountain, Yukon Territory, Canada; Journal of Geochemical Exploration, v. 36, p. 73-102.

McKenzie, C.B.
1986: Geology and mineralization of the Chetwynd deposit, southwestern Newfoundland, Canada, in Proceedings Gold '86, an International Symposium on the Geology of Gold, (ed.) A.J. Macdonald; Konsult International, Toronto, Ontario, p. 137-148.

Morin, J. and Downing, D.A. (comp.; ed.)
1984: Gold-silver deposits and occurrences in Yukon Territory; Exploration and Geological Services Division, Yukon, Indian and Northern Affairs Canada, Open File, 1:2 000 000 scale map with marginal notes and tables.

Mosier, D.L., Sato, T., and Singer, D.A.
1985: Grade and tonnage model of Creede epithermal veins; in 1985, Geologic Characteristics of Sediment- and Volcanic-hosted Disseminated Gold Deposits – Search for an Occurrence Model, (ed.) E.W. Tooker; United States Geological Survey, Bulletin 1646, p. 146-149.

Mutschler, F.E. and Mooney, T.C.
1993: Precious-metal deposits related to alkalic igneous rocks: provisional classification, grade-tonnage data and exploration frontiers; in Mineral Deposit Modeling, (ed.) R.V. Kirkham, W.D. Sinclair, R.I. Thorpe, and J.M. Duke; Geological Association of Canada, Special Paper 40, p. 479-520.

Naito, K., Matsuhisa, Y., Izawa, E., and Takaoka, H.
1993: Oxygen isotopic zonation of hydrothermally altered rocks in the Hishikari gold deposit, southern Kyushu, Japan: its implications for mineral prospecting; Resource Geology Special Issue, no. 14, p. 71-84.

Nakamura, H., Sumi, K., Karagiri, K., and Iwata, T.
1970: The geological environment of the Matsukawa geothermal area, Japan; Geothermics (Special Issue 2), v. 2, pt. 1, p. 221-231.

Nicholson, K.
1989: Early Devonian geothermal systems in northeast Scotland: exploration targets for epithermal gold; Geology, v. 17, p. 568-571.

*Panteleyev, A.
1986: A Canadian Cordilleran model for epithermal gold-silver deposits; Geoscience Canada, v. 13, p. 101-111.
*1991: Gold in the Canadian Cordillera -- a focus on epithermal and deeper environments; British Columbia Ministry of Energy, Mines and Petroleum Resources, Paper 1991-4, p. 163-212.

Parrish, R.R. and Coleman, M.E.
1990: A model of middle Eocene extension and strike-slip faulting for the Canadian Cordillera and Pacific Northwest (abstract); Geological Association of Canada, Program with Abstracts, v. 15, p. A101.

Pride, M.J.
1986: Description of the Mt. Skukum volcanic complex in southern Yukon; in Yukon Geology, v. 1, (ed.) J.A. Morin and D.S. Emond; Exploration and Geological Services Division, Yukon, Indian and Northern Affairs Canada, p. 148-160.

Pride, M.J. and Clark, G.S.
1985: An Eocene Rb-Sr isochron for rhyolite plugs, Mt. Skukum area, Yukon Territory; Canadian Journal of Earth Sciences, v. 22, p. 1747-1753.

Rye, R.O., Bethke, P.M., and Wasserman, M.D.
1992: The stable isotope geochemistry of acid sulfate alteration; Economic Geology, v. 87, p. 225-262.

Schroeter, T.G.
1982: Toodoggone River (94E); in Geological Fieldwork 1981, British Columbia Ministry of Energy, Mines and Petroleum Resources, Paper 1982-1, p. 122-133.
1985: Toodoggone River area; in Geological Fieldwork 1984, British Columbia Ministry of Energy, Mines and Petroleum Resources, Paper 1985-1, p. 291-298.
1986: Brief studies of selected gold deposits in southern British Columbia; in Geological Fieldwork 1985, British Columbia Ministry of Energy, Mines and Petroleum Resources, Paper 1986-1, p. 15-22.

Shen, K. and Sinclair, A.J.
1982: Preliminary results of a fluid inclusion study of Sam Goosly deposit, Equity Mines Ltd., Houston; Geological Fieldwork 1981, British Columbia Ministry of Energy, Mines and Petroleum Resources, Paper 1982-1, p. 229-233.

Shen, K., Champigny, N., and Sinclair, A.J.
1982: Fluid inclusion and sulphur isotope data in relation to genesis of the Cinola Gold deposit, Queen Charlotte Islands, B.C.; in Geology of Canadian Gold Deposits, (ed.) R.W. Hodder and W. Petruk; Proceedings of the CIM Gold Symposium, September 1980, The Canadian Institute of Mining and Metallurgy, Special Volume 24, p. 255-257.

Siddeley, G. and Araneda, R.
1986: The El Indio-Tambio gold deposits, Chile; in Proceedings Gold '86, an International Symposium on the Geology of Gold, (ed.) A.J. Macdonald; Konsult International, Toronto, Ontario, p. 445-456.

Sillitoe, R.H.
1989: Gold deposits in western Pacific island arcs; the magmatic connection; Economic Geology, Monograph No. 6, p. 274-291.
*1993: Epithermal models; genetic types, geometrical controls and shallow features; in Mineral Deposit Modeling, (ed.) R.V. Kirkham, W.D. Sinclair, R.I. Thorpe, and J.M. Duke; Geological Association of Canada, Special Paper 40, p. 403-417.

Sillitoe, R.H. and Bonham, H.F., Jr.
1984: Volcanic landforms and ore deposits; Economic Geology, v. 79, p. 1286-1298.
1990: Sediment-hosted gold deposits; distal products of magmatic hydrothermal systems; Geology, v. 18, p. 157-161.

Simmons, S.F. and Christenson, B.W.
1994: Origins of calcite in a boiling geothermal system; American Journal of Science, v. 294, p. 361-400.

Slack, J.F.
1980: Multistage vein ores of the Lake City district, western San Juan mountains, Colorado; Economic Geology, v. 75, p. 963-991.

Spycher, N.F. and Reed, M.H.
1989: Evolution of a Broadlands-type epithermal ore fluid along alternative P-T paths: implications for the transport and deposition of base, precious, and volatile metals; Economic Geology, v. 84, p. 328-359.

Stoffregen, R.E.
1987: Genesis of acid-sulfate alteration and Au-Cu-Ag mineralization at Summitville, Colorado; Economic Geology, v. 82, p. 1575-1591.

Symonds, R.P., Rose, W.I., Reed, M.H., Lichte, F.E., and Finnegan, D.L.
1987: Volatilization, transport and sublimation of metallic and non-metallic elements in high temperature gases at Merapi Volcano, Indonesia; Geochimica et Cosmochimica Acta, v. 51, p. 2083-2101.

Taylor, B.E.
1987: Stable isotope geochemistry of ore-forming fluids; in Stable Isotope Geochemistry of Low Temperature Fluids, (ed.) T.K. Kyser; Short Course Handbook Volume 13, Mineralogical Association of Canada, p. 337-445.
1992: Degassing of H_2O from rhyolite magma during eruption and shallow intrusion, and the isotopic composition of magmatic water in hydrothermal systems; Geological Survey of Japan, Report No. 279, p. 190-194.

Taylor, H.P., Jr.
1973: O^{18}/O^{16} evidence for meteoric-hydrothermal alteration and ore deposition in the Tonopah, Comstock Lode, and Goldfield Mining Districts, Nevada; Economic Geology, v. 68, p. 747-764.

Thiersch, P. and Williams-Jones, A.E.
1990: Paragenesis and ore controls of the Shasta Ag-Au deposit, Toodoggone River area, British Columbia; in Geological Fieldwork 1989; British Columbia Ministry of Energy, Mines and Petroleum Resources, Paper 1990-1, p. 315-321.

Thomson, M.L., Mastalerz, M., Sinclair, A.J., and Bustin, R.M.
1992: Fluid source and thermal history of an epithermal vein deposit, Owen Lake, central British Columbia: evidence from bitumen and fluid inclusions; Mineralium Deposita, v. 27, p. 219-225.

Tolbert, R.S. and Froc, N.V.
1988: Geology of the Cinola gold deposit, Queen Charlotte Islands, B.C., Canada; in Major Gold-Silver Deposits of the Northern Canadian Cordillera, Society of Economic Geologists, Field Guide and Notebook, p. 15-40.

*Tooker, E.W. (ed.)
1985: Geologic characteristics of sediment- and volcanic-hosted disseminated gold deposits – search for an occurrence model; United States Geological Survey, Bulletin 1646, 150 p.

Vivian, G., Morton, R.D., Changkakoti, A., and Gray, J.
1987: Blackdome Eocene epithermal Ag-Au deposit, British Columbia, Canada – nature of ore fluids; Transactions of the Institution of Mining and Metallurgy (Section B; Applied Earth Science), v. 96, p. B9-B14.

Vulimiri, M.R., Tegart, P., and Stammers, M.A.
1986: Lawyers gold-silver deposits, British Columbia; The Canadian Institute of Mining and Metallurgy, Special Volume 37, p. 191-201.

Walton, L.
1986: Textural characteristics of the Venus vein and implications for ore shoot distribution; in Yukon Geology, v. 1 -1984; Exploration and Geological Services Division, Yukon, Indian and Northern Affairs Canada, p. 67-71.

Walton, L. and Nesbitt, B.E.
1986: Evidence for late stage Au-galena mineralization in the Venus arsenopyrite-pyrite-quartz vein, southwest Yukon (abstract); Program with Abstracts, Geological Association of Canada-Mineralogical Association of Canada, Annual Meeting, Ottawa, 1986, v. 11, p. 141.

White, D.E., Muffler, L.J.P., and Truesdell, A.H.
1971: Vapor-dominated hydrothermal systems compared with hot-water systems; Economic Geology, v. 66, p. 75-97.

*White, N.C. and Hedenquist, J.W.
1990: Epithermal environments and styles of mineralization; variations and their causes, and guidelines for exploration; in Epithermal Gold Mineralization of the Circum-Pacific; Geology, Geochemistry, Origin and Exploration, II; (ed.) J. Hedenquist, N.C. White, and G. Siddeley; Journal of Geochemical Exploration, v. 36, p. 445-474.

Williams, A.E. and McKibben, M.A.
1989: A brine interface in the Salton Sea geothermal system, California; fluid geochemical and isotopic characteristics; Geochimica et Cosmochimica Acta, v. 53, p. 1905-1920.

Wojdak, P.J. and Sinclair, A.J.
1984: Equity Silver silver-copper-gold deposit; alteration and fluid inclusion studies; Economic Geology, v. 79, p. 969-990.

Wood, D.G., Porter, R.G., and White, N.C.
1990: Geological features of some Paleozoic epithermal gold occurrences in northeastern Queensland, Australia; in Epithermal Gold Mineralization of the Circum-Pacific; Geology, Geochemistry, Origin and Exploration, II; (ed.) J. Hedenquist, N.C. White, and G. Siddeley; Journal of Geochemical Exploration, v. 36, p. 413-443.

Wrucke, C.T. and Armbrustmacher, T.J.
1975: Geochemical and geologic relations of gold and other elements at the Gold Acres open-pit mine, Lander County, Nevada; United States Geological Survey, Professional Paper 860, 27 p.

Zhang, X., Nesbitt, B.E., and Muehlenbachs, K.
1989: Gold mineralization in the Okanagan Valley, southern British Columbia; fluid inclusion and stable isotope studies; Economic Geology, v. 84, p. 410-424.

15.2 QUARTZ-CARBONATE VEIN GOLD

François Robert

INTRODUCTION

This subtype of gold deposits consists of simple to complex quartz-carbonate vein systems associated with brittle-ductile shear zones and folds in deformed and metamorphosed volcanic, sedimentary, and granitoid rocks. In these deposits, gold occurs in veins or as disseminations in immediately adjacent altered wall rocks, and is generally the only or the most significant economic commodity. The veins occur in structural environments characterized by low- to medium-grade metamorphic rocks and brittle-ductile rock behavior, corresponding to intermediate depths within the crust, and by compressive tectonic settings. Deposits of this type have commonly been referred to as mesothermal gold-quartz vein deposits, but they in fact encompass both mesothermal and hypothermal classes as initially defined by Lindgren (1933).

Quartz-carbonate vein gold deposits are widely spread throughout Canada and they occur principally in the following geological areas (Fig. 15.2-1): the greenstone belts of the Superior, Churchill, and Slave provinces, the oceanic terranes of the Canadian Cordillera, and the turbiditic Meguma terrane and the ophiolitic Baie Verte district in the Appalachians. The largest concentration of these deposits occurs in the greenstone belts of the south-central Superior Province.

Typical Canadian examples of such deposits, located on Figure 15.2-1, include: Goldenville, Nova Scotia; Sigma-Lamaque, O'Brien, and Casa-Berardi, Quebec; Kerr Addison, Macassa, Dome, Hollinger-McIntyre, Campbell Red Lake, and MacLeod-Cockshutt, Ontario; San Antonio,

Robert, F.
1996: Quartz-carbonate vein gold; in Geology of Canadian Mineral Deposit Types, (ed.) O.R. Eckstrand, W.D. Sinclair, and R.I. Thorpe; Geological Survey of Canada, Geology of Canada, no. 8, p. 350-366 (also Geological Society of America, The Geology of North America, v. P-1).

<hr>

[1] *Editorial note:* Within this volume, in a companion account of gold deposits hosted by iron-formation (subtype 15.3), Lupin is presented as the type example of stratiform iron-formation hosted gold deposits and has been interpreted to be a syngenetic deposit.

Manitoba; Star Lake, Saskatchewan; Giant Yellowknife, Camlaren, and Lupin[1], Northwest Territories; Bralorne-Pioneer and Cariboo Gold Quartz-Island Mountain, British Columbia. Other examples throughout the world include the following deposits or districts: Mother Lode and Grass Valley, California; Alaska-Juneau, Alaska; Homestake, South Dakota; Mt. Charlotte, Victory, Norseman, and Bendigo-Ballarat, Australia; Ashanti and Prestea, Ghana; and Passagem, São Bento and Crixas, Brazil.

IMPORTANCE

Quartz-carbonate vein deposits account for approximately 80% of the production from lode gold deposits in Canada (Fig. 15.2-2). The Canadian Shield, and the Superior Province in particular, contains the most significant deposits

and accounts for more than 85% of the gold production from quartz-carbonate veins in Canada. Total production from the main geological areas of occurrence in Canada is given in Table 15.2-1.

SIZE AND GRADE OF DEPOSITS

Quartz-carbonate vein gold deposits display a wide range of sizes (Fig. 15.2-2), which can vary as a function of the price of gold, as it is possible in almost every case to selectively mine the higher grade portions of the deposits at times of lower gold prices, and lower grade material as well at times of higher prices. Deposits of Superior Province are the largest, typically containing between 6 and 60 t of gold to a maximum of 1000 t, those of Churchill Province between 5 and 10 t, and those of the Meguma terrane, less

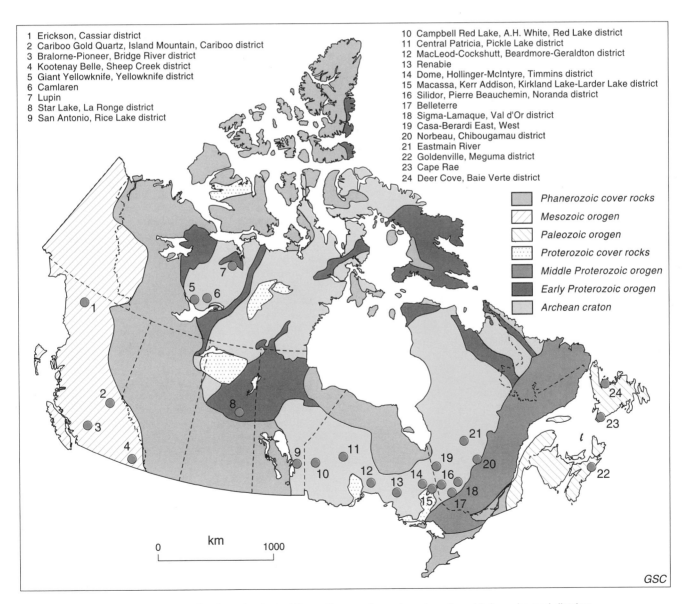

1 Erickson, Cassiar district
2 Cariboo Gold Quartz, Island Mountain, Cariboo district
3 Bralorne-Pioneer, Bridge River district
4 Kootenay Belle, Sheep Creek district
5 Giant Yellowknife, Yellowknife district
6 Camlaren
7 Lupin
8 Star Lake, La Ronge district
9 San Antonio, Rice Lake district

10 Campbell Red Lake, A.H. White, Red Lake district
11 Central Patricia, Pickle Lake district
12 MacLeod-Cockshutt, Beardmore-Geraldton district
13 Renabie
14 Dome, Hollinger-McIntyre, Timmins district
15 Macassa, Kerr Addison, Kirkland Lake-Larder Lake district
16 Silidor, Pierre Beauchemin, Noranda district
17 Belleterre
18 Sigma-Lamaque, Val d'Or district
19 Casa-Berardi East, West
20 Norbeau, Chibougamau district
21 Eastmain River
22 Goldenville, Meguma district
23 Cape Rae
24 Deer Cove, Baie Verte district

Phanerozoic cover rocks
Mesozoic orogen
Paleozoic orogen
Proterozoic cover rocks
Middle Proterozoic orogen
Early Proterozoic orogen
Archean craton

km
0   1000

GSC

Figure 15.2-1. Distribution of selected Canadian quartz-carbonate vein gold deposits and districts.

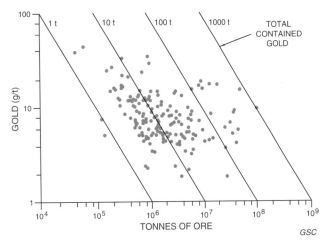

Figure 15.2-2. Tonnage versus grade diagram for Canadian quartz-carbonate vein gold deposits.

Table 15.2-1. Approximate gold content and typical grade and tonnage figures for individual quartz-carbonate vein deposits in the three main geological gold-producing areas in Canada.

	Appalachian Province	Canadian Shield	Canadian Cordillera
Total production (t)	66	6000	360
Deposit grade (g/t)	10	8	14
Deposit tonnage (t)	5×10^5	6×10^6	2×10^6

than 3 t (Table 15.2-1). Typical tonnage and grade of quartz-carbonate vein deposits are a few million tonnes of ore at a grade of 6 to 10 g/t gold (Fig. 15.2-2).

GEOLOGICAL FEATURES

Geological setting

At the regional scale, quartz-carbonate vein gold deposits occur in two contrasting geological environments: deformed clastic sedimentary terranes and deformed volcano-plutonic terranes containing diverse volcanic assemblages of island-arc and oceanic affinities. Despite lithological and structural differences (see below), these two types of environments share the following characteristics: greenschist to locally lower amphibolite metamorphic facies, brittle-ductile nature of the deformation, and geological structures recording compressional to transpressional tectonic settings.

Quartz-carbonate vein gold deposits in these environments tend to occur in clusters, or districts, and they are by far more abundant in volcano-plutonic terranes than in clastic sedimentary terranes. Both types of environments are present in a number of districts, in which they are separated by major fault zones. However, in such cases auriferous quartz-carbonate veins preferentially occur in the volcano-plutonic domains. Key characteristics and examples of these two geological environments are presented below.

Clastic sedimentary terranes

Clastic sedimentary terranes mineralized with quartz-carbonate veins are not very common in Canada but, where present, they typically occupy extensive areas. These terranes include the Meguma terrane, Nova Scotia (Fig. 15.2-3), the "Yellowknife basin" in the Slave Province, and sedimentary rocks of the Sheep Creek district and of the Barkerville terrane in the Cariboo district, both in British Columbia.

Most clastic sedimentary terranes are characterized by important thicknesses of well-bedded turbidites consisting of greywacke, mudstone, shale, and minor conglomerate. In the Meguma terrane (Fig. 15.2-3), the turbidite sequence consists of vein-bearing quartz-rich greywacke and inter-bedded slate of the Goldenville Formation and overlying thinly laminated slate of the Halifax Formation (Graves and Zentilli, 1982). Some sequences, such as the Contwoyto Formation in the Slave Province, also contain significant proportions of interbedded iron-formation and mafic volcanic rocks. The presence of quartzite and/or limestone in the Cariboo (Sutherland-Brown, 1957) and Sheep Creek districts (Matthews, 1953) are indicative of continental margin environments. Clastic sedimentary sequences contain only small proportions of intrusive rocks, most of which form large, postfolding dioritic to granitic bodies such as the Devonian granodiorites and monzogranites in the Meguma terrane (Fig. 15.2-3).

Gold-bearing clastic sedimentary sequences are invariably folded, and commonly in a complex manner. Folds range from open to isoclinal, and may be accompanied by a penetrative axial plane cleavage. In many cases, younger faults cut the folds at moderate to high angles. The Meguma terrane is characterized by a series of shallowly plunging, northeast-to east-northeast-trending upright folds which are cut by northwest-striking faults and intruded by Devonian granites (Fig. 15.2-3). Most sequences have been metamorphosed to the greenschist facies, and in some regions, such as in the Contwoyto Lake area, to the lower and middle amphibolite facies.

Volcano-plutonic terranes

Volcano-plutonic terranes are the most important hosts to vein gold mineralization in Canada. They are represented by the abundant Precambrian greenstone belts of the Canadian Shield and by the Phanerozoic island arc-oceanic assemblages of the Canadian Cordillera and the Appalachians. Representative districts include: Baie Verte, Newfoundland; Val d'Or, Cadillac, and Casa-Berardi, Quebec; Larder Lake, Kirkland Lake, Timmins, Beardmore-Geraldton district, and Red Lake, Ontario; Rice Lake, Manitoba; La Ronge, Saskatchewan; and Coquihalla, Bridge River, and Cassiar, British Columbia (Fig. 15.2-1).

Mineralized volcano-plutonic terranes form elongate belts bounded by, or transected by, crustal-scale fault zones. These belts typically comprise contrasting geological domains, which may include clastic sedimentary sequences, separated from the volcano-plutonic domains by the major fault zones. This is the case at Val d'Or (Fig. 15.2-4) and Beardmore-Geraldton (Fig. 15.2-5), where volcano-plutonic terranes to the north are separated from turbidite sequences to the south by the Larder Lake-Cadillac and Barton Bay fault zones, respectively. In other districts, such as Bridge River, major faults may separate contrasting volcanic assemblages: the Fergusson thrust fault separates the oceanic Bridge River Group from the Cadwallader Group of island arc affinity (Fig. 15.2-6; Leitch, 1990).

Volcano-plutonic terranes are lithologically more diverse than clastic sedimentary sequences. Volcanic supracrustal rocks dominate and typically include basaltic tholeiitic domains of oceanic affinity and mafic to felsic,

tholeiitic to calc-alkaline domains of island arc affinity. Ultramafic rocks are volumetrically important in some Archean terranes where they form komatiitic volcanic domains. In Phanerozoic terranes, ultramafic rocks occur mostly as serpentinite bodies along fault zones, as in the Bridge River district (Fig. 15.2-6), and may represent remnant ophiolite sequences. Narrow belts of clastic sedimentary rocks are also present in many volcano-plutonic terranes and include both flysch-like and molasse-like facies. The flysch-like facies consist of greywacke-mudstone

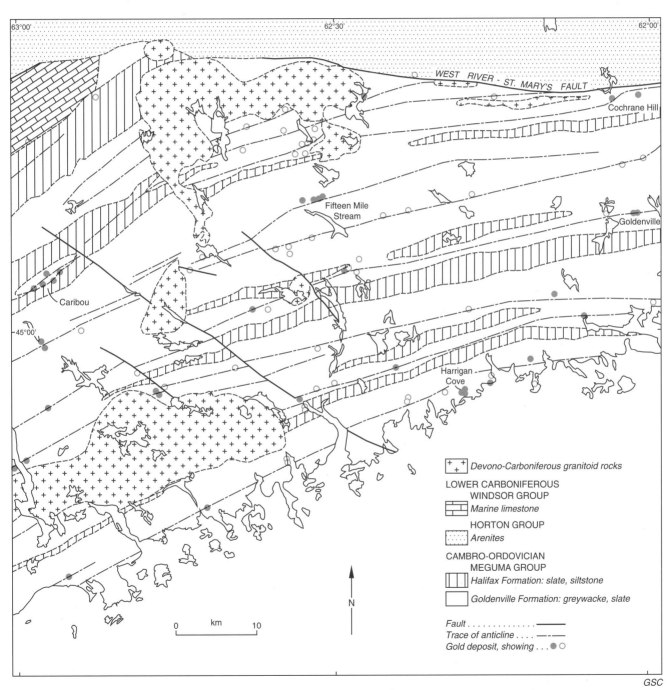

Figure 15.2-3. Simplified geological map of the eastern portion of the Meguma terrane, Nova Scotia, showing the distribution of quartz-carbonate vein gold deposits. (modified from McMullen et al., 1987)

353

with locally abundant conglomerate and iron-formation, as represented by the Cadillac Group at Val d'Or (Fig. 15.2-4) and the Northern, Central, and Southern Metasedimentary Belts at Beardmore-Geraldton (Fig. 15.2-5). Fluvial-alluvial sequences of polymictic conglomerate, arenite, and sandstone, referred to as Timiskaming-type in Superior Province, are representative of the molasse-like facies and are present along major fault zones and uncomformably overlie volcanic rocks in many Precambrian districts such as Kirkland Lake, Rice Lake, and La Ronge. In the Bridge River district, ribbon chert and argillites overlie basalts of the oceanic Bridge River Complex (Fig. 15.2-6).

In contrast to clastic sedimentary sequences, volcanoplutonic terranes contain abundant associated intrusive rocks, including batholiths, stocks, sills, and dykes, emplaced at several stages during their volcanic and tectonic evolution. Early, synvolcanic intrusions include gabbro sills and dykes and subvolcanic diorite-tonalite plutons such as the Bourlamaque pluton at Val d'Or (Fig. 15.2-4) and the Bralorne intrusions at Bridge River (Fig. 15.2-6). Syn- to late tectonic intrusions evolve from commonly porphyritic diorite-tonalite stocks and dykes, to monzonitic to syenitic plutons, to late granitic batholiths.

Superimposed tectonic fabrics and folds in many volcanoplutonic terranes indicate complex structural evolutions linked with the history of associated major fault zones. In many areas, a dominant episode of compressional deformation, involving thrusting, folding, and development of upright penetrative fabrics subparallel to major faults, is followed by transcurrent deformation largely localized along the major faults (Card, 1990; Leitch, 1990). In addition to first-order major faults, these terranes are characterized by abundant higher-order subsidiary shear zones and faults, subparallel to the regional trend, any of which may host auriferous quartz-carbonate veins. Metamorphic grade is greenschist in most volcano-plutonic terranes but reaches lower amphibolite in some districts such as Red Lake, Ontario.

Distribution of quartz-carbonate vein districts and deposits

A large number of quartz-carbonate vein gold districts, especially those in volcano-plutonic terranes, are spatially associated with crustal-scale fault zones, which are generally regarded as the major conduits for auriferous fluids. This association is particularly well illustrated by gold

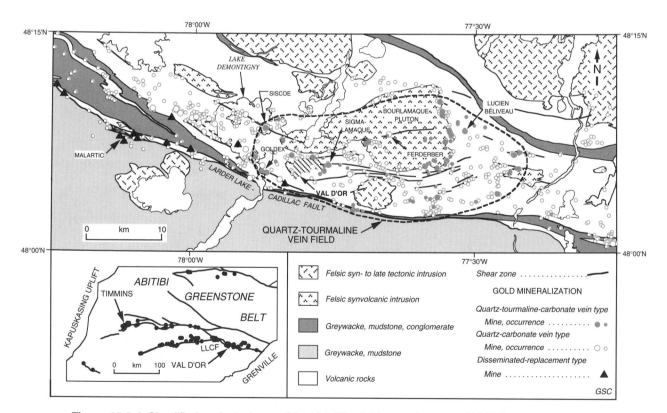

Figure 15.2-4. Simplified geological map of the Val d'Or district, southeastern Abitibi Subprovince, showing the distribution of the major vein gold deposits. In contrast to the widely distributed quartz-carbonate veins, quartz-tourmaline-carbonate veins occur in a well defined field. The inset shows the distribution of gold deposits and major fault zones within the Abitibi Subprovince; LLCF = Larder Lake-Cadillac fault (modified from Robert, 1994).

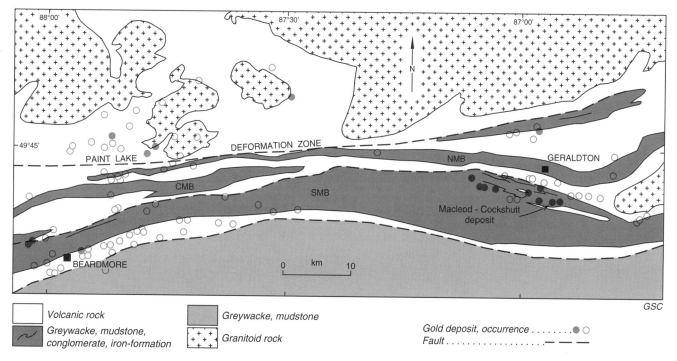

Figure 15.2-5. Simplified geological map of the Beardmore-Geraldton district, Ontario, showing the distribution of quartz-carbonate vein gold deposits; SMB, CMB, and NMB: Southern, Central, and Northern Metasedimentary Belts, respectively. (modified from Ontario Department of Mines, 1966)

deposits of the Abitibi greenstone belt (inset, Fig. 15.2-4). Within districts, however, auriferous veins are in fact more closely associated with smaller subsidiary structures adjacent to major faults, resulting in a dispersion of deposits away from such faults, as in the Val d'Or district (Fig. 15.2-4).

Within volcano-plutonic terranes, quartz-carbonate veins may occur in any rock type present within a district, and deposits typically consist of simple to complex networks of veins and related shear zones. They are most common in parts of the districts that are dominated by mafic volcanic rocks, as in the Red Lake, Yellowknife, and Cassiar districts. Vein deposits also occur in areas dominated by iron-formation-bearing clastic sedimentary belts such as in the Beardmore-Geraldton district (Fig. 15.2-5), and in large felsic plutons as illustrated by the Bourlamaque pluton at Val d'Or (Fig. 15.2-4). These different lithological associations are considered further in a subsequent section.

In clastic sedimentary terranes not adjacent to volcano-plutonic terranes, the distribution of gold districts does not show any recurring pattern and appears to reflect features specific to the host sequence. In the Meguma terrane, for example, gold districts are located at the crests of gently- and doubly-plunging anticlines and occur mostly within the Goldenville Formation (Fig. 15.2-3), whereas the distribution of deposits in the Contwoyto Formation, Northwest Territories, is controlled by that of folded iron-formation (Lhotka and Nesbitt, 1989). Mineralized veins may occur in fold hinges as in the Camlaren deposit in the Slave Province or in the Goldenville deposit (Fig. 15.2-7), or in postfolding veins parallel to fold axial planes as at the MacLeod-Cockshutt deposit (Fig. 15.2-8) or in oblique faults as in the Sheep Creek and Cariboo districts.

Age of host rocks and mineralization

Volcanic and sedimentary host rocks to quartz-carbonate vein gold deposits in Canada range in age from Archean to Jurassic. However, most veins occur in rocks of four main age groups: Late Archean, Early Proterozoic, Cambrian-Ordovician, and Triassic-Jurassic. Of these four groups, rocks of Late Archean age have yielded most of the Canadian gold production from deposits of this type (Table 15.2-2).

In a large number of volcano-plutonic terranes, field and geochronology studies show that the gold-bearing veins formed relatively late in the local structural evolution, after folding of supracrustal rocks and emplacement of the syn- to late tectonic intrusions. At Val d'Or, the Sigma-Lamaque vein system (Fig. 15.2-9) cuts a 2685 ± 2 Ma tonalite stock and a swarm of 2694 ± 2 Ma feldspar porphyry dykes that have both intruded 2705 ± 2 Ma volcanic rocks (Wong et al., 1991). Deposits in the Kirkland Lake and Timmins districts, hosted in 2725-2700 Ma volcanic rocks, postdate Timiskaming sedimentation, bracketed between 2680 and 2676 Ma, and the intrusion of 2673 +6/-2 Ma albitite dykes at Hollinger-McIntyre (Corfu, 1993). In the Red Lake district, gold mineralization is bracketed between 2720 and 2700 Ma, corresponding to the last stages of tectonism and plutonism, and is much younger than the volcanism, which lasted from 3000 to 2730 Ma (Corfu and Andrews, 1987). Similar young relative ages are indicated for the Bralorne-Pioneer deposit (Fig. 15.2-6): quartz-carbonate veins are hosted by 270 ± 5 Ma diorite-tonalite and coeval volcanic rocks, but they cut albitite dykes dated at 91.4 ± 1.4 Ma (Leitch, 1990). Thus, in most documented cases, quartz-carbonate veins are significantly younger than the host

355

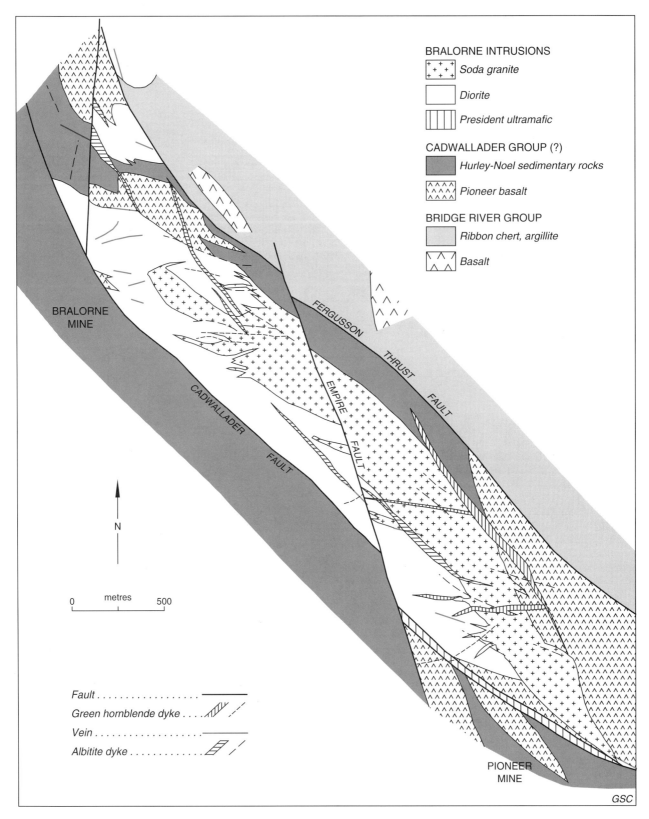

Figure 15.2-6. Simplified geological map of the Bralorne-Pioneer deposit, Bridge River deposit, British Columbia (modified from Leitch, 1990).

volcanic sequences and emplaced more or less synchronously with late magmatic activity within, and adjacent to the greenstone belts during the late Archean.

In clastic sedimentary terranes, two distinct relative ages of vein formation are recognized: (1) prefolding, such as in the sedimentary strata of the Meguma terrane of Nova Scotia (Fig. 15.2-7; Graves and Zentilli, 1982); and (2) postfolding, associated with fractures and faults oblique to fold axial surfaces, such as in the Cariboo and Sheep Creek districts in British Columbia (Matthews, 1953; Sutherland-Brown, 1957).

The absolute ages of quartz-carbonate vein deposits are not well constrained. In southern Abitibi greenstone belt, direct dating of hydrothermal rutile, scheelite, and muscovite by U-Pb, Sm-Nd, and ^{40}Ar-^{39}Ar techniques, respectively, give ages 50-80 Ma younger than any known

Table 15.2-2. Age distribution of host rocks to quartz-carbonate vein deposits and respective gold endowment.

Age	Examples	Contained gold (t)
Archean	Greenstone belts of the Superior and Slave provinces	6000
Proterozoic	Churchill and Grenville provinces	150
Cambrian to Ordovician	Cariboo, B.C.; Meguma terrane, N.S.; Baie Verte, Nfld.	100
Triassic to Jurassic	Cassiar and Bridge River districts, Canadian Cordillera	150

plutonic rock in the area (Corfu, 1993). At Val d'Or, rutile and scheelite ages of ~2600 Ma from quartz-tourmaline-carbonate veins at the Sigma deposit conflict with the 2682 Ma age of a hydrothermal zircon from the same sets of veins (Claoué-Long et al., 1990). The significance of such "young" ages is still unclear.

In the Canadian Cordillera, the age of the Bralorne-Pioneer deposit is bracketed between ~90 and ~85 Ma by premineral albitite dykes and intra- to postmineral hornblende-bearing dykes (Leitch, 1990). The K/Ar ages of vein-related white micas suggest mineralization ages of ~130 Ma in the Cassiar district (Sketchley et al., 1986) and ~140 Ma in the Cariboo district (Andrew et al., 1983). Similar Lower Cretaceous mineralization has also been documented along the Mother Lode gold belt in California (Bohlke and Kistler, 1986).

In some districts, there is growing evidence for the existence of multiple generations of auriferous quartz-carbonate veins. In the Rice Lake district, Brommecker et al. (1989) have documented two generations of gold-bearing quartz-carbonate veins related to two distinct deformation increments. At Val d'Or, late quartz-tourmaline-carbonate veins crosscut dykes and are typically not deformed, whereas earlier quartz-carbonate veins are overprinted by deformation and commonly cut by dykes (Robert, 1994).

Host rock associations

In general, quartz-carbonate veins occur in any rock type present in a given district. However, there are a number of recurring deposit-scale lithological associations which are in part reflected in the geometric and/or hydrothermal characteristics of the deposits. These different lithological

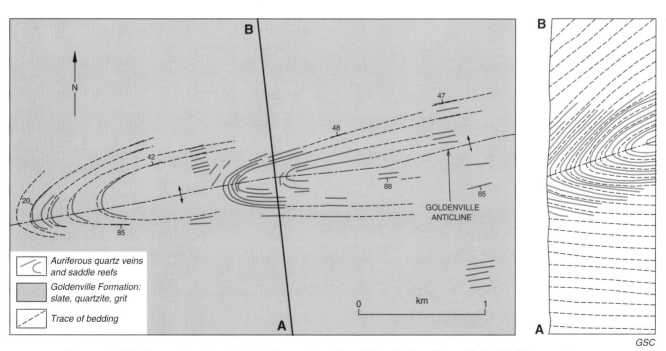

GSC

Figure 15.2-7. Generalized geological plan and section of the Goldenville gold district, Meguma terrane, Nova Scotia (modified from Boyle, 1979).

357

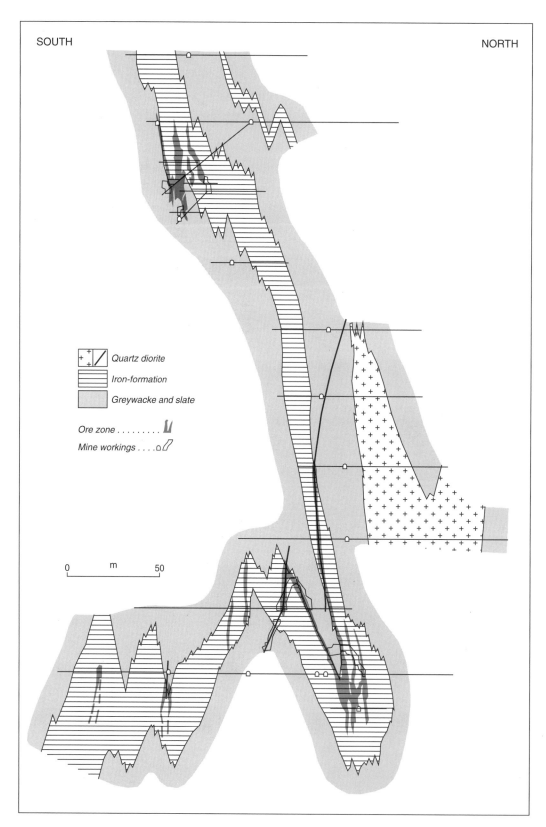

SOUTH

NORTH

Quartz diorite

Iron-formation

Greywacke and slate

Ore zone . . . .

Mine workings . . . . .

0   m   50

Figure 15.2-8. Cross-section through the MacLeod-Cockshutt deposit, Beardmore-Geraldton district (adapted from Horwood and Pye, 1955).

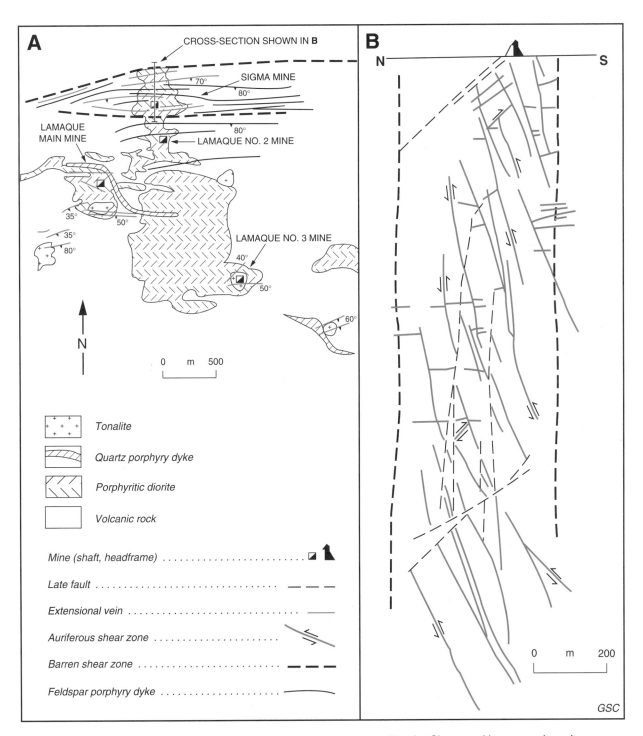

Figure 15.2-9. A) Simplified geological map of the area surrounding the Sigma and Lamaque deposits. **B)** Simplified vertical cross-section through the Sigma mine, showing the configuration of the shear zone and vein network (adapted from Robert and Brown, 1986a).

associations are best regarded as different facies, or styles, of quartz-carbonate vein deposits. They reflect variations in structural and chemical controls exerted by the host lithology on the development of the vein networks.

Volcanic-hosted quartz-carbonate vein deposits are the most common. They occur most commonly in mafic volcanic rocks and associated ultramafic rocks and are represented by the Belleterre, Kerr Addison, Campbell Red Lake, Giant Yellowknife, and Erickson deposits. Characteristics common to this category of deposits include relatively wide, highly schistose host shear zones and wide haloes of carbonate alteration (fuschsite-bearing if hosted in ultra-mafic rocks), reflecting both the ductile and the Fe-Mg-rich nature of the host rocks. Several deposits of this group are centered on intrusive complexes comprising stocks, irregu-lar bodies, and dykes of diorite, tonalite, and syenite, which are commonly porphyritic. This is the case at the Sigma-Lamaque (Fig. 15.2-9A), Macassa, Dome, Hollinger-McIntyre, and Bralorne-Pioneer (Fig. 15.2-6) deposits, which display relatively complex vein and shear zone patterns. Other deposits, represented by the San Antonio and Norbeau mines, occur in laterally extensive differenti-ated tholeiitic gabbro sills. They consist of relatively complex vein networks which are largely confined to the most differentiated, quartz-bearing or granophyric units within the sills. Veins may be confined to such units because of their more competent nature and because their Fe-rich nature is favourable for gold precipitation. Volcanic-hosted deposits include many of the largest Canadian quartz-carbonate vein deposits. Some deposits of this subtype also have the greatest vertical extent, reaching 2 km or more in several mines, including Sigma (Fig. 15.2-9B).

Another group of deposits is *tonalite-hosted* and occurs in large diorite-tonalite and monzonite plutons within volcano-plutonic terranes. Examples include the Ferderber and other deposits in the Bourlamaque pluton at Val d'Or (Fig. 15.2-4), the Silidor and Pierre Beauchemin deposits in the Flavrian pluton at Noranda, and the Star Lake deposit and pluton in the La Ronge belt. The host intrusion may also lie immediately outside greenstone belts, as at Renabie. Deposits of this type are characterized by rela-tively simple geometries and the quartz-carbonate veins and host shear zones are spatially associated with mafic dykes present in these intrusions.

Iron-formation-hosted quartz carbonate veins also form an important group of deposits in both clastic sedimentary sequences and volcano-plutonic terrane, represented by the Central Patricia, MacLeod-Cockshutt (Fig. 15.2-8), and Lupin deposits. Orebodies in such deposits are within zones that contain abundant quartz-carbonate veins and that are generally restricted to the iron-formation layers. The veins in all cases postdate folding of the sedimentary layers and, in a number of cases, they are parallel to the axial planes of the folds (Fig. 15.2-8).

Finally, other deposits are *turbidite-hosted*. In these, veins either occur in fold hinges as at Goldenville (Fig. 15.2-7) and at Camlaren (Boyle, 1979), or in fractures and faults cutting the folds at a moderate to high angle, as in the Cariboo and Sheep Creek districts. These deposits lack obvious spatial relationships to intrusive rocks and are characterized by poorly developed alteration haloes. In some districts, specific sedimentary units are preferen-tially mineralized, such as the Upper Nugget and Upper

Navada quartzites in the Sheep Creek district (Matthews, 1953), or the Rainbow Formation in the Island Mountain deposit (Sutherland-Brown, 1957).

In several districts within volcano-plutonic terranes, there is one particular setting of quartz-carbonate veins which dominates, despite the presence of other rock types. For example, nearly all vein deposits in the La Ronge district occur within granitoid intrusions, whereas those in the Beardmore-Geraldton district are associated with iron-formation (Fig. 15.2-5).

Form and structure

Quartz-carbonate vein gold deposits consist of networks of veins and related host structures. An important charac-teristic of a large number of vein deposits, especially in volcano-plutonic terranes, is their significant vertical extent, which exceeds 1 km in several deposits, and 2 km in a few deposits listed above. The networks display simple to complex geometries involving single to multiple sets of veins and host structures (Hodgson, 1989). They comprise veins in one or more of the following structural settings: (1) in faults and shear zones; (2) in extensional fractures and stockwork zones, including breccias; and (3) in associa-tion with folds. As illustrated by the Sigma-Lamaque deposit at Val d'Or, a large number of networks combine veins in shear zones and in spatially associated extensional fractures (Fig. 15.2-9B). Veins and their different settings are described below. Vein networks in volcanic-hosted deposits commonly display complex geometries, especially those centred in intrusive complexes such as Bralorne-Pioneer (Fig. 15.2-6) and Sigma-Lamaque (Fig. 15.2-9B), whereas those in tonalite-hosted deposits generally consist of a single set of mineralized structures.

Veins in faults and shear zones

Faults and shear zones probably represent the most common host structures to quartz-carbonate veins, and they are a component of almost every gold deposit. Veins hosted by these types of structures occur principally in volcanic-dominated terranes, where they are found in practically every rock type. The nature of the host shear zones ranges from ductile to most commonly brittle-ductile, correlating in part with the metamorphic grade of the host rocks (Colvine, 1989). These shear zones have moderate to steep dips, and can be traced for several hundred metres to a few kilometres along strike and down dip. They are typically high-angle reverse to reverse-oblique shear zones, and less commonly strike-slip.

The mineralized shear zones may occur individually, as parallel sets, or may form anastomosing, conjugate, or more complex arrays (Poulsen and Robert, 1989). These shear zones are generally discordant to the stratigraphic layering but, in a number of cases, they parallel bedding planes or intrusive contacts (such as along dykes), reflecting the influence of strength anisotropy on their development.

Quartz-carbonate veins in shear zones and faults, commonly referred to as shear veins, typically form tabular to lenticular bodies within the central parts of brittle-ductile shear zones, either parallel, or slightly oblique, to the host structure (Hodgson, 1989; Poulsen and Robert, 1989). The veins range in thickness from a few tens of centimetres to a few metres and may reach a few hundred metres in their longest dimension. Mineralized shear veins or portions of veins commonly occur at splays and intersections of shear zones, at bends in the general trend of the host structure, as well as at the intersection of the shear zone with a specific rock type.

Shear veins in shear zones are typically laminated (Fig. 15.2-10A). Laminations are defined by thin septa and slivers of altered and foliated wall rocks, incorporated into the vein by multiple-opening episodes. In several deposits, individual quartz-carbonate laminae are also bounded by striated slip surfaces, in some cases with hydrothermal slickenlines indicating vein development in active shear zones. With increasing proportion and thickness of wall rock slivers, laminated veins may also grade into sheeted veinlet zones.

In a number of deposits, shear veins display some degree of folding and boudinage due to postvein displacement along the host shear zone or to subsequent folding of the entire shear zone (Poulsen and Robert, 1989).

Veins in extensional fractures and stockwork zones

Veins in extensional fractures, or extensional veins, stockwork zones, and hydrothermal breccias occur principally in volcano-plutonic terranes and are present in a significant number of deposits. They are not as common as shear veins and represent a major source of ore in only a small proportion of deposits.

Extensional veins may form arrays of planar to sigmoidal veins within shear zones or at frontal and lateral terminations of shear veins (Robert, 1994), or form sets of regular tabular bodies (Fig. 15.2-10B) extending outside shear zones in less deformed rocks, such as the sub-horizontal extensional veins of the Sigma-Lamaque deposit (Fig. 15.2-9B). They also occur as sets of en echelon veins in relatively competent host lithologies such as small intrusions of intermediate to felsic composition. In most cases, extensional veins are spatially associated with shear veins and they have relatively shallow dips, which are consistent with the reverse to reverse-oblique movements along the associated shear zone.

Extensional veins within shear zones and stockwork zones are typically a few centimetres thick and a few metres long, whereas those outside shear zones are commonly several tens of centimetres thick and a few hundred metres in their longest dimensions. At the Sigma-Lamaque deposit, subhorizontal extensional veins, less than one metre thick, commonly occupy areas as great as 5000 m^2 in extent (Robert and Brown, 1986a). The internal structure of extensional veins contrasts with that of shear veins and is commonly characterized by mineral fibres at high angles to vein walls (Fig. 15.2.-10B), as well as by crack-seal and open-space filling textures.

Stockwork zones are important in a number of deposits; at San Antonio in the Rice Lake district, for example, they constituted a large proportion of the ore mined. Stockworks consist of several sets of extensional veins (Fig. 15.2-10C), which can grade into hydrothermal breccias in areas of intense veining. They are preferentially developed in competent lithologies, such as the granophyric facies of the differentiated gabbro sill hosting the San Antonio deposit.

Other types of hydrothermal breccias also occur along shear veins: they include "jigsaw-puzzle" breccias, characterized by angular fragments of altered wall rock in a fine grained matrix of quartz and/or tourmaline, and by fault breccias, composed of crushed and rotated vein and wall rock fragments in a dominantly hydrothermal matrix.

Veins associated with folds

Veins associated with folds probably represent the least common structural setting of quartz-carbonate veins. Veins in such settings occur almost exclusively in folded clastic sedimentary rocks, in either volcano-plutonic or clastic sedimentary terranes.

Quartz-carbonate veins are associated with folds ranging from those of regional scale, as in the Meguma terrane (Fig. 15.2-3), to deposit-scale asymmetric folds, as in the MacLeod-Cockshutt deposit (Fig. 15.2-8). Veins display diverse geometric and age relationships to the folds. They may be folded along with their host rocks, as in the case of bedding-parallel veins in the Meguma terrane (Fig. 15.2-10D), which occur in anticlinal hinge areas where they are typically stacked and saddle-shaped (Fig. 15.2-7). Veins may also be syn- to late folding and be either parallel to axial plane cleavage in hinge zones, as at MacLeod-Cockshutt (Fig. 15.2-8), or in extensional veins perpendicular to fold axes (AC joints), as is the case in the Cariboo district (Sutherland-Brown, 1957). In other cases, laminated quartz veins occur in fractures and faults cutting obliquely across fold axial surfaces as at the Lupin deposit (Lhotka and Nesbitt, 1989) and in the Sheep Creek district (Matthews, 1953).

Ore and gangue mineralogy

Ore mineralogy

In most quartz-carbonate vein deposits, as at Sigma-Lamaque, gold mineralization occurs in both the veins and the adjacent altered wall rocks, in varying proportions. The bulk of the gold occurs within the veins in turbidite-hosted deposits but within altered wall rocks in iron-formation-hosted deposits. In most cases, gold is intimately associated with sulphide minerals, both in the veins and altered wall rocks. The dominant sulphide mineral is pyrite, or arsenopyrite in sediment-hosted deposits, commonly accompanied by variable, but minor amounts of sphalerite, chalcopyrite, pyrrhotite, and galena. Trace amounts of molybdenite are also present in a number of deposits. The sulphide contents of the veins rarely exceed 5 volume per cent; within laminated veins, sulphide minerals are commonly distributed along thin, altered wall rock slivers, which thus indirectly control the distribution of gold within the veins.

The main ore mineral in most deposits is native gold, which typically contains some silver. Gold-to-silver ratios of the ore range from 5:1 to more than 9:1, and cluster around a ratio of ~9:1, distinct from that of most epithermal veins (see Introduction, "Lode gold"). Gold typically occurs as coatings on, or as inclusions and fracture-fillings within, sulphide grains, as well as isolated grains and fracture fillings in quartz. Other significant ore minerals in quartz-carbonate veins are tellurides, mostly petzite and

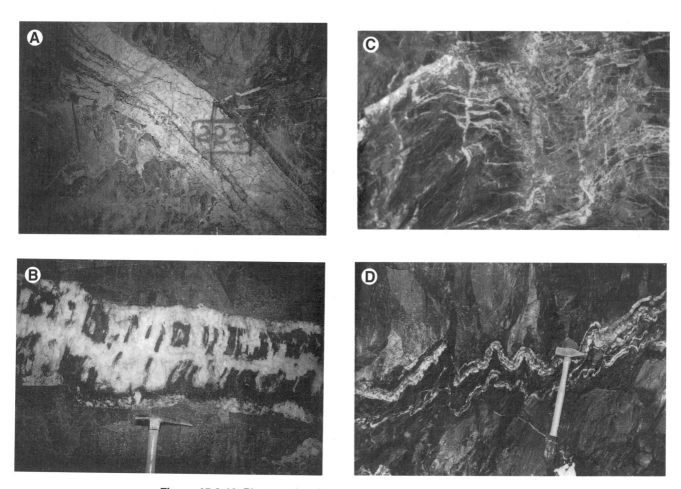

Figure 15.2-10. Photographs of typical quartz-carbonate vein features.

A) Shear zone-hosted laminated shear vein; Lucien Béliveau deposit, Val d'Or district. The fine dark laminae within the vein, marked by tourmaline, are slip surfaces. Field of view ~5 m. Photo by F. Robert. GSC 1994-398

B) Subhorizontal quartz (white)-tourmaline (black) extensional vein; Sigma mine, Val d'Or, Quebec. Subvertical tourmaline fibres are cut by a quartz-rich ribbon in the center of the vein, indicating repeated opening episodes. Hammer for scale. Photo by F. Robert. GSC 1994-399

C) Stockwork from the San Antonio mine, Bissett, Manitoba, consisting of shallowly dipping sigmoidal veins and subvertical veins. Due to increasing vein abundance towards its core, the stockwork grades into a hydrothermal breccia. Field of view ~2 m. Photo by K.H. Poulsen. GSC 1995-024

D) Folded veins in slate at their contact with overlying greywacke on the crest of an anticline, Tangier district, Meguma terrane. Hammer for scale. Photo by A.L. Sangster. GSC 204390-S

calaverite, which are particularly abundant in deposits associated with felsic stocks such as Macassa (Thompson et al., 1950) and Sigma-Lamaque (Robert and Brown, 1986b).

Gangue mineralogy

The most common gangue minerals in the vein deposits considered here are quartz and carbonate. Quartz typically accounts for more than 85% of the vein fillings. Carbonates, including calcite, dolomite, or ankerite in various combinations, typically comprise less than 10-15% of the vein fillings. Veins at the Campbell Red Lake deposit, which are dominated by dolomite and ferro-dolomite, represent a notable exception (Andrews et al., 1986). Other generally minor constituents of the veins include albite, chlorite, and white mica. Tourmaline and scheelite are also present in minor amounts in many quartz-carbonate veins. Tourmaline is particularly abundant in veins in the Val d'Or district, where it may represent up to 15-20 volume per cent of the vein fillings (Robert and Brown, 1986b).

Host rock composition exerts some influence on the accessory gangue mineralogy of the veins. Arsenopyrite rather than pyrite is the dominant vein and altered wall rock sulphide mineral in deposits hosted by sedimentary rocks, such as Lupin and those of the Meguma terrane. The composition of carbonate minerals in the veins also reflects that of the host lithology: the Fe and Mg contents of Ca-carbonates increase proportionally with the Fe and Mg contents of the host rocks. Fuschite normally occurs in veins which are in the vicinity of altered ultramafic rocks.

Quartz-carbonate vein deposits typically lack vertical mineralogical zoning, despite their significant vertical extent. A notable exception is the Sigma-Lamaque deposit, in which the tourmaline-pyrite assemblage gives way in some veins to a pyrrhotite-chlorite-biotite assemblage at depths in excess of 1.6 km (Robert and Brown, 1986b). In general, pyrite is the dominant sulphide mineral in deposits hosted by greenschist grade rocks, whereas pyrrhotite dominates in deposits hosted by amphibolite grade rocks (Colvine, 1989).

Hydrothermal alteration

Wall rock hydrothermal alteration around auriferous quartz-carbonate veins varies in scale, intensity, and mineralogy as a function of host rock composition. Several fundamental types of alteration can be distinguished and these generally combine to form zoned alteration haloes at the vein or the deposit scales. In most documented cases, alteration assemblages have been superimposed on previously metamorphosed rocks, as is the case at Bralorne-Pioneer (Leitch, 1990) and at Sigma-Lamaque (Robert and Brown, 1986b). Two documented exceptions include the Campbell Red Lake and adjoining A.H. White (Dickenson) deposits, where wall rock alteration either predated or was synchronous with amphibolite grade metamorphism (Andrews et al., 1986), and the Eastmain River deposit in northern Quebec, where wall rock alteration is interpreted to have taken place during amphibolite grade metamorphism (Couture and Guha, 1990).

Alteration types

The main types of alteration around quartz-carbonate veins include carbonatization, sulphidation, alkali metasomatism, chloritization, and silicification (Boyle, 1979). Carbonatization is the most common and most extensive type of alteration. Zones of carbonate alteration around individual veins and structures commonly coalesce to envelope the entire orebody. This type of alteration involves progressive replacement of Ca-, Fe-, and Mg-silicates by carbonate minerals and is characterized by additions of CO_2, accompanied by release of Al and Si, fixed in other alteration minerals or in veins. The amounts of introduced carbonates depend, in part, on the amount of Ca, Fe, and Mg present in the host lithology.

Sulphidation of wall rocks is common around veins and, in most cases, is restricted to their immediate proximity. Pyrite is the most common sulphide, followed by pyrrhotite, mostly present in amphibolite grade rocks. Arsenopyrite is also common around veins hosted by clastic sedimentary rocks. Sulphides generally comprise less than 10% of the altered rocks, except in oxide facies iron-formation, in which they make up as much as 75% of the altered rocks, as at McLeod-Cockshutt (Horwood and Pye, 1955).

Sodium and potassium metasomatism is observed in proximity to most quartz-carbonate veins. Potassium metasomatism is the most common and typically consists of sericitization of chlorite and plagioclase; fuchsite, rather than sericite, is generally present in altered ultramafic rocks, and K-feldspar and biotite are alteration products in a few deposits. Sodium metasomatism results largely in the formation of albite, and in some cases of paragonite. Chloritization of amphibole, biotite, and pyroxene (at constant Fe and Mg), commonly accompanies incipient carbonatization. In some deposits, intense chloritization may be accompanied by addition of Fe and Mg to the rock. A distinction should be made between hydrothermal chlorite considered here and chlorite produced by metamorphism of the host rocks. Silicification, sensu stricto, i.e. the addition of silica, has been documented mostly in clastic sedimentary rocks (Boyle, 1979). A more common form of silicification in mafic and ultramafic host rocks, due to silica release from carbonatization reactions, is a local increase in the abundance of quartz, either as quartz-flooding of the rock matrix or as abundant quartz veinlets.

Gold is commonly enriched in intensely altered rocks adjacent to quartz-carbonate veins. In many cases, as at Sigma, these altered zones reach economic grades (Robert and Brown, 1986b). In fact, a significant proportion of the extracted gold in several deposits is derived from altered rocks adjacent to veins.

Alteration zoning patterns

The above different types of alteration commonly combine to form zoned alteration envelopes around veins or deposits (Roberts, 1987). The resulting zoning patterns, summarized in Table 15.2-3, result largely from progressive carbonatization of wall rocks and accompanying alkali metasomatism.

Table 15.2-3. Idealized alteration zoning patterns (from least to most altered assemblages) around quartz-carbonate veins and deposits in igneous host rocks of different compositions. Note that not all zones are necessarily present around a given vein. Adapted from Roberts (1987).

rock composition	alteration zone	serpentine	talc	amphibole	epidote	chlorite	albite	quartz	sericite	calcite	dolomite	pyrite
ultramafic	unaltered[1]	X	X	X		X						
	chloritic[1]		X			X	X	X			X	
	chloritic[2]					X	X	X			X	
	carbonate[1]						X	X	X		X[2]	X
	carbonate[2]						X	X			X[2]	X
mafic	unaltered			X	X	X	X	X				
	chloritic					X	X	X		X		
	carbonate[1]						X	X	X		X[3]	X
	carbonate[2]						X	X			X[3]	X
Intermediate	unaltered			X	X	X	X	X	X			
	chloritic					X	X	X	X	X		
	carbonate[1]						X	X	X		X[3]	X
	carbonate[2]						X	X			X[3]	X

[1] Mineral assemblages of unaltered rocks are taken here as the most commonly observed greenschist assemblages.
[2] Siderite and magnesite may also be present.
[3] Ferroan dolomite and ankerite are the dominant carbonate minerals in most cases.

In igneous wall rocks of ultramafic to intermediate composition, outer alteration zones are characterized by replacement of metamorphic amphibole, epidote, and/or serpentine by calcite±dolomite and chlorite; those minerals are accompanied by talc±tremolite in ultramafic rocks and albite in mafic to intermediate rocks (Table 15.2-3). With increasing intensity of alteration and proximity to veins, chlorite-calcite assemblages are replaced by dolomite-white mica assemblages with or without pyrite. Inner alteration assemblages consist of ankerite-albite-pyrite assemblages; magnesite and siderite are also present in Mg- and Fe-rich igneous host rocks. In general, the iron content of carbonate minerals increases towards the mineralized zones.

Veins in clastic sedimentary rocks typically lack well defined alteration envelopes. Where present, they tend to be narrow and are characterized by replacement of chlorite and biotite by carbonates, white mica, and albite, and by formation of arsenopyrite. Where veins intersect iron-formation, the alteration is typically controlled by bedding and laminations: for example, layers of magnetite are selectively altered and replaced by sulphides, generally pyrite, over distances as great as several decimetres on either side of a vein.

DEFINITIVE CHARACTERISTICS

Quartz-carbonate vein gold deposits consist of simple to complex vein and shear zone networks with significant vertical extents, hosted by rocks in deformed volcano-plutonic terranes, and less commonly in deformed clastic sedimentary terranes. The deposits occur in districts spatially associated with large-scale fault zones. The veins occupy shear zones, faults, stockwork zones, and extensional fractures, or are associated with folds: they are generally discordant, at least in part, to lithological units. The veins are composed mainly of quartz, with less abundant carbonate and pyrite. Commonly associated minerals include tourmaline, scheelite, fuchsite, and arsenopyrite. Hydrothermal alteration of wall rocks is dominated by carbonatization, and accompanied by alkali metasomatism and sulphidation of the rocks immediately adjacent to the veins.

GENETIC MODELS

In contrast to many other deposit types, there is no real consensus on the origin of quartz-carbonate veins in deformed terranes and, as a result, a number of genetic models have been proposed for their formation (Roberts, 1987; Kerrich, 1989). Studies of fluid inclusions and hydrothermal alteration in several deposits point to a relatively uniform fluid composition and temperature, irrespective of their occurrence in volcano-plutonic or clastic sedimentary terranes (Kerrich and Wyman, 1990). The auriferous fluids are typically CO_2-bearing (5-15 mol % $CO_2 \pm CH_4$), low-salinity fluids, at 300-350°C, which underwent phase separation in a number of deposits. Differences between districts in the Sr, Pb, C, and O isotope compositions of the auriferous fluids contrast with the relatively uniform bulk fluid composition and indicate multiple source regions for these fluid components, including sources external to, and underneath, the host supracrustal sequences (Kerrich, 1989). However, such isotopic tracers do not allow unequivocal discrimination of the nature and origin of the fluids.

Among all the genetic models proposed for quartz-carbonate veins, the orthomagmatic model has historically been the most commonly advocated (e.g., Emmons, 1937). According to this model, gold and the hydrothermal fluids are derived from ascending felsic magmas generated during tectonism and metamorphism. A variation on this model involves derivation of the gold from the host supracrustal sequences by their interaction with the magma and associated hydrothermal fluids.

In the last two decades, a number of fluid-source models, based largely on fluid inclusion and isotopic tracer studies, have also been proposed and reviewed by Roberts (1987), Kerrich (1989), and others. In the metamorphic model, gold is considered to be leached from the underlying supracrustal rocks by a metamorphic fluid released during prograde metamorphism and focused into shear zones and related dilational zones. A variation on this model has been suggested by Graves and Zentilli (1982) for the origin of the folded veins of the Meguma terrane by which pore fluids, released by greenschist metamorphism during incipient folding and cleavage development, induced hydraulic fracturing and transported locally-derived gold and other vein constituents into these fractures. Nesbitt and Muehlenbachs (1989) developed a model involving deep circulation of meteoric waters in the vicinity of major fault zones for quartz-carbonate vein deposits of the Canadian Cordillera. In the mantle degassing/granulitization model, upward streaming of mantle-derived CO_2 is thought to induce dehydration and granulitization of the lower crust, possibly accompanied by magma generation; the resulting H_2O-CO_2 fluids, leaching gold from the lower crust, rise to higher crustal levels along major shear zones, where gold and other components are deposited.

In light of the recent recognition that many quartz-carbonate vein gold districts occur at transpressive accretionary plate margins, many authors relate the formation of these deposits to accretionary processes (e.g. Kerrich and Wyman, 1990). In this model, fluids are generated by thermal re-equilibration and metamorphism of subducted material following cessation of subduction. Such deep fluids, which may dissolve gold and other vein components anywhere along their path, are thought to be channelled upwards along crustal-scale faults.

RELATED DEPOSIT TYPES

A number of gold deposits that are primarily of quartz-carbonate vein type, contain orebodies typical of the disseminated-replacement subtype of gold deposits (see subtype 15.4), which suggests a possible genetic link between the two subtypes. In the Cariboo district, for example, both quartz-carbonate veins and pyrite replacement (manto) orebodies in limestone were mined (Sutherland-Brown, 1957); the Campbell Red Lake-Dickenson deposit, apart from more abundant quartz-carbonate vein orebodies, also includes sulphidic orebodies of the East South "C" type (Andrews et al., 1986). In the Cariboo district, quartz-carbonate veins clearly overprint pre-existing pyrite replacement orebodies (Robert and Taylor, 1990) and the two styles of ore are not related to the same hydrothermal event. However, in most hybrid gold deposits, the temporal and possible genetic relationships between different styles of orebodies are not clearly established.

A similar problem exists for iron-formation-hosted gold deposits of the stratiform type (see subtype 15.3): the relationships are not clearly established between finely disseminated gold in cherty sulphide-banded iron-formation and quartz-carbonate veins, with which at least some gold is spatially associated. In contrast, iron-formation-hosted gold deposits of the nonstratiform type simply represent a subset of the quartz-carbonate vein deposits considered here.

EXPLORATION GUIDELINES

Because quartz-carbonate vein gold deposits are typically sulphide-poor deposits, geophysical methods commonly fail to reveal their presence. As a result, exploration must be based heavily on geological criteria, as reviewed by Hodgson et al. (1982).

At the regional scale, portions of volcano-plutonic terranes containing significant volumes of mafic volcanic rocks and a major fault zone, especially along terrane boundaries, should be considered favourable. At the district- and mining property-scale, exploration should focus on shear zones and faults subsidiary to, and distributed around, major fault zones. Emphasis should be placed on segments of shear zones intersecting or following favourable host rocks such as small felsic intrusions and dykes, iron-formations, and iron-rich igneous rocks. Favourable segments of shear zones could also be selected on the basis of splays or deflections of the overall trend of the shear zone and, in mafic to ultramafic lithologies, on the basis of mapping the distribution of the different carbonate minerals along carbonatized shear zones and units, using simple mineral staining techniques.

Geophysical methods can be used directly, for example to identify shear zones and faults, or indirectly for selection of favorable target areas. For example, the abrupt loss of magnetic signature along a magnetic unit, such as iron-formation, serpentinized ultramafic rock, or iron-rich gabbro, may indicate the presence of a zone of carbonatization or sulphidation related to gold mineralization.

In glaciated areas such as the Canadian Shield, heavy mineral concentrates in basal till, as well as surficial till geochemistry, can be used to outline mineralized areas along major shear zones (DiLabio, 1982).

ACKNOWLEDGMENTS

Thanks are extended to Howard Poulsen for enlightening discussions on lode gold deposits, and to Howard Poulsen and Roger Eckstrand for their critical reviews of this document.

REFERENCES

References with asterisks (*) are considered to be the best source of general information on this deposit subtype.

Andrew, A., Godwin, C.I., and Sinclair, A.J.
1983: Age and genesis of Cariboo gold mineralization determined by isotopic methods; British Columbia Ministry of Energy, Mines and Petroleum Resources, Geological Fieldwork, Paper 1983-1, p. 305-313.

*Andrews, A.J., Hugon, H., Durocher, M., Corfu, F., and Lavigne, M.J.
1986: The anatomy of a gold-bearing greenstone belt: Red Lake, northwestern Ontario, Canada; in Proceedings of Gold '86, an International Symposium on the Geology of Gold Deposits, (ed.) A.J. Macdonald; Toronto, p. 3-22.

Bohlke, J.K. and Kistler, R.W.
1986: Rb-Sr, K-Ar and stable isotope evidence for the ages and sources of fluid components of gold-bearing quartz veins in the Northern Sierra Nevada Foothills metamorphic belt, California; Economic Geology, v. 81, p. 296-322.

*Boyle, R.W.
1979: The geochemistry of gold and its deposits; Geological Survey of Canada, Bulletin 280, 584 p.

Brommecker, R., Poulsen, K.H., and Hodgson, C.J.
1989: Preliminary report on the structural setting of gold at the Gunnar mine in the Beresford Lake area, Uchi Subprovince, southeastern Manitoba; in Current Research, Part C; Geological Survey of Canada, Paper 89-1C, p. 325-332.

Card, K.D.
1990: A review of the Superior Province of the Canadian Shield, a product of Archean accretion; Precambrian Research, v. 48, p. 99-156.

Claoué-Long, J.C., King, R.W., and Kerrich, R.
1990: Archaean hydrothermal zircon in the Abitibi greenstone belt: constraints on the timing of gold mineralization; Earth and Planetary Science Letters, v. 98, p. 109-128.

*Colvine, A.C.
1989: An empirical model for the formation of Archean gold deposits: products of final cratonization of the Superior Province, Canada; in The Geology of Gold Deposits: the Perspective in 1988, (ed.) R.R. Keays, W.R.H. Ramsay, and D.I. Groves; Economic Geology, Monograph 6, p. 37-53.

Corfu, F.
1993: The evolution of the southern Abitibi greenstone belt in light of precise U-Pb geochronology; Economic Geology, v. 88, p. 1323-1340.

Corfu, F. and Andrews, A.J.
1987: Geochronological constraints on the timing of magmatism, deformation and gold mineralization in the Red Lake greenstone belt, northwestern Ontario; Canadian Journal of Earth Sciences, v. 24, p. 1302-1320.

Couture, J.-F. and Guha, J.
1990: Relative timing of emplacement of an Archean lode-gold deposit in an amphibolite terrane: the Eastmain River deposit, northern Quebec; Canadian Journal of Earth Sciences, v. 27, p. 1621-1636.

DiLabio, R.W.
1982: Drift prospecting near gold occurrences at Onaman River, Ontario and Oldham, Nova Scotia; in Geology of Canadian Gold Deposits, (ed.) W. Petruk and R.W. Hodder; The Canadian Institute of Mining and Metallurgy, Special Volume 24, p. 261-266.

*Emmons, W.H.
1937: Gold Deposits of the World; McGraw-Hill, New York, 562 p.

Graves, M.C. and Zentilli, M.
1982: A review of the geology of gold in Nova Scotia; in Geology of Canadian Gold Deposits, (ed.) R.W. Hodder and W. Petruk; The Canadian Institute of Mining and Metallergy, Special Volume 24, p. 233-242.

*Hodgson, C.J.
1989: The structure of shear-related vein-type gold deposits: a review; Ore Geology Reviews, v. 4, p. 231-273.

Hodgson, C.J., Chapman, R.S.G., and MacGeehan, P.J.
1982: Application of exploration criteria for gold deposits in Superior Province of the Canadian Shield to gold exploration in the northern Cordillera; in Precious Metals in the Northern Cordillera, (ed.) A.A. Levinson; The Association of Exploration Geochemists, Special Publication Number 10, p. 173-207.

Horwood, H.C. and Pye, E.G.
1955: Geology of Ashmore Township; Ontario Department of Mines Annual Report, v. LX, pt. V, 1951, 105 p.

*Kerrich, R.
1989: Geochemical evidence on the sources of fluids and solutes for shear zone hosted mesothermal Au deposits; in Mineralization and Shear Zones, (ed.) J.T. Bursnall; Geological Association of Canada, Short Course Notes, v. 6, p. 129-197.

Kerrich, R. and Wyman, D.
1990: Geodynamic setting of mesothermal gold deposits: as association with accretionary tectonic regimes; Geology, v. 18, p. 882-885.

Leitch, C.H.B.
1990: Bralorne: a mesothermal, Shield-type vein gold deposit of Cretaceous age in southern British Columbia; The Canadian Institute of Mining and Metallurgy Bulletin, v. 83, p. 53-80.

Lhotka, P.G. and Nesbitt, B.E.
1989: Geology of unmineralized and gold-bearing iron-formation, Contwoyto Lake-Point Lake region, Northwest Territories, Canada; Canadian Journal of Earth Sciences, v. 26, p. 46-64.

*Lindgren, W.
1933: Mineral Deposits; McGraw-Hill Book Co., New York, 930 p.

Matthews, W.H.
1953: Geology of the Sheep Creek camp; British Columbia Department of Mines, Bulletin, no. 31, 94 p.

McMullen, J.M., Richardson, G.G., and Goodwin, T.A.
1987: Preliminary gold compilation maps of the Meguma Terrane; Nova Scotia Department of Mines and Energy, Open File maps 86-049 to 86-056, scale 1:100 000.

Nesbitt, B.E. and Muehlenbachs, K.
1989: Geology, geochemistry, and genesis of mesothermal lode gold deposits of the Canadian Cordillera: evidence for ore formation from evolved meteoric water; in The Geology of Gold Deposits: the Perspective in 1988, (ed.) R.R. Keays, W.R.H. Ramsay, and D.I. Groves; Economic Geology, Monograph 6, p. 553-563.

Ontario Department of Mines
1966: Tashota-Geraldton sheet; Ontario Department of Mines, Map 2102, scale 1:253 440.

Poulsen, K.H. and Robert, F.
1989: Shear zones and gold: practical examples from southern Canadian Shield; in Mineralization and Shear Zones, (ed.) J.T. Bursnall; Geological Association of Canada, Short Course Notes, v. 6, p. 239-266.

Robert, F.
1994: Vein fields in gold districts: the example of Val d'Or, southeastern Abitibi; in Current Research 1994-C, Geological Survey of Canada, p. 295-302.

Robert, F. and Brown, A.C.
1986a: Archean gold-bearing quartz veins at the Sigma mine, Abitibi greenstone belt, Quebec: Part I. Geologic relations and formation of the vein system; Economic Geology, v. 81, p. 578-592.
1986b: Archean gold-bearing quartz veins at the Sigma mine, Abitibi greenstone belt, Quebec: Part II. Vein paragenesis and hydrothermal alteration; Economic Geology, v. 81, p. 593-616.

Robert, F. and Taylor, B.E.
1990: Structural evolution and gold remobilization at the Mosquito Creek Gold Mine, Cariboo district, British Columbia; Geological Association of Canada-Mineralogical Association of Canada, Joint Annual Meeting, Vancouver, 1990, Program with Abstracts, v. 15, p. A112.

*Roberts, R.G.
1987: Ore deposit models #11. Archean lode gold deposits; Geoscience Canada, v. 14, p. 37-52.

Sketchley, D.A., Sinclair, A.J., and Godwin, C.I.
1986: Early Cretaceous gold-silver mineralization in the Sylvester allochthon, near Cassiar, north-central British Columbia; Canadian Journal of Earth Sciences, v. 23, p. 1455-1458.

Sutherland-Brown, A.
1957: Geology of the Antler Creek area, Cariboo district, British Columbia; British Columbia Department of Mines, Bulletin no. 38, 105 p.

Thompson, J.E., Charlewood, G.H., Griffin, K., Hawley, J.E., Hopkins, H., MacIntosh, C.G., Orgizio, S.P., Perry, O.S., and Ward, W.
1950: Geology of the main ore zone at Kirkland Lake; Ontario Department of Mines, v. 57, pt. 5, p. 54-196.

Wong, L., Davis, D.W., Krogh, T.E., and Robert, F.
1991: U-Pb zircon and rutile chronology of Archean greenstone formation and gold mineralization in the Val d'Or region, Quebec; Earth and Planetary Science Letters, v. 104, p. 325-336.

15.3 IRON-FORMATION-HOSTED STRATABOUND GOLD

J.A. Kerswill

INTRODUCTION

Gold deposits hosted by iron-formation are characterized by: (1) a close association between native gold and iron sulphide minerals; (2) the presence of gold-bearing quartz veins and/or shear zones; (3) structural complexity of the host terranes; and (4) paucity of lead and zinc in the ores. Two principal varieties of iron-formation-hosted gold deposits can be defined, based on the dominant style of gold distribution (Kerswill, 1986, 1993): stratiform and non-stratiform (or vein type). Some deposits have characteristics of both varieties and thus have a hybrid character.

In the vein-type deposits, gold hosted by iron-formation is restricted to late structures (quartz veins and/or shear zones) and/or iron sulphide-rich zones adjacent to such structures. Ore is confined to discrete, commonly small shoots separated by barren (gold- and sulphide-poor) iron-formation, typically of oxide facies. These nonstratiform ores are essentially a variety of the mesothermal quartz-carbonate vein deposits that are described elsewhere (see subtype 15.2).

Deposits of the stratiform-type can be subdivided into those occurring within sediment-dominated settings and those within mixed volcanic-sedimentary settings. In the former, gold is uniformly disseminated in thin, but laterally extensive units of cherty pyrrhotite-rich iron-formation that are conformably interlayered with sulphide- and oxide-poor iron-formation and pelitic sedimentary rocks in portions of turbidite basins relatively distant from felsic volcanic centers. In the deposits within mixed settings, gold is uniformly disseminated in thin, but laterally extensive units of cherty sulphide-iron-formation that are associated with carbonate-iron-formation and black carbonaceous shale relatively close to volcanic centres.

Gold is the principal commodity in all deposits and occurs in the free native form. Silver is recovered with the gold from all deposits, but the gold:silver ratio is variable (see below).

Examples of vein-type deposits include: the North ore zone of the Hard Rock and MacLeod-Cockshutt properties in the Geraldton camp, Ontario (Horwood and Pye, 1951; Macdonald and Fyon, 1986); the Central Patricia mine and portions of the Pickle Crow mine near Pickle Lake, Ontario (Thomson, 1938); the Cullaton Lake B-zone in the Northwest Territories (Page, 1981; Sethu Raman et al., 1986; Miller, 1992); a number of deposits in Western Australia, including the Hill 50, Nevoria, and Water Tank Hill mines (Phillips et al., 1984); several deposits in Zimbabwe, including the Lennox mine (Foster, 1989); and probably the São Bento mine in Brazil (Mosley and Hofmeyr, 1986). Recently discovered iron-formation-hosted gold deposits in the George Lake (Olson, 1989; Chandler and Holmberg, 1990; Padgham, 1990) and Meliadine (Miller et al., 1993) areas of the Northwest Territories also appear to be vein type.

Important examples of the sediment-hosted stratiform-type deposits include: the Lupin mine, Northwest Territories (Gardiner, 1986; Kerswill, 1986; Lhotka, 1988; Bullis, 1990), the Jardine deposit, Montana, U.S.A. (Hallager, 1984; Cuthill et al., 1990), and probably the Homestake mine, South Dakota, U.S.A. (Nelson, 1986; Caddey et al., 1991). Iron-formation-hosted gold mineralization in the Russell Lake area (Bunner, 1988), Northwest Territories, also appears to be of this type.

Examples of stratiform deposits in mixed volcanic-sedimentary settings include the Morro Velho (Ladeira, 1980; Vieira et al., 1991b) and Cuiabá (Vial, 1988; Vieira et al., 1991a) mines in Brazil, and the Agnico-Eagle mine in Quebec (Barnett et al., 1982; Wyman et al., 1986; Dubé et al., 1991).

Several Canadian deposits have characteristics of both the vein and stratiform subtypes. These include the Wedge Lake deposit in the La Ronge Domain, Saskatchewan (Netolitsky, 1986) and the Musselwhite (Hall and Rigg, 1986) and Dona Lake deposits in Ontario (Cohoon, 1986). However, these appear to be dominantly vein type.

IMPORTANCE

Combined total world-wide production and reserves for all iron-formation-hosted gold deposits exceed 3000 t. Much of the production has come from a few world-class deposits.

Nine significant Canadian deposits (Table 15.3-1) within this class account for about 220 t of contained gold (production plus reserves) or about 5% of the lode gold category. Deposits of the vein-type have not been large producers in Canada. Central Patricia produced more than 19 t gold and the North ore zone at Geraldton produced more than 15 t gold. However, at the global scale (Fig. 15.3-1), Hill 50 and São Bento can be considered major deposits.

Kerswill, J.A.
1996: Iron-formation-hosted stratabound gold; in Geology of Canadian Mineral Deposit Types, (ed.) O.R. Eckstrand, W.D. Sinclair, and R.I. Thorpe; Geological Survey of Canada, Geology of Canada, no. 8, p. 367-382 (also Geological Society of America, The Geology of North America, v. P-1).

[1] *Editorial note:* Within this volume, in a companion account of quartz-carbonate vein gold deposits (subtype 15.2), Lupin is presented as an example of a quartz-carbonate vein gold deposit and has been interpreted as epigenetic.

Table 15.3-1. Grades and tonnages of selected Canadian iron-formation-hosted gold deposits.

Deposit	Type	Tonnage (t)	Grade (g/t)	Contained gold (t)	Au:Ag
Central Patricia	Vein	1 568 780	12.34	19.30	10.7
Cullaton B-zone	Vein	321 870	17.14	5.50	14.1
North ore zone, Geraldton	Vein	3 175 200	5.14	16.20	29.9
Pickle Crow	Vein	199 580	17.14	3.40	8.6
Agnico-Eagle (Joutel)	Stratiform	4 031 780	6.51	25.50	5.0
Lupin	Stratiform	9 080 000	10.75	97.58	6.2
Dona Lake	Hybrid	1 179 360	9.26	10.90	??
Musselwhite	Hybrid	4 200 000	9.60	40.32	9.1
Wedge Lake	Hybrid	544 320	6.17	3.40	??
?? = unknown					

The Lupin mine, a stratiform deposit, is one of Canada's largest gold producers and has an average annual production of about 6 t. The Homestake mine is one of the largest gold producers in the world, and its total gold production since 1876 has been in excess of 1100 t gold. Indeed, production at the Homestake mine in 1993 was 13 t, (447 600 ounces), its highest annual output since 1971. Morro Velho is the largest lode deposit in South America and has produced more than 310 t gold since 1834.

SIZE AND GRADE OF DEPOSITS

Tonnage and grade figures for nine Canadian deposits are presented in Table 15.3-1 and selected deposits are plotted in Figure 15.3-1. It is noteworthy that Lupin and Agnico-Eagle, the only stratiform deposits and the only deposits currently in production, contribute more than half the total contained gold in Canadian banded iron-formation-hosted deposits.

GEOLOGICAL FEATURES

All iron-formation-hosted deposits are characterized by a strong association between native gold and iron sulphide minerals, the presence of gold-bearing quartz veins, the occurrence of deposits in structurally complex terranes, and lack of lead and zinc enrichment in the ores.

Vein-type deposits

Gold in vein-type deposits is restricted to late structures or to sulphide-iron-formation adjacent to the veins and/or shear zones. The ores occur in either sediment- or volcanic-dominated portions of greenstone belt terranes. All deposits are structurally controlled, occurring particularly in fold hinges (Fig. 15.3-2A), and most are hosted by rocks of relatively low metamorphic grade. Oxide-iron-formation is the dominant type of iron-formation associated with gold. Pyrite and/or pyrrhotite clearly replace other pre-existing iron-rich minerals (Fig. 15.3-3A). Arsenic-bearing minerals are common, but not always present; where they are present, a strong positive correlation generally exists between gold and arsenic. Ores are relatively silver-poor with gold:silver

ratios characteristically greater than 8.0 (Table 15.3-1). Intrusions of feldspar porphyry are spatially associated with a number of the deposits and may contain shear- or vein-related mineralization similar in style to that hosted by the nearby banded iron-formation (Fig. 15.3-2A).

Stratiform-type deposits

Setting

In sediment-hosted stratiform deposits such as Lupin, Jardine, and Homestake, gold is restricted to Algoma sulphide-iron-formation occurring within portions of greenstone terranes dominated by clastic sedimentary rocks (mostly turbidites) or, locally, to quartz veins that crosscut such banded iron-formation. Pelitic sedimentary rocks are commonly interbedded with the gold-bearing sulphide-iron-formation. Clearly identifiable products of volcanism do not occur within the orebodies, but volcanic rocks are typically interbedded with basinal clastic sedimentary rocks at the regional scale.

The sedimentary rocks that host these deposits are typically deformed, with at least local domains characterized by tight to isoclinal folding. Several generations of folds have been recognized, but major regional scale faults ("breaks") similar to those associated with many vein-type deposits have yet to be identified near the deposits.

Granitoid bodies of variable size and age have intruded the supracrustal rocks near deposits of this type. A locally pegmatitic tourmaline-bearing peraluminous two-mica granite occurs within the late Archean Contwoyto batholith to the north of Lupin. Quartz-feldspar porphyry intrusions have not been recognized at Lupin or Jardine, although Tertiary dykes, sills, and local breccias are abundant at Homestake.

Metamorphic grade at these deposits ranges from middle greenschist to lower amphibolite facies. Homestake is well within greenschist facies (staurolite-bearing clastic metasedimentary rocks, i.e., knotted schists, diagnostic of amphibolite facies metamorphism, occur several kilometres northeast of the mine). Jardine is upper greenschist grade, and Lupin occurs at the greenschist to amphibolite grade transition.

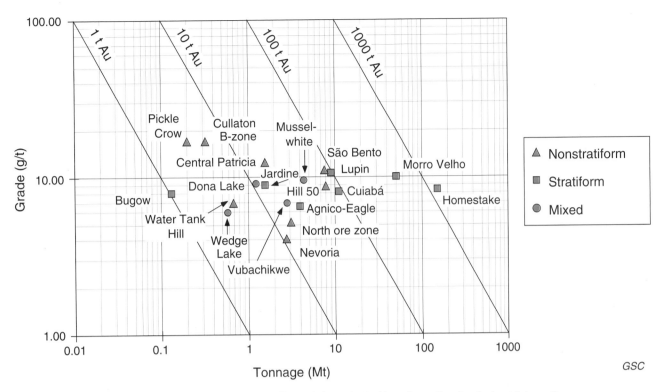

Figure 15.3-1. Grade-tonnage diagram for selected iron-formation-hosted gold deposits.

In deposits in mixed volcanic-sedimentary settings, gold occurs in pyritic sulphide-iron-formation that is associated with felsic pyroclastic rocks in the upper portions of volcanic cycles (Agnico-Eagle and Cuibá) or in pyrrhotite-rich sulphide-iron-formation interbedded with lapa seca (banded to massive quartz-carbonate rock) and rhyolitic tuff in a pelite-dominated environment (Morro Velho). Carbonate-iron-formation, other carbonate-bearing lithologies, and carbonaceous shales occur within the mine sequences. Metamorphic grade is lower greenschist at all three deposits.

Age of host rocks and mineralization

The supracrustal rocks that host the iron-formation at Lupin, Jardine, Morro Velho, Cuibá, and Agnico-Eagle are late Archean (2.6-2.8 Ga.), but Homestake occurs in rocks of Proterozoic age (1.9-2.1 Ga.; DeWitt et al., 1986; Redden et al., 1987).

Structure

The rocks that host the stratiform deposits, as well as the orebodies themselves, are deformed, principally by folding. In the Homestake mine area two principal sets of folds have been recognized regionally (Dewitt et al, 1986). Early northeast-trending isoclinal folds were refolded by northwest-trending isoclinal folds. At the mine itself, Caddey et al. (1991) identified two major deformational events which include six periods of folding. Early F_{1b} open to isoclinal folds that trend northwesterly and plunge southeasterly overprint late F_{1a} sheath folds of similar trend and plunge. At Lupin, the ore is confined to a tightly folded Z-shaped

portion (Fig. 15.3-2B) of a larger doubly-plunging structure that was produced by two generations of folding (Relf, 1989). Three periods of deformation have been identified in the Jardine area; an early period of isoclinal folding was followed by northwest-trending and northeast-trending folding events. Caddey et al. (1991) have identified an anastomosing set of late shear zones (middle D_{1b}) in the Homestake mine which were synchronous with retrogressive metamorphism and emplacement of gold-bearing quartz veins. Auriferous quartz veins in other stratiform deposits are also related to later stages of deformation.

Orebodies

Deposits are stratiform by definition, but in all cases, particularly at Homestake, the original geometry of orebodies has been obscured by folding. However, lateral or down-plunge extents of orebodies are tens to hundreds of times greater than their thicknesses.

In both sediment-hosted deposits and those occurring within mixed volcanic-sedimentary settings, gold is concentrated in several discrete units of sulphide-iron-formation that are conformably interlayered with barren silicate- and/or carbonate-iron-formation (Fig. 15.3-3B). Gold is, for the most part, relatively uniformly disseminated throughout the sulphide-iron-formation of individual orebodies, although the late quartz veins contain modest amounts of coarse (visible) gold. Arsenic is a significant component in all sediment-hosted deposits (Fig. 15.3-3C) except those in the Russell Lake area (Bunner, 1988), but is less common in deposits in mixed settings. Indeed, it is possible to identify two principal types of ore in sediment-hosted stratiform deposits on the basis of arsenic content.

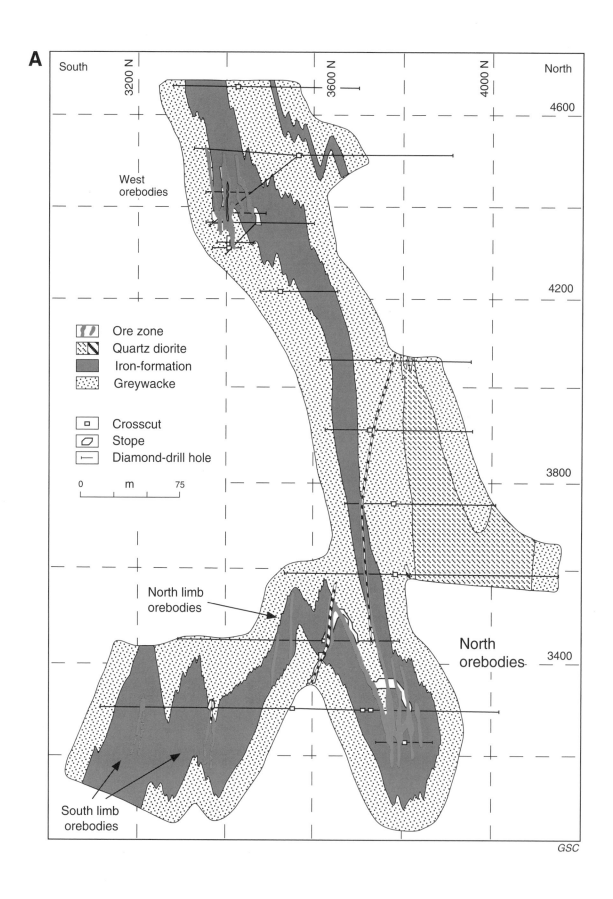

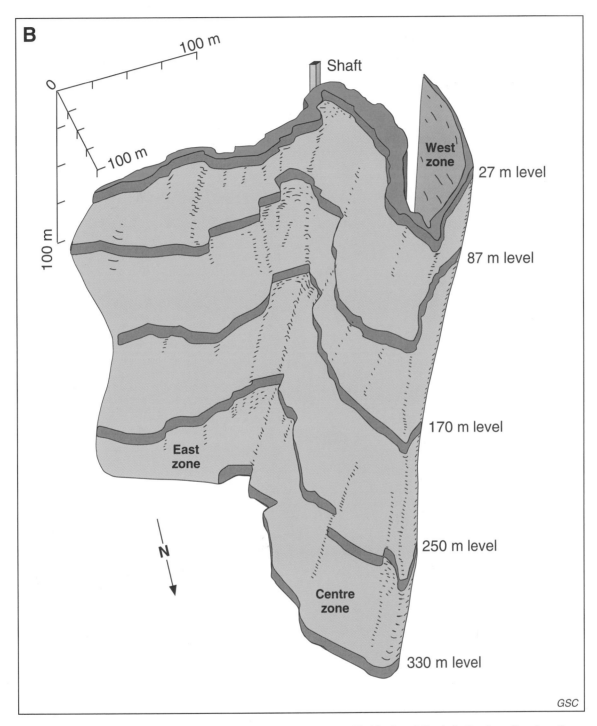

Figure 15.3-2. A) Vertical north-south section along 17400E, MacLeod-Cockshutt mine showing the distribution of several ore zones, including the North ore zone (after Horwood and Pye, 1951). The nonstratiform ore is confined to structurally controlled sulphide-rich shoots that are separated by much barren oxide-iron-formation. **B)** Lupin orebody, Northwest Territories; isometric view of upper levels as defined by assay data using a cutoff grade of 6.86 g/t or 0.2 troy oz./ton (after the 1982 Annual Report of Echo Bay Mines Ltd.). The continuity of gold distribution is one of the most remarkable features of the Lupin deposit and is a critical constraint in modelling genesis. The strike extent of the stratiform orebody exceeds 800 m. Ore reserves defined at the end of 1990 extended to a depth of greater than 1000 m. Average width of the orebody varies from about 2.5 m in the West zone to greater than 20 m in fold-thickened portions of the Centre zone. The average width of ore in the Centre zone is about 10 m.

Arsenic-rich sulphide-iron-formation occurs in areas immediately adjacent to late quartz veins or shear zones at Lupin (Fig. 15.3-3D) and Homestake (Caddey et al., 1991), and appears to be similarly controlled at Jardine (Seager, 1944). Arsenic-poor sulphide-iron-formation is more widely distributed and is the principal ore type in all deposits.

At the Homestake mine, production has come from nine elongate zones or "ledges" confined to the Homestake Formation, and each zone contains a large number of discrete orebodies. Sulphide-iron-formation that forms these orebodies constitutes approximately 3% of the total volume of the Homestake Formation. The Lupin mine occurs within the Lupin ore unit and, in at least the upper portions of the deposit, can be viewed as a single orebody with mineralization essentially continuous throughout the deposit (Fig. 15.3-2B). The limits of the Lupin orebody coincide with a marked decrease in the proportion of auriferous sulphide-iron-formation relative to barren silicate-iron-formation and clastic sedimentary rocks in the ore unit.

Although stratiform orebodies are tightly to isoclinally folded and quartz veins are locally abundant, the distributions of gold and sulphur are not obviously controlled by either the folds or the veins. At Lupin, the arsenic-rich sulphide-iron-formation adjacent to late quartz veins does not consistently contain more gold or sulphur than arsenic-poor sulphide-iron-formation further from the veins. Furthermore, in much of the deposit, mesobands of consistently auriferous pyrrhotite-rich iron-formation can be traced around fold hinges and along fold limbs without significant changes in thickness or sulphur or gold contents for tens of metres. Such mesobands are clearly not restricted to the vicinity of quartz veins or to areas of closely spaced quartz veins. The spacing between quartz veins in the Lupin orebody varies from less than one metre to greater that 10 m and averages about 4 m.

Caddey et al. (1991) reported a strong spatial association between gold-rich sulphide-iron-formation (arsenic-poor as well as arsenic-rich varieties) and late structures (shear zones and quartz veins) at Homestake. Recent work at Lupin by Bullis et al. (1992) has resulted in the discovery of narrow gold- and sulphide-rich haloes immediately adjacent to several late quartz veins in lower grade portions of the Lupin orebody at depth. However, fieldwork by the writer indicates that most, but clearly not all, of the ore on the deeper levels at Lupin is of the stratiform-type and similar to that on the upper levels of the mine.

Alteration

Alteration related to deposition of the stratiform ores is not clearly defined, largely because it is difficult to consistently determine whether individual minerals are products of isochemical metamorphism of auriferous chemical sedimentary rocks, or of metasomatism associated with formation of the late quartz veins. Both processes have undoubtedly affected the rocks at all deposits. Chlorite-rich alteration envelopes occur immediately adjacent to the late quartz veins at Homestake, Lupin, and Jardine, but chlorite unrelated to vein formation is also abundant in sulphide-banded iron-formation relatively distant from veins. Carbonate minerals are present at some stratiform deposits. The widespread Mg-rich siderite (sideroplesite) at Homestake is generally not thought to be a product of pervasive carbonatization associated with formation of the quartz veins. At Agnico-Eagle, however, carbonate-rich tuffs and agglomerates, as well as abundant ferroan dolomite have been interpreted as products of late carbonatization coeval with gold concentration (Wyman et al., 1986). The origin of the Lapa Seca at Morro Velho is controversial.

Mineralogy

In sediment-hosted stratiform deposits, pyrrhotite is ubiquitous in cherty sulphide-iron-formation and is closely associated with gold. Pyrite is present in some cases, but appears to be either vein-related or a late alteration product after pyrrhotite. As noted previously, arsenic-bearing sulphide minerals occur in most of these deposits, but are spatially related to late quartz veins and/or shear zones and appear to be later than pyrrhotite. Pyrrhotite content of sulphide-iron-formation averages between 10 and 15 modal per cent at Lupin and around 8 modal per cent at Homestake (Caddey et al., 1991). The modal abundance of arsenic-bearing minerals varies from greater than 50 per cent immediately adjacent to quartz veins to less than one per cent several tens of centimetres away from veins. There is no consistent correlation between gold and arsenic at Lupin or Homestake, because much of the ore is arsenic-poor. Löllingite occurs with arsenopyrite at Lupin and a number of other occurrences in the Contwoyto Lake area, but is rare or unreported at Homestake and Jardine. Non-sulphide minerals associated with gold-rich iron-formation include quartz, chlorite, siderite, grunerite, garnet, hornblende, and hedenbergite. At Lupin, hornblende and chlorite are more abundant in gold-rich iron-formation than in

Figure 15.3-3. A) Replacement of oxide iron-formation by sulphide minerals, principally pyrite, adjacent to late quartz vein, Solomons Pillars property, Geraldton area (Colvine et al., 1988, photograph by J. Macdonald). GSC 203652-I **B)** Sulphide-rich iron-formation at Lupin: underground photograph illustrating stratiform character of arsenic-poor variety of sulphide-iron-formation. Note alternating units of pyrrhotite-rich sulphide-iron-formation (gold-bearing) (light grey-white, banded), garnetiferous silicate-banded iron-formation (barren) (dark grey, banded), and pelitic sedimentary rock (barren) (dark grey, massive). Individual units of sulphide-iron-formation can be followed for tens of metres without significant change in thickness, sulphide content, or gold content. Dimensions of the photographed area approximately 2 m by 3 m. GSC 1993-170H **C)** Arsenic- and pyrrhotite-rich variety of sulphide-iron-formation at Lupin, occurring immediately adjacent to a late quartz vein (Q). Note megacrysts of arsenopyrite-löllingite-pyrrhotite that appear to overgrow banded pyrrhotite. Scale bar adjacent to the polished slab is one centimetre. GSC 1995-025 **D)** Arsenic-poor and arsenic-rich varieties of sulphide-banded iron-formation at Lupin (underground photograph). Note restriction of arsenic-rich ore to zones immediately adjacent to a late quartz vein and the decrease in both the abundance and size of the arsenic-bearing megacrysts with distance from the vein. The dimensions of the photographed area approximately 2 m by 3 m. GSC 1993-170I

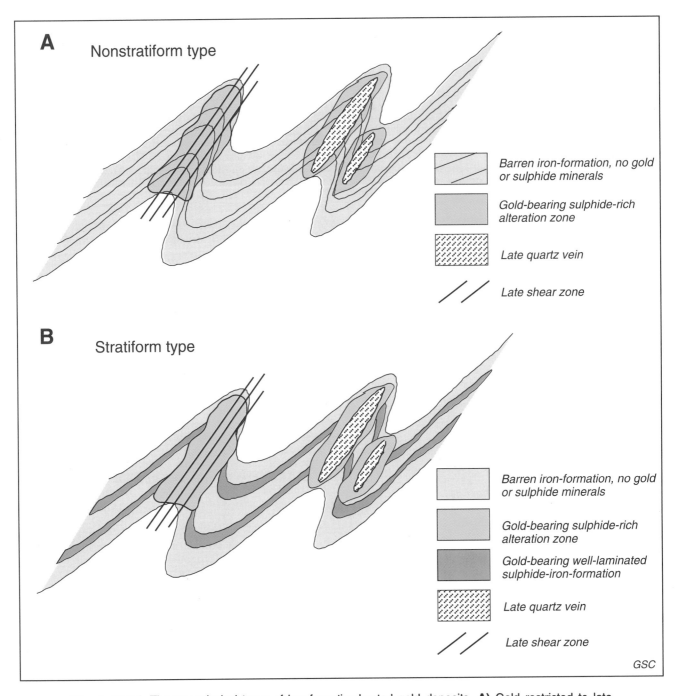

Figure 15.3-4. The two principal types of iron-formation-hosted gold deposits. **A)** Gold restricted to late structures or to sulphide-iron-foundation immediately adjacent to such structures. Examples are many and include the North ore zone, Geraldton, Ontario, Central Patricia and portions of Pickle Crow, Pickle Lake, Ontario, numerous deposits in Western Australia, including Hill 50, Nevoria, and Water Tank Hill, numerous deposits in Zimbabwe, including the Lennox mine, and probably São Bento in Brazil and the Cullaton B-zone, Northwest Territories. **B)** Gold occurs in thin, but laterally continuous units of well laminated sulphide-iron-formation as well as in sulphide-rich alteration zones adjacent to late structures. Examples are few but include Lupin, Northwest Territories; Jardine, Montana; and probably Homestake, South Dakota; all within sediment-dominated settings, as well as Morro Velho and Cuiabá, Brazil, and probably Agnico-Eagle, Quebec, all within mixed sedimentary-volcanic terranes.

gold-poor iron-formation, whereas garnet and grunerite are more abundant in gold-poor iron-formation. However, hornblende is also an essential mineral in barren silicate-iron-formation, not only in the Lupin and Jardine orebodies, but also throughout the Lupin and Jardine areas. Carbonaceous material is locally abundant at Lupin, Homestake, and Jardine.

In stratiform deposits occurring within mixed volcanic-sedimentary settings, as at Agnico-Eagle and Cuiabá, gold is associated with pyrite (characteristically greater than 20 modal per cent), and both pyrrhotite and arsenopyrite are rare. Pyritic iron-formation at these deposits is interlayered with carbonate-banded iron-formation. At Morro Velho gold occurs with pyrrhotite, pyrite, and arsenopyrite.

Gold is present as the native metal and is silver-rich compared to that in nonstratiform ores hosted by iron-formation. Gold-silver ratios for Lupin and Agnico-Eagle, as well as for Homestake, Jardine, Morro Velho, and Cuiabá, fall within the relatively restricted range of 4.0 to 6.5.

Ore textures

The sulphide-rich iron-formation in sediment-hosted stratiform deposits is typically well laminated. All components of the iron-formation, including the iron sulphide minerals, occur in clearly defined layers (Fig. 15.3-3B). Sulphidation textures, which are common in vein-type deposits in iron-formation (Fig. 15.3-3A) and which demonstrate replacement of iron oxide, carbonate, or silicate minerals by iron sulphide minerals, are lacking in the stratiform pyrrhotite-rich ores. By contrast, arsenic-bearing sulphide minerals in arsenic-rich sulphide-iron-formation adjacent to quartz veins or shear zones appear to have overgrown or replaced pyrrhotite (Fig. 15.3-3C,D). At Lupin, gold most commonly is in direct contact with pyrrhotite, but also occurs interstitially to silicate minerals in sulphide-iron-formation. In the arsenic-rich ore adjacent to quartz veins at Lupin, significant gold occurs along arsenopyrite-löllingite grain boundaries in complex megacrysts of pyrrhotite, arsenopyrite, and löllingite. Minor gold occurs along fractures in the sulphides.

The stratiform ores in mixed volcanic-sedimentary settings range from well laminated to relatively massive. Gold in the Cuiabá ore is in well laminated, cherty sulphide-iron-formation and is intimately associated with both a microcrystalline variety of pyrite and a more coarsely crystalline euhedral variety of pyrite. At Agnico-Eagle, much of the gold also occurs in well laminated, cherty sulphide-rich iron-formation. At least some of this gold is associated with a distinctive fine grained euhedral form of pyrite, and occurs along intragranular fractures, as 3-5 μm particles embedded within pyrite, and as small veinlets as much as 10 μm in length (Wyman et al., 1986). The sulphide-rich ore at Morro Velho, though banded, is typically relatively massive.

There is considerable textural evidence in some stratiform deposits that gold-rich sulphide-iron-formation was present before metamorphism and deformation. For example, thin bands of arsenic-poor sulphide-iron-formation at Lupin are commonly more tightly folded than the interlayered units of silicate-iron-formation and clastic sedimentary rocks, suggesting that the sulphide-iron-formation was present before folding. Also at Lupin, the

ubiquitous occurrence in arsenic-poor sulphide-iron-formation of pyrrhotite in lens-shaped mosaics in which individual pyrrhotite grains meet at triple junctions, the presence of chalcopyrite at such triple junction positions, a greater sulphide grain size in more deformed portions of the deposit, and an increase of not only discordant pyrrhotite veinlets, but also of nonmagnetic pyrrhotite in the more highly metamorphosed portions of the deposit, are compatible with the presence of sulphide-banded iron-formation before deformation and metamorphism. At both Lupin and Homestake coarse clots to pods of massive pyrrhotite are common in sulphide-banded iron-formation immediately adjacent to the late quartz veins, and the chlorite- and arsenide-rich alteration zones adjacent to late quartz veins appear to overprint well-laminated arsenic-poor sulphide-iron-formation.

DEFINITIVE CHARACTERISTICS

The essential features of stratiform and nonstratiform iron-formation-hosted gold deposits are illustrated in Figure 15.3-4. Table 15.3-2 provides a detailed comparison between nonstratiform and stratiform sediment-hosted ores. Of the characteristics listed for nonstratiform deposits, numbers 1, 5, 9, and 12 are the most definitive. In other words, these deposits contain sulphide-rich, nonstratiform orebodies in which the sulphide minerals clearly replace other iron-rich minerals, typically magnetite in barren oxide-iron-formation, within or immediately adjacent to late structures. The most definitive characteristics of stratiform, Lupin-like, deposits are numbers 1, 3, 4, 5, 9, 12, and 13. In other words, these deposits contain well-laminated units of laterally continuous cherty, pyrrhotite-rich sulphide-iron-formation that are conformably interlayered with barren silicate- and/or carbonate-iron-formation and clastic sedimentary rocks. Furthermore, such ores lack obvious sulphidation textures and are not clearly controlled by late structures.

Many of the characteristics of selected deposits are summarized in Table 15.3-3. Figure 15.3-5 is an idealized stratigraphic section through the Lupin ore unit. This has been drawn to scale and illustrates inter-relationships among the different types of iron-formation, clastic sedimentary rocks, late quartz veins, alteration zones, and gold that may be typical of stratiform sediment-hosted deposits.

GENETIC MODELS

Numerous genetic models or working hypotheses have been proposed to best explain the critical features of gold deposits hosted by iron-formation. These fall into two main categories discussed below.

Syngenetic models

All components of a deposit (Fe, Si, Ca, S, As, Au, Ag, Cu, C, CO_2, W, etc.) were deposited from hydrothermal fluids during chemical sedimentation or early diagenesis. Localized remobilization of certain components (Si, Ca, As, W, etc.) during metamorphism and/or deformation in an essentially closed system is called upon to account for the typically vein-controlled distribution of some components.

Table 15.3-2. Comparison of characteristics of iron-formation-hosted gold deposits.

Features common to all deposits
1. Very strong spatial association between native gold and iron sulphide minerals 2. Gold-bearing quartz-rich veins and/or shear zones are present and locally abundant 3. Deposits occur in structurally complex settings 4. Ores contain only background contents of lead and zinc
Features diagnostic of nonstratiform deposits
1 Deposits are nonstratiform 2 Gold is commonly not restricted to sulphide-iron-formation or veins that crosscut iron-formation 3 Sulphide-iron-formation does not occur in laterally continuous units 4 Sulphide-iron-formation is not well laminated; iron sulphide minerals are commonly massive 5 Distributions of iron sulphide minerals and gold are clearly controlled by veins and/or late structures 6 Orebodies are typically less deformed than associated rocks 7 Iron sulphide minerals tend to be relatively undeformed and unmetamorphosed 8 Deposits not restricted to, but most abundant in greenschist facies 9 Sulphidation textures are common 10 Orebody scale alteration exists 11 Alteration products generally similar to those in "mesothermal vein" gold deposits 12 Oxide-iron-formation is typically the principal iron-formation lithology in the deposit 13 Pyrite is commonly the dominant iron sulphide mineral 14 Arsenic, if present, is characteristically directly correlated with gold 15 Silver contents of gold grains are typically low (Au-to-Ag ratios greater than 8.0) 16 Deposits are relatively common, generally small, and relative to stratiform deposits, are difficult to evaluate and mine
Features diagnostic of sediment-hosted (Lupin-like) stratiform deposits
1 Deposits are stratiform 2 Gold mostly restricted to iron-formation or to veins that crosscut sulphide-iron-formation 3 Sulphide-iron-formation occurs in several thin, but laterally continuous units that are conformably interlayered with barren silicate-iron-formation and/or carbonate-iron-formation and clastic sedimentary rocks 4 Sulphide-iron-formation is well laminated and chert-rich; iron sulphide minerals are typically finely layered 5 Distributions of iron sulphide minerals and gold are not clearly controlled by veins and/or late structures 6 Orebodies are as deformed or more deformed than associated rocks 7 Iron sulphide minerals show effects of deformation and metamorphism 8 Deposits occur in both greenschist and amphibolite facies terranes 9 Sulphidation textures are absent 10 Lack of orebody-scale alteration; localized vein-related alteration does occur 11 Vein-related alteration is commonly atypical of "mesothermal vein" gold deposits 12 Oxide-iron-formation is lacking in the deposits, irrespective of metamorphic grade 13 Pyrrhotite is typically the dominant iron sulphide mineral; in some cases early pyrrhotite has been replaced by pyrite 14 Arsenic is generally abundant adjacent to late quartz veins, but is not well correlated with gold 15 Silver contents of gold grains are moderately high (Au-to-Ag ratios ~ 3.0-7.0) 16 Deposits are rare, can be very large, and are less difficult to evaluate and mine than are the nonstratiform deposits

From the late 1960s through the early 1980s this essentially exhalative model was favoured for many gold deposits hosted by iron-formation (Sawkins and Rye, 1974; Hutchinson and Burlington, 1984). There is still some support for this model, particularly among those who have spent considerable time working with stratiform deposits (Nelson, 1986; Vial, 1988; Ladeira, 1991; Vieira et al., 1991a, b). Modern unequivocal examples of significant syngenetic gold concentrations lend further support to this model. Such concentrations occur within base metal-rich volcanic-associated massive sulphide deposits accumulating at Axial Seamount, within base metal-poor sulphide concentrations in sediment-covered portions of the Gorda Ridge (Escanaba Trough), and in subaerial hot springs associated with active geothermal zones in New Zealand and the western United States.

Epigenetic models

Some components of iron-formation were deposited during chemical sedimentation (Fe, Ca, some Si and CO_2, etc.), but others related to ore formation (S, Au, Ag, Cu, As, W, some Si and CO_2, etc.) were added during vein-related hydrothermal activity associated with much later deformation, metamorphism, and/or magmatism. Sulphidation of relatively Fe-rich host rocks adjacent to shear zones and/or veins is viewed as the principal ore-forming process. Extensive carbonatization may have preceded sulphidation in some deposits.

This model was favoured by most economic geologists up until the late 1960s and is currently advocated by many for most, if not all, gold deposits hosted by banded iron-formation (Phillips et al., 1984; Macdonald and Fyon, 1986; Colvine et al., 1988). Several potential sources of the

Table 15.3-3. Summary of characteristic features of selected iron-formation-hosted gold deposits.

Features	L	RL	J	H	C	MV	AE	NZ	CP	PC	SB	CL	DL	M
Type of deposit														
Stratiform	+	+	+	?	+	+	?	-	-	-	?	?	?	?
Probably stratiform	-	-	-	+	-	-	+	-	-	-	?	?	?	?
Nonstratiform	-	-	-	?	-	-	?	+	+	+	?	?	?	?
Probably nonstratiform	-	-	-	?	-	-	?	-	-	-	+	+	+	+
Sediment-hosted	+	+	+	+	-	-	-	+	+	-	-	+	-	-
Mixed volcanic-sedimentary	-	-	-	-	+	+	+	-	-	+	+	-	+	+
Iron-rich sedimentary rock														
Sulphide-iron-formation	+	+	+	+	+	+	+	+	+	+	+	+	+	+
Oxide-iron-formation	-	-	-	-	-	-	-	+	+	+	+	+	+	+
Carbonate-iron-formation	-	-	-	+	+	-	+	?	?	?	+	-	-	?
Silicate-iron-formation	+	+	+	+	-	-	-	-	-	-	-	-	+	+	-
Lapa Seca	-	-	-	-	-	+	-	-	-	-	-	-	-	-
Ore host														
Sulphide-iron-formation	+	+	+	+	+	+	+	+	+	+	+	+	+	+
Clastic sedimentary rocks	-	-	-	-	-	-	-	?	?	-	-	-	-	-
Mafic volcanic rocks	-	-	-	-	-	-	-	-	-	+	-	-	-	-
Felsic intrusions	-	-	-	-	-	-	-	+	-	+	-	-	-	-
Age														
Archean	+	+	+	-	+	+	+	+	+	+	+	?	+	+
Proterozoic	-	-	-	+	-	-	-	-	-	-	-	?	-	-
Associated minerals														
Pyrrhotite	++	+	++	++	+	++	+	+	++	++	?	++	++	++
Pyrite	+	++	+	+	++	+	++	++	-	++	++	+	+	?
Arsenopyrite	++	-	++	++	+	++	+	+	++	+	++	+	-	+
Löllingite	++	-	-	-	-	-	-	-	-	-	-	-	-	+
Quartz														
as chert	++	++	+	++	++	+	++	+	+	+	+	++	+	+
as veins	+	+	+	++	+	+	+	+	+	+	+	++	?	+
Chlorite	+	+	+	++	?	?	?	?	?	?	?	+	?	+
Hornblende	+	+	+	-	?	?	?	?	?	?	?	?	?	?
Grunerite	+	+	+	++	?	?	?	?	?	?	?	+	?	+
Fe-carbonate	-	-	?	++	++	++	++	++	?	?	++	++	-	-
Metamorphic grade														
Greenschist (G)	+	-	+	+	+	+	+	+	+	+	+	+	+	-
Amphibolite (A)	+	+	-	-	-	-	-	-	-	-	-	-	-	+
G-A transition	+	-	-	-	-	-	-	-	-	-	-	-	-	-
Granulite	-	-	-	-	-	-	-	-	-	-	-	-	-	-
Felsic intrusions														
Abundant within deposit	-	+	?	+	?	?	?	+	+	+	?	-	?	+
Abundant near deposit	+	+	+	+	?	?	+	+	+	+	?	-	?	+
Sulphidation textures	-	-	-	-	-	-	-	+	+	+	+	+	+	+
Au-to-Ag ratio (Au/Ag)	6.2	3.6	6.0	5.0	6.0	5.0	5.0	29.9	10.7	8.6	9.0	14.1	?	9.1

Explanation: L Lupin; RL Russell Lake; J Jardine; H Homestake; C Cuiabá; MV Morro Velho; AE Agnico-Eagle; NZ North ore zone, Geraldton; CP Central Patricia; PC Pickle Crow; SB São Bento; CL Cullaton B-zone; DL Dona Lake; M Musselwhite

Associated minerals: ++ Abundant; + Present in significant amounts; - Rare or not present; ? Unknown
Other categories: + Present; - Not present; ? Unknown

epigenetic ore-forming fluids have been proposed. Some suggest the fluids were metamorphically derived during devolatilization reactions associated with prograde metamorphism of deeper crustal rocks. Others believe the fluids were derived from late felsic magmas. A direct contribution from the mantle has also been proposed.

The epigenetic model is clearly appropriate for the vein (nonstratiform) type of iron-formation-hosted deposits. Such deposits are structurally controlled and related to late sulphidation of iron-formation that was initially gold- and sulphur-poor.

The genesis of stratiform ores is controversial, but this writer considers that synsedimentary concentration of gold, silver, and copper during deposition of sulphide-iron-formation

on or just below the seafloor best accounts for many of the critical features of these deposits (Kerswill, 1993). The conformable interlayering of discrete units of gold-rich sulphide-iron-formation with barren silicate-iron-formation and clastic sedimentary rocks, the finely laminated character of the ores, their remarkable continuity, the strong positive correlation between gold and sulphur, the lack of clear evidence for structural control, and an absence of sulphidation textures suggest that both gold and sulphur were primary components of the deposits. Arsenic, tungsten, and much silica were, however, probably introduced during formation of the late quartz veins. Mobilization of synsedimentary gold during subsequent metamorphism and deformation, followed by its deposition in structurally favourable sites, were probably not essential for the genesis

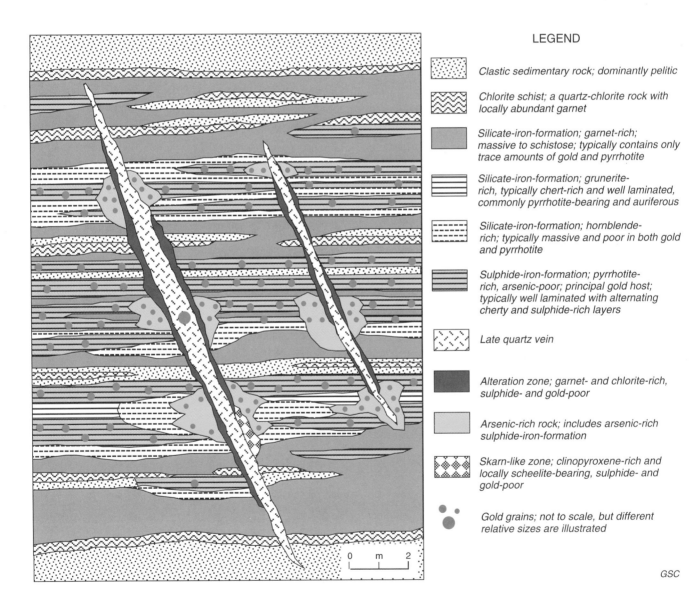

LEGEND

:::::::: *Clastic sedimentary rock; dominantly pelitic*

Chlorite schist; a quartz-chlorite rock with locally abundant garnet

Silicate-iron-formation; garnet-rich; massive to schistose; typically contains only trace amounts of gold and pyrrhotite

Silicate-iron-formation; grunerite-rich, typically chert-rich and well laminated, commonly pyrrhotite-bearing and auriferous

Silicate-iron-formation; hornblende-rich; typically massive and poor in both gold and pyrrhotite

Sulphide-iron-formation; pyrrhotite-rich, arsenic-poor; principal gold host; typically well laminated with alternating cherty and sulphide-rich layers

Late quartz vein

Alteration zone; garnet- and chlorite-rich, sulphide- and gold-poor

Arsenic-rich rock; includes arsenic-rich sulphide-iron-formation

Skarn-like zone; clinopyroxene-rich and locally scheelite-bearing, sulphide- and gold-poor

Gold grains; not to scale, but different relative sizes are illustrated

0   m   2

GSC

Figure 15.3-5. Idealized stratigraphic section through the Lupin ore unit showing the distributions of gold, sulphide-iron-formation, several varieties of silicate-iron-formation, chlorite schist, clastic sedimentary rocks, late quartz veins, arsenic-rich rock, chlorite- and garnet-rich alteration zones, and skarn-like zones.

Table 15.3-4. Comparison between iron-formation-hosted gold deposits in sediment dominate settings and skarn gold deposits.

	IRON-FORMATION GOLD	SKARN GOLD
Presence of calc-silicate minerals (garnet±hornblende±clinopyroxene)	Yes	Yes
Direct correlation between distribution of calc-silicate minerals and distributions of gold and sulphur	No	Yes
Typical compositions of diagnostic calc-silicate minerals (garnet and clinopyroxene)	Pyralspite, Hedenbergite	Ugrandite, Diopsidic
Significant ore occurs within an intrusion	No	Common
Calc-silicate minerals are zoned about an intrusion and/or major structure	No	Common
Much of the ore is discordant rather than stratiform	No	Yes
Much of the ore is clearly controlled by late structures and/or lithological contacts	No	Yes
Ores contain significant concentrations of bismuth and/or tellurium	No	Common
Presence of locally abundant tungsten as scheelite	Common	Uncommon
Presence of abundant magnetite in the ore	No	Common
Strong direct correlation between gold and arsenic	No	Common
Silicate mineral assemblages most commonly associated with ore	Prograde	Retrograde
Apparent timing of gold and sulphur introduction	Early	Late
Inferred genesis of calc-silicate minerals most commonly associated with ore	Metamorphic	Metasomatic

of the stratiform ores. Some gold in stratiform deposits may represent new gold introduced during formation of the late quartz veins rather than gold that has just been remobilized.

However, epigenetic models for the genesis of stratiform deposits continue to receive support. For example, recent work at the Homestake mine has led Caddey et al. (1991) to abandon the syngenetic model of Nelson (1986) in favour of epigenetic introduction of gold during reverse movement on high-angle shears synchronous with late quartz vein emplacement. Lhotka (1988) and Lhotka and Nesbitt (1989) have concluded from their investigations that Lupin is an epigenetic deposit in which all the gold, sulphur, and arsenic were introduced by the late quartz veins. Bullis (1990) also proposed a model for Lupin that is similar to that of Lhotka (1988). In the author's opinion much of the available evidence for stratiform deposits is most consistent with a syngenetic model.

RELATED DEPOSIT TYPES

Vein or nonstratiform type

Iron-formation-hosted gold deposits of the vein or nonstratiform type are similar to mesothermal gold deposits as discussed elsewhere in this volume (see subtype 15.2).

Stratiform type

Stratiform gold deposits, though possessing some of the features of mesothermal vein deposits and/or stratabound sulphide schist deposits, are more similar to some of the types of stratiform deposits discussed below.

Sedimentary-exhalative massive sulphides

Stratiform iron-formation-hosted deposits in sediment-dominated settings are similar to sedimentary exhalative (Sedex) deposits in that both types are stratiform,

sediment-hosted, well laminated, and sulphide-rich. However, there are major differences in the metal contents; sedimentary exhalative deposits are base metal-rich and gold-poor, whereas the stratiform gold deposits are base metal-poor but gold-rich. The two deposit types may have formed in similar geological settings, but from hydrothermal fluids of different character; reduced, acidic, and relatively high temperature for sedimentary exhalative deposits, but reduced, alkaline, and only moderate temperature for stratiform gold ores. Sedimentary exhalative deposits range in age from Proterozoic to the present. Homestake, the largest of the stratiform sediment-hosted gold deposits, is Proterozoic, but Lupin and Jardine are Archean. As noted previously, the auriferous pyrrhotite-rich but lead- and zinc-poor sulphide accumulations in the Escanaba Trough may be similar to Lupin and other Lupin-like deposits.

Volcanic-hosted massive sulphides

Stratiform gold deposits of the mixed volcanic-sedimentary subtype are somewhat similar to volcanic-associated sulphide deposits. Numerous ancient, and several modern, deposits of the latter type contain significant gold, but the stratiform gold deposits are not enriched in lead and zinc and contain only minor concentrations of copper.

Iron-rich sedimentary rocks (Algoma type)

Both types of iron-formation-hosted gold deposits occur within Algoma-type banded iron-formation and thus share similarities with it. The sulphide-iron-formation in the stratiform deposits is probably a true sulphide facies iron-formation and was most likely deposited from hydrothermal fluids during chemical sedimentation or early diagenesis. However, most Algoma-type iron deposits contain abundant magnetite and are poor in both sulphur and gold. In the case of nonstratiform deposits, the sulphide-rich iron-formation within and adjacent to late veins and/or shear zones is a product of structurally and chemically controlled alteration processes related to metamorphism and/or deformation, and was clearly not originally a primary chemical sediment.

Skarns

Some features of stratiform iron-formation-hosted gold deposits in sediment-dominated settings are characteristic of reduced skarn deposits that are rich in tin and/or tungsten. Calc-silicate assemblages occur in both, pyrrhotite is the principal iron sulphide, ore units are typically thin, but laterally continuous, and ores are commonly restricted to beds hosted by clastic sedimentary rocks. However, the banded iron-formation-hosted deposits are only superficially similar to gold-rich skarns. Significant differences between these two deposit types are presented in Table 15.3-4. Unequivocal evidence for the pervasive metasomatism associated with skarn deposits is lacking in stratiform gold deposits. Indeed, evidence for skarn-like metasomatism is typically restricted to narrow gold- and sulphide-poor alteration zones immediately adjacent to late quartz veins. Most of the mineralogical and chemical features of the stratiform ores away from veins are adequately explained by prograde metamorphism of variably mixed clastic and chemical sediments containing different amounts of aluminous clays, Fe-rich silicate and/or carbonate minerals, abundant quartz, Fe-monosulphide minerals and gold.

EXPLORATION GUIDES

A number of largely empirical exploration guidelines that may be useful in the search for additional iron-formation-hosted gold deposits can be proposed. These include general guides for either stratiform- or vein-type targets, as well as specific guides for each type. Comparison between the guidelines and the characteristics of a target may help one determine whether the target is more likely to contain stratiform or nonstratiform mineralization. Early recognition of the most probable deposit type should permit more effective evaluation of the prospect. Although stratiform deposits can be very large and are more easily evaluated and mined than vein-type deposits, the favourable grade and tonnage characteristics of some nonstratiform deposits (see Fig. 15.3-1) make this deposit type an attractive target as well.

The most useful empirical exploration guide for any iron-formation-hosted gold deposit is the presence of abundant iron sulphide minerals (pyrrhotite and/or pyrite). These minerals are in some cases relatively uniformly disseminated in stratiform units, and in other cases are restricted to alteration zones adjacent to late structures. Units of well laminated, cherty, pyrrhotite-rich banded iron-formation are particularly favourable for the occurrence of stratiform ores in sediment-dominated target areas. Auriferous pyrrhotite-rich iron-formation in such deposits is typically interbedded with silicate- and/or carbonate-bearing iron-formation and pelitic sedimentary rocks. Carbonate-iron-formation may be a useful exploration guide for stratiform mineralization in terranes at relatively low metamorphic grade (Homestake, Morro Velho, Cuiabá, and Agnico-Eagle). Calc-silicate-bearing

iron-formation may be the metamorphosed equivalent of carbonate-iron-formation within higher grade terranes (Lupin, Jardine, Bugow, and Musselwhite). Sulphidic, but gold-poor, black shales are consistently associated with stratiform iron-formation-hosted gold deposits in mixed volcanic-sedimentary settings (Agnico-Eagle, Cuiabá, and Morro Velho). Such rocks are not characteristic of sediment-hosted deposits, but pyrrhotite-rich gold-poor graphitic mudstone is a significant component of the Poorman Formation, which underlies the Homestake Formation at Homestake. Despite the apparent preferred occurrence of iron-formation-hosted gold ores in Archean rocks, the Proterozoic age of Homestake, the largest stratiform deposit, suggests that exploration, particularly for stratiform ores, need not be restricted to Archean terranes.

Oxide-iron-formation is a negative indicator for stratiform gold mineralization, but such rocks are the typical host of nonstratiform deposits. However, the presence of oxide-iron-formation does not mean that an area or region is unfavourable for the occurrence of stratiform mineralization. Although lateral transitions from sulphide-iron-formation to oxide-iron-formation have not been recognized in stratiform deposits, oxide-iron-formation is not uncommon in the vicinity of stratiform mineralization. Indeed, oxide-iron-formation is widespread near the Lupin mine and hosts numerous occurrences of the nonstratiform type. This suggests that the presence of nonstratiform mineralization may be useful in identifying areas with potential for stratiform mineralization.

Late quartz veins and/or shear zones are present in most known examples of iron-formation-hosted gold deposits. Thus they provide an obvious exploration guide, particularly if concentrations of iron sulphide minerals are spatially and genetically associated with the veins. Complex folds are characteristic of many iron-formation-hosted gold deposits and may be viewed as a positive feature in evaluating exploration targets. In nonstratiform deposits, the distributions of gold-bearing veins and sulphide-rich zones are commonly controlled by fold structures. In stratiform deposits, structural repetition and thickening of sulphide-iron-formation in fold hinges can significantly increase ore tonnage. However, the distributions of gold and sulphur in these deposits are not obviously controlled by such structures. This suggests that in the case of targets of possibly stratiform nature, fold limbs as well as fold hinges require testing.

Arsenic-bearing minerals can be viewed as useful empirical ore guides, because late quartz veins with associated arsenic constitute a volumetrically important component in many nonstratiform deposits and most sediment-hosted stratiform deposits (Bugow and SP in the Russell Lake area, Northwest Territories, are notable exceptions). However, in stratiform deposits, sulphur is consistently more reliable than arsenic as a guide for gold. This is largely because much stratiform mineralization is arsenic-poor. In some nonstratiform deposits, such as the North ore zone at Geraldton and the São Bento mine, gold is closely associated with arsenopyrite and the positive correlation between gold and arsenic is very strong. In many cases, the presence of abundant arsenic can be linked to a relatively late hydrothermal event in a setting containing abundant sedimentary rocks.

Zones or lithologies anomalously rich in nonsulphide minerals that may be products of alteration associated with gold mineralization can be viewed as possible ore hosts or as indicators of nearby mineralization. Such nonsulphide minerals include chlorite, carbonate, biotite, and sericite. The dolomite-rich Lapa Seca that is closely associated with the sulphide ore at Morro Velho is an excellent exploration guide.

The consistently high silver content of gold grains in stratiform ores, and the apparent ubiquitous presence of minor chalcopyrite in association with pyrrhotite in sediment-hosted stratiform ores, suggest that silver and copper may be useful pathfinders for stratiform mineralization. The significant difference in gold-to-silver ratios between stratiform (Au:Ag typically between 3.0 and 7.0) and nonstratiform (Au:Ag typically greater than 8.0) deposits indicates that variations in this ratio may be helpful in distinguishing styles or types of mineralization. A general consistency of gold-silver ratios within individual deposits suggests that anomalous ratios may help identify samples that should be reassayed.

Magnetic surveys are particularly useful in exploration for iron-formation-hosted gold deposits. The abundant pyrrhotite in occurrences of the sediment-hosted stratiform-type produces positive magnetic anomalies that generally can be distinguished from the background response of sulphide- and oxide-poor silicate-iron-formation, as well as from positive anomalies associated with the presence of oxide-iron-formation or mafic dykes. The replacement of magnetite by pyrite and/or pyrrhotite in vein deposits produces subtle negative magnetic anomalies associated with gold mineralization. Electromagnetic, self-potential, induced polarization, and resistivity surveys may prove useful in detecting sulphide concentrations.

The recognition that syngenetic processes may have controlled the distributions of gold and sulphur in stratiform iron-formation-hosted gold deposits has considerable exploration significance. The principal implication is that exploration for such deposits should involve documenting and evaluating features of the primary depositional environment, particularly the distribution of sulphide-bearing iron-formation. In some cases, the detailed work involved in delineating and tracing particular units of sulphide-iron-formation is best attempted after potential problems related to the interpretation of complex fold patterns and metamorphic overprint have been resolved. Exploration for nonstratiform deposits should emphasize structural and metamorphic features.

SELECTED BIBLIOGRAPHY

References with asterisks (*) are considered to be the best source of general information on this deposit subtype.

Barnett, E.S., Hutchinson, R.W., Adamcik, A., and Barnett, R.
1982: Geology of the Agnico-Eagle gold deposit, Quebec; in Precambrian Sulphide Deposits, H.S. Robinson Memorial Volume, (ed.) R.W. Hutchinson, C.D. Spence, and J.M. Franklin; Geological Association of Canada, Special Paper 25, p. 403-426.

*Boyle, R.W.
1979: The geochemistry of gold and its deposits (together with a chapter on geochemical prospecting for the element); Geological Survey of Canada, Bulletin 280, 584 p.

Bullis, H.R.
1990: Geology of the Lupin Deposit, N.W.T.; in Mineral Deposits of the Slave Province, Northwest Territories, 8th International Association on the Genesis of Ore Deposits Symposium, Field Trip Guidebook, Field Trip 13, (ed.) W.A. Padgham and D. Atkinson; Geological Survey of Canada, Open File 2168, p. 115-125.

Bullis, H.R., Hureau, R.A., and Penner, B.D.
1992: Controls of gold and sulphide distribution at Lupin, N.W.T.; in Exploration Overview 1992, Northwest Territories, Mining, Exploration and Geological Investigations, (ed.) J.A. Brophy; Northwest Territories Geology Division, Department of Indian and Northern Affairs Canada, p. 18.

Bunner, D.P.
1988: The geologic setting and geochemistry of auriferous iron-formations on the Bugow Property, Russell Lake area, District of Mackenzie, Northwest Territories; MSc. thesis, Carleton University, Ottawa, Ontario, 189 p.

*Caddey, S.W., Bachman, R.L., Campbell, T.J., Reid, R.R., and Otto, R.P.
1991: The Homestake gold mine, an Early Proterozoic iron-formation-hosted gold deposit, Lawrence County, South Dakota; United States Geological Survey, Bulletin 1857-J, 67 p.

Chandler, T.E. and Holmberg, H.W.
1990: Geology of gold deposits at George Lake area, Back River region, N.W.T., an exploration overview; in Proceedings, American Institute of Mining, Metallurgical and Petroleum Engineers Regional Meeting, Lead, South Dakota, p.361-370.

Cohoon, G.A.
1986: Gold in an iron-formation: the Dona Lake deposit; The Northern Miner Magazine, v. 1, no. 8, p. 16-20.

Colvine, A.C., Fyon, J.A., Heather, K.B., Marmont, S., Smith, P.M., and Troop, D.G.
1988: Archean lode gold deposits in Ontario: Part I, a depositional model, Part II, a genetic model; Ontario Geological Survey, Miscellaneous Paper 139, 136 p.

Cuthill, J., Oliver, D., and Hofer, W.
1990: The structure of the Jardine gold deposit, Jardine, Montana; in Proceedings, American Institute of Mining, Metallurgical and Petroleum Engineers Regional Meeting, Lead, South Dakota, p. 353-360.

DeWitt, E., Redden, J.A., Burack Wilson, A., and Buscher, D.
1986: Mineral resource potential and geology of the Black Hills National Forest, South Dakota and Wyoming; United States Geological Survey, Bulletin 1580, 135 p.

Dubé, L.M., Hubert, C., Brown, A.C., and Simard, J.M.
1991: The Telbel orebody of the Agnico-Eagle mine in the Joutel area of the Abitibi greenstone belt, Quebec, Canada: a stratabound, gold-bearing massive siderite deposit with early diagenetic pyritization; in Brazil Gold '91, The Economics, Geology, Geochemistry and Genesis of Gold Deposits, (ed.) E.A. Ladeira; A.A. Balkema, Rotterdam, Brookfield, p. 493-498.

Ford, R.C.
1988: Comparative geology of gold-bearing Archean iron-formation, Slave Structural Province, Northwest Territories; MSc. thesis, University of Western Ontario, London, Ontario, 233 p.

Foster, R.P.
1989: Archean gold mineralization in Zimbabwe: implications for metallogenesis and exploration; in The Geology of Gold Deposits: the Perspective in 1988, (ed.) R.R. Keays, W.R.H. Ramsay, and D.I. Groves; Economic Geology, Monograph 6, p. 54-70.

Gardiner, J.J.
1986: Structural geology of the Lupin gold mine, Northwest Territories; MSc. thesis, Acadia University, Wolfville, Nova Scotia, 206 p.

Hall, R.S. and Rigg, D.M.
1986: Geology of the West Anticline Zone, Musselwhite Prospect, Opapimiskan Lake, Ontario, Canada; in Proceedings of Gold '86, an International Symposium on the Geology of Gold Deposits, (ed.) A.J. Macdonald; Toronto, 1986, p. 124-136.

Hallager, W.S.
1984: Geology of gold-bearing metasediments near Jardine, Montana; in Proceedings Symposium Gold '82, Geological Society of Zimbabwe, Special Publication 1, A.A. Balkema, Rotterdam, The Netherlands, p. 191-218.

*Horwood, H.C. and Pye, E.G.
1951: Geology of Ashmore Township; Ontario Department of Mines Annual Report, v. LX, pt. V, 105 p.

*Hutchinson, R.W. and Burlington, J.L.
1984: Some broad characteristics of greenstone belt gold lodes; in Proceedings of Gold '82 Symposium, (ed.) R.P. Foster; Geological Society of Zimbabwe, Special Publication 1, A.A. Balkema, Rotterdam, The Netherlands, p. 339-372.

Kerswill, J.A.
1986: Gold deposits hosted by iron-formation in the Contwoyto Lake area, Northwest Territories; in Poster Volume for Gold '86, an International Symposium on the Geology of Gold Deposits, (ed.) A.M. Chater; Toronto, 1986, p. 82-85.
*1993: Models for iron-formation-hosted gold deposits; in Mineral Deposit Modeling, (ed.) R.V. Kirkham, W.D. Sinclair, R.I. Thorpe, and, J.M. Duke; Geological Association of Canada, Special Paper 40, p. 171-199.

Ladeira, E.A.
1980: Metallogenesis of gold at the Morro Velho mine, and in the Nova Lima District, Quadrilatero Ferrifero, Minas Gerais, Brazil; PhD. thesis, University of Western Ontario, London, Ontario, 272 p.
1991: Genesis of gold in the Quadrilátero Ferrífero: a remarkable case of permanency, recycling and inheritance - a tribute to Djalma Guimarães, Pierre Routhier and Hans Ramberg; in Brazil Gold '91, The economics, geology, geochemistry and genesis of gold deposits, (ed.) E.A. Ladeira; A.A. Balkema, Rotterdam, Brookfield, p. 11-30.

Lhotka, P.G.
1988: Geology and geochemistry of gold-bearing iron-formation in the Contwoyto Lake-Point Lake region, Northwest Territories, Canada; PhD. thesis, University of Alberta, Edmonton, Alberta, 265 p.

Lhotka, P.G. and Nesbitt, B.E.
1989: Geology of unmineralized and gold-bearing iron formation, Contwoyto Lake–Point Lake region, Northwest Territories, Canada; Canadian Journal of Earth Sciences, v. 26, p. 46-64.

Macdonald, A.J. and Fyon, J.A.
1986: Sulphidation - the key to gold mineralization in banded iron-formation; in Poster Volume for Gold '86, an International Symposium on the Geology of Gold Deposits, (ed.) A.M. Chater; Toronto, 1986, p. 96-97.

Miller, A.R.
1992: Gold metallogeny, Churchill structural province; Geological Survey of Canada, Open File 2484, p. 157-160.

Miller, A.R., Balog, M.J., and Gochnauer, K.
1993: Contrasting oxide iron-formation-hosted lode gold deposit types in the Meliadine trend, Rankin Inlet Group, Churchill Province, with emphasis on alteration assemblages; in Exploration Overview 1993, Northwest Territories, Mining, Exploration and Geological Investigations, (ed.) S.P. Goff; Northwest Territories Geology Division, Indian and Northern Affairs Canada, p. 41-42.

Mosley, G. and Hofmeyr, P.K.
1986: The geology and mineralogy of the São Bento gold deposit, Minas Gerais, Brazil; in Extended Abstracts, Geocongress '86, Geological Society of South Africa, Johannesburg, 1986, p. 321-324.

*Nelson, G.
1986: Gold mineralization at the Homestake gold mine, Lead, South Dakota; in Gold in the Western Shield, Proceedings of a Symposium Held in Saskatoon, September, 1985, (ed.) L.A. Clark; The Canadian Institute of Mining and Metallurgy, Special Volume 38, p. 347-358.

Netolitzky, R.K.
1986: An exploration review of the Weedy Lake, Tower Lake and Wedge Lake gold deposits, Saskatchewan; in Gold in the Western Shield, Proceedings of a Symposium Held in Saskatoon, September, 1985, (ed.) L.A. Clark; The Canadian Institute of Mining and Metallurgy, Special Volume 38, p. 229-252.

Olson, R.A.
1989: Geology of the gold-bearing zones at George Lake area, Back River region, N.W.T. (extended abstract); in Exploration Overview, 1989, Northwest Territories Geology Division, Department of Indian and Northern Affairs, Yellowknife, p. 46-47.

Padgham, W.A.
1990: The Slave Province, an overview; in Mineral Deposits of the Slave Province, Northwest Territories, Field Trip 13, 8th International Association on the Genesis of Ore Deposits Field Trip Guidebook, (ed.) W.A. Padgham, and D. Atkinson; p. 1-40.

Page, C.
1981: The B-zone deposit, Cullaton Lake, District of Keewatin, N.W.T.; in Proceedings of the Gold Workshop, (ed.) R.D. Morton; Yellowknife, Northwest Territories, December, 1979, p. 323-347.

*Phillips, G.N., Groves, D.I., and Martyn, J.E.
1984: An epigenetic origin for Archaean banded iron-formation hosted gold deposits; Economic Geology, v. 79, p. 162-171.

Redden, J.A., Peterman, Z.E., and DeWitt, E.
1987: U-Th-Pb zircon and monazite ages and preliminary interpretation of Proterozoic tectonism in the Black Hills, South Dakota (abstract); Geological Association of Canada, Program with Abstracts, Annual General Meeting, Saskatoon, Saskatchewan, May 1987.

Relf, C.
1989: Archean deformation of the Contwoyto Formation metasediments, western Contwoyto Lake area, Northwest Territories; in Current Research, Part C; Geological Survey of Canada, Paper 89-1C, p. 95-105.

*Sawkins, F.J. and Rye, D.M.
1974: Relationship of Homestake-type gold deposits to iron-rich Precambrian sedimentary rocks; Institution of Mining and Metallurgy, Transactions, Section B, v. 83, p. B56-B59, (also discussion in Transactions, Section B, v. 84, 1975, p. B37-B38).

*Seager, G.F.
1944: Gold, arsenic and tungsten deposits of the Jardine-Crevasse Mountain District, Park County, Montana; State of Montana Bureau of Mines and Geology, Memoir 23, 110 p.

Sethu Raman, K., Kruse, J., and Tenney, D.
1986: Geology, geophysics and geochemistry of the Cullaton B-Zone gold deposit, Northwest Territories; in Gold in the Western Shield, Proceedings of a Symposium Held in Saskatoon, September, 1985, (ed.) L.A. Clark; The Canadian Institute of Mining and Metallurgy, Special Volume 38, p. 307-321.

Thomson, J.E.
1938: The Crow River area, Ontario; Ontario Department of Mines, v. XLVII, pt. III, 65 p.

Vial, D.S.
1988: Geology of the Cuiaba gold mine, Quadrilatero Ferrifero, Minas Gerais, Brazil: in Extended Abstracts, Poster Programme, Volume 1, Bicentennial Gold 88, Melbourne, Geological Society of Australia Inc., Abstracts Volume, no. 23, p. 134-136.

Vieira, F.W.R., Filho, M.C., de Faria Fonseca, J.T., Pereira, A., de Oliveira, G.A.I., and Clemente, P.L.C.
1991a: Excursion to the Cuiabá gold mine, Minas Gerais, Brazil; in Gold Deposits Related to Greenstone Belts in Brazil-Deposit Modeling Workshop, Part A-Excursions, (ed.) C.H. Thorman, E.A. Ladeira, and D.C. Schnabel; United States Geological Survey, Bulletin 1980-A, p. A75-A86.

Vieira, F.W.R., Lisboa, L.H.A., Chaves, J.L., de Oliveira, G.A.I., Clemente, P.L.C., and de Oliveira, R.L.
1991b: Excursion to the Morro Velho gold mine, Minas Gerais, Brazil; in Gold Deposits Related to Greenstone Belts in Brazil-Deposit Modeling Workshop, Part A-Excursions, (ed.) C.H. Thorman, E.A. Ladeira, and D.C. Schnabel; United States Geological Survey, Bulletin 1980-A, p. A63-A74.

Wyman, D.A., Kerrich, R., and Fryer, B.J.
1986: Gold mineralization overprinting iron-formation at the Agnico-Eagle deposit, Quebec, Canada: mineralogical, microstructural and geochemical evidence; in Proceedings of Gold '86, an International Symposium on the Geology of Gold, (ed.) A.J. Macdonald; Toronto, 1986, p. 108-123.

15.4 DISSEMINATED AND REPLACEMENT GOLD

K.H. Poulsen

INTRODUCTION

In all geological environments, gold deposits are commonly composed of either vein or disseminated ores, or combinations of the two. The term disseminated refers to ores in which veins are minor and gold is "finely dispersed in host rocks of variable composition where little or no fabric control on mineralization is apparent, at least at the hand specimen scale" (Romberger, 1986). Although this term is commonly applied to deposits inferred to have formed at shallow to intermediate crustal depths in younger terranes (Carlin-type, mantos, gold-rich volcanic-associated massive sulphides, etc.), there are also many important Canadian deposits in metamorphic terranes, particularly in Precambrian and younger greenstone belts (sensu lato), that consist dominantly of disseminated ores for which the origins are more obscure. Disseminated and replacement gold deposits comprise mainly stratabound, auriferous bodies of disseminated to massive sulphides, commonly pyritic, that are hosted either by micaceous and/or aluminous schists, derived from tuff and volcanic sandstone, or by carbonate-clastic sedimentary rocks; spatial associations with granitoid rocks are common. They have low contents of base metal sulphides and commonly less silver than gold. Important Canadian examples (Fig. 15-1) include the Hemlo deposit in Ontario, QR and Equity Silver deposits in British Columbia, and the Hope Brook deposit in the Chetwynd area, Newfoundland (Table 15.4-1). The Archean Big Bell and Sons of Gwalia deposits in Western Australia and the Late Proterozoic-Paleozoic Haile, Brewer, and Ridgeway deposits in South Carolina, U.S.A. are of similar type. Sediment-associated deposits such as Island Mountain, British Columbia and Ketza River and Brewery Creek, Yukon Territorry have broadly similar characteristics. The last of these is analogous to "Carlin-like" deposits in the U.S.A.

IMPORTANCE

Sulphide-rich disseminated and replacement gold deposits represent approximately 5% of Canada's historical gold production and reserves but, with significant production from three mines on the Hemlo deposit, they account for approximately 25% of current output.

SIZE AND GRADE

The disseminated gold deposits have a similar size range as other subtypes (Fig. 15-1). From the giant Hemlo deposit (three mines, 600 t gold) to the smaller Hope Brook deposit (50 t gold), the stated total deposit size is dependent on the number and size of individual orebodies that constitute the deposit. It is not uncommon for deposits of this subtype to comprise several lenticular orebodies.

GEOLOGICAL FEATURES

Setting

Disseminated and replacement gold deposits occur in host rocks of both volcanic and sedimentary derivation. This includes tuffaceous metavolcanic rocks in the Precambrian greenstone belts and Phanerozoic arc terranes, as well as clastic and carbonate sedimentary rocks such as those found in the deformed passive margins of ancestral North America. The best Archean examples of the volcanic-associated type occur in the Superior Province at Hemlo in Ontario (Fig. 15.4-2); the sulphide orebodies in the Red Lake district, Ontario (Madsen, Campbell-Dickenson) and Beattie, Quebec are similar in many respects. The Hope Brook deposit (Fig. 15.4-3A), hosted by Late Precambrian to Paleozoic La Poile Group in the Chetwynd district, Newfoundland, is the most directly analogous deposit in younger terranes, but the early Proterozoic MacLellan deposit in Manitoba possesses some similar attributes. Similar younger deposits include the Tertiary Equity Silver (Fig. 15.4-3B) and Mesozoic QR deposits in British Columbia, which occur in the volcanic and volcaniclastic rocks of accreted island arc terranes. Sulphide "replacement" orebodies such as those at Island Mountain, British Columbia and Ketza River, Yukon Territory are examples of sediment-associated deposits of this type. The former is hosted by the Late Precambrian to Paleozoic Barkerville carbonate and clastic rocks in the Kootenay terrane of the Cariboo district, and the latter by similar Cambrian strata in the Cassiar terrane. The Brewery Creek deposit in Yukon Territory also occurs in Paleozoic clastic rocks of the continental miogeocline.

Deposits of this type have notable similarities and significant differences in their geological settings at a district scale (Table 15.4-1). In most cases, the deposits occur in linear belts containing a diversity of lithological units with subparallel contacts (Fig. 15.4-2, 15.4-3). Mafic host rocks are regionally important at MacLellan deposit, Red Lake district, and QR deposit; are present at Hemlo and Cariboo districts, but are rare in the sequences at Hope Brook and Equity Silver deposits. Where they are present and well preserved, the mafic units are interpreted to be volcanic flows. Felsic rocks at these deposits can be ascribed to both volcanic and sedimentary origins. The host rocks at the Madsen deposit, Red Lake district, consist of intermediate

Poulsen, K.H.
1996:   Disseminated and replacement gold; in Geology of Canadian Mineral Deposit Types, (ed.) O.R. Eckstrand, W.D. Sinclair, and R.I. Thorpe; Geological Survey of Canada, Geology of Canada, no. 8, p. 383-392 (also Geological Society of America, The Geology of North America, v. P-1).

to felsic volcaniclastic rocks; at Hemlo and Chetwynd districts, both volcaniclastic rocks and wacke are present. In the Cariboo district, quartz wacke and phyllite are dominant.

Intrusions form a significant proportion of the rocks in most of the districts containing disseminated gold deposits (Table 15.4-1). These take the form of stocks and dykes, ranging from mafic to felsic composition and from pre- to post-tectonic timing. Granodioritic plutons and associated dykes are present at Hemlo (Cedar Lake Pluton), where they have an inferred pre- to synkinematic timing (Fig. 15.4-2), whereas the granodioritic Faulkenham Lake stock at Madsen deposit, and the Chetwynd Granite at Hope Brook, are both postkinematic. Pre- to synkinematic mafic dykes are common at Hemlo and Hope Brook deposit. At Equity Silver, the deposit lies adjacent to a quartz-monzonite stock containing subeconomic porphyry style Cu-Mo mineralization and the ore zones are cut by unaltered dacite dykes (Fig. 15.4-4C). The Brewery Creek deposit is related to Late Cretaceous felsic dykes and QR to quartz monzonite porphyry, and a deeply buried intrusion is inferred at Ketza River on the basis of existing hornfelsed rocks. No intrusive activity can conclusively be linked to the Cariboo district deposits.

Regional dynamothermal metamorphism of low to medium grade has affected the rocks in all districts that contain deposits of this subtype. Middle to upper greenschist metamorphic conditions are inferred at Cariboo district and Hope Brook deposit; the Red Lake district; MacLellan, and Hemlo deposits occur in rocks at the transition from greenschist to amphibolite facies. Where rocks of the upper greenschist and amphibolite facies are present, the presence of diagnostic minerals, such as cordierite and andalusite at Red Lake, and co-existing sillimanite and kyanite at Hemlo, indicates that metamorphism was of low to moderate pressure.

Structure

In each of the districts cited, the rocks were penetratively deformed during regional metamorphism (Table 15.4-1), and this has resulted in at least one generation of tectonic fabrics that overprints the main lithological units. In most cases, a strong foliation, amplified in discrete fault zones (Fig. 15.4-2, 15.4-3A), strikes subparallel to the regional lithological trend. In most cases, minor folds have been noted to be contemporaneous with foliation, and the transposition of bedding into parallelism with foliation is

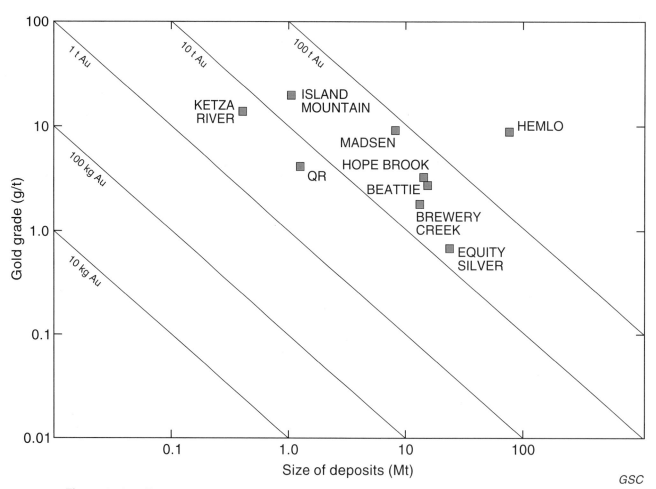

GSC

Figure 15.4-1. Tonnage-grade plot of selected Canadian disseminated and replacement gold deposits.

Table 15.4-1. Geological characteristics of selected Canadian disseminated and replacement gold deposits.

Deposit	Size	Age	Host rocks	Structure	Intrusions	Ore mineralogy	Gangue	Alteration
volcanic-associated:								
Hemlo	600 t Au	Archean	felsic tuff, wacke	intense E-W foliation	feldspar porphyry dykes	pyrite, molybdenite, sphalerite, arsenopyrite, stibnite, cinnabar, realgar	quartz, muscovite, biotite, microcline, barite, V-muscovite	kyanite, sillimanite, staurolite
Hope Brook	50 t Au	Paleozoic	wacke-tuff	intense NE-SW foliation	quartz-feldspar porphyry	pyrite, chalcopyrite, molybdenite	quartz, sericite	andalusite, paragonite, pyrophyllite
Equity Silver	25 t Au	Tertiary	Cretaceous lapilli tuff, tuff, conglomerate, sandstone	N-S folds	quartz monzonite stock; gabbro-monzonite complex	arsenopyrite, pyrite, chalcopyrite, tetrahedrite, sphalerite	quartz, chlorite, sericite	andalusite, pyrophyllite, corundum, tourmaline, scorzalite
QR	7 t Au	Mesozoic	Triasssic-Jurassic basalt, tuff, breccia, and siltstone	thrust and normal faults	early Cretaceous diorite	pyrite, chalcopyrite, sphalerite; rare arsenopyrite, galena	epidote	carbonate, epidote, chlorite, calcite
Beattie	50 t Au	Archean	basalt flows, breccias	E-W foliation, local faults	porphyritic syenite stock and dykes	pyrite, arsenopyrite; trace chalcopyrite, sphalerite, galena, molybdenite	quartz, microcline	microcline
Madsen	84 t Au	Archean	felsic tuff	intense NE-SW foliation	adjacent quartz porphyry	pyrite, pyrrhotite, arsenopyrite, sphalerite, chalcopyrite, molybdenite	quartz, biotite, chlorite	andalusite, cordierite, garnet
MacLellan	10 t Au	Early Proterozoic	basalt, siltstone	intense E-W foliation	none	pyrite, pyrrhotite, arsenopyrite, galena, sphalerite	quartz, biotite	andalusite, staurolite, sillimanite
sediment-associated:								
Island Mountain	22 t Au	Mesozoic (?)	Paleozoic limestone beds in wacke	intense NW-SE foliation	none	pyrite, arsenopyrite; minor sphalerite-galena	quartz	sericite(?)
Ketza River	5.3 t Au	Mesozoic	Cambrian dolomitic limestone in argillite-quartzite hornfels	intersecting steep faults	inferred buried stock	pyrrhotite, pyrite, arsenopyrite; minor sphalerite, galena	siderite, calcite	silicification
Brewery Creek	27.5 t Au	Mesozoic	Paleozoic argillite, sandstone, and barite	adjacent normal faults	quartz monzonite porphyry	pyrite, arsenopyrite; stibnite in late veins; local realgar	quartz	sericite

an attribute of all the districts (Alldrick, 1983; Andrews et al., 1986; O'Brien, 1987; Muir and Elliot, 1987). Such transposition accounts for the "straightness" of the belts (e.g., Fig. 15.4-2) and is largely responsible for obscuring the primary relationships between the ore deposits and their host rocks. Linear fabrics, such as the axes of asymmetric minor folds and mineral and shape lineations, are also characteristic of these areas and are of consistent orientation within a district. Fold hinges and lineations with shallow plunge are present at Cariboo district, whereas moderate to steep plunges have been noted at Hemlo and Red Lake.

Nature and composition of orebodies

Disseminated and replacement gold orebodies are commonly stratabound at the scale of a district (Fig. 15.4-2, 15.4-3). This is attributable to the fact that they occur within, and along the strike of, well defined lithotectonic packages of rocks. Furthermore, they commonly occur at contacts between distinctive lithological units or solely within a particular unit. The lenticular to tabular shape of most orebodies is such that they are geometrically concordant with their host rocks (Fig. 15.4-3, 15.4-4). Individual deposits commonly comprise several subparallel orebodies

that are arranged in a stacked fashion or along strike from one another. The long axes of orebodies (e.g., Fig. 15.4-5) are commonly parallel to other linear fabrics in a district. Ores within these deposits are sulphide-bearing, commonly schistose rocks in which the proportions of sulphides and the nature of the silicate hosts differ from orebody to orebody and from deposit to deposit (Table 15.4-1).

The Hemlo ores (Fig. 15.4-4A, 15.4-5) contain on average 8% pyrite (Harris, 1989), and occur principally in three rock types (Kuhns et al., 1986; Walford et al., 1986; Burk et al., 1986), namely quartz-microcline±barite±molybdenite schist, quartz-muscovite schist, and biotite schist. The ore types contain a wide variety of ore minerals (Table 15.4-1) that reflect chemical enrichment of Au, Mo, Sb, As, Hg, Tl, V, and Ba (Harris, 1989). The mercury minerals are located centrally within the "A" orebody where the ore is thickest (Fig. 15.4-5) and, although they occur mainly in a late paragenesis, their distribution suggests a cogenetic relationship with the gold orebody.

The Hope Brook ores are composed almost entirely of fine grained quartz with minor sericite and about 5% pyrite; chalcopyrite is a minor but common constituent (McKenzie, 1986). They occur within a wider zone of strong silicification in the structural hanging wall of a barren pyritic zone, the "pyritic cap" (Fig. 15.4-4B).

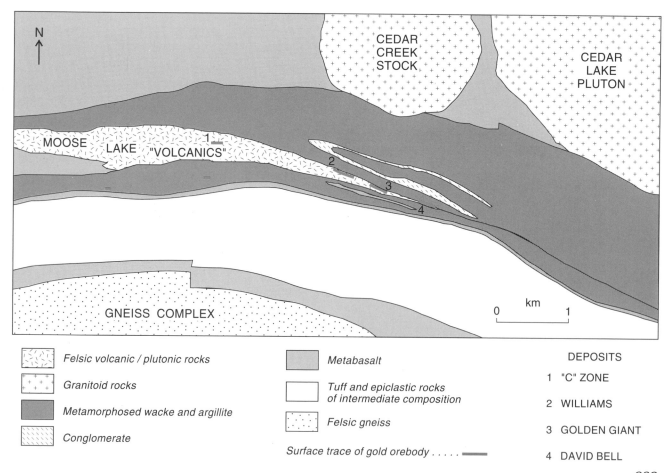

Figure 15.4-2. Geological setting of the Hemlo gold deposits (adapted after Patterson, 1985).

GSC

At Equity Silver silver-copper-gold-antimony ores occur in three separate orebodies (Fig.15.4-3B). The ores of the Main zone and its northern extension, the Waterline zone, are composed of pyrite-chalcopyrite-tetrahedrite and arsenopyrite finely disseminated in the matrix of volcanic breccia and closely associated with andalusite, blue scorzalite, tourmaline, and dumortierite (Wojdak and Sinclair, 1984). The Southern Tail ores (Fig. 15.4-3B, 15.4-4C) also are composed mainly of pyrite, arsenopyrite, chalcopyrite, and tetrahedrite, but are coarser grained fracture fillings and are related to sericitic alteration that overprints aluminous alteration.

Horwood (1940) described the Madsen orebodies as "lenses of sheared tuff with sulphide mineralization". He noted an early barren phase of massive pyrite overprinted by "orebodies" containing "variable quantities of pyrite, pyrrhotite, arsenopyrite, sphalerite, chalcopyrite, magnetite and gold".

The MacLellan ores are quartz-biotite-sulphide schists containing four successive generations of veinlets (Gagnon and Sampson, 1989). Gold and silver occurs predominantly in third generation quartz-arsenopyrite veinlets whereas veinlets of the fourth generation contain quartz, galena, sphalerite, pyrite, and arsenopyrite.

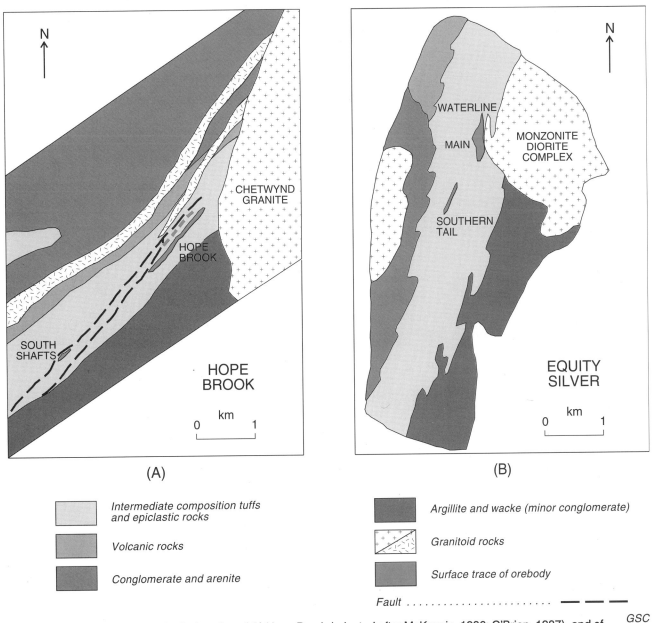

(A)

(B)

Intermediate composition tuffs and epiclastic rocks

Volcanic rocks

Conglomerate and arenite

Argillite and wacke (minor conglomerate)

Granitoid rocks

Surface trace of orebody

Fault . . . .

GSC

Figure 15.4-3. Geological setting of **A)** Hope Brook (adapted after McKenzie, 1986; O'Brien, 1987), and of **B)** Equity Silver (after Cyr et al., 1984).

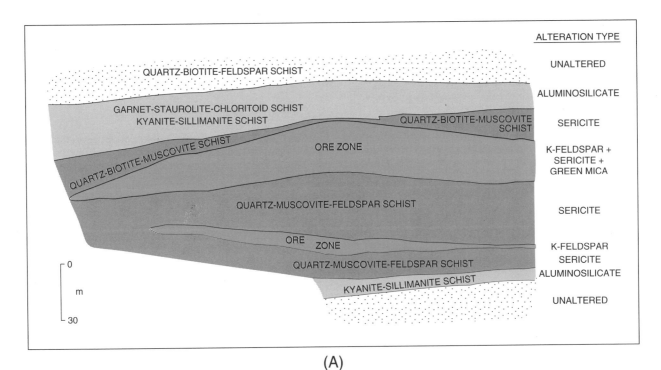

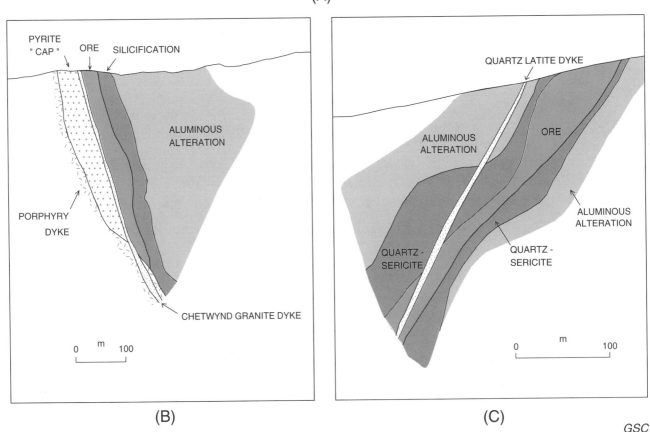

Figure 15.4-4. Distribution of alteration minerals in, and adjacent to, disseminated and replacement gold orebodies, which are shown in red: **A)** Plan of the Golden Giant mine, Hemlo (after Kuhns, 1986), **B)** Cross-section of the Hope Brook deposit, Chetwynd (after McKenzie, 1986), **C)** Cross-section of the Southern Tail zone of the Equity Silver deposit (after Wojdak and Sinclair, 1984).

GSC

The immediate host of the sulphide ores of the Cariboo district is siliceous, sericitic, dolomitic limestone. Ore lenses are disseminated to massive accumulations of fine grained pyrite, and lesser arsenopyrite, sphalerite, galena, and pyrrhotite (Alldrick, 1983). Similar pyrite-pyrrhotite-arsenopyrite ores at Ketza River are hosted by dolomitic limestone.

Alteration

All deposits in this category occur in regionally metamorphosed terranes in which minerals that are normally attributable to hydrothermal alteration are also part of the mineral assemblage of metamorphosed, but unaltered, rocks. This is particularly true because, in some cases, the deposits occur in rocks that are of an original pelitic composition. Nonetheless, anomalous abundances of some minerals, coupled with enrichment and depletion of selected trace elements, have been noted at the various deposits. These are likely the products of hydrothermal alteration, although their direct relationship to gold deposition has rarely been demonstrated. Unlike the quartz-carbonate vein gold deposits that occur in the same types of metamorphic terranes, the disseminated deposits are not noted for abundant carbonate alteration; the QR deposit is an exception (Melling and Watkinson, 1988). They do, however, display evidence of sericitic alteration like those of the vein type, but also show a notable spatial association with rocks that contain peraluminous mineral assemblages (Table 15.4-1) or in some cases, potassic alteration.

Sericitic alteration is a common feature of most deposits of this type. The sericite, or muscovite and/or biotite in higher grade metamorphic assemblages, occurs with quartz in mineral assemblages containing few phases, and in abundances that preclude formation by metamorphism of an unaltered protolith. For example, quartz-muscovite schists at Hemlo (Fig. 15.4-4A) are devoid of Na, Mg, and Ca (Kuhns et al., 1986) and are almost certainly hydrothermally altered. Similarly, the Hope Brook orebody (Fig. 15.4-4B) is enclosed in a silicified zone composed entirely of quartz and minor sericite (McKenzie, 1986).

Potassic alteration, in the form of abundant microcline, in part barium-rich, occurs in quartz-microcline rocks that host and envelop the ore (Fig.15.4-4A) at Hemlo (Kuhns et al., 1986; Harris, 1989). It is also an important constituent of other smaller, disseminated deposits in volcanic rocks, such as Beattie, Lac Shortt, and Bachelor Lake in the Abitibi belt of Quebec.

Aluminous mineral assemblages are distinctive features of the rocks adjacent to many of the volcanic-associated gold deposits of this type. A 200 m wide zone of aluminous alteration (Fig. 15.4-4B), containing abundant andalusite, pyrophyllite, paragonite, and local alunite,

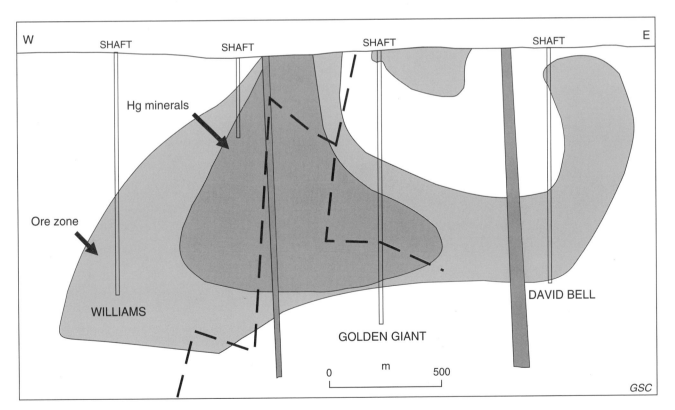

Figure 15.4-5. Vertical longitudinal projection (looking north) of the westward-plunging Hemlo "A" orebody (pink) through the Williams, Golden Giant, and David Bell mines, illustrating the distribution of mercury minerals (red) (after Harris, 1989). Diabase dykes (grey) crosscut the orebody.

occupies the structural hanging wall of the Hope Brook deposit (McKenzie, 1986; Stewart, 1990). Units of kyanite-sillimanite schist flank the Hemlo orebody (Fig. 15.4-4A), and the ore zones at Equity Silver are enveloped by a zone (Fig. 15.4-4C) containing abundant andalusite and minor pyrophyllite (Wojdak and Sinclair, 1984). Ferromagnesian aluminous minerals also occur in significant proportions adjacent to several of the deposits (Table 15.4-1). Chloritoid is notable at Hemlo, staurolite at Hemlo and Madsen, and cordierite at Hope Brook and Madsen (Durocher, 1983; McKenzie, 1986).

The distribution and significance of the alteration types with respect to ore is, in most cases, difficult to assess. This is primarily because of the uncertainty in distinguishing mineral assemblages that are attributable solely to iso-chemical metamorphism of unaltered pelitic rocks from those that are products of high temperature metasomatism or of metamorphism of hydrothermally altered rocks. The possibility that ore has been dislocated from its wall rocks by faults during postore deformation is an added complication at some deposits. Nonetheless, zoning patterns have been documented at some deposits (Fig. 15.4-4), and they illustrate the importance of sericitic, silicic, potassic, and aluminous alteration. Potassic and sericitic alteration at Hemlo and in association with the southern Tail Zone at Equity Silver are closer to ore than the peripheral alumi-nous type, whereas at Hope Brook sericite occurs through-out the aluminous alteration zone which envelops silicified rocks, but which is thickest in the structural hanging wall of the deposit.

The sericitic and aluminous alteration observed at many of the volcanic-associated deposits in this category are similar, in many respects, to those ascribed to seafloor alteration (Mathieson and Hodgson, 1984). Also, the altera-tion pattern is comparable to that of certain porphyry deposits and epithermal gold deposits of acid-sulphate type (Schmidt, 1985; Ririe, 1990).

Definitive characteristics

As a group, these deposits possess the following definitive characteristics:

1) they are sulphidic gold deposits in which ore distribu-tion is not dictated by vein quartz;

2) they are commonly stratabound at the district and deposit scale and are commonly hosted by clastic rocks of volcanic and/or sedimentary origin;

3) with few exceptions, granitoid rocks, both as dykes and stocks, are present in the ore environment;

4) with few exceptions, they have low contents of base metals (less than one per cent combined metal) and gold contents exceeding those of silver; arsenopyrite is a common constituent.

5) orebodies in volcanic environments are closely associ-ated with zones of potassic alteration or zones of silici-fication enclosed by aluminous alteration; sericitic alteration is ubiquitous.

GENETIC MODEL

Although the deposits of this type conform to a reasonably coherent descriptive model, they do not necessarily all have a common genesis. The failure to establish consistent genetic models for these deposits stems mainly from the difficulty in distinguishing the effects of ore-forming proc-esses from those of regional metamorphism and deforma-tion in the absence of significant veins that could be used as tectonic markers. The most commonly proposed models for these deposits are of two types: those that relate the deposits directly to deformation and metamorphism, and those that view the deposits to be pretectonic entities (e.g., porphyry-related mantos, epithermal, or volcanic exhalative) which, in some cases, were merely redistrib-uted by overprinting deformation and metamorphism.

The historically accepted genetic model for these depos-its is that of "replacement" of wall rocks during deforma-tion and metamorphism (e.g., Cooke, 1946). At one time or another, such a model has been applied to all of the deposits described above. In its modern application (e.g., Phillips, 1985; Colvine et al., 1988), their formation is considered to be identical in timing, and in fluid source, to that of "mesothermal" quartz-carbonate vein deposits, the only difference being one of depositional setting. The dissemi-nated ores of the type at Madsen and Hemlo are considered to represent a deeper, hotter, and more ductile depositional environment than that of quartz vein deposits. A point in favour of such a model is that, in many cases, the dissemi-nated ores strongly resemble vein-type ores in their bulk chemical composition, particularly the associated trace metal suite (As, Sb, Te) and gold:silver ratios. Furthermore, many gold mines contain both types of orebodies; the controversial East South "C" orebodies in the Campbell-Dickenson deposit at Red Lake are well studied examples in which sulphide ores at deep levels grade into vein ores at higher structural levels in the same deposit (Mathieson and Hodgson, 1984), and the Cariboo deposits are known for both their vein and "replacement" ores. Therefore, such dynamothermal replacement models do not treat disseminated greenstone gold deposits as a subtype separate from quartz-carbonate veins, but rather as a vari-ation in a unified model of "mesothermal and hypothermal" lode gold deposits, the differences being attributable to crustal level of ore deposition.

Alternative models portray disseminated and replace-ment gold deposits as a separate class of pretectonic depos-its of epigenetic origin. Such models typically are analogous to those for porphyry and epithermal deposits in younger geological environments. Porphyry and intrusion-related models are attractive for at least some of the disseminated deposits identified here, particularly if one considers the inferred relationships among porphyry deposits, overlying epithermal deposits and peripheral gold-bearing mantos in younger terranes. For example, McKenzie (1986) used the gold-copper-barium geochemical association and the devel-opment of extensive advanced argillic alteration to argue in favour of an epithermal model for the Hope Brook deposit. Similarly the Au-Hg-Sb-As metal association at Hemlo suggests an epithermal affiliation, whereas the high Mo concentrations and the presence of substantial potassic alteration favour a porphyry affiliation (Schmidt, 1985; Kuhns, 1986). A key unresolved question in this regard is whether the Moose Lake "Volcanics" at Hemlo (Fig. 15.4-2), a unit of fine grained quartz-phyric felsite, is indeed of extrusive origin or merely a deformed hypabyssal intrusion. The presence of Cu-Mo stockwork mineralization in granodiorite adjacent to, and of approximately the same age as, Au-Ag-Cu-Sb ore at Equity Silver also strongly favours a genetic relationship between the two. Although strongly deformed, the Island Mountain "replacement orebodies" resemble other mantos (e.g., Ketza River) in both form and setting (F. Robert and B.E. Taylor, pers. comm., 1988) and the Brewery Creek deposit is broadly similar to Carlin-type and Carlin-like deposits in the western U.S.A. which, in turn, locally display clear relationships to intrusions.

RELATED DEPOSIT TYPES

Auriferous volcanic-associated massive sulphide deposits

There are several superficial similarities in the geological characteristics of disseminated and replacement gold deposits in volcanic rocks with auriferous volcanic-associated massive sulphide deposits (see subtype 6.3). Although they share common aluminosilicate alteration and a primarily sulphidic nature, the case for volcanic exhalative origins for deposits of the former type (e.g., Valliant and Bradbrook, 1986) is, however, not particularly strong. In no cases (e.g., Hemlo, Hope Brook, Equity Silver, etc.) have exhalative units been convincingly identified, and related intrusions tend to be post- rather than synvolcanic in timing. A common point in the genesis of the deposits discussed here and auriferous massive sulphides may lay, however, in the fact that, like the stockwork zones beneath exhalative deposits, they are subsurface zones of veining and replacement. One distinction between auriferous volcanogenic massive sulphide deposits (subtype 6.3) and those of "disseminated and replacement" type may be their composition: the former tend to have lower gold:silver ratios than the latter.

EXPLORATION GUIDES

On a regional scale, disseminated gold deposits occur at major lithological contacts which mark a distinctive change in volcanic and sedimentary facies. Furthermore, these contacts are, in part, of deformational origin in that they commonly coincide with zones of intense layer transposition. At a local scale, the presence of aluminous mineral assemblages in rocks in which they are not normally expected may be a useful exploration guide. The sulphide contents of many of these deposits are sufficient to produce geophysical responses and, owing to the disseminated nature of the sulphides, induced polarization methods should be the most effective. Lithogeochemical anomalies for trace elements such as Sb, As, and Hg have locally been shown to correlate with orebodies of this type (Durocher, 1983; Kuhns, 1986; Harris, 1989), but no single element, or suite of elements, is diagnostic of all of the deposits of this subtype. For those deposits like Hemlo, in which potassic alteration is important, gamma-ray spectrometry has proven to be a useful tool for mapping hydrothermal alteration.

REFERENCES

References with asterisks (*) are considered to be the best source of general information on this deposit subtype.

Alldrick, D.
1983: The Mosquito Creek Mine, Cariboo gold belt; in Geological Fieldwork, 1982; British Columbia Ministry of Energy, Mines and Petroleum Resources, Paper 1983-1, p. 98-112.

Andrews, A.J., Hugon, H., Durocher, M., Corfu, F., and Lavigne, M.J.
1986: The anatomy of a gold-bearing greenstone belt: Red Lake, northwestern Ontario, Canada; in Proceedings of Gold '86, an International Symposium on the Geology of Gold, (ed.) A.J. Macdonald; Toronto, 1986, p. 3-22.

Burk, R., Hodgson C.J., and Quartermain, R.A.
1986: The geological setting of the Teck-Corona Au-Mo-Ba deposit, Hemlo, Ontario, Canada; in Proceedings of Gold '86, an International Symposium on the Geology of Gold, (ed.) A.J. Macdonald; Toronto, 1986, p. 311-326.

Colvine, A.C., Fyon, J.A., Heather, K.B., Marmont, S., Smith, P.M., and Troop, D.G.
1988: Archean lode gold deposits in Ontario; Ontario Geological Survey, Miscellaneous Paper 139, 136 p.

Cooke, H.C.
1946: Canadian lode gold areas; Canada Department of Mines and Resources, Economic Geology Series, No. 15, 86 p.

Cyr, J.B., Pease, R.B., and Schroeter, T.G.
1984: Geology and mineralization at Equity Silver mine; Economic Geology, v. 79, p. 947-968.

Durocher, M.E.
1983: The nature of hydrothermal alteration associated with the Madsen and Starratt-Olsen gold deposits, Red Lake area; in The Geology of Gold in Ontario; Ontario Geological Survey, Miscellaneous Paper 110, p. 111-140.

Gagnon, J.E. and Sampson, I.M.
1989: Geology, mineralization and alteration in the MacLellan Au-Ag deposit, Lynn Lake, Manitoba; in Manitoba Energy and Mines, Minerals Division, Report of Field Activities, 1989, p. 13-15.

*Harris, D.C.
1989: The mineralogy and geochemistry of the Hemlo gold deposit, Ontario; Geological Survey of Canada, Economic Geology Report 38, 88 p.

Horwood, C.
1940: Geology and mineral deposits of the Red Lake area; Ontario Department of Mines, 49th Annual Report, v. 44, pt. 2, 231 p.

*Kuhns, R.J.
1986: Alteration styles and trace element dispersion associated with the Golden Giant deposit, Hemlo, Ontario, Canada; in Proceedings of Gold '86, an International Symposium on the Geology of Gold, (ed.) A.J. Macdonald; Toronto, 1986, p. 340-354.

Kuhns, R.J., Kennedy, P., Cooper, P., Brown, P., Mackie, B., Kusins, R., and Friesen, R.
1986: Geology and mineralization associated with the Golden Giant deposit, Hemlo, Ontario, Canada; in Proceedings of Gold '86, an International Symposium on the Geology of Gold, (ed.) A.J. Macdonald; Toronto, 1986, p. 327-339.

Mathieson, N.A. and Hodgson, C.J.
1984: Alteration, mineralization, and metamorphism in the area of the East South "C" ore zone, 24th level of the Dickenson mine, Red Lake, northwestern Ontario; Canadian Journal of Earth Sciences, v. 21, p. 35-54.

McKenzie, C.B.
1986: Geology and mineralization of the Chetwynd deposit, southwestern Newfoundland, Canada; in Proceedings of Gold '86, an International Symposium on the Geology of Gold, (ed.) A.J. Macdonald; Toronto, 1986, p. 137-148.

Melling, D.R. and Watkinson, D.H.
1988: Alteration of fragmental basaltic rocks: the Quesnel River gold deposit, central British Columbia; in Geological Fieldwork, 1987, British Columbia Ministry of Energy, Mines and Petroleum Resources, Paper 1988-1, p. 335-347.

Muir, T.L. and Elliott, C.G.
1987: Hemlo tectono-stratigraphic study, District of Thunder Bay; in Summary of Field Work and Other Activities, 1987, by the Ontario Geological Survey; Ontario Geological Survey, Miscellaneous Paper 137, p. 117-129.

O'Brien, B.H.
1987: The lithostratigraphy and structure of the Grand Bruit-Cinq Cerf area (parts of 11/O/9 and 11/O/16), southwestern Newfoundland; in Current Research (1987); Newfoundland Department of Mines and Energy, Mineral Development Division, Report 87-1, p. 311-334.

Patterson, G.C.
1985: Exploration history and field stop descriptions of the Hemlo area; in Gold and Copper-zinc Metallogeny within Metamorphosed Greenstone Terrain Hemlo-Manitouwadge-Winston Lake Ontario, Canada, (ed.) R.H. McMillan and D.J. Robinson; Geological Association of Canada and The Canadian Institute of Mining and Metallurgy, p. 66-86.

*Phillips, G.N.
1985: Interpretation of Big Bell/Hemlo-type deposits: precursors, metamorphism, melting and genetic constraints; Transactions of the Geological Society of South Africa, v. 88, p. 159-174.

Ririe, G.T.
1990: A comparison of alteration assemblages associated with Archean gold deposits in Western Australia and Paleozoic gold deposits in the southeast United States; Canadian Journal of Earth Sciences, v. 27, p. 1560-1576.

Romberger, S.B.
1986: Ore deposits #9: disseminated gold deposits; Geoscience Canada, v. 13, p. 23-31.

*Schmidt, R.G.
1985: High-alumina hydrothermal systems in volcanic rocks and their significance to mineral prospecting in the Carolina Slate Belt; United States Geological Survey, Bulletin 1562, 59 p.

Stewart, P.W.
1990: A shear zone-hosted and metamorphosed acid-sulphate gold deposit, Hope Brook, Newfoundland; Geological Association of Canada-Mineralogical Association of Canada, Program with Abstracts, v. 15, p. A125.

Walford, P., Stephens, J., Skrecky, G., and Barnett, R.
1986: The geology of the "A" Zone, Page-Williams Mine, Hemlo, Ontario, Canada; in Proceedings of Gold '86, an International Symposium on the Geology of Gold, (ed.) A.J. Macdonald; Toronto, 1986, p. 362-378.

Wojdak, P.J. and Sinclair, A.J.
1984: Equity Silver silver-copper-gold deposit: alteration and fluid inclusion studies; Economic Geology, v. 79, p. 969-990.

Valliant, R.I. and Bradbrook, C.J.
1986: Relationship between stratigraphy, faults and gold deposits, Page-Williams Mine, Hemlo, Ontario, Canada; in Proceedings of Gold '86, an International Symposium on the Geology of Gold, (ed.) A.J. Macdonald; Toronto, 1986, p. 355-361.

Authors' addresses

J.A. Kerswill
Geological Survey of Canada
601 Booth Street
Ottawa, Ontario
K1A 0E8

K.H. Poulsen
Geological Survey of Canada
601 Booth Street
Ottawa, Ontario
K1A 0E8

François Robert
Geological Survey of Canada
601 Booth Street
Ottawa, Ontario
K1A 0E8

B.E. Taylor
Geological Survey of Canada
601 Booth Street
Ottawa, Ontario
K1A 0E8

Printed in Canada

16. CLASTIC METASEDIMENT-HOSTED VEIN SILVER-LEAD-ZINC

16. CLASTIC METASEDIMENT-HOSTED VEIN SILVER-LEAD-ZINC

G. Beaudoin and D.F. Sangster

INTRODUCTION

Silver-lead-zinc vein districts are commonly associated with major fault zones in clastic metasedimentary terranes; individual veins occur in a variety of lithologies ranging in age from Proterozoic to Cenozoic. Silver-lead-zinc veins are a late feature in the tectonic evolution of orogens. Classical examples are the Kokanee Range (British Columbia), Keno Hill (Yukon Territory), Coeur d'Alène (U.S.A.), Příbram (Czechoslovakia), and the Harz Mountains and Freiberg (Germany) (Fig. 16-1).

IMPORTANCE

Silver-lead-zinc veins constitute one of the largest silver resources in the world with the Coeur d'Alène district being the world's largest silver district; in Europe they have been mined since the Middle Ages. In Canada, deposits of this type are of diminishing economic importance.

SIZE AND GRADE OF DEPOSITS

Metal production in a single district ranges up to 30 kt Ag, 7 Mt Pb, and 3 Mt Zn. In terms of contained metals, a large silver-lead-zinc vein district (such as Coeur d'Alène) compares with a large Zn-Pb-Ag sedimentary-exhalative

deposit or a large Zn-Pb Mississippi Valley-type district. Grades are highly variable as they may be biased by selective mining methods. Examples follow:

Kokanee Range – district total: 10.4 Mt grading 5.1% Pb, 4.8% Zn, and 251 g/t Ag;
  Silvana: 3.8 kt grading 5.8% Pb, 5.1% Zn, and 515 g/t Ag;
  Bluebell: 4.8 Mt grading 4.8% Pb, 4.8% Zn, and 46 g/t Ag;
Keno Hill – district total: 4.54 Mt grading 6.8% Pb, 4.6% Zn, and 1412 g/t Ag;
  Husky – production to 1984: 3.6 kt grading 3.96% Pb, 0.27% Zn, and 1450 g/t Ag;
Coeur d'Alène – production to 1965: >100 Mt grading 6.2% Pb, 1.9% Zn, and 193 g/t Ag.

Silver-lead-zinc vein districts are characterized by Pb/(Pb+Zn) ratios ranging from 0.51 to 0.72 and (Agx100)/((Agx100)+Pb) ratios ranging from 0.22 to 0.63. In Ag-Pb-Zn space (Fig. 16-2A), lead-zinc skarns are characterized by high Zn contents, whereas porphyry copper and epithermal vein districts have higher Ag, compared with silver-lead-zinc vein districts. In Ag-Pb-Au space, carbonate replacement and manto deposits are enriched in Au relative to Ag-Pb-Zn vein districts (Fig. 16-2B).

GEOLOGICAL FEATURES

Geological setting

The classical silver-lead-zinc vein districts are in two orogens: the Cordilleran Orogen of North America and the Variscan Orogen of Europe (Fig. 16-1). The districts are in metasedimentary terranes typically dominated by thick

Beaudoin, G. and Sangster, D.F.
1996: Clastic metasediment-hosted vein silver-lead-zinc; in Geology of Canadian Mineral Deposit Types, (ed.) O.R. Eckstrand, W.D. Sinclair, and R.I. Thorpe; Geological Survey of Canada, Geology of Canada, no. 8, p. 393-398 (also Geological Society of America, The Geology of North America, v. P-1).

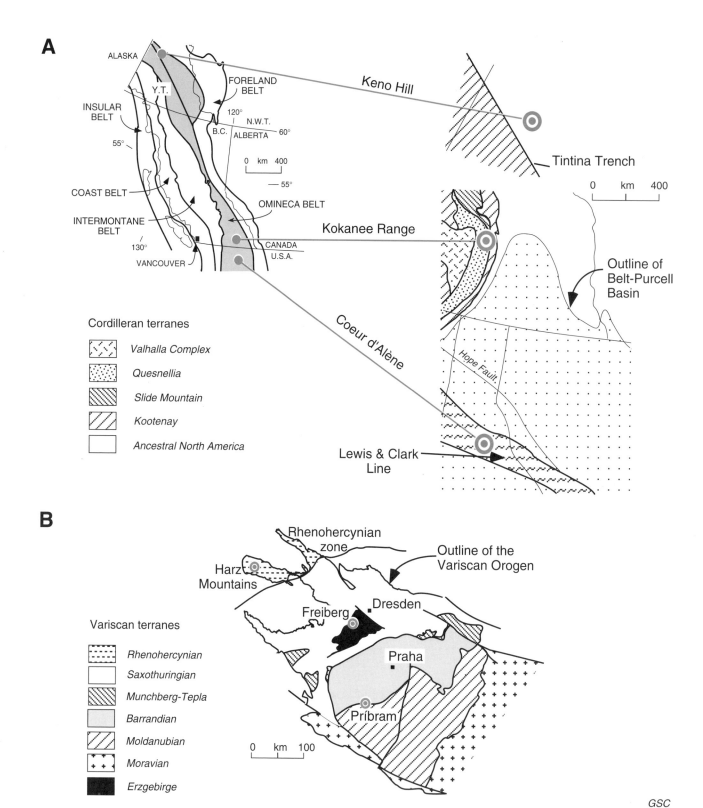

Figure 16-1. Location of six classical silver-lead-zinc vein districts in: **A)** the Cordilleran Orogen of North America; **B)** the Variscan Orogen of Europe (from Beaudoin and Sangster, 1992; reproduced from Economic Geology, 1992, v. 87, p. 1007).

and monotonous sequences of fine- to medium-grained clastic rocks with minor carbonate, mafic volcanic, and tuff units. The sedimentary basins, initiated as epicratonic embayments on passive margins, or in continental, oceanic, or back-arc marginal basins, are commonly part of large Pb-Zn metallogenic provinces containing large sedimentary-exhalative deposits (e.g. Selwyn Basin and Belt-Purcell Basin). These basins have typically been deformed, metamorphosed, and intruded by igneous rocks. The latter comprise zoned monzonitic to syenitic plutons (Coeur d'Alène), syn- to late orogenic granodioritic batholiths (Kokanee Range, Příbram), gabbronorite (Harz Mountains), and postorogenic dioritic to granitic plutons (Harz Mountains, Keno Hill). The intrusions can be classified as I, S, and within-plate types with alkaline to calc-alkaline affinities (Beaudoin and Sangster, 1992).

Age of host rocks and mineralization

Silver-lead-zinc veins are found in sedimentary, volcanic, or plutonic rocks ranging in age from Proterozoic to Eocene. Although the veins, traditionally, have been genetically related to the intrusion of granitic plutons or batholiths, recent geochronological data have demonstrated a significant age difference between intrusion and mineralization. The latter has been shown (Beaudoin and Sangster, 1992) to be younger(-) or older(+) than those intrusions which, traditionally, have been considered to be genetically related to mineralization: ~-110 Ma (Kokanee Range); ~-170 Ma (Harz Mountains); >+770 Ma (Coeur d'Alène). These large differences in ages have resulted in a re-assessment of the relationship between intrusions and Ag-Pb-Zn veins.

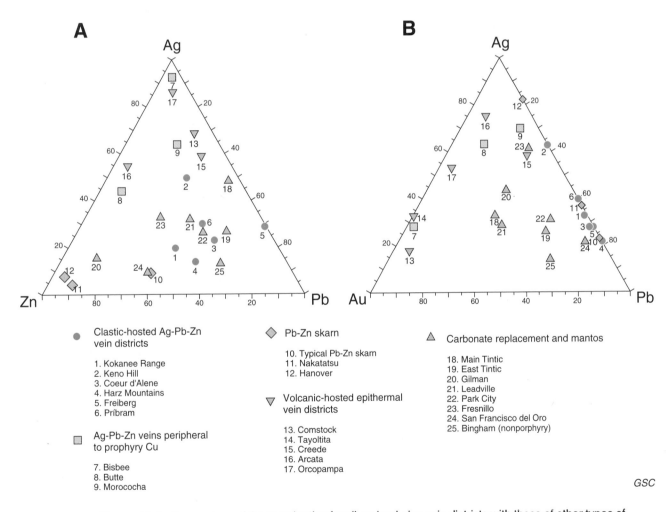

A

B

Clastic-hosted Ag-Pb-Zn vein districts

1. Kokanee Range
2. Keno Hill
3. Coeur d'Alene
4. Harz Mountains
5. Freiberg
6. Příbram

Ag-Pb-Zn veins peripheral to prophyry Cu

7. Bisbee
8. Butte
9. Morococha

Pb-Zn skarn

10. Typical Pb-Zn skarn
11. Nakatatsu
12. Hanover

Volcanic-hosted epithermal vein districts

13. Comstock
14. Tayoltita
15. Creede
16. Arcata
17. Orcopampa

Carbonate replacement and mantos

18. Main Tintic
19. East Tintic
20. Gilman
21. Leadville
22. Park City
23. Fresnillo
24. San Francisco del Oro
25. Bingham (nonporphyry)

GSC

Figure 16-2. Comparison of the metal ratios for silver-lead-zinc vein districts with those of other types of deposits mined for Ag, Pb, or Zn; **A)** Ag-Pb-Zn ternary space; note Tayoltita (14) not plotted on A, zinc data not available; **B)** Ag-Pb-Au ternary space (from Beaudoin and Sangster, 1992; reproduced from Economic Geology, 1992, v. 87, p. 1009).

Associated structures

Vein districts were formed late in the tectonic evolution of their orogens and three of these districts are located at terrane boundaries (Beaudoin and Sangster, 1992). The Kokanee Range is in the upper plate of the Valhalla metamorphic core complex, which was unroofed during Eocene extension of the Cordilleran Orogen (Parrish et al., 1988). The Erzgebirge gneiss hosting the Freiberg district forms the lower plate of a low angle extensional shear zone with the veins occurring in a conjugate set of shear and tension fractures cutting the gneiss dome (Matte et al., 1990). Mineralization in the Příbram district is in structures subsidiary to the Central Bohemian shear zone, a major, dextral transpression fault zone at the boundary between the Barrandian and Moldanubian terranes (Zák and Dobeš, 1991).

In the Selwyn Basin, mineralized faults are conjugate to the Mayo Lake shear zone, a dextral transcurrent fault zone (Lynch, 1989a). Mineralization in the Harz Mountains is in sets of parallel and wedge-shaped oblique, dextral strike-slip, and normal faults cutting late Carboniferous Variscan thrusts and isoclinal folds (Hannak, 1981). The Coeur d'Alène district is contained within the Lewis and Clark line, a major intracontinental plate boundary that was initiated as Early Proterozoic synsedimentary faults that remained active to the Holocene, including a major event of dextral strike-slip deformation in the Late Cretaceous (Wallace et al., 1990).

Form of deposits

The veins occur as massive lenticular bodies as much as several metres thick in fractures or in fault zones, or occur as stockwork structures up to 80 m wide in shear zones. The structures containing the veins range from small tension fractures to large fault zones with several kilometres of strike length. Mineralization in an individual district can occur within a large vertical interval ranging from 350 m in Keno Hill to 2300 m in Coeur d'Alène.

Mineralogy and ore textures

Galena and sphalerite are commonly associated with minor pyrite, chalcopyrite, and a diverse and complex suite of sulphosalt minerals, mainly tetrahedrite but also including minor amounts of pyrargyrite, stephanite, bournonite, acanthite, and native silver. Gangue is typically composed of siderite and/or quartz, and lesser amounts of dolomite or calcite. Textures are typically coarse grained and drusy; sphalerite may be characterized by rhythmic banding. In the Kokanee Range and Keno Hill districts, calcite and/or dolomite are typically late stage minerals associated with flooding of the hydrothermal system by meteoric water. Postmineralization deformation, shearing, and brecciation in some districts has obliterated most primary textures.

Silver-lead-zinc veins have long been the subject of mineralogical zoning studies. In some cases, the mineral zones were formed by multiple hydrothermal events or a telescoped single event, rather than by regular zoning about a single point (Kutina, 1963). In the Keno Hill district, Lynch (1989b) documented a mineralogical zonation similar to classical zoning models.

Alteration

Typically, hydrothermal alteration is restricted to the vicinity of the veins and extends as much as a few metres into the wall rocks. The alteration is commonly phyllic, characterized by sericitization, silicification, and pyritization of the wall rocks. In the Coeur d'Alène district, however, large zones of hydrothermal bleaching are typical.

Geochemistry

The following is summarized from a literature review of the six classical metasediment-hosted silver-lead-zinc vein districts (Beaudoin and Sangster, 1992).

Hydrothermal fluids are characterized by temperatures near 250-300°C and salinities ranging from 0 to 26 wt.% NaCl equivalent; CO_2 is abundant in fluid inclusions in some districts. As many as three hydrothermal fluids have been identified, in some districts, from oxygen and hydrogen isotopic studies: 1) a deep-seated fluid of metamorphic origin characterized by high temperature isotopic exchange with crustal rocks; 2) a second fluid which may originally have been of meteoric origin, but which has undergone a long history of isotope exchange with upper crustal rocks; and 3) meteoric water that commonly dominates the waning stage of the hydrothermal system.

Sulphur isotope compositions are usually zoned and correlate with the local country rocks. Carbon isotope compositions in carbonates may be either heterogeneous ($-14 < \delta^{13}C < 0$), reflecting a variety of local carbon sources, or homogeneous ($-8 < \delta^{13}C < -5$), reflecting deep-seated carbon sources. Lead isotope compositions typically form linear arrays and identify primarily upper crustal sources for the lead. Other minor lead sources comprise depleted upper mantle and lower crustal Pb reservoirs.

DEFINITIVE CHARACTERISTICS

Silver-lead-zinc veins are characterized by their mineralogy, metal ratios, and local phyllic alteration. The veins are in faults and fractures that are commonly associated with deep crustal breaks at terrane boundaries and are hosted by monotonous sequences of clastic rocks deposited in basins within various tectonic settings and which have been intruded by granitic to gabbroic plutonic rocks. Mineralization occurs late in the tectonic evolution of an orogen and may be associated with the extensional collapse of the orogen.

GENETIC MODEL

Three genetic models have been recently proposed for the formation of metasediment-hosted silver-lead-zinc veins. The magmatic differentiation model (Tischendorf and Förster, 1990) holds that chacophile elements such as Pb, Zn, and Ag, concentrated in the intercumulate fluid phase of an accumulating crystal pile, are expelled, along with mafic and ultrapotassic magmas, to form coeval silver-lead-zinc veins and dykes.

The magmatic-hydrothermal model proposes that heat, derived from coeval crystallizing intrusions, drives the hydrothermal system at temperatures between 250° to 300°C (Möller et al., 1984; Lynch et al., 1990; Criss and Fleck, 1990). Mineralogical, isotope, and other zonations are developed around the intrusion. In the Keno Hill district, Lynch et al. (1990) considered graphitic-rich host rocks to be an important genetic feature, buffering hydrothermal fluids to high CO_2 partial pressures and providing an important metal source.

A third, and our preferred, genetic model involves deep-seated, metamorphic hydrothermal fluids that are channelled along deep crustal faults to higher crustal levels where mineral precipitation takes place as a result of mixing with upper crustal hydrothermal fluids and local boiling. The deep crustal faults are first-order fluid channels which directed fluids from deep-seated sources into subsidiary related structures. Mineralization occurred in these secondary structures, from dilute to saline fluids, at depths as great as 6 km and at temperatures around 250° to 300°C (Beaudoin and Sangster, 1992).

RELATED DEPOSIT TYPES

Silver-lead-zinc veins share several geological characteristics with Ag-Pb-Zn carbonate replacement and manto deposits. Replacement of limestone by massive sulphide bodies along fractures exists in some silver-lead-zinc vein districts. A genetic link between carbonate replacement and manto deposits and nearby intrusions has been suggested, based on zonation from skarn to chimneys, spatial association, and geochemical data (Haynes and Kesler, 1988; Megaw et al., 1988). Carbonate replacement and manto deposits formed at shallow depths, under pressures ranging from 0.3 to 0.8 kbar (Megaw et al., 1988), compared with silver-lead-zinc veins (6 km and 1.6 kbar). Silver-lead-zinc veins appear to be a type of ore deposit distinct from carbonate replacement and manto deposits, but no single deposit scale feature is distinctive; a district scale comparison is required.

EXPLORATION GUIDES

All six classical Ag-Pb-Zn vein districts (Fig. 16-1) were discovered centuries or decades ago in surface outcrop, thereby precluding good examples of recent exploration guides. Development in these areas has been largely directed toward extension of known ore veins. However, in spite of these limitations, some suggested guidelines follow:

1. Regions containing a late, major crustal fault cutting a clastic metasedimentary terrane would be regarded as favourable for vein silver-lead-zinc deposits.

2. A favourable metasedimentary terrane should be part of a Pb-Zn metallogenic province containing sedimentary-exhalative Pb-Zn deposits.

3. Subsidiary faults to the major crustal fault commonly host ore shoots.

4. Boyle (1965) reported that, in the Keno Hill area, analyses of residual soil along traverses across known mineralized vein faults gave broad anomalies with strong contrast for Ag, Pb, Zn, Sb, As, and Mn. Subsequent experience, however, showed that ore shoots in the area yield poor soil geochemical expressions (Watson, 1986).

5. Because Ag-Pb-Zn veins occur in clastic sedimentary rocks commonly containing abundant graphite and/or disseminated pyrite, electromagnetic surveys have had limited success in locating new veins. Faults are well located using EM surveys, but oreshoots give a poor response.

6. In the Keno Hill area, Watson (1986) reported that drilling fences across suspected faults or segments of known faults has been the most successful exploration tool and is credited with the discovery of several deposits. The drill used was of the rotary percussion-type; cuttings were recovered using an air-flush system.

SELECTED BIBLIOGRAPHY

References marked with asterisks (*) are considered to be the best sources of general information on this deposit type.

*Beaudoin, G. and Sangster, D.F.
1992:   A descriptive model for silver-lead-zinc veins in clastic metasedimentary terranes; Economic Geology, v. 87, p. 1005-1021.

Boyle, R.W.
1965:   Geology, geochemistry, and origin of the lead-zinc-silver deposits of the Keno Hill area, Yukon Territory; Geological Survey of Canada, Bulletin 311, 302 p.

Criss, R.E. and Fleck, R.J.
1990:   Oxygen isotope map of the giant metamorphic-hydrothermal system around the northern part of the Idaho batholith, U.S.A.; Applied Geochemistry, v. 5, p. 641-655.

Hannak, W.W.
1981:   Genesis of the Rammelsberg ore deposit near Goslar/Upper Harz, Federal Republic of Germany; in Handbook of Strata-bound and Stratiform Ore Deposits, Volume 9, (ed.) K.H. Wolf; Elsevier Scientific Publishing Co., Amsterdam-Oxford-New York, p. 551-642.

Haynes, F.M. and Kesler, S.E.
1988:   Compositions and sources of mineralizing fluids for chimney and manto limestone-replacement ores in Mexico; Economic Geology, v. 83, p. 1985-1992.

Kutina, J.
1963:   The distinguishing of the monoascendent and polyascendent origin of associated minerals in the study of the zoning of the Příbram ore veins; in Problems of Postmagmatic Ore Deposition Symposium, Prague, Geological Survey of Czechoslovakia, v. 1, p. 200-206.

Lynch, J.V.G.
1989a: Hydrothermal zoning in the Keno Hill Ag-Pb-Zn vein system, Yukon: a study in structural geology, mineralogy, fluid inclusions, and stable isotope geochemistry; PhD. thesis, University of Alberta, Edmonton, Alberta, 190 p.
1989b: Large-scale hydrothermal zoning reflected in the tetrahedrite-freibergite solid solution, Keno Hill Ag-Pb-Zn district, Yukon; Canadian Mineralogist, v. 27, p. 383-400.

Lynch, J.V.G., Longstaffe, F.J., and Nesbitt, B.E.
1990:   Stable isotopic and fluid inclusion indications of large-scale hydrothermal paleoflow, boiling, and fluid mixing in the Keno Hill Ag-Pb-Zn district, Yukon Territory, Canada; Geochimica et Cosmochimica Acta, v. 54, p. 1045-1059.

Matte, P., Maluski, H., Rajlich, P., and Franke, W.

1990: Terrane boundaries in the Bohemian Massif: result of large-scale Variscan shearing; Tectonophysics, v. 177, p. 151-170.

Megaw, P.K., Ruiz, J., and Titley, S.R.

1988: High-temperature, carbonate-hosted Ag-Pb-Zn(Cu) deposits of northern Mexico; Economic Geology, v. 83, p. 1856-1885.

Möller, P., Morteani, G., and Dulski, P.

1984: The origin of the calcites from Pb-Zn veins in the Harz Mountains, Federal Republic of Germany; Chemical Geology, v. 45, p. 91-112.

Parrish, R.R., Carr, S.D., and Parkinson, D.L.

1988: Eocene extensional tectonics and geochronology of the southern Omineca Belt, British Columbia and Washington; Tectonics, v. 7, p. 181-212.

Tischendorf, G. and Förster, H.-J.

1990: Acid magmatism and related metallogenesis in the Erzgebirge; Geological Journal, v. 25, p. 443-454.

Wallace, C.A., Lidke, D.J., and Schmidt, R.G.

1990: Faults of the central part of the Lewis and Clark line and fragmentation of the Late Cretaceous foreland basin in west-central Montana; Geological Society of America Bulletin, v. 102, p. 1021-1037.

Watson, K.W.

1986: Silver-lead-zinc deposits of the Keno Hill-Galena Hill area, Central Yukon, in Yukon Geology, Volume 1 (ed.) J.A. Morin and D.S. Emond; Indian and Northern Affairs, Canada, Exploration Geological Services Division, p. 83-88.

Zák, K. and Dobeš, P.

1991: Stable isotopes and fluid inclusions in hydrothermal deposits: the Příbram ore region; Rozpravy Československé Akadamie Věd, Praha, v. 101, no. 5, 109 p.

Author's addresses

G. Beaudoin
Département de géologie et de
  génie géologique
Université Laval
Québec, Québec
G1K 7P4

D.F. Sangster
Geological Survey of Canada
601 Booth Street
Ottawa, Ontario
K1A 0E8

Printed in Canada

17. VEIN COPPER

17. VEIN COPPER

R.V. Kirkham and W.D. Sinclair

INTRODUCTION

Vein copper deposits include various vein-type deposits in which copper is the dominant metal. The deposits are structurally controlled and occur in faults, fault systems, and vein-breccia zones; replacement zones in associated country rocks are also present. They are typically small, but are highly varied in both size and grade. Although vein copper deposits occur in association with many different host rocks and in diverse geological settings, two main subtypes are recognized, which are based on the associated intrusive rocks; the deposits do not necessarily occur in these rocks in all cases, but they are likely genetically related to them. The first subtype consists of vein copper deposits associated with mafic intrusive rocks; these have been referred to as 'Churchill type' after the Churchill Copper (Magnum) deposit in British Columbia (Kirkham, 1973). Other examples of this subtype include the Davis-Keays and Bull River deposits in British Columbia; Bruce Mines, Crownbridge, Ethel Copper, and some veins in the Cobalt district, Ontario; Icon-Sullivan, Quebec; and deposits in the East Arm of Great Slave Lake (e.g., Susu Lake) and Coppermine River (e.g., Copper Lamb) areas, Northwest Territories (Fig. 17-1).

The second subtype consists of vein copper deposits associated with intermediate to felsic intrusions, including some intrusions related to porphyry copper deposits. Canadian examples of this subtype include the Alwin copper deposit and copper-gold deposits of the Rossland

Kirkham, R.V. and Sinclair, W.D.
1996:   Vein copper; in Geology of Canadian Mineral Deposit Types, (ed.) O.R. Eckstrand, W.D. Sinclair, and R.I. Thorpe; Geological Survey of Canada, Geology of Canada, no. 8, p. 399-408 (also Geological Society of America, The Geology of North America, v. P-1).

camp, British Columbia; and the copper-gold deposits of the Chibougamau and Opemiska mining camps, Quebec (Fig. 17-1). Foreign examples include vein copper deposits of Magma, Arizona; Maria, Mexico; 'plutonic' copper veins in Chile such as Tamaya (Sillitoe, 1992); and polymetallic veins such as those of the Morococha and Quiruvilca districts, Peru, and the Ashio, Akenobe, and Osarizawa districts in Japan. The copper veins at Butte, Montana, many of which were mined individually, are also included although collectively they have many of the characteristics of porphyry copper deposits and, in some cases, were mined by bulk mining methods.

IMPORTANCE

Although not of major importance in Canada at present, vein copper deposits were historically important as the source of the first copper produced in Canada, in 1847 at Bruce Mines, Ontario. They have been an important source of copper production in the Chibougamau and Opemiska mining camps, Quebec, which together have produced more than 60 Mt of ore grading about 2% Cu and 1-2 g/t Au (Table 17-1). Significant amounts of copper were also produced from the gold-rich veins of the Rossland area, British Columbia. Overall, vein copper deposits account for approximately 3% of Canada's copper production and less than 2% of copper reserves.

Historically, vein copper deposits have been important locally in various parts of the world, such as Cornwall and Devon, England where copper production in first half of the nineteenth century amounted to 40% of world output (Dines, 1956). In the latter part of the nineteenth century, 'plutonic' veins in Chile were a major source of world copper production (Sillitoe, 1992). Current production of copper from vein deposits, however, is small compared to production from other types of deposits.

SIZE AND GRADE OF DEPOSITS

Vein copper deposits are relatively small, typically ranging from tens of thousands to a few million tonnes of ore, except for a few mining camps with more than 10 Mt, and the Butte deposits which have produced several hundred million tonnes of ore (Table 17-1). Copper grades are typically 1 to 3%, although some deposits contain greater than 10% (e.g., Maria, Mexico). Grade-tonnage relationships for the various selected vein copper deposits and districts are shown in Figure 17-2.

GEOLOGICAL FEATURES

Geological setting

Vein copper deposits occur in diverse tectonic environments. Churchill-type deposits apparently occur in extensional tectonic settings, particularly in Proterozoic sedimentary basins intruded by diabase and gabbro (e.g., Churchill Copper, British Columbia and Bruce Mines, Ontario; Kirkham, 1973) (Fig. 17-3). No mafic rocks have been documented in the immediate vicinity of the Icon-Sullivan deposit, but the deposit is truncated on its

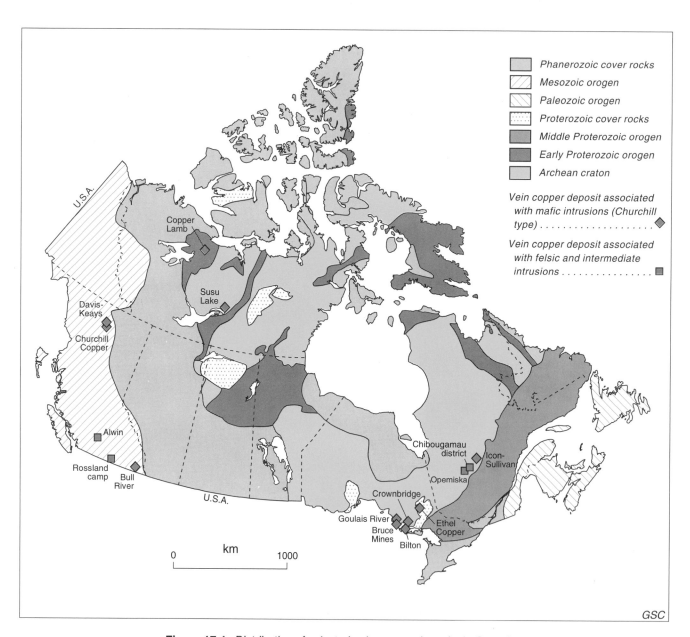

Figure 17-1. Distribution of selected vein copper deposits in Canada.

Table 17-1. Production/reserves of selected Canadian and foreign vein copper deposits.

Deposit	Production/reserves	Comments/references
Vein copper deposits associated with mafic intrusions (Churchill type)		
Canadian deposits:		
Churchill Copper (Magnum), British Columbia	0.6 Mt; 2.9% Cu	Production 1970-75, plus reserves; British Columbia Ministry of Energy, Mines and Petroleum Resources, Minfile.
Davis-Keays, British Columbia	1.9 Mt; 3.65% Cu	Proven and probable reserves; Canadian Mines Handbook 1990-91, p. 335.
Bull River, British Columbia	0.5 Mt; 1.5% Cu 13.5 g/t Ag 0.27 g/t Au	Production 1972-74; British Columbia Ministry of Energy, Mines and Petroleum Resources, Minfile.
Copper Lamb, Northwest Territories	5000 t; "high grade bornite ore"	Reserves; Kindle (1972).
Susu Lake, Northwest Territories	0.13 Mt; 0.95% Cu	Reserves; Thorpe (1972).
Bruce Mines, Ontario	0.4 Mt; 3% Cu	Production, 1847-1921; Shklanka (1969); Cu grade is approximate.
Crownbridge, Ontario	0.4 Mt; 2% Cu	Reserves; The Northern Miner, 29 October, 1964.
Bilton, Ontario	0.5 Mt; 1.7% Cu	Reserves; Pearson et al. (1985).
Goulais River, Ontario	0.2 Mt; 2.35% Cu 0.26 g/t Ag	Reserves; Pearson et al. (1985).
Ethel Copper, Ontario	7700 t; 1.2% Cu 10 g/t Ag 0.3 g/t Au	Production 1962-1967; Shklanka (1969).
Icon-Sullivan, Quebec	1.4 Mt; 2.9% Cu	Production 1967-1975; Canadian Mines Handbook 1977-1978, p. 61.
Vein copper deposits associated with felsic and intermediate intrusions		
Canadian deposits:		
Alwin (OK), British Columbia	1.2 Mt; 2.2% Cu 13 g/t Ag 0.2 g/t Au	Intermittent production 1916-1982, plus reserves; British Columbia Ministry of Energy, Mines and Petroleum Resources, Minfile. Gold grade based on historical production.
Rossland camp, British Columbia	5.6 Mt; 1% Cu 16 g/t Au 21 g/t Ag	Production 1894-1941; Gilbert (1948).
Chibougamau district, Quebec	40 Mt; 1.5-2% Cu 1-2 g/t Au	Production from Cu-Au veins in the Chibougamau district 1955-1992; Gobeil and Racicot (1984) and Canadian Mines Handbooks, 1984-1992.
Merrill Island (Canadian Merrill), Chibougamau, Quebec	1.2 Mt; 2.25% Cu 0.46 g/t Au	Production 1958-1967; Gobeil and Racicot (1984).
Copper Rand, Chibougamau, Quebec	16.3 Mt; 1.81% Cu 2.3 g/t Au	Production 1960-1989, plus reserves; Blais (1990).
Portage, Chibougamau, Quebec	6.5 Mt; 1.75% Cu 2.7 g/t Au	Production 1960-1989, plus reserves; Blais (1990).
Devlin, Chibougamau, Quebec	1.4 Mt; 2.25% Cu	Reserves; Campbell Resources Annual Report, 1983.
Corner Bay, Chibougamau, Quebec	1.5 Mt; 4% Cu 0.34 g/t Au 14 g/t Ag	Reserves; Bertoni and Vachon (1984).
Opemiska, Quebec	23.4 Mt; 2.2% Cu 1.6 g/t Au 9 g/t Ag	Production 1953-1978; Watkins and Riverin (1982) and Canadian Mines Handbooks, 1979-1992.
Foreign deposits:		
Butte, Montana	296 Mt; 2.5% Cu 68 g/t 0.3 g/t Au Ag	Production 1880-1964; Meyer et al. (1968); includes some production from lower grade, porphyry-type zones.
Magma, Arizona	12.4 Mt; 5.69% Cu 66.2 g/t Ag 1.1 g/t Au	Production 1911-1964; includes some production from replacement deposits in limestone; Hammer and Peterson (1968).
Maria, Mexico	0.47 Mt; 12.8% Cu 0.25% Mo 62 g/t Ag	Mineable reserves; Reuss and Ollivier (1992).
Dalcoath Lode, Cornwall, England	0.35 Mt; 6-7.5% Cu	Production 1750-1905; Dines (1956); significant amounts of Sn were also produced from lower, Cu-poor parts of the vein system.
Quiruvilca district, Peru	>10 Mt; Cu-Pb-Zn-Ag ore	Production 1789-1990; Bartos (1990).
Akenobe, Japan	17 Mt; 1.1% Cu 2.0% Zn 20 g/t Ag 0.4% Sn	Production 1935-1986; Shimizu and Kato (1991); In also present (T. Nakamura, pers. comm., 1994).

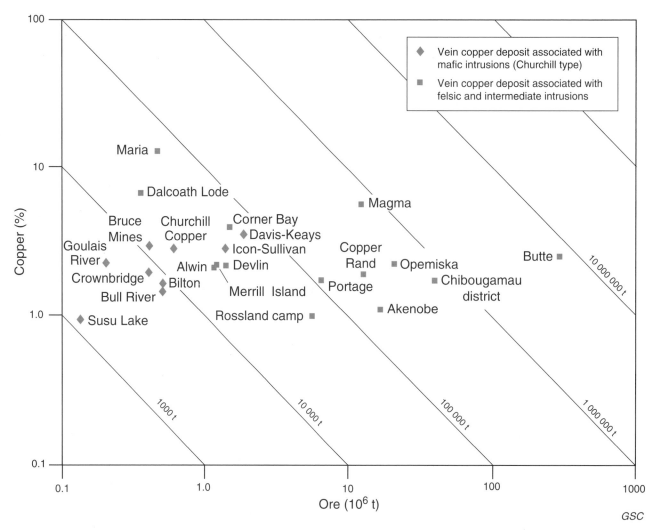

Figure 17-2. Grade versus tonnage diagram for vein copper deposits (data from Table 17-1).

Figure 17-3. Cliff face in the Gataga River area, northeastern British Columbia showing a typical diabase dyke swarm cutting Proterozoic sedimentary rocks; Churchill-type veins occur in the same or related fault and fracture systems as the dykes. GSC 1994-707 and 1994-708 (composite)

southeast side by a steep northeast-trending fault related to the Grenville Front; mafic rocks may have been present southeast of the mine area but have been removed or obscured by Grenvillian metamorphism and deformation.

Felsic and intermediate intrusion-associated vein copper deposits characteristically occur in subduction-related contintental- and island-arc settings, typically in areas of high-level felsic and intermediate intrusions and especially those associated with porphyry copper deposits (e.g., Alwin, British Columbia; Butte, Montana; Magma, Arizona; and Morococha, Peru). Although the vein copper-gold deposits of the Chibougamau and Opemiska mining camps are mainly in differentiated mafic intrusive rocks, the deposits are closely associated with, and probably genetically related to, felsic plutonic rocks (Sinclair et al., 1994; Robert, 1994). The copper-gold deposits of the Rossland camp, British Columbia are associated with monzonitic plutonic rocks (Dunne and Höy, 1992).

Age of host rocks and ore

Vein copper deposits and their host rocks range in age from Archean to Recent. The ores are epigenetic and in many cases are much younger than their host rocks.

Associated rocks

Country rocks for vein copper deposits are diverse and depend on the particular geological settings in which the deposits occur. Churchill-type deposits typically occur in clastic sedimentary rocks and mafic igneous rocks such as diabase or gabbro (e.g., Churchill Copper and Bruce Mines). Vein copper deposits associated with felsic and intermediate intrusions occur in a wide variety of host rocks, including layered mafic intrusive rocks (e.g., Chibougamau and Opemiska deposits), felsic plutonic rocks (e.g., Alwin, British Columbia; Butte, Montana), metamorphic and sedimentary rocks (e.g., Magma, Arizona), and volcanic rocks (e.g., Quiruvilca, Peru; Akenobe, Japan).

Form and size of deposits

Deposits range from simple veins to anastomosing and reticulate veins (Fig. 17-4), vein sets, vein breccia, local stockworks, and horsetails. Individual veins and vein sets can be tens to hundreds to thousands of metres long and from less than one metre to tens of metres wide. Vertical extent of Churchill-type veins in northeastern British Columbia ranges from at least 150 m for the Churchill Copper deposit (Fig. 17-5 and 17-6) to more than 500 m for the Eagle vein system of the Davis-Keays deposit (Preto, 1971). The vertical extent of vein copper deposits associated with felsic and intermediate intrusions ranges from 250 m for the Alwin deposit (W.J. McMillan, 1972) to nearly 1400 m for the Anaconda vein system at Butte (Meyer et al., 1968). The copper-gold deposits of the Chibougamau camp are massive to disseminated, lenticular sulphide zones that occur within steeply-dipping shear zones several kilometres long, hundreds of metres wide, and extending to depths of 1000 m or more (Guha et al., 1983; Archambault et al., 1984).

Individual high-grade ore shoots within veins are structurally controlled, and typically occur where veins change attitude, as at the Churchill Copper deposit (Carr, 1971).

At the Icon-Sullivan deposit, high-grade massive sulphide ore was localized in the vicinity of structural terraces or "rolls" in the flat-lying host rocks (Troop and Darcy, 1973). Distribution of ore minerals within ore shoots is typically irregular and may be disseminated, banded, patchy, or massive; disseminated to massive ore minerals may be present in adjacent altered host rocks.

Mineralogy

In most Churchill-type vein copper deposits, chalcopyrite is the principal ore mineral; bornite, tetrahedrite, covellite, and galena are present in some deposits in minor amounts. Pyrite is the main associated gangue mineral; others include pyrrhotite, quartz, calcite, dolomite, ankerite, and hematite.

The principal ore minerals in vein copper deposits associated with felsic and intermediate intrusion-associated vein copper deposits are more varied and include chalcopyrite, bornite, chalcocite, enargite, tetrahedrite-tennantite, bismuthinite, molybdenite, sphalerite, native gold, and

Figure 17-4. Typical anastomosing quartz-carbonate-chalcopyrite-pyrite-specularite vein (Churchill type) exposed in outcrop beside Highway 17 about 40 km west of Iron Bridge, Ontario. GSC 202516E

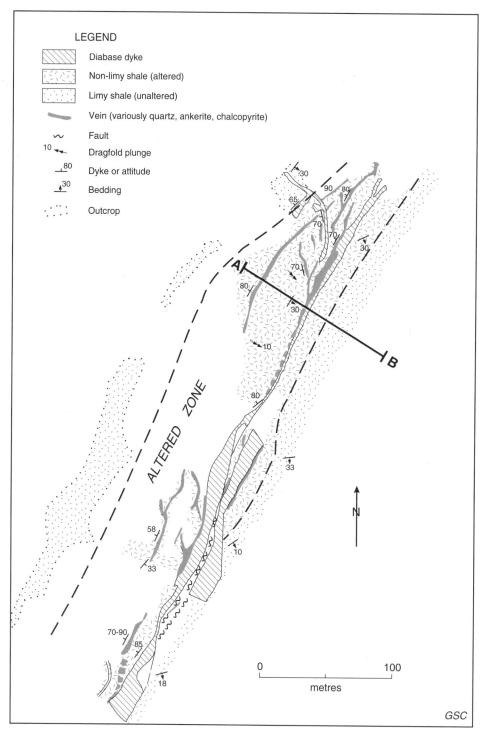

Figure 17-5. Surface geological map of the north part of the Churchill Copper (Magnum) vein system (adapted from Carr, 1971).

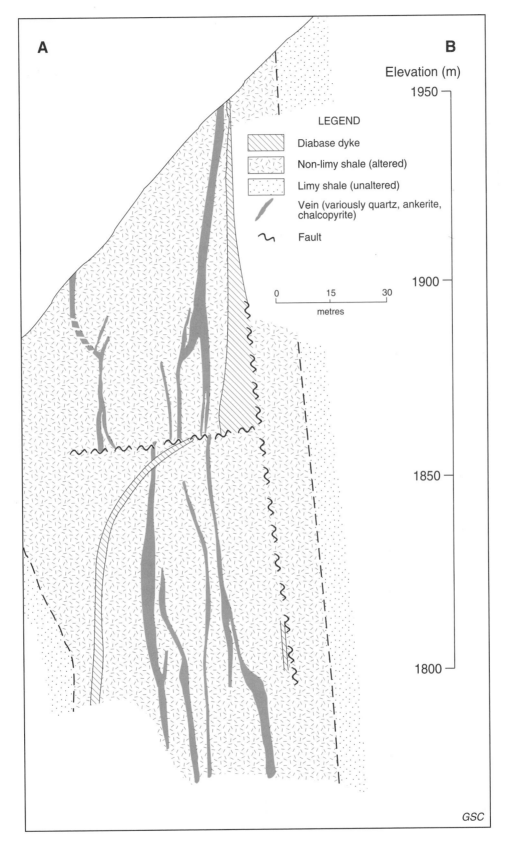

Figure 17-6. Vertical cross-section A-B of the Churchill Copper (Magnum) vein system (adapted from Carr, 1971). Location of the cross-section is shown in Figure 17-5.

LEGEND

Diabase dyke

Non-limy shale (altered)

Limy shale (unaltered)

Vein (variously quartz, ankerite, chalcopyrite)

Fault

electrum. Associated gangue minerals include pyrite, pyrrhotite, magnetite, hematite, quartz, K-feldspar, epidote, calcite, ankerite, siderite, chlorite, sericite, and clay minerals.

Zoning

Zoning of minerals or metals may be present in large deposits or districts associated with felsic and intermediate intrusions. In the Rossland camp, three zones were defined by Thorpe (1967): 1) a central zone of massive pyrrhotite veins with minor chalcopyrite; 2) an intermediate zone with a diverse sulphide mineralogy dominated mainly by pyrrhotite and chalcopyrite; and 3) an outer zone comparable to the intermediate zone, except for the addition of galena, tetrahedrite, boulangerite, and abundant sphalerite. The copper grade and copper-to-gold ratio of the Rossland deposits increased outward from the central zone. Three crudely concentric zones have also been delineated at Butte: 1) a central copper zone which includes a molybdenite zone at depth; 2) an intermediate zone of copper with minor zinc; and 3) a peripheral zone of silver, zinc, and manganese (Sales, 1914; Meyer et al., 1968). In the Quiruvilca district, Peru, copper-rich deposits form a central core with lead-zinc-silver zones at the periphery (Bartos, 1990).

Alteration

Principal types of alteration in Churchill-type vein copper deposits appear to be carbonatization and silicification. At Churchill Copper, limy strata have been altered to non-limy rocks by decalcification which has left abundant graphite in the host rocks and ankerite as metacrysts and replacement zones; pyrite is also present in places as seams and disseminated zones that are roughly conformable to bedding in the host strata (Carr, 1971; Preto, 1971). In the vein copper deposits north of Lake Huron, such as Bruce Mines and Crownbridge, breccia fragments in the veins are typically silicified and hematitized (Pearson, 1979).

Sericitization and chloritization are the principal alteration types in vein copper deposits associated with felsic to intermediate intrusions. Granodioritic rocks that border the Alwin copper vein, for example, have been intensely altered to sericite and clay minerals; chlorite is present in places and massive pods of epidote occur locally adjacent to ore zones (W.J. McMillan, 1972). The copper-gold veins of the Chibougamau camp occur in extensively sericitized anorthositic rocks with abundant siderite, ankerite, and magnetite in some of the deposits. Chlorite, chloritoid, and actinolite occur in close proximity to some ore zones (Eckstrand, 1963; Allard, 1976).

DEFINITIVE CHARACTERISTICS

Churchill-type deposits are characterized by structural control, occurrence in predominantly extensional tectonic settings (rift and failed rift zones) and association with mafic intrusive rocks. Vein copper deposits associated with felsic and intermediate intrusions are also characterized by structural control, and are distinguished from Churchill-type deposits by their association with high level, felsic and intermediate intrusive rocks, particularly those associated with porphyry copper deposits, and by their occurrence in predominantly subduction-related continental- and island-arc settings.

GENETIC MODEL

Vein copper deposits are a diverse and relatively understudied group of deposits for which no single, well constrained genetic model exists. For Churchill-type deposits, Kirkham (1973) proposed a model based the occurrence of the deposits in regions of crustal extension where passive upwelling of mafic magmas has occurred, as reflected by numerous mafic dykes. In such environments, water entrained in country rocks, or meteoric water that migrated to depth, could have been heated by the mafic magmas. The heated waters selectively extracted copper from the ascending magmas and/or the country rocks and subsequently deposited the copper in cooler, favourable structural environments at higher levels in the crust.

Genetic models for vein copper deposits associated with felsic and intermediate intrusions reflect either prograding or retrograding evolution of the deposits. Prograding evolution in base metal lode deposits is indicated by a paragenetic sequence of vein-filling minerals that reflects a temporal increase in sulphidation state (i.e., later deposition of minerals with increasing sulphur-to-metal contents), accompanied by pervasive wall rock alteration characterized by increasing hydrogen ion activity (i.e., strong sericitic and/or advanced argillic alteration) (Bartos, 1989). Prograde deposits form at moderate to deep levels in the crust (>3000 m below the paleosurface), mainly from magmatic-hydrothermal fluids expelled from the associated intrusions, but also by deep hypogene leaching and enrichment of copper-bearing protore by meteoric waters (e.g., Butte; Brimhall, 1979). According to Bartos (1989), vein copper deposits associated with porphyry copper deposits are typically prograde.

Retrograding evolution is indicated by temporal decrease in sulphidation state of vein-filling minerals and wall rock alteration characterized by decreasing hydrogen ion activity (i.e., maximum intensity of alteration occurs prior to significant base metal deposition). Retrograde deposits form mainly at shallow to moderate levels in the crust (approximately 200 to 3000 m below the paleosurface) from hydrothermal systems that have had a significant input of ground water. Such deposits tend to have a high component of lead and zinc, and in some cases grade outward to precious metal deposits (Bartos, 1989).

Deposits such as Alwin and copper-gold veins in the Chibougamau area are closely related to porphyry copper deposits, although they may have formed in different structural settings and may be superimposed on, or peripheral to, the porphyry deposits. The Alwin deposit, for example, occurs in the core of the Guichon Creek batholith in the vicinity of the Highland Valley porphyry copper deposits. A close association between the copper-gold deposits of the Chibougamau mining camp and felsic dykes of the Chibougamau pluton is evident, and preliminary observations suggest the deposits may be related to low-grade, porphyry-type copper-molybdenum deposits (Sinclair et al., 1994; Robert, 1994). The copper-gold veins are in strongly foliated, sericitic shear zones, however, and have a complex history that includes deformation and remobilization (e.g., Guha et al., 1983; Tessier and Hodgson, 1994).

The origin of the Opemiska veins is less certain as no porphyry deposits have been identified in the area. They could be related to the nearby Opemiska pluton (Duquette, 1970), which would account for the presence of minor tungsten and molybdenum, but alternative hypotheses involving leaching of metals from the associated host rocks (Derry and Folinsbee, 1957; Watkins and Riverin, 1982) and metamorphism of volcanic-associated vein-type deposits (R.H. McMillan, 1972) have been proposed.

RELATED DEPOSIT TYPES

Churchill-type vein copper deposits are probably related to silver-cobalt-arsenic veins at Cobalt and Thunder Bay, Ontario and the Camsell River area, Northwest Territories (see subtype 14.1, "Arsenide vein silver-cobalt"). Deposits related to vein copper deposits associated with felsic and intermediate intrusions include high sulphidation gold-copper veins (e.g., El Indio, Chile), copper-bearing skarn and manto deposits, and porphyry copper deposits.

Also related to vein copper deposits associated with felsic and intermediate intrusions are polymetallic vein and replacement deposits such as Tintic, Utah; Central City, Colorado; and Cerro de Pasco, Peru. Other related deposits include precious metal veins which grade into base metal veins and vein copper deposits at depth, as in the Banská Štiavnica-Hodruša district, Carpathian Mountains, Slovakia (Štohl and Lexa, 1993) and iron-copper-mercury veins, such as the Droždiak and Hrubá veins of the Rudňany district, Slovakia, which are currently mined primarily for iron but which have produced signficant amounts of copper, silver, and mercury (Varček, 1967).

EXPLORATION GUIDES

Exploration guidelines for vein copper deposits include the following:

1. Rifted Proterozoic sedimentary successions characterized by diabase dykes and gabbro bodies may contain copper veins of the Churchill type.

2. Ore shoots may be localized along dilational bends within veins, and high grade sulphide shoots may cut lower grade sulphide-quartz-carbonate parts of veins as in the Icon-Sullivan deposit (Troop and Darcy, 1973).

3. Vein copper deposits associated with felsic and intermediate intrusions may be mineralogically zoned, on both vein and district scales.

4. Primary geochemical dispersion aureoles in host rocks (mainly Cu) are likely to be limited in Churchill-type copper veins but may be more extensive in vein copper deposits associated with felsic and intermediate intrusions. Secondary dispersion halos in overburden and stream sediments may help identify target areas at regional and local scales.

5. Electromagnetic surveys can be used to trace favourable structures such as faults or fracture zones, and may help outline areas of high concentrations of sulphides in veins.

ACKNOWLEDGMENTS

K.H. Poulsen, V.A. Preto, and R.I. Thorpe reviewed the manuscript and provided constructive comments.

SELECTED BIBLIOGRAPHY

References marked with asterisks (*) are considered to be the best sources of general information on this deposit type.

Allard, G.O.
1976: Doré Lake Complex and its importance to Chibougamau geology and metallogeny; Québec Ministère des Richesses Naturelles, DP-368, 446 p.

Archambault, G., Guha, J., Tremblay, A., and Kanwar, R.
1984: Implications of the geomechanical interpretation of the Copper Rand deposit on the Dore Lake shear belt; in Chibougamau - Stratigraphy and Mineralization, (ed.) J. Guha and E.H. Chown; The Canadian Institute of Mining and Metallurgy, Special Volume 34, p. 300-318.

Bartos, P.J.
*1989: Prograde and retrograde base metal lode deposits and their relationship to underlying porphyry copper deposits; Economic Geology, v. 84, p. 1671-1683.
1990: Metal ratios of the Quiruvilca mining district, northern Peru; Economic Geology, v. 85, p. 1629-1644.

Bertoni, C.H. and Vachon, A.
1984: The Corner Bay deposit: a new copper discovery in the Doré Lake Complex; in Chibougamau - Stratigraphy and Mineralization, (ed.) J. Guha and E.H. Chown; The Canadian Institute of Mining and Metallurgy, Special Volume 34, p. 319-328.

Blais, A.
1990: Copper Rand mine; in Litho-tectonic Framework and Associated Mineralization of the Eastern Extremity of the Abitibi Greenstone Belt; (ed.) J. Guha, E.H. Chown, and R. Daigneault; Guidebook for Field Trip 3, 8th International Association on the Genesis of Ore Deposits Symposium; Geological Survey of Canada, Open File 2158, p. 63-69.

Brimhall, G.H., Jr.
1979: Lithologic determination of mass transfer mechanisms of multiple-stage porphyry copper mineralization at Butte, Montana: vein formation by hypogene leaching and enrichment of potassium-silicate protore; Economic Geology, v. 74, p. 556-589.

*Carr, J.M.**
1971: Geology of the Churchill Copper deposit; The Canadian Institute of Mining and Metallurgy Bulletin, v. 64, no. 710, p. 50-54.

Derry, D.R. and Folinsbee, J.C.
1957: Opemiska copper mines; in Structural Geology of Canadian Ore Deposits, v. II (Congress Volume); The Canadian Institute of Mining and Metallurgy, Montreal, Quebec, p. 430-441.

Dines, H.G.
1956: The metalliferous mining region of south-west England, volume 1; Memoir of the Geological Survey of Great Britain, London, 508 p.

Dunne, K.P.E. and Höy, T.
1992: Petrology of pre to syntectonic Early and Middle Jurassic intrusions in the Rossland Group, southeastern British Columbia (82F/SW); in Geological Fieldwork 1991, (ed.) B. Grant and J.M. Newell; British Columbia Ministry of Energy, Mines and Petroleum Resources, Paper 1992-1, p. 9-19.

*Duquette, G.**
1970: Archean stratigraphy and ore relationships in the Chibougamau district; Quebec Department of Natural Resources, Special Paper 8, 16 p.

Eckstrand, O.R.
1963: Crystal chemistry of chlorite; PhD. thesis, Harvard University, Boston, Massachusetts, 128 p. (includes a section on chloritic alteration at the Copper Rand mine, Chibougamau district, northern Quebec).

Gilbert, G.
1948: Rossland camp; in Structural Geology of Canadian Ore Deposits (Jubilee Volume); The Canadian Institute of Mining and Metallurgy, Montreal, Quebec, p. 189-196.

Gobeil, A. and Racicot, D.

1984: Chibougamau: histoire et minéralisations; in Chibougamau - Stratigraphy and Mineralization, (ed.) J. Guha and E.H. Chown; The Canadian Institute of Mining and Metallurgy, Special Volume 34, p. 261-270.

*Guha, J. and Chown, E.H. (ed.)

1984: Chibougamau - Stratigraphy and Mineralization; The Canadian Institute of Mining and Metallurgy, Special Volume 34, 534 p.

Guha, J., Archambault, G., and Leroy, J.

1983: A correlation between the evolution of mineralizing fluids and the geomechanical development of a shear zone as illustrated by the Henderson 2 Mine, Quebec; Economic Geology, v. 78, p. 1605-1618.

*Hammer, D.F. and Peterson, D.W.

1968: Geology of the Magma mine area, Arizona; in Ore Deposits in the United States, 1933-1967 (The Graton-Sales Volume), (ed.) J.D. Ridge; The American Institute of Mining, Metallurgical, and Petroleum Engineers, Inc., New York, p. 1282-1310.

Kindle, E.D.

1972: Classification and description of copper deposits, Coppermine River area, District of Mackenzie; Geological Survey of Canada, Bulletin 214, 109 p.

Kirkham, R.V.

1973: Tectonism, volcanism and copper deposits; in Volcanism and Volcanic Rocks, (ed.) I.F. Ermanovics; Geological Survey of Canada, Open File 164, p. 130-151.

McMillan, R.H.

1972: Petrology, geochemistry and wall rock alteration at Opemiska - a vein copper deposit crosscutting a layered Archean ultramafic-mafic sill; PhD. thesis, University of Western Ontario, London, Ontario, 169 p.

McMillan, W.J.

1972: OK (Alwin) mine; in Geology, Exploration and Mining in British Columbia, 1972; British Columbia Department of Mines and Petroleum Resources, p. 153-157.

*Meyer, C., Shea, E.P., Goddard, C.G., Jr., Zeihen, L.G., Guilbert, J.M., Miller, R.N., McAleer, J.F., Brox, G.B., Ingersoll, R.G., Jr., Burns, G.J., and Wigal, T.

1968: Ore deposits at Butte, Montana; in Ore Deposits in the United States, 1933-1967 (The Graton-Sales Volume), (ed.) J.D. Ridge; The American Institute of Mining, Metallurgical, and Petroleum Engineers, Inc., New York, p. 1373-1416.

Pearson, W.N.

1979: Copper metallogeny, north shore region of Lake Huron, Ontario; in Current Research, Part A; Geological Survey of Canada, Paper 79-1A, p. 289-304.

Pearson, W.N., Bretzlaff, R.E., and Carrière, J.J.

1985: Copper deposits and occurrences in the north shore region of Lake Huron, Ontario; Geological Survey of Canada, Paper 83-28, 34 p.

Preto, V.A.

1971: Lode copper deposits of the Racing River-Gataga River area; in Geology, Exploration and Mining in British Columbia, 1971; British Columbia Department of Mines and Petroleum Resources, p. 75-107.

Reuss, R.J. and Ollivier, F.J.

1992: Mina Maria: a new mine in Mexico's Cananea district; Mining Engineering, v. 44, no. 11, p. 1323-1329.

Robert, F.

1994: Timing relationships between Cu-Au mineralization, dykes, and shear zones in the Chibougamau camp, northeastern Abitibi Subprovince, Quebec; in Current Research 1994-C; Geological Survey of Canada, p. 287-294.

Sales, R.H.

1914: Ore deposits at Butte, Montana; American Institute of Mining Engineers, Transactions, v. 46, p. 4-106.

Shimizu, M. and Kato, A.

1991: Roquesite-bearing tin ores from the Omodani, Akenobe, Fukoko and Ikuno polymetallic vein-type deposits in the Inner Zone of southwestern Japan; Canadian Mineralogist, v. 29, p. 207-215.

Shklanka, R.

1969: Copper, nickel, lead and zinc deposits in Ontario; Ontario Department of Mines, Mineral Resources Circular No. 12, 394 p.

Sillitoe, R.H.

1992: Gold and copper metallogeny of the central Andes - past, present and future exploration objectives; Economic Geology, v. 87, p. 2205-2216.

Sinclair, W.D., Pilote, P., Kirkham, R.V., Robert, F., and Daigneault, R.

1994: A preliminary report of porphyry Cu-Mo-Au and shear zone-hosted Cu-Au deposits in the Chibougamau area, Quebec; in Current Research 1994-C; Geological Survey of Canada, p. 303-309.

Štohl, J. and Lexa, J.

1993: Banská Štiavnica-Hodruša ore district; in Field Trip Guide, Field Conference on Plate Tectonic Aspects of Alpine Metallogeny in the Carpatho-Balkan Region and International Union of Geological Sciences/UN Educational, Scientific and Cultural Organization Deposit Modeling Workshop, May, 1993; Hungarian Geological Survey, Budapest, p. 8-16.

Tessier, A.C. and Hodgson, C.J.

1994: Syn-tectonic auriferous quartz-carbonate vein-type orbodies formed by metamorphic remobilization of Au and Cu from pre-metamorphic sulphide lenses at the Portage mine, Chibougamau, Quebec: implications for crustal recycling of metals and the origin of metal provinces; Geological Association of Canada-Mineralogical Association of Canada Annual Meeting, Waterloo, Ontario, Program with Abstracts, v. 19, p. A111.

Thorpe, R.I.

1967: Controls on hypogene sulphide zoning, Rossland, British Columbia; PhD. thesis, University of Wisconsin, Madison, Wisconsin, 141 p.

1972: Mineral exploration and mining activities, mainland Northwest Territories, 1966 to 1968 (excluding the Coppermine River area); Geological Survey of Canada, Paper 70-70, 204 p.

Troop, A.J. and Darcy, G.

1973: Geology of the Icon Sullivan Joint Venture copper deposit, Quebec; The Canadian Institute of Mining and Metallurgy Bulletin, v. 66, no. 729, p. 89-95.

Varček, C.

1967: Ore deposits of the West Carpathians; Guidebook for Excursion 24AC, 23rd International Geological Congress (1968); Geological Survey of Czechoslovakia, 48 p.

Watkins, D.H. and Riverin, G

1982: Geology of the Opemiska copper-gold deposits at Chapais, Quebec; in Precambrian Sulphide Deposits, (ed.) R.W. Hutchison, C.D. Spence, and J.M. Franklin; The Geological Association of Canada, Special Paper 25, p. 427-446.

Authors' addresses

R.V. Kirkham
Geological Survey of Canada
100 West Pender Street
Vancouver, B.C.
V6B 1R8

W.D. Sinclair
Geological Survey of Canada
601 Booth Street
Ottawa, Ontario
K1A 0E8

Printed in Canada

18. VEIN-STOCKWORK TIN, TUNGSTEN

18. VEIN-STOCKWORK TIN, TUNGSTEN

W.D. Sinclair

INTRODUCTION

Vein-stockwork deposits of tin and tungsten occur in a wide variety of structural styles that include individual veins, multiple vein systems, vein and fracture stockworks, breccias, and replacement zones in altered wall rocks adjacent to veins. The deposits generally occur in or near granitic intrusions which have been emplaced at relatively shallow levels (1 to 4 km) in the Earth's crust. The associated intrusions are highly fractionated and typically enriched in lithophile elements such as Rb, Li, Be, Sn, W, Mo, Ta, Nb, U, Th, and REEs, and volatile elements such as F and B.

Hydrothermal alteration of wall rocks associated with vein-stockwork tin and tungsten deposits is commonly greisen-type alteration that is characterized by Li-, F-, and/or B-bearing minerals such as topaz, fluorite, tourmaline, and various F- and/or Li-rich micas. Vein-stockwork deposits with extensively greisenized wall rocks that contain disseminated tin and/or tungsten minerals have been referred to as "greisen" deposits (Shcherba, 1970; Taylor, 1979; Reed, 1986). Such deposits are included in this review as vein-stockwork deposits; "greisen" is used primarily in reference to a type of alteration rather than as a type of deposit.

Canadian examples of vein-stockwork deposits of tin and tungsten include Grey River, Newfoundland (tungsten); East Kemptville, Nova Scotia (tin, copper, zinc); Mount Pleasant-North Zone (tin) and Burnthill (tungsten) in New Brunswick; Regal Silver (tungsten, tin) and Red Rose (tungsten) in British Columbia; and Kalzas, Yukon

Territory (tungsten)(Fig. 18-1). Important foreign examples include deposits in Cornwall, England (tin, copper); the Erzgebirge region of central Europe (tin, tungsten); the Oruro, Llallagua, and Potosi districts (tin, silver), Chojlla (tungsten, tin), and Chambillaya (tungsten) in Bolivia; Aberfoyle (tin, tungsten), Ardlethan (tin), and Mount Carbine (tungsten) in Australia; Xihuashan, China (tungsten); Panasqueira, Portugal (tungsten, tin); and San Rafael (tin, copper), Palca Once (tungsten), and Pasto Bueno (tungsten, copper, lead, silver) in Peru.

IMPORTANCE

Until recently, vein-stockwork deposits accounted for little production of tin and tungsten in Canada. In the past, most tin was produced as a byproduct from the Sullivan sediment-hosted zinc-lead-silver deposit, and tungsten production has been primarily from skarn and porphyry deposits. Since 1986, however, production of tin has been dominated by the East Kemptville deposit in Nova Scotia; from 1986 to January, 1992 when it closed, the East Kemptville mine produced approximately 20 000 t of tin, and minor quantities of copper and zinc, from about 17 Mt of ore mined (Bourassa, 1988; Canadian Mines Handbooks 1990-93). Production of tungsten from vein-stockwork deposits includes 1540 t of WO_3 from the Red Rose mine in British Columbia (Sutherland Brown, 1960), about 24 t of wolframite concentrates from Burnthill, New Brunswick (MacLellan et al., 1990), and 3.3 t of WO_3 from the Regal Silver mine in British Columbia (Mulligan, 1984).

Worldwide, vein-stockwork deposits have been important sources of both tin and tungsten, although in recent years their importance as a source of tin has declined. In 1989, world tin production was estimated at 217 500 t (Amlôt, 1990), much of which was derived from placer deposits in southeast Asia and Brazil, and from carbonate-replacement deposits in Australia and China; no more than about 10 to 15% of tin production was derived from

Sinclair, W.D.
1996:   Vein-stockwork tin, tungsten; in Geology of Canadian Mineral Deposit Types, (ed.) O.R. Eckstrand, W.D. Sinclair, and R.I. Thorpe; Geological Survey of Canada, Geology of Canada, no. 8, p. 409-420 (also Geological Society of America, The Geology of North America, v. P-1).

vein-stockwork deposits, primarily in Bolivia, Peru, England, and Russia. On the other hand, of the estimated world production in 1989 of 49 000 t of tungsten metal (Maby, 1990), more than half came from vein-stockwork deposits in China, Russia, Kazakhstan, Portugal, Bolivia, and Peru.

SIZE AND GRADE OF DEPOSITS

Many vein deposits of tin and tungsten are relatively small, on the order of tens of thousands to hundreds of thousands of tonnes of ore; however, deposits that consist of multiple veins or stockworks may contain millions to tens of millions of tonnes. Grades in vein deposits typically range from 0.5 to 2% Sn, and from 0.3 to 1.5% WO_3, although much higher grades occur locally. Stockwork deposits that can be mined using bulk mining methods have grades as low as 0.165% Sn (e.g. East Kemptville) or 0.1% WO_3 (e.g. Mount Carbine).

Production and/or reserves for Canadian and some important foreign deposits are given in Table 18-1. Grade-tonnage relationships are shown in Figure 18-2.

GEOLOGICAL FEATURES
Geological setting

Vein-stockwork deposits of tin and tungsten range in age from Archean to Tertiary, although a considerable proportion are late Paleozoic, Mesozoic, or Cenozoic. In Canada,

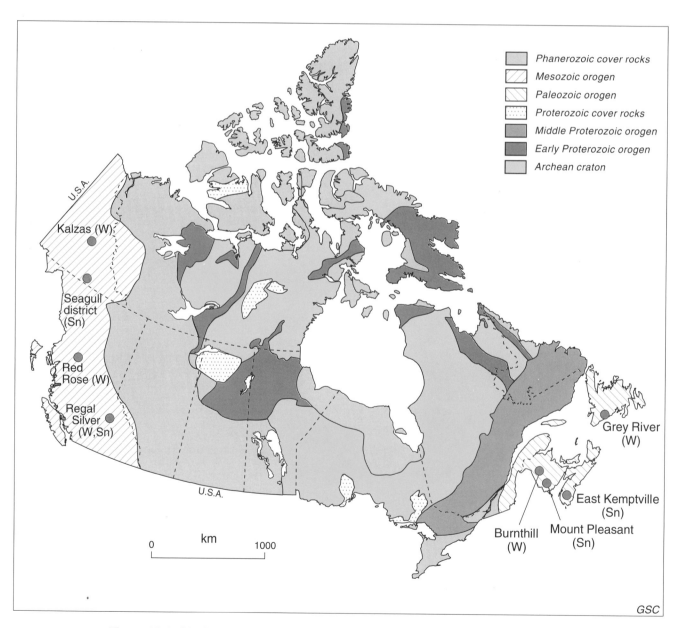

Figure 18-1. Distribution of selected vein-stockwork tin and tungsten deposits in Canada.

Table 18-1. Production/reserves of selected Canadian and foreign vein-stockwork tin, tungsten deposits.

Deposit	Production/reserves	Comments/references
Canadian deposits:		
East Kemptville, Nova Scotia	a) 56 Mt; 0.165% Sn b) 27.9 Mt; 0.182% Sn, 0.075% Cu 0.180% Zn	Preproduction reserves (Moyle, 1984). Proven and probable reserves, Dec. 31, 1989 (Canadian Mines Handbook 1990-91).
Mount Pleasant-North zone, New Brunswick	5.1 Mt; 0.79% Sn	Reserves-North Zone (The Northern Miner, 6 November, 1989).
Seagull district, Yukon Territory	None reported	Numerous tin-bearing veins are associated with the Seagull batholith (Mato et al., 1983).
Burnthill, New Brunswick	4 Mt; 0.12% WO_3	Reserves (R.M. Crosby, pers. comm., in MacLellan et al., 1990); minor tin also present.
Grey River, Newfoundland	0.52 Mt; 1.09% WO_3	Assumed plus possible reserves (Higgins, 1985).
Red Rose, British Columbia	0.12 Mt; 1.4% WO_3	Production 1942-1943, 1952-1954, plus reserves (Sutherland Brown, 1960).
Regal Silver, British Columbia	3.3 t, 100% WO_3	Production 1930-1954 (Mulligan, 1984); Sn (in stannite) also present.
Kalzas, Yukon Territory	None reported	Wolframite and minor cassiterite occur in sheeted quartz veins and stockworks over an area 1500 m by 1000 m (Lynch, 1989).
Foreign deposits:		
Aberfoyle, Australia	1.6 Mt; 0.84% Sn, 0.28% WO_3	Production 1931-1962, plus reserves (Kingsbury, 1965).
Ardlethan, Australia	9 Mt; 0.5% Sn	Production plus reserves (Taylor and Pollard, 1986).
Baal Gammon, Australia	3 Mt; 0.3% Sn, 1.2% Cu, 46 g/t Ag, 50 g/t In	Drill-indicated reserve (McKinnon and Seidel, 1988).
Bolivar, Bolivia	0.8 Mt; 1% Sn, 274 g/t Ag, 15% Zn, 1.3% Pb	Proven and probable reserves (Mining Journal, March 15, 1991).
San Rafael, Peru	~2 Mt; 1.5-2% Sn, ~1.5% Cu	Production 1970-1988 (United States Bureau of Mines annual reports).
Akenobe, Japan	17 Mt; 0.4% Sn, 1.1% Cu, 2.0% Zn, 20 g/t Ag	Production 1935-86 (Shimizu and Kato,1991); In also present.
Wheal Jane, Cornwall, England	5 Mt; 1.20% Sn	Reserves (Mining Magazine, November, 1971).
South Crofty, Cornwall, England	3.85 Mt; 1.55% Sn	Reserves (Sutphin et al., 1990).
Geevor, Cornwall, England	5 Mt; ~0.65% Sn	Production 1911-1983 (Mount, 1985).
Kelapa Kampit, Indonesia	2 Mt; 1.20% Sn	Reserves (Omer-Cooper et al., 1974).
Dzhida district, Russia	11 Mt; 0.43% WO_3	In situ resource (Anstett et al., 1985).
Xihuashan, China	~20 Mt; 0.25% WO_3	Production 1959-1983, plus reserves (based on data provided by mine staff, 1983);Mo, Sn, Bi, and Cu recovered as byproducts.
Hemerdon, Cornwall, England	42 Mt; 0.18% WO_3, 0.025% Sn	Reserves (Skillings' Mining Review, 7 January, 1984).
Panasqueira, Portugal	~31 Mt; 0.3% WO_3, 0.02% Sn	Production 1934-1981 (Smith, 1979; McNeil, 1982) plus reserves (Anstett et al., 1985).
Mount Carbine, Australia	35 Mt; 0.1% WO_3	Production 1972-1986 (Australian Mineral Industry Reviews for 1972-1986) plus reserves (Roberts, 1988).
Chicote Grande, Bolivia	21.2 Mt; 0.43% WO_3	Reserves (Willig and Delgado, 1985).
Chojlla, Bolivia	~8 Mt; 0.45% WO_3, ~0.4% Sn	Production plus reserves (Valenzuela,1979; Willig and Delgado, 1985).
Chambillaya, Bolivia	~4 Mt; 0.6% WO_3	Production (Valenzuela, 1979; Willig and Delgado, 1985).
Palca Once, Peru	1.5 Mt; 1.34% WO_3	Reserves (Willig and Delgado, 1985).
Pasto Bueno, Peru	1.1 Mt; 0.44% WO_3	Reserves (Willig and Delgado, 1985); Cu, Pb, and Ag have also been produced.

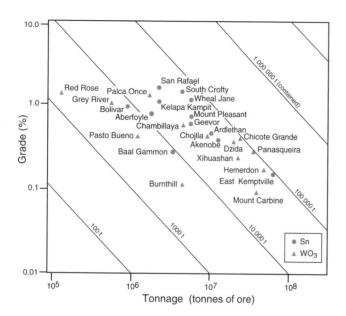

Figure 18-2. Grade versus tonnage diagram for vein-stockwork tin-tungsten deposits (Sn and WO_3 data are from Table 18-1).

important deposits of late Paleozoic age occur in the Appalachian Orogen; lesser deposits of Mesozoic age are present in the Cordilleran Orogen.

The granitic rocks associated with vein-stockwork tin and tungsten deposits are generally discordant to both local and regional structures and are either late orogenic or anorogenic. For example, the East Kemptville deposit is associated with granitic rocks that were emplaced between 360 and 370 Ma ago in metasedimentary rocks of the Meguma Terrane (Kontak and Chatterjee, 1992); the granitic rocks therefore postdate regional metamorphism associated with accretion of the Meguma Terrane onto North America, which has been dated at 405-390 Ma (Muecke et al., 1988).

The tectonic settings in which the granitic rocks have been emplaced are characteristically continental: they include continental collision belts, such as southeast Asia (Permo-Triassic), western North America (Mesozoic), eastern North America (Devonian), and the Variscan Belt of central Europe (late Paleozoic); inner continental arcs, such as Bolivia-Peru (Tertiary) and Burma-Thailand (Tertiary); and continental rift zones, such as northern Bolivia (Permo-Triassic), southern China (Cretaceous), and Nigeria (Jurassic).

On a more local scale, vein-stockwork tin and tungsten deposits are associated with granitic intrusions emplaced at relatively shallow depths ranging from about 1 to 4 km. Some of the intrusions are subvolcanic (e.g. Mount Pleasant). The intrusions range from cusp-shaped or irregular protrusions on batholiths to small cupolas and subvolcanic stocks. The deposits occur close to the contact zones of the intrusions, and are hosted to varying degrees in the granitic rocks themselves, or in associated sedimentary, volcanic, metamorphic, or older intrusive rocks.

Form of deposits

Vein-stockwork deposits are structurally controlled and take their form accordingly. Deposits may consist of single veins or narrow vein systems (e.g. Grey River, Aberfoyle); subparallel vein systems, also referred to in some cases as "sheeted veins" (e.g. Burnthill, Mount Carbine, Xihuashan); stockworks of interconnecting veins and fractures (e.g. East Kemptville); and breccias (e.g. Ardlethan). The different styles of mineralization are represented schematically in Figure 18-3. Individual veins range from less than 1 cm to several metres wide, but most are on the order of 10 to 20 cm wide. Veins may bend, branch, or pinch out over tens to hundreds of metres both laterally and vertically. Some vein systems in sheeted vein and stockwork deposits are hundreds of metres wide and more than a thousand metres long. Breccias are highly variable in size and shape; they are commonly subvertical and pipe-like, but can also be dyke-like to irregular in shape.

Structural control may not be obvious in some deposits with extensive greisen alteration. At East Kemptville, for example, mineralized alteration zones range from 1 cm to 20 m wide. The larger zones are irregular in shape and controlling fractures have been obscured by alteration.

Ore minerals and distribution

Cassiterite is the principal tin ore mineral, although stannite and other tin sulphides are present in some deposits. Wolframite is the main tungsten mineral; however, scheelite can also be present, and in some deposits is more abundant than wolframite (e.g. Red Rose). Grain size of the ore minerals ranges from fine to coarse; wolframite crystals in some quartz veins are as much as 10 to 20 cm long. Associated minerals include molybdenite, bismuthinite, chalcopyrite, sphalerite, pyrite, pyrrhotite, hematite, arsenopyrite, tourmaline, topaz, fluorite, muscovite, beryl, lepidolite, zinnwaldite, biotite, chlorite, quartz, K-feldspar, albite, and clay minerals.

Within some individual veins, the distribution of the main ore minerals is systematic, but in others it is irregular or random to complex. Some veins, for example, display comb textures indicating sequential deposition from the wall rocks inward; cassiterite commonly occurs at, or close to, the vein walls, and is succeeded inward by wolframite, although the reverse also occurs. Other veins are banded and appear to have undergone repeated opening of the vein and deposition of ore minerals. Crosscutting veins and fractures typical of many stockwork deposits also indicate that mineralization occurred in multiple stages (Fig. 18-4).

Ore composition and zoning

The compositions of vein-stockwork tin-tungsten ores vary widely (Table 18-1). Most deposits consist predominantly of either tin or tungsten, with minor amounts of the other; a notable exception is the Chojlla deposit in Bolivia, which has produced approximately equal amounts of tin and tungsten. In addition to tin and tungsten, other metals such as copper, lead, zinc, molybdenum, bismuth, silver, and indium may be present in economically significant amounts (Table 18-1).

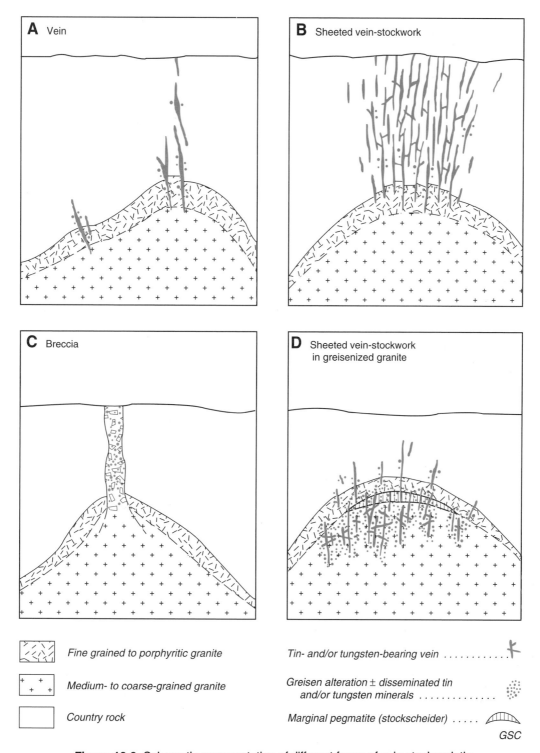

Fine grained to porphyritic granite

Medium- to coarse-grained granite

Country rock

Tin- and/or tungsten-bearing vein . . . .

Greisen alteration ± disseminated tin
    and/or tungsten minerals . . . .

Marginal pegmatite (stockscheider) . . . . .

GSC

Figure 18-3. Schematic representation of different forms of vein-stockwork tin-tungsten deposits.

Figure 18-4. Cassiterite- and arsenopyrite-bearing veins (dark) cutting fractured and highly altered granite, Mount Pleasant, North zone. GSC 204880B

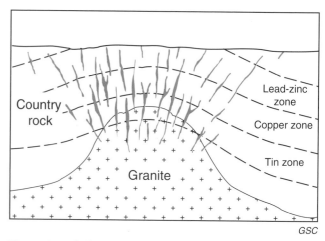

Figure 18-5. Schematic diagram illustrating regional zonation of metals in vein deposits of Cornwall and Devon (after Dines, 1934).

Metal zoning is common in many vein-stockwork tin and tungsten deposits and districts. In the tin-copper deposits of Cornwall, for example, tin (±tungsten) veins occur closest to, and commonly within, associated granite bodies; copper and lead-zinc veins are distributed above and outward from the tin veins (Fig. 18-5). Zoning also occurs within individual veins in Cornwall; the Dalcoath lode, for example, consisted of mainly copper in the upper part and tin in the lower part (Dines, 1956). Other deposits or districts which display comparable metal zoning include the San Rafael deposit, Peru (Clark et al., 1983), the Akenobe deposit, Japan (Sato and Akiyama, 1980), and the Herberton tin field, Australia (Blake and Smith, 1970; Taylor and Steveson, 1972). In some deposits, however, zonation of metals can be complex and highly varied, as in the tin-silver deposits of Bolivia (Turneaure, 1960).

Many vein tungsten deposits in southern China display vertical zonation with respect to vein morphology and distribution of ore minerals. The general model for these deposits (Fig. 18-6) consists of five zones. The uppermost part of a deposit, zone I, consists of parallel veinlets less than 1 cm wide which are closely to widely spaced. Veinlets in zone II are 1 to 10 cm wide, and are subparallel; branching or merging is common. In zones III and IV, vein width increases from 10 to more than 50 cm; vein density decreases as vein width increases. The lowermost part of a deposit, zone V, consists of branching or merging veins and veinlets that pinch out with depth. Zone V veins and veinlets are commonly rooted in granite. Tungsten occurs primarily as wolframite, and is most abundant in zones III and IV, and to a lesser extent in zone II. Cassiterite distribution overlaps that of tungsten in zones II, III, and the upper part of zone VI, and commonly extends into zone I. Chalcopyrite and sphalerite occur mainly in zones III and IV; molybdenite occurs in zones IV and V.

Zonation in stockwork deposits is less pronounced, if present at all. In the Hub deposit, Czech Republic, for example, the ratio of tin to tungsten increases slightly with depth (Stemprok, 1986). Copper, zinc, and molybdenum are associated with tin and tungsten in the upper part of the deposit, but are rare or absent in the lower part.

Metals may also be weakly dispersed in zones surrounding vein-stockwork tin-tungsten deposits. At East Kemptville, a zone of anomalous tungsten, due to wolframite and scheelite in veinlets and fractures in metasedimentary rock, is limited to 200 m from the contact of the metasedimentary rock with greisenized, tin-bearing granite; zones of copper, tin, and zinc, respectively, extend increasingly further, for as much as 600 m from the contact (G.J.A. Kooiman, pers. comm., 1991).

Associated granitic rocks

Granitic rocks associated with vein-stockwork tin deposits are commonly referred to as tin granites or "specialized" granites. Tin granites range from peraluminous to peralkaline in composition. Compared to normal granites, such as the low-Ca granite in the compilation by Turekian and Wedepohl (1961), tin granites are characterized by high contents of SiO_2, and alkali, lithophile, and high field strength elements such as Rb, Li, Be, Ga, REEs, Y, Sn, Ta, Nb, U, W, Mo, and volatiles such as F and B; they are typically depleted in CaO, MgO, TiO_2, total FeO, Sr, and Ba. The average composition of specialized granites, as compiled by Tischendorf (1977), is given in Table 18-2; the average compositions of low-Ca granite and A- and S-type granites are included for comparison. The compositions of tin granites from East Kemptville and Mount Pleasant are also given in Table 18-2. Tin granites have many characteristics in common with A- and S-type granites, but appear to be more highly fractionated. Granites associated with vein-stockwork tungsten deposits are similar to tin granites; an example is the granite associated with the Burnthill tungsten deposit (Table 18-2).

Many granites associated with tin-tungsten deposits were emplaced in multiple stages of intrusion that represent progressive degrees of fractionation. The deposits are in most cases related to specific intrusive phases that are among the most highly fractionated. Such phases are also characterized by a variety of textures that are the result of interaction between the granitic magma and aqueous, ore-bearing fluids. For example, concentration of aqueous fluids at the top of a cupola can result in pegmatitic zones distinguished by feldspar crystals that have grown from the contact of the cupola inward; these are referred to in European literature as "stockscheider". Comb quartz layers and other unidirectional solidification textures that are typical of many fluid-saturated felsic intrusions associated with porphyry molybdenum-tungsten deposits may also be present (cf. Shannon et al., 1982; Kirkham and Sinclair, 1988). Sudden or rapid loss of aqueous fluids from granitic melts can cause undercooling (or quenching) due to the rapid decrease in fluid pressure, resulting in the formation of aplitic, porphyritic, and micrographic or granophyric textures (cf. Fenn, 1986); the development of such textures within the upper marginal portions of small cupolas is typical of many tin-tungsten granites, for example, Mount Pleasant (Sinclair et al., 1988).

Tin and tungsten granites are also distinguished by their accessory minerals. In the eastern Transbaikal region of Russia, tin and tungsten deposits are associated with granites containing ilmenite and monazite, whereas granites containing sphene and allanite are barren (Ivanova and Butuzova, 1968). Ishihara (1977, 1981) found that most tin and tungsten deposits worldwide are associated with what he referred to as ilmenite-series granites, which contain ilmenite but are devoid of magnetite, and that none (or few) are related to magnetite-series granites, which can contain both magnetite and ilmenite. Other accessory minerals in various tin-tungsten granites include tourmaline, topaz, fluorite, apatite, xenotime, andalusite, and cordierite.

Alteration

Hydrothermal alteration assemblages associated with vein-stockwork tin and tungsten deposits are characterized by F-, Li-, and/or B-rich minerals such as topaz, fluorite, tourmaline, lepidolite, zinnwaldite, and F- and Li-rich muscovite and biotite. Altered rock containing these minerals is commonly referred to as greisen (Shcherba, 1970). Other minerals associated with greisen-type alteration include albite, microcline, chlorite, and quartz. Cassiterite, wolframite, chalcopyrite, sphalerite, pyrite, and other sulphide minerals may be disseminated in greisen-altered rock.

Alteration margins along individual veins or fractures range from narrow selvages one centimetre wide or less to broader zones as much as several metres wide. Zonal distribution of alteration assemblages, both lateral and vertical, is also present in some deposits; in general, zones containing abundant topaz and/or tourmaline occur closest to veins or fractures and are surrounded by zones containing muscovite (or sericite) and chlorite.

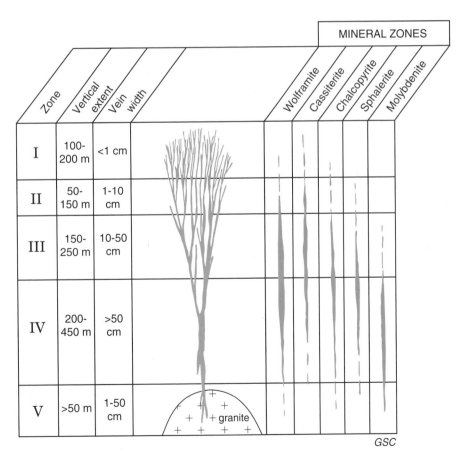

Figure 18-6. Vertical zonation of a typical tungsten vein deposit in southeast China (after Gu, 1982; Li, 1993).

Table 18-2. Composition of granites associated with vein-stockwork tin and tungsten deposits.

	Specialized granites (Tischendorf, 1977)	Low-Ca granite (Turekian and Wedepohl, 1961)	Average A-type granite (Whalen et al., 1987)	Average S-type granite (Whalen et al., 1987)	Biotite monzogranite, East Kemptville (Boyle, 1990)	Granite II, Mount Pleasant (Sample 87-11; Sinclair and Kooiman, 1990)	Biotite microgranite, Burnthill (Unit 3b, average of 8 analyses in MacLellan et al., 1990)
Major elements (weight %)							
SiO_2	73.38 ± 1.39	74.2	73.81	70.27	75.81	75.9	77.00
TiO_2	0.16 ± 0.10	0.20	0.26	0.48	0.15	0.04	0.08
Al_2O_3	13.97 ± 1.07	13.6	12.40	14.10	13.01	12.9	12.69
Fe_2O_3	0.80 ± 0.47	2.0*	1.24	0.56	0.99	0.0	1.04*
FeO	1.10 ± 0.47	n.d.**	1.58	2.87	1.07	1.5	n.d.
MnO	0.045 ± 0.040	0.05	0.06	0.06	0.03	0.03	0.07
MgO	0.47 ± 0.56	0.17	0.20	1.42	0.24	0.02	0.18
CaO	0.75 ± 0.41	0.71	0.75	2.03	0.54	0.70	0.50
Na_2O	3.20 ± 0.61	3.48	4.07	2.41	2.94	3.3	3.78
K_2O	4.69 ± 0.68	5.01	4.65	3.96	4.62	4.92	4.50
P_2O_5	-	0.14	0.04	0.15	0.15	0.0	0.01
F	0.37 ± 0.15	0.09	n.d.	n.d.	0.17	0.55	0.10
Trace elements (ppm)							
Rb	550 ± 200	170	169	217	439	823	507
Sr	n.d.	100	48	120	31	10	16
Li	400 ± 200	40	n.d.	n.d.	118	220	n.d.
Be	13 ± 6	3	n.d.	n.d.	7	n.d.	n.d.
Ga	n.d.	17	24.6	17	n.d.	26	n.d.
Nb	n.d.	21	37	12	14	59	35
Ce	n.d.	92	137	64	n.d.	147	46
Y	n.d.	40	75	32	30	164	85
U	n.d.	3.0	5	4	16	n.d.	21
Zr	n.d.	175	528	165	103	94	72
Sn	30 ± 15	3	n.d.	n.d.	17	22	19
W	7 ± 3	2.2	n.d.	n.d.	10	12	3.3
Cu	n.d.	10	2	11	5	10	n.d.
Mo	4 ± 2	1.3	n.d.	n.d.	3	7	n.d.
Pb	n.d.	19	n.d.	n.d.	12	64	n.d.
Zn	n.d.	39	120	62	75	68	26

*Total Fe reported as Fe_2O_3.   ** n.d. = no data

Alteration associated with stockwork deposits can be present on a broad scale, forming extensive areas of continuous alteration over distances of tens to hundreds of metres. This style of alteration is common in the upper parts of granitic intrusions, as in the classic "greisen" deposits of the Erzgebirge in central Europe (Baumann, 1970) and at East Kemptville (Fig. 18-7, 18-8).

DEFINITIVE CHARACTERISTICS

The principal characteristics of vein-stockwork tin and tungsten deposits are structural control (veins, fractures); association with highly fractionated or specialized granitic intrusions, particularly ilmenite-series granites; and greisen-type alteration.

GENETIC MODEL

The association of vein-stockwork tin and tungsten deposits with granitic rocks has long been recognized and a genetic relationship is generally accepted. A brief historical review of ideas concerning the origin of tin deposits was provided by Tischendorf (1977).

Granitic rocks associated with vein-stockwork tin and tungsten deposits occur in a variety of tectonic settings and range from peraluminous to peralkaline in composition. However, virtually all the genetically-related granites are highly fractionated and formed from magmas rich in silica, alkali elements, water, and other volatiles, such as fluorine and boron, and lithophile elements, especially tin and/or tungsten. Although a crustal source of the magmas is generally accepted, disagreement exists concerning the source of the metals. On one hand, workers in southern China (e.g. Xu and Zhu, 1988) consider enrichment of tin and tungsten in the source rocks to be an important factor in the generation of tin- and tungsten-bearing granitic magmas. On the other hand, according to Lehmann et al. (1990), enrichment of tin in the igneous rocks of the Bolivian tin belt was due to magmatic processes rather than anomalous tin in the source rocks.

Whatever the source of the metals, granites associated with vein-stockwork tin and tungsten deposits probably originate as fluid-undersaturated, anatectic melts at depths of 15 to 20 km, or more (Strong, 1981). In the case of S-type granites, these melts are derived in an orogenic setting from mature sedimentary rocks (Chappell and White, 1974); the melts which form A-type granites are

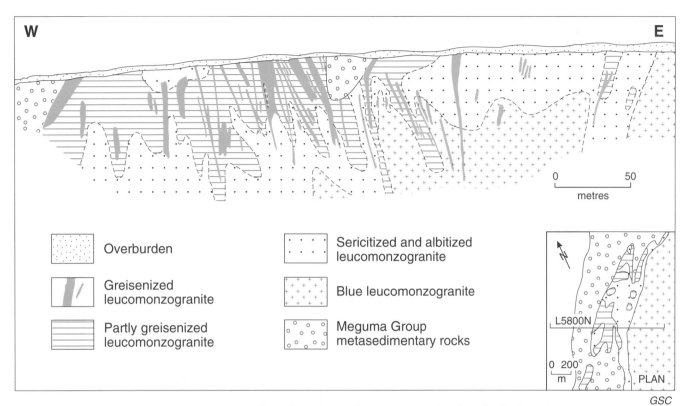

Figure 18-7. Cross-section L5800 of the East Kemptville tin deposit showing distribution of tin-bearing, greisen-altered zones; inset map shows plan view of surface and location of cross-section (after Richardson, 1988).

Figure 18-8. Photograph of the East Kemptville tin deposit showing tin-bearing greisen (grey, partly obscured by talus) and granite (light grey) below the subhorizontal contact of the granite with Meguma Group metasedimentary rocks (dark grey). GSC 1991-426

derived in an anorogenic setting by partial melting of felsic granulite residue from which granite has been produced previously (Collins et al., 1982; Whalen et al., 1987). When these melts are emplaced at higher levels in the crust and begin to crystallize, they become saturated in water and other volatiles such as chlorine, fluorine, boron, and carbon dioxide. Aqueous fluids, into which these volatile elements plus tin, tungsten, and other metals are strongly partitioned, subsequently separate from the crystallizing magma.

The style of mineralization (i.e. vein versus stockwork) depends on the magnitude of pressure developed within the crystallizing magma by the metal-rich aqueous fluids and the confining pressure of the surrounding country rock (mainly lithostatic pressure plus tensile strength of the country rock, including previously crystallized magma). If the confining pressure of the surrounding rocks is low, metal-bearing fluids can escape from the magma, primarily into faults and fractures to form vein deposits (Fig. 18-3A). Moderate to high confining pressure, combined with high fluid pressure in the magma, can result in extensive fracturing of the country rocks and the development of sheeted veins and stockworks (Fig. 18-3B); extremely high fluid pressure in the magma results in explosive failure of the country rocks and the formation of breccia pipes (Fig. 18-3C). If the confining pressure is high and/or fluid pressure in the magma is low, fracturing will be limited. In this case, the fluids will be trapped mainly within the crystallizing magma, leading to the formation of marginal pegmatites or stockscheider. The trapped fluids will also react with the magma and previously crystallized granite to form extensive zones of greisen alteration (Fig. 18-3D).

The magnitude of pressure developed in the crystallizing magma, and consequently the style of mineralization, is affected by the relative abundance of boron and fluorine in the mineralizing system. According to Pollard et al. (1987), the higher solubility of water in boron-bearing magmas compared to fluorine-bearing magmas leads to higher fluid pressures during the crystallization of the residual magma. Thus, boron-rich systems result dominantly in breccia, stockwork, and vein deposits, whereas fluorine-rich systems lead to the formation of deposits characterized more by extensive alteration (greisenization) and disseminated mineralization.

RELATED DEPOSIT TYPES

Vein-stockwork tin-tungsten deposits are the result of magmatic-hydrothermal activity related to the crystallization of felsic, peraluminous to peralkaline granitic rocks. Porphyry tungsten-molybdenum and porphyry tin deposits are associated with similar granitic rocks and have comparable styles of mineralization; the relationship between porphyry deposits and large vein-stockwork deposits is probably gradational. Deposits of lithium, beryllium, niobium, and tantalum associated with peraluminous to subalkaline rare-metal granites (Pollard, 1989) are also close relatives, but do not share the same degree of structural control.

Pipe deposits, which occur in the upper parts of granitic intrusions, also represent a different style of mineralization. These deposits are typically 1 to 2 m across, hundreds of metres long, and in many cases have sinuous shapes that appear unrelated to any obvious structural control. Such pipes occur in the Zaaiplaats tin deposit, South Africa (Strauss, 1954; Pollard et al., 1989) and in the tungsten deposits of Wolfram Camp, Australia (Blanchard, 1947).

Carbonate-replacement and skarn tin deposits, and some skarn tungsten deposits, are also associated with granitic rocks similar in composition to those associated with vein-stockwork tin and tungsten deposits.

EXPLORATION GUIDES

Exploration guidelines for vein-stockwork tin and tungsten deposits include the following:

1) Associated granitic rocks: vein-stockwork tin and tungsten deposits are associated with postorogenic to anorogenic granites that are typically enriched in lithophile elements such as Sn, W, Li, Rb, Be, Ga, REEs, Y, Ta, Nb, U, and Mo, and volatiles such as F and B.

2) Alteration: hydrothermal alteration typically consists of greisen-type assemblages characterized by F-, Li-, and/or B-rich minerals such as topaz, fluorite, tourmaline, and F- and/or Li-bearing micas such as lepidolite and zinnwaldite.

3) Structural control: faults and fractures are major ore controls in these deposits.

4) Zoning: tin and tungsten may be zoned relative to base metals at both regional (district) and local (deposit) scales.

5) Geochemical approaches: primary dispersion aureoles in host rocks (Sn, W, Cu, Zn, Pb, Rb, Li, F, B), secondary dispersion halos in overburden, and heavy minerals (cassiterite, wolframite, scheelite, topaz, tourmaline) in stream sediments help identify target areas at both regional and local scales.

6) Geophysical approaches: radiometric surveys may be useful for identifying tin and tungsten granites that are enriched in U, Th, and K. Magnetic surveys can be used to outline potential tin or tungsten granites of low magnetic response that are primarily ilmenite-series granites; magnetic surveys can also identify associated zones of magnetite- or pyrrhotite-bearing hornfels. Gravity surveys can be used to outline hidden granites in host rocks of contrasting density.

ACKNOWLEDGMENT

D.R. Boyle reviewed the manuscript and provided numerous constructive comments.

SELECTED BIBLIOGRAPHY

References with asterisks (*) are considered to be the best source of general information on this deposit type.

Amlôt, R.
1990: Tin; in Metals and Minerals Annual Review - 1990, Mining Journal Ltd., London, United Kingdom, p. 43-44.
Anstett, T.F., Bleiwas, D.I., and Hurdelbrink, R.J.
1985: Tungsten availability - market economy countries; United States Bureau of Mines, Information Circular 9025, 51 p.
Baumann, L.
1970: Tin deposits of the Erzgebirge; Transactions of the Institution of Mining and Metallurgy, v. 79, p. B68-B75.

*Beus, A.A. (ed.)
1986:   Geology of tungsten; International Geological Correlation Programme, Project 26 'MAWAM', UN Educational Scientific Cultural Organization, Earth Sciences, v. 18, 280 p.

Blake, D.H. and Smith, J.W.
1970:   Mineralogical zoning in the Herberton tinfield, North Queensland, Australia; Economic Geology, v. 65, p. 993-997.

Blanchard, R.
1947:   Some pipe deposits of eastern Australia; Economic Geology, v. 42, p. 265-304.

Bourassa, A.
1988:   Tin; in Canadian Minerals Yearbook 1987, Review and Outlook; Energy, Mines and Resources Canada, Mineral Report 36, p. 64.1-64.11.

Boyle, D.R.
1990:   The East Kemptville polymetallic domain, in Mineral Deposits of New Brunswick, (ed.) D.R. Boyle; 8th International Association on the Genesis of Ore Deposits Symposium, Guidebook, Field Trip 2; Geological Survey of Canada, Open File 2157, p. 88-132.

Chappell, B.W. and White, A.J.R.
1974:   Two contrasting granite types; Pacific Geology, v. 8, p. 173-174.

Clark, A.H., Palma, V.V., Archibald, D.A., Farrar, E., Arenas, M.J., and Robertson, R.C.R.
1983:   Occurrence and age of tin mineralization in the Cordillera Oriental, southern Peru; Economic Geology, v. 78, p. 514-520.

Collins, W.J., Beams, S.D., White, A.J.R., and Chappell, B.W.
1982:   Nature and origin of A-type granites with particular reference to southeastern Australia; Contributions to Mineralogy and Petrology, v. 80, p. 189-200.

Dines, H.G.
1934:   The lateral extent of the ore-shoots in the primary depth zones of Cornwall; Transactions of the Royal Geological Society of Cornwall, v. XVI, pt. XI, p. 279-296.
1956:   The metalliferous mining region of south-west England; Memoir of the Geological Survey of Great Britain (two volumes), London, 795 p.

Fenn, P.M.
1986:   On the origin of graphic granite; American Mineralogist, v. 71, p. 325-330.

Gu Juyun
1982:   Morphological zoning of the vein-type tungsten deposits in southern China; in Tungsten Geology, China, (ed.) Hepworth, J.V. and Yu Hong Zhang; UN Economic and Social Commission for Asia and the Pacific/Regional Mineral Resources Development Centre (Indonesia), Bandung, Indonesia, p. 269-278.

Higgins, N.C.
1985:   Wolframite deposition in a hydrothermal vein system: the Grey River tungsten prospect, Newfoundland, Canada; Economic Geology, v. 80, p. 1297-1327.

*Hosking, K.F.G.
1979:   Tin distribution patterns; Bulletin of the Geological Society of Malaysia, no. 11, p. 1-70.

Ishihara, S.
1977:   The magnetite-series and ilmenite-series granitic rocks; Mining Geology, v. 27, p. 293-305.
*1981:   The granitoid series and mineralization; in Economic Geology, Seventy-fifth Anniversary Volume, 1905-1980, (ed.) B.J. Skinner; p. 458-484.

Ivanova, G.F. and Butuzova, Y.G.
1968:   Distribution of tungsten, tin and molybdenum in the granites of Eastern Transbaykaliya; Geochemistry International, v. 5, no. 3, p. 572-583.

Kingsbury, C.J.R.
1965:   Cassiterite and wolframite veins of Aberfoyle and Story's Creek; in Geology of Australian Ore Deposits, Australasian Institute of Mining and Metallurgy, p. 506-511.

Kirkham, R.V. and Sinclair, W.D.
1988:   Comb quartz layers in felsic intrusions and their relationship to porphyry deposits; in Recent Advances in the Geology of Granite-related Mineral Deposits, (ed.) R.P. Taylor and D.F. Strong; The Canadian Institute of Mining and Metallurgy, Special Volume 39, p. 50-71.

Kontak, D.J. and Chatterjee, A.K.
1992:   The East Kemptville tin deposit, Yarmouth County, Nova Scotia: a Pb-isotope study of the leucogranite and mineralized greisens - evidence for a 366 Ma metallogenetic event; Canadian Journal of Earth Sciences, v. 29, p. 1180-1196.

Lehmann, B., Ishihara, S., Michel, H., Miller, J., Rapela, C., Sanchez, A., Tistl, M., and Winkelmann, L.
1990:   The Bolivian tin province and regional tin distribution in the central Andes: a reassessment; Economic Geology, v. 85, p. 1044-1058.

Li Yidou
1993:   Polytype model for tungsten deposits and vertical structural zoning model for vein-type tungsten deposits in South China; in Mineral Deposit Modeling, (ed.) R.V. Kirkham, W.D. Sinclair, R.I. Thorpe, and J.M. Duke; Geological Association of Canada, Special Paper 40, p. 555-568.

Lynch, J.V.G.
1989:   Hydrothermal alteration, veining, and fluid inclusion characteristics of the Kalzas wolframite deposit, Yukon; Canadian Journal of Earth Sciences, v. 26, p. 2106-2115.

Maby, M.
1990:   Tungsten; in Metals and Minerals Annual Review - 1990, Mining Journal Ltd., London, p. 67-69.

MacLellan, H.E., Taylor, R.P., and Gardiner, W.W.
1990:   Geology and geochemistry of Middle Devonian Burnthill Brook granites and related tin-tungsten deposits, York and Northumberland counties, New Brunswick; New Brunswick Department of Natural Resources and Energy, Minerals and Energy Division, Mineral Resource Report 4, 95 p.

Mato, G., Ditson, G., and Godwin, C.I.
1983:   Geology and geochronometry of tin mineralization associated with the Seagull batholith, south-central Yukon Territory; The Canadian Institute of Mining and Metallurgy Bulletin, v. 76, no. 854, p. 43-49.

McKinnon, A. and Seidel, H.
1988:   Tin; in Register of Australian Mining 1988/89, (ed.) R. Louthean; Resource Information Unit Ltd., Subiaco, Western Australia, p. 197-204.

McNeil, M.
1982:   Panasqueira-the largest mine in Portugal; World Mining, December, 1982, p. 52-55.

Mount, M.
1985:   Geevor mine: a review; in High Heat Production (HHP) Granites, Hydrothermal Circulation and Ore Genesis; Papers presented at the High Heat Production (HHP) Granites, Hydrothermal Circulation and Ore Genesis Conference, organized by the Institution of Mining and Metallurgy and held in St. Austell, Cornwall, England, September 22-25, 1985, p. 221-238.

Moyle, J.E.
1984:   Development and construction begins at East Kemptville, North America's only primary tin mine; Mining Engineering, April 1984, p. 335-336.

Muecke, G.K., Elias, P., and Reynolds, P.H.
1988:   Hercynian/Alleghanian overprinting of an Acadian Terrane: $^{40}Ar/^{39}Ar$ studies in the Meguma Zone, Nova Scotia; Chemical Geology, v. 73, p. 153-167.

Mulligan, R.
*1975:   Geology of Canadian tin occurrences; Geological Survey of Canada, Economic Geology Report 28, 155 p.
*1984:   Geology of Canadian tungsten occurrences; Geological Survey of Canada, Economic Geology Report 32, 121 p.

Omer-Cooper, W.R.B., Hewitt, W.V., and van Hees, H.
1974:   Exploration for cassiterite-magnetite-sulphide veins on Belitung, Indonesia; in 4th World Conference on Tin, Kuala Lumpur, 1974, Volume Two: Prospecting and Mining; International Tin Council, London, p. 95-117.

Pollard, P.J.
1989:   Geologic characteristics and genetic problems associated with the development of granite-related deposits of tantalum and niobium; in Lanthanides, Tantalum and Niobium, (ed.) P. Möller, P. Cerný and F. Saupé; Society for Geology Applied to Mineral Deposits, Special Publication No. 7, Springer-Verlag, Berlin, p. 240-256.

Pollard, P.J., Pichavant, M., and Charoy, B.
1987:   Contrasting evolution of fluorine- and boron-rich tin systems; Mineralium Deposita, v. 22, p. 315-321.

Pollard, P.J., Taylor, R.G., and Tate, N.M.
1989:   Textural evidence for quartz and feldspar dissolution as a mechanism of formation for Maggs pipe, Zaaiplaats tin mine, South Africa; Mineralium Deposita, v. 24, p. 210-218.

Reed, B.L.
1986:   Descriptive model of Sn greisen deposits; in Mineral Deposit Models, (ed.) D.P. Cox and D.F. Singer; United States Geological Survey, Bulletin 1693, p. 70.

Richardson, J.M.
1988:   Field and textural relationships of alteration and greisen-hosted mineralization at the East Kemptville tin deposit, Davis Lake complex, southwest Nova Scotia; in Recent Advances in the Geology of Granite-Related Mineral Deposits, (ed.) R.P. Taylor and D.F. Strong; The Canadian Institute of Mining and Metallurgy, Special Volume 39, p. 265-279.

Roberts, R.
1988: Tungsten; in Register of Australian Mining 1988/89, (ed.) R. Louthean; Resource Information Unit Ltd., Subiaco, Western Australia, p. 319-322.

Sato, N. and Akiyama, Y.
1980: Structural control of the Akenobe tin-polymetallic deposits, southwest Japan; in Granitic Magmatism and Related Mineralization, (ed.) S. Ishihara and S. Takenouchi; The Society of Mining Geologists of Japan, Mining Geology Special Issue, no. 8, p. 175-188.

Shannon, J.R., Walker, B.M., Carten, R.B., and Geraghty, E.P.
1982: Unidirectional solidification textures and their significance in determining relative ages of intrusions at the Henderson mine, Colorado; Geology, v. 10, p. 293-297.

*Shcherba, G.N.
1970: Greisens; International Geology Review, v. 12, p. 114-150, 239-259.

Shimizu, M. and Kato, A.
1991: Roquesite-bearing tin ores from the Omodani, Akenobe, Fukoko and Ikuno polymetallic vein-type deposits in the Inner Zone of southwestern Japan; Canadian Mineralogist, v. 29, p. 207-215.

Sinclair, W.D. and Kooiman, G.J.A.
1990: The Mount Pleasant tungsten-molybdenum and tin deposits; in Mineral Deposits of New Brunswick and Nova Scotia, (ed.) D.R. Boyle, 8th International Association on the Genesis of Ore Deposits Symposium, Guidebook, Field Trip 2; Geological Survey of Canada, Open File 2157, p. 78-87.

Sinclair, W.D., Kooiman, G.J.A., and Martin, D.A.
1988: Geological setting of granites and related tin deposits in the North Zone, Mount Pleasant, New Brunswick; in Current Research, Part B; Geological Survey of Canada, Paper 88-1B, p. 201-208.

Smith, A.
1979: Mining at Panasqueira mine, Portugal; Institution of Mining and Metallurgy Transactions, v. 88, p. A108-A115.

Stemprok, M.
1986: Tungsten deposits of central Europe, in Geology of Tungsten, (ed.) A.A. Beus; International Geological Correlation Programme, Project 26, 'MAWAM', UN Educational, Scientific and Cultural Organization, Earth Sciences, v. 18, p. 79-87.

Strauss, C.A.
1954: The geology and mineral deposits of the Potgietersrus tinfields; Geological Survey of South Africa, Memoir 46, 241 p.

Strong, D.F.
1981: A model for granophile mineral deposits; Geoscience Canada, v. 8, p. 155-161.

Sutherland Brown, A.
1960: Geology of the Rocher Deboule Range; British Columbia Department of Mines and Petroleum Resources, Bulletin no. 43, 78 p.

Sutphin, D.M., Sabin, A.E., and Reed, B.L.
1990: International Strategic Minerals Inventory summary report – tin; United States Geological Survey, Circular 930-J, 52 p.

*Taylor, R.G.
1979: Geology of Tin Deposits; Elsevier, Amsterdam, 544 p.

Taylor, R.G. and Pollard, P.J.
1986: Recent advances in exploration modelling for tin deposits and their application to the Southeast Asian environment; GEOSEA V Proceedings, Volume 1, Geological Society of Malaysia, Bulletin 19, p. 327-347.

Taylor, R.G. and Steveson, B.G.
1972: An analysis of metal distribution and zoning in the Herberton tinfield, North Queensland; Economic Geology, v. 67, p. 1234-1240.

* Tischendorf, G.
1977: Geochemical and petrographic characteristics of silicic magmatic rocks associated with rare-element mineralization; in Metallization Associated with Acid Magmatism, Volume 2, (ed.) M. Stemprok, L. Burnol, and G. Tischendorf; Geological Survey (Czechoslovakia), Prague, p. 41-96.

Turekian, K.K. and Wedepohl, K.H.
1961: Distribution of the elements in some major units of the earth's crust; Geological Society of America Bulletin, v. 72, p. 175-192.

Turneaure, F.S.
1960: A comparative study of major ore deposits of central Bolivia. Part II; Economic Geology, v. 55, p. 574-606.

Valenzuela, S.R.
1979: Geology of the main wolfram mines in Bolivia; Primary Tungsten Association, Bulletin no. 7, June, 1979, p. 4-8.

Whalen, J.B., Currie, K.L., and Chappell, B.W.
1987: A-type granites: geochemical characteristics, discrimination and petrogenesis; Contributions to Mineralogy and Petrology, v. 95, p. 407-419.

Willig, C.D. and Delgado, J.
1985: South America as a source of tungsten; in Tungsten: 1985, Proceedings of the Third International Tungsten Symposium, Madrid, May 1985, MPR Publishing Services, Shrewsbury, England, p. 58-85.

Xu Keqin and Zhu Jinchu
1988: Time-space distribution of tin/tungsten deposits in South China and controlling factors of mineralization; in Geology of Tin Deposits, (ed.) C.S. Hutchison; Springer-Verlag, Berlin, p. 265-277.

Author's address

W.D. Sinclair
Geological Survey of Canada
601 Booth Street
Ottawa, Ontario
K1A 0E8

Printed in Canada

19. PORPHYRY COPPER, GOLD, MOLYBDENUM, TUNGSTEN, TIN, SILVER

19.1 Copper ($\pm$ gold, molybdenum, silver, rhenium)

19.2 Copper-molybdenum ($\pm$ gold, silver)

19.3 Copper-gold ($\pm$ silver)

19.4 Gold ($\pm$ silver, copper, molybdenum)

19.5 Molybdenum ($\pm$ tungsten, tin)

19.6 Tungsten-molybdenum ($\pm$ bismuth, tin)

19.7 Tin ($\pm$ tungsten, molybdenum, silver, bismuth, indium)

19.8 Tin-silver ($\pm$ tungsten, copper, zinc, molybdenum, bismuth)

19.9 Silver ($\pm$ gold, zinc, lead)

19. PORPHYRY COPPER, GOLD, MOLYBDENUM, TUNGSTEN, TIN, SILVER

R.V. Kirkham and W.D. Sinclair

INTRODUCTION

Porphyry deposits are large, low- to medium-grade deposits in which hypogene ore minerals are primarily structurally controlled and which are spatially and genetically related to felsic to intermediate porphyritic intrusions (Kirkham, 1972). The large size and structural control (e.g., veins, vein sets, stockworks, fractures, 'crackled zones', and breccia pipes) serve to separate porphyry deposits from genetically-related

Kirkham, R.V. and Sinclair, W.D.
1996: Porphyry copper, gold, molybdenum, tungsten, tin, silver; in Geology of Canadian Mineral Deposit Types, (ed.) O.R. Eckstrand, W.D. Sinclair, and R.I. Thorpe; Geological Survey of Canada, Geology of Canada, no. 8, p. 421-446 (also Geological Society of America, The Geology of North America, v. P-1).

(e.g., some skarns, high-temperature mantos, breccia pipes, peripheral mesothermal ("intermediate", "transitional") veins, epithermal precious-metal deposits) and unrelated deposit types.

Supergene minerals may be developed in enriched zones in porphyry deposits by weathering of primary sulphides. Such zones typically have much higher copper grades, thereby enhancing the possibility of economic exploitation. Oxidization of porphyry deposits can also reduce sulphide contents of gold zones, thus improving extraction of gold by heap-leach methods.

The metal contents and petrogenetic associations of porphyry deposits are diverse. In this paper, nine subtypes with different metal contents are grouped into one large "deposit type". Definition of subtypes is based on the simple principle that metals essential to the economics of the deposit define the subtype; byproduct and potential byproduct metals are listed in parentheses. For deposits with subeconomic grades and tonnages, projections would have to be made as to probable coproduct and byproduct metals, assuming that the deposits were economic. Other types of porphyry deposits might be identified in the future and changes in metal prices could affect the groupings. For example, some deposits, such as Bingham in Utah, contain economically important amounts of molybdenum and gold as well as copper, so that a Cu-Mo-Au grouping might be appropriate for this as well as other deposits (Table 19-1) (c.f., Kesler, 1973; Sinclair et al., 1982; Cox and Singer, 1988). Another possible subtype might be porphyry tungsten deposits, for example large, bulk-mineable vein-stockwork tungsten deposits such as Mount Carbine, Australia (35 Mt grading 0.08% W) and Hemerdon, England (42 Mt grading 0.14% W and 0.025% Sn)(see Type 18, Vein-stockwork tin, tungsten). Important subtypes that are well established include **porphyry copper** and **copper-molybdenum** deposits associated typically with calc-alkaline intrusive rocks (Titley and Hicks, 1966; Lowell and Guilbert, 1970; Sillitoe, 1972, 1973, 1986; Hollister, 1974; Gustafson and Hunt, 1975; Sutherland Brown, 1976; Gustafson, 1978; McMillan and Panteleyev, 1980; Titley and Beane, 1981; Titley, 1982, 1993; Guilbert, 1986; Guilbert and Park, 1986; McMillan, 1991); **porphyry copper-gold** deposits, many of which, but not all, are associated with alkaline intrusive rocks (Barr et al., 1976; Mutschler et al., 1985; Imai et al., 1992; Mutschler and Mooney, 1993; Sillitoe, 1993b; Müller and Groves, 1993; MacDonald and Arnold, 1993, 1994; Müller et al., 1994; Carlile and Mitchell, 1994; Meldrum et al., 1994; Perelló, 1994); **porphyry gold** (Vila and Sillitoe, 1991; Sillitoe, 1993b; Fraser, 1993; Richards and Kerrich, 1993); and **porphyry molybdenum** deposits, including high grade deposits associated with fluorine-rich, high silica granites, and lower grade deposits associated with low-fluorine granitic and granodioritic rocks (Wallace et al., 1968; Woodcock and Hollister, 1978; Sillitoe, 1980; White et al., 1981; Westra and Keith, 1981; Mutschler et al., 1981; Theodore and Menzie, 1984; Carten et al., 1988b, 1993; Huang et al., 1988). Other subtypes, which are not as well established, include **porphyry tungsten-molybdenum** deposits (Liu, 1981; Noble et al. 1984; Kooiman et al., 1986; Sinclair, 1986; Kirkham and Sinclair, 1988); **porphyry tin and tin-silver** deposits (Sillitoe et al., 1975; Grant et al., 1977, 1980; Guan et al., 1988; Richardson, 1988; Villalpando, 1988; Huang and Zhang, 1989),

and especially **porphyry silver** deposits (e.g., Real de Angeles, Mexico – no known intrusive rocks but hornfels at depth; Pearson et al., 1988). Linkages (gradations of one subtype to another), distinctions, and petrogenetic associations of porphyry subtypes are not well documented; nevertheless, we believe that several porphyry deposit metal subtypes exist (as listed above). In the following sections, we give examples of subtypes, describe their characteristics and suggest possible linkages between them.

Examples of the various subtypes are listed below.

Porphyry copper:
Canadian – Island Copper, Bethlehem, Lornex, Valley Copper, Bell Copper, and Granisle, British Columbia; Queylus and Gaspé Copper (Copper Mountain), Quebec.
Foreign – Butte, Montana; Bingham, Utah; Morenci, Arizona; Cananea and La Caridad, Mexico; Cerro Colorado, Panama; Toquepala, Cuajone, and Michiquillay, Peru; Chuquicamata, El Teniente, La Escondida, and El Salvador, Chile; Atlas and Sipalay, Philippines; Sar Cheshmeh, Iran; Kounrad and Almalyk, Kazakhstan; Recsk, Hungary; Bor, former Yugoslavia; Haib, Namibia; Malanjkhand, India; Tongkuangyu, Dexing, and Yulong, China.

Porphyry copper-molybdenum:
Canadian – Brenda, Highmont, and Berg, British Columbia; McLeod Lake, Quebec.
Foreign – Mount Tolman, Washington; Cumo, Idaho; Esperanza, Sierrita, and Mineral Park, Arizona; Mocoa, Colombia; Copaquire, Chile; Nuggetty Gully, Australia.

Porphyry copper-gold:
Canadian – Copper Mountain, Afton, Mount Polley, Fish Lake, Mount Milligan, and Galore Creek, British Columbia; Casino, Yukon Territory.
Foreign – Tanama, Puerto Rico; Bajo de la Alumbrera, Argentina; Dizon, Guinaoang, and Lepanto Far Southeast, Philippines; Grasberg and Batu Hijau, Indonesia; Panguna, Papua New Guinea; Goonumbla (Endeavour), Australia; Skouries, Greece.

Porphyry gold:
Canadian – Lac Troilus, Quebec; possibly Sulphurets (Snowfield zone), British Columbia; possibly Brewery Creek and Dublin Gulch, Yukon Territory; possibly Moss Lake, Ontario.
Foreign – Fort Knox, Alaska; Cripple Creek, Colorado; possibly Metates, Mexico; possibly Kori Kollo, Bolivia; and Marte, Lobo, Refugio, and Alebarán (Cerro Casale), Chile. Porgera and Ladolam, Papua New Guinea and Kelian, Indonesia are telescoped systems with epithermal mineralization superimposed on porphyry deposits.

Porphyry molybdenum:
Canadian – Boss Mountain, Endako, Glacier Gulch, Kitsault, and Adanac, British Columbia; Red Mountain, Yukon Territory; Setting Net Lake, Ontario.
Foreign – Quartz Hill, Alaska; Thompson Creek, Idaho; Pine Grove and Mount Hope, Nevada; Climax and Henderson, Colorado; Questa, New Mexico; Malmbjerg, Greenland; Jinduicheng, China.

Table 19-1. Production/reserves of selected Canadain and foreign porphyry deposits.

Deposit	Production/Reserves	Comments/References
19.1 Porphyry copper deposits		
Canadian deposits:		
Valley Copper, British Columbia	692 Mt; 0.414% Cu 0.0069% Mo	Measured and indicated reserves (Northern Miner, April 27, 1992, p. 5)
Lornex, British Columbia	545 Mt; 0.42% Cu 0.015% Mo	Production plus reserves (Lornex Mining Corporation Ltd., Annual Reports 1972-78; Northern Miner, March 24, 1977, p. 23)
Island Copper, British Columbia	345 Mt; 0.42% Cu 0.017% Mo 0.19 g/t Au 1.4 g/t Ag	Mill feed, 1971 to end of 1993 (Perelló et al., 1995)
Granisle, British Columbia	125 Mt; 0.44% Cu	Production plus reserves (Northern Miner, January 18, 1973, p. 3)
Gaspé Copper (Copper Mountain zone), Quebec	153 Mt; 0.37% Cu 0.02% Mo	Reserves (Kirkham et al., 1982)
Foreign deposits:		
Bingham, Utah	2068 Mt; 0.78% Cu 0.032% Mo 0.29 g/t Au 2.38 g/t Ag	Production 1904-1972 (James, 1978) plus estimated open pit reserves (Argall, 1981); Pt, Pd, and Se are also recovered as byproducts
Butte, Montana	2083 Mt; 0.85% Cu	Miller, 1973; World Mining, February, 1977, p. 66-67; Engineering and Mining Journal, January 1990, p. c65.
Morenci-Metcalf, Arizona	1822 Mt; 0.77% Cu	Production and reserves (Engineering and Mining Journal, January 1990, p. c16; United States Bureau of Mines, Minerals Yearbook, 1983, v. 2)
Cananea, Mexico	1633 Mt; 0.7% Cu ~0.006% Mo	Geologic reserves (Bushnell, 1988)
Cerro Colorado, Panama	2210 Mt; 0.6% Cu 0.009% Mo 0.07 g/t Au 4.7 g/t Ag	Reserves (Linn et al., 1981)
Chuquicamata, Chile	10 838 Mt; 0.56% Cu 0.024% Mo	Total reserves (Sutulov, 1977; Ambrus, 1978)
El Teniente, Chile	8350 Mt; 0.68% Cu 0.027% Mo	Total reserves (Sutulov, 1977; Ambrus, 1978)
La Escondida, Chile	1873 Mt; 1.56% Cu ~ 0.021% Mo	Total reserves (Ojeda, 1986)
Atlas(Frank, Biga, Carmen), Philippines	898 Mt; 0.46% Cu 0.25% g/t Au	Reserves (Saegart and Lewis, 1977)
Sipalay, Philippines	740 Mt; 0.49% Cu 0.015% Mo 0.05 g/t Au 1.5 g/t Ag	Production (Sillitoe and Gappe, 1984)
Sar Chesmeh, Iran	1200 Mt; 1.2% Cu 0.03% Mo	Reserves (Mining Journal, June 14, 1991, p. 454)
Kounrad, Kazakhstan	225 Mt; 0.6% Cu	Reserves (Strishkov, 1984)
Malanjkhand, India	789 Mt; 0.83% Cu 0.004% Mo 0.2 g/t Au 6.0 g/t Ag	Estimated reserves (Sikka et al., 1991)
Dexing, China	1500 Mt; 0.43% Cu 0.017% Mo 0.16 g/t Au 1.9 g/t Ag	Reserves (Chen and Li, 1990)

Table 19-1. (cont.)

Deposit	Production/Reserves	Comments/References
Tongkuangyu, China	380 Mt 0.67% Cu	Reserves (mine staff, pers. comm., 1988)
Goonumbla (Endeavour 26), Australia	166 Mt; 0.74% Cu 0.12 g/t Au 1.7 g/t Ag	Possible reserves (Jones, 1985)
19.2 Porphyry copper-molybdenum deposits		
Canadian deposits:		
Brenda, British Columbia	227 Mt; 0.16% Cu 0.039% Mo	Production plus reserves (Mining Journal, June 6, 1980, p. 475; Brenda Mines Limited Annual Report)
Highmont, British Columbia	122 Mt; 0.26% Cu 0.026% Mo	Reserves (Canadian Mining Journal, May 1980, p. 89-90)
Berg British Columbia	363 Mt; 0.40% Cu 0.05% Mo	Reserves (Northern Miner, April 15, 1976, p. 3)
McLeod Lake, Quebec	38 Mt; 0.44% Cu 0.05% Mo 0.048 g/t Au	Reserves (Northern Miner, November 26, 1990, p. 19)
Foreign deposits:		
Mount Tolman, Washington	1180 Mt; 0.084% Cu 0.0552% Mo	Drill-indicated resource at 0.03% Mo cutoff (company data, 1980)
Cumo, Idaho	907 Mt; 0.06% Cu 0.06% Mo 0.008% W 2.06 g/t Ag	Possible resource at 0.03% Mo cutoff (D. Baker, pers. comm., 1985)
Mineral Park, Arizona	91 Mt; 0.39% Cu 0.043% Mo	Total production (company official, April, 1985)
Esperanza, Arizona	123 Mt; 0.38% Cu 0.026% Mo	Combined production and reserves (United States Bureau of Mines, Minerals Yearbook, 1980, Arizona, p. 9)
Sierrita, Arizona	579 Mt; 0.32% Cu 0.031% Mo	Mining Journal, January 23, 1970, p. 74
La Caridad, Mexico	1274 Mt; 0.42% Cu 0.038% Mo	Reserves (Mining Magazine, May, 1987, p. 370-381)
Mocoa, Columbia	254 Mt; 0.4% Cu 0.056% Mo	Reserves (Mining Journal, November 30, 1984, p. 379)
Copaquire, Chile	50 Mt; 0.2% Cu 0.13% Mo	Reserves (Lowell, 1974)
Nugetty Gully (Marble Bar), Australia	102 Mt; 0.152% Cu 0.105% Mo	Reserves (Lowell, 1978)
19.3 Porphyry copper-gold deposits		
Canadian deposits:		
Copper Mountain (Ingerbelle), British Columbia	235 Mt; 0.45% Cu 0.16 g/t Au	Production plus reserves (R.V. Kirkham, unpub. compilation)
Afton, British Columbia	32 Mt; 1.09% Cu 0.79 g/t Au	Production plus reserves (Afton Operating Corp., company official,1991)
Mount Polley , British Columbia	230 Mt; 0.25% Cu 0.343 g/t Au	Preliminary reserve estimate (Northern Miner, July 22, 1991, p. 3)
Fish Lake, British Columbia	1148 Mt; 0.22% Cu 0.411 g/t Au	Estimated reserves (Northern Miner,December 14, 1992, p. 3)
Mount Milligan British Columbia,	298 Mt 0.22% Cu 0.45 g/t Au	Total size (Engineering and Mining Journal, April, 1992, p. 32-WW)
Galore Creek, British Columbia	113 Mt; 1.06% Cu 0.45 g/t Au 8.6 g/t Ag	Indicated reserves in Central Zone (Northern Miner, October 1, 1990, p. 5)
Casino, Yukon Territory	162 Mt; 0.37% Cu 0.039% Mo 0.48 g/t Au	Reserves (Mining Review, 1992)

Deposit	Production/Reserves	Comments/References
Foreign deposits:		
Tanama, Puerto Rico	126 Mt; 0.64% Cu 0.002% Mo 0.7 g/t Au	Reserves (Cox, 1985; D. Cox, pers. comm., 1985)
Bajo de la Alumbrera Argentina	494 Mt; 0.53% Cu ~0.007% Mo 0.68 g/t Au	Reserves (Northern Miner, February 14, 1994, p. 2)
Dizon, Philippines	187 Mt; 0.355% Cu 0.746 g/t Au 2.0 g/t Ag	Estimated reserves (Imai et al., 1992)
Lepanto Far Southeast, Philippines	356 Mt; 0.73% Cu 1.24 g/t Au	Reserves (Concepcion and Cinco, 1989)
Grasberg, Indonesia	a) 1000 Mt; 1.4% Cu 1.8 g/t Au 3.9 g/t Ag	Reserves (MacDonald and Arnold, 1993)
	b) 4000 Mt; 0.6% Cu 0.64 g/t Au	Resource; 675 Mt mineable at 1.45% Cu and 1.87 g/t Au (van Leeuwen, 1994, p. 64)
Batu Hijau, Indonesia	335 Mt; 0.8% Cu 0.7 g/t Au	Inferred resource (Mining Journal, September 11, 1992, p. 179)
Ok Tedi, Papua New Guinea	377 Mt; 0.7% Cu 0.01% Mo 0.66 g/t Au	Estimated reserves (Goldie, 1990)
Panguna, Papua New Guinea	691 Mt; 0.4% Cu 0.47 g/t Au	Mining Journal, April 13, 1990, p. 307
Goonumbla, Australia	55 Mt; 1.07% Cu 0.56 g/t Au	Open pit and underground mineable reserves (Heithersay et al., 1990),

19.4 Porphyry gold deposits

Deposit	Production/Reserves	Comments/References
Canadian deposits:		
Moss Lake, Ontario	75 Mt; 1.06 g/t Au	Reserves (Northern Miner, October 5, 1992, p. 18)
Young-Davidson and Matachewan Consolidated, Ontario	8.7 Mt; 3.39 g/t Au 0.94 g/t Ag	Combined production, 1934-1956 (Sinclair, 1982)
Lac Troilus, Quebec	79.5 Mt; 0.11% Cu 1.3 g/t Au 1.38 g/t Ag	Geological reserves (Metall Mining Corporation, poster presentation, April, 1994)
Sulphurets (Snowfield zone) British Columbia	7.75 Mt; 2.8 g/t Au	Geological inventory (Newhawk Gold Mines Ltd., 1985)
Brewery Creek, Yukon Territory	16 Mt; 1.63 g/t Au	Reserves (Mining Journal, March 25, 1994, p. 212)
Dublin Gulch, Yukon Territory	90 Mt; 1.2 g/t Au	Speculative resource data (Northern Miner, February 21, 1994, p. 15)
Foreign deposits:		
Fort Knox, Alaska	158 Mt; 0.823 g/t Au	Reserves (Northern Miner, March 29, 1993, p. B5)
Montana Tunnels, Montana	61 Mt; 0.96 g/t Au 12.0 g/t Ag 0.67% Zn 0.28% Pb	Diatreme breccia; geological reserve (Sillitoe et al., 1985)
Zortman-Landusky, Montana	100 Mt; 0.62 g/t Au	Proven and probable reserves with significant silver; epithermal characteristics but structural control and widespread potassic alteration (Wampler, 1993)
Cripple Creek, Colorado	81.6 Mt; 1.13 g/t Au	Veins and diatreme breccias; bulk mineable reserves and resources end of 1992; previous mining from high-grade veins about 600 t gold at average grade about 30 to 60 g/t Au at about 15 g/t Au cutoff (J.A. Pontius, Cripple Creek Field Guide, Pikes Peak Mining Company, March 1993, 14 p; Thompson et al., 1985)

Table 19-1. (cont.)

Deposit	Production/Reserves	Comments/References
Las Cristinas, Venezuela	189 Mt; 1.269 g/t Au	Reserves (Northern Miner, April 11, 1994, p. B1); Cu present
Kori Kollo, Bolivia	53 Mt; 2.32 g/t Au 14.5 g/t Ag	Reserves, sulphide zone (Columba C. and Cunningham, 1993)
Marte, Chile	44 Mt; 1.4 g/t Au 0.06% Cu 0.004% Mo	Geological resources (company official, April, 1990)
Lobo, Chile	80 Mt; 1.6 g/t Au 0.1% Cu 0.001% Mo	Geological resources (company official, April, 1990)
Refugio, Chile	216 Mt; 0.88 g/t Au <3 g/t Ag 0.04% Cu	Geological reserves, Verde deposit at 0.5 g/t Au cutoff; additional reserves present in Pancho (100 Mt) and Guanco (30 Mt) deposits which contain 0.5 to 1.5 g/t Au and <0.1% Cu (Flores V., 1994).
Porgera, Papua New Guinea	51.5 Mt; 7.9 g/t Au	Reserves (Richards and Kerrich, 1993)
Ladolam (Lihir), Papua New Guinea	188 Mt; 3.6 g/t Au	Mineable resource at 1.6 g/t Au cutoff, epithermal deposit superimposed on low-grade porphyry deposit (Mining Journal, April 15, 1994, p. 266)
Kidston, Australia	36 Mt; 1.28 g/t Au	Breccia pipe; mineable reserve at 1.22:1 waste to ore ratio (Engineering and Mining Journal, May 1992, p. 12-WW)
19.5 Porphyry molybdenum deposits		
Canadian deposits:		
Endako, British Columbia	a) 100 Mt; 0.084% Mo b) 182 Mt; 0.0785% Mo	Production 1965 to May 1980 Reserves, December, 1979 (Kirkham et al., 1982).
Glacier Gulch, British Columbia	92 Mt; 0.178% Mo 0.05% W ~0.03% Cu	Geological reserve (Kirkham et al., 1982)
Boss Mountain, British Columbia	a) 5.5 Mt; 0.2165% Mo b) 6.7 Mt; 0.13% Mo c) 51.5 Mt; 0.06% Mo	Production 1967 to December, 1979 (Kirkham et al., 1982) Estimated mineable reserve March 26, 1981 (Kirkham et al., 1982) Subeconomic resource March 26, 1981 (Kirkham et al., 1982)
Kitsault, British Columbia	95 Mt; 0.112% Mo	Open pit reserves (Northern Miner, March 4, 1981, p. A1, & A29)
Adanac, British Columbia	152 Mt; 0.063% Mo	Reserves (Christopher and Pinsent, 1982)
Setting Net Lake, Ontario	90 Mt; 0.054% Mo	Inferred resource based on six diamond drillholes (Northern Miner, 1978, p. 17)
Foreign deposits:		
Quartz Hill, Alaska	1700 Mt; 0.136% Mo	Reserves (Nokleberg et al., 1987)
Thompson Creek, Idaho	300 Mt; 0.108% Mo ~0.005% W	Preproduction reserves (E. Wozniak pers. comm., 1981)
Mount Hope, Nevada	451 Mt; 0.09% Mo	Reserves (J. Erickson, pers. comm., April 20, 1993 and Exxon Annual Report)
Pine Grove, Nevada	113 Mt; 0.18% Mo	Geological reserves (company official pers. comm., 1985)
Climax, Colorado	907 Mt; 0.240% Mo 0.02% W	Geological reserves (Carten et al., 1993); W grade from S.W. Hobbs (pers. comm., 1985)
Henderson, Colorado	727 Mt; 0.171% Mo	Geological reserves (Carten et al., 1993)

Deposit	Production/Reserves	Comments/References
Questa, New Mexico	277 Mt; 0.144% Mo	Mineable reserves (Carten et al., 1993)
Malmbjerg, Greenland	136 Mt; 0.138% Mo 0.016% W	Reserves at 0.096% Mo cutoff (Schassberger, 1977, quoted in Geyti and Thomassen, 1984)
Jinduicheng, China	130 Mt; 0.20% Mo	Estimated open pit reserves (R.V. Kirkham, pers. comm., 1986)
Malala, Indonesia	100 Mt; 0.084% Mo ~0.02% Cu ~0.01% W	Geological resource (van Leeuwen et al., 1994)
19.6 Porphyry tungsten-molybdenum deposits		
Canadian deposits:		
Logtung, Yukon Territory	162 Mt; 0.10% W 0.03% Mo	Geological reserves (Noble et al., 1984)
Mount Pleasant, New Brunswick	22.5 Mt; 0.21% W 0.10% Mo 0.08% Bi	Drill-indicated resource in Fire Tower zone (Parrish and Tully, 1978)
Foreign deposits:		
Xinglokeng, China	78 Mt; 0.18% W 0.024% Mo	Tonnage estimate from Anstett et al. (1985); W, Mo grades from Liu (1981)
19.7 Porphyry tin deposits		
Canadian deposits:		
Mount Pleasant, New Brunswick	5.1 Mt; 0.79% Sn	Reserves in North zone tin deposits (Northern Miner, November 6, 1989, p. A2)
East Kemptville, Nova Scotia	a) 56 Mt; 0.165% Sn b) 27.9 Mt; 0.182% Sn 0.075% Cu 0.18% Zn	Preproduction reserves (Moyle, 1984) Proven and probable reserves, December 31, 1989 (Canadian Mines Handbook 1990-91, p. 385)
Foreign deposits:		
Altenberg, Germany	60 Mt; 0.3% Sn	Production plus reserves (mine staff, pers. comm., 1993)
Ardlethan, Australia	9 Mt; 0.45% Sn	Production plus reserves (Taylor and Pollard, 1986)
Taronga, Australia	48.2 Mt; 0.145% Sn	Reserves (McKinnon and Seidel, 1988)
Yinyan, China	~50-100 Mt? ; 0.46% Sn	Size of deposit (Lin, 1988)
19.8 Porphyry tin-silver deposits		
Foreign deposits:		
Cerro Rico (Potosi), Bolivia	828 Mt; 0.3-0.4% Sn 150-250 g/t Ag; 0.3% Sn	Reserves (Suttill, 1988)
Llallagua, Bolivia	80 Mt	Reserves (Grant et al., 1980)
19.9 Possible porphyry silver deposits		
Foreign deposits:		
Real de Angeles, Mexico	85 Mt; 75 g/t Ag 1.0% Pb 0.92% Zn	Reserves (Pearson et al., 1988); Cd also present

Porphyry tungsten-molybdenum:
Canadian – Logtung, Yukon Territory; Mount Pleasant, New Brunswick (Fire Tower zone).
Foreign – Xinglokeng, China.

Porphyry tin:
Canadian - possibly Mount Pleasant, New Brunswick (North zone); East Kemptville, Nova Scotia.
Foreign - Altenberg, Germany; Taronga and Ardlethan, Australia; Yinyan, China; Khingan, Russia.

Porphyry tin-silver:
Foreign - Cerro Rico (Potosi), Chorolque, and Llallagua, Bolivia.

Porphyry silver:
Foreign - possibly Real de Angeles, Mexico.

The distribution of selected porphyry deposits in Canada is shown in Figure 19-1.

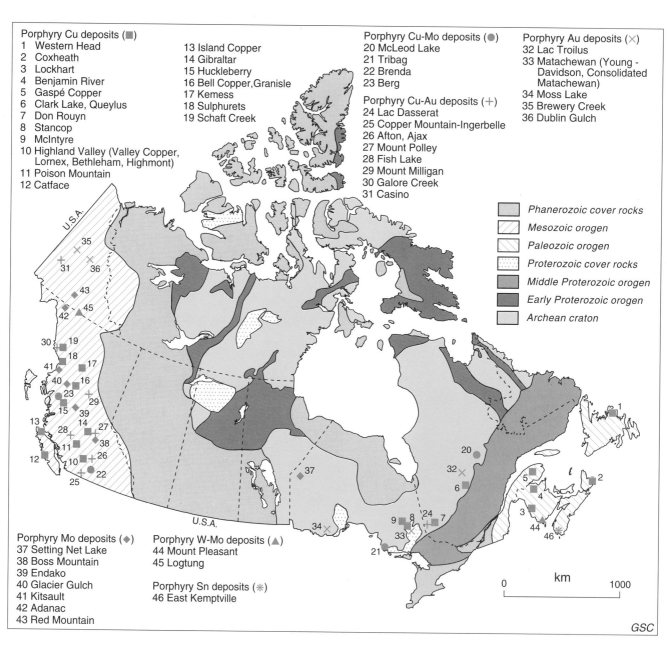

Figure 19-1. Distribution of selected porphyry deposits in Canada.

IMPORTANCE

Porphyry deposits are the world's most important source of copper, molybdenum, and rhenium, and are major sources of gold, silver, and tin; other byproduct metals include tungsten, platinum, palladium, and selenium. They account for about 50 to 60% of world copper production, although less than 50% of Canadian copper production is from porphyry deposits. This is primarily because of important Canadian copper production from copper-nickel ores at Sudbury and from numerous volcanogenic massive sulphide deposits scattered across the country, as well as significant production from skarn and vein deposits. About 60% of Canadian copper reserves are in porphyry deposits, largely in the Cordillera (Fig. 19-1), but they include a considerable amount of low-grade copper resources that are currently subeconomic. Porphyry deposits account for more than 99% of both Canadian and world molybdenum production and reserves. In the past few years, porphyry copper-gold (e.g., Grasberg; MacDonald and Arnold, 1993, 1994) and porphyry gold deposits have been recognized as increasingly important sources of gold (Vila and Sillitoe, 1991). At present, no porphyry gold, tungsten-molybdenum, or tin deposits are in production in Canada.

SIZE AND GRADE OF DEPOSITS

Porphyry deposits are large and typically contain hundreds of millions of tonnes of ore, although they range in size from tens of millions to billions of tonnes; grades for the different metals vary considerably but generally average less than one per cent. Grade-tonnage relationships for various selected porphyry deposit subtypes are shown in Figure 19-2 and grade relationships between various metals are shown in Figure 19-3.

In **porphyry copper** deposits, copper grades range from 0.2% to more than 1% Cu (Fig. 19-2A); molybdenum content ranges from approximately 0.005 to about 0.03% Mo (Fig. 19-2B). Gold contents range from 0.004 to 0.35 g/t (Fig. 19-2C), and silver from 0.2 to 5 g/t (Fig. 19-2D). Rhenium is also a significant byproduct from some porphyry copper deposits; at Island Copper, for example, rhenium is extracted from molybdenite concentrates that typically contain greater than 1000 ppm Re.

Copper grades in **porphyry copper-gold** deposits are comparable to those of the porphyry copper subtype (Fig. 19-2A), but gold contents are higher (>0.4 to 2.0 g/t) (Fig. 19-2C). Sillitoe (1993b) suggested that porphyry copper deposits should contain 0.4 g/t Au to be called gold rich. Figure 19-3B, however, shows that a 1:1 ratio (i.e., 1 g/t Au to 1% Cu) could be useful in defining gold-rich porphyry copper deposits at higher copper grades. Although the number of deposits in this class is limited, deposits such as Grasberg in Indonesia, with a resource greater than 1 billion tonnes grading 1.4% Cu and 1.8 g/t Au, indicate that porphyry copper-gold deposits can contain not only major copper, but also major gold resources.

Porphyry gold deposits contain 0.8 to 2.0 g/t Au in deposits that range in size from about 30 Mt to greater than 200 Mt of ore (Fig. 19-2C). Lac Troilus, Quebec, which is probably an intensely deformed porphyry deposit (Fraser, 1993), contains about 79.5 Mt of material grading 1.3 g/t Au, 1.38 g/t Ag and 0.11% Cu that might be mineable in two open pits.

Grade-tonnage relationships for **porphyry molybdenum** deposits show that the very large and rich Climax and Henderson deposits in Colorado are end members of a spectrum of molybdenum-bearing deposits, most of which have lower molybdenum grades and/or tonnages (Fig. 19-2B) (Carten et al., 1993). Limited data are available for tungsten and tin grades in most porphyry molybdenum deposits (Table 19-1), but some deposits, such as Climax, have produced small amounts of tungsten and tin.

Copper and molybdenum contents indicate that a continuum may exist between porphyry copper and porphyry molybdenum deposits (Fig. 19-3A). End member deposits are abundant and important economically (e.g., Guilbert and Park, 1986, p. 427), but some deposits have intermediate copper and molybdenum contents, suggesting that porphyry copper deposits with minor or no molybdenum grade to porphyry molybdenum deposits with negligible copper contents (e.g., Westra and Keith, 1981) (Fig. 19-3A). Similarly, a continuum may exist between porphyry molybdenum and porphyry tungsten-molybdenum and tungsten deposits (Table 19-1), although more data are required to substantiate such a relationship. These examples illustrate some of the difficulties in making sharp distinctions between different porphyry deposit subtypes and one reason for viewing porphyry deposits as a single large class of deposits characterized by diverse metal contents with gradational boundaries between metal subtypes. More work is required to document metal contents within porphyry deposits.

GEOLOGICAL FEATURES

Geological setting

Porphyry deposits occur in close association with epizonal and mesozonal, felsic to intermediate intrusions. Possible exceptions are some porphyry gold deposits such as Porgera, Papua New Guinea and QR, British Columbia that show a close association with small alkaline mafic intrusions emplaced at very shallow depths (Richards and Kerrich, 1993; A. Panteleyev, pers. comm., 1994). Intrusions related to porphyry deposits show a wide range in compositions and petrogenetic associations (Fig. 19-4, 19-5) and occur in a variety of tectonic settings. For example, **porphyry copper** deposits typically occur in the root zones of andesitic stratovolcanoes in subduction-related, continental-arc and island-arc settings (Mitchell and Garson, 1972; Sillitoe, 1973, 1988a; Sillitoe and Bonham, 1984) (Fig. 19-6). **Porphyry copper-gold** deposits, such as those associated with Triassic and Lower Jurassic silica-saturated, alkaline intrusions in British Columbia, formed in an island-arc setting, but possibly during periods of extension; Ladolam, a bulk-tonnage epithermal gold deposit with an early porphyry stage of mineralization, formed in the Tabar–Feni alkaline island-arc during late stage rifting related to spreading in the adjacent Manus Basin (Moyle et al., 1990; McInnes and Cameron, 1994). Grasberg and Porgera formed in a continental-island-arc collisional zone during or immediately following subduction (MacDonald and Arnold, 1993, 1994; Richards and Kerrich, 1993). **Porphyry gold** deposits of Tertiary age in the Maricunga belt in Chile appear to have formed in a continental-arc setting along strike to the north from major porphyry copper deposits of the same general age (Sillitoe, 1992, 1993b).

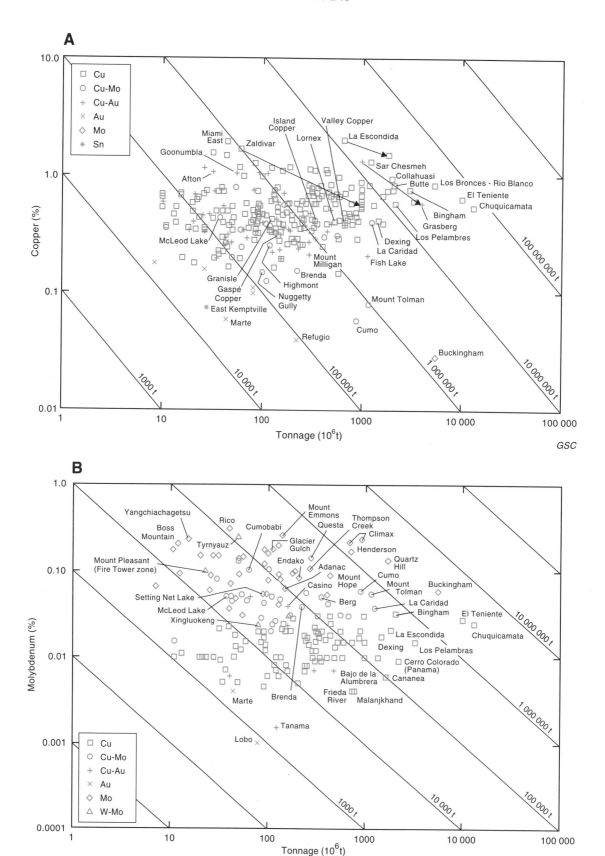

Figure 19-2. Grade versus tonnage relationships for Canadian and foreign porphyry deposits (data from Table 19-1 and other sources): **A)** copper grades and tonnages (arrows link smaller, higher grade zones to the overall deposits in which they occur); **B)** molybdenum grades and tonnages; **C)** gold grades and tonnages; **D)** silver grades and tonnages.

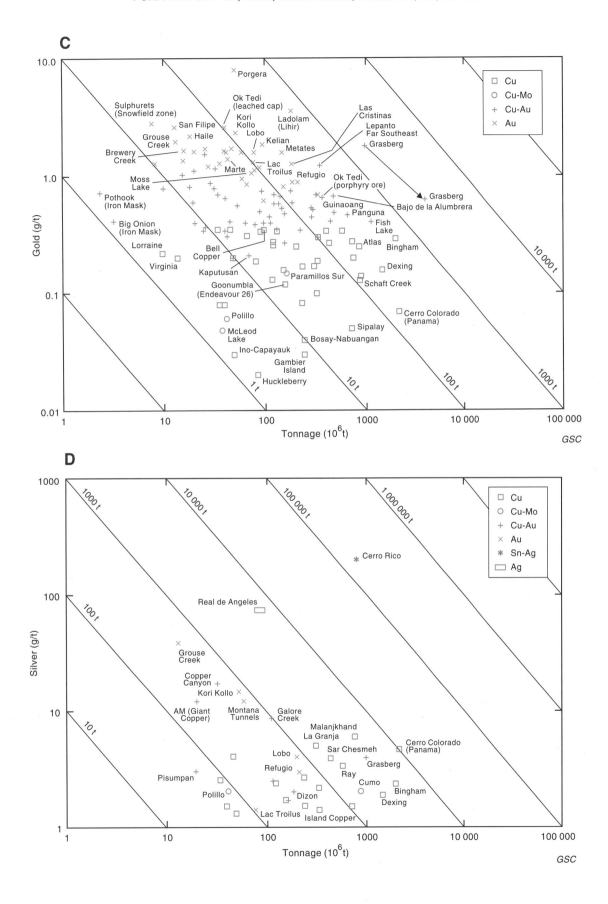

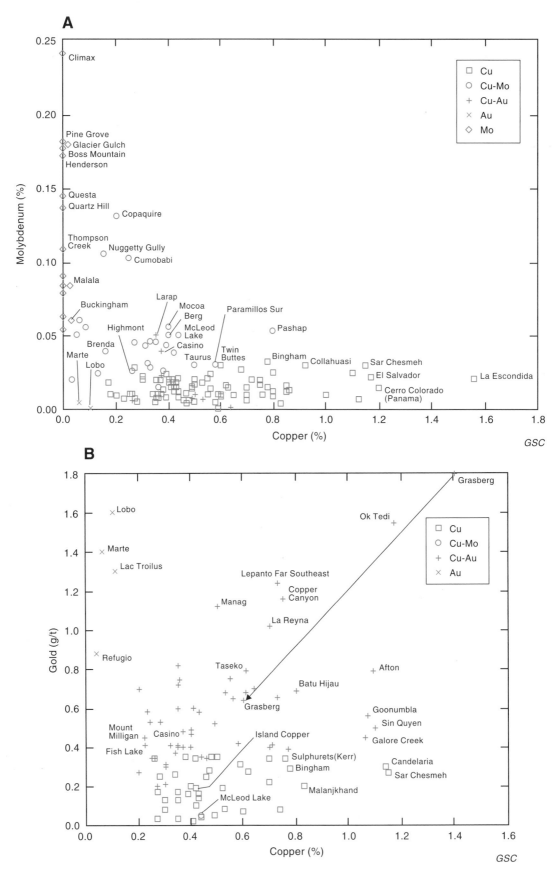

Figure 19-3. Metal ratios in Canadian and foreign porphyry deposits: **A)** molybdenum versus copper (see also Westra and Keith, 1981); **B)** gold versus copper.

Porphyry molybdenum deposits are typically associated with anorogenic or A-type granites that have been emplaced in continental settings, particularly rift or extensional environments (Fig. 19-6). The Climax and Henderson deposits, for example, are genetically related to small cupolas (small plugs and stocks) of a regional batholith emplaced during active extension in the Rio Grande rift (Bookstrom, 1981; Carten et al., 1988b, 1993). The Questa deposit farther south in New Mexico also formed during active extension and bimodal volcanism along the Rio Grande rift system (Leonardson et al., 1982; Lipman, 1988; Johnson et al., 1989; Meyer and Foland, 1991). The Pine Grove porphyry molybdenum deposit is likewise associated with bimodal igneous rocks that were emplaced during regional extension in the Basin and Range Province of Nevada (Keith et al., 1986; Keith and Shanks, 1988). Other porphyry molybdenum deposits appear to have formed during extension in areas adjacent to strike-slip faults (e.g., northern Cordillera – Quartz Hill, Adanac, Casmo, and Mount Haskins) (Fig. 19-6). A few deposits, such as Mount Pleasant, New Brunswick and Questa, New Mexico, are associated with high-silica rhyolites and granites that formed in continental calderas (Lipman, 1988; McCutcheon, 1990; McCutcheon et al., in press). For most porphyry deposits, however, the depth of erosion is such that caldera settings are conjectural (e.g., Lipman, 1984).

Some **porphyry molybdenum** deposits, along with **porphyry tungsten-molybdenum** and **porphyry tin** deposits, formed in areas of great continental thickness related to collisional tectonic settings, although the deposits generally postdate the collision event. Porphyry tin deposits in Bolivia, in particular, are related to S-type peraluminous intrusions that were emplaced above deep levels of a Benioff Zone (Ishihara, 1981; Kontak and Clark, 1988; Lehmann, 1990).

Schematic tectonic settings of deposits are illustrated in Figure 19-6. Details of each setting and related controls on magma generation, composition, and emplacement conceivably had a major influence on the size, metal contents, and nature of individual deposits. However, exceptions to typical settings, such as the Tribag and Jogran porphyry copper (molybdenum, tungsten) deposits in Ontario that apparently are related to a continental rift environment (Kirkham, 1973; Norman and Sawkins, 1985), and the Malmbjerg porphyry molybdenum deposit in East Greenland that is related to the Iceland mantle plume, indicate that individual porphyry deposits can occur in diverse and unique settings.

Age of host rocks and ore

Most porphyry deposits are Triassic or younger, but individual deposits range in age from approximately 3.0 Ga to Recent. Examples of Precambrian deposits include McIntyre and Setting Net Lake, Ontario; Clark Lake, Queylus, McLeod Lake, and Lac Dasserat, Quebec; Malanjkhand, India; Haib, Namibia; Tongkuangyu, China; and Nuggetty Gully, Australia. Porphyry mineralization is

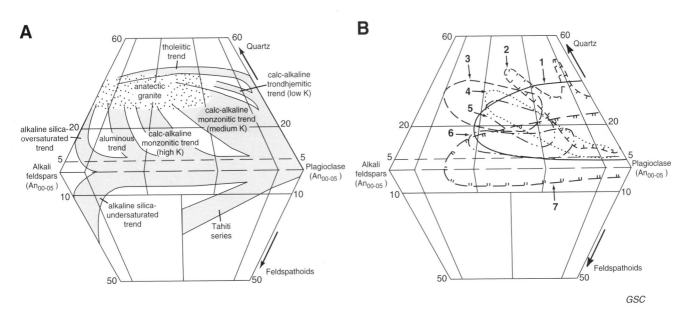

GSC

Figure 19-4. **A)** QAPF diagram showing schematic relationships between igneous rock suites (after Lameyre and Bowden, 1982). **B)** Same as A) with data from granitic rocks associated with porphyry deposits. The various trends are as follows: 1 – General field for porphyry copper deposits (R.V. Kirkham, unpub. compilation, 1985); 2 – Guichon batholith (McMillan et al., 1985); 3 – General field for most porphyry tungsten-molybdenum and tin deposits (R.V. Kirkham, unpub. compilation, 1985); 4 – Cornelia pluton, Ajo (from Kesler et al., 1975); 5 – General field for most porphyry copper-molybdenum deposits (R.V. Kirkham, unpub. compilation, 1985); 6 – Bingham-Last Chance stocks (from Kesler et al., 1975); 7 – General field for porphyry copper-gold deposits (data from Preto, 1972; Barr et al., 1976; Perelló, 1994). Boundaries of the general fields are tentative and need further refinement.

characteristically superimposed on older host rocks, although in virtually all cases, contemporaneous, genetically-related intrusions are present (Kirkham, 1971; Shannon et al., 1982; Carten et al., 1988a; Kirkham and Sinclair, 1988). Sillitoe and Bonham (1984) and Sillitoe (1988a) suggested that many porphyry copper deposits formed in the root zones of andesitic stratovolcanoes.

Associated structures

At the scale of ore deposits, associated structures, such as veins, vein sets, stockworks, fractures, 'crackled zones', and breccia pipes (Fig. 19-7) are of fundamental importance. In large, complex, economic porphyry deposits, mineralized veins and fractures have a very high density. Orientations of mineralized structures can be related to local stress environments around the tops of plutons or can reflect regional stress conditions (Rehrig and Heidrick, 1972; Heidrick and Titley, 1982; Titley et al., 1986; Carten et al., 1988a). Where they are superimposed in a large volume of rock, the combination of mineralized structures results in higher grade zones and the characteristic large size of porphyry deposits (e.g., Carten et al., 1988a).

Regional structures are thought to be important in controlling the distribution of porphyry deposits; for example, the Rio Grande rift system in the western United States is the locus for porphyry molybdenum deposits (Bookstrom, 1981). The West Fault along strike of the Eocene porphyry copper belt in northern Chile, from El Salvador in the south to beyond Collahausi in the north, was active both during and following porphyry emplacement and hydrothermal activity (Baker and Guilbert, 1987). Also within this belt, cross-structures apparently controlled the distribution of individual deposits such as Quebrada Blanca and Collahausi (Rosario – Ujina) (Sillitoe, 1992). The major strike-slip fault system in the northern part of the Philippine island arc system, similar to the West Fault in northern Chile, was probably also a control on the location of major magmatic and hydrothermal centres, which might be localized in areas that are pull-apart structures at dilational bends. In many districts, however, perhaps because of intense alteration and multiple intrusions, regional structural control is obscure.

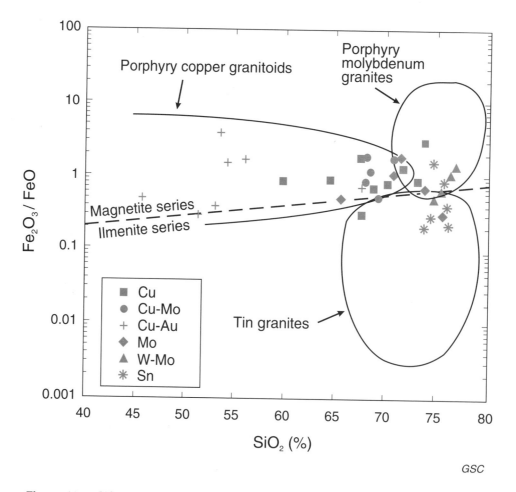

GSC

Figure 19-5. SiO_2-Fe_2O_3/FeO variation diagram for granitic rocks related to porphyry copper, molybdenum, and tin deposits; the fields for porphyry copper granitoids, porphyry molybdenum granitoids, and tin granites are generalized from Lehmann (1990).

Form of deposits

The overall form of individual porphyry deposits is variable and includes irregular, oval, solid, or "hollow" cylindrical and inverted cup shapes (Sutherland Brown, 1969; James, 1971; McMillan and Panteleyev, 1980). Orebodies may occur separately or overlap and, in some cases, are stacked one on top of the other (Wallace et al., 1968; White et al., 1981; Carten et al., 1988a). Individual orebodies measure hundreds to thousands of metres in all three dimensions.

Orebodies are characteristically zoned, with barren cores and crudely concentric metal zones that are surrounded by barren pyritic halos with or without peripheral veins, skarns, replacement manto zones, and epithermal precious-metal deposits (Einaudi, 1982; Sillitoe, 1988a, b; Jones, 1992) (Fig. 19-8). Complex, irregular ore and alteration patterns are due, in part, to the superposition or overlap of mineral and alteration zones of different ages.

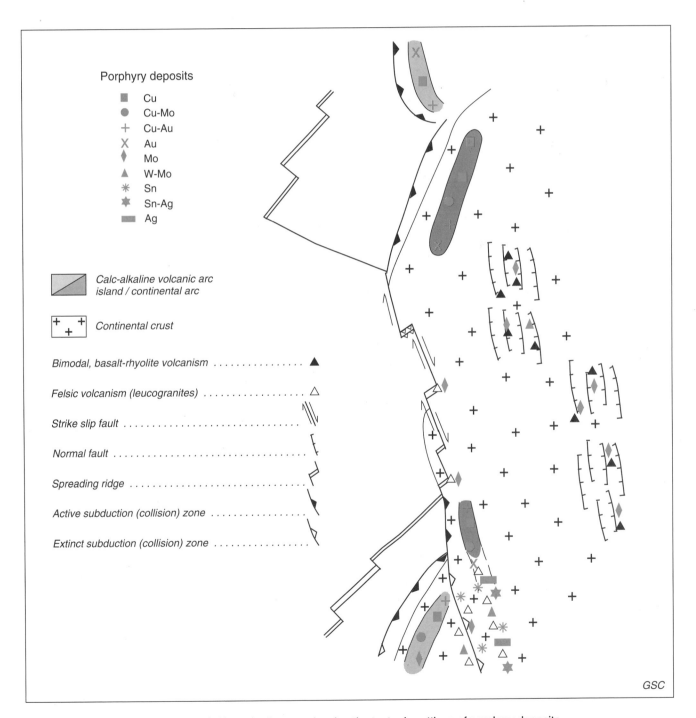

Figure 19-6. Schematic diagram showing the tectonic settings of porphyry deposits.

Mineralogy

The mineralogy of porphyry deposits is highly varied, although pyrite is typically the dominant sulphide mineral in porphyry copper, copper-molybdenum, copper-gold, gold, and silver deposits, reflecting the fact that large amounts of sulphur were added to the deposits. In porphyry deposits of the more lithophile elements, i.e., tin, tungsten, and molybdenum, the overall sulphur and sulphide mineral contents are lower. Principal ore and associated minerals of the different porphyry deposit subtypes are as follows:

Porphyry copper and copper-molybdenum deposits: Principal ore minerals are chalcopyrite, bornite, chalcocite, tennantite, enargite, other copper sulphides and sulphosalts, molybdenite, and electrum; associated minerals include pyrite, magnetite, quartz, biotite, K-feldspar, anhydrite, muscovite, clay minerals, epidote, and chlorite.

Porphyry copper-gold deposits: Principal ore minerals are chalcopyrite, bornite, chalcocite, tennantite, other copper minerals, native gold, electrum, and tellurides; associated minerals include pyrite, arsenopyrite, magnetite, quartz, biotite, K-feldspar, anhydrite, epidote, chlorite, scapolite, albite, calcite, fluorite, and garnet.

Porphyry gold deposits: Principal ore minerals are native gold, electrum, chalcopyrite, bornite, and molybdenite; associated minerals include pyrite, magnetite, quartz, biotite, K-feldspar, muscovite, clay minerals, epidote, and chlorite.

Figure 19-7. A) Quartz-molybdenite stockwork, Roundy Creek porphyry molybdenum deposit, British Columbia. GSC 202871-I; (from Soregaroli and Sutherland Brown, 1976; reprinted with permission of The Canadian Institute of Mining, Metallurgy and Petroleum); **B)** Erratic, discontinuous veins, lenses, and disseminations of chalcopyrite (white) in intensely altered monzonite, Copper Mountain-Ingerbelle porphyry copper-gold deposit, British Columbia. GSC 203886-G; (from Kirkham and Sinclair, 1984); **C)** Intermineral porphyry dyke cutting fine grained porphyry, and magnetite- and chalcopyrite-bearing quartz veins, Granisle porphyry copper deposit, British Columbia. GSC 201531-M (from Kirkham and Sinclair, 1984).

Porphyry molybdenum deposits: Principal ore minerals are molybdenite, scheelite, wolframite, cassiterite, bismuthinite, and native bismuth; associated minerals include magnetite, quartz, K-feldspar, biotite, muscovite, clay minerals, fluorite, and topaz.

Porphyry tungsten-molybdenum deposits: Principal ore minerals are scheelite, wolframite, molybdenite, cassiterite, stannite, bismuthinite, and native bismuth; other minerals include pyrite, arsenopyrite, loellingite, quartz, K-feldspar, biotite, muscovite, clay minerals, fluorite, and topaz.

Porphyry tin and tin-silver deposits: Principal ore minerals are cassiterite, tetrahedrite, argentite, stannite, wolframite, chalcopyrite, sphalerite, franckeite, cylindrite, teallite, molybdenite, bismuthinite, other sulphides and sulphosalts, native silver, and native bismuth; associated minerals include pyrite, arsenopyrite, loellingite, quartz, K-feldspar, biotite, muscovite, clay minerals, fluorite, and topaz.

Porphyry silver deposits: Principal ore minerals are freibergite, stephanite, acanthite, sphalerite, and galena; associated minerals include arsenopyrite, pyrrhotite, pyrite, adularia, quartz, fluorite, and calcite.

Alteration

Hydrothermal alteration is extensive and typically zoned both on a deposit scale and around individual veins and fractures. In many porphyry deposits, alteration zones on a deposit scale consist of an inner potassic zone characterized by biotite and/or K-feldspar (±amphibole±magnetite±anhydrite) and an outer zone of propylitic alteration that consists of quartz, chlorite, epidote, calcite, and, locally, albite associated with pyrite (Fig. 19-9). Zones of phyllic alteration (quartz+sericite+pyrite) and argillic alteration (quartz+illite+pyrite±kaolinite±smectite±montmorillonite±calcite) may be part of the zonal pattern between the potassic and propylitic zones, or can be irregular or tabular, younger zones superimposed on older alteration and sulphide assemblages (e.g., Ladolam; Moyle et al., 1990).

Economic sulphide zones are most closely associated with potassic alteration, as demonstrated by Carson and Jambor (1974) for several porphyry copper (±molybdenum) deposits. Sodic alteration (mainly as secondary albite) is associated with potassic alteration in some porphyry copper-gold deposits, such as Copper Mountain and Ajax, British Columbia (Preto, 1972; Barr et al., 1976; Ross et al., 1995). Albitic alteration partly overlaps potassic alteration and copper zones on the north side of the Ingerbelle deposit at Copper Mountain. At the Ajax deposit, highest copper grades occur near, but not in, the most intensely altered albitic rocks. Eaton and Setterfield (1993) indicated that the low grade Nasivi 3 porphyry copper deposit in the centre of the shoshonitic Tavua caldera, adjacent to the epithermal Emperor gold mine in Fiji, contains an albitic, copper-bearing core surrounded by peripheral propylitic alteration and overprinted by younger phyllic alteration. Sodic-calcic alteration (oligoclase+quartz+sphene+apatite± actinolite±epidote) has been documented in the deep root zones beneath, and peripheral to, potassically altered porphyry copper deposits at Yerington and Ann-Mason, Nevada (Carten, 1986; Dilles and Einaudi, 1992).

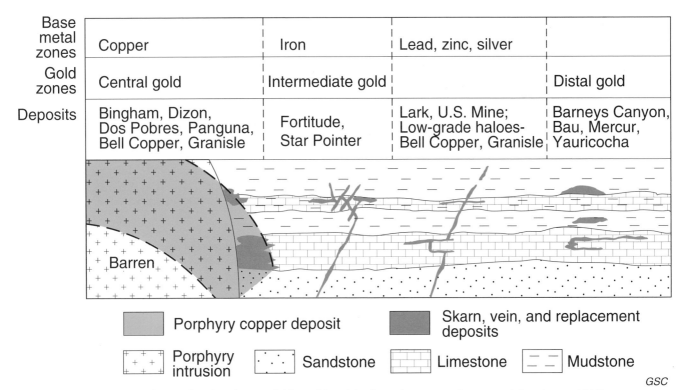

Base metal zones	Copper	Iron	Lead, zinc, silver	
Gold zones	Central gold	Intermediate gold		Distal gold
Deposits	Bingham, Dizon, Dos Pobres, Panguna, Bell Copper, Granisle	Fortitude, Star Pointer	Lark, U.S. Mine; Low-grade haloes- Bell Copper, Granisle	Barneys Canyon, Bau, Mercur, Yauricocha

Porphyry copper deposit

Skarn, vein, and replacement deposits

Porphyry intrusion   Sandstone   Limestone   Mudstone

GSC

Figure 19-8. Generalized zoning model for gold-enriched porphyry copper systems (after Jones, 1992).

Alteration mineralogy is controlled in part by the composition of the host rocks. In mafic host rocks with significant iron and magnesium, biotite (±lesser hornblende) is the dominant alteration mineral in the potassic alteration zone, whereas K-feldspar dominates in more felsic rocks. In carbonate-bearing host rocks, calc-silicate minerals such as garnet and diopside are abundant.

Alteration mineralogy is also controlled by the composition of the mineralizing system. In more oxidized environments, minerals such as pyrite, magnetite (±hematite), and anhydrite are common, whereas pyrrhotite is present in more reduced environments. Fluorine-rich systems, such as those related to many porphyry tin and tungsten-molybdenum deposits, and some porphyry molybdenum deposits, commonly contain fluorine-bearing minerals as part of the alteration assemblages. At Mount Pleasant, for example, potassic alteration is rare and the principal alteration associated with the tungsten-molybdenum deposit consists of quartz, topaz, fluorite, and sericite, and the surrounding propylitic alteration consists of chlorite+sericite (Kooiman et al., 1986). Similarly, alteration in some low-grade tin deposits in Australia (e.g., Ardlethan) grades out from a central zone of quartz+topaz to zones of sericite and chlorite±carbonate (Scott, 1981). Siems (1989) suggested that lithium silicate alteration (e.g., lithium-rich mica and tourmaline, with associated fluorite), which accompanies tin, tungsten, and molybdenum in some granite-related deposits, is analogous to potassic alteration in porphyry copper and molybdenum deposits.

Phyllic alteration zones are not present in all porphyry deposits. In many deposits in which they are present, however, phyllic alteration is superimposed on earlier potassic alteration assemblages (Carson and Jambor, 1979) (Fig. 19-9). At Chuquicamata in Chile, for example, a zone of intense phyllic alteration extends to depth in the core of the deposit and is superimposed on earlier potassic alteration and small amounts of associated copper sulphides with low copper grades. This phyllic zone contains higher than average copper grades and associated arsenic-bearing copper minerals and molybdenite.

Advanced argillic (high sulphidation) and adularia-type (low sulphidation) epithermal alteration zones with associated precious-metal deposits occur above or near several porphyry copper and copper-molybdenum deposits. These alteration zones, in places, show a marked telescoping of older potassic and younger epithermal alteration (Fig. 19-8, 19-9; Sillitoe, 1990, 1993a, b; Moyle et al., 1990; Vila and Sillitoe, 1991; Setterfield et al., 1991; Eaton and Setterfield, 1993; Richards and Kerrich, 1993). The advanced argillic assemblages include illite, quartz, alunite, natroalunite, pyrophyllite, diaspore, and a high pyrite content. Adularia assemblages, with quartz, sericite, and clay minerals, have lower pyrite contents. Sillitoe (1993a) suggested that advanced-argillic or high-sulphidation-type epithermal systems can occur in spatial association with porphyry copper, copper-molybdenum, copper-gold, and gold deposits, but not with porphyry molybdenum deposits. Adularia- or low-sulphidation-type epithermal systems probably form from more dilute ore fluids and may or may not occur on the peripheries of porphyry systems. Furthermore, Sillitoe (1993a) suggested that base-metal-rich epithermal deposits form from more concentrated NaCl brines and, similar to porphyry deposits, are parts of magmatic-hydrothermal systems.

DEFINITIVE CHARACTERISTICS

The following features serve to distinguish porphyry deposits from other types of deposits: large size, widespread alteration, structurally-controlled ore minerals superimposed on pre-existing host rocks, distinctive metal associations, and spatial, temporal, and genetic relationships to porphyritic epizonal and mesozonal intrusions.

GENETIC MODEL

The most applicable model for porphyry deposits is a magmatic-hydrothermal one, or variations thereon, in which the ore metals were derived from temporally and genetically-related intrusions (Fig. 19-10, 19-11). Large polyphase hydrothermal systems developed within and above genetically-related intrusions and commonly interacted with meteoric fluids (and possibly seawater) on their tops and peripheries. During the waning stages of hydrothermal activity, the magmatic-hydrothermal systems collapsed inward upon themselves and were replaced by waters of dominantly meteoric origin. Redistribution, and possibly further concentration of metals, occurred in some deposits during these waning stages.

Variations of the magmatic-hydrothermal model for porphyry deposits, commonly referred to as the "orthomagmatic" model, have been presented by such authors as Burnham (1967, 1979), Phillips (1973), and Whitney (1975, 1984). These authors envisaged felsic and intermediate magma emplacement at high levels in the crust and border

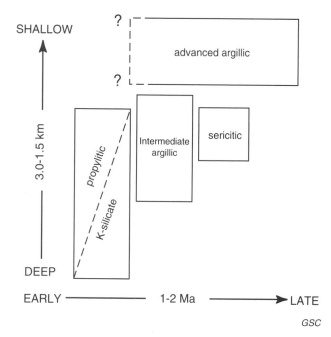

Figure 19-9. Schematic time-depth relations of principal alteration types in gold-rich porphyry copper systems and other types of porphyry deposits (after Sillitoe, 1993b).

zone crystallization along the walls and roof of the magma chamber. As a consequence of this crystallization, super-saturation of volatile phases occurred within the magma, resulting in separation of volatiles due to resurgent, or second, boiling. Ore metals and many other components were strongly partitioned into these volatile phases, which became concentrated in the carapace of the magma chamber (Christiansen et al., 1983; Candela and Holland, 1986; Manning and Pichavant, 1988; Candela, 1989; Cline and Bodnar, 1991; Heinrich et al., 1992). When increasing fluid pressures exceeded lithostatic pressures and the tensile strength of the overlying rocks, fracturing of these rocks occurred, permitting rapid escape of hydrothermal fluids into newly created open space. A fundamental control on ore deposition was the pronounced adiabatic cooling of the ore fluids due to their sudden expansion into the fracture and/or breccia systems, thus the importance of structural control on ore deposition in porphyry deposits. Aplitic and micrographic textures in granitic rocks associated with porphyry deposits are the result of pressure-quench crystallization related to the rapid escape of the ore fluids (Shannon et al., 1982; Kirkham and Sinclair, 1988).

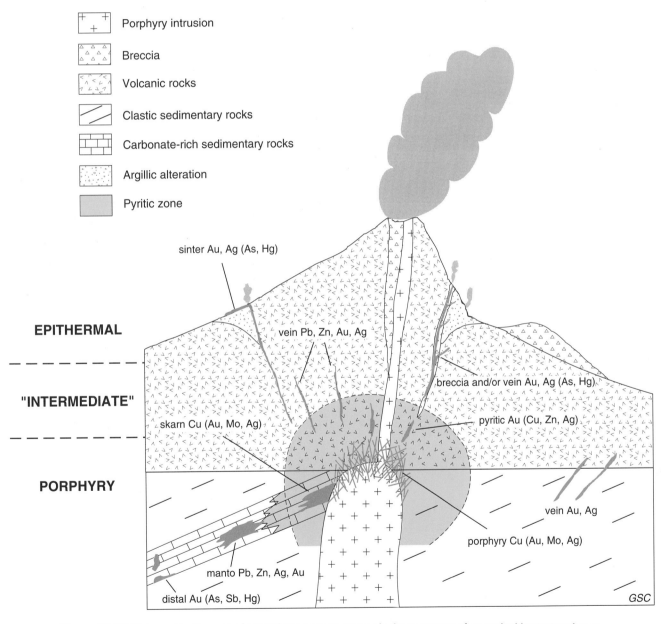

Figure 19-10. Schematic diagram of a porphyry copper system in the root zone of an andesitic stratovolcano showing mineral zonation and possible relationship to skarn, manto, "mesothermal" or "intermediate" precious metal and base metal vein and replacement, and epithermal precious-metal deposits.

Some modification of the above orthomagmatic model is required for at least some, if not most, porphyry deposits, in view of studies by Shannon et al. (1982), Carten et al. (1988a), and Kirkham and Sinclair (1988). These authors concluded that, in several deposits, the underlying genetically-related intrusions were largely liquid in their carapaces until ore formation was essentially complete. Kirkham and Sinclair (1988) suggested that crystallization deep within a batholithic magma chamber (Fig. 19-11) could have been the cause of resurgent boiling, rather than local border zone crystallization as envisaged by Burnham (1967, 1979), Whitney (1975, 1984), and Carten et al. (1988b). According to this model, volatiles that streamed through large volumes of magma, stripping it of its metal content, accumulated in small cupolas at the top of the magma chambers. These volatile-rich, ore-forming fluids would have lowered the liquidus temperature of the magmas in the cupolas, keeping them largely liquid during the ore-forming process. Areas where these ore-forming fluids accumulated in cupolas of siliceous intrusions associated with some porphyry molybdenum, copper-molybdenum, and tungsten-molybdenum deposits are indicated by abundant comb quartz layers (Shannon et al., 1982; Carten et al., 1988a; Kirkham and Sinclair, 1988). Such a model is consistent with the sequence of erupted products from large-volume ash-flow tuff eruptions – that is, early high-silica eruptive products with few crystals followed by more mafic eruptive products rich in crystals (Hildreth, 1979, 1981; Smith, 1979; Keith et al., 1986; Keith and Shanks, 1988). Similarities in chemical characteristics of siliceous intrusions associated with the Quartz Hill porphyry molybdenum deposit in Alaska and the Bishop Tuff in California (Hudson et al., 1981) indicate that the magmas responsible for the Quartz Hill deposit could have been similar to those that produced the Bishop Tuff.

Carten et al. (1993) suggested an interesting alternative for high-grade porphyry molybdenum deposits, namely that volatiles (F, Cl, S, CO_2) released from underlying saturated mafic magmas are responsible for stripping metals from the overlying felsic magmas. Keith and Shanks (1988) suggested that the Pine Grove porphyry molybdenum deposit in Utah formed from a large volume of silicic magma with a low molybdenum content. Similarly, calculations by Westra (1978) for the porphyry copper deposit at Ely, Nevada and by Heithersay et al. (1990) for the porphyry copper-gold deposits at Goonumbla, Australia, indicated that very large volumes of magma, much greater than that in the exposed intrusions, were required for the formation of these deposits.

Wall rocks of the intrusions and deposits are not considered to be viable sources for the metals in porphyry deposits. Perhaps the most convincing argument against a wall rock source for metals is the strong, universal petrogenetic and temporal association of deposits of specific metals with intrusions of specific compositions and petrogenesis. With the exception of some gold deposits, such as Porgera in Papua New Guinea, no known significant porphyry-type deposits are related to gabbros or more mafic rocks, suggesting that heat engine models for genesis of porphyry deposits have little or no relevance. Furthermore, the metal content of most porphyry deposits is related to one or more specific phase(s) of intrusion, as at Henderson, Colorado, where two of the eleven identified phases, the Seriate and the Henderson stocks, together provided an estimated 62% of the molybdenum in the deposit (Carten et al., 1988a). At Bingham, Utah the early Last Chance augite monzonite intrusion has no known significant associated mineralization, although it was emplaced at a time when a scavenging heat engine should have been most effective; on the other hand, the subsequent quartz monzonite phases of the Bingham stock and the related small, but not insignificant, latite porphyry phases (Wilson, 1978) have huge amounts of associated metals. Another example is the Battle Mountain district in Nevada where, at essentially the same place in the Earth's crust at different times, a porphyry molybdenum deposit, and a porphyry copper deposit with related gold-rich skarn zones, were formed (Theodore et al., 1982, 1992; Kirkham, 1985). Such evidence indicates strongly that input of metal-rich magmatic-hydrothermal fluids was essential for the formation of these deposits.

RELATED DEPOSIT TYPES

Deposits genetically related to porphyry deposits include skarns, high-temperature mantos, mesothermal and epithermal veins, and probably many economically important bulk-tonnage epithermal gold and silver deposits. However, because epithermal and porphyry deposits tend to form at different depths and times, and are not necessarily in contact with each other, this latter relationship has been difficult to demonstrate. An interesting case in which a porphyry deposit instantaneously became an epithermal deposit as a result of the catastrophic collapse of the overlying volcano, is the Ladolam deposit on Lihir Island in the alkaline Tabar-Feni arc in Papua New Guinea (Moyle et al., 1990). Sillitoe (1988b, 1993a) pointed out the association of some porphyry deposits with large epithermal precious metal deposits. The metal contents of associated skarns, mantos, and veins differ from district to district,

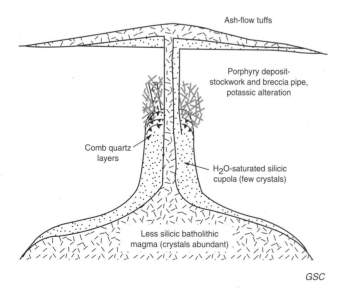

Ash-flow tuffs

Porphyry deposit-
stockwork and breccia pipe,
potassic alteration

Comb quartz
layers

H_2O-saturated silicic
cupola (few crystals)

Less silicic batholithic
magma (crystals abundant)

GSC

Figure 19-11. Schematic diagram of a crystallizing batholithic mass with an overlying volatile-saturated cupola and related ash-flow tuffs illustrating the environment of formation of porphyry deposits (modified from Kirkham and Sinclair, 1988).

but are probably, in some way related to the metal contents of associated porphyry deposits (cf., Jones, 1992). Panteleyev (1991) documented deposits that can be considered "transitional" or "intermediate" between deeper porphyry and shallower epithermal systems (e.g., Fig. 19-10). However, in some areas, as noted by Sillitoe (1992), no clear distinction can be drawn between porphyry deposits and epithermal environments.

EXPLORATION GUIDES

Several features of porphyry deposits conducive to exploration are related to their large size. Metal, mineral, and alteration patterns tend to be large, concentric, and zoned, thus yielding useful clues to areas with exploration potential. Large pyritic halos, for example, may be used to delineate the extent of the deposits, and also the intensity and complexity of the hydrothermal system.

On a regional scale, the presence of epizonal to mesozonal felsic to intermediate porphyritic intrusions, especially if accompanied by large pyritic alteration zones, indicate that the area could be prospective for porphyry deposits. Porphyry copper and copper-molybdenum deposits are relatively abundant in island-arc and continental-arc volcanic terranes; porphyry molybdenum deposits are relatively abundant in subaerial areas of crustal extension with bimodal mafic and felsic magmatism (Fig. 19-6). The tectonic settings of other subtypes of porphyry deposits are less well understood.

Porphyry deposits tend to have large geochemical dispersion halos; consequently, reconnaissance stream sediment and soil geochemical surveys have been effective exploration tools in many parts of the world. Careful study and interpretation of leached cappings have also been used to differentiate between barren and mineralized deposits, some with major supergene-enriched ores (Blanchard, 1968; Anderson, 1982).

Induced polarization surveys have been useful in outlining sulphide distribution in porphyry deposits, and magnetic surveys have been used to outline porphyry copper and copper-gold deposits with abundant hydrothermal magnetite, and pyrrhotite- and/or magnetite-bearing hornfels zones around porphyry-related intrusive rocks. Conversely, some deposits are characterized by magnetic lows due to the destruction of magnetite in phyllic alteration zones. Recently, ground and airborne gamma-ray spectrometry surveys have been used to outline potassic alteration zones closely related to copper-gold zones in the Mount Milligan deposit in central British Columbia (R. Shives, pers. comm., 1994).

ACKNOWLEDGMENTS

The authors have benefitted from discussions with many people over many years and from many informative visits to porphyry deposits. A.G. Galley, A. Douma, J.J. Carrière, D.F. Garson, and K. Ross helped compile data used in this paper; D.F. Garson also produced several computer plots used to make diagrams. W. Kornelson, C. Plant, and L. O'Neill kindly helped with word processing the manuscript. A. Panteleyev reviewed the manuscript and made constructive suggestions.

SELECTED BIBLIOGRAPHY

References with asterisks (*) are considered to be the best source of general information on this deposit type.

Ambrus, J.
1978: Chile; in International Molybdenum Encyclopaedia 1778-1978, Volume I, Resources and Production, (ed.) A. Sutulov; Intermet, Santiago, Chile, p. 54-85.

Anderson, J.A.
1982: Characteristics of leached capping and appraisal; in Advances in Geology of the Porphyry Copper Deposits, Southwestern North America, (ed.) S.R. Titley; University of Arizona Press, Tucson, Arizona, p. 275-296.

Anstett, T.F., Bleiwas, D.I., and Hurdelbrink, R.J.
1985: Tungsten availability – market economy countries; United States Bureau of Mines, Information Circular 9025, 51 p.

Argall, G.O.
1981: Takeovers shake U.S.A. mining companies; World Mining, May, p. 56-59.

Ayres, L.D., Averill, S.A., and Wolfe, W.J.
1982: An Archean molybdenite occurrence of possible porphyry type at Setting Net Lake, northwestern Ontario, Canada; Economic Geology, v. 77, p. 1105-1119.

Baker, R.C. and Guilbert, J.M.
1987: Regional structural control of porphyry copper deposits in northern Chile (abstract); Geological Society of America, Abstracts with Programs, v. 19, no. 7, p. 578.

Barr, D.A., Fox, P.E., Northcote, K.E., and Preto, V.A.
1976: The alkaline suite porphyry deposits: a summary; in Porphyry Deposits of the Canadian Cordillera, (ed.) A. Sutherland Brown; The Canadian Institute of Mining and Metallurgy, Special Volume 15, p. 359-367.

Blanchard, R.
1968: Interpretation of leached outcrops; Nevada Bureau of Mines, Bulletin 66, 196 p.

Bookstrom, A.A.
1981: Tectonic setting and generation of Rocky Mountain porphyry molybdenum deposits; in Relations of Tectonics to Ore Deposits in the Southern Cordillera, (ed.) W.R. Dickinson and W.D. Payne; Arizona Geological Society Digest, v. 14, p. 251-226.

Burnham, C.W.
1967: Hydrothermal fluids at the magmatic stage; in Geochemistry of Hydrothermal Ore Deposits, (ed.) H.L. Barnes; Holt, Rinehart and Winston Inc., New York, p. 34-76.
1979: Magma and hydrothermal fluids; in Geochemistry of Hydrothermal Ore Deposits, 2nd edition, (ed.) H.L. Barnes; Wiley Interscience, New York, p. 71-136.

Bushnell, S.E.
1988: Mineralization at Cananea, Sonora, Mexico, and the paragenesis and zoning of breccia pipes in quartzofeldspathic rock; Economic Geology, v. 83, p. 1760-1781.

Candela, P.A.
1989: Calculation of magmatic fluid contributions to porphyry-type ore system: predicting fluid inclusion chemistries; Geochemical Journal, v. 23, p. 295-305.

Candela, P.A. and Holland, H.D.
1986: A mass transfer model for copper and molybdenum in magmatic hydrothermal systems: the origin of porphyry-type ore deposits; Economic Geology, v. 81, no. 1, p. 1-19.

Carlile, J.C. and Mitchell, A.H.G.
1994: Magmatic arcs and associated gold and copper mineralization in Indonesia; Journal of Geochemical Exploration, v. 50, p. 91-142.

Carson, D.J.T. and Jambor, J.L.
1974: Mineralogy, zonal relationships and economic significance of hydrothermal alteration at porphyry copper deposits, Babine Lake area, British Columbia; The Canadian Institute of Mining and Metallurgy, Bulletin, v. 76, no. 742, p. 110-133.
1979: The occurrence and significance of phyllic overprinting at porphyry copper-molybdenum deposits (abstract); The Canadian Institute of Mining and Metallurgy, v. 72, no. 803, p. 78.

Carten, R.B.
1986: Sodium-calcium metasomatism: chemical, temporal, and spatial relationships at the Yerington, Nevada, porphyry copper deposit; Economic Geology, v. 81, p. 1495-1519.

*Carten, R.B., Geraghty, E.P., and Walker, B.M.
1988a: Cyclic development of igneous features and their relationship to high-temperature hydrothermal features in the Henderson porphyry molybdenum deposit, Colorado; Economic Geology, v. 83, p. 266-296.

Carten, R.B., Walker, B.M., Geraghty, E.P., and Gunow, A.J.
1988b: Comparison of field-based studies of the Henderson porphyry molybdenum deposit, Colorado with experimental and theoretical models of porphyry systems; in Recent Advances in the Geology of Granite-related Mineral Deposits, (ed.) R.P. Taylor and D.F. Strong; The Canadian Institute of Mining and Metallurgy, Special Volume 39, p. 351-366.

*Carten, R.B., White, W.H., and Stein, H.J.
1993: High-grade, granite-related molybdenum systems: classification and origin: in Mineral Deposit Modeling, (ed.) R.V. Kirkham, W.D. Sinclair, R.I. Thorpe, and J.M. Duke; Geological Association of Canada, Special Paper 40, p. 521-554.

Carter, N.C.
1981: Porphyry copper and molybdenum deposits, west-central British Columbia; British Columbia Ministry of Energy, Mines and Petroleum Resources, Bulletin 64, 150 p.

Chen Chucai and Li Guansheng
1990: Dexing copper mine; Mining Magazine, v. 162, p. 287-288.

Christiansen, E.H., Burt, D.M., Sheridan, M.F., and Wilson, R.T.
1983: The petrogenesis of topaz rhyolites from the western United States; Contributions to Mineralogy and Petrology, v. 83, p. 16-30.

Christopher, P.A. and Pinsent, R.
1982: Geology of the Ruby Creek and Boulder Creek area near Atlin (104N/11W); British Columbia Ministry of Energy, Mines and Petroleum Resources, notes to accompany Preliminary Map 52, 10 p.

Cline, J.S. and Bodnar, R.J.
1991: Can economic porphyry copper mineralization be generated by a typical calc-alkaline melt?; Journal of Geophysical Research, v. 96, p. 8113-8126.

Columba C., M. and Cunningham, C.G.
1993: Geologic model for the mineral deposits of the La Joya district, Oruro, Bolivia; Economic Geology, v. 88, p. 701-708.

Concepcion, R.A. and Cinco, J.C., Jr.
1989: Geology of Lepanto Far Southeast gold-rich porphyry copper deposit, Mankayan, Benguet, Philippines (abstract); 28th International Geological Congress, Washington, D.C., Abstracts, v. 1, p. 319-320.

Cox, D.
1985: Geology of the Tanama and Helecho porphyry copper deposits and vicinity, Puerto Rico; United States Geological Survey, Professional Paper 1327, 57 p.

Cox, D.P. and Singer, D.A.
1988: Distribution of gold in porphyry copper deposits; United States Geological Survey, Open File Report 88-46, 22 p.

Dilles, J.H. and Einaudi, M.T.
1992: Wall-rock alteration and hydrothermal flow paths about the Ann-Mason porphyry copper deposit, Nevada – a 6-km vertical reconstruction; Economic Geology, v. 87, p. 1963-2001.

Eaton, P.C. and Setterfield, T.N.
1993: The relationship between epithermal and porphyry hydrothermal systems within the Tavua Caldera, Fiji; Economic Geology, v. 88, p. 1053-1083.

Einaudi, M.T.
1982: Description of skarns associated with porphyry copper plutons: in Advances in Geology of the Porphyry Copper Deposits, (ed.) S.R. Titley; The University of Arizona, Press, Tuscon, Arizona, p. 139-183.

Flores V., R.
1994: Precious metal deposits of the Refugio area, northern Chile; Society for Mining, Metallurgy and Exploration, Inc., for Society of Mining Engineers Annual Meeting, Albuquerque, New Mexico, February 14-17, 1994, Pre-print 94-109, 9 p.

Fraser, R.J.
1993: The Lac Troilus gold-copper deposit, northwestern Quebec: a possible Archean porphyry system; Economic Geology, v. 88, p. 1685-1699.

Geyti, A. and Thomassen, B.
1984: Molybdenum and precious metal mineralization at Flannefjeld, southeast Greenland; Economic Geology, v. 79, p. 1921-1929.

Goldie, R.
1990: Ok Tedi: a copper-gold porphyry emplaced in a compressional environment; The Gangue (Newsletter of the Mineral Deposits Division of the Geological Association of Canada), issue no. 33, October, 1990, p. 10-14.

Grant, J.N., Halls, C., Sheppard, S.M.F., and Avila, W.
1980: Evolution of the porphyry tin deposits of Bolivia; in Granitic Magmatism and Related Mineralization, (ed.) S. Ishihara and S. Takenouchi; Mining Geology Special Issue, no. 8, The Society of Mining Geologists of Japan, p. 151-173.

Grant, N., Halls, C., Avila, W., and Avila, G.
1977: Igneous geology and the evolution of hydrothermal systems in some sub-volcanic tin deposits of Bolivia; Geological Society of London, Special Volume 7, p. 117-126.

Guan Xunfan, Shou Yongqin, Xiao Jinghua, Lian Shuzhao, and Li Jinmao
1988: A new type of tin deposit - the Yinyan porphyry tin deposit in China; in Geology of Tin Deposits in Asia and the Pacific, (ed.) C.S. Hutchison; Springer-Verlag, Berlin, New York, p. 487-494.

Guilbert, J.M.
1986: Recent advances in porphyry base metal deposit research; in Geology and Metallogeny of Copper Deposits, Proceedings of the Copper Symposium, 27th International Geological Congress, Moscow, 1984, p. 196-208.

Guilbert, J.M. and Park, C.F., Jr.
1986: The Geology of Ore Deposits; W.H. Freeman, New York, 985 p.

Gustafson, L.B.
1978: Some major factors of porphyry copper genesis; Economic Geology, v. 73, p. 600-607.

*Gustafson, L.B. and Hunt, J.P.
1975: The porphyry copper deposit at El Salvador, Chile; Economic Geology, v. 70, p. 857-912.

Heidrick, T.L. and Titley, S.R.
1982: Fracture and dike patterns in Laramide plutons and their structural and tectonic implication: American Southwest; in Advances in Geology of the Porphyry Copper Deposits, Southwestern North America, (ed.) S.R. Titley; University of Arizona Press, Tucson, Arizona, p. 73-91.

Heinrich, C.A., Ryan, C.G., Mernach, T.P., and Eadington, P.J.
1992: Segregation of ore metals between magmatic brine and vapor: a fluid inclusion study using PIXE microanalysis; Economic Geology, v. 87, p. 1566-1583.

Heithersay, P.S., O'Neill, W.J., van der Helder, P., Moore, C.R., and Harbon, P.G.
1990: Goonumbla porphyry copper district – Endeavour 26 North, Endeavour 22 and Endeavour 27 copper-gold deposits; in Geology of the Mineral Deposits of Australia and Papua New Guinea, (ed.) F.E. Hughes; The Australasian Institute of Mining and Metallurgy, Melbourne, Australia, p. 1385-1398.

Hildreth, W.
1979: The Bishop Tuff: evidence for the origin of compositional zonation in silicic magma chambers; in Ash-flow Tuffs; Geological Society of American, Special Paper 180, p. 43-75.
1981: Gradients in silicic magma chambers: implications for lithospheric magmatism; Journal of Geophysical Research, v. 86, p. 10153-10192.

Hollister, V.F.
1974: Regional characteristics of porphyry copper deposits of South America; Society of Mining Engineers of American Institute of Mining, Metallurgical and Petroleum Engineers, Transactions, v. 255, p. 45-53.

Huang Dianhao, Wu Chengyu, and Nie Fengjun
1988: Geological features and genesis of the Jinduicheng porphyry molybdenum deposit, Shaanxi Province, China; Chinese Journal of Geochemistry, v. 7, p. 136-147.

Huang Xunde and Zhang Dingyuan
1989: Geochemical zoning pattern of the Yinyan tin deposit; Journal of Geochemical Exploration, v. 33, p. 109-119.

Hudson, T., Arth, J.G., and Muth, K.G.
1981: Geochemistry of intrusive rocks associated with molybdenite deposits, Ketchikan Quadrangle, southeastern Alaska; Economic Geology, v. 76, p. 1225-1232.

Imai, A., Muyco, J.D., Domingo, E.G., Almeda, R.L., Villones, R.I., Jr., Yumul, G.P., Jr., Damasco, F.V., Baluda, R.P., Malihan, T.D., Datuin, R.T., Punongbayan, R.S., Listanco, E.L., and Santos, R.A.
1992: Porphyry copper and gold mineralization in the Philippines; in Mineral Deposits of Japan and the Philippines; 29th International Geological Congress Field Trip Guide Book, Volume 61, Field Trip C36, p. 1-30.

Ishihara, S.
1981: The granitoid series and mineralization; in Economic Geology Seventy-fifth Anniversary Volume, 1905-1980, (ed.) B.J. Skinner; Economic Geology Publishing Co., p. 458-484.

James, A.H.
1971:   Hypothetical diagrams of several porphyry copper deposits; Economic Geology, v. 66, p. 43-47.

James, L.P.
1978:   The Bingham copper deposits, Utah, as an exploration target: history and pre-excavation geology; Economic Geology, v. 73, p. 1218-1227.

Johnson, C.M., Czamanske, G.K., and Lipman, P.W.
1989:   Geochemistry of intrusive rocks associated with the Latir volcanic field, New Mexico, and contrasts between evolution of plutonic and volcanic rocks; Contributions to Mineralogy and Petrology, v. 103, p. 90-109.

Jones, B.K.
1992:   Application of metal zoning to gold exploration in porphyry copper systems; Journal of Geochemical Exploration, v. 43, p. 127-155.

Jones, G.J.
1985:   The Goonumbala porphyry copper deposits, New South Wales; Economic Geology, v. 80, p. 591-613.

Keith, J.D. and Shanks, W.C., III
1988:   Chemical evolution and volatile fugacities of the Pine Grove porphyry molybdenum and ash-flow tuff system, southwestern Utah; in Recent Advances in the Geology of Granite-related Mineral Deposits, (ed.) R.P. Taylor and D.F. Strong; The Canadian Institute of Mining and Metallurgy, Special Volume 39, p. 402-423.

Keith, J.D., Shanks, W.C., III, Archibald, D.A., and Farrar, E.
1986:   Volcanic and intrusive history of the Pine Grove porphyry molybdenum system, southwestern Utah; Economic Geology, v. 81, p. 553-577.

Kesler, S.E.
1973:   Copper, molybdenum and gold abundances in porphyry copper deposits; Economic Geology, v. 68, p. 106-112.

Kesler, S.E., Jones, L.M., and Walker, R.L.
1975:   Intrusive rocks associated with porphyry copper mineralization in island arc areas; Economic Geology, v. 70, p. 515-526.

Kirkham, R.V.
1971:   Intermineral intrusions and their bearing on the origin of porphyry copper and molybdenum deposits; Economic Geology, v. 66, p. 1244-1250.

1972:   Porphyry deposits; in Report of Activities, Part B: November 1971 to March 1972; Geological Survey of Canada, Paper 72-1, Part B, p. 62-64.

1973:   Tectonism, volcanism and copper deposits; in Volcanism and Volcanic Rock; Geological Survey of Canada, Open File 164, p. 129-151.

1985:   Tectonic and petrochemical control on distribution and metal contents of granite-related molybdenum deposits (extended abstract); in Granite-related Mineral Deposits, (ed.) R.P Taylor and D.F. Strong; The Canadian Institute of Mining and Metallurgy Conference, Halifax, September, 1984, p. 165-168.

Kirkham, R.V. and Sinclair, W.D.
1984:   Porphyry copper, molybdenum, tungsten in Canadian Mineral Deposit Types: a Geological Synopsis, (ed.) O.R. Eckstrand; Geological Survey of Canada, Economic Geology Report 36, p. 50-52.

1988:   Comb quartz layers in felsic intrusions and their relationship to origin of porphyry deposits; in Recent Advances in the Geology of Granite-related Mineral Deposits, (ed.) R.P. Taylor and D.F. Strong; The Canadian Institute of Mining and Metallurgy, Special Volume 39, p. 50-71.

Kirkham, R.V., McCann, C., Prasad, N., Soregaroli, A.E., Vokes, F.M., and Wine, G.
1982:   Molybdenum in Canada, part 2: MOLYFILE – an index-level computer file of molybdenum deposits and occurrences in Canada; Geological Survey of Canada, Economic Geology Report 33, 208 p.

Kontak, D.J. and Clark, A.H.
1988:   Exploration criteria for tin and tungsten mineralization in the Cordillera Oriental of southeastern Peru; in Recent Advances in the Geology of Granite-related Mineral Deposits, (ed.) R.P. Taylor and D.F. Strong; The Canadian Institute of Mining and Metallurgy, Special Volume 39, p. 157-169.

Kooiman, G.J.A., McLeod, M.J., and Sinclair, W.D.
1986:   Porphyry tungsten-molybdenum orebodies, polymetallic veins and replacement bodies, and tin-bearing greisen zones in the Fire Tower zone, Mount Pleasant, New Brunswick; Economic Geology, v. 81, p. 1356-1373.

Lameyre, J. and Bowden, P.
1982:   Plutonic rock types series: discrimination of various granitoid series and related rocks; Journal of Volcanology and Geothermal Research, v. 14, p. 169-186.

Lehmann, B.
1990:   Metallogeny of tin; Lecture Notes in Earth Sciences, v. 32, Springer-Verlag, Berlin, 211 p.

Leonardson, R.W., Dunlop, G., Starquist, V.L., Bratton, G.P., Meyer, J.W., Osborne, L.W., Atkin, S.A., Molling, P.A., Moore, R.F., and Olmore, S.D.
1982:   Preliminary geology and molybdenum deposits at Questa, New Mexico; The Genesis of Rocky Mountain Ore Deposits: Changes with Time and Tectonics; Proceeding of Denver Region Exploration Geologists Society Symposium, November 1982, p. 151-155.

Lin Guiqing
1988:   Geological characteristics of the ignimbrite-related Xiling tin deposit in Guangdong Province; in Geology of Tin Deposits in Asia and the Pacific, (ed.) C.S. Hutchison; Springer-Verlag, Berlin, New York, p. 495-506.

Linn, K.O., Wieselmann, E.A., Galay, I., Harvey, J.J.T., Tufino, G.F., and Winfield, W.D.B.
1981:   Geology of Panama's Cerro Colorado porphyry copper deposit; Mineral and Energy Resources, v. 24, no. 6, p. 1-14.

Lipman, P.W.
1984:   The roots of ash flow calderas in western North America: windows into the tops of granitic batholiths; Journal of Geophysical Research, v. 89, p. 8801-8841.

1988:   Evolution of silicic magma in the upper crust: the mid-Tertiary Latir volcanic field and its cogenetic granitic batholith, northern New Mexico, U.S.A.; Transactions of the Royal Society of Edinburgh; Earth Sciences, v. 79, p. 265-288.

Lipman, P.W. and Sawyer, D.A.
1985:   Mesozoic ash-flow caldera fragments in southeastern Arizona and their relation to porphyry copper deposits; Geology, v. 13, p. 652-656.

Liu Wengzhang
1981:   Geological features of mineralization of the Xingluokeng tungsten (molybdenum) deposit, Fujian Province; in Tungsten Geology, China, (ed.) J.V. Hepworth and Yu Hong Zhang, UN Economic and Social Commission for Asia and the Pacific/Regional Mineral Resources Development Centre, Bandung, Indonesia, p. 339-348.

Lowell, J.D.
1974:   Three new porphyry copper mines for Chile?; Mining Engineering, v. 26, no. 11, p. 22-28.

1978:   Porphyry model; in International Molybdenum Encyclopaedia 1778-1978, Volume I - Resources and Production, (ed.) A. Sutulov; Intermet Publications, Santiago, Chile, p. 261-270.

*Lowell, J.D. and Guilbert, J.M.
1970:   Lateral and vertical alteration-mineralization zoning in porphyry ore deposits; Economic Geology, v. 65, p. 373-408.

MacDonald, G.D. and Arnold, L.C.
1993:   Intrusive and mineralization history of the Grasberg deposit Irian Jaya, Indonesia; for presentation at the Society of Mining Engineers Annual Meeting Reno, Nevada, February 15-18, 1993, Society for Mining, Metallurgy, and Exploration, Inc., Preprint number 93-92, p. 1-10.

1994:   Geological and geochemical zoning of the Grasberg Igneous Complex, Irian Jaya, Indonesia; Journal of Geochemical Exploration, v. 50, p. 143-178.

Manning, D.A.C. and Pichavant, M.
1988:   Volatiles and their bearing on the behaviour of metals in granitic systems; in Recent Advances in the Geology of Granite-related Mineral Deposits, (ed.) R.P. Taylor and D.F. Strong; The Canadian Institute of Mining and Metallurgy, Special Volume 39, p. 13-24.

McCutcheon, S.R.
1990:   The Mount Pleasant caldera: geological setting of associated tungsten-molybdenum and tin deposits; in Mineral Deposits of New Brunswick and Nova Scotia, (ed.) D.R. Boyle; 8th International Association on the Genesis of Ore Deposits Symposium, Field Trip Guidebook, Geological Survey of Canada, Open File 2157, p. 73-77.

McCutcheon, S.R., Anderson, H.E., and Robinson, P.T.
in press: Stratigraphy and eruptive history of the Late Devonian Mount Pleasant caldera complex, Canadian Appalachians; Geological Magazine.

McInnes, B.I.A. and Cameron, E.M.
1994:   Carbonated alkaline hybridizing melts from a sub-arc environment: mantle wedge samples from the Tabar-Lihir-Targa-Feni arc, Papua New Guinea; Earth and Planetary Science Letters, v. 122, p. 125-144.

McKinnon, A. and Seidel, H.
1988:   Tin; in Register of Australian Mining, 1988/89, (ed.) R. Louthean; Resource Information Unit Ltd., Subiaco, Western Australia, p. 197-204.

McMillan, W.J.
1991:   Porphyry deposits in the Canadian Cordillera; in Ore Deposits, Tectonics and Metallogeny in the Canadian Cordillera, British Columbia Geological Survey Branch, Paper 1991-4, p. 253-276.

McMillan, W.J. and Panteleyev, A.
1980: Ore deposit models - 1. Porphyry copper deposits; Geoscience Canada, v. 7, p. 52-63.

McMillan, W.J., Newman, K., Tsang, L., and Sanford, G.
1985: Geology and ore deposits of the Highland Valley camp; Geological Association of Canada, Field Guide and Reference Manual Series, no. 1, 121 p.

Meldrum, S.J., Aquino, R.S., Gonzales, R.I., Burke, R.J., Suyadi, A., Irianto, B., and Clarke, D.S.
1994: The Batu Hijau porphyry copper-gold deposit, Sumbawa Island, Indonesia; Journal of Geochemical Exploration, v. 50, p. 203-220.

Meyer, J. and Foland, K.A.
1991: Magmatic-tectonic interaction during early Rio Grande rift extension at Questa, New Mexico; Geological Society of America Bulletin, v. 103, p. 993-1006.

Miller, R.N.
1973: Production history of the Butte district and geological function, past and present; in Guidebook for the Butte Field Meeting of Society of Economic Geologists, (ed.) R.N. Miller; August 18-21, 1973, p. F1-F10.

Mitchell, A.H. and Garson, M.S.
1972: Relationship of porphyry copper and circum-Pacific tin deposits to palaeo-Benioff zones; Institution of Mining and Metallurgy, Transaction, v. 81, p. B10-25.

Moyle, A.J., Doyle, B.J., Hoogvliet, H., and Ware, A.R.
1990: Ladolam gold deposit, Lihir Island; in Geology of the Mineral Deposits of Australia and Papua New Guinea, (ed.) F.E. Hughes; The Australasian Institute of Mining and Metallurgy, Melbourne, Australia, p. 1793-1805.

Moyle, J.E.
1984: Development and construction begins at East Kemptville, North America's only primary tin mine; Mining Engineering, April 1984, p. 335-336.

Müller, D. and Groves, D.I.
1993: Direct and indirect associations between potassic igneous rocks, shoshonites and gold-copper deposits; Ore Geology Reviews, v. 8, p. 383-406.

Müller, D., Heithersay, P.S., and Groves, D.I.
1994: The shoshonite porphyry Cu-Au association in the Goonumbla district, N.S.W., Australia; Mineralogy and Petrology, v. 50, p. 299-321.

Mutschler, F.E. and Mooney, T.C.
1993: Precious-metal deposits related to alkaline igneous rocks – provisional classification, grade-tonnage data and exploration frontiers; in Mineral Deposit Modeling, (ed.) R.V. Kirkham, W.D. Sinclair, R.I. Thorpe, and J.M. Duke; Geological Association of Canada, Special Paper 40, p. 479-520.

Mutschler, F.E., Griffin, M.E., Stevens, D.S., and Shannon, S.S., Jr.
1985: Precious metal deposits related to alkaline rocks in the North American Cordillera – an interpretive review; Transactions of the Geological Society of South Africa, v. 88, p. 355-377.

Mutschler, F.E., Wright, E.G., Ludington, S., and Abbott, J.T.
1981: Granite molybdenite systems; Economic Geology, v. 76, p. 874-897.

Noble, S.R., Spooner, E.T.C., and Harris, F.R.
1984: The Logtung large tonnage, low-grade W (scheelite)-Mo porphyry deposit, south-central Yukon Territory; Economic Geology, v. 79, p. 848-868.

Nokleberg, W.J., Bundtzen, T.K., Berg, H.C., Brew, D.A., Grybeck, D., Robinson, M.S., Smith, T.E., and Yeend, W.
1987: Significant metalliferous lode deposits and placer districts of Alaska; United States Geological Survey, Bulletin 1786, 104 p.

Norman, D.I. and Sawkins, F.J.
1985: The Tribag breccia pipes: Precambrian Cu-Mo deposits, Batchawana Bay, Ontario; Economic Geology, v. 80, p. 1593-1621.

Ojeda F., J.M.
1986: Escondida porphyry copper deposit, II Región, Chile: exploration drilling and current geological interpretation; in Papers Presented at the Mining Latin/Minería Latinoamerican Conference; Institute of Mining and Metallurgy, Meeting, November 17-19, Santiago, Chile, p. 299-318.

Panteleyev, A.
1981: Berg porphyry copper-molybdenum deposit; British Columbia Ministry of Energy, Mines and Petroleum Resources, Bulletin 66, 158 p.
1991: Gold in the Canadian Cordillera – a focus on epithermal and deeper environments; in Ore Deposits, Tectonics and Metallogeny in the Canadian Cordillera; British Columbia Ministry of Energy, Mines and Petroleum Resources, Paper 1991-4, p. 163-212.

Parrish, I.S. and Tully, J.V.
1978: Porphyry tungsten zones at Mt. Pleasant, N.B.; The Canadian Institute of Mining and Metallurgy Bulletin, v. 71, no. 794, p. 93-100.

Pearson, M.F., Clark, K.F., and Porter, E.W.
1988: Mineralogy, fluid characteristics, and silver distribution at Real de Angeles, Zacatecas, Mexico; Economic Geology, v. 83, p. 1737-1759.

Perelló, J.A.
1994: Geology, porphyry Cu-Au, and epithermal Cu-Au-Ag mineralization of the Tombulialato district, North Sulawesi, Indonesia; Journal of Geochemical Exploration, v. 50, p. 221-256.

Perelló, J.A., Fleming, J.A., O'Kane, K.P., Burt, P.D., Clarke, G.A., Himes, M.D., and Reeves, A.T.
1995: Porphyry Copper-gold-molybdenum Mineralization in the Island Copper Cluster, Vancouver Island; The Canadian Institute of Mining and Metallurgy, Special Volume 46, p. 214-238.

Phillips, W.J.
1973: Mechanical effects of retrograde boiling and its probable importance in the formation of some porphyry ore deposits; Institution of Mining and Metallurgy Transactions, v. B82, p. 90-98.

Preto, V.
1972: Geology of Copper Mountain; British Columbia Department of Mines and Petroleum Resources, Bulletin 59, 87 p.

Rehrig, W.A. and Heidrick, T.L.
1972: Regional fracturing in Laramide stocks of Arizona and its relationship to porphyry copper mineralization; Economic Geology, v. 67, p. 198-213.

Richards, J.P. and Kerrich, R.
1993: The Porgera gold mine, Papua New Guinea: magmatic hydrothermal to epithermal evolution of an alkalic-type precious metal deposit; Economic Geology, v. 88, p. 1017-1052.

Richardson, J.M.
1988: Field and textural relationships of alteration and greisen-hosted mineralization at the East Kemptville tin deposit, Davis Lake complex, southwest Nova Scotia; in Recent Advances in the Geology of Granite-related Mineral Deposits, (ed.) R.P. Taylor and D.F. Strong; The Canadian Institute of Mining and Metallurgy, Special Volume 39, p. 265-279.

Ross, K.V., Godwin, C.I., Bond, L., and Dawson, K.M.
1995: Geology, alteration and mineralization of the Ajax East and Ajax West deposits, southern Iron Mask Batholith, Kamloops, British Columbia; in The Canadian Institute of Mining and Metallurgy, Special Volume 46, p. 565-580.

Saegart, W.E. and Lewis, D.E.
1977: Characteristics of Philippine porphyry copper deposits and summary of current production and reserves; American Institute of Mining and Metallurgy, Transactions, v. 262, p. 199-208.

Scott, K.M.
1981: Wall-rock alteration in disseminated tin deposits, southeastern Australia; Proceedings of the Australasian Institute of Mining and Metallurgy, no. 280, December, p. 17-28.

Setterfield, T.N., Eaton, P.C., Rose W.J., and Sparks, R.S.J.
1991: The Tavua Caldera, Fiji: a complex shoshonitic caldera formed by concurrent faulting and downsagging; Journal of the Geological Society, v. 148, p. 115-127.

Shannon, J.R., Walker, B.M., Carter, R.B., and Geraghty, E.P.
1982: Unidirectional solidification textures and their significance in determining relative ages of intrusions at the Henderson mine, Colorado; Geology, v. 19, p. 293-297.

Siems, P.L.
1989: Lithium silicate alteration of tin granites: an analog of potassium silicate alteration in porphyry copper and molybdenite deposits (abstract); The Geological Society of America, Abstracts with Programs, v. 21, no. 5, p. 143.

Sikka, D.G., Petruk, W., Nehru, C.E., and Zhang, Z.
1991: Geochemistry of secondary copper minerals from Proterozoic porphyry copper deposit, Malanjkhand, India; Ore Geology Reviews, v. 6, p. 257-290.

Sillitoe, R.H.
1972: A plate tectonic model for the origin of porphyry copper deposits; Economic Geology, v. 67, p. 184-197.
1973: The tops and bottoms of porphyry copper deposits; Economic Geology, v. 68, p. 700-815.
1980: Types of porphyry molybdenum deposits; Mining Magazine, June 1980, p. 550-551.
1986: Space-time distribution, crustal setting and Cu/Mo ratios of Central Andean porphyry copper deposits: metallogenic implications; Geology and Metallogeny of Copper Deposits, Proceedings of the Copper Symposium, 27th International Geological Congress, Moscow, 1984, p. 235-250.

Sillitoe, R.H. (cont.)

1988a: Ores in volcanoes; in Proceedings of the Seventh Quadrennial International Association on the Genesis of Ore Deposits Symposium, (ed.) E. Zachrisson; E. Schweizerbart'sche Verlagsbuchhandlung, Stuttgart, p. 1-10.

1988b: Gold and silver deposits in porphyry systems; in Bulk Mineable Precious Metal Deposits of the Western United States, Symposium Proceedings, April 6-8, Reno, Nevada, 1987, p. 233-257.

1990: Gold-rich porphyry copper deposits of the circum-Pacific region – an updated overview; in Proceedings (v. 2), The Australasian Institute of Mining and Metallurgy, Pacific Rim 90 Congress, Gold Coast, Queensland, Australia, May, 1990, p. 119-126.

*1992: Gold and copper metallogeny of the central Andes – past, present, and future exploration objectives; Economic Geology, v. 87, p. 2205-2216.

1993a: Epithermal models: genetic types, geometrical controls and shallow features; in Mineral Deposit Modeling, (ed.) R.V. Kirkham, W.D. Sinclair, R.I. Thorpe, and J.M. Duke; Geological Association of Canada, Special Paper 40, p. 403-417.

*1993b: Gold-rich porphyry copper deposits: geological model and exploration implications; in Mineral Deposit Modeling, (ed.) R.V. Kirkham, W.D. Sinclair, R.I. Thorpe, and J.M. Duke; Geological Association of Canada, Special Paper 40, p. 465-478.

Sillitoe, R.H. and Bonham, H.F., Jr.

1984: Volcanic landforms and ore deposits; Economic Geology, v. 79, p. 1286-1298.

*Sillitoe, R.H. and Gappe, I.M., Jr.

1984: Philippine porphyry copper deposits: geologic setting and characteristics; Committee for Co-ordination of Joint Prospecting of Mineral Resources in Asian Offshore Areas (CCOP), Technical Report 14, 89 p.

Sillitoe, R.H., Grauberger, G.L., and Elliott, J.E.

1985: A diatreme-hosted gold deposit at Montana Tunnels, Montana; Economic Geology, v. 80, p. 1707-1721.

Sillitoe, R.H., Halls, C., and Grant, J.N.

1975: Porphyry tin deposits in Bolivia; Economic Geology, v. 70, p. 913-927.

Sinclair, A.J., Drummond, A.D., Carter, N.C., and Dawson, K.M.

1982: A preliminary analysis of gold and silver grades of porphyry-type deposits in western Canada; in Precious Metals in the Northern Cordillera, (ed.) A.A. Levinson; The Association of Exploration Geochemists, p. 157-172.

Sinclair, W.D.

1982: Gold deposits of the Matachewan area, Ontario; in Canadian Gold Deposits, (ed.) R.W. Hodder and W. Petruk; The Canadian Institute of Mining and Metallurgy, Special Volume 24, p. 83-93.

*1986: Molybdenum, tungsten and tin deposits and associated granitoid intrusions in the northern Canadian Cordillera and adjacent parts of Alaska; in Mineral Deposits of Northern Cordillera, (ed.) J.A. Morin; The Canadian Institute of Mining and Metallurgy, Special Volume 37, p. 216-233.

Smith, R.L.

1979: Ash-flow magmatism; in Ash-flow Tuffs; Geological Society of America, Special Paper 180, p. 5-27.

Soregaroli, A.E. and Sutherland Brown, A.

1976: Characteristics of Canadian Cordilleran molybdenum deposits; in Porphyry Deposits of the Canadian Cordillera, (ed.) A. Sutherland Brown; The Canadian Institute of Mining and Metallurgy, Special Volume 15, p. 417-431.

Strishkov, V.V.

1984: The copper industry of the U.S.S.R.: problems, issues and outlook; United States Bureau of Mines, Mineral Issues, 80 p.

Sutherland Brown, A.

1969: Mineralization in British Columbia and the copper and molybdenum deposits; The Canadian Institute of Mining and Metallurgy, v. 72, p. 1-15.

*Sutherland Brown, A. (ed.)

1976: Porphyry Deposits of the Canadian Cordillera; The Canadian Institute of Mining and Metallurgy, Special Volume 15, 510 p.

Suttill, K.R.

1988: Cerro Rico de Potosi; Engineering and Mining Journal, March, 1988, p. 50-53.

Sutulov, A.

1977: Chilean copper resources said to be world's largest; American Metal Market, August 4, p. 18-19.

Taylor, R.G. and Pollard, P.J.

1986: Recent advances in exploration modelling for tin deposits and their application to the Southeast Asian environment; in Regional Conference on the Geology and Mineral Resources of Southeast Asia V Proceedings, v. 1, Geological Society of Malaysia, Bulletin 19, p. 327-347.

*Taylor, R.P. and Strong, D.F. (ed.)

1988: Recent Advances in the Geology of Granite-related Mineral Deposits; The Canadian Institute of Mining and Metallurgy, Special Volume 39, 445 p.

Theodore, T.G. and Menzie, W.D.

1984: Fluorine-deficient porphyry molybdenum deposits in the western North American Cordillera; Proceedings of the Sixth Quadrennial International Association on the Genesis of Ore Deposits Symposium, E. Schweizerbart'sche Verlagsbuchhandlung, Stuttgart, p. 463-470.

Theodore, T.G., Blake, D.W., and Kretschmer, E.L.

1982: Geology of the Copper Canyon porphyry copper deposits, Lander County, Nevada; in Advances in the Geology of the Porphyry Copper Deposits, (ed.) S.R. Titley; The University of Arizona Press, Tucson, Arizona, p. 543-550.

Theodore, T.G., Blake, D.W., Loucks, T.A., and Johnson, C.A.

1992: Geology of the Buckingham Stockwork molybdenum deposit and surrounding area, Lander County, Nevada; United States Geological Survey, Professional Paper 798-D, p. D1-D307.

Thompson, T.B., Trippel, A.D., and Dwelley, P.C.

1985: Mineralized veins and breccias of the Cripple Creek district, Colorado; Economic Geology, v. 80, p. 1669-1688.

*Titley, S.R. (ed.)

1982: Advances in Geology of the Porphyry Copper Deposits – Southwestern North America; The University of Arizona Press, Tucson, Arizona, 560 p.

Titley, S.R.

1993: Characteristics of porphyry copper occurrence in the American southwest; in Mineral Deposit Modeling, (ed.) R.V. Kirkham, W.D. Sinclair, R.I. Thorpe, and J.M. Duke; Geological Association of Canada, Special Paper 40, p. 433-464.

Titley, S.R. and Beane, R.E.

1981: Porphyry copper deposits; in Economic Geology Seventy-fifth Anniversary Volume, 1905-1980, (ed.) B.J. Skinner; Economic Geology Publishing Co., p. 214-269.

Titley, S.R. and Hicks, C.L. (ed.)

1966: Geology of the Porphyry Copper Deposits, Southwestern North America; The University of Arizona Press, Tucson, Arizona, 287 p.

Titley, S.R., Thompson, R.C., Haynes, F.M., Manske, S.L., Robison, L.C., and White, J.L.

1986: Evolution of fractures and alteration in the Sierrita-Esperanza hydrothermal system, Pima County, Arizona; Economic Geology, v. 81, p. 343-370.

van Leeuwen, T.M.

1994: 25 years of mineral exploration and discovery in Indonesia; Journal of Geochemical Exploration, v. 50, p. 13-90.

van Leeuwen, T.M., Taylor, R., Coote, A., and Longstaffe, F.J.

1994: Porphyry molybdenum mineralization in a continental collision setting at Malala, northwest Sulawesi, Indonesia; Journal of Geochemical Exploration, v. 50, p. 279-315.

*Vila, T. and Sillitoe, R.H.

1991: Gold-rich porphyry systems in the Maricunga gold-silver belt, northern Chile; Economic Geology, v. 86, p. 1238-1260.

Villalpando, B.A.

1988: The tin ore deposits of Bolivia; in Geology of Tin Deposits in Asia and the Pacific, Selected Papers from the International Symposium on the Geology of Tin Deposits held in Nanning, China, October 26-30, 1984, p. 201-215.

*Wallace, S.R., Muncaster, N.K., Jonson, D.C., Mackenzie, W.B., Bookstrom, A.A., and Surface, V.A.

1968: Multiple intrusion and mineralization at Climax, Colorado; in Ore Deposits of the United States, 1933-1967 (Graton-Sales volume), (ed.) J.D. Ridge; American Institute of Mining, Metallurgical, and Petroleum Engineers, Inc., New York, p. 605-640.

Wampler, P.

1993: Geology, hydrothermal alteration, and geographical information system analysis of the Zortman gold mine, Montana (extended abstract); in Integrated Methods in Exploration and Discovery, (ed.) S.B. Romberger and D.I. Fletcher; Conference Program and Extended Abstracts, Golden, Colorado, April, 1993, p. AB 124-125.

Westra, G.

1978: Porphyry copper genesis at Ely, Nevada: in Papers on Mineral Deposits of Western North America, (ed.) J.D. Ridge; 5th International Association on the Genesis of Ore Deposits Quadrennial Symposium Proceedings, Volume II, Nevada Bureau Mines and Geology, Report 33, p. 127-140.

Westra, G. and Keith, S.B.

1981: Classification and genesis of stockwork molybdenum deposits; Economic Geology, v. 76, p. 844-873.

*White, W.H., Bookstrom, A.A., Kamilli, R.J., Ganster, M.W., Smith, R.P., Ranta, D.E., and Steininger, R.C.
1981: Character and origin of Climax-type molybdenum deposits; <u>in</u> Economic Geology Seventy-fifth Anniversary Volume, 1905-1980, (ed.) B.J. Skinner, Economic Geology Publishing Co., p. 270-316.

Whitney, J.A.
1975: Vapour generation in a quartz monzonite magma: a synthetic model with application to porphyry copper deposits; Economic Geology, v. 70, p. 346-358.
1984: Volatiles in magmatic systems; <u>in</u> Fluid-mineral Equilibria in Hydrothermal Systems; Reviews in Economic Geology, v. 1, p. 155-175.

Wilson, J.C.
1978: Ore fluid-magma relationships in a vesicular quartz latite porphyry dike at Bingham, Utah; Economic Geology, v. 73, p. 1287-1307.

Woodcock, J.R. and Hollister, V.F.
1978: Porphyry molybdenite deposits of the North American Cordillera; Mineral Science and Engineering, v. 10, p. 3-18.

Authors' addresses

R.V. Kirkham
Geological Survey of Canada
100 West Pender Street
Vancouver, B.C.
V6B 1R8

W.D. Sinclair
Geological Survey of Canada
601 Booth Street
Ottawa, Ontario
K1A 0E8

20. SKARN DEPOSITS

20. SKARN DEPOSITS

INTRODUCTION

Skarn deposits are abundant, variable, and economically important. They are a principal global source of tungsten, a major source of copper, and an important source of iron, molybdenum, zinc, and gold.

Skarn is an assemblage of dominantly calcium and magnesium silicates typically formed in carbonate-bearing rocks as a result of regional and thermal metamorphism, and by metasomatic replacement. Regional and stratiform metamorphic skarn deposits include, for example, skarn iron deposits that were derived from iron-rich sedimentary and volcanic rocks by recrystallization, isochemical metamorphism, and bimetasomatism. Recrystallization, in particular, results in upgrading the quality of ore for concentration, beneficiation, and metallurgical recovery by increasing grain size of the ore minerals. The term "skarn"

was first applied to the calc-silicate gangue associated with some Swedish iron ores of this type (Geijer and Magnusson, 1952). It is not normally used for skarn-type mineral assemblages produced by regional metamorphism of pre-existing deposits, for example highly metamorphosed lithofacies of iron-formation (Gross, 1968; "Skarn iron", subtype 20.4).

Thermal metamorphism of calcareous rocks by adjacent plutons causes a bimetasomatic exchange of ions between dissimilar lithologies, e.g., limestone and pelite, in addition to recrystallization of limestone. The resultant calc-silicate hornfels and marble is subsequently converted to anhydrous prograde skarn under the metasomatic influence of hot hydrothermal fluids emanating from the adjacent crystallizing pluton. Most economic concentrations of ore minerals occur during the cooling of the hydrothermal system, coincident with the onset of retrograde alteration. In rare instances, existing mineral deposits are converted to skarn deposits by metamorphism, as proposed by Sangster et al. (1990) for Meat Cove and Lime Hill, Nova Scotia; Johnson et al. (1990) for Franklin Furnace, New Jersey; Gemmell et al. (1992) for Aguilar, Argentina; and Hodgson (1975) for Broken Hill, Australia.

Most skarn deposits consist of metallic ore minerals and skarn silicates as gangue, and form as a result of magmatic hydrothermal processes, as inferred from their ubiquitous association with intrusive rocks. The skarn itself may be classified according to its calcic or magnesian mineral assemblage, derived from its limestone or dolostone host rock, respectively. Skarn deposits, i.e. skarns which contain economic concentrations of minerals, on the other hand, are most usefully classified according to the dominant contained economic metal. On this basis, the skarn subtypes zinc-lead-silver, copper, gold, iron, and tungsten can be distinguished and are described separately in the following accounts.

REFERENCES

Geijer, P. and Magnusson, N.H.
1952: The iron ores of Sweden; 19th International Geological Congress, Algiers, 1952, v. 2, p. 477-499.

Gemmell, G.B., Zantop, H., and Meinert, L.D.
1992: Genesis of the Aguilar zinc-lead-silver deposit, Argentina: contact metasomatic versus sedimentary exhalative; Economic Geology, v. 87, p. 2085-2112.

Gross, G.A.
1968: Iron ranges of the Labrador geosyncline; Volume III in Geology of Iron Deposits in Canada; Geological Survey of Canada, Economic Geology Report 22, 179 p.

Hodgson, C.J.
1975: The geology and geological development of the Broken Hill lode in the New Broken Hill Consolidated mine, Australia; Part II, mineralogy; Geological Society of Australia, Journal, v. 22, p. 33-50.

Johnson, C.A., Rye, D.M., and Skinner, B.J.
1990: Petrology and stable isotope geochemistry of the metamorphosed zinc-iron-manganese deposit at Sterling Hill, New Jersey; Economic Geology, v. 85, p. 1133-1161.

Sangster, A.L., Justino, M.F., and Thorpe, R.I.
1990: Metallogeny of the Proterozoic marble-hosted zinc occurrences at Lime Hill and Meat Cove, Cape Breton Island, Nova Scotia; in Mineral Deposit Studies in Nova Scotia, Volume 1, (ed.) A.L. Sangster; Geological Survey of Canada, Paper 90-8, p. 31-66.

20.1 SKARN ZINC-LEAD-SILVER

K.M. Dawson

IDENTIFICATION

Zinc-lead-silver skarn deposits are distinguished from other skarn deposits by their distinctive Mn- and Fe-rich mineralogy and their common occurrence along structural pathways and lithological contacts at some distance from intrusive contacts. Plutons may occur at distances of several kilometres from the deposit, or may not be exposed. The metamorphic aureole centred on the skarn is less extensively developed than in W and Cu skarn deposits. Related intrusions are variable in size, composition, and depth of emplacement. Johannsenitic (i.e., Mn- and Fe-rich) pyroxene is more abundant than andraditic garnet in prograde skarn, and manganiferous actinolite, epidote, ilvaite, and chlorite are common minerals in retrograde skarn.

Dawson, K.M.
1996: Skarn zinc-lead-silver; in Geology of Canadian Mineral Deposit Types, (ed.) O.R. Eckstrand, W.D. Sinclair, and R.I. Thorpe; Geological Survey of Canada, Geology of Canada, no. 8, p. 448-459 (also Geological Society of America, The Geology of North America, v. P-1).

A continuum is recognized that extends from endoskarn and reaction skarn at or near the intrusive contact, through exoskarn to more stratigraphically or structurally controlled manto and chimney deposits with progressively higher sulphide and lower calc-silicate gangue contents with increasing distance from the intrusion, passing ultimately to carbonate-hosted veins with manganese-rich silicate and carbonate gangue. Meinert (1992), in acknowledging some problems in Zn skarn classification, noted that most large skarn deposits contain both skarn-rich and skarn-poor ore in a variety of geometric settings, including mantos and chimneys. In this context, deposits identified in the literature as "replacements" with little or no calc-silicate gangue, e.g., Bluebell, British Columbia; Gilman, Colorado; and Tintic, Utah are recognized as genetically related to skarn Zn-Pb deposits and intrusions, even if neither are exposed.

Significant Canadian Zn-Pb skarns and mantos (Fig. 20.1-1) include the previously producing deposits Sa Dena Hes at Mount Hundere, Yukon Territory and Bluebell, Mineral King, and Jersey in southeastern British Columbia, and the developed manto deposits at Quartz Lake, Yukon Territory; Midway, British Columbia; and Prairie Creek, Northwest Territories. Small deposits occur adjacent to the Cassiar, Seagull, and Mount Billings batholiths and Flat River stock in the northern Canadian Cordillera, and on Vancouver Island, British Columbia.

Some large foreign Zn-Pb skarn deposits include Santa Eulalia, Naica, San Martin, and Velardeña, Mexico; the Central Mining district, New Mexico; the Central Colorado Mineral Belt, Colorado; Stantrg, Yugoslavia; El Mochito, Honduras; Nikolaevskoe, Primor'ye, Russia; Kamioka and Nakatatsu, Japan; Yeonhua, Korea; and Shuikoushan, China.

IMPORTANCE

Significant production of Ag, Pb, and Zn has been obtained from several skarn and replacement deposits in southeastern British Columbia. Production has taken place from only one of numerous base metal skarns in the northern Canadian Cordillera, but significant reserves have been developed in several others (Table 20.1-1). Skarn deposits of Zn, Pb, and Ag are common throughout the world, but statistics on production from skarn deposits relative to other base metal deposit types are not readily available. The more silver-rich deposits are important sources of the

world's silver, with the bulk of such ore coming from replacement deposits in limestone peripheral to areas of skarn formation.

SIZE OF DEPOSIT

On a global basis, most large skarn deposits mined underground for predominantly Zn and Pb contain 1 to 10 Mt of Zn+Pb metal in orebodies that range from 3 to 90 Mt at relatively high average grades of 10 to 15% Zn+Pb, with Zn usually dominant (Table 20.1-1, Fig. 20.1-2). The large Ag-rich skarn deposits of northern Mexico constitute a subgroup of deposits with greater than 10 Mt contained Zn+Pb metal. Most base metal skarns contain 30 to 300 g/t Ag, but Ag-rich deposits, whose average grade commonly exceeds 500 g/t Ag, are often mined for their Ag content alone. Copper may be recovered from base metal skarns, at grades averaging 0.2 to 2% Cu. Tungsten, gold, cadmium, and tin are present in small amounts in several Cordilleran deposits.

Skarn deposits and orebodies in British Columbia and Yukon Territory range from less than 1 t to about 8 Mt of ore. Tonnage and grade figures are not available for the numerous skarn occurrences in the northern Cordillera, but these occurrences are generally small. Size, grade, and other characteristics of Zn-Pb-Ag skarns worldwide have been tabulated by Einaudi et al. (1981).

GEOLOGICAL FEATURES

Geological setting

Base metal skarns in the North American Cordillera are hosted by the same Upper Proterozoic to mid-Paleozoic shelf sedimentary rocks that host W skarns, except on Vancouver Island where host rocks are Paleozoic and lower Mesozoic oceanic arc-type volcanic-carbonate sequences. Mexican and Central American deposits are hosted by a Jura-Cretaceous transgressive carbonate-clastic overlap assemblage. Asian, Australian, South American, European, and Russian deposits are hosted by dominantly Paleozoic limestones developed on continental margins. Host sedimentary strata in the North American Cordillera are underlain by basement lithologies dominated by thick, craton-derived clastic assemblages which, in the southern Cordillera, overlie crystalline Precambrian rocks. Associated granitoid rocks commonly are late orogenic to postorogenic, but may be synorogenic.

In the northern Canadian Cordillera, Zn-rich skarn deposits occur in belts in cratonal host rocks adjacent to the Mount Billings batholith in southeastern Yukon Territory and the Flat River and smaller stocks in southwestern Northwest Territories. A second belt of Ag-Pb-Zn manto and skarn deposits, e.g., Midway and Ketza River, occurs 100 km to the west within equivalents of North American shelf sedimentary rocks in the displaced Cassiar cratonal terrane, adjacent to the Cassiar and Seagull batholiths and small satellitic intrusions. In the southeastern Canadian Cordillera, skarn, replacement, and vein deposits of the Salmo, Riondel-Ainsworth and Slocan districts are adjacent to the synorogenic Nelson batholith and younger intrusions, but are hosted by carbonate strata of both Kootenay and Quesnellia terranes. The Mineral King deposit was formed along the margin of the North American craton, in dolostone of the Mount Nelson Formation in the

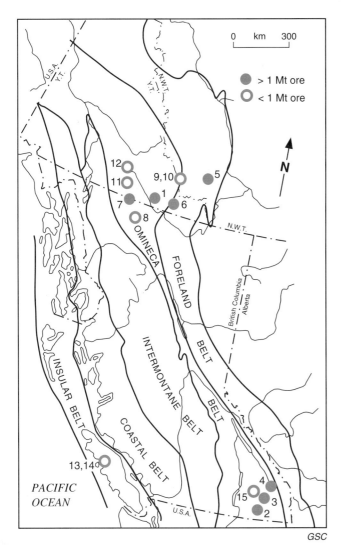

Figure 20.1-1. Significant Zn-Pb-Ag skarn and replacement deposits of the Canadian Cordillera. Numbered symbols correspond to deposits listed in Table 20.1-1.

449

upper part of the Purcell Assemblage, and no associated intrusion is exposed (Fyles, 1960). In contrast to late orogenic to postorogenic deposits of the northern Cordillera, these deposits are more deformed and metamorphosed (Dawson et al., 1991).

Age of host rocks, associated rocks, and mineralization

Host rocks for Zn-Pb-Ag skarns are the metamorphosed equivalents of limestone and calcareous to carbonaceous pelites: skarn, calc-silicate hornfels, schist, and marble. On

Table 20.1-1. Significant Canadian and foreign Pb-Zn-Ag skarns and mantos.

No.	Deposit/ District/ Location	Type	Size (Mt)	Zn (%)	Pb (%)	Cu (%)	Ag (g/t)	Au (g/t)	Other	Skarn minerals	References
Canadian deposits											
1	Sa Dena Hes; Mount Hundere, Y.T.	skarn, manto	4.9 (R)	12.7	4	n.d.	60	n.d.	tet	gross, diop, act, qtz, fluor	Indian and Northern Affairs Canada, 1992
2	Jersey, Emerald; Salmo, B.C.	skarn, manto	7.7 (P)	3.5	1.65	n.d.	3.1	n.d.	W 0.23%, As	px, gar, dol, cal, trem, qtz, chl, wol, oliv, serp, musc	Höy, 1982; Whishaw, 1954
3	Bluebell; Riondel, B.C.	manto	4.8 (P)	6.3	5.2	minor	45	n.d.		sid, ank, minn, dic, cal, qtz, kneb	Ohmoto and Rye, 1970; Höy, 1980
4	Mineral King; B.C.	manto	2.1 (P)	4.12	1.76	n.d.	25	n.d.		dol, qtz, bar	Fyles, 1960
5	Prairie Creek, N.W.T.	manto	3.75 (R)	14.7	13	0.5	202	n.d.	Hg, W	si, cal	Northern Miner, 28/02/1994
6	Quartz Lake, Y.T.	manto	1.5 (R)	6.6	5.5	minor	102	n.d.	As, Sb	qtz, dol, sid, ank	Vaillancourt, 1982
7	Midway; B.C.	manto	1.2 (R)	9.6	7	minor	410	minor	Sn 0.1%, Sb,As,Bi	qtz, ser, cal, sid, rut, trem, ep	Bradford and Godwin, 1988
8	Magno, D; Cassiar district, B.C.	skarn, manto	0.5 (R)	4.4	5.3	minor	168	1.4	Sn to 1.5%	px, gar, act, rhodoc, chl	C. Bloomer, unpub. data, 1981
9	Lucky Lake, N.W.T.	skarn	0.36 (est.)	6.2	2.5	minor	14	n.d.	W	diop, gar, act	Aho, 1969
10	Roy, N.W.T.	manto	0.15 (R)	3.1	5.2	0.1	147	1.4	Sn 0.13% As, B	qtz, cal	V. Kukor, unpub. data, 1982
11	Silver Hart, Y.T.	skarn	0.1 (R)	3.8	1.4	minor	958	n.d.	Sb, As, W	gar, diop, ves, qtz, cal	E. Buhlmann, pers. comm., 1988; Abbott, 1983
12	Tintina Silver, Y.T.	skarn	0.1 (R)	10	6	minor	690	n.d.	Sb, As, W	diop, gar, qtz, ilv, trem	Northern Miner, 31/01/1980; Morin et al., 1977
13	Zip, B.C.	skarn	0.08 (est.)	12.5	3.7	1.7	64	n.d.		ep, px, gar, chl	Gunning, 1932
14	Caledonia, B.C.	skarn	0.07 (P)	7.45	0.6	6.04	704	0.01		ep, gar, act, ser	GCNL 221, 1981; Webster et al., 1992
15	Piedmont, B.C.	skarn, manto	0.005 (P)	13	4.7	0.1	124	tr	As, Bi, Cd, Sb	gar, px, cp, biot	Webster et al., 1992
Foreign deposits											
16	Santa Eulalia, Mexico	skarn, manto, chimney	50 (P+R)	3	2	0.1	125	tr	Sn, V, Hg, Sb, As, W	gar, joh-hed, ep, chl, qtz, fluor, mgt	Hewitt, 1968; Megaw et al., 1988
17	Providencia-Concepcion del Oro, Mexico	skarn, chimney, manto	25 (P+R)	0.6	1	2	30	0.6	Sb, As, Hg	gar, trem, wol, mgt, scap, diop	Buseck, 1966
18	Naica, Mexico	skarn, chimney	21 (P+R)	3.8	4.5	0.4	150	0.3	Mo, Hg, W to 0.12%	gar, wol, ves, diop, trem	Erwood et al., 1979; Ruiz et al., 1986
19	San Martin, Mexico	skarn, chimney	21 (P+R)	5.3	0.6	1.24	146	0.7	As, Mo	gar, hed, trem, act, wol, ves, ep, chl	Megaw et al., 1988; White, 1980
20	Charcas, Mexico	skarn	15 (P+R)	8	2.5	0.5	140	n.d.	Sb, As, Hg, Sn	gross, hed, wol, ep, diop, trem, ilv, qtz, dat, dan, axin, tourm	Megaw et al., 1988

a global basis, Upper Proterozoic to Cretaceous host rocks are intruded by Paleozoic to mid-Tertiary granitoid batholiths, stocks, and dykes. Most skarn deposits in the Canadian and American Cordillera have developed preferentially in limestone beds of Upper Proterozoic to Upper Paleozoic cratonal or pericratonic sedimentary sequences. Skarns on Vancouver Island occur in Upper Triassic, volcanic arc-related limestone of accreted Wrangellia.

Zinc-lead skarns of the Central Mining district, New Mexico are hosted by Carboniferous shelf limestone (Meinert, 1987). Skarn deposits of northern Mexico developed in Lower Cretaceous limestones within a Jura-Cretaceous transgressive carbonate-clastic assemblage that overlaps Paleozoic cratonal sedimentary terranes (Campa and Coney, 1983). Zinc-lead-silver skarns in the northern Canadian Cordillera, like W skarns, have formed preferentially in

No.	Deposit/ District/ Location	Type	Size (Mt)	Zn (%)	Pb (%)	Cu (%)	Ag (g/t)	Au (g/t)	Other	Skarn minerals	References
21	Velardeña, Mexico	skarn, manto, chimney	15 (P+R)	5	4	2.5	175	0.5	As to 12%	grand, ves, wol, ep, diop-hed, spec	Gilmer et al., 1986; Megaw et al., 1988
22	Catorce, Mexico	skarn, chimney	10 (P+R)	6	10	minor	80	minor	Sb, As, Bi, Hg	gar, px, ep	Megaw et al., 1988
23	Zimapan district; La Negra, El Monte, Mexico	skarn	10 (P+R)	2.5	1.2	0.65	150	n.d.	W, Te, B, As, Sb	hed, and-gross, wol, bors	Dawson, 1985; Megaw et al., 1988
24	Central Mineral district; N. Mexico: Groundhog, Hanover, etc.	skarn	18 (P+R)	14	2	1	96	n.d.		joh-hed, and, bust, ilv, cum, amph, chl	Meinert, 1987
25	Gilman, Colorado	manto, chimney	11.7 (P)	8.5	1.5	0.9	228	1.7	Mo, As, Sb, Ba, F	mn-ank, sid, rhodoc, si, dol	Beaty et al., 1990
26	Leadville, Colorado	manto, skarn	23.8 (P)	3	4.2	0.2	320	3.7	Sb, Bi, W, Cd, Te	jasp, dol, sid, qtz, musc, bar, fluor	Beaty et al., 1990; Thompson and Arehart, 1990
27	Lark, Bingham Canyon, Utah	manto, skarn	39.2 (P)	1.9	4.7	0.93	106.2	1.85		wol, diop, qtz, cal	Atkinson and Einaudi, 1978
28	Tintic district, Utah	manto, chimney	17.2 (P)	1.2	5.9	0.9	485	4.86		si, mn-carb, ser, py, bar, dol	Morris, 1968
29	Park City, Utah	manto	13.1 (P)	4.5	8.7	0.38	556.3	2.3		qtz, rhodoc, hmt, cal, mn-cal, rhodon	Barnes and Simos, 1968
30	Mozumi, Maruyama, Tochibora: Kamioka district, Japan	skarn	90 (P+R)	5	0.7	n.d.	30	n.d.	Sn, Mo, W	hed, ep, gar, chl, ser, hmt, cal, qtz	Sakurai and Shimazaki, 1993
31	Nakayama, Hitokata, etc.; Nakatatsu district, Japan	skarn	16.2 (P)	5.5	0.4	0.1	31	n.d.	Sn, Bi, Te	hed-joh, and, wol, bust, qtz, rhodoc, act, chl, rhodon	Kano and Shimizu, 1992
32	El Mochito, Honduras	skarn, chimney	7.1 (P)	8	4.2	minor	128	n.d.		and, hed, bust, amph, ilv, chl, fluor	Shultz and Hamann, 1977
33	Stantrg, Yugoslavia	skarn	12.5 (P)	3.8	8.6	0.2	140	n.d.		hed, and, ilv, amph	Forgan, 1950
34	Yeonhua I & II, S. Korea	skarn	9.6 (P)	6.6	3	minor	minor	n.d.		hed, and, bust, rhodon, chl	Yun, 1979
35	Nikolaevskoe; Primor'ye, Russia	skarn	40 (P+R)	n.d.	n.d.	0.2-0.5%	30-50	n.d.	Sn, Bi, In	gar, hed, wol, axin, dat, fluor, act, ep	V.V. Ratkin, pers. comm., 1994
36	Shuikoushan, China	skarn	1.5 (P)	20	17	n.d.	224	n.d.		gar, px, ep, chl, zeol	Hsieh, 1950
37	Tienpaoshan, China	skarn	3 (P)	6	5	1.8	minor	n.d.		px, ep, fluor	Hsieh, 1950

Abbreviations:
Minerals: act = actinolite, amph = amphibole, and = andradite, ank = ankerite, axin = axinite, bar = barite, biot = biotite, bors = borospurrite, bust = bustamite, cal = calcite, carb = carbonate, chl = chlorite, cp = chalcopyrite, cum = cummingtonite, dan = danburite, dat = datolite, dic = dickite, diop = diopside, dol = dolomite, ep = epidote, fluor = fluorite, gar = garnet, grand = grossular-andradite, gross = grossularite, hed = hedenbergite, hmt = hematite, ilv = ilvaite, jasp =jasperoid, joh = johannssenite, kneb = knebelite, minn = minnesotaite, mgt = magnetite, mn = manganiferous, musc = muscovite, oliv = olivine, px = pyroxene, py = pyrite, qtz = quartz, rhodoc = rhodocrosite, rhodon = rhodonite, rut = rutile, scap = scapolite, ser = sericite, serp = serpentine, si = silica, sid = siderite, spec = specularite, tet = tetrahedrite, tourm = tourmaline, trem = tremolite, ves = vesuvianite, wol = wollastonite, zeol = zeolite.
Others: P = production, R = reserves, est. = estimate, n.d. = no data, GCNL = George Cross Newsletter, Vancouver.

Lower Cambrian shelf limestone, but have also developed in a broader age range of host rocks than W skarns, and, in addition, occur in some cases in regionally metamorphosed host rocks (Dawson and Dick, 1978).

Intrusive rocks associated with Zn-Pb-Ag skarns commonly are calc-alkaline felsic to intermediate batholiths, stocks, dykes, and sills, but also span a wide range of compositions from high-silica leucogranite and topaz granite, and also syenite plutons, through dioritic dykes. Small quartz monzonite stocks are most common. Intrusions show a broader compositional and morphological range and depth of emplacement than those associated with W and Cu skarns, from deep-seated batholiths to shallow dyke-sill intrusive complexes. In the Canadian Cordillera, both 'S-type' and 'I-type' granitoids of Chappell and White (1974) are associated with base metal skarns. Topaz-bearing peraluminous granite stocks and dykes are associated with some Zn-rich skarns in the northern Canadian Cordillera and Mexico. Stocks and dykes have in some cases been altered adjacent to mineralized exoskarn and endoskarn to assemblages containing sericite, epidote, clay, quartz, and fluorite. In almost all cases, zinc skarns occur distal to their associated igneous rocks.

Ore deposition was penecontemporaneous with associated intrusive rocks: mid-Cretaceous to Eocene in the North American Cordillera, except Jurassic in southern British Columbia and Eocene to Oligocene in northeastern Mexico; Mesozoic in eastern Asia; Permo-Triassic in Australia and Russia; and Tertiary in Europe.

Form of deposit and associated structures, zoning, and distribution of ore minerals

Zinc-lead-silver skarns are similar to the W skarns in that they commonly form in the thermal metamorphic aureole at the contact between granitoid intrusions and calcareous sedimentary rocks, but the contact metamorphic aureole is less extensive than that associated with deeper seated W skarns. Zinc-rich exoskarns have in many cases developed

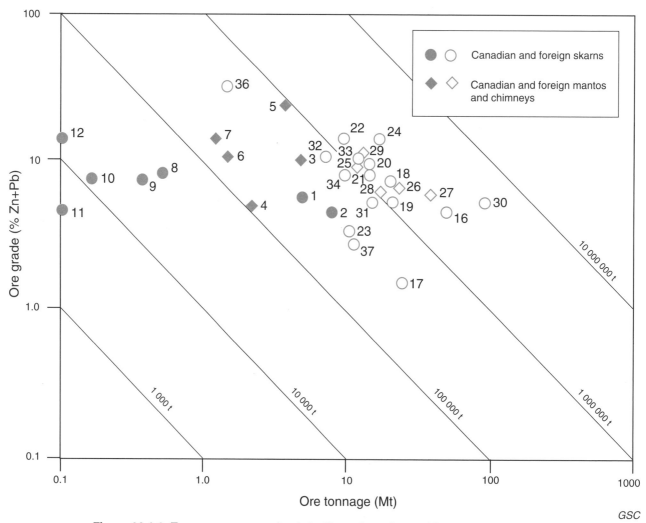

Figure 20.1-2. Tonnage versus grade of significant Canadian and foreign Zn-Pb-Ag skarn and replacement deposits. Numbered symbols correspond to deposits listed in Table 20.1-1.

GSC

along lithological contacts and structural pathways in unmetamorphosed rocks distant from the intrusive contact, and in places, e.g., Santa Eulalia (Hewitt, 1968), pass distally into essentially concordant mantos and discordant chimneys significantly higher in ore tonnage and grade than the skarn. The distinctive morphology, abrupt changes in ore grade and mineralogy, and paucity of calc-silicate gangue that characterize these distal members of the Zn-Pb skarn system have contributed to their classification as a distinct deposit type, i.e. "high temperature carbonate-hosted massive sulfide ores" of Titley (1993), "high-temperature carbonate-hosted Ag-Pb-Zn(Cu) deposits" of Megaw et al. (1988), and "carbonate-hosted sulfide deposits" of Beaty et al. (1990). Zinc-lead skarn deposits form a structurally zoned continuum that includes a) relatively small contact skarns adjacent to batholiths; b) endoskarns and extensive structurally and stratigraphically controlled exoskarns within, adjacent to, and at some distance from, stock contacts; c) skarn and replacement bodies following dykes and branching out along bedding and faults; d) 'rootless' exoskarn and replacement bodies in structural and stratigraphic traps distant from a probable or unknown igneous source; and e) most distal of all, carbonate-hosted vein deposits with calc-silicate and (or) manganese-silicate gangue (Einaudi et al., 1981).

Zinc-lead-silver skarns characteristically contain johannsenitic (manganiferous) pyroxene and subordinate amounts of andraditic garnet as prograde skarn minerals. Vesuvianite, bustamite, and wollastonite are present in some deposits. Iron-rich sphalerite predominates over galena in the ratio of about 3:2. Pyrrhotite or pyrite are common; magnetite, chalcopyrite, and arsenopyrite are less common. Retrograde minerals are Mn-rich actinolite, epidote, chlorite, ilvaite, calcite, and siderite; also rhodonite, fluorite, and quartz. The prograde mineral assemblage in Zn-rich skarns in Yukon Territory is distinctive in its abundant epidote, rare andradite, and common intergrowth of the two pyroxenes, Mn-hedenbergite and diopside (Dick, 1979).

Contact skarns and exoskarns close to the intrusive contact commonly are high in Fe, Cu, and W, and rarely, in Au. Zinc tends to occur inwards of Pb at intermediate distances from the intrusion. Skarns and replacements distant from the intrusive contact are rich in Mn, Ag, and Pb and, less commonly, in Sb and As. Zonal metal distribution patterns in some deposits have been modified and overprinted by retrograde mineral assemblages. In the large Zn-Pb-Ag(Cu, W, Au) skarn deposit at Naica, Chihuahua, Mexico, gold is concentrated with tungsten and copper in prograde skarn close to rhyolite dyke contacts (Clark et al., 1986). In the Midway and YP Ag-Pb-Zn (Au, Cu, Sn) manto deposits near the Yukon Territory-British Columbia border, Au is concentrated commonly with massive sulphide ores rich in Fe, Cu, and Zn, rather than with more distal Pb- and Ag-rich zones (K.M. Dawson, unpub. data, 1994). A similar trend is manifested as district-wide zonation at Ketza River, Yukon Territory where gold-rich pyrrhotite-chalcopyrite manto orebodies apparently grade outward over 1 to 3 km to Ag-Pb-Zn replacement deposits (Cathro, 1990). Both proximal Au-rich and distal Ag-rich zones probably are underlain by the same unexposed Cretaceous stock (Abbott, 1986).

The largest production from a Canadian Zn-Pb-Ag skarn deposit was attained from the Bluebell mine at Riondel, British Columbia between 1895 and 1971 (Fig. 20.1-1). A geological sketch map of the district is given in Figure 20.1-3, and a schematic block diagram of the orebodies in Figure 20.1-4. These enigmatic deposits have been described as 'vein replacements' by several authors, including Ohmoto and Rye (1970), Ransom (1977), Höy (1980), and Sangster (1986), and previously as 'syngenetic'

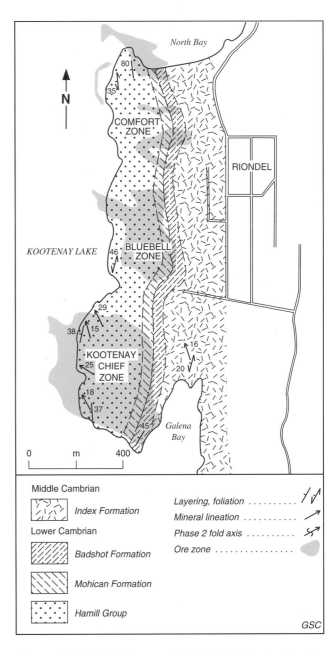

Figure 20.1-3. Sketch map of Riondel Peninsula showing distribution of formations and Bluebell mine orebodies, projected to surface (after Höy, 1980). Similar Zn-Pb-Ag replacement deposits of the Ainsworth district lie 4 km to the west, adjacent to the contact of the mid-Jurassic Nelson batholith.

453

by Sangster (1970). Features supporting relatively early mineralization include: essentially stratiform morphology and finely laminated texture of orebodies, consistent metal zoning with Pb stratigraphically higher than Zn, and apparent coincidence of some mineralization with predeformational brecciation and dolomitization. Features that argue more strongly for epigenesis include: alignment of ore shoots with steeply-dipping, crosscutting faults; lateral replacement of marble adjacent to these steep structures, including marble of the structurally overlying Mohican Formation; a distinctive skarn mineral assemblage that includes prograde knebelite (Fe, Mn-olivine) and retrograde minnesotaite (Fe-talc), Fe-, Mg-, and Mn-carbonates, chlorite, calcite, and quartz; fluid inclusion and stable isotopic data supporting a relatively high temperature, deeply seated origin (Ohmoto and Rye, 1970); and coarsely crystalline sulphide and gangue minerals in cavities formed after metamorphism and deformation. Lead isotopic data cited by Andrew et al. (1984) support either a primary replacement origin, as in the similar Ainsworth and Slocan districts, or a remobilized stratiform origin, both under the influence of the Nelson batholith. A

younger, Eocene age of epigenetic mineralization (59 ± 3 Ma) is indicated by K-Ar data from vein muscovite cutting a gabbro dyke (Beaudoin et al., 1992).

Significant production was also obtained from two probable skarn orebodies, the Jersey and Emerald Zn-Pb mines in the Salmo district of southeastern British Columbia. These deposits are hosted by carbonate units of the lower Cambrian Laib Formation of the Kootenay Arc, and the Jersey has been grouped with the H.B., Remac, and Duncan Lake deposits as "concordant" (Fyles, 1966), "Remac type" (Sangster, 1970), and "sedimentary exhalative" (Sangster, 1986). The Jersey is adjacent to the Emerald, Feeney, and Dodger W skarn orebodies, within the same anticlinal structure and underlain by the same granitoid stock and related calc-silicate skarn horizon that hosts the W skarns. The geology of the Emerald Tungsten camp is given in Figure 20.1-5. The relationship of the Jersey to the Dodger in section, is shown in Figure 20.1-6. Both the ore mineral assemblage, which includes sphalerite with 6% Fe, pyrrhotite, arsenopyrite, and scheelite in addition to galena and pyrite, and the gangue assemblage of dolomite, calcite, quartz, muscovite, chlorite, tremolite, wollastonite, olivine,

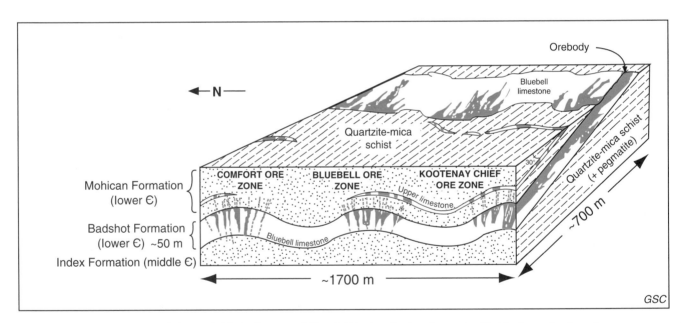

Figure 20.1-4. Schematic block diagram of Riondel Peninsula showing major Zn-Pb-Ag orebodies of the Bluebell mine. The westward-dipping inverted Cambrian succession lies within the lower limb of the Riondel nappe (Phase 1). The host Bluebell limestone (Badshot Formation) occupies the western limb of a Phase 2 northerly trending antiform. Replacement orebodies are localized by Phase 3 warps in earlier folds and by steeply dipping cross-fractures, both trending westerly to northwesterly. After Ohmoto and Rye (1970) and Höy (1980).

Figure 20.1-5. Geology of the Emerald Tungsten camp, Salmo, British Columbia, showing locations of skarn Zn-Pb and W deposits. Geology after Fyles and Hewlett, 1959 (modified after Webster et al., 1992, Fig. 2-2-5).

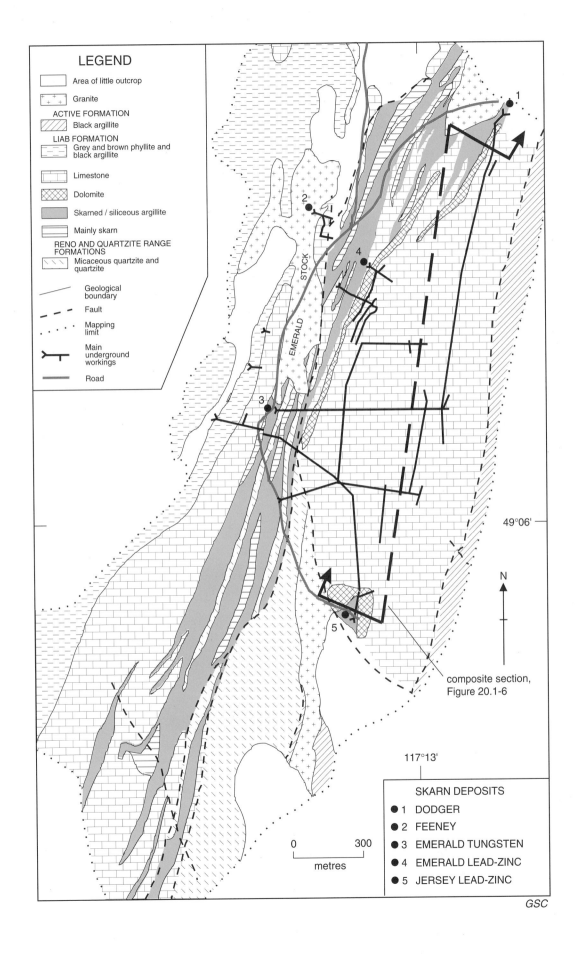

LEGEND

Area of little outcrop

Granite

ACTIVE FORMATION
Black argillite

LIAB FORMATION
Grey and brown phyllite and
black argillite

Limestone

Dolomite

Skarned / siliceous argillite

Mainly skarn

RENO AND QUARTZITE RANGE
FORMATIONS
Micaceous quartzite and
quartzite

Geological
boundary

Fault

Mapping
limit

Main
underground
workings

Road

EMERALD STOCK

49°06'

117°13'

N

composite section,
Figure 20.1-6

0   300
metres

SKARN DEPOSITS

● 1 DODGER
● 2 FEENEY
● 3 EMERALD TUNGSTEN
● 4 EMERALD LEAD-ZINC
● 5 JERSEY LEAD-ZINC

GSC

and serpentine (Whishaw, 1954; Bradley, 1970) are more representative of a Zn-Pb skarn deposit than is the simpler mineralogy of the Remac deposit to the south. The relationship of these probable Zn-Pb skarns to adjacent intrusions and consanguineous stratiform Zn-Pb deposits remains unclear. Production from the combined Jersey and Emerald Zn-Pb mines between 1914 and 1972 was 7.68 Mt of ore grading 3.49% Zn, 1.65% Pb, 3.08 g/t Ag, and 0.23% W (Höy, 1982).

The Sa Dena Hes mine at Mount Hundere, near Watson Lake, Yukon Territory is an example of a large Zn-Pb skarn without an exposed igneous source (Table 20.1-1). Several tabular skarn orebodies are concordant with bedding in a deformed Upper Proterozoic-Lower Cambrian limestone-phyllite sequence (Abbott, 1981). Distribution of abundant prograde actinolite-hedenbergite-grossularite and retrograde quartz-fluorite skarn assemblages is apparently unrelated to small aplite and diabase dykes, but may reflect the influence of a buried pluton (Dawson and Dick, 1978). Two additional large deposits, which formed at a distance from a probable or unknown igneous source, are Quartz Lake in southeastern Yukon Territory and Prairie Creek (Cadillac) near the South Nahanni River in Northwest Territories. The Quartz Lake deposit is a stratabound pyritic replacement of limestone in an Upper Proterozoic-Lower Cambrian

conglomeratic quartzite-limestone-argillite shelf sequence (Gabrielse and Blusson, 1969; Morin, 1981). The ore mineral assemblage of pyrite, dark sphalerite, and galena, and lesser arsenopyrite, boulangerite, tetrahedrite, and chalcopyrite, and the gangue assemblage of quartz, dolomite, siderite, and ankerite, support an epigenetic origin as a skarn related to small, barely unroofed Cretaceous plutons in the district. Recent exploration drilling of the developed Prairie Creek prospect, an extensive Zn-Pb-Ag vein in a shear zone in dolostone and shale of the Middle Devonian Arnica Formation (Thorpe, 1972) has revealed a stratabound orebody at depth that more than doubled known reserves (Table 20.1-1). Consistently high Zn+Pb and elevated Cu values in ore, and lesser amounts of W and Hg (Jonasson and Sangster, 1975) support an igneous source, but the closest known plutons are exposed 80 km to the west.

Mineralogy

Principal ore minerals: sphalerite, galena (in part argentiferous), Ag sulphosalts.

Other opaque minerals: chalcopyrite, pyrrhotite, magnetite, arsenopyrite, pyrite, Pb sulphosalts, tetrahedrite, scheelite.

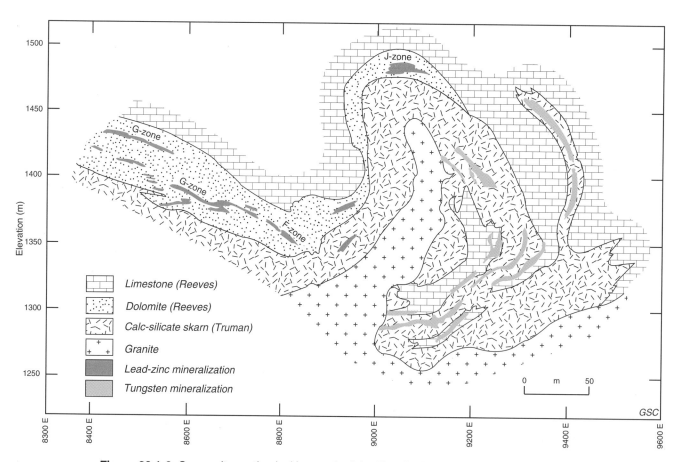

Figure 20.1-6. Composite section looking north of the East Dodger tungsten mine and Jersey lead-zinc mine, Salmo, British Columbia. Approximate line of composite section shown in Figure 20.1-5. Modified after Thompson (1973).

Prograde skarn minerals: Mn-rich pyroxene (johannsenitic hedenbergite), andraditic garnet, vesuvianite, wollastonite, bustamite, epidote, diopside, olivine (knebelite).

Retrograde skarn minerals: manganoan varieties of actinolite, epidote, chlorite, siderite, and ilvaite, and also calcite, quartz, fluorite, rhodonite, axinite, talc.

DEFINITIVE CHARACTERISTICS

1. Manganese- and iron-rich mineralogy.
2. Occurrence along stratigraphic contacts and structural pathways, including dyke contacts.
3. Occurrence at a distance from intrusive contacts.
4. Limited size of metamorphic aureole in comparison with other skarn types.
5. Abundant sulphide deposition beyond metamorphic aureole, as replacement of carbonate rock.
6. Pyroxene dominant over garnet in prograde skarn.
7. Pronounced structurally-controlled zoning of deposit morphology and metal zoning, with proximal skarns rich in Cu and W, and distal skarns, mantos, and veins rich in Mn, Ag, and Pb.

GENETIC MODEL

In the genetic model for Pb-Zn-Ag skarns, limestone and calcareous pelite adjacent to a cooling and crystallizing batholith, stock, or dyke-sill complex undergoes isochemical thermal metamorphism to marble and calc-silicate hornfels. Metasomatism is initiated by the separation of a magmatic-hydrothermal fluid released by crystallization and hydrofracturing of the pluton, and this aqueous fluid may combine with metamorphic waters to transport Fe, Mg, Si, Al, and base metal cations and sulphur. Metasomatic fluids pass along pluton and dyke contacts, stratigraphic contacts, fold hinges, fractures, and other permeable pathways and infiltrate the surrounding wall rocks. Marble and calc-silicate hornfels are replaced by prograde exoskarn, but some exoskarn and most sulphide replacements extend along structural pathways beyond the limits of the metamorphic aureole into limestone. Simultaneously, epidote-rich endoskarns form in the pluton and in pelitic hornfels by introduction of calcium from the host rock, and wollastonite forms in marble.

Skarn close to the pluton is rich in andraditic garnet, but clinopyroxene is the dominant prograde mineral. More distal assemblages include pyroxenoids, i.e., rhodonite, wollastonite, and bustamite, and prograde hydrous minerals, i.e., epidote and chlorite. Johannsenitic pyroxene becomes depleted successively in Mg, then Fe, and enriched in Mn along the strike of structural conduits away from the intrusion, towards the marble front away from fluid channelways, and at higher elevations in the system. Calc-silicate compositions are controlled mainly by the composition of the host rocks, the oxidation state, and the temperature of the system. Exoskarns may pass continuously beyond the thermal aureole into sulphide-rich and calc-silicate-poor concordant to discordant skarn bodies termed mantos and chimneys, respectively, and most distally to Ag-Pb-Zn vein deposits that contain quartz, carbonate, and minor calc-silicate gangue.

Late in the prograde stage of skarn formation, the evolving ore fluid is enriched in sulphur, manganese, and ferric iron, and sulphide and magnetite deposition commences. Copper- and zinc-rich skarn assemblages form closest to the intrusive contact, together with W and Au, if present. Greater deposition of Pb, Ag, and Mn occurs at a distance from the intrusion, in conjunction with Pb- and Ag-sulphosalt minerals. The main period of sulphide deposition follows skarn formation and is accompanied by hydrous alteration of early skarn minerals and hydrolytic alteration of intrusions. Early calcium-rich silicates are altered to calcium-depleted, Fe- and Mn-enriched silicates, carbonates, and iron oxides. Distal skarns and replacements are affected less by hydrous alteration than proximal skarns. Some peripheral mantos, chimneys, and veins are formed at this late stage.

Most skarn minerals can be enriched in manganese, including pyroxene, garnet, amphibole, pyroxenoid, olivine, ilvaite, chlorite, and serpentine. Meinert (1987) described a systematic increase in the pyroxene:garnet ratio and the manganese content of pyroxene along the path of hydrothermal fluid flow at the Groundhog deposit, New Mexico. Evolutionary stages for skarns in general are outlined in "Skarn tungsten" (deposit subtype 20.5; Fig. 20.5-3) and "Skarn copper" (deposit subtype 20.2; Fig. 20.2-7).

RELATED DEPOSIT TYPES

Massive sulphide bodies with little or no accompanying calc-silicate gangue minerals, in addition to the mantos, chimneys, and carbonate-hosted veins described above, are commonly termed "replacement deposits" and not widely recognized as distal members of a base metal skarn system, particularly if no related intrusion is recognized. In discussing the skarn deposits of northern Mexico, Megaw et al. (1988) made the important point that many zinc skarn districts grade outward from intrusion-associated to intrusion-free ores, therefore those districts lacking known intrusion relationships may not have been traced to their ends. The important Gilman and Leadville deposits of the Colorado Mineral Belt are recognized by Beaty et al. (1990) and Titley (1993) as mantos similar to the manto and skarn orebodies of northern Mexico, as described by Megaw et al. (1988). Similar mantos that occur at Pioche and Eureka, Nevada; Tombstone, Arizona; Tintic and Park City, Utah; U.S. and Lark mines, Bingham, Utah; and Magdalena, New Mexico are described by Titley (1993).

Incompletely explored Zn-Pb skarn districts may have only one or a few of the characteristic skarn and manto zones exposed, in addition to an unexposed intrusion. The presence of skarn minerals anywhere in the system, such as garnet- and manganese-rich pyroxene and amphibole, will definitively distinguish these deposits from other types of Zn-Pb-Ag deposits such as Mississippi Valley-type deposits, which they may resemble.

Zinc-rich skarns that have formed as peripheral parts of some W-Cu skarns in the northern Canadian Cordillera are relatively rich in W, but deficient in Pb and Ag (Dick, 1979). Base metal skarns adjacent to the Seagull batholith, a F- and B-rich leucogranite pluton in south-central Yukon Territory, are enriched in Sn, B, Be, and F to the extent that several are predominantly Sn skarns.

EXPLORATION GUIDES

The recognition of any of the following features within a prospective terrain may lead to the discovery of a Zn-Pb-Ag skarn deposit.

1. Relatively thick pure and impure limestone beds, interbedded with shale or pelite, and intruded by granitoid plutons of generally small dimensions, e.g., dykes and sills.

2. Stockwork fractures, faults, and porphyry dykes along and adjacent to a pluton/limestone contact.

3. Structural traps in carbonate host rock, particularly the intersection of anticlines and domes by structural conduits such as dykes, faults, and fault intersections.

4. Presence of calc-silicate skarn minerals, particularly garnet, pyroxene, and amphibole in association with base metal assemblages. Purplish-black Mn-rich gossan developed over calc-silicate and carbonate gangue minerals rich in Mn and Fe.

5. Pyroxene:garnet ratios and manganese contents of pyroxene decrease, and magnesium content of pyroxene increases, systematically towards the intrusive contact and hydrothermal conduit, enabling one to identify proximal and distal zones of large systems (Meinert, 1987).

6. Small base metal sulphide occurrences lacking apparent skarn affinity may be part of a major magmatic-hydrothermal system. Such deposits may be hosted by carbonate-pelite rocks distant from an intrusive source which may not be exposed and which may lack a thermal metamorphic aureole; consequently, these deposits may have little or no associated calc-silicate gangue. These deposits may, in places, include Ag-rich Zn-Pb veins and 'rootless' conformable and discordant Fe-Zn-Pb replacement bodies.

7. Metal zoning in the skarn assemblage shows enrichment away from an intrusive contact or hydrothermal conduit in the general order Fe-Cu-Zn-Pb-Ag-Sb-As.

REFERENCES

References marked with asterisks (*) are considered to be the best sources of general information on this deposit type.

Abbott, J.G.
1981:   A new geological map of Mt. Hundere and the area north; in Yukon Geology and Exploration 1979-80, Indian and Northern Affairs Canada, p. 45-50.
1983:   Silver-bearing veins and replacement deposits of the Rancheria district; in Yukon Exploration and Geology 1983, Indian and Northern Affairs Canada, Exploration and Geological Services Division, Whitehorse, p. 34-44.
1986:   Epigenetic mineral deposits of the Ketza-Seagull district, Yukon; Yukon Geology, v. 1, p. 56-66.

Aho, A.E.
1969:   Base metal province of Yukon; The Canadian Institute of Mining and Metallurgy, Bulletin, April, 1969, p. 71-83.

Andrew, A., Godwin, C.I., and Sinclair, A.J.
1984:   Mixing line isochrons: a new interpretation of galena lead isotope data from southeastern British Columbia; Economic Geology, v. 79, p. 919-932.

Atkinson, W.W., Jr. and Einaudi, M.T.
1978:   Skarn formation and mineralization in the contact aureole at Carr Fork, Bingham, Utah; Economic Geology, v. 73, p. 1326-1365.

Barnes, M.P. and Simos, J.G.
1968:   Ore deposits of the Park City district with a contribution on the Mayflower lode; in Ore Deposits of the United States, 1933-1967, The Graton-Sales Volume, (ed.) J.D. Ridge; v. II, American Institute of Mining, Metallurgical and Petroleum Engineers, Inc., New York, p. 1102-1126.

*Beaty, D.W., Landis, G.P., and Thompson, T.B.
1990:   Carbonate-hosted sulfide deposits of the central Colorado mineral belt: introduction, general discussion, and summary; in Carbonate-Hosted Sulfide Deposits of the Central Colorado Mineral Belt, (ed.) D.W. Beaty, G.P. Landis, and T.B. Thompson; Economic Geology, Monograph 7, p. 1-18.

Beaudoin, G., Roddick, J.C., and Sangster, D.F.
1992:   Eocene age for Ag-Pb-Zn-Au vein and replacement deposits of the Kokanee Range, southeastern British Columbia; Canadian Journal of Earth Sciences, v. 29, p. 3-14.

Bradford, J.A. and Godwin, C.I.
1988:   Midway silver-lead-zinc manto deposit, northern British Columbia; in Geological Fieldwork, 1987, British Columbia Ministry of Energy Mines and Petroleum Resources, Paper 1988-1, p. 353-360.

Bradley, O.E.
1970:   Geology of the Jersey lead-zinc mine, Salmo, British Columbia; in Lead-Zinc Deposits in the Kootenay Arc, Northeastern Washington and Adjacent British Columbia, (ed.) A.E. Weissenborn; State of Washington Department of Natural Resources, Bulletin no. 61, p. 89-98.

Buseck, P.R.
1966:   Contact metasomatism and ore deposition: Concepcion del Oro, Mexico; Economic Geology, v. 61, p. 97-136.

Campa, M.F. and Coney, P.J.
1983:   Tectono-stratigraphic terranes and mineral resource distribution in Mexico; Canadian Journal of Earth Sciences, v. 20, p. 1040-1051.

Cathro, M.S.
1990:   Gold, silver and lead deposits of the Ketza River district, Yukon: Preliminary results of field work; in Mineral Deposits of the Northern Canadian Cordillera, Yukon-Northeastern British Columbia (Field Trip 14), (ed.) J.G. Abbott and R.J.W. Turner; 8th International Association on the Genesis of Ore Deposits Symposium, Ottawa, Field Trip Guidebook, Geological Survey of Canada, Open File 2169, p. 269-282.

Chappell, B.W. and White, A.J.R.
1974:   Two contrasting granite types; Pacific Geology, v. 8, p. 173-174.

Clark, K.F., Megaw, P.K.M., and Ruiz, J. (ed.)
1986:   Lead-zinc-silver carbonate-hosted deposits of northern Mexico; Guidebook for Field and Mine Excursions, November 13-17, 1986, Society of Economic Geologists, 329 p.

Dawson, K.M.
1985:   MDD field trip to replacement deposits of Mexico, February 18-28, 1985; in The Gangue, Mineral Deposits Division Newsletter, Geological Association of Canada, no. 21, May, 1985, p. 5-6.

*Dawson, K.M. and Dick, L.A.
1978:   Regional metallogeny of the northern Cordillera: tungsten and base metal-bearing skarns in southeastern Yukon and southwestern Mackenzie; in Current Research, Part A; Geological Survey of Canada, Paper 78-1A, p. 287-292.

Dawson, K.M., Panteleyev, A., Sutherland Brown, A., and Woodsworth, G.J.
1991:   Regional metallogeny, Chapter 19 in Geology of the Cordilleran Orogen in Canada, (ed.) H. Gabrielse and C.J. Yorath; Geological Survey of Canada, Geology of Canada, no. 4, p. 707-768 (also Geological Society of America, The Geology of North America, v. G-2).

*Dick, L.A.
1979:   Tungsten and base metal skarns in the northern Cordillera; in Current Research, Part A; Geological Survey of Canada, Paper 79-1A, p. 259-266.

*Einaudi, M.T., Meinert, L.D., and Newberry, R.J.
1981:   Skarn deposits; Economic Geology Seventy-fifth Anniversary Volume, 1905-1980, (ed.) B.J. Skinner; p. 317-391.

Erwood, R.J., Kesler, S.E., and Cloke, P.L.
1979:   Compositionally distinct saline hydrothermal solutions, Naica mine, Chihuahua, Mexico; Economic Geology, v. 74, p. 95-108.

Forgan, C.G.
1950:   Yugoslavia - ore deposits of the Stantrg lead-zinc mine; 18th International Geologic Congress, London, 1948, Report, pt. 7, p. 290-301.

Fyles, J.T.
1960:   Windermere, Mineral King (Sheep Creek Mines Limited), British Columbia; Minister of Mines, Annual Report, 1959, p. 74-89.

Fyles, J.T. (cont.)

1966: Lead-zinc deposits in British Columbia; in Tectonic History and Mineral Deposits of the Western Cordillera; The Canadian Institute of Mining and Metallurgy, Special Volume No. 8, p. 231-238.

Fyles, J.T. and Hewlett, C.G.

1959: Stratigraphy and structure of the Salmo lead-zinc area; British Columbia Department of Mines, Bulletin 41, 162 p.

Gabrielse, H. and Blusson, S.L.

1969: Geology of Coal River map-area, Yukon Territory and District of Mackenzie (95D); Geological Survey of Canada, Paper 68-38, 22 p.

Gilmer, A.L., Clark, K.F., Hernandez, C.I., Conde, J.C., and J.I. Figueroa S.

1986: Geological and mineralogical summary of metalliferous deposits in the Santa Maria dome, Velardena, Durango; in Lead-Zinc-Silver Carbonate-Hosted Deposits of Northern Mexico, (ed.) K.F. Clark, P.K.M. Megaw, and J. Ruiz; Guidebook for Field and Mine Excursions, November 13-17, 1986, Society of Economic Geologists, p. 143-153.

Gunning, H.C.

1932: H.P.H. group, Nahwitti Lake, Vancouver Island, British Columbia; Geological Survey of Canada, Summary Report, 1931, p. 36A-45A.

Hewitt, W.P.

1968: Geology and mineralization of the main mineral zone of the Santa Eulalia district, Chihuahua, Mexico; Society of Mining Engineers American Institute of Mining, Metallurgical and Petroleum Engineers Transactions, v. 241, p. 228-260.

Höy, T.

1980: Geology of the Riondel area, central Kootenay Arc, southeastern British Columbia; British Columbia Ministry of Energy, Mines and Petroleum Resources, Bulletin 73, 89 p.

1982: Stratigraphic and structural setting of stratabound lead-zinc deposits in southeastern B.C.; The Canadian Institute of Mining and Metallurgy, Bulletin, v. 75, no. 840, p. 114-134.

Hsieh, C.Y.

1950: China-note on the lead, zinc and silver deposits in China; 18th International Geologic Congress, London, 1948, Report, pt. 7, p. 380-399.

Indian and Northern Affairs Canada

1992: Yukon Exploration and Geology 1991; Exploration and Services Division, Indian and Northern Affairs Canada, Whitehorse, Yukon, 42 p.

Jonasson, I.R. and Sangster, D.F.

1975: Variations in mercury content of sphalerite from some Canadian sulphide deposits; Geochemical Exploration 1974, Association of Exploration Geochemists, Special Publication No. 2, p. 324-325.

Kano, T. and Shimizu, M.

1992: Mineral deposits and magmatism in the Hida and Hida Marginal Belts, central Japan; Field Trip C 20 Guidebook, 29th International Geological Congress, Kyoto, 41 p.

*Megaw, P.K.M., Ruiz, J., and Titley, S.R.

1988: High-temperature, carbonate-hosted Ag-Pb-Zn (Cu) deposits of northern Mexico; Economic Geology, v. 83, p. 1856-1885.

*Meinert, L.D.

1987: Skarn zonation and fluid evolution in the Groundhog mine, Central District, New Mexico; Economic Geology, v. 82, no. 3, p. 523-545.

1992: Skarns and skarn deposits; Geoscience Canada, v. 19, no. 4, p. 145-162.

Morin, J.A.

1981: The McMillan deposit - a stratabound lead-zinc-silver deposit in sedimentary rocks of upper Proterozoic age; in Yukon Geology and Exploration 1979-80, Indian and Northern Affairs, p. 105-109.

Morin, J.A., Sinclair, W.D., Craig, D.B., and Marchand, M.

1977: Eagle; Tintina Silver Mines Limited; in Mineral Industry Report 1976, EGS 1977-1, Indian and Northern Affairs Canada, Whitehorse, p. 199-203.

Morris, H.T.

1968: The main Tintic mining district, Utah; in Ore Deposits of the United States, 1933-1967, The Graton-Sales Volume, (ed.) J.D. Ridge; v. II, American Institute of Mining, Metallurgical and Petroleum Engineers, Inc., New York, p. 1043-1073.

Ohmoto, H. and Rye, R.O.

1970: The Bluebell mine, British Columbia. I. Mineralogy, paragenesis, fluid inclusions, and the isotopes of hydrogen, oxygen and carbon; Economic Geology, v. 65, p. 417-437.

Ransom, P.W.

1977: An outline of the geology of the Bluebell mine, Riondel, B.C.; in Lead-zinc Deposits of Southeastern British Columbia, (ed.) T. Höy; Geological Association of Canada Annual Meeting, 1977, Fieldtrip Guidebook No. 1, p. 44-51.

Ruiz, J., Sweeney, R., and Palacios, H.

1986: Geology and geochemistry of Naica, Chihuahua, Mexico; in Lead-zinc-silver Carbonate-hosted Deposits of Northern Mexico, (ed.) K.F. Clark, P.K.M. Megaw, and J. Ruiz; Guidebook for Field and Mine Excursions, November 13-17, 1986, Society of Economic Geologists, p. 169-178.

Sakurai, W. and Shimazaki, H.

1993: Exploration of blind skarn deposits based on the mineralization model of the Kamioka mine, Gifu Prefecture, Central Japan; Resource Geology, Special Issue, no. 16, p. 141-150.

Sangster, D.F.

1970: Metallogenesis of some Canadian lead-zinc deposits in carbonate rocks; Geological Association of Canada, Proceedings, v. 22, p. 27-36.

1986: Classification, distribution and grade-tonnage summaries of Canadian lead-zinc deposits; Geological Survey of Canada, Economic Geology Report 37, 68 p.

Shultz, J. and Hamann, R.

1977: El Mochito: a good mine taxed to the hilt; Engineering and Mining Journal, November, 1977, p. 166-183.

Thompson, R.I.

1973: Invincible, East Dodger (82F/SW-234); in Geology, Exploration and Mining in British Columbia 1973, British Columbia Department of Mines and Petroleum Resources, p. 54-57.

Thompson, T.B. and Arehart, G.B.

1990: Geology and the origin of ore deposits in the Leadville district, Colorado: part I. Geologic studies of orebodies and wall rocks; in Carbonate-hosted Sulfide Deposits of the Central Colorado Mineral Belt, (ed.) D.W. Beaty, G.P. Landis, and T.B Thompson; Economic Geology Monograph 7, p. 130-154.

Thorpe, R.I.

1972: Mineral exploration and mining activities, mainland Northwest Territories, 1966 to 1968 (excluding Coppermine River area); Geological Survey of Canada, Paper 70-70, p. 130-139.

Titley, S.R.

1993: Characteristics of high temperature carbonate-hosted massive sulfide ores in southeastern United States and northeastern Mexico; in Mineral Deposit Modeling, (ed.) R.V. Kirkham, W.D. Sinclair, R.I. Thorpe, and J.M. Duke; Geological Association of Canada, Special Paper 40, p. 433-464.

Vaillancourt, P. de G.

1982: Geology of pyrite-sphalerite-galena concentrations in Proterozoic quartzite at Quartz Lake, southwestern Yukon; in Yukon Exploration and Geology 1982, Indian and Northern Affairs Canada, Exploration and Geological Services, Whitehorse, p. 73-77.

Webster, I.C.L., Ray, G.E., and Pettipas, A.R.

1992: An investigation of selected mineralized skarns in British Columbia; in Geological Fieldwork 1991, British Columbia Geological Survey Branch, Paper 1992-1, p. 235-252.

Whishaw, Q.G.

1954: The Jersey lead-zinc deposit, Salmo, B.C.; Economic Geology, v. 49, no. 5, p. 521-529.

White, L.

1980: Mining in Mexico (San Martin); Engineering and Mining Journal, November, 1980, v. 181, p. 63-194.

Yun, S.

1979: Geology and skarn ore mineralization of the Yeonhua-Ulchin zinc-lead mining district, S.E. Tagbaggsan region, Korea; PhD. thesis, Stanford University, Stanford, California, 184 p.

20.2 SKARN COPPER

20.2a Copper skarns not associated with porphyry copper deposits
20.2b Copper skarns associated with porphyry copper deposits

K.M. Dawson and R.V. Kirkham

INTRODUCTION

Skarn copper deposits are characterized by their large size relative to tungsten, lead-zinc, and gold skarns; association with both porphyry copper- and nonporphyry copper-related intrusions; and relatively oxidized, gold-rich mineral assemblages developed close to intrusive contacts with carbonate rocks. Principal commodities are Cu and Mo, but Fe, Au, and Ag are common byproducts, and Zn, Pb, W, and Bi are common minor constituents. Two principal divisions of copper skarns are recognized, i.e., those not associated with porphyry copper deposits (subtype 20.2a) and those associated with porphyry copper deposits (subtype 20.2b). Two additional subdivisions are recognized, based on the composition of the plutons, i.e. calc-alkaline and alkaline (Table 20.2-1). Significant Canadian and foreign examples of the above are: non-porphyry-associated: Phoenix, British Columbia; Whitehorse Copper Belt, Yukon Territory; Sayak I, Kazakhstan, Concepcion del Oro, Mexico; and Tongling, Lower Yangtze, China; porphyry-associated calc-alkaline: Gaspé Copper, Quebec; Twin Buttes, Arizona, U.S.A.; and Gold Coast, Ok Tedi district, Papua New Guinea; and porphyry-associated alkaline: Ingerbelle and Galore Creek, British Columbia and Larap, Philippines. Distribution of principal skarn copper deposits in Canada is shown in Figure 20.2-1.

IMPORTANCE

Skarn copper deposits account for approximately 5% of Canada's copper production and less than 5% of its reserves. Copper skarns are perhaps the world's most abundant skarn type. Significant deposits are mined around the world; in the porphyry copper province of southwestern United States, in the Ural Mountains of Russia, Kazakhstan, and in the Lower Yangtze River area, China, skarn deposits are a major source of copper.

Dawson, K.M. and Kirkham, R.V.
1996:   Skarn copper; in Geology of Canadian Mineral Deposit Types, (ed.) O.R. Eckstrand, W.D. Sinclair, and R.I. Thorpe; Geological Survey of Canada, Geology of Canada, no. 8, p. 460-476 (also Geological Society of America, The Geology of North America, v. P-1).

SIZE AND GRADE OF DEPOSIT

Tonnages and grades of significant Canadian and foreign skarn copper deposits are given in Table 20.2-2. Size, grade, and some geological characteristics of global skarn copper deposits were tabulated by Einaudi et al. (1981), and of porphyry copper-related skarns by Einaudi (1982). Copper-bearing skarns associated with, and forming part of, porphyry copper deposits, which include some of the world's largest skarn deposits, range in size from 50 to greater than 320 Mt of skarn ore. They average about 100 Mt at a grade of 1% Cu. In addition to skarn copper ore, variable amounts of porphyry-type stockwork and disseminated copper ore are hosted within adjacent intrusions and other rocks. Canadian porphyry deposit-related copper skarn deposits average about 90 Mt of ore grading between 0.4 and 1.0% Cu. Large porphyry deposit-related copper skarns, shown as circles on Figure 20.2-2, contain more than 1 Mt of Cu metal. Copper skarns developed adjacent to intrusions with no related porphyry copper deposits, shown as square symbols on Figure 20.2-2, are smaller, i.e. generally between 2 and 30 Mt of ore, but higher in copper grade, i.e., in the 1.5% to 2.5% Cu range. Canadian examples of this skarn type, such as Phoenix, British Columbia and Whitehorse Copper Belt, Yukon Territory have produced significant copper. Limited production and reserve data indicate that the Tongling and adjacent skarn districts in the Lower Yangtze River area in China combined contain greater than 10 Gt of ore grading about 1% Cu.

GEOLOGICAL FEATURES
Geological setting

Porphyry deposit-related copper skarns commonly occur in continental marginal belts, where miogeoclinal calcareous sediments have been intruded by epizonal calc-alkaline granodioritic to quartz monzonitic magnetite-bearing, I-type plutons. Large deposits in the southwestern United States occur in such a setting, in association with subduction-related magmatic arcs. Deposits in similar settings are located in Mexico, Peru, Russia, China, and Japan. Many of these deposits occur within evolved magmatic arcs which developed on continental crust. Timing of skarn-related intrusion is commonly postorogenic.

Table 20.2-1. Mineralogy of three skarn copper subtypes.

Plutonic asociation of skarn Cu	Prograde skarn minerals	Retrograde skarn minerals
Mineralized calc-alkaline porphyry Cu stocks	andraditic garnet, diopsidic pyroxene, wollastonite	actinolite, tremolite, chlorite, montmorillonite, quartz, calcite
Unmineralized stocks, unrelated to porphyry Cu	grandite garnet, diopsidic pyroxene, wollastonite, magnetite, epidote	epidote, actinolite, tremolite, hornblende, chlorite, quartz, calcite
Mineralized alkaline porphyry Cu-Au stocks	andraditic garnet, diopsidic pyroxene, orthoclase, biotite, albite, epidote, sphene, vesuvianite	chlorite, actinolite, epidote, calcite, albite, scapolite, zeolite, hematite

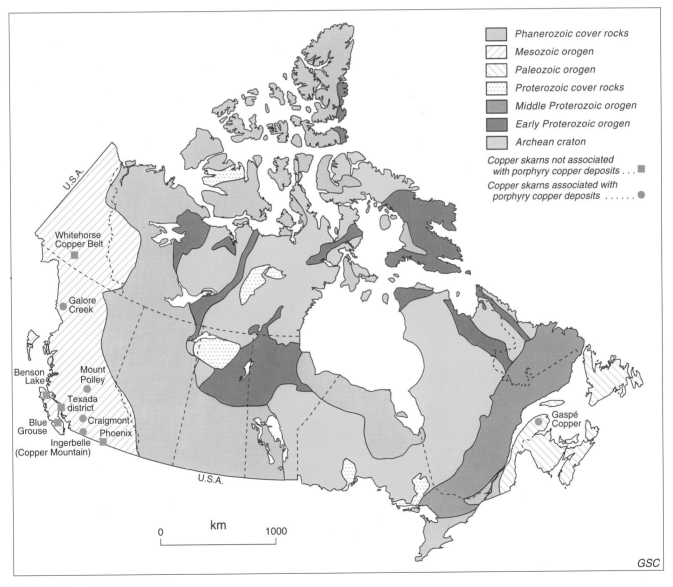

Figure 20.2-1. Distribution of principal skarn copper deposits in Canada.

Table 20.2-2. Tonnages and grades of significant Canadian and foreign skarn copper deposits.

Deposit, location	Size (million tonnes)	Average Grade %Cu	Average Grade g/t Au	Average Grade g/t Ag	Average Grade other	Comments P = production R = reserves
Canadian deposits						
1. Ingerbelle (Copper Mountain), Princeton, British Columbia	240 (skarn 216)	0.40	0.16	0.63		about 90% is skarn P (1972-1988) + R (Similco, Mines Ltd., internal publication, 1989)
2. Stikine Copper, Galore Creek, British Columbia	113 (skarn 90)	1.06	0.4	7.7		about 80% is skarn R (Allen et al., 1976)
3. Craigmont, Merritt, British Columbia	35	1.13	tr.	tr.	19.6% Fe 0.12% Mo	P (1961-1982)+ R (BC MINFILE, 1988; Morrison, 1980)
4. Phoenix, Greenwood, British Columbia	34	0.8	1.1	15	Fe, Pb, Zn	P (1893-1976) + R (Church, 1986)
5. Mount Polley, British Columbia (Cariboo Bell)	48 (skarn 10)	0.38	0.55	4.5	Fe	about 20% is skarn. R (Imperial Metals Corp., internal publication, 1991)
6. Whitehorse Copper Belt, Yukon Territory	10.1	1.4	1.0	8.8	Mo	P (1898-1982) + R (Watson, 1984; Meinert, 1986)
7. Prescott, Marble Bay, etc., Texada Island, British Columbia	21.2	0.2	0.1	1.9	Fe	P (1885-1976) + R (Ettlinger and Ray, 1988)
8. Old Sport, Coast Copper, Benson Lake, British Columbia	2.7	1.5	1.4	4.0	33% Fe	P (1962-1971) + R (Meinert, 1986)
9. Blue Grouse, Cowichan Lake, British Columbia	0.25	3.0	tr.*	11	Fe, Zn	P (1917-1960), Fyles (1955), Ray and Webster (1991)
10. Gaspé Copper (Mines Gaspé), Murdochville, Quebec	67	1.45	--	10 (E zone)	Mo, Zn, W, Bi	A,B,C,E, skarn zones) P+R (Allcock, 1982; Wares and Williams-Jones, 1993)
Foreign deposits						
11. Carr Fork, Utah, U.S.A.	400 (skarn 80)	2.2	0.6	12		P+R; 20% is skarn (Cameron and Garmoe, 1987)
12. Twin Buttes, Arizona, U.S.A.	400 (skarn 320)	0.7	--	6.0	0.03% Mo	P+R; 80% is skarn (Barter and Kelly, 1982)
13. Santa Rita, New Mexico, U.S.A.	350 (skarn 70)	0.9	0.2	--		P+R; 20% is skarn (Nielson, 1970)
14. Ely District, Nevada, U.S.A.	339 (skarn 68)	0.7	0.3	0.9		P+R; 20% is skarn (James, 1976)
15. Ertsberg, Irian Jaya, Indonesia	52.6	2.4	0.8	8.0		P+R (MacDonald and Arnold, 1993; Meinert, 1987)
16. Gold Coast, Ok Tedi district, Papua-New Guinea	54	1.53	1.6	--	Mo	R; Gold Coast skarn (Hewitt et al., 1980)
17. Larap, Philippines	17	0.42	0.3	--	0.09% Mo 22% Fe W, U, Bi	P+R (Sillitoe and Gappe, 1984)
18. Rosita, Honduras	4	1.6	0.9	6.0	Fe	P+R (Bevan, 1973)
19. Concepcion del Oro, Mexico	2	2.0	1.6	--	Fe, Zn, Pb	P+R (Buseck, 1966)
20. Memé, Haiti	1.2	2.5	--	--	Fe, Mo	P+R (Kesler, 1968)
21. Yaguki, Japan	1.2	0.8	3.0	156	Fe, W, Co, Bi	P+R (Shimazaki, 1969)
22. Red Dome, Australia	13.8	0.46	2.0	4.6	1% Zn	P+R (Torrey et al., 1986)
23. Morococha, Peru	>10	2.4	--	123.5	1.5% Zn 0.5% Pb 20% Fe	P+R (estimated) (Petersen, 1965)
24. Mission Arizona, U.S.A.	400 (skarn 320)	0.8	--	9.6	0.019% Mo 0.024% Pb	P+R; 80% is skarn (Einaudi et al., 1981)
25. Christmas Arizona, U.S.A.	100	0.7	--	--		P+R; 100% skarn (Perry, 1968)
26. El Tiro (Silver Bell), Arizona, U.S.A.	50 (skarn 7.5)	0.8	--	--	0.8% Zn	P+R; 15% is skarn (Graybeal, 1982)
27. Kasaan Peninsula, Alaska, U.S.A.	4.3	2	--	440	50% Fe	P+R (Green et al., 1989)
28. Zackly, Alaska, U.S.A.	1.25	2.7	6.0	30		P+R (Nokleberg et al., 1988)

*tr. = trace

Porphyry deposit-related copper skarns in the Canadian Cordillera differ from the deposits cited above in several ways: their accreted oceanic-volcanic arc setting, i.e. Quesnellia and Stikinia, is in contrast to craton marginal settings in southwestern United States, Mexico, and Russia; limestone host rocks are scarce, resulting in less abundant calc-silicate gangue; and they are associated with alkaline, mainly monzonitic intrusions, rather than intermediate to felsic calc-alkaline plutons (Table 20.2-1). Other porphyry deposit-related copper skarns are found in similar oceanic-island arc settings in the Philippines, Papua New Guinea, Honduras, and Haiti (Table 20.2-2). Craigmont mine at Merritt, British Columbia is a copper-magnetite skarn associated with the border phase of the calc-alkaline Guichon Creek batholith, whose younger intrusive phases host the large Highland Valley (Valley Copper, Lornex) porphyry district. Although the border phase is unmineralized, Craigmont is classified in this paper as a porphyry-associated skarn copper deposit.

Copper skarns not associated with porphyry copper deposits occur typically in oceanic-island arc settings, within interstratified limestone-volcanic flow-volcaniclastic sequences (e.g., Phoenix, Greenwood district), but also occur in continental marginal carbonate strata (e.g., Concepcion del Oro, Zacatecas, Mexico). Associated intrusions are generally intermediate to mafic in composition. The skarn mineral assemblages tend to be iron-rich and to share characteristics, including accreted arc and rifted continental marginal settings, with calcic iron skarns (Einaudi et al., 1981).

Age of host rocks and mineralization

Host rocks in which skarn copper deposits have been preferentially formed include pure and impure limestone, dolostone, calcareous sedimentary rocks, thermally metamorphosed equivalents of these lithologies, and also calcic metavolcanic and meta-intrusive rocks. Host rocks range from Precambrian to Cenozoic in age, and are predominantly Phanerozoic. The host rocks of North American skarn copper deposits are predominantly Paleozoic cratonic carbonate strata in the western United States and the Appalachian Orogen, Late Triassic oceanic-island arc volcano-sedimentary sequences in the northern Cordillera, and Cretaceous shelf carbonate strata in Mexico and Central America. Skarn host rocks are dominantly Late Paleozoic to Tertiary oceanic-arc assemblages in Japan, Philippines, and Papua New Guinea, and Paleozoic craton marginal beds in Australia and Russia.

Age of skarn mineralization is mainly Mesozoic or younger, and penecontemporaneous with the associated intrusion. Mineralization age is dominantly early Tertiary in the western United States, Mexico, and Central and South America; Early to mid-Jurassic in British Columbia; Late Jurassic and Cretaceous in Alaska, Yukon Territory, and China; Paleozoic in the Appalachians, Australia, and Russia; and Tertiary in Japan, Philippines, and Papua New Guinea.

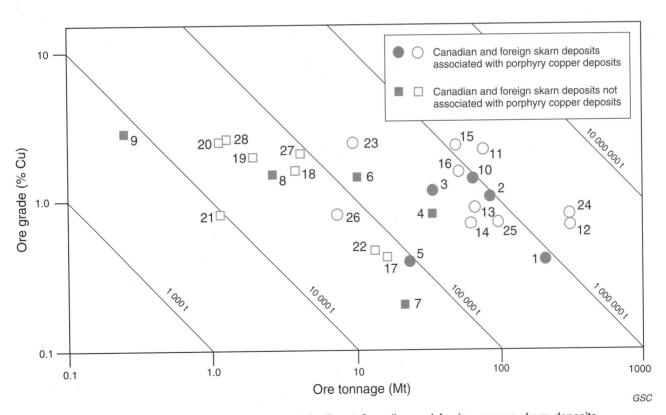

GSC

Figure 20.2-2. Grade versus tonnage of significant Canadian and foreign copper skarn deposits. Numbers correspond to deposits listed in Table 20.2-1.

Form of deposit and associated structures, zoning, and distribution of ore minerals

The morphology of calc-alkaline porphyry copper-related skarn copper deposits reflects the relatively high level of emplacement of associated felsic porphyry stocks, with resultant less extensive development and lower metamorphic grade of thermal aureoles compared with those surrounding deeper seated batholiths with associated tungsten-rich skarns. Multiple intrusive events contributed to the characteristically intense fracturing, brecciation, and breccia pipe formation. The relatively great extent and thickness of the skarns reflects the high fracture porosity and low pressure of the shallow porphyry system. Skarn has been formed close to intrusive contacts as irregular, tabular, vein-like to peneconcordant contact skarns or exoskarns with patchy, massive, and disseminated ore mineral assemblages.

Alteration in calc-alkaline porphyry stocks that evolved from early potassium silicate to late sericite assemblages parallels skarn alteration in carbonate wall rocks: in limestone an oxidized prograde assemblage of ferric iron-rich andradite garnet and ferrous iron-poor clinopyroxene is developed; pyroxene hornfels is altered to actinolite and biotite along sulphide-magnetite veinlets; pyrite-chalcopyrite-magnetite near plutonic contacts passes gradationally to bornite-chalcopyrite-wollastonite near marble contacts, reflecting the trend with time, and towards the periphery of the system, to a decrease in total iron and an increase in oxidation-sulphidation states (Burt, 1972; Einaudi et al., 1981). Sphalerite, pyrrhotite, tennantite, and galena are enriched in peripheral parts of the skarn system. Voluminous hydrous silicate-carbonate alteration of skarn and silica-pyrite alteration of limestone develops contemporaneously with late sericite-clay alteration of the stock. Retrograde skarn assemblages include actinolite, carbonates, clay, silica, iron oxides, and sulphides, and lesser amounts of chlorite, epidote, and talc. Low vein temperatures (<350°C, Roedder, 1971), influx of oxidized groundwater (Sheppard and Taylor, 1974), and peripheral deposition of base and precious metals represent the final stages of a long-lived sulphur-rich hydrothermal system. Zonation of sulphide and calc-silicate skarn assemblages in a large porphyry copper skarn system is exemplified by the Carr Fork deposit, Bingham mining district, Utah (Fig. 20.2-3A, B).

The Mines Gaspé copper deposits at Murdochville, Quebec are hosted by a folded, carbonate-rich Lower Paleozoic sequence of the Gaspé-Connecticut Valley synclinorium and are associated with porphyritic epizonal intrusions of a more deeply seated Devonian plutonic complex (Procyshyn et al., 1989). Several skarn and stockwork vein deposits occur within a broad aureole of calc-silicate hornfels enclosing the Copper Mountain porphyritic biotite granodiorite plug and the Porphyry Mountain diatreme breccia pipe and dyke-sill swarm (Wares and Williams-Jones, 1993). A map of the Mines Gaspé area and a cross-section are given in Figures 20.2-4A and 20.2-4B.

Thermal metamorphism preceded development of skarn and stockwork ores of four types: 1) the most economically significant deposits are the relatively early massive to disseminated pyrrhotite-chalcopyrite-cubanite replacements of discrete limestone beds enveloped, in part, by actinolite-rich skarn assemblages, that include the E-zone deposits with aggregate mining reserves of 6.7 Mt grading 2.8% Cu and 10 g/t Ag (Wares and Williams-Jones, 1993); 2) tabular skarn orebodies of disseminated to semimassive sulphides that replaced limestone; deposits of this type in the B and C horizons at Needle Mountain contained a premining reserve of 47 Mt grading 1.48% Cu (Wares and Williams-Jones, 1993). The skarn assemblage of andradite, salite, quartz, actinolite, calcite, epidote, and K-feldspar replaced earlier-formed metamorphic minerals (Allcock, 1982). Garnet/pyroxene ratios and Fe-contents of garnets decrease away from the Copper Mountain and Porphyry Mountain plutons (Murphy, 1986), supporting an intrusive source of metasomatic fluids; 3) coevally with skarn, porphyry-style disseminated and stockwork Cu- and Mo-bearing veinlets were deposited in fractured potassic calc-silicate hornfels, adjacent to the Copper Mountain plug (i.e., Copper Mountain zone) and in the Needle Mountain "A" zone; these two zones combined total 222 Mt grading 0.42% Cu and 0.02% Mo (Wares and Williams-Jones, 1993); and 4) retrograde sulphide-quartz-actinolite-fluorite-anhydrite veins formed in potassic hornfels as the final stage of mineralization in the Copper Mountain zone.

Skarn copper deposits not associated with porphyry copper deposits are distinguished from porphyry copper-related skarns mainly by lack of stockwork and disseminated Cu and Mo sulphides in the intrusion and other rocks, and also by smaller size, more massive nature, and higher Cu grade. They are associated with calcic iron skarns with which they have several features in common, including tectonic setting, composition of intrusions, skarn morphology, and mineralogy. Prograde skarn minerals include garnet intermediate in composition between grossularite and andradite, diopsidic pyroxene, wollastonite, magnetite, and epidote. Examples of coexisting calcic copper and iron skarns are found in the Texada Island, Benson Lake, and Greenwood districts of British Columbia (Tables 20.2-1 and 20.2-2). This type of skarn copper deposit lacks the intense stockwork fracturing typical of porphyry deposits, resulting in the restriction of retrograde mineral assemblages to vug fillings, contact zones, and widely spaced, structurally controlled fluid conduits. Retrograde skarn assemblages commonly include actinolite, tremolite, epidote, chlorite, quartz, and calcite; significant concentrations of base metals and precious metals are associated with the retrograde skarn assemblages.

In the Whitehorse Copper Belt, Yukon Territory, thirty-two Cu, Fe (Mo, Au, Ag) skarn deposits occur in both calcareous and dolomitic units of the upper Triassic Lewes River Group, a back-arc succession of arkosic clastic and carbonate rocks. Skarns are localized mainly along the irregular western contact of the Whitehorse batholith (Fig. 20.2-5A), a composite calc-alkaline granodioritic pluton with a dioritic margin (Morrison, 1981) that postorogenically intruded accreted Stikinia in mid-Cretaceous time (Dawson et al., 1991). The silicate mineralogy of individual skarn deposits is largely a function of the skarn protolith (Morrison, 1981). Skarns formed from a limestone protolith (e.g., War Eagle, Fig. 20.2-5B) contain abundant andraditic garnet, less abundant iron-rich (up to Hd_{37}; Meinert, 1986) pyroxene, wollastonite, and vesuvianite, and variable amounts of the retrograde alteration products actinolite, epidote, and chlorite. Skarns with a dolomitic

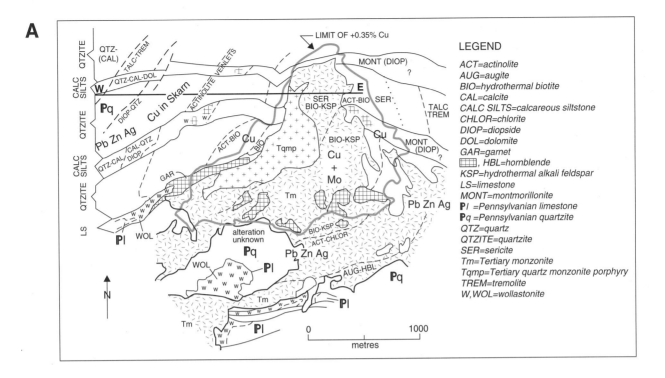

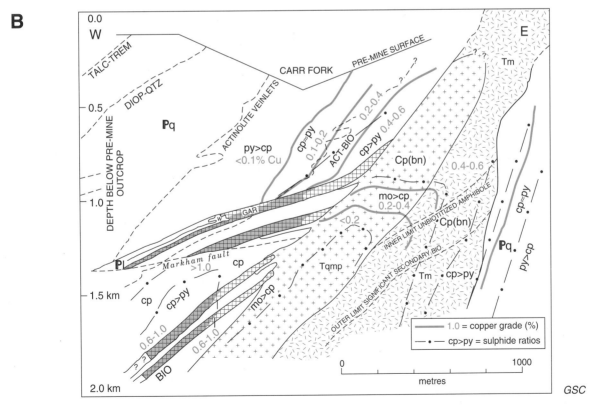

Figure 20.2-3. Mineral zoning and alteration in the Bingham mining district, Utah, U.S.A. After Atkinson and Einaudi (1978) and Einaudi (1982, Fig. 7.14, 7.15). **A)** Alteration in igneous and sedimentary rocks at surface, and metal zonation relative to central quartz monzonite porphyry stock. **B)** Cross-section W-E (looking north) of the Carr Fork skarn copper deposit, western contact zone of the Bingham stock, illustrating alteration, copper grades, and sulphide mineral ratios.

protolith (e.g., Arctic Chief, Fig. 20.2-5C) contain abundant prograde magnesian minerals such as diopside and for-steritic olivine, less abundant andradite, and retrograde phlogopite, brucite, serpentine, and talc. The sulphide minerals are associated mainly with retrograde alteration assemblages: chalcopyrite and pyrite preferentially with actinolite and chlorite; and bornite and chalcocite with epidote. The rare micaceous copper mineral valleriite (4(Fe,Cu)S·3(Mg,Al)(OH)$_2$) is restricted to retrograde alteration

assemblages in magnesian, i.e. dolomitic, rocks. The highest average Au and Ag grades occur where massive sulphide and retrograde alteration assemblages coexist in skarn (Meinert, 1986).

In British Columbia, some porphyritic members of the distinctive Copper Mountain alkaline plutonic suite, which are associated with porphyry Cu-Au deposits, also developed a distinctive type of skarn copper deposit. Synvolcanic syenite, monzonite, and diorite porphyry plutons intruded

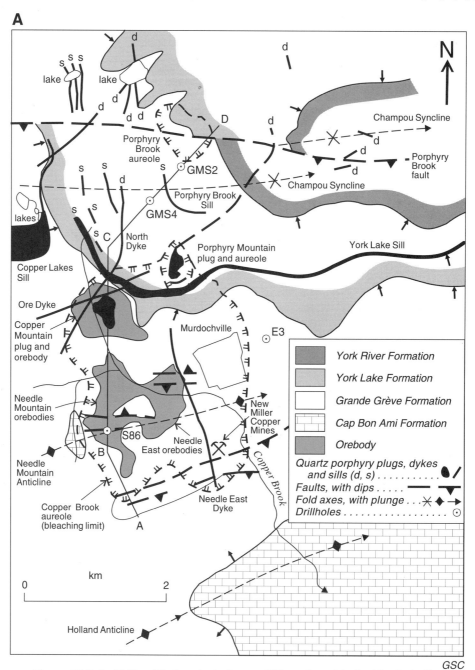

Figure 20.2-4. A) Simplified geological map of Mines Gaspé in the Murdochville, Quebec area, after Allcock (1982, Fig. 2). The 150 m contour of the Copper Mountain plug is included to show its shape. Section ABCD is shown in Figure 4B. **B)** Section along line ABCD in Figure 4A, after Allcock (1982, Fig. 3). Deep drillholes are numbered. Distributions of minerals are given.

andesitic volcanic rocks, calcareous volcaniclastic rocks, and minor carbonate strata of the Quesnellia and Stikinia terranes prior to their accretion to North America in middle Jurassic time (Dawson et al., 1991). Similarities between the alkaline Copper Mountain plutonic suite and their host volcanic rocks of the eastern Nicola Assemblage suggest that they are subvolcanic equivalents (Woodsworth et al., 1991). The genetic relationship of alkaline to coeval calc-alkaline plutonic suites, and the relationship of both to the Early Jurassic tectonic regime, is not well understood. Skarn ore, which constitutes 20% (Mount Polley) to 80% (Galore Creek) of total reserves (Table 20.2-2), consists mainly of porphyry-style disseminations, vein stockworks, and breccia fillings of chalcopyrite, bornite, pyrite, and magnetite plus massive sulphide-magnetite replacements of calcareous volcanic and sedimentary units.

At the Ingerbelle mine, Princeton, British Columbia, Sutherland Brown et al. (1971), Preto (1972), Macauley (1973), and Fahrni et al. (1976) recognized skarn-like ore and gangue mineral zonation relative to the contact of the composite Lost Horse diorite-monzonite-syenite stock with agglomerate, tuff, tuff breccia, and sedimentary rocks of the Upper Triassic Nicola Group. Early biotite hornfels was overprinted by prograde albite-epidote-chlorite±andradite±diopsidic pyroxene±sphene; both stock and prograde skarn have been flooded by retrograde albite, K-feldspar, scapolite, calcite, and hematite. Chalcopyrite-bornite ore, about 90% of which is located in andesitic volcanic rocks, follows contacts, apophyses, and dykes of the Lost Horse intrusion. Elevated contents of Pt (about 0.15 ppm) and Pd (about 3.0 ppm) are present in sulphide concentrates from the adjacent Copper Mountain porphyry Cu-Au deposit (L.J. Hulbert, pers. comm., 1991).

At Galore Creek in northwestern British Columbia, prograde skarn copper ore, which was developed in calcareous pyroclastic, volcaniclastic, and shoshonitic (high K) volcanic rocks adjacent to contacts of pseudoleucite-phyric syenite dykes, constitutes about 80% of total ore reserves (Allen et al., 1976). Pervasive alteration assemblages of orthoclase and biotite in, or adjacent to, potassic rocks give way to calcic assemblages of zoned, anisotropic andradite, diopside, Fe-rich epidote, and vesuvianite plus chalcopyrite, bornite, pyrite, and magnetite and minor amounts of chalcocite, sphalerite, and galena, in calcareous host rocks (Fig. 20.2-6A-C). Retrograde assemblages include anhydrite, chlorite, sericite, calcite, gypsum, and fluorite.

About 20% of ore reserves at Mount Polley, in the Cariboo region of south-central British Columbia, are in copper skarn (Table 20.2-2) that developed in calcareous crystal and lapilli tuff of the Upper Triassic Takla Group adjacent to diorite, monzodiorite, and monzonite porphyry intrusions (Hodgson et al., 1976). Most of the ore is contained within several types of hydrothermal breccias (Fraser, 1994). Potassic alteration assemblages of orthoclase and biotite in, and adjacent to, the intrusions grade outward from the contacts of these intrusions and breccia contacts to prograde skarn assemblages of diopside-andradite-magnetite and peripheral propylitic assemblages of epidote-pyrite-albite. The bulk of the sulphide ore, i.e. chalcopyrite, bornite, and lesser amounts of pyrite, is associated with the retrograde assemblage carbonate, chlorite, zeolite, prehnite, epidote, and hematite.

Alkaline porphyry copper-related skarn deposits, like many calc-alkaline ones, are associated with small stocks emplaced at high levels in comagmatic volcanic sequences, with resultant intense fracturing, brecciation, and breccia pipe formation. Like calcic magnetite skarns, they are characterized by epidote-diopside-garnet endoskarn

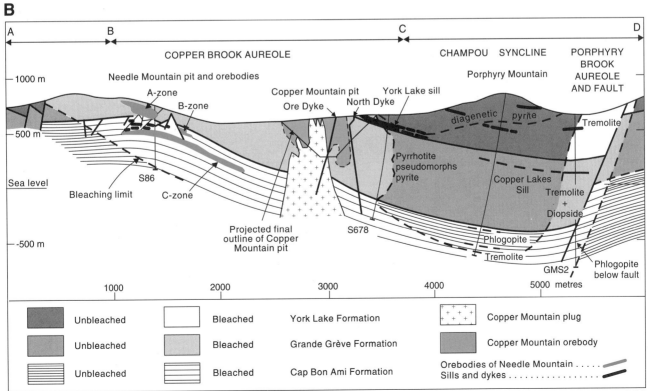

GSC

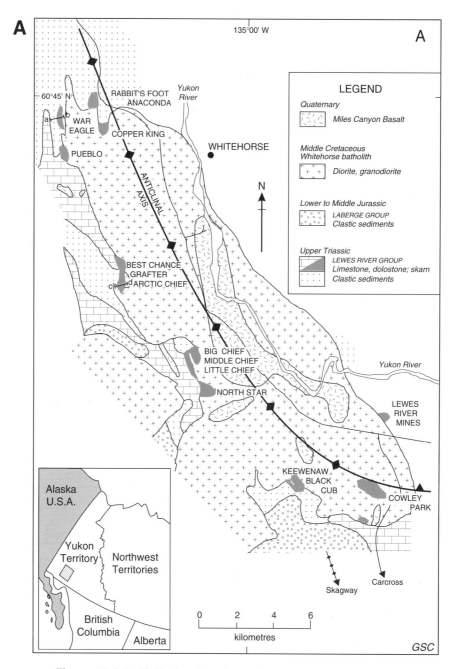

Figure 20.2-5. A) Regional geology of the Whitehorse Copper Belt; vertical cross-sections a-b and c-d are shown in Figures 20.2-5B and 20.2-5C. After Kindle (1964) and Tenney (1981). **B)** Geological cross-section (a-b) of the War Eagle calcic skarn copper deposit, looking north; cpy = chalcopyrite, bn = bornite, mt = magnetite, mo = molybdenite. After Morrison (1981) and Watson (1984). **C)** Geological cross-section (c-d) of the Arctic Chief magnesian skarn copper deposit, looking north. After Morrison (1981) and Watson (1984).

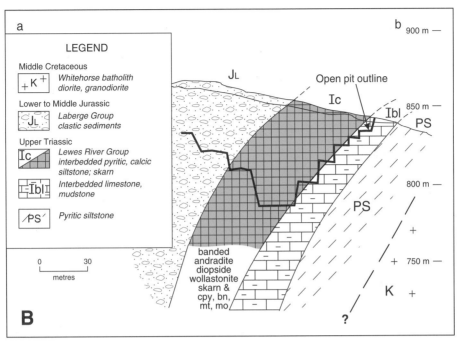

a

LEGEND

Middle Cretaceous

+K+ *Whitehorse batholith diorite, granodiorite*

Lower to Middle Jurassic

JL *Laberge Group clastic sediments*

Upper Triassic

Ic *Lewes River Group interbedded pyritic, calcic siltstone; skarn*

Ibl *Interbedded limestone, mudstone*

PS *Pyritic siltstone*

0 ____ 30
metres

B

b — 900 m

JL

Open pit outline

Ic

Ibl

PS

— 850 m

— 800 m

PS

K

— 750 m

banded andradite diopside wollastonite skarn & cpy, bn, mt, mo

?

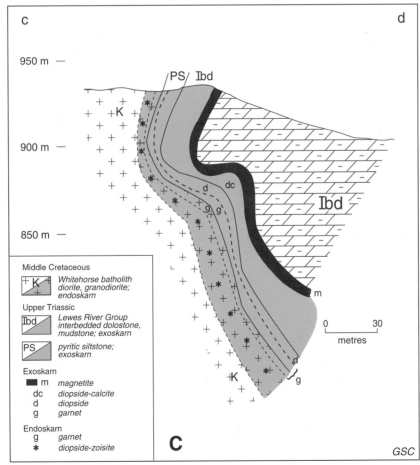

c   d

950 m —

PS / Ibd

+K+

900 m —

Ibd

850 m —

d   dc

g g

m

0 ____ 30
metres

d

g

+K+

Middle Cretaceous

+K+ *Whitehorse batholith diorite, granodiorite; endoskarn*

Upper Triassic

Ibd *Lewes River Group interbedded dolostone, mudstone; exoskarn*

PS *pyritic siltstone; exoskarn*

Exoskarn

m *magnetite*
dc *diopside-calcite*
d *diopside*
g *garnet*

Endoskarn

g *garnet*
* *diopside-zoisite*

C

GSC

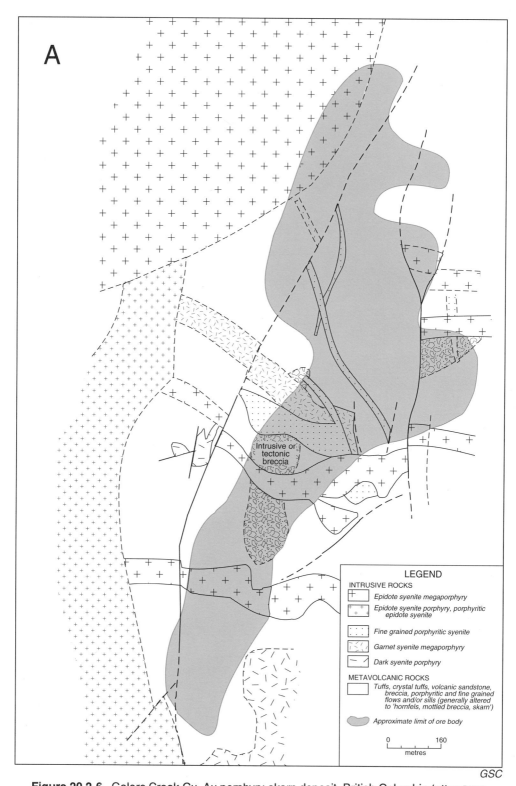

A

Intrusive or
tectonic
breccia

LEGEND

INTRUSIVE ROCKS

Epidote syenite megaporphyry

Epidote syenite porphyry, porphyritic
epidote syenite

Fine grained porphyritic syenite

Garnet syenite megaporphyry

Dark syenite porphyry

METAVOLCANIC ROCKS

Tuffs, crystal tuffs, volcanic sandstone,
breccia, porphyritic and fine grained
flows and/or sills (generally altered
to 'hornfels, mottled breccia, skarn')

Approximate limit of ore body

0   160
          metres

GSC

Figure 20.2-6. Galore Creek Cu, Au porphyry-skarn deposit, British Columbia (after Allen et al., 1976) **A)** generalized geology of central zone; **B)** sulphide and magnetite distribution, central zone, 610 m level. Disseminated magnetite occurs in a belt to the west and north, pyrite is mainly to the east; Bn = bornite. **C)** K-feldspar, garnet, and diopside distribution, 610 m level within the central zone. Zones rich in K-feldspar and biotite parallel the syenite dyke contacts; garnet is abundant toward the north end of the breccia pipe, and diminishes away from its contacts.

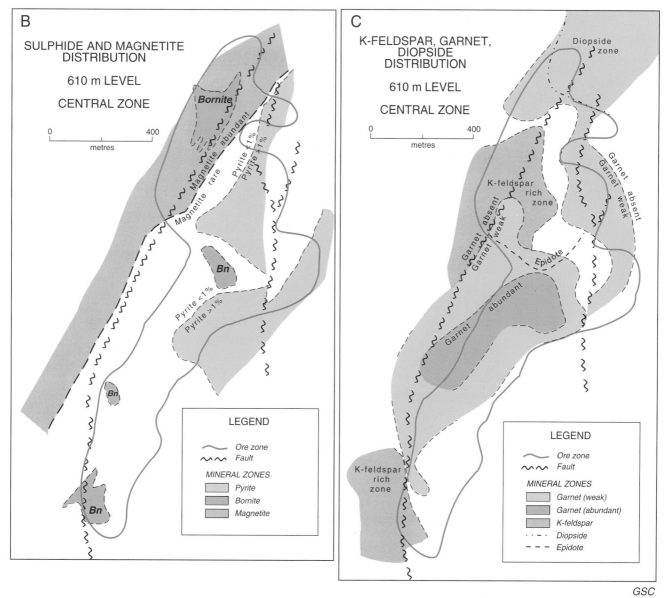

Table 20.2-3. Examples of Canadian and selected foreign skarn copper deposits, classified by plutonic association.

Pluton composition	Mineralization in pluton	
	Unmineralized intrusion related	**Porphyry-type mineralization**
Calc-alkaline	Greenwood (Phoenix) British Columbia	Gaspé Copper, Quebec
	Whitehorse Copper Belt, Yukon Territory	Craigmont, Merritt, British Columbia
	Old Sport, Coast Copper (Benson Lake) British Columbia	Ely, Nevada, U.S.A.
	Prescott, Marble Bay, Texada Island, British Columbia	Santa Rita, Arizona, U.S.A.
	Yaguki, Japan	
	Sayak 1, Kazakhstan	Ertsberg, Irian Jaya, Indonesia
	Conception del Oro, Mexico	Gold Coast (Ok Tedi), Papua New Guinea
	Rosita, Honduras	
	Memé, Haiti	
Alkaline		Galore Creek (Stikine Copper) British Columbia
		Ingerbelle (Copper Mountain), Princeton, British Columbia
		Larap, Philippines

within the intrusion, extensive prograde and retrograde potassic (K-feldspar, biotite) and sodic (albite, scapolite) metasomatic alteration, plus a ferric iron-rich, magnesium-poor bulk skarn composition. The most notable compositional characteristics are overall deficiency in silica, elevated contents of Au, Ag, and platinum group elements in the ore (Mutschler and Mooney, in press), and lack of peripheral Pb and Zn sulphides. Like their associated porphyry deposits, alkaline porphyry copper-related skarn deposits are readily distinguished from calc-alkaline porphyry-related skarn deposits by elevated Ag contents, i.e., mean of 4.15 g/t versus 1.66 g/t (Sinclair et al., 1982).

Mineralogy

Principal ore minerals: chalcopyrite, bornite, magnetite, molybdenite, electrum, gold, silver.

Subsidiary ore minerals: platinum group elements, sphalerite, galena, scheelite, Au and Ag tellurides, Ag sulphosalts, tetrahedrite, cobaltite, wittichenite, bismuth, bismuthinite.

Other opaque minerals: pyrite, pyrrhotite, arsenopyrite, marcasite, hematite.

DEFINITIVE CHARACTERISTICS

The definitive characteristics of the various skarn copper subtypes are summarized in Table 20.2-4.

GENETIC MODEL

Copper, molybdenum, associated metals, sulphur, and cations other than calcium were derived mainly from the associated pluton by the separation of an orthomagmatic hydrothermal fluid. All types of skarn copper deposits have the following three main evolutionary stages in common; these stages are similar to those for skarn deposits in general, which are represented in Figure 20.2-7:

Stage 1: Isochemical thermal metamorphism of limestone to marble, calcareous pelite to calc-silicate hornfels, and clastic and volcaniclastic rocks to biotite hornfels accompanies the intrusion of magma into upper levels of the crust (Fig. 20.2-7A).

Stage 2: The main stage of prograde skarn formation begins as the crystallizing magma releases hydrothermal fluid through conduits developed by hydrofracturing of the crystallized outer portion of the pluton and hornfels. Some mixing of this hydrothermal fluid with metamorphic fluids can occur, but the degree to which metals and sulphur, derived by the metamorphic dehydration of the host rock, contribute to the skarn assemblage is not known, but is probably minor. Fluids infiltrate the host rock along plutonic and dyke contacts, fractures, breccias, and lithological contacts, to react with either the calcareous host rock, marble, or calc-silicate hornfels to form prograde anhydrous skarn. Endoskarn forms in the pluton by introduction of calcium from the host rock. Calcic exoskarn forms in limestone and calcareous sedimentary rocks, and magnesian exoskarn forms in

dolostones. The mineral assemblage of the resultant skarn is influenced by the composition, temperature, and oxidation state of the system. Stage 2 anhydrous skarn is deficient in copper and sulphur, and later skarn stages are enriched in ferric iron and sulphur and depleted in magnesium. Deposition of most magnetite and sulphide minerals starts late in this stage (Fig. 20.2-7B).

Stage 3: The main period of sulphide deposition follows the termination of prograde anhydrous skarn development and accompanies the start of hydrous alteration of early skarn minerals and associated intrusive rocks. Subsequent cooling of the system and influx of meteoric water initiates hydrous retrograde alteration of early calcium-rich silicates to silicates more depleted in calcium, plus iron oxides, carbonates, and plagioclase. Processes occurring at this stage include hydrolytic alteration of rocks and deposition of most sulphides, including some redistribution of earlier-formed sulphides. The assemblage of opaque minerals associated with hydrous retrograde alteration reflects higher oxidation and sulphidation states and lower temperatures than do early prograde opaque minerals (Fig. 20.2-7C).

RELATED DEPOSIT TYPES

1. Skarn gold: Most types of skarn copper deposits are enriched in gold to some degree and, with increasing Au content, will grade into gold skarn deposits. Skarn copper proximal to a copper mineralized porphyry stock at Copper Canyon, Nevada passes laterally to Au-Ag skarns at the Lower and Upper Fortitude and Minnie-Tomboy deposits (Blake et al., 1984; Wotruba et al., 1988).

2. Calc-alkaline and alkaline porphyry copper deposits: Significant porphyry copper deposits with related Cu skarn deposits include Grasberg and Ertsberg, Irian Jaya, Indonesia (MacDonald and Arnold, 1993), and Mount Fubilan and Gold Coast, Ok Tedi district, Papua New Guinea (Hewitt et al., 1980). Several large skarn copper deposits associated with porphyry copper deposits in southwestern United States have been documented by Einaudi (1982).

3. Skarn iron: Copper skarns associated with unmineralized stocks unrelated to porphyry copper deposits share a common plutonic association and tectonic setting with, and may pass laterally into, calcic magnetite, magnetite-copper, and magnetite-gold skarns. Canadian examples include Phoenix, Whitehorse Copper Belt, Benson Lake, and Texada Island.

EXPLORATION GUIDES

Copper skarns associated with calc-alkaline porphyry copper deposits

Geological guides

a) Extensive skarn and hornfels zones adjacent to exposed copper-bearing porphyry stocks, or overlying buried ones.

b) Relatively thick, pure and impure limestone beds in a craton-marginal setting.

c) Stockwork fractures, breccias, and breccia pipes adjacent to porphyry copper stocks.

d) Potassic and sericitic alteration of stocks that contain disseminated Cu- and Mo- sulphides.

Geochemical guides

a) In general, geochemical prospecting techniques, i.e. media, elements analyzed for, sample spacing, etc. are similar to those employed in porphyry Cu exploration.

b) Hydromorphic dispersion from blind orebodies commonly causes proximal anomalies in Mo, Pb, W, and Au, and distal anomalies in Cu, Zn, Ag, As, and Sb.

Geophysical guides

a) Plutonic contacts and common magnetite association with skarn generate strong magnetic anomalies.

b) Relatively massive skarn sulphide mineralization peripheral to disseminated porphyry-style copper-iron sulphides generates an enhanced induced potential and electromagnetic response.

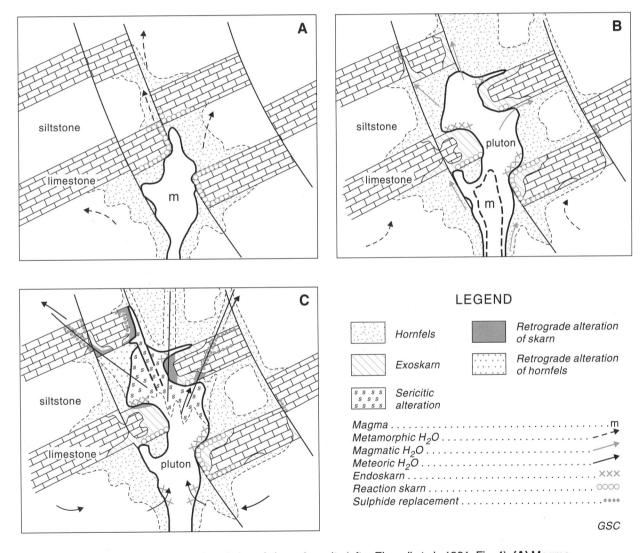

Figure 20.2-7. Stages of evolution of skarn deposits (after Einaudi et al., 1981, Fig. 4). **(A)** Magma emplacement; dehydration; thermal metamorphism; reaction skarn formation. **(B)** Crystallization and magmatic fluid separation; stratigraphically and structurally controlled exoskarn formation; endoskarn formation locally in pluton; peripheral replacement. **(C)** Cooling, meteoric water influx; retrograde alteration of pluton, hornfels and skarn; structurally and stratigraphically controlled sulphide-silica-carbonate replacement.

Table 20.2-4. Definitive characteristics of skarn copper deposits.

SKARN COPPER SUBTYPE	DEFINITIVE CHARACTERISTICS
Skarn Cu associated with calc-alkaline porphyry Cu-related intrusions	Disseminated Cu, Mo mineralization in altered intrusion; high fracture permeability; breccia pipes; thick and extensive skarn; low Au grade; late stage sericite-clay and silica-pyrite alteration; peripheral Zn, Pb mineralization.
Skarn Cu associated with unmineralized nonporphyry Cu-related intrusions	Lack of disseminated Cu, Mo mineralization in relatively small, mafic, unaltered intrusion; massive sulphide orebody relatively rich Cu, Au, and Ag; restricted zones of retrograde alteration.
Skarn Cu associated with alkaline porphyry Cu-Au stocks	Low Mo content in intrusion; low silica in total system; high Au, Ag, and PGE content of ore; sodic and potassic alteration of skarn and intrusion; intrusive breccias and breccia pipes; high magnetite content; limestone host rocks minor to absent

Copper skarns associated with unmineralized stocks unrelated to porphyry copper deposits

Geological guides

a) In the northern part of the North American Cordillera, all examples of this subtype, including the important districts of Greenwood, British Columbia and Whitehorse Copper Belt, Yukon Territory are hosted by Upper Triassic-Jurassic accreted oceanic-arc terranes.

b) Irregularities in pluton-limestone contacts, particularly re-entrants and troughs, tend to serve as structural control.

c) Highest copper and precious metal concentrations frequently occur within structurally controlled zones that have the most intense retrograde alteration.

Geochemical guides

a) Guides are the same as for calc-alkaline porphyry Cu-related skarns, with the addition of Co to the element suite analyzed for, particularly in the case of magnetite-rich Cu skarns.

Geophysical guides

a) The almost ubiquitous association of copper sulphides with magnetite yields strong magnetic anomalies.

b) Lack of disseminated or stockwork sulphides in associated stocks reduces induced potential response.

c) Massive morphology and high chalcopyrite content relative to that of porphyry copper-related skarns generates a strong electromagnetic response.

Skarn copper associated with alkaline porphyry copper deposits

Geological guides

a) Silica-deficient, magnetite-rich composite alkaline intrusions.

b) Limestone deficient, alkaline to subalkaline members of volcanic, volcaniclastic, and clastic sedimentary assemblages in accreted Upper Triassic to Lower Jurassic oceanic-arc terranes in the Canadian Cordillera.

c) Magnetite breccias, breccia pipes, and dykes, with and without associated copper sulphides.

d) Extensive potassic and sodic alteration of the pluton, skarn, and hornfels; sericitic alteration weak to absent; distal Zn and Pb mineralization weak to absent.

Geochemical guides

a) Guides the same as for calc-alkaline porphyry Cu-related skarns, except for a reduced Mo anomaly over the stock and skarn.

Geophysical guides

a) Disseminated magnetite-rich mineral assemblage in the stock demonstrates higher magnetic susceptibility than that of calc-alkaline stocks.

SELECTED BIBLIOGRAPHY

References marked with asterisks (*) are considered to be the best sources of general information on this deposit type.

Allcock, J.B.
1982: Skarn and porphyry copper mineralization at Mines Gaspé, Murdochville, Quebec; Economic Geology, v. 77, p. 971-999.

Allen, D.G., Panteleyev, A., and Armstrong, A.T.
1976: Galore Creek; in Porphyry Deposits of the Canadian Cordillera, (ed.) A. Sutherland Brown; The Canadian Institute of Mining and Metallurgy, Special Volume 15, p. 402-414.

Atkinson, W.W., Jr. and Einaudi, M.T.
1978: Skarn formation and mineralization in the contact aureole at Carr Fork, Bingham, Utah; Economic Geology, v. 73, p. 1326-1365.

Barter, C.F. and Kelly, J.L.
1982: Geology of Twin Buttes mineral deposit, Pima mining district, Pima County, Arizona; Chapter 20 in Advances in Geology of the Porphyry Copper Deposits, Southwestern North America, (ed.) S.R. Titley; University of Arizona Press, Tucson, Arizona, p. 407-432.

Bevan, P.A.
1973: The Rosita mine - a brief history and geological description; The Canadian Institute of Mining and Metallurgy, Bulletin, v. 66, no. 736, p. 80-84.

Blake, D.W., Wotruba, P.R., and Theodore, T.G.
1984: Skarn environment at the Tomboy-Minnie gold deposits, Lander County, Nevada; Arizona Society Digest, v. 15, p. 67-71.

Burt, D.M.
1972: Mineralogy and geochemistry of Ca-Fe-Si skarn deposits; PhD. thesis, Harvard University, Cambridge, Massachusetts, 256 p.

Buseck, P.R.
1966: Contact metasomatism and ore deposition: Concepcion del Oro, Mexico; Economic Geology, v. 61, p. 97-136.

Cameron, D.E. and Garmoe, W.J.
1987: Geology of skarn and high-grade gold in the Carr Fork mine, Utah; Economic Geology, v. 82, p. 1219-1333.

Church, B.N.
1986: Geological setting and mineralization in the Mount Attwood-Phoenix area of Greenwood mining camp; British Columbia Ministry of Energy, Mines and Petroleum Resources, Paper 1986-2, 65 p.

Dawson, K.M., Panteleyev, A., Sutherland Brown, A., and Woodsworth, G.J.
1991: Regional Metallogeny, Chapter 19 in Geology of the Cordilleran Orogen in Canada, (ed.) H. Gabrielse and C.J. Yorath; Geological Survey of Canada, Geology of Canada, no. 4, p. 707-768 (also Geological Society of America, The Geology of North America, v. G-2).

*Eckstrand, O.R. (ed.)
1984: Canadian mineral deposit types: a geological synopsis; Geological Survey of Canada, Economic Geology Report 36, 86 p.

*Einaudi, M.T.
1982: Description of skarns associated with porphyry copper plutons; Chapter 7 in Advances in Geology of the Porphyry Copper Deposits, Southwestern North America, (ed.) S.R. Titley; University of Arizona Press, Tucson, Arizona, p. 139-184.

*Einaudi, M.T., Meinert, L.R., and Newberry, R.J.
1981: Skarn deposits; in Seventy-fifth Anniversary Volume, 1905-1980, (ed.) D.J. Skinner; Economic Geology, p. 317-391.

Ettlinger, A.D. and Ray, G.E.
1988: Gold-enriched skarn deposits of British Columbia; in Geological Fieldwork, 1987, British Columbia Ministry of Energy, Mines and Petroleum Resources, Paper 1988-1, p. 263-279.

Fahrni, K.C., Macauley, T.N., and Preto, V.A.G.
1976: Copper Mountain and Ingerbelle; in Porphyry Deposits of the Canadian Cordillera, (ed.) A. Sutherland Brown; The Canadian Institute of Mining and Metallurgy, Special Volume 15, p. 368-375.

Fraser, T.M.
1994: Hydrothermal breccias and associated alteration of the Mount Polley copper-gold deposit (93A/12); in Geological Fieldwork, 1993, (ed.) B. Grant and E. Newell; British Columbia Geological Survey Branch, Paper 1994-1, p. 259-267.

Fyles, J.T.
1955: Geology of the Cowichan Lake area, Vancouver Island, British Columbia; British Columbia Department of Mines, Bulletin, no. 37, p. 54-57.

Graybeal, F.T.
1982: Geology of the El Tiro ore deposit, Silver Bell mining district, Arizona; in Advances in Geology of the Porphyry Copper Deposits, Southwestern North America, (ed.) S.R. Titley; University of Arizona Press, Tucson, Arizona, p. 487-505.

Green, C.B., Bundtzen, T.K., Peterson, R.J., Seward, A.F., Deagen, J.R., and Burton, J.E.
1989: Alaska's mineral industry, 1988; Alaska Division of Geological and Geophysical Surveys, Special Report 43, p. 67.

Hewitt, W.V., Leekie, J.F., Lorraway, R., Moule, K., Rush, P.M., Seegers, H.J., and Tarr, G.L.
1980: Ok Tedi project - geological and principal ore deposits; BHP Technical Bulletin, v. 24, no. 1, p. 3-10.

Hodgson, C.J., Bailes, R.J., and Verzosa, R.S.
1976: Cariboo-Bell; in Porphyry Deposits of the Canadian Cordillera, (ed.) A. Sutherland Brown; The Canadian Institute of Mining and Metallurgy, Special Volume 15, p. 388-396.

James, L.B.
1976: Zoned alteration in limestone at porphyry copper deposits, Ely, Nevada; Economic Geology, v. 71, p. 488-512.

Kesler, S.E.
1968: Contact-localized ore formation at the Memé mine, Haiti; Economic Geology, v. 63, p. 541-552.

Kindle, E.D.
1964: Copper and iron resources, Whitehorse Copper Belt, Yukon Territory; Geological Survey of Canada, Paper 63-41, 46 p.

Macauley, T.N.
1973: Geology of the Ingerbelle and Copper Mountain deposits at Princeton, B.C.; The Canadian Institute of Mining and Metallurgy, Bulletin, v. 66, no. 732, p. 105-112.

MacDonald, G.D. and Arnold, L.C.
1993: Intrusive and mineralization history of the Grasberg deposit, Irian Jaya, Indonesia; Society for Mining, Metallurgy and Exploration Inc., Annual Meeting, Reno, Nevada, preprint no. 93-92, 10 p.

Meinert, L.D.
1986: Gold in skarns of the Whitehorse Copper Belt, southern Yukon; in Yukon Geology, v. 1, Exploration and Geological Services, Division, Yukon, Indian and Northern Affairs Canada, p. 19-43.
1987: Gold in skarn deposits - a preliminary overview; Proceedings of the 7th Quadrennial International Association on the Genesis of Ore Deposits Symposium, E. Schweizerbart'sche Verlagsbuchhandlung, Stuttgart, p. 363-374.

Morrison, G.W.
1980: Stratigraphic control of Cu-Fe skarn ore distribution and genesis at Craigmont, British Columbia; The Canadian Institute of Mining and Metallurgy Bulletin, v. 73, no. 820, p. 109-123.
1981: Setting and origin of skarn deposits in the Whitehorse, Copper Belt, Yukon; PhD. thesis, University of Western Ontario, London, Ontario, 306 p.

Murphy, R.K.
1986: Geochemistry of garnets and pyroxenes from the skarn ore bodies, Mines Gaspé, Quebec; MSc. thesis, University of Missouri, Columbia, Missouri, 191 p.

Mutschler, F.E. and Mooney, T.C.
1993: Precious metal deposits related to alkaline igneous rocks-provisional classification, grade-tonnage data and exploration frontiers; in Mineral Deposit Modelling, (ed.) R.V. Kirkham, W.D. Sinclair, R.I. Thorpe, and J.M. Duke; Geological Association of Canada, Special Paper 40, p. 479-520.

Nielson, R.L.
1970: Mineralization and alteration in calcareous rocks near the Santa Rita stock, New Mexico; New Mexico Geological Society Guidebook, 21st Field Conference, p. 133-139.

Nokleberg, W.J., Bundtzen, T.K., Berg, H.C., Brew, D.A., Grybeck, D., Robinson, M.S., Smith, T.E., and Yeend, W.
1988: Metallogeny and major mineral deposits of Alaska; United States Geological Survey, Open File Report 88-73, 87 p.

Perry, D.V.
1968: Genesis of the contact rocks at the Christmas mine, Gila County, Arizona; PhD. thesis, University of Arizona, Tucson, Arizona, 229 p.

Petersen, U.
1965: Regional geology and major ore deposits of Peru; Economic Geology, v. 60, no. 3, p. 407-476.

Preto, V.A.
1972: Geology of Copper Mountain; British Columbia Department of Mines and Petroleum Resources, Bulletin 59, 87 p.

Procyshyn, E.L., Bernard, P., and Duquette, G.
1989: The copper ore deposits at Mines Gaspé, Murdochville, Quebec; in Mineral Deposits Associated with Siluro-Devonian Plutonism in the New Brunswick-Gape Sector, (ed.) E.L. Procyshyn, W.D. Sinclair, and A.E Williams-Jones; Geological Association of Canada-Mineralogical Association of Canada Joint Annual Meeting, Montreal '89, Excursion Guidebook B5, p. 80-104.

Ray, G.E. and Webster, C.L.
1991: An overview of skarn deposits; in Ore Deposits, Tectonics and Metallogeny in the Canadian Cordillera; British Columbia Ministry of Energy, Mines and Petroleum Resources, Paper 1991-4, p. 213-252.

Roedder, E.
1971: Fluid inclusion studies on the porphyry-type ore deposits of Bingham, Utah, Butte, Montana and Climax, Colorado; Economic Geology, v. 66, p. 98-120.

Sheppard, S.M.F. and Taylor, H.P.
1974: Hydrogen and oxygen isotope evidence for the origins of water in the Boulder batholith and the Butte ore deposits, Montana; Economic Geology, v. 69, p. 926-946.

Shimazaki, H.
1969: Pyrometasomatic copper and iron ore deposits of the Yaguki mine, Fukushima Prefecture, Japan; University of Tokyo Faculty Section Journal, sec. 2, v. 17, p. 317-350.

Sillitoe, R.H. and Gappe, I.M., Jr.
1984: Philippine porphyry copper deposits: geologic setting and characteristics; United Nations Economic and Social Commission for Asia and the Pacific, Committee for Co-ordination of Joint Prospecting for Mineral Resources in Asian Offshore Areas, Technical Publication 14, 89 p.

Sinclair, A.J., Drummond, A.D., Carter, N.C., and Dawson, K.M.
1982: A preliminary analysis of gold and silver grades of porphyry-type deposits in western Canada; in Precious Metals in the Northern Cordillera, (ed.) A.A. Levinson; Symposium Proceedings, Association of Exploration Geochemists, Vancouver, April 1981, p. 157-182.

Sutherland Brown, A., Cathro, R.J., Panteleyev, A., and Ney, C.S.
1971: Metallogeny of the Canadian Cordillera; in The Canadian Institute of Mining and Metallurgy, Bulletin, v. 64, no. 709, p. 37-61, and Transactions, v. 74, p. 121-145.
Tenney, D.
1981: The Whitehorse Copper Belt: mining, exploration and geology (1967-1980); Department of Indian and Northern Affairs, Geology Section, Yukon, Bulletin 1, 29 p.
Torrey, C.E., Karjalainen, H., Joyce, P.J., Erceg, M., and Stevens, M.
1986: Geology and mineralization of the Red Dome (Mungana) gold skarn deposit, north Queensland, Australia; in Proceedings of Gold '86, An International Symposium on the Geology of Gold, (ed.) A.J. MacDonald; Toronto, Ontario, p. 504-517.
Wares, R.P. and Williams-Jones, A.E.
1993: Porphyry copper-skarn mineralization at Mines Gaspé, Quebec; Exploration and Mining Geology, v. 2, no. 4, p. 414-415.

Watson, P.H.
1984: The Whitehorse Copper Belt - a compilation; Exploration and Geological Services Division, Yukon, Indian and Northern Affairs Canada, Open File, map, scale 1:25 000 with marginal notes.
Woodsworth, G.J., Anderson, R.G., and Armstrong, R.L.
1991: Plutonic regimes, Chapter 15 in Geology of the Cordilleran Orogen in Canada, (ed.) H. Gabrielse and C.J. Yorath; Geological Survey of Canada, Geology of Canada, no. 4, p. 491-531 (also Geological Society of America, The Geology of North America, v. G-2).
Wotruba, P.R., Benson, R.G., and Schmidt, K.W.
1988: Geology of the Fortitude gold-silver skarn deposit, Copper Canyon, Lander County, Nevada; in Bulk Mineable Precious Metal Deposits of the Western United States, (ed.) R.W. Schafer, J.J. Cooper, and P.G. Vikre; Geological Society of Nevada, Reno, Nevada, p. 159-171.

20.3 SKARN GOLD

K.M. Dawson

IDENTIFICATION

Skarns which contain sufficient gold to be mined economically for that commodity alone, herein termed 'Au skarns', have several common characteristics distinct from other types of Au-bearing skarns. Characteristics of four additional subtypes of skarn deposits, from which byproduct or coproduct Au is recovered, as proposed by Meinert (1988), are outlined below, with examples. Another approach to classification of productive gold-bearing skarn deposits was taken by Orris et al. (1987). They subdivided deposits with 1 ppm gold or more into "gold skarns" and "byproduct-gold skarns", depending upon the primary commodity recovered. A similar approach was taken by Theodore et al. (1991). Tonnage, grade, and subtype classification for significant Canadian and foreign gold skarn deposits are listed in Table 20.3-1. Grades versus tonnages are plotted in Figure 20.3-1. The locations of significant deposits in the Canadian Cordillera are shown in Figure 20.3-2.

Gold skarns, relative to other Au-bearing skarns, in addition to having a relatively high Au grade, are rich in As, Bi, and Te, are deficient in base metals, are dominated by reduced and Fe-rich gangue minerals, including Fe-rich pyroxene and lesser grandite garnet, have a higher clastic component in their host rocks, and are associated with

intrusions of more mafic character. Canadian Cordilleran examples, including Hedley, Quesnel River, Dividend-Lakeview, and Tillicum Mountain, are hosted by mainly Upper Triassic volcanic arc lithologies of the accreted Quesnellia terrane, and are associated with Lower to mid-Jurassic dioritic intrusions. Large foreign examples include the Fortitude mine at Battle Mountain, Nevada (Wotruba et al.,1986); McCoy Creek mine, Lander County, Nevada (Lane, 1987); and the Crown Jewel at Buckhorn Mountain, Washington (Hickey, 1992).

Gold-bearing skarns associated with porphyry Cu deposits, compared to other gold-bearing skarns, are relatively large, low in Au grade, and rich in andraditic garnet, diopsidic pyroxene, disseminated Cu sulphides, magnetite, and hematite. Gold occurs with sulphides either in prograde skarn or in zones of intense retrograde alteration. The large copper-gold skarns associated with alkalic porphyry deposits at Ingerbelle (Princeton) and Galore Creek, British Columbia are the most significant Canadian examples (see subtype 20.2, "Skarn copper"). Large foreign deposits of this type include Carr Fork, Bingham, Utah (Cameron and Garmoe, 1987); the Ely district, Nevada (James, 1976); Ok Tedi, Papua New Guinea (Davies et al., 1978); and the Red Dome or Mungana mine at Chillagoe, Queensland, Australia (Torrey et al., 1986; Ewers and Sun, 1988).

Copper-gold skarns are distinguished from porphyry Cu-Au skarns mainly by the lack of disseminated Cu-Mo minerals in the generally more mafic associated intrusion, and also by the smaller size, more massive nature, and higher Au grade of the orebodies. Copper-gold skarns in the North American Cordillera are associated, to varying degrees, with calcic Fe skarns. In the Intermontane Belt of British Columbia, magnetite is only a common component

Dawson, K.M.
1996: Skarn gold; in Geology of Canadian Mineral Deposit Types, (ed.) O.R. Eckstrand, W.D. Sinclair, and R.I. Thorpe; Geological Survey of Canada, Geology of Canada, no. 8, p. 476-489 (also Geological Society of America, The Geology of North America, v. P-1).

Table 20.3-1. Tonnage, grade, and subtype of significant Canadian and foreign gold and gold-rich skarn deposits, listed in decreasing order of contained Au. Skarn subtypes are after Meinert (1988).

	Name	Location	Ore (10⁶t)	Average grade g/t Au	Subtype	Reference source
Canadian deposits						
1	Hedley (Nickel Plate)	B.C.	8.4	7.3	Au	Ray et al., 1993
2	Greenwood district	B.C.	31.8	1.1	Cu-Au	Church, 1986
3	Ingerbelle	B.C.	216	0.16	porCu-Au*	Fahrni et al., 1976
4	Whitehorse Copper Belt	Y.T.	~10	1.0	Cu-Au	Meinert, 1986
5	Ketza River district	Y.T.	0.7	13	Au	Canamax Resources Inc., Annual Report, 1987
6	Tillicum Mountain district	B.C.	2.9	2.7	Au	Esperanza Explorations Ltd., 1989
7	Quesnel River (QR)	B.C.	1.0	6.5	Au	Fox et al., 1987
8	Benson Lake (Coast Copper)	B.C.	2.7	1.4	Cu-Au	Sangster, 1969
9	Marble Bay (Texada Iron)	B.C.	0.3	7.9	Fe-Au	Ettlinger and Ray, 1988
10	Banks I. (Tel, etc.)	B.C.	0.14	16.4	Au	Ettlinger and Ray, 1988
11	Tasu (Wesfrob)	B.C.	~18	0.1	Fe-Au	Sutherland Brown, 1968
12	Dividend-Lakeview	B.C.	0.11	4.5	Au	McKechnie, 1964
13	Galore Creek	B.C.	90	0.4	porCu-Au	Allen et al., 1976
14	Mount Polley	B.C.	10	0.55	porCu-Au	Imperial Metals Corp., Annual Report, 1991
Foreign deposits						
15	Carr Fork (Bingham)	Utah, U.S.A.	400	0.6	porCu-Au	Cameron and Garmoe, 1987
16	Ok Tedi (Mt. Fubilan)	Papua New Guinea	265	0.65	porCu-Au	Davies et al., 1978
17	Ely district	Nevada, U.S.A.	339	0.3	porCu-Au	James, 1976
18	Fortitude	Nevada, U.S.A.	10.3	6.9	Au	Wotruba et al., 1986
19	Santa Rita	New Mexico, U.S.A.	350	0.2	porCu-Au	Neilson, 1970
20	Bisbee	Arizona, U.S.A.	120	0.5	porCu-Au	Bryant and Metz, 1966
21	Red Dome	Australia	15	2.6	porCu-Au	Ewers and Sun, 1988
22	Larap	Philippines	20	1.2	Fe-Au	Frost, 1965
23	Salsigne	France	1.5	13	Au	Reynolds, 1965
24	Christmas	Arizona, U.S.A.	80	0.2	porCu-Au	Koski and Cook, 1982
25	McCoy Creek	Nevada, U.S.A.	8.6	1.7	Au	Lane, 1987
26	Minnie-Tomboy	Nevada, U.S.A.	3.9	2.8	Au	Blake et al., 1984
27	Mission	Arizona, U.S.A.	100	0.1	porCu-Au	Einaudi, 1982
28	Continental	New Mexico, U.S.A.	47	0.2	porCu-Au	Einaudi, 1982
29	Copper Canyon	Nevada, U.S.A.	10	0.8	porCu-Au	Blake et al., 1978
30	Naica	Mexico	20	0.4	Zn-Pb-Au	Clark et al., 1986
31	Zackly	Alaska, U.S.A.	1.25	6	Cu-Au	Nokleberg et al., 1988
32	Mount Biggenden	Australia	0.5	15	Au	Clarke, 1969
33	Cable	Montana, U.S.A.	1	6	Au	Earll, 1972
34	Yaguki	Japan	1.2	3	Cu-Au	Shimazaki, 1969
35	Concepcion del Oro	Mexico	25	0.6	Cu-Au	Megaw et al., 1988
36	El Mochito	Honduras	15.5	0.1	Zn-Pb-Au	Shultz and Hamann, 1977
37	La Luz	Nicaragua	16	4.1	porCu-Au	Sillitoe, 1983
38	Bau	Malaysia	2.4	7.2	Au	Wolfenden, 1965
39	Crown Jewel	Washington, U.S.A.	6.5	5.6	Au	Hickey, 1992
40	Siana	Philippines	5.4	5.1	Au	Orris et al., 1987
*porCu-Au = porphyry copper-gold						

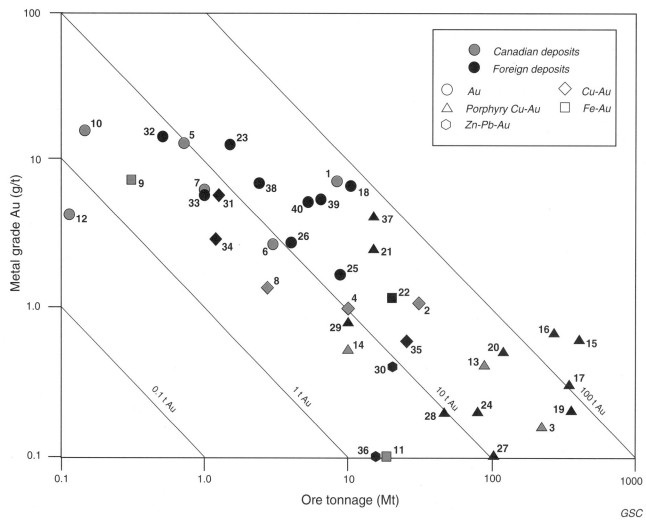

Figure 20.3-1. Grade versus tonnage of significant Canadian and foreign gold skarn and gold-rich skarn deposits. Numbers correspond to deposits listed in Table 20.3-1.

of the Cu-Au skarn mineral assemblage, whereas in the Insular Belt Cu-Au skarns are associated with large calcic magnetite skarn deposits. Almost all deposits of this kind in the northern North American Cordillera, including those in the important districts of Greenwood, British Columbia and the Whitehorse Copper Belt, Yukon Territory, as well as smaller deposits in the Insular Belt, associated with the Nelson batholith in southeastern British Columbia, and in Alaska, are hosted by Upper Triassic to Lower Jurassic carbonate-clastic-volcanic assemblages within accreted oceanic-arc terranes. Foreign examples include Yaguki, Japan (Shimaziki, 1969) and Concepcion del Oro, Mexico (Megaw et al., 1988)

Iron-gold skarns are associated with large calcic magnetite skarns mined primarily for their Fe content, but with significant byproduct Au and Cu. Gold, with Co and As, is concentrated with erratically distributed Fe and Cu sulphides, rather than with the Fe oxides. Canadian examples include Merry Widow (Ettlinger and Ray, 1989), Texada Iron (Lake, Paxton, Prescott, Yellow Kid; Ettlinger and Ray,

1988), and Tasu (Sutherland Brown, 1968); and the Oro Denoro and Emma deposits of the Greenwood district (Church, 1986). Foreign deposits include Larap, Philippines (Frost, 1965) and Nabesna and Rambler deposits, Alaska (Nokleberg et al., 1988).

Zinc-lead skarns and related replacement deposits are more commonly enriched in Ag than in Au. However, three northern Cordilleran Ag-Pb-Zn skarn and replacement deposits, i.e., Midway, YP, and Roy, contain important minor values in Au, mainly in proximal skarn and mantos. Two British Columbia Au skarns, the Dividend-Lakeview and Banks Island, contain substantial amounts of Pb and

Figure 20.3-2. Significant skarn gold deposits of the Canadian Cordillera. Numbered symbols correspond to those in Figure 20.3-1 and Table 20.3-1.

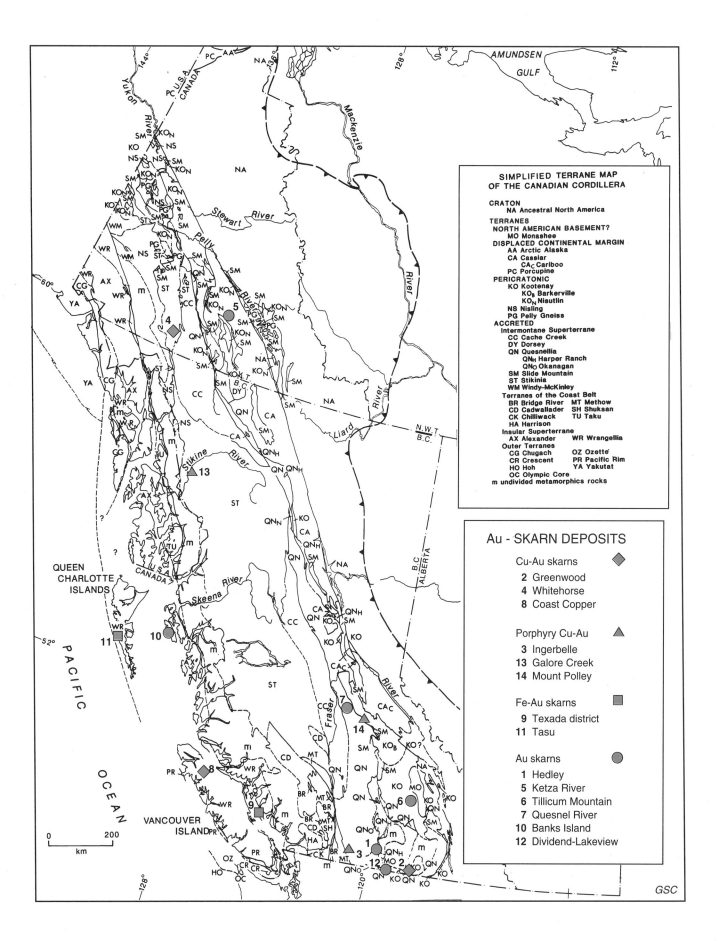

SIMPLIFIED TERRANE MAP
OF THE CANADIAN CORDILLERA

CRATON
  NA Ancestral North America
TERRANES
  NORTH AMERICAN BASEMENT?
    MO Monashee
  DISPLACED CONTINENTAL MARGIN
    AA Arctic Alaska
    CA Cassiar
      CA$_C$ Cariboo
    PC Porcupine
  PERICRATONIC
    KO Kootenay
      KO$_B$ Barkerville
      KO$_N$ Nisutlin
    NS Nisling
    PG Pelly Gneiss
  ACCRETED
    Intermontane Superterrane
      CC Cache Creek
      DY Dorsey
      QN Quesnellia
        QN$_H$ Harper Ranch
        QN$_O$ Okanagan
      SM Slide Mountain
      ST Stikinia
      WM Windy–McKinley
    Terranes of the Coast Belt
      BR Bridge River   MT Methow
      CD Cadwallader   SH Shuksan
      CK Chilliwack   TU Taku
      HA Harrison
    Insular Superterrane
      AX Alexander   WR Wrangellia
    Outer Terranes
      CG Chugach   OZ Ozette
      CR Crescent   PR Pacific Rim
      HO Hoh   YA Yakutat
    OC Olympic Core
    m undivided metamorphics rocks

Au - SKARN DEPOSITS

Cu-Au skarns ◆
  2 Greenwood
  4 Whitehorse
  8 Coast Copper

Porphyry Cu-Au ▲
  3 Ingerbelle
  13 Galore Creek
  14 Mount Polley

Fe-Au skarns ■
  9 Texada district
  11 Tasu

Au skarns ●
  1 Hedley
  5 Ketza River
  6 Tillicum Mountain
  7 Quesnel River
  10 Banks Island
  12 Dividend-Lakeview

GSC

479

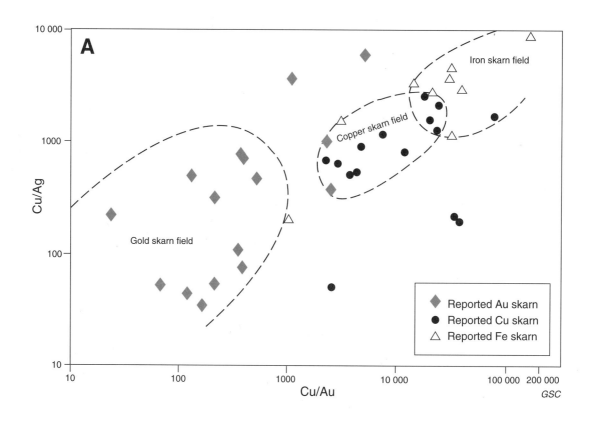

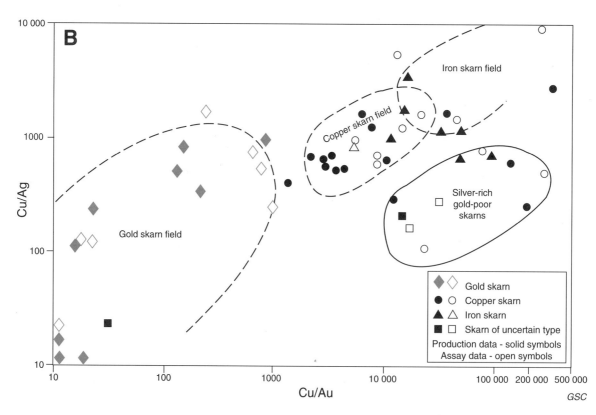

Zn. Large foreign gold-rich Zn-Pb-Ag skarn and replacement deposits include Naica, Mexico (Clark et al., 1986) and El Mochito, Honduras (Shultz and Hamann, 1977).

A method of classifying gold and gold-bearing skarns on the basis of Cu:Au and Cu:Ag ratios, proposed by Ettlinger and Ray (1989), broadly differentiates gold skarns from gold-bearing copper and iron skarns. Figure 20.3-3A, a plot of Cu:Au versus Cu:Ag for a database of 40 international skarn deposits, and Figure 20.3-3B, the same plot of 54 precious metal-enriched British Columbia skarns, show the same well differentiated fields, with gold skarns having Cu:Au values less than 1000, copper skarns 2000-25 000, and iron skarns 20 000-160 000. In addition, in Figure 20.3-3B, silver-rich, gold-poor skarns with affinities for Zn-Pb skarns plot in a separate field.

IMPORTANCE

Meinert (1988) estimated that production from all known Au skarn deposits totals about 1000 t Au. This represents about 1% of Boyle's (1979) estimate of total historical Au production up to 1975. Statistics on global Au production from skarns relative to other types of Au deposits are not readily available. However, Ray and Webster (1991) noted that precious metal-enriched skarns in British Columbia have produced about 10% of the world total: 49 skarn deposits have produced 95 t Au and 342 t Ag. Significant production has been attained from four active and several previously-producing Au skarns in the Canadian Cordillera. The economic significance of Au skarn deposits in the Cordillera has been emphasized by the redevelopment, in 1987, of the large Nickel Plate deposit at Hedley as an open pit mine, commencement of production, in 1987, of the Ketza River and Tillicum Mountain deposits, and exploration activity centred on numerous skarn Au prospects, as summarized by Ettlinger and Ray (1988). High-grade gold skarn deposits at Fortitude and McCoy, Copper Canyon district, Nevada, and Crown Jewel, Buckhorn district, Washington have been documented by Wotruba et al. (1986) and Hickey (1992), respectively.

SIZE OF DEPOSIT

Meinert (1988) has compiled data on 69 globally-distributed Au-bearing skarn deposits, as reproduced in Table 20.3-2.

The six Au skarns cited for the Canadian Cordillera in Table 20.3-1 range widely in tonnage and Au grade, but the calculated average of 2.6 Mt of 9.3 g/t Au is similar to the global average. The four Canadian Cu-Au skarns are larger than the global average, but lower in Au grade, at a calculated average of 11 Mt of 1.1 g/t Au. Data for 25 selected Au- and Ag-enriched skarns of the world tabulated by

Table 20.3-2. Average tonnage and Au grade for Au skarns and Au-bearing skarn subtypes.

Skarn type/ subtype	No.	Mt ore (average)	g/t Au (average)
Au	14	4.6	10.6
Fe-Au	14	1.6	4.4
Cu-Au	19	2.1	2.5
Porphyry Cu-Au	14	153	0.5
Zn-Pb-Au	8	3.8	0.6

Ettlinger and Ray (1989, their Table 1) yield a calculated average of 7.45 Mt at a grade of 2.6 g/t Au. Grades and tonnages of 62 globally distributed Au-bearing skarns with average grades of at least 1 g/t Au have been compiled by Orris et al. (1987) and two subtypes have been proposed: (1) Au skarns, exploited primarily for Au (median grade 6.8 g/t Au); and (2) byproduct Au skarns, mined primarily for their base metal content (median grade 3.4 g/t Au). All Au-bearing skarns have a median size of 400 000 t, and a median grade of 5 g/t Au. Theodore et al. (1991) employed a larger global database and calculated median grades and tonnages for 40 Au skarn deposits as 8.6 g/t Au, 5.0 g/t Ag, and 213 000 t respectively; and for 50 byproduct Au skarn deposits as 3.7 g/t Au, 37 g/t Ag, and 330 000 t respectively.

GEOLOGICAL FEATURES

Geological setting

Gold skarn deposits are found in two principal tectonic settings: continental shelves and oceanic or arc volcanic-sedimentary assemblages. Gold skarns in the southwestern United States and Australia occur commonly in Paleozoic to lower Mesozoic shelf and basinal sedimentary rocks of cratonal origin. Related calc-alkaline plutonism in the United States is late Laramide and in Australia, Carboniferous. Two Canadian Au skarns, i.e., Ketza River, Yukon Territory and Banks Island, British Columbia have similar settings in displaced, rather than autochthonous, cratonal terranes associated with Cretaceous granitoid bodies. The other four Canadian deposits, hosted by oceanic arc carbonate-clastic-volcanic lithologies of accreted Quesnellia, are representative of the other principal setting for Au skarn deposits in allochthonous or accreted rocks with a significant marine volcanic component. Intrusions associated with Au skarns in the Canadian Cordillera constitute a distinctive suite of calc-alkaline to late orogenic timing. At Hedley, British Columbia, the quartz dioritic and gabbroic intrusions are enriched in iron, depleted in total alkalis and silica and have low F_2O_3/FeO ratios, i.e., are reduced, relative to intrusions associated with other types of skarn deposits (Ray and Webster, 1991).

The tectonic setting of porphyry Cu-Au skarns is continental margin orogenic belts, which have been intruded by calc-alkaline granodioritic to quartz monzonitic stocks, as exemplified by the deposits in the southwestern United States, i.e., Utah, Arizona, Nevada, and New Mexico. Relatively

Figure 20.3-3. (A) Plot showing Cu/Au versus Cu/Ag ratios of 40 international skarn deposits. Gold, copper, and iron skarn types are outlined in discrete fields (after Ettlinger and Ray, 1989, Fig. 50). **(B)** Plot showing Cu/Au versus Cu/Ag ratios of 54 precious metal-enriched skarns in British Columbia. Skarn fields correspond to those in Figure 20.3-3A, with the addition of silver-rich, gold-poor skarns with zinc-lead skarn affinities (after Ettlinger and Ray, 1989, Fig. 51).

few deposits of this type, e.g., Ok Tedi, Papua New Guinea and Ingerbelle, British Columbia are known from oceanic-island arc settings and their accreted equivalents. Copper-gold skarns associated with barren stocks unrelated to porphyry deposits, on the other hand, have tectonic settings similar to those of calcic Fe skarns in oceanic island arcs, their accreted equivalents and rifted continental margins (Einaudi et al., 1981). The limited data on the Au-rich subtype of Zn-Pb skarns indicate they have a tectonic setting similar to other Zn-Pb skarns, i.e., they occur in cratonal sediments at continental margins and are associated with synorogenic to late orogenic calc-alkaline stocks.

Age of host rocks, associated rocks, and mineralization

Host rocks for deposits of the Au skarn subtype are metamorphosed equivalents of carbonate rocks which occur either as relatively thick and pure limestones in a miogeoclinal setting or as carbonate units interbedded with clastic and volcaniclastic rocks, tuffites, and flows in an oceanic-island arc setting. Host rock ages range from Cambrian to Miocene, but are dominantly Paleozoic in Australia, Russia, and the southwestern United States, and Late Triassic in British Columbia. Plutonic rocks associated with Au skarns range

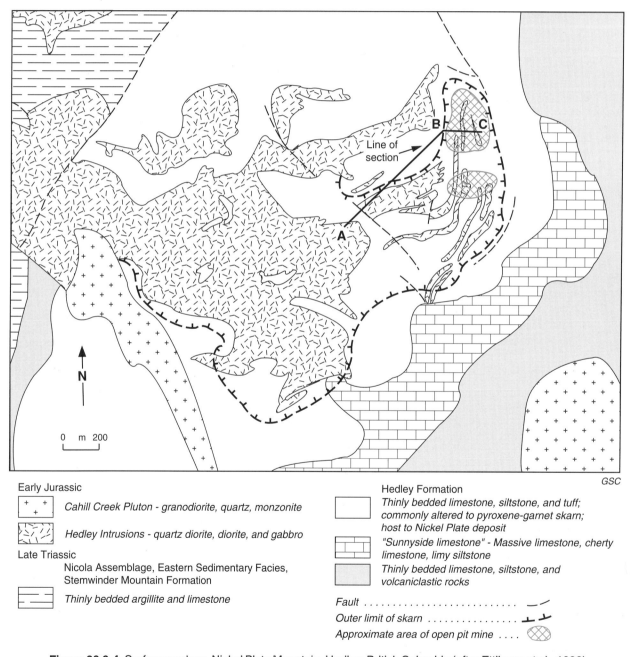

GSC

Early Jurassic

| + + + | Cahill Creek Pluton - granodiorite, quartz, monzonite |

| 〰 | Hedley Intrusions - quartz diorite, diorite, and gabbro |

Late Triassic

Nicola Assemblage, Eastern Sedimentary Facies, Stemwinder Mountain Formation

| ☰ | Thinly bedded argillite and limestone |

Hedley Formation

| ☐ | Thinly bedded limestone, siltstone, and tuff; commonly altered to pyroxene-garnet skarn; host to Nickel Plate deposit |

| ▥ | "Sunnyside limestone" - Massive limestone, cherty limestone, limy siltstone |

| ▨ | Thinly bedded limestone, siltstone, and volcaniclastic rocks |

Fault . . . .

Outer limit of skarn . . . .

Approximate area of open pit mine . . . .

Figure 20.3-4. Surface geology, Nickel Plate Mountain, Hedley, British Columbia (after Ettlinger et al., 1992). Section A-B-C shown in Figure 20.3-5.

in age from Early Paleozoic in Australia to Miocene in the Philippines (Meinert, 1988), and in composition from diorite, gabbro, and syenodiorite in British Columbia to rhyolite porphyry at Bau, Malaysia (Wolfenden, 1965). A general correlation between the global distributions of gold skarns and porphyry copper deposits was noted by Ray et al. (1990). The majority of Au skarn deposits are associated with relatively small, mafic to intermediate plutons. Meinert (1992) noted the association of most high-grade gold skarns with reduced, i.e., ilmenite-bearing, Fe^{+3}/Fe^{+2} less than 0.75, diorite-granodiorite plutons and dyke-sill complexes. Age of mineralization is penecontemporaneous with intrusion, and is dominantly Early Tertiary in southwestern United States, Early to mid-Jurassic in British Columbia, and Paleozoic in Australia and Russia.

Large Cu skarns, all enriched in part in Au, are associated with some porphyry Cu plutons emplaced in continental margin carbonate strata. The largest group of porphyry copper-related gold-rich skarns, associated with the Laramide porphyry Cu province of southwestern United States, is hosted by dominantly Paleozoic cratonal sedimentary rocks. Other notable districts occur in Mexico, Peru, Russia, and Japan. At Ingerbelle, British Columbia, an atypical alkalic (Na-K-Ca) porphyry Cu-Au skarn is hosted by Upper Triassic arc-related andesitic to basaltic volcaniclastic rocks and flows adjacent to alkalic stocks (Preto, 1972; Fahrni et al., 1976) of the Early Jurassic Copper Mountain suite.

Copper-gold skarns and related calcic Fe-Au skarns are hosted by Paleozoic and Mesozoic limestones interbedded with clastic, volcaniclastic, and tholeiitic to calc-alkaline volcanic rocks in several tectonic settings, including oceanic, island-arc, and back arc. The formation of Cu-Au and Fe-Au skarns in the Canadian Cordillera accompanied the emplacement of both pre-accretionary dioritic Early Jurassic plutons and postaccretionary granodioritic mid- to Late Cretaceous plutons.

Form of deposit and associated structures, zoning, and distribution of ore minerals

The morphology of Au skarns is not known to be distinctive. Gold commonly accompanies the dominant sulphide mineral, usually arsenopyrite, pyrrhotite, pyrite, or chalcopyrite, in both prograde skarn and retrograde alteration zones. Both stratigraphic control on the development of

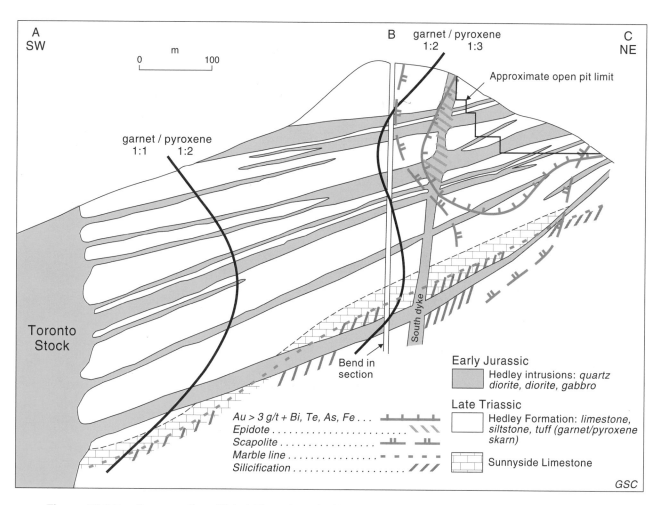

Figure 20.3-5. Cross-section, Nickel Plate deposit, Hedley, British Columbia. Skarn assemblages and sulphide-Au mineralization after Ettlinger et al. (1992).

483

exoskarn by bedding in the carbonate-clastic rocks, and structural control by intrusive contacts, faults, fractures, and folds are common. No consistent zonal distribution of ore minerals has been recognized, but prograde calcic skarn assemblages typically are zoned toward the intrusive contact from marble, through wollastonite and diopside-hedenbergite, to andradite.

The Nickel Plate skarn Au orebodies at Hedley, British Columbia are semiconformable, tabular sulphide zones developed near the skarn-marble boundary. Host limestone of the Upper Triassic Nicola Group is interbedded with argillite and siltstone and has been intruded by sills and some dykes of Late Triassic to Early Jurassic diorite and gabbro that have yielded zircon U/Pb ages between 219 and 194 Ma (Ray et al., 1993). Alternating layers of garnet-rich and diopside-hedenbergite-rich prograde skarn follow bedding. Gold, together with anomalous amounts of Bi, Te, and Co, is concentrated with arsenopyrite, pyrrhotite, and pyrite in the latest stage, a retrograde quartz-calcite-epidote-sulphide assemblage deposited near the skarn-marble boundary (Billingsley and Hume, 1941; Ettlinger and Ray, 1988, 1989; Ettlinger et al., 1992) (Fig. 20.3-4, 20.3-5, and Table 20.3-1).

The morphologies of gold-rich porphyry Cu skarns reflect a relatively high level of emplacement of the associated intrusive rocks, and resultant intense fracturing, brecciation, and breccia pipe formation. Typically extensive development of thick skarn units reflects the high fracture permeability of the porphyry Cu system. The calc-silicate assemblage of ferric iron-rich andradite and ferrous iron-poor clinopyroxene reflects the relatively oxidized environment, as in Au-deficient Cu skarns associated with porphyry Cu deposits (Einaudi et al., 1981). Most skarns are part of a large zoned system with proximal garnet-rich and distal Zn-Pb-Ag-rich zones, e.g., Fortitude mine at Battle Mountain, Nevada (Theodore and Blake, 1978). Some porphyry Cu-Au skarns, such as Carr Fork, Utah have high concentrations of Au within zones of localized intense retrograde alteration (Cameron and Garmoe, 1987), but gold occurs generally in low concentrations and is recoverable only as a byproduct.

The large disseminated Cu-Au-Ag deposits at Copper Mountain, British Columbia have been classified by Sutherland Brown et al. (1971) as complex porphyry-type deposits of the alkaline suite, and deposits at adjacent Ingerbelle as skarn deposits gradational to a porphyry. Dolmage (1934), Preto (1972), and Macauley (1973) noted

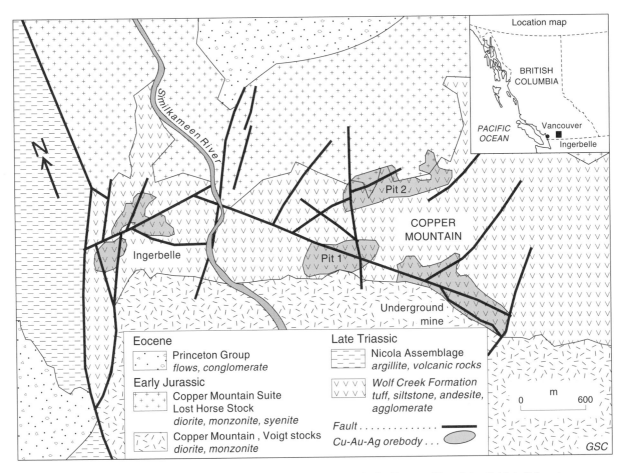

Figure 20.3-6. Geology and Cu-Au-Ag deposits of the Ingerbelle-Copper Mountain district, Princeton, British Columbia (after Fahrni et al., 1976).

the pyrometasomatic characteristics of the Copper Mountain and Ingerbelle deposits. Fahrni et al. (1976) emphasized the zonal distribution of most of the following gangue assemblages at Ingerbelle relative to the contact of the Lost Horse diorite-monzonite-syenite stock: early biotite hornfels overprinted by main stage sulphides plus prograde albite-epidote-chlorite±andradite±diopsidic pyroxene±sphene; and both prograde skarn and stock flooded by retrograde albite, K-feldspar, scapolite, calcite, and hematite. The gold-rich chalcopyrite-bornite ore, about 90% of which occurs in andesitic volcanic host rocks, is distributed along the contacts of the Lost Horse intrusion. These mineral assemblages are interpreted to represent the superposition of silica-deficient, alkali- and alumina-rich magnesian skarn assemblages upon biotite hornfels, developed in an andesite protolith adjacent to an alkalic intrusion. The local geology and distribution of orebodies are given in Figure 20.3-6 after Fahrni et al. (1976). Gold-rich alkalic porphyry copper skarn deposits at Galore Creek (Allen et al., 1976) and Mount Polley (Fraser, 1994), British Columbia are described in "Skarn copper", subtype 20.2.

Copper-gold skarns associated with barren stocks unrelated to porphyry deposits have several features in common with calcic Fe-Au skarns, including tectonic setting, composition of intrusions, and skarn morphology and mineralogy (see deposit subtype 20.2, "Skarn copper"). The highest

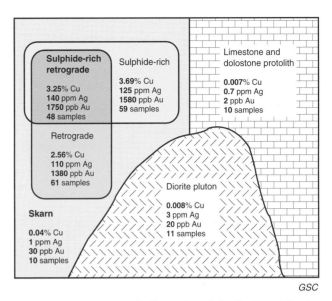

Figure 20.3-8. Schematic illustration of distribution of Cu, Ag, and Au in rock and alteration types in the Whitehorse Copper Belt (after Meinert, 1986).

Au contents occur in zones of coincident high sulphide content and intense retrograde alteration of skarn (Meinert, 1988). Examples of coexisting Cu-Au and Fe-Au skarns can be found in the Texada Island, Benson Lake, and Greenwood districts of British Columbia (Table 20.3-1). Calcic Fe-Au skarns, e.g., Tasu (Wesfrob) and Texada Iron, commonly are large, but Au is in low overall concentration and erratically distributed within Cu- and S-rich zones. Gold and cobalt may, in some cases, be concentrated in the early formed sulphides, chalcopyrite, pyrite, and pyrrhotite, or in later retrograde alteration assemblages with sulphides, magnetite, amphibole, epidote, and ilvaite (Meinert, 1984; Ettlinger and Ray, 1989). The Crown Jewel Au skarn at Buckhorn Mountain, Washington is distant from a proximal Cu-Fe skarn in a large system also zoned in proximal garnet-rich and distal pyroxene-rich assemblages (Hickey, 1992).

A reconnaissance study of the Cu-Au-Ag skarns of the Whitehorse Copper Belt, Yukon Territory by Meinert (1986) illustrates the distribution of Au and its association with retrograde alteration. Copper, iron (molybdenum, gold, silver) skarn deposits are developed in both dolomitic and calcareous carbonate units of the Upper Triassic Lewes River Group, a back-arc succession of arkosic clastic and carbonate rocks. Some 32 skarn deposits and occurrences are localized mainly along the irregular western contact of the Whitehorse batholith (Fig. 20.3-7), a composite calc-alkaline granodiorite pluton with a diorite margin (Morrison, 1981) that postorogenically intruded accreted Stikinia in mid-Cretaceous time (Dawson et al., 1991).

Skarns formed from a limestone protolith at Whitehorse contain abundant andraditic garnet, hedenbergitic pyroxene, wollastonite, vesuvianite, and variable amounts of the retrograde alteration products actinolite, epidote, and chlorite. Skarns with a dolomitic protolith contain abundant prograde magnesian minerals, such as diopside and forsteritic olivine plus andradite, and retrograde phlogopite, brucite, talc, and serpentine. Copper,

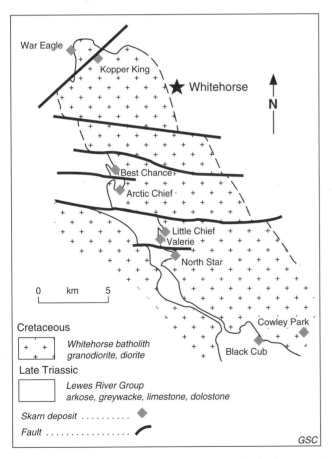

Figure 20.3-7. Simplified regional geological map of Whitehorse Copper Belt showing principal Cu-Au-Ag skarn deposits (after Meinert, 1986).

gold, and silver contents in unaltered carbonate rock are low; in diorite they are about equivalent to crustal averages; and in unmineralized skarn are slightly enriched over protolith values (Fig. 20.3-8). Both sulphide-rich skarn and retrograde-altered skarn are enriched in Cu, Au, and Ag, as would be expected, but highest average Au and Ag grades occur where massive sulphides and retrograde alteration coexist in skarn (Meinert, 1986).

Most Zn-Pb skarns are rich in Ag, but deficient in Au. Where Au is present in significant amounts, it is commonly concentrated with sulphide-rich proximal ore and gangue mineral assemblages either close to a known intrusive contact or in the core of the deposit (as deduced from mineral zonation when no intrusion is exposed). Examples of Au-rich Zn-Pb-Ag skarns from the northern Canadian Cordillera and northern Mexico are given in the account for deposit subtype 20.1, "Skarn zinc-lead-silver".

Mineralogy

Principal ore minerals: gold, electrum, gold tellurides.

Subsidiary ore minerals: chalcopyrite, bornite, sphalerite, galena, molybdenite, scheelite, silver, silver tellurides, silver sulphosalts, tetrahedrite, bismuth, bismuthinite, bismuth tellurides, wittichenite, cobaltite, gersdorffite.

Other opaque minerals: pyrrhotite, pyrite, arsenopyrite, marcasite, hematite, magnetite. Typical prograde and retrograde skarn minerals are listed in Table 20.3-3.

DEFINITIVE CHARACTERISTICS

Skarn deposit classification may be based on various features, such as dominant economic metals, morphology, host rock type and composition, temperature and redox conditions of formation, and tectonic setting. Many modern authors prefer a descriptive classification that first emphasizes dominant economic metals, and is modified by genetic, compositional, morphological, and other factors. This approach is followed in the skarn descriptions in this volume, and adapted to include in the "skarn Au" category gold-rich subtypes of other principal types of skarn deposits. Although gold-rich skarn subtypes possess definitive characteristics as given below, the principal features are similar to those of other skarn deposits, e.g. skarn copper, skarn iron, etc.

1. Gold skarns: ore assemblage rich in As, Bi, and Te; calcic skarn mineral assemblage low in Mn; Fe- and Al-rich pyroxene more abundant than grandite garnet; abundant clastic and volcaniclastic components in host rocks; association with relatively Fe-rich, I-type, ilmenite-bearing subalkalic to calcalkalic intrusions.

2. Copper-gold skarns associated with porphyry Cu deposits: disseminated Cu and Mo minerals in altered intrusion; high fracture permeability; relatively thick and extensive skarn; low Au grade; those in the southwestern United States have a craton margin setting.

3. Copper-gold skarns associated with unmineralized plutons unrelated to porphyry deposits: lack of disseminated Cu-Mo minerals in unaltered intrusion; Au concentrated in sulphide-rich retrograde alteration zones; orebodies more massive and higher in Au grade than Cu-Au skarns associated with porphyry deposits.

4. Iron-gold skarns: high Au, Co, and As contents associated with sulphide-rich zones; calcic magnetite-rich skarn shows gradation distally to Cu and Cu-Au skarn; oceanic-island arc setting.

5. Zinc-lead-silver-gold skarns: high Ag content; Ag concentrated distally to Au±Cu, W-rich proximal skarn; craton margin setting.

Table 20.3-3. Prograde and retrograde skarn minerals.

Prograde skarn minerals	Retrograde skarn minerals
(a) Au: andraditic garnet, hedenbergitic pyroxene, wollastonite, vesuvianite, scapolite	actinolite, hornblende, chlorite, epidote
(b) porphyry Cu: andraditic garnet, diopsidic pyroxene, wollastonite	actinolite, chlorite, montmorillonite, quartz, calcite
(c) Cu: grandite garnet (intermediate between grossular and andradite), wollastonite, diopsidic pyroxene, magnetite, epidote	epidote, actinolite, hornblende, chlorite
(d) Fe: grandite garnet, diopsidic to johannsenitic pyroxene, magnetite, epidote	amphibole, chlorite, ilvaite, epidote
(e) Zn-Pb: johannsenitic pyroxene, andraditic garnet, vesuvianite, wollastonite, bustamite.	manganoan actinolite, epidote, chlorite, siderite and ilvaite; and calcite, quartz, fluorite

GENETIC MODEL

The role of characteristic geological features such as clastic-rich host rocks, tectonic setting, and mafic plutons in the formation of Au skarns is not well understood. Most Au and Au-bearing skarns in the Canadian Cordillera are associated with suites of pre-accretionary subalkaline to alkaline plutons comagmatic with the volcanic component of an accreted arc assemblage. Gold may have been contributed from the marine volcanic, volcaniclastic, and clastic sedimentary host rocks of Quesnellia. Plutons of the same alkaline suite, associated elsewhere with Au- and Ag-rich porphyry Cu deposits, may represent regional enrichment of Au in the relatively undifferentiated synorogenic magma.

In a global context, the compositions of host rocks and intrusions related to Au-rich skarns range widely and compositions of gangue minerals are not well documented. Whether or not these skarns are mineralogically and genetically distinctive remains to be established. The genesis of Au-rich subtypes of porphyry Cu, Cu, Fe, and Zn-Pb skarns does not differ substantially from that of the principal skarn types. Genesis of these principal skarn types is covered elsewhere in this volume (e.g., see deposit subtype 20.2, Skarn copper) and has been discussed in a summary by Einaudi et al. (1981).

RELATED DEPOSIT TYPES

Most Au and Au-bearing skarns in the Canadian Cordillera are associated with a unique suite of dominantly subalkaline to alkaline, Late Triassic to mid-Jurassic granitoid plutons which, elsewhere in the same Quesnellia host rock sequences, contain significant porphyry Cu-Au-Ag deposits. Examples of deposits related to these intrusions and host rocks include the Ingerbelle and Galore Creek porphyry Cu-Au-Ag skarns (Allen et al., 1976), and Au skarns such as the Nickel Plate (Ettinger et al., 1992). The Au vein deposits at Rossland (Wilson et al., 1990; Höy and Andrew, 1991) and near Nelson, British Columbia (Second Relief; Ettlinger and Ray, 1989) possess some mineralogical characteristics of Au skarns, and may be related. Skarn deposits of several types are enriched in Au, particularly in S-rich and intensely retrograde altered zones. Porphyry Cu (Mo) skarns and Cu skarns are enriched in coproduct and byproduct Au to the greatest degree, whereas W, Sn, and Mo skarns are poorest in Au.

EXPLORATION GUIDES

General guidelines for the discovery of the principal types of skarn deposits, covered elsewhere in this volume, also apply to Au-rich subtypes of those deposits. Most skarns, except those of W, Mo, and Sn, are potentially rich in Au, and some specific guides to Au-rich skarns are as follows:

1. Gold skarns contain anomalous amounts of Bi, Te, and Ag, in addition to Cu and As.

2. Gold-rich parts of calcic Fe skarns also are enriched in Cu, Co, and As, in addition to S.

3. Gold is concentrated in S-rich parts of most skarns, particularly calcic Cu and Fe skarns.

4. Gold and silver are concentrated in intensely retrograde altered zones of calcic Cu and porphyry Cu skarns.

5. Gold is, in some cases, concentrated in proximal parts of Zn-Pb skarn and replacement deposits that are enriched in Cu, W, and Fe.

General exploration guides to discovery of Au skarns are:

1. Gold skarns may be discovered in either craton marginal or oceanic-island arc settings, commonly in interbedded carbonate-clastic-volcaniclastic sequences and adjacent to relatively small dioritic plutons.

2. In the Canadian Cordillera, most Au skarns are hosted by accreted sedimentary and volcanic assemblages of Quesnellia where these rocks have been intruded by Late Triassic to Middle Jurassic, mainly subalkaline to alkaline plutons that are comagmatic with the volcanic rocks. Good potential exists for discovery of Au-rich skarns adjacent to known porphyry Cu-Au-Ag deposits hosted by and associated with this plutonic suite.

REFERENCES

References marked with asterisks (*) are considered to be the best sources of general information on this deposit type.

Allen, D.G., Panteleyev, A., and Armstrong, A.T.
1976:   Galore Creek; in Porphyry Deposits of the Canadian Cordillera, (ed.) A. Sutherland Brown; The Canadian Institute of Mining and Metallurgy, Special Volume 15, p. 402-414.

Billingsley, P. and Hume, C.G.
1941:   The ore deposits of Nickel Plate Mountain, Hedley, B.C.; The Canadian Institute of Mining and Metallurgy, Bulletin, v. 44, p. 524-590.

Blake, D.W., Theodore, T.G., and Kretschmer, E.L.
1978:   Alteration and distribution of sulphide mineralization at Copper Canyon, Lander County, Nevada; Arizona Geological Society Digest, v. 11, p. 67-78.

Blake, D.W., Wotruba, P.R., and Theodore, T.G.
1984:   Zonation in the skarn environment at the Tomboy-Minnie gold deposits, Lander County, Nevada; Arizona Geological Society Digest, v. 15, p. 67-72.

Boyle, R.W.
1979:   The geochemistry of gold and its deposits; Geological Survey of Canada, Bulletin 280, 584 p.

Bryant, D.G. and Metz, H.E.
1966:   Geology and ore deposits of the Warren mining district; in Geology of the Porphyry Copper Deposits, Southwestern North America, (ed.) S.R. Titley and C.L. Hicks; University of Arizona Press, Tucson, Arizona, p. 189-204.

Cameron, D.E. and Garmoe, W.J.
1987:   Geology of skarn and high-grade gold in the Carr Fork mine, Utah; Economic Geology, v. 82, p. 1219-1333.

Church, B.N.
1986:   Geological setting and mineralization in the Mount Attwood-Phoenix area of the Greenwood mining camp; British Columbia Ministry of Energy, Mines and Petroleum Resources, Paper 1986-2, 65 p.

Clark, K.F., Megaw, P.K.M., and Ruiz, J.
1986:   Lead-zinc-silver carbonate-hosted deposits of northern Mexico; Society of Economic Geologists, Guidebook for Field and Mine Excursions, November 13-17, 1986, 329 p.

Clarke, D.E.
1969:   Geology of the Mount Biggenden gold and bismuth mine and environs; Queensland Geological Survey, Report no. 32, 16 p.

Davies, H.L., Howell, W.J.S., Fardon, R.S.H., Carter, R.J., and Bumstead, E.D.
1978:   History of the Ok Tedi porphyry copper prospect, Papua New Guinea; Economic Geology, v. 73, no. 5, p. 796-809.

Dawson, K.M., Panteleyev, A., Sutherland Brown, A., and Woodsworth, G.J.
1991:   Regional metallogeny, Chapter 19 in Geology of the Cordilleran Orogen in Canada, (ed.) H. Gabrielse and C.J. Yorath; Geological Survey of Canada, Geology of Canada, no. 4, p. 707-768 (also Geological Society of America, The Geology of North America, v. G-2).

Dolmage, V.

1934: Geology and Ore Deposits of Copper Mountain, British Columbia; Geological Survey of Canada, Memoir 171, 69 p.

Earli, F.M.

1972: Mines and mineral deposits of the southern Flint Creek Range, Montana; Montana Bureau of Mines and Geology, Bulletin 84, 54 p.

*Einaudi, M.T.

1982: Descriptions of skarns associated with porphyry copper plutons; in Advances in Geology of the Porphyry Copper Deposits: Southwestern North America, (ed.) S.R. Titley; University of Arizona Press, Tucson, p. 139-183.

*Einaudi, M.T., Meinert, L.D., and Newberry, R.J.

1981: Skarn deposits in Seventy-fifth Anniversary Volume, 1905-1980, (ed.) B.J. Skinner; Economic Geology, p. 317-391.

Ettlinger, A.D. and Ray, G.E.

1988: Gold-silver enriched skarn deposits of British Columbia in Geological Fieldwork 1987; British Columbia Ministry of Energy, Mines and Petroleum Resources, Paper 1988-1, p. 263-279.

*1989: Precious metal enriched skarns in British Columbia: an overview and geological study; British Columbia Ministry of Energy, Mines and Petroleum Resources, Paper 1989-3, 128 p.

Ettlinger, A.D., Meinert, L.D., and Ray, G.E.

1992: Gold skarn mineralization and evolution of fluid in the Nickel Plate deposit, Hedley District-British Columbia; Economic Geology, v. 87, p. 1541-1565.

Ewers, G.R. and Sun, S.-S.

1988: Genesis of the Red Dome deposit, Northeast Queensland; in Bicentennial Gold 88, Extended Abstracts Oral Programme, (comp.) A.D.T. Goode and L.I. Bosma; Geological Society of Australia Abstract Series, no. 22, p. 110-115.

Fahrni, K.C., Macauley, T.N., and Preto, V.A.G.

1976: Copper Mountain and Ingerbelle; in Porphyry Deposits of the Canadian Cordillera, (ed.) A. Sutherland Brown; The Canadian Institute of Mining and Metallurgy, Special Volume 15, p. 368-375.

Fox, P.E., Cameron, R.S., and Hoffman, S.J.

1987: Geology and soil geochemistry of the Quesnel River gold deposit, British Columbia; in GEOEXPO 86-Exploration in the North American Cordillera, (ed.) I.L. Elliott and B.W. Smee; The Association of Exploration Geochemists, Proceedings Volume from Symposium, Vancouver, May 12-14, 1986, p. 61-71.

Fraser, T.M.

1994: Hydrothermal breccias and associated alteration at the Mount Polley copper-gold deposit (93A/12); in Geological Fieldwork 1993, (ed.) B. Grant and E. Newell; British Columbia Geological Survey Branch, Paper 1994-1, p. 259-267.

Frost, J.E.

1965: Controls of ore deposition for the Larap mineral deposits, Camarines Norte, Philippines; PhD. thesis, Stanford University, Stanford, California, 173 p.

Hickey, R.J., III

1992: The Buckhorn Mountain (Crown Jewel) gold skarn deposit, Okanogan County, Washington; Economic Geology, v. 87, p. 125-141.

Höy, T. and Andrew, K.P.E.

1991: Geology of the Rossland-Trail area, southeastern British Columbia; British Columbia Ministry of Energy, Mines and Petroleum Resources, Open File Map 1991-2.

James, L.P.

1976: Zoned alteration in limestone at porphyry copper deposits, Ely, Nevada; Economic Geology, v. 71, p. 488-512.

Koski, R.A. and Cook, D.S.

1982: Geology of the Christmas porphyry copper deposit: Gila County, Arizona; in Advances in Geology of the Porphyry Copper Deposits, Southwestern North America, (ed.) S.R. Titley; University of Arizona Press, Tucson, Arizona, 560 p.

Lane, M.L.

1987: Geology and mineralization of the McCoy skarn, Lander County, Nevada; in Program and Abstracts, Northwest Miners Association Annual Meeting, Spokane, Washington, p. 28.

Macauley, T.N.

1973: Geology of the Ingerbelle and Copper Mountain deposits at Princeton, B.C.; The Canadian Institute of Mining and Metallurgy, Bulletin, v. 66, no. 732, p. 105-112.

McKechnie, N.D.

1964: Gem, Dividend-Lakeview; British Columbia Minister of Mines, Annual Report, 1963, p. 65-67.

Megaw, P.K.M., Ruiz, J., and Titley, S.R.

1988: High-temperature, carbonate-hosted Ag-Pb-Zn(Cu) deposits of northern Mexico; Economic Geology, v. 83, no. 8, p. 1856-1885.

Meinert, L.D.

1984: Mineralogy and petrology of iron skarns in western British Columbia, Canada; Economic Geology, v. 79, p. 869-882.

*1986: Gold in skarns of the Whitehorse Copper Belt, southern Yukon; in Yukon Geology, v. 1, Exploration and Geological Services Division, Yukon, Indian and Northern Affairs Canada, p. 19-43.

*1988: Gold in skarn deposits – a preliminary overview; in Proceedings of the 7th Quadrennial International Association on the Genesis of Ore Deposits Symposium, (ed.) E. Zachrisson; Schweizerbart'sche Verlagsbuchhandlung, Stuttgart, Germany, p. 363-374.

1992: Skarns and skarn deposits; Geoscience Canada, v. 19, no. 4, p. 145-162.

Morrison, G.W.

1981: Setting and origin of skarn deposits in the Whitehorse Copper Belt, Yukon; PhD. thesis, University of Western Ontario, London, Ontario, 306 p.

Neilson, R.L.

1970: Mineralization and alteration in calcareous rocks near the Santa Rita stock, New Mexico; New Mexico Geological Society Guidebook, 21st Field Conference, p. 133-139.

Nokleberg, W.J., Bundtzen, T.K., Berg, H.C., Brew, D.A., Grybeck, D., Robinson, M.S., Smith, T.E., and Yeend, W.

1988: Metallogeny and major mineral deposits of Alaska; United States Geological Survey, Open File Report 88-73, 87 p.

Orris, G.J., Bliss, J.D., Hammarstrom, J.M., and Theodore, T.G.

1987: Description and grades and tonnages of gold-bearing skarns; United States Geological Survey, Open File Report 87-273, 50 p.

Preto, V.A.

1972: Geology of Copper Mountain; British Columbia Department of Mines and Petroleum Resources, Bulletin 59, 87 p.

Ray, G.E. and Webster, I.C.L.

1991: An overview of skarn deposits; in Ore Deposits, Tectonics and Metallogeny in the Canadian Cordillera, British Columbia Ministry of Energy, Mines and Petroleum Resources, Paper 1991-4, p. 213-252.

Ray, G.E., Ettlinger, A.D., and Meinert, L.D.

1990: Gold skarns: their distribution, characteristics and problems in classification; in Geological Fieldwork 1989; British Columbia Ministry of Energy, Mines and Petroleum Resources, Paper 1990-1, p. 237-246.

Ray, G.E., Webster, I.C.L., Dawson, G.L., and Ettlinger, A.D.

1993: A geological overview of the Hedley gold skarn district, southern British Columbia (92H); in Geological Fieldwork 1992; British Columbia Ministry of Energy, Mines and Petroleum Resources, Paper 1993-1, p. 269-279.

Reynolds, D.G.

1965: Geology and mineralization of the Salsigne gold mine, France; Economic Geology, v. 60, p. 772-791.

Sangster, D.F.

1969: The contact metasomatic magnetite deposits of southwestern British Columbia; Geological Survey of Canada, Bulletin 172, 85 p.

Shimazaki, H.

1969: Pyrometasomatic copper and iron ore deposits of the Yaguki mine, Fukushima Prefecture, Japan; University of Tokyo Faculty Section Journal, sec. 2, v. 17, p. 317-350.

Shultz, J. and Hamann, R.

1977: El Mochito: a good mine taxed to the hilt; Engineering and Mining Journal, November, 1977, p. 166-183.

Sillitoe, R.H.

1983: Low-grade gold potential of volcano-plutonic arcs; in Proceedings of Society of Mining Engineers of American Institute of Mining, Metallurgical and Petroleum Engineers, Precious Metals Symposium, (ed.) V.E. Kral; Sparks, Nevada, 1980, Nevada Bureau of Mines and Geology Report 36, p. 62-68.

Sutherland Brown, A.

1968: Geology of the Queen Charlotte Islands, British Columbia; British Columbia Department of Mines and Petroleum Resources, Bulletin No. 54, 226 p.

Sutherland Brown, A., Cathro, R.J., Panteleyev, A., and Ney, C.S.

1971: Metallogeny of the Canadian Cordillera; The Canadian Institute of Mining and Metallurgy Bulletin, v. 64, no. 709, p. 37-61.

Theodore, T.G. and Blake, D.W.

1978: Geology and geochemistry of the West orebody and associated skarns, Copper Canyon porphyry copper deposits, Lander County, Nevada; United States Geological Survey, Professional Paper 798-C, 85 p.

Theodore, T.G., Orris, G.J., Hammarstrom, J.M., and Bliss, J.D.

1991: Gold-bearing skarns; United States Geological Survey, Bulletin 1930, 61 p.

Torrey, C.E., Karjalainen, H., Joyce, P.J., Erceg, M.,
and Stevens, M.
1986:   Geology and mineralization of the Red Dome (Mungana) gold skarn deposit, North Queensland, Australia; in Proceedings of Gold '86, An International Symposium of the Geology of Gold, (ed.) A.J. Macdonald; Toronto, 1986, p. 504-517.

Wilson, G.C., Rucklidge, J.C., and Kilius, L.R.
1990:   Sulfide gold content of skarn mineralization at Rossland, British Columbia; Economic Geology, v. 85, p. 1252-1259.

Wolfenden, E.B.
1965:   Bau mining district, West Sarawak, Malaysia; Part 1, Bau; Geological Survey of Malaysia (Borneo region), Bulletin 7, pt. 1, 147 p.

Wotruba, P.R., Benson, R.G., and Schmidt, K.W.
1986:   Battle Mountain describes the geology of its Fortitude gold-silver deposit at Copper Canyon; Mining Engineering, v. 38, p. 495-499.

20.4 SKARN IRON

20.4a Contact metasomatic (associated with intrusive rocks)
20.4b Stratiform in metamorphic terrane

G.A. Gross

INTRODUCTION

The term "skarn" is used here as a lithological term, without specific genetic connotations, to designate a large group of mineral deposits that have a significant proportion of gangue rock composed of calcium and magnesium silicate minerals that formed by metamorphic and metasomatic processes in a diversity of geological settings and terranes. Skarn iron ore deposits have developed in many different metallogenes[1] and minerogenic environments, but commonly formed along the contacts of igneous intrusions with limestone, dolomite, and shale or volcanic rocks by metasomatic processes; by metamorphism and alteration of sequences of sedimentary, volcanic, and intrusive rocks; and in metallogenetic environments controlled principally by tectonic and structural features (Einaudi et al., 1981; Meinert, 1993).

Skarn-type iron deposits are usually composed of complex mineral assemblages consisting of pyroxene, amphibole, epidote, garnet, biotite, and chlorite, with associated magnetite, hematite, siderite, titaniferous magnetite, pyrite, pyrrhotite, and chalcopyrite, that are developed in the host rocks by metamorphic and metasomatic replacement processes. Skarn-type deposits as a group are important sources of iron, copper, tungsten, lead, zinc, molybdenum, and tin and some may contain significant amounts of other ferrous, nonferrous, and precious metals. They are also sources of graphite, asbestos, wollastonite, magnesite, phlogopite, talc, and fluorite (Gross, 1967a; Einaudi et al., 1981; Meinert, 1983, 1993).

Skarn deposits that provide significant resources of iron belong to two main subtypes:

1. contact metasomatic or replacement skarn deposits developed along the margins of igneous intrusions in association with limestone, dolomite, argillaceous sediments, and mafic volcanic rocks; and

2. stratiform skarn deposits hosted in sequences of highly metamorphosed sediments and volcanic rocks, that may have no evident or demonstrated association with intrusive rocks, and that may have developed by metamorphism of iron-bearing protolithic rocks of unknown origin.

The term "skarn iron" is usually not used or recommended for stratiform metalliferous deposits in which the nature and kinds of protolithic rocks bearing calcium and

Gross, G.A.
1996:   Skarn iron; in Geology of Canadian Mineral Deposit Types, (ed.) O.R. Eckstrand, W.D. Sinclair, and R.I. Thorpe; Geological Survey of Canada, Geology of Canada, no. 8, p. 489-495 (also Geological Society of America, The Geology of North America, v. P-1).

[1]   The term metallogene is used to designate sequences of interrelated geological processes associated with prominent sedimentary, igneous, metamorphic, and/or structural features in a geological domain that have produced concentrations of elements and minerals in deposits of significant size and type. More than one type of mineral deposit may be developed in a single metallogene (Gross, 1977).

magnesium silicate "skarn type" mineral assemblages have been positively identified or demonstrated, as in the case of metamorphosed lithofacies of iron-formation or other iron-bearing sedimentary and volcanic rocks.

Examples of intrusion-associated, contact metasomatic skarn deposits in Canada include Tasu, Iron Hill, and Texada Island in British Columbia. Foreign examples include Cornwall, Pennsylvania; Iron Springs, Utah; Iron Hat, California; Tayeh, China; and Sarbai, Sokolovsk, and Magnitnaya in the former U.S.S.R.

Canadian examples of stratiform skarn deposits in metamorphic terranes include the Marmora deposit, Ontario; the Hilton mine, Quebec; and probably many others derived from iron-formations in which primary features and the nature of the protolithofacies has been destroyed by metamorphism or is not clearly defined. Foreign examples include the Marcona deposits in Peru and numerous deposits in northern Sweden which may have been derived from ferruginous sediments, including highly metamorphosed iron-formations.

IMPORTANCE

Iron ore production from skarn deposits has probably now fallen to less than 2% of total world production, and some skarn deposits are only mined because copper and other byproduct minerals are recovered in the processing of the ore. The greatest production has been from the large skarn deposits in the former U.S.S.R., China, Peru, and U.S.A. Although production of iron ore from other skarn deposits has been relatively small, the mines developed in them have had an important impact on the economy of local communities in Canada, China, and many other countries.

Past production of iron ore from skarn deposits in Canada was generally less than one million tonnes of concentrate per year, from deposits containing less than 30 Mt of ore and with grades ranging from 35 to 50% Fe. Production of iron ore concentrate from the skarn deposits in southwestern British Columbia was developed for an export market and, although small by world standards for iron ore (less than 2 Mt per year) was an important factor in the regional economy. For example, the total production from British Columbia in 1964 from the larger mine areas,

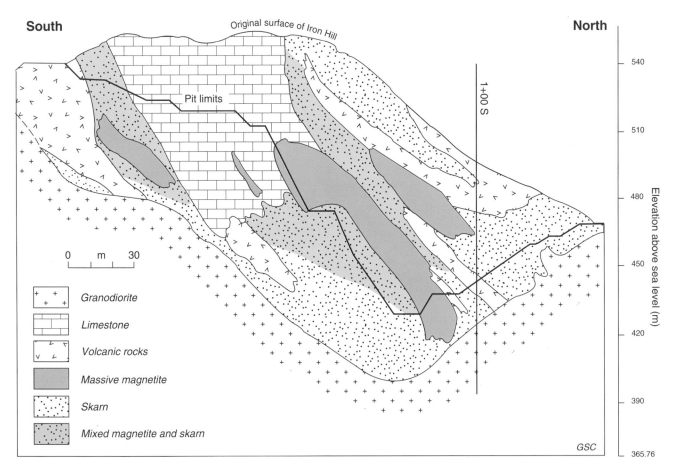

Figure 20.4-1. Geology in section of the Iron Hill iron deposit, Vancouver Island (after Sangster, 1969).

Texada Mines Ltd., Zeballos Iron Mines, Jedway Iron Ore, Empire Development, Coast Copper Co., and Brynnor Mines, was about 1.7 Mt of magnetite concentrate, which was recovered from about 3 Mt of crude ore containing 30 to 54% Fe. The iron content of the concentrate ranged from 54.1 to 62.56% Fe, and the SiO_2 content from 3.3 to 8.4%.

Production of iron ore concentrate on the west coast of Canada began on Texada Island in 1952. Most of the larger mine areas continued production of less than 0.5 Mt of magnetite concentrate per year for periods of about 10 years from the early 1950s to the late 1960s.

Copper was a prime product with the iron ore concentrate at the Coast Copper Co. mine, and was recovered as a byproduct in the concentration of the skarn iron ore from some of the deposits in the region. The copper content ranged from 0.2 to 1.3% at Tasu, formerly the Wesfrob mine, on Moresby Island, Queen Charlotte Islands, and production of copper with the iron ore concentrate was an important factor in extending production in this mine area.

Wesfrob Mines at Tasu Harbour began production in 1967 at a rate of about one million tonnes per year of crude iron ore, and in 1983 produced more than 490 000 t of magnetite concentrate grading 63.3% natural iron (69.1% dry analysis) from about 850 000 t of ore mined. Production from the mine ceased in late 1983.

The Marmora deposit in Hastings County, Ontario is typical of stratiform iron-rich skarns in the Grenville Province of eastern Canada. It was first detected by an airborne geophysical survey in 1949 by the Geological Survey of Canada and the Ontario Department of Mines, below a 30-40 m thick capping of Paleozoic limestone. The first shipment of pellets was made in 1955 following the removal of 20 Mt of limestone and the normal production rate during mining operations was about 0.5 Mt of pellets a year. A total of 1.126 Mt of waste rock was removed in 1959 and 0.313 Mt of concentrate grading 66.25% Fe was obtained from 762 785 t of ore treated in the mill.

SIZE AND GRADE OF DEPOSITS

The largest skarn iron deposits in the world, such as Sokolovsk and Sarbai in the former U.S.S.R., contain 1000 Mt of ore, but most in North America have less than 50 Mt, and grade from 35 to 50% Fe. Proven reserves in most of the contact metasomatic deposits in southern British Columbia were less than 20 Mt and ore from a cluster of small deposits was concentrated at central mills. Individual deposits ranged in size from less than 2 to 10 Mt of crude ore grading 30 to 54% Fe. The Tasu skarn deposit on Moresby Island was probably the largest in the British Columbia coastal region. It produced about 21 Mt of ore from 1914 to 1983 at recovered grades of about 40% Fe, 0.29% Cu, 2.4 g/t Ag, and 0.064 g/t Au.

The initial drilling program at the Marmora stratiform skarn iron deposit indicated about 20 Mt of ore grading 35 to 37% Fe within the projected limits of an open pit mine extending to a depth of about 150 m. Production from 1955 to 1978 was about 28 Mt of ore at a grade of 42.8% Fe.

GEOLOGICAL FEATURES

Geological setting

Contact metasomatic skarn deposits are commonly formed in volcanic arc and rifted continental margin tectonic settings at the contacts of felsic or mafic intrusions with carbonate, calcareous clastic, and volcanic rocks. The more complex stratiform-type skarn deposits occur in highly metamorphosed terranes such as the Grenville Province in eastern Canada where deposits of known sedimentary and replacement origin have been subjected to several stages of tectonic deformation and metamorphism with considerable remobilization of the iron and major constituents.

Age of host rocks

Skarn mineralization occurs in rocks of all ages, but is probably most abundant in Mesozoic and younger tectonic belts.

Form of deposits

Skarn deposits vary greatly in form and mineralogy. Contact metasomatic deposits range from irregular, massive and disseminated patches, to veins and dyke-like masses (Fig. 20.4-1, Iron Hill). Stratiform skarn deposits are more uniform, tabular, massive, or layered conformable bodies. Mineral distribution in both subtypes is patchy to uniform and textures vary greatly. The patchy and irregular to erratic distribution of magnetite in skarn deposits is typical, and generally the whole mass, composed of high grade pods, lenses, stringers, and disseminated ore, is mined and processed to produce a magnetite concentrate.

Mineralogy

The ore minerals in iron-rich skarns are magnetite, hematite, martite, and goethite, with minor amounts of pyrite, pyrrhotite, and chalcopyrite. They are associated with gangue mineral assemblages consisting of calc-silicate skarn minerals, pyroxene, amphibole, chlorite, epidote, calcite, dolomite, siderite, and variable amounts of apatite, alkali feldspar, biotite, garnet, and quartz. Skarn ores may vary from massive bodies of nearly pure magnetite to disseminated zones with less than 10% iron oxide. Skarn ore containing as little as 15% Fe has been mined successfully, but the average iron content of most of the ore mined is 35 to 50% Fe or greater.

Because of the great diversity in genetic processes that form skarn iron deposits, they often contain recoverable amounts of other ferrous, nonferrous, and precious metals, as well as tin, tungsten, molybdenum, titanium, phosphate, and a wide variety of other elements which reflect the composition and petrology of the associated intrusions.

Distribution of ore and gangue minerals and relationships to host rocks

Skarn deposits differ greatly in their mineral assemblages, textures, internal structures, and composition, depending on fundamental differences in their tectonic setting, nature of the host rocks and kinds of associated intrusions, and the metamorphic and alteration processes that produced them. Meinert (1983) pointed out that underlying the variations in size, texture, and mineralogy of skarn deposits are common patterns of 1) early initial isochemical metamorphism, 2) multiple intermediate stages of metasomatism, and 3) late retrograde alteration.

Contact metasomatic skarns in the coastal regions of British Columbia are composed mainly of garnet (andradite-grossularite), pyroxene (diopside-hedenbergite), epidote, and magnetite. Chalcopyrite, pyrite, pyrrhotite, and arsenopyrite are locally abundant (Sangster, 1969). The skarn deposits of this area are irregular masses that vary in composition and mineralogy depending on the associated host rocks. Most of the deposits have replaced volcanic rocks near contacts with limestone, some occur entirely within limestone, and rarely in the intrusive rocks. Stocks adjacent to the deposits are usually intermediate in composition, but range from gabbro to quartz monzonite. Folds and faults and the presence of limestone or other calcareous rocks are important controls for the development of skarn deposits.

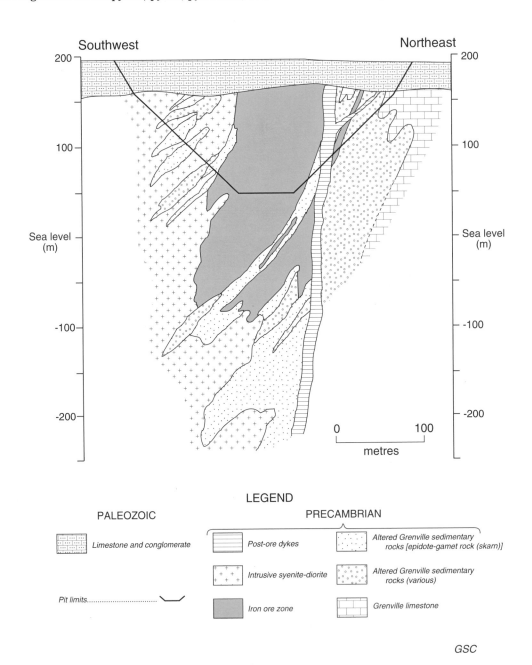

Figure 20.4-2. Geology in section of the Marmora iron mine, Hastings County, Ontario (after Rose, 1958).

Numerous small skarn deposits of all types in the southwestern part of the Grenville Province in Ontario and Quebec were studied extensively by Rose (1958, 1969) and Gross (1967a). Many of the deposits in this highly metamorphosed terrane are stratiform lenses of skarn developed in metasedimentary and metavolcanic rocks, gneisses, and schists.

The Marmora deposit lies within a belt of metasedimentary rocks composed of crystalline limestone, amphibolite, granite gneiss, and quartzitic gneiss that have been intruded by rocks of the gabbro-anorthosite-diorite group, by later granite and syenite, and finally by diabase dykes. The metasedimentary rocks in the mine strike northnortheast and dip 60°W. Marble predominates on the east side and considerable impure metaquartzite is distributed in the magnetite skarn and with the grey syenite and diorite intrusive masses which cut the skarn ore zone in the southwestern side and deeper parts of the mine (Fig. 20.4-2).

The ore deposit consists of a number of thin sinuous magnetite-rich zones, 15 to 30 m wide, composed of skarn rock impregnated with magnetite, that are roughly parallel to the strike and dip of the metasediments and form a large tabular mass about 130 m thick and 800 m long that extends in depth to 250 m. Most of the skarn, especially in the eastern part of the deposit, consists of medium- to fine-grained dark green pyroxene and amphibole and variable amounts of epidote, garnet, chlorite, and talc. Epidote-garnet-pyroxene skarn is more prevalent and coarser grained in the western part of the deposit near the syenite intrusions (Fig. 20.4-3).

Magnetite is mostly fine- to medium-grained and is disseminated in granoblastic textures with the skarn minerals, or forms thin lenses, shoots, and stringers in the skarn zone. Pyrite, pyrrhotite, and traces of chalcopyrite in

fractures in the magnetite skarn were most abundant in the western part of the deposit where they formed less than 5% of the ore.

GENETIC MODEL

Genetic processes for skarn deposits may vary greatly even in a single metallogene, and the complexities in the processes involved have been shown by studies of hydrothermal alteration in the Iron Hat skarn deposit in California by Hall et al. (1988). Major factors that influence the nature and distribution of skarn deposits are: 1) tectonic setting and geological history of an area; 2) depth of emplacement of intrusive rocks, mineral replacement, or metamorphism; 3) temperature of intrusions and wall rocks; 4) composition and petrogenesis of the associated igneous intrusions; and 5) composition of the sedimentary and associated volcanic rocks that were intruded.

Every skarn deposit is developed under a unique set of conditions and by a specific sequence of genetic processes which are difficult to discern in the final complex mineral assemblage.

Typical contact metasomatic skarn development takes place within the metamorphic aureoles developed around the margins of intrusive stocks where there is extensive circulation of hydrothermal solutions highly charged with metals derived by alteration and metamorphism of the wall rocks, and by cooling and differentiation of the adjacent magma. Hydrothermal solutions ranging in temperature to as great as 600°C may produce wall rock alteration and skarn mineral development at several stages or intervals during the formation of a deposit. In the course of wall rock alteration and skarn development, the hydrothermal solutions react most extensively with the carbonate rocks, resulting in an increase in their pH, silicification of the wall rock, development of silicate skarn minerals, and deposition of magnetite and other associated oxide and sulphide minerals. Three episodes of skarn formation in calcareous wall rocks adjacent to granite intrusions and two alteration events in the associated plutons have been documented for the Iron Hat skarn deposits in California (Hall et al., 1988).

Stratiform skarn deposits were probably derived from metasedimentary rocks originally rich in iron, or from mafic magmatic rocks in which the iron was mobilized to some extent during subsequent metamorphism. The genesis of skarn and igneous metamorphic deposits was discussed at some length by Park and MacDiarmid (1970), and Einaudi et al. (1981) and Meinert (1993) have given extensive consideration to the genesis of skarn deposits in general.

The large stratiform iron deposits of northern Sweden originally referred to as skarn iron ores are now considered by many to be highly metamorphosed sedimentary rocks and iron-formations, in which there has been considerable mobilization of the iron and of the magnesium and calcium constituents of the skarn minerals (Frietsch, 1973, 1974, 1977, 1978, 1979, 1980a, b, 1982a, b). The mineral assemblages and textures of many of the highly metamorphosed

Figure 20.4-3. Banded magnetite ore with amphibole, pyroxene, garnet, and epidote gangue minerals, Marmora deposit, Ontario (GSC 152352 1-7-58; from Gross, 1967a).

lithofacies of iron-formation in the Mount Wright and northeast regions of the Grenville Province in Canada (Gross, 1968) are very similar to those found in some of the stratiform skarn deposits.

RELATED DEPOSIT TYPES

Deposits related to skarn iron deposits may be classified into two principal groups. The first group includes syngenetic mineral concentrations in: a) the associated igneous intrusions, such as the banded and injected ilmenite and titaniferous magnetite deposits in anorthosite, gabbro, syenite, and granite; and b) the associated sedimentary and volcanic rocks, such as iron-formation and metalliferous sediments (Gross, 1968, 1986; Frietsch, 1973, 1974, 1977, 1978, 1980a, b, 1982a, b).

The second group includes skarn deposits of many different types and compositions that developed by replacement and alteration processes in the contact aureoles of intrusions, and consist of pegmatite dykes, veins, and disseminated mineralization in skarn zones.

Deposits of the second group are usually classified on the basis of the principal metals recovered and include iron-rich skarns as described above, copper-bearing skarn deposits, (e.g., Tasu, British Columbia); tungsten skarns, (e.g., MacMillan Pass, Yukon Territory); molybdenum skarns, (e.g., Little Boulder Creek, Idaho); zinc-lead-bearing skarns, (e.g., Naica, Mexico); tin-bearing skarns (e.g., Moina, Tasmania, and Pingyung in northern Kwangtung Province, China). Some deposits of this group associated with pegmatite dykes contain niobium, tantalum, uranium, rutile, and zirconium.

Skarn-related iron deposits also occur in volcanic breccias, volcanic necks, and calderas in association with explosive and intrusive breccia cemented by hematite, magnetite, siderite, pyrite, pyrrhotite, copper-bearing sulphides, uranium oxide, gold, silver, and rare-earth elements. The Olympic Dam deposits in Southern Australia are considered to be outstanding examples of skarn-related mineralization associated with intrusive and volcanic breccia (Kennedy, 1988).

A wide variety of nonmetallic mineral resources occur in skarn environments, including wollastonite, feldspar, mica, brucite, magnesite, talc, serpentinite, tremolite, spodumene, amblygonite, apatite, and graphite.

EXPLORATION GUIDES

Skarn deposits are most likely to be found in the contact zones and alteration aureoles around intrusive stocks, especially where these are associated with carbonate and calcareous clastic sediments, and volcanic rocks. Skarn mineral zones are usually indicated by a) hydrothermal alteration in the potential host rocks, especially in carbonate and mafic rocks that react readily to neutralize solutions of low pH; b) by evidence of silicification, chloritization, epidotization, and/or albitization in host rocks; and c) volatile constituents in the host rocks that could have been introduced by alteration processes.

Highly metamorphosed and deformed supracrustal sequences that contain iron-rich and calcareous strata are favourable host rocks for the development of stratiform skarn deposits.

Magnetic anomalies in highly metamorphosed terrane, even of a diffuse and irregular configuration, should be examined as they may indicate metalliferous zones that have been highly deformed, and where considerable mobilization of metals and rock constituents has taken place.

Many skarn deposits are too small to give geochemical anomalies, or they do not contain conspicuous marker elements that can be detected easily by geochemical prospecting methods. However, these methods may be used successfully in conjunction with detailed studies of host rock petrology and alteration aureoles.

Studies of host rock petrology may provide useful indicators of rock alteration and skarn development.

SELECTED BIBLIOGRAPHY

References marked with asterisks (*) are considered to be the best sources of general information on this deposit type.

*Einaudi, M.T., Meinert, I.D., and Newberry, R.J.
1981:   Skarn deposits; in Economic Geology, Seventy-fifth Anniversary Volume, 1905-1980, (ed.) B.J. Skinner; p. 317-391.

Frietsch, R.
1966:   Berggrund och malmer i Svappavaarafältet, Norra Sverige; English summary: Geology and ores of the Svappavaara area, northern Sweden; with 5 plates in a separate folder; Sveriges Geologiska Undersökning, ser. C, no. 604, Stockholm, 282 p.

1973:   Precambrian iron ores of sedimentary origin in Sweden; in Genesis of Precambrian Iron and Manganese Deposits; Proceedings Kiev Symposium 1970, UN Educational, Scientific and Cultural Organization, 1973, Earth Sciences, 9, p. 77-82.

1974:   The occurrence and composition of apatite with special reference to iron ores and rocks in northern Sweden; Sveriges Geologiska Undersökning, ser. C, no. 694, 49 p.

1977:   The iron ore deposits in Sweden; in The Iron Ore Deposits of Europe and Adjacent Areas, (ed.) A. Zitzmann; Bundesanstalt fur Geowissenschaften und Rohstoffe, Hanover, Germany, p. 279-293.

1978:   On the magmatic origin of iron ores of the Kiruna type; Economic Geology, v. 73, p. 478-485.

1979:   Petrology of the Kurravaara area northeast of Kiruna, northern Sweden; Sverige Geologiska Undersökning, ser. C, no. 760, 81 p.

1980a: Precambrian ores of the northern part of Norrbotten County, northern Sweden; Guidebook to excursions 078 A+C, Part 1 (Sweden), 26th International Geological Congress, Paris 1980, Geological Survey of Finland, Espoo, 1980, 35 p.

1980b: The ore deposits of Sweden; Geological Survey of Finland, Bulletin 306, 20 p.

1982a: A model for the formation of the non-apatitic iron ores, manganese ores and sulphide ores of central Sweden; Sveriges Geologiska Undersökning, ser. C, no. 795, 41 p.

1982b: Alkali metasomatism in the ore-bearing metavolcanics of central Sweden; Sveriges Geologiska Undersökning, ser. C, no. 791, 54 p.

1985:   The Lannavaara iron ores northern Sweden; Sveriges Geologiska Undersökning, ser. C, no. 807, 55 p.

Gross, G.A.
*1965: General geology and evaluation of iron deposits; Volume I in Geology of Iron Deposits in Canada; Geological Survey of Canada, Economic Geology Report 22, 181 p.

*1967a: Iron deposits in the Appalachian and Grenville regions of Canada; Volume II in Geology of Iron Deposits in Canada; Geological Survey of Canada, Economic Geology Report 22, 111 p.

1967b: Iron deposits of the Soviet Union; The Canadian Institute of Mining and Metallurgy, Bulletin, v. 60, no. 668, p. 1435-1440.

1968:   Iron Ranges of the Labrador geosyncline; Volume III in Geology of Iron Deposits in Canada; Geological Survey of Canada, Economic Geology Report 22, 179 p.

1977:   Metallogenetic evolution of the Canadian Shield; in Volume 2, Correlation of the Precambrian, (ed.) A.V. Sidorenko; Publishing Office "Nauka", Moscow, p. 274-293.

1986:   The metallogenetic significance of iron-formation and stratafer rocks; Journal of the Geological Society of India, v. 28, no. 2-3, p. 92-108.

Hall, D.L., Cohen, L.H., and Schiffman, P.
1988: Hydrothermal alteration associated with the Iron Hat iron skarn deposit, Eastern Mojave Desert, San Bernardino County, California; Economic Geology, v. 83, p. 568-587.

Kennedy, A.
1988: Olympic Dam Project; Mining Magazine, November 1988, p. 330-344.

Landergren, S.
1948: On the geochemistry of Swedish iron ores and associated rocks; Sveriges Geologiska Undersökning, ser. C, no. 496, 182 p.

Lang, A.H., Goodwin, A.M., Mulligan, R., Whitmore, D.R.E., Gross, G.A., Boyle, R.W., Johnston, A.G., Chamberlain, J.A., and Rose, E.R.
1970: Economic minerals of the Canadian Shield; Chapter V, Geology and Economic Minerals of Canada; Economic Geology Report No. 1, fifth edition, Department of Energy, Mines and Resources, Ottawa, Canada, p. 152-226.

Meinert, L.D.
1983: Variability of skarn deposits: guides to exploration; in Revolution in the Earth Sciences - Advances in the Past Half-century, (ed.) S.J. Boardman; Kendall/Hunt Publishing Company, Iowa, p. 301-316.

Meinert, L.D. (cont.)
1993: Skarns and skarn deposits; Geoscience Canada, v. 19, no. 4, p. 145-162.

Oftedahl, C.
1958: A theory of exhalative-sedimentary ores; Geologiska Föreningens i Stockholm Förhandlingar, no. 492, Band 80, Häfte 1, p. 1-19.

Parak, T.
1975: The origin of the Kiruna iron ores: Sveriges Geologiska Undersökning, ser. C, no. 709, 209 p.

Park, C.F. and MacDiarmid, R.A.
1970: Ore Deposits, Second Edition; W.H. Freeman and Company, San Francisco, 522 p.

Rose, E.R.
1958: Iron deposits of eastern Ontario and adjoining Quebec; Geological Survey of Canada, Bulletin 45, 120 p.
1969: Geology of titanium and titaniferous deposits of Canada; Geological Survey of Canada, Economic Geology Report 25, 176 p.

*Sangster, D.F.
1969: The contact metasomatic magnetite deposits of southwestern British Columbia; Geological Survey of Canada, Bulletin 172, 86 p.

20.5 SKARN TUNGSTEN

K.M. Dawson

IDENTIFICATION

Tungsten skarns are typically coarse grained assemblages of ore and calc-silicate gangue minerals that form commonly in the thermal aureole at the contact between felsic, calc-alkaline intrusive and calcareous sedimentary rocks. Scheelite, commonly with pyrrhotite and either chalcopyrite or molybdenite, is unevenly distributed throughout a prograde calc-silicate assemblage of mainly hedenbergitic pyroxene and grossular-andradite-almandine garnet, and a hydrous retrograde assemblage of mainly hornblende and biotite. Significant Canadian examples include Canada Tungsten, Northwest Territories; Salmo district, British Columbia; and MacMillan Tungsten, Yukon Territory. Some large foreign deposits include King Island, Tasmania; Sangdong, South Korea; Tyrnyauz and Vostok-2, Russia; Shizhuyuan, China; and Bishop district, California.

Dawson, K.M.
1996: Skarn tungsten; in Geology of Canadian Mineral Deposit Types, (ed.) O.R. Eckstrand, W.D. Sinclair, and R.I. Thorpe; Geological Survey of Canada, Geology of Canada, no. 8, p. 495-502 (also Geological Society of America, The Geology of North America, v. P-1).

IMPORTANCE

Almost 100% of Canadian tungsten production, which at one time constituted about 5% of the annual world production, has been derived from skarn deposits. Skarns are the second most important type of economic tungsten deposit after vein-stockwork deposits, and account for an estimated 30% of total world production.

SIZE OF DEPOSIT

The bulk of the world's skarn tungsten production and reserves is accounted for by relatively few deposits with greater than 10 000 t contained metal (Fig. 20.5-1). A few open pit mines, operated during World War II, produced ore with grades below 0.4% WO_3, but most underground operations require average grades of at least 0.4% WO_3. Remote areas, such as the northern Canadian Cordillera, require large scale mining of ore with an average grade of 1% WO_3 or greater. Tonnages and grades of significant Canadian and foreign skarn tungsten deposits are given in Figure 20.5-1 and Table 20.5-1. Size, grade, and other characteristics of global tungsten skarns have been tabulated by Einaudi et al. (1981).

GEOLOGICAL FEATURES

Geological setting

Tungsten skarns in the North American Cordillera are commonly localized in the thermal aureole of Mesozoic plutons that have discordantly intruded Paleozoic cratonal shelf carbonate-pelite sequences. In the Omineca Belt of the eastern Canadian Cordillera, tungsten-rich skarns are hosted by North American miogeoclinal shelf limestone-pelite assemblages and their displaced equivalents. Several developed prospects with large reserves define an arcuate belt which includes Canada Tungsten (Cantung) mine and flanks Selwyn Basin on the east and northeast (Fig. 20.5-2).

Tungsten skarns are emplaced in a generally deeper, higher temperature, and more reduced environment than Cu- and Zn-rich skarns, as deduced from extensive thermal aureoles, coarse grained intrusive textures, presence of aplites and pegmatites, lack of breccias and hydrothermal alteration, low ferric:ferrous ratios in calc-silicate gangue, and abundant carbon and pyrite in host rocks. A typical example of a pluton associated with a reduced tungsten skarn is the MacMillan Tungsten (Mactung) stock (Dick and Hodgson, 1982), a coarse grained K-feldspar megacrystic quartz monzonitic member of the calc-alkaline Selwyn plutonic suite (Anderson, 1983). An oxidized subtype of tungsten skarn, exemplified by King Island, Tasmania, formed in noncarbonaceous or hematitic rocks at lesser depths (Newberry, 1979).

Age of host rocks, associated rocks, and mineralization

Host rocks for tungsten skarns are the contact metamorphosed equivalents of relatively pure limestones and calcareous to carbonaceous pelites: skarn, calc-silicate hornfels, and biotite-pyrite hornfels. On a global basis, host rocks ranging in age from late Proterozoic to early Mesozoic have been intruded by granitoid plutons that are dominantly Paleozoic to late Mesozoic. Exceptions include the Fostung (Ontario) and Yxsjöberg (Sweden) deposits, which are associated with Proterozoic granite plutons, and the San Alberto (Mexico) deposit, which is associated with a Tertiary pluton. Cordilleran skarns have developed preferentially in the lowest thick limestone bed of an Upper Proterozoic to Carboniferous cratonal or pericratonic sedimentary sequence. A typical Canadian Cordilleran setting is Cambrian shelf limestone, underlain by and interbedded with pelite and carbonaceous shale, and intruded by a postorogenic mid-Cretaceous quartz monzonite stock.

Intrusive rocks associated with tungsten skarns are calc-alkaline felsic stocks, plutons, or batholiths. Quartz monzonite is most common; quartz diorite is least common. Newberry and Swanson (1986) noted that most granitoids associated with scheelite skarns in the western United States show features characteristic of the 'I-type granites' of Chappell and White (1974). Tungsten skarn-related granitoid plutons in the northern Canadian Cordillera, on the other hand, are both 'S-type' and 'I-type' (Anderson, 1983). Plutons commonly are coarse grained, porphyritic, and unaltered, but border phases are in some cases locally

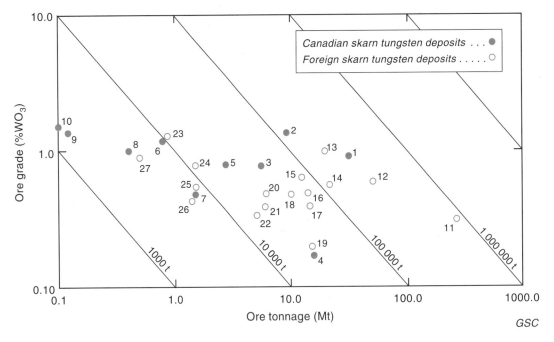

Figure 20.5-1. Grade versus tonnage of significant Canadian and foreign skarn tungsten deposits. Numbers correspond to deposits listed in Table 20.5-1.

argillized, greisenized, or tourmalinized. Stockwork quartz+scheelite+molybdenite veins are not extensive, but are more abundant in the intrusive rock than in the skarn. Porphyritic dykes are rarely associated with W skarns, but aplite and pegmatite dykes are common. Breccia pipes, intrusive and shatter breccias, and other features indicative of high levels of emplacement are absent.

Ore deposition is penecontemporaneous with emplacement of associated intrusive rocks: dominantly Paleozoic in Europe, the former Soviet Union, and Australia; Late Mesozoic in eastern Asia; and Late Jurassic to Early Tertiary, but mainly Cretaceous, in the North American Cordillera. A few deposits, as cited above, are Proterozoic in age.

Table 20.5-1. Tonnage and grade of significant Canadian and foreign tungsten skarn deposits, listed in decreasing order of contained tungsten metal.

Deposit	Tonnage; grade	Comments/references
Canadian deposits		
1. MacMillan Tungsten (Mactung), Yukon Territory	32 Mt; 0.92% WO$_3$	Reserves (Atkinson and Baker, 1986)
2. Canada Tungsten (Cantung), Northwest Territories	9 Mt; 1.42% WO$_3$	Production plus reserves (Mathieson and Clark, 1984)
3. Ray Gulch, Yukon Territory	5.44 Mt; 0.82% WO$_3$	Reserves (Lennan, 1986)
4. Fostung, Ontario	16.2 Mt; 0.23% WO$_3$	Reserves (Ginn and Beecham, 1986)
5. Risby, Yukon Territory	2.7 Mt; 0.81% WO$_3$	Reserves (The Northern Miner, 8 July, 1982, p. 30)
6. Lened, Northwest Territories	about 0.75 Mt; 1.2% WO$_3$	Reserves (Glover and Burson, 1986)
7. Salmo district, British Columbia	about 1.5 Mt; 0.5% WO$_3$	Production 1943-1973 (Mulligan, 1984)
8. Bailey, Yukon Territory	0.4 Mt; 1.0% WO$_3$	Reserves (D.I.A.N.D., 1981)
9. Baker, Northwest Territories	about 0.12 Mt; 1.4% WO$_3$	Reserves (S. Bartlett, pers. comm., 1986)
10. Clea, Yukon Territory	about 0.1 Mt; 1.5% WO$_3$	Reserves (Godwin et al., 1980)
Foreign deposits		
11. Shizhuyuan, China	170 Mt; 0.33% WO$_3$	Reserves, about 2/3 is skarn (Zhang, 1980)
12. Tyrnyauz, Russia	50.8 Mt; 0.6% WO$_3$	In situ resources (Anstett et al., 1986)
13. Sangdong, South Korea	about 20 Mt; 1.0% WO$_3$	Production 1916-1985, plus reserves (Yih and Wang, 1979; Anstett et al., 1986)
14. Vostok-2, Russia	22 Mt; 0.58% WO$_3$	In situ resources (Anstett et al., 1986)
15. King Island, Tasmania	13 Mt; 0.65% WO$_3$	Production 1911-1972, plus reserves (Danielson, 1975)
16. Uludag, Turkey	14.5 Mt; 0.5% WO$_3$	Reserves (Karahan et al., 1980)
17. Brejui/Barra Verde/Boca de Lage, Brazil	about 15 Mt; 0.4% WO$_3$	Production 1943-1985, plus reserves (Willig and Delgado, 1985)
18. Pine Creek, California	about 10 Mt; 0.5% WO$_3$	Production 1918-1977 (Yih and Wang, 1979)
19. Indian Springs, Nevada	15.8 Mt; 0.2% WO$_3$	Reserves (Nevada State Journal, 05 October 1969)
20. Brown's Lake, Montana	about 6 Mt; 0.5% WO$_3$	Production 1953-1958, plus reserves (J.E. Elliott, pers. comm., 1985)
21. Mill City district, Nevada	about 6 Mt; 0.4% WO$_3$	Production 1925-1958, plus reserves (J.E. Elliott, pers. comm., 1985)
22. Yxsjöberg, Sweden	5 Mt; 0.3-0.4% WO$_3$	Production 1938-1963, plus reserves (Hübner, 1971)
23. Salau, France	0.85 Mt; 1.48% WO$_3$	Production 1972-1977, plus reserves (Reymond, 1981)
24. Los Santos, Spain	1.5 Mt; 0.8% WO$_3$	Reserves (Billiton Espanola, internal report, 1987)
25. San Alberto, Mexico	about 1.5 Mt; 0.55% WO$_3$	Reserves (W. Gruenweg, pers. comm., 1986)
26. Osgood Mountains, Nevada	1.4 Mt; 0.45% WO$_3$	Production 1942-1955 (Hotz and Willden, 1964)
27. Strawberry, California	0.5 Mt; 0.9% WO$_3$	Production 1942-1966, plus reserves (Nokleberg, 1981)

Form of deposit and associated structures, zoning, and distribution of ore and gangue minerals

Tungsten skarns commonly form as essentially stratiform exoskarns tens to hundreds of metres away from an intrusive contact and are continuous for as much as hundreds of metres along a lithological, e.g., carbonate-pelite, contact. Tungsten skarns also form as semiconcordant to discordant bodies immediately adjacent to an intrusive contact, either as early reaction skarn, metasomatic exoskarn, or as retrograde alterations of both. They form less commonly as endoskarn, either as replacements of the intrusive rock itself or of xenoliths, roof pendants, and screens within a plutonic border phase. In both prograde and retrograde skarn assemblages, essentially stratiform replacement textures predominate over discordant, vein-like morphology. Abundant fractures, like other indications of forceful emplacement of the intrusion, are lacking. Vein skarns are

rare. The structurally and lithologically controlled evolutionary stages and morphologies of a typical skarn are shown in Figure 20.2-7 (in subtype 20.2, "Skarn copper").

Scheelite commonly occurs in an essentially stratiform almandine-hedenbergite exoskarn assemblage that metasomatically overprints and replaces earlier metamorphic calc-silicate hornfels. Scheelite also may be erratically redistributed in association with structurally-controlled retrograde amphibole-biotite alteration. Scheelite, commonly with pyrrhotite and either chalcopyrite or molybdenite, is unevenly distributed throughout the prograde and retrograde calc-silicate mineral assemblages. Mineral zoning may be well developed and typically consists of sphalerite deposited peripherally to scheelite-rich skarn. Pyrrhotite and biotite-amphibole-rich skarn assemblages are structurally controlled, and retrograde alteration products of garnet and pyroxene-rich skarn and pyrite occupy late structures.

Prior to its 1986 closure, the Canada Tungsten (Cantung) mine at Tungsten, Northwest Territories (Fig. 20.5-2) was the largest tungsten producer in the western world. Other significant Canadian production has been obtained from four skarn deposits in the Salmo district of southeastern British Columbia (Fig. 20.5-1 and 20.5-2). At Cantung, three high-grade scheelite skarn orebodies are localized in the limbs of an overturned anticline composed of the Lower Cambrian 'Swiss Cheese limestone' and 'Ore limestone' host rocks which are underlain by phyllite and overlain by quartzite and dolomite (Blusson, 1968). A schematic geological cross-section through the mine area (Fig. 20.5-3) shows the E-zone orebody developed in the lower limb adjacent to the shallowly dipping contact of a mid-Cretaceous quartz monzonite stock. This deposit, which had initial reserves of 4.2 Mt of 1.6% WO$_3$, is unusually large and high grade. Zonal distribution of skarn facies and relative W contents of the E-zone orebody are shown in Figure 20.5-4.

Mineralogy

Principal ore mineral: scheelite.

Opaque minerals: chalcopyrite, sphalerite, molybdenite, pyrrhotite, pyrite (late), magnetite, bismuth, bismuthinite.

Prograde skarn minerals: pyroxene (hedenbergite-diopside), garnet (grossularite-andradite-almandine); calcite, dolomite, quartz, vesuvianite, wollastonite.

Retrograde skarn minerals: hornblende, biotite, plagioclase, epidote, sphene, chlorite, actinolite, apatite.

DEFINITIVE CHARACTERISTICS

1. Stratiform skarn morphology dominates over discordant and vein-like forms.

2. Extensive thermal aureoles are commonly developed in shelf carbonate-pelite host rocks.

3. Associated intrusions are coarse grained, unaltered, porphyritic, felsic, calc-alkaline granitoid stocks.

4. Associated dykes are pegmatite and aplite, rarely porphyry.

5. Breccias, breccia pipes, intense stockwork fractures in the intrusion and other high level structures are generally absent.

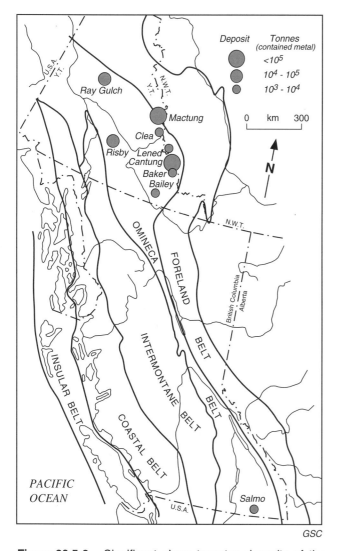

Figure 20.5-2. Significant skarn tungsten deposits of the Canadian Cordillera.

6. Deposits of reduced subtype have low ferric:ferrous ratios in skarn, and pyrite and carbon are present in their host rocks; deposits of oxidized subtype have high ferric:ferrous ratios, and hematite is present in their host rocks.

GENETIC MODEL

Tungsten, associated metals, sulphur, and cations other than calcium were derived mainly from the associated pluton by the separation of a magmatic-hydrothermal fluid. Constituents of opaque minerals, deposited abundantly during the cooling of the system, were derived mainly from an evolved hydrothermal fluid and, to a lesser extent, from the alteration of early minerals. The contribution of metals and sulphur from host and basement rocks by the late stage influx of meteoric waters, relative to associated plutonic rocks, is not known. Isochemical thermal metamorphism of limestone to marble and calcareous pelite to calc-silicate hornfels preceded the main stage of prograde metasomatic reactions. Hydrothermal fluids, released by crystallization and hydrofracturing of the pluton, infiltrated stratigraphic and structural channelways to react with calcareous host rock and calc-silicate hornfels to form prograde skarn. Chemical components of the system are of both local and exotic derivation: endoskarn forms in the pluton by introduction of calcium from the host rock, magnesian exoskarn forms in dolostones, and calcic exoskarn forms in limestones. The mineral assemblage of the resultant skarn is influenced by the composition, temperature, and oxidation state of the system. A typical calcic assemblage would include contemporaneously developed almandine-hedenbergite-scheelite exoskarn in calc-silicate hornfels, wollastonite-vesuvianite distal metamorphic skarn in marble, and hornblende-pyroxene-biotite-plagioclase endoskarn in the pluton and

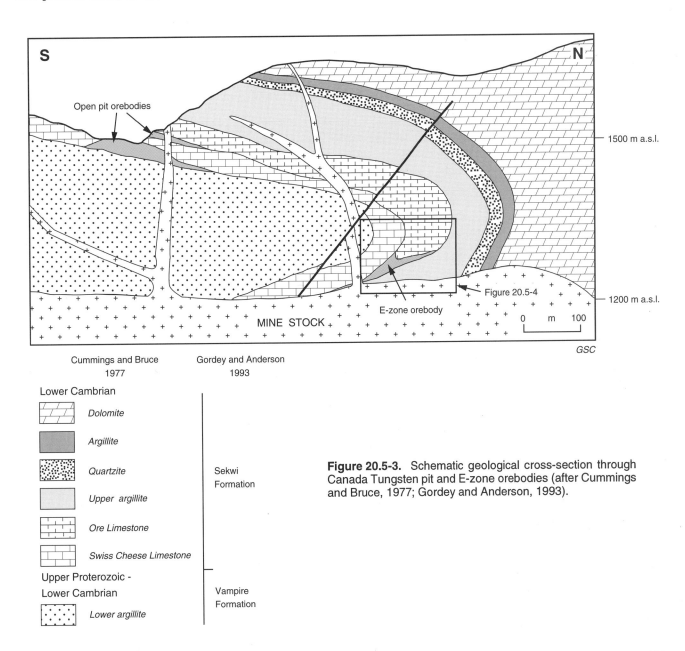

Cummings and Bruce
1977

Gordey and Anderson
1993

Lower Cambrian

Dolomite

Argillite

Quartzite   Sekwi
                   Formation
Upper argillite

Ore Limestone

Swiss Cheese Limestone

Upper Proterozoic -
Lower Cambrian   Vampire
                        Formation
Lower argillite

Figure 20.5-3. Schematic geological cross-section through Canada Tungsten pit and E-zone orebodies (after Cummings and Bruce, 1977; Gordey and Anderson, 1993).

499

in pelitic hornfels. Early anhydrous skarn is deficient in tungsten and sulphides, late stage prograde calcic skarn is enriched in ferric iron, manganese, sulphides, and scheelite. The main period of deposition of opaque minerals followed the termination of prograde skarn development, and accompanied the start of hydrous alteration of early skarn minerals and associated intrusive rocks. Subsequent cooling of the system and influx of meteoric water initiated hydrous retrograde alteration of early, calcium-rich silicates to calcium-depleted silicates, iron oxides, carbonates, and plagioclase. Processes occurring at this stage may have included hydrolytic alteration of associated intrusive rocks, deposition of most sulphides, and either primary deposition of scheelite or its redistribution with both depletion and upgrading. The suite of opaque minerals associated with hydrous retrograde assemblages reflects higher oxidation and sulphidization states and lower temperatures than conditions associated with early prograde opaque minerals. Evolutionary stages for skarns in general are given in Figure 20.2-7.

RELATED DEPOSIT TYPES

Some skarn tungsten deposits in the northern Canadian Cordillera are associated with quartz-molybdenite-scheelite stockworks in the adjacent intrusion, but none of these stockworks approach economic proportions as porphyry molybdenum-tungsten deposits. Molybdenum- and tungsten-bearing veinlets are confined mainly to altered border phases of the plutons, but in some cases overprint the adjacent scheelite-rich calc-silicate skarns. Several Mo-W stockwork-skarn deposits of this type, hosted by Lower Paleozoic limestones of the Cassiar Terrane in northern British Columbia and southern Yukon Territory, include the Logtung, Mount Haskin, and Boya prospects (Dawson et al., 1991; see Type 19, "Porphyry copper, gold, molybdenum, tungsten, tin, silver"). A large foreign example is the Shizhuyuan composite W skarn, W-Bi-Mo greisen-stockwork deposit in southern Hunan, China (Yang, 1982).

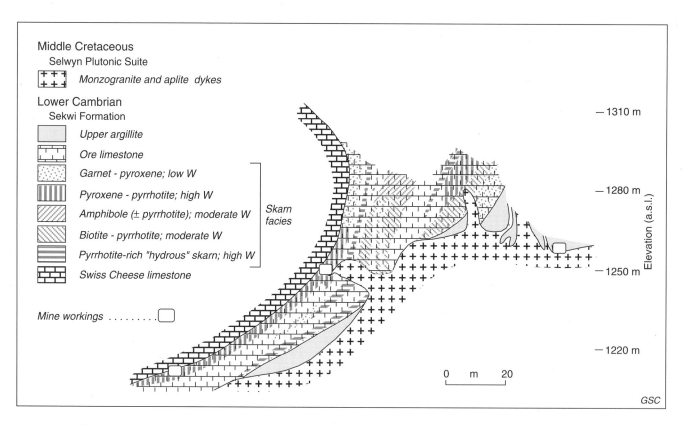

Figure 20.5-4. Cross-section 1215W through the Cantung E-zone orebody, lying to the west of the schematic section in Figure 20.5-3 (after Mathieson and Clark, 1984). Note anhydrous skarn along the Ore limestone unit in the footwall and hanging wall, and hydrous, in part pyrrhotite-rich, skarn facies in the axial zone (after Mathieson and Clark, 1984).

EXPLORATION GUIDES

Geological guides

The following geological features within a prospective terrain are favourable exploration guides:

a) Extensive hornfels zone adjacent to an exposed pluton, or overlying a buried one.

b) Relatively thick, pure and impure limestone beds.

c) Shallowly dipping pluton-limestone contacts.

d) Structural and stratigraphic traps in carbonate-pelite host rocks.

e) Irregularities in pluton-limestone contact, particularly reentrants and troughs.

f) Stockwork fracturing along pluton-limestone contact.

Geochemical guides

In regional geochemical soil and silt sampling surveys, samples of conventional size are less effective in detecting the dominantly particulate scheelite than are bulk samples of alluvium collected from principal drainages and reduced to a heavy mineral concentrate.

The elements Cu, Mo, and Zn, generally associated with W in skarns, may be used as pathfinders in regional geochemical surveys. Waters draining from plutons related to W skarns may be enriched in F.

Geophysical guides

Shallowly buried plutons, which may have controlled the development of W skarns, may be detected readily by airborne magnetic surveys due to their characteristically annular magnetic high surrounding a central low.

Multispectral satellite imagery may be selected to enhance colour anomalies associated with oxidized pyritic hornfels developed extensively in pelitic host rocks adjacent to the less extensive W skarn.

Prospecting guides

The effectiveness of the ultraviolet lamp as a scheelite prospecting tool is improved markedly by first locating, in daylight, prospective terrain, such as the base of a talus slope below a granite-limestone contact. The area should then be systematically prospected with the lamp during darkness, and any scheelite marked for follow-up examination in daylight.

REFERENCES

References marked with asterisks (*) are considered to be the best sources of general information on this deposit type.

Anderson, R.L.
1983: Selwyn plutonic suite and its relationship to tungsten skarn mineralization, southeast Yukon and District of Mackenzie; in Current Research, Part B; Geological Survey of Canada, Paper 83-1B, p. 151-163.

Anstett, T.F., Bleiwas, D.I., and Hurdelbrink, R.J.
1986: Tungsten availability - market economy countries; United States Bureau of Mines, Information Circular 9025, 51 p.

Atkinson, D. and Baker, D.J.
1986: Recent developments in the geologic picture of Mactung; in Mineral Deposits of Northern Cordillera, (ed.) J.A. Morin; The Canadian Institute of Mining and Metallurgy, Special Volume 37, p. 234-244.

Blusson, S.L.
1968: Geology and tungsten deposits near the headwaters of Flat River, Yukon Territory and southwestern District of Mackenzie, Canada; Geological Survey of Canada, Paper 67-22, 77 p.

Chappell, B.W. and White, A.J.R.
1974: Two contrasting granite types; Pacific Geology, v. 8, p. 173-174.

Cummings, W.W. and Bruce, D.E.
1977: Canada Tungsten - change to underground mining and description mine-mill practices; The Canadian Institute of Mining and Metallurgy, Bulletin, v. 70, no. 784, p. 94-101.

Danielson, M.J.
1975: King Island scheelite deposits; in Economic Geology of Australia and Papua New Guinea; Australasian Institute of Mining and Metallurgy, Monograph Series No. 5, p. 592-598.

Dawson, K.M., Panteleyev, A., Sutherland Brown, A., and Woodsworth, G.J.
1991: Regional metallogeny, Chapter 19; in Geology of the Cordilleran Orogen in Canada, (ed.) H. Gabrielse and C.J. Yorath; Geological Survey of Canada, Geology of Canada, no. 4, p. 707-768 (also Geological Society of America, The Geology of North America, v. G-2).

D.I.A.N.D.
1981: Pat (Bailey) summary; in Yukon Geology and Exploration 1979-80; Department of Indian Affairs and Northern Development (Canada), Northern Affairs Program, Exploration and Geological Services Division, Whitehorse, Yukon Territory, p. 140.

***Dick, L.A. and Hodgson, C.J.**
1982: The Mactung W-Cu (Zn) contact metasomatic and related deposits of the northeastern Canadian Cordillera; Economic Geology, v. 77, p. 845-867.

***Einaudi, M.T., Meinert, L.D., and Newberry, R.J.**
1981: Skarn deposits; in Economic Geology, Seventy-fifth Anniversary Volume 1905-1980, (ed.) B.J. Skinner, p. 317-391.

Ginn, R.M. and Beecham, A.W.
1986: The Fostung scheelite deposit, Espanola, Ontario; Canadian Geology Journal of The Canadian Institute of Mining and Metallurgy, v. 1, no. 1, p. 46-54.

Glover, J.K. and Burson, M.J.
1986: Geology of the Lened tungsten skarn deposit, Logan Mountains, Northwest Territories; in Mineral Deposits of Northern Cordillera, (ed.) J.A. Morin; The Canadian Institute of Mining and Metallurgy, Special Volume 37, p. 255-265.

Godwin, C.I., Armstrong, R.L., and Tompson, K.M.
1980: K-Ar and Rb-Sr dating and the genesis of tungsten at the Clea tungsten skarn property, Selwyn Mountains, Yukon; The Canadian Institute of Mining and Metallurgy, Bulletin, v. 73, no. 821, p. 90-93.

Gordey, S.P. and Anderson, R.G.
1993: Evolution of the northern Cordilleran miogeocline, Nahanni map area (105I), Yukon and Northwest Territories; Geological Survey of Canada, Memoir 428, 214 p.

Hotz, P.E. and Willden, R.
1964: Geology and mineral deposits of the Osgood Mountains quadrangle, Humboldt County, Nevada; United States Geological Survey, Professional Paper 431, 128 p.

Hübner, H.
1971: Molybdenum and tungsten occurrences in Sweden; Sveriges Geologiska Undersökning, Series Ca, no. 46, 29 p.

Karahan, S., Demirci, A., and Atademir, R.
1980: Turkish tungsten producer redesigns for efficiency; World Mining, v. 33, no. 10, p. 46-51.

Lennan, W.B.
1986: Ray Gulch tungsten skarn deposit, Dublin Gulch area, central Yukon; in Mineral Deposits of Northern Cordillera, (ed.) J.A. Morin; The Canadian Institute of Mining and Metallurgy, Special Volume 37, p. 245-254.

Mathieson, G.A. and Clark, A.H.

1984:   The Cantung E-zone scheelite skarn orebody, Tungsten, Northwest Territories: a revised genetic model; Economic Geology, v. 79, no. 5, p. 883-901.

*Mulligan, R.

1984:   Geology of Canadian tungsten occurrences; Geological Survey of Canada, Economic Geology Report 32, 121 p.

Newberry, R.J.

1979:   Systematics in W-Mo-Cu skarn formation and tungsten deposition in the Sierra Nevada: an overview; Geological Society of America, Abstracts with Programs, v. 11, p. 486.

Newberry, R.J. and Swanson, S.E.

1986:   Scheelite skarn granitoids: an evaluation of the roles of magmatic source and process; Ore Geology Reviews, v. 1, p. 57-81.

Nokleberg, W.J.

1981:   Geologic setting, petrology and geochemistry of zoned tungsten-bearing skarns at the Strawberry mine, Central Sierra Nevada, California; Economic Geology, v. 76, p. 111-133.

Reymond, M.

1981:   Anglade: scheelite mining in the snow-capped French Pyrénées; Mining Magazine, v. 145, no. 5, November 1981, p. 356-361.

Willig, C.D. and Delgado, J.

1985:   South America as a source of tungsten; in Tungsten: 1985, Proceedings of the Third International Tungsten Symposium, Madrid, 1985; MPR Publishing Services Ltd., Shrewsbury, England, p. 58-85.

Yang Chaoqun

1982:   Mineralization of the composite greisen-stockwork-skarn type W (scheelite and wolframite) Bi-Mo deposit of Shizhuyuan, Dongpo, southern Hunan, China; Tungsten Geology Symposium, Jiangxi, China, Proceedings, UN Economic and Social Commission for Asia and the Pacific, Regional Mineral Resources Development Centre, Bangdung, Indonesia, p. 503-520.

Yih, S.W.H. and Wang, C.T.

1979:   Tungsten: sources, metallurgy, properties and applications; Plenum Press, New York, 500 p.

Zhang Fumin

1980:   Brief introduction of metal mines in the People's Republic of China; in Proceedings, Fourth Joint Meeting, The Mining and Metallurgical Institute of Japan-American Institute of Mining, Metallurgical and Petroleum Engineers, 1980, Tokyo, Special Session A and B, p. 121.

Authors' addresses

K.M. Dawson
Geological Survey of Canada
100 West Pender Street
Vancouver, British Columbia
V6B 1R8

G.A. Gross
Geological Survey of Canada
601 Booth Street
Ottawa, Ontario
K1A 0E8

R.V. Kirkham
Geological Survey of Canada
100 West Pender Street
Vancouver, British Columbia
V6B 1R8

Printed in Canada

21. GRANITIC PEGMATITES

21. GRANITIC PEGMATITES

W.D. Sinclair

INTRODUCTION

Pegmatites are holocrystalline rocks typically composed of igneous rock-forming minerals that are, in part, very coarse grained, although some are extremely varied in grain size (Jahns, 1955). They are commonly granitic in composition, consisting mainly of quartz, feldspar, and mica, but more mafic varieties composed of olivine, pyroxene, and plagioclase also occur. Mafic pegmatites, however, have negligible economic significance and are not considered in this review.

A general classification scheme for granitic pegmatites based on their environments of formation and mineralogical features as suggested by Ginsburg (1984) and modified by Černý (1990, 1991b), is summarized in Table 21-1. Abyssal and muscovite class pegmatites are commonly mineralogically simple. Abyssal pegmatites consist mainly of quartz and feldspar, and are generally barren or poorly mineralized with regard to rare elements such as niobium, tantalum, rare-earth elements, yttrium, and beryllium. Muscovite class pegmatites contain extensive reserves of mica and feldspar and are in some cases enriched in uranium and rare-earth elements. Many of the pegmatites of the northeastern and central Grenville zones, Ontario likely belong to the muscovite class, but others contain significant rare element minerals and for these a transitional classification between muscovite class and rare element pegmatites is probably more appropriate (Černý, 1990). Miarolitic (gem-bearing) pegmatites are extremely rare in Canada.

The emphasis in this review is on rare element pegmatites, which are mineralogically complex and typically enriched in lithophile elements and rare metals such as beryllium, lithium, rubidium, cesium, tin, tantalum, niobium, rare-earth elements, and uranium. They also contain industrial minerals such as feldspar and mica. Rare element pegmatites have been subdivided by Černý (1990, 1991b) into rare-earth, beryl, complex, albite-spodumene, and albite types; in this review, however, they are considered as a single group.

Canadian examples of rare element pegmatites include the Tanco pegmatite, Manitoba (tantalum, lithium, cesium); uranium-bearing pegmatites of the Bancroft area, Ontario; and lithium-bearing pegmatites in the Preissac-Lacorne area, Quebec and the Yellowknife area, Northwest Territories. Notable foreign examples include the tin-spodumene belt, North Carolina, U.S.A. (lithium); the Manono and Kitotolo pegmatites, Zaïre (lithium, tin); the Bikita pegmatite, Zimbabwe (lithium, cesium); the Kamativi pegmatite, Zimbabwe (tin); the Uis pegmatite field, Namibia (tin); the Greenbushes pegmatite, Australia (tantalum, niobium, tin, lithium); and the pegmatite fields of the Afghan Hindukush (lithium, tantalum, cesium).

IMPORTANCE

Pegmatite deposits account for nearly all historical Canadian production of tantalum, cesium, and lithium, and contain most of the known reserves of these commodities. The Tanco pegmatite, for example, contains the largest known concentration of pollucite $[(Cs,Na)_2(Al_2Si_4)O_{12} \cdot H_2O]$ and was the world's largest tantalum producer in the 1970s (Crouse et al., 1984). Production of uranium from pegmatites in the Bancroft area has been of minor importance and present reserves are limited. Feldspar and mica have been produced in the past from numerous pegmatites in Ontario and Quebec, but production in recent years has been sporadic and of minor economic significance.

Sinclair, W.D.
1996:   Granitic pegmatites; in Geology of Canadian Mineral Deposit Types, (ed.) O.R. Eckstrand, W.D. Sinclair, and R.I. Thorpe; Geological Survey of Canada, Geology of Canada, no. 8, p. 503-512 (also Geological Society of America, The Geology of North America, v. P-1).

Table 21-1. Classification of pegmatite deposits (from Ginsburg, 1984; Černý, 1990, 1991b).

Pegmatite class	Environment of formation	Metamorphic facies of host rocks	Relationship to parent granites	Economic minerals
Miarolitic (gem-bearing)	~1-2 kbar	Greenschist	Within or peripheral to subvolcanic granitic plutons	Quartz crystals, beryl, topaz, tourmaline
Rare-element	~2-4 kbar	Lower amphibolite (Abukuma-type)	Peripheral to granitic intrusions	Spodumene, amblygonite, petalite, lepidolite, pollucite, beryl, columbite-tantalite, microlite, wodginite, uraninite, cassiterite, xenotime, gadolinite
Muscovite	~5-8 kbar	Upper amphibolite (Barrovian-type)	No obvious association with granitic intrusions in many cases	Muscovite, feldspar, uraninite
Abyssal	~4-9 kbar	Granulite (Barrovian- to Abukuma-type)	May be associated with migmatitic granite	Feldspar, quartz

On a world scale, pegmatites have been a major source of beryllium, lithium, cesium, tantalum, muscovite mica, and feldspar, and a minor source of uranium, yttrium, rare-earth elements, tin, and tungsten. Miarolitic pegmatites are an important source of gemstones such as beryl (emerald), topaz, and tourmaline.

SIZE AND GRADE OF DEPOSITS

Pegmatite deposits that have been commercially exploited range in size from thousands to millions of tonnes. Deposits in Ontario and Quebec that were mined in the past for feldspar and sheet mica were small and these are no longer in production for the most part. In other parts of the world, however, production from small pegmatite deposits is significant, particularly for beryllium and tantalum in Brazil and in many African and Asian countries.

Examples of pegmatite deposits in Canada that are significant for either their production or reserves of tantalum, cesium, lithium, and uranium are listed in Table 21-2 and shown in Figure 21-1; important foreign examples are included in Table 21-2 for comparison. Grade-tonnage relationships are shown in Figure 21-2. Quantitative data are unavailable for extensive pegmatite fields with sizable reserves of rare metals in Afghanistan (Hindukush), northern Australia (Finnis River), central Finland and Sweden, Russia, Ukraine, and China.

GEOLOGICAL FEATURES

Geological setting

Pegmatites and associated host rocks throughout the world range in age from early Precambrian to Tertiary. In Canada, the majority of commercially interesting pegmatites are Late Archean (Kenoran) or Late Proterozoic (Grenvillian) in age; some pegmatites are associated with Phanerozoic intrusive rocks but are of only minor commercial significance. Most pegmatites occur in orogenic belts, although the type of pegmatite formed differs according to the nature of its geological setting. Abyssal class pegmatites

typically occur in migmatitic rocks of upper amphibolite to granulite facies metamorphism. Muscovite class pegmatites occur in slightly lower grade Barrovian-type metamorphic terranes, mainly amphibolite facies. For both abyssal and muscovite class pegmatites, the host rocks represent deeply eroded root zones of orogenic belts. Rare element pegmatites occur in less deeply eroded Abukuma-type metamorphic terranes, generally of cordierite-amphibolite facies. They are commonly peripheral to larger granitic plutons that, in many cases, represent the parental granite from which the pegmatite was derived. The Late Archean pegmatites of the Superior Province are typically localized along deep fault systems which in many areas coincide with major metamorphic and tectonic boundaries. For example, rare element pegmatites associated with the Ghost Lake batholith in northwestern Ontario occur within thrust-faulted and, in places, migmatized rocks of the Sioux Lookout Terrane, which forms the boundary between the Winnipeg River and Wabigoon subprovinces (Breaks and Moore, 1992).

Form of deposits

Many pegmatites occur as dyke-like or lenticular bodies but they range considerably in both shape and size. Pegmatites in high grade metamorphic rocks form irregular, tabular to ellipsoidal bodies that are typically conformable to the foliation of the host rocks. Some pegmatites in lower grade metamorphic rocks are conformable with the host rocks, but others occupy discordant, crosscutting structures such as tension faults. Pegmatites formed within larger granitic bodies have bulbous to highly irregular shapes.

Most pegmatites range in size from a few metres to hundreds of metres long and from 1 cm to several hundred metres wide, although a few pegmatites are much larger. The Tanco pegmatite, for example, is a subhorizontal body 1440 m long, as much as 820 m wide, and slightly more than 100 m thick. The main FI pegmatite dyke in the Yellowknife district is more than 2000 m long and averages about 8 m wide. Pegmatite dykes on the Quebec Lithium property in the Preissac-Lacorne district are as much as

Table 21-2. Production/reserves of selected Canadian and foreign pegmatite deposits.

Deposit	Production/reserves	Comments/references
Canadian deposits		
Tanco mine, Manitoba	a) 1.9 Mt; 0.216% Ta_2O_5 b) 6.6 Mt; 2.76% Li_2O 　(in spodumene + petalite) c) 0.3 Mt; 23.3% Cs_2O d) 0.8 Mt; 0.20% BeO	Reserves (Crouse et al., 1984)
Preissac-Lacorne area, Quebec	19 Mt; 1.25% Li_2O	Production plus reserves from the (former) Quebec Lithium property (Flanagan, 1978)
FI (J.M.-Lit), Yellownknife district, N.W.T.	13.9 Mt; 1.19% Li_2O	"Identified paramarginal resources" (Lasmanis, 1978)
Thor (Echo), Yellowknife district, N.W.T.	8.4 Mt; 1.5% Li_2O	"Identified paramarginal resources" (Lasmanis, 1978)
Violet, Herb Lake area, Manitoba	5.9 Mt; 1.2% Li_2O	Reserves (Williams and Trueman, 1978)
Nama Creek, Georgia Lake area, Ontario	3.9 Mt; 1.06% Li_2O	Reserves, North and South zones (Pye, 1965)
Lac la Croix, Ontario	1.4 Mt; 1.3% Li_2O	Reserves (Mulligan, 1965)
Madawaska mine, (formerly Faraday mine), Bancroft district, Ontario	4.5 Mt; 0.09% U_3O_8	Production 1957-1964 and 1976-1982 (Carter and Colvine, 1985)
Foreign deposits		
Tin-spodumene Belt, North Carolina	a) 26 Mt; 1.5% Li_2O	Measured and indicated reserves, Kings Mountain, Foote Mineral Co. (Kunasz, 1982)
	b) 30.5 Mt; 1.5% Li_2O	Reserves, Bessemer City, Lithium Corporation of America (Company news release, 1976)
Bikita, Zimbabwe	10.8 Mt; 3.0% Li_2O	Reserves (Wegener, 1981)
Kamativi, Zimbabwe	100 Mt; 0.114% Sn, 0.603% Li_2O	"Maximum inferrable reserves of a single pegmatite" (Bellasis and van der Heyde, 1962)
Uis, Namibia	87 Mt; 0.134% Sn	Mineable plus possible reserves (Mining Magazine, November, 1983, p. 291)
Greenbushes, Australia	a) 28 Mt; 0.114% Sn, 0.043% Ta_2O_5, 0.031% Nb_2O_5	Underground reserves (Knight and Wallace, 1982)
	b) 33.5 Mt; 2.55% Li_2O	Proven and probable reserves (Knight, 1986)
Manono-Kitotolo, Zaire	35 Mt; 1.3% Li_2O	Reserves proved by systematic exploration (Evans, 1978)
Minas Gerais and Ceara states, Brazil	106 Mt; 0.04% BeO	Estimated in situ ore (Soja and Sabin, 1986)

600 m long and 30 m wide. The Bikita pegmatite in Zimbabwe is 1.8 to 2.1 km long and 300 m wide (Martin, 1964). The main pegmatite at Greenbushes, Australia is more than 2 km long and as wide as 230 m (Hatcher and Bolitho, 1982).

Internal structure

Many pegmatites are unzoned and relatively uniform in both composition and texture; however, rare element pegmatites in particular can have complex internal structures. In the 1940s, geologists of the United States Geological Survey devised a system to describe these internal structures (Cameron et al., 1949). Although revisions to this system have been suggested (e.g. Černý, 1982a; Norton, 1983), it is still widely used. According to this system, internal units in complex pegmatites consist of a sequence of zones, mainly concentric, which conform roughly to the shape of the pegmatite, and differ in mineral assemblages and textures. From the margin inward, these zones consist of a border zone, a wall zone, intermediate zones, and a core zone. The border zone is thin, averaging a few centimetres wide, and typically aplitic in texture. The border zone in some pegmatites is metasomatic in part, but in many cases it represents a chilled margin of the pegmatite. The wall zone is wider and coarser grained than the border zone and marks the beginning of coarse crystallization characteristic of pegmatites. The wall zone consists mainly of quartz, feldspar, and muscovite, although lesser biotite, apatite, tourmaline, beryl, and garnet may also be present. Intermediate zones, where present, are more complex

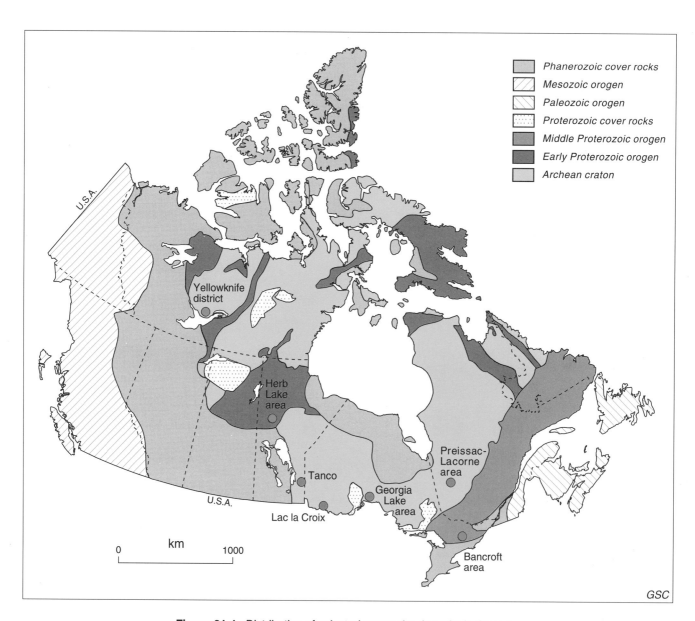

Figure 21-1. Distribution of selected pegmatite deposits in Canada.

mineralogically and contain a variety of economically important minerals such as sheet mica, beryl, spodumene, amblygonite, lepidolite, columbite-tantalite, and cassiterite. In the intermediate zones of some pegmatites, individual crystals of quartz, feldspar, mica, apatite, beryl, tourmaline, spodumene, and other minerals may be metres or even tens of metres in maximum dimension. The core zone consists mainly of quartz, either as solid masses or as euhedral crystals. Not all of the above zones are necessarily present in every pegmatite; however, zones that do occur are generally in the sequence described. In addition to these concentric zones, other internal features such as replacement bodies and fracture fillings may be present.

At the Tanco pegmatite, nine internal zones and an exomorphic zone of metasomatic alteration have been documented (Crouse et al., 1984). The distribution of these zones is shown in Figure 21-3 and their compositional characteristics are summarized in Table 21-3. Some of the zones (2, 4, 5, 6, 7, 8) are considered to have formed mainly by primary crystallization, whereas others (1, 3, 9) have features that suggest they are metasomatic in origin (Černý, 1982b, 1989a). However, this classification is oversimplified; features indicative of partial metasomatism are present in the primary zones (Černý, 1982b) and recent studies of the saccharoidal albite unit (zone 3) indicate that it is primary rather than replacive in origin (London, 1986; Thomas and Spooner, 1988). Minerals of commercial interest in the various zones are indicated in Table 21-3. The most recent mining activity at Tanco has been centred mainly on the saccharoidal albite zone (for tantalum) and the upper intermediate zone (for spodumene).

Regional zoning

Some pegmatites associated with granitic intrusions, particularly rare element pegmatites, are distributed in zonal patterns around such intrusions. In general, the pegmatites most enriched in rare metals and volatile components are located farthest from the intrusions (Trueman and Černý, 1982; Černý, 1989b); this relationship is shown schematically in Figure 21-4.

An example of horizontal zoning that fits this pattern occurs in the Ross Lake area, Northwest Territories, where five zones peripheral to the Redout Lake granite were defined by Rowe (1952), and subsequently modified and described in greater detail by Hutchinson (1955), Meintzer (1987), and Wise (1987). According to these authors, zone I (closest to the Redout Lake granite) contains giant pegmatites characterized by graphic granite; in zone II, pegmatites contain graphic granite and beryl; pegmatites of zone III contain beryl but not graphic granite; zone IV pegmatites contain beryl and niobium- and tantalum-bearing minerals; and pegmatites of zone V are characterized by spodumene and rare grains of columbite.

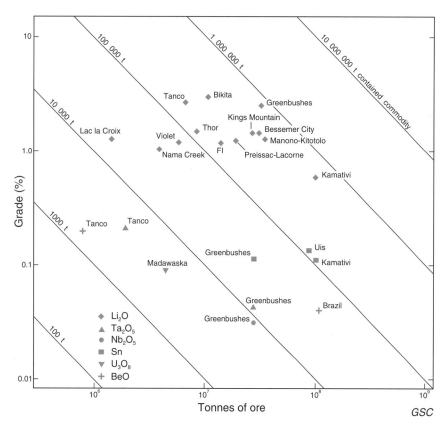

Figure 21-2. Grade versus tonnage diagram for pegmatite deposits (data are from Table 21-2).

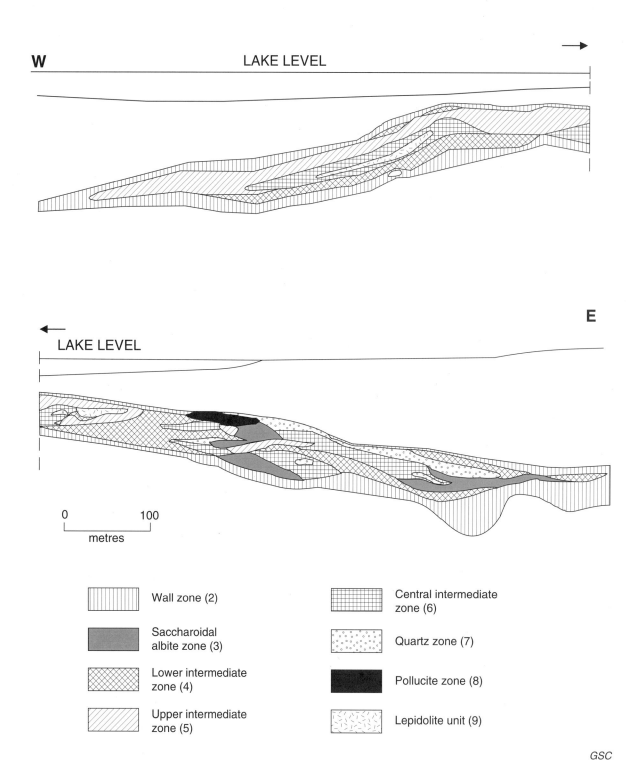

W LAKE LEVEL

LAKE LEVEL E

0   100

metres

	Wall zone (2)		Central intermediate zone (6)
	Saccharoidal albite zone (3)		Quartz zone (7)
	Lower intermediate zone (4)		Pollucite zone (8)
	Upper intermediate zone (5)		Lepidolite unit (9)

GSC

Figure 21-3. Representative longitudinal west-east section through the Tanco pegmatite, Manitoba (modified from Crouse et al., 1984). The host rock amphibolite is not patterned; the exomorphic zone and border unit are relatively thin and are not shown. The main zones that have been mined are shown in red; these are the saccharoidal albite zone (tantalum) and the upper intermediate zone (spodumene).

Table 21-3. Mineral zones in the Tanco pegmatite (from Crouse et al., 1984).

Zone	Main constituents*	Minor to rare constituents*
Exomorphic zone	Biotite, tourmaline, holmquistite	Arsenopyrite
1. Border unit	Albite, quartz	Tourmaline, apatite, biotite, beryl, triphylite
2. Wall zone	Albite, quartz, muscovite, Li-muscovite, microcline-perthite	**Beryl**, tourmaline
3. Saccharoidal albite zone	**Albite**, quartz, muscovite	Muscovite, **Ta-oxide minerals, beryl**, apatite, tourmaline, cassiterite, ilmenite, zircon-hafnon, sulphides
4. Lower intermediate zone	Microcline-perthite, albite, quartz, spodumene, amblygonite	Li-muscovite, lithiophilite, lepidolite, petalite, Ta-oxide minerals
5. Upper intermediate zone	**Spodumene, quartz, amblygonite**	Pollucite, lithiophilite, microcline-perthite, albite, Li-muscovite, petalite, eucryptite, Ta-oxide minerals
6. Central intermediate zone	**Microcline-perthite**, quartz, **albite**, muscovite	Lithiophilite, apatite, spodumene
7. Quartz zone	**Quartz**	Spodumene, amblygonite
8. Pollucite zone	**Pollucite**	Quartz, spodumene, petalite, muscovite, lepidolite, albite, microcline, apatite
9. Lepidolite unit	**Li-muscovite, lepidolite**, microcline-perthite	Albite, quartz, **beryl, Ta-oxide minerals**, cassiterite, zircon-hafnon

* Minerals outlined in **bold type** occur in economic or potentially economic quantities in the zones indicated.

Chemical composition

The chemical compositions of most pegmatites are similar to those of differentiated granitic rocks with respect to major elements, except that pegmatites tend to have lower total Fe, MgO, and CaO, and higher Al_2O_3 (Černý, 1991b). Rare element pegmatites, however, display extreme fractionation and enrichment in lithophile elements such as Li, Rb, Cs, Tl, Be, Nb, Ta, and Ga, and volatiles such as B, F, and P (Černý et al., 1985). Some rare element pegmatites, for example, contain as much as 2% Li_2O, 1% Rb_2O or more, 1.5% Cs_2O, 0.8% B_2O_3, and 1% F (Černý, 1982a). Pegmatites in high grade metamorphic rocks have high contents of Ca, Ba, Sr, Fe, Mn, Ti, and, in some cases, B, F, and rare-earth elements, but the content of rare metals in these pegmatites is low.

DEFINITIVE CHARACTERISTICS

Pegmatites are recognized by their granitic composition and by their highly variable grain size, including extremely coarse crystals. Rare element pegmatites, found in medium grade metamorphic terranes, contain dispersed rare metal-bearing minerals.

GENETIC MODEL

Pegmatites are generally considered to form by primary crystallization from a volatile-rich, siliceous melt (e.g. Jahns, 1955; Jahns and Burnham, 1969; Černý, 1982b, 1991b; London, 1990, 1992). In the case of rare element pegmatites, these melts are related to highly differentiated granitic magmas and represent strongly fractionated residual melts rich in silica, alumina, alkali elements, water and other volatiles, lithophile elements, and rare metals. According to Černý (1991a), the lithology of the source rocks for these melts is a major control on the ultimate composition of subsequently formed rare element pegmatites: undepleted upper crustal lithologies result in peraluminous granites that give rise to pegmatites enriched in lithium, cesium, and tantalum (e.g. Tanco), whereas depleted lower crustal rocks generate metaluminous to peralkaline granites that are parental to pegmatites enriched in niobium and rare-earth elements (including yttrium). Muscovite and abyssal class pegmatites, which occur in high grade metamorphic terranes, crystallized from melts that likely resulted from partial melting of their host rocks. These anatectic melts were also

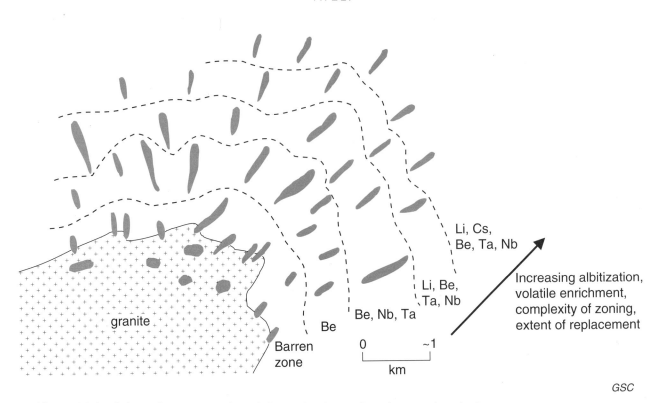

Figure 21-4. Schematic representation of the regional zonation of pegmatites (red) around a granite intrusion (modified from Trueman and Černý, 1982).

siliceous and volatile-rich, but the pegmatites they produced are not as highly fractionated and have no appreciable content of lithophile elements and rare metals.

According to Jahns and Burnham (1969), crystallization of pegmatite-producing melts takes place mainly under closed-system conditions, from the contacts of the pegmatite inward to produce concentric mineral zones. Some of these zones are sufficiently enriched in rare elements to be of commercial interest. Progressive evolution of a coexisting supercritical aqueous phase during this crystallization facilitates the growth of large crystals and provides a means to concentrate elements not easily incorporated in silicate minerals. This aqueous phase can react with earlier-formed minerals at various stages of pegmatite formation to produce metasomatic zones that are enriched in lithophile elements and rare metals. Fracture fillings may also form at various stages, and represent intermittent open-system conditions that probably occur briefly during pegmatite crystallization. As a further modification of this model, London (1990, 1992) has shown that highly fractionated pegmatites of the Tanco type crystallized largely from homogeneous melts enriched in B, P, F, and Li, and extremely enriched in H_2O. He also suggested that many "metasomatic" units are possibly primary, and that separation of aqueous fluid may, in fact, be very late in the consolidation history of pegmatites.

RELATED DEPOSIT TYPES

Pegmatites appear to represent a transitional phase between granitic intrusions and quartz veins. For example, zones of pegmatitic texture occur in several types of granite-related deposits, such as "stockscheider" associated with tin- and tungsten-bearing stockworks and greisens, and pegmatitic zones in felsic intrusions associated with porphyry copper and porphyry molybdenum deposits. Such pegmatitic zones, however, generally do not host significant mineralization.

Geochemical characteristics of tin- and tungsten-bearing granites (e.g. "specialized" granites of Tischendorf, 1977) and of felsic intrusions associated with porphyry molybdenum deposits resemble those of fertile granites that generate rare element pegmatites (Černý and Meintzer, 1988); intrusions associated with most porphyry copper deposits, however, are more mafic in composition and are substantially different geochemically.

Peraluminous to subalkaline rare metal granites with associated lithium, beryllium, niobium, and tantalum, as well as tungsten and tin mineralization (Pollard, 1989), are the closest relatives to pegmatite deposits. Some rare metal granites display pegmatitic cupolas that suggest an origin from pegmatitic melts that did not separate from their plutonic parent (Černý, 1992).

EXPLORATION GUIDES

Exploration guidelines for rare element pegmatites (Trueman and Černý, 1982; Černý, 1989b, 1991c) include the following:

1. Geological setting: rare element pegmatites typically occur in rock suites of medium grade Abukuma-type metamorphic facies, along fault systems and lithological boundaries, or closely associated with anorogenic granitoid plutons.

2. Regional zoning: identification of zonal patterns of pegmatite distribution can help isolate specific areas of interest.

3. Fractionation: mineral assemblages and chemistry of individual minerals in pegmatites indicate fractionation levels and economic potential.

4. Geochemical approaches: primary dispersion aureoles in host rocks (e.g. Li, Rb, Cs, Be, B), secondary dispersion halos in overburden, and light plus heavy minerals in stream sediments (e.g. beryl, spodumene, tourmaline, columbite-tantalite) help identify target areas at both regional and local scales.

5. Geophysical approaches: radiometric surveys may be useful for identifying parent granites and/or associated pegmatites that are enriched in U and Th. Gravity surveys can be used to outline pegmatites in host rocks of contrasting density.

ACKNOWLEDGMENT

P. Černý reviewed the paper and provided many constructive comments.

SELECTED BIBLIOGRAPHY

References marked with asterisks (*) are considered to be the best sources of general information on this deposit type.

Bellasis, J.W.M. and van der Heyde, C.
1962: Operations at Kamativi Tin Mines Ltd., Southern Rhodesia; Chamber of Mines Journal, v. 4, no. 9 (Sept.), p. 36-41.

Breaks, F.W. and Moore, J.M., Jr.
1992: The Ghost Lake batholith, Superior Province of northwestern Ontario: a fertile, S-type, peraluminous granite - rare-element pegmatite system; The Canadian Mineralogist, v. 30, p. 835-875.

***Cameron, E.N., Jahns, R.H., McNair, A., and Page, L.R.**
1949: Internal structure of granitic pegmatites; Economic Geology, Monograph 2, 115 p.

Carter, T.R. and Colvine, A.C.
1985: Metallic mineral deposits of the Grenville Province, southeastern Ontario; The Canadian Institute of Mining and Metallurgy Bulletin, v. 78, no. 875, p. 95-106.

Černý, P.
1982a: Anatomy and classification of granitic pegmatites; in Granitic Pegmatites in Science and Industry, (ed.) P. Černý; Mineralogical Association of Canada, Short Course Handbook, v. 8, p. 1-39.
1982b: The Tanco pegmatite at Bernic Lake, southeastern Manitoba; in Granitic Pegmatites in Science and Industry, (ed.) P. Černý; Mineralogical Association of Canada, Short Course Handbook, v. 8, p. 527-543.
*1982c: Granitic Pegmatites in Science and Industry, (ed.) P. Černý; Mineralogical Association of Canada, Short Course Handbook, v. 8, 555 p.
1989a: Characteristics of pegmatite deposits of tantalum; in Lanthanides, Tantalum and Niobium, (ed.) P. Möller, P. Černý, and F. Saupé; Society for Geology Applied to Mineral Deposits, Special Publication No. 7, Springer-Verlag, Berlin, p. 192-236.
1989b: Exploration strategy and methods for pegmatite deposits of tantalum; in Lanthanides, Tantalum and Niobium, (ed.) P. Möller, P. Černý, and F. Saupé; Society for Geology Applied to Mineral Deposits, Special Publication No. 7, Springer-Verlag, Berlin, p. 274-302.
1990: Distribution, affiliation and derivation of rare-element pegmatites in the Canadian Shield; Geologische Rundschau, Band 79, Heft 2, p. 163-226.
1991a: Fertile granites of Precambrian rare-element pegmatite fields: is geochemistry controlled by tectonic setting or source lithologies?; Precambrian Research, v. 51, p. 429-468.
*1991b: Rare-element pegmatites, Part I: anatomy and internal evolution of pegmatite deposits; Geoscience Canada, v. 18, no. 2, p. 49-67.

Černý, P. (cont)
*1991c: Rare-element pegmatites, Part II: regional to global environments and petrogenesis; Geoscience Canada, v. 18, no. 2, p. 68-81.
1992: Geochemical and petrogenetic features of mineralization in rare-metal granitic pegmatites in the light of current research; Applied Geochemistry, v. 7, p. 393-416.

Černý, P. and Meintzer, R.E.
1988: Fertile granites in the Archean and Proterozoic fields of rare-element pegmatites: crustal environment, geochemistry and petrogenetic relationships; in Recent Advances in the Geology of Granite-Related Mineral Deposits, (ed.) R.P. Taylor and D.F. Strong; The Canadian Institute of Mining and Metallurgy, Special Volume 39, p. 170-207.

Černý, P., Meintzer, R.E., and Anderson, A.J.
1985: Extreme fractionation in rare-element granitic pegmatites: selected examples of data and mechanisms; The Canadian Mineralogist, v. 23, p. 381-421.

Crouse, R.A., Černý, P., Trueman, D.L., and Burt, R.O.
1984: The Tanco pegmatite, southeastern Manitoba; in The Geology of Industrial Minerals in Canada, (ed.) G.R. Guillet and W. Martin; The Canadian Institute of Mining and Metallurgy, Special Volume 29, p. 169-176.

Evans, R.K.
1978: Lithium reserves and resources; Energy, v. 3, p. 379-385.

Flanagan, J.T.
1978: Lithium deposits and potential of Quebec and Atlantic provinces, Canada; Energy, v. 3, p. 391-398.

Ginsburg, A.I.
1984: The geological condition of the location and the formation of granitic pegmatites; Proceedings of the 27th International Geological Congress, v. 15 (Non-metallic Mineral Ores), VNU Science Press, Utrecht, The Netherlands, p. 245-260.

Hatcher, M.I. and Bolitho, B.C.
1982: The Greenbushes pegmatite, south-west Western Australia; in Granitic Pegmatites in Science and Industry, (ed.) P. Černý; Mineralogical Association of Canada, Short Course Handbook, v. 8, p. 513-525.

Hutchinson, R.W.
1955: Regional zonation of pegmatites near Ross Lake, District of Mackenzie, Northwest Territories; Geological Survey of Canada, Bulletin 34, 50 p.

Jahns, R.H.
1955: The study of pegmatites; in Economic Geology, Fiftieth Anniversary Volume, (ed.) A.M. Bateman; p. 1025-1130.

***Jahns, R.H. and Burnham, C.W.**
1969: Experimental studies of pegmatite genesis: I. A model for the derivation and crystallization of granitic pegmatites; Economic Geology, v. 64, p. 843-864.

Knight, N.D.
1986: Lithium; in Australian Mineral Industry Annual Review for 1983, Bureau of Mineral Resources, Geology and Geophysics, Canberra, Australia, p. 181-182.

Knight, N.D. and Wallace, D.A.
1982: Tantalum and niobium; in Australian Mineral Industry Annual Review for 1980, Bureau of Mineral Resources, Geology and Geophysics, Canberra, Australia, p. 253-256.

Kunasz, I.
1982: Foote Mineral Company - Kings Mountain operation; in Granitic Pegmatites in Science and Industry, (ed.) P. Černý; Mineralogical Association of Canada, Short Course Handbook, v. 8, p. 505-511.

Lasmanis, R.
1978: Lithium resources in the Yellowknife area, Northwest Territories, Canada; Energy, v. 3, p. 399-407.

London, D.
1986: Magmatic-hydrothermal transition in the Tanco rare element pegmatite: evidence from fluid inclusions and phase-equilibrium experiments; The American Mineralogist, v. 71, p. 376-395.
*1990: Internal differentiation of rare-element pegmatites; a synthesis of recent research; in Ore-Bearing Granite Systems; Petrogenesis and Mineralizing Processes, (ed.) H.J. Stein and J.L. Hannah; Geological Society of America, Special Paper 246, p. 35-50.
1992: The application of experimental petrology to the genesis and crystallization of granitic pegmatites; The Canadian Mineralogist, v. 30, p. 499-540.

Martin, H.J.
1964: The Bikita tinfield; Southern Rhodesia Geological Survey Bulletin, no. 58, p. 114-132.

Martin, R.F. and Černý, P. (ed.)
*1992: Granitic pegmatites; The Canadian Mineralogist, v. 30, p. 497-954.

Meintzer, R.E.

1987: The mineralogy and geochemistry of the granitoid rocks and related pegmatites of the Yellowknife pegmatite field, Northwest Territories; PhD. thesis, University of Manitoba, Winnipeg, Manitoba, 708 p.

Mulligan, R.

1965: Geology of Canadian lithium deposits; Geological Survey of Canada, Economic Geology Report 21, 131 p.

Norton, J.J.

1983: Sequence of mineral assemblages in differentiated granitic pegmatites; Economic Geology, v. 78, p. 854-874.

Pollard, P.J.

1989: Geologic characteristics and genetic problems associated with the development of granite-related deposits of tantalum and niobium; in Lanthanides, Tantalum and Niobium, (ed.) P. Möller, P. Černý, and F. Saupé; Society for Geology Applied to Mineral Deposits, Special Publication No. 7, Springer-Verlag, Berlin, p. 240-256.

Pye, E.G.

1965: Georgia Lake area; Ontario Department of Mines, Geological Report No. 31, 113 p.

Rowe, R.B.

1952: Pegmatite mineral deposits of the Yellowknife-Beaulieu River region, Northwest Territories; Geological Survey of Canada, Paper 52-8, 36 p.

Soja, A.A. and Sabin, A.E.

1986: Beryllium availability - market economy countries; United States Bureau of Mines, Information Circular 9100, 19 p.

Thomas, A.V. and Spooner, E.T.C.

1988: Occurrence, petrology and fluid inclusion characteristics of tantalum mineralization in the Tanco granitic pegmatite, southeastern Manitoba; in Recent Advances in the Geology of Granite-related Mineral Deposits, (ed.) R.P. Taylor and D.F. Strong; The Canadian Institute of Mining and Metallurgy, Special Volume 39, p. 208-222.

Tischendorf, G.

1977: Geochemical and petrographic characteristics of silicic magmatic rocks associated with rare-element mineralization; in Metallization Associated with Acid Magmatism, v. 2, (ed.) M. Stemprok, L. Burnol, and G. Tischendorf; Geological Survey (Czechoslovakia), Prague, p. 41-96.

Trueman, D.L. and Černý, P.

1982: Exploration for rare-element granitic pegmatites; in Granitic Pegmatites in Science and Industry, (ed.) P. Černý; Mineralogical Association of Canada, Short Course Handbook, v. 8, p. 463-493.

Wegener, J.E.

1981: Profile on Bikita - processed petalite the new priority; Industrial Minerals, v. 165, p. 51-53.

Williams, C.T. and Trueman, D.L.

1978: An estimation of lithium resources and potential of northwestern Ontario, Manitoba and Saskatchewan, Canada; Energy, v. 3, p. 409-413.

Wise, M.A.

1987: Geochemistry and crystal chemistry of Nb, Ta and Sn minerals from the Yellowknife pegmatite field; PhD. thesis, University of Manitoba, Winnipeg, Manitoba, 368 p.

Author's address

W.D. Sinclair
Geological Survey of Canada
601 Booth Street
Ottawa, Ontario
K1A 0E8

Printed in Canada

22. KIRUNA/OLYMPIC DAM-TYPE IRON, COPPER, URANIUM, GOLD, SILVER

22. KIRUNA/OLYMPIC DAM-TYPE IRON, COPPER, URANIUM, GOLD, SILVER

S.S. Gandhi and R.T. Bell

INTRODUCTION

Deposits of this type are characterized by an abundance of magnetite and/or hematite in tabular bodies (Kiruna), breccia-fillings (Olympic Dam), veins, disseminations and replacements, all in continental, dominantly felsic volcano-plutonic settings in a late tectonic or anorogenic environment. The host rocks include all volcanic lithofacies ranging from proximal, epizonal feeder systems to distal volcaniclastic sediments, and related epizonal plutons. The deposits range from essentially monometallic (Fe) to polymetallic ($Fe+Cu\pm U\pm Au\pm Ag\pm REEs$). Large monometallic deposits have been mined for iron alone, but in polymetallic deposits associated metals are at the main economic interest.

The largest and most important examples occur outside Canada; these include the classic magnetite-apatite-actinolite deposits of the Kiruna district in northern Sweden and in southeastern Missouri, U.S.A. and the giant Olympic Dam Cu-U-Au-Ag-REE deposit in South Australia (Table 22-1). The breccia-hosted polymetallic Sue-Dianne deposit in the Northwest Territories is small and subeconomic, but it provides the best Canadian example of this deposit type (Fig. 22.1). All of these deposits are Proterozoic in age.

Large Phanerozoic deposits, mineralogically similar to the Kiruna deposits, occur in the circum-Pacific region (Chile, Mexico, China) and in Iran and Turkey. They are associated with intermediate to felsic volcanic and related plutonic rocks in orogenic settings and in relatively stable cratonic regions. Smaller occurrences of this type occur in the North American Cordillera, e.g. in the Tatoosh pluton of Mount Rainier, Washington, U.S.A., and in the Iron Mask batholith near Kamloops, British Columbia.

IMPORTANCE

In Canada, mineable deposits of this type have not yet been found, and the known prospects are too small to form a significant mineral resource.

In Sweden, China, Chile, Mexico, U.S.A., and Iran, deposits of this type have been mined for iron, and form a small but significant proportion of world iron resources. The Olympic Dam deposit, discovered in 1975, is mined for copper, gold, silver, and uranium. It rivals the largest porphyry copper deposits in terms of contained copper, and forms the largest single resource of uranium in the world. Some of the Chinese deposits have been mined for iron and copper.

SIZE AND GRADE OF DEPOSITS

In Canada, the Sue-Dianne deposit has drill-indicated resources of 8 million tonnes averaging 0.8% Cu, approximately 100 ppm U, and locally significant values in Au (Gandhi, 1989). Other polymetallic prospects of this type in the Great Bear Magmatic Zone are smaller, e.g. Mar, Damp, and Fab Main (Gandhi, 1989, 1994).

Gandhi, S.S. and Bell, R.T.
1996:   Kiruna/Olympic Dam-type iron, copper, uranium, gold, silver; in Geology of Canadian Mineral Deposit Types, (ed.) O.R. Eckstrand, W.D. Sinclair, and R.I. Thorpe; Geological Survey of Canada, Geology of Canada, no. 8, p. 513-522 (also Geological Society of America, The Geology of North America, v. P-1).

Table 22-1. Deposits of Kiruna/Olympic Dam-type in Proterozoic volcano-plutonic settings.

Deposit characteristics / District/region	Kiruna-type deposits — Tabular & pipe-like bodies, dykes, and veins of magnetite±apatite±actinolite; essentially monometallic	Olympic Dam-type deposits — Breccia (one or more stages); with hematite-magnetite matrix and Cu±U±Au±Ag±REE	Age (Ma)
Gawler Ranges, South Australia [1]		Olympic Dam *, Acropolis, Oak Dam, Wirrda Well	1600-1590
Kiruna and Bergslagen districts, Sweden [2]	Kiirunavaara *, Luossavaara, Rektorn, Haukivaara, Nukutusvaara, Lappmalmen, Painirova, Mertainen, Gruvberget, Grängesberg		1890-1880
St. Francois Mountains, Missouri, U.S.A. [3]	Pea Ridge, Pilot Knob, Bourbon, Camel's Hump, Katz Spring, Iron Mountain	Boss-Bixby †	1450-1350
Great Bear Magmatic Zone, N.W.T., Canada [4]	Echo Bay, Contact Lake, Terra, Nod, Fab North, Ketcheson Lake, Blanchet Island, Labelle Peninsula, Regina Bay	Sue-Dianne, Mar, Fab Main, Damp	1865-1855

* Estimated 2000 million tonnes.
† Multiple orebodies; mineralization mainly as disseminated copper in a felsic pluton (pers. comm., C.R. Allen, Cominco American Resources Incorporated, 1994).
[1] Roberts and Hudson, 1983; Paterson et al., 1986; Oreskes and Einaudi, 1990; Reeve et al., 1990; Parker, 1990; Hitzman et al., 1992.
[2] Geijer, 1930; Magnusson, 1970; Parák, 1975; Frietsch et al., 1979; Lundberg and Smellie, 1979; Lundstrom and Papunen, 1986.
[3] Emery, 1968; Panno and Hood, 1983; Hagni and Brandom, 1988; Kisvarsanyi and Kisvarsanyi, 1989; Marikos et al., 1989.
[4] Badham and Morton, 1976; Badham, 1978; Hildebrand, 1986; Gandhi, 1989, 1992, 1994.

Most foreign deposits are in the 100 to 500 million tonne range and average 45 to 65% iron (Cox and Singer, 1986, p. 172-174; Hitzman et al., 1992), but the Kiirunavaara and Olympic Dam deposits each contain 2 billion tonnes of ore. The Kiruna district has several other deposits (Table 22-1), and together with the Kiirunavaara, they constitute more than 3.4 billion tonnes of iron ore resource (Frietsch et al., 1979). Kiruna-type deposits also occur in the Bergslagen district in central Sweden, the largest being the Grängesberg deposit. In southeastern Missouri, some 30 deposits are known, and the larger ones among them (containing more than 100 million tonnes) are listed in Table 22-1. The aggregate resource of the Missouri district exceeds one billion tonnes of iron (Kisvarsanyi and Kisvarsanyi, 1989).

The larger Chilean deposits are El Romeral, El Algarrobo, and Los Colorados in the northern part of the country, and the El Laco deposits are in the Chilean Altiplano (Frutos and Oyarzùn, 1975; Bookstrom, 1977; Oyarzùn and Frutos, 1984). The Mexican examples are the Cerro de Mercado group of deposits in the west central part of the country (Lyons, 1988). Tonnage figures are not available for the individual Chinese iron deposits, but on the district level they approach the magnitude for those in other countries (Research Group on Porphyrite Iron Ore of the Middle-Lower Yangtze Valley, 1977; Li and Kuang, 1990; Xu, 1990). Occurrences in the North American Cordillera,

exemplified by those in Washington and British Columbia (Fiske et al., 1963; Cann, 1979), are too small to be of economic interest. In Iran, the major deposits are the Chador Malu, Chogurt, and Seh Chahoon, located in the Bafq district in the central part of the country (Förster and Knittel, 1979; Förster and Jafarzadeh, 1984, 1994; Förster, 1990). The Murdere and Miskel deposits in Turkey are smaller examples (Helvaci, 1984).

The polymetallic Olympic Dam deposit averages 1.6% Cu, 0.05% U, 3.5 g/t Ag, 0.6 g/t Au, approximately 35% Fe, and contains notable amounts of rare-earth elements, in particular 0.2% La and 0.3% Ce (Reeve et al., 1990). Minerals carrying these metals are intimately associated with iron oxides, the proportion of which ranges from 30% to more than 80% in the deposit.

GEOLOGICAL FEATURES
Geological setting

The deposits are associated with continental felsic volcano-plutonic complexes that have been developed in late- and post-orogenic tectonic settings. In Sweden, this felsic magmatism occurred at the end of the Svecofennian orogeny ca. 1900 Ma (Gaál and Gorbatschev, 1987; Skiöld, 1987; Cliff et al., 1990). The granite-rhyolite terrane of southeastern

Missouri is part of a major, early to middle Proterozoic crustal zone that extends through central Labrador and south Greenland to Scandinavia, and is characterized by episodic felsic magmatism (Bickford et al., 1986; Gower et al., 1990; Gandhi and Bell, 1993). The Olympic Dam and other related deposits in South Australia are spatially related to a granite-rhyolite suite that formed in a stable cratonic environment, after the early Proterozoic Sleaford orogeny, and the Kimban orogenic and postorogenic events that occurred during the interval 1900 to 1650 Ma

(Fanning et al., 1988; Mortimer et al., 1988; Johnson and Cross, 1991; Creaser and Cooper, 1993). They are concealed under some 300 m of flat-lying late Proterozoic and early Paleozoic strata, and were discovered by drill testing of coincident magnetic and gravity anomalies (Paterson et al., 1986; Reeve et al., 1990). The Great Bear Magmatic Zone in the northwestern Canadian Shield was formed during the period 1870-1840 Ma, after the Wopmay orogeny, which culminated ca. 1900 Ma (Hoffman, 1980; Hildebrand et al., 1987).

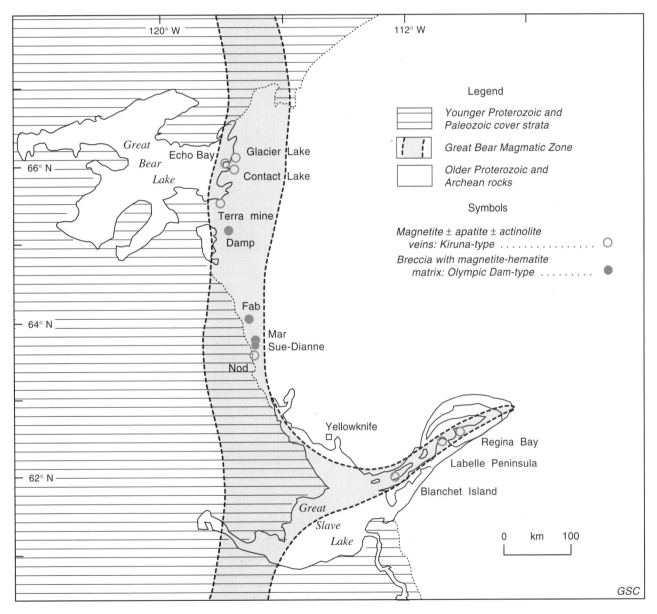

Figure 22-1. Areal extent of the Great Bear Magmatic Zone and location of selected occurrences of magnetite-rich veins and breccia-fillings, Great Bear Lake and Great Slave Lake area, Northwest Territories. Outline of the magmatic zone in the area covered by younger strata and lakes is based on the continuation of the magnetic anomalies, which are diagnostic of the zone in its exposed part, as noted from the regional airborne magnetic survey maps (Gandhi, 1994).

The deposits in Chile, Mexico, China, Iran, and Turkey are Phanerozoic in age. El Laco in Chile, Cerro de Mercado in Mexico, and the Iranian deposits were formed in stable cratonic environments and are closely associated with rhyolitic flows (Frutos and Oyarzùn, 1975; Förster and Jafarzadeh, 1984; Lyons, 1988). The host rocks in Mexico are part of the Tertiary Sierra Madre Occidental volcanic province, which is the world's largest continuous field of rhyolitic rocks. Other deposits in Chile and those in China are, however, in orogenic zones of Jurassic-Cretaceous age, and are associated with basalt-andesite-dacite volcanic sequences and related intrusions that are products of subduction of oceanic plates beneath continents (Oyarzùn and Frutos, 1984; Xu, 1990). In this generally compressive regime, however, extension may have played some role from time to time. For example the Chilean deposits, numbering more than fifty, occur along a large north-trending fracture zone which, according to Oyarzùn and Frutos (1984), developed at an early extensional stage, and controlled the emplacement of the host volcanic and plutonic rocks. Large granitic intrusions were emplaced during later compression. The Turkish deposits are in early Paleozoic metamorphosed volcanic strata (Helvaci, 1984).

Age of host rocks and mineralization

The deposits range in age from 1900 Ma to recent. The age range of the Proterozoic deposits is given in Table 22-1. Older examples are not known. This may reflect the prerequisite of a thick, and laterally extensive crust for the formation of such deposits. It is noteworthy in this regard that globally the rate of crustal growth and thickening peaked in early Proterozoic time (West, 1980).

Geochronological constraints are reasonably good for the host rock/sequences based on radiometric dates and fossil records, but the precise time of mineralization in many of the deposits is less well constrained because of uncertainties inherent in isotopic studies of datable minerals in the deposits, mainly uraninite/pitchblende and apatite. Field relations and available geochronological data, however, point to the formation of the deposits essentially coeval with, or within a few million years after, the emplacement of the host rocks, as is well established for the Kiirunavaara deposit in Sweden (Cliff et al., 1990) and the Olympic Dam and Acropolis deposits in South Australia (Fanning et al., 1988; Mortimer et al., 1988; Johnson and Cross, 1991; Creaser and Cooper, 1993), and for the younger deposits, such as the Infracambrian deposits of Iran (Förster and Jafarzadeh, 1984, 1994), Oligocene deposits of Mexico (Lyons, 1988), and late Pliocene deposits of El Laco in Chile (Oyarzùn and Frutos, 1984).

Associated structure

Block faulting, breccia development, and minor structural adjustments in the host rocks are common, but apparently neither major tectonic disturbance nor orogenic deformation occurred in these rocks immediately prior to mineralization. In some cases, major lineaments have been regarded as controlling factors, e.g. in the case of the Olympic Dam deposit (Parker, 1990).

Relation of ore to host rocks

The ore bodies generally have sharply defined boundaries, and are commonly discordant to or cut the host volcanic, volcaniclastic, and plutonic rocks, except for some of the stratabound and stratiform deposits. Large tabular bodies in the Kiruna district are essentially concordant with the host volcanic sequence, but details of the contacts reveal some discordances as well as brecciation and veining in the host rock. In breccia deposits, the clasts are fragments of the host rocks, and the matrix is rich in the iron oxides. Replacement of wall rocks and fragments is locally intensive in some deposits as in case of the Olympic Dam deposit (Oreskes and Einaudi, 1990), but commonly it is not extensive, and is lacking in many deposits.

Form of deposits

The deposits occur in many forms, ranging from stratabound tabular to discordant breccia zones and veins. The largest tabular deposit is the Kiirunavaara deposit, which has a strike length of 4 km and a thickness of 100 m. It dips steeply eastwards to a depth of 1.5 km, and forms a prominent ridge. Some deposits display features indicative of extrusion of iron oxide-rich melt, e.g. El Laco and Cerro de Mercado deposits (Park, 1961, 1972; Henriquez and Martin, 1978; Lyons, 1988). Other Chilean deposits are steeply dipping lenticular bodies, with associated breccia-fillings and veins. The deposits in southeastern Missouri and Iran are mostly discordant plug-like or irregular bodies with associated brecciated zones, although there are some concordant zones in the volcanic strata (Förster and Jafarzadeh, 1984, 1994; Kisvarsanyi and Kisvarsanyi, 1989). The largest complex of breccia zones is at the Olympic Dam deposit where multiple brecciation is evident (Reeve et al., 1990). The Chinese iron oxide deposits include massive irregular bodies, veins, and breccia-fillings near the boundary of subvolcanic plutons, disseminations of coarse crystals magnetite in volcanic rocks and subvolcanic intrusions, replacement bodies, and exhalative sedimentary iron-formation in a volcanic sequence (Li and Kuang, 1990). Breccia bodies in the Great Bear Magmatic Zone are lobate, circular, and tabular, as exemplified by the Sue-Dianne, Mar, and Damp deposits, respectively. They are discordant to the host volcanic strata, and contain abundant fragments of the volcanic rocks. Their vertical extent is not fully explored. In addition to these there are numerous veins in the magmatic zone, most of which are located close to the margins of quartz monzonitic intrusions (Gandhi and Prasad, 1982; Hildebrand, 1986).

Ore composition

Two compositional varieties are common, but gradations between them also occur, forming a broad spectrum of ore composition. The most common variety is the magnetite-apatite-actinolite ore typical of the Kiruna deposits themselves. The ore is essentially monometallic, although traces of copper are not uncommon. The other compositional variety is typified by the polymetallic Olympic Dam deposit, in which the dominant iron oxide is hematite, and the ore contains significant amounts of Cu, U, Au, Ag, and REEs. The gradational varieties include magnetite-rich ore carrying significant amounts of Cu and U, and traces of Au, Ag, Co, Ni, and Bi, as seen in deposits of the Great Bear Magmatic Zone. Fluorite occurs in some of the deposits.

Alteration

Hematite, chlorite, epidote, and albite are common, although individual deposits may have only one or two of these minerals. In some deposits, mostly the small ones, alteration may be negligible. In addition to the primary alteration, some of the deposits have been affected by secondary alteration. Fractures cutting some of the deposits have secondary concentrations of pitchblende that yield much younger radiometric ages, e.g. the Sue-Dianne deposit (Gandhi, 1994). Deposits subjected to paleoweathering may have undergone oxidation of magnetite to hematite, and acquired supergene enrichment of uranium and copper. The deposits in South Australia were exposed prior to the deposition of the middle Proterozoic continental siliciclastic

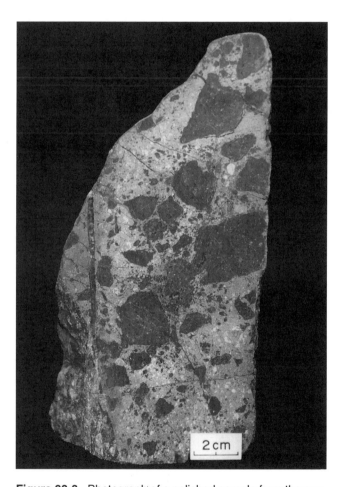

Figure 22-2. Photograph of a polished sample from the central part of the Mar deposit, southern Great Bear magmatic zone (Fig. 22-1), showing fragments of a feldspar porphyritic rhyodacite volcanic unit (dark grey) in magnetite matrix (medium to light grey). Note highly reflective coarse crystals and aggregates (white) in the mass of finer crystal aggregate of magnetite. Younger, veinlets contain quartz with some epidote. GSC 2054271

sediments and the Paleozoic cover strata, hence they are likely to have been affected by such supergene alterations, as suggested by Oreskes and Einaudi (1990) and observed by the writers, but opinions differ regarding the importance of such alterations.

Texture and mineralogy

In the monometallic Kiruna-type deposits, relative proportions of magnetite, apatite, and actinolite vary considerably. Magnetite is dominant overall, although apatite and/or actinolite form high concentrations locally in some deposits (Parák, 1975). Two or more generations of magnetite are found in many of the deposits; the earlier one occurs commonly as coarse grained, euhedral crystal aggregates, and the later ones as smaller grains. In some parts of the deposits, magnetite is altered in part, along grain boundaries and cracks, or wholly, to hematite (martite). Some of the deposits are rich in hematite, and display stratification, e.g. some of the deposits in the Kiruna, Missouri, and Mexican districts. Apatite occurs as euhedral crystals, and actinolite forms aggregates of acicular crystals. Copper sulphides, where present, form interstitial grains or aggregates. Pyrite occurs as small euhedral to subhedral crystals. The proportions of other minerals, such as calcite and quartz, are negligible. Cobalt and nickel arsenides occur in some of the veins in the Great Bear Magmatic Zone.

In the polymetallic Olympic Dam deposit, hematite predominates over magnetite, and forms 30 to 70% of the ore as breccia matrix. More than one generation of hematite is present. It occurs as fine grained aggregates and laths. Copper sulphides, pitchblende, and coffinite are intimately associated with hematite as disseminations, stringers, and aggregates (Roberts and Hudson, 1983; Oreskes and Einaudi, 1990; Reeve et al., 1990). Gold occurs locally in some of the deposits, and is commonly associated closely with copper minerals. Chalcopyrite, bornite, chalcocite, pitchblende, coffinite, and native gold are minerals of prime value, and fluorite, pyrite, bastnaesite, florencite, quartz, sericite, and barite are accessory minerals.

In the breccia deposits of the Great Bear Magmatic Zone, pitchblende, coffinite, and sulphides form aggregates, disseminations, veins, and stringers in the magnetite-specularite matrix. Fragments of volcanic rocks are common, e.g. in the Mar deposit (Fig. 22-2). Some vein-type deposits in this district display zones of coarse apatite and actinolite crystals that have grown inward from the walls (Gandhi and Prasad, 1982).

The extrusive character of some deposits, such as El Laco and Cerro de Mercado, is reflected in their lava flow-like character and prismatic crystals that reflect rapid cooling (Park, 1961, 1972; Henriquez and Martin, 1978; Lyons, 1988). Dykes and intrusive breccias, indicative of injection of iron-rich ore 'melt', are associated with deposits of the Kiruna district, and dendritic magnetite and miniature diapir-like apatite concentrations also reflect their near surface intrusive or extrusive character (Geijer, 1930; Frietsch, 1978; Nyström and Henriquez, 1994). Replacement textures have been reported in some of the Chinese deposits, and have been referred to as 'skarnoid' (Li and Kuang, 1990). Primary disseminations of magnetite occur

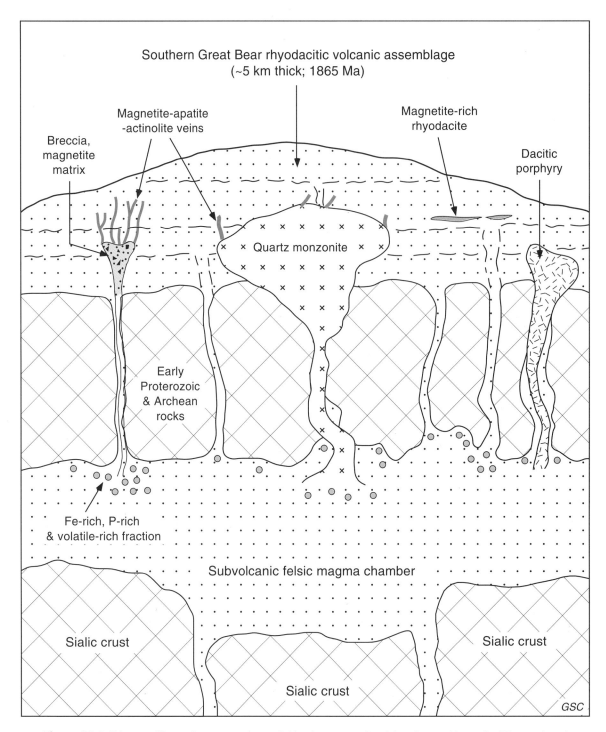

Figure 22-3. Diagram illustrating a genetic model for the magnetite-rich veins and breccia-fillings related to the Great Bear magmatic activity (Gandhi, 1994).

in host igneous intrusions, some parts of which contain this mineral well in excess of accessory amounts, viz., in the range of 5 to 10%.

DEFINITIVE CHARACTERISTICS

Deposits of Proterozoic age have some definitive characteristics as listed below. Many of the Phanerozoic examples do not conform to all of these features, reflecting a wide range of geological environments in which deposits of this type can form.

i) High concentrations of magmatic or hydrothermally deposited iron are present.

ii) Host rocks are commonly felsic to intermediate volcanic rocks and related epizonal plutons, which are meta- or peraluminous and mildly to highly alkaline in character.

iii) The deposits are coeval with the volcano-plutonic activity that formed the host rocks, and are younger than 1900 Ma.

iv) Copper, gold, silver, uranium, and rare-earth elements occur in variable amounts in association with magnetite and hematite.

v) Hematitization, epidotization, carbonatization, and soda metasomatism of the host rocks are common primary alteration features of the deposits.

GENETIC MODEL

Processes of formation of this diverse group of deposits are magmatic fractionation and hydrothermal activity, the latter in some cases involving meteoric waters. Opinions differ as to the relative importance of these and in details of mineralization for individual deposits; but there is a consensus that iron, the most abundant metal in these deposits, is derived from the same magma source that formed the host volcanic and plutonic rocks. For many deposits compelling textural and structural evidence points to crystallization from an iron-rich, volatile-charged magmatic fraction or 'melt' that intruded and/or extruded, and cooled like magma (Geijer, 1930; Park, 1961, 1972; Frietsch, 1978; Henriquez and Martin, 1978; Lyons, 1988; Nyström and Henriquez, 1994). Magnetite-rich deposits can be expected to form from such fractions at high temperatures in subsurface environments, and hematite-rich deposits at relatively lower temperatures in near surface and subaerial environments. The proportion of magmatic water in such fractions is conjectural. Mixing with oxidized meteoric waters would favour deposition of hematite. Formation of the hematite-rich Olympic Dam deposit has been attributed to magmatogenic hydrothermal solutions in the near surface environment of a volcano-plutonic complex, characterized by repeated fault-related brecciation, phreatomagmatic explosions, epiclastic sedimentation, and chemical corrosion (Reeve et al., 1990). Hematite (or specularite) can also develop by partial or complete alteration of earlier formed magnetite due to postdepositional changes in a mineralizing system. For some of the hematite-rich stratabound deposits, there are differences of opinion as to whether hematite was deposited as a primary iron oxide from hydrothermal solution, formed as a replacement of magnetite and other minerals, or was precipitated in a marine exhalative-sedimentary environment. Replacement may be due to the action of hydrothermal fluids, or of oxidized ground waters if a deposit was subjected to weathering processes.

Apatite is common in magnetite-rich deposits, and phosphorus-bearing minerals (florencite, xenotime, and monazite) occur in the Olympic Dam deposit. The role of phosphorus in lowering the crystallization temperature of magnetite in magma, and thus facilitating the generation and transport of an iron-rich magmatic fraction has been strongly emphasized by many workers (Geijer, 1930; Frietsch, 1978; Lyons, 1988). Volatiles would further enhance enrichment of iron in a residual magmatic fraction. Violent escape of volatiles would create space and conditions favourable for forceful injection or passive ingress of an iron-rich fluid, and for deposition of iron oxide. This process is manifested on a small scale as veins and breccia fillings in the roof zones and along the margins of epizonal quartz monzonite plutons in the Great Bear Magmatic Zone (Fig. 22-3). Differentiation of the plutons in situ may be adequate to explain the formation of these small occurrences (Gandhi, 1994). Large deposits, on the other hand, were most likely developed by differentiation in larger magma chambers at depth, which were the source of the volcanic rocks, subvolcanic plutons, and iron-rich fractions. Transportation of the iron-rich fractions to higher levels took place via their ascent in a pulse or pulses, either in association with silicate fractions or separately, to tectonically prepared zones or in outlets created by explosive discharge of contained volatiles. Among the volatiles, F, Cl, and H_2O are considered the most important ones. With an increase in proportion of magmatic and/or meteoric water, the iron-rich fraction passes into hydrothermal solution, and forms deposits with hydrous phases as is the case with the Olympic Dam deposit. Magnetite-rich deposits, on the other hand, show a paucity of hydrous minerals that suggests low water/rock ratio and a limited role of meteoric water. Furthermore, sulphides and carbonates are present in only small amounts, reflecting the paucity of S, CO, and CO_2 in the mineralizing fluids. The system thus differs from the porphyry copper system, which involves development of a carapace in the roof zone of a pluton and a second boiling, prior to mineralization (Burnham and Ohmoto, 1980). The porphyry copper system has a relatively greater water/rock ratio, lower abundance of iron, lower oxygen fugacity, and higher sulphur pressure. The differences, however, become obscure in the case of alkaline porphyry gold-copper deposits, which have associated pods and veins of magnetite and apatite (±amphibole).

The ultimate source of iron and mechanisms for its concentration into mineralizing fluids are topics of considerable speculation and debate, as can be expected for a group of such diverse deposits. The sources that have been invoked are calc-alkaline mafic and/or intermediate and felsic magmas (Geijer, 1930; Frietsch, 1978; Oyarzùn and Frutos, 1984; Hildebrand, 1986), alkaline and peralkaline magmas (Förster and Jafarzadeh, 1984, 1994; Förster, 1990), and pre-existing sedimentary iron deposits (Park, 1972). The mechanisms suggested include liquid immiscibility of the silicate and iron-phosphorus melts (Badham and Morton, 1976), crystallization differentiation (Geijer, 1930; Frietsch, 1978), volatile transfer (Lyons, 1988), hydrothermal concentration (Bookstrom, 1977; Hildebrand, 1986; Oreskes and Einaudi, 1990; Reeve et al., 1990) and mixing of a relatively reduced fluid with an oxidized fluid derived from a mafic or mixed mafic and felsic volcanic provenance C. Haynes et al., 1995) in the case of magmatic sources; and high grade metamorphism and assimilation of iron-rich rocks by magma in the case of a sedimentary source. In the authors' opinion, the most likely scenario is differentiation and volatile transfer in a dacitic to rhyolitic magma chamber at depth.

RELATED DEPOSIT TYPES

Alkaline porphyry Cu-Au deposits have veins and irregular pods of magnetite±apatite±actinolite(amphibole), and many skarn deposits are rich in magnetite and contain copper and other metals, hence both can be considered as related deposit types. Felsic volcanic-associated uranium deposits are indirectly related to the Kiruna/Olympic Dam-type in that they too occur in the continental, felsic-dominated, volcano-plutonic terrane, commonly in the late tectonic and post-tectonic environments.

Some iron oxide-rich breccias and lensoid bodies in metasedimentary rocks are mineralogically and texturally comparable with the deposits of Kiruna/Olympic Dam-type. They are, however, in settings that lack the evidence of large scale coeval magmatic activity, which is regarded here as an important feature of this deposit type. For example, breccias in the middle Proterozoic, folded sediments of the Wernecke Supergroup in Yukon Territory contain Cu, U, Co, Ag, and Au, and resemble those of the Olympic Dam deposit in many respects. There is no field evidence for any major igneous event to which the Wernecke breccias can be directly linked, but a deep seated magmatic hydrothermal source has been invoked for the formation and mineralization of the breccias (Hitzman et al., 1992; Thorkelson and Wallace, 1993, 1994). On the other hand, an evaporite diapir model, based on comparison with the Copperbelt in Zaïre and the Flinders Ranges in South Australia, has been proposed by Bell (1989). It requires dissolution and removal of the evaporites, but evidence for the presence of evaporites in or beneath the Wernecke Supergroup is weak.

Lensoid bodies of magnetite-apatite±amphibole, mineralogically similar to the Kiruna deposits, occur in some folded metasedimentary beds of early Proterozoic age, e.g. in the basement of southern Great Bear Magmatic Zone (Gandhi, 1994) and in the Singhbhum district in eastern India (Sarkar, 1984). They are essentially metamorphic segregations of the constituents deposited with the enclosing sediments.

EXPLORATION GUIDES

Important geological guides are the features listed as "Definitive characteristics". Of these, the extensional tectonic environment is of fundamental importance, especially if accompanied by voluminous felsic volcanism. There is indeed an empirical observation that continental felsic or bimodal (basalt-rhyolite) suites formed during the interval 2000 to 1400 Ma are very important, as found in the Proterozoic terranes of the Canadian, Baltic, and Australian shields. The most important geophysical guides are coincident magnetic and gravity anomalies.

REFERENCES

References marked with asterisks (*) are considered to be the best sources of general information on this deposit type.

Badham, J.P.N.
1978:   Magnetite-apatite-amphibole-uranium and silver-arsenide mineralizations in lower Proterozoic igneous rocks, East Arm, Great Slave Lake, Canada; Economic Geology, v. 73, no. 8, p. 1474-1491.

Badham, J.P.N. and Morton, R.D.
1976:   Magnetite-apatite intrusions and calc-alkali magmatism, Camsell River, N.W.T.; Canadian Journal of Earth Sciences, v. 13, p. 348-354.

Bell, R.T.
1989:   A conceptual model for development of megabreccias and associated mineral deposits in Wernecke Mountains, Canada, Copperbelt, Zaïre, and Flinders Range, Australia; in Uranium Resources and Geology of North America; Proceedings of a Technical Committee Meeting, Saskatoon, Canada, September 1987; International Atomic Energy Agency, Vienna, TECDOC-500, p. 149-169.

Bickford, M.E., Van Schmus, W.R., and Zietz, I.
1986:   Proterozoic history of the midcontinent region of North America; Geology, v. 14, p. 492-496.

*Bookstrom, A.A.
1977:   The magnetite deposits of El Romeral, Chile; Economic Geology, v. 72, no. 6, p. 1101-1130.

Burnham, W.C. and Ohmoto, H.
1980:   Late stage processes of felsic magmatism; Mining Geology Special Issue, no. 8, Society of Mining Geologists of Japan, Tokyo, p. 1-11.

Cann, R.M.
1979:   Geochemistry of magnetite and the genesis of magnetite-apatite lodes in the Iron Mask batholith, British Columbia; M.Sc. thesis, University of British Columbia, Vancouver, British Columbia, 196 p.

Cliff, R.A., Rickard, D., and Blake, K.
1990:   Isotopic systematics of the Kiruna magnetite ores, Sweden; Part I. Age of the ore; Economic Geology, v. 85, no. 8, p. 1770-1776.

*Cox, D.P. and Singer, D.M. (ed.)
1986:   Mineral deposit models; United States Geological Survey, Bulletin 1693, p. 172-174.

Creaser, R.A. and Cooper, J.A.
1993:   U-Pb geochronology of middle Proterozoic felsic magmatism surrounding the Olympic Dam Cu-U-Au-Ag and Moonta Cu-Au-Ag deposits, South Australia; Economic Geology, v. 88, no. 1, p. 186-197.

Emery, J.A.
1968:   Geology of the Pea Ridge iron ore body; in Ore Deposits of the United States, 1933-1967, Graton-Sales Volume, (ed.) J.D. Ridge; American Institute of Mining and Metallurgical Engineers, New York, p. 359-369.

Fanning, C.M., Flint, R.B., Parker, A.J., Blissett, A.H.,
and Ludwig, K.R.
1988:   Refined Proterozoic evolution of the Gawler Range craton, South Australia, through U-Pb geochronology; Precambrian Research, v. 40/41, p. 363-386.

Fiske, R.S., Hopson, C.A., and Waters, A.C.
1963:   Geology of Mount Rainier National Park, Washington; United States Geological Survey, Professional Paper 444, 93 p.

Förster, H.
1990:   Scapolite-analcite-bearing magnetite ore from Seh Chahoon, central Iran; Program with Abstracts, Eighth International Association on the Genesis of Ore Deposits Symposium, August 1990, Ottawa, Canada, p. A196.

Förster, H. and Jafarzadeh, A.
1984:   The Chador Malu iron ore deposit (Bafq district, central Iran) - magnetite filled pipes; Neues Jahrbuch für Geologie und Paläontologie Abhandlungen, v. 168, no. 2/3, p. 524-534.
1994:   The Bafq mining district in central Iran – a highly mineralized infracambrian volcanic field; Economic Geology, v. 89, no. 8, p. 1697-1721.

Förster, H. and Knittel, U.

1979: Petrographic observations on a magnetite deposit at Mishdovan, central Iran; Economic Geology, v. 74, no. 6, p. 1485-1510.

*Frietsch, R.

1978: On the magmatic origin of iron ores of the Kiruna type; Economic Geology, v. 73, no. 4, p. 478-485.

*Frietsch, R., Papunen, H., and Vokes, F.M.

1979: The ore deposits in Finland, Norway and Sweden - a review; Economic Geology, v. 74, p. 975-1001.

Frutos, J. and Oyarzùn, J.

1975: Tectonic and geochemical evidence concerning the genesis of El Laco magnetite lava flow deposits, Chile; Economic Geology, v. 70, no. 5, p. 988-990.

Gaál, G. and Gorbatschev, R.

1987: An outline of the Precambrian evolution of the Baltic Shield; Precambrian Research, v. 35, p. 15-32.

Gandhi, S.S.

1989: Rhyodacite ignimbrites and breccias of the Sue-Dianne and Mar Cu-Fe-U deposits, southern Great Bear magmatic zone, Northwest Territories; in Current Research, Part C; Geological Survey of Canada, Paper 89-1C, p. 263-273.

1992: Magnetite-rich breccia of the Mar deposit and veins of the Nod prospect, southern Great Bear magmatic zone, Northwest Territories; in Current Research, Part C; Geological Survey of Canada, Paper 92-1C, p. 237-249.

*1994: Geological setting and genetic aspects of mineral occurrences in the southern Great Bear magmatic zone, Northwest Territories; in Studies of Rare-metal Deposits in the Northwest Territories, (ed.) W.D. Sinclair and D.G. Richardson; Geological Survey of Canada, Bulletin 475, p. 63-96.

*Gandhi, S.S. and Bell, R.T.

1993: Metallogenic concepts to aid exploration for the giant Olympic Dam-type deposits and their derivatives; Proceedings of the 8th Quadrennial International Association on the Genesis of Ore Deposits Symposium, Ottawa, Canada, August 12-18, 1990; (ed.) Yvon T. Maurice; E. Schweizerbart'sche Verlagsbuchhandlung, Stuttgart, Germany, p. 787-802.

Gandhi, S.S. and Prasad, N.

1982: Comparative petrochemistry of two cogenetic monzonitic laccoliths and genesis of associated uraniferous actinolite-apatite-magnetite veins, east arm of Great Slave Lake, District of Mackenzie; in Uranium in Granites, (ed.) Y.T. Maurice; Geological Survey of Canada, Paper 81-23, p. 81-90.

Geijer, P.

1930: The iron ores of Kiruna type: geographical distribution, geological characters, and origin; Sveriges Geologiska Undersökning, ser. C, no. 367, 39 p.

Gower, C.F., Ryan, B., and Rivers, T.

1990: Mid-Proterozoic Laurentia-Baltica: an overview of its geological evolution and a summary of contributions made in this volume; in Mid-Proterozoic Laurentia-Baltica, (ed.) C.F. Gower, B. Ryan, and T. Rivers; Geological Association of Canada, Special Paper 32, p. 1-20.

Hagni, R.D. and Brandom, R.T.

1988: Comparison of the Boss-Bixby, Missouri and Olympic Dam, South Australia ore deposits, and potential for these deposits in the midcontinent region; in Proceedings, North American Conference on Tectonic Control of Ore Deposits and the Vertical and Horizontal Extent of Ore Systems, (ed.) G. Kisvarsanyi and S.K. Grant; University of Rolla, Missouri, p. 333-335.

Haynes, D.W., Cross, K.C., Bills, R.T., and Reed, M.H.

1995: Olympic Dam ore genesis: a fluid mixing model; Economic Geology, v. 90, no. 2, p. 281-307.

Helvaci, C.

1984: Apatite-rich iron deposits of the Avnik (Bingöl) region, southeastern Turkey; Economic Geology, v. 79, no. 2, p. 354-371.

Henriquez, F. and Martin, R.F.

1978: Crystal-growth textures in magnetite flows and feeder dykes, El Laco, Chile; Canadian Mineralogist, v. 16, pt. 4, p. 581-589.

Hildebrand, R.S.

1986: Kiruna-type deposits: their origin and relationship to intermediate subvolcanic plutons in the Great Bear magmatic zone, northwest Canada; Economic Geology, v. 81, no. 3, p. 640-659.

Hildebrand, R.S., Hoffman, P.F., and Bowring, S.A.

1987: Tectono-magmatic evolution of the 1.9-Ga Great Bear magmatic zone, Wopmay orogen, northwestern Canada; Journal of Volcanology and Geothermal Research, v. 32, p. 99-118.

*Hitzman, M.W., Oreskes, N., and Einaudi, M.T.

1992: Geological characteristics and tectonic setting of Proterozoic iron oxide (Cu-U-Au-REE) deposits; Precambrian Research, v. 58, p. 241-287.

Hoffman, P.F.

1980: Wopmay Orogen: a Wilson cycle of Early Proterozoic age in the northwest of the Canadian Shield; in Continental Crust and its Mineral Deposits, (ed.) D.W. Strangway; Geological Association of Canada, Special Paper 20, p. 523-549.

Johnson, J.P. and Cross, K.C.

1991: Geochronological and Sm-Nd isotopic constraints on the genesis of the Olympic Dam Cu-U-Au-Ag deposit, South Australia; in Proceedings of the 25th Anniversary Meeting of the Society for Geology Applied to Mineral Deposits: 'Source, Transport, and Deposition of Ore Minerals', (ed.) M. Pagel and J.L. Leroy; A.A. Balkema, Rotterdam, p. 395-400.

*Kisvarsanyi, G. and Kisvarsanyi, E.B.

1989: Precambrian geology and ore deposits of the southeast Missouri iron metallogenic province; in Olympic Dam-type Deposits and Geology of Middle Proterozoic Rocks in the St. Francois Mountains Terrane, Missouri, (ed.) V.M. Brown, E.B. Kisvarsanyi, and R.D. Hagni; Society of Economic Geologists, Guidebook Series, v. 4, p. 1-54.

Li, Wenda and Kuang, Fuxiang

1990: The geology and geochemistry of porphyrite iron deposits in the Nanjing-Wuhu area, southeast China; Chinese Journal of Geochemistry, v. 9, no. 1, p. 1-26.

Lundberg, B. and Smellie, J.

1979: Painirova and Mertainnen iron ores: two deposits of Kiruna ore type in northern Sweden; Economic Geology, v. 74, no. 8, p. 1131-1152.

Lundstrom, I. and Papunen, H. (ed.)

1986: Mineral deposits of southwestern Finland and the Bergslagen Province, Sweden; 7th International Association on the Genesis of Ore Deposits Symposium Excursion Guide No. 3, Sveriges Geologiska Undersökning, Luleå, Sweden, ser. C, no. 61, 44 p.

Lyons, J.I.

1988: Volcanogenic iron oxide deposits, Cerro de Mercado and vicinity, Durango, Mexico; Economic Geology, v. 83, no. 8, p. 1886-1906.

Magnusson, N.H.

1970: The origin of the iron ores in central Sweden and the history of their alterations; Sveriges Geologiska Undersökning, ser. C, no. 643, pt. 1 & 2, 127 p. & 364 p.

Marikos, M.A., Nuelle, L.M., and Seeger, C.M.

1989: Geology of the Pea Ridge mine; in Olympic Dam-type Deposits and Geology of Middle Proterozoic Rocks in the St. Francois Mountains Terrane, Missouri; (ed.) by V.M. Brown, E.B. Kisvarsanyi, and R.D. Hagni; Society of Economic Geologists, Guidebook Series, v. 4, p. 41-54.

Mortimer, G.E., Cooper, J.A., Paterson, H.L., Cross, K., Hudson, G.R.T., and Uppill, R.K.

1988: Zircon U-Pb dating in the vicinity of the Olympic Dam Cu-U-Au deposit, Roxby Downs, South Australia; Economic Geology, v. 83, no. 4, p. 694-709.

Nyström, J.O. and Henriquez, F.

1994: Magmatic features of iron ores of the Kiruna type in Chile and Sweden: ore textures and magnetite geochemistry; Economic Geology, v. 89, no. 4, p. 820-839.

Oreskes, N. and Einaudi, M.T.

1990: Origin of rare earth element-enriched hematite breccias at the Olympic Dam Cu-U-Au-Ag deposit, Roxby Downs, South Australia; Economic Geology, v. 85, no. 1, p. 1-28.

Oyarzùn, J. and Frutos, J.

1984: Tectonic and petrological frame of the Cretaceous iron deposits of north Chile; Mining Geology, v. 34, no. 1, p. 21-31.

Panno, S.V. and Hood, W.C.

1983: Volcanic stratigraphy of the Pilot Knob iron deposits, Iron County, Missouri; Economic Geology, v. 78, no. 5, p. 972-982.

*Parák, T.

1975: The origin of Kiruna iron ores; Sveriges Geologiska Undersökning, ser. C, no. 709, 209 p.

Park, C.F.

1961:   A magnetite "flow" in northern Chile; Economic Geology, v. 56, no. 2, p. 431-441.

1972:   The iron ore deposits of the Pacific basin; Economic Geology, v. 67, no. 2, p. 339-349.

Parker, A.J.

1990:   Gawler Craton and Stuart Shelf - Regional Geology and Mineralization; in Geology of the Mineral Deposits of Australia and Papua New Guinea, (ed.) F.E. Hughes; Australasian Institute of Mining and Metallurgy, Melbourne, Australia, Monograph 14, p. 999-1008.

Paterson, H.L., Dalgarno, C.R., Esdale, D.J., and Tonkin, D.

1986:   Basement geology of the Stuart Shelf region, South Australia; Olympic Dam/Stuart Shelf, 8th Australian Geological Convention, Excursion Guidebook A1, Geological Society of Australia, p. 30.

*Reeve, J.S., Cross, K.C., Smith, R.N., and Oreskes, N.

1990:   Olympic Dam copper-uranium-gold-silver deposit; in Geology of the Mineral Deposits of Australia and Papua New Guinea, (ed.) F.E. Hughes; The Australasian Institute of Mining and Metallurgy, Melbourne, Australia, Monograph 14, p. 1009-1035.

*Research Group on Porphyrite Iron Ore of the Middle-Lower Yangtze Valley

1977:   Porphyrite iron ore - a genetic model of a group of iron ore deposits in andesitic volcanic area; Acta Geologica Sinica, v. 51, no. 1, p. 1-19.

Roberts, D.E. and Hudson, G.R.T.

1983:   The Olympic Dam copper-uranium-gold deposit, Roxby Downs, South Australia; Economic Geology, v. 78, no. 5, p. 799-822 (see also discussions by I.P. Youles and reply in Economic Geology, v. 79, p. 1941-1945).

Sarkar, S.C.

1984:   Geology and ore mineralization in the Singhbhum copper-uranium belt, eastern India; Jadavpur University, Calcutta, 263 p.

Skiöld, T.

1987:   Implications of new U-Pb zircon chronology to early Proterozoic crustal accretion in northern Sweden; Precambrian Research, v. 38, p. 147-164.

Thorkelson, D.J. and Wallace, C.A.

1993:   Development of Wernecke breccia in Slats Creek (106D/16) map area, Wernecke Mountains; in Yukon Exploration and Geology, 1992; Exploration and Geological Services, Yukon, Indian and Northern Affairs Canada, p. 77-87.

1994:   Geological setting of mineral occurrences in Fairchild Lake map area (106C/13), Wernecke Mountains; in Yukon Exploration and Geology, 1992; Exploration and Geological Services, Yukon, Indian and Northern Affairs Canada, p. 79-92.

West, G.F.

1980:   Formation of continental crust; in The Continental Crust and its Mineral Deposits, (ed.) D.W. Strangway; Geological Association of Canada, Special Paper 20, p. 117-148.

Xu, Zhigang

1990:   Mesozoic volcanism and volcanogenic iron-ore deposits in eastern China; Geological Society of America, Special Paper 237, 46 p.

Authors' addresses

R.T. Bell
2430 Garmil Crescent
North Gower, Ontario
Canada
K0A 2T0

S.S. Gandhi
Geological Survey of Canada
601 Booth Street
Ottawa, Ontario
Canada
K1A 0E8

Printed in Canada

23. PERALKALINE ROCK-ASSOCIATED RARE METALS

23. PERALKALINE ROCK-ASSOCIATED RARE METALS

D.G. Richardson and T.C. Birkett

INTRODUCTION

Commodities associated with peralkaline rocks are diverse and include a variety of metals and industrial rocks and minerals. The elements of economic interest in peralkaline rock-associated rare metal deposits include zirconium (Zr), niobium (Nb), beryllium (Be), uranium (U), thorium (Th), tantalum (Ta), rare-earth elements (REEs), yttrium (Y), and gallium (Ga). Commonly several elements are concentrated in a deposit.

Peralkaline rocks are characterized by a molar excess of alkali elements (Na_2O+K_2O) over aluminum (Al_2O_3). Mineralogically, this chemical distinction is commonly manifested in the presence of alkali amphiboles and pyroxenes. Aluminous minerals such as topaz and biotite, which are more typically associated with peraluminous rocks (i.e. molar $CaO+K_2O+Na_2O>Al_2O_3$), are generally absent. Peralkaline rocks span the range of silica saturation, from granites through syenites to feldspathoid-bearing undersaturated rocks. Deposits of rare metals in peralkaline rocks occur in all rock types without regard for silica activity. Silica-saturated to silica-undersaturated rocks concentrate similar suites of elements, although some of the silica-undersaturated, carbonatite-associated rocks of the Kola Peninsula also contain large deposits of apatite.

The mineralizing processes within peralkaline igneous rocks can be subdivided conceptually into magmatic and metasomatic end members. In nature, however, these systems commonly pass without interruption from magmatic to postmagmatic (hydrothermal) conditions. Three Canadian examples illustrate the behaviour of peralkaline systems. The Strange Lake deposit, Quebec-Labrador, and the Mann #1 occurrence, Labrador can be described as end member magmatic deposits, whereas the Thor Lake deposits, Northwest Territories, cross the magmatic-metasomatic transition. The locations of these deposits are shown on Figure 23-1.

Important foreign examples of deposits associated with peralkaline rocks include the Zr-Y deposits of the Ilimaussaq complex in southwestern Greenland, the giant apatite, Nb and Zr deposits of the Khibiny and Lovozero complexes in the Kola Peninsula of Russia, the Y-Zr-REE Brockman deposit, Western Australia, the Kvanefjeld and Motzfeldt Centre deposits in southern Greenland, deposits in apogranites in Russia, and numerous deposits in Saudi Arabia.

The Kipawa deposit (Quebec, Fig. 23-1) is a small Y-Zr deposit in a highly deformed and partly metasomatized alkaline complex which includes peralkaline granites, as well as carbonate rocks of indeterminate origin. Intense deformation and high grade metamorphism have obscured the origin of the mineralization.

In terms of end members, magmatic deposits consist of identifiable igneous units; the principal rare metal-bearing minerals are typically disseminated throughout the ore-forming unit and represent essential constituents that crystallized with their host rocks. Hydrothermal alteration associated with the deposits, if present, is generally late deuteric and local in nature; characteristically there are no extensive zones of alteration surrounding the deposits.

Richardson, D.G. and Birkett, T.C.
1996: Peralkaline rock-associated rare metals; in Geology of Canadian Mineral Deposit Types, (ed.) O.R. Eckstrand, W.D. Sinclair, and R.I. Thorpe; Geological Survey of Canada, Geology of Canada, no. 8, p. 523-540 (also Geological Society of America, The Geology of North America, v. P-1).

Metasomatic deposits occur in or near peralkaline igneous rocks, and are superimposed on pre-existing rocks. The deposits are related to the cooling and fluid release of intrusions; hydrothermal alteration may be extensive.

IMPORTANCE

Peralkaline rocks contain large resources of rare metals, including Nb, Ta, Be, Zr, Y, and REEs. These resources are generally undeveloped, except in Russia where about 2300 t of Nb_2O_5 are produced annually from the mining of loparite in the Lovozero complex, Kola Peninsula (I.G. Argamakov, pers. comm., 1992). In comparison, the equivalent of about 20 000 t of Nb_2O_5 was produced in market-economy countries in 1991 (Cunningham, 1991).

During the period 1957 to 1964, the Bokan Mountain deposit in Alaska produced a small amount of uranium. No significant production of rare metals has taken place from deposits of this class in Canada, although deposits such as Strange Lake and Thor Lake represent important potential sources of Be, Y, Nb, Ta, and Zr. Related deposits of fluorite in, and surrounding, peralkaline granites in Newfoundland have been a significant source of fluorite production in Canada (3.43 Mt of 70-95% CaF_2 concentrate from 1933 to 1978; Collins and Strong, 1985).

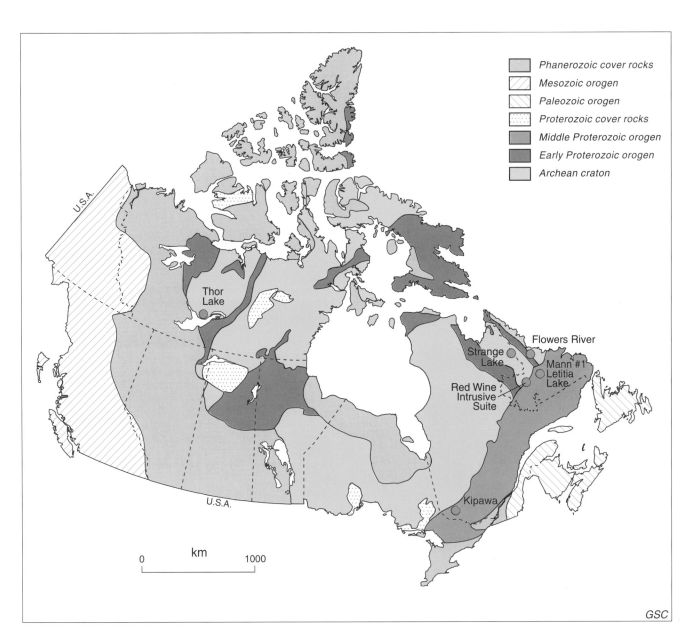

Figure 23-1. Locations of selected Canadian peralkaline rock-associated rare metal deposits/occurrences discussed in the text.

SIZE AND GRADE OF DEPOSITS

Peralkaline rock-associated rare metal deposits range widely in size, from less than one million tonnes to hundreds of millions of tonnes. Grades of Nb, Ta, Be, Y, and REEs are generally less than 1%; Zr typically ranges from 1 to 5%. The sizes and grades of Canadian peralkaline rock-associated rare metal deposits and selected other deposits worldwide are listed in Tables 23-1 and 23-2, respectively.

Grade-tonnage relationships of peralkaline rock-associated rare metal deposits are shown in Figure 23-2.

GEOLOGICAL FEATURES

Geological setting

Magmatic and metasomatic rare metal deposits generally occur within specific phases of peralkaline complexes. Drysdall et al. (1984), Miller (1989), and Pollard (1989a, b) have all noted that in deeper level environments, rare metal deposits associated with peralkaline rocks are commonly most closely associated with small satellite phases of the batholiths (e.g., Ghurayyah, Saudi Arabia); in some cases they occur within relatively fine grained marginal facies varieties of large plutons (e.g., Jabal Tawlah, Saudi Arabia).

Table 23-1. Production/reserves of selected Canadian peralkaline rare metal deposits/occurrences.

DEPOSIT	PRODUCTION/RESERVES/GRADE	COMMENTS/REFERENCES
Flowers River Igneous Suite, Labrador Lat. 55°35' N Long. 61°05' W	None	0.05 to 0.24% Y_2O_3 and 0.67 to 2.84% ZrO_2 occur in peralkaline aphyric quartz-poor lithic ash-flow tuffs and breccias of the Nuiklavik volcanics (Miller, 1993, 1994).
Mann # 1 Showing, Letitia Lake Group, Labrador Lat. 54°14' N Long. 62°25' W	1.8 Mt; 0.35-0.40 BeO, 0.24% Nb_2O_5	Volcanic-subvolcanic complex containing several Be and Nb occurrences. At Mann #1, beryllium minerals include barylite and eudidymite; niobium minerals are niobophyllite and pyrochlore. Limited drilling indicates irregular mineralization is contained in a lenticular oversaturated peralkaline body (2.0 km by 0.5 km) to a vertical depth of 60 m (Evans and Dujardin, 1961; Miller, 1987, 1988)
Red Wine Intrusive Suite, Labrador Lat. 54°05' N Long. 62°35' W	None; however, disseminated REE mineralization is reported in eudialyte near the boundaries of the complex	Eudialyte occurs within high-level, silica-undersaturated peralkaline phases of the North and South Red Wine plutons (Currie, 1976; Curtis and Currie, 1981; Hill and Miller, 1991).
Strange Lake/ Lac Brisson peralkaline complex Labrador-Quebec Lat. 56°18' N Long. 62°05' W	Open pit mineable reserves: 52 Mt; 2.93% ZrO_2, 0.31% Y_2O_3, 0.38% Nb_2O_5, 0.54% REEs, 0.08% BeO High grade zone within reserve (tonnage not published to date) contains 3.25% ZrO_2, 0.66% Y_2O_3, 0.56% Nb_2O_5	Highest grade mineralization occurs in a later stock and associated zoned pegmatite-aplite lens situated in the central part of a Middle Proterozoic arfvedsonite-aegirine peralkaline granite (Miller, 1985, 1986, 1990; Zajac et al., 1984).
Kipawa syenite gneiss complex, Quebec Lat. 46°48' N Long. 78°30' W	No tonnage reported but grades of ≥ 0.1% Y_2O_3, 0.5-1.2% ZrO_2 occur in a continuous zone 1300 m long and 30 m to 100 m wide	Mineralized zone consists of three higher grade, crudely defined subzones, termed the eudialyte, mosandrite, and britholite subzones. These subzones appear to be conformable with lithological contacts in the alkalic complex (Allan, 1990, 1992). Textural relations demonstrate that amphibolite-grade metamorphism, deformation, and possibly emplacement of an igneous alkaline protolith were roughly coeval (Currie and van Breemen, 1994).
Thor Lake, Northwest Territories Lat. 62°06' N Long. 112°35' W	North T- zone: 0.46 Mt; 1.11% BeO, 0.17% Y_2O_3, 0.58% Nb_2O_5 and 0.51 Mt; 0.28% $(REE)_2O_3$ 200 000 t of polylithionite South T-zone: 1.13 Mt; 0.62% BeO, 0.1% Y_2O_3, 0.2% $(REE)_2O_3$, 0.46% Nb_2O_5 Lake zone: 64 Mt; 0.04% Ta_2O_5, 0.57% Nb_2O_5, 1.99% $(REE)_2O_3$, 4.73% ZrO_2	The T-zone deposits are greisenized/metasomatized deposits hosted by oversaturated peralkaline members of the Blatchford Lake Intrusive Complex. The Lake zone deposit represents mineralization associated with a large metasomatized/altered breccia zone (Černý and Trueman, 1985; Highwood Resources, Limited Annual Report 1987; Pedersen and LeCouteur, 1990).

Table 23-2. Production/reserves of selected foreign peralkaline rare metal deposits/occurrences.

DEPOSIT	PRODUCTION/RESERVES/GRADE	COMMENTS/REFERENCES
Thabayadiotsa, Pilanesberg, South Africa	Estimated reserves: 13.5 Mt; 0.7% $(REE)_2O_3$ + ThO_2	The Pilanesberg alkaline complex consists of volcanic rocks intruded by a sequence of green and white foyaites, red syenite, and tinguaite disposed in concentric rings around a central core of red foyaite. Disseminations, irregular veinlets and sheet-like subzones rich in britholite occur on or near the contact of tinguaite with green trachytoidal foyaite (Lurie, 1986).
Brockman, Western Australia	Indicated resources: 8.97 Mt; 1.027% ZrO_2, 0.116% Y_2O_3, 0.437% Nb_2O_5, 0.036% Ta_2O_5, 0.038% HfO_2, 0.01% Ga, 0.105% $(REE)_2O_3$	Extremely fine grained minerals (< 20 micrometres) containing Zr, Hf, Y, and Ga in an ash-flow tuff (Niobium Tuff). Bastnaesite (± parisite and synchysite) and bertrandite occur in late-stage calcite veins. Mineralization is thought to be the result of alteration and remobilization of magmatic precursor minerals such as columbite and zircon by F-rich deuteric solutions that were retained in the tuff unit. Silicification and muscovitization are associated with this alteration (Industrial Minerals, February 1990; Ramsden et al., 1993).
Pocos de Caldas Alkaline Complex, Osamu Utsumi mine, Brazil	Reasonably assured reserves: 25.7 Mt; 0.0847% U_3O_8 Estimated (inferred) ores: 22.7 Mt; 0.11% MoO_3 21.3 Mt; 0.81% ZrO_2	Reserves, delineated to a depth of 370 m, are contained in a zone 1240 m long by 440 m wide, in the south-central portion of the largest known peralkaline complex in Brazil (1000 km²). Zr-, Mo-, and U-bearing minerals (pitchblende, uraninite, coffinite, phosphuranylite, and uranothorianite) are disseminated in tuffaceous phonolites and foyaites. Uranium occurs mainly in refractory minerals, especially zircon. Genetically, mineralization is closely related to postmagmatic hydrothermal activity. Associated alteration products include clay (kaolinite and halloysite), sericite, pyrite, and fluorite (Loureiro and Dos Santos, 1988).
Pajarito Mountain, Otero County, New Mexico, U.S.A.	Known resources: 2.4 Mt; 0.18% Y_2O_3, 1.2% ZrO_2	Eudialyte contained in Proterozoic peralkaline riebeckite granite-quartz syenite complex (Mining Engineering, 1989; Mariano, 1989).
Bokan Mountain (Kendrick Bay), Alaska, U.S.A.	Tonnage mined from the Ross-Adams deposit: 89 000 t; 1% U_3O_8, 3% ThO_2	Thorium and REE-rich rock containing uranothorite, uraninite, and generally < 2% sulphides, is associated with a Late Jurassic peralkaline granite ring-dyke complex. Wall rock alteration within and adjacent to orebodies consists of pervasive hydrothermal albite and lesser amounts of chlorite, fluorite, calcite, quartz, sericite, and tourmaline. Specular hematite is present in the outer-distal parts of the ore zone. The pod-like Ross-Adams deposit appears to have formed during regional faulting synchronous with magma crystallization and subsequent hydrothermal events (Thompson, 1988).
Lovozero complex, Russia	Tonnage figures not published	Unique 650 km² (surface area) layered funnel (lopolithic) shaped multiphase intrusive massif of peralkaline composition with a determined depth of 6-8 km. Identified intrusive phases include: Phase I - pegmatoid poikilitic and porphyritic nepheline and hydrosodalite syenites; Phase II - loparite-bearing lujavrite-foyaite-urtite; Phase III - eudialyte lujavrites and murmanite (lovozerite) lujavrites; Phase IV - veins and dykes of porphyritic murmanite-bearing lujavrite.
Phase II rocks: Stratified loparite-bearing lujavrite-foyaite-urtite series	Reserves are very large, on the order of billions of tonnes; 0.30% Nb_2O_5, 0.8-1.5% $(REE)_2O_3$	Loparite-bearing (Phase II) rocks comprise 75% of the volume (1650 m thickness) of the massif. They consist of numerous (>203), persistent, comparatively thin, rhythmically alternating layers which in turn have been subdivided into 65 units, all of which dip gently (8-16°) towards the centre of the massif. Economic concentrations of loparite are generally found in the basal portion (1-2 m) of each of the 65 units; however, significant concentrations can be found as thin (0.4-0.5 m) layers in the upper parts of the units. With depth there is a marked increase in the number and thickness of ore horizons (i.e. 3.7 m) and grade (Vlasov, 1966; Ginzburg and Fel'dman, 1977; Kogarko, 1987).
Phase III rocks: Eudialyte lujavrites, murmanite lovozerite lujavrites	Eudialyte lujavrites: 3.45% (Zr,Hf)O_2, 0.28% (Nb,Ta)$_2O_5$, 0.30 $(REE)_2O_3$ Eudialyte-rich zones: 1.37% TiO_2, 6.26-8.68% (Zr, Hf)O_2, 0.39-0.93% (Nb,Ta)$_2O_5$, 1.01-1.68% $(REE)_2O_3$ Lovozerite lujavrites; 1.7-2.4% ZrO_2, 0.17-0.33% (Nb,Ta)$_2O_5$, 0.14-0.39% $(REE)_2O_3$	Rhythmically layered eudialyte-bearing lujavrite sill-like bodies, (150 to 500 m thick) with steep sided/vertical contacts, are contained within Phase II rocks. The lenticular, almost monomineralic, eudialyte-rich zones are nest-like, and range in thickness from 1 to 75 m in the basal portion of the eudialyte-bearing lujavrites. The ore horizons are made up of 50-80% (modal) euhedral eudialyte crystals (Vlasov, 1966; Kogarko, 1987, 1990). Porphyritic lovozerite lujavrites occur in a 1 km² area (Vlasov, 1968).

DEPOSIT	PRODUCTION/RESERVES/GRADE	COMMENTS/REFERENCES
Midyan region, northwestern Saudi Arabia		
Ghurayyah	440 Mt; 8.6% Zr (zircon), 2.2% Nb (columbite-tantalite-pyrochlore), 0.13% Y (synchysite)	Disseminated Nb, Ta, Sn, Y, Th, U, and Zr minerals contained in a 0.9 km in diameter peralkaline microgranite with steeply dipping contacts that has intruded metavolcanic and metasedimentary rocks. Wall rocks show little alteration (Drysdall et al., 1984; Jackson, 1986).
Jabal Tawlah	Diamond drill indicated reserves to a depth of 65 m: 6.4 Mt; 0.34% Nb, 0.52% Y, 3.73% Zr	Disseminated Nb-Ta (columbite), Y-HREE (gagarinite, fergusonite, xenotime, and yttrium fluorite), Th (thorite), and Zr (zircon) contained in a composite sill up to 80 m thick with a strike length of 320 m. Wall rocks are in places silicified, feldspathized, and mineralized but elsewhere are apparently unaltered (Drysdall et al., 1984; Drysdall and Douch, 1986)
Hijaz region, central Saudi Arabia		
Jabal Sa'id	Potential reserves: 58 Mt; 0.33% Y, 0.108% Nb, 0.104% Ce, >1.9% Zr	Disseminated Nb, Ta, Sn, REE, Y, Th, U, and Zr minerals (zircon, thorite, bastnaesite, synchysite-(Y), monazite, thorian uraninite, pyrochlore) concentrated in the apical portion of a prominently layered, 150 m thick, aplite-pegmatite zone that extends for approximately 2.4 km along the contact between altered peralkaline microgranite and metavolcanic country rocks (Drysdall et al., 1984; Hackett, 1986; Jackson, 1986).
Umm al Birak	Reserve potential to 150 m below surface: High grade zone only - 6.6 Mt; 0.16% Nb, 0.51% Zr	Disseminated Nb, Ta, Sn, W, REE, Y, Th, and Zr minerals (zircon, monazite, bastnaesite, pyrochlore, scheelite, thorite) contained in a porphyritic, albite-microcline alkali microgranite stock (measuring 700 m by 400 m) and minor veins and pegmatites that are intrusive into metavolcanic rocks (Drysdall et al., 1984; Jackson, 1986).
Jabal Hamra	Reserve potential to 100 m below surface: 18 Mt; 0.17% Nb, 0.34% Ce, 0.16% Y, 1.33% Zr	Disseminated Nb, Ta, Sn, REE, Y, Th, and Zr minerals (monazite, bastnaesite, zircon, uraninite) contained in a 300 m long by 100 m wide, cresent-shaped stock of pervasively silicified and cataclasized silexite (rock composed essentially of quartz + hematite + alkali feldspar) (Drysdall et al., 1984; Jackson and Douch,1986).
Greenland alkaline complexes		
Ilimaussaq intrusion:		Ellipsoidal, 8 km by 14 km, approximately 1000 m thick differentiated peralkaline intrusion of nepheline syenite, naujaite, and lujavrite. In the southern part of the intrusion, stratified red, white, and black kakortokites occur among the lujavrites (these have a maximum thickness of 400 m in an area covering 35 km^2).
Kringlerne	Measured mineral resource: 14.5 Mt; 6% ZrO_2, 19.5 Mt; 0.2% Y_2O_3 6.8 Mt; 3.0% $(REE)_2O_3$ 27.9 Mt; 0.2% Nb_2O_5	Estimation of the resources contained only in the eudialyte-rich (extremely high grade) red kakortokite igneous cumulates in the Kangerluarsuk fiord area of the complex (Sørensen, 1992; Kalvig and Appel, 1994).
Agpat	Estimated resources: 30 Mt; 1% ZrO_2, 0.1% Y_2O_3	Resource estimate based on examination by Highwood Resources of various types of lujavrite found at three sites adjacent to the Tunugdliarfik fiord (Sørensen, 1992).
Kvanefjeld	Reasonably assured resources to a depth of 200 m: 794 Mt; 0.00034% U 110 Mt; 0.36% Zr 73 Mt; 0.07% Nb 141 Mt; 0.8% REEs 131 Mt; 0.08% Y	The Kvanefjeld area (3 km^2 in size) on the northwestern margin of the Ilimaussaq intrusion has high concentrations of U, Th, Y (± REEs) in sheared and metasomatized sheets and masses of medium- to coarse-grained lujavrite, near the contact with overlying sheared volcanics/gabbros. Radioactive minerals are steenstrupine (a uranium-rich variety of monazite) and thorite. Zirconium appears to be enriched in the lower lujavrite levels. Niobium minerals, related to either a late water/volatile-rich magma or hydrothermal solutions, occur in veins and sheared masses located near the roof of the intrusion or along sheared xenolith contacts (Sørensen et al., 1974; Kunzendorf et al., 1982).
Igaliko Nepheline Syenite:		
Motzfeldt Centre	Indicated mineable resources in microsyenite: 80 Mt; 0.4-1.0% Nb_2O_5, 0.01-0.03% Ta_2O_5 and 1-2% ZrO_2 Additional resources in zones of altered microsyenite: 50 Mt; 0.03-0.1% Ta_2O_5	Peralkaline sheets of microsyenite, pegmatites, and hydrothermal alteration zones (characterized by the presence of hematitic alteration) are enriched in U-Nb-Ta-Zr-REEs contained in fine grained pyrochlore and subordinate amounts of columbite (Tukiainen, 1988; Sørensen, 1992; Kalvig and Appel, 1994).

Peralkaline-rock associated REE deposits generally occur in cratonic, anorogenic settings. As noted by Pollard (1989a), the localization of individual igneous complexes in anorogenic regions is frequently controlled by regional scale fault systems, and the complexes are in many cases characterized by the presence of bimodal (basaltic and rhyolitic) magmatism. In shallow level environments, these plutons are commonly controlled by ring fractures, and intrude contemporaneous volcanic rocks within shallow level ring complexes (Pollard, 1989a). In some instances, both peralkaline and peraluminous magmas are spatially related (e.g. Arabian and Benin-Nigerian shields).

Form of deposits

Magmatic deposits assume the form of the host intrusive body. They may be irregular, following the contacts with country rocks or earlier intruded units, but can also be tabular, such as the rhythmically-layered kakortokite unit of the Illimaussaq complex.

Metasomatic rare metal deposits range in form from veins and stockworks to irregular replacement zones in which minerals are typically fine grained and widely disseminated, usually localized within the upper contact zones of small, steep-sided plutons, or within dome-shaped protuberances or cupolas (e.g., Jabal Sa'id, Saudi Arabia) or elongate, ridge-like structures on the upper surface of larger plutons (e.g., REE-rich zone, Motzfeldt Centre deposit, Igaliko complex, Greenland).

Subvolcanic and volcanic analogues of these intrusive deposits are also recognized. Jackson (1986) and Miller (1989) have suggested that some Nb-Zr-REE-bearing pegmatitic-aplite dykes and veins in Labrador and Saudi Arabia represent mineralization associated with vented roof zones of peralkaline intrusive bodies. In near surface environments, emplacement of late intrusive phases of magmatic complexes is in many cases controlled by ring fractures. Commonly, fine grained disseminated rare-metal mineralization occurs in contemporaneous volcanic rocks within ring complexes [e.g., Nb-Zr mineralization associated with peralkaline granites of the Nigerian ring complexes (Bowden, 1985; Bowden et al., 1987); REE+Th mineralization at the Thabayadiotsa deposit, Pilanesberg, South Africa (Lurie, 1986); and the Y-REE occurrences in the Flowers River Igneous Suite (Miller, 1993, 1994)].

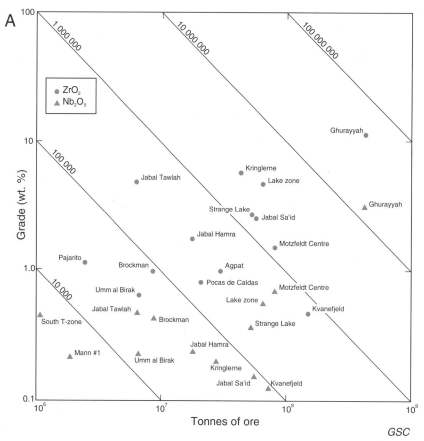

Figure 23-2. Grade-tonnage relationships for peralkaline rare metal deposits. Grades of the deposits are expressed in weight per cents. The diagonal lines indicate the quantity, in tonnes, of the contained commodity in the deposit. Data from Tables 23-1 and 23-2.

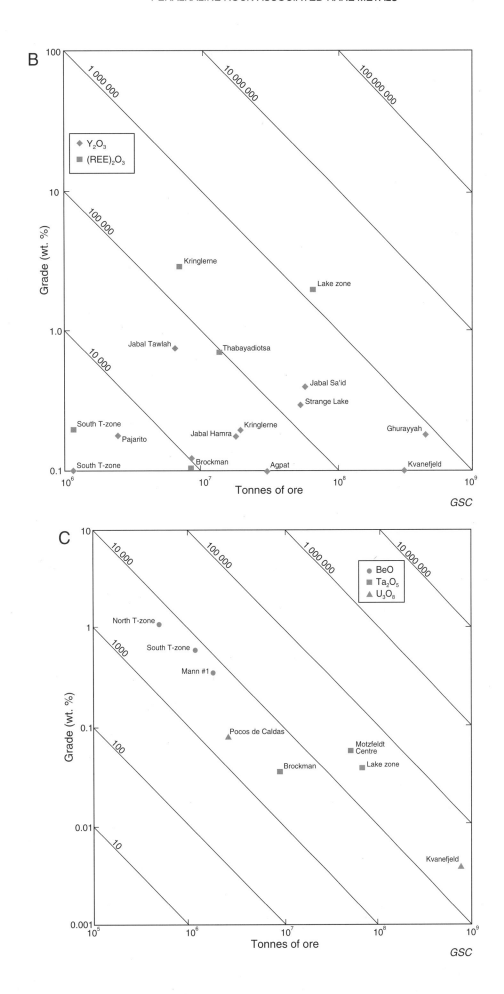

Table 23-3. Formulae of some of the unusual minerals mentioned in the text.

Mineral name	Formula
Silicates:	
allanite-(Ce)	$(Ce,Ca)_2(Al,Fe^{3+})_3(SiO_4)_3(OH)$
aenigmatite	$Na_2Fe^{2+}_5TiSi_6O_{20}$
armstrongite	$CaZrSi_6O_{15} \cdot 3H_2O$
astrophyllite	$(K,Na)_3Fe^{2+}_7Ti(Si_8O_{24})(O,OH,F)_7$
barylite	$BaBe_2Si_2O_7$
bertrandite	$Be_4Si_2O_7(OH)_2$
calcium catapleiite	$CaZrSi_3O_9 \cdot 2H_2O$
coffinite	$U(SiO_4)_{1-x}(OH)_{4x}$
chkalovite	$Na_2BeSi_2O_6$
dalyite	$K_2ZrSi_6O_{15}$
elpidite	$Na_2ZrSi_6O_{15} \cdot 3H_2O$
eudialyte	$Na_4(Ca,Ce)_2(Fe^{2+},Mn,Y)ZrSi_8O_{22}(OH,Cl)_2$
eudidymite	$NaBeSi_3O_7(OH)$
gadolinite (Ca-rich)	$Be_2(Ca,REE,Fe)_3Si_2O_{10}$
gittinsite	$CaZrSi_2O_7$
helvite	$(Fe,Mn,Zn)_4(BeSiO_4)_3S$
kainosite	$Ca_2(Y,Ce)_2Si_4O_{12}(CO_3) \cdot H_2O$
leifite	$Na_2(Si,Al,Be)_7(O,OH,F)_{14}$
lovozerite	$Na_2Ca(Zr,Ti)Si_6(O,OH)_{18}$
milarite	$K_2Ca_4Al_2Be_4Si_{24}O_{60} \cdot H_2O$
mosandrite (rinkite)	$(Na,Ca,Ce)_3Ti(SiO_4)_2F$
murmanite	$Na_2(Ti,Nb)_2Si_2O_9 \cdot nH_2O$
narsarsukite	$Na_2(Ti,Fe^{3+})Si_4(O,F)_{11}$
niobophyllite	$(K,Na)_3(Fe^{2+},Mn)_6(Nb,Ti)_2Si_8(O,OH,F)_{31}$
phenakite	Be_2SiO_4
polylithionite	$KLi_2AlSi_4O_{10}(OH,F)_2$
thorite	$(Th,Fe,Y,P,Ca)SiO_4$
titanite	$CaTiSiO_5$
uranothorite	$(U,Th)SiO_4$
vlasovite	$Na_2ZrSi_4O_{11}$
zircon	$ZrSiO_4$
Oxides:	
aeschynite-(Nd)	$(Nd,Ce,Ca)(Ti,Nb)_2(O,OH)_6$
columbite	$(Fe,Mn,Ti,Ta)Nb_2O_6$
fergusonite-(Y)	$(Y,Ce,U,Th)(Nb,Ta,Ti)O_4$
loparite	$(Na,Ca,REE)_2(Ti,Nb)_2O_6$
pyrochlore	$(Na,Ca)_2Nb_2O_6(OH,F)$
uraninite (pitchblende)	UO_2
uranothorianite (thorian uraninite)	$(U,Th)O_2$
Carbonates/fluorides:	
bastnaesite-(Ce)	$(Ce,La)(CO_3)F$
gagarinite -(Y)	$NaCaY(F,Cl)_6$
parisite-(Ce)	$Ca(Ce,La)_2(CO_3)_3F_2$
synchysite-(Ce)	$Ca(Ce,La)(CO_3)_2F$
roentgenite-(Ce)	$Ca_2(Ce,La)_3(CO_3)_5F_3$
Phosphates:	
britholite	$(Ca,Ce,Y)_5(SiO_4,PO_4)_3(OH,F)$
monazite-(Ce)	$(La,Ce,Nd,Th)PO_4$
phosphuranylite	$Ca(UO_2)_3(PO_4)_2(OH)_2 \cdot 6H_2O$
xenotime-(Y)	YPO_4

Age of deposits

Peralkaline igneous complexes with rare metal deposits range in age from Proterozoic to Tertiary, although a large number are mid-Proterozoic (1400 to 1000 Ma) in age (Currie and Gittins, 1993). Deposits in the Arabian Shield are related to late Proterozoic, post-tectonic peralkaline granite complexes that range in age from about 610 Ma to 510 Ma (Stoeser, 1986). Castor (1990) postulated that most of the world's Proterozoic REE deposits are associated with an axial zone of anorogenic magmatism that existed on a Proterozoic supercontinent. Anorogenic peralkaline complexes and associated rare metal deposits in Nigeria are primarily Jurassic in age (Kinnaird et al., 1985). The Lake zone at Thor Lake has been dated at 2094 ± 10 Ma (Sinclair et al., 1994).

Mineralogy

In most rare metal deposits, the principal Nb- and Ta-bearing minerals are columbite-tantalite and pyrochlore, although loparite is the source of the Nb and REE currently recovered from the Lovozero complex. In many deposits, Zr occurs mainly in eudialyte, which in some cases contains REEs. In addition, rare metal deposits associated with alkalic rocks can also contain a wide variety of unusual or rare minerals. Formulae of some of the unusual minerals found in peralkaline rock-associated rare metal deposits are given in Table 23-3.

The magmatic deposits of the Illimaussaq intrusion of southern Greenland contain alkali silicate minerals with essential components of the ore metals, including chkalovite, eudialyte, and vlasovite, and oxides and phosphates such as pyrochlore and monazite.

Alteration

As noted by Bowden (1985), Kinnaird (1985), Jackson (1986), Bowden et al. (1987), and Pollard (1989a, b), despite differences in primary magma compositions, the distribution of alteration and textural zones associated with rare metal deposits are similar, with a medium- to coarse-grained microcline-rich zone at the deepest levels succeeded upwards by an intermediate fine grained albite-rich zone and an upper pegmatitic zone of greisenization. In the case of peralkaline systems, greisenization is rare and alteration is dominated instead by albitization (sodium metasomatism). For example, albite-rich zones host the mineralization at Bokan Mountain, U.S.A. (Thompson, 1988) and at Thor Lake, Northwest Territories (Trueman et al., 1988). Pollard (1989a) noted that regardless of whether zones of intense albitization are magmatic or metasomatic, they generally occur in the top 50 m, and the grade of mineralization decreases gradually, laterally and vertically away from the apex of the intrusion. The development of U-Nb-Ta deposits and associated alteration at Pocos de Caldas, Brazil, and at Kvanefjeld and Motzfeldt Centre, Greenland seems to be related to the remobilization of rare metals by volatile-rich, deuteric fluids and their concentration in the carapace of the related intrusion. Primary geochemical haloes in the rocks hosting peralkaline rare metal deposits are typically absent, but, if present, are generally not extensive. The typical textural features of rare metal peralkaline rock-hosted deposits are illustrated and described in Table 23-4.

GENETIC MODEL

The formation of magmatic peralkaline rare metal deposits is related to igneous differentiation of intrusive complexes under closed conditions. Rare metal concentrations in the magmas increase through crystal fractionation, possibly supplemented by vapour-phase transport. Layering, such

Table 23-4. Schematic representation and summary of textural features typically associated with peralkaline and peraluminous rare metal deposits (compiled from: Bowden, 1985; Kinnaird, 1985; Jackson, 1986; Bowden et al., 1987; Pollard, 1989a, b).

TEXTURAL ZONE	SPATIAL LOCATION	THICKNESS	DESCRIPTION OF POSSIBLE FEATURES
Marginal pegmatite facies (stockscheider)	Upper cupola/apex/ upper contact zone of pluton	Variable thickness. If present, however, from few centimetres to >5 m. Thickness decreases away from apex of intrusion.	- Development of subhorizontal quartz-feldspar pegmatite interlayered with fine grained granite. - Development of downward branching (plumose) alkali feldspar in matrix of fine grained granite. - Presence of "brain rock" - narrow (<1 cm) alternating layers of microgranite and oriented quartz. - Development of layering parallel to upper contact of the main granite (i.e., individual quartz-rich and amphibole-rich zones within a network matrix of K-feldspar).
Fine grained granite	Upper contact zone/roof zone of pluton	In large plutons, zone is typically 10-100 m thick, but may be dominant rock type in small isolated plutons	- Presence of segregations of coarse grained pegmatite as much as a metre in diameter. - Miarolitic cavities are a common feature and are frequently infilled with hydrothermal minerals. - Presence of granophyric quartz-feldspar intergrowths within a network of coarser grained quartz-feldspar-mica pegmatite.
Medium- to coarse-grained ±porphyritic	Deeper levels of pluton	May possibly extend 200-300 m below pluton apex	- Development of microcline "porphyroblasts" adjacent to incipient fractures.

as that seen at the Lovozero complex, Russia and Kringlerne, Greenland, forms by crystal accumulation and by injection of residual magma into a semiconsolidated crystal mush. Pegmatites and magmatic-hydrothermal fluids are also formed from residual magmas. The pegmatites share the mineralogy of the surrounding rocks, but commonly have higher concentrations of rare metals (e.g., Strange Lake, Quebec-Labrador).

Disseminated Nb-Zr-Be-U-Th-REE-bearing minerals in pegmatite-aplite dykes and adjacent rocks in Labrador (Flowers River, Mann #1) and South Africa (Thabayadiotsa) are thought to represent subvolcanic- and volcanic-hosted mineralization associated with nearby peralkaline intrusive bodies (Lurie, 1986; Miller 1987, 1994).

The concentration of F (and possibly of CO_2) is fundamental in the development of magmatic and metasomatic peralkaline rock-associated rare metal deposits. Fluorine expands the temperature interval between liquidus and solidus, lowers viscosity of magmas, and, through complexing,

assists in the transport of elements, including those of economic interest. Deposits form through relatively simple concentration of elements during crystallization (Strange Lake, Ilimaussaq intrusion), possibly with some transport of material by fluids (supra-solidus alteration).

In some deposits, such as Thor Lake, mineralization results from the upward migration of postmagmatic fluids rich in F and CO_2, probably in excess of H_2O. The timing and nature of fluids exsolved from peralkaline magmas probably exercise a fundamental control on the subsequent mineralization.

Pollard (1989b) suggested that rare metals may be concentrated through scavenging by postmagmatic, metasomatic fluids, and the mineralization at Thor Lake is a good illustration of the potential for transport and concentration of rare metals by metasomatic processes. Fluids not related to the igneous event (for example groundwaters) are not generally considered to be involved in the mineralizing processes.

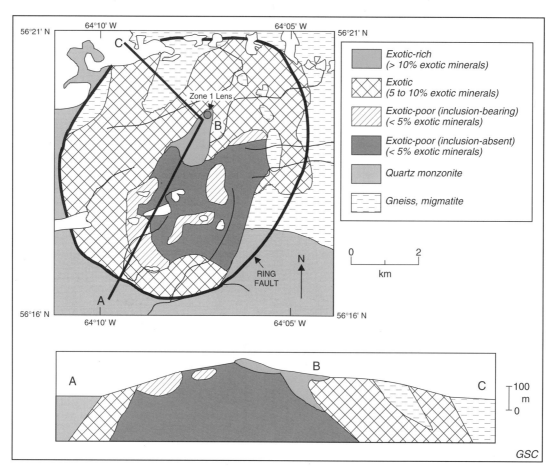

Figure 23-3. Generalized geology and south-north cross-section A-B-C of the Strange Lake peralkaline complex (after Miller, 1990; Hill and Miller, 1991). The term 'exotic' in the legend is used to define the main mineralogical phases of the Strange Lake Granite. These phases are distinguished on the basis of the modal abundances of an unusual suite of minerals that includes the Zr-bearing minerals zircon, gittinsite, armstrongite, vlasovite, elpidite, catapleiite, and dalyite, together with pyrochlore, kainosite, and gadolinite, as well as titanite, fluorite, monazite, thorite, narsarsukite, astrophyllite, and aenigmatite. The 'exotic-poor' granite is further subdivided on the basis of whether or not it contains host rock inclusions.

DEFINITIVE CHARACTERISTICS

Peralkaline rock-associated rare metal deposits have several common features, including:

1. Elevated contents of rare metals, such as Nb, Zr, Y, REEs, U, and Th, and of the volatiles F and CO_2.

2. Rare-earth elements are usually contained in oxides or silicates of Nb, Ti, Zr, Al, Be, and Na; Ca-phosphates and Ca-fluorocarbonates, such as bastnaesite; or yttrium members such as synchysite (Möller, 1989c). Niobium is primarily contained in pyrochlore, rather than columbite-tantalite.

3. Associated intrusive, subvolcanic, and volcanic rocks are peralkaline in composition (molar proportions Na_2O+K_2O/Al_2O_3 >1), and they may be either silica-undersaturated or silica-saturated.

4. Distinction between magmatic and metasomatic deposits is clear in cases in which extensive hydrothermal alteration activity has been inhibited (e.g. Strange Lake, Kringlerne, Illimaussaq intrusion, Greenland), but becomes arbitrary in cases in which the mineralizing system has passed continuously from magmatic to hydrothermal conditions (e.g., Thor Lake deposits, Northwest Territories; Motzfeldt Centre, Igaliko complex, Greenland).

SYNOPSIS OF CANADIAN DEPOSITS AND OCCURRENCES

Strange Lake deposit

Geological setting

The Strange Lake deposit, located on the Newfoundland-Quebec border, is part of a post-tectonic, peralkaline granite complex dated at 1240 ± 2 Ma (R.R. Miller, L.M. Heaman, and T.C. Birkett, unpub. data, 1994), which has intruded along the contact between older gneisses and monzonite of the Churchill Province of the Canadian Shield. The complex is subcircular, and consists of generally concentric, high level granitic intrusions bounded by sharp contacts with the country rock (Zajac et al., 1984; Miller, 1985, 1986, 1990). Ring faults, at or near the contact of the alkalic complex, dip outward at low to moderate angles (20-35°) (Miller, 1985). At the geometric centre of the complex is a small (approximately 1.5 km^2) stock of medium grained, generally nonporphyritic granite 'exotic-rich' granite on Fig. 23-3), with very high overall values of zirconium, niobium, and yttrium. Rooted within this medium grained granite stock are dykes of aplite-pegmatite that contain very high values of rare metals. The principal deposit outlined to date is the 'Zone 1 lens', a flat-lying aplite-pegmatite dyke located just north and east of the granite stock (Miller, 1990). This dyke is as much as 20 m thick and has a surface expression of 0.75 km^2. The basal portion of the dyke is a fine grained aplitic rock with flow-aligned phenocrysts which impart a directional fabric (Miller, 1986, 1990). Overlying the fine grained portion is a coarse grained, generally massive pegmatite of broadly similar composition. Occurrences of similar aplite-pegmatite dykes are known in other parts of the Strange Lake alkalic complex, but have not been systematically explored.

Mineralogy

Early hypersolvus granites at Strange Lake contain sodium-rich minerals such as narsarsukite, aenigmatite, astrophyllite, sodic amphibole and pyroxene, elpidite, leifite, and vlasovite, in addition to pyrochlore, fluorite, quartz, and feldspar. In later subsolvus granites, calcium-rich phases appear, such as gittinsite, calcium-catapleiite, armstrongite, kainosite, titanite, milarite, pyrochlore, prehnite, and an unusual calcium-rich member of the gadolinite group. Magnesium is generally present in clay minerals.

The potential ore minerals at Strange Lake include Be- and Y-bearing minerals such as the gadolinite group Ca-rich silicate minerals, an unnamed Ca-Y silicate and kainosite; the Zr-bearing mineral gittinsite; and Nb-bearing minerals such as pyrochlore and titanite.

Genetic model

Magmatic evolution within the Strange Lake complex led to enrichment of Ca and Mg, as well as of the elements of interest. Whether or not concentration of these elements in the apical portions of the pluton was assisted by volatile transport is not clear. Within the deposit, reaction of earlier formed minerals with magmatic and postmagmatic fluids has led to local metasomatic reaction and pseudomorphous replacement of earlier formed minerals. The absence of veins within and around the deposit is noteworthy.

Mineralization of the Letitia Lake Group and Flowers River Igneous Suite, Labrador

Mann #1 occurrence, Letitia Lake Group

The Mann #1 occurrence, in the Letitia Lake area of Labrador, has been described by Evans and Dujardin (1961), Thomas (1981), and Miller (1987, 1989). Mineralization is present in a variety of rare Be-, Nb-, and Y-bearing minerals, including barylite, eudidymite, niobophyllite, and pyrochlore (Miller, 1987, 1988). According to Miller (1988, 1989), high grade Nb-Be mineralization at the Mann #1 occurrence is contained in: 1) subvolcanic veins (i.e., aegirine-feldspar veins and albite-rich felsic veins) associated with fine- to medium-grained, massive, equigranular peralkaline aegirine-riebeckite±quartz syenite; and 2) disseminations in near-vent peralkaline trachytes (i.e., banded feldspar-riebeckite volcanic rocks and massive aegirine-feldspar volcanic rocks) that overlie, and are probably coeval with, the peralkaline syenite. Although most Nb-Be showings in the Letitia Lake area are associated with the emplacement of the peralkaline syenite, significant mineralization has not been discovered near all of the mapped syenites (Miller, 1988). Miller (1988) also noted that peralkaline granites in the area, which have similar mineralogy to the syenite, except that quartz and astrophyllite are more abundant and riebeckite and aegirine are less abundant, do not appear to host mineralization, although they are enriched in many of the rare metals.

Flowers River Igneous Suite

The Mid-Proterozoic Flowers River Igneous Suite includes the Nuiklavik felsic volcanic rocks and associated predominantly peralkaline granites. The Nuiklavik volcanic rocks, approximately 340 m thick and exposed in several partially eroded nested calderas, have been subdivided into five major units: 1) basal tuff; 2) amphibole-bearing porphyry; 3) lower crystal-rich ash flow; 4) crystal-poor and quartz-phyric ash-flow; and 5) upper ash-flow (Miller, 1993). Anomalous Zr and Y (2000-12 000 ppm Zr, 400 to 2000 ppm Y) are found in the crystal-poor and quartz-phyric ash-flow unit (Abdel-Rahman and Miller, 1994). Miller (1993) documented the presence of a thin sequence (about 4 m thick) of mineralized volcanic rocks (300-1900 ppm Y and >4000 ppm Zr) over an area of 14 km^2. In one locality, this unit is 32 m thick. Although mineralization is associated with Na-depleted peralkaline volcanic rocks, trace element data suggest that REE mineralization is magmatic in origin (Miller, 1994).

Thor Lake deposits

Geological setting

The Thor Lake rare metal deposits are centrally located within the Aphebian Blatchford Lake Intrusive Complex, which has intruded metasedimentary rocks of the Archean Yellowknife Supergroup. The complex consists of a western series of gabbroic, granitic, and syenitic rocks; to the east, these rocks are cut by a larger subcircular body of peralkaline granite (Grace Lake Granite) that encloses a central syenite body (Thor Lake Syenite) (Davidson 1978, 1981, 1982). Gravity studies by Birkett et al. (1994) suggest that the peralkaline rocks form a subhorizontal, relatively thin (extending to a depth of 1.5 to 1 km) lobe that appears to be centred over a subsided block of Yellowknife Supergroup metasedimentary rocks. Pinckston (1989) and Pinckston and Smith (1991) identified nepheline-bearing rocks as the youngest phase at Blatchford Lake, and have suggested they were emplaced as a ring complex.

The six Thor Lake mineral deposits shown in Figure 23-4 (Fluorite zone, Lake zone, R zone, S zone, North T-zone, and South T-zone), which at one time may have been contiguous, are separated by vertical faults (Trueman, 1986). Although the R, S, and Fluorite zones represent important mineralized areas, only the Lake Zone and both the North and South T-zone deposits are of economic interest (Schiller, 1985). Schematic geological cross-sections of the Be-Y-rich North T-zone are shown in Figure 23-5.

The T-zone deposits form a peralkaline pegmatite system which was emplaced as an intrusive, dyke-like body. In contrast, the Lake zone deposit consists of disseminated Nb-, Ta-, and Zr-bearing minerals contained in a large core of hydrothermally altered syenite breccia or pseudobreccia.

The T-zone and Lake zone deposits are linked by an albitized wall zone, which envelopes the T- and Lake zones. The two portions, however, are quite distinct.

Ore compositions and zoning of ore

The Thor Lake deposits display large-scale zonation, with Be, REEs, and Y concentrated in the northern portions of the system (the T-zone deposits) and Ta and Zr concentrated in the southern part, (the Lake zone). Trueman et al. (1988) and Pedersen and LeCouteur (1990) have documented more detailed patterns of metal distribution at Thor Lake, including:

1. The five Be-enriched zones outlined in the North T-zone show a trend to phenakite enrichment upward and a concomitant decrease in the proportion of Be as bertrandite and the minor Be-bearing minerals gadolinite and helvite.

2. Phenakite mineralization is richest in the North T-zone and values decrease downward and southward in the South T-zone where bertrandite (±gadolinite and helvite) predominate. Beryllium is not concentrated in the Lake zone; however, anomalous values have been detected along its eastern contact.

3. Yttrium, present in xenotime and Th-Y silicates, occurs in the central portion of the Lake zone and to a lesser degree in the central and lower portions of the North T-zone, where it may or may not be associated with Be mineralization (Fig. 23-4 and 23-5).

4. Cerium and lanthanum, present in the REE fluorocarbonates bastnaesite, parisite, synchysite, and roentgenite, attain their highest values in the upper North T-zone where they form a discrete upper quartz-bastnaesite zone (Fig. 23-5). This zone contains 60 000 t grading approximately 8% rare-earth oxides (REOs) (Sinclair et al., 1992).

5. Niobium (columbite and pyrochlore) follows a similar pattern to Be and Y, showing extreme enrichment in the North T-zone and decreasing southward into the Lake zone. Schiller (1985) reported that a portion of the Wall zone of the North T-zone contains a separate zone grading 1.0% Nb_2O_5 and 0.05% Ga.

6. In the Lake zone, Ta and Nb, contained in ferrocolumbite, pyrochlore group minerals, and aeschynite group minerals, are accompanied by enrichment of LREE (allanite-(Ce), monazite-(Ce), bastnaesite-(Ce)) and Y (fergusonite-(Y), xenotime, zircon).

7. Zirconium, almost absent in the T-zone deposits, attains significant concentrations in the Lake zone (>3.5% Zr in zircon).

Detailed descriptions of the zones, subzones, and mineralogies of the significant Thor Lake deposits are provided in Table 23-5.

Genetic model

The T-zone deposit is thought to have evolved by magmatic crystallization upward and inward from the walls. The main stage of mineralization was formed by the reaction of the residual liquids of the system (possibly with the addition of more fluids from below) with the already-solidified

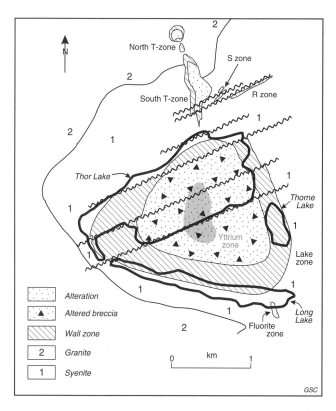

Figure 23-4. General geology of the Thor Lake area, showing the location of the North and South T-, Lake-, R-, S- and Fluorite zones, after Trueman et al. (1988). Area colored in red denotes approximate location of the high grade yttrium zone in the Lake zone deposit as determined by diamond drilling (Highwood Resources Limited, Annual Report, 1987). The heavy black lines outline lakes.

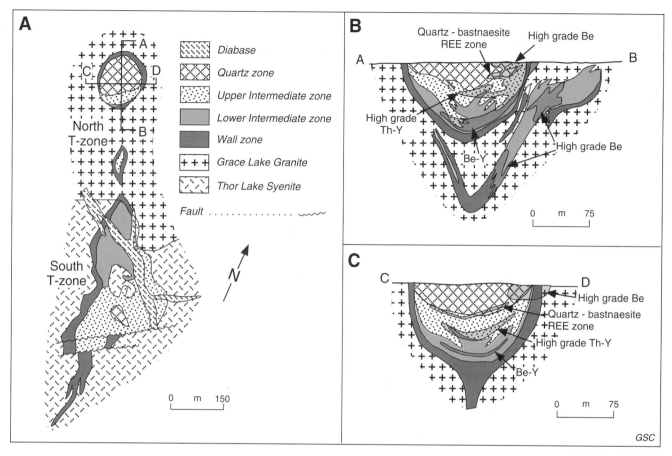

Figure 23-5. Plan of the T-zone deposit showing the major rock types **A)**; cross-sections AB **B)** and CD **C)** are from the North T-zone deposit, modified from Trueman et al. (1988). Areas shaded red denote zones of high-grade Be, high grade Th-Y, Be-Y, and quartz-bastnaesite REE mineralization (after Trueman et al., 1984; Schiller, 1985; and Pedersen and LeCouteur, 1990).

minerals. The progressive development and upward migration of the various zones, is thought to be analogous to the formation of a chromatogram, where various chemical compounds are separated and then subsequently deposited into a distinct zonal pattern. Alteration patterns and mineralization at Thor Lake, which include relic granite and syenite protoliths; albitites; microclinites; quartz-fluorite-polylithionite-phenakite greisens; and a massive quartz core, are indicative of superimposed potassic, sodic, and Fe-rich greisen metasomatism on an original peralkaline pegmatite (Pinckston and Smith, 1988, 1991; Trueman et al. 1988; and Trueman, 1989).

Uranium-lead zircon and monazite geochronology by Sinclair et al. (1994) suggests that Thor Lake mineralization substantially postdates the peralkaline phases of the Blatchford Lake Intrusive Complex (the Grace Lake Granite and Thor Lake Syenite). The presence of REE-bearing minerals such as zircon, cerianite-(Ce), britholite-(Ce), and thorite in the upper parts of the as yet undated, late stage undersaturated nepheline syenite body below the Thor Lake deposits led Pinckston (1989) and Pinckston and Smith (1991) to suggest that Thor Lake mineralization may be related to the crystallization and cooling of this syenite.

However, an outstanding problem with this hypothesis concerns the derivation of the relatively siliceous wall zone rocks that envelope the deposits (Birkett et al., 1990).

RELATED DEPOSIT TYPES

Felsic anorogenic rocks commonly evolve to compositions enriched in the rare elements without regard to peralkalinity or silica activity. The gradation of composition between peralkaline and peraluminous rocks commonly results in members of the two classes being closely related in space and time. Thus lithophile element mineralization associated with peraluminous and metaluminous granitic rocks, such as Sn-W vein and stockwork, porphyry Mo-W, and pegmatitic and vein-hosted Li-Ta-Nb-Be deposits, are related to the peralkaline rock-associated deposits (see Types 18, 19, and 21, respectively). The Sn-Nb (columbite type) mineralization associated with peraluminous biotite granites in the 1250 km long belt of ring complexes extending across Niger and Nigeria (Kinnaird et al., 1985; Pollard, 1989a), as well as certain Sn-W deposits associated with peraluminous to metaluminous, postorogenic granites

Table 23-5. Summary of geology/mineralogy of the T-zone and Lake zone deposits, Thor Lake, Northwest Territories (compiled from: Černý and Trueman, 1985; Trueman et al., 1988; Pedersen and LeCouteur, 1990; and Trueman, 1990).

DEPOSIT/ DIMENSIONS	HOST ROCK	ALTERATION/ MINERALIZED ZONE	DESCRIPTION	MINERALOGY
T-ZONE DEPOSITS 1 km long, 275 m wide, 150 m deep	**North T-zone** hosted by Grace Lake Granite	QUARTZ ZONE (QZ)	Monomineralic coarse grained quartz core as much as 35 m thick. Gradational with UIZ.	Patchy zones of green fluorite and separate zone of honey yellow sphalerite found near footwall boundary.
	South T-zone hosted by both Grace Lake Granite and Thor Lake Syenite	UPPER INTERMEDIATE ZONE (UIZ)	Transitional with QZ and LIZ and boundaries between these units are indistinct. Significant Be mineralization is contained in UIZ, along with some Y enrichment. Four pronounced subzones have been recognized.	Subzone 1 - Quartz-mica/quartz-mica feldspar/quartz feldspar Massive lenses and anastomosing stringers of euhedral polylithionite in a quartz matrix. Pink albite common, green fluorite and carbonate are common accessories. Subzone 2 - Quartz-feldspar-mica-magnetite ± phenakite Abundant phenakite intergrown with quartz, and large randomly oriented tabular crystals of magnetite and biotite. Fluorite and bastnaesite-group minerals are accessories. Subzone 3 - Quartz-bastnaesite Quartz with "brick-red" bastnaesite group fluorocarbonates (bastnaesite-(Ce), parisite-(Ce), synchysite-(Ce), roentgenite-(Ce) with accessory fluorite. Subzone 4 - Quartz-thorite-feldspar Lower half of UIZ, anastomosing stringers and blebs of fine grained mixtures of "chocolate-brown" thorium minerals (thorite, Th-Y silicates) in dark grey, very fractured quartz. This subzone is highly enriched in Y and Be. Beryllium minerals include phenakite, lesser amounts of bertrandite, and minor gadolinite. Xenotime, columbite, and dark purple fluorite are common.
		LOWER INTERMEDIATE ZONE (LIZ)	Displays gradational contact with UIZ, and with WZ breccias. Abundant granite-syenite xenoliths are present, and exhibit varying degrees of alteration. Zone is enriched in Be, Y, and Nb. Five pronounced subzones have been recognized.	Subzone 1 - Quartz-biotite/chlorite-feldspar Upper boundary of LIZ, massive lenses and anastomosing stringers of euhedral biotite in quartz, with accessory fluorite and columbite. Subzone 2 - Biotite/chlorite-feldspar-carbonate quartz Found only in south T-zone, mineralogy is similar to LIZ subzone 1, but the subzone contains 25% grey dolomite. Subzone 3 - Quartz-biotite/chlorite-feldspar (altered granite) and biotite/chlorite-feldspar± quartz (altered syenite) Mineralogy similar to LIZ subzone 1, relict xenoliths of granite and syenite exhibit minor to complete alteration. Subzone 4 - Quartz-magnetite-biotite/chlorite Forms fairly continuous brownish black "envelope" between UIZ and the main LIZ alteration. Highly siliceous, rich in magnetite, and locally in Nb (columbite). Subzone 5 - Quartz, feldspar, ± magnetite Gradational with LIZ subzone 4, similar to UIZ Subzone 1, but less siliceous and is significantly enriched in Be (phenakite).
		WALL ZONE (WZ)	Forms outer feldspathic shell of the T-zone. Outer contact with host granite and syenite is sharp, but the inner brecciated boundary with the LIZ is gradational. Niobium is found in columbite, and Ga is enriched in feldspars. Can be subdivided into three distinct subunits.	Subunit 1 - Feldspar-breccia Found along inner boundary of WZ, partly to completely albitized large K-feldpar crystals in matrix of quartz and quartz-magnetite. Subunit 2 - Microclinite, albitite Central core of WZ, light pink coarse microcline partly to completely replaced by cleavelandite. Subunit 3 - Banded aplite, albitite Fine grained outer margin of WZ, contains disseminated fluorite and mafic accessories.
LAKE ZONE Triangular in shape (covers an area of about 2 km², has been delineated to a depth of 335 m	Hosted by Thor Lake Syenite	CORE ZONE BRECCIA	Core zone is surrounded by the wall zone.	Not a breccia sensu stricto; what appear to be syenite fragments often display diffuse boundaries, grading back into unaltered protolith. Core zone assemblage consists of crystals and fragments of K-feldspar, all in a matrix of fine- to coarse-grained mica and lesser magnetite and amphibole. The matrix contains fluorite, columbite, uraninite, bastnaesite, parisite, synchysite, xenotime, monazite-(Ce), apatite, allanite-(Ce), fergusonite- (Y), aeschynite- (Nd), and zircon.
		WALL ZONE	Zone consists of albitite, similar to that enveloping the T-zone deposits.	Outer part consists of fine grained albite and quartz; inner part is characterized by a diverse suite of rocks rich in mafic minerals such as aegirine, biotite, and Fe and Ti oxides. Inner part also contains fluorite, quartz, and coarse fragments of perthitic K-feldpar, and displays evidence of extensive albitization.

within the eastern Arabian Shield (Jackson, 1986; Du Bray et al., 1988), are examples of spatially and temporally related peralkaline and peraluminous deposits.

Mineralization associated with fluorine-rich silicic lava flows (e.g., Spor Mountain Be deposit) and subvolcanic equivalents (e.g., Mexican fumarolic Sn deposits; Burt and Sheridan, 1987; Ramsden et al., 1993) are linked through the essential involvement of F in volcanic and subvolcanic peralkaline rare metal deposits. Similarly, the Fe-Cu-U-Au-REE Olympic Dam deposit of southern Australia (Oreskes and Einaudi, 1990) and the Fe-REE-apatite deposits at Kiruna, Sweden are associated with felsic alkalic rocks rich in F (see deposit Type 22, Kiruna-Olympic Dam Fe-Cu-U). These deposits exhibit extensive metasomatism (Frietsch, 1978, 1989; Hauck et al., 1989; Oreskes and Einaudi, 1990). The iron-rich nature of these deposits could be due to the involvement of groundwaters in the hydrothermal system.

Mineralogical, geochemical, and petrogenetic similarities also exist between carbonatites and peralkaline rock-associated rare metal deposits. For example, Bowden (1985) noted that from a mineralogical point of view, there are many parallels, including the presence of sphalerite, REE minerals, zircon, complex titanium-zirconium silicate minerals, pyrochlore, columbite, and uranium and thorium minerals (see Type 24).

EXPLORATION GUIDES

Exploration guidelines for peralkaline rock-associated rare metal deposits include the following:

1. Broad-scale features: deposits occur principally in anorogenic tectonic settings in which magma generation and intrusion are associated with crustal extension (Pollard, 1989a, b). Mineralized plutons typically form steep-sided, ovoid to elongate bodies that range from less than one square kilometre to several tens of square kilometres, which generally constitute the final intrusive phases of multiple-staged batholiths (Pollard, 1989b). In the subvolcanic environment, these plutons are in many cases controlled by ring fractures, and have commonly intruded contemporaneous volcanic rocks within shallow level ring complexes. In deeper level environments, mineralized plutons generally form small satellite phases to the main pluton, and in some cases form relatively fine grained, marginal facies varieties of large plutons.

2. Mineralogy: undersaturated peralkaline intrusion-related deposits are characterized by the presence of colourful, relatively rare alkali-rich minerals, such as sodalite (dark blue), eudialyte (pink/red), acmite (brown/green), alkali-amphiboles (blue/black), rinkolite (red brown/yellow brown), and gadolinite (green/brown-black).

In the Be-bearing metasomatic deposits, the prominent minerals are quartz and fluorite, whereas the primary beryllium minerals (phenakite, beryl, bertrandite, and helvite) are inconspicuous and easily overlooked. At Thor Lake, the beryllium potential of the T-zone deposit was not recognized until 1978 when phenakite was identified.

The presence of U and Th in these deposits commonly results in fluorite being purple.

3. Geochemical approaches: peralkaline rocks associated with mineralization typically have high Rb/Sr values, and are anomalously enriched in Zr, Zn, Nb, Y, Th, U, LREE and HREE, F, Be, and Pb (Bowden, 1985). These elements provide strong contrasts to regional background concentrations. Geochemical surveys of lake water or till can effectively locate the peralkaline rocks likely to host deposits. In the Gardar igneous province in southern Greenland, the Nb content in the 0.1 mm fraction of stream sediments, collected at a reconnaissance scale (1 sample per 6.25 km^2) proved most effective in defining potentially economic mineralization associated with peralkaline and carbonatitic intrusive complexes (Steenfelt, 1991). Indicator minerals in heavy mineral concentrates from stream sediments include pyrochlore, chrysoberyl, helvite, and euclase.

The Strange Lake deposit was discovered in 1979 as a result of exploration by the Iron Ore Company of Canada following regional lake water and sediment surveys that had revealed anomalous fluorine, uranium, and lead values (Geological Survey of Canada, 1979). Subsequent tracing of glacially transported mineralized boulders, initially recognized some 20 km from their source, led to the discovery of the deposit (Miller, 1985, 1986). Mapping of Quaternary deposits and geochemical studies have determined that anomalies associated with the Strange Lake deposit extend for 25 to 40 km in a down-ice direction from the mineralized source (McConnell et al., 1984; McConnell and Batterson, 1987; Batterson, 1989).

Although not applicable in Canada, the leaves of hickory trees have been found to concentrate REEs to an extraordinary degree and can serve as a biogeochemical indicator of REE-Y mineralization (Möller, 1989b).

4. Geophysical approaches: in Canada and Greenland, airborne radiometric (K, U, Th) surveys at 1 km line-spacing have proven effective in delineating peralkaline-alkaline intrusions and their mineral deposits (Batterson and LeGrow, 1986; Steenfelt, 1991). According to A.N. Mariano (Geological Survey of Canada Logan Club, oral presentation, February 24, 1994) the presence of positive airborne radiometric anomalies and negative aeromagnetic signatures characterize the Strange Lake, Kipawa, and Pajarito Mountain deposits.

Because most REE (Y) ores are somewhat radioactive, largely owing to the co-presence of Th, and to a lesser extent U, radiometric surveys may be useful for identifying parent granites, actual REE deposits, and other relevant features associated with mineralization (e.g., alteration, structural controls). A 1971 reconnaissance scale airborne radiometric survey carried out by the Geological Survey of Canada detected both U and Th anomalies over the area containing the Thor Lake deposits (Davidson, 1982). A subsequent detailed airborne radiometric/magnetic/VLF survey in 1988 by the Geological Survey of Canada of a 500 km^2 area at a line spacing of 250 m identified: 1) the major lithologies of the Blatchford Lake Intrusive Complex; 2) the Thor Lake deposits; and 3) potential sites of additional REE mineralization (Charbonneau and Legault, 1994).

537

Gravity surveys can be used to outline REE deposits in host rocks of contrasting density. Modelling of gravity data obtained over the Lake zone deposit by Highwood Resources Limited in 1983, suggests that the deposit has the form of an easterly-plunging cone (Trueman et al., 1988).

ACKNOWLEDGMENTS

R.R. Miller's careful review and numerous suggestions have greatly improved the manuscript. Discussions with W.D. Sinclair have focused and clarified many ideas. A.N. Mariano provided insight into the importance of peralkaline rocks to rare metal mineralization.

SELECTED BIBLIOGRAPHY

References with asterisks (*) are considered to be the best source of general information on this deposit type.

Abdel-Rahman, A.M. and Miller, R.R.
1994:   Extreme Na-depletion in peralkaline volcanic rocks of the Flowers River cauldron complex, Labardor; in Program with Abstracts, Geological Association of Canada-Mineralogical Association of Canada, Joint Annual Meeting, Waterloo, Ontario, May, 1994, v. 19, p. A1.

Allan, J.M.
1990:   Kipawa zirconium-yttrium property, Quebec; The Canadian Institute of Mining and Metallurgy Bulletin, v. 83, no. 935, p. 81.
1992:   Geology and mineralization of the Kipawa zirconium-yttrium property, Quebec; Exploration and Mining Geology, v. 1, p. 283-295.

Batterson, M.J.
1989:   Glacial dispersal from the Strange Lake alkalic complex, northern Labrador; in Drift Prospecting, (ed.) R.N.W. DiLabio and W.B. Coker; Geological Survey of Canada, Paper 89-20, p. 31-40.

Batterson, M.J. and LeGrow, P.
1986:   Quaternary exploration and surficial mapping in the Letitia Lake area, Labrador; in Current Research (1986), Report 86-1, Newfoundland Department of Mines and Energy, Mineral Development Division, p. 257-265.

Birkett, T.C. and Miller, R.R.
1990:   The role of hydrothermal processes in the granite-hosted Zr, Y, REE deposit at Strange Lake, Quebec/Labrador: evidence from fluid inclusions: Discussion; Geochimica et Cosmochimica Acta, v. 55, p. 3443-3445.

Birkett, T.C. and Richardson, D.G.
1988:   Comparative metallogeny of the Strange Lake and Thor Lake rare metal deposits; in Program with Abstracts, Geological Association of Canada-Mineralogical Association of Canada-Canadian Society of Petroleum Geologists, Joint Annual Meeting, St. John's, Newfoundland, May 1988, v. 13, p. A10.

Birkett, T.C., Miller, R.R., Roberts, A.C., and Mariano, A.N.
1992:   Zirconium-bearing minerals of the Strange Lake Intrusive Complex, Quebec-Labrador; Canadian Mineralogist, v. 30, p. 191-205.

Birkett, T.C., Richardson, D.G., and Sinclair, W.D.
1994:   Gravity modelling of the Blatchford Lake Intrusive Suite, Northwest Territories; in Studies of Rare Metal Deposits in the Northwest Territories, (ed.) W.D. Sinclair and D.G. Richardson; Geological Survey of Canada, Bulletin 475, p. 5-16.

Birkett, T.C., Sinclair, W.D., and Richardson, D.G.
1990:   Origin of the Thor Lake rare metals deposit; in Program with Abstracts, 8th International Association on the Genesis of Ore Deposits Symposium, Ottawa, Ontario, August 1990, p. A254-255.

Bowden, P.
1985:   The geochemistry and mineralization of alkaline ring complexes in Africa (a review); Journal of African Earth Sciences, v. 3, p. 17-39.

Bowden, P., Black, R., Martin, R.F., Ike, E.C., Kinnaird, J.A., and Batchelor, R.A.
1987:   Niger-Nigerian alkaline ring complexes: a classic example of African Phanerozoic anorogenic mid-plate magmatism; in Alkaline Igneous Rocks, (ed.) J.G. Fitton and B.G.J. Upton; Geological Society Special Publication No. 30, p. 357-379.

Burt, D.M. and Sheridan, M.F.
1987:   Types of mineralization related to fluorine-rich silicic lava flows and domes; Geological Society of America, Special Paper 212, p. 103-109.

Castor, S.B.
1990:   Rare earth deposits and Proterozoic anorogenic magmatism; in Program with Abstracts, 8th International Association on the Genesis of Ore Deposits Symposium Ottawa, Ontario, August 1990, p. A255-256.

Cerny, P. and Trueman, D.L.
1985:   Polylithionite from the rare metal deposits of the Blatchford Lake alkaline complex, N.W.T., Canada; American Mineralogist, v. 70, p. 1127-1134.

Charbonneau, B.W. and Legault, M.I.
1994:   Interpretation of airborne geophysical data for the Thor Lake area, Northwest Territories; in Studies of Rare Metal Deposits in the Northwest Territories, (ed.) W.D. Sinclair and D.G. Richardson; Geological Survey of Canada, Bulletin 475, p. 17-31.

Collins, C.J.
1984:   Genesis of the St. Lawrence fluorite deposits; in Mineral Deposits of Newfoundland - a 1984 Perspective; Newfoundland Department of Mines and Energy, Mineral Development Division, Report 84-3, p. 164-170.

Collins, C.J. and Strong, D.F.
1985:   A fluid inclusion and trace element study of fluorite veins associated with the peralkaline St. Lawrence Granite, Newfoundland; in Recent Advances in the Geology of Granite - related Mineral Deposits (ed.) R.P. Taylor and D.F. Strong; The Canadian Institute of Mining and Metallurgy, Special Volume 39, p. 291-302.

Cunningham, L.D.
1991:   Columbium (niobium) and tantalum; in 1991 Minerals Yearbook, Volume 1, Metals and Minerals, United States Bureau of Mines, p. 467-500.

Currie, K.L.
1973:   The Red Wine-Letitia alkaline province of Labrador; in Report of Activities, Part A: April to October, 1972; Geological Survey of Canada, Paper 73-1, Part A, p. 138.
*1976:   The alkaline rocks of Canada; Geological Survey of Canada, Bulletin 239, 228 p.
1985:   An unusual peralkaline granite near Lac Brisson, Quebec-Labrador; in Current Research, Part A; Geological Survey of Canada, Paper 85-1A, p. 73-80.

Currie, K.L. and Gittins, J.
1993:   Preliminary report on peralkaline silica-undersaturated rocks in the Kipawa syenite gneiss complex, western Quebec; in Current Research, Part E; Geological Survey of Canada, Paper 93-1E, p. 197-205.

Currie, K.L. and van Breemen, O.
1994:   Tectonics and age of the Kipawa syenite-complex, western Quebec; in Program with Abstracts, Geological Association of Canada-Mineralogical Association of Canada, Joint Annual Meeting, Waterloo, Ontario, May 1994, v. 19, p. A25.

Curtis, L.W. and Currie, K.L.
1981:   Geology and petrology of the Red Wine alkaline complex, central Labrador; Geological Survey of Canada, Bulletin 294, 61 p.

Davidson, A.
1978:   The Blatchford Lake Intrusive Suite: an Aphebian akaline plutonic complex in the Slave Province, Northwest Territories; in Current Research, Part A; Geological Survey of Canada, Paper 78-1A, p. 119-127.
1981:   Petrochemistry of the Blatchford Lake Complex, District of Mackenzie; Geological Survey of Canada, Open File 764, report and two maps, scale 1:50 000.
1982:   Petrochemistry of the Blachford Lake Complex near Yellowknife, Northwest Territories; in Uranium in Granites, (ed.) Y.T. Maurice; Geological Survey of Canada, Paper 81-23, p. 71-79.

Drysdall, A.R. and Douch, C.J.
1986:   Nb-Th-Zr mineralization in microgranite-microsyenite at Jabal Tawlah, Midyan region, Kingdom of Saudi Arabia; Journal of African Earth Sciences, v. 4, p. 275-288.

*Drysdall, A.R., Jackson, N.J., Ramsay, C.R., Douch, C.J., and Hackett, D.
1984:   Rare element mineralization related to Precambrian alkali granites in the Arabian Shield; Economic Geology, v. 79, p. 1366-1377.

Du Bray, E.A., Elliott, J.E., and Stuckless, J.S.
1988:   Proterozoic peraluminous granites and associated Sn-W deposits, Kingdom of Saudi Arabia; in Recent Advances in the Geology of Granite - related Mineral Deposits, (ed.) R.P. Taylor and D.F. Strong; The Canadian Institute of Mining and Metallurgy, Special Volume 39, p. 142-156.

Evans, E.L. and Dujardin, R.A.
1961: A unique beryllium deposit in the vicinity of Ten Mile Lake, Seal Lake area, Labrador; Geological Association of Canada Proceedings, v. 13, p. 45-51.

Frietsch, R.
1978: On the magmatic origin of iron ores of the Kiruna type; Economic Geology, v. 73, p. 478-485.
1989: The Kiruna ores; in Abstracts with Programs, Geological Society of America, Annual Meeting, St. Louis, Missouri, November 1989, v. 21, no. 6, p. A33.

Geological Survey of Canada
1979: Regional lake sediment and water geochemical reconnaissance data, Labrador; Geological Survey of Canada, Open File 559, 9 maps.

Gerasimovsky, V.I., Volkov, V.P., Kogarko, L.N., and Polyakov, A.I.
1974: Alkaline provinces - IV.2 Kola Peninsula; in The Alkaline Rocks, (ed.) H. Sorensen; Wiley, London, p. 206-221.

Ginzburg, A.I. and Fel'dman, L.G.
1977: Deposits of tantalum and niobium; in Ore Deposits of the USSR - Volume III, (ed.) V.I. Smirnov; Pitman Publishing, London, p. 372-424.

Hackett, D.
1986: Mineralization aplite-pegmatite at Jabal Sa'id, Hijaz region, Kingdom of Saudi Arabia; Journal of African Earth Sciences, v. 4, p. 257-267.

Hauck, S.A., Hinze, W.J., Kendall, E.W., and Adams, S.S.
1989: A conceptual model of the Olympic Dam Cu-U-Au-REE-Fe deposit: a comparison of central North American and South Australian terranes; in Abstracts with Programs, Geological Society of America, Annual Meeting, St. Louis, Missouri, November 1989, v. 21, no. 6, p. A33.

Hill, J.D.
1981: Geology of the Flowers River area, Labrador; Newfoundland Department of Mines and Energy, Report 81-6, 40 p.
1982: Geology of the Flowers River-Notakwanon River area, Labrador; Newfoundland Department of Mines and Energy, Report 82-6, 138 p.

Hill, J.D. and Miller, R.R.
1991: A review of Middle Proterozoic epigenetic felsic magmatism in Labrador; in Mid-Proterozoic Laurentia-Baltica, (ed.) C.F. Gower, T. Rivers, and B. Ryan; Geological Association of Canada, Special Paper 38, p. 417-431.

Jackson, N.J.
1986: Mineralization associated with felsic plutonic rocks in the Arabian Shield; Journal of African Earth Sciences, v. 4, p. 213-227.

Jackson, N.J. and Douch, C.J.
1986: Jabal Hamra REE-mineralized silexite, Hijaz region, Kingdom of Saudi Arabia; Journal of African Earth Sciences, v. 4, p. 269-272.

Kalvig, P. and Appel, P.W.U.
1994: Greenlandic mineral resources for use in advanced materials; Industrial Minerals, April 1994, p. 45-51.

Kinnaird, J.A.
1985: Hydrothermal alteration and mineralization of the alkaline anorogenic ring complexes of Nigeria; Journal of African Earth Sciences, v. 3, no. 1/2, p. 229-251.

Kinnaird, J.A., Bowden, P., Ixer, R.A., and Odling, N.W.A.
1985: Mineralogy, geochemistry and mineralization of the Ririwai complex, northern Nigeria; Journal of African Earth Sciences, v. 3, p. 185-222.

Kogarko, L.N.
1987: Alkaline rocks of the eastern part of the Baltic Shield (Kola Peninsula); in Alkaline Igneous Rocks, (ed.) J.G. Fitton and B.G.J. Upton; Geological Society Special Publication No. 30, p. 531-544.
1990: Ore-forming potential of alkaline magmas; in Alkaline Igneous Rocks and Carbonatites, (ed.) A.R. Woolley and M. Ross; Lithos, v. 26, p. 167-175.

Kunzendorf, H., Nyegaard, P., and Nielsen, B.L.
1982: Distribution of characteristic elements in the radioactive rocks of the northern part of Kvanefjeld, Ilimaussaq intrusion, south Greenland; Grønlands Geologiske Undersøgelse, Rapport Nr. 109, 32 p.

Loureiro, F.E.L. and Dos Santos, R.C.
1988: The intra-intrusive uranium deposits of Pocos de Caldas, Brazil; Ore Geology Reviews, v. 3, p. 227-240.

Lurie, J.
1986: Mineralization of the Pilanesberg Alkaline Complex; in Mineral Deposits of Southern Africa, (ed.) C.R. Anhaeusser and S. Maske; Geological Society of South Africa, v. II, p. 2215-2228.

*Mariano, A.N.
1989: Economic geology of rare earth elements; in Geochemistry and Mineralogy of Rare Earth Elements,(ed.) B.R. Lipin and G.A. McKay; Reviews in Mineralogy, v. 21, p. 309-336.

McConnell, J.W. and Batterson, M.J.
1987: The Strange Lake Zr-Y-Nb-Be-REE deposit, Labrador: a geochemical profile in till, lake and steam sediments and water; Journal of Geochemical Exploration, v. 29, p. 105-127.

McConnell, J.W., Vanderveer, D.G., Batterson, M.J., and Davenport, P.H.
1984: Geochemical orientation studies and Quaternary mapping around the Strange Lake deposit, northern Labrador; Newfoundland Department of Mines and Energy, Paper 84-1, p. 98-102.

Miller, R.R.
1985: Geology of the Strange Lake alkalic complex and the associated Zr-Y-Nb-Be-REE mineralization; in Granite - related Mineral Deposits (Geology, Petrogenesis and Tectonic Setting), (ed.) R.P. Taylor and D.F. Strong; The Canadian Institute of Mining and Metallurgy, Montreal, p. 193-196.
1986: Geology of the Strange Lake alkalic complex and the associated Zr-Y-Nb-Be-REE mineralization; in Current Research (1986), Report 86-1, Newfoundland Department of Mines and Energy, Mineral Development Division, p. 11-19.
1987: The relationship between Mann-type Nb-Be mineralization and felsic peralkaline intrusives, Letitia lake project, Labrador; in Current Research (1987), Report 87-1, Newfoundland Department of Mines and Energy, Mineral Development Division, p. 83-91.
*1988: Yttrium (Y) and other rare metals (Be, Nb, REE, Ta, Zr) in Labrador; in Current Research (1988), Report 88-1, Newfoundland Department of Mines and Energy, Mineral Development Division, p. 229-245.
*1989: Rare metal targets in insular Newfoundland; in Current Research (1989), Report 89-1, Newfoundland Department of Mines and Energy, Geological Survey Branch, p. 171-179.
1990: The Strange Lake pegmatite-aplite-hosted rare metal deposit, Labrador; in Current Research, Report 90-1, Newfoundland and Labrador Department of Mines and Energy, Geological Survey Branch, p. 171-182.
1991: Preliminary evaluation of rare metal targets in insular Newfoundland; in Current Research (1991), Report 91-1, Newfoundland Department of Mines and Energy, Geological Survey Branch, p. 327-334.
1992: Preliminary report of the stratigraphy and mineralization of the Nuiklavik volcanic rocks of the Flowers River Igneous Suite; in Current Research (1992), Report 92-1, Newfoundland Department of Mines and Energy, Geological Survey Branch, p. 251-258.
1993: Rare metal mineralization in the Nuiklavik volcanic rocks of the Flowers River Igneous Suite; in Current Research (1993), Report 93-1, Newfoundland Department of Mines and Energy, Geological Survey Branch, p. 363-371.
1994: Extreme Na-depletion in the peralkaline volcanic rocks of the Middle Proterozoic Flowers River cauldron complex, Labrador; in Current Research (1994), Report 94-1, Newfoundland Department of Mines and Energy, Geological Survey Branch, p. 233-246.

Mining Engineering
1989: Molycorp, Apache Indians to develop yttrium/zirconium deposit; Mining Engineering, v. 41, no. 7, p. 515.

Möller, P.
*1989a: Prospecting for rare-earth element deposits; in Lanthanides, Tantalum and Niobium; (ed.) P. Möller, P. Černý, and F. Saupé; Society for Geology Applied to Mineral Deposits, Special Publication No. 7, p. 263-265.
*1989b: REE(Y), Nb, and Ta enrichment in pegmatities and carbonatite-alkalic rock complexes; in Lanthanides, Tantalum and Niobium; (ed.) P. Möller, P. Černý, and F. Saupé; Society for Geology Applied to Mineral Deposits, Special Publication No. 7, p. 103-144.
*1989c: Rare earth mineral deposits and their industrial importance; in Lanthanides, Tantalum and Niobium; (ed.) P. Möller, P. Černý, and F. Saupé; Society for Geology Applied to Mineral Deposits, Special Publication No. 7, p. 171-188.

Oreskes, N. and Einaudi, M.T.
1990: Origin of rare earth element enriched hematite breccias at the Olympic Dam Cu-U-Au-Ag deposit, Roxby Downs, South Australia; Economic Geology, v. 85, p. 1-28.

Pedersen, J.C. and LeCouteur, P.C.
1990: The Thor Lake beryllium-rare metal deposits, Northwest Territories; in Mineral Deposits of the Slave Province, Northwest Territories; (ed.) W.A. Padgham and D. Atkinson; Geological Survey of Canada, Open File 2168, p. 128-136.

Pinckston, D.R.
1989:   Mineralogy of the Lake Zone deposit, Thor Lake, Northwest Territories; MSc. thesis, University of Alberta, Edmonton, Alberta, 155 p.

Pinckston, D.R. and Smith, D.G.W.
1988:   Mineralogy of the Lake Zone, Thor Lake, N.W.T.: final report for DIAND contract YK-87-87-026; Indian and Northern Affairs Canada, EGS-1988-5, 12 p.
1991:   Mineralogy and petrogenesis of the Lake Zone, Thor Lake rare metals deposit, N.W.T., Canada; Indian and Northern Affairs Canada, EGS-1991-5, 25 p.

Pollard, P.J.
*1989a: Geologic characteristics and genetic problems associated with the development of granite-hosted deposits of tantalum and niobium; in Lanthanides, Tantalum and Niobium, (ed.) P. Möller, P. Cerny, and F. Saupé; Society for Geology Applied to Mineral Deposits, Special Publication No. 7, p. 240-256.
*1989b: Geochemistry of granites associated with tantalum and niobium mineralization; in Lanthanides, Tantalum and Niobium; (ed.) P. Möller, P. Cerny, and F. Saupé; Society for Geology Applied to Mineral Deposits, Special Publication No. 7, Springer-Verlag, Berlin, p. 145-168.
1989c: Exploration for granite-hosted tantalum deposits: an approach via district analysis; in Lanthanides, Tantalum and Niobium, (ed.) P. Möller, P. Cerny, and F. Saupé; Society for Geology Applied to Mineral Deposits, Special Publication No. 7, Springer-Verlag, Berlin, p. 266-273.

Ramsden, A.R., French, D.H., and Chalmers, D.I.
1993:   Volcanic-hosted rare metals deposit at Brockman, Western Australia; Mineralium Deposita, v. 28, p. 1-12.

Salvi, S. and Williams-Jones, A.E.
1990:   The role of hydrothermal processes in the granite-hosted Zr, Y, REE deposit at Strange Lake, Quebec/Labrador: evidence from fluid inclusions; Geochimica et Cosmochimica Acta, v. 54, p. 2403-2418.

Schiller, E.A.
1985:   Beryllium-geology, production and uses; Mining Magazine, v. 152, no. 4, p. 317-322.

Sinclair, W.D., Hunt, P.A., and Birkett, T.C.
1994:   U-Pb zircon and monazite ages of the Grace Lake Granite, Blatchford Lake Intrusive Suite, Slave Province, Northwest Territories; in Radiogenic Age and Isotopic Studies: Report 8; Geological Survey of Canada, Current Research 1994-F, p. 15-20.

***Sinclair, W.D., Jambor, J.L., and Birkett, T.C.**
1992:   Rare earths and the potential for rare-earth deposits in Canada; Mining and Exploration Geology, v. 1, p. 265-281.

***Sørensen, H.**
1992:   Agpaitic nepheline syenites: a potential source of rare elements; Applied Geochemistry, v. 7, p. 417-427.

Sørensen, H., Rose-Hansen, J., Nielsen, B.L., Lovborg, L., Sørensen, E., and Lundgaard, L.,
1974:   The uranium deposit at Kvanefjeld, the Ilimaussaq intrusion, South Greenland: geology, reserves and beneficiation; Grønlands Geologiske Undersøgelse, Rapport Nr. 60, 54 p.

Steenfelt, A.
1991:   High-technology metals in alkaline and carbonatitic rocks in Greenland: recognition and exploration; Journal of Geochemical Exploration, v. 40, p. 263-279.

Stoeser, D.B.
1986:   Distribution and tectonic setting of plutonic rocks of the Arabian Shield; Journal of African Earth Sciences, v. 4, p. 21-46.

Taylor, R.G. and Pollard, P.J.
1988:   Pervasive hydrothermal alteration in tin-bearing granites and implications for the evolution of ore-bearing magmatic fluids; in Recent Advances in the Geology of Granite - related Mineral Deposits, (ed.) R.P. Taylor and D.F. Strong; The Canadian Institute of Mining and Metallurgy, Special Volume 39, p. 86-95.

Thomas, A.
1981:   Geology along the southwestern margin of the Central Mineral Belt, Labrador; Newfoundland Department of Mines and Energy, Mineral Development Division, Report 81-4, 40 p.

Thompson, T.B.
1988:   Geology and uranium-thorium mineral deposits of the Bokan Mountain granite complex, southeastern Alaska; Ore Geology Reviews, v. 3, p. 193-210.

Tremblay-Clark, P. and Kish, L.
1978:   Le district radioactif de Kipawa; Ministère de l'Energie et des Ressources de Québec, DPV 579, 28 p.

Trueman, D.L.
1986:   The Thor Lake rare metals deposits, Northwest Territories, Canada; in 7th Industrial Minerals International Congress, Industrial Minerals Papers - Volume 1, (ed.) G.M. Clarke and J.B. Griffiths; Monaco, April 1986, p. 127-131.
1989:   Geology of the T-Zone rare metal deposit at Thor Lake, N.W.T.; in Program with Abstracts, Geological Association of Canada-Mineralogical Association of Canada, v. 14, p. A1.
1990:   Thor Lake rare metal deposits, Northwest Territories; The Canadian Institute of Mining and Metallurgy Bulletin, v. 83, no. 935, p. 81.

Trueman, D.L., Pedersen, J.C., and de St. Jorre, L.
1984:   Geology of the Thor Lake beryllium deposits: an update; in Contributions to the Geology of the Northwest Territories, v. 1, Indian and Northern Affairs Canada, EGS 1984-6, p. 115-120.

Trueman, D.L., Pedersen, J.C., de St. Jorre, L., and Smith, D.G.W.
1988:   The Thor Lake, N.W.T. rare metal deposits; in Recent Advances in the Geology of Granite - related Mineral Deposits, (ed.) R.P. Taylor and D.F. Strong; The Canadian Institute of Mining and Metallurgy, Special Volume 39, p. 280-290.

Tukiainen, T.
1988:   Niobium-tantalum mineralization in the Motzfeldt Centre of the Igaliko nepheline syenite complex, south Greenland; in Mineral Deposits within the European Community, (ed.) J. Boissonnas and P. Omenetto; Society for Geology Applied to Mineral Deposits, Special Publication No. 6, p. 230-246.

Vlasov, K.A.(ed.)
1966:   Geochemistry and Mineralogy of Rare Elements and Genetic Types of their Deposits, Volume 3: Genetic Types of Rare-Element Deposits, Institute of Mineralogy, Geochemistry and Crystal Geochemistry, Academy of Sciences of the Union of Soviet Socialist Republics, Moscow, 860 p. (translation by Israel Program for Scientific Translations, Jerusalem, 1968, 916 p.)

***Woolley, A.R.**
1987:   Alkaline rocks and carbonatites of the world: Part 1. North and South America; British Museum (Natural History), London, 216 p.

Zajac, I.S., Miller, R.R., Birkett, T.C., and Nantel, S.
1984:   The Strange Lake deposit, Quebec-Labrador; The Canadian Institute of Mining and Metallurgy Bulletin, v. 77, p. 60.

Authors' addresses

T.C. Birkett
SOQUEM
2600, boul. Laurier
Tour Belle Cour, bureau 2500
Sainte-Foy, Quebec
G1V 4M6

D.G. Richardson
Geological Survey of Canada
601 Booth Street
Ottawa, Ontario
K1A 0E8

Printed in Canada

24. CARBONATITE-ASSOCIATED DEPOSITS

D.G. Richardson and T.C. Birkett

INTRODUCTION

Carbonatite-associated deposits include a variety of mineral deposits that occur both within and in close spatial association with carbonatites and related alkalic silicate rocks. Carbonatite-associated deposits are mined for rare-earth elements (REEs), niobium, iron, copper, apatite, vermiculite, and fluorite. Byproducts include barite, zircon or baddeleyite, tantalum, uranium, and in the unique Palabora carbonatite of South Africa, platinum group elements, silver, and gold. In some complexes, calcite-rich carbonatite is mined as a source of lime to produce Portland cement, and in Europe, carbonatites have provided lime and iron for hundreds of years (Dawson, 1974; Deans, 1978; Bowden, 1985).

Carbonatites are igneous rocks which contain at least 50% modal carbonate minerals, mainly calcite, dolomite, ankerite, or sodium- and potassium-bearing carbonates (nyerereite and gregoryite). Other minerals commonly present include diopside (in early carbonatites – e.g., Bond zone, Oka, Quebec), sodic pyroxenes or amphiboles, phlogopite, apatite, and olivine. A large number of rare or exotic minerals also occur in carbonatites. Definitions of rock names used in describing carbonatite-associated deposits are provided in Table 24-1 and chemical formulae of some less common minerals are given in Table 24-2.

Carbonatites occur mainly as intrusive bodies of generally modest dimensions (as much as a few tens of square kilometres), and to a lesser extent as volcanic rocks (flows and derived deposits), which are associated with a wide range of alkali silicate rocks (syenites, nepheline syenites, nephelinites, ijolites, urtites, pyroxenites, etc.) (Bowden, 1985; Barker, 1989). Although carbonatites are invariably associated with alkalic rocks, the inverse relationship does not necessarily hold (Möller, 1989b). Carbonatites are generally surrounded by an aureole of metasomatically altered rocks called fenites produced by reaction of country rock with peralkaline fluids released from the carbonatite complex (Morogan, 1994).

Carbonatite-associated deposits can be subdivided into magmatic and metasomatic types. Magmatic deposits are formed through processes associated with the crystallization of carbonatites, whereas metasomatic deposits form by the reaction of fluids released during crystallization with pre-existing carbonatite or country rocks. Spatially distinct niobium and rare-earth element (REE) mineralization at the Niobec mine, Quebec, are examples of magmatic and metasomatic mineralization, respectively.

The Aley (British Columbia); Oka, Crevier, and Niobec (Quebec); and Argor, Manitou Islands, Lackner Lake, and Nemegosenda Lake (Ontario) deposits are Canadian carbonatite-associated deposits (Fig. 24-1). Significant foreign carbonatite-associated deposits include Palabora (South Africa), Kovdor (Russia), Siilinjärvi (Finland), Sarfartôq (Greenland), Tapira and Jacupiranga (Brazil), the Mountain Pass bastnaesite deposit (California), the Kangankunde Hill monazite deposit (Malawi), Bayan Obo (People's Republic of China), and Amba Dongar (India).

IMPORTANCE

Carbonatite-associated deposits contain the majority of the known reserves of niobium in the world. The Niobec mine near Chicoutimi, Quebec, accounts for approximately 10% of western world niobium production, and is the only underground niobium mine (Scales, 1989). Production from this deposit makes Canada the world's second largest

Richardson, D.G. and Birkett, T.C.
1996:   Carbonatite-associated deposits; in Geology of Canadian Mineral Deposit Types, (ed.) O.R. Eckstrand, W.D. Sinclair, and R.I. Thorpe; Geological Survey of Canada, Geology of Canada, no. 8, p. 541-558 (also Geological Society of America, The Geology of North America, v. P-1).

Table 24-1. Definitions of rock names used in describing carbonatite complexes (compiled from: Sørensen, 1977; Bates and Jackson, 1987; Le Bas, 1987; Woolley and Kempe, 1989; Rock, 1991; Sage, 1991; A.N.Mariano, pers. comm., 1994).

bebedourite	Diopside-salite $(Ca(Mg,Fe)Si_2O_6)$ pyroxenite with low acmite and hedenbergite contents. The pyroxenite may contain biotite, with accessory perovskite, apatite, and titanomagnetite (from the classic locality – the Bebedoure Mountains of Salitre, Minas Gerais, Brazil).
magnesiocarbonatite/ dolomitic carbonatite/ beforsite	Magnesiocarbonatite contains >50% carbonate minerals and has $MgO>FeO + Fe_2O_3 + MnO$; beforsite = hyabyssal dolomitic carbonatite.
fenite	A rock that has been altered by alkali (Na/K) metasomatism.
ferrocarbonatite	A ferroan calcite to ankerite calcite carbonatite that is strongly enriched in some or all of REEs, Ba, Mn, Fe, Zn, F, and U, with possible lower limits of 1.0% MnO, 5000 ppm REEs, and 5000 ppm Ba. This rock type is very rare, but is the main source for mineralization of these elements in carbonatites. Ferrocarbonatites usually have $FeO + Fe_2O_3 + MNO>MgO$
glimmerite/biotitite	Igneous rock composed almost entirely of bitoite $\pm$ calcite $\pm$ zeolite $\pm$ magnetite $\pm$ REEs.
ijolite	A nepheline-pyroxene rock with a nepheline content between 30 and 70%. Rocks containing more than 70% nepheline are classified as **urtite** and those with less than 30% as **melteigite**. Some specimens may contain significant amounts of biotite in place of pyroxene. Potassium feldspar content is 10% or less and those rocks with 10% or less nepheline are classified as **pyroxenite**. **Nephelinite** is a fine grained or porphyritic extrusive or hypabyssal rock, of basaltic character, but primarily composed of nepheline and pyroxene (especially titanaugite), and lacking feldspar.
jacupirangite	An ultramafic plutonic rock that is part of the ijolite series, and is composed chiefly of titanaugite and magnetite, with a smaller amount of nepheline. Considered to be a nepheline-bearing clinopyroxenite.
malignite	A melanocratic nepheline syenite. In general, nepheline, pyroxene, and potassium feldspar occur in roughly equal proportions. The potassium feldpsar content must exceed 10% or the rock is classified as belonging to the ijolite suite. Both the nepheline and pyroxene content must exceed 10% or the rock would be classified with the syenites. This rock group is transitional between the ijolites and overlaps the syenitic rock group.
natrocarbonatite	Fine grained carbonatite lava composed of sodium-potassium-calcium carbonates (i.e., predominantly nyerereite and gregoryite).
phoscorite/foskorite/ kamaforite	Magmatic rock consisting predominatly of three major rock-forming minerals – olivine (forsterite), magnetite, and apatite. Accessory phases include variable amounts of calcite, baddelyite, and sulphides. Variations in the relative proportions of the three essential mineral components results in a variety of rock types in the phoscorite series, including: forsteritite (rock composed almost entirely of forsterite olivine), magnetitite (rock composed almost entirely of magnetite), nelsonite (magnetite-apatite [oxide-phosphate] rock), and apatitite (rock composed almost entirely of apatite).
rauhaugite	A carbonatite that contains ankerite or dolomite.
silicocarbonatite	A sövite-type of carbonatite rock containing 50% or more oxide and silicate minerals. Where silicate or oxide mineralogy exceeds 90%, various other rock names are applied (i.e. ijolite, pyroxenites, etc.).
sövite/calciocarbonatite	Coarse grained carbonatite rock composed of 50% or more calcite that displays adcumulate texture. If it is medium- or fine-grained the rock is a micro-sövite. Various mineralogical modifiers are used to classify the sövite (e.g., apatite-magnetite sövite, olivine-amphibole sövite, etc.). An alvikite is medium- to fine-grained type of sövite enriched in incompatible elements such as REEs, Ba, Mn, Zn, and commonly Nb. The boundary between sövite and alvikite is uncertain but may be set at 0.4% MnO, 1500 ppm Ba, and 200 ppm REEs. The calcite of an alvikite is slightly ferroan. Calciocarbonatites contain >80% CaO.
tinguaite	A textural variety of phonolite (extrusive equivalent of nepheline syenite), typically found in dykes, and characterized by conspicuous acicular crystals of acmite arranged in radial or criss-cross patterns in the groundmass. The phenocrysts are equigranular alkali feldspar and nepheline.
melilite series/melilitolite	A group of extrusive/plutonic, usually olivine-free, mafic rocks composed of melilite and augite or other mafic minerals that comprise more than 90% of the rock. This rock type may contain minor amounts of feldspathoids, apatite, calcite, anatase, and sometimes plagioclase and phlogopite. The melilite series includes **uncompahgrite**. **Ultramafic lamprophyres** correspond broadly to melilites and melilite-nephelinites, but are considerably enriched in volatile elements (H_2O, CO_2, F, Cl) and large-ion lithophile elements (K, Rb, Ba, Mg, Cr, Ni, etc.).

Table 24-2. Formulae of some of the less common minerals mentioned in the text.

Mineral name	Formula
Silicates	
allanite	$(Ce,Ca,Y)_2(Al,Fe^{+3})_3(SiO_4)_3(OH)$
eckermannite	$Na_3(Mg,Fe^{+2})_4AlSi_8O_{22}(OH)_2$
kimzeyite	$Ca_3(Zr,Ti)_2(Si,Al,Fe^{+3})_3O_{12}$
niocalite	$(Ca,Nb)_{16}Si_8(O,OH,F)_{36}$
melanite (titanian andradite)	$Ca_3(Fe,Ti)(SiO_4)_3$
melilite	$(Na,Ca)_2(Mg,Al)(Si,Al)_2O_7$
monticellite	$CaMgSiO_4$
taeniolite	$K,Li,Mg_2Si_4O_{10}F_2$
Oxides	
aeschynite	$(LREE,Ca,Fe,Th)(Ti,Nb)_2(O,OH)_6$
baddeleyite	ZrO_2
brookite	TiO_2
columbite	$(Fe,Mn)(Nb,Ta)_2O_6$
latrappite	$(Ca,Na)(Nb,Ti,Fe)O_3$
loparite	$(Ce,Na,Ca)_2(Ti,Nb)_2O_6$
lueshite	$NaNbO_3$
fersmite	$(Ca,Na)(Nb,Ta,Ti)_2(O,OH,F)_6$
perovskite	$CaTiO_3$
pyrochlore	$(Na,Ca,Ce)_2(Nb,Ta,Ti)_2O_6(OH,F)$
uranothorianite	$(U,Th)O_2$
Carbonates/fluorides	
ancylite	$SrCe(CO_3)_2(OH) \cdot H_2O$
burbankite	$(Na,Ca,Sr,Ba,LREE)_6(CO_3)_5$
bastnaesite	$(LREE)(CO_3)F$
cordylite	$Ba(LREE,Ca,Sr)_2(CO_3)_3F_2$
gregoryite	$(Na,K)_2CO_3$
huanghoite	$Ba,LREE,(CO_3)_2F_2$
nyerereite	$(Na,K)_2Ca(CO_3)_2$
parisite	$(LREE)_2Ca(CO_3)_3F_2$
synchysite	$Ca(LREE)(CO_3)_2F$
Phosphates	
britholite	$(Ca,Ce)_5(SiO_4,PO_4)_3(OH,F)$
florencite	$(LREE)Al_3(PO_4)_2(OH)_6$
monazite	$(LREE,Y)PO_4$
xenotime	YPO_4

niobium producer, after Brazil. Apatite and magnetite, and byproduct baddeleyite or zircon, are also derived from magmatic carbonatite deposits. The Palabora carbonatite is unique in that it hosts large economic concentrations of copper (Palabora Mining Company Limited, Mine Geological and Mineralogical Staff, 1976; Eriksson, 1989).

Carbonatite-associated deposits account for a significant portion of world REE production and contain most known reserves. The Bayan Obo orebody is the world's largest known REE deposit, with published reserves of 37 Mt of contained rare-earth oxides (REO) (Möller, 1989a). Over the past few decades, Mountain Pass has consistently been the world's leading producer of REE concentrates (Möller, 1989a). The central metasomatic beforsite core of the St. Honoré (Niobec) carbonatite contains bastnaesite and monazite in veinlets. These rocks contain approximately 2% rare-earth oxides (REOs) to a depth of 460 m (E. Denommé, pers. comm., 1991). The fine grained nature of the REE mineralization has, however, been an impediment to the production of a REE concentrate (Sinclair et al., 1992).

A substantial inferred resource of phosphate (apatite) is present in the Aley carbonatite, British Columbia (Table 24-3). The Jacupiranga carbonatite in Brazil is unique in that it represents one of the lowest grade commercial phosphate deposits (5.3% P_2O_5), and is the only carbonatite in Brazil from which phosphate is recovered from fresh rock rather than from a weathered residuum (Gomes et al., 1990). Phoscorites contained within alkali silicate-carbonatite complexes are also a major source of phosphate and form the basis of domestic phosphate production in South Africa (Notholt, 1980). The Kovdor (Russia) phoscorite has produced iron ore (magnetite) and phosphate since 1962. Canadian examples of phoscorite occur at the Lackner Lake alkalic complex, northern Ontario (Sage, 1988a), and at the Howard Creek carbonatite, Blue River area, British Columbia (A.N. Mariano, pers. comm., 1994). Phoscorite deposits generally have grades of 6 to 10% P_2O_5, but the product can be beneficiated to high-grade concentrates of 36-40% P_2O_5 at relatively low cost (Krauss et al., 1984).

Other commodities produced more rarely from carbonatite-associated deposits include: fluorite (e.g., Amba Dongar, India; Okorusu, Namibia; and Mato Preto, Paraná State, Brazil), vermiculite (e.g., Palabora, South Africa; Kovdor, Russia), vanadium (e.g., Magnet Cove Complex, U.S.A.), and baddeleyite (e.g., Palabora and Kovdor). Some carbonatite deposits are enriched in U and Th, but low U/Th ratios generally make them uneconomic (Le Bas, 1987).

SIZE AND GRADE OF DEPOSITS

Many carbonatite-associated deposits are relatively small, on the order of tens of thousands to hundreds of thousands of tonnes. However, significant production of phosphate, niobium, and REOs is derived from larger, higher grade (>5.0% P_2O_5; >0.6% Nb_2O_5; >5.0% REOs) deposits in Brazil, Canada, and South Africa, which vary greatly in size and grade (4 Mt to 5000 Mt grading >5% to 15% P_2O_5; 0.05% to >0.6% Nb_2O_5; 6 to >25% REOs; 3 to 4% TiO_2; and 30 to >50% CaF_2). Published grades and tonnages of Canadian and world carbonatite-associated deposits are presented in Tables 24-3 and 24-4, respectively. The grade-tonnage relationships of these deposits are shown in Figure 24-2.

GEOLOGICAL FEATURES

Geological setting

Of the more than 330 alkali silicate-carbonatite complexes presently known, most occur in relatively stable, intraplate areas (Le Bas, 1987). The regional distribution of these complexes is controlled by major tectonic features. About one-half of known carbonatites are located in topographic highs or domes, which can be from tens of kilometres to thousands of kilometres in diameter, and are bounded by zones of crustal-scale faulting (Woolley, 1989). Other major controls on carbonatite emplacement are major faults, anorogenic rifts, and the intersections of major faults. The Niobec deposit, for example, is located along the northern border of the Saguenay-Lac St. Jean graben, a branch of the St. Lawrence rift zone (Gagnon, 1981). A few carbonatites are found near plate margins and may be linked with orogenic activity or plate separation, and an even smaller number (i.e., Canary Islands and Cape Verde Islands) are aligned with oceanic fracture/fault zones and appear to be emplaced in hybrid oceanic/continental lithospheric rocks (Bowden, 1985; Woolley, 1989; B.A. Kjarsgaard, pers. comm., 1994). Because carbonatites are generally localized in clusters or provinces that display episodic magmatic activity, Woolley (1989) has postulated that the physical and/or chemical properties of lithospheric plates exert some control on their location and genesis.

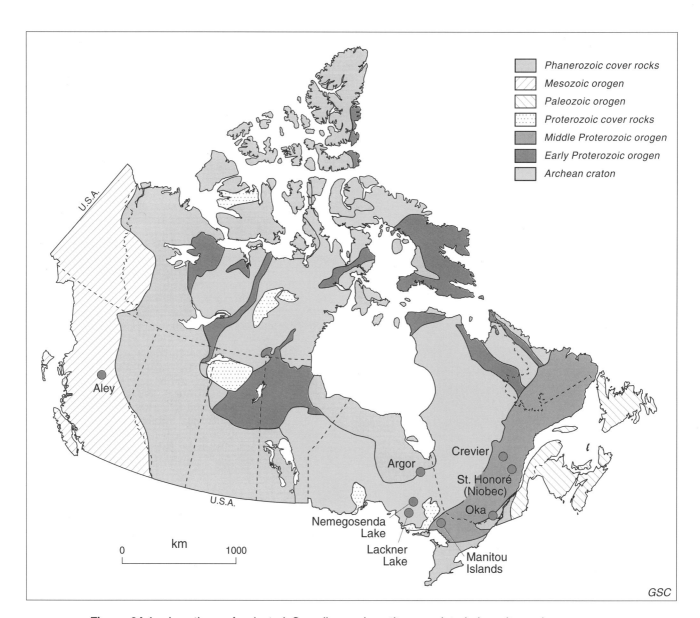

Figure 24-1. Locations of selected Canadian carbonatite-associated deposits and occurrences discussed in the text.

Age

According to Woolley's (1989) compilation of carbonatite ages (Fig. 24-3), there has been a gradual increase in carbonatite magmatism with time, and the dates fall into groups that generally correspond to major orogenic and tectonic events. Identified age groups include: 1) a mid-Proterozoic group (1800-1550 Ma) that corresponds to the Hudsonian and Svecokarelian orogenies of North America and Europe, respectively; 2) a mid- to late Proterozoic group corresponding to the Grenville orogeny (peak at ~1100 Ma); 3) a group between 750 Ma and 500 Ma which includes early Caledonian dates from both northern Europe and North America, as well as some African localities; and 4) a major period starting at 200 Ma which is perhaps associated with the breakup of Pangaea (Woolley, 1989). A few carbonatites of

Table 24-3. Production and reserves of selected Canadian carbonatite-associated deposits and occurrences.

DEPOSIT	PRODUCTION/RESERVES/GRADE	COMMENTS/REFERENCES
Aley property, British Columbia Lat. 56°27'N Long. 123°45'W	Inferred open pit reserves: 20 Mt; 0.7% Nb_2O_5 Potential resources: 5 Gt; 3-5% P_2O_5	Eight Nb-bearing zones have been identified in the dolomitic carbonatite core, which consists of 80-95% dolomite and 5-15% apatite. Also minor REE-bearing zones (Pell, 1986, 1987; Mäder, 1987; Faulkner et al., 1992).
Argor, Ontario Lat. 50°45'N Long. 81°01'W	Diamond drill indicated reserves: 56.2 Mt; 0.52% Nb_2O_5	Deposit also has potential for byproduct apatite and possibly zircon. Columbite is related to fracturing and hematite alteration (Woolley, 1987; Sage, 1988b).
Lackner Lake, Ontario, Lat. 47°47'N Long. 83°06'W	Drill indicated/inferred reserves in zones 3, 4, 6, & 8 to a maximum depth of 243 m: 111 Mt; 0.23% Nb_2O_5	Pyrochlore mineralization occurs within nepheline syenite and mafic silicate phases as: 1) fine grains or irregular clusters in the ijolites and malignites (3 & 4 zones); 2) finely disseminated grains in massive titaniferous magnetite-apatite bodies (6 zone); and 3) scattered grains within ijolitic malignites (8 zone). Two dyke-like carbonatite bodies host part of the No. 8 zone (Woolley, 1987; Sage, 1988a, 1991).
Nemegosenda Lake, Ontario Lat. 48°00'N Long. 83°06'W	Drill indicated reserves/resources to a depth of 180 m: 18.1 Mt; 0.47% Nb_2O_5	Nb, present as pyrochlore, is found in all rock types, but appears to be concentrated in fenites and in plugs and breccia zones within the marginal zone of fenitization. There is a correlation of radioactivity with Nb content (Sage, 1976, 1987; Woolley, 1987).
Manitou Islands, Ontario Lat. 46°15'N Long. 79°35'W	Drill indicated reserves: 3.26 Mt; 0.627% Nb_2O_5; 0.032% U_3O_8	Five uranian pyrochlore deposits are confined to an aegirine-potassic feldspar fenite zone. Brownish crystals of uranian pyrochlore range from about 0.01 mm to 3.5 mm. The aegirine-potassic feldspar fenite also contains a few per cent magnetite, pyrite, and pyrrhotite (Lumbers, 1971; Currie, 1976; Woolley, 1987).
St. Honoré (Niobec), Quebec Lat. 48°33'N Long. 71°04'W	Total past production: 9.6 Mt; avg. 0.66% Nb_2O_5 Proven and probable reserves at Dec. 31, 1992: 10.1 Mt; 0.65% Nb_2O_5	Pyrochlore in two major zones in the southern section of the nearly circular St. Honoré complex. Production data from Canadian Mines Handbook 1977-78 to 1993-94; reserves from same source 1993-94.
Crevier-Lagorce townships, Quebec Lat. 49°29'N Long. 72°45'W	Drill indicated reserves: 15.2 Mt; 0.189% Nb_2O_5, 0.020% Ta_2O_5 (Société Quebecoise d'Exploration Minière Annual Report 1981-82)	Mineralization is in two zones: 1) disseminated apatite, magnetite, pyrochlore, betafite (uranpyrochlore), and chalcopyrite in carbonatite and carbonate-biotite-melanosyenite; 2) disseminated fine grained pyrochlore in porphyritic pegmatitic nepheline syenite dykes.
Oka carbonatite, Quebec Lat. 45°30'N Long. 74°01'W	Drill indicated reserves of all zones: 112.7 Mt; 0.44% Nb_2O_5 23.8 Mt; 0.20-0.50% REOs	Drilling indicated six potentially exploitable zones of niobium mineralization, consisting primarily of pyrochlore, niocalite, perovskite, and latrappite. Hydrothermal REE mineralization is in arcuate inward-dipping tabular bodies associated with cone-sheet fractures (Gold et al., 1967, 1986; Woolley, 1987; Mineral Policy Sector, 1989).

Table 24-4. Production and reserves of selected foreign carbonatite-associated deposits.

DEPOSIT	PRODUCTION/RESERVES/GRADE	COMMENTS/REFERENCES
Siilinjärvi, Finland	Inferred resources to a depth of 150 m: 470 Mt; 4% P_2O_5	Phosphate mineralization is confined to Archean glimmerite-carbonatite rocks. Locally apatite comprises 25-30% of the carbonatite (Notholt, 1979, 1980; Puustinen and Kauppinen, 1989).
Kovdor, Kola Peninsula, Russia	Indicated reserves 700 Mt; 6-7% P_2O_5	Economically important apatite-forsterite phoscorites surround an elongated core magnetite "ore complex" (as much as 50% magnetite and 16% apatite). Chalcopyrite is a persistent but minor accessory in the Kovdor carbonatites. Vermiculite in an olivine-phlogopite unit is mined from the central pyroxenite core (Deans, 1966; Notholt, 1980; Ilyin, 1989; Harben and Bates, 1990a).
Mountain Pass, California, U.S.A.	Proven reserves at end of 1986: 36.3 Mt; 7.67% REOs; 20-25% $BaSO_4$ Indicated reserves: 90.7 Mt; 5% REOs	REE mineralization is contained in a north-trending tabular carbonatite body (Sulphide Queen). The principal REE-bearing minerals are bastnaesite and parisite. The ore mined consists of 40-75% calcite, 15-50% barite, and 5-15% bastnaesite/parisite (Olsen et al., 1954; Woolley, 1987; Mariano 1989a, b; Möller, 1989a).
Magnet Cove Complex, Arkansas, U.S.A.	Estimated resources at the Christy deposit: 11.7 Mt; 0.0513% Nb 10.8 Mt; 0.83% V 10 Mt; 2.0% Ti	About 4700 t of rutile concentrate were produced 1934 to 1944 from veins hosted by altered phonolite/trachyte. The Christy Ti-V-Nb deposit is hosted in an interbedded chert-shale horizon (Arkansas Novaculite) adjacent to the Magnet Cove complex. It may have been formed by carbonatite-derived alkali-rich fluids that infiltrated the Arkansas Novaculite (Woolley, 1987; Flohr, 1994).
Powderhorn Complex (Iron Hill), Colorado, U.S.A.	Measured and indicated reserves: 223 Mt; 4.0% TiO_2 Inferred reserves: 123 Mt; 4.0% TiO_2	Perovskite and magnetite occur as irregular lens-like bodies and as accessory minerals in pyroxenite. Although the orebody is reported to contain 12% TiO_2, much of this is bound in augite, magnetite, mica, and leucoxene. The recoverable Ti mineral is perovskite (50% TiO_2) which comprises 8% of the orebody (Fantel et al., 1986; Woolley, 1987).
Sarfartôq, Greenland	Indicated reserve: 500 Mt; 3.5% P_2O_5 Drill indicated reserves of high grade Nb zone: 300 000 t; 10% Nb_2O_5	Apatite-bearing carbonatite sheets 2-30 m thick, and as much as 200-500 m in strike length, are situated within the outer carbonatite core. The veins and shear zones of the marginal zone are enriched in Nb, U, and LREEs contained in pyrochlore. The high grade pyrochlore zone is a semimassive replacement body (Woolley, 1987; Secher, 1989; Kalvig and Appel, 1994).
Qaqarssuk, Greenland	Drill indicated reserves: 4.0 Mt; 0.5% Nb_2O_5 3.4 Mt; 3.5-6% P_2O_5	Coarse grained apatite occurs in early calcitic and dolomitic carbonatite, which also contains minor phlogopite and magnetite. Pyrochlore is present in late stage calcitic carbonatites and in glimmerites, which locally contain zones consisting almost entirely of apatite-magnetite-pyrochlore. Lanthanides are locally present (Knudsen, 1989a, b; Kalvig and Appel, 1994).
Jacupiranga, São Paulo, Brazil	Estimated reserves: 200 Mt; 5.30% P_2O_5	Phosphate mineralization, as apatite, is generally contained in a carbonatite body that forms part of a complex of ultramafic rocks, ijolite, nepheline syenite, and carbonatite (Woolley, 1987; Born, 1989a; Gomes et al., 1990).
Ipanema, São Paulo, Brazil	Indicated reserves: 25 Mt; 7.5% P_2O_5	Apatite mineralization occurs primarily in glimmerite bodies and in a zone, 50-200 m wide, in enveloping fenites (Ulbrich and Gomes, 1981; Born, 1989b; Mariano, 1989b; Gomes et al., 1990).
Anitápolis, Santa Catarina, Brazil	Indicated reserves: 186 Mt; 4% P_2O_5	Apatite-rich calcitic carbonatite (sövite) forms a small plug in the complex, but is also present as dykes and veins (Ulbrich and Gomes, 1981; Woolley, 1987; Gomes et al., 1990).
Tapira, Minas Gerais, Brazil	Proven resources: 114 Mt; 1.2% Nb_2O_5 241 Mt; 8% P_2O_5	A pyroxenite phase is micaceous and rich in apatite and perovskite. A sövite core is rich in pyrochlore and apatite and is surrounded by an outer annulus composed of bebedourite and glimmerite (Woolley, 1987; Mariano, 1989b; Gomes et al., 1990; Eby and Mariano, 1992).

DEPOSIT	PRODUCTION/RESERVES/GRADE	COMMENTS/REFERENCES
Mato Preto, Paraná State, Brazil	Indicated resources: 4.3 Mt; 58% CaF_2	A small body consists of a feldspathic carbonatite breccia and a ferruginous carbonatite with uneconomic REEs, Th, and P. A larger sövite body contains abundant hydrothermal fluorite and pyrite veins (Woolley, 1987; Mariano, 1989b; Gomes et al., 1990).
Palabora, South Africa	Existing reserves in main (PMC) pit: 600 Mt; 7% P_2O_5 286 Mt; 0.69% Cu Apatite concentrate contains 0.5% REOs and total recoverable reserves are estimated at 2.16 Mt REOs. Vermiculite and baddeleyite are also mined.	Vermiculite ore grading 22% has been mined since 1963. Production was about 89 500 t/a. in the 1960s, but has been increased to about 180 000. Very coarse grained, locally pegmatitic phoscorite consists largely of serpentine-magnetite and apatite. Its minerals of economic interest are apatite, copper sulphides, and baddeleyite, as well as magnetite (most has >4 wt.% TiO_2). Banded carbonatite averages 6.5% P_2O_5 and also contains copper sulphides, magnetite, baddeleyite, and uranothorianite. A central carbonatite core has higher copper (1 wt.%), magnetite with lower Ti, and a lower apatite content (average 4.5% P_2O_5). Minor Ni, Au, Pt group metals, Ag, and Se are also recovered from the Cu ores (Palabora Mining Company Limited, Mine Geological and Mineralogical Staff, 1976; Clarke, 1981; De Jager, 1989; Eriksson, 1989; Mariano, 1989b; Verwoerd, 1989; Harben and Bates, 1990b).
Panda Hill (Mbeya), Tanzania	Drill indicated reserves: 113 Mt; 0.3% Nb_2O_5 High grade zone: 3.4 Mt; 0.79% Nb_2O_5	Disseminated pyrochlore-apatite and magnetite are in a sövite plug. The carbonatite has well developed flow structures accentuated by apatite-rich streaks and pyrochlore stringers. In places, pyrochlore is partially replaced by columbite (Bowden, 1985; van Straaten, 1989).
Gakara-Karonge, Burundi	United States Bureau of Mines estimated reserves: 907 t contained REOs. Intermittent mining occurred 1950 to 1981.	A stockwork and REE veins of massive bastnaesite, and lesser monazite, quartz, goethite, and barite, occur in four major zones within the Lake Tanganyika Rift. No associated carbonatite is known (van Wambeke, 1977; Roskill Information Services Limited, 1988; Mariano, 1989a).
Okorusu, Namibia	Drill indicated reserves: 7.9 Mt; 50% CaF_2	Hydrothermal fluorite mineralization occurs in sedimentary rocks, in part metasomatically altered, along the southern rim of the complex. Economic mineralization is a coarse mixture of fluorite and quartz, with minor calcite and apatite, in steeply-dipping fractures and vein-like fracture-filled breccia zones (Gittins, 1966; Kilgore et al., 1986).
Kangankunde Hill, Malawi	United States Bureau of Mines estimated the total monazite reserve contains 269 000 t REOs.	An ankeritic-dolomite carbonatite unit contains an average of 7% monazite. Other REE minerals associated with quartz are hydrothermal. Difficulty in separating monazite from waste rock has prevented the deposit from being exploited (Deans, 1978; Roskill Information Services Limited, 1988; Mariano, 1989a).
Bayan Obo, Inner Mongolia, People's Republic of China	Proven reserves: 37 Mt of REOs; 1 Mt of Nb (avg. grades: 6% REOs; 0.10% Nb)	This is the largest known REE deposit. It consists mainly of bastnaesite, monazite, parisite, xenotime, aeschynite, and allanite, and associated dolomite, fluorite, quartz, calcite, and apatite, in lenticular bodies of magnetite-hematite ore. REE-Nb-iron mineralization is considered to have a carbonatite, mantle-derived origin (Argall, 1980; O'Driscoll, 1988; Drew et al., 1989a, b; 1990; Mariano, 1989a; Philpotts et al., 1989; Zhongxin et al., 1992)
Amba Dongar, India	11.6 Mt; 30% CaF_2	Fluorite occurs principally around the outer margin of a calcitic carbonatite ring dyke as hydrothermal quartz-fluorite veins and replacements of the late stage dolomitic carbonatite plugs (locally also fluorite-bearing). The ore is associated with fenites along the contact between carbonatite and country rocks (Gittins, 1966; Deans, 1978; Mariano, 1989b).

Archean age are known (e.g., Lac Shortt carbonatite in the northwestern Abitibi region of Quebec, which has a U-Pb zircon age of 2691 +5/-3 Ma; A. Joanisse, pers. comm., 1994); however, Proterozoic and, in particular, Mesozoic and younger ages are much more abundant. Veizer et al. (1992) have cautioned that the latter interpretation may be misleading since the temporal distribution of carbonatites could in fact be a function of their preservation. Because the preservation of carbonatites decreases exponentially with increasing age of the crustal segments (i.e. Archean and earliest Proterozoic carbonatites would be eroded or buried by later sedimentation), these carbonatites would not be accounted for in Woolley's compilation, and therefore, the temporal distribution of carbonatites actually involves a combination of orogenic activity and erosional dispersal. Regardless of how the temporal distribution of carbonatites is interpreted, there does not appear to be a correlation between age of carbonatite intrusion and economic potential (Sage, 1986).

Relationships of ore to host rocks

Carbonatites may consist of a number of intrusive phases with different textural and mineralogical characteristics. Early phases typically consist mainly of calcite, do not contain peralkaline pyroxenes or amphiboles, except as overprinting phases, and contain associated apatite+magnetite±pyrochlore mineralization (e.g., Panda Hill and Palabora) (Bowden, 1985; Flohr, 1994). Later phases, which may contain dolomite, ankerite, and siderite, in addition to calcite, are commonly enriched in pyrochlore. At Niobec, the principal host rock for the niobium deposit is coarse grained dolomitic carbonatite, which is surrounded, in part, by barren calcitic carbonatite (Fig. 24-4, 24-5) (Fortin-Bélanger, 1981; Gagnon, 1981; Vallée and Dubuc, 1981). Many very late stage carbonatites contain only trace, or no, Nb mineralization, and are enriched in primary REE-bearing minerals (A.N. Mariano, pers. comm., 1994). This is the case at Niobec, where the core of the dolomitic carbonatite contains a zone rich in REEs (Fig. 24-4).

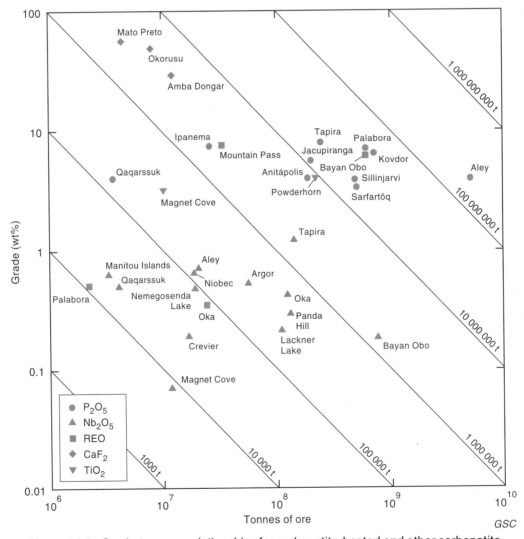

Figure 24-2. Grade-tonnage relationships for carbonatite-hosted and other carbonatite-associated deposits. Data from Tables 24-1 and 24-2.

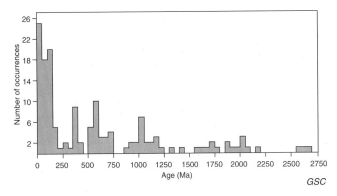

Figure 24-3. Frequency distribution diagram of carbonatite age dates (from Woolley, 1989).

In most complexes, iron-rich (ferro) carbonatites occur in minor quantities as thin brown dykes and veins with chilled margins against, and crosscutting relations to, earlier carbonatites. In some cases, late stage ferrocarbonatite magmas were highly enriched in volatiles (and REEs, F, Ba, U, Th). The ferrocarbonatites and associated minerals deposited by these volatile-rich magmas may replace earlier calcite and dolomite carbonatites (e.g., REE±U mineralization at both Niobec and Sarfartôq), or form dykes and veins in the roof zones of carbonatites (e.g. fluorite veins at Amba Dongar and Mato Preto). At Mountain Pass and Kangankunde Hill, however, bastnaesite, parisite, and monazite appear to have crystallized with calcite, barite, and dolomite as primary igneous minerals in a late stage carbonatite (Mariano, 1989a, b).

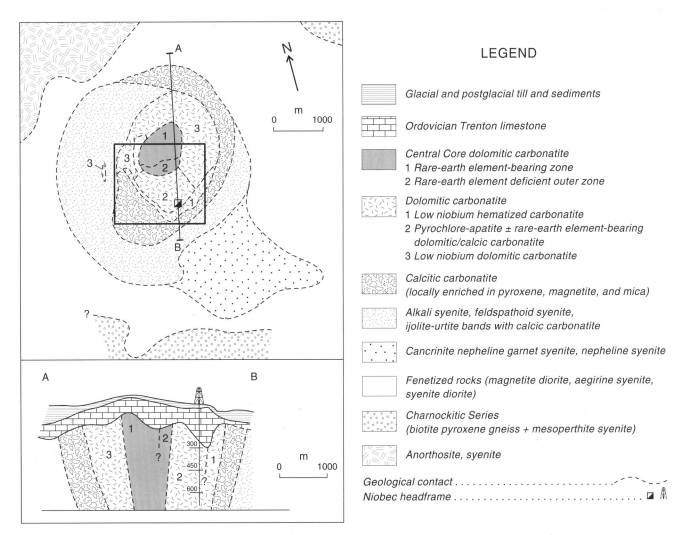

LEGEND

Glacial and postglacial till and sediments

Ordovician Trenton limestone

Central Core dolomitic carbonatite
1 Rare-earth element-bearing zone
2 Rare-earth element deficient outer zone

Dolomitic carbonatite
1 Low niobium hematized carbonatite
2 Pyrochlore-apatite ± rare-earth element-bearing dolomitic/calcic carbonatite
3 Low niobium dolomitic carbonatite

Calcitic carbonatite
(locally enriched in pyroxene, magnetite, and mica)

Alkali syenite, feldspathoid syenite, ijolite-urtite bands with calcic carbonatite

Cancrinite nepheline garnet syenite, nepheline syenite

Fenetized rocks (magnetite diorite, aegirine syenite, syenite diorite)

Charnockitic Series
(biotite pyroxene gneiss + mesoperthite syenite)

Anorthosite, syenite

Geological contact . . . .
Niobec headframe . . . .

Figure 24-4. Subsurface geology of the St. Honoré carbonatite complex below Ordovician Trenton limestone cover, and schematic north-south cross-section A-B (vertical exaggeration = 9) (after Gagnon, 1981; Thivierge et al., 1983). Area of Figure 24-5 is indicated by the bold outline.

Other carbonatite-associated deposits are hosted in a range of felsic to ultramafic alkali silicate plutonic rocks, the most important of which include: 1) the apatite-mica pyroxenite and apatite-forsterite-magnetite phoscorites of Palabora; 2) the perovskite-rich biotite pyroxenites of the Powderhorn Complex (Verwoerd, 1989); and 3) vermiculite deposits at Palabora, contained in hyrothermally altered pyroxenitic phases of ultramafic ring complexes.

Form of deposit

Mineralized carbonatites are found in lava flows and tephra, plugs, cone sheets, dykes, and rare sills, but never in large homogenous plutons (Barker, 1989). Economic mineralization, however, is generally associated with plutonic carbonatitic phases, not with lavas. Magmatic carbonatite deposits generally occur in small (3-5 km) plug-like and crescentric bodies in composite plutons with

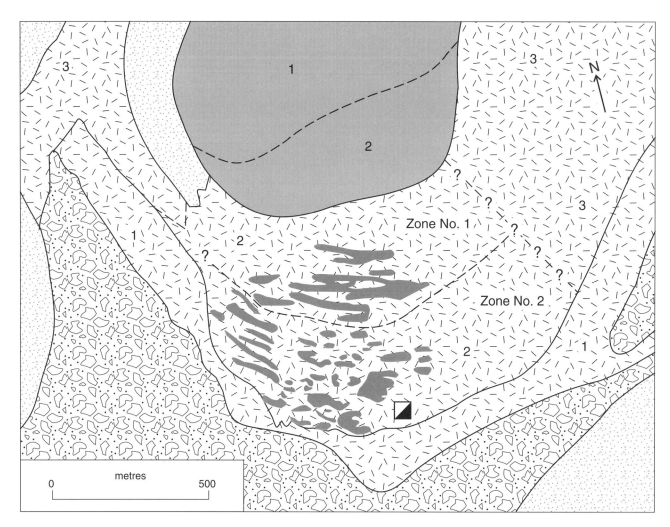

Figure 24-5. Vertical projection to surface of geology and ore lenses (solid red) containing Nb_2O_5 >0.5% from the 300 m and 450 m levels (after Gagnon, 1981). Zone 2 lenses, in the southern portion of the complex, differ from those in Zone 1 in that they: 1) are generally of higher grade; 2) have a more pinch-and-swell shape; 3) are contained in more calcitic carbonatite; 4) contain niobium almost exclusively in sodic pyrochlore, not columbite; 5) lack the paleosurficial ferruginous alteration layer associated with the carbonatite-Trenton contact; and 6) contain abundant biotite and a lower content of magnetite. Legend same as Figure 24-4, except that well-defined geological contacts are shown as solid lines, poorly-defined or gradational contacts are shown as dashed lines, and uncertain contacts are designated by dashed lines with question marks.

coeval silica-undersaturated mafic and/or ultramafic rocks (Notholt, 1980; Dawson and Currie, 1984; Barker, 1989). Mineralization is commonly related to magmatic layering and flow structures within the host rocks, defined by parallelism of xenoliths and by trains of apatite, oxides, and mafic silicates. The deposits are commonly groups of lenses or irregular ore shoots that, in plan, have crescent-shaped or annular forms. In section, these deposits generally have steep dips, parallel to the walls of the intrusive complex, and may extend to great depths. The niobium zones at Niobec are irregular, steeply dipping, east-striking lenses that are 12 to 60 m wide and of variable length. The largest mineralized lens has a maximum length of 305 m (Fig. 24-5). The lenses occur in two distinct zones which collectively occupy an area measuring 610 m x 760 m to a depth of at least 440 m (Gagnon and Gendron, 1981; Scales, 1989).

Metasomatic carbonatite deposits typically have the form of: 1) dykes and dilatant veins of ankerite or dolomite, with or without calcite; 2) thin hydrothermal veins; 3) stockworks; and 4) replacement bodies rich in calcite and dolomite or ankerite (Barker, 1989). Veins and dykes, which locally form radial or annular patterns, commonly crosscut consanguineous fenitized alkaline lithologies and adjacent country rocks (Gomes et al., 1990).

Phoscorites occur as: 1) stocks, stockworks, ring dykes, and linear dykes of widely variable thickness (i.e. from centimetres to hundreds of metres) that form a distinct component of the alkaline-carbonatite complex (e.g. Palabora); and 2) inclusions of various sizes and shapes in carbonatites (Yegorov, 1993).

Mineralogy and ore compositions

Ore minerals of magmatic carbonatite deposits include: 1) pyrochlore; 2) rare-earth fluorocarbonates or phosphates (e.g., bastnaesite, parisite, monazite) which are present as major rock-forming components with barite and strontianite; 3) perovskite-magnetite±loparite±lueshite±pyrochlore (e.g., Oka, Quebec, and Prairie Lake, Ontario – A.N. Mariano, pers. comm., 1994); and 4) apatite. The host rocks are typically calcite or dolomite carbonatite which contain accessory hematite, magnetite, apatite, zircon, allanite, sphene, and biotite. At Niobec, the principal ore minerals are finely disseminated pyrochlore and columbite; associated minerals include apatite, biotite, pyrite, pyrrhotite, zircon, ilmentite, sphalerite, barite, and traces of hydrocarbons (Gagnon, 1981).

The multiple dyke and vein complexes comprising late stage REE-rich metasomatic carbonatite deposits typically show strong enrichment in the light rare-earth elements (LREEs), Ba, and Sr, lack appreciable magnetite and pyrochlore, and contain quartz, sphene, zircon, allanite, sulphides, garnet, (±)perovskite, and fluorite as accessory phases (van Wambeke, 1977; Bowden, 1985; Le Bas, 1987, 1989). If phosphate is abundant in the carbonatite, REEs are typically concentrated in monazite; if phosphate is absent, economic minerals include bastnaesite, parisite, synchysite, ancylite, and britholite (Mariano, 1989a). At Niobec, the REEs occur mainly in bastnaesite and, to a minor extent, in monazite (Sinclair et al., 1992). Fluorite

mineralization, which is largely independent of REE mineralization, occurs in the roof zones of carbonatite complexes (Le Bas, 1987, 1989). The Mato Preto, Okorusu, and Amba Dongar vein deposits each consist of a relatively coarse mixture of fluorite and quartz with minor calcite, apatite, barite, magnetite±galena±pyrite±chalcopyrite±pyrrhotite (Kilgore, 1986; Mariano, 1989b; Gomes et al., 1990). Barite, which is present in nearly all carbonatite complexes, may be magmatic or metasomatic (Mariano, 1989b). Magmatic barite is typically associated with calcite, dolomite, bastnaesite, parisite, and strontianite.

Phoscorite deposits consist of apatite, forsterite olivine (serpentine), and magnetite, with subordinate amounts of calcite, vermiculite, baddeleyite, pyroxene, biotite, and phlogopite (Notholt, 1980).

Alteration

Carbonatites and related intrusions are typically surrounded by alkali metasomatized country rocks, referred to as fenites. Most fenite alteration haloes are characterized by desilicification, addition of ferric (Fe^{3+}) iron, sodium, potassium, and anomalous concentrations of LREEs, large ion lithophile (LIL), and other incompatible elements (Sage, 1991). Proximal to alkali silicate-carbonatite complexes, fenite aureoles are characterized by: 1) pervasive alteration/replacement of pre-existing minerals by alkali feldspars and sodic pyroxenes and amphiboles; 2) recrystallization of quartz and feldspar; and 3) partial recrystallization and incomplete assimilation of country rocks (Gomes et al., 1990). With increased distance from the intrusions, the distribution of fenite alteration is controlled by structural features (i.e., faults and fractures) (Sage, 1991). Although the extent of the fenite surrounding the St. Honoré (Niobec) complex has not been determined, it is characterized by: 1) presence of numerous red or green carbonate veinlets; 2) blue sodic amphiboles that have been altered to aegirine; 3) sericitization of feldspars in the vicinity of carbonate veinlets; and 4) the conversion of oligoclase-andesine to albite (Fortin-Bélanger, 1981).

Both sodic (Na-rich) and potassic (K-rich) fenites can occur around a single alkali silicate-carbonatite complex (e.g., Amba Dongar). Sodic fenites are composed of albite, alkali feldspar, aegirine to aegirine-augite, magnesio-arfvedsonite to eckermannite, and biotite; whereas potassic fenite mineralogy is characterized by orthoclase, microcline, biotite-phlogopite, and iron-titanium oxides (Bowden, 1985). Le Bas (1989) has suggested that fenites are depth zone-related, with potassic fenitization characterizing high level emplacement, and sodic fenitization developing only at deeper levels.

Fenites enveloping carbonatites are distinct from the similar-looking syenitic fenites formed around ijolitic rocks, which consist of much higher temperature minerals (Le Bas, 1989; Morogan, 1994). The potassic fenite aureoles developed around pyroxenites and ijolites are commonly smaller than those around carbonatites (Dawson and Currie, 1984), and are characterized by extensive phlogopitization, feldspathization, and amphibolitization (Bowden, 1985; Gomes et al., 1990).

In addition to fenitization, late stage carbonatitic fluids are responsible for postmagmatic autometasomatic alteration and redistribution of incompatible elements. For example, REEs in magmatic fluorocarbonates may be leached and precipitated, with other minerals, in veins in the country rock (Möller, 1989b). Another common alteration feature produced by late stage fluids is the development of a distinctive biotite-rich rock commonly referred to as biotitite or glimmerite (Sage, 1991). Prominent glimmerite zones occur within the various lithostructural units of the Oka Complex (Gold et al., 1986). Considerable local Nb enrichment along contact zones, fractures, and faults can also be produced by mobilization of Nb by late alkaline fluids (e.g., Sarfartôq; Möller, 1989b). The development of rödberg, a low temperature, iron-rich hematite-dolomite-sövite, formerly mined for iron at the Fen complex in south Norway, has also been ascribed to late autometasomatic processes.

Supergene alteration of biotite and phlogopite leads to the development of economic concentrations of vermiculite (Harben and Bates, 1990b).

DEFINITIVE CHARACTERISITICS

Magmatic carbonatite deposits are igneous rocks which form small, volumetrically limited but integral portions of carbonatite complexes. They are generally contained in small (<3 to 4 km in diameter) rounded plug-like and crescent-shaped bodies or nested cone sheets that intrude larger bodies of felsic to ultramafic alkali silicate rocks (i.e., feldspathoid-bearing nephelinites, phonolites, nepheline syenites, urtite-ijolite-melteigite series, melilite/melilitolite-series rocks) (Barker, 1989; Sage, 1991). Deposits are characterized by: 1) a suite of incompatible rare elements and volatile components (i.e., Sr, Ba, REE(Y), Nb(Ta), P, ±Th) (Möller, 1989b); and 2) distinct mantle-derived isotopic signatures (i.e., $^{87}Sr:^{86}Sr$ <0.706, $\delta^{13}C$ = -1 to -9‰ PDB; $\delta^{18}O$ = +6 to +12‰ SMOW) (Barker, 1989; Möller, 1989b; Philpotts et al., 1989).

REE-bearing carbonatite deposits are strongly enriched in LREEs. Chondrite-normalized rare-earth patterns are characteristically steep and Eu anomalies are minor or absent (Bowden, 1985; Möller, 1989b).

Different types of mineralization are associated with the various textural and mineralogical varieties of carbonatite that commonly occur in a given complex. Early sövite phases, which generally form the largest carbonatite bodies in most complexes, are usually enriched in apatite and magnetite±pyrochlore (Bowden, 1985; Barker, 1989). Late stage magnesium- and iron-rich carbonatites contain deposits of various commodities including: fluorite, REE-fluorocarbonates, U and Th in silica-rich zones, and barite vein and microvein complexes (Le Bas, 1989).

Those carbonatite-associated deposits that are not hosted in carbonatites may be contained in distinct zones of phoscorite, pyroxenite, and ijolite-urtite that are components of steeply dipping alkaline ring complexes (Notholt, 1979, 1980).

The alkali silicate-carbonatite complexes that host carbonatite-associated deposits are usually surrounded by sodic and/or potassic fenitized aureoles of metasomatically altered rocks (Bowden, 1985; Le Bas, 1987, 1989; Barker, 1989).

GENETIC MODEL

Carbonatite-associated deposits are genetically related to host alkali silicate-carbonatite complexes. There are three competing hypotheses for the formation of carbonatites and related alkalic rocks: 1) carbonatite and alkaline magmas represent two distinct primary magmas; 2) carbonatites form by immiscible separation from silica-undersaturated carbonated nephelinitic magmas after prolonged differentiation at shallow crustal levels; and 3) carbonatites represent residual melts derived by fractional crystallization of a parent carbonated alkaline silicate magma (Barker, 1989; Möller, 1989b; Wyllie, 1989).

In general, the emplacement of carbonatites represents the final intrusive phase of most alkali silicate-carbonatite complexes. According to Barker (1989), the emplacement sequence of a typical complex, from periphery to core (i.e., oldest to youngest) is: nepheline syenite to nepheline-clinopyroxene rocks to carbonatite. Kapustin (1986) listed the sequence as dunite and/or clinopyroxenite to melilite-bearing rocks to nepheline-clinopyroxene rocks, nepheline syenite, and finally carbonatite. Kapustin's sequence is observable at Powderhorn (U.S.A.) and Tapira, Salitre I and II, Serra Negra, and Jacupiranga in Brazil (A.N. Mariano, pers. comm., 1994). The mechanisms controlling magma chemistry and the emplacement sequence are largely unknown; however, there are strong indications of magmatic differentiation in early-formed rock units with a corresponding progressive increase in alkalinity as successive units form (e.g., emplacement sequence at Mountain Pass: shonkinite (with no peralkaline pyriboles) – syenite (with Na-amphibole) – alkali granite – carbonatite bodies, dykes, and veins; Olsen et al., 1954; DeWitt et al., 1987). Although limited radiometric age studies have been performed on the different rock units comprising some South American alkali silicate-carbonatite complexes (i.e., Chiriguelo and Cerro Sarambi, Amamby and Concepción provinces, Paraguay, and Catalão I and II, Goiás Province, Brazil; Eby and Mariano, 1992), the temporal development of rock units, and carbonatite phases comprising worldwide alkali silicate-carbonatite complexes is not well understood.

Fractional crystallization and related differentiation of carbonatite magmas results in the various types of carbonatite deposits. The earliest crystals to precipitate from a carbonate melt are usually strontium-rich calcites that are accompanied by apatite, magnetite, phlogopite, zircon, sulphides (pyrrhotite), olivine, pyrochlore±perovskite cumulates (Le Bas, 1989; Möller, 1989a). The apatite+magnetite+pyrochlore association in magmatic deposits results from early crystallization of apatite which removes P_2O_5 and F from the magma, lowers niobium solubility, and results in the crystallization of pyrochlore (Mariano, 1989b). Rare-earth elements in deposits of apatite+magnetite+pyrochlore typically do not form fluorocarbonate minerals, but instead are incorporated into apatite and calcite. Recent studies of physical properties of carbonatite magmas have shown that carbonatite magmas have very low viscosities (similar to that of water) and heats of fusion, and high thermal diffusivities as compared to the more common silicate magmas (Treiman and Schedl, 1983; Treiman, 1989). Thus crystal settling is rapid, the grain size of cumulate phases is small, and flow systems (convective or other) are rapid and turbulent (Treiman, 1989).

Further fractionation of carbonatite magma can produce dolomitic/magnesiocarbonatites and ferrocarbonatites (beforsites and rauhaugites), and residual F-REE-Sr-Ba-U-Th-rich fluid (Le Bas, 1987, 1989). The late fluids promote partial replacement of calcite by ankerite or dolomite and form late stage veins (Bowden, 1985; Le Bas, 1987, 1989; Barker, 1989; Möller, 1989a). Mineralization associated with the intrusion of late stage ferrocarbonatites that bear ankerite or ankeritic dolomite, and the introduction of associated end-stage metasomatic volatile phases, include REEs, fluorite, U, and Th, and is in many cases accompanied by silicification (Le Bas, 1987, 1989). Late stage F-enriched volatile carbonatitic phases also concentrate Ba and SO_4^{2-} as incompatible components (Le Bas, 1989). Andersen (1984) has postulated that the hematite contained in rödberg, unlike the iron in more typical carbonatite-associated magnetite-apatite deposits, formed by reactions involving a combination of both deuteric fluids and groundwater convecting in hydrothermal cells set up by the intrusion of the carbonatite and other intrusive phases. The degree to which meteoric fluids in the country rocks hosting other carbonatite-associated deposits have been involved in the formation of metasomatic deposits has not been determined.

Although vein and microvein metasomatic mineralization is usually spatially and temporally associated with late stage carbonatitic phases, in the presence of local structural controls, this mineralization may be distally deposited and direct evidence of a carbonatite association is obscured (e.g., Gakara-Karonge and Bayan Obo) (Mariano, 1989a, b).

Late stage fluids in some cases result in the alteration or remobilization of Nb and REEs in carbonatite-associated deposits. For example, although pyrochlore within the Niobec deposit was magmatic, columbite may have formed by metasomatism of pyrochlore (replacement of calcium in pyrochlore by iron; Gagnon, 1981). In contrast, however, the presence of karstic features at Niobec led Sage (1988b) to suggest that columbite replacement of pyrochlore may have been a supergene phenomenon. Fluid inclusion data obtained from calcite and dolomite within the dolomitic beforsite of the REE-zone at Niobec have provided evidence of postmagmatic hydrothermal activity and confirmed the mobility of REE in aqueous solutions (Heinritzi et al., 1989).

The evidence for either a hydrothermal or magmatic origin for the sulphides (bornite, chalcopyrite, and cubanite) in the Palabora carbonatite, which constitute a unique deposit, remains equivocal (Eriksson, 1989). Also, the origin of many carbonatite-associated phoscorite deposits and their relationship to carbonatites remains unclear (Yegorov, 1993). Nevertheless, whether they result from liquid immiscibility or from crystal accumulation, these rocks, although commonly affected by postmagmatic processes, are of magmatic origin (Yegorov, 1993).

RELATED DEPOSIT TYPES

The ore-forming environment of the giant Olympic Dam Cu-U-Au-Ag-REE deposit in southern Australia (see Type 22) has many similarities with carbonatite-associated deposits, including: 1) location in extensional environments, in late orogenic and anorogenic settings; 2) REE mineralization (i.e., bastnaesite, florencite, monazite, xenotime, and britholite) is associated with extensive hydrothermal alteration and metasomatism; 3) evidence that REE, and possibly Cu mineralization may in part be contributed from alkaline mafic/ultramafic magmas; and 4) apatite and fluorite as principal nonmetallic minerals (Gandhi and Bell, 1990; Oreskes and Einaudi, 1990). However, unlike carbonatite deposits which have a mantle source, the Olympic Dam deposit is thought to be a hydrothermal breccia complex formed in a subvolcanic environment as an integral part of a major crustal melting event triggered by mantle plume activity beneath continental crust (J.P. Johnson, unpub. data, Geological Survey of Canada, Minerals Colloquium, Ottawa, 1994).

The relationship between carbonatite and megacryst-bearing, olivine-rich and micaceous diamondiferous kimberlites (see subtype 25.1) has been in dispute for several decades. Both deposit types are characterized by: 1) the presence of high temperature calcite; 2) mantle isotopic signatures; 3) enrichment of the trace elements Cr, Ni, V, La, Li, Nb, Sr, and Y; and 4) similar morphology (i.e., deep-seated cyclindrical plug-like bodies) (Sage, 1983). Although a close kimberlite-carbonatite association can be observed in certain localities (e.g., Premier mine, South Africa; Barker, 1989), a worldwide relationship cannot be clearly established, and, except for the fact that both deposit types are magmatic and of mantle origin, there is no obvious genetic connection. However, the rare occurrence of diamonds in what were formerly called 'central complex' kimberlites (also known as alnöites/damtjernites/ultramafic lamprophyres), which commonly form part of alkali-silicate carbonatite complexes, is noteworthy (e.g., Île Bizard, Quebec; Alnö, Sweden) (Rock, 1991).

Mineralogical, geochemical, and petrogenetic similarities also exist between carbonatite deposits and peralkaline rock-associated rare metal deposits described in Type 23 of this volume. Most notable of these is the layered ijolitic (nepheline-aegirine-apatite-sphene) rocks of the very large (1327 km²) Khibiny ultramafic alkaline complex in the Kola Peninsula of Russia, which contains the largest known accumulation of monomineralic apatite rock in the world (Notholt, 1980; Ilyin, 1989; Mariano, 1989a). The relationship between the recently discovered carbonatite/silicocarbonatite stock, located along the eastern periphery of the Khibiny intrusion, and the apatite-bearing silica-undersaturated rocks remains unclear (Kogarko, 1987; Ilyin, 1989).

Weathering of carbonatite-associated deposits, may, under favourable conditions, produce: residual apatite, pyrochlore, and anatase deposits; supergene REE deposits; and, as previously noted, supergene vermiculite from the conversion of phlogopite/biotite (for further details, see subtype 4.3, "Residual carbonatite-associated deposits").

EXPLORATION GUIDES

Exploration guidelines for carbonatite-associated deposits include the following:

Broad scale features

1. Individual carbonatites usually contain more than one mineral of potential economic interest. Therefore, each carbonatite should be considered as a multicommodity exploration target and should be thoroughly evaluated for Nb, REEs, U, P, and fluorite (Sage, 1986).

2. Carbonatites occur as irregular, centrally located, rounded masses or continuous to semicontinuous concentric rings and dykes, commonly within more extensive alkali-silicate complexes. These complexes themselves are notable for their relatively small size (surface area usually does not exceed 50 km^2), steeply dipping or vertical walls, and that most are of undetermined depth (Notholt, 1980). The alkaline rocks have variable compositions, felsic members include nepheline-syenites, mafic members, ijolites, ultramafic peridotites, mica pyroxenites, and jacupirangites (Notholt, 1980; Ulbrich and Gomes, 1981; Gomes et al., 1990).

3. Extrusive pyroclastic carbonatites can be laterally extensive. For example, the Mount Grace extrusive carbonatite in the Omineca Belt of the Canadian Cordilleran averages 2-5 m in thickness and has been traced and extrapolated for at least 100 km along the northwestern margin of the Frenchman Cap Dome (Pell, 1987; Pell and Höy, 1989).

4. Because carbonatite deposits are commonly contained in plug-like intrusive bodies or ring structures, many are associated with annular topographic features. In the absence of appreciable overburden, the characteristic circular patterns are still readily discernable on aerial photographs, even in the case of those having very subdued relief (Gold et al., 1967).

5. Because deposits associated with carbonatitic rocks are usually contained in specific lithologies, definition of igneous contacts among the various intrusive phases is an important exploration tool.

6. The presence of the characteristic carbonatite fenite facies may prove useful as a general guide in the detection of carbonatites and associated mineralization. Fenitization is generally characterized by desilicification; however, in certain REE-rich magnesiocarbonatites and ferrocarbonatites (e.g., Mountain Pass and Kangankunde Hill), mineralization is associated with enrichment in silica (Woolley and Kempe, 1989).

7. Both magmatic (e.g., Mountain Pass, U.S.A.) and metasomatic (e.g., Wigu Hill, Tanzania; Itapirapuá, Brazil; Adrounedj, Mali) REE-rich carbonatite deposits are low in P, Ti, Zr, and Nb relative to earlier carbonatite rock units (A.N. Mariano, Geological Survey of Canada, Logan Club presentation, February 24, 1994).

Geochemical approaches

Elements associated with carbonatite-associated deposits (i.e., Nb±Ta, Ba, Sr, U, Th, LREEs, Ti, F, and P) are usually contained in distinct, chemically resistant minerals (e.g., pyrochlore, monazite, perovskite) that can be detected through regional studies of heavy mineral suites from unconsolidated sediments. In western Greenland, for example, multi-element analysis of the fine fraction of stream sediments and U, F, and conductivity determinations of waters, collected at a very low density of 1 sample/30 km^2, successfully delineated the Sarfartôq and Qaqarssuk carbonatites (Steenfelt, 1991). A reconnaissance scale (1 sample /10 km^2) stream sediment survey completed in the vicinity of the Magnet Cove alkali silicate-carbonatite complex determined that REEs, Ti, and F in the 0.075 mm fraction were the best indicators for the presence of carbonatite (Sadeghi and Steele, 1989).

Geophysical approaches

Magnetic, radiometric, and gravimetric techniques can all be applied to carbonatite deposits. Because carbonatites commonly contain appreciable magnetite, they generally have a higher magnetic susceptibility than their host rocks, and appear on aeromagnetic maps as small-diameter (about 5 km) high-intensity, circular to elliptical positive magnetic anomalies (Gold et al., 1967; Sage, 1986). Associated magnetite-bearing alkalic rock complexes are commonly characterized by less intense, larger, circular to elliptical anomalies (Sage, 1986). However, because strong alkalinity associated with fenitization stabilizes iron in the ferric state due to the "alkali ferric-iron effect" (Carmichael and Nicholls, 1967), magnetite usually does not form in the fenite aureole, and these rocks have a negative or low magnetic response. The magnetite-poor fenites and carbonatitic phases of the magmatic Fen and Mountain Pass deposits display negative aeromagnetic responses (Saether, 1958; A.N. Mariano, pers. comm., 1994). Metasomatic REE-bearing carbonatite deposits may or may not have positive magnetic expression, depending on the presence of appreciable magnetite (A.N. Mariano, Geological Survey of Canada, Logan Club presentation, February 24, 1994).

Exposed carbonatites that contain radioactive minerals (e.g. thorian pyrochlore, monazite) have positive radiometric responses. The lack of penetration of the conventionally measured radiation (K, U, Th) detracts from the usefulness of radiometric techniques in areas of thick soil development or glacial overburden (Gold et al., 1967). Reconnaissance radiometric surveys in Canada (5 km line spacing, 120 m ground clearance survey over the 0.4 km^2 Allan Lake carbonatite) and Greenland (3 km line spacing, 100 m ground clearance over the 10 km^2 Sarfartôq and 15 km^2 Qaqarssuk carbonatites) successfully detected the carbonatites or their associated dispersion trains (Steenfelt, 1987, 1991; Ford et al., 1988).

The location of carbonatite deposits in regionally extensive fault zones, rift valleys, and crustal warps is fortuitous as these large scale structural features appear as pronounced linear features on regional gravity anomaly maps. At the deposit or intrusive complex scale, if the density contrast between country rocks and the lithologies comprising the alkaline-carbonatite complexes is sufficient, deep-rooted carbonatite bodies will be characterized by steep, high gradient gravity anomalies.

Case studies of the application of integrated exploration techniques to Canadian carbonatites have been documented for Niobec (Vallée and Dubuc, 1970), Allen Lake (Ford et al., 1988), and Oka (Gold et al., 1967).

ACKNOWLEDGMENTS

The careful reviews by B.A. Kjarsgaard and A.N. Mariano, and editing by W.D. Sinclair and R.I. Thorpe have greatly improved the manuscript.

SELECTED BIBLIOGRAPHY

References marked with asterisks (*) are considered to be the best sources of general information on this deposit type.

Andersen, T.
1984:   Secondary processes in carbonatites: petrology of rödberg (hematite-calcite-dolomite carbonatite) in the Fen central complex, Telemark (south Norway); Lithos, v. 17, p. 227-245.

Argall, G.O.
1980:   Three iron ore bodies of Bayan Obo; World Mining, January, p. 38-41.

***Barker, D.S.**
1989:   Field relations of carbonatites; in Carbonatites - Genesis and Evolution, (ed.) K. Bell; Unwin Hyman, London, p. 38-69.

Bates, R.L. and Jackson, J.A.
1987:   Glossary of Geology, Third Edition; American Geological Institute, Alexandria, Virginia, 788 p.

Bergeron, A.
1981:   Petrography and geochemistry of the Crevier igneous alkaline complex and the metasomatized country rocks; in The St. Honoré and Crevier Niobium - Tantalum Deposits and Related Alkalic Complexes, Lac St. Jean, Quebec, The Canadian Institute of Mining and Metallurgy, Excursion Guidebook 18, p. 37-38.

Bonneau, J.
1981:   The Crevier alkaline igneous complex and associated niobium-tantalum-uranium mineralization; in The St. Honoré and Crevier Niobium - Tantalum Deposits and Related Alkalic Complexes, Lac St. Jean, Quebec, The Canadian Institute of Mining and Metallurgy, Excursion Guidebook, p. 29-35.

Born, H.
1989a: The Jacupiranga apatite deposit, São Paulo, Brazil; in Phosphate Deposits of the World, Volume 2, Phosphate Rock Resources, (ed.) A.J.G. Notholt, R.P. Sheldon, and D.F. Davidson; International Geological Correlation Programme, Project 156-Phosphorites, p. 111-115.

1989b: The Ipanema phosphate deposit São Paulo, Brazil; in Phosphate Deposits of the World, Volume 2, Phosphate Rock Resources, (ed.) A.J.G. Notholt, R.P. Sheldon, and D.F. Davidson; International Geological Correlation Programme, Project 156-Phosphorites, p. 116-119.

***Bowden, P.**
1985:   The geochemistry and mineralization of alkaline ring complexes in Africa (a review); Journal of African Earth Sciences, v. 3, no. 1/2, p. 17-39.

Carmichael, I.S.E. and Nicholls, S.J.
1967:   Iron-titanium oxides and oxygen fugacities in volcanic rocks; Journal of Geophysical Research, v. 72, p. 4665-4687.

Chao, E.C.T., Back, J.M., and Minkin, J.A.
1992:   Host-rock controlled epigenetic, hydrothermal metasomatism origin of the Bayan Obo REE-F-Nb ore deposit, Inner Mongolia, P.R.C.; Applied Geochemistry, v. 7, p. 443-458.

Clarke, G.
1981:   The Palabora complex-triumph over low grade ores; Industrial Minerals, October 1981, p. 45-62.

Currie, K.L.
1976:   The alkaline rocks of Canada; Geological Survey of Canada, Bulletin 239, 228 p.

Dawson, K.R.
1974:   Niobium (columbium) and tantalum in Canada; Geological Survey of Canada, Economic Geology Report 29, 157 p.

***Dawson, K.R. and Currie, K.L.**
1984:   Carbonatite-hosted deposits; in Canadian Mineral Deposit Types, a Geological Synopsis, Geological Survey of Canada, Economic Geology Report 36, p. 48-49.

Deans, T.
1966:   Economic mineralogy of African carbonatites; in Carbonatites, (ed.) O.F. Tuttle and J.C. Gittins; John Wiley and Sons, New York, p. 385-413.

1978:   Mineral production from carbonatite complexes, a world review; in Proceedings of the First International Symposium on Carbonatites, Pocos de Caldas, Brazil, 1976; Departamento Nacional da Produção, Rio de Janeiro, p. 123-133.

De Jager, D.H.
1989:   Phosphate resources in the Palabora igneous complex, Transvaal, South Africa; in Phosphate Deposits of the World, Volume 2, Phosphate Rock Resources, A.J.G. Notholt, R.P. Sheldon, and D.F. Davidson; International Geological Correlation Programme, Project 156-Phosphorites, p. 267-272.

DeWitt, E., Kwak, L.M., and Zartman, R.E.
1987:   U-Th-Pb and $^{40}Ar/^{39}Ar$ dating of the Mountain Pass carbonatite and alkalic igneous rocks, S.E. California; in Abstracts with Programs, Geological Society of America, v. 19, no. 7, p. 642.

Drew, L.J., Qingrun, M., and Weijun, S.
1989a: Geological setting of iron-niobium-rare-earth orebodies at Bayan Obo, Inner Mongolia, China, and a proposed regional model; in USGS Research on Mineral Resources - 1989, Program and Abstracts, Fifth Annual V.E. McKelvey Forum on Mineral and Energy Resources, United States Geological Survey, Circular 1035, p. 14-15.

1989b: Observations on regional geology and alkali metasomatism associated with iron-niobium-rare earths ore bodies at Bayan Obo, Inner Mongolia, China; in Abstracts Volume 1 of 3, 28th International Geological Congress, Washington, D.C., p. 1-416.

1990:   The Bayan Obo iron-rare-earth-niobium deposits, Inner Mongolia, China; Lithos, v. 26, no. 1/2, p. 43-65

Eby, G.N. and Mariano, A.N.
1992:   Geology and geochronology of carbonatites and associated alkaline rocks peripheral to the Paraná Basin, Brazil-Paraguay; Journal of South American Earth Sciences, v. 6, no. 3, p. 207-216.

Eriksson, S.C.
1989:   Phalaborwa: a saga of magmatism, metasomatism, and miscibility; in Carbonatites - Genesis and Evolution, (ed.) K. Bell; Unwin Hyman, London, p. 221-254.

Fantel, R.J., Buckingham, D.A., and Sullivan, D.E.
1986:   Titanium mineral availability - market economy countries: a minerals availability appraisal; United States Bureau of Mines, Information Circular 9061, 28 p.

Faulkner, T., Meyers, R., Malott, M-L., Wilton, P., and Melville, D.
1992:   1991 producers and potential producers, mineral and coal; British Columbia Ministry of Energy, Mines and Petroleum Resources, Open File 1992-1 (includes MINFILE data for 373 major occurrences).

Flohr, M.J.K.
1994:   Titanium, vanadium and niobium mineralization and alkali metasomatism from the Magnet Cove Complex, Arkansas; Economic Geology, v. 89, p. 105-130.

Ford, K.L., Delabio, R.N.W., and Rencz, A.N.
1988:   Geological, geophysical and geochemical studies around the Allan Lake carbonatite, Algonquin Park, Ontario; Journal of Geochemical Exploration, v. 30, p. 99-121.

Fortin-Bélanger, M.
1981: The annular carbonatite complex of St. Honoré and associated niobium mineralization: petrographic and geochemical study; in The St. Honoré and Crevier Niobium - Tantalum Deposits and Related Alkalic Complexes, Lac St. Jean, Quebec, Canadian Institute of Mining and Metallurgy, Excursion Guidebook, p. 19-20.

Gagnon, G.
1981: The St. Honoré carbonatite complex and associated niobium deposits; in The St. Honoré and Crevier Niobium - Tantalum Deposits and Related Alkalic Complexes, Lac St. Jean, Quebec, The Canadian Institute of Mining and Metallurgy, Excursion Guidebook, p. 4-16.

Gagnon, G. and Gendron, L.A.
1981: Geology and current developement of the St. Honoré niobium columbium deposits; in The St. Honoré and Crevier Niobium - Tantalum Deposits and Related Alkalic Complexes, Lac St. Jean, Quebec, The Canadian Institute of Mining and Metallurgy, Excursion Guidebook, p. 18.

Gandhi, S.S. and Bell, R.T.
1990: Metallogenic concepts to aid exploration for the giant Olympic Dam-type deposits and their derivatives; in Program with Abstracts, 8th Symposium of the International Association on the Genesis of Ore Deposits, Ottawa, 1990, p. A-7.

Gauthier, A.
1981: Geochemical, petrographic and mineralogical study of rare-earth zones of the St. Honoré carbonatite; in The St. Honoré and Crevier Niobium - Tantalum Deposits and Related Alkalic Complexes, Lac St. Jean, Quebec, Canadian Institute of Mining and Metallurgy, Excursion Guidebook, p. 21.

Gittins, J.C.
1966: Summaries and bibliographies of carbonatite complexes; in Carbonatites, (ed.) O.F. Tuttle and J.C. Gittins; John Wiley and Sons, New York, p. 417-570.

Gold, D.P.
1972: The Monteregian Hills: ultra-alkaline rocks and the Oka carbonatite complex; 26th International Geological Congress, Montreal, 1972, Guidebook, Excursion B-11, 47 p.

Gold, D.P., Eby, G.N., Bell, K., and Vallée, M.
1986: Carbonatites, diatremes, and ultra-alkaline rocks in the Oka area, Quebec; Geological Association of Canada, Mineralogical Association of Canada, Canadian Geophysical Union, Joint Annual Meeting, Ottawa '86, Field Trip 21: Guidebook, 51 p.

Gold, D.P., Vallée, M., and Charette, J.-P.
1967: Economic geology and geophysics of the Oka alkaline complex, Quebec; The Canadian Institute of Mining and Metallurgy, Bulletin, v. 60, no. 666, p. 1131-1144.

Gomes, C.B., Ruberti, E., and Morbidelli, L.
1990: Carbonatite complexes from Brazil: a review; Journal of South American Earth Sciences, v. 3, no. 1, p. 51-63.

Harben, P.W. and Bates, R.L.
1990a: Phosphate rock; in Industrial Minerals Geology and World Deposits; Industrial Minerals Division, Metal Bulletin Plc., London, p.190-204.
1990b: Vermiculite; in Industrial Minerals Geology and World Deposits; Industrial Minerals Division, Metal Bulletin Plc., London, p. 295-298.

Heinrich, E.W.
1966: The Geology of Carbonatites; Rand McNally and Company, Chicago, 553 p.

Heinritzi, F., Williams-Jones, A.E., and Wood, S.A.
1989: Fluid inclusions in calcite and dolomite of the REE-Zone in the St. Honoré carbonatite complex, Quebec; Program with Abstracts, Geological Association of Canada/Mineralogical Association of Canada, v. 14, p. A20-A21.

Ilyin, A.V.
1989: Apatite deposits in the Khibiny and Kovdor alkaline igneous complexes, Kola Peninsula, northwestern USSR; in Phosphate Deposits of the World, Volume 2, Phosphate Rock Resources, (ed.) A.J.G. Notholt, R.P. Sheldon, and D.F. Davidson; International Geological Correlation Programme, Project 156-Phosphorites, p. 485-493.

Kalvig, P. and Appel, P.W.U.
1994: Greenlandic mineral resources for use in advanced materials; Industrial Minerals, April 1994, p. 45-51.

Kapustin, Y.L.
1986: The origin of early calcitic carbonatites; International Geology Review, v. 28, p. 1031-1044.

Kilgore, C.C., Kraemer, S.R., and Bekkala, J.A.
1986: Fluorspar availability - market economy countries and China: A mineral availability appraisal; United States Bureau of Mines, Information Circular 9060, 57 p.

Knudsen, C.K.
1989a: Pyrochlore group minerals from the Qaqarssuk carbonatite complex; in Lanthanides, Tantalum and Niobium, (ed.) P. Möller, P. Cerny, and F. Saupé; Society for Geology Applied to Mineral Deposits, Special Publication No. 7, p. 80-99.
1989b: Apatite mineralization in the Qaqarssuk carbonatite complex, southern West Greenland; in Phosphate Deposits of the World, Volume 2, Phosphate Rock Resources, (ed.) A.J.G. Notholt, R.P. Sheldon, and D.F. Davidson; International Geological Correlation Programme, Project 156-Phosphorites, p. 84-86.

Kogarko, L.N.
1987: Alkaline rocks of the eastern part of the Baltic Shield (Kola Peninsula); in Alkaline Igneous Rocks; (ed.) J.G. Fitton and B.G.J. Upton, Geological Society Special Publication No. 30, p. 531-544.

Krauss, V.H., Saam, H.G., and Schmidt, H.W.
1984: International strategic minerals inventory summary report - phosphate; United States Geological Survey, Circular 930-C, 41 p.

Laplante, R.
1981: Nb-Ta-U mineralization study from the Crevier alkaline igneous complex, Roberval County, Lac St. Jean, Quebec; in The St. Honoré and Crevier Niobium - Tantalum Deposits and Related Alkalic Complexes, Lac St. Jean, Quebec, The Canadian Institute of Mining and Metallurgy, Excursion Guidebook, p. 39.

Le Bas, M.J.
*1987: Nephelinites and carbonatites; in Alkaline Igneous Rocks, (ed.) J.G. Fitton and B.G.J. Upton; Geological Society Special Publication No. 30, p. 53-83.
*1989: Diversification of carbonatites; in Carbonatites - Genesis and Evolution, (ed.) K. Bell; Unwin Hyman, London, p. 428-447.

Lumbers, S.B.
1971: Geology of the North Bay area, Districts of Nipissing and Parry Sound; Ontario Department of Mines, Geological Report 94, 104 p.

Mäder, U.K.
1987: The Aley carbonatite complex, northern Rocky Mountains, British Columbia; in Geological Fieldwork, 1986, British Columbia Ministry of Energy, Mines and Petroleum Resources, Paper 1987-1, p. 283-288.

Mariano, A.N.
*1989a: Economic geology of rare earth elements; in Geochemistry and Mineralogy of Rare Earth Elements, (ed.) B.R. Lipin and G.A. McKay; Mineralogical Society of America, Reviews in Mineralogy, v. 21, p. 309-337.
*1989b: Nature of economic mineralization in carbonatites and related rocks; in Carbonatites - Genesis and Evolution, (ed.) K. Bell; Unwin Hyman, London, p. 149-176.

Mineral Policy Sector
1989: Canadian mineral deposits not being mined in 1989; National Mineral Inventory, Department of Energy, Mines and Resources, Canada, Mineral Bulletin MR 223, National Mineral Inventory QUE 102 to QUE 104, QUE 106 and QUE 107.

Möller, P.
1989a: Rare earth mineral deposits and their industrial importance; in Lanthanides, Tantalum and Niobium; (ed.) P. Möller, P. Cerny, and F. Saupé; Society for Geology Applied to Mineral Deposits, Special Publication No. 7, p. 171-188.
*1989b: REE(Y), Nb, and Ta enrichment in pegmatites and carbonatite-alkalic rock complexes; in Lanthanides, Tantalum and Niobium; (ed.) P. Möller, P. Cerny, and F. Saupé; Society for Geology Applied to Mineral Deposits, Special Publication No. 7, p. 103-144.

Morogan, V.
1994: Ijolite versus carbonatite as sources of fenitization; Terra Nova, v. 6, no. 2, p. 166-176.

Morteani, G.
1989: Prospection for niobium-rich alkaline rocks; in Lanthanides, Tantalum and Niobium; (ed.) P. Möller, P. Cerny, and F. Saupé; Society for Geology Applied to Mineral Deposits, Special Publication No. 7, p. 309-320.

Notholt, A.J.G.

*1979:   The economic geology and development of igneous phosphate deposits in Europe and the USSR; Economic Geology, v. 74, p. 339-350.

*1980:   Igneous apatite deposits: mode of occurrence, economic development and world resources; in Fertilizer Mineral Potential in Asia and the Pacific, (ed.) R.P. Sheldon and W.C. Burnett; Proceedings of the Fertilizer Raw Materials Resources Workshop, August 20-24, 1979, p. 263-285.

O'Driscoll, M.

1988:   Rare earths: enter the dragon; Industrial Minerals, no. 254, November, p. 21-55.

Olsen, J.C., Shawe, D.R., Pray, L.C., and Sharp, W.N.

1954:   Rare-earth mineral deposits of the Mountain Pass District, San Bernardino County, California, United States Geological Survey, Professional Paper 261, 75 p.

Ontoyev, D.O.

1990:   The problem of the origin of the Bayan Obo complex iron-rare earth deposit, China; International Geology Review, v. 32, no. 10, p. 988-996.

Oreskes, N. and Einaudi, M.T.

1990:   Origin of rare earth element enriched hematite breccias at the Olympic Dam Cu-U-Au-Ag deposit, Roxby Downs, South Australia; Economic Geology, v. 85, p. 1-28.

Palabora Mining Company Limited, Mine Geological and Mineralogical Staff

1976:   The geology and economic deposits of copper, iron and vermiculite in the Palabora igneous Complex: a brief review; Economic Geology, v. 71, p. 177-192.

Pell, J.

1986:   Carbonatites in British Columbia: the Aley property (NTS 94B/5); in Geological Fieldwork, 1985, British Columbia Ministry of Energy, Mines and Petroleum Resources, Paper 1986-1, p. 275-277.

1987:   Alkaline ultrabasic rocks in British Columbia; carbonatites, nepheline syenites, kimberlites, ultramafic lamprophyres and related rocks; British Columbia Ministry of Energy, Mines and Petroleum Resources, Geological Survey Branch, Open File 1987-17, 109 p.

1990:   High-tech metals in British Columbia; British Columbia Ministry of Energy, Mines and Petroleum Resources, Geological Survey Branch, Information Circular 1990-19, 27 p.

Pell, J. and Höy, T.

1989:   Carbonatites in a continental margin environment - the Canadian Cordillera; in Carbonatites - Genesis and Evolution, (ed.) K. Bell; Unwin Hyman, London, p. 200-220.

Philpotts, J.A., Tatsumoto, M., Wang, K., and Fan, P-F.

1989:   Petrography, chemistry, age and origin of the rare-earth iron deposit at Bayan Obo, China, and implications of Proterozoic iron ores in earth evolution; in United States Geological Survey, Research on Mineral Resources - 1989, Program and Abstracts, Fifth Annual V.E. McKelvey Forum on Mineral and Energy Resources, United States Geological Survey, Circular 1035, p. 53-55.

Puustinen, K. and Kauppinen, H.

1989:   The Siilinjärvi carbonatite complex, eastern Finland; in Phosphate Deposits of the World, Volume 2, Phosphate Rock Resources, (ed.) A.J.G. Notholt, R.P. Sheldon, and D.F. Davidson; International Geological Correlation Programme, Project 156-Phosphorites, p. 394-397.

Rock, N.M.S.

1991:   Lamprophyres; Blackie and Sons Ltd., Glasgow and London, Van Nostrand Reinhold, New York, 285 p.

Rose, E.R.

1979:   Rare earth prospects in Canada; The Canadian Institute of Mining and Metallurgy, Bulletin, v. 72, p. 110-116.

Roskill Information Services Limited

1988:   The economics of rare earths and yttrium - Seventh Edition 1988; Roskill Information Services Limited, London, 356 p.

Sadeghi, A. and Steele, K.F.

1989:   Use of stream sediment elemental enrichment factors in geochemical exploration for carbonatite and uranium, Arkansas, U.S.A.; Journal of Geochemical Exploration, v. 32, p. 279-286.

Saether, E.

1958:   The alkaline rock province of the Fen area in southern Norway; Det Kongelige Norske Videnskabers Selskabs, Skrifter, 1957, no. 1, 158 p.

Sage, R.P.

1976:   No. 13 - Carbonatite-alkaline complexes; in Summary of Field Work, 1976, (ed.) V.G. Milne, W.R. Cowan, K.D. Card, and J.A. Robertson; Ontario Division of Mines, Miscellaneous Paper 67, p. 56-79.

*1983:   Literature review of alkalic rocks - carbonatites; Ontario Geological Survey, Open File Report 5436, 277 p.

1986:   Alkalic rock complexes-carbonatites of northern Ontario and their economic potential; PhD. thesis, Carleton University, Ottawa, Ontario, 355 p.

1987:   Geology of carbonatite-alkalic rock complexes in Ontario: Nemegosenda Lake Alkalic Rock Complex, District of Sudbury; Ministry of Northern Development and Mines, Ontario Geological Survey, Study 34, 132 p.

1988a: Geology of carbonatite-alkalic rock complexes in Ontario: Lackner Lake alkalic rock complex, District of Sudbury; Ministry of Northern Development and Mines, Ontario Geological Survey, Study 32, 141 p.

1988b: Geology of carbonatite-alkalic rock complexes in Ontario: Argor carbonatite complex, District of Cochrane; Ministry of Northern Development and Mines, Ontario Geological Survey, Study 41, 90 p.

1991:   Alkalic rock, carbonatite and kimberlite complexes of Ontario, Superior Province; Chapter 18 in Geology of Ontario, (ed.) P.C. Thurston, H.R. Williams, R.H. Sufcliffe, and G.M. Stott; Ontario Geological Survey, Special Volume 4, pt. 1, p. 683-709.

Samoilov, V.S.

1991:   The main geochemical features of carbonatites; Journal of Geochemical Exploration, v. 40, p. 251-262.

Scales, M.

1989:   Niobec - one-of-a-kind mine; Canadian Mining Journal, June, p. 43-46.

Secher, K.

1989:   Phosphate resources in the Sarfartôq carbonatite complex, southern West Greenland; in Phosphate Deposits of the World, Volume 2, Phosphate Rock Resources, (ed.) A.J.G. Notholt, R.P. Sheldon, and D.F. Davidson; International Geological Correlation Programme, Project 156-Phosphorites, p. 87-89.

Sinclair, W.D., Jambor, J.L., and Birkett, T.C.

1992:   Rare earths and the potential for rare-earth deposits in Canada; Exploration and Mining Geology, v. 1, p. 265-281.

Sørensen, H.

1977:   Glossary of alkaline and related rocks; in The Alkaline Rocks, (ed.) H. Sørensen; John Wiley & Sons, p. 558-577.

Steenfelt, A.

1987:   Geochemical mapping and prospecting in Greenland: a review of results and experiences; Journal of Geochemical Exploration, v. 29, p. 183-205.

1991:   High-technology metals in alkaline and carbonatitic rocks in Greenland: recognition and exploration; Journal of Geochemical Exploration, v. 40, p. 263-279.

Thivierge, S., Roy, D.-W., Chown, E.H., and Gauthier, A.

1981:   Evolution of the St. Honoré carbonatite after intrusion; in The St. Honoré and Crevier Niobium - Tantalum Deposits and Related Alkali Complexes, Lac St. Jean, Quebec, The Canadian Institute of Mining and Metallurgy, Excursion Guidebook, p. 23.

1983:   Evolution du complexe alcalin de St.-Honoré (Québec) après sa mise en place; Mineralium Deposita, v. 18, p. 267-283.

Treiman, A.H.

1989:   Carbonatite magma: properties and processes; in Carbonatites - Genesis and Evolution, (ed.) K. Bell; Unwin Hyman, London, p. 149-176.

Treiman, A.H. and Schedl, A.

1983:   Properties of carbonatite magma and processes in carbonatite magma chambers; Journal of Geology, v. 91 p. 437-447.

Tuttle, O.F. and Gittins, J.C.

1966:   Carbonatites; John Wiley and Sons, London, 591 p.

Ulbrich, H.H.G.L. and Gomes, C.B.

1981:   Alkaline rocks from continental Brazil; Earth Science Reviews, v. 17, p. 135-154.

Vallée, M. and Dubuc, F.

1970: The St-Honoré Carbonatite Complex, Quebec; The Canadian Institute of Mining and Metallurgy, Transactions, v. 73, p. 346-356.

1981: The St. Honoré carbonatite complex, Quebec; in The St. Honoré and Crevier Niobium - Tantalum Deposits and Related Alkalic Complexes, Lac St. Jean, Quebec, The Canadian Institute of Mining and Metallurgy, Excursion Guidebook, p. 17.

van Straaten, P.

1989: Nature and structural relationships of carbonatites from southwest and west Tanzania; in Carbonatites - Genesis and Evolution, (ed.) K. Bell; Unwin Hyman, London, p. 177-199.

Van Wambeke, L.

1977: The Konge rare earth deposits, Republic of Burundi: new mineralogical-geochemical data and origin of the mineralization; Mineralium Deposita, v. 12, no. 3, p. 373-380.

Veizer, J., Bell, K., and Jansen, S.L.

1992: Temporal distribution of carbonatites; Geology, v. 20, p. 1147-1149.

Verwoerd, W.J.

1989: Genetic types of ore deposits associated with carbonatites; in Abstracts Volume 3 of 3, 28th International Geological Congress, Washington, D.C., p. 3-295.

Woolley, A.R.

*1987: Alkaline rocks and carbonatites of the world: Part 1. North and South America, British Museum (Natural History), London, 216 p.

*1989: The spatial and temporal distribution of carbonatites; in Carbonatites - Genesis and Evolution, (ed.) K. Bell; Unwin Hyman, London, p. 149-176.

Woolley, A.R. and Kempe, D.R.C.

*1989: Carbonatites: nomenclature, average chemical compositions, and element distribution; in Carbonatites - Genesis and Evolution, (ed.) K. Bell; Unwin Hyman, London, p. 1-14.

Wyllie, P.J.

1989: Origin of carbonatites: evidence from phase equilibrium studies; in Carbonatites - Genesis and Evolution, (ed.) K. Bell; Unwin Hyman, London, p. 500-545.

Yegorov, L.S.

1993: Phoscorites of the Maymecha-Kotuy ijolite-carbonatite association; International Geology Review, v. 35, no. 4, p. 346-358.

Zhongxin, Y., Ge, B., Chenyu, W., Zhongqin, Z., and Xianjiang, Y.

1992: Geological features and genesis of the Bayan Obo REE ore deposit, Inner Mongolia, China; Applied Geochemistry, v. 7, p. 429-442.

Authors' addresses

T.C. Birkett
SOQUEM
2600, boul. Laurier
Tour Belle Cour, bureau 2500
Sainte-Foy, Quebec
G1V 4M6

D.G. Richardson
Geological Survey of Canada
601 Booth Street
Ottawa, Ontario
K1A 0E8

Printed in Canada

25. PRIMARY DIAMOND DEPOSITS

25.1 Kimberlite-hosted diamond
25.2 Lamproite-hosted diamond

25. PRIMARY DIAMOND DEPOSITS

INTRODUCTION

Diamonds are lithologically widely distributed, and are found in unconsolidated and consolidated sediments (placers and paleoplacers), various igneous rock types of deep-seated origin (kimberlite, orangeite, lamproite, alnoite, aillikite, picritic monchiquite, alkali basalt), high pressure mantle xenoliths, high pressure metamorphic rocks, and also meteorites and their impact structures. Of these, only diamond-bearing kimberlite, orangeite, and lamproite, plus associated placers and paleoplacers, are economically viable. Prior to 1960, more than 80% of all diamonds were derived from secondary deposits; by 1990, this figure was less than 25% (Levinson et al., 1992).

Diamond is the only mineral commodity extracted from kimberlite- or lamproite-hosted deposits. Diamonds are subdivided into industrial, near-gem, and gem quality stones. However, they are also described as being either 'cuttable' or 'industrial' (Levinson et al., 1992). Based on 1992 world production figures, approximately 50% by weight of a total production of 105 Mc (where Mc = million metric carats; c = metric carat = 0.2 g) was industrial grade, the remainder being cuttable. Industrial grade stones are used for a variety of purposes, but compete with synthetically produced industrial diamonds (estimated 1993 production 450-500 Mc; G.T. Austin, pers. comm., 1994).

Only primary diamond deposits are discussed here. These have been subdivided into two groups on the basis of their host rocks, which are either kimberlites or lamproites. In addition to their host rock differences, these deposits also differ in morphology, mineralogy, and other respects. These differences between the two types are discussed in the summary accounts that follow.

25.1 KIMBERLITE-HOSTED DIAMOND[1]

B.A. Kjarsgaard

IDENTIFICATION

In kimberlite-hosted deposits, diamonds occur mainly as sparsely dispersed, mantle-derived xenocrysts and diamondiferous mantle xenoliths in the kimberlite matrix. Economic quantities of diamond are mainly found in kimberlite diatremes. Kimberlites with preserved crater facies rocks are much rarer, but are in a few cases important high grade and high tonnage deposits.

The best examples of diamond-bearing kimberlites in Canada are several pipes in the Lac de Gras field, Northwest Territories. Grades established during current drilling and bulk sampling associated with pipe evaluation include A-154 south pipe (450 c/100 t), Misery pipe (419 c/100 t), Panda and Koala pipes (95 c/100 t), Leslie pipe (33 c/100 t) and the Fox pipe (27 c/100 t). These grades (as well as per cent gem quality stones) from preliminary samples are similar to those at producing mines (see below). Exploration and grade establishment also continues in the Attawapiskat and Kirkland Lake fields, Ontario and the Fort à la Corne and Candle Lake fields in Saskatchewan. Published grades for pipes from the Kirkland Lake and Fort à la Corne fields range from 1-23 c/100 t (Brummer et al., 1992; Northern Miner, 1995). These and other Canadian kimberlite localities are shown in Figure 25.1-1.

World class examples of kimberlite-hosted diamond deposits include the Orapa (67-130 c/100 t) and Jwaneng (154 c/100 t) pipes in Botswana, the Venetia (120 c/100 t) and Premier (35 c/100 t) pipes in South Africa, and the Mir (200 c/100 t) and Udachnaya (100 c/100 t) pipes in Yakutia.

IMPORTANCE

There is no past or current production of diamonds from kimberlite-hosted deposits in Canada. If diamond grade and stone quality from large bulk samples currently being extracted from the Lac de Gras pipes match results from smaller samples taken previously (see above), production decisions could be made by late 1996, with mining commencing in 1998. Diamond is an important mineral commodity, with many uses, including gemstones, abrasives, semiconductors, scientific instruments, surgical instruments, machine cutting tools, and drill bits. Before 1980, all production of diamonds was derived from kimberlite and related placer and paleoplacer deposits. Currently, this has decreased to about 65% (by weight; 93% by market value); the remainder is derived from lamproite and related placer deposits.

SIZE AND GRADE OF DEPOSIT

In kimberlite pipes, the grades and qualities of diamond vary considerably. Approximately 1% of all kimberlite pipes worldwide are economic. There are about 5000 kimberlites worldwide; fifty were mined at some time or another, twenty are active, and fifteen are major producers. The viability of any deposit is dependent upon a number of variables, including stone quality, stone size, grade (c/100 t), tonnage, extraction method (open pit versus underground), and processing costs, as well as local tax structure, environmental legislation, and infrastructure. An important economic parameter utilized is average US$/carat, determined from large (5000+ carats) parcel(s) of stones. The highly variable character of producing mines is illustrated by the ranges of the following parameters: size (1-150 ha), grade (4-600 c/100 t), average carat value (10-400 US$/c). Typical grades of economic kimberlites are listed throughout this paper.

In simple terms, deposit size is related to the erosional level of the pipe, coupled with its original shape (see Fig. 25.1-2A, B) and geology. The maximum long axis for near-surface craters is 1.5-2 km and their surface extent ranges from 200-40 ha. Examples include M1, Botswana (216 ha); Mwadui, Tanzania (146 ha); Pioneerskaya, Arkhangel, Russia (40 ha); and Orapa (106 ha). Very large kimberlites occasionally form due to multiple pipe coalescence; examples include the Jwaneng (52 ha) and Premier (32 ha) triple pipes and the Udachnaya (22 ha) and Frank Smith-Weltevreden, (South Africa, 8 ha) double pipes. Usually, however, at upper diatreme levels the maximum long axis is no more than 700 m. Examples of large diatremes include the Zarnitsa, Siberia (25 ha) and Letseng, Lesotho (16 ha, but the ore zone is only 4 ha in area) pipes. Due to the downward tapering of the diatreme, at root zone level (see Fig. 25.1-2A) diameters may only be tens of metres (e.g., Kimberley and De Beers mines, South Africa).

Grade (c/100 t) combined with stone value (US$/c) (Fig. 25.1-3A) illustrates the ore value for a number of economic pipes worldwide. An approximate 'in ground value' in US$billion to 120 m depth (Fig. 25.1-3B) can be calculated from deposit size to 120 m (Mt) combined with ore value (US$/t). Figure 25.1-3B illustrates that most economic pipes have 'values' of US$0.5-5 billion, exceptionally rich pipes have 'values' of US$10-17 billion. In practice, many pipes are mined to 1 km depth or more and with this increased tonnage theoretical mine 'values' can be upwards of $75 billion. The life spans for individual mines range from 25 years to more than 100 years (e.g. the Kimberley area Dutoitspan, Bulfontein, and Wesselton mines)

Kjarsgaard, B.A.
1996: Kimberlite-hosted diamond; in Geology of Canadian Mineral Deposit Types, (ed.) O.R. Eckstrand, W.D. Sinclair, and R.I. Thorpe; Geological Survey of Canada, Geology of Canada, no. 8, p. 560-568 (also Geological Society of America, The Geology of North America, v. P-1).

[1] Two types of kimberlite have been recognized (Wagner, 1914; Smith, 1983): 'basaltic' or 'Group I' kimberlite, hereafter termed kimberlite, and; 'micaceous' or 'Group II' kimberlite, hereafter termed orangeite (Mitchell, 1991). These latter rocks have only been recognized in South Africa, and therefore will not be discussed in greater detail.

GEOLOGICAL FEATURES

Setting and associated structure

Kimberlites are restricted to continental shield areas and are not associated with rift valleys. Economic kimberlites are found within Archean (>2.5 Ga) cratons. Kimberlites generally occur in clusters of two to twenty pipes; a kimberlite field (being approximately 50 km in diameter), consists of one to a number of separate kimberlite clusters of similar age. Kimberlite provinces consist of one or more fields. The Yakutia kimberlite province consists of twenty kimberlite fields; magmatism occurred in five distinct episodes from the Late Ordovician to the Late Jurassic. The initiation of kimberlite magmatism is deep seated, and correlation of this magmatism with hotspots or plate tectonic processes (transform fault extensions, subduction zones, etc.) has not been satisfactorily demonstrated on a worldwide basis. No viable theory exists which can predict the location of kimberlite fields within a craton. However, at the scale of a kimberlite field, individual pipes are believed to be located upon linear or arcuate trends related to major crustal fracture zones. These structural features provide an easily exploitable route for the ascent of deep-seated kimberlite magmas (Mitchell, 1991). The worldwide distribution of kimberlite pipes in relation to Archean cratons is shown in Figure 25.1-4.

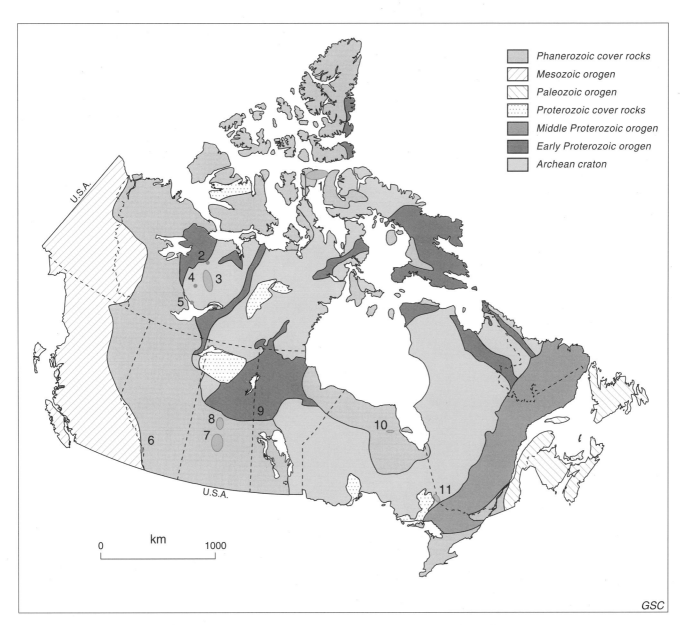

Figure 25.1-1. Location map of kimberlites in Canada: 1) Somerset Island field; 2) Ranch Lake; 3) Lac de Gras field; 4) Cross Lake; 5) Dry Bones Bay; 6) Crossing Creek; 7) Fort à la Corne field; 8) Candle Lake field; 9) Snow Lake-Wekusko; 10) Attawapiskat field; 11) Kirkland Lake/Timiskiming fields.

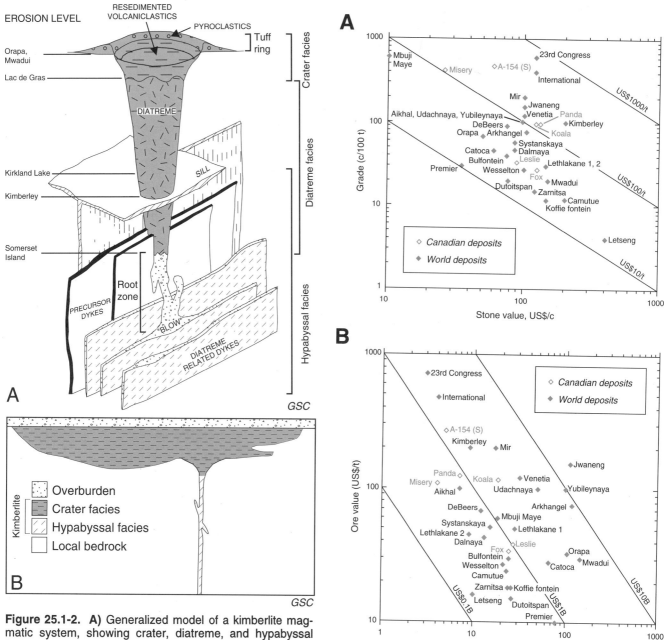

Figure 25.1-2. A) Generalized model of a kimberlite magmatic system, showing crater, diatreme, and hypabyssal facies rocks. Crater facies rocks consist of resedimented volcaniclastic and pyroclastic rocks; diatreme facies rocks consist dominantly of tuffisitic breccias; hypabyssal facies rocks are found in the root zone of the diatreme and consist of dykes, blows (enlarged dykes), and sills. Also shown are the present erosion levels of some representative economic and Canadian kimberlites (adapted from Mitchell, 1986), **B)** Generalized model of a kimberlite crater in which diatreme facies are absent. Thin (metre scale) hypabyssal feeder dykes are not necessarily observed. These craters are dominated by a wide variety of different pyroclastic and resedimented volcaniclastic rock types. Examples of this type of kimberlite crater include Mbuji Maye, Zaire; and Fort à la Corne, Saskatchewan (adapted from Meyer de Stadhelhofen, 1963 and Lehnert-Thiel et al., 1992).

Figure 25.1-3. A) Kimberlite ore value (US$/t) as determined by grade (c/100t) multiplied by diamond value (US$/c) for a number of economic kimberlite pipes worldwide (labelled filled diamonds; data from Janse, 1993) as compared to the Northwest Territories pipes (labelled open diamonds; data from various press releases). **B)** 'In ground kimperlite pipe value' (in US$B to a depth of 120 m) as determined by deposit size (Mt to a depth of 120 m) multiplied by average value per tonne (US$/t) for a number of economic kimberlite pipes worldwide (labelled filled diamonds; data from Janse, 1993) as compared to the Northwest Territories pipes (labelled open diamonds; ore value from Figure 25.1-3A; tonnage to 120 m depth estimated by the author).

Age of host rocks and diamond

Intrusion ages of economic kimberlite pipes range from the Mesoproterozoic (Middle Proterozoic) to the Middle Eocene (Table 25.1-1). The presence of diamonds (and associated kimberlite indicator minerals) in the Witwatersrand Conglomerates (ca. 2.9-2.7 Ga) is taken as evidence for kimberlite volcanism of Archean age. Consistent with these Archean ages are syngenetic diamond inclusions which have been dated at or have model ages of 3.3-0.6 Ga, inferred to be the formation ages of the diamonds. Examples in which both kimberlite host rock age and diamond formation age have been determined illustrate that the diamonds are 3.2 Ga to 1 Ma older than the host rocks.

Relationship of diamond to host rock

Age determinations on diamonds and their kimberlite hosts (see above) are consistent with other evidence suggesting that kimberlite-derived diamonds (specifically macro-diamonds: >1 mm) are xenocrysts. Kimberlites act as transportation agents only, bringing diamonds or diamond-bearing mantle xenoliths from within the diamond stability field (>4.5 GPa or 150 km depth) to the surface. In general, diamonds are disseminated throughout the kimberlite host, although 'intact' diamond-bearing mantle xenoliths are also found. Diamond inclusion silicate minerals and silicate mineral assemblages in diamond-bearing mantle xenoliths indicate that macro-diamonds can be of either eclogitic (E-type) or peridotitic (P-type) paragenesis. Diamond inclusion studies illustrate that the proportion of E-type to P-type stones at different mines is variable. At Wesselton the diamond population consists of 2% E-type and 98% P-type stones, whereas at Orapa, 85% of the stones are E-type and only 15% are P-type (Gurney, 1989). Grades reported for diamondiferous mantle xenoliths have been extrapolated to suggest that mantle source rocks are moderately to highly diamondiferous; inferred grades range from 0.5-650 c/100 t for peridotitic mantle to 17-37 000 c/100 t for eclogitic mantle.

Form of deposit and diamond distribution

In kimberlites which have preserved crater facies rocks, two distinct types of craters have been recognized. The most common type consists of resedimented volcaniclastic and rare pyroclastic rocks that overlie diatreme facies kimberlite (e.g., Mwadui and Orapa; see Fig. 25.1-2A). Crater walls dip inward at angles ranging from 25° to 75°. Craters of the other type are extremely rare and have only recently been recognized. These consist mainly of pyroclastic kimberlite with associated resedimented volcaniclastics (Fig. 25.1-2B). Contacts are horizontal to shallowly (0° to 35°) dipping. Diatreme facies rocks are absent and the feeder dyke(s) are

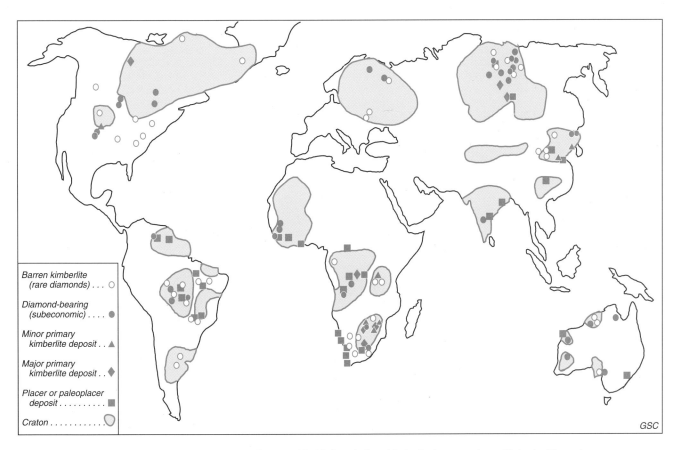

Figure 25.1-4. Distribution of kimberlites worldwide in relationship to Archean cratons. (Adapted from Janse, 1984; Atkinson, 1989; Gurney, 1989; Helmstaedt, 1993; and Janse, (pers. comm., 1994).

Table 25.1-1. Radiometric ages of selected economic kimberlites worldwide, plus additional Canadian examples. Data sources are from the bibliography and cited references of this paper (Ages indicated by * from L. Heaman and B.A. Kjarsgaard, unpub. data).

	Age (Ma)	Localities
Middle Eocene	52 55-50	Lac de Gras, Canada Mwadui, Tanzania
Late Cretaceous	71 73* 84 93 95	Mjubi Maye, Zaire Lac de Gras, Canada Dutoitspan, South Africa Orapa, Botswana Kimberley pool, South Africa
Early Cretaceous	99* 99*	Fort à la Corne, Canada Somerset Island, Canada
Late Jurassic	155*	Kirkland Lake, Canada
Early Triassic	250-235 250-240	Jwaneng, Botswana Crossing Creek, Canada
Late Devonian	361	Mir, Yakutia
Mesoproterozoic	1180	Premier, South Africa

not always observed. Examples include the Mbuji Maye pipe in Zaire, and kimberlite pipes in the Fort à la Corne field, Saskatchewan.

The concentrations of diamonds in craters facies kimberlite are in some cases enhanced due to weathering (denudation of tuffs, resulting in a residual concentration of resistant minerals) and/or associated sedimentary reworking (fluvial, lacustrine, marine processes). Exceptionally high diamond grades of 660 c/100 t are reported for the Mbuji Maye kimberlite, although more than 95% of the stones are industrial quality. At Mwadui, it has been inferred that the resedimented volcaniclastic rocks were richer in diamonds (both grade and stone size) than the diatreme facies rocks below. Similarly, the surface (crater) grade at Orapa is 132 c/100 t, almost twice the reported subsurface (diatreme) grade of 64-69 c/100 t. The grade of the Jwaneng crater is 154 c/100t.

Usually kimberlites manifest themselves as cone-shaped diatremes (see Fig. 25.1-2A), with steeply dipping (75°-85°) country rock contacts. Diatreme facies rocks consist mainly of tuffisitic kimberlite breccias, which are relatively uniform compared to crater or hypabyssal facies rocks. Although a diatreme can be comprised of ten to twenty separate, identifiable types of kimberlite, grades are relatively uniform throughout the diatreme. This is envisaged as a result of mixing processes involved in diatreme formation (Mitchell, 1991). Grades reported from mined kimberlite diatremes are highly variable, i.e. 4 c/100 t (Letseng), 60 c/100 t (Systanskaya, Yakutia), 120 c/100 t (Venetia, South Africa), 400 c/100 t (International, Yakutia). Very high grades may be encountered in weathered diatreme facies kimberlite at the surface. Wagner (1914) observed at the Premier pipe that surface grades (highly weathered kimberlite) were initially more than 150 c/100 t but after four years of mining were less than 50 c/100 t and

after 15 years of mining were 32 c/100 t (the present day grade). Although official grades are unavailable for the Russian pipes, at Mir surficial diatreme facies kimberlite grades of 300-400 c/100 t are inferred to be much higher than subsurface diatreme facies rock which grades 60-200 c/100 t.

With increasing depth, kimberlite diatremes grade into root zones (see Fig. 25.1-2A), consisting of multiple intrusions of kimberlite forming dykes, blows (enlarged dykes), and sills. Pipes which have been eroded to the level of the root zone are usually not economic (Mitchell, 1991). However, extraction of root zone kimberlite may be viable if mining continues after the diatreme facies rocks are exhausted (e.g., pipes in the Kimberley cluster, South Africa). In the root zone of a kimberlite pipe, several distinct intrusive phases of hypabyssal kimberlite occur. Grades of different intrusive phases within a root zone are variable: e.g., 1.6 - 17.8 c/100 t at Dutoitspan and less than 10 c/100 t to more than 40 c/100 t for the W3 kimberlite at Wesselton.

Alteration

Porous kimberlite crater and diatreme facies rocks are highly susceptible to alteration by weathering processes after emplacement. This alteration leads to the development of 'yellow' and 'blue' ground, whose properties (e.g., resistivity) can be used in exploration programs. Diamonds, however, are not affected by these surficial weathering processes and therefore surface grades may be much higher (see previous discussion of Premier and Mir) due to kimberlite volume loss. During transport by the kimberlite magma from the mantle source area to the surface, diamond is removed from its stability field and may undergo partial or complete resorption. Diamond can be converted to either graphite or a C-O gas species (CO, CO_2), depending upon magma fO_2 and reaction kinetics (P-T dependent). At low fO_2 diamond is very stable. It has been suggested that resorption of octahedral macro-diamonds to stones with tetrahexahedroid morphology implies a weight loss on the order of 45 to 60% (Gurney, 1989). Kimberlite-derived magnesian-ilmenite compositions are utilized as a monitor of redox conditions to indicate the potential for diamond preservation. In general (if the kimberlite is diamond-bearing) ilmenites with low Fe^{3+}/Fe^{2+} ratios (i.e., low fO_2) are associated with higher diamond contents, whereas diamonds are not found in association with high Fe^{3+} (i.e. high fO_2) ilmenites that are low in MgO.

Ore mineralogy

Diamond is the only 'ore' mineral extracted from the kimberlite host. The associated minerals are discussed below.

DEFINITIVE CHARACTERISTICS

Kimberlite is a volatile-rich ultrabasic rock that has an enriched incompatible (Sr, Zr, Hf, Nb, REEs) and compatible (Ni, Cr, Co) element signature similar to, but distinct from, lamproites. Kimberlite often appears hybrid in nature, as they may contain mantle xenoliths, xenocrysts, and macrocrysts (large crystals 1-20 cm in size), plus crustal xenoliths in a matrix crystallized from kimberlite melt.

The following definition of kimberlite has been adapted from Clement et al. (1984) and Mitchell (1986). Kimberlites are CO_2- and H_2O-rich ultrabasic rocks that have a distinctive inequigranular texture due to the presence of large, rounded, anhedral macrocrysts (i.e., megacrysts and xenocrysts) plus euhedral to subhedral phenocrysts set in a finer grained groundmass. The macrocryst suite of minerals includes minerals derived from disaggregated mantle xenoliths plus olivine (the essential macrocryst), Mg-ilmenite, Ti-Cr-pyrope garnet, clinopyroxene, phlogopite, enstatite, and zircon of the megacryst suite. Primary matrix minerals include second generation euhedral olivine phenocrysts/microphenocrysts, and one or more of the following: spinels, ilmenite, perovskite, monticellite, apatite, phlogopite-kinoshitalite$_{ss}$ mica, carbonates, and serpentine. Primary groundmass microcrystalline diopside has been observed only in crustally contaminated rocks. Commonly, macrocrysts and both early- and late-formed matrix minerals (e.g., monticellite) are replaced by deuteric serpentine and calcite.

The diverse mineralogy and associated mineral chemistry of kimberlites are reflections of the unusual major and trace element composition of these rocks. In this respect, combined petrographic, mineral chemistry, and whole-rock geochemical studies can usually discriminate kimberlites from other rock types of similar mineralogy (e.g., alnoite, aillikite, and other lamprophyres) and magmatic style (Mitchell, 1986). Chemical zoning trends observed in minerals such as phlogopite (plots of Al_2O_3-FeO and Al_2O_3-TiO_2) and spinel (reduced and oxidized spinel prism plots) can be particularly useful in constraining the identification of an unknown rock type (Mitchell, 1986).

Peridotite and eclogite xenoliths, plus minerals derived from their disaggregation are also observed in kimberlites. Eclogite xenoliths are characterized by pyrope-almandine garnet and omphacitic pyroxene, as well as accessory rutile, kyanite, corundum, coesite, and diamond. Peridotite xenoliths are olivine-rich with variable amounts of orthopyroxene, clinopyroxene, spinel, and garnet plus accessories (e.g., phlogopite, amphibole, rutile, and diamond).

Kimberlite typically occurs as small (<1 km diameter), steep walled (75°-85° dips), carrot-shaped diatremes occurring in clusters. Complex root zones consisting of hypabyssal kimberlite are found at the base of the diatreme. Large (to 2 km) craters are rarely preserved. Kimberlite-filled craters have shallowly dipping (0°-75°) contacts and may resemble vents formed by hydrovolcanic processes (see also subtype 25.2, "Lamproite-hosted diamond"). The greatest potential for diamonds is found in pipes with preserved diatreme and/or crater facies kimberlite. Unfortunately, these rocks are highly susceptible to alteration and weathering, and either do not outcrop (e.g., under lakes) or form poor outcrops in low or swampy ground.

GENETIC MODEL

Thermobarometric calculations on mineral assemblages from diamondiferous mantle xenoliths and polymineralic diamond inclusions are consistent with diamond existing in regions of the mantle at depths greater than 150 km. It is rarely possible to establish precise constraints on the formation of macro-diamonds; igneous, metamorphic, and metasomatic origins have all been suggested. On the basis of carbon isotope studies of eclogitic diamonds, it is inferred

that the carbon for at least some of these stones originated at or near the Earth's surface and was transported into the mantle via subduction processes. In contrast to eclogite paragenesis diamonds, peridotitic diamonds have a restricted range of carbon isotopic compositions, consistent with a juvenile (mantle) source of the carbon. Macrodiamonds are transported from the mantle to the surface by kimberlite magmas (see Fig. 25.1-5).

Kimberlites occur in a restricted tectonic setting and are observed only in ancient continental shield regions older than 1.5 Ga (Clifford, 1966). The most favourable tectonic environment for kimberlite pipes is a thick, old craton with low heat flow values; economic kimberlites are restricted to Archean cratons (>2.5 Ga; see Fig. 25.1-4). The initiation of kimberlite magmatism is deep seated, and magma generation is poorly understood. Correlation of this magmatism with hotspots or plate tectonic processes (transform fault extensions, subduction zones) cannot be satisfactorily demonstrated on a worldwide basis. Kimberlite magmas are thought to form by the partial melting of carbonated peridotite source regions (Eggler, 1989). However, Ringwood et al. (1993) have proposed an alternate model in which kimberlite magma is generated by partial melting in the transition zone (400-650 km depth). Ultra-high pressure majorite garnets that occur as inclusions in diamonds (Moore and Gurney, 1985) and in mantle xenoliths (Haggerty and Sautter, 1990) are consistent with kimberlite magma formation at depths of at least 300 km.

The range in diamond contents of kimberlites is dependent upon the amount of diamond-bearing mantle material entrained by the ascending magma, the proportions of various mantle lithologies (eclogite and peridotite; eclogite often contains higher modal diamond content) sampled and the degree to which resorption and mechanical sorting of this entrained material occurs during transport to the surface. Kimberlites probably ascend through the mantle at substantial velocities (10-30 kilometres per hour; Eggler, 1989) by crack propagation processes. Near the surface, vent velocities of several hundred kilometres per hour may be possible, due to rapid CO_2 degassing from the magma. Highly explosive, near surface volcanism is consistent with the formation of kimberlite diatremes and craters as well as the entrainment of large amounts of angular crustal material. This can cause dilution of grade which in some cases is significant. Crater and diatreme facies kimberlite contain the highest diamond grades, hypabyssal rocks generally have low diamond tenors.

RELATED DEPOSIT TYPES

Lamproite (see subtype 25.2, "Lamproite-hosted diamond") and orangeite form the only other important primary diamond deposits with established economic potential. Associated with primary deposits are secondary (placer and paleoplacer) deposits. The distribution patterns of important secondary deposits closely mimics primary distribution (i.e., closely associated with stable cratonic nuclei; Gurney, 1989). Diamonds in these deposits are inferred to be dominantly kimberlite-derived, although in South Africa orangeite-derived diamonds are also important.

The alluvial diamond deposits of Sierra Leone and along the Zaire-Angola border region, as well as the marine terrace deposits in Namaqualand and southwest Africa-Namibia, are examples of economic secondary deposits.

The Namibian eolian deposits are thought to be reworked marine terrace deposits. Diamonds from placer deposits in general have a very high proportion of gem-quality stones; this improvement is thought to be due to the preferential breakage of inferior crystals (Gurney, 1989). Extreme secondary enrichment of diamonds (grades of 1000 c/m³) have been reported from favourable trap sites in placers and paleoplacers.

The primary sources for some secondary diamond fields remain unknown. Secondary diamond deposits (of unknown source) which have yielded significant quantities of diamonds are located in Brazil, India, southeast Australia, China, and western Transvaal, South Africa (Gurney, 1989). In the Great Lakes region of North America, more than 80 diamonds of as much as 21 carats in weight (most are less than 1 carat) have been recovered from glacial drift. Although the primary source of these stones is unknown (Brummer, 1978), they are possibly derived from kimberlites of the Kirkland Lake or Attawapiskat fields. Many 'sourceless' lone diamonds have been recovered from diverse regions of the globe, but this is not surprising in light of the inherent hardness and chemical stability of diamond. These diamonds, however, must have had a primary source, likely a kimberlite or lamproite.

EXPLORATION GUIDES

Economic kimberlites are found in old (>2.5 Ga) stable cratons characterized by thick crust and low geothermal gradients. Various methods are used to locate kimberlites, depending upon local conditions: i.e., type of country rock, climate, and overburden. The main exploration techniques used are: 1) indicator mineral sampling (heavy mineral separates from stream sediment sampling, soil sampling, and till sampling); 2) remote sensing (LANDSAT, airphoto interpretation); 3) geophysical surveys (magnetic, gravity, electrical, radiometric, seismic profiling); and 4) geochemical. Biogeochemical methods have also been utilized. Atkinson (1989) provided a recent general summary of exploration techniques.

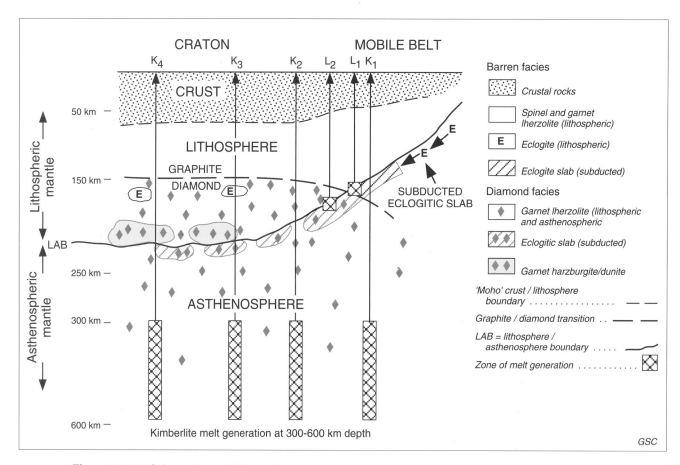

Figure 25.1-5. Schematic model illustrating magma source regions and the relationship between the ascent of these magmas and diamond source rocks in an Archean craton and surrounding mobile belt. Salient points on this diagram are described in the figure key. Kimberlites (asthenospheric source), depending upon the nature of the mantle they ascend through may contain: no diamonds (path K1); diamonds from lithospheric and asthenospheric garnet lherzolite (path K2); diamonds predominantly from eclogites, plus minor contributions from garnet harzburgite/dunite and garnet lherzolite (path K3); diamonds predominantly from garnet harzburgite/dunite, plus minor contributions from eclogite and garnet lherzolite (path K4). Lamproite (lithospheric source: see subtype 27.2) ascent routes are shown as path L1 (barren) and L2 (diamondiferous). Modified from Mitchell (1991).

Because kimberlites are rare rocks which generally form poor outcrops, exploration methods (like those for many other ore deposits) must be capable of finding a hidden target. While the geophysical signature of kimberlites is not unique, it is unusual and can be discerned by low-level aeromagnetic and EM surveys. Numerous kimberlite pipes in the Arkhangel region (Russia) and the central Kalahari in Botswana were located by aeromagnetic surveys. This technique is most effective in areas with a uniform and low magnetic background. Combined aeromagnetic and electromagnetic surveys have been used with a high success rate in the Lac de Gras field, Northwest Territories. A recent, comprehensive review of geophysical techniques as applied to kimberlite exploration can be found in Urquart and Hopkins (1993).

The unique mineralogical signature of kimberlites enables the application of indicator mineral sampling exploration techniques. The identification of resistant minerals that can indicate the potential presence of a kimberlite has been widely and successfully applied as an exploration technique in South Africa, Yakutia, and Canada. However, it is extremely important to note that these so-called kimberlite indicator minerals are also found in many other rock types that do not contain diamonds. Kimberlite indicators include minerals derived from the kimberlite (spinel, olivine, ilmenite, and perovskite), all the macrocryst minerals (olivine, spinel, low-Cr Ti-pyrope, Mg-ilmenite, Cr-diopside, enstatite, and zircon), as well as minerals from disaggregated mantle xenoliths (olivine, enstatite, Cr-diopside, chrome pyrope garnet, Cr-spinels, pyrope-almandine garnet, omphacitic pyroxene, and diamond). In Canada, application of the indicator mineral method to stream sediment sampling is problematic due to Quaternary glaciation. However, success in locating kimberlite pipes has been obtained by esker and till sampling in the Lac de Gras, and Kirkland Lake areas. A combination of alluvial and stream sediment sampling coupled with ground magnetics was utilized in the discovery of pipes at Attawapiskat.

Because diamond is a rare mineral in kimberlite, (0-1.4 ppm), a subset of the indicator minerals, termed 'diamond indicators' is used to indicate the potential presence of diamond in these rocks. This is based on studies of silicates and oxide inclusions in diamond and minerals from diamond-bearing mantle xenoliths (Gurney, 1989). Specific diamond indicator minerals (with xenolith-paragenesis type in brackets) include subcalcic Cr-pyrope (garnet-bearing harzburgite/dunite source), Cr-pyrope garnet (garnet-bearing lherzolite source), high-Cr-Mg chromite (chromite-bearing harzburgite/dunite source), and high-Na-Ti pyrope-almandine garnet (eclogite source). It is important to note that these minerals (and xenoliths) are not definitive of kimberlite volcanism, as they can be observed in other rock types of deep-seated origin (e.g., ultramafic lamprophyres). Furthermore, these minerals are not an infallible indicator of the presence of diamond in kimberlite; the Skerring (Australia) and Zero (South Africa) kimberlites both contain subcalcic Cr-pyrope garnet, but lack diamonds (Gurney, 1989).

ACKNOWLEDGMENTS

Critical reviews by Barbara Scott-Smith and Roger Mitchell on an early version of the manuscript greatly improved the content. Comments by R.I. Thorpe, O.R. Eckstrand, D.F. Sangster, and A.N. LeCheminant on later versions were quite helpful. The final version was capably reviewed by A.J.A. Janse, whose comments clarified a number of points.

REFERENCES

References with asterisks (*) are considered to be the best source of general information on this deposit type. All kimberlite conference proceedings volumes are excellent sources of information.

*Ahrens, L.H., Dawson, J.B., Duncan, A.R., and Erlank, A.J. (ed.)
1975: Proceedings of the First International Kimberlite Conference; Physics and Chemistry of the Earth, v. 9, 940 p.

*Atkinson, W.J.
1989: Diamond exploration philosophy, practice, and promises: a review; in Proceedings of the Fourth International Kimberlite Conference, Volume 2. Kimberlites and Related Rocks: Their Mantle/Crust Setting, Diamonds and Diamond Exploration, (ed.) J. Ross; Geological Society of Australia, Special Publication 14, Blackwell Scientific Publications, Oxford, 1986, p. 1075-1107.

Brummer, J.J.
1978: Diamonds in Canada; The Canadian Mining and Metallurgical Bulletin, v. 71, no. 798, p. 64-79.

Brummer, J.J., MacFadyen, D.A., and Pegg, C.C.
1992: Discovery of kimberlites in the Kirkland Lake area northern Ontario, Canada. Part II: kimberlite discoveries, sampling, diamond content, ages and emplacement; Exploration and Mining Geology, v. 1, no. 4, p. 351-370.

Clement, C.R., Skinner, E.M.W., and Scott-Smith, B.H.
1984: Kimberlite re-defined; Journal of Geology, v. 92, p. 223-228.

Clifford, T.N.
1966: Tectono-metallogenic units and metallogenic provinces of Africa; Earth and Planetary Science Letters, v. 1, p. 421-434.

*Dawson, J.B.
1980: Kimberlites and Their Xenoliths; Springer Verlag, Berlin, 252 p.

Eggler, D.H.
1989: Kimberlites: how do they form?; in Proceedings of the Fourth International Kimberlite Conference, Volume 1. Kimberlites and Related Rocks: Their Composition, Occurrence, Origin and Emplacement, (ed.) J. Ross; Geological Society of Australia, Special Publication 14, Blackwell Scientific Publications, Oxford, 1986, p. 323-342.

*Glover, J.E. and Groves, D.I. (ed.)
1980: Kimberlites and diamonds; Publication #5, Geology Department/ Extension Services, The University of Western Australia, Perth, Australia, 133 p.

*Glover, J.E. and Harris, P.G. (ed.)
1984: Kimberlite occurrence and origin; Publication #8, Geology Department/Extension Services, The University of Western Australia, Perth, Australia, 298 p.

*Gurney, J.J.
1989: Diamonds; in Proceedings of the Fourth International Kimberlite Conference, Volume 1. Kimberlites and Related Rocks: Their Composition, Occurrence, Origin and Emplacement, (ed.) J. Ross; Geological Society of Australia, Special Publication 14, Blackwell Scientific Publications, Oxford, 1986, p. 935-965.

Haggerty, S.E. and Sautter, V.
1990: Ultradeep (greater than 300 km), ultramafic upper mantle xenoliths; Science, v. 248, p. 993-996.

*Helmstaedt, H.H.
1993: Natural diamond occurrences and tectonic setting of "primary" diamond deposits; in Diamonds: Exploration, Sampling and Evaluation, (ed.) P. A. Sheahan and A. Chater; Short Course Proceedings, Prospectors and Developers Association of Canada, Toronto, March 27, 1993, p. 1-72.

Janse, A.J.A.

1984: Kimberlites - where and when; in Kimberlite Occurrence and Origin, (ed.) J.E. Glover and P.G. Harris; Publication #8, Geology Department/Extension Services, The University of Western Australia, Perth, Australia, p. 19-62.

1993: The aims and economic parameters of diamond exploration; in Diamonds: Exploration, Sampling and Evaluation, (ed.) P.A. Sheahan and A. Chater; Short Course Proceedings, Prospectors and Developers Association of Canada, Toronto, March 27, 1993, p. 173-184.

Kornprobst, J. (ed.)

*1984a: Proceedings of the Third International Kimberlite Conference, Volume 1. Kimberlites I: Kimberlites and Related Rocks; Developments in Petrology 11A, Elsevier, Amsterdam, 466 p.

*1984b: Proceedings of the Third International Kimberlite Conference, Volume 2. Kimberlites II: The Mantle and Crust-Mantle, Relationships; Developments in Petrology 11A, Elsevier, Amsterdam, 393 p.

Lehnert-Thiel, K., Loewer, R., Orr, R.G., and Robertshaw, P.

1992: Diamond-bearing kimberlites in Saskatchewan, Canada: the Fort a la Corne case history; Exploration and Mining Geology, v. 1, no. 4, p. 391-403.

Levinson, A.A., Gurney, J.J., and Kirkley, M.B.

1992: Diamond sources and production: past, present and future; Gems and Gemology, v. 28, no. 4, p. 234-254.

Meyer, H.O.A. and Boyd, F.R. (ed.)

*1979a: Proceedings of the Second International Kimberlite Conference, Volume 1. Kimberlites, Diatremes and Diamonds: Their Geology, Petrology and Geochemistry; American Geophysical Union, Washington, D.C., 400 p.

*1979b: Proceedings of the Second International Kimberlite Conference, Volume 2. The Mantle Sample: Inclusions in Kimberlites and Other Volcanics, American Geophysical Union, Washington, D.C., 424 p.

Meyer de Stadhelhofen, C.

1963: Les Breches kimberlitique du Terratoire du Bakwanga (Congo); Archives de Science, v. 16, fasc. 1, p. 87-144.

*Mitchell, R.H.

1986: Kimberlites: Mineralogy, Geochemistry, and Petrology; Plenum Press, New York, 442 p.

*1991: Kimberlites and lamproites: primary sources of diamond; Geoscience Canada, v. 18, no. 1, p. 1-16.

Moore, R.O. and Gurney, J.J.

1985: Pyroxene solid solutions in garnets included in diamond; Nature, v. 318, p. 553-555.

*Nixon, P.H. (ed.)

1987: Mantle Xenoliths; J. Wiley and Sons, Toronto, 844 p.

Northern Miner

1995: Uranurz, partners drill-test Saskatchewan kimberlite field; The Northern Miner, Toronto, vol. 81, no. 22, p. 1, 2.

Ringwood, A.E., Kesson, S.E., Hibberson, W., and Ware, N.

1993: Origin of kimberlites and related magmas; Earth and Planetary Science Letters, v. 113, p. 521-538.

Ross, J. (ed.)

*1986a: Proceedings of the Fourth International Kimberlite Conference, Volume 1. Kimberlites and Related Rocks: Their Composition, Occurrence, Origin and Emplacement; Geological Society of Australia, Special Publication 14, Blackwell Scientific Publications, Oxford, 646 p.

*1986b: Proceedings of the Fourth International Kimberlite Conference, Volume 2. Kimberlites and Related Rocks: Their Mantle/Crust Setting, Diamonds and Diamond Exploration; Geological Society of Australia, Special Publication 14, Blackwell Scientific Publications, Oxford, 1986.

*Sheahan, P. and Chater, A. (ed.)

1993: Diamonds: Exploration, Sampling and Evaluation; Short Course Proceedings, Prospectors and Developers Association of Canada, Toronto, March 27, 1993, 384 p.

Smith, C.B.

1983: Pb, Sr, and Nd isotopic evidence for sources of Cretaceous kimberlite; Nature, v. 304, p. 51-54.

*Urquhart, W.E.S. and Hopkins, R.

1993: Exploration geophysics and the search for diamondiferous diatremes; in Diamonds: Exploration, Sampling and Evaluation, (ed.) P. Sheahan and A. Chater; Short Course Proceedings, Prospectors and Developers Association of Canada, Toronto, March 27, 1993, p. 249-287.

Wagner, P.A.

1914: The Diamond Fields of South Africa; Transvaal Leader, Johannesburg, South Africa, 347 p.

25.2 LAMPROITE-HOSTED DIAMOND

B.A. Kjarsgaard

IDENTIFICATION

In lamproites, as in kimberlites, diamond occurs as sparsely dispersed xenocrysts in the matrix. Economic quantities of diamonds are found mainly in lamproite pyroclastic rocks, but rarely also in dykes. Diamonds have yet to be found in lamproite lavas. Viable lamproite diamond deposits are all hosted by vent facies olivine lamproite tuffs; examples include the Argyle AK1 mine, Australia and the Majhgawan mine, India.

No lamproites have yet been found in Canada, although they are known in the U.S.A. (e.g., Leucite Hills, Wyoming; Prairie Creek, Arkansas).

Kjarsgaard, B.A.

1996: Lamproite-hosted diamond; in Geology of Canadian Mineral Deposit Types, (ed.) O.R. Eckstrand, W.D. Sinclair, and R.I. Thorpe; Geological Survey of Canada, Geology of Canada, no. 8, p. 568-572 (also Geological Society of America, The Geology of North America, v. P-1).

IMPORTANCE

Initially, the Argyle deposit was considered to be kimberlitic (Atkinson et al., 1984). The importance of lamproite as a diamond host rock has only been recognized since 1984, a result of the landmark studies of Scott-Smith et al. (1984, 1989). They determined that some 'anomalous' diamond-bearing kimberlites are actually lamproites (e.g. Prairie Creek; Majhgawan mine; Kapamba, Zambia). The Argyle AK1 mine in Australia, discovered in 1979, is of vast economic importance. This deposit currently produces just over one third (by weight) of all diamonds mined (38.4 Mc/a), 400% more than the most productive kimberlite mine (Jwaneng, Botswana; 9.4 Mc/a). In value, however, Argyle stones represent only 7% of world production, as 95% of the diamonds mined are either industrial grade or poor quality gemstones.

In Canada, rocks of the lamproite type are unknown and there is no past or current production of diamonds from this deposit type.

SIZE AND GRADE OF DEPOSIT

The Argyle AK1 mine has a surface area of approximately 46 ha. The reported grade for 1987 for a variety of tuffs ranged from 100 to 680 c/100 t (Grice and Boxer, 1990). Hyalo-olivine lamproite lapilli tuffs at the Majhgawan mine (9 ha surface area), have diamond grades of 8 to 15 c/100 t (Scott-Smith, 1989). The Prairie Creek vent has a surface area of 27 ha, and grades approximately 13 c/100 t. At present, this deposit is not mined (it is now a state park), but is being re-evaluated. Approximately 100 000+ carats

of diamond were mined from 1907 to 1933 (Waldman and Meyer, 1992). The Ellendale 4 vent, Australia (surface area 84 ha) grades 3 to 25 c/100 t, but is a sub-economic deposit.

Established grades for diamond-bearing lamproites (<1-680 c /100 t) are similar to those of kimberlites. However, the exceptionally high grade of the Argyle AK1 mine is anomalous with respect to other lamproites. Typical reported grades for other diamondiferous olivine lamproites range from <1 to 30 c /100 t, lower than most economic kimberlites. The economic viability of any diamond-bearing lamproite is dependent upon a variety of factors, such as grade, tonnage, average $/carat etc. (see also subtype 25.1, "Kimberlite-hosted diamond").

GEOLOGICAL FEATURES

Setting and associated structure

Diamond-bearing lamproites occur in a wide variety of geological and tectonic settings. This precludes the formulation of a universal model constraining the geotectonic setting in which they were emplaced. The following diamondiferous lamproites illustrate this variety. The Argyle and Ellendale lamproites are in Mesoproterozoic mobile belts at the margins of the Archean/Paleoproterozoic

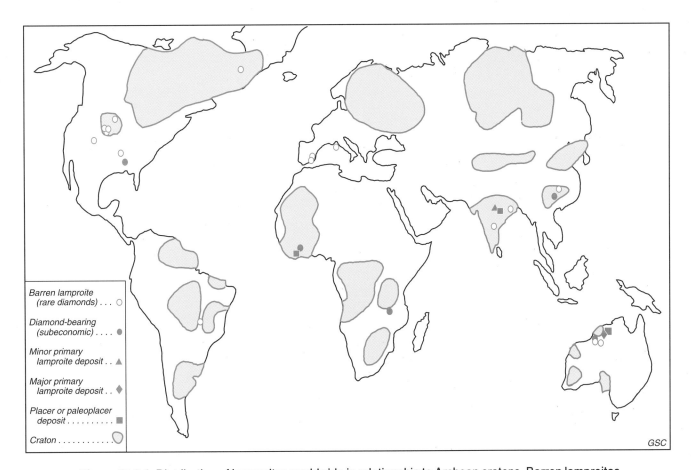

Barren lamproite
(rare diamonds) . . . ○

Diamond-bearing
(subeconomic) . . . . ●

Minor primary
lamproite deposit . . ▲

Major primary
lamproite deposit . . ◆

Placer or paleoplacer
deposit . . . . ■

Craton . . . . ◯

GSC

Figure 25.2-1. Distribution of lamproites worldwide in relationship to Archean cratons. Barren lamproites in Antarctica are not shown. Localities from Mitchell and Bergman (1991), Helmstaedt (1993), and Janse (pers. comm., 1994).

Table 25.2-1. Radiometric ages of diamondiferous lamproites worldwide. Data from Mitchell and Bergman (1991).

	Age (Ma)	Location
Tertiary	22-18	Ellendale, Australia
Cretaceous	106-97	Prairie Creek, U.S.A.
Late Triassic	≈220	Kapamba, Zambia
Mesoproterozoic	1170-1140 1178-1126 1455-1150	Majhgawan, India Argyle, Australia Bobi, Ivory Coast

(Early Proterozoic) Kimberley block. Their emplacement appears to have been strongly controlled by major fracture zones that represent lithospheric lines of weakness. The Prairie Creek vent (which lies well off craton) was emplaced near the intersection of the Reelfoot rift and the southeastern edge of the Phanerozoic Ouachita orogenic belt (marginal to the 1.7-1.6 Ga Mazatzal-Pecos structural province). The Kapamba lamproites are located in the Luangwa graben, an extension of the East African Rift (Scott-Smith et al., 1989). This graben occurs in the Proterozoic Irumide and Mozambique tectonic belts, to the south of the Archean Tanzanian craton. The Bobi dykes (Ivory Coast) have intruded Archean-Paleoproterozoic granitic basement of the West African Shield. Lamproites are also found within stable Archean cratons (e.g., Leucite Hills, Wyoming craton). The worldwide distribution of lamproites is shown in Figure 25.2-1.

Age

Intrusion ages of the lamproite host rocks span the range from the Mesoproterozoic to the Late Pleistocene. Typical radiometric ages of diamondiferous lamproites are listed in Table 25.2-1. As for kimberlites, diamonds in lamproites are interpreted to have formed during the Early Archean to the Proterozoic, previous to their entrainment in the lamproite host magma.

Relationship of diamond to host rock

Diamonds in lamproites are considered to be xenocrysts (see "Age" above) derived from regions of the mantle within the diamond stability field and brought to the surface by lamproite magmas. Diamonds are disseminated throughout the host, and also found in diamond-bearing mantle xenoliths. Mineral inclusions in diamond from the Ellendale, Argyle, and Prairie Creek lamproites indicate that macro-diamonds (>1 mm) are of both eclogite and peridotite paragenesis.

Form of deposit and diamond distribution

For lamproites, subeconomic to commercial quantities of diamond have been found mainly associated with olivine lamproite vents. Figure 25.2-2A illustrates the typical form of a champagne glass-shaped lamproite vent (e.g. Ellendale field vents, Prairie Creek). Figure 25.2-2B illustrates the funnel-shaped form of the lamproite vents observed at

Ellendale 9 vent

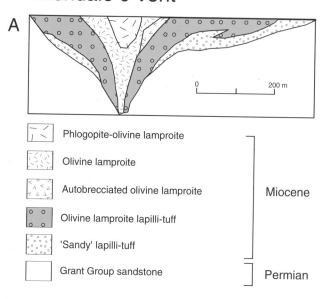

	Phlogopite-olivine lamproite	
	Olivine lamproite	
	Autobrecciated olivine lamproite	Miocene
	Olivine lamproite lapilli-tuff	
	'Sandy' lapilli-tuff	
	Grant Group sandstone	Permian

Argyle AK1 vent

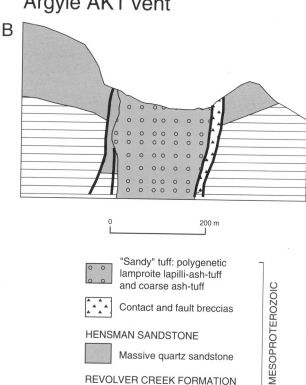

	"Sandy" tuff: polygenetic lamproite lapilli-ash-tuff and coarse ash-tuff	
	Contact and fault breccias	
	HENSMAN SANDSTONE	
	Massive quartz sandstone	MESOPROTEROZOIC
	REVOLVER CREEK FORMATION	
	Sandstone, siltstone, and shale	
	Fault	GSC

Figure 25.2-2. A) Simplified cross-section of the champagne-glass shaped Ellendale 9 lamproite vent (adapted from Jacques et al., 1986); **B)** Simplified cross-section of the funnel-shaped Argyle lamproite vent (adapted from Jacques et al., 1986).

Argyle and Majhgawan. At Argyle, Ellendale, Majhgawan, and Prairie Creek, the earliest tuffaceous phases appear to have the highest diamond tenors (Mitchell, 1991). Diamonds are also found in economic quantities in olivine lamproite dykes (e.g., Bobi; the Lissadell Road dykes associated with the Argyle mine), although lamproite lavas are nondiamondiferous. Diamond grades can be highly variable in vent facies rocks. For example, at Argyle the sandy tuffs (pyroclastic rocks with as much as 60% xenocrystic quartz) contain as much as 680 c/100 t, but tuffs rich in juvenile clasts have much lower grades (100 c/100 t; Deakin and Boxer, 1989). In contrast, at Ellendale diamond grades increase from the earliest contaminated 'sandy' lapilli-tuff (1-4 c/100 t) to later uncontaminated olivine lamproite lapilli tuffs (3-30 c /100 t; Jacques et al., 1986).

Alteration

During transport by the lamproite magma from the mantle source area to the surface, diamond is removed from its stability field and may undergo graphitization or partial to complete resorption (conversion to CO_2). Although H_2O-rich lamproite magmas are thought to have higher intrinsic oxygen fugacities (i.e., high diamond resorption potential) than CO_2-rich kimberlite magmas (Mitchell, 1991), this is an oversimplification. Lamproite diamond populations from Argyle (high degree of resorption and graphitization, irregular stone shapes) and Ellendale (no or low degrees of resorption, preserved crystal shapes) illustrate this variability.

Ore mineralogy

Diamond is the only 'ore' mineral extracted from the lamproite host. The associated minerals are discussed below.

DEFINITIVE CHARACTERISTICS

Lamproite is defined as an ultrapotassic ($K_2O/Na_2O>3$), peralkaline ($[Na+K]/Al>1$) and typically perpotassic ($K>Al$) rock, ranging from ultrabasic to intermediate (37-64 wt.% SiO_2) in composition. These rocks have enriched incompatible (Rb, Ba, Sr, Zr, Hf, Ti, P, Nb, REEs, Y, Th, U) and compatible (Ni, Cr, Co, V, Sc) element abundances similar to, but distinct from, kimberlites. Lamproites are characterized by the occurrence of at least one of the following: olivine, leucite, richterite, diopside, and sanidine ($\pm$ glass).

The following explicit definition of lamproite is adopted from Scott-Smith and Skinner (1984) and Mitchell and Bergman (1991). Lamproites contain the following typomorphic minerals (5-90 vol.%): Al-poor, Ti-rich phlogopite phenocrysts; poikilitic groundmass Ti-rich tetraferriphlogopite; Ti-K-richterite; forsteritic olivine; Al-Na-poor diopside; Fe^{3+}-rich leucite; Fe^{3+}-rich sanidine. Not all of the above phases are required for a rock to be termed a lamproite, as one or two minerals may be modally dominant and the others subordinate or absent. Lamproites are divided into five petrographic groups, based upon the modal dominance of olivine, leucite, richterite, diopside, and sanidine. Accessory phases include apatite, perovskite, Mg-chromite, Mg-Ti-chromite, Mg-Ti-magnetite, potassian barian titanates (priderite and jeppeite), and potassian zirconian or titanian silicates (wadeite, davanite, and shcherbakovite). Other typomorphic accessories include armalcolite, ilmenite, and enstatite. Common alteration and secondary phases include analcite (replacing leucite and/or sanidine), barite, quartz, TiO_2-polymorphs, Ba-rich zeolites, chlorite, and carbonates. Megacrysts of Ti-Cr-pyrope and Mg-ilmenite (typical of kimberlites) are very rare to absent in lamproites. Mantle peridotite and eclogite xenoliths, plus minerals derived from their disaggregation are also (rarely) observed in these rocks.

The diverse mineralogy and associated mineral chemistry of lamproites are reflections of the unusual major and trace element compositions of these rocks. In this respect, combined petrographic, mineral chemistry, and whole-rock geochemical studies can discriminate lamproites from other rock types of similar mineralogy (e.g., leucite basanites, potassic alkali basalts, kamafugites, minettes, ultramafic lamprophyres, kimberlites, and orangeites) and magmatic style (Mitchell and Bergman, 1991). Chemical zoning trends observed in specific minerals such as phlogopite (plots of Al_2O_3-FeO and Al_2O_3-TiO_2) and spinel (reduced and oxidized spinel prism plots) are particularly useful in arriving at an identification of an unknown rock type (Mitchell and Bergman, 1991).

Lamproites occur as extrusive, subvolcanic and hypabyssal rocks. Ponded magma which has formed a lava lake, and pyroclastic rocks are the most common form of lamproite. Lava flows are rare. Lamproite volcanism is similar in style to alkali basaltic volcanism and small magma volumes preclude the formation of large stratovolcanoes and calderas. In contrast to kimberlite volcanism, lamproites do not form diatremes or root zones, but rather vents and dyke-lava lake systems (see Fig. 25.2-2A, B). The greatest potential for diamondiferous lamproites is in olivine-phyric tuffs in champagne glass- or funnel-shaped vents.

GENETIC MODEL

Lamproites occur in a wide variety of tectonic and geological settings that may be either on or off Archean cratons. Specifically, all economic and near-economic lamproites are found in Proterozoic terranes. These magmas have commonly intruded crust that overlies lithospheric mantle affected by earlier subduction or rifting events (Mitchell and Bergman, 1991). Lamproite magmas are derived by partial melting of ancient, enriched (metasomatized, i.e., amphibole-, phlogopite- and apatite-bearing) upper mantle sources (lithospheric) which have previously been depleted in Na, Al, and Ca (leaving a residuum of harzburgitic composition). The source region for diamond-bearing olivine lamproites must be at a depth of more than 150 km; these magmas sample diamond-bearing mantle en route to the surface, transporting the ore as xenocrysts and diamond-bearing xenoliths. Vent-filling, olivine lamproite pyroclastic rocks contain the highest diamond grades. The formation of these vents is believed to be a result of hydrovolcanic processes (resulting in maars and tuff rings).

RELATED DEPOSIT TYPES

Important secondary (placer) diamond deposits are associated with primary lamproite diamond deposits. These are shown on Figure 25.2-1 and include the Smoke and Limestone Creek placers downstream from the Argyle pipe (with grades lower than the pipe; 70-140 c /100 t) the Bow River placer east of Argyle (27 c /100 t) and the Sequéla placers, Ivory Coast (derived from the Bobi dykes).

571

EXPLORATION GUIDES

Economic lamproites are found within Proterozoic terranes, which are usually adjacent to stable cratonic blocks characterized by thick crust and low geothermal gradients. Exploration methodologies used to locate lamproites are the same as those for kimberlites (see subtype 25.1; "Exploration guides"). However, it is noteworthy that the responses to these surveys and tests are not necessarily the same as those obtained for kimberlites (Atkinson, 1989).

The two most successfully used lamproite exploration techniques are stream sampling for indicator minerals and low-level airborne surveys. The Ellendale and Argyle lamproites were initially discovered by stream sampling; follow-up aeromagnetic work located additional vents in the Ellendale area. In contrast to kimberlites, airborne electromagnetic surveys do not appear to be particularly effective; no additional pipes were found in the Ellendale field by EM methods. As in kimberlites, diamonds are rare in lamproites, and thus concentrates are examined to identify associated minerals that belong to the 'lamproite indicator mineral suite'. This suite is similar to, yet distinct from, the kimberlite indicator mineral suite and has a finer grain size. Lamproite indicators include minerals from disaggregated mantle xenoliths (diamond, Cr-Al-Mg-Fe spinels, Cr-pyrope garnet, pyrope-almandine garnet, olivine, and rarely Cr-diopside and orthopyroxene) and lamproite liquidus phases (Cr-rich spinels, olivine, Ti-rich phlogopite and richterite, diopside, priderite, wadeite, perovskite). Megacrysts of Ti-Cr-pyrope and Mg-ilmenite (typical of kimberlites) are very rare to absent in lamproites.

As in the case of kimberlites, the specific compositions of certain indicator minerals can be used to assess diamond potential. Unfortunately, the well defined method of estimating diamond grade from extensive studies of low calcium, high chromium pyrope garnets in kimberlites (Gurney, 1989) is not reliably applicable to lamproites; studies based on kimberlite indicator minerals at both Argyle and Prairie Creek resulted in forecasts of low diamond potential due to the lack or near absence of such garnets.

ACKNOWLEDGMENTS

Critical reviews by Barbara Scott-Smith and Roger Mitchell on an early version of the manuscript greatly improved the scientific content. Comments by R.I. Thorpe, O.R. Eckstrand, D.F. Sangster, and A.N. LeCheminant on later versions were quite helpful. The final version was capably reviewed by A.J.A. Janse.

SELECTED BIBLIOGRAPHY

References with asterisks (*) are considered to be the best source of general information on this deposit type.

*Atkinson, W.J.
1989:   Diamond exploration philosophy, practice, and promises: a review; Proceedings of the Fourth International Kimberlite Conference, Volume 2. Kimberlites and Related Rocks: Their Mantle/Crust Setting, Diamonds and Diamond Exploration, (ed.) J. Ross; Geological Society of Australia, Special Publication 14, Blackwell Scientific Publications, Oxford, 1989, p. 1075-1107.

Atkinson, W.J., Hughes, F.E., and Smith, C.B.
1984:   A review of the kimberlitic rocks of Western Australia; Proceedings of the Third International Kimberlite Conference, Volume 1. Kimberlites I: Kimberlites and Related Rocks, (ed.) J. Kornprobst; Developments in Petrology 11A, Elsevier, Amsterdam, 1984, p. 195-224.

Deakin, A.S. and Boxer, G.L.
1989:   Argyle AK1 diamond size distribution; the use of fine diamonds to predict the occurrence of commercial size diamonds; Proceedings of the Fourth International Kimberlite Conference, Volume 2. Kimberlites and Related Rocks: Their Mantle/Crust Setting, Diamonds and Diamond Exploration, (ed.) J. Ross; Geological Society of Australia, Special Publication 14, Blackwell Scientific Publications, Oxford, 1989, p. 1117-1122.

Grice, J.D. and Boxer, G.L.
1990:   Diamonds from Kimberly, Western Australia; The Mineralogical Record, v. 21, p. 559-564.

*Gurney, J.J.
1989:   Diamonds; in Proceedings of the Fourth International Kimberlite Conference, Volume 2. Kimberlites and Related Rocks: Their Mantle/Crust Setting, Diamonds and Diamond Exploration, (ed.) J. Ross; Geological Society of Australia, Special Publication 14, Blackwell Scientific Publications, Oxford, 1989, p. 935-965.

*Helmstaedt, H.H.
1993:   Natural diamond occurrences and tectonic setting of "primary" diamond deposits; in Diamonds: Exploration, Sampling and Evaluation; (ed.) P.A. Sheahan and A. Chater, Prospectors and Developers Association of Canada, Short Course Proceedings, Toronto, March 27, 1993, p. 1-72.

*Jacques, A.L., Lewis, J.D., and Smith, C.B.
1986:   The kimberlites and lamproites of Western Australia; Geological Survey of Western Australia, Bulletin, no. 132, 268 p.

*Mitchell, R.H.
1989:   Aspects of the petrology of kimberlites and lamproites: some definitions and distinctions; in Kimberlites and Related Rocks, (ed.) J. Ross, A.L. Jacques, J. Ferguson, D.N. Green, S.Y. O'Reilly, R.V. Danchin, and A.J.A. Janse, Geological Society of Australia, Special Publication 14, Blackwell Scientific Publications, Oxford, 1989, p. 1-45.
*1991:   Kimberlites and lamproites: primary sources of diamond; Geoscience Canada, v. 18, no. 1, p. 1-16.

*Mitchell, R.H. and Bergman, S.C.
1991:   Petrology of Lamproites; Plenum Press, New York, 447 p.

Scott-Smith, B.H.
1984:   A new look at Prairie Creek, Arkansas; in Proceedings of the Third International Kimberlite Conference, Volume 1. Kimberlites I: Kimberlites and Related Rocks, (ed.) J. Kornprobst; Developments in Petrology 11A, Elsevier, Amsterdam, 1984, p. 255-284.
1989:   Lamproites and kimberlites in India; Neves Jahrbuch für Mineralogie, v. 161, p. 193-225.

Scott-Smith, B.H. and Skinner, E.M.W.
1984:   Diamondiferous lamproites; Journal of Geology, v. 92, p. 433-438.

Scott-Smith, B.H., Skinner, E.M.W., and Loney, P.E.
1989:   The Kapamba lamproites of the Luangwa valley, Eastern Zambia; Proceedings of the Fourth International Kimberlite Conference, Volume 1. Kimberlites and Related Rocks: Their Composition, Occurrence, Origin and Emplacement, (ed.) J. Ross, Geological Society of Australia, Special Publication 14, Blackwell Scientific Publications, Oxford, 1989, p. 189-205.

Waldman, M. and Meyer, H.O.A.
1992:   Great expectations in North America; Diamonds International, May/June, p. 42-48.

Author's address

B.A. Kjarsgaard
Geological Survey of Canada
601 Booth Street
Ottawa, Ontario
K1A 0E8

26. MAFIC INTRUSION-HOSTED TITANIUM-IRON

26.1. Anorthosite-hosted titanium-iron
26.2. Gabbro-anorthosite-hosted iron-titanium

26. MAFIC INTRUSION-HOSTED TITANIUM-IRON

G.A. Gross

INTRODUCTION

Large ilmenite and titaniferous magnetite deposits are hosted in massive and layered intrusive complexes–dominantly ilmenite in Proterozoic anorthosite (subtype 26.1); and titaniferous magnetite in gabbro and leucogabbro (formerly termed gabbro-anorthosite; e.g. Wager and Brown, 1968) (subtype 26.2). Deposits of both subtypes include irregular discordant masses in layered or massive intrusions, and concordant oxide-rich layers produced during fractional crystallization. The principal ore minerals are oxides of iron and titanium: ilmenite ($FeTiO_3$), hemo-ilmenite (a solid solution of $FeTiO_3$-Fe_2O_3), magnetite (Fe_3O_4), and titaniferous magnetite. The term "titaniferous magnetite" refers to granular aggregates and exsolution intergrowths consisting of ilmenite, magnetite, hematite, and titanomagnetite (a solid solution of Fe_3O_4-Fe_2TiO_4).

Gross, G.A.
1996: Mafic intrusion-hosted titanium-iron; *in* Geology of Canadian Mineral Deposit Types, (ed.) O.R. Eckstrand, W.D. Sinclair, and R.I. Thorpe; Geological Survey of Canada, Geology of Canada, no. 8, p. 573-582 (<u>also</u> Geological Society of America, The Geology of North America, v. P-1).

The iron- and titanium-rich deposits are classified as two subtypes on the basis of the principal ore minerals and the petrology of the host intrusions. The proportions of the principal ore minerals vary from ilmenite-dominant in anorthosite host rocks to titaniferous magnetite-dominant in gabbro and leucogabbro host rocks. The dominant mineralogy determines whether deposits are of interest as resources of titanium and iron or mainly of iron (Gross, 1965, 1967a).

Subtype 26.1 deposits consist mainly of ilmenite and hemo-ilmenite with minor titaniferous magnetite, and form massive irregular discordant intrusions or layered bodies hosted in massif anorthosite. Important examples are Lac Tio (Lac Allard), Degrosbois, Lac des Pins Rouges, St. Urbain, and Ivry (Morin anorthosite) in Quebec, Canada; Tellnes and Egersund in Norway; and Ilmen Mountains in the former U.S.S.R.

Subtype 26.2 deposits consist mainly of titaniferous magnetite and minor ilmenite and complex Fe-Ti oxide mineral assemblages hosted in layered and/or massive intrusions of leucogabbro, gabbro, norite, and rocks of intermediate composition. Examples include Magpie Mountain, St. Charles, Lac Doré complex, Kiglapait, Newboro Lake, and Lodestone Mountain in Canada; Smaalands-Taberg in Sweden; Bushveld Igneous Complex in South Africa; Kachkanar and Kusinskoye in the former U.S.S.R.; Tahawus and Iron Mountain in the United States.

Deposits of both subtypes provide resources of titanium, vanadium, and iron. Some deposits contain important quantities of apatite (Gross, 1967a; von Gruenewaldt, 1993).

IMPORTANCE

The Lac Tio deposit near Lac Allard, Quebec (Bergeron, 1972) is the only titanium-iron deposit (subtype 26.1) being mined in Canada at present. Mining was started in this area in 1951, and currently about 800 000 t of TiO_2 and 600 000 t of iron are produced annually from the processing of approximately 2 million tonnes of ilmenite ore (Harben and Bates, 1990). Production from Lac Tio accounts for nearly 25% of the world production of titanium oxide (Adams, 1994). High quality iron metal and TiO_2 are coproducts recovered from titaniferous slag produced from the ilmenite ore of subtype 26.1 deposits. Iron ore concentrates in which the titanium content has been reduced to 1% or less have been produced from subtype 26.2 deposits. Other titanium-iron deposits hosted in mafic rocks are mined in Norway (subtype 26.1) and in Russia (subtype 26.2). Deposits of subtype 26.2 have been minor sources of iron ore in Canada in the past, and have been a substantial source of iron ore in the former U.S.S.R.

Titanium dioxide powder is a nontoxic, white pigment used in paint, plastics, rubber, and paper. Titanium metal, resistant to corrosion and with a high strength-to-weight ratio, is used in the manufacture of aerospace and marine components. Significant changes are taking place throughout the world with respect to the kinds and sources of raw materials used for the production of titanium oxide and metal. For example, environmental regulations in many countries make production of hard-rock-derived ilmenite impossible because of the large acid requirements. About 95% of the total titanium mineral production, from both primary magmatic deposits and heavy mineral placer deposits, is used in the production of titanium dioxide. About 20% of the world production of titanium oxide is recovered in the processing of rutile (TiO_2) derived from beach sands (see Adams, 1994).

SIZE AND GRADE OF DEPOSITS

Ilmenite deposits of subtype 26.1 rarely host more than 300 Mt of ore; they contain from 10 to 45% TiO_2, from 32 to 45% Fe, and less than 0.2% V. The ratios of Fe:Ti are usually about 2, and the contents of Cu, Cr, Mn, and Ni commonly range from 0.05 to 0.2% for each element. Sulphide minerals and apatite are present in low and variable amounts. Ore treatment processes currently in use for the production of titanium dioxide require ore concentrates that contain at least 45% TiO_2 (pure ilmenite contains 52.7% TiO_2).

The two largest iron-titanium deposits of subtype 26.1 which are hosted in anorthositic intrusive rocks are the Lac Tio deposit in Canada and the Tellnes deposit in

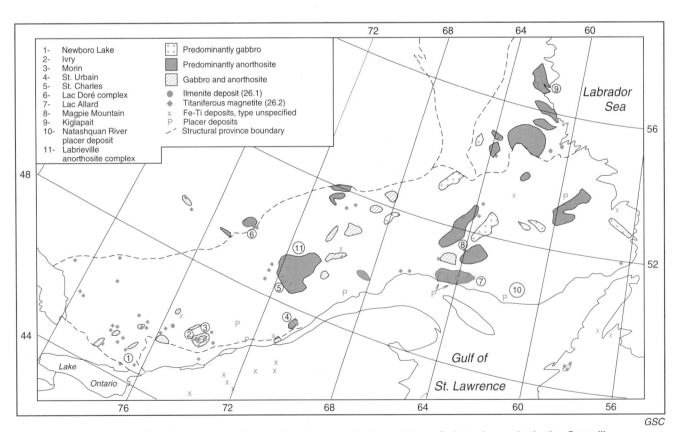

Figure 26-1. Titaniferous magnetite and ilmenite deposits hosted in mafic intrusive rocks in the Grenville Province and Eastern Canada. Important deposits mentioned in text are labelled.

southwestern Norway. Lac Tio is a flat-lying irregular tabular intrusive mass, 1100 m long and 1000 m wide, which is estimated to contain more than 125 million tonnes of ore averaging 32% TiO_2 as ilmenite and 36% FeO. The high-grade ore contains as much as 75% ilmenite and 20% hematite. The Tellnes deposit is about 2800 m long, 400 m wide, and at least 350 m deep. Estimated reserves are 300 million tonnes of ore averaging 18% TiO_2 as ilmenite, 2% magnetite, and 0.25% sulphides.

Titaniferous magnetite deposits of subtype 26.2 range in size from one million tonnes to more than 1000 million tonnes. They usually contain from 20 to 45% iron and from 2 to 20% TiO_2. Ratios of Fe:Ti range from 40:1 to 2:1 and are commonly about 5:1. The average content of V is about 0.25%, Cr is present in trace amounts, and the content of P_2O_5 is variable, but usually less than 7.1%.

GEOLOGICAL FEATURES

Despite the geological and economic importance of iron and titanium deposits hosted in mafic intrusions, few comprehensive reviews are available. Gross (1967a) and Rose (1969) provided geological descriptions and analytical data for deposits being mined and many of the deposits of possible economic importance known in Canada at the time. More recent reviews of the characteristics of titanium ores can be found in Korneliussen et al. (1985), and of anorthosite-hosted deposits in Ashwal (1993).

Geological setting

Ilmenite and titaniferous magnetite deposits associated with anorthosite and gabbro are widely distributed in the Grenville Province and in many other tectonic belts of North America and the world. Both types of intrusive complexes are typically associated with granitoid gneisses, granulites, schists, amphibolites, quartzites, and skarn rocks of deep crustal settings but some occur in greenschist-facies terranes.

Deposits of subtype 26.1 are hosted worldwide in anorthosite; intrusions of the Grenville Province are typical (Fig. 26-1). Most of the deposits form discordant dykes, sills, and stock-like masses in the host anorthositic rocks; others are layered concentrations of Fe-Ti oxides within anorthosite or gabbro, concordant to layering in the host and to the internal fabric of late stage intrusions.

Subtype 26.2 deposits are hosted worldwide in mafic layered and massive intrusions, and are also widely distributed in the Grenville Province (Fig. 26-1). The layered deposits generally form concordant, laterally continuous magnetite-rich layers measuring centimetres to metres thick. Deposits in massive intrusions usually consist of disseminated titaniferous magnetite. Deposits of subtype 26.2 also include massive discordant stock-like bodies of Fe-Ti oxide in layered deposits, as at Newboro Lake in Canada. The host intrusive complexes are typically differentiated and include gabbro, leucogabbro, diorite, diabase, gabbro-diorite, and quartz monzonite.

Concentrations of metallic oxide minerals in both subtypes 26.1 and 26.2 are conspicuously developed in four styles:

1. disseminated syngenetic metal oxides in the host rocks;

2. irregular to conformable autointrusions which have sharp to indistinct or gradational borders with earlier phases of the host anorthosite and gabbro, and were emplaced during the lithification and cooling of the host intrusive rocks;

3. late stage dykes and intrusions transecting the lithified host anorthosite and gabbro complexes;

4. in the skarn rock and alteration zones at the contact of the host intrusions and wall rocks.

Ages of host rocks and ore

Anorthositic host rocks to deposits of subtype 26.1 that have been dated in Canada are Proterozoic in age. These anorthosite complexes range in age from 1.65 Ga (Mealy Mountains; Emslie and Hunt, 1990) to 1.01 Ga (Labrieville; Owens et al., 1994). Major anorthosite-hosted deposits such as Lac Tio and lesser deposits such as St. Urbain, and Ivry and Degrosbois in the Morin anorthosite complexes occur within a much more restricted period with ages ranging from the 1.16 Ga Morin anorthosite (Doig, 1991) through the 1.06 Ga Havre-Saint-Pierre intrusion (van Breemen and Higgins, 1993).

In most cases the precise timing of the Fe-Ti oxide mineralization relative to the crystallization ages of the host anorthosite and gabbro rocks is not known specifically because suitable data are not available. The crystallization age of the Tellnes deposit (southern Norway), 920 ± 2 Ma, is measurably younger than the crystallization age of the host anorthosite, 930 ± 4 Ma (Duchesne et al., 1993), whereas crystallization ages for the Sybille deposit and host anorthosite (Wyoming) are indistinguishable within error at 1434 ± 1 Ma by the uranium-lead method (Scoates and Chamberlain, 1993).

Host rocks to deposits of subtype 26.2 in Canada do not appear to be restricted in time. Ages of crystallization range from 2727 ± 1.3 Ma for the P3 ferric pyroxenite member of the Lac Doré complex (Mortensen, 1993; U-Pb zircon) through the 1305 Ma Kiglapait intrusion (DePaolo, 1985), which contains massive titanomagnetite layers, to the ~540 Ma Sept-Îles intrusion, which contains local concentrations of titaniferous magnetite and ilmenite (Higgins and Doig, 1981).

Form of deposits and relation to host rocks

Generalizations on the form and relationships of these deposits to host rocks are tenuous because of the many variations from deposit to deposit in the host rocks, mineralogy, and geological settings. Nevertheless the two groupings used herein may be of use for discussion, research, and exploration purposes. Both types of Fe-Ti oxide deposits occur in two general forms: massive lenses, dykes, sills, and irregular intrusions; and stratiform, layered, concordant, or irregular bodies. The Fe-Ti oxide minerals may be disseminated and interstitial to the silicate minerals or occur as massive aggregates separated from them. Deposits of subtype 26.1 of economic interest for the recovery of TiO_2 and iron metal are massive irregular intrusions. Deposits of subtype 26.2 are predominantly stratiform and layered. In some cases (e.g. Tahawus and Iron Mountain) attributes of both forms are combined in a single intrusive complex.

Figure 26-2. A) Irregular dyke (width approximately 5-10 m) of ilmenite (black) in anorthosite, northeast side of Lac Tio deposit, Lac Allard, Quebec. GSC 112341-A; **B)** Inclusions (width approximately 1-5 m) of anorthosite (light grey) in massive ilmenite (black) in Lac Tio deposit, Lac Allard, Quebec. GSC 152327

Ilmenite deposits of subtype 26.1, are typically massive discordant intrusive bodies in anorthositic host rocks (Fig. 26-2A), but some also occur as conformable layers within late stage gabbroic, troctolitic, and dioritic intrusions in anorthosite. Some of the Fe-Ti oxide masses, especially along their borders with the host rocks, have local fragmented or brecciated structures, show evidence of plucking and stoping of the enclosing rocks, and contain abundant xenoliths of anorthosite and xenocrysts of plagioclase derived from anorthosite (Fig. 26-2B). Both massive and disseminated ores are found within a single intrusion (Fig. 26-3A, 26-3B). The massive discordant intrusions of Fe-Ti oxide range in shape from sinuous dyke-like forms to irregular equidimensional masses.

Layered stratiform deposits of subtype 26.2 hosted in gabbro and leucogabbro usually contain layers of disseminated titaniferous magnetite which alternate with layers of feldspar and mafic silicate minerals (Fig. 26-4). Individual layers range in thickness from centimetres to metres. Lateral continuity of oxide-rich layers in large intrusions may be in the order of several thousand metres.

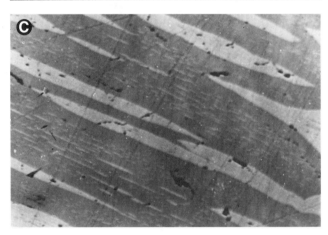

Figure 26-3. A) Massive and disseminated ilmenite (black) in anorthosite. Width of field of view approximately 1 m. GSC 112341-U; **B)** Coarse grained ilmenite, Lac Tio deposit, Lac Allard, Quebec. Width of field of view approximately 1 m. GSC 152330; **C)** Photomicrograph showing hematite (light grey) exsolved from ilmenite (dark grey) Lac Tio deposit, Lac Allard, Quebec. (Width of field of view ~2 mm). GSC 1995-020

Figure 26-4. Layered anorthositic gabbro with disseminated magnetite near Newboro Lake, Ontario. GSC 152354

Table 26-1. Partial compositions of: I) typical ilmenite-hematite ore at Lac Allard (subtype 26.1), II) average of three titaniferous magnetite deposits at Magpie Mountain (subtype 26.2), and III) Massive titaniferous magnetite at Chaffey mine (subtype 26.2) (in weight per cent).

	I) Lac Allard	II) Magpie Mountain	III) Chaffey mine
Fe	40.2	43.69	50.23
TiO_2	35.0	10.9	9.8
V_2O_5	0.3	0.17	NA
MgO	3.2	5.78	NA
FeO	28.3	NA	NA
Fe_2O_3	26.2	NA	NA
P_2O_5	0.015	0.085	NA
Al_2O_3	3.1	11.57	5.65
SiO_2	3.7	5.9	7.1

After: I) Hatch and Cuke, 1956;
II) Rose, 1969;
III) Unpublished analyzes, GSC
NA = not analysed.

Ore mineralogy, composition, and texture

The proportions of the common ore minerals, ilmenite, hemo-ilmenite, titaniferous magnetite, titanomagnetite, and magnetite vary greatly from one deposit or deposit type to another. The complex exsolution textures and mineral relationships that indicate mineral paragenesis and sequence of crystallization vary greatly and appear to be distinctive for individual deposits.

The principal ore minerals in deposits of subtype 26.1 are ilmenite, hemo-ilmenite and their exsolution intergrowths, and titanomagnetite. They are associated with plagioclase, pyroxene, olivine, garnet, biotite, apatite, ulvöspinel, quartz, hornblende, rutile, and pyrrhotite which are present in varying proportions. Hemo-ilmenite, the principal ore mineral at the Lac Tio and Tellnes deposits (subtype 26.1) hosted in anorthosites, is typically equigranular with coarse exsolution lamellae of hematite that constitute as much as 30 mole per cent of the grains (Fig. 26-3C). A second set of very fine exsolution lamellae of ilmenite is commonly developed within the broad hematite lamellae. The forms of earlier titanomagnetite grains can be recognized where the diagnostic trellis lamellae of ilmenite are still preserved along the {111} planes of the host magnetite.

Some parts of the Lac Allard ilmenite deposits contain 8 to 10% fluorapatite (Gross, 1967a). Ilmenite-apatite occurrences (nelsonites) have been reported in many anorthosites (Kolker, 1982). Some of the anorthosite-hosted Fe-Ti oxide deposits contain minor rutile, sapphirine, corundum, sillimanite, and graphite (Ashwal, 1993).

The principal ore minerals in deposits of subtype 26.2 are titanomagnetite, and other varieties of titaniferous magnetite and ilmenite which occur as discrete grains and as exsolution intergrowths in various proportions in magnetite. They are associated with plagioclase (commonly labradorite), olivine, pyroxene, and small amounts of apatite, titanite (sphene), rutile, spinel, biotite, pyrite, chalcopyrite, and pyrrhotite.

Mineralogy and texture are important factors to be considered in assessing potential resources that might be recovered from Fe-Ti oxide deposits. Massive ilmenite deposits of subtype 26.1, mined for the production of titanium oxide and iron metal, are usually coarse (<1 cm) equigranular aggregates of ilmenite with minor titaniferous magnetite. The large titaniferous magnetite deposits of greatest interest as potential sources of iron ore consist of titaniferous magnetite, magnetite, and minor ilmenite in coarse, discrete grains that have a minimum of exsolution textures and intergrowths of Fe-Ti minerals. Material of this kind is amenable to processing and can provide concentrates of relatively pure magnetite that contain less than one per cent titanium.

Partial data on the composition of the Lac Allard (subtype 26.1), Magpie Mountain (subtype 26.2), and Chaffey mine (subtype 26.2) Fe-Ti oxide deposits are shown in Table 26-1, for the purpose of illustrating the concentrations of the main ore elements for mineral processing.

Alteration

Some aspects of mineral alteration are considered in the section on genetic models.

CANADIAN EXAMPLES

Subtype 26.1

Lac Allard deposits

The ilmenite deposits of the Lac Allard area, located 40 km north of Havre-Saint-Pierre on the north shore of the Gulf of St. Lawrence, are described briefly to illustrate some of the typical geological features of subtype 26.1 deposits. Six deposits ranging in size from one million to several hundred million tonnes are located near the east border of a large oval-shaped anorthosite intrusion that extends west from the Romaine River for more than about 150 km (Fig. 26-1). Uranium-lead zircon dating of the southwest lobe of the Havre-Saint-Pierre anorthosite intrusion indicates that it is 1062 ± 4 Ma old and the parallelism of magmatic and solid-state foliation with the adjacent Abbe-Huard lineament suggest that anorthosite parental magmas rose along this shear zone. The anorthosite complex of this area formed as part of a widespread magmatic event between 1.09-1.05 Ga that included extensive intrusion of anorthosite in the Grenville Province. This magmatic event was a manifestation of deeper mantle processes that probably were not coupled to, but coincided with, the tectonic regime of the overlying crust and the late stage convergent tectonics in the southwestern Grenville Province (van Breemen and Higgins, 1993).

The host intrusion at Lac Allard is composed of almost pure plagioclase anorthosite, associated with leucogabbro, norite, ilmenite-rich anorthosite, and gabbro. The most abundant variety of anorthosite has less than 5% mafic minerals. It is medium- to coarse-grained, porphyritic in places, and varies from light to dark grey pinkish to greenish brown. Most of it is massive, but some parts are distinctly layered and foliated. Protoclastic textures are present throughout most of the intrusion. The composition of the plagioclase ranges from An_{40} to An_{52}. The mafic minerals consist mainly of hypersthene with minor clinopyroxene, ilmenite, amphibole, and biotite.

Noritic and other mafic phases commonly are layered and foliated. Some form dykes and sills that have intruded the anorthosite. Gabbro and diorite dykes cut gneiss and granitic rocks adjacent to anorthosite. The border zones between anorthosite and the host gneissic rocks consist of a group of hypersthene-bearing hybrid rocks with altered phases of anorthosite, and dykes of syenite, granite, and pegmatite that cut the anorthosite and gabbro. Some of the pegmatite dykes are zoned and cut granite, massive ilmenite, and anorthosite; these may be the youngest intrusions of the region.

The ilmenite-magnetite norite phases of the Lac Allard anorthosite complex were intruded during late stages of consolidation of the anorthosite and consist of elongated lenticular sheets that are several kilometres long, a kilometre or less in width, and dip steeply east. They are composed of plagioclase and 50 to 60% mafic minerals including hypersthene, apatite, ilmenite, and magnetite, and 30% or less of iron and titanium oxide minerals that are interstitial to the other minerals. Most of the norite is banded or gneissic and the oxide minerals are most abundant in the lower parts of the bands. Grain size varies from 1 to 20 mm. Ilmenite and magnetite occur as discrete grains and ilmenite is usually the most abundant. The ilmenite grains contain exsolved lamellae of hematite but magnetite grains appear to be homogeneous and contain from 0.75 to 3.3% TiO_2 in solid solution. The content of fluorapatite associated with the oxides ranges from 8 to 10%.

The Lac Allard ilmenite deposits, clustered in the eastern part of the anorthosite intrusive complex, are massive, medium- to coarse-grained dykes, sills, lenses, and irregular bodies. The Lac Tio ilmenite deposit, a good example of subtype 26.1 deposits, is the largest and best known in the area, and is an irregular tabular mass, about 1100 m long and 1000 m wide, with a maximum thickness of 100 m and surface relief of about 130 m. It forms a basin-like or open synformal body of ilmenite in anorthosite. Typical ore structures, textures, xenoliths of anorthosite, and typical dykes and intrusions of ilmenite and hematite in anorthosite are illustrated in Figures 26-2 and 26-3.

Lac Tio orebody is composed of a complex interconnected maze of sills and dykes of hemo-ilmenite. Inclusions and xenoliths of anorthosite range from single feldspar crystals to angular and irregular masses several metres in size. Vertical boundaries of the ore mass are usually sharp and well defined, but horizontal boundaries are irregular and indistinct. Massive ore grades to crudely banded disseminated ore and anorthosite along many margins of the ilmenite mass. Small stringers and dykes of massive ilmenite extend from the main ilmenite mass through the marginal zones into the anorthosite.

The massive ilmenite ore at Lac Tio consists of 88 to 97% combined Fe and Ti oxides (Fig. 26-3A). It is equigranular, coarse grained, dull black to brownish black, and has a glistening jet black surface where freshly broken. Ilmenite grain size ranges overall from 5 to 15 mm, but crystals are fairly uniform in size in local patches several metres in diameter (Fig. 26-3B). The ilmenite that is located close to fault zones is, in most cases, much finer grained, dense, and extremely hard. Ilmenite grains are uniform in composition; they contain microscopic blades and lenses of exsolved hematite which constitute about 15% of the ore (Fig. 26-3C). A small amount of magnetite is present in the ore. Pyrite, pyrrhotite, and chalcopyrite form veins or interstitial fillings around iron-titanium oxide mineral grains.

The ratios of iron to titanium throughout the orebody are remarkably uniform. The ore contains an average of 32% TiO_2 and 36% iron. The grade of the ore can be estimated to within 2% from its specific gravity, which varies from 4.46 to 4.9. The composition of typical ilmenite-hematite ore reported by Hatch and Cuke (1956) is given in Table 26-1.

Subtype 26.2

Magpie Mountain deposit

One of the largest massive titaniferous magnetite deposits known is located 3 km west of the St. Jean River and 200 km northeast of Sept-Îles, Quebec. The steeply dipping tabular masses of magnetite form the crests of ridges that rise 180 m to 300 m above the surrounding terrane. The magnetite bodies are surrounded by anorthosite zones up to 30 m thick and the composite mass has been intruded into granite and granite gneiss.

Four separate deposits are intersected by two major steeply dipping reverse faults (see also description in Rose, 1969, p. 103, 104). The magnetite is medium- to coarse-grained and contains ilmenite in very fine, exsolved blades. The tabular masses consist mainly of magnetite and 15% plagioclase and pyroxene. The largest of the four deposits is 3500 m long, 300 m wide, and is reported to contain 737 Mt of material that contains 45.7% iron, 10.8% TiO_2, and 7.45% silica. Resources of potential ore suitable for open-pit mining in the four deposits are estimated to be greater than 1500 Mt. The average composition for three deposits of titaniferous magnetite reported by Rose (1969) is given in Table 26-1.

Newboro Lake deposit

This deposit is typical of many banded and layered leucogabbro intrusions in the Grenville Province that contain sufficient magnetite and titaniferous magnetite to be of interest as possible sources of iron ore, and that have granular textures which make them amenable to concentration of the iron and separation of the titanium. The leucogabbro that hosts the Newboro Lake deposit is part of a differentiated intrusive complex of layered gabbro, leucogabbro, monzonite, and migmatite, and layered titaniferous magnetite-bearing gabbro. The Newboro Lake iron deposit as now defined consists of magnetite-rich zones in dark grey to green, ophitic-textured, layered leucogabbro in which distinct banding is developed by the separation of variable proportions of andesine and labradorite feldspar, augite, ferroaugite, hornblende, biotite, titaniferous magnetite, ilmenite, and accessory apatite (Fig. 26-4).

The massive titaniferous magnetite intrusions in which the previous Chaffey and Matthews mines were located are enclosed in a layered magnetite-rich zone in the leucogabbro host rock which is about 100 m wide and 1000 m long and forms the Newboro Lake deposit. It contains at least 50 million tonnes of potential ore within the possible limits of an open pit mine. It has an average grade of 26.7% iron and about 6% titanium dioxide. A magnetite concentrate containing 51% iron, was produced by grinding the crude ore to 28 mesh, and 80% of the iron was recovered with a crude ore:concentrate ratio of 2.4:1.

Some layers of magnetite are essentially free of inclusions, but in others ilmenite has exsolved along the octahedral planes of the magnetite and Mg-Fe spinel has exsolved along the cubic planes. Ilmenite with exsolved lamellae of hematite (Fig. 26-3C) forms discrete grains intermixed with magnetite. Titanite (sphene) associated with calcite is present in some parts of the magnetite-rich zones, and in fractures in the gabbro. Small amounts of pyrite are disseminated in some of the magnetite-rich layers. Analyses of higher grade material from the Chaffey mine are shown in Table 26-1.

DEFINITIVE CHARACTERISTICS OF ORE

1. Massive and layered ilmenite and hemo-ilmenite deposits (subtype 26.1) are hosted in anorthosites. Layered and massive concentrations of titanomagnetite, titaniferous magnetite, magnetite, and ilmenite (subtype 26.2) are hosted in differentiated mafic layered and massive intrusions.

2. Subtype 26.1 deposits are massive irregular to tabular bodies and disseminated masses of coarse grained ilmenite containing blades of exsolved hematite, pure ilmenite, and titaniferous magnetite hosted in massive or layered anorthosite and leucogabbro intrusive complexes, stocks, and sills.

3. Typical subtype 26.1 deposits contain from 20 to 40% titanium and 25 to 45% iron with Fe/Ti ratios of about 2:1, and 100 million tonnes or less mineable ore.

4. Subtype 26.2 deposits consist of layered disseminated concentrations and massive irregular to tabular intrusions of titaniferous magnetite, titanomagnetite, magnetite, and ilmenite. These minerals are distributed as discrete grains, and as granular and exsolution intergrowths. The host silicate phases include gabbro, gabbroic anorthosite, and other differentiated intrusive complexes ranging in composition from gabbro, through norite, quartz monzonite, to syenite.

5. The iron content in subtype 26.2 deposits ranges from 20 to 45%; TiO_2 from 2 to 20%; Fe:Ti ratios vary from 40:1 to 2:1 and are commonly about 5:1; the content of P_2O_5 varies to a maximum of about 8% and the content of V, Cu, Ni, Cr, and Mn may vary greatly, but the average for each element is about 0.25% or less.

6. As a group, subtype 26.2 deposits vary greatly in composition, mineralogy, and physical characteristics, but individual deposits are fairly uniform.

7. The mafic-hosted titanium-iron deposits of both subtypes vary greatly in character and composition depending on the kinds of associated host intrusions, the stage of differentiation and oxygen potential in the magma from which they were derived, tectonic setting, and mobilization of elements during metamorphism (cf. Yoder, 1968).

8. They are important as sources of titanium oxide and high quality iron metal that are recovered as coproducts, and as resources of iron ore concentrate in which the titanium content can be reduced to one per cent or less.

RELATED DEPOSIT TYPES

The iron-titanium oxide deposits described here as subtypes 26.1 and 26.2 are the largest and most common type of metallic oxide occurrences hosted in mafic intrusive rocks. They occur in layered intrusions (e.g. Bushveld Complex; Cameron, 1970; Naldrett and Cabri, 1976) that may also host deposits which contain chromium (see Type 28, "Mafic/ultramafic-hosted chromite") and platinum group elements (see subtype 27.2, "Magmatic platinum group elements") (Wardle, 1987). Significant concentrations of vanadium (Rose, 1973), chromium, apatite, nickel and copper sulphides, and platinum group elements occur in or are associated with the Fe-Ti oxide deposits and their host rocks.

Concentrations of magnetite in zoned ultramafic complexes of southeastern Alaska and in the gabbro-ultramafic belt of the Ural Mountains of Russia have been compared by Taylor and Noble (1969). The large Kachkanar and Kusinskoye Fe-Ti deposits on the east slope of the Urals, classified in this paper with other deposits hosted in gabbro and gabbro-anorthosite rocks as subtype 26.2 deposits, were reported to be hosted mainly in the marginal zones of a complex of gabbroic intrusions (see Gross, 1967b; Sokolov, 1970) which is intruded by numerous ultramafic bodies.

The anorthosites themselves are potential sources of alumina. Many of the anorthosite and gabbro intrusions provide decorative and building stone products of high quality and beauty.

Large placer deposits of economic significance that contain ilmenite, magnetite, rutile, and zircon are developed in drainage systems and on beaches in terranes that have prominent anorthosite and gabbro intrusions, with iron and titanium oxide deposits. The concentrations of black sand in the delta area of the Natashquan River on the north shore of the St. Lawrence River (Fig. 26-1) are typical examples of heavy mineral placer deposits derived from a gabbro-anorthosite terrane.

EXPLORATION GUIDELINES

1. Both layered and massive ilmenite deposits, subtype 26.1, are commonly hosted in anorthosite*. Titaniferous magnetite deposits, subtype 26.2, are commonly hosted in gabbroic intrusive complexes.

2. Massive ilmenite and hemo-ilmenite deposits, subtype 26.1, commonly have a distinctive negative magnetic anomaly, or irregular patterns of negative and positive anomalies that mark erratic polarization in segments of the deposits.

3. Intrusive rocks bearing significant concentrations of Fe-Ti oxide are characterized by high positive magnetic anomalies that show broad, smooth profiles or patterns.

4. Iron and titanium oxide deposits and the mafic intrusive rocks which host them have higher gravity anomalies than the surrounding granitic and gneissic rocks.

5. Iron-titanium oxide minerals in stream sediments can be used as effective markers or tracers in exploration for ilmenite and magnetite deposits.

6. Ilmenite deposits of subtype 26.1 appear to be best developed in anorthosite intrusions located along deep-seated fault zones and fracture systems as developed at the margins of major tectonic provinces and belts. In Canada, for example, the best deposits are associated with intrusive complexes along the St. Lawrence River lineament near the southeast margin of the Grenville Province (Gross, 1977).

7. The host intrusive complexes commonly consist of a number of differentiated phases of mafic rock that range in composition from anorthosite, through gabbro and norite to diorite and syenite.*

8. Ilmenite deposits (subtype 26.1) are associated with anorthosite intrusions in which the Fe:Ti ratios in the disseminated metal oxides are less than 3, usually about 2.

9. Titaniferous magnetite deposits (subtype 26.2) are commonly associated with the magnesian, labradorite phases of mafic intrusions, or igneous phases related to them. The Fe:Ti ratios in their metallic oxide minerals vary from 40:1 to 2:1 and are commonly about 5:1.

10. Titaniferous magnetite deposits of subtype 26.2 are most commonly developed in:

    a) the gabbroic phases near the margins of gabbro intrusive stocks;

    b) in the upper stratigraphic parts of mafic layered intrusions; and

    c) in the gabbro-diorite stocks, dykes, and sills which are associated with major gabbro intrusions.

* *Editorial note*, from T. Birkett, pers. comm., 1994: Massif anorthosites, at least those of the Grenville Province (see also Hill, 1988 for Nain area), are enveloped in jotunite. As well, jotunites form septa within many anorthosites, possibly defining "cells" within the larger complexes. Anorthosite-hosted Fe-Ti ores are invariably associated with jotunite. In the past, jotunites have been called gabbro-anorthosite, ferrodiorite, monzodiorite, gabbroic anorthosites, and oxide-apatite-gabbronorites. Jotunites are orthopyronene-bearing rocks of mainly monzodioritic composition, commonly containing appreciable Fe-Ti oxide and apatite. Jotunites are distinguished from minor products of crystallization of anorthosite by their higher concentrations of Ti, Fe, and especially P.

GENETIC MODELS FOR MAFIC INTRUSION-HOSTED TITANIUM-IRON DEPOSITS

J.S. Scoates

The titanium-iron deposits that are associated with Proterozoic anorthosites and layered mafic intrusions are clearly late products of the crystallization history of individual intrusions. Brecciation of ore-hosting anorthosite and truncation of structural elements in anorthosite are clear evidence for late intrusion of the ore-forming magmas in many subtype 26.1 deposits. Conformable layers in small intrusions in anorthosite and in large, mafic layered intrusions throughout the world indicate an origin by crystal settling and accumulation on the floors of magma chambers for subtype 26.2 deposits and parts of subtype 26.1 deposits.

Both subtypes of deposits require extensive periods of prior plagioclase crystallization to concentrate Fe and Ti in residual magmas, and variations in the oxidation state of the magmas (monitored by the intensive parameter - oxygen fugacity) to promote the formation of the titanium-iron deposits. Hemo-ilmenite deposits (subtype 26.1) require relatively more oxidizing conditions of formation compared to the more reduced titanomagnetite deposits (subtype 26.2).

Evidence is lacking for the presence of hydrous fluids during formation of the Ti-Fe deposits, although CO_2-dominant fluids were likely present. The preserved primary mineral assemblages are typified by anhydrous mineralogies. Hydrous minerals are always late and volumetrically minor ($<<1\%$) or definitely related to crosscutting monzonitic or granitic intrusions. The presence of grain-boundary graphite and CO_2-rich inclusions in apatite from anorthosites indicates that the very small amounts of fluids associated with anorthosites were probably CO_2-dominated.

The genesis of the discordant, massive Fe-Ti oxide deposits associated with Proterozoic anorthosites is the least understood of the deposit types. Two end-member genetic models are currently under consideration: (1) remobilization of Fe-Ti oxide-rich cumulates, and (2) formation of an Fe-Ti-oxide-rich, silica-poor immiscible melt. The remobilization mechanism involves the intrusion of dense, solidified Fe-Ti-oxide-rich cumulates into cracks or fractures within the host anorthosite (Bateman, 1951; Hammond, 1952; Ashwal, 1982, 1993). A similar remobilization mechanism, but also involving magma mixing, has been proposed for the Tellnes deposit of Norway (Wilmart et al., 1989). In this scenario, a noritic magma crystallized Fe-Ti oxides which concentrated at the bottom of the chamber, and plagioclase which concentrated at the top of the chamber. Before complete solidification, the chamber was tapped and Fe-Ti oxide cumulates were injected into a dyke that already contained a fractionated monzonitic melt.

A liquid immiscibility origin for the massive ores has been proposed for most of the large deposits in Quebec for many years (Hargraves, 1962; Anderson, 1966; Lister, 1966; Philpotts, 1966, 1967). Liquid immiscibility, the separation of a single magma into two distinctive liquid phases, is most likely to occur in systems with bulk compositions high in total iron, TiO_2, P_2O_5, and high ratios of Fe^{3+}/Fe^{2+} (oxidized conditions) (Naslund, 1983). Rocks of these compositions are found throughout Proterozoic anorthosite complexes and are referred to as ferrodiorites, monzonorites, jotunites, and oxide-apatite-rich gabbronorites. These Fe-Ti-P-enriched rocks are considered to have formed as residual liquids following extensive crystallization of plagioclase to produce the associated anorthosites.

Experimental support for the liquid immiscibility mechanism is derived from the observation in the system magnetite-fluorapatite that an immiscible eutectic melt with a composition of approximately two-thirds by volume magnetite and one-third apatite can separate from a silicate melt (Philpotts, 1967), although the temperatures of the experiments were geologically unreasonable (1420°C). Liquid immiscibility may be appropriate for the production of small apatite-rich oxide deposits, referred to as nelsonites (Kolker, 1982), but the majority of the major deposits are apatite-poor. If liquid immiscibility is to remain a reasonable option in the formation of titanium-iron deposits, then an additional suitable flux must be found associated with the massive ores, because the melting temperatures of pure Fe-Ti oxides are unrealistically high.

Titanium-free oxide liquids do exist, as exemplified by the magnetite lava flows of El Laco, Chile (Park, 1961; Henriquez and Martin, 1978), and by the experiments of Weidner (1982), which show that graphite and C-O fluids flux oxide liquids to temperatures below 1000°C. However, there is limited evidence for the existence of Fe-Ti oxide melts. Recent experimental work shows that graphite does not stabilize Ti in oxide liquids (Lindsley and Philipp, 1993), and thus the mechanism required for the generation of apatite-poor, Ti-bearing immiscible melts remains elusive.

The origin of conformable Fe-Ti oxide-rich layers in layered intrusions is more straightforward than that for the discordant massive intrusions. The conformable layers represent the overproduction of Fe-Ti oxides in a progressively crystallizing magma, mainly in response to local variations in oxygen fugacity (Morse, 1980). Prior to the cumulus arrival of magnetite and/or ilmenite in a magma, protracted crystallization of plagioclase will enrich the residual magma in Fe, Ti, (and V), and increase the density of this residual melt. The prominent titanomagnetite layers in the Kiglapait intrusion of Labrador (Morse, 1969, 1980) and the Bushveld Igneous Complex of South Africa (Willemse, 1970; Reynolds, 1985) occur relatively high in the stratigraphic sections of these intrusions, and require crystallization from magnetite-supersaturated liquids.

The compositions of Fe-Ti oxides in both hemo-ilmenite-rich and titanomagnetite-rich ores can undergo substantial modification during cooling by both intra- and intercrystalline reaction and exchange. During slow cooling, the titanium component in titanomagnetite may be exsolved by oxidation to form either discrete lamellae of ilmenite in magnetite, or granular exsolutions of ilmenite around magnetite grains, a process called oxy-exsolution (Buddington and Lindsley, 1964):

$$6Fe_2TiO_4 + O_2 = 6FeTiO_3 + 2Fe_3O_4$$
in magnetite   ilmenite   magnetite

This reaction may be facilitated by the presence of a CO_2-rich fluid, and can occur to very low temperatures (400-500°C). As a result, titanomagnetite grains can purge themselves entirely of the original titanium component, and the resultant ore mineralogy and texture is one of interlocking discrete grains of magnetite and ilmenite. In addition, at relatively high temperatures, exchange of

titanium and iron between individual grains of magnetite and ilmenite can occur according to the following equilibrium reaction, which proceeds to the right with decreasing temperature:

$$Fe_2TiO_4 + Fe_2O_3 = FeTiO_3 + Fe_3O_4$$
in magnetite   in ilmenite   ilmenite   magnetite

This produces magnetite and ilmenite grains that will approach their end-member compositions as cooling proceeds.

Oxidation of ilmenite-rich deposits can result in the alteration of ilmenite to rutile. Associated alteration of silicate and Fe-Ti oxide minerals always postdates the formation of the deposits.

ACKNOWLEDGMENTS

Special thanks are expressed to R.I. Thorpe, B.R. Frost, T.C. Birkett, and C.W. Jefferson for thorough reviews and suggestions. G.A. Gross wishes to acknowledge J. Scoates for additional information on the dating of host rocks and review of some of the recent literature. Word processing assistance was provided by C.M. Plant and L.C. O'Neill. Figure 26-1 was prepared by Jeff Werner.

SELECTED BIBLIOGRAPHY

References with asterisks (*) are considered to be the best source of general information on this deposit type.

*Adams, R.
1994:   The world market for TiO_2 feedstocks; Minerals Industry International, Bulletin of the Institution of Mining and Metallurgy, no. 1016, p. 9-16.

Allard, G.D.
1976:   Doré Lake Complex and its importance to Chibougamau geology and metallogy; Quebec ministère des richesses naturelles, Direction Générale des Mines, DP-368, 446 p., map scale 1:100 000 .

Anderson, A.T., Jr.
1966:   Mineralogy of the Labrieville anorthosite, Quebec; American Mineralogist, v. 51, p. 1671-1711.

Ashwal, L.D.
1982:   Mineralogy of mafic and Fe-Ti oxide-rich differentiates of the Marcy anorthosite massif, Adirondacks, New York; American Mineralogist, v. 67, p. 14-27.
*1993: Anorthosites; Springer-Verlag, Berlin, 422 p.

*Bateman, A.M.
1951:   The formation of late magmatic oxide ores; Economic Geology, v. 46, p. 404-426.

*Bergeron, M.
1972:   Quebec Iron and Titanium Corporation ore deposit at Lac Tio, Quebec; Guidebook, Excursion B-09, Twenty-fourth Session, International Geological Congress, Canada, 8 p.

Buddington, A.F. and Lindsley, D.H.
1964:   Iron-titanium oxide minerals and synthetic equivalents; Journal of Petrology, v. 5, p. 310-357.

*Cameron, E.N.
1970:   Compositions of certain coexisting phases in the eastern part of the Bushveld Complex; The Geological Society of South Africa, Special Publication 1, p. 40-58.

DePaolo, D.J.
1985:   Isotopic studies of processes in mafic magma chambers: I. The Kiglapait intrusion, Labrador; Journal of Petrology, v. 26, p. 925-951.

Doig, R.
1991:   U-Pb zircon dates of Morin anorthosite suite rocks, Grenville Province, Quebec; Journal of Geology, v. 99, p. 729-738.

Duchesne, J.C., Scharer, U., and Wilmart, E.
1993:   A 10 Ma period of emplacement for the Rogaland anorthosites, Norway: evidence from U-Pb ages; Terra Nova, v. 5, p. 64.

Emslie, R.F. and Hunt, P.A.
1990: Ages and petrogenetic significance of igneous mangerite-charnockite suites associated with massif anorthosites, Grenville Province; Journal of Geology, v. 98, p. 213-231.

Gross, G.A.
*1965: General geology and evaluation of iron deposits; Volume I in Geology of Iron Deposits in Canada, Geological Survey of Canada, Economic Geology Report 22, 181 p.
*1967a: Iron deposits in the Appalachian and Grenville Regions of Canada; Volume II in Geology of Iron Deposits in Canada, Geological Survey of Canada, Economic Geology Report 22, 111 p.
1967b: Iron deposits of the Soviet Union; The Canadian Mining and Metallurgical Bulletin, December, v. 60, p. 1435-1440.
*1977: Metallogenetic evolution of the Canadian Shield; in Correlation of the Precambrian, Volume 2, (ed.) A.V. Sidorenko; Publishing Office "Nauka", Moscow, p. 274-293.

*Hammond, P.
1952: Allard Lake ilmenite deposits; Economic Geology, v. 47, p. 634-649.

Harben, P.W. and Bates, R.L.
1990: Titanium and zirconium minerals; in Industrial Minerals Geology and World Deposits; Industrial Minerals Division, Metal Bulletin Plc, London, p. 282-294.

*Hargraves, R.B.
1962: Petrology of the Allard Lake anorthosite suite, Quebec; in Petrologic Studies: a Volume in Honor of A.F. Buddington, (ed.) A.E.J. Engel, H.L. James, and B.F. Leonard; Geological Society of America, p. 163-189.

*Hatch, G.G. and Cuke, N.H.
1956: Iron operations of the Quebec Iron and Titanium Corporation; Transactions, The Canadian Institute of Mining and Metallurgy, v. 59, p. 359-362.

Henriquez, F. and Martin, R.F.
1978: Crystal-growth textures in magnetite flows and feeder dykes, El Laco, Chile; Canadian Mineralogist, v. 16, p. 581-589.

Higgins, M.D. and Doig, R.
1981: The Sept Iles anorthosite complex: field relationships, geochronology, and petrology; Canadian Journal of Earth Sciences, v. 18, p. 561-573.

Hill, J.D.
1988: Alkalic to transitional ferrogabbro magma associated with Paleohelikian anorthositic plutons in the Flowers River area, southeastern Nain igneous complex, Labrador; Contributions to Mineralogy and Petrology, v. 99, p. 113-125.

Kolker, A.
1982: Mineralogy and geochemistry of Fe-Ti oxide and apatite (nelsonite) deposits and evaluation of the liquid immiscibility hypothesis; Economic Geology, v. 77, p. 1146-1158.

*Korneliussen, A., Geis, H.P., Gierth, E., Krause, H., Robins, B., and Schott, W.
1985: Titanium ores: an introduction to a review of titaniferous magnetite, ilmenite and rutile deposits in Norway; Norges Geologiske Undersøkelse Bulletin, v. 402, p. 7-23.

Lindsley, D.H. and Philipp, J.R.
1993: Experimental evaluation of Fe-Ti oxide magmas; EOS, v. 74, p. 337.

*Lister, G.F.
1966: The composition and origin of selected iron-titanium deposits; Economic Geology, v. 61, p. 275-310.

Morse, S.A.
*1969: The Kiglapait layered intrusion, Labrador; Geological Society of America, Memoir 112, 146 p.
*1980: Kiglapait mineralogy 2: Fe-Ti oxide minerals and the activites of oxygen and silica; Journal of Petrology, v. 21, p. 685-719.

*Mortensen, J.K.
1993: U-Pb geochronology of the eastern Abitibi Subprovince. Part 1: Chibougamau-Matagami-Joutel region; Canadian Journal of Earth Sciences, v. 30, p. 11-28.

*Naldrett, A.J. and Cabri, L.J.
1976: Ultramafic and related mafic rocks: their classification and genesis with special reference to the concentration of nickel sulphides and Platinum-group elements; Economic Geology, v. 71, p. 1131-1158.

Naslund, H.R.
1983: The effect of oxygen fugacity on liquid immiscibility in iron-bearing silicate melts; American Journal of Science, v. 283, p. 1034-1059.

Owens, B.E., Dymek, R.F., Tucker, R.D., Brannon, J.C., and Podosek, F.A.
1994: Age and radiogenic isotopic composition of a late- to post-tectonic anorthosite in the Grenville Province: the Labrieville massif, Quebec; Lithos, v. 31, p. 189-206.

Park, C.F., Jr.
1961: A magnetite "flow" in northern Chile; Economic Geology, v. 56, p. 431-441.

Philpotts, A.R.
1966: Origin of anorthosite-mangerite rocks in southern Quebec; Journal of Petrology, v. 7, p. 1-64.
1967: Origin of certain iron-titanium oxide and apatite rocks; Economic Geology, v. 62, p. 303-315.

*Reynolds, I.M.
1985: The nature and origin of titaniferous magnetite-rich layers in the Upper Zone of the Bushveld Complex: a review and synthesis; Economic Geology, v. 80, p. 1089-1108.

Rose, E.R.
*1969: Geology of Titanium and Titaniferous Deposits of Canada; Geological Survey of Canada, Economic Geology Report 25, 177 p.
*1973: Geology of vanadium and vanadiferous occurrences of Canada; Geological Survey of Canada, Economic Geology Report 27, 130 p.

Scoates, J.S. and Chamberlain, K.R.
1993: The duration of anorthositic magmatism in the 1.43 Ga Laramie anorthosite complex, Wyoming; Geological Society of America, Abstracts with Programs, v. 25, p. A446.

Sokolov, G.A.
1970: Iron ore deposits of the Union of Soviet Socialist Republics; in Survey of World Iron Ore Resources, Occurrence and Appraisal; United Nations, New York, p. 381-410.

Taylor, H.P., Jr. and Noble, J.A.
1969: Origin of magnetite in the zoned ultramafic complexes of southeastern Alaska; in Magmatic Ore Deposits: a Symposium, (ed.) H.D.B. Wilson; Economic Geology, Monograph 4, p. 209-230.

*van Breemen, O. and Higgins, M.D.
1993: U-Pb zircon age of the southwest lobe of the Havre-Saint-Pierre anorthosite complex, Grenville Province, Canada; Canadian Journal of Earth Sciences, v. 30, p. 1453-1457.

von Gruenewaldt, G.
1993: Ilmenite-apatite enrichments in the Upper Zone of the Bushveld Complex: a major titanium-rock phosphate resource; Economic Geology, v. 35, p. 987-1000.

*Wager, L.R. and Brown, G.M.
1968: Layered Igneous Rocks; W. H. Freeman and Co., New York, 588 p.

*Wardle, R.J.
1987: Platinum-group-element potential in Labrador; in Current Research (1987), Newfoundland Department of Mines and Energy, Mineral Development Division, Report 87-1, p. 211-223.

Weidner, J.R.
1982: Iron-oxide magmas in the system Fe-C-O; Canadian Mineralogist, v. 20, p. 555-566.

*Willemse, J.
1970: The vanadiferous magnetic iron ore of the Bushveld igneous complex; in Magmatic Ore Deposits: a Symposium, (ed.) H.D.B. Wilson; Economic Geology, Monograph 4, p. 187-208.

Wilmart, E., Demaiffe, D., and Duchesne, J.C.
1989: Geochemical constraints on the genesis of the Tellnes ilmenite deposit, southwest Norway; Economic Geology, v. 84, p. 1047-1056.

*Yoder, H.S., Jr.
1968: Experimental studies bearing on the origin of anorthosite; New York State Museum and Science Service, Memoir 18, p. 13-22.

Author's address

G.A. Gross
Geological Survey of Canada
601 Booth Street
Ottawa, Ontario
K1A 0E8

27. MAGMATIC NICKEL-COPPER-PLATINUM GROUP ELEMENTS

27.1 Nickel-copper sulphide

27.1a Astrobleme-associated nickel-copper

27.1b Rift- and continental flood basalt-associated nickel-copper

27.1c Komatiite-hosted nickel

27.1d Other tholeiitic intrusion-hosted nickel-copper

27.2 Magmatic platinum group elements

27. MAGMATIC NICKEL-COPPER-PLATINUM GROUP ELEMENTS

O.R. Eckstrand

A broad group of deposits containing nickel-copper-platinum group elements (PGE) occur as sulphide segregations associated with a variety of mafic and ultramafic magmatic rocks. Among such deposits, two main subtypes are distinguishable. In the first (27.1 "Nickel-copper sulphide"), nickel and copper are the main economic commodities, contained in sulphide-rich ores that are associated with differentiated mafic sills and stocks and ultramafic (komatiitic) volcanic flows and sills. The second subtype (27.2 "Magmatic platinum group elements") is mined principally for PGEs, which are associated with sparsely dispersed sulphides in medium to large, typically layered mafic-ultramafic intrusions.

Eckstrand, O.R.
1996: Magmatic nickel-copper-platinum group elements; in Geology of Canadian Mineral Deposit Types, (ed.) O.R. Eckstrand, W.D. Sinclair, and R.I. Thorpe; Geological Survey of Canada, Geology of Canada, no. 8, p. 583 (also Geological Society of America, The Geology of North America, v. P-1).

27.1 NICKEL-COPPER SULPHIDE

O.R. Eckstrand

INTRODUCTION

Nickel-copper sulphide deposits are sulphide concentrations that occur in certain mafic and/or ultramafic intrusions or volcanic flows. Nickel is the main economic commodity, copper may be either a coproduct or byproduct, and platinum group elements (PGEs) are usual byproducts. Other commodities recovered in some cases include gold, silver, cobalt, sulphur, selenium, and tellurium. These metals are associated with sulphides, which generally make up more than 10% of the ore. The locations of Canadian deposits are shown in Figure 27.1-1.

The mafic and ultramafic magmatic bodies that host the ores are diverse in form and composition, and can be subdivided into the following four subtypes:

(27.1a) an astrobleme-associated sill-like mafic intrusion that contains ores in which Ni:Cu is approximately 1:1 (Sudbury, Ontario is the only known example).

(27.1b) rift- and continental flood basalt-associated mafic sills and dyke-like bodies, in which Ni:Cu ratios of the related ores may be either somewhat greater or less than 1 (Noril'sk-Talnakh, Russia; Duluth Complex, Minnesota; Crystal Lake intrusion, Ontario; possibly Jinchuan, China).

(27.1c) komatiitic volcanic flows and related intrusions, which have ores with Ni:Cu ratios that are commonly greater than 10, but less in some cases (Thompson, Manitoba; Expo Ungava and Marbridge, Quebec; Langmuir, Ontario; Kambalda and Agnew, Australia; Pechenga, Russia; Shangani, Trojan, and Hunter's Road, Zimbabwe; Kabanga, Tanzania).

(27.1d) other tholeiitic intrusions, in which the Ni:Cu ratios of the ores are commonly in the range 2 to 3 (Lynn Lake, Manitoba; Giant Mascot, British Columbia; Kotalahti, Finland; Råna, Norway; Selebi-Pikwe, Botswana).

IMPORTANCE

As a group, magmatic nickel-copper sulphide deposits have accounted for most of the world's past and current production of nickel. International reserves of magmatic sulphide nickel remain large, though they are exceeded by those of lateritic nickel deposits, the only other significant source of nickel.

Eckstrand, O.R.
1996: Nickel-copper sulphide; *in* Geology of Canadian Mineral Deposit Types, (ed.) O.R. Eckstrand, W.D. Sinclair, and R.I. Thorpe; Geological Survey of Canada, Geology of Canada, no. 8, p. 584-605 (*also* Geological Society of America, The Geology of North America, v. P-1).

Sudbury, the sole known example of the astrobleme subtype (27.1a) constitutes the world's largest nickel-producing camp as measured by total past production plus reserves. It accounts for about two-thirds to three-quarters of Canada's current nickel production. As a deposit type, it is the second most important Canadian producer of copper, and the only producer of PGEs and cobalt.

The rift- and continental flood basalt-associated subtype (27.1b) includes Noril'sk in Russia and probably Jinchuan in China, the second and third largest nickel-producing camps in the world. The undeveloped, low-grade Great Lakes Nickel deposit is the only example known in Canada, and is similar to the much larger, but also undeveloped, low grade deposits in the Duluth mafic complex in Minnesota.

The komatiitic subtype (27.1c) is the third most important type in the world. Proterozoic komatiitic deposits of the Thompson Nickel Belt in Manitoba account for one quarter to one third of current nickel production in Canada. Archean komatiitic deposits at Kambalda and elsewhere in Western Australia yield most of that country's produced nickel. Several small nickel mines in the Abitibi greenstone belt of Ontario and Quebec are also Archean komatiitic deposits.

Nickel deposits of other tholeiitic affiliation (subtype 27.1d) have been significant producers in the past; but only Selebi-Pikwe in Botswana has continued to operate recently.

SIZE AND GRADE OF DEPOSITS

Grades and tonnages of some of the more significant Canadian and foreign nickel deposits are listed in Table 27.1-1, and illustrated in Figure 27.1-2.

Most nickel sulphide deposits consist of several closely adjacent, but discrete orebodies, therefore the definition of "deposit" is rather arbitrary. Individual orebodies may contain from a few hundred thousand to a few million tonnes of ore, and in some instances tens of millions of tonnes of ore. Mining grades are generally about 1 to 3% Ni, but may be higher in some small deposits. Noteworthy exceptions are some of the ore zones in the Talnakh camp of the Noril'sk area, where substantial orebodies average several per cent Ni and greater than 20% Cu.

GEOLOGICAL FEATURES

All subtypes of magmatic nickel sulphide deposits have some general similarities. For example, the host intrusions in all cases are either mafic or ultramafic in composition. In addition, most deposits occur as sulphide concentrations toward the base of their magmatic host bodies. Furthermore, all subtypes of nickel sulphide ores usually consist mainly of the simple sulphide assemblage pyrrhotite-pentlandite-chalcopyrite, either as massive sulphides,

Table 27.1-1. Size and grade of nickel-copper sulphide deposits (production + reserves).

No.	Subtype	Deposit	Age	Size (Mt)	Ni%	Cu%	Reference
Canadian deposits							
1	27.1a	Sudbury (total), Ontario	Proterozoic	1648	1.20	1.03	Canadian Mines Handbook 91-92; Naldrett, 1994
2	27.1b	Great Lakes Nickel, Ontario	Proterozoic	45.6	0.183	0.344	Great Lakes Nickel Ltd., 1976 Annual Report
3	27.1c	Thompson Ni Belt (INCO), Manitoba	Proterozoic	89	2.5	0.13	INCO Prospectus, 1968; Naldrett, 1994
4	27.1c	Amax Area 1 ("Nose"), Manitoba	Proterozoic	7.3	1.33	na	Roth (1975)
5	27.1c	Manibridge, Manitoba	Proterozoic	1.409	2.55	0.27	Coats and Brummer (1971)
6	27.1c	Bucko, Manitoba	Proterozoic	2.5	2.23	0.17	Falconbridge Review, 1991
7	27.1c	Bowden, Manitoba	Proterozoic	80	0.6	na	Northern Miner, 1970-08-06
8	27.1c	Raglan deposits (6), Ungava, Quebec	Proterozoic	18.5	3.13	0.88	Northern Miner, 1992-11-02
9	27.1c	Expo Ungava, Quebec	Proterozoic	6.3	0.86	1.01	Northern Miner, 1992-11-02
10	27.1c	Texmont, Ontario	Archean	3.19	0.93	na	Coad (1979)
11	27.1c	Langmuir (No. 1 & 2), Ontario	Archean	1.6	2.09	0.08	Coats (1982)
12	27.1c	Marbridge, Quebec	Archean	0.774	2.82	0.1	Brett et al. (1976)
13	27.1c	Alexo Mine, Ontario	Archean	0.057	3.58	na	Shklanka (1969)
14	27.1c	Redstone, Ontario	Archean	1.22	2.39	0.09	Barrie et al. (1993)
15	27.1c	Dumont, Quebec	Archean	150	0.5	na	Duke (1986)
16	27.1c	Gordon Lake, Ontario	Archean	1.07	1.62	0.68	Coats (1982)
17	27.1c	Shebandowan, Ontario	Archean	15	1.5	1	Coats (1982)
18	27.1c	Namew Lake, Manitoba	Proterozoic	2.6	2.44	0.9	Canadian Minerals Yearbook, p. 45.2
19	27.1d	Montcalm, Ontario	Archean	3.56	1.44	0.68	Barrie and Naldrett (1989)
20	27.1d	St. Stephen (3 zones), New Brunswick	Devonian	1	1.05	0.53	Pactunç (1986)
21	27.1d	Macassa, Limerick Township, Ontario	Proterozoic	1.8	0.91	0.26	Northern Miner, 1971-10-14
22	27.1d	Lynn Lake, Manitoba	Proterozoic	20.151	1.023	0.535	Pinsent (1980)
23	27.1d	Giant Mascot, British Columbia	Cretaceous	2.05	1.4	0.5	Coats (1982)
24	27.1d	Canalask, Yukon Territory	Triassic	0.5	1.68	0.04	NMI 115F/15Ni001 **
25	27.1d	Wellgreen, Yukon Territory	Triassic	0.669	2.04	1.42	Hulbert et al. (1988)
26	27.1d(?)	Lorraine, Quebec	Archean	0.661	0.39	0.91	NMI 031M/07Cu002 **
Foreign deposits							
27	27.1b	Noril'sk-Talnakh district (Russia)	Triassic	555	2.7	2.07*	DeYoung et al. (1985); Naldrett (1994)
28	27.1b	Jinchuan (China)	Proterozoic	515	1.06	0.67	Chen and Mingliang (1987); Naldrett (1994)
29	27.1b	Duluth Complex (Minnesota)	Proterozoic	4000	0.2	0.66	Listerud and Meineke (1977)
30	27.1c	Pechenga (Russia)	Proterozoic	36	1	0.4*	DeYoung et al. (1985)
31	27.1c	Kambalda district (Australia)	Archean	48	3.6	0.25*	DeYoung et al. (1985); Naldrett (1994)
32	27.1c	Agnew (Australia)	Archean	46.764	2.08	0.1*	Billington (1984)
33	27.1c	Windarra district (Australia)	Archean	13.161	1.45	na	DeYoung et al. (1985)
34	27.1c	Mt. Keith (Australia)	Archean	270	0.6	na	DeYoung et al. (1985)
35	27.1c	Hitura (Finland)	Proterozoic	12.3	0.56	0.16	DeYoung et al. (1985)
36	27.1c	Shangani (Zimbabwe)	Archean	22	0.71	na	DeYoung et al. (1985)
37	27.1c	Trojan (Zimbabwe)	Archean	20.35	0.68	na	DeYoung et al. (1985)
38	27.1c	Hunter's Road (Zimbabwe)	Archean	30	0.7	na	DeYoung et al. (1985)
39	27.1c	Kabanga (Tanzania)	Proterozoic	11.7	1.72	0.26	Northern Miner 1993-03-08, p. 14
40	27.1d	Monchegorsk (Russia)	Proterozoic	47	0.7	0.4	Coates (1982)
41	27.1d	Kotalahti (Finland)	Proterozoic	23.2	0.7	0.3	DeYoung et al. (1985)
42	27.1d	Selebi-Pikwe (Botswana)	Archean	49.444	1.04	1.12	DeYoung et al. (1985)

* Cu grade approximate
** National Mineral Inventory file, Natural Resources Canada
na = not available

sulphide-matrix breccias, or disseminations of sulphides. Nickel-copper sulphide ores of any of the subtypes that have undergone tectonic remobilization have been converted to similar-appearing sulphide-matrix breccias.

However, the subtypes differ significantly in their geological-tectonic settings and in the geometric form and style of differentiation of the host magmatic bodies. They differ also in that the magmatic hosts in most subtypes are intrusions, but in the komatiitic subtype most are volcanic flows. Furthermore the ores of the various subtypes show some differences in composition, most noticeably in their Ni:Cu ratios. A general review of magmatic nickel sulphide deposits has been given by Naldrett (1989a).

Astrobleme-associated nickel-copper (Sudbury camp): subtype 27.1a

The Sudbury Igneous Complex is unique in a number of respects, including the exceptional concentration of associated nickel deposits. Perhaps its most unique characteristic

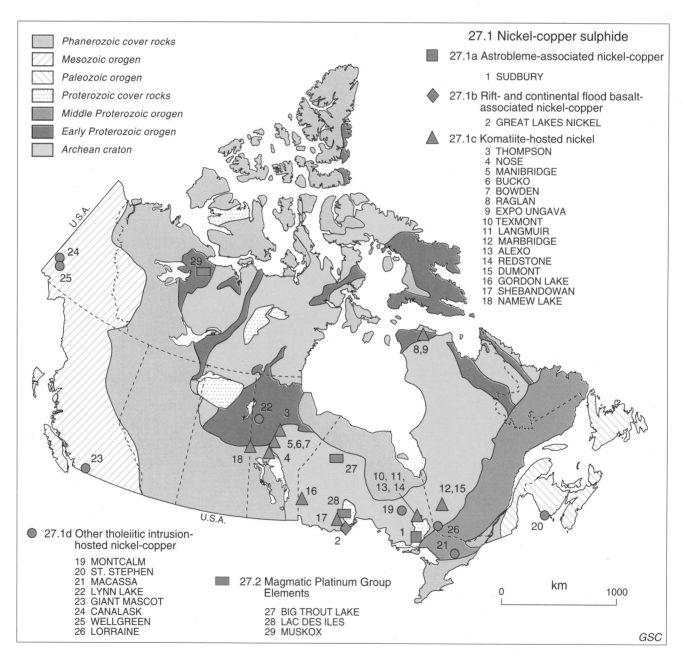

Figure 27.1-1. Locations of Canadian nickel deposits or districts. Numbers for nickel deposits refer to those listed in Table 27.1-1.

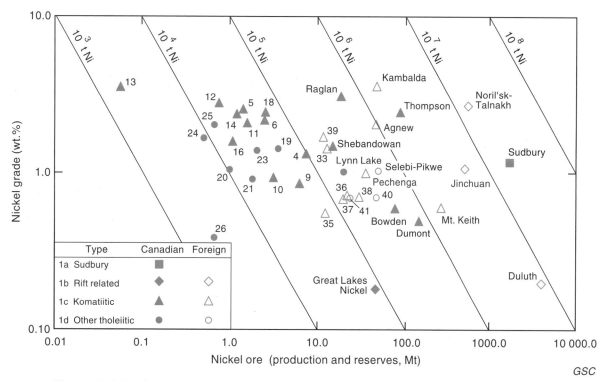

Figure 27.1-2. Grades versus tonnages of selected Canadian and foreign nickel deposits or districts. Numbers refer to the deposits listed in Table 27.1-1.

is the evidence for an associated shock metamorphic event, which is now widely believed to have been a meteoritic impact. Nevertheless, though other deposits of this type have not yet been identified, there is no apparent geological reason why they could not exist.

Geological setting

A comprehensive account of the geology of the Sudbury camp is given in the volume edited by Pye et al. (1984). Ages cited for units discussed in the following are from Krogh et al. (1984) in that volume.

The Sudbury Igneous Complex (SIC; 1850 ± 1 Ma) and its nickel-copper ores (Fig. 27.1-3) occur near the southern limit of the Archean Superior Province craton which is overlain by the Proterozoic Huronian Supergroup (Card et al., 1984). The northern margin or "North Range" of the basin-shaped Sudbury Igneous Complex cuts Archean rocks, mainly migmatitic tonalitic gneisses (Levack Gneiss, 2711 Ma) and anorogenic granitic plutons (Cartier Granite, about 2680 Ma). The southern margin or "South Range" has intruded lower Huronian mafic and felsic metavolcanic rocks (about 2450 Ma) and felsic stocks (about 2388 Ma). A little more than ten kilometres to the southeast of the complex lies the faulted Grenville Front, the northwestern margin of the mid-Proterozoic Grenville Province.

The complex is overlain by the basin-shaped Early Proterozoic Whitewater Group, a conformable sequence consisting, in upward succession, of a heterolithic breccia (Onaping Formation), a carbonaceous and pyritic argillite (Onwatin Formation), and a proximal turbidite sequence

(Chelmsford Formation). The base of the Onaping Formation has been intruded by the underlying granophyre of the complex.

The Sudbury Igneous Complex outcrops as a crude oval ring about 65 km long and 25 km across, representing a basin- or funnel-shaped intrusion (Fig. 27.1-3). The inward dip of the complex averages about 30° along the North Range, and 45 to 60° along the South Range. The complex consists of two main layer-like units (Naldrett and Hewins, 1984), the Lower Zone and the Upper Zone. The Lower Zone (0.5-2.5 km thick) comprises mainly norite and gabbronorite that grades upward into a quartz- and oxide-rich gabbro. The Upper Zone (1-2.5 km thick), which is made up mainly of granophyre, has an abruptly transitional contact with the underlying quartz-rich gabbro.

The Sudbury Igneous Complex has no obvious small-scale layering (Naldrett and Hewins, 1984), and only the Lower Zone shows a relatively weak, simple differentiation trend. Quartz-bearing gabbronorite is the most common rock type of the Lower Zone, and is composed of plagioclase-orthopyroxene cumulates, cumulus or intercumulus clinopyroxene, and intercumulus quartz, quartz-feldspar micrographic intergrowth, biotite, and Fe-Ti oxides. Grain size increases from fine grained at the base to medium- to coarse-grained. Transition to the overlying quartz- and oxide-rich gabbro involves loss of orthopyroxene and appearance of cumulus titaniferous magnetite and apatite. Gabbro passes upward into the relatively homogeneous granophyre which consists of about two-thirds micrographic intergrowth of quartz and feldspar, together with plagioclase and lesser amounts of mafic minerals.

At the base of the Lower Zone is the "Sublayer", which contains the nickel-copper sulphide deposits (Souch et al., 1969; Pattison, 1979). Its contact with the Lower Zone norite is reportedly sharp in some places, but is gradational in others. The Sublayer has two facies; (1) Contact Sublayer (generally less than 200 m thick) consists of discontinuous gabbronoritic lenses along the basal contact of the Sudbury Igneous Complex, which grade into (2) Offset Sublayer that constitutes apophyses of mainly quartz diorite which project outward into the footwall rocks. The Contact Sublayer

(Souch et al., 1969; Pattison, 1979; Naldrett et al., 1984) is a gabbronorite that is characterized by the presence of nickel-copper sulphides and xenoliths of wall rock and mafic and ultramafic rocks of nonlocal origin. It typically consists of a fine grained assemblage of zoned plagioclase laths, subophitic hypersthene and augite, minor amounts of primary biotite and hornblende, and widely varying amounts of quartz. Although closely similar in composition, in the lithology of its xenoliths, and in the presence of sulphides,

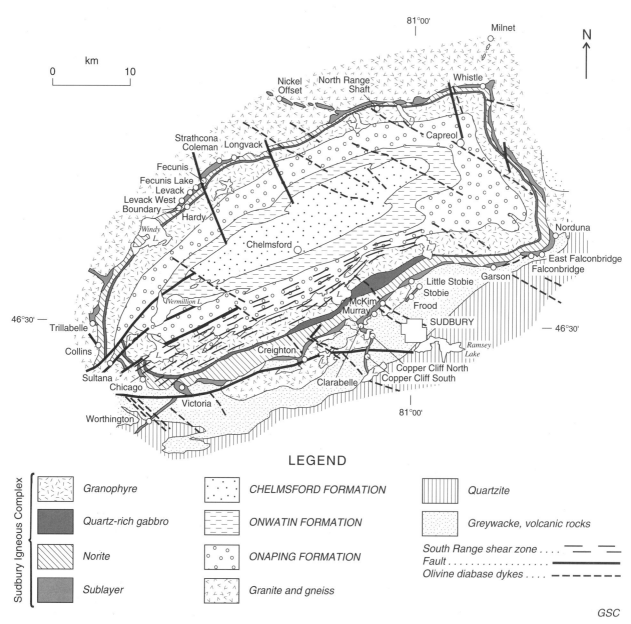

LEGEND

Sudbury Igneous Complex

Granophyre	CHELMSFORD FORMATION	Quartzite
Quartz-rich gabbro	ONWATIN FORMATION	Greywacke, volcanic rocks
Norite	ONAPING FORMATION	South Range shear zone ....
Sublayer	Granite and gneiss	Fault ....
		Olivine diabase dykes ....

GSC

Figure 27.1-3. Geological map of the Sudbury Igneous Complex, showing locations of some of the more important nickel-copper mines and deposits (after Naldrett, 1989d; Sublayer after Pattison, 1979, and Naldrett et al., 1984; South Range shear zone after Shanks and Schwerdtner, 1991).

the quartz dioritic Offset Sublayer has distinct mineralogy, and consists mainly of quartz, plagioclase, biotite, and hornblende, the last commonly pseudomorphic after pyroxene.

The composition of the Sudbury Igneous Complex reflects a strong influence by crustal rocks, quite likely the immediate footwall rocks (see "Genetic model"). This is indicated by the high silica and potash contents, the LREE enrichment (Naldrett, 1984), and Os isotopic ratios (Walker et al., 1991).

Features related to the intrusion of the Sudbury Igneous Complex (Dressler, 1984) include a contact metamorphic halo as much as 1.2 km in width, and a set of shock-induced structures generally attributed to a meteoritic impact. The shock features include shatter cones, and planar dislocations and kink bands in certain rock-forming minerals; these are found in footwall rocks and in clasts in the Onaping Formation and the granophyre. In addition, the complex is underlain by a quasi-conformable "Footwall Breccia" of variable thickness that is leucocratic and prominent in the North Range, but more obscure in the South Range due to its mafic nature. Outside the Footwall Breccia and around the entire periphery of the complex are irregular veins and masses of Sudbury Breccia that are known to be present at least 80 km outward from the complex. Clast compositions in both breccias suggest an essentially in situ genesis. The North Range Footwall Breccia is a highly variable unit whose clasts consist largely of Levack Gneiss set in a fine grained granoblastic matrix of plagioclase, quartz, biotite, amphibole, and pyroxene. Sudbury Breccia has clasts of all sizes (commonly as large as tens of metres), mainly consisting of wall rocks. Typically the finest matrix material is a dark rock flour (Dressler, 1984).

The Sudbury basin owes its elliptical shape to northwestward directed ductile thrusting probably related to the Penokean Orogeny (Rousell, 1984; Shanks and Schwertner, 1991; Milkereit et al., 1992; Cowan and Schwertner, 1994). A prominent element of this deformation is the east-northeast-trending South Range shear zone that transects the southern part of the Sudbury basin.

Nature of ore deposits

Five types of ore zones can be distinguished, based on the host rocks, controlling structures, and ore composition. The first four of these are in or adjacent to the Sublayer and comprise pyrrhotite-dominated sulphides with Ni:Cu ratios near 1. The fifth type occurs in footwall rocks and consists mainly of copper sulphides, highly enriched in PGEs.

In the South Range contact type (Souch et al., 1969), the ore typically occurs in noritic Sublayer, which occupies a depression in the base of the Sudbury Igneous Complex and plunges down the basal contact (Fig. 27.1-4A), as in the Creighton, Murray, and Little Stobie 1 mines. The ore generally has a Ni:Cu ratio greater than 1, and is zoned. The lowest zone consists of massive sulphide ore that contains angular wall rock fragments ("inclusion massive sulphide"), and stringers of sulphide that project into the underlying footwall rocks. This ore grades upward into a sulphide-matrix breccia with an increasing amount of norite in the matrix, and numerous pyroxenite and peridotite inclusions ("gabbro-peridotite inclusion

sulphide"). Above this, the amount of sulphide matrix diminishes in the inclusion-rich Sublayer ("ragged disseminated sulphide").

Typical ores of the North Range type (Pattison, 1979; Coats and Snajdr, 1984), such as the Levack and Strathcona ore zones, occur in noritic Sublayer and underlying Footwall Breccia, which together occupy embayments that bulge downward from the base of the Lower Zone norite into the underlying Levack Gneiss (Fig. 27.1-4B). The ores are hosted mainly in the Footwall Breccia (considered by some authors to be a facies of the Sublayer). They consist of massive stringers and lenses of pyrrhotite-dominated Ni-Cu sulphides (Ni:Cu generally greater than 1) that are commonly concentrated in depressions at the base of the Footwall Breccia host, and are oriented subparallel to the dip of the Sudbury Igneous Complex. Some ore zones, such as the Deep Ore zone at Strathcona, extend downward 200 m into the Levack Gneiss footwall as lenses and stringers of massive sulphide. North Range ores are generally zoned, with Ni:Cu ratios decreasing downward from the complex; at Strathcona Ni:Cu goes from 3:1 in noritic Sublayer to 2:1 in the Footwall Breccia to 1:1 in the Deep Ore zone.

The "offset" type of ores is hosted in the quartz dioritic "offsets" (Fig. 27.1-4C), the dyke-like facies of Sublayer that projects outward from the Sudbury Igneous Complex and penetrates the footwall rocks (Pattison, 1979; Cochrane, 1984; Grant and Bite, 1984). The offsets typically follow dykes of Sudbury Breccia, are steeply dipping, and are either subradial or subparallel to the basal contact of the complex. The Copper Cliff and Frood-Stobie offsets contain the most important of these ores. The ore mineralization generally has a Ni:Cu ratio close to 1, and occurs in two main forms: as disseminated sulphide "blebs" in lenticular zones of inclusion-rich quartz diorite located centrally in the offset, or as sheaths of sulphide bleb-bearing quartz diorite or sulphide-matrix breccia along the margins of the offset.

The Falconbridge, East Falconbridge, and Garson mines contain ore zones of the fault-related type (Souch et al., 1969; Owen and Coats, 1984). All are associated with near-vertical faults that cut the South Range Lower Zone norite and adjacent Huronian footwall mafic metavolcanic rocks of the Stobie Formation (Fig. 27.1-4D). Ores are of two types: "contorted schist inclusion sulphide" within the shear zones constitute the Main zone; and "inclusion massive sulphide" occurs as discontinuous lenses ("Southwall ores") in adjacent metavolcanic rocks.

Ores of the deep copper vein type (Fig. 27.1-4B) occur within the footwall, as much as 500 m below the Sublayer. The Deep Copper zone at Strathcona (Abel et al., 1979; Li et al., 1992; Money, 1993; Morrison et al., 1994) is hosted mainly in masses of thermally metamorphosed Sudbury Breccia within Levack Gneiss. The ore consists of anastamosing veins and stringers of massive sulphides. The minerals are mainly chalcopyrite and cubanite, and include minor amounts of pentlandite, magnetite, millerite, and pyrrhotite. Platinum and palladium are more highly enriched than in any other type of ore at Sudbury, but nickel values are low (Ni:Cu <<1). The massive veins are asymmetrically zoned, with pentlandite and pyrrhotite occupying the footwall side of the veins.

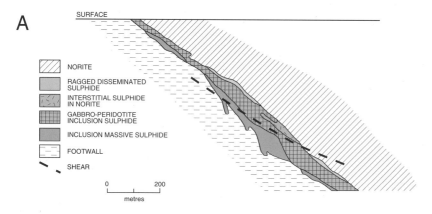

A

NORITE

RAGGED DISSEMINATED
SULPHIDE

INTERSTITIAL SULPHIDE
IN NORITE

GABBRO-PERIDOTITE
INCLUSION SULPHIDE

INCLUSION MASSIVE SULPHIDE

FOOTWALL

SHEAR

SURFACE

0   200
metres

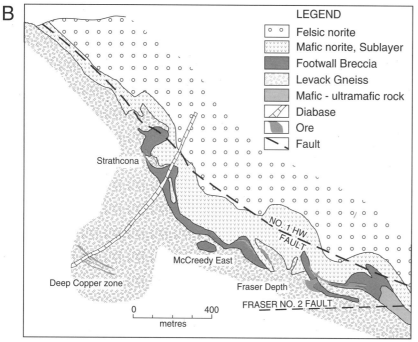

B

LEGEND

Felsic norite

Mafic norite, Sublayer

Footwall Breccia

Levack Gneiss

Mafic - ultramafic rock

Diabase

Ore

Fault

Strathcona

Deep Copper zone

McCreedy East

Fraser Depth

NO. 1 HW FAULT

FRASER NO. 2 FAULT

0   400
metres

Figure 27.1-4. Typical deposits of the Sudbury Igneous Complex:
A) Cross-section through the Murray mine, South Range, looking west (after Souch et al., 1969).
B) Cross-section through the Strathcona, McCreedy East, and Fraser mines, North Range, looking east (after Coats and Snajdr, 1984). Strathcona Deep zone lies off this section, in the Levack Gneiss, stratigraphically about 100 m below the Strathcona Main zone.
C) Plan of the Copper Cliff South mine in the Copper Cliff offset (after Cochrane, 1984).
D) Plan of the Falconbridge mine, 4525 level (after Owen and Coats, 1984).

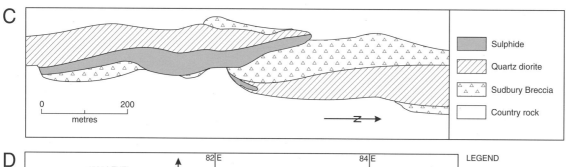

C

Sulphide

Quartz diorite

Sudbury Breccia

Country rock

0   200
metres

N →

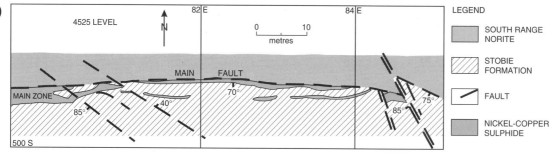

D

4525 LEVEL

N

82 E   84 E

0   10
metres

MAIN   FAULT
70°

MAIN ZONE

85°   40°

85°   75°

500 S

LEGEND

SOUTH RANGE NORITE

STOBIE FORMATION

FAULT

NICKEL-COPPER SULPHIDE

GSC

590

Table 27.1-2. Compositions of typical Sudbury ores.

Deposit	Type	no.*	Ni	Cu	Co	Pt	Pd	Rh	Ru	Ir	Os	Au	Ni:Cu	Pd:Ir
Levack West	NR	21	5.73	3.72	0.16	1150	1250	186	60	7	22	150	1.54	178
Strathcona Main	NR	18	3.40	1.20	0.14	413	380	20	12	7	4	78	2.83	54
Strathcona Deep	FW	8	4.00	2.30	0.13	750	701	16	4	4	3	112	1.74	175
Strathcona Deep Copper	FW	32	1.97	28.27	0.03	4719	5213	<0.16	<2	0.11	<0.53	296	0.07	47 000
Little Stobie 1	SR	23	3.83	4.41	0.19	1930	2120	119	123	62	29	862	0.87	34
Falconbridge	FR	23	5.35	1.52	0.22	546	381	287	225	144	40	174	3.52	2.6

* no. of samples
Ni, Cu, Co in wt. %; Pt, Pd, Rh, Ru, Ir, Os, Au in ppb
NR = North Range, FW=Footwall, SR=South Range, FR=Fault-related
Data for Levack West, Little Stobie 1, and Falconbridge from Naldrett et al. (1982); data for Strathcona zones from Li et al. (1992).

Ore composition and zoning

Typical compositions of ore samples from several Sudbury deposits are given in Table 27.1-2. Data shown for Levack West and Strathcona Main are representative of the main North Range type, and those for Little Stobie 1 and Falconbridge are typical of two types found in the South Range.

In the North Range, ores are compositionally zoned from the mafic norite of the Sudbury Igneous Complex downward through the Footwall Breccia into the Levack Gneiss (Li et al., 1992). Figure 27.1-4B in conjunction with Table 27.1-2 illustrate the downward increase of Cu, Pt, Pd, and Au contents and decrease of Rh, Ru, Ir, and Os contents exhibited by the Main, Deep, and Deep Copper zones within the Strathcona mine. Enrichment of Cu, Pt, and Pd is extreme in the Copper and Deep Copper zones. Some South Range deposits exhibit similar zoning, e.g., the Lindsley deposit (Binney et al., 1994).

Mineralogy

The mineral assemblages of the Contact Sublayer, Offset Sublayer, and Footwall Breccia, which are the main hosts to ore, have been described above. The main ores (Naldrett, 1984) consist largely of the assemblage pyrrhotite (dominant, both hexagonal and monoclinic), pentlandite, chalcopyrite, pyrite, and magnetite. Bornite occurs locally in copper-rich zones. South Range ores are characterized by higher arsenic content, expressed as the arsenides niccolite and maucherite, and the sulpharsenides gersdorffite and cobaltite. The copper-rich vein ores deep in the footwall are dominated by chalcopyrite and cubanite, and contain lesser pentlandite, magnetite, and pyrrhotite. A considerable number of platinum-group minerals have been identified, the most abundant of which are michenerite (PdBiTe), moncheite (PtTe$_2$), and sperrylite (PtAs$_2$).

Rift- and continental flood basalt-associated nickel-copper: subtype 27.1b

The two best known examples of deposits of this type, the Noril'sk-Talnakh deposits of northern Siberia, and deposits in the Duluth Complex, Minnesota are associated with large continental flood basalt provinces. In Canada, this type is represented only by the Great Lakes Nickel deposit in the Crystal Lake intrusion, which is probably a northern extension of the Duluth Complex. The world-class Jinchuan deposit in China may also belong to this type. Though broadly similar in tectonic setting, the geology of these three districts is rather different in many respects, and they are described separately.

Noril'sk-Talnakh camp

The Permo-Triassic Noril'sk-Talnakh deposits (Fig. 27.1-5) occur near the northwestern margin of the Siberian Platform (Naldrett, 1989a; Simonov et al., 1994). The margin has a crystalline Proterozoic basement overlain by Riphean molassic strata, Lower Paleozoic marine dolostones, argillites, and sandstones, and Devonian marls and evaporites. Carboniferous shallow water limestones and continental sedimentary rocks that include coal measures are capped by a 3.5 km thickness of traps and tuff of the Permo-Triassic Siberian Flood Basalt Province, one of the largest in the world. Large mafic masses underlying the traps are indicated by aeromagnetic anomalies (Rempel, 1994). Near the base of the traps and related to them are small mafic-ultramafic hypabyssal intrusions (248 Ma; Likhachev, 1994) with which the Noril'sk-Talnakh nickel-copper-PGE deposits are associated. These sill-like intrusions occur along the regional north-northeast-trending Noril'sk-Kharayelakh fault, and have intruded strata from mid-Devonian to lower Triassic in age (Fig. 27.1-5).

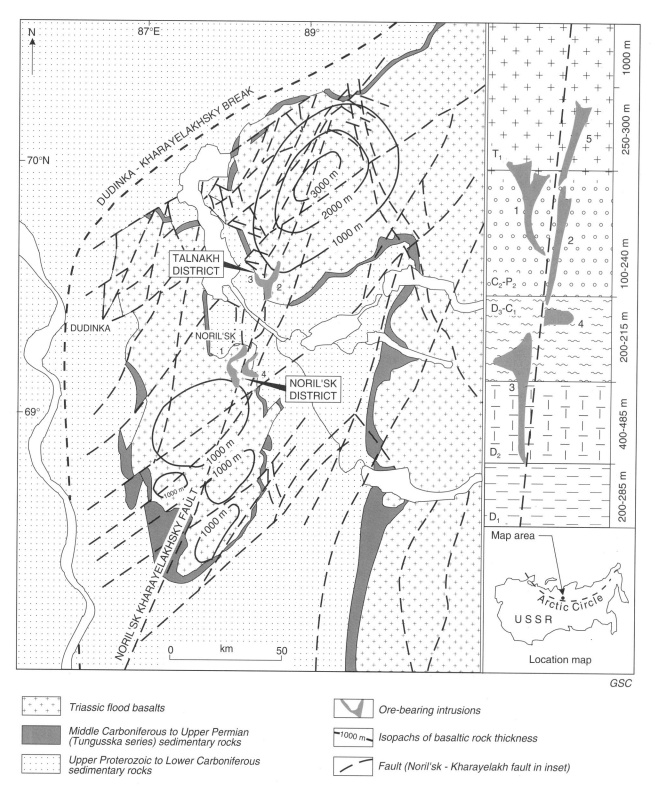

Figure 27.1-5. Regional geology of the Noril'sk-Talnakh area, showing the vertically projected locations of ore-bearing intrusions: 1 – Noril'sk I; 2 – Talnakh; 3 – Kharayelakh; 4 – Chernogorsky; 5 – (sic, not identified, not shown in plan) D = Devonian, C = Carboniferous, P = Permian, T = Triassic. The inset shows the stratigraphic position of the intrusions (after Likhachev, 1994).

The Noril'sk-Talnakh ore-bearing intrusions (Likhachev, 1994; Zen'ko and Czamanske, 1994) are mostly sill-like with known dimensions as great as 15 km in length, 0.5 to 2 km in width, and 50 to 300 m in thickness (Fig. 27.1-6). A typical intrusion consists of a highly differentiated main sill that passes from olivine-bearing, melanocratic gabbro-dolerites near the base upward through the succession of mafic units shown in Figure 27.1-7 to leucocratic gabbro-dolerites near the top. This sequence is considered by some to result from multiple magmatic pulses of different compositions. Olivine and plagioclase exhibit compositions ranging from Fo_{85} to Fo_{60}, and An_{95} to An_{40}, respectively (Likhachev, 1994). An exceptionally thick hornfels halo, as great as 250 m in thickness above and 100 m below, has affected the surrounding wall rocks.

The three main types of ore mineralization are disseminated, massive, and "copper" ores. The disseminated ores consist of droplets, schlieren, and fine sulphide veinlets dispersed through picritic, taxitic, and contact gabbro-dolerite in the lower portions of the intrusions, and form sheet-like conformable orebodies that in some cases exceed 40 m in thickness. Massive ores are also sheet-like bodies, but they undulate along the basal contacts of the intrusions, and commonly depart from the contacts to intrude the underlying metasedimentary footwall rocks. These ores are the most important type in the Talnakh camp, and form massive sulphide bodies as large as 1.5 km long, several hundreds of metres wide, and several tens of metres thick (Distler, 1994). The bulk of these ores have a uniform composition (pyrrhotite-dominated chalcopyrite-pentlandite, Cu:Ni=0.8), but some portions exhibit striking compositional zoning which gives rise to zones having extreme copper- and PGE-enrichment. The resulting succession of sulphide assemblages ranges from pyrrhotite-chalcopyrite-pentlandite, through decrease in the proportion of pyrrhotite and increase in copper sulphide, to one consisting almost entirely of cubanite, mooihoekite, and talnakhite. The "copper" ores are of two types; one consists of disseminated veinlets of copper-rich sulphides that form a halo around the periphery of massive ore in both sedimentary and intrusive rocks; while the other comprises breccia ores with a copper sulphide matrix that occur in the foundered roofs of some intrusions (Fig. 27.1-7).

Noril'sk ores are probably the richest in PGEs of all known Ni-Cu sulphide (subtype 27.1) deposits. The ordinary massive pyrrhotite-chalcopyrite-pentlandite ores reported by Zientek et al. (1994) and Distler (1994) contain 0.6-3 g/t Pt and 4-13 g/t Pd. The PGE contents increase with

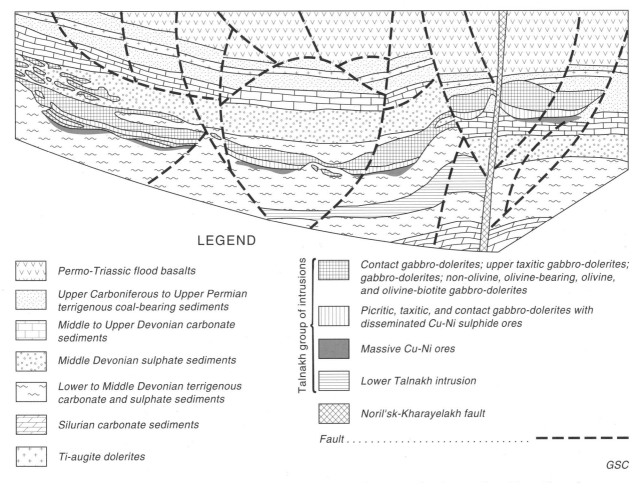

LEGEND

- ∨∨∨ Permo-Triassic flood basalts
- Upper Carboniferous to Upper Permian terrigenous coal-bearing sediments
- Middle to Upper Devonian carbonate sediments
- Middle Devonian sulphate sediments
- Lower to Middle Devonian terrigenous carbonate and sulphate sediments
- Silurian carbonate sediments
- +∙+ Ti-augite dolerites

Talnakh group of intrusions
- Contact gabbro-dolerites; upper taxitic gabbro-dolerites; gabbro-dolerites; non-olivine, olivine-bearing, olivine, and olivine-biotite gabbro-dolerites
- Picritic, taxitic, and contact gabbro-dolerites with disseminated Cu-Ni sulphide ores
- Massive Cu-Ni ores
- Lower Talnakh intrusion

- Noril'sk-Kharayelakh fault
- Fault . . . . - - - - - -

GSC

Figure 27.1-6. Sckematic cross-section through Talnakh camp, showing stratigraphic setting of ore-bearing intrusions and locations of sulphide ores (after Duzhikov et al., 1992).

Cu grade, achieving their highest values in massive sulphides that have >25% Cu, namely 5-24 g/t Pt and 30-140 g/t Pd.

Duluth Complex-related copper-nickel mineralization

The Duluth Complex in Minnesota is an arcuate mass of mafic intrusions more than 225 km in length that constitute part of a major 1.1 Ga intracontinental rift structure known as the Mid-Continent Rift system (Green, 1983). The complex is closely associated with the overlying tholeiitic flood basalts (North Shore Volcanics); together they have a combined thickness of 15 km. South of Thunder Bay in Ontario, the Crystal Lake gabbro is a small intrusion that appears to be a northern satellite of the Duluth Complex.

Both contain similar, low-grade disseminated copper-nickel mineralization, as yet undeveloped, but representing significant resources.

The Duluth Complex as a whole dips gently southeastward, and is made up of an extensive cap of layered anorthositic rocks and a suite of younger troctolitic intrusions that forms the basal portion of the complex (Weiblen and Morey, 1980; Weiblen, 1982; Naldrett, 1989a). The intrusions also include dunite, peridotite, pyroxenite, gabbro, and norite. The base of the complex lies in contact with Archean granite and greenstone, and Lower Proterozoic graywacke, slate, and iron-formation. The lower parts of the troctolitic masses are characterized by abundant xenoliths, which include gabbroic and anorthositic rocks, but also Lower Proterozoic metasedimentary rocks.

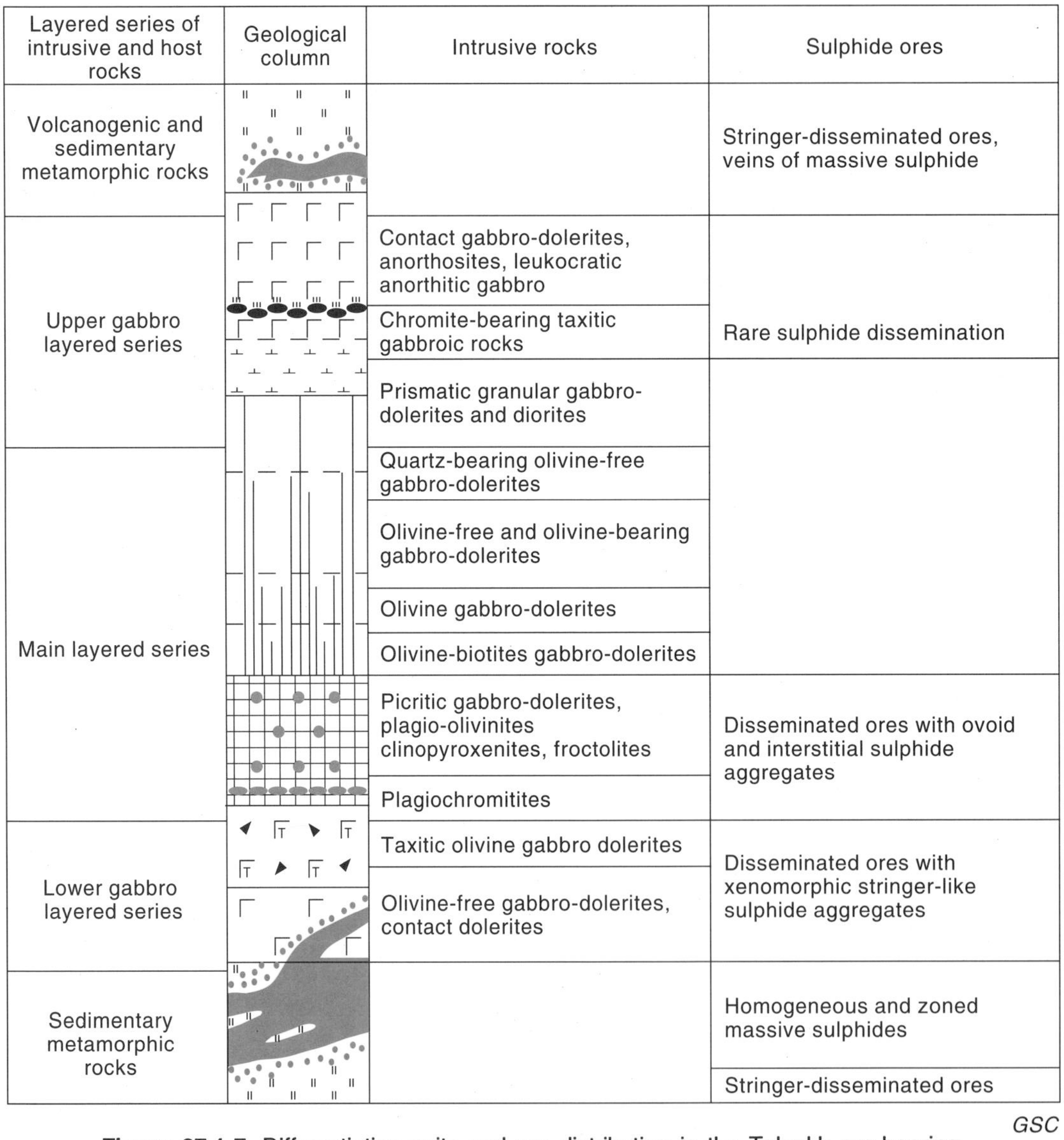

Layered series of intrusive and host rocks	Geological column	Intrusive rocks	Sulphide ores
Volcanogenic and sedimentary metamorphic rocks			Stringer-disseminated ores, veins of massive sulphide
Upper gabbro layered series		Contact gabbro-dolerites, anorthosites, leukocratic anorthitic gabbro	Rare sulphide dissemination
		Chromite-bearing taxitic gabbroic rocks	
		Prismatic granular gabbro-dolerites and diorites	
Main layered series		Quartz-bearing olivine-free gabbro-dolerites	
		Olivine-free and olivine-bearing gabbro-dolerites	
		Olivine gabbro-dolerites	
		Olivine-biotites gabbro-dolerites	
		Picritic gabbro-dolerites, plagio-olivinites clinopyroxenites, froctolites	Disseminated ores with ovoid and interstitial sulphide aggregates
		Plagiochromitites	
Lower gabbro layered series		Taxitic olivine gabbro dolerites	Disseminated ores with xenomorphic stringer-like sulphide aggregates
		Olivine-free gabbro-dolerites, contact dolerites	
Sedimentary metamorphic rocks			Homogeneous and zoned massive sulphides
			Stringer-disseminated ores

GSC

Figure 27.1-7. Differentiation units and ore distribution in the Talnakh ore-bearing intrusions (after Distler, 1994).

Copper-nickel mineralization in the Duluth Complex (Ripley, 1986) occurs as numerous zones (combined in the composite grade/tonnage data shown in Table 27.1-1), mainly in the basal portions of two of the troctolitic intrusions (South Kawishiwi, Partridge River). The mineralization consists principally of disseminated pyrrhotite, chalcopyrite, cubanite, and pentlandite. Massive sulphides are significant in only a few of the mineralized zones. The mineralization generally occurs within 100 m of the basal contact, but some zones are 300 to 400 m above the contact. The mineralized zones are typified by abundant metasedimentary xenoliths, textural heterogeneity of the host rock, and variable rock composition.

The Crystal Lake intrusion in Ontario (Geul, 1970; Eckstrand et al., 1989b; Cogulu, 1990, 1993), though much smaller, bears many similarities to the Duluth Complex. It is dyke-like in plan, about 700 m wide, and canoe-shaped in cross-section. Internal layering is trough-shaped. The intrusion comprises a basal contact gabbro, a lower zone of olivine gabbro, an overlying cyclically layered zone of anorthosite-troctolite, and an upper zone of olivine gabbro. The lower zone contains the Great Lakes Nickel deposit (see Table 27.1-1) which consists of pyrrhotite-chalcopyrite-cubanite-pentlandite mineralization that is disseminated within olivine gabbro. Like the mineralized troctolite of the Duluth Complex, the olivine gabbro has heterogeneous texture, and abundant cognate and hornfelsed metasedimentary xenoliths.

Jinchuan camp

The Jinchuan nickel-copper sulphide deposits are located in Gansu province, north central China in a northwesterly trending, uplifted and faulted belt that forms the southwestern margin of the Sino-Korean platform (Chai and Naldrett, 1992a, b). The deposits are hosted by a dyke-like ultramafic body that has intruded Lower Proterozoic migmatite, gneiss, and marble. Chai and Naldrett (1992a) have suggested that the intrusion may have been a feeder for continental rift-associated basaltic volcanism.

Rock types range from dunite to olivine pyroxenite, but the matrix and disseminated nickel-copper sulphide ores occur mainly in dunite and peridotite in the lower part of the intrusion. The main ore minerals are pyrrhotite, pentlandite, violarite, chalcopyrite, and cubanite.

Komatiite-hosted nickel subtype 27.1c

Geological setting

Naldrett (1989a) has recognized two types of komatiite-hosted nickel deposits, the first found in Archean greenstone belts, and the second in rifted, mainly Lower Proterozoic, continental margins.

Komatiites in Archean greenstone belts

The greatest concentration of nickel deposits within Archean greenstone belts is in Western Australia, in the northerly trending, 800 km long Norseman-Wiluna greenstone belt in the eastern part of the Yilgarn craton (Marston et al., 1981; Groves et al., 1984; Lesher, 1989). Most of the deposits (Kambalda, Agnew, Mt. Keith, Nepean, Scotia) are confined to a central fault-bounded rift zone as much as 200 km wide, characterized by abundant komatiites and sulphidic cherts thought to represent a deep marine environment. The sequence of nickel ore-bearing spinifex-textured komatiitic peridotite flows in the Kambalda dome (Fig. 27.1-8) is believed to be correlative with similar ore-bearing komatiites (Gemuts and Theron, 1975) throughout an extensive surrounding part of this central rift zone. On either side of the central rift, the greenstone sequences comprise platformal suites of basalts, shallow-water volcaniclastic rocks, and oxide facies iron-formation, and contain a few komatiitic nickel deposits (Windarra and Forrestania districts).

Most of the Canadian Archean nickel deposits are found in the Abitibi greenstone belt of the Superior province (e.g., Langmuir, Redstone, Marbridge, Texmont, Alexo; Coad, 1979; Barnes, 1985; Barrie et al., 1993). Those in the Shaw dome area south of Timmins are associated with a discontinuous horizon of spinifex-textured komatiitic flows (probably 2707 Ma; Corfu, 1993) at the base of a volcanic cycle which comprises an ultramafic-mafic-felsic succession. Other nickel deposit-bearing komatiites in the Timmins area are considered to be part of the same horizon. Furthermore, in both the Kambalda and Shaw dome-Timmins areas, most of the ore-bearing komatiites are directly underlain by sulphidic sediments (Coad, 1979; Lesher, 1989). Many believe these sediments are the source of sulphur that became incorporated in the komatiitic magmas and gave rise to the ores.

In both the Norseman-Wiluna and Abitibi greenstone belts, some larger, but subeconomic, low grade disseminated nickel sulphide deposits occur in thicker and larger dunitic sills of komatiitic affiliation. These are generally separate from the areas in which extrusive komatiite-hosted nickel deposits occur.

Komatiites in rifted (Proterozoic) continental margins

The most important komatiitic nickel deposits associated with rifted continental margins are those in the early Proterozoic Thompson Nickel Belt of northern Manitoba (Zurbrigg, 1963; Peredery et al., 1982). These deposits are associated with peridotitic lenses that typically occur in sulphidic metapelites (biotite schist) of a sedimentary-volcanic shelf sequence (Ospwagan Group) on the northwestern margin of the Archean Superior province craton (Bleeker, 1990). The supracrustal rocks include siltstone, sandstone, quartzite, shale, phyllite, dolomite, iron-formation, and pillowed basalts. Subsequent Hudsonian (early Proterozoic) collisional deformation has given rise to the pronounced linearity of the belt, and juxtaposed it with upper amphibolite and granulite gneisses (Kisseynew Group) of the Trans-Hudson orogen to the northwest. Several periods of folding and amphibolite facies metamorphism have pervasively reworked the original ore-host relationships in many deposits.

The komatiitic nickel deposits of the early Proterozoic Cape Smith belt in the Ungava Peninsula of northern Quebec, like those of Thompson, are also related to the rifted continental margin of the Superior Province craton (Baragar and Scoates, 1981). The deposits occur in a Hudsonian fold-and-thrust belt in a series of peridotitic lenses along a particular sediment-volcanic contact in an allochthonous, recumbently folded sequence of marine

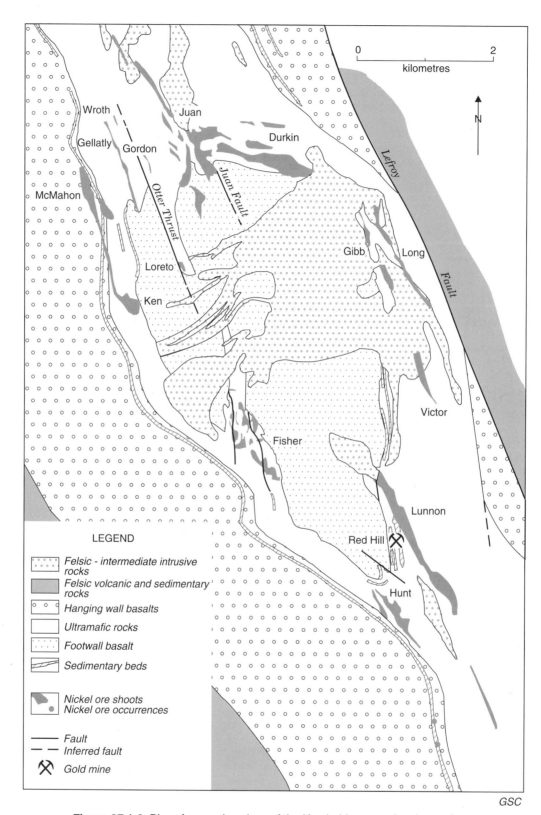

LEGEND

Felsic - intermediate intrusive rocks

Felsic volcanic and sedimentary rocks

Hanging wall basalts

Ultramafic rocks

Footwall basalt

Sedimentary beds

Nickel ore shoots
Nickel ore occurrences

Fault

Inferred fault

Gold mine

GSC

Figure 27.1-8. Plan of general geology of the Kambalda camp showing surface projections of orebodies (after Gresham and Loftus-Hills, 1981).

Table 27.1-3. Comparison of komatiitic hosts and nickel ores.

Deposit	MgO% (rock)	FeO/FeO+MgO (rock)	Olivine composition (%Fo)	Type of pyroxene	Ni:Cu	Reference
Langmuir (Archean, Ontario)	48.5	0.24	na	na	16	Green and Naldrett, 1981
Kambalda (Archean, Australia)	43	0.16-0.20	94	cpx	13	Marston et al., 1981; Lesher, 1989
Thompson (Proterozoic, Manitoba)	44	0.15	90	opx>>cpx	14	Peredery, 1979; Peredery et al., 1982; Paktunç, 1984
Pipe 2 (Proterozoic, Manitoba)	na	0.25	na	opx>>cpx	12	Peredery, 1979; Peredery et al., 1982
Katiniq (Proterozoic, Quebec)	36	0.31	na	cpx	4.3	Barnes et al., 1982

na = not available
Fo = forsterite, cpx = clinopyroxene, opx = orthopyroxene

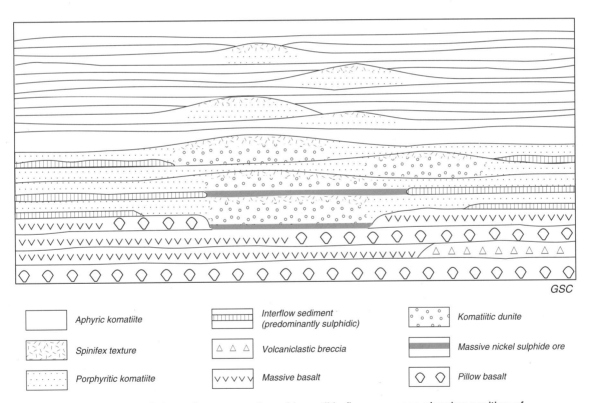

Aphyric komatiite	Interflow sediment (predominantly sulphidic)	Komatiitic dunite
Spinifex texture	Volcaniclastic breccia	Massive nickel sulphide ore
Porphyritic komatiite	Massive basalt	Pillow basalt

Figure 27.1-9. Schematic cross-section of komatiitic flow sequence showing position of basal and hanging wall orebodies, and graben-like depressions in which basal ores are typically located (after Lesher, 1989).

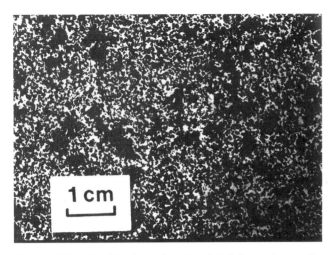

Figure 27.1-10. Matrix sulphides (white) enclose relict cumulus olivine (black) in serpentinized komatiitic peridotite, Alexo mine, Timmins, Ontario. GSC 203887

shales and komatiitic basalts (St-Onge and Lucas, 1990). The peridotites were originally regarded as intrusions (Barnes et al., 1982), but an extrusive origin now seems more likely (Barnes and Barnes, 1990), and they appear to form part of the lower Chukotat volcanic sequence (Lesher et al., 1991).

Nature of ore deposits

Most komatiite-hosted nickel sulphide deposits occur at or near the base of an extrusive sequence, and are typically in the lowest, thickest, and most magnesian of the flows. Barrie et al. (1993) have noted that MgO contents (marginal chilled zone, anhydrous) of ore-bearing komatiites are 20 to 35% whereas those of barren komatiites are 10 to 25%.

The ores comprise conformable lenses of mineralization concentrated within, and near the basal contact of the layer-like or lenticular peridotitic host unit. Most ore bodies occur in the lowermost flow unit, but some, generally smaller, are in overlying units (Fig. 27.1-9). The mineralization includes matrix (continuous network of sulphides enclosing olivine crystals) and massive sulphides, and in many cases, disseminated sulphides. The common sequence of sulphides from the base upward is massive, matrix, and disseminated. The ores commonly occur in local depressions of one kind or another in the base of the host unit. At Kambalda, many of these depressions resemble fault-bounded troughs (Fig. 27.1-9), but their mode of formation remains unclear (Lesher, 1989). Similarly, the Langmuir deposits in the Timmins area of Ontario are controlled by basal trough-like structures (Green and Naldrett, 1981). In the less structurally disrupted deposits of the Cape Smith Belt, the ore-bearing basal depressions appear to be simple embayments, which Lesher (1989) has interpreted as thermal erosion channels (<u>see</u> "Genetic model").

The host rocks and ores differ from district to district in the magnesian character of the host rocks, and the Ni:Cu ratios of the ores (Table 27.1-3), and may also show variation within a single district. On a district comparison basis, MgO content of the ultramafic hosts correlates

roughly with Ni:Cu ratio of the ores. Thompson appears unique in that the peridotitic host rocks are harzburgites rather than wehrlites as they are in most other cases.

Metamorphism and alteration

Komatiitic ores and their hosts have generally been subjected to regional metamorphism at grades ranging from prehnite-pumpellyite (e.g., Alexo, east of Timmins, Ontario) to upper amphibolite (e.g., Thompson, Manitoba). In general, this has profoundly affected silicate mineralogy and texture, whereas sulphide-rich ores have simply recrystallized with a new fabric but with little change in the original pyrrhotite-pentlandite-pyrite assemblage (Lesher, 1989). At low metamorphic grades, peridotitic hosts become serpentinized but typically retain relict cumulus olivine and intercumulus matrix sulphide textures (Fig. 27.1-10). At intermediate grades (lower amphibolite), the new metamorphic minerals may form intimate intergrowths with sulphides that seriously hamper efficient mechanical recovery of nickel sulphides. At these metamorphic grades, carbonate (magnesite-dolomite) alteration is common. This destroys primary textures, and partially or completely converts pyrrhotite-pentlandite assemblages to pyrite-millerite (NiS). In leaner mineralization (<5% sulphides), serpentinization results in replacement of pentlandite by heazlewoodite (Ni_3S_2) or awaruite (~$FeNi_3$), and of pyrrhotite by magnetite (Eckstrand, 1975), rendering metallurgical recovery more complex.

Effects of deformation

Because sulphide concentrations (especially massive sulphides) are structurally incompetent, and because the ores generally occur at basal contacts where there is a competency contrast with wall rocks, later quasiconformable faulting has commonly been localized within the ore zones. This has in certain deposits (e.g. Kambalda, Langmuir) produced sheet-like zones of mobilized sulphide-matrix breccia ores which in some cases extend beyond the host ultramafic unit.

Deformation of ores and host rocks at the Thompson mine has been extreme, and little remains of their original physical relationships. The parent harzburgitic sill(s) has been stretched and dismembered into a horizon of ultramafic blocks and boudins that are enclosed in the sulphide-rich pelitic schist unit that hosts the ores (Bleeker, 1990). The pyrrhotite-dominated nickel sulphide ores are zones in the pelitic schist that contain abundant quasiconformable lenses and stringers of massive sulphides which enclose numerous wall rock inclusions. Other deposits and their hosts in the Thompson Nickel Belt (Pipe, Birchtree, Soab, Manibridge, Bucko, Bowden) show similar though less extreme effects of deformation. In general, most komatiitic sulphide-rich ores have undergone some deformation, typically resulting in sulphide-matrix breccias containing wall rock clasts (Fig. 27.1-11).

Mineralogy

The main ore minerals at Kambalda in order of decreasing abundance are pyrrhotite, pentlandite, pyrite, magnetite, chromite, millerite, and violarite. This ore assemblage is representative for most komatiitic deposits.

Figure 27.1-11. Suphide-matrix (white) breccia ore with clasts of mafic metavolcanic wall rocks (black), Shebandowan, Ontario. GSC 203886-F

Original minerals in komatiites comprised mainly olivine, clinopyroxene, glass, and chromite (Lesher, 1989). However all komatiitic rocks have been metamorphosed and/or altered to assemblages that include combinations of lizardite, antigorite, tremolite, chlorite, talc, magnesite, dolomite, magnetite, and ferrochromite. The particular assemblage depends on the composition of the ultramafic precursor, metamorphic conditions, and the extent and intensity of carbonate metasomatism.

Other tholeiitic intrusion-hosted nickel-copper: subtype 27.1d

Geological setting

A diverse group of mostly small nickel-copper deposits are associated with tholeiitic, commonly stock-like, intrusions that occur in a variety of terranes. These include Lynn Lake (Pinsent, 1980), Giant Mascot (Aho, 1956), Montcalm (Barrie and Naldrett, 1989), and St. Stephen (Paktunc, 1989) in Canada. The most notable foreign examples are Monchegorsk in Russia (Gorbunov et al., 1985), Kotalahti in Finland (Gaál, 1980; Papunen and Koskinen, 1985), and Selebi-Pikwe in Botswana (Table 27.1-1). Ages range from Archean to Mesozoic, and the settings include Archean greenstone belts, and Proterozoic and Phanerozoic, weakly to strongly metamorphosed, sedimentary-volcanic fold belts. The intrusions are generally small, for example the Lynn Lake pluton is 1.5 km by 3.5 km in surface plan. They are strongly differentiated, generally ranging in composition from peridotite to quartz diorite. Most of the intrusions appear to have been deformed and metamorphosed.

Nature of ore deposits

The nickel-copper ores are mainly associated with the mafic and ultramafic facies of the intrusions, rarely with the intermediate or felsic facies. This is the case at Kotalahti where most ore zones occur in peridotites and pyroxenites, and only a few are in gabbro (Papunen and Koskinen, 1985). At St. Stephen, the main ores are hosted in troctolite

and olivine gabbro (Paktunc, 1989). The ores typically form lenticular shoots, such as those in the Lynn Lake deposit (Pinsent, 1980) where many of the orebodies are subvertically oriented cigar-shaped zones of mineralization within similarly oriented masses of peridotite, gabbro, and amphibolite. All the three usual types of mineralization, massive, matrix, and disseminated, are present in most deposits. The massive type is commonly sulphide-matrix breccia or massive veins that result from deformation of sulphide-rich ore, as for example at Lynn Lake and Montcalm. The less sulphide-rich ores commonly consist of lenticular or ribbon-like zones of disseminated sulphide blebs and stringers.

Ore composition

The sulphide minerals consist of the usual pyrrhotite-pentlandite-chalcopyrite-pyrite assemblage. Pyrrhotite is more abundant than in ores of the other subtypes, reflecting the generally lower bulk content of Ni (mostly <5 wt.%), Cu, and PGEs in sulphides of this subtype. Nickel:copper ratios typically lie in the range 3:1 to 1:1.

The Voisey Bay Ni-Cu deposit near Nain in Labrador, discovered in 1994, adds a significant new example to the tholeiitic intrusion-hosted type of nickel deposits (27.1d). It consists of a lens of massive sulphide up to 100 m thick which contains a minimum of 27 Mt grading 3.6% Ni, 2.17% Cu and 0.15% Co. The lens is hosted in a melatroctolitic dyke-like body that may be an apophysis of a troctolitic member (Reid Brook intrusion) of the anorogenic Nain Plutonic Suite (1350-1290 Ma). The deposit is located near the 1800 Ma collisional boundary between the Archean Nain Province and the Paleoproterozoic Churchill Province (Ryan et al., 1995).

DEFINITIVE CHARACTERISTICS

General

1. All magmatic sulphide deposits are associated with mafic and/or ultramafic igneous rocks. In most subtypes of these deposits, these bodies are intrusions, but in the case of the komatiitic subtype (27.1c) they are extrusive flows.

2. The sulphide ores are commonly segregated at or near the base of the hosting magmatic body, and consist of the assemblage pyrrhotite-pentlandite-chalcopyrite(-pyrite-magnetite).

Astrobleme-associated nickel-copper: subtype 27.1a

1. The Sudbury deposits are uniquely characterized by the associated presence of shock metamorphic features in the wall rocks to the intrusion, attributed by most geologists to meteoritic impact. Most ores that extend into footwall rocks are hosted by breccias which are considered shock or impact related.

2. The ores are contained in the Sublayer, a unit at the base of the Sudbury Igneous Complex that is characterized by the presence of nickel-copper sulphides and ultramafic xenoliths of nonlocal origin. The Sublayer

consists of a discontinuous noritic basal contact unit and quartz dioritic apophyses ("Offsets") that project into footwall rocks.

Rift- and continental flood-basalt-associated nickel-copper: subtype 27.1b

1. These deposits are associated with mafic-ultramafic intrusions that are related to continental rifting and flood basalt suites.
2. At Noril'sk-Talnakh, the richest ores are separate apophyses of massive sulphide that lie below the associated differentiated sill.
3. At Jinchuan, the ore consists of matrix and disseminated sulphides in dunitic and peridotitic facies of a dyke-like ultramafic host intrusion.

Komatiite-hosted nickel: subtype 27.1c

1. The ores are typically hosted by Precambrian komatiitic volcanic flows.
2. The orebodies are generally lenses of matrix and massive sulphides that occupy depressions in the floors of the flows.
3. These are the most nickel-rich ores of all subtypes; the more magnesian komatiitic hosts contain ores that commonly have Ni:Cu ratios >10.

Other tholeiitic intrusion-hosted nickel-copper: subtype 27.1d

1. The ores are associated with well differentiated, generally stock-like tholeiitic ultramafic-mafic intrusions of a wide range of ages and tectonic settings.

GENETIC MODEL

Genetic models for all subtypes of nickel sulphide deposits require (1) the generation of a sulphide melt associated with a mafic or ultramafic silicate magma, and (2) its accumulation into economic concentrations of nickel-copper sulphide. If such a sulphide melt is brought into equilibrium with a sufficiently large volume of silicate magma containing normal background levels of nickel, copper, and PGEs, the preference of these metals for the sulphide phase (i.e., their chalcophile tendency) will produce ore grade concentrations in the melt. Experiment-based numerical modelling involving partition coefficients and "R factors" (ratio of silicate magma to sulphide melt) has demonstrated that the melt should acquire the approximate Ni, Cu, and PGE contents that are observed in magmatic sulphide deposits (Campbell and Naldrett, 1979; Naldrett, 1989a, b; Fleet et al., 1993). Furthermore, variations in the contents and ratios of base and precious metals in these sulphide deposits can be attributed to differences in the proportions (R factors) of silicate and sulphide melts that came into equilibrium, and the compositions of the original silicate melts.

An important corollary of the silicate-sulphide equilibration just described is that the resulting silicate mass should be depleted of some of its original metal content.

This depletion has been identified in a number of komatiitic hosts containing nickel deposits (Thompson and Naldrett, 1984; Duke, 1986; Lesher, 1989).

It is generally considered that Ni, Cu, and PGEs were present in the original magmas in quantities sufficient to produce the sulphide ores in which these metals became concentrated. Whether there was also enough sulphur in the original magma to produce the sulphide ores is not clear; in fact this seems doubtful. Evidence from a number of magmatic nickel deposits such as Thompson (Eckstrand et al., 1989a), Kambalda, and other Archean komatiitic deposits, Ungava, Duluth Complex (Ripley, 1986), Crystal Lake (Eckstrand et al., 1989b), and St. Stephen (Paktunc, 1989) suggests that nickeliferous sulphide melts were generated by contamination of the mafic magma with sulphur from crustal rocks, generally sulphidic sediments. The evidence is based on isotopic data (S, Sr, Nd, Os), Se/S ratios, and the presence of xenolithic material. Mechanisms of contamination that have been suggested include assimilation of sulphidic wall rock and metasomatic migration of sulphur out of the wall rock, into the magma. Consistent with this hypothesis of a crustal source of sulphur, is the fact that virtually all nickel sulphide deposits occur in rocks underlain by continental crust.

However, some deposits have characteristics that contradict crustal contamination. In the Noril'sk deposits, although S isotopic evidence reflects a crustal source of sulphur (Grinenko, 1985), Os isotopes indicate a mantle source (Walker et al., 1994). Abitibi komatiitic flows that host nickel deposits show little isotopic evidence of crustal contamination (Barrie et al., 1993). Cases such as these could possibly be explained by invoking uncontaminated mantle-derived magmas that were affected in the crustal regime only by the addition of sulphur through metasomatism.

Basal massive and matrix sulphides suggest early separation (liquation) of sulphide melt out of the magma, and its gravitational settling. In contrast, internal zones of disseminated sulphides more likely result from separation when silicate crystallization was more advanced, and settling of sulphide droplets was impeded (Lesher, 1989).

When a large body of sulphide melt accumulates and solidifies, it will undergo fractional crystallization, and, analogous to silicate magmas, produce crystallized sulphides of a range of compositions. Naldrett et al. (1994b) have proposed that this mechanism operated in the case of the Sudbury sulphide ores of the North Range (Fig. 27.1-4B). It has given rise to a sequence of ore zones that range from pyrrhotite-rich and Cu-poor at the base of the Lower zone (representing early crystallized monosulphide solid solution) to Cu-Pt-Pd-rich and lacking in pyrrhotite well down in the footwall (representing late residual liquids). The same mechanism could explain the dramatic compositional zoning of copper, platinum, and palladium content in massive sulphide ore at Talnakh (Naldrett et al., 1994c; Zientek et al., 1994).

The different subtypes of deposits differ in detail with regard to modes of emplacement of the magmatic hosts and the accumulation of the sulphide ores.

The Sudbury Igneous Complex and its ores (subtype 27.1a) are considered to have been emplaced at the site of a major explosive event, probably a meteoritic impact, which is the most favored hypothesis, or possibly a volcanic explosion (Naldrett, 1984). Shock features identified in footwall rocks indicative of peak pressures of 20 GPa (Müller-Mohr, 1992) appear to require a meteoritic impact origin. The composition of the Sudbury Igneous Complex indicates a large crustal component, and the hypotheses proposed to explain this range from strongly contaminated, mantle-derived basaltic magma (Naldrett, 1984) to a completely locally derived melt resulting from meteoritic impact (Golightly, 1994; Grieve, 1994). An intermediate hypothesis proposes that the Upper zone granophyre represents impact melt, and the Lower zone norite is a highly contaminated mantle-derived magma (Chai and Eckstrand, 1994).

In other respects, however, the genesis of the Sudbury ores seems explicable by normal terrestrial processes. After recognition of the Sublayer and of its significance as the main ore-bearing unit (Souch et al., 1969), it came to be regarded as a separate intrusion at the base of the complex. However, whether the Sublayer is older or younger than the main Irruptive remains controversial because of apparently conflicting evidence. It seems possible that the Sublayer simply represents the chilled basal border zone of the complex in which sulphides and dense xenoliths have accumulated gravitationally from the overlying Lower zone (Golightly, 1994).

Rift- and continental flood basalt-associated nickel-copper deposits (subtype 27.1b) are associated with major magmatic provinces. In the case of Noril'sk-Talnakh, the Noril'sk-Kharayelakh fault zone was the main conduit for this magmatism, and gave rise to the thick and geochemically varied trap sequence and sill-like intrusions. Though differing degrees of crustal contamination have affected the flows, the ore-bearing intrusions and ores seem curiously devoid of crustal influence (Czamanske et al., 1994), except that sulphur in the ores ($\delta^{34}S$ = 10-12‰) must clearly have come in large part from a crustal source that has not yet been identified.

Many komatiitic nickel deposits (subtype 27.1c) have been shown to contain sulphur that was most likely derived from the sulphide in sediments (Groves et al., 1979; Eckstrand et al., 1989a). Lesher (1989) has proposed that the sulphur was gained by melting and assimilation of sulphidic sediments under the main channel of komatiite flow whose extrusion temperature was significantly greater than its solidus, 1200°C.

The komatiitic magmas themselves were derived by significant partial melting of mantle material. Their great abundance in the Archean and Lower Proterozoic, and subsequent virtual disappearance is believed to reflect the greater thermal flux and lesser thickness of the Archean crust.

RELATED DEPOSIT TYPES

The Fe-rich immiscible sulphide melts from which nickel sulphide deposits originate can only exist at relatively high temperature (greater than ~900°C); this limits the possible host magmas to those of Mg-rich mafic and ultramafic composition, whose minimum solidus temperatures would be about 1000°C. In such primitive (i.e., high temperature) magmas, the only available elements that will preferentially enter the coexisting sulphide melt in sufficient quantity to produce ore grade concentrations are Ni, Cu, Co, PGEs, and Au. These are the very elements that occur in magmatic nickel-copper sulphide (subtype 27.1) and magmatic platinum group element (subtype 27.2) deposits. Consequently the apparent lack of other, related magmatic sulphide deposits is not surprising.

Magmatic nickel arsenide-chromite occurrences have been reported in peridotitic massifs in Spain and Morocco (Leblanc et al., 1990). These deposits also contain Cu, Co, PGEs, and Au, and appear similar in many respects to the nickel-copper sulphide subtype (27.1), but are rare, and seem to have little economic significance.

Mafic-ultramafic-hosted chromite deposits (Type 28) occur in similarly primitive magmatic rocks, in some cases even in the same magmatic complex. However chromite deposits do not require formation of an immiscible sulphide phase, a process that is fundamental for the genesis of magmatic sulphide deposits.

EXPLORATION GUIDES

The features of interest as guides to exploration can be subdivided into those applicable at different scales ranging from regional scale, through camp scale, to deposit scale.

Regional

- Magmatic nickel-copper sulphide deposits are universally associated with mafic and ultramafic magmatic rocks.

- These are intrusions associated with flood basalt provinces and/or continental rift regimes in the case of subtype 27.1b. The late Proterozoic Franklin diabase-gabbro suite on Victoria Island may have potential for rift-flood basalt-related nickel deposits similar to those of the Noril'sk-Talnakh region of Russia (Jefferson et al., 1993).

- Nickel deposits of the komatiitic subtype (27.1c) occur in Archean greenstone belts, and Proterozoic sedimentary-volcanic fold belts.

- The ore-related highly differentiated tholeiitic intrusions of subtype 27.1d can occur in a variety of terranes.

Camp scale

- Selection of the most favorable intrusion or flow sequence may be aided by recognizing depletion of Ni, Cu, or PGEs in the prospective host rock (Fedorenko, 1994). Such depletion should ideally indicate the equilibration of the magma with a sulphide melt that may have resulted in an economic nickel sulphide deposit.

- Chemical indications of sulphide contamination that could potentially generate economic nickel sulphide concentrations would include: (1) $\delta^{34}S$ values distinctly different from those of the mantle; (2) Se/S ratios much lower than those of the mantle; and (3) anomalously high Zn contents in associated chromite, indicative of the assimilation of Zn-bearing sulphidic metasediments.

Deposit scale

- The preferred site for nickel sulphide ores is at the stratigraphic base of the hosting mafic-ultramafic intrusion or flow sequence. In the case of komatiitic flows, orebodies may also occur at the base of flows higher in the sequence.

- At Sudbury and Noril'sk, some orebodies lie a few metres to hundreds of metres below the associated intrusion. These were the sites of original emplacement of the sulphide ores.

- The thickened parts of komatiitic sequences is where komatiitic ores are most commonly found.

- Nickel deposits that have been deformed, particularly by strike-parallel faults, have, in some cases, been mobilized along the faults well beyond the lateral extent of the associated host intrusion or flow (e.g., Falconbridge mine, Gordon Lake mine, Redstone mine).

ACKNOWLEDGMENTS

The author gratefully acknowledges a critical review by L.J. Hulbert, and useful comments by R.I. Thorpe. B. Williamson assembled the tonnage and grade data.

SELECTED BIBLIOGRAPHY

References marked with asterisks (*) are considered to be the best sources of general information on this deposit type.

Abel, M.K., Buchan, R., Coats, C.J.A., and Penstone, M.E.
1979: Copper mineralization in the Footwall Complex, Strathcona mine, Sudbury, Ontario; Canadian Mineralogist, v. 17, p. 275-285.

Aho, A.E.
1956: Geology and genesis of ultrabasic nickel-copper-pyrrhotite deposits at the Pacific Nickel property, southwestern British Columbia; Economic Geology, v. 51, p. 444-481.

Baragar, W.R.A. and Scoates, R.F.J.
1981: The Circum-Superior Belt: a Proterozoic plate margin?; in Precambrian Plate Tectonics, (ed.) A. Kröner; Elsevier, Amsterdam, p. 297-330.

Barnes, S.-J.
1985: The petrography and geochemistry of komatiite flows from the Abitibi greenstone belt and a model for their formation; Lithos, v. 18, p. 241-270.

*Barnes, S.-J. and Barnes, S.J.
1990: A new interpretation of the Katiniq nickel deposit; Economic Geology, v. 85, p. 1269-1272.

Barnes, S.J., Coats, C.J.A., and Naldrett, A.J.
1982: Petrogenesis of a Proterozoic nickel sulfide-komatiite association: the Katiniq sill, Ungava, Quebec; Economic Geology, v. 77, p. 413-429.

Barrie, C.T. and Naldrett, A.J.
1989: Geology and tectonic setting of the Montcalm Gabbroic Complex and Ni-Cu deposit, western Abitibi Subprovince, Ontario, Canada; in Magmatic Sulphides - The Zimbabwe Volume, (ed.) M.D. Prendergast and M.J. Jones; Institution of Mining and Metallurgy, London, p. 151-164.

Barrie, C.T., Ludden, J.N., and Green, T.H.
1993: Geochemistry of volcanic rocks associated with Cu-Zn and Ni-Cu deposits in the Abitibi subprovince; Economic Geology, v. 88, p. 1341-1358.

Billington, L.G.
1984: Geological review of the Agnew nickel deposit, Western Australia; in Sulfide Deposits in Mafic and Ultramafic Rocks, (ed.) D.L. Buchanan and M.J. Jones; Institution of Mining and Metallurgy, Special Publication, p. 43-54.

Binney, W.P., Poulin, R.Y., Sweeny, J.M., and Halladay, S.H.
1994: The Lindsley Ni-Cu-PGE deposit and its geological setting; in The Sudbury-Noril'sk Symposium, (ed.) A.J. Naldrett, P.C. Lightfoot, and P. Sheahan; Ontario Geological Survey, Special Publication 5, p. 91-103.

Bleeker, W.
1990: Thompson area - general geology and ore deposits; in Geology and Mineral Deposits of the Flin Flon and Thompson Belts, Manitoba, (ed.) A.G. Galley, A.H. Bailes, E.C. Syme, W. Bleeker, J.J. Macek, and T.S. Gordon; International Association on the Genesis of Ore Deposits, Guide Book No. 10, Geological Survey of Canada, Open File 2165, p. 93-136.

Brett, P.R., Jones, R.E., Leuner, W.R., and Latulippe, M.
1976: LaMotte Township; Quebec Department of Natural Resources, Geological Report 160, 158 p.

Campbell, I.H. and Naldrett, A.J.
1979: The influence of silicate:sulfide ratios on the geochemistry of magmatic sulfides; Economic Geology, v. 74, p. 1503-1505.

Card, K.D., Gupta, V.K., McGrath, P.H., and Grant, F.S.
1984: The Sudbury Structure; its regional geological and geophysical setting; in The Geology and Ore Deposits of the Sudbury Structure, (ed.) E.G. Pye, A.J. Naldrett, and P.E. Giblin; Ontario Geological Survey, Special Volume 1, p. 25-43.

Chai, G. and Eckstrand, O.R.
1994: Rare earth element characteristics and origin of the Sudbury igneous complex; Chemical Geology, v. 113, p. 221-244.

Chai, G. and Naldrett, A.J.
1992a: The Jinchuan ultramafic intrusion: cumulate of high-Mg basaltic magma; Journal of Petrology, v. 33, p. 277-303.
*1992b: Characteristics of Ni-Cu-PGE mineralization and genesis of the Jinchuan deposit, northwest China; Economic Geology, v. 87, p. 1475-1495.

Chen, J.Y. and Mingliang, J.
1987: Jin Chuan nickel; Engineering and Mining Journal, v. 188, no. 9 (September), p. 44-57.

*Coad, P.R.
1979: Nickel sulphide deposits associated with ultramafic rocks of the Abitibi belt and economic potential of mafic-ultramafic intrusions; Ontario Geological Survey, Study 20, 84 p.

Coats, C.J.A.
1982: Geology and nickel sulfide deposits of the Raglan area, Ungava, Quebec; ministère de l'Énergie et des Ressources, Quèbec, GM-40480, 121 p.

Coats, C.J.A. and Brummer, J.J.
1971: Geology of the Manibridge nickel deposit, Wabowden, Manitoba; Geological Association of Canada, Special Paper No. 9, p. 155-165.

*Coats, C.J.A. and Snajdr, P.
1984: Ore deposits of the North Range, Onaping-Levack area, Sudbury; in The Geology and Ore Deposits of the Sudbury Structure, (ed.) E.G. Pye, A.J. Naldrett, and P.E. Giblin; Ontario Geological Survey, Special Volume 1, p. 327-346.

Cochrane, L.B.
1984: Ore deposits of the Copper Cliff offset; in The Geology and Ore Deposits of the Sudbury Structure, (ed.) E.G. Pye, A.J. Naldrett, and P.E. Giblin; Ontario Geological Survey, Special Volume 1, p. 347-359.

Cogulu, E.H.
1990: Mineralogical and petrological studies of the Crystal Lake Intrusion, Thunder Bay, Ontario; Geological Survey of Canada, Open File 2277, 15 p.
1993: Mineralogy and chemical variations of sulphides from the Crystal Lake Intrusion, Thunder Bay, Ontario; Geological Survey of Canada, Open File 2749, 17 p.

Corfu, F.
1993: The evolution of the southern Abitibi greenstone belt in light of precise U-Pb geochronology; Economic Geology, v. 88, p. 1323-1340.

Cowan, E.J. and Schwerdtner, W.M.
1994: Fold origin of the Sudbury Basin; Ontario Geological Survey, Special Publication 5, p. 45-55.

Czamanske, G.K., Wooden, J.L., Zientek, M.L., Fedorenko, V.A., Zen'ko, T.E., Kent, J., King, B.-S.W., Knight, R.J., and Siems, D.F.
1994: Geochemical and isotopic constraints on the petrogenesis of the Noril'sk-Talnakh ore-forming system; in The Sudbury-Noril'sk Symposium, (ed.) A.J. Naldrett, P.C. Lightfoot, and P. Sheahan, Ontario Geological Survey, Special Publication 5, p. 313-341.

*DeYoung, J.H., Jr., Sutphin, D.M., Werner, A.B.T., and Foose, M.P.
1985: International Strategic Minerals Inventory summary report - nickel; United States Geological Survey, Circular 930-D, 62 p.

Distler, V.V.
1994: Platinum mineralization of the Noril'sk deposits; in The Sudbury-Noril'sk Symposium, (ed.) A.J. Naldrett, P.C. Lightfoot, and P. Sheahan; Ontario Geological Survey, Special Publication 5, p. 243-260.

Dressler, B.O.
1984: The effects of the Sudbury Event and the intrusion of the Sudbury igneous complex on the footwall rocks of the Sudbury Structure; in The Geology and Ore Deposits of the Sudbury Structure, (ed.) E.G. Pye, A.J. Naldrett, and P.E. Giblin; Ontario Geological Survey, Special Volume 1, p. 97-136.

Duke, J.M.
1986: Petrology and economic geology of the Dumont sill: an Archean intrusion of komatiitic affinity in northwestern Quebec; Geological Survey of Canada, Economic Geology Report 35, 56 p.

Duzhikov, O.A., Distler, V.V., Strunin, B.M., Mkrtychyan, A.K., Sherman, M.L., Sluzhenikin, S.S., and Lurje, A.M.
1992: Geology and metallogeny of sulfide deposits of Noril'sk region, U.S.S.R.; Society of Economic Geologists, Special Publication No. 1, 242 p.

Eckstrand, O.R.
1975: The Dumont serpentinite: a model for control of nickeliferous opaque mineral assemblages by alteration reactions in ultramafic rocks; Economic Geology, v. 70, p. 183-201.

Eckstrand, O.R., Cogulu, E.H., and Scoates, R.F.J.
1989b: Magmatic Ni-Cu-PGE mineralization in the Crystal Lake layered intrusion, Ontario, and the Fox River sill, Manitoba; in Workshop on the Applicability of Gold and Platinum-group Element Models in Minnesota, (ed.) G.B. Morey; Minnesota Geological Survey, p. 45-46.

Eckstrand, O.R., Grinenko, L.N., Krouse, H.R., Paktunç, A.D., Schwann, P.L., and Scoates, R.F.J.
1989a: Preliminary data on sulphur isotopes and Se/S ratios, and the source of sulphur in magmatic sulphides from the Fox River Sill, Molson Dykes, and Thompson nickel deposits, northern Manitoba; in Current Research, Part C; Geological Survey of Canada, Paper 89-1C, p. 235-242.

Fedorenko, V.A.
1994: Evolution of magmatism as reflected in the volcanic sequence of the Noril'sk region; in The Sudbury-Noril'sk Symposium, (ed.) A.J. Naldrett, P.C. Lightfoot, and P. Sheahan; Ontario Geological Survey, Special Publication 5, p. 171-183.

Fleet, M.E., Chryssoulis, S.L., Stone, W.E., and Weisener, C.G.
1993: Partitioning of platinum-group elements and Au in the Fe-Ni-Cu-S system: experiments on the fractional crystallization of sulfide melt; Contributions to Mineralogy and Petrology, v. 115, p. 36-44.

Gaál, G.
1980: Geological setting and intrusion tectonics of the Kotalahti nickel-copper deposit, Finland; Geological Society of Finland, Bulletin, v. 52, p. 101-128.

Gemuts, I. and Theron, A.C.
1975: The Archaean between Coolgardie and Norseman - stratigraphy and mineralization; in Economic Geology of Australia and Papua New Guinea, I. Metals, (ed.) C.L. Knight; Australasian Institute of Mining and Metallurgy, Monograph 5, p. 66-74.

Geul, J.J.C.
1970: Devon and Pardee Townships and the Stuart Location; Ontario Department of Mines, Geological Report 87, 52 p.

Golightly, J.P.
1994: The Sudbury Igneous Complex as an impact melt: evolution and ore genesis; in The Sudbury-Noril'sk Symposium, (ed.) A.J. Naldrett, P.C. Lightfoot, and P. Sheahan; Ontario Geological Survey, Special Publication 5, p. 105-117.

Gorbunov, G.I., Yakovlev, Yu. N., Goncharov, Yu.V., Gorlov, V.A., and Tel'nov, V.A.
1985: The nickel areas of the Kola Peninsula; Geological Survey of Finland, Bulletin 333, p. 41-109.

Grant, R.W. and Bite, A.
1984: Sudbury quartz diorite offset dikes; in The Geology and Ore Deposits of the Sudbury Structure, (ed.) E.G. Pye, A.J. Naldrett, and P.E. Giblin; Ontario Geological Survey, Special Volume 1, p. 275-300.

Green, A.H. and Naldrett, A.J.
1981: The Langmuir volcanic peridotite-associated nickel deposits: Canadian equivalents of the Western Australian occurrences; Economic Geology, v. 76, p. 1503-1523.

Green, J.C.
1983: Geologic and geochemical evidence for the nature and development of the Middle Proterozoic (Keweenawan) Midcontinent rift of North America; Tectonophysics, v. 94, p. 413-437.

*Gresham, J.J. and Loftus-Hills, G.D.
1981: The geology of the Kambalda nickel field, Western Australia; Economic Geology, v. 76, p. 1373-1416.

Grieve, R.A.F.
1994: An impact model of the Sudbury structure; in The Sudbury-Noril'sk Symposium, (ed.) A.J. Naldrett, P.C. Lightfoot, and P. Sheahan; Ontario Geological Survey, Special Publication 5, p. 119-132.

Grinenko, L.I.
1985: Sources of sulfur of the nickeliferous and barren gabbro-dolerite intrusions of the northwest Siberian platform; International Geology Review, v. 27, p. 695-708.

Groves, D.I., Barrett, F.M., and McQueen, K.G.
1979: The relative roles of magmatic segregation, volcanic exhalation and regional metamorphism in the generation of volcanic-associated nickel ores of Western Australia; Canadian Mineralogist, v. 17, p. 319-336.

*Groves, D.I., Hudson, D.R., Marston, R.J., and Ross, J.R. (ed.)
1981: A Special Issue on Nickel Deposits and their Host Rocks in Western Australia; Economic Geology, v. 76, no. 6, p. 1289-1783.

Groves, D.I., Lesher, C.M., and Gee, R.D.
1984: Tectonic setting of sulfide nickel deposits of the Western Australian shield; in Sulfide Deposits in Mafic and Ultramafic Rocks, (ed.) D.L. Buchanan and M.J. Jones; Institution of Mining and Metallurgy, Proceedings of IGCP Projects 161 and 91, Third Nickel Sulphide Field Conference, Perth, Western Australia, May 23-25, 1982, p. 1-13.

Hulbert, L.J., Duke, J.M., Eckstrand, O.R., Lydon, J.W., Cabri, L.J., and Irvine, T.N.
1988: Geological environments of the platinum group elements; Geological Survey of Canada, Open File 1440, 148 p.

Jefferson, C.W., Chandler, F.W., Hulbert, L.J., Smith, J.E.M., Fitzhenry, K., and Powis, K.
1993: Assessment of mineral and energy resource potential in the Laughland Lake terrestrial area and Wager Bay marine area, N.W.T.; Geological Survey of Canada, Open File 2659, 48 p.

Krogh, T.E., Davis, D.W., and Corfu, F.
1984: Precise U-Pb zircon and baddeleyite ages for the Sudbury area; in The Geology and Ore Deposits of the Sudbury Structure, (ed.) E.G. Pye, A.J. Naldrett, and P.E. Giblin; Ontario Geological Survey, Special Volume 1, p. 431-446.

Leblanc, M., Gervilla, F., and Jedwab, J.
1990: Noble metals segregation and fractionation in magmatic ores from Ronda and Beni Bousera lherzolite massifs (Spain, Morocco); Mineralogy and Petrology, v. 42, p. 233-248.

*Lesher, C.M.
1989: Komatiite-associated nickel sulfide deposits; in Ore Deposition Associated with Magmas, (ed.) J.A. Whitney and A.J. Naldrett; Reviews in Economic Geology, v. 4, p. 45-101.

Lesher, C.M., Thacker, J.L., Thibert, F., Tremblay, C., and Dufresne, M.W.
1991: Physical volcanology of Proterozoic komatiitic peridotites in Chukotat Group, Cape Smith Belt, New Quebec; Geological Association of Canada-Mineralogical Association of Canada Annual Meeting, Toronto, Ontario, May 27-29, 1991, Program with Abstracts, v. 16, p. A74.

*Li, C., Naldrett, A.J., Coats, C.J.A., and Johannessen, P.
1992: Platinum, palladium, gold, and copper-rich stringers at the Strathcona mine, Sudbury: their enrichment by fractionation of a sulfide liquid; Economic Geology, v. 87, p. 1584-1598.

Likachev, A.P.
1994: Ore-bearing intrusions of the Noril'sk region; in The Sudbury-Noril'sk Symposium, (ed.) A.J. Naldrett, P.C. Lightfoot, and P. Sheahan; Ontario Geological Survey, Special Publication 5, p. 185-201.

Listerud, W.H. and Meineke, D.G.
1977: Mineral resources of a portion of the Duluth Complex and adjacent rocks in St. Louis and Lake counties, northeastern Minnesota; Minnesota Department of Natural Resources, Division of Minerals, Report 93, 74 p.

Marston, R.J., Groves, D.I., Hudson, D.R., and Ross, J.R.
1981: Nickel sulfide deposits in Western Australia: a review; Economic Geology, v. 76, p. 1330-1363.

Milkereit, B., Green, A., and the Sudbury Working Group
1992: Deep geometry of the Sudbury structure from seismic reflection profiling; Geology, v. 20, p. 807-811.

Money, D.P.

1993: Metal zoning in the Deep Copper Zone 3700 level, Strathcona mine, Ontario; Exploration and Mining Geology, v. 2, p. 307-320.

Morrison, G.G., Jago, B.C., and White, T.L.

1994: Footwall mineralization of the Sudbury Igneous Complex; Ontario Geological Survey, Special Publication 5, p. 57-64.

Müller-Mohr, V.

1992: Sudbury Project: (5) new investigations on Sudbury Breccia; in Papers Presented to the International Conference on Large Meteorite Impacts and Planetary Evolution, (ed.) B.O. Dressler and V.L. Sharpton; Lunar Planetary Institute, Contribution No. 790, p. 53.

Naldrett, A.J.

1984: Summary, discussion, and synthesis; in The Geology and Ore Deposits of the Sudbury Structure, (ed.) E.G. Pye, A.J. Naldrett, and P.E. Giblin; Ontario Geological Survey, Special Volume 1, p. 533-569.

*1989a: Magmatic Sulfide Deposits; Clarendon Press-Oxford University Press, New York-Oxford, 186 p.

1989b: Sulfide melts: crystallization temperatures, solubilities in silicate melts, and Fe, Ni, and Cu partitioning between basaltic magmas and olivine; in Ore Deposition Associated with Magmas, (ed.) J.A. Whitney and A.J. Naldrett; Reviews in Economic Geology, v. 4, p. 5-20.

*1989c: Ores associated with flood basalts; in Ore Deposition Associated with Magmas, (ed.) J.A. Whitney and A.J. Naldrett; Reviews in Economic Geology, v. 4, p. 103-118.

1989d: Contamination and the origin of the Sudbury structure and its ores; in Ore Deposition Associated with Magmas, (ed.) J.A. Whitney and A.J. Naldrett; Reviews in Economic Geology, v. 4, p. 119-134.

1994: The Sudbury-Noril'sk Symposium, an overview; Ontario Geological Survey, Special Publication 5, p. 3-8.

Naldrett, A.J. and Hewins, R.H.

1984: The main mass of the Sudbury igneous complex; in The Geology and Ore Deposits of the Sudbury Structure, (ed.) E.G. Pye, A.J. Naldrett, and P.E. Giblin; Ontario Geological Survey, Special Volume 1, p. 235-251.

Naldrett, A.J., Asif, M., Gorbachev, N.S., Kunilov, V.Ye., Stekhin, A.I., Fedorenko, V.A., and Lightfoot, P.C.

1994c: The composition of the Ni-Cu ores of the Oktyabr'sky deposit, Noril'sk region; in The Sudbury-Noril'sk Symposium, (ed.) A.J. Naldrett, P.C. Lightfoot, and P. Sheahan; Ontario Geological Survey, Special Publication 5, p. 357-371.

Naldrett, A.J., Hewins, R.H., Dressler, B.O., and Rao, B.V.

1984: The contact sublayer of the Sudbury igneous complex; in The Geology and Ore Deposits of the Sudbury Structure, (ed.) E.G. Pye, A.J. Naldrett, and P.E. Giblin; Ontario Geological Survey, Special Volume 1, p. 253-274.

Naldrett, A.J., Innes, D.G., Sowa, J., and Gorton, M.P.

1982: Compositional variations within and between five Sudbury ore deposits; Economic Geology, v. 77, p. 1519-1534.

*Naldrett, A.J., Lightfoot, P.C., and Sheahan, P. (ed.)

1994a: The Sudbury-Noril'sk Symposium; Ontario Geological Survey, Special Publication 5, 423 p.

Naldrett, A.J., Pessaran, A., Asif, M., and Li, C.

1994b: Compositional variation in the Sudbury ores and prediction of the proximity of footwall copper-PGE orebodies; in The Sudbury-Noril'sk Symposium, (ed.) A.J. Naldrett, P.C. Lightfoot, and P. Sheahan; Ontario Geological Survey, Special Publication 5, p. 133-143.

Owen, D.L. and Coats, C.J.A.

1984: Falconbridge and East mines; in The Geology and Ore Deposits of the Sudbury Structure, (ed.) E.G. Pye, A.J. Naldrett, and P.E. Giblin; Ontario Geological Survey, Special Volume 1, p. 371-378.

Paktunç, A.D.

1984: Petrogenesis of ultramafic and mafic rocks of the Thompson Nickel Belt, Manitoba; Contributions to Mineralogy and Petrology, v. 88, p. 348-353.

1986: St. Stephen mafic-ultramafic intrusion and related nickel-copper deposits, New Brunswick; in Current Research, Part A; Geological Survey of Canada, Paper 86-1A, p. 327-331.

1989: Petrology of the St. Stephen intrusion and the genesis of related nickel-copper sulfide deposits; Economic Geology, v. 84, p. 817-840.

Papunen, H. and Koskinen, J.

1985: Geology of the Kotalahti nickel-copper ore; Geological Survey of Finland, Bulletin 333, p. 229-240.

*Papunen, H. and Vorma, A.

1985: Nickel deposits in Finland, a review; Geological Survey of Finland, Bulletin 333, p. 123-143.

*Pattison, E.F.

1979: The Sudbury sublayer: its characteristics and relationships with the main mass of the Sudbury Irruptive; Canadian Mineralogist, v. 17, p. 257-274.

Peredery, W.V.

1979: Relationship of ultramafic amphibolites to metavolcanic rocks and serpentinites in the Thompson belt, Manitoba; Canadian Mineralogist, v. 17, p. 187-200.

Peredery, W.V. and Geological staff

1982: Geology and nickel sulfide deposits of the Thompson belt, Manitoba; in Precambrian Sulphide Deposits, (ed.) R.W. Hutchinson, C.D. Spence, and J.M. Franklin; Geological Association of Canada, Special Paper No. 25, p. 165-209.

Pinsent, R.H.

1980: Nickel-copper mineralization in the Lynn Lake Gabbro; Manitoba Department of Energy and Mines, Economic Geology Report ER79-3, 138 p.

*Pye, E.G., Naldrett, A.J., and Giblin, P.E. (ed.)

1984: The geology and ore deposits of the Sudbury Structure; Ontario Geological Survey, Special Volume 1, 603 p.

Rempel, G.G.

1994: Regional geophysics at Noril'sk; in The Sudbury-Noril'sk Symposium, (ed.) A.J. Naldrett, P.C. Lightfoot, and P. Sheahan; Ontario Geological Survey, Special Publication 5, p. 147-160.

*Ripley, E.M.

1986: Genesis of Cu-Ni sulfide mineralization in the Duluth Complex; in Metallogeny of Basic and Ultrabasic Rocks (Regional Presentations), (ed.) S.S. Augustithis; Athens, Theophrastus, p. 391-414.

Roth, J.

1975: Exploration of the southern extension of the Manitoba Nickel Belt; The Canadian Institute of Mining and Metallurgy Bulletin, v. 68, no. 761, p. 73-80.

Rousell, D.H.

1984: Structural geology of the Sudbury Basin; in The Geology and Ore Deposits of the Sudbury Structure, (ed.) E.G. Pye, A.J. Naldrett, and P.E. Giblin; Ontario Geological Survey, Special Volume 1, p. 83-95.

Ryan, B., Wardle, R.J., Gower, C.F., and Nunn, G.A.G.

1995: Nickel-copper-sulphide mineralization in Labrador: the Voisey Bay discovery and its exploration implications; Newfoundland Department of Natural Resources, Geological Survey Branch, Current Research, Report 95-1, p. 177-204.

Schklanka, R. (ed.)

1969: Copper, nickel, lead and zinc deposits of Ontario; Ontario Department of Mines, Circular No. 12, 394 p.

Shanks, W.S. and Schwerdtner, W.M.

1991: Structural analysis of the central and southwestern Sudbury Structure, Southern Province, Canadian Shield; Canadian Journal of Earth Sciences, v. 28, p. 411-430.

Simonov, O.N., Lul'ko, V.A., Amosov, Yu.N., and Salov, V.M.

1994: Geological structure of the Noril'sk region; in The Sudbury-Noril'sk Symposium, (ed.) A.J. Naldrett, P.C. Lightfoot, and P. Sheahan; Ontario Geological Survey, Special Publication 5, p. 161-170.

*Souch, B.E., Podolsky T., and Geological Staff

1969: The sulfide ores of Sudbury; their particular relationship to a distinctive inclusion-bearing facies of the Nickel Irruptive; in Magmatic Ore Deposits, (ed.) H.D.B. Wilson; Economic Geology Monograph 4, Economic Geology Publishing Company, p. 252-261.

St-Onge, M.R. and Lucas, S.B.

1990: Evolution of the Cape Smith Belt; early Proterozoic continental underthrusting, ophiolite obduction, and thick-skinned folding; in The Early Proterozoic Trans-Hudson Orogen of North America, (ed.) J.F. Lewry and M.R. Stauffer; Geological Association of Canada, Special Paper 37, p. 313-351.

Thompson, J.F.H. and Naldrett, A.J.

1984: Sulphide-silicate reactions as a guide to Ni-Cu-Co mineralization in central Maine, U.S.A.; in Sulphide Deposits in Mafic and Ultramafic Rocks, (ed.) D.L. Buchanan and M.J. Jones; Institution of Mining and Metallurgy, London, p. 103-113.

Walker, R.J., Morgan, J.W., Hanski, E., and Smolkin, V.F.

1994: The role of the Re-Os isotope system in deciphering the origin of magmatic sulphide ores: a tale of three ores; in The Sudbury-Noril'sk Symposium, (ed.) A.J. Naldrett, P.C. Lightfoot, and P. Sheahan; Ontario Geological Survey, Special Publication 5, p. 343-355.

Walker, R.J., Morgan, J.W., Naldrett, A.J., Li, C., and Fassett, J.D.
1991:   Re-Os isotope systematics of Ni-Cu sulfide ores, Sudbury igneous complex, Ontario; evidence for a major crustal component; Earth and Planetary Science Letters, v. 105, p. 416-429.

Weiblen, P.W.
1982:   Keweenawan intrusive igneous rocks; in Geology and Tectonics of the Lake Superior Basin, (ed.) R.J. Wold and W.J. Hinze; Geological Society of America, Memoir 156, p. 57-82.

Weiblen, P.W. and Morey, G.B.
1980:   A summary of the stratigraphy, petrology, and structure of the Duluth Complex; in The Jackson Volume, (ed.) A.J. Irving and M.A. Dungan; American Journal of Science, v. 280-A, pt. 1, p. 88-133.

Zen'ko, T.E. and Czamanske, G.K.
1994:   Spatial and petrologic aspects of the intrusions of the Noril'sk and Talnakh ore junctions; in The Sudbury-Noril'sk Symposium, (ed.) A.J. Naldrett, P.C. Lightfoot, and P. Sheahan; Ontario Geological Survey, Special Publication 5, p. 263-281.

Zientek, M.L., Likhachev, A.P., Kunilov, V.E., Barnes, S.-J., Meier, A.L., Carlson, R.R., Briggs, P.H., Fries, T.L., and Adrian, V.M.
1994:   Cumulus processes and the composition of magmatic ore deposits: examples from the Talnakh district, Russia; in The Sudbury-Noril'sk Symposium, (ed.) A.J. Naldrett, P.C. Lightfoot, and P. Sheahan, Ontario Geological Survey, Special Publication 5, p. 373-392.

*Zurbrigg, H.F.
1963:   Thompson mine geology; Canadian Institution of Mining and Metallurgy, Transactions, v. LXVI, p. 227-236.

27.2 MAGMATIC PLATINUM GROUP ELEMENTS

C. Tucker Barrie

INTRODUCTION

The platinum group elements: Os, Ir, Ru, Rh, Pt, and Pd (collectively referred to as PGEs) are group VIIIA transition elements with strong siderophile and chalcophile characters. Although they are concentrated in a variety of geological settings, the principal PGE-dominant ores are associated with mafic-ultramafic intrusions. Platinum group element ores are rare: only a dozen mafic-ultramafic intrusions world-wide have economic or near-economic concentrations.

There are two principal deposit types: reef-type or stratiform PGE deposits, such as the Merensky Reef and UG-2 chromitite layer of the Bushveld Complex, South Africa, and the J-M Reef of the Stillwater Complex, Montana, are the most important. Canadian examples include the Big Trout Lake and Muskox occurrences (Fig. 27.2-1). A second type is termed here the "supersolidus intrusion breccia" type (SIB type), and is exemplified by the Lac des Iles deposit, Ontario (Fig. 27.2-1).

SIZE AND GRADE OF DEPOSITS

Nine-tenths of the PGEs mined are recovered from PGE-dominant ores (Table 27.2-1), with the bulk of the remainder recovered from magmatic Ni-Cu or alluvial deposits (Naldrett, 1989). Individual deposits in the Bushveld Complex (Fig. 27.2-2) have several millions to many tens

of millions of tonnes of mined ore plus mineable reserves, whereas the Lac des Iles deposit has a geological reserve of approximately 6 million tonnes (Table 27.2-1). As precious metals, the PGEs have comparable values to gold, and the economics of PGE deposits are broadly analogous to those of gold-only deposits, with ore grades from 8 to 20 g/t combined PGE+Au (Table 27.2-1). Platinum group element deposits differ from gold deposits in that each of the PGEs are valued separately, with values that range from one-third that of gold for Pd, to more than three times that of gold for Rh. Platinum (which generally has one to two times the value of gold) and palladium make up 75% to 90% of the PGEs present in magmatic ore deposits, so the Pt to Pd ratio (Pt:Pd) has a profound effect on the economic value of the deposit. Rhodium constitutes 8% of the PGEs in the UG-2 chromite ore of the Bushveld Complex, the world's largest Rh reserve (Buchanan, 1979).

GEOLOGICAL FEATURES

Reef- and supersolidus intrusion breccia-type PGE deposits share a number of geological features, but they contrast with each other in several important respects. The geology and mineral deposits of the Bushveld and Stillwater complexes are presented here to highlight the features of reef-type deposits, and the Lac des Iles Complex and its deposit are presented to represent the supersolidus intrusion breccia-type mineralization.

Geological setting

The Bushveld Complex (Fig. 27.2-2A), one of the world's great repositories for PGEs, chromite, and magnetite (+vanadium+titanium), is a 240 km by 400 km, Mid-Proterozoic layered intrusion within Early Proterozoic

Barrie, C.T.
1996:   Magmatic platinum group elements; in Geology of Canadian Mineral Deposit Types, (ed.) O.R. Eckstrand, W.D. Sinclair, and R.I. Thorpe; Geological Survey of Canada, Geology of Canada, no. 8, p. 605-614 (also Geological Society of America, The Geology of North America, v. P-1).

Transvaal siliciclastic rocks and Late Archean granitoid plutons. It comprises mafic and ultramafic cumulates around a cloverleaf-shaped perimeter, and three granitic suites in its central portion. The cumulate rocks are well-layered, and form a stratigraphic section more than 7 km thick known as the Rustenburg Layered Suite. The cumulate rocks are divided into four zones: the Lower zone of olivine-bronzite-chromite cumulates; the Critical zone of plagioclase-pyroxene cumulates, with local cumulus olivine and chromite; the Main zone of plagioclase-pyroxene cumulates; and the Upper zone of plagioclase-pyroxene-Fe-Ti oxide (+apatite+Fe-rich olivine+hornblende) cumulates (Fig. 27.2-2B).

The Merensky Reef and UG-2 chromitite layers, the major reef-type PGE deposits of the complex, are within the lower 500 m of the Critical zone. Other types of PGE mineralization are found in the Platreef in the northeastern limb of the complex, and in dunite pipes that cut the Lower and Critical zones. The Platreef, where the lower Critical zone is in contact with country rocks, comprises lower grade mineralization within altered, mafic and ultramafic cumulates that contain xenoliths of Transvaal Group dolomite, and has agmatitic and back-veined textures with granitoid rocks. The dunite pipes contain mineralization that is locally of high grade and in which much of the PGE ore is present as metals, Fe-alloys, and arsenides (Sharpe, 1985; Naldrett, 1989).

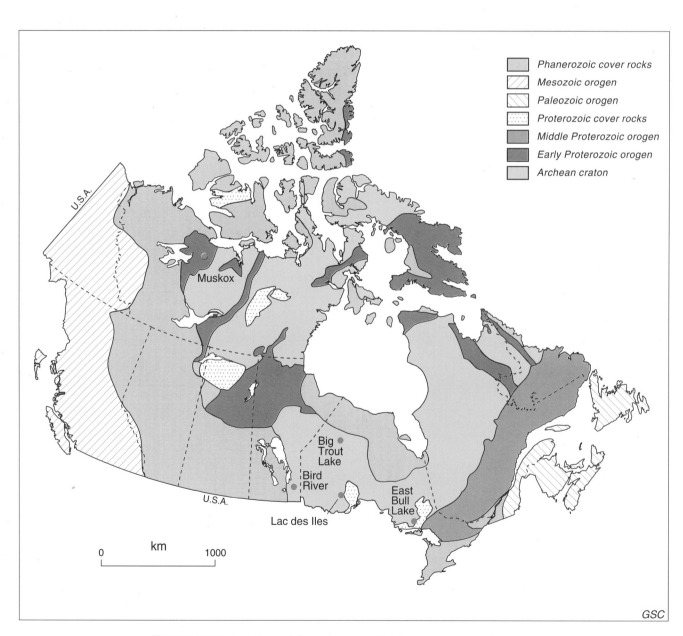

Figure 27.2-1. Locations of Canadian deposits of platinum group elements.

The Stillwater Complex is a 48 km (strike) by 2-7 km (across strike), Late Archean mafic-ultramafic intrusion that intruded ca. 3.3 Ga granitoid and metasedimentary rocks. Its cumulate stratigraphy has been divided into three zones: the Basal Contact zone of pyroxene-plagioclase cumulates with abundant inclusions of hornfelsed wall rock and low grade Ni-Cu sulphide locally; the Ultramafic zone of olivine-bronzite-chromite cumulates, which is overlain by the Banded zone of plagioclase-pyroxene-olivine (+chromite) cumulates. More fractionated cumulates are probably present up-section but are covered by Phanerozoic rocks. The principal PGE reef of the complex, J-M Reef, is located approximately 500 m above the base of the Banded zone, whereas a second lower grade reef, the G-Chromite layer, is found about 500 m below the top of the Ultramafic zone. Minor Pd-rich PGE mineralization such as the Picket Pin occurrence is present as transgressive and semiconformable zones in the anorthositic upper sections of the Banded zone (Czamanske and Zientek, 1985).

The Lac des Iles intrusion (Fig. 27.2-3A) is a 10 km by 2-4 km, Late Archean mafic-ultramafic intrusion, one of at least three broadly similar mafic-ultramafic intrusions in the region, all found within coeval, or nearly coeval granitoid rocks. It comprises a well layered, funnel-shaped northern ultramafic zone of olivine-pyroxene cumulates which dip moderately to steeply toward the centre; a southern ultramafic zone of similar composition but with a less well defined cumulate stratigraphy; a more southerly gabbro zone with plagioclase-pyroxene cumulates, mafic pegmatites, and intrusion breccia zones; and hornblendite and pegmatitic gabbroic rocks along the central western margin of the complex. Supersolidus intrusion breccia-type PGE mineralization occurs within the gabbro zone in and near the "Robie zone", and minor stratiform PGE occurrences are present in the northern ultramafic zone (Brugmann et al., 1989; Sutcliffe et al., 1989).

Age of host rocks and mineralization

The majority of PGE deposits and their hosting mafic-ultramafic intrusions or intrusive complexes are Precambrian (Stillwater, Munni Munni, Lac des Iles, Bird River, Big Trout Lake at 2.7 Ga; Great Dyke and East Bull Lake at 2.45 Ma; Bushveld at 2.05 Ga, Muskox at 1.3 Ga); the Tertiary Skaergaard Intrusion is an exception. For the reef-type deposits, the host intrusions are generally >200 Ma younger than the country rock, reflecting a stable cratonic setting at the time of emplacement. In contrast, the Lac des Iles intrusion and supersolidus intrusion breccia-type mineralization are nearly coeval with adjacent granites; and regional granitoid rocks are only ca. 30 Ma older.

Form of deposits and relationship of ore to host rocks

Reef-type mineralization is stratiform, but there are differences in the degree of conformity with adjacent cumulate layers from deposit to deposit. The Merensky Reef comprises a layer one to two metres thick that is continuous over hundreds of kilometres within the lower Critical zone. In detail, however, it is disrupted by numerous syn- and postmagmatic faults, and by "potholes": circular depressions from a few metres to several thousand metres in

Table 27.2-1. Magmatic platinum-group element deposits.

	Grade (g/t) PGE+Au	Pt:Pd	Geological reserves (t x 10⁶)	Reference
WORLD **Reef-type**				
Bushveld Complex, South Africa				Buchanan, 1979
Merensky Reef	8.1	2.4:1	2160	
UG-2 Chromite	8.7	1.2:1	3700	
Platreef	7.3	0.9:1	1700	
Stillwater Complex, Montana				Zientek, 1993
J-M Reef	18.8	0.3:1	421	
G chromite	2.4	0.4:1	3.4	
Great Dyke, Zimbabwe	4.7	1.4:1	1680	Naldrett, 1989
Munni Munni, Australia	2.9	0.6:1	25*	Barnes et al., 1990
Skaergaard, E. Greenland	~2	0.1:1	–	Bird et al., 1991
CANADA **Reef-type**				
Muskox Intrusion, Northwest Territories	~1	0.1:1	–	Hulbert et al., 1988
Bird River Sill, Manitoba	~0.6	0.5:1	–	Scoates et al., 1989
Big Trout Lake, Ontario	~2	1:1	–	Borthwick, 1984
Supersolidus mixing-type				
Lac des Iles, Ontario	5.4	0.14:1	6.7	Natural Resources Canada Mining Industry Quarterly Report, Fall 1993
East Bull Lake, Ontario	2.5	0.3:1	–	Peck et al., 1993
* Geological resource – = no published data				

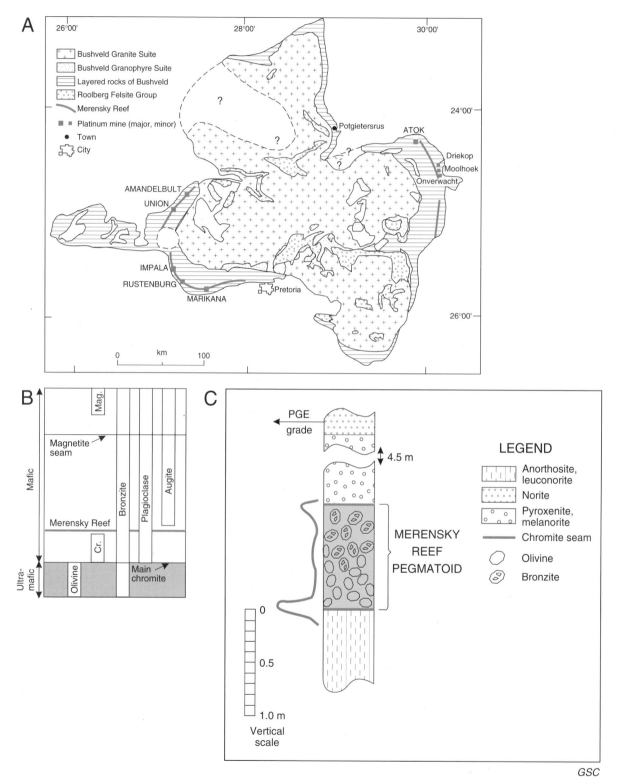

Figure 27.2-2. A) Geology of the Bushveld Complex, South Africa (after Campbell et al., 1983). **B)** Stratigraphy and cumulus phases in the Bushveld Complex (after Campbell et al., 1983) Cr.= chromite, Mag.= magnetite. **C)** Lithological section of the Merensky Reef at the Union mine, Bushveld Complex (after Naldrett, 1989). The PGEs are concentrated at the base and top of the Merensky Reef.

diameter and as much as to 30 m deep where the reef transects footwall cumulate stratigraphy (Campbell, 1986). The Merensky Reef corresponds to a marked increase in whole rock initial strontium isotopic ratios across stratigraphy (Kruger and Marsh, 1982), and represents a separate cyclic unit as defined by mineral chemistry (Naldrett, 1989). The J-M Reef of the Stillwater Complex is more erratic. It constitutes a 1-5 m thick zone within a 4-25 m thick "reef zone" of olivine-rich, plagioclase-olivine cumulates in the lower Banded zone (Barnes and Naldrett, 1986). The reef zone and many cumulate layers at Stillwater Complex, have less well defined stratigraphic boundaries than the cumulate stratigraphy of the Bushveld Igneous Complex (Czamanske and Zientek, 1985).

Supersolidus intrusion breccia-type mineralization at Lac des Iles (Fig. 27.2-3B) forms irregularly-shaped zones from several square metres to 50 m by 500 m (Robie zone), within lithologically complex gabbros and pyroxenite dykes. Mineralization cuts cumulate layering and is associated with pegmatitic mafic and ultramafic dykes as much as several metres thick. It is also associated with complex intrusion breccias that have mutual crosscutting relationships between gabbroic and pyroxenitic rocks, and gabbro-pyroxenite-hornblendite pegmatite dykes and apophyses (Macdonald, 1987a). Minor leucotonalite dykes related to wall rock granitoid bodies cut supersolidus intrusion breccia-type mineralization, with textures that indicate intrusion prior to complete crystallization of the gabbroic rocks.

Nature of the ore, distribution of ore minerals, and ore textures

In detail, Merensky Reef ore varies from bronzite-olivine pegmatoid cumulate, with thin chromitite seams at the top and bottom (Union mine: Fig. 27.2-2C), to bronzite-olivine-chromite pegmatoid and cumulate (Rustenberg and Atok mines), to bronzite cumulate with thin chromitite seams (Marikana mine: Naldrett, 1989). Intercumulus phases are predominantly plagioclase and bronzite; sulphides and the PGE minerals constitute 3-7 volume per cent (generally as much as 2 wt.% S), and in places are concentrated at the base or top of the reef. Coarse pegmatitic textures are present locally and may contain quartz, graphite, apatite, hornblende, and other hydrous silicates with appreciable chlorine contents (Ballhaus and Stumpfl, 1986). The UG-2 chromitite, from one to two and half metres thick, has 60 to 90 modal per cent chromite, and the remainder consists of bronzite, plagioclase, sulphides, PGE minerals, oxides, and biotite (Naldrett, 1989). The UG-2 and other PGE-bearing chromitite layers of the Bushveld Igneous Complex have much lower total sulphide contents than the Merensky Reef. The J-M Reef at Stillwater comprises a relatively olivine- and sulphide-rich "layer" or layers, from one to three metres thick, within the reef-zone. The PGE-rich layers rarely overlap where more than one are present in the same area (Radeke and Vian, 1986). The J-M Reef has cumulus, amoeboid-shaped olivine, oikocrystic bronzite locally, and intercumulus plagioclase, sulphide (as much as 2 wt.% S), and minor phlogopite, apatite, and chromite. "Mixed rock" reef ore contains additional pegmatitic pods of plagioclase and olivine.

In addition to intrusion breccia and pegmatitic dykes and apophyses, the supersolidus intrusion breccia-type mineralization at Lac des Iles has a variety of textures on a scale of decimetres to metres and which are common to metasomatic rocks in porphyry molybdenum systems. These include pegmatitic comb layering, orbicular textures, miarolitic enclosures of pegmatitic quartz gabbro with crystal growth directed inward, and "brain-rock" (Macdonald, 1987a). Similar textures are found in supersolidus intrusion breccia-type PGE mineralization at the margins of Huronian mafic intrusions in central Ontario, particularly the East Bull Lake and Shakespeare-Dunlop intrusions. There the mineralization is characterized by rounded anorthositic nodules, dendritic and pegmatitic mafic pods, and intrusion breccias (Peck et al., 1993).

Ore mineralogy and zonation

In reef- and supersolidus intrusion breccia-type settings, the PGEs are found almost exclusively in sulphides and PGE minerals: chalcopyrite, pentlandite, pyrrhotite, pyrite, braggite ((Pt, Pd, Ni)S), cooperite (PtS), laurite (RuS_2), Pt-Fe and PGE alloys, and PGE arsenides and antimonides. Kinloch (1982) documented mineralogical (and compositional) zonation on a scale of tens of kilometres for the Merensky and UG-2 reefs, and reported zonation near reef pothole depressions, where normal reef mineralogy gives way to Pt-Fe alloys at the pothole floor. Chromitite reefs are relatively enriched in Os, Ir, and Ru, and depleted in sulphur, in comparison to the Merensky and J-M reefs, and this is reflected in their ore mineralogy (Naldrett, 1989). The Pd-rich mineralization at Lac des Iles has vysotskite (PdS) as an important constituent (Watkinson and Dunning, 1979).

Alteration

Reef-type PGE deposits are commonly within near-pristine cumulate and pegmatitic rocks, but they may have deuteric mineral assemblages confined to a zone that is within metres of the reef horizon and that extends for tens of metres to kilometres along strike. Pegmatite-rich areas of the reefs may have intercumulus biotite, chlorite, amphibole, serpentine, talc, and graphite associated with intercumulus sulphide, and interstitial plagioclase may be sericitized (Ballhaus and Stumpfl, 1986; Boudreau et al., 1986; Mathez et al., 1989). At Lac des Iles, the central Robie zone is characterized by moderate deuteric (uralitic) alteration with an irregular distribution, and is gradational with less-altered gabbroic rocks to the west and east (Watkinson and Dunning, 1979; Brugmann et al., 1989).

DEFINITIVE CHARACTERISTICS

Reef-type and supersolidus intrusion breccia-type deposits share the following features:

- Located within primitive, medium to large mafic-ultramafic intrusions of tholeiitic (+ultramafic) affinity;

- Cumulate layering in host intrusion;

- Pegmatitic textures within largely cumulate rocks;

- Platinum group element enrichment, accompanied by minor sulphide with recoverable Ni and Cu contents, or by significant chromite;

- Presence of minor hydrous silicates locally.

Reef-type deposits have the following additional characteristics:

- Host intrusion emplaced into stable cratonic settings;
- Host intrusion with primitive and boninitic chill compositions that may have been sulphur undersaturated and PGE-enriched prior to emplacement.
- Evidence for magma influxes during formation of cumulus stratigraphy: the recurrence of high temperature phases such as olivine and/or chromite; an increase in Mg number in mafic phases; or a significant change in radiogenic isotope initial ratios in whole rocks;

Supersolidus intrusion breccia-type deposits are characterized by the following features in addition to the five shared features above:

- Textural evidence for mingling of two distinct magmas after formation of cumulus stratigraphy under supersolidus conditions;
- Significant metasomatic enrichment of PGEs in complex zones of intrusion breccia;
- Volatile input from adjacent granitic intrusions or partially melted wall rock.

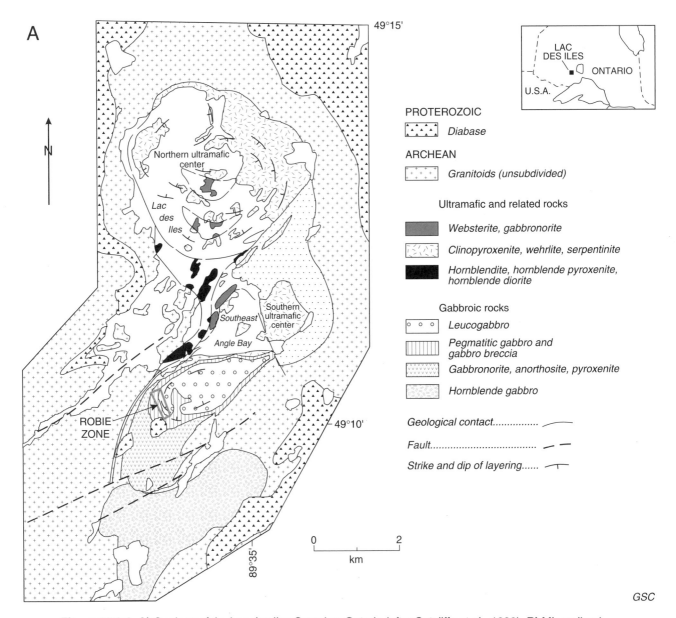

Figure 27.2-3. A) Geology of the Lac des Iles Complex, Ontario (after Sutcliffe et al., 1989). **B)** Mineralized outcrop 200 m south of the Robie zone (after Sutcliffe et al., 1989). The PGEs are concentrated in and at the margins of the pegmatitic gabbro dyke as indicated.

GENETIC MODELS

Considering the characteristics listed above, the genesis of PGE-dominant deposits must involve a balance of magmatic and volatile-related processes. Campbell et al. (1983) outlined a model for reef-type deposits in which magma mixing was considered as the fundamental process for enrichment. This model, and subsequent refinements, are summarized as follows (Fig. 27.2-4): 1) a large, density-stratified magma chamber with cumulus layers forming at the base is injected by a new pulse of hot, primitive magma as a turbulent plume; 2) the newly injected magma rises to a density level equal to its own and spreads out laterally, mixing turbulently with entrained resident magma; 3) during turbulent mixing, minor amounts of immiscible sulphide liquid precipitate due to mixing (Naldrett and von Gruenewaldt, 1989) or to a decrease in temperature (Haughton et al., 1974); 4) the sulphide mixes with a proportionally large volume of silicate magma (high "R" factor: silicate magma-to-sulphide ratio) and efficiently scavenges PGEs from the silicate magma due to their chalcophile affinity (sulphide-silicate partition coefficients of approximately 10^5); 5) with further cooling and crystallization, the mixed magma layer, containing PGE-enriched sulphide, crystals, and liquid, becomes more dense than underlying liquid and descends (as downspouts) to the base of the intrusion, forming a loosely-packed, PGE-rich orthocumulate layer, the PGE reef (Naldrett, 1989). An alternative model, similar to the Campbell et al. (1983) model in most respects, has mixing internal to a single, fractionating magma rather than between resident and new primitive magmas (Hoatson and Keays, 1989).

Pegmatitic textures and hydrous minerals common to PGE reefs may be products of: 1) the trapping of excess volatiles migrating upward from lower cumulates by relatively abundant intercumulus liquid within the orthocumulate reef; and 2) recrystallization of the reef cumulate phases in the presence of volatile-rich intercumulus liquid under supersolidus conditions (Naldrett, 1989).

Brugmann et al. (1989) proposed a model for the supersolidus intrusion breccia-type mineralization at Lac des Iles, using zone refining as the principal mechanism of PGE enrichment. Their model involves: 1) formation of plagioclase-pyroxene cumulates; 2) injection of fractionated pyroxenite dykes enriched in PGEs; 3) partial remelting of the plagioclase-pyroxene cumulates under supersolidus conditions due to volatile-rich magma fluxing, with the partial melt phase progressively enriched in incompatible elements, including Au, Pt, Pd, and volatiles, and the oxide-bearing residuum retaining Ir and Os; 4) precipitation of PGE-enriched sulphide from the volatile-rich liquid due to

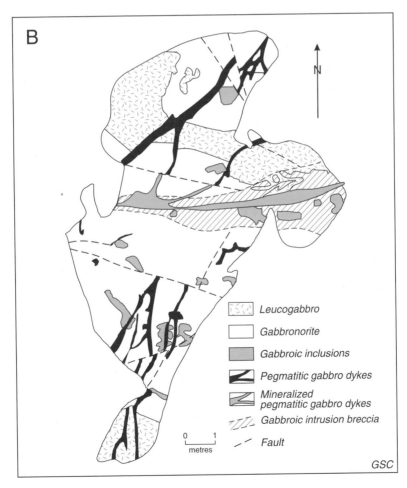

Figure 27.2-3. (cont.)

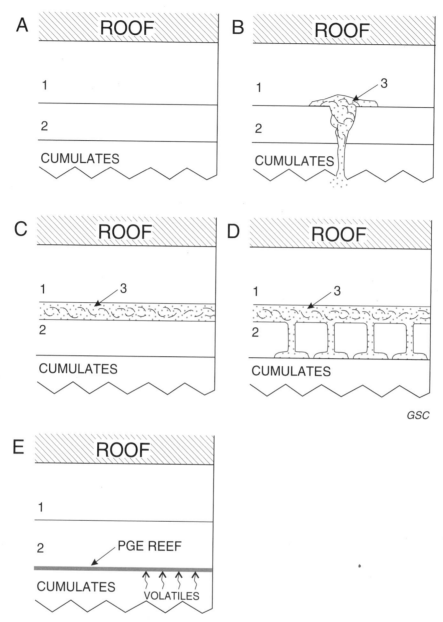

Figure 27.2-4. Model for the formation of the Merensky and J-M reefs. **A)** Initial, density-stratified magma chamber, with cumulates forming at base. **B)** Influx of new, hot primitive magma "3", as turbulent plume, rises to a level equal to its own density between layers 1 and 2. **C)** Primitive magma mixes turbulently at its own density level, entraining adjacent magma. Minor amounts of immiscible sulphide globules form due to magma mixing processes or to temperature decrease, and scavenge PGEs from silicate magma. **D)** With cooling, new turbulently-mixed magma layer becomes more dense than underlying layer and descends, probably as downspouts, through underlying layer, forming new cumulate layer at base. **E)** PGE-rich reef forms at the base as a stratiform, orthocumulate layer, traps volatiles migrating upward from underlying cumulate layers under supersolidus conditions, and crystallizes, with pegmatitic textures developed locally. Modified after Naldrett (1989).

sulphur saturation. Prior PGE enrichment occurred by magma mixing processes within the ultramafic magma (Brugmann et al., 1989; Sutcliffe et al., 1989). The volatiles responsible for pegmatite and hornblendite development may have been derived from adjacent, contemporaneous granitoid rocks during tectonism.

In summary, genetic models for both reef-type and supersolidus intrusion breccia-type PGE mineralization involve two processes: an initial enrichment in PGEs in magmatic sulphide by magma mixing, and a secondary enrichment or redistribution of the PGEs by volatiles or a volatile-enriched magma under supersolidus conditions. Efficient magma mixing in large intrusions in density-stratified layers that have very high silicate magma-to-sulphide ratios adequately accounts for the extreme PGE enrichment in the Merensky and J-M reefs; volatile-related processes account for local, minor redistribution. Magma mixing plays a subordinate role for supersolidus intrusion breccia-type mineralization, and volatile-induced zone refining under supersolidus conditions can adequately explain most of the geological features and the PGE enrichment.

RELATED DEPOSIT TYPES

Reef-type PGE deposits are most similar to chromite, Fe-Ti oxide, and gold (e.g., Skaergaard: Bird et al., 1991) deposits concentrated in stratiform cumulate layers in large mafic-ultramafic intrusions. They are also akin to magmatic sulphide Ni-Cu-PGE deposits, and to PGE deposits in transgressive dunite pipes that cut cumulate layering in the Bushveld Igneous Complex (Schiffries, 1982). Supersolidus intrusion breccia-type deposits are distantly analogous to Cu-PGE-rich metasomatic apophyses and breccias found in mafic alkaline porphyry Cu-Au deposits (Mutschler et al., 1985). Other related deposit types are PGE-enriched footwall quartz sulphide veins (e.g., Sudbury: Li et al., 1992); magmatic PGE enrichments in intrusions with alkaline affinity (e.g., Coldwell Complex, Ontario: Good and Crocket, 1994; Tulameen Alaska-type intrusion, British Columbia: Nixon et al., 1990); and unconformity-related U-PGE-Au deposits (e.g., Nicholson Bay, Saskatchewan: Hulbert et al., 1988).

EXPLORATION GUIDES

Geological and geochemical criteria that indicate favourable environments for magmatic PGE deposits are: 1) the presence of a large, layered mafic-ultramafic intrusion, particularly in a stable cratonic setting where the country rock is >200 Ma older than the intrusion; 2) evidence for mixing of primitive magma with more fractionated magma within the magma chamber, such as the recurrence of cumulus olivine or chromite within a thick plagioclase-pyroxene cumulate sequence in a well-layered intrusion, or complex metasomatic and intrusion breccia textures; 3) presence of sulphide- and/or oxide-bearing, stratiform or transgressive pegmatitic mafic rocks, within or at the margin of cumulate rocks of the intrusion; 4) determination of PGE-enriched, S-undersaturated, primitive (e.g., mantle-derived, ultramafic or tholeiitic) or boninitic chilled marginal rocks (e.g., Hamlyn and Keays, 1986); and 5) determination of metal contents in magmatic sulphides

that indicate a primitive character (high Ni/Cu, Ir/Pd) and a high R factor (R = silicate magma-to-sulphide ratio: Campbell et al., 1983; Naldrett, 1989; Barnes, 1990).

ACKNOWLEDGMENTS

I thank L. Hulbert, J. Macdonald, M. Zientek, K. Dawson, and B. Williamson for help in compilation, R.I. Thorpe for editorial comments, and A.J. Naldrett and his students for continued inspiration regarding magmatic sulphide deposits.

SELECTED BIBLIOGRAPHY

References marked with asterisks (*) are considered to be the best sources of general information on this deposit type.

Ballhaus, C.G. and Stumpfl, E.F.
1986:   Sulfide and platinum mineralization in the Merensky reef: evidence from hydrous silicates and fluid inclusions; Contributions to Mineralogy and Petrology, v. 94, p. 193-204.

Barnes, S.-J.
1990:   The use of metal ratios in prospecting for platinum-group element deposits in mafic and ultramafic intrusions; Journal of Geochemical Exploration, v. 37, p. 91-99.

Barnes, S.J. and Naldrett, A.J.
1986:   Geochemistry of the J-M Reef of the Stillwater Complex, Minneapolis Adit area II. Silicate mineral chemistry and petrogenesis; Journal of Petrology, v. 27, p. 791-825.

Barnes, S.J., McIntyre, J.R., Nisbet, B.W., and Williams, C.R.
1990:   Platinum group element mineralisation in the Munni Munni complex, Western Australia; Mineralogy and Petrology, v. 42, p. 141-164.

Bird, D.K., Brooks, C.K., Gannicott, R.A., and Turner, P.A.
1991:   A gold-bearing horizon in the Skaergaard intrusion, east Greenland; Economic Geology, v. 86, p. 1083-1092.

Borthwick, A.A.
1984:   The geology and geochemistry of the Big Trout Lake Complex, northwestern Ontario; MSc. thesis, University of Toronto, Toronto, Ontario, 200 p.

Boudreau, A.E., Mathez, E.A., and McCallum, I.S.
1986:   Halogen geochemistry of the Stillwater and Bushveld complexes: evidence for transport of platinum-group elements by Cl-rich fluids; Journal of Petrology, v. 27, p. 967-986.

Brugmann, G.E., Naldrett, A.J., and Macdonald, A.J.
1989:   Magma mixing and constitutional zone refining in the Lac des Iles Complex, Ontario: genesis of platinum group element mineralization; Economic Geology, v. 84, p. 1557-1573.

Buchanan, D.L.
1979:   Platinum metal production from the Bushveld Complex and its relationship to world markets; Bureau for Mineral Studies, University of Witwatersrand, Johannesburg, South Africa, Report #4, 31 p.

Campbell, I.H.
1986:   A fluid dynamic model for the potholes of the Merensky Reef; Economic Geology, v. 81, p. 1118-1125.

Campbell, I.H., Naldrett, A.J., and Barnes, S.J.
1983:   A model for the origin of the platinum group rich sulfide horizons in the Bushveld and Stillwater complexes; Journal of Petrology, v. 24, p. 133-165.

*Czamanske, G.K. and Zientek, M.L. (ed.)
1985:   The Stillwater Complex, Montana: geology and guide; Montana Bureau of Mines and Geology, Special Publication 92, 396 p. and maps.

Economic Geology
*1985:   A Special Issue Devoted to the Bushveld Complex; Economic Geology, v. 80, no. 4, p. 803-1211.
*1986:   A Third Issue Devoted to Platinum Deposits; Economic Geology, v. 81, no. 5, p. 1045-1285.

Energy, Mines and Resources Canada
1993:   Mineral Industry Quarterly Report (Fall); Ottawa, Ontario, 59 p.

Good, D.J. and Crocket, J.H.
1994: Genesis of the Marathon Cu-platinum-group element deposit, Port Coldwell alkalic complex, Ontario: a mid-continent rift-related magmatic sulfide deposit; Economic Geology, v. 89, p. 131-150.

Hamlyn, P.R. and Keays, R.R.
1986: Sulfur saturation and second stage melts: application to the Bushveld platinum-metal deposits; Economic Geology, v. 81, p. 1431-1445.

Haughton, D.R., Roeder, P.L., and Skinner, B.J.
1974: Solubility of sulfur in mafic magmas; Economic Geology, v. 69, p. 451-467.

Hoatson, D.M. and Keays, R.R.
1989: Formation of platiniferous sulfide horizons by crystal fractionation and magma mixing in the Munni Munni layered intrusion, west Pilbara block, Western Austraia; Economic Geology, v. 84, p. 1775-1804.

*Hulbert, L.M., Duke, J.M., Eckstrand, O.R., Lydon, J.W., Scoates, R.F.J., Cabri, L.J., and Irvine, T.N.
1988: Geological environments of the platinum group elements; Geological Survey of Canada, Open File 1440, 148 p.

Kinloch, E.D.
1982: Regional trends in the platinum-group element mineralogy of the Critical Zone of the Bushveld Complex, South Africa; Economic Geology, v. 77, p. 1328-1347.

Kruger, F.J. and Marsh, J.S.
1982: Significance of Sr^{87}/Sr^{86} ratios in the Merensky cyclic unit of the Bushveld Complex; Nature, v. 298, p. 53-55.

Li, C., Naldrett, A.J., Coats, C.J.A., and Johannssen, P.
1992: Platinum, palladium, gold, and copper-rich stringers at the Strathcona mine, Sudbury: their enrichment by fractionation of a sulfide liquid; Economic Geology, v. 87, p. 1584-1598.

Macdonald, A.J.
1987a: Platinum-group element mineralisation and the relative processes of magmatic and deuteric processes: field evidence from the Lac des Iles deposit, Ontario, Canada; in Geo-platinum 87 - Conference Proceedings, (ed.) H.M. Pritchard, P.J. Potts, J.F.W. Bowles, and S.J. Cribb; Elsevier, London, U.K., p. 215-236.

*1987b: Ore deposit models #12: the platinum group element deposits: classification and genesis; Geoscience Canada, v. 14, p. 155-166.

Mathez, E.A., Dietrich, V.J., Holloway, J.R., and Boudreau, A.E.
1989: Carbon distribution in the Stillwater Complex and evolution of vapor during crystallization of Stillwater and Bushveld magmas; Journal of Petrology, v. 30, p. 153-173.

Mutschler, F.E., Griffin, M.E., Stevens, D.S., and Shannon, S.S.
1985: Precious metal deposits related to alkaline rocks in the North American Cordillera - an interpretive review; Transactions of the Geological Society of South Africa, v. 88, p. 355-377.

*Naldrett, A.J.
1989: Magmatic Sulfide Deposits; Oxford University Press, Oxford, England, 186 p.

Naldrett, A.J. and von Gruenewaldt, G.
1989: Association of platinum-group elements with chromitite in layered intrusions and ophiolite complexes; Economic Geology, v. 84, p. 180-187.

Nixon, G.T., Cabri, L.J., and Laflamme, J.H.
1990: Platinum-group-element mineralization in lode and placer deposits associated with the Tulameen Alaskan-type Complex, British Columbia; Canadian Mineralogist, v. 28, p. 503-535.

Peck, D.C., James, R.S., and Chubb, P.T.
1993: Geological environmnets for PGE-Cu-Ni mineralization in the East Bull Lake gabbro-anorthosite intrusion, Ontario; Exploration and Mining Geology, v. 2, p. 85-104.

Radeke, L.D. and Vian, R.W.
1986: A three-dimensional view of mineralization in the Stillwater J-M reef; Economic Geology, v. 81, p. 1187-1195.

Schiffries, C.M.
1982: The petrogenesis of a platiniferous dunite pipe in the Bushveld Complex: infiltration metasomatism by a chloride solution; Economic Geology, v. 77, p. 1439-1453.

Scoates, R.F.J., Williamson, B.L., Eckstrand, O.R., and Duke, J.M.
1989: Stratigraphy of the Bird River Sill and its chromitiferous zone, and preliminary geochemistry of the chromitite layers and PGE-bearing units, Chrome property, Manitoba; Geological Survey of Canada, Open File 2213, p. 69-82.

*Sharpe, M.R.
1985: Bushveld Complex - Excursion Guidebook Geocongress '86; Institute for Geological Research on the Bushveld Complex, University of Pretoria, Republic of South Africa, 143 p.

Sutcliffe, R.H., Sweeney, J.M., and Edgar, A.D.
1989: The Lac des Iles Complex, Ontario: petrology and platinum-group element mineralization in an Archean mafic intrusion; Canadian Journal of Earth Sciences, v. 26, p. 1408-1427.

Watkinson, D.H. and Dunning, G.
1979: Geology and platinum-group mineralization, Lac des Iles complex, northwestern Ontario; Canadian Mineralogist, v. 17, p. 453-462.

Zientek, M.L.
1993: Mineral resource appraisal for locatable minerals: the Stillwater Complex; in Mineral Resource Assessment of the Absaroka-Beartooth Study Area, Custer and Gallatin National Forests, Montana, (ed.) J.M. Hammarstrom, M.L. Zientek, and J.E. Elliott; United States Geological Survey, Open File Report 93-207, p. F1-F83.

Authors' addresses

C.T. Barrie
Geological Survey of Canada
601 Booth Street
Ottawa, Ontario
K1A 0E8

O.R. Eckstrand
Geological Survey of Canada
601 Booth Street
Ottawa, Ontario
K1A 0E8

Printed in Canada

28. MAFIC/ULTRAMAFIC-HOSTED CHROMITE

J.M. Duke

Chromite is mined almost exclusively from massive to semimassive accumulations in ultramafic or mafic igneous rocks. Eluvial and alluvial deposits derived by the erosion of such rocks account for a small fraction of total production. Significant, but as yet unexploited, resources reside in laterites such as those of the Ramu River deposit in Papua New Guinea. Hard rock chromite deposits are normally assigned to one of two classes on the basis of deposit geometry, petrological character, and tectonic setting. Stratiform deposits are sheet-like accumulations of chromite that occur in layered ultramafic to mafic igneous intrusions. Podiform deposits are irregular but fundamentally lenticular chromite-rich bodies that occur within Alpine peridotite or ophiolite complexes. The Alpine peridotite or ophiolitic affiliation is an essential part of the latter definition: this has led to some confusion because some deformed stratiform deposits have morphological similarities to podiform deposits.

Chromite is the only mineral extracted from the ore[1] and, depending upon its physical and chemical properties, may be directly reduced to ferrochrome or used in its natural state for refractories, foundry sands, glass pigments, or the manufacture of chemicals. Until relatively recently, chromium-rich chromite (i.e., Cr/Fe >2.8) was required for the production of ferrochrome and was referred to as "metallurgical grade". Similarly, aluminum-rich chromite (i.e., Al_2O_3 >20%) was used for manufacturing refractory bricks and was designated as "refractory grade". Iron-rich chromite was largely restricted to use for chemicals and pigments, and was therefore referred to as

Duke, J.M.
1996:   Mafic/ultramafic-hosted chromite; in Geology of Canadian Mineral Deposit Types, (ed.) O.R. Eckstrand, W.D. Sinclair, and R.I. Thorpe; Geological Survey of Canada, Geology of Canada, no. 8, p. 615-616 (also Geological Society of America, The Geology of North America, v. P-1).

[1] The significant exception is the UG2 stratiform chromite deposit of the Bushveld Complex which is mined for platinum group elements: however, the chromite is not currently being recovered.

"chemical grade". Technological advances have meant that ferrochrome is now made routinely from chromite with substantially lower chromium to iron ratios (e.g., Cr/Fe = 1.5), and iron-rich and chromium-rich materials are used in refractory applications. Therefore, the traditional metallurgical, refractory, and chemical grades are more properly designated as high-chromium, high-aluminum, and high-iron, respectively. Representative composition of chromite concentrates from Canadian and foreign deposits are given in Table 28-1.

Table 28-1. Representative compositions of chromite concentrates from Canadian and foreign chromite deposits.

	Stratiform deposits							Podiform deposits				
	1	2	3	4	5	6	7	8	9	10	11	12
SiO_2	1.37	-	0.12	0.22	-	0.34	0.21	0.55	0.30	0.09	0.24	1.18
TiO_2	0.72	0.67	0.44	0.43	0.35	-	-	-	0.11	0.19	0.15	-
Al_2O_3	15.70	12.50	15.70	16.90	11.80	14.00	8.70	14.40	14.70	14.20	8.12	32.30
Cr_2O_3	42.50	47.00	47.10	47.80	55.00	56.20	56.30	50.70	53.40	54.50	59.00	35.40
$FeO(t)$	31.20	27.40	25.10	21.30	17.80	16.60	22.20	19.90	14.90	15.70	13.60	13.40
MnO	0.65	0.20	-	0.16	0.20	0.44	0.21	-	-	0.13	0.14	-
MgO	6.65	8.70	11.10	12.10	12.60	11.50	11.40	13.10	16.70	14.20	18.20	17.40
CaO	0.66	-	0.53	0.04	-	-	-	0.50	0.44	0.01	0.10	0.14
Total	99.45	96.47	100.09	98.95	97.75	99.08	99.02	99.15	100.55	99.02	99.55	99.82

1. Bird River Sill, Chrome claims (Bateman, 1943).
2. Kemi (Soderholm and Inkinen, 1982) - average of 25 samples.
3. Bushveld, Steelpoort Seam (Cameron and Desborough, 1969).
4. Stillwater Complex, H. Chromitite, Mouat mine (Thayer, 1964)
5. Great Dyke (Slatter, 1980) - average of 40 samples.
6. Selukwe, Selukwe Peak mine (Cotterill, 1969) - average of 5 samples
7. Campo Formoso, Campinhos Mine (Hedlund et al., 1974).
8. Sterrett mine, Quebec (Dennis, 1931) - average of 2 analyses.
9. Caribou mine, Quebec (Dennis, 1931) - average of 2 analyses.
10. Chrome mine, Cyprus (Greenbaum, 1977).
11. Molodehnoe deposit, Kempirsai, Kazakhstan (Smirnov, 1978).
12. Masinloc mine, Zambales, Philippines (Stoll, 1958).not determined

28.1 STRATIFORM CHROMITE

J.M. Duke

INTRODUCTION

Stratiform chromite deposits are sheet-like accumulations of chromite that occur in layered ultramafic to mafic igneous intrusions.

The best examples of stratiform chromite deposits in Canada occur in the Bird River Sill in southeastern Manitoba and in the Big Trout Lake intrusion in northwestern Ontario. Other intrusions in Canada with chromitite layers include the Muskox complex in the Northwest Territories, the Lac des Montagnes body in Quebec, and the Puddy Lake and Crystal Lake intrusions in Ontario (Fig. 28.1-1).

The Bushveld Complex in southern Africa contains the world's most important stratiform chromite deposits in terms of both production and reserves. Other important deposits include Kemi in Finland, Campo Formoso and the Jacurici Valley in Brazil, the Great Dyke and the Selukwe Complex in Zimbabwe, and Andriamena in Madagascar.

IMPORTANCE

There is no past or current production from stratiform chromite deposits in Canada apart from approximately 6000 t of material which was mined at Puddy Lake, Ontario in the 1930s. The deposits of the Bird River Sill constitute a significant strategic resource of chromite. Stratiform deposits account for 45% of world chromite production and 95% of reserves.

SIZE AND GRADE OF DEPOSITS

Productive stratiform chromite deposits are typically very large as indicated by the following data on reserves: Bushveld Complex - 1.1 billion tonnes, mostly grading 42 to 45% Cr_2O_3; Great Dyke - 113 million tonnes grading 26 to 51% Cr_2O_3; Kemi - 59 million tonnes grading 26% Cr_2O_3; Selukwe - 11.5 million tonnes grading 47% Cr_2O_3; Campo Formoso - 17 million tonnes grading 17 to 21% (DeYoung et al., 1984). These figures include resources categorized as "R1E" under the United Nations classification: that is, reliable estimates of economically exploitable material in known deposits. The Great Dyke and, in particular, the Bushveld Complex contain substantial additional resources of marginally economic material.

Duke, J.M.
1996:   Stratiform chromite; in Geology of Canadian Mineral Deposit Types, (ed.) O.R. Eckstrand, W.D. Sinclair, and R.I. Thorpe; Geological Survey of Canada, Geology of Canada, no. 8, p. 617-620 (also Geological Society of America, The Geology of North America, v. P-1).

The Bird River Sill is the only Canadian stratiform chromite deposit for which resource estimates have been published. Measured and indicated resources on two properties amount to 3.6 million tonnes grading 21% Cr_2O_3 (Bannatyne and Trueman, 1982), contained in the 3 m thick Upper Main chromitite (Fig. 28.1-2). Measured resources of about 60 million tonnes grading 4.6% Cr_2O_3 are contained in a 20 m thick zone including the Upper Main as well as several thinner chromitite layers (Watson, 1985).

GEOLOGICAL FEATURES

Geological setting

Stratiform chromite deposits occur in large, layered intrusions which are commonly differentiated into a lower ultramafic zone and an upper mafic zone. The intrusions fall into two broad categories with respect to morphology. The first includes essentially tabular bodies which were emplaced as sill-like intrusions in which igneous layering is conformable to the floor. Examples include Kemi, Campo Formoso, Stillwater Complex, Bird River Sill, and Big Trout Lake. Bodies of the second category, which includes Bushveld Complex, Great Dyke, and Muskox Intrusion, comprise one or more funnel-shaped intrusions in which the layering dips at a shallow angle towards the centre giving a synclinal cross-section. Although some chromitite-bearing intrusions, including the Bushveld Complex (480 x 380 km) and the Great Dyke (530 x 6 km), are very large, significant productive deposits also occur in much smaller bodies such as those at Kemi (15 x 2 km) and Campo Formoso (40 x 1 km).

The intrusions which host stratiform chromite deposits occur in a variety of tectonic settings. The Bushveld Complex, Great Dyke, and Muskox Intrusion are unmetamorphosed and were emplaced into stable cratonic settings. The Kemi and Campo Formoso intrusions are prekinematic and occur at the unconformable contact between Archean granitic basement and overlying, mainly sedimentary Proterozoic supracrustal rocks. The Bird River Sill and Big Trout Lake body are synvolcanic intrusions in Archean greenstone belt settings.

Age of host rocks and mineralization

Stratiform chromite deposits are syngenetic with their host intrusions, and the economically significant deposits are Archean or early Proterozoic in age. Radiometric ages are 2050 Ma for the Bushveld Complex, 2444 Ma for Kemi, 2461 Ma for the Great Dyke, 2701 Ma for Stillwater Complex, and 2745 Ma for Bird River Sill. Dating of crosscutting rocks indicate minimum ages of 3420 Ma for Selukwe and 2000 Ma for Campo Formoso.

Form of deposits

Most stratiform chromite deposits comprise laterally extensive chromite-rich layers which, despite local irregularities, are generally conformable to, and form an integral part of, the igneous layering that characterizes such intrusions. The individual chromitite layers range from less than 1 cm to more than 1 m in thickness, but their lateral extent is measured in kilometres or tens of kilometres. Chromitite may be interlayered with a variety of rock types including dunite, peridotite, orthopyroxenite, anorthosite, and norite, and may occur at various stratigraphic levels within the host layered intrusion. However, because chromite in the most primitive rocks tends to be the most chromium-rich, the immediate host rocks of economically significant chromitites are peridotites (e.g., Kemi, Campo Formoso, Selukwe, Stillwater Complex, Bird River Sill) or, less commonly, pyroxenites (e.g., Bushveld).

Orebodies may comprise discrete layers of massive chromitite, as in the Bushveld Complex and the Kemi deposit, or a number of closely spaced chromitite layers separated by ultramafic rock, as in the mines of the Campo Formoso district of Brazil. The bulk of the chromite production of the Bushveld Complex comes from two layers in particular: the Steelpoort seam (also called the LG6, Magazine, or Main seam) which is from 0.6 m to 1.3 m thick where mined, and the F chromitite which is 1.3 m thick. The Kemi mine in Finland exploits a very interesting chromitite layer. Over much of the 15 km length of the Kemi intrusion, the layer is a few centimetres to a metre or so thick. However, over a strike length of 4.5 km in the widest part of the intrusion, there are several successive swellings where the layer attains thicknesses of 30 to 90 m, and it is these swellings that constitute the orebodies. The most economically important orebodies at Campo Formoso occur

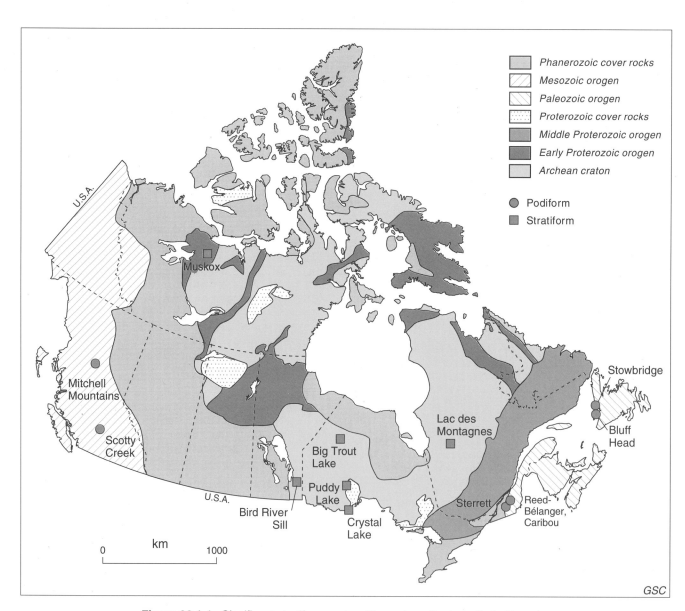

Figure 28.1-1. Significant stratiform and podiform chromite deposits in Canada.

618

Figure 28.1-2. Upper Main Group chromitite layers, Bird River Sill, Manitoba. This unit averages 3 m in thickness and comprises two chromite cumulate layers (dark, left and centre) and a chromite-olivine cumulate layer (right), separated by layers of olivine-chromite cumulate. Stratigraphic top is to the right (after Williamson, 1990). GSC Photo 204887-F

in the No. 4 seam, which averages 9 m in thickness and includes two massive chromitites and several layers of disseminated and net-textured chromite in serpentinite.

The Bird River Sill contains at least 35 individual chromitite layers which have been assigned to six groups (Scoates, 1983), and the three layers that make up the Upper Main Group were the focus of early exploration (Fig. 28.1-2).

Ore mineralogy and textures

Stratiform ores are made up of massive, "net-textured" and disseminated chromite. The massive ores may be described lithologically as 'chromitite' or as 'chromite cumulates' in cumulus terminology. Typical massive chromite ore textures are illustrated in Figure 28.1-3. The massive chromitites at Kemi and Bird River Sill contain 50 to 70 modal per cent chromite and are texturally orthocumulate rocks. The Bushveld Complex ores contain more than 90 modal per cent chromite and are texturally adcumulate rocks. Hulbert and Von Gruenewaldt (1985) have argued persuasively that the high proportion of chromite in the Bushveld Complex chromitites results from postcumulus sintering of a chromite orthocumulate precursor. The distinction between orthocumulates, mesocumulates, and adcumulates is reflected directly in the ore grade and also partly determines whether the ore must be beneficiated prior to smelting: it is therefore of great economic significance. The so-called "net-" or "chain-texture" chromite ores are typically chromite-olivine cumulates, (chromite peridotites or olivine chromitites in lithological terminology), and contain 20 to 50 modal per cent chromite (Fig. 28.1-3C). Disseminated chromite occurs also in peridotite, pyroxenite, or anorthosite and, although it does not constitute ore in its own right, may be recovered where it is interlayered with massive or net-textured chromitites.

Chromite, as noted above, is the only ore mineral. Olivine, orthopyroxene, plagioclase, and clinopyroxene are the most common associated magmatic minerals. These minerals are subject to alteration and serpentine, tremolite, chlorite, magnetite, talc, and carbonate are common secondary minerals.

Ore composition

The compositions of chromite concentrates from some stratiform deposits are given in Table 28-1. These illustrate the importance of texture in determining the grade of the deposit, as noted above. For example, although the Cr_2O_3 content of Bushveld Complex chromite is the same as that at Kemi and at Stillwater (about 47%), the ore grade is almost twice as high (42 to 45%, as compared with 26% at Kemi and 20 to 24% at Stillwater). The relatively low Cr_2O_3 and high iron content of Bird River Sill concentrates probably reflects the modification of primary compositions during regional metamorphism. The chromite concentrates from the Great Dyke, Selukwe, and Campo Formoso deposits are relatively chrome-rich and are similar in composition to those from most podiform deposits.

DEFINITIVE CHARACTERISTICS

The definitive characteristics of stratiform deposits are:

1. Conformity of chromitite seams to overall igneous layering in differentiated mafic-ultramafic intrusions.
2. Occurrence of cumulus igneous textures.

GENETIC MODEL

Stratiform chromite deposits clearly formed by the fractional crystallization of mafic or ultramafic magmas. The chromitites are igneous cumulates, and their formation is a special case of the larger problem of the origin of layered igneous rocks. The occurrence of chromitites in which chromite is the only cumulus mineral is problematic because chromite is not expected to be the sole liquidus phase during the normal course of fractional crystallization of mafic or ultramafic magma. A number of mechanisms have been suggested to give rise to the special circumstances that cause chromite to be the only liquidus phase. These include an increase in oxygen fugacity (Ulmer, 1969; Cameron and Desborough, 1969), changes in total pressure (Cameron, 1980), and contamination of the magma through assimilation of salic country rocks (Irvine, 1975). However, the most popular model currently is that chromite precipitation is caused by mixing of primitive and fractionated magmas (Irvine, 1977; Irvine and Sharpe, 1986).

RELATED DEPOSIT TYPES

Podiform chromite deposits (see deposit subtype 28.2) are closely related to the stratiform chromite deposits and probably form by similar processes. The association of chromite with magmatic platinum group element deposits (see deposit subtype 27.2) is well known. The association is two-fold. Firstly, the intrusions hosting the most prominent PGE "reefs" (Bushveld Complex, Stillwater Complex, and Great Dyke) also contain significant stratiform chromite deposits, normally at lower levels in the igneous stratigraphy. Secondly, chromite may be an important constituent of the PGE ores. The best example is the UG2 chromitite layer in the Bushveld Complex which is mined for PGEs:

the iron-rich chromite tailings are stockpiled for possible future use. The Merensky Reef in the Bushveld Complex is characterized by two thin seams of iron-rich chromitite whereas chromite is present in amounts of 1 to 2% in the J-M Reef in the Stillwater Complex. Platinum group elements are also spatially associated with chromitite layers in the Ultramafic Series of the Bird River Sill (Scoates et al., 1987). Mafic intrusion-hosted iron-titanium deposits (see deposit Type 26) are in many respects analogous to stratiform chromite deposits, the difference being that the former segregated from less primitive magmas which gave rise to less chrome-rich oxide phases.

EXPLORATION GUIDES

1. Stratiform chromite deposits occur in layered, differentiated, sill-like or funnel-shaped, mafic-ultramafic intrusions.

2. The deposits are igneous layers and so the presence of layering indicates that the necessary igneous processes occurred in the intrusion.

3. Economically significant chromitite seams are most commonly associated with peridotite and pyroxenite in the lower ultramafic parts of the intrusions.

4. Disseminated chromite is often a conspicuous mineral in the host rocks of the chromitite or chromite-rich seams.

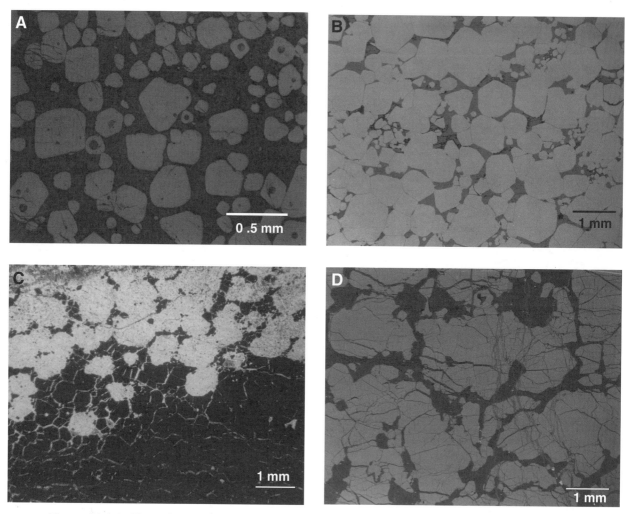

Figure 28.1-3. Photomicrographs of chromitite ore textures. **A)** Chromite orthocumulate, Upper Main Chromitite Group, Bird River Sill, Manitoba. The presence of spherical silicate inclusions in chromite is common in Bird River chromitites. Reflected light. GSC 1995-193C **B)** Fine grained massive chromite from the stratiform A chromitite zone of the Stillwater Complex, Montana. The chromite in this chromite mesocumulate exhibits euhedral to subhedral boundaries against intercumulus silicate minerals (dark grey) but shows the effects of mutual interference against other chromite grains. Reflected light. GSC 1995-193B **C)** Massive chromite from the stratiform LG6 chromitite, Bushveld Complex, South Africa. Primary cumulate textures have been virtually obliterated by postcumulus recrystallization. "Net-textured" olivine-chromite cumulate is shown in the upper part of the photograph. Transmitted light. GSC 1995-193C **D)** Coarse grained, massive chromite from the podiform deposit at the Sterrett mine, St. Cyr, Quebec. Note the extremely irregular shapes of the chromite anhedra. Reflected light. GSC 1995-193D

28.2 PODIFORM (OPHIOLITIC) CHROMITE

J.M. Duke

INTRODUCTION

Podiform chromite deposits were defined by Thayer (1964) as irregular but fundamentally lenticular chromite-rich bodies that occur within Alpine peridotite or ophiolite complexes. Thus they generally occur in orogenic settings.

Podiform chromite deposits occur in ophiolitic terranes in Canada. Significant deposits occur in the Appalachian orogen in Quebec (e.g., Reed-Belanger, Caribou, and Sterrett mines) (Fig. 28.2-1) and Newfoundland (e.g., Bluff Head and Stowbridge deposit; Snelgrove, 1934) and in the Cordillera of British Columbia (e.g., Scotty Creek and Mitchell Mountain; Whittaker and Watkinson, 1984) and Yukon Territory.

The most significant podiform chromite deposits worldwide are those of the Kempirsai Massif in the southern Urals in Kazakhstan, the Bater-Martanesh district of Albania, the Zambales ophiolite of the Philippines, the Cuttack district of India, and the Guleman district in Turkey.

IMPORTANCE

There is no current production from podiform chromite deposits in Canada, although about 250 000 t were mined from 1894 to 1949, mainly in Quebec, but also to a limited extent in British Columbia and Newfoundland. Indeed, at the turn of the century, Canada ranked fourth in world chromite production after Russia, Turkey, and New Caledonia. Podiform deposits account for about 55% of current world chromite production but only for about 5% of reserves. The small share of reserves partly reflects the difficulty in exploring and delineating these deposits.

SIZE AND GRADE OF DEPOSITS

Individual podiform chromite bodies are small; they range from a few tens to a few millions of tonnes, but those greater than 1 million tonnes are rare. The Coto orebody of the Masinloc mine in the Phillipines is reputed to be the largest known body and it contains an estimated 13 million tonnes of high-Al ore with 36.5% Cr_2O_3, 31% Al_2O_3, and Cr:Fe = 2.2. Most important mines exploit a number of pods. For example, the Kavak mine in Turkey includes 21 orebodies totalling about 2 million tonnes grading 28 to 30% Cr_2O_3 from which a 51% Cr_2O_3 concentrate with Cr:Fe = 3.2 is

Duke, J.M.
1996: Podiform (ophiolitic) chromite; in Geology of Canadian Mineral Deposit Types, (ed.) O.R. Eckstrand, W.D. Sinclair, and R.I. Thorpe; Geological Survey of Canada, Geology of Canada, no. 8, p. 621-624 (also Geolocit Society of America, The Geology of North America, v. P-1).

produced (Ergunalp, 1980). The Forty Years of the Kazak SSR deposit in Kazakhstan includes at least 23 orebodies totalling about 90 million tonnes with an average grade of 50% Cr_2O_3.

The following are the resources plus past production of some Canadian podiform deposits: Reed-Belanger, 1.1 million tonnes grading 7 to 14% Cr_2O_3; Sterrett, 180 000 t grading 18%, and Caribou, 60 000 t grading 27%. Concentrates of the Quebec ores typically graded about 48% Cr_2O_3 with Cr:Fe = 2.4 to 2.8, (Table 28-1).

GEOLOGICAL FEATURES

Geological setting

The ideal ophiolite succession comprises, in descending order, marine sediments, pillowed basaltic lavas, sheeted diabase dykes, noncumulate and cumulate mafic rocks, ultramafic (mainly dunitic) cumulates, and ultramafic (mainly harzburgite) tectonites. Most ophiolites are clearly allochthonous and represent obducted fragments of oceanic lithosphere. The tectonites are thought to represent the residual, partially melted mantle, and their contact with the ultramafic cumulates has been interpreted as the boundary between the mantle and crust (i.e., the petrological Moho). The interpretation of the uppermost ultramafic rocks as cumulates has been challenged by Nicolas and Prinzhofer (1983) who argued that these also represent residua from partial melting. In any case, chromite deposits are most abundant in the uppermost part of the tectonized harzburgite, but also occur in the lower part of the cumulate succession. For example, Cassard et al. (1981) estimated that 90% of the chromite deposits of the Massif du Sud in New Caledonia occur within the upper 1000 m of the harzburgite, with the balance occurring in the overlying dunites. In the Oman ophiolite, although chromitites occur as far as 12 km beneath the cumulate sequence, they are concentrated in the uppermost 1.5 km and the largest bodies are within 500 m of the cumulates (Brown, 1980).

Age of host rocks and mineralization

Podiform chromite deposits that occur in the mantle tectonite are younger than their host harzburgite whereas those in the cumulate sequence are presumably syngenetic with their host rocks. In either case, the deposits predate obduction of the ophiolite. The absolute ages of almost all podiform chromite deposits fall within the Phanerozoic, including those of the Philippines (Tertiary), Albania (Jurassic), Turkey (Paleozoic), and Kazakhstan (Paleozoic). The only significant deposits of Precambrian age are those of the Ingessana Hills in Sudan which are believed to be Proterozoic. The deposits in the Canadian Appalachians are Cambro-Ordovician; those in the Cordillera are mainly Permo-Triassic.

Form of deposits and host rocks

Podiform chromite deposits are generally lenticular bodies of massive to heavily disseminated chromite. Tabular, rod-shaped and irregular bodies are also observed. Cassard et al. (1981) classified the deposits into concordant, subconcordant, and discordant types according to their orientation with respect to the penetrative fabric of the enclosing peridotite. The discordant bodies tend to be irregular or rod-shaped bodies and contain relict magmatic textures. The concordant bodies are tabular and, less commonly, rod-shaped. The spectrum of discordant to concordant deposits represents a range of least deformed through most deformed. The ores display mainly tectonite textures. The immediate host rocks are in almost all cases dunite or serpentinized dunite. Even those in the tectonized harzburgite are normally surrounded by a thin envelope of dunite. The high-Al chromitites of the Camaguey district of Cuba are associated with troctolites.

Ore mineralogy and textures

Chromite is the only ore mineral in podiform deposits. Olivine, orthopyroxene, plagioclase, and clinopyroxene are the principal associated primary minerals, with serpentine, magnetite, chlorite, tremolite, talc, and carbonate being common secondary minerals.

Podiform chromite ores display a bewildering range of textures from those which are clearly magmatic through to features that result from either ductile or brittle deformation. Features characteristic of layered igneous rocks are common in some deposits and include cumulate textures, modal grading, and grain size grading (Greenbaum, 1977; Brown, 1980; Burgath and Weiser, 1980; Hock and Friedrich, 1985). Nodular textures are a common and striking feature of podiform deposits and comprise loosely packed, ellipsoidal chromite nodules 5 to 20 mm in diameter in a dunite matrix. In some cases the nodules are cored by skeletal chromite which Greenbaum (1977), in conjunction

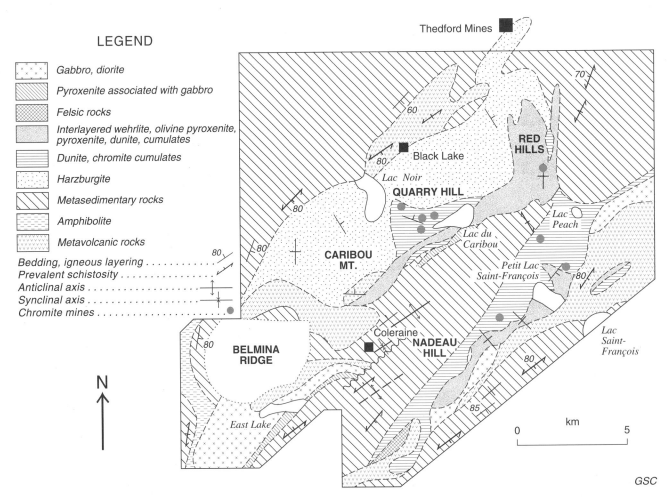

Figure 28.2-1 Location of significant chromite deposits in the Thetford Mines ophiolite complex (after Kacira, 1982; Marcotte, 1980).

with other evidence, suggested had a magmatic origin. Textures attributable to tectonism include lineation or foliation of chromite and olivine, stretching, boundinage and "pull apart" of chomite grains, brecciation, and mylonitization. The chromite in massive ore tends to occur as coarse grained, interlocking anhedra; grain sizes on the order of 5 to 10 mm are common (Fig. 28.1-3D).

Chemical composition of ores

Selected ore compositions are included in Table 28.1-1. By comparison with stratiform chromite ores, most podiform ores are richer in chrome and MgO, and poorer in iron and titanium. The occurrence of distinct Cr-rich and Al-rich orebodies is a noteworthy feature of some podiform chromite districts. The Al-rich deposits appear to be associated in part with feldspar-bearing rocks including feldspathic peridotite, troctolite, and gabbro, and they are presumably derived from less primitive magmas than the Cr-rich bodies.

DEFINITIVE CHARACTERISTICS

The occurrence of massive or semimassive chromite in an ophiolitic succession or in Alpine-type peridotite is diagnostic. Nodular textures are also characteristic of podiform deposits.

GENETIC MODEL

There is little doubt that podiform chromite deposits formed initially by magmatic segregation, but the precise nature of the process is controversial. An origin by fractional crystallization of basaltic magma has long been accepted for the chromitites within the ultramafic cumulates of ophiolite successions. The origin of chromitites in harzburgite tectonites has been more problematical. Some authors have suggested that they are outliers of the cumulate succession that were incorporated in the tectonite either by gravitational sinking of dense autoliths (Dickey, 1975) or by infolding of the lowermost cumulate layers (Greenbaum, 1977), but neither hypothesis is widely accepted. An alternative view is that the chromitites as well as the harzburgites are the refractory residua of partial melting of mantle lherzolite (Dickey and Yoder, 1972). The current consensus is that the chromitites segregated from basaltic magma in small chambers (Neary and Brown, 1979) or steeply inclined cavities in the conduit system (Lago et al., 1982) through which the magma moved en route to the crust. In any event, the primary form and textures of the deposits have been variably modified by deformation during both flow in the mantle and obduction.

RELATED DEPOSIT TYPES

The genesis of podiform chromite deposits may involve similar magmatic segregation processes as those which give rise to stratiform chromite ores (see deposit subtype 28.1).

Many significant ultramafic-hosted asbestos deposits (see deposit Type 11) also occur in the tectonized harzburgite portions of ophiolitic sequences and, in some cases, are spatially associated with podiform chromites.

EXPLORATION GUIDES

1. Podiform chromite deposits occur in ophiolitic sequences and are most common in the uppermost 500 to 1000 m of the tectonite member.

2. The podiform bodies in the tectonite commonly occur within an envelope of dunite a few centimetres to a few metres in thickness.

3. Podiform bodies which are conformable to the fabric of the harzburgite (i.e., concordant deposits) are normally the largest, the most regular in shape, and the highest grade deposits.

ACKNOWLEDGMENTS

D.H. Watkinson and C.T. Barrie critically reviewed the manuscript.

SELECTED BIBLIOGRAPHY

References with asterisks (*) are considered to be the best source of general information on this deposit type.

Bannatyne, B.B. and Trueman, D.L.
1982:   Chromite reserves and geology of the Bird River Sill, Manitoba; Manitoba Energy and Mines, Geological Services, Open File Report OF82-1, 73 p.

Bateman, J.D.
1943:   Bird River chromite deposits, Manitoba; The Canadian Institute of Mining and Metallurgy Transactions, v. 46, p. 154-183.

Brown, M.
1980:   Textural and geochemical evidence for the origin of some chromite deposits in the Oman ophiolite; in Ophiolites, Proceedings of the International Ophiolite Symposium, Cyprus, 1979, (ed.) A. Panayiotou; p. 714-721.

Burgath, K. and Weiser, T.
1980:   Primary features and genesis of Greek podiform chromite deposits; in Ophiolites, Proceedings of the International Ophiolite Symposium, Cyprus, 1979, (ed.) A. Panayiotou; p. 675-690.

Cameron, E.N.
1980:   Evolution of the Lower Critical Zone, central sector of the Eastern Bushveld Complex, and its chromite deposits; Economic Geology, v. 75, p. 845-871.

Cameron, E.N. and Desborough, G.A.
1969:   Occurrence and characteristics of chromite deposits - Eastern Bushveld Complex; Economic Geology, Monograph 4, p. 23-40.

Cassard, D., Rabinovicz, M., Nicholas, A., Moutte, J., Leblanc, M., and Prinzhofer, A.
1981:   Structural classification of chromite pods in southern New Caledonia; Economic Geology, v. 76, p. 805-831.

Cotterill, P.
1969:   The chromite deposits of Selukwe, Rhodesia; Economic Geology, Monograph 4, p. 154-186.

Dennis, B.T.
1931:   The chromite deposits of the eastern townships of the province of Quebec; Quebec Department of Mines, Annual Report Part D, 106 p.

DeYoung, J.H., Lee, M.P., and Lipin, B.R.
1984:   International Strategic Minerals Inventory Summary Report - Chromium; United States Geological Survey, Circular 930-B, 41 p.

Dickey, J.S.
1975:   A hypothesis of origin for podiform chromite deposits; Geochimica et Cosmochimica Acta, v. 71, p. 1061-1074.

Dickey, J.S. and Yoder, H.S.
1972:   Partitioning of chromium and aluminum between clinopyroxene and spinel; Carnegie Institution of Washington Yearbook, v. 71, p. 384-392.

*Duke, J.M.
1983:   Magmatic segregation deposits of chromite; Geoscience Canada, v. 10, no. 1, p. 15-24.

Ergunalp, F.
1980: Chromite mining and processing at Kavak mine, Turkey; Institution of Mining and Metallurgy Transactions, v. 89, p. A179-A184.

Greenbaum, D.
1977: The chromitiferous rocks of the Troodos ophiolite complex, Cyprus; Economic Geology, v. 72, p. 1175-1194.

Hedlund, D.C., Moriera, P., Pinto, A., da Silva, J., and Souza, G.
1974: Stratiform chromitite at Campo Formoso, Bahia, Brazil; United States Geological Survey, Journal of Research, v. 2, p. 551-562.

Hock, M. and Friedrich, G.
1985: Structural features of ophiolitic chromitites in the Zambales Range, Luzon, Philippines; Mineralium Deposita, v. 20, p. 290-301.

Hulbert, L.J. and Von Gruenewaldt, G.
1985: Textural and compositional features of chromite in the Lower and Critical Zones of the Bushveld Complex south of Potgietersrus; Economic Geology, v. 80, p. 872-895.

Irvine, T.N.
1975: Crystallization sequences in the Muskox intrusion and other layered intrusions. II. Origin of chromitite layers and similar deposits of other magmatic ores; Geochimica et Cosmochimica Acta, v. 39, p. 991-1020.
1977: Origin of chromitite layers in the Muskox intrusion and other stratiform intrusions: a new interpretation; Geology, v. 5, p. 273-277.

Irvine, T.N. and Sharpe, M.R.
1986: Magma mixing and the origin of stratiform oxide ore zones in the Bushveld and Stillwater complexes; in Metallogeny of Basic and Ultrabasic Rocks, (ed.) M.J. Gallagher, R.A. Ixer, C.R. Neary, and H.M. Pritchard; The Institution of Mining and Metallurgy, London, p. 183-198.

***Kacira, N.**
1982: Chromite occurrences of the Canadian Appalachians; The Canadian Institute of Mining and Metallurgy, v. 75, no. 837, p. 73-82.

Lago, B.L., Rabinowicz, M., and Nicholas, A.
1982: Podiform chromite ore bodies: a genetic model; Journal of Petrology, v. 23, p. 103-125.

***Marcotte, R.**
1980: Gîtes et Indices de chromite au Québec; Ministére de l'energie et des ressources, Open File 724, 58 p.

Neary, C.R. and Brown, M.
1979: Chromites from the Al'ays complex, Saudi Arabia and the Semail complex, Oman; Institute of Applied Geology, Kingdom of Saudi Arabia, Bulletin 3, p. 193-205.

Nicholas, A. and Prinzhofer, A.
1983: Cumulative or residual origin for the transition zone in ophiolites: structural evidence; Journal of Petrology, v. 24, p. 188-206.

Scoates, R.F.J.
1983: A preliminary stratigraphic examination of the ultramafic zone of the Bird River Sill; in Manitoba Department of Energy and Mines, Report of Field Activities 1983, p. 70-83.

Scoates, R.F.J., Eckstrand, O.R., and Cabri, L.J.
1987: Interelement correlation, stratigraphic variation and distribution of PGE in the Ultramafic Series of the Bird River Sill; in Geoplatinum 87, (ed.) H.M. Pritchard, P.J. Potts, J.F.W. Bowles, and S.J. Cribb; Elsevier Applied Science, London, p. 239-249.

Slatter, D. de L.
1980: The composition of Zimbabwean chromium ores and the derivation of chemical and physico-chemical ratings for smelting the ores to high-carbon ferrochromium; Institute of Mining Research, University of Zimbabwe, Report C193, 79 p.

Smirnov, V.I.
1978: Ore Deposits of the U.S.S.R.; v. 1, Pitman Publications, London, United Kingdom, 352 p.

Snelgrove, A.K.
1934: Chromite deposits of Newfoundland; Geological Survey of Newfoundland, Bulletin 1, 26 p.

Soderholm, K. and Inkinen, O.
1982: The Tornio layered intrusion - a recently discovered intrusion with chromitite horizons in northern Finland; Bulletin of the Geological Society of Finland, v. 54, no. 1-2, p. 15-24.

Stoll, W.C.
1958: Geology and petrology of the Masinloc chromite deposit, Zambales, Luzon, Philippine Islands; Geological Society of America Bulletin, v. 69, p. 419-448.

Thayer, T.P.
1964: Geologic features of podiform chromite deposits; in Methods of Prospection for Chromite, (ed.) R. Woodtli; Organization for Economic Co-operation and Development, Paris.

Ulmer, G.C.
1969: Experimental investigation of chromite spinels; Economic Geology, Monograph 4, p. 114-131.

Watson, D.F.
1985: Chromite reserves of the Bird River Sill; Manitoba Energy and Mines, Geological Services, Open File Report OF85-8, 22 p.

Whittaker, P.J.
1986: Chromite deposits in Ontario; Ontario Geological Survey, Study 55, 97 p.

Whittaker, P.J. and Watkinson, D.H.
1984: Genesis of chromitite from the Mitchell Range, central British Columbia; The Canadian Mineralogist, v. 22, p. 161-172.

Williamson, B.
1990: Geology of the Bird River Sill at the Chrome property, southeast Manitoba; Geological Survey of Canada, Open File 2067, 26 p.

Author's address

J.M. Duke
Geological Survey of Canada
601 Booth Street
Ottawa, Ontario
K1A 0E8

Printed in Canada

INDEX

APPENDIX

Index to mineral deposits shown on Figure 2 (*in pocket*)

NEWFOUNDLAND

1 Wabana Fe
2 Foxtrap pyrophyllite
3 Kelligrews River Mn
4 St. Lawrence fluorite
5 Grey River W
6 Hope Brook (Chetwynd) Au,Cu
7 Cape Ray Au
8 Bishop Fe,Ti,V
9 Strickland Pb,Zn
10 Fischell's Brook gypsum
11 Flat Bay gypsum
12 Romaines Brook gypsum
13 York Harbour Cu,Zn
14 Glover Island Au
15 Tulk's Zn,Pb,Cu,Ag,Au
16 Tulk's East Zn,Pb,Cu,Ag,Au
17 Skidder Cu,Zn
18 Buchans Zn,Pb,Cu,Ag,Au,Cd,barite
19 Duck Pond Cu,Zn,Pb,Ag,Au
20 Boundary Cu,Zn
21 Great Burnt Lake Cu
22 Beaver Brook Sb
23 Lake Bond Zn,Cu
24 Gullbridge Cu
25 Hand Camp Cu,Au,Ag
26 Point Leamington Zn,Cu
27 First Pond magnesite
28 Pilleys Island pyrite,Cu
29 Little Bay Cu
30 Whalesback Cu
31 Colchester Cu
32 Rendell-Jackman (Hammerdown) Au
33 Betts Cove Cu
34 Nugget Pond Au
35 Tilt Cove Cu
36 Rambler Cu,Au,Zn,Ag,Cd
37 Advocate asbestos
38 Daniels Harbour Zn
39 Aillik Bay (Makkovik) U,Mo
40 Kitts U
41 Michelin U
42 Ten Mile Lake Be
43 Wabush Lake Fe
44 Schefferville Fe
45 Strange Lake Y,Zr,REEs
46 Voisey Bay Ni,Cu,Co

NOVA SCOTIA

1 Meat Cove Zn,Cd,Ge,Ag
2 Dingwall gypsum
3 Cheticamp gypsum
4 Lake Ainslie barite,fluorite
5 Little Narrows gypsum
6 River Denys gypsum
7 Lime Hill Zn
8 Yava Pb,Ag
9 Lake Enon celestite
10 Stirling (Mindamar) Zn,Cu,Pb,Ag,Au

11 Antigonish gypsum
12 Lochaber Lake Cu
13 Dort's Cove andalusite
14 Lower Seal Harbour Au
15 Goldboro Au
16 Cochrane Hill Au
17 Goldenville Au
18 Fifteen Mile Stream Au
19 Tangier Au
20 Mooseland Au
21 Moose River Au
22 Caribou Au
23 Gays River Pb,Zn
24 Milford gypsum
25 Oldham Au
26 Waverley Au
27 Montague Au
28 East Kemptville Sn,Cu,Zn,Ag
29 Millett Brook U
30 Windsor gypsum
31 Cheverie gypsum
32a Magnet Cove barite,Ag,Pb,Cu,Zn
32b Walton gypsum
33 Londonderry Fe
34 Pugwash salt
35 Nappan salt

NEW BRUNSWICK

1 Dorchester Cu
2 Dorchester barite
3 Hillsborough gypsum
4 Havelock gypsum
5a Plumweseep potash
5b Millstream potash
6 Salt Springs potash,salt
7 Cape Spencer Au
8 St. Stephen Ni,Cu
9 Mount Pleasant W,Mo,Sn,Bi,Cu,Pb,
Zn,Ag,In
10 Lake George Sb
11 Connell Mountain Cu,Mo
12 Woodstock Mn,Fe
13 Sisson Brook W,Cu,Mo
14 Burnt Hill W,Sn
15 Mount Costigan Zn,Pb,Ag
16 Chester Zn,Pb,Cu,Ag,Au
17 Maliseet Mountain Ni,Cu,Bi,Co,Ag
18 Half Mile Lake Zn,Pb,Cu
19 Stratmat Zn,Pb,Cu,Ag
20 Heath Steele Zn,Cu,Pb,Ag,Cd
21 Captain North Extension Zn,Pb,Ag
22 Key Anacon Zn,Pb,Cu,Ag
23 Flat Landing Brook Zn,Pb,Cu,Ag
24 Brunswick No. 6 Zn,Pb,Cu,Ag,Cd
25 Brunswick No. 12 Zn,Pb,Cu,Ag,Cd
26 Nepisiguit Zn,Pb,Cu
27 Canoe Landing Lake Zn,Cu,Pb,Au,Ag
28 Wedge Cu,Ag

29 Restigouche Zn,Pb,Ag,Cu,Au
30 Murray Brook Cu,Au
31 Caribou Pb,Zn,Cu
32 Armstrong Brook Zn,Pb,Cu,Ag,Au
33 Tetagouche Falls Mn
34 Beresford Copper Cu
35 Nigadoo Zn,Pb,Ag,Cu,Cd
36 Keymet Zn,Pb,Cu,Ag
37 Elmtree Au
38 Turgeon Cu,Zn
39 Nash Creek Zn,Pb,Ag
40 Benjamin River Cu,Mo

QUEBEC

1a Dauphin salt
1b Grosse-Île salt
2a Havre-Aubert salt,potash
2b Île d'Entrée salt
3 Mid-Patapedia Cu,Ag
4 Pekan Cu
5 Gaspé Copper Cu,Mo,Ag,Se,Te,Bi
6 Madeleine Cu,Ag
7 Natashquan Ti,Fe,zircon,garnet
8 Doran U
9 Lac Tio Ti,Fe
10 Magpie Mountain Ti,Fe,V,Cr
11 Lac Kachiwiss U
12 La Blache-Hervieux-Shmoo lakes Ti,Fe,V
13 Lac Brûlé Ti,Fe
14 La Hache Ti,Fe,phosphate
15 K-Nuts Ti,Fe
16 Niobec Nb
17 St. Charles Fe,Ti,V,phosphate
18 Ménard Township wollastonite
19 Roberval Ti,Fe
20 Crevier Nb,Ta
21a Bédard phlogopite
21b Parent phlogopite
21c Siscoe phlogopite
22 Lamy phlogopite
23 Chasseur Suzorite phlogopite
24 St. Urbain Ti,Fe
25a Tétreault Zn,Pb,Au,Ag
25b Montauban Au,Ag,Zn,Pb
26 Portneuf Mo,mica
27 Cranbourne asbestos
28 Golden Age asbestos
29 Carey Canadian asbestos,talc
30 National-Flintkote-Pennington asbestos
31a British Canadian asbestos
31b Black Lake asbestos
31c Normandy-Vimy Ridge-Penhale asbestos
31d Bell-King-Beaver-Johnson asbestos
31e Aylmer asbestos
32 St. Adrien Mountain asbestos
33 Nicolet asbestos
34 Jeffrey asbestos
35 Lili-St. Cyr-Steel Brook asbestos,Cr
36 Derogan (Melbourne) asbestos

37 Reed-Bélanger Cr
38 Montréal Cr
39 Chaudière River Au
40 Mount St. Sebastien Mo
41 Clinton Cu,Zn,Ag
42a Cupra d'Estrie Cu,Zn,Ag,Pb,Au,Cd
42b Solbec Cu,Zn,Pb,Ag,Au,Cd
43 Weedon Cu,Zn,Ag,Au
44 Lingwick Zn,Cu,Ag
45 Moulton Hill (Aldermac) Zn,Pb,Cu,Ag,Au
46 Eustis Cu
47 Suffield Zn,Cu,Pb,Ag,Au
48 Huntingdon Cu,Au,Ag
49 Oka Nb,REEs
50a Pin-Rouge Lake Ti,Fe
50b Tamara-Drummond Ti,Fe
51 Desgrosbois Ti,Fe
52 Kilmar magnesite
53 Eastern Asbestos asbestos
54 Hull (Forsythe) Fe
55 Hilton (Bristol) Fe
56 Portage-du-Fort magnesite
57 Clarendon graphite
58 Calumet Zn,Pb,Ag,Au,Cu
59 Leman-Axe Lake U
60 Capri U
61 Renzy Lake Ni,Cu
62 Houdet Fe
63 Kipawa Y,Zr
64 Belleterre Au,Ag
65 Kelly Lake Cu,Ni
66 Lorraine Cu,Ni
67 Russian Kid (Bordulac) Au,W
68a Magusi River Zn,Cu,Au,Ag
68b Hébécourt (New Insco) Cu,Au,Ag
69 Beattie Au
70 Francoeur Au
71 Aldermac Cu,Au,Ag
72 Wasamac Au
73 Eldrich Au
74 Elder Au
75 Don Rouyn Cu,Mo
76a Durbar Au
76b Silidor Au
76c Astoria Au
76d Stadacona Au
76e Senator Rouyn Au
76f Powell Rouyn Au
76g Chadbourne Au
77a Amulet F Cu,Zn,Au,Ag
77b Amulet C Cu,Zn,Au,Ag
77c Amulet A-Lower A Cu,Zn,Au,Ag
77d Millenbach Cu,Zn,Au,Ag
77e Corbet Cu,Zn,Au,Ag
78a Lac Dufault (Norbec) Cu,Zn,Au,Ag,Cd
78b East Waite Cu,Zn,Au,Ag
78c Waite Amulet Cu,Zn,Au,Ag
78d Vauze Cu,Zn,Ag,Au
78e Ansil Cu,Zn,Au,Ag
79 Destor (Thurbois) Au
80 Mobrun Zn,Cu,Ag,Au
81 Gallen (West Macdonald) Zn,Cu,Au,Ag
82 Donalda Au
83a Horne Cu,Au,Ag,Se,Te
83b Quemont Cu,Zn,Au,Ag
83c Joliet Cu,Au
84 McWatters Au
85a Hosco Au
85b Héva Au
86 Mic Mac Au

87 Doyon (Silverstack) Au
88 Bousquet No. 1 Au,Cu
89 Bousquet No. 2-LaRonde Au,Ag,Cu
90 O'Brien Au
91a Anglo American (Cadillac Moly) Mo,Bi
91b Preissac Mo,Bi
92 Cadillac-Univex Fe
93 Marbridge Ni,Cu
94 Lacorne Mo,Bi
95 Canadian Malartic Au
96a Barnat Au
96b East Malartic Au
97 Malartic Hygrade Au
98 Camflo Au
99 Malartic Gold Fields Au
100a Marban Au
100b Norlartic Au
101 Callaghan Au
102 Kiena Au
103 Wesdome Au
104 Siscoe Au
105 Goldex Au
106 Sullivan Consolidated Au
107 Sigma-Lamaque Au
108 Manitou-Barvue Zn,Ag,Cu,Au,Pb
109 East Sullivan Cu,Zn,Ag
110 Dunraine Cu
111 Louvicourt asbestos
112a New Pascalis-South Au
112b Perron-Pascalis North Au
113 Louvem Zn,Cu,Cd,Au,Ag
114 Aur Resources (Louvicourt) Cu,Zn,Au,Ag
115 Abitibi Copper Cu,Au,Ag
116 Bevcon-Buffadison Au
117 Zulapa Ni,Cu
118 Chimo Au
119 Vauquelin (Nordeau) Fe,Au
120 Croinor-Pershing Au
121 Coniagas Zn,Ag,Pb
122 Atlas Fe
123 Vendome (Consolidated Mogador) Zn,Cu,Pb,Ag,Au
124 Quebec Lithium Li
125a Barvue Zn,Ag
125b Consolidated Pershcourt Zn,Ag
126 Frebert Zn,Ag
127 Trinity Fe
128 Conigo Cu,Ag
129 Dumont Ni
130 Normétal Cu,Zn,Au,Ag
131 Golden Pond Au
132 Selbaie (Detour) Zn,Cu,Ag,Au
133 Estrades Zn,Cu,Pb,Au,Ag
134 Agnico Eagle-Telbel Au
135a Joutel Cu,Zn
135b Poirier Cu,Zn,Ag
135c Explo-Zinc Zn,Cu,Ag
136 Douay Au
137 Phelps Dodge-La Gauchetière Township Cu,Zn,Ag
138 Chabouillé Lake Cu,Ni,Ag
139 New Hosco Cu,Zn,Ag,Au
140a Mattagami Lake Zn,Cu,Ag,Au,Cd
140b Orchan Zn,Cu,Ag,Au,Cd
140c Norita Zn,Cu,Ag,Au
140d Isle Dieu Zn,Cu,Ag,Au
141 Abitibi Asbestos asbestos
142 Sleeping Giant Au
143 Chesbar Fe
144 Nicobi Lake Ni,Cu

145 Bachelor Lake Au
146 Lac Shortt Au
147 Joe Mann-Meston Lake Au
148 Opemiska Cu,Au,Ag
149 Norbeau Au
150 Campbell Chibougamau-Merrill Island Cu,Ag
151 Lac Doré Cu,Au
152 Copper Rand Cu,Au
153a Cedar Bay Cu,Au
153b Copper Cliff Cu,Fe
154 Henderson-Portage Island Cu,Au
155 Devlin Cu
156a R-2 Cu,Mo
156b Corner Bay Cu,Au,Ag
157 Lemoine Cu,Zn,Ag,Au
158 Kellogg (Bischoff) Fe,Ti,V
159 Asbestos Island asbestos
160 Roberge Lake asbestos
161 Icon Cu,Ag
162 Lessard Cu,Zn,Ag,Au
163 Troïlus Au,Cu
164 Albanel Minerals-Canso East Fe
165 McLeod Lake Cu,Mo
166 Eastmain Au
167 Indicator magnesite
168 Pambrun Lake Ti,Fe
169 Gagnon Fe
170 Fire Lake Fe
171 Lac Croche kyanite
172 Lac Knife graphite
173 Mount Wright Fe
174 Sakami Lake U
175 Duncan Range Fe
176 Grande Rivière de la Baleine E Fe
177 Grande Rivière de la Baleine D Fe
178 Grande Rivière de la Baleine A Fe
179 Ruby Lake-Nancy Island Zn,Pb
180 Dieter Lake U
181 Schefferville Fe
182 Eclipse Fe
183 Retty Lake Cu,Ni,PGEs
184 Franelle Cu
185 Strange Lake Y,Zr,REEs
186 Lac Otelnuk Fe
187 Lac de l'Hématite Fe
188 Marymac Cu,Ni
189 Old Red Hill-Gossan Hill Fe
190 Boylen Zn,Cu,Pb,Au,Ag
191 Leslie No. 2 Cu,Ni
192 Partington Fe,S
193 Lac Irony Fe
194 Soucy No. 1 Cu,Zn,Au,Ag
195 Prud'homme No. 1 Cu,Zn,Au,Ag
196a Lac Berard Fe
196b Lac Mannic Fe
197 Hopes Advance Bay 3 Cu,Ni
198 Hopes Advance Bay 1 and 1N Cu,Ni
199 Ford Lake Fe
200a Morgan Range Fe
200b Lac Morgan Fe
201 Payne Bay Fe
202 Kyak Bay Fe
203 Yvon Lake Fe
204 Expo Ungava Ni,Cu
205 Raglan (Donaldson) Ni,Cu
206 Raglan (Katiniq) Ni,Cu
207 Asbestos Hill asbestos
208 Raglan (Cross Lake) Ni,Cu
209 Kenty Lake Ni,Cu,PGEs

ONTARIO

1 Ojibway salt
2 Chatham salt
3 Sarnia salt
4 Warwick salt
5 Adelaide salt
6 Goderich salt
7 Drumbo salt
8 Hagersville gypsum
9 Caledonia gypsum
10 Matthews Fe,Ti
11 Seeleys Bay wollastonite
12 Portland (Victoria Graphite) graphite
13 Olympus vermiculite
14 Timmins graphite
15 Kirkham graphite
16 Olden Township wollastonite
17 Addington Au
18 Henderson-Conley talc
19 Marmora Fe
20 Moira River wollastonite
21 Methuen Ti
22 Blue Mountain nepheline syenite
23 Macassa Ni,Cu,Co
24 Madawaska (Faraday) U
25 Wilbermere Lake fluorite,U
26 Harcourt graphite
27 Craigmont corundum
28 Renprior Zn,Pb
29 Haley Mg
30 Cummings tremolite,talc,phlogopite
31 Todd graphite
32 Graphite Lake graphite
33 Bissett Creek graphite
34 Mattawa Fe,Ti,V
35 Crocan Lake kyanite
36 Manitou Islands Nb,U
37 Angus Fe,Ti
38 Keeley-Frontier Ag,Co
39 Cobalt Ag,Co,Ni,Cu
40 Casey-Cobalt Ag,Co,Ni,Cu
41 Sherman Fe
42 Temagami Cu,Au,Ag,Ni,Co,PGEs
43 Cummings Lake Fe
44 Golden Rose Au
45 Moose Mountain Fe
46 North Range Ni,Cu,PGEs,Au
47 Victor Ni,Cu,PGEs,Au
48 Maclennan Ni,Cu,PGEs,Au
49 Street Township garnet
50 Wanapitei kyanite
51a Norduna Ni,Cu,PGEs,Au
51b Falconbridge East Ni,Cu,PGEs,Co,Au,Se
51c Falconbridge Ni,Cu,PGEs,Co,Au,Se
52a Garson Ni,Cu,PGEs,Au
52b Kirkwood Ni,Cu,PGEs
53a Blezard Ni,Cu,PGEs,Au
53b Lindsley Ni,Cu,PGEs
53c Mount Nickel Ni,Cu,PGEs,Au
53d Little Stobie Ni,Cu,PGEs,Au
53e Frood-Stobie Ni,Cu,PGEs,Co,Au
54a McKim Ni,Cu,PGEs,Co,Au
54b Murray Ni,Cu,PGEs
54c Copper Cliff North-Clarabelle Ni,Cu,PGEs,Au
54d Copper Cliff No. 2 Ni,Cu,PGEs,Au
54e Copper Cliff Ni,Cu,PGEs,Au
54f Copper Cliff No. 1 Ni,Cu,PGEs,Au
54g Copper Cliff South Ni,Cu,PGEs,Au
55a North Star Ni,Cu,PGEs,Au
55b Creighton Ni,Cu,PGEs,Au
56a Lockerby Ni,Cu,PGEs,Au

56b Ellen Ni,Cu,PGEs,Co,Se,Au
56c Crean Hill Ni,Cu,PGEs,Co,Se,Au
56d Victoria Ni,Cu,PGEs
57 Totten Ni,Cu,PGEs,Co
58a McCreedy West Ni,Cu,PGEs
58b Onaping Ni,Cu,PGEs
58c Craig Ni,Cu,PGEs
58d Boundary Ni,Cu,PGEs,Au
58e Hardy Ni,Cu,PGEs,Au
58f Levack West Ni,Cu,PGEs,Co,Au
58g Levack Ni,Cu,PGEs,Co,Au
59a North Ni,Cu,PGEs,Au
59b Fecunis Ni,Cu,PGEs,Co,Au
59c Fraser Ni,Cu,PGEs
59d McCreedy East Ni,Cu,PGEs
59e Strathcona Ni,Cu,PGEs,Co
59f Coleman Ni,Cu,PGEs,Au
59g Longvack South Ni,Cu,PGEs,Au
59h Longvack Ni,Cu
59i Big Levack Ni,Cu,PGEs,Au
60 Errington Cu,Zn,Pb,Ag,Au
61 Vermilion Lake Cu,Zn,Pb,Ag,Au
62 Agnew Lake U
63 Spanish River Cu,Au,Ag
64 Shakespeare Ni,Cu
65 Fostung W,Mo
66 Pater Cu,Au,Ag
67 Elliot Lake-Nordic zone U,Y
68 Elliot Lake-Quirke zone U,Y
69 Bruce Mines Cu,Ag,Au
70 Coppercorp Cu
71a Tribag mine Cu,Mo,Ag,Au,W
71b Tribag-East Breccia zone Cu,Mo,Ag,Au,W
72 Renner Ni,Cu
73 Lackner Lake Nb,phosphate,Fe
74 Nemegosenda Nb
75 Shunsby Cu,Zn
76 Orofino (Swayze) Au
77 Stackpool Fe
78 Reeves asbestos
79 Penhorwood talc
80 Texmont Ni
81 Midlothian asbestos
82 Siscoe (Miller Lake O'Brien) Ag,Co
83a Young-Davidson Au,W
83b Matachewan Consolidated Au,W
84a Macassa Au
84b Kirkland Lake Au
84c Teck-Hughes Au
84d Lake Shore Au
84e Sylvanite Au
84f Wright-Hargreaves Au
84g Toburn Au
85 Adams Fe
86 Victoria Creek Au
87 Upper Canada Au
88a Omega Au
88b Raven River Au
89a Kerr Addison Au
89b Chesterville Au
90a Holt-McDermott Au
90b Matawasaga Au
90c Lightning zone Au
91 Matheson asbestos
92 Hedman serpentine
93 Munro asbestos
94 Ross Au
95 St. Andrews Goldfields Au
96 Clavos Au
97 Langmuir Ni
98 Redstone Ni
99 Porcupine-Southgate magnesite,talc

100a Dome Au
100b Preston Au
101a Paymaster Au
101b Ankerite Au
101c Delnite Au
101d Aunor Au
102a McIntyre Au,Cu,Mo,W
102b Hollinger Au
102c Vipond-Crown Au
102d Coniaurum Au
103a Pamour Au
103b Hoyle Pond Au
104 Owl Creek Au
105a Hallnor Au
105b Broulan Au
106 Reef Au
107 Bell Creek Au
108 Kidd Creek Cu,Zn,Ag,Pb,Cd,Se,In
109 Kam Kotia Cu,Zn,Ag
110 Montcalm Ni,Cu
111 Detour Lake Au,Ag,Cu
112 Argor Nb
113 Martison Lake phosphate, Nb
114 Cargill phosphate
115 Renabie Au
116a Magino Au
116b Kremzar Au
117 Lakemount Ni,Cu,PGEs
118a MacLeod-Helen-Sir James Fe
118b Josephine-Lucy Fe
119 Surluga Au
120 Magnacon (Mishibishu Lake) Au
121a Geco Cu,Zn,Ag,Pb,Cd
121b Willecho Zn,Cu,Ag,Pb
121c Willroy Cu,Zn
121d Big Nama Creek Zn,Cu,Ag
122a Golden Giant Au,Mo
122b Teck-Corona Au,Mo
122c Page-Williams Au,Mo
123 Anaconda Cu,PGEs
124 Coubran Lake Fe,Cu,Ti
125 Winston Lake (Zenmac) Zn,Cu
126 Georgia Lake Li
127 Jean Lake Li
128 Nama Creek-Conway Li
129 Leitch Au
130 Magnet Consolidated Au
131 Brookbank Au
132 Marshall Lake Cu,Zn,Ag,Cu
133a McLeod-Cockshutt Au
133b Mosher Au
133c Hard Rock Au
133d Little Long Lac Au
134 Can-Fer Fe
135 Juneau Lake Ni,Cu
136 Obonga Lake Ni,Cu
137a Lyon Lake-Creek Zone Zn,Cu,Ag,Pb,Au
137b Sturgeon Lake Zn,Cu,Ag,Pb,Au
138 Mattabi Zn,Cu,Ag,Pb,Au
139 Lac des Iles PGEs,Au,Ni,Cu
140 Anderson Lake Mo
141 Great Lakes Nickel Ni,Cu,PGEs
142 Shebandowan Ni,Cu,PGEs,Co
143 North Coldstream Cu,Au,Ag
144 Moss Lake Au
145 Hammond Reef Au
146 Atikokan Iron Fe,Cu
147 Steep Rock Fe
148 Lac La Croix Li
149 Seine Bay-Bad Vermilion Lake Ti
150 Northrock (Grassy Portage Bay) Cu
151 Emo Ni,Cu,Co

152 Duport (Cameron Island) Au
153 Cameron Lake Au
154 Maybrun Cu
155 Kenbridge Ni,Cu
156 Pidgeon Molybdenum Mo
157 Goldlund Au
158 Reynar Lake-Almo Lake Ni,Cu
159 Gordon Lake Ni,Cu,PGEs,Co
160 Madsen Red Lake Au
161a Howey Au
161b Hasaga Au
162a Cochenour-Willans Au
162b McKenzie Red Lake Au
163a Campbell Red Lake Au
163b Dickenson Au
164 Griffith Fe
165 South Bay (Uchi Lake) Zn,Cu,Ag
166 McCombe (Root Lake) Li
167 Golden Patricia Au
168a Thierry Cu
168b Kapkichi Lake Cu,Ni
169 Central Patricia Au
170 Dona Lake Au
171 Pickle Crow Au
172 Norton Lake Ni,Cu,PGEs
173 Opapimiskan Lake (Musselwhite) Au
174 Setting Net Lake Mo
175 Berens River Au,Ag,Pb,Zn
176 Big Trout Lake Cr
177 Lingman Lake Au

MANITOBA

1 Sunbeam-Waverley Au
2 Bernic Lake (Tanco) Ta,Li,Cs,Sn, Be,Rb,Ga
3 Buck-Coe-Pegli Li
4 Bird Lake Cr
5 Dumbarton-Maskwa West Ni,Cu
6 Page Cr
7 Chrome Group Cr
8 Euclid Lake Cr
9 Irgon Li
10 Spot Li
11 San Antonio-New Forty-Four Au
12 Silver Plains gypsum
13 Amaranth gypsum
14 Neepawa Fe
15 St. Lazare potash
16 Canamax potash
17 Gypsumville gypsum,anhydrite
18 Little Stull Lake Au
19 Gods Lake Au
20 Moak Ni
21 Mystery Lake South Ni
22 Birchtree Ni,Cu
23 Thompson Ni,Cu,Co,PGEs
24 Pipe Ni
25 Hambone Ni
26 Soab Ni
27 Bucko and Bowden lakes Ni
28 Resting Lake Ni,Cu
29 Manibridge Ni,Cu
30 Minago River Ni
31 Dyce Siding (Sylvia zone) Cu,Zn
32 Violet Li
33 Lit Li
34 Stall Lake Cu,Zn,Ag,Au
35 Osborne Lake Cu,Zn
36 Anderson Lake Cu,Ag,Au

37 Ghost Lake Zn,Cu,Pb,Ag,Au
38 Chisel Lake Zn,Ag,Au,Cu,Pb
39 Nor-Acme Au
40 Squall Lake Au
41 Wim Cu,Ag,Au
42 Dickstone Cu,Zn,Ag,Au
43 Reed Lake Cu,Zn
44 Namew Lake Ni,Cu,PGEs
45 Morgan Lake Zn,Au,Cu,Ag
46 Century Au
47 Pine Bay Cu
48 Centennial Cu,Zn,Ag,Au
49 Westarm Cu,Zn,Ag,Au
50 Schist Lake Cu,Zn,Ag,Au
51 Flin Flon Cu,Zn,Ag,Au,Cd,Se,Te
52 Embury (Trout Lake) Cu,Zn
53 Tartan Lake Au
54 Vamp Lake Au,Ag,Cu,Zn
55 Puffy Lake Au
56 Sherridon Cu,Zn,Ag,Au
57 Bob Lake Cu,Zn,Ag,Au
58 Jungle Lake Cu,Zn
59 Ruttan Cu,Zn,Ag,Au
60 MacBride Lake Zn,Cu,Ag,Au
61 Farley Lake Au
62 Lasthope Au
63 Burnt Timber (BT) Au
64 MacLellan (Agassiz) Au
65 Lynn Lake Ni,Cu
66 Fox Cu,Zn,Ag,Au

SASKATCHEWAN

1 Rocanville potash
2 Esterhazy potash,salt
3 Sybouts Lake sodium sulphate
4 Horseshoe Lake sodium sulphate
5 Frederick Lake sodium sulphate
6 Kalium-Belle Plaine potash,salt
7 Chaplin Lake sodium sulphate
8 Snakehole Lake sodium sulphate
9 Ingebrigt Lake sodium sulphate
10 Vincent Lake sodium sulphate
11 Alsask Lake sodium sulphate
12 Porcupine Prime potash
13 Watrous potash
14 Lanigan potash
15 Allan-Colonsay potash
16 Saskatoon potash
17 Vanscoy-Cory potash
18 Whiteshore Lake sodium sulphate
19 Unity salt
20 Bigstone Lake Cu,Zn
21 McIlvenna Bay Cu,Zn,Au,Ag
22 Coronation Cu,Zn,Ag,Au
23 Fon Zn
24 Callinan Cu,Zn,Au,Ag
25 Flin Flon Cu,Zn,Ag,Au,Cd,Se,Te
26 Wildnest Lake Cu,Zn
27 Shotts Lake Cu
28 Mokoman Lake Cu
29 Seabee Au
30 Anglo-Rouyn Cu,Ag,Au
31 Nemeiben Lake Ni,Cu
32 Elizabeth Lake Cu
33 Contact Lake (Bakos-Pap zones) Au
34 North Lake Au
35 Ivy Ni,Cu
36a Star Lake Au
36b Jolu Au

36c Jasper Au
37 Tower Lake East Au
38 Komis Au
39 Wedge Lake Au
40 Weedy Lake Au
41 McKenzie Zn,Cu
42 Pollon Lake (Deep Bay) graphite
43 George Lake Zn,Pb
44 Key Lake U
45 Maw Y,REEs
46 McArthur River U
47 Cigar Lake U
48 Midwest Lake U
49 Dawn Lake U
50 McClean Lake U
51a Rabbit Lake U
51b Horseshoe U
51c Raven U
52a Collins Bay U
52b Eagle Point U
53 Cluff Lake U,Ni
54 Gunnar U
55 Eldorado (Fay-Ace-Verna) U
56a Box Au
56b Athona Au
57 Axis Lake Ni,Cu,Co
58 Pluto Bay U,Mo

ALBERTA

1 Spionkop Creek-Yarrow Creek Cu,Ag
2 Kananaskis gypsum
3 Metisko Lake sodium sulphate
4 Lindbergh salt
5 Duvernay salt
6 Fetherstonehaugh gypsum
7 Fort Saskatchewan salt
8 Clearwater gypsum
9 Fort McMurray salt
10 Athabaska gypsum
11 Peace Point gypsum

BRITISH COLUMBIA

1 St. Eugene Ag,Pb,Zn,Au
2 Sullivan Pb,Zn,Ag,Cd,Sn
3 Bluebell Zn,Pb,Ag,Cd,Cu
4 Jackson Pb,Zn,Ag,Cd,Au,Cu
5 Lucky Jim Zn,Ag,Pb,Cd,Au
6 Aylwin Creek (Rockland) Cu,Au,Ag
7 Tillicum Au
8 Ymir Au
9 Jack Pot Pb,Zn
10a H.B. Zn,Pb,Ag,Cd
10b Jersey Zn,Pb,Ag,Cd
11 Emerald-Dodger W
12 Reeves MacDonald Zn,Pb,Ag,Cd
13 Rossland Au,Ag,Cu
14 Red Mountain (Giant) Mo,Au
15 Basic Ni
16 Mastodon Ni
17 Phoenix Cu,Au,Ag
18 Lexington Cu,Au
19 Mother Lode Cu,Au,Ag
20 Old Nick Ni
21 Beaverdell Ag,Zn,Pb
22 Blizzard U
23 Carmi Mo,U
24 Hydraulic Lake (Tye) U

25 Crystal Peak garnet
26 Hedley Mascot-Nickel Plate Au
27 Brenda Cu,Mo
28 Siwash North (Elk) Au
29 Primer Cu
30 Axe Cu
31 Copper Mountain-Ingerbelle Cu,Au,Ag
32 Lodestone Mountain Fe,V,PGEs
33 Canam Cu,Mo,Ag,Au
34 Carolin (Ladner Creek) Au
35 Pride of Emory Ni,Cu,Co
36 Harrison Gold Au
37 Seneca Zn,Cu,Pb,Au,Ag
38 Gem Mo
39 Pacific Talc talc
40 H talc,magnesite
41 Rawhide talc,magnesite
42 Brittania Cu,Zn,Au,Ag,Pb,Cd
43 Gambier Island Cu,Mo,Ag,Au
44 Mineral Hill wollastonite
45 Lara Au,Ag,Zn,Cu,Pb
46 Sunro (Jordan River) Cu,Ag,Au
47 Texada Fe,Cu,Au,Ag
48 Hi Ho Cu
49 OK Cu,Mo
50 Mount Washington Cu,Mo,Au
51 Westmin (Myra,Lynx,H-W) Zn,Cu,Pb,Ag,Au
52 Iron Hill Fe
53 Brynnor Fe
54 Catface Cu
55 Indian Chief Cu,Ag,Au
56 Zeballos Camp (Privateer) Au,Ag,Cu,Zn,Pb
57 Zeballos Iron (Ford) Fe
58 Iron Crown Fe
59 Kingfisher-Merry Widow Fe
60 Coast Copper (Old Sport) Cu,Fe,Au,Ag
61 Island Copper Cu,Mo,Au,Re
62 Hushamu Lake (Expo) Cu,Mo,Au
63 Red Dog Cu,Au
64 Fish Lake Cu,Au,Au
65 Spokane Cu,Au,Ag
66 Buzzer Cu,Mo
67 Northair-Van Silver Au,Ag,Pb,Zn
68 Owl Creek Cu,Mo
69 Bralorne-Pioneer Au
70 Congress Au
71 Empire Mercury Hg
72 Poison Mountain Cu,Mo,Au
73 Blackdome Mountain Au
74 Golden Eagle Hg
75 Maggie Cu,Mo
76 Cache Creek Zeolite zeolites
77a Trojan Cu
77b Krain Cu,Mo
78 Alwyn (OK) Cu,Ag,Au
79a Valley Copper-Lornex Cu,Mo
79b Bethlehem-JA Cu,Mo
79c Highmont Cu,Mo
79d Minex-Ann No. 1 Cu,Mo
80 Craigmont Cu,Fe,Ag,Au
81 Rey Cu
82 Glen Iron Fe
83 Afton Cu,Au,Ag
84a Ajax Cu
84b Evening Star Cu,Au
84c Rainbow Cu,Mo,Ag
84d Victor Cu,Au,Ag
85 Nan Fe
86 Falkland gypsum
87 Colby Zn,Pb
88 River Jordan Zn,Pb
89 Wigwam Pb,Zn
90 Big Ledge Zn
91 Trout Lake Mo
92a Mineral King Zn,Pb,Ag,Cd,Cu,barite
92b Duncan Pb,Zn
93 Lussier River-Truroc gypsum
94 Windermere (Western Gypsum) gypsum
95 Mount Brussilof magnesite
96 Brisco barite
97 Monarch Ag,Pb,Zn,Cd
98 Regal Silver-Snowflake Ag,Pb,Zn,Sn,W
99 J & L Au,Ag,Zn,Pb
100 Goldstream Cu,Zn,Ag
101 Bend Pb,Zn
102 Ruddock Creek Zn,Pb
103 CK Zn,Pb,Ag
104 Harper Creek Cu
105 Rexspar U,fluorite
106a Homestake Ag,Zn,Pb,Cu,barite
106b Rea (Samatosum) Ag,Au,Zn,Pb,Cu
107 Chu Chua Cu,Zn,Ag,Au,Co
108 Boss Mountain Mo
109 Frasergold Au
110 Verity Nb,Ta
111 Canoe mica
112 Eaglet fluorite
113 Mount Polley (Cariboo-Bell) Cu,Au,Ag
114 QR Au
115 Gibraltar Cu,Mo
116 Fraser River Au
117 Cariboo area Au
118a Mosquito Creek-Aurum Au
118b Cariboo Gold Quartz Au
119 Pinchi Lake Hg
120 Endako Mo
121 Capoose Ag,Au
122 Equity Ag,Cu,Au,Sb,As
123 Silver Queen Au,Ag,Zn,In,Ge
124 Poplar Cu,Mo,Au
125 Lucky Ship Mo
126 Nanika Cu
127 Berg Cu,Mo,Ag
128 Whiting Creek Mo,Cu
129 Huckleberry Cu,Mo
130 Ox Lake Cu,Mo,Au,Ag
131 Red Bird (CAFB) Mo
132 Surf Inlet-Pugsley Au,Ag,Cu
133 Jedway Fe,Cu
134 Jib Fe
135 Tasu Fe,Cu,Au,Ag
136 Cinola Au
137 Surf Point Au
138 Banks Island Au
139 Ecstall Zn,Cu,Ag,Au
140 Iron Mountain Fe
141 Kelly Creek Cu,Ag
142 JB Mo
143 Serb Creek Mo
144 Louise Lake Cu,Au,Mo
145 Glacier Gulch Mo,W
146 Big Onion Cu
147 Dome Mountain Au
148 Topley Richfield Au,Ag
149 Granisle Cu,Mo,Au,Ag
150 Bell Copper Cu,Au,Ag
151 Fireweed Ag,Zn,Pb
152 Morrison Cu,Mo
153 Dorothy Cu,Mo
154 Misty Cu
155 Kwanika Cu
156 Germanson River area Au
157 Lorraine Cu,Au
158 Rondah Cu
159 Mount Milligan Cu,Au
160 Wolf Ag,Pb,Zn
161 Lonnie Nb,Zr
162 Aley Nb
163 Beveley Zn,Pb,Ag
164 Robb Lake Pb,Zn
165 Cirque Zn,Pb,Ag
166 Driftpile Pb,Zn
167 Magnum Cu
168 Davis-Keays Cu
169 Al Au
170 Lawyers Au
171 Golden Neighbour Au,Ag
172 Baker Au
173 Kemess Cu,Au
174 Sustut Cu
175 Sping Cu
176 Red Cu,Ag
177 Thomlinson Mountain Mo
178 Bonanza Cu
179 Ajax Mo
180 Bell Moly Mo
181 Kitsault Mo
182 Roundy Creek Mo
183 Tidewater Mo
184 Hidden Creek (Anyox) Cu,Au,Ag
185 Double Ed Cu,Zn
186 Maple Bay Cu
187 Dolly Varden Ag,Pb,Cu,Au
188 Toric Ag,Pb,Zn
189 Georgia Au
190 Porter-Idaho Ag,Pb,Zn
191 Red Mountain Au
192 Silbak-Premier Au,Ag,Zn,Pb,Cu,Cd
193 Silver Butte Au
194 Big Missouri Au,Ag,Zn,Pb
195 Martha Ellen Au
196 Granduc Cu,Au,Ag
197 Doc Au
198 Max Fe,Cu
199 Brucejack Au
200 Kerr Cu,Au
201 Eskay Creek Au
202 E & L Ni,Cu
203 Stonehouse (Reg) Au
204 Bronson Slope Au,Ag,Cu,Mo
205 Snip Au,Cu
206 Copper Canyon Cu,Au
207 Galore Creek Cu,Au,Ag
208 Schaft Creek Cu,Mo,Au,Ag
209 Red-Chris Cu,Au
210 Gnat Lake (June, Stikine) Cu
211 Eagle Cu,Mo,Ag
212 Letain asbestos
213 Kutcho Creek Cu,Zn,Ag,Au
214 Sulphur Creek barite
215 Liard (Tam) fluorite
216 Dease Creek Au
217 Thibert Creek Au
218 McDame Creek Au
219 Mount Haskin Mo
220 Erickson Au
221 Cassiar asbestos,jade(nephrite)
222 Storie (Casmo) Mo
223 Magno Ag,Pb,Zn
224 Kuhn (Windy) W,Mo
225 Ewen Barite barite
226 Midway Ag,Zn,Pb,Sn
227 Golden Bear (Muddy Lake) Au

228 Mount Ogden Mo
229 Erickson-Ashby Ag,Pb,Zn
230 Polaris-Taku Au,Ag,Cu
231 Tulsequah Chief Zn,Cu,Pb,Cd,Au,Ag
232 Atlin Au
233 Adanac Mo
234 O'Connor gypsum,anhydrite
235 Windy-Craggy Cu,Co,Au

YUKON TERRITORY

1 Mel Zn,Pb,barite
2 McMillan (Quartz Lake) Zn,Pb,Ag
3 Bailey W,Cu
4 Sa Dena Hes (Mount Hundere) Zn,Pb,Ag
5 Logan Zn,Ag
6 Silver Hart Ag,Zn,Pb
7 Logtung W,Mo
8 Logjam Ag,Au,Zn,Pb
9 JC Sn
10 Red Mountain Mo
11 Livingstone Creek area Au
12 War Eagle Cu,Au,Ag,Mo
13 Little Chief Cu,Au,Ag
14 Cowley Park Cu,Au,Ag,Mo
15 Venus Ag,Au,Pb,Zn
16 Becker-Cochran Sb
17 Skukum Creek (Mount Reid) Au
18 Mount Skukum Au
19 Kane Ag,Pb
20 Hopkins Cu,Au
21 Bullion Creek-Sheep Creek Au
22 Burwash Creek Au
23 Wellgreen Ni,Cu,Co,PGEs
24 Casino Cu,Mo,Au
25 Cash Cu,Mo
26 Nucleus Au
27 Minto Cu,Au,Ag
28 Williams Creek Cu,Au,Ag
29 Tinta Hill Zn,Pb,Cu,Ag,Au
30 Mount Freegold Au
31 Mount Nansen Au,Ag,Zn,Pb
32 Clear Lake Zn,Pb,Ag,barite
33 Faro Zn,Pb,Ag
34a Grum Zn,Pb,Ag
34b Vangorda Zn,Pb,Ag
35 Dy Zn,Pb,Ag
36 Swim Zn,Pb,Ag
37 Grew Creek Au
38 Risby W
39 Groundhog Ag,Au,Zn,Pb,Cd
40 Ketza Au
41 Eagle (Tintina) Ag,Zn,Pb
42 Kudz Ze Kayah (ABM) Zn,Cu,Ag,Au
43 Matt Berry Ag,Pb,Zn,Sb,Cd
44 Howards Pass Zn,Pb
45 Anniv Zn,Pb
46 Moose barite
47 Tea barite
48 Jason Zn,Pb,Ag,barite
49 Tom Zn,Pb,Ag,barite

50 Mactung W,Cu
51 Cathy barite
52 Goz Creek Zn
53 Snake River (Crest) Fe
54 Craig Zn,Pb,Ag
55 Vera (Rusty Mountain) Ag,Zn,Pb
56 Val Ag,Pb,Zn
57 Blende Zn,Pb
58 Marg Cu,Zn,Pb,Ag,Au
59 Clark Ag,Pb,Zn
60 Keno Hill-Galena Hill Ag,Pb,Zn
61 Peso-Rex Ag,Pb
62 Ray Gulch (Mar) W
63 Dublin Gulch Au
64 Hart River Cu,Zn,Pb,Ag,Au
65 Mayo area Au
66 Clear Creek Au
67 Zeta Ag,Sn
68 Brewery Creek Au
69 Klondike Au
70 Lucky Joe Cu,Au,Ag
71 Sixtymile River area Au
72 Caley asbestos
73 Clinton Creek asbestos
74 Shell Creek Fe
75 Marn Au,Ag,Cu,W
76 Rein barite
77 Alto Fe
78 Fish River Fe,P

NORTHWEST TERRITORIES

1 Nastapoka Islands Fe
2 Innetalling Island Fe
3 Haig Inlet Fe
4 East Korok Fe
5 Chorkbak Inlet Fe
6 Maltby Lake Fe
7 Meliadine River (Discovery) Au
8 Rankin Inlet Ni,Cu,PGEs
9 Cache Au
10 Turquetil Lake Au
11 Heninga Lake Zn,Cu,Ag,Au
12 McConnell River (Ice) Fe
13a Cullaton Au
13b Shear Lake Au
14 Lac Cinquante U
15 Ferguson Lake Cu,Ni
16 Baker Lake U,Mo
17 Meadowbank (Third Portage Lake) Au
18 Kiggavik (Lone Gull) U
19 Nickel King (Thye Lake) Ni,Cu,Co,Ag
20 Pine Point Zn,Pb
21 Great Slave Reef Zn,Pb
22 Lens Li
23 Thor Lake Nb,Ta,Be,Y,Zr,REEs
24a Moose Li,Ta
24b Elk Li,Ta
25 Thor (Echo) Li
26 Bullmoose Lake Au,W
27 Kennedy Lake (Indian Mountain Lake) Zn,Cu,Ag,Cd

28 Sunrise Lake Zn,Pb,Cu,Ag,Au
29 Bear Zn,Pb,Ag,Au
30 Gordon Lake (Mahe) Au
31 VO Li
32 Thompson-Lundmark Au,Ag
33 Hidden Lake Li
34a Jake-Paint Li
34b Ann Li
35 Big (Murphy) Li
36 Nite Li
37a Giant Yellowknife-Royal Oak Au,Ag
37b Supercrest (Akaitcho) Au
37c Con-Rycon Au,Ag
37d Negus Au,Ag
38 Discovery Au
39 Nicholas Lake Au
40 Sue-Dianne Cu,U
41 Indin Lake (Kim) Au
42 Colomac (Hydra) Au
43a Tundra (Fat) Au
43b Salmita Au
43c Tundra Au,Ag
44 MacKay Lake (Deb) Zn,Cu,Ag
45 Lac de Gras (Point Lake) diamond
46 Musk Zn,Cu,Pb,Ag
47 Yava Zn,Cu,Pb,Ag,Au
48 George Lake Au
49 Hackett River Zn,Pb,Cu,Ag,Au
50 Lupin Au
51 Gondor Zn,Pb,Cu,Ag
52 Ren Au
53 Izok Lake Zn,Cu,Pb,Ag
54 Takijuq Lake (Hood River) Cu,Zn,Ag
55 Terra (Silver Bear) Ag,Cu,Bi,Co,U
56a Eldorado U,Ag,Cu,Co,Ni,Pb,Ra,Po
56b Echo Bay Ag,Cu
57 Mountain Lake U
58 Wreck Lake (Dot 47) Cu,Ag
59 June Cu,Ag
60 Muskox Cr,Cu,Ni,PGEs
61 Arcadia-Coronation Gulf Au
62 High Lake Cu,Zn,Au,Ag
63 Turner Lake Au
64 Pistol Lake Au
65 Borealis West Fe
66 Borealis East Fe
67 Eqe Bay Fe
68 Mary River Fe
69 Nanisivik Zn,Pb,Ag,Cd,Ge
70 Polaris Zn,Pb
71 Eclipse Pb,Zn
72 Wrigley Pb,Zn,Ag,Ge
73 Prairie Creek Zn,Pb,Cu,Ag,Cd
74 Ram-Rod Zn,Pb,Ag
75 Cantung W,Cu,Bi
76 Lened W,Cu
77 Howards Pass Zn,Pb
78 Coates Lake Cu,Ag
79 Jay Cu,Ag
80 Bear-Twit Zn,Pb,Ag
81 Gayna River Zn,Pb